STATISTICAL METHODS

Combined Edition (Volumes I & II)

STATISTICAL METHODS

N. G. DAS

Formerly,

Professor-in-charge, Department of Statistics,

St. Xavier's College, Calcutta

and

Lecturer, Goenka College of Commerce & Business Administration, Calcutta

Tata McGraw Hill Education Private Limited

NEW DELHI

McGraw-Hill Offices

New Delhi New York St Louis San Francisco Auckland Bogotá Caracas Kuala Lumpur Lisbon London Madrid Mexico City Milan Montreal San Juan Santiago Singapore Sydney Tokyo Toronto

Tata McGraw-Hill

Published by Tata McGraw Hill Education Private Limited,
7 West Patel Nagar, New Delhi 110 008

Sixth reprint 2012
RQZLCRAHRQBXL

ISBN 13: 978-0-07-008327-1
ISBN 10: 0-07-008327-4

Managing Director: *Ajay Shukla*

General Manager: Publishing—SEM & Tech Ed: *Vibha Mahajan*
Sponsoring Editor: *Shalini Jha*
Sditorial Researcher: *Smruti Snigdha*
Senior Copy Editor: *Dipika Dey*
Production Executive: *Suneeta S Bohra*

General Manager: Marketing—Higher Education & School: *Michael J Cruz*
Product Manager—SEM & Tech Ed: *Biju Ganesan*

Controller—Production: *Rajender P Ghansela*
Asst. General Manager—Production: *B L Dogra*

Published by Tata McGraw Hill Education Private Limited,
7 West Patel Nagar, New Delhi 110 008, typeset at Bukprint India, B-180A, Guru Nanak Pura, Laxmi Nagar-110 092 and printed at Adarsh Printers, Delhi 110 032

Cover: A P Offset

The McGraw-Hill Companies

CONTENTS

PREFACE

Statistics: The whole concept of this mathematical science revolves around *data*. Thus, **Statistical Methods** deal with techniques for collecting, analysing and drawing conclusions from data. Today, the subject has applications, wide and varied, spanning from disciplines like natural and social sciences, humanities, government and business to fields like economics and criminology.

PURPOSE

This book will serve the purpose of a text to the students of UG and PG courses such as B.Sc. (Statistics), B.Sc. (Mathematics), M.Sc. (Statistics), B.A. (Economics), B.Com, B.B.A, and M.B.A. wherever statistics features as a fundamental paper. It presents the mathematics underlying elementary statistical methods in as simple a manner as possible. Detailed solution of numerous questions (both theoretical and numerical) from C.A. (Intermediate), I.C.W.A. (Intermediate and Final), M.Com, B.Sc. (Economics Honours), B.Com (Honours and Pass), Diplomas in Social Welfare and Management examinations of various universities have been incorporated to make the book more versatile and hence, will be of considerable value to these students as well.

FEATURES

The structuring of this book facilitates students to understand statistical methods as efficiently as possible. It succeeds at simplifying the subject and makes it easy even to an average student. A brief discussion of important theoretical aspects has been provided in a simple language. Special notes and techniques of working out numerical problems, with elaborate steps enhance the understanding of concepts. Chapter-end Exercises with Answers have been provided to enable students confirm their mastery over the subject. The classic and wholesome approach to Statistics covering all the major concepts in a simple and lucid style helps the readers gain proficiency in the subject.

In order to make the book more user-friendly, the following important features have been incorporated.

- ❑ *Theory and Solved Examples presented in a simple, lucid and easy-to-understand manner*
- ❑ *Around 1800 Solved and Unsolved Problems*
- ❑ *Wide Spectrum of Problems ranging from simple to difficult level with detailed stepwise solutions*
- ❑ *Relevant Tables and Figures supporting theory*
- ❑ *A complete list of Statistical Formulae and Results*
- ❑ *Special Supplement including Solved Problems and Additional Exercises with Answers*

- *Rich appendices includes Area under Standard Normal Curve, Significant Points of Chi Square, Significant points of t, Significant points of F, Value of Exp(-m), Random Numbers*

The major highlight of this book is the inclusion of Solved Examination Questions of Calcutta University (C.U.), Burdwan University (B.U.), Institute of Chartered Accountants of India (C.A.), Institute of Cost and Works Accountants of India (I.C.W.A.), Diploma in Social Welfare (D.S.W.) of Calcutta University, Diploma in Management (D.M.) of Calcutta University and West Bengal Higher Secondary Council (W.B.H.S.).

ORGANISATION

The coherent arrangement of material in this book makes it a good reference source for statistical methods. Spanning over 19 chapters, this comprehensive book offers a lucid presentation of theory interspersed with explicitly solved examples. **Chapter 1** deals with Collection of Data, Classification and Tabulation. Charts and Diagrams and Useful Mathematical Devices are enumerated in **Chapters 2 and 3.** Graphical ways of summarizing a group of data are dealt with in **Chapter 4** on Frequency distributions. **Chapter 5** on Measures of Central Tendency covers ways of describing the central position of a frequency distribution for a group of data. Measures of Dispersion are elucidated in **Chapter 6**. **Chapter 7** focuses on Moments, Skewness and Kurtosis whereas **Chapter 8** explains Curve Fitting and Method of Least Squares. **Chapter 9** and **10** discuss concepts of Correlation, Regression and Interpolation. **Chapter 11** provides a detailed exposition of the Theory of Probability. Theoretical Distributions – Binomial, Poisson, Hypergeometric and Normal are clarified in **Chapter 12.** Sampling Theory, Estimation and Test of Significance are explained in **Chapters 13** and **14** respectively. **Chapters 15, 16 and 17** focus on Analysis of Variance, Time Series and Index Numbers. **Chapters 18 and 19** simplify concepts of Vital Statistics and Statistical Quality Control. •

ACKNOWLEDGEMENTS

Thanks are due to The Institute of Cost and Works Accountants of India and the Institute of Chartered Accountants of India for the kind permission accorded to make use of their questions. Also, a note of acknowledgement to all those who have been directly or indirectly associated with the book.

Every effort has been made to make this text as error-free as possible. However, feedback from readers is always welcome.

1 COLLECTION OF DATA CLASSIFICATION AND TABULATION

1.1 MEANING OF 'STATISTICS'

Statistics is a familiar word now-a-days. It has two different meanings. In the plural sense, the word implies a set of numerical figures, usually obtained by measurement or counting. These are also collectively known as *data*. In the singular sense, 'Statistics' refers to the subject of scientific activity which deals with the theories and methods of collection, analysis and interpretation of such data.

Example 1.1 *Distinguish between 'Statistics as data' and 'Statistics as tool'. Illustrate your answer with examples.* [C.U., M.Com. '66]

Or,

What are the two senses in which the term Statistics is generally used?

[B.U., B.A.(Econ) '73]

Solution The word *Statistics* is used in two different senses. Firstly, it refers to a collection of numerical figures or data obtained from the fields of human activity. For example, we frequently speak of Statistics of Price, Statistics of Food, Statistics of Educational Institutions, and so on, denoting thereby a set of facts and figures on the concerned topic. Secondly, the word is used to imply the subject of scientific activity dealing with the theories and methods of collection, analysis and interpretation of such data. It may be mentioned that unlike Physics, Chemistry, Biology, Geology etc., Statistics is not a separate branch of science. It is a scientific method or tool, like Mathematics, and is applied for handling numerical data in all branches of science, commerce, trade or humanities, where measurement or counting is possible. Heaps of apparently inconsistent data, resulting from a complex combination of many known or unknown factors, are analysed by applying statistical tools and techniques in order to find out the underlying pattern of variation. For example, the monthly sales figures of a company over several years, when plotted on a graph paper may show haphazard fluctuations; but when the time series data (see Chapter 14) are analysed, they may reveal important seasonal and cyclical movements superimposed on trend, which may be used with advantage for forecasting future sales.

Example 1.2 *What are the characteristics and limitations of Statistics?*

[C.A., Nov. '70, '72, '76]

Solution *Characteristics* 1. In Statistics, all available information are expressed in quantitative terms. In the study of 'variables' numerical figures arise as a direct consequence of measurement, e.g. age in years—28, 34, 27, etc. If the character studied is an 'attribute', the number of cases in each category gives rise to figures, e.g. defective articles 23, non-defectives 389. Sometimes in the study of attributes like intelligence, scores or marks are assigned to certain suitably framed questions, and intelligence is then indicated by total scores, say 53, 48, etc. Even 'dispersion', 'skewness', 'probability', business activity, etc. are expressed numerically.

2. Statistics must be aggregates of facts. A single fact, say, 672 lakh metric tons production of coal in India during the year 1965, is of no interest to a statistician. But if production figures for several consecutive years are available, they may be statistically analysed to draw important conclusions.

3. Statistics must be related to some field of inquiry. In other words, statistical data will be collected with a definite object in view. Stray collection of figures will be of no use.

4. Statistics should be capable of being related to each other, so that some cause-and-effect relationship can be established. In regression analysis, these are useful for 'prediction'.

5. Statistics are affected by a multiplicity of causes. In all fields of inquiry, the observed data are the result of a large number of factors, each of which contributes to the final figure. It is for the statistician to split up the effects due to component factors and study them.

6. Exactness cannot be guaranteed. No estimates or forecasts or statistical decisions can be taken to be exact. They are all subject to certain amount of inaccuracy whose limits may be stated in terms of probability. Almost in all cases, estimates and conclusions are arrived at on the basis of samples, and so it cannot be expected that they will lead to exact results.

Limitations 1. Statistics is applicable only to quantitative data. This cannot be used to study such events which cannot be expressed numerically.

2. Statistics can be used to analyse only collective matters, and not individual events.

3. Statistical decisions are applicable only on the average and in the long run. They may not hold in a particular case.

4. Statistical methods, if not applied in the proper perspective of the collected data, may lead to false conclusions. These should therefore be handled with utmost care and by experts only.

5. Statistical data must be uniform, in the sense that they should be subject to a stable causal system. There should not be much change in the group of factors responsible for variation in the data.

Example 1.3 *Illustrate with suitable examples the use of statistical methods in commerce, business and industry.*

[C.A., May '69 & '73; Nov. '70 & '77; I.C.W.A., June '73; C.U., M.Com., '74]

Solution With the gradual industrialisation and consequent expansion of trade, businessmen can no longer rely on the old system of hit-or-miss methods, or leave their future on chances. They have now to proceed on scientific principles, prepare themselves for competitive markets and plan their business accordingly. The traders and manufacturers have therefore to depend on a variety of factors, e.g. past sales records, present labour conditions and prices of raw materials used, prices of similar products already available in the market, etc. All these factors are statistically taken account of before fixing the price of a new commodity, so that it may find a suitable place in the market.

The following are some of the statistical techniques applied in Commerce, Business and Industry:

1. The various statistical measures of 'average', 'dispersion', 'skewness', 'correlation', etc. are generally useful for bringing out the main characteristics of available records. Their calculation enables comparison between different time periods or geographical areas.

2. 'Time series' analysis helps in isolating the components of variation noticed in the chronological records of sales, profits, prices, production, etc. These components—'trend', 'seasonal' and 'cyclical'—are used for future planning and forecasts.

3. 'Index numbers' of various types are constructed by the use of statistical methods. Index number of price, Index number of production, Index number of cost of living, Index number of business activity, etc. are so important in business and commerce that they have come to be known as 'business barometers'.

4. 'Regression' analysis enable establishment of relationship between various economic factors, e.g. input and output, production and sales, demand and per capita income. 'Regression equation' of a dependent variable upon one or several independent variables also helps in assessing the impact of each of the several factors, and may be used for forecasting.

5. Various sampling methods, e.g. 'random sampling', 'stratified sampling' etc. are available for conducting business and market surveys with the least amount of time, labour and money. These may also be used in test audit for checking the accuracy of records.

6. 'Statistical Quality Control' is now an indispensable tool for controlling the quality of manufactured products and in the maintenance of quality during production. 'Sampling inspection' is used for decision to accept or reject lots of manufactured product.

7. 'Vital statistics' and 'demography' are useful for the calculation of mortality rates in a community classified by age, sex, profession etc. These rates, or probabilities of death, are used in insurance business for fixing the rates of premium payable at different ages.

8. In the recruitment of right type of personnel for a job, the aptitude and efficiency of candidates are statistically determined by using 'test scores'.

1.2 VARIABLE AND ATTRIBUTE

The character of statistical information collected from a group of individuals or objects, is of two types—quantitative and qualitative. Information about the ages of a group of men is quantitative, because age is expressed in numbers, say 29 years, 43.5 years, etc. Religion of a group of men is qualitative, because religion cannot be stated is numerical terms; e.g. either Hindu or Buddhist or Christian etc. Age is a quantitative character and religion is a qualitative character. The quantitative character is technically called *variable* and the qualitative character is called *attribute.* A variable takes different 'values' and these values can be measured numerically in suitable units. An attribute cannot be measured, but can only be classified under different heads or 'categories'.

Example 1.4 *Explain with illustration the distinction between (i) an attribute and a variable, and (ii) a continuous variable and a discrete variable.*

[C.U., B.A.(Econ) '72; M.Com. '69, '74; I.C.W.A., June '74]

Solution Variable is a quantitative character of an individual or item and its values can always be measured. A variable whose values depend on chance and cannot be predicted is called a *random variable* or *variate*. Statistical theories deal with random variables only, and hence very often the term 'variable' is used in practice. Variates are of two types–(i) Continuous, and (ii) Discrete or Discontinuous.

A variate is said to be *continuous,* when it can take any value within a specified interval. For example, height of an individual may have any value between, say, 58 inches and 72 inches. It may be 65.2 inches, 65.23 inches, or even 65.22709 inches, provided we could measure the height so precisely. Hence, a continuous variate can take an infinite number of values within a given interval, however small it may be. Height, weight, temperature, density etc. are examples of a continuous variate.

A *discrete* variate, however can take only some isolated values. The number of children per family, the number of workers in a factory, the price of a cinema ticket sold during a week, etc. are examples of this type. The number of children per family can take only the values 0, 1, 2, 3, ..., i.e. whole numbers; it cannot take any value, e.g. a fractional value. Similarly, the price of a cinema ticket has only some specified values.

Example 1.5 *Classify each of the following characters as either an attribute, or a discrete variable, or a continuous variable: —Family size, Family income (per month), Mother tongue of a student, size of agricultural holding of a family, Division obtained (by a student) at the Higher Secondary Examination.* [B.U., B.A.(Econ) '73]

Solution (i) Family size—Discrete variable. Because it can take only some isolated values. The family size (i.e. number of members) may be either 1, or 2, or 3, or 4, etc. It cannot take any value. For example, we cannot speak of 2.73 members in a family.

(ii) Family income (per month)—Discrete variable. It can take values like Rs 874.27 P., or Rs. 874.28 P., etc. but not any value. For example, family income cannot take a value between the above two, because income in fractions of a Paisa is impossible.

(iii) Mother tongue—Attribute. It cannot be measured, but can only be classified as Bengali, Hindi, Tamil, etc.

(iv) Size of agricultural holding—Continuous variable. It can take any value within a given range, say 47.325697... acres. The sizes of two holdings may differ by any amount (theoretically), say, even by fractions of a square inch.

(v) Division obtained by a student—Attribute. Although it is expressed in numbers, e.g. Division 1, or 2, or 3, Division is not actually measured. It is only a means of classification into one of the categories.

1.3 PRIMARY DATA AND SECONDARY DATA

Statistical data may be of two types, viz. 'primary' and 'secondary'. *Primary data* are those which are collected for a specific purpose directly from the field of enquiry, and hence are original in nature. Such data are published by authorities who themselves are responsible for their collection. *Secondary data* are such numerical information which have previously been collected by some agency for one purpose and are merely compiled from that source for use in a different connection. In fact, data collected by someone when used by another, or collected for one purpose when used for another, will be called Secondary data. The same data are primary in the hands the collecting authority, but are secondary in the hands of another. For example, the census figures published by the Registrar General of India will be primary data, while the same data contained in any other publication will be called secondary data.

Sometimes, the sources of data are also classified as primary or secondary, depending on the type of data available therefrom. The authority who collects the data from the field of enquiry is known as *Primary source,* while some agency publishing the data originally collected by another is known as *Secondary source.*

The nature, scope and objects of a statistical investigation should be taken into account for deciding whether the data are to be collected originally or whether the available secondary data are to be utilised. It is generally preferable to make use of primary data from several standpoints—(i) Such data usually show detailed information and a description regarding the definition of the terms used. (ii) Very often, a note on the method of collection and any approximations used are also available, so that while using these data it can be decided in advance how much reliance can be placed on these figures. (iii) Secondary data usually contain errors due to transcription, rounding etc. and hence are hardly reliable. Inspite of all the advantages of primary data, secondary data are used when either due to limitations of time and money at the disposal of the investigator the data cannot be collected directly, or it becomes necessary to compare the data collected over a period of time, or utmost accuracy is not essential.

Publications containing Primary Data:

(i) "Reserve Bank of India Bulletin" issued *monthly* by the Reserve Bank of India, Bombay.

(ii) "Annual Report of the Chief Inspector of Mines in India", issued *annually* by the office of the Chief Inspector of Mines, Dhanbad.

(iii) "Indian Textile Bulletin", issued *monthly* by the Textile Commissioner, Bombay.

(iv) "Monthly Coal Bulletin", issued *monthly* by the Office of the Chief Inspector of Mines, Dhanbad.

Publications containing Secondary Data:

(i) "Statistical Abstract of the Indian Union", issued *annually* by Central Statistical Organisation (C.S.O.), New Delhi.

(ii) "Monthly Abstract of Statistics", issued by C.S.O.

Example 1.6 *Describe the various methods of collecting primary data and comment on their relative advantages.*

[I.C.W.A., Jan. '72; C.U., M.Com. '68 & '71; C.A., May '66 & '68, Nov. '69; D.S.W., Nov. '73]

Solution The following methods are generally used for collection of primary data:

(a) Direct personal observation,
(b) Indirect oral investigation,
(c) Questionnaires sent by mail,
(d) Schedules sent through investigators.

In *'direct personal observation'*, the investigator collects the requisite information personally through observation or by measurement. For example, in order to study the conditions of students residing in Calcutta hostels, the investigator meets the students in their hostels and collects necessary data after a personal study. The method is time-consuming and costly, but yields very accurate results. It is therefore suitable for such studies when the field of enquiry is small.

In *'indirect oral investigation'*, data are collected through indirect sources. Persons who are likely to have information about the problem, are interrogated and on the basis of their answers, factual data have to be compiled. Most of the Commissions of Enquiry or Committees appointed by Government collect primary data by this method. The accuracy of the method depends largely upon the type of persons interviewed and hence these persons have to be selected very carefully.

In the *'mailed questionnaire method'*, the most important instrument is the questionnaire. This contains a set of questions, relevant to the subject of enquiry, answers whereto are expected to yield the requisite information. Printed questionnaires are sent by mail to a selected list of persons, with the request to return them duly filled in. Supplementary instructions regarding the definitions of terms used and methods of filling up the forms, should also accompany the questionnaire. The method is cheap and expeditious, and a large area can be covered within a limited cost. Two principal disadvantages of the method are—the low degree of reliability of the collected data, and a large number of non-responses.

'Schedules sent through investigators' is the most widely used method of collection of primary data. Here paid investigators (also called 'enumerators') are employed for data collection. The investigators carry with them printed 'schedules' specially designed for the purpose, interview the people concerned, and fill up the schedules themselves on the spot, based on answers received from the informants. The method is very popular and yields satisfactory results. Much of the accuracy of the collected data however depends on the ability and tactfulness of investigators, who are given special training as to how they should elicit the correct information through friendly discussions. This method is adopted during the decennial census of population in this country.

Example 1.7 *Define "Secondary Data." State the chief sources, and point out the dangers involved in their use and what precautions are necessary before using them.* [C.A., Nov. '67 & '71, May '73; I.C.W.A., Dec. '73]

Solution Secondary data are such statistical information which have already been collected by someone for his own purpose and are now available for use by others for their purposes. Indeed, numerical data which are not gathered directly from the field of enquiry, but are merely compiled from other sources, are referred to as secondary data. The chief sources of such data are:

(i) Official publications of State and Central Governments, Foreign Governments, and international bodies, such as U.N.O., I.L.O. etc.

(ii) Publications and reports of trade associations, Chambers of Commerce, co-operative societies etc.

(iii) Journals and magazines published by private agencies, e.g. Commerce, Capital, etc.

(iv) Reports of Committees and Commissions of Enquiry.

(v) Unpublished reports prepared by research scholars, labour and trade unions, etc.

One must be very careful in using secondary data, because of their many limitations. For example, the definition and coverage of the terms, and the units of measurement may not quite agree with those required by the user. The basis of classification of such data may also be different. Sometimes the available secondary data may be mere estimates and not facts. The method employed for the collection of data may not be satisfactory. Secondary data in most cases contain errors due to transcription, rounding etc.

Detailed scrutiny of secondary data is therefore extremely essential before putting them into use. The user must satisfy himself with (i) the scope and object of enquiry for which the data were originally collected, (ii) method of collection, (iii) time and area covered by the data, (iv) precise definitions of the terms used, (v) extent of accuracy of the data, and (vi) the integrity and experience of the authority who collected these data.

1.4 POPULATION (OR UNIVERSE) AND SAMPLE

The totality of statistical information on a particular character, from all members covered by an enquiry, is called *population* or *universe*; e.g. population of income, population of registered factories in West Bengal, etc. The *sample* is a selected part of the population and is used to throw light on the population characteristics. In a socio-economic investigation covering the inhabitants of Calcutta city, if the monthly income figures of each and every Calcuttan could be obtained, the whole collection of figures of income would constitute the 'population of income.' This, however, is impracticable, and only a few thousand inhabitants would ordinarily be interviewed and their income figures obtained. These few thousand figures of income would comprise a 'sample of income', based on which the average income is estimated.

1.5 COMPLETE ENUMERATION (OR CENSUS) AND SAMPLE SURVEY

Example 1.8 *Distinguish between Census method and Sample Survey method.*
[C.A., Nov. '75 & '76; May '70; D.S.W., Nov. '69]

Solution There are two ways of collection of statistical data–(i) *Census* or *Complete enumeration*, and (ii) *Sample survey.* In census, information is collected from each and every member of the population, whereas sample survey refers to the study of population on the basis of information obtained from a sample. Census of population (human) in India and other countries of the world is an example of complete enumeration. The 'test audit' covering only a small number of entries, which is applied to verify the accuracy of all the entries in books of accounts, is an example of sample survey method. The same method is employed in estimating the crop yield of a country, or in counting the number of trees in a forest.

The census method, since the entire population is investigated, would require a large number of enumerators and other staff involving considerable time, money and labour. It is therefore generally applied for intensive investigations or when the number of members in the population is not large. Although the method is expected to yield accurate information, in practice the results are influenced by large non-sampling errors, whose magnitudes cannot be ascertained.

The sample survey method involving only a small part of the population, has in comparison with the census method, greater flexibility as regards intensity of data collection and the degree of accuracy of the final results, depending on the amount of time and money available for the purpose. Here sampling errors no doubt influence the final conclusions, but non-sampling errors can be kept to a minimum because of the possibility of deployment of highly trained personnel. On the other hand, the limits of sampling error can be stated in terms of probability. Sample survey method is suitable for all situations, but in some cases census method cannot be applied.

Example 1.9 *Discuss the advantages of Sampling methods over the Census method of collecting statistical information.*
[I.C.W.A., Jan. '65; C.U., B.A.(Econ) '69, '73; B.U., B.A.(Econ) '67; C.A., Nov. '74 & '76]

Solution Sample survey method has definite advantages over census method of collecting statistical information from several standpoints:

1. Sample survey takes *less time, labour and money* than is necessary in complete enumeration, because only a part of the whole population is investigated. In most cases, time and money available for the completion of an investigation are limited, and as such it becomes necessary to employ sample survey method.

2. There is *greater scope* in sample survey than in census. In the former, only a small group of skilled investigators is employed for collection of information. Although the cost per investigator is larger in sample survey than in census, because of the specialized training to be imparted to these personnel, the amount of information collected by each is much larger. As

such, a larger geographical area can be covered and more intensive data collection can be made in sample survey with the same amount of money spent in census.

3. Sample survey yields *more precise results* than census, because of the deployment of specially trained investigators, and the possibility of better control and supervision over them. Census necessitates a large army of investigators all of whom cannot be expected to have the same standards of skill, intelligence and efficiency. Sample survey no doubt introduces sampling error, but non-sampling errors are practically absent here. On the other hand, although in the census method, there is no question of sampling error (because the whole population is investigated), yet large magnitudes of non-sampling errors affect the accuracy of results more severely than the sampling errors introduced in a sample survey.

4. An extra advantage of sample survey method is that the *magnitude of error is known.* Sample survey depends on the laws of probability, and as such the magnitude of sampling errors can be theoretically calculated. Since non-sampling errors do not follow any definite probability law, no such estimate of error is possible in census method.

5. In some cases, *complete enumeration is not feasible* and sample survey method is the only way. For example, a rice merchant can not afford to examine every single grain of rice he purchases. He has to depend only on a sample, based on which he forms an idea about the quality of rice in the whole consignment.

Because of these advantages, random sampling and various other sampling techniques are getting very popular now-a-days. A great limitation of sample survey method is that it yields results for the concerned phenomena as a whole or its broad subdivisions, but not for small sectors. Therefore, if time and cost are not important factors or the population is not very large, census method will be more appropriate than sample survey method.

1.6 STATISTICAL ENQUIRY

Example 1.10 *Describe the different stages of a Statistical enquiry.*

[C.U., M.Com. '70; C.A., May '77]

Solution *'Statistical enquiry'* refers to some investigation wherein relevant information is collected, analysed and interpreted by the application of statistical methods. Broadly, the different stages of a statistical enquiry are as follows :

1. *Planning the Enquiry*—In any scientific investigation, planning is essential. Without proper planning the collected data may not be found suitable for the purpose and considerable time and money will be wasted. The following points should be considered at the planning stage:—(i) *Object and scope of the enquiry*—The first and most essential thing is to prepare a clear-cut statement of the purpose of the enquiry. This will determine its scope. (ii) *Source of data, whether primary or secondary*—Usually primary data are collected during statistical enquiries, and as such the method of collection must be decided in advance. This depends on the time and money available for the investigation. (iii) *Type of enquiry, whether census or sample survey*—Because of the many advantages of sample survey over census, most statistical enquiries use the former. The type of sampling to use, e.g. random, stratified, or multi-stage etc. and the selection of the actual sample, has to be finalised. (iv) *Definition of statistical units*—It is extremely essential that the units be accurately and clearly defined, and the same definitions be strictly adhered to during the course of the whole investigation. (v) *Degree of accuracy*—The standard of accuracy of the results and of the final conclusions derived therefrom must be determined in advance on the basis of available resources. Absolute accuracy is neither possible nor very necessary. The enquiry should be so planned as to obtain the maximum accuracy within the limited cost.

2. *Collection of Data*—Numerical data form the raw material for any statistical enquiry, and hence the collection of accurate data is the most important part in the whole investigation. The

method of collection depends to a large extent on the nature, object and scope of the enquiry on one hand and the availability of time and money on the other. For the collection of primary data, one of the following methods may be used—(i) Direct personal observation, (ii) Indirect oral investigation, (iii) Sending postal questionnaires, and (iv) Sending schedules through investigators. Of these, the first gives fairly accurate results; but it is very costly and has a limited scope. The third method has a very large coverage, but suffers from the disadvantage of a high proportion of non-response or incomplete returns. Collection of data through enumerators equipped with schedules is the most widely used method. This involves organisational problems associated with the selection of suitable enumerators and training of these personnel for tactful collection of data.

3. *Editing the Returns and Presentation of Data*—Soon after the collection of data begins, arrangements should be made to receive and scrutinize the completed forms in batches. These returns usually contain many inconsistencies, errors and omissions. The defective returns should be immediately referred back for correction. When all returns have been received and checked, the information is classified and tabulated. Usually many tables of various types have to be prepared and in large-scale enquiries, it is not possible to complete the work manually. Specially developed machines, e.g. electric tabulators, may then be employed for the purpose, where the work of sorting and tabulation is simultaneously done and at a great speed. However, when the enquiry does not involve a large number of returns, manual sorting and tabulation is resorted to. Besides tabular presentation, numerous charts and diagrams are also prepared.

4. *Analysis of Data and Interpretation of Results*—Statistical analysis is a method of abstracting significant facts from the large mass of numerical data collected during the enquiry. The process includes determination of various statistical measures, the estimation of statistical constants, and subsequent tests of significance. The techniques for analysis require specialised knowledge in statistical theories. The results of analysis are then interpreted and conclusions are drawn.

5. *Preparation of Report*—After the work of the enquiry has been completed, it is necessary that a detailed report is published. This report should give a complete description of all the stages of the enquiry, definitions of the terms used, coverage, degree of reliability of the data, and final results of the enquiry. Charts and tables are also included in the report. The report is helpful in planning any future enquiries on a related problem.

Example 1.11 *What do you mean by a questionnaire? State the essential points to be observed in drafting a good questionnaire.*

Describe carefully the merits and defects of the mailed questionnaire method of investigation and the precautions necessary in using this method of investigation.

[I.C.W.A., July '72, Dec. '73; C.A., Nov. '73 & May '71, '75 '76]

Solution The 'questionnaire' is a proforma containing a sequence of questions relevant to a statistical enquiry. It is used for collection of primary data from individual persons through their responses to the set of questions. The questionnaire is used in two ways. Very often, questionnaires are sent by post to individual informants who are required to write the answers against each question and return the completed proforma again by post. The method is very popular and is usually referred to as the 'mailed questionnaire method' of investigation. Sometimes, however, the enumerators (i.e. people employed for collection of data) themselves carry the questionnaires and note down the answers while interviewing the informants. This method is employed during the census of population.

The drafting of a good questionnaire requires utmost skill, and the success of any investigation depend to a large extent on tactful drafting of the questionnaire. The following points should be observed in drafting a questionnaire:

1. The questionnaire should be as short as possible. Many questions may arise during an investigation. But if all are included, the questionnaire will become unduly lengthy with the

consequence that the respondents (i.e. persons who are required to answer them) will feel bored and reluctant to answer all the questions.

2. The individual questions should be, simple, unambiguous and precise. Lengthy questions cause irritation, resulting in careless and inaccurate replies. Complicated questions should be split up into several smaller parts which can be easily answered by the respondents. Explanations and definitions of some of the terms used in the questionnaire must therefore accompany each proforma.

3. If possible, questions should be so set as to elicit only two possible definite answers—'yes' or 'no'.

4. Questions affecting the pride and sentiments of informants must not be asked. Questions on personal matters like income and property should be avoided as much as possible.

5. Questions should be capable of yielding objective answers, avoiding matters of opinion.

6. The units in which the information is to be collected should be clearly and precisely mentioned in the questionnaire.

7. The arrangement of questions in the proforma should be such as to have an easy and systematic flow of answers in turn. Questions should not skip back and forth from one topic to another.

8. After the questionnaire has been devised, it is desirable to try it on a few individuals. The procedure, which is known as *pilot survey,* is useful in detecting the shortcomings of the questionnaire, so that necessary modifications may be made before it is used in the actual enquiry.

Merits—(i) Mailed questionnaire method of investigation is the least expensive. Since the enumerators do not have to travel to the informants' places, the requisite information can be collected only at a fraction of the cost needed in the other methods. (ii) For the same reason, a vast area can be covered by the enquiry. (iii) The information can also be gathered at a comparatively short period of time; because the questionnaires can be sent out simultaneously to many informants. (iv) The method is useful for collecting information on such matters which the informants may not like to disclose in the presence of enumerators.

Defects—(i) Mailed questionnaire method is not very satisfactory due to the low proportion of response. In many cases, the informants do not care to take the trouble and the completed proformae are not returned. (ii) The quality of the collected information is also very poor. Many respondents do not carefully fill in the questionnaires, and haphazard answers are given. Incomplete returns are also not rare. This necessitates large follow-up correspondence. (iii) The respondents may not fully understand the meaning of some of the questions (inspite of the accompanying instructions) resulting in the low accuracy of the collected information.

Precautions—People have their own whims and feelings of pride and prejudice. Unless compelled to submit returns under an Act of law, they are usually reluctant to disclose any information or take unnecessary trouble. Much persuasion is therefore necessary to elicit information in such form as to be really useful for valid statistical analysis. The following precautions should be taken: (i) The purpose of the enquiry should be clearly explained and the informants requested for co-operation. They should be assured of complete secrecy of their identity and that the outcome of the investigation will in no way harm them. (ii) Utmost care should be taken in drafting the questionnaire. The questions should be simple and unambiguous, so that respondents do not find it difficult to answer them. Questions on personal matters should be avoided as much as possible. (iii) It should be made clear that no postage need be paid by the respondents and that the investigator is fully responsible for the cost of postage. All that is necessary is to fill up the proforma and post it as early as possible. (iv) Reminders in the form of courteous letters should be sent to persons who fail to return the completed questionnaires. Sometimes telegraphic or express reminders are also helpful. (v) As soon as each return is received, it should be scrutinized for any inconsistencies or anomalies. If necessary, some of the returns should be referred back immediately for correction.

1.7 USEFUL TERMS

(1) Law of Statistical Regularity

The Law of Statistical Regularity states that a moderately large number of items selected at random from a larger group (technically called 'population') are almost sure, on the whole, to possess the characteristics of the larger group. In other words, if a random sample (Chapter 13) of large size is taken, the sample will to a fairly accurate degree exhibit the characteristics of the entire population. Two points should be kept in mind—(i) The sample must be drawn 'at random'. This means that the items constituting the sample are obtained by giving equal chance of selection to all the items in the population. (ii) The number of items included in the sample should be sufficiently large. The larger the size of the sample, the more representative will the sample be of the population.

Note that the Law of Statistical Regularity does not guarantee that the large random sample must reflect all the characteristics of the population. A part can never represent the whole in toto. The law only states about the tendency of the sample to contain the characteristics of the population. If 10,000 persons are selected at random from a certain city and their income figures are obtained, these data are likely to form a miniature model of the income structure for the city. We are almost sure that the average of these 10,000 figures will be close to the average income of all the inhabitants, i.e. the true per capita income for the whole city. Again, if data from another random sample of 20,000 people were available, the average calculated from this larger sample is expected to be closer to the true average, thus reducing the error with increase in the size of sample. It is because of the operation of this law that sufficiently accurate estimates about the population are obtained by sample survey, replacing the costly and laborious method of complete enumeration. Insurance companies insure against death, fire, etc., banks and commercial firms can undertake risks in business, and gamblers bet money on the basis of this law.

(2) Statistical Unit

Example 1.12 *Define a Statistical Unit and explain what should be the essential requisites of a good statistical unit.* [C.A., May '73. '74 & '75]

Solution *Statistical Unit* is that thing in terms of which requisite information is collected and expressed during a statistical inquiry. For effective comparison and valid conclusions, it is necessary that the statistical unit is suitably chosen. A mistake in the choice of the appropriate unit is more harmful than mistakes in the collection of data.

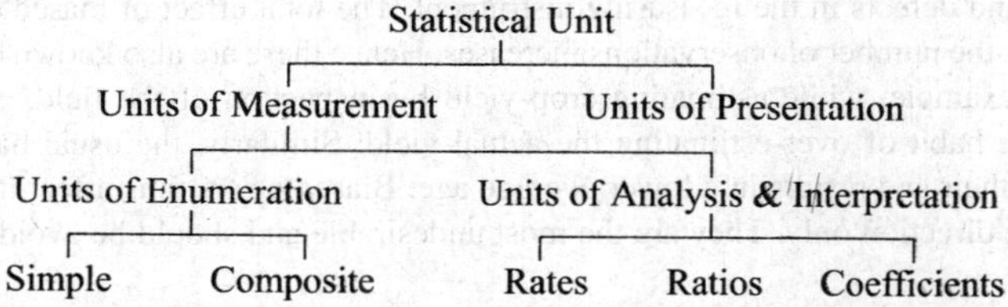

Examples: 'Simple' units—yard, ton, etc.; 'Composite' units—manday, kilowatt-hour, passenger-mile, etc.; Units of Analysis and Interpretation—maund per acre (yield), per capita per annum (income), number per thousand (birth rate).

The requisites of a good statistical unit are:

1. It should be clear and unambiguous. For instance, during the census of population in India, 'household' is often employed as a unit. Whether all persons living in the same house, or those having a common kitchen, or only blood relations living in the same house would comprise the household, has to be clearly defined before the collection of data begins.

2. It should be rigid, easy to ascertain and homogeneous. In the collection of data for estimation of crop-yield, all enumerators should be told specifically whether per acre yield figures are to be given in maunds or tons; and if in tons, whether metric ton or imperial ton.

3. It should be suitable for the inquiry. In a survey of industries, volume of production may be stated either in tons, or in number, or even in rupees. If the survey covers different types of industries, naturally 'rupees' should be the appropriate unit.

4. It should be stable. If production of electric fans in India over several years is compared, the 'number' of fans produced would be a better unit than their values in 'rupees'. The former unit is more stable than the latter, which depends on the price per fan varying from year to year.

(3) Statistical Error

Example 1.13 *Define Statistical Error. How is it different from a mistake? Write an essay on the different types of statistical errors and how they are measured.*

[C.A., Nov. '71]

Solution In statistical terminology, the word 'error' is used in a special sense. Error shows the extent to which the observed value of a quantity exceeds (or falls short of) the true value.

Error = Observed value – True value.

It may be positive, negative or zero. Since statistical data are obtained either by measurement or by observation, it is never possible to attain absolute accuracy either in the collected data or in the final results. Inaccuracies are bound to occur. For example, data on income, for obvious reasons, are hardly accurate. Measurements on any continuous variable can never be exact, e.g. heights measured upto the nearest tenths of an inch. Personal prejudices of investigators, or faulty methods of collection of data also affect the accuracy. Again, it is not possible to assure absolute accuracy while making estimates about the population based on the analysis of samples. Such inaccuracies arising in the course of statistical investigations are called '*statistical errors*'. Statistical errors should not, however, be confused with 'mistakes'. Mistakes are caused by carelessness, e.g. following a wrong method of calculation, computational mistakes, or clerical lapses. Statistical errors are inherent in any statistical enquiry, and enter the process unconsciously. They can never be eliminated.

Statistical errors may be classified into

(a) Biased error, (b) Unbiased error.

Biased errors are those which arise due to personal bias or prejudices of investigators and informants, and defects in the measuring instrument. The total effect of biased errors goes on increasing, as the number of observations increases. Hence these are also known as 'cumulative errors'. For example, while estimating crop-yield by inspection of the field, an investigator may have the habit of over-estimating the actual yield. Similarly, the usual habit of women understating their age results in a lower average age. Biased errors thus affect the accuracy of results in one direction only. They are the most undesirable and should be avoided as much as possible.

Unbiased errors are those which enter into a statistical enquiry due to chance causes. Since they are sometimes in one direction and sometimes in the other, unbiased errors are compensating. With increase in the number of observations, the total effect of unbiased errors therefore diminishes. Unbiased errors follow the law of statistical regularity, and hence the magnitude of such errors can be predicted based on probability considerations. For example, while measuring heights of persons to the nearest inch, some recorded values are smaller, while some others are larger, than the true heights. The total of the observations do not therefore, differ much from the actual total. The larger the number of observations, the smaller will be the difference between the two. Thus unbiased errors can be minimised by taking a large number of observations.

(4) Statistical Fallacies

All statistical data and conclusions derived from their analysis must be interpreted in their proper perspective and by experts only. Mistakes in interpretation often give rise to wrong and absolutely baseless propositions being imputed into the results. Such erroneous conclusions reached on a misinterpretation of data are known as *Statistical Fallacies*. The chief causes of statistical fallacies are:

(a) The characteristics of a group may be wrongly assumed to hold in respect of each individual member. Example : (i) "The per capita national income in India increased from Rs 300 in 1950 to Rs 900 in 1970, and so the prosperity of Indians has increased three times" (*C.A., May '72*). (ii) Conversation between patient and doctor: "Patient—'What are the chances of getting better? Doctor—'One hundred per cent. Medical records show that 9 out of every 10 die of the disease you have. Yours is the tenth case I have treated; the others all died. Statistics are statistics; so you are bound to get well'."

(b) An unrepresentative sample may be analysed and the results may be generalised to cover the whole population. Example: "80% of the people who die of cancer are found to be smokers in a hospital, and so it may be concluded that smoking causes cancer" (*C.A., May '66*). In order to justify the statement, it is necessary to examine the percentage of smokers amongst patients dying of other diseases and from all causes. Other factors like sex, age-group, etc. of the patients should also be considered. It may turn out that smokers among patients dying from another cause is greater than 80%, or that cancer patients resort to smoking habits, being aware of the fatality of the disease.

(c) The characteristics may not be properly defined as to make the data comparable. Example: (i) "In a war, 40% of the people of a State had to go to the field where 10% of them died; 12% of the people who remained at home died during the period. As the death-rate among those who did not go to the field was higher than those who went to the field, the conclusion was that joining the war was safer than remaining at home" (*C.U., M.Com. '71*). Here the composition of people in the two categories–those who went to the field, and those who remained at home—has been completely ignored. The risks of death at various ages, and for males and females are different. Healthy adult people have gone to the field, and the old and infirm have been left at home, whose risk of death from natural causes is high. (ii) "The average cost of production was Rs 1.50 in 1960 and Rs 1.75 in 1961, and so the factory has become inefficient" (*C.A., May '70*). The cost may have increased because of increase in prices of raw materials, but still the overall profits may have gone up.

(d) A cause-and-effect relationship may be wrongly assumed, which may not hold for all values of the variables. Example: "Since we had more than double the normal rainfall during the current year, we can expect a record yield in rice" (*C.A., May '72*). Obviously, rainfall increases rice yield to a limited extent; excessive rains may spoil the standing crop in the field.

(e) Sometimes, data transformed into percentages may show a misleading result. Example: "The production of a factory increased by 10% from 1972 to 1973, and by 15% during the next year; so production increased by 25% over the two-year period 1972–74." Here the productions at the beginning of the two periods were not the same, and hence the percentages cannot be added.

1.8 CLASSIFICATION

Example 1.14 *Define 'classification' and explain the various ways of classification adopted in Statistics.* [C.A., Nov. '68]

Solution *Classification* is the process of arranging the collected statistical information under different categories or classes according to some common characteristic possessed by the individual members. Statistical data collected during the course of an investigation are so varied that it is not possible to appreciate, even after a careful study, the true significance of the figures, unless they are arranged properly. To make the data really useful, they must be *classified* or grouped into homogeneous categories, so that the like will go with the like and the unlike with the unlike. Classification prepares the ground for enabling comparison and analysis by instituting a logical and orderly arrangement of data. For example, during the population census, apart from the number of members in each family, various other information, e.g. age, sex, occupation etc. of all people in the country are collected. The total population is then classified according to sex into males and females; according to age groups 0–10 years, 10–20 years, etc.; according to livelihood into agricultural classes, production other than cultivation, commerce, transport, etc. If such classifications are not made, it will not be possible to analyse the data and reveal their true significance from the heaps of material collected during the population census.

Types of Classification

Broadly, there are four types of classification:

1. *On qualitative basis*—Classification of the total population of a country on the basis of sex, religion, occupation, etc. belongs to this type. This is also known as *classification by attributes.*

2. *On quantitative basis*—Classification of the total population according to age, or of industries according to the number of persons employed, etc. are included in this type. Here, the basis of classification is some variable, and hence this is also known as *classification by variables.*

3. *On time basis*—Some statistical data are arranged in order of their time of occurrence. Production of a factory may be shown by weeks, months, quarters or years. Statistical data classified according to time are known as *time series* (Chapter 14).

4. *On geographical basis*—The total population of a country may be classified by States or districts, or exports of a particular commodity from India may be classified by the country to which exported. The basis of classification in such cases is by geographical regions.

1.9 TABULATION

Tabulation may be defined as the logical and systematic organisation of statistical data in rows and columns, designed to simplify the presentation and facilitate comparisons. The advantages of tabulation are—(i) Tabulation enables the significance of data readily understood, and leaves a lasting impression than textual presentation. (ii) It facilitates quick comparison of statistical data shown between rows and columns. (iii) Errors and omissions can be readily detected when data are tabulated. (iv) Repetition of explanatory terms and phrases can be avoided, and the concise tabular form clearly reveals the characteristics of data.

Example 1.15 *Define a Statistical Table and state the essentials of a good table.*
[C.A., Nov. '68 & '69, May '74 & '75; ICWA, June '73 & '74]

Solution *Statistical Table* is a systematic arrangement of quantitative data under appropriate heads in rows and columns. After the data have been collected, they should be tabulated, i.e. put in the form of a table, so that the whole information can be had at a glance. There is no ideal method of tabulation. Skill in tabulation is generally attained after years of experience. Nevertheless, the following general rules should be followed while tabulating statistical information:

1. A table must contain a title. This title should be clear and convey in as few words as possible, the contents of the table. Columns should similarly have definite and, at the same time, sufficiently comprehensive headings.

2. When several tables are used, they should be numbered, e.g. Table 1, Table 2, etc., to facilitate future reference. If there are many columns, say more than four, they should also be serially numbered.

3. Units of measurement must be clearly shown in the column headings, e.g. price (Rs), height (inches), weight (lbs.), etc.

4. The table should be well-balanced in length and breadth. When the table is unusually long, a separate coarse grouping should be tried. If, however, this causes any serious loss of information, some breaks or extra spaces should be left at equal distances apart.

5. The arrangement of items in the table should have a logical sequence. Time series data, if any, must be arranged chronologically.

6. Columns of figures which are directly comparable should be kept as close as possible. If percentages or averages appear in a table, they should be placed near the figures to which they relate. Large numbers should be rounded and mentioned in thousands, lakhs or millions

7. Light and heavy rulings may be used to distinguish between sub-columns and main columns. Double rulings may also be used instead of heavy ones.

8. Totals of columns may be shown at the bottom of the table. In cases where row totals are useful, they should also be shown.

9. If the tabulated data have been compiled from another source, this source should be mentioned in the foot-note.

10. Items which are in any way different from the rest, e.g. estimated figures, revised figures, etc. should be marked with asterisk or numbers and explanatory notes given in the foot-note.

Types of Tabulation

In very general terms, there are two types of tabulation–(i) *Simple tabulation,* and (ii) *Complex tabulation*. A simple table contains data in respect of one characteristic only, information relating to other characteristics being left out. A complex table contains figures relating to several characteristics. Tabulation may also be *Single, Double, Treble* or *Manifold.* A single tabulation is one that answers one or more groups of independent questions. Double, Treble or Manifold tabulation shows the subdivision of a total according to two, three or many characteristics respectively and answers as many mutually dependent questions.

Example 1.16 *What are the different parts of a table?*
[C.A., May '65; I.C.W.A., July '71 & Jan. '74]

Solution The different parts of a table are:

(i) *Title*—This is a brief description of the contents and is shown at the top of the table.

(ii) *Stub*—The extreme left part of the table, where descriptions of rows are shown, is called Stub.

(iii) *Caption* and *Box-head*—The upper part of the table which shows the description of columns and sub-columns is called Caption. The whole of the upper part, including caption, units of measurement and column-numbers, if any, is called Box-head.

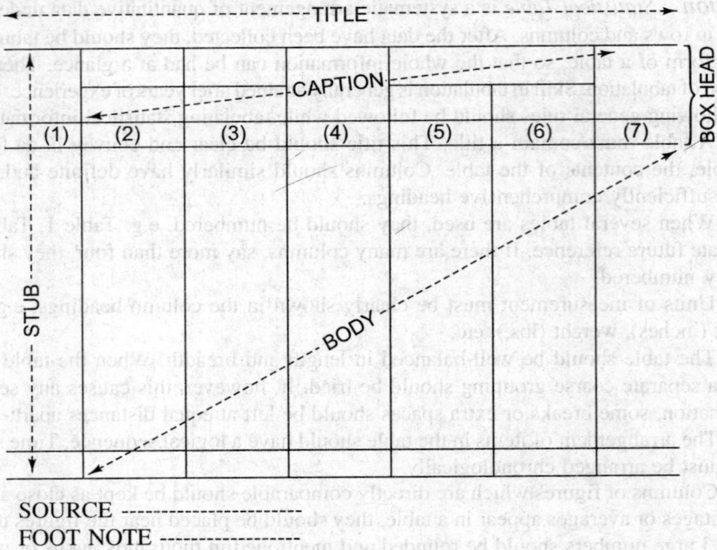

Fig. 1.1 Different Parts of Table

(iv) *Body*—It is that part of the table which shows the figures.

(v) *Footnote*—This is the part below the Body, where the source of data and any explanations are shown.

Example 1.17 *Prepare a blank table showing the average wages of males and females classified into two age-groups of eighteen years and over, and under eighteen years in five industries at two different dates.* [C.U., B.A.(Econ) '73; M.Com. '63]

Solution

Table 1.1 Average (Monthly) Wages (Rs) in Five Industries

Industry	*As on... (date)*				*As on... (date)*			
	Under 18 years		18 years & over		Under 18 years		18years & over	
	Male	Female	Male	Female	Male	Female	Male	Female
1.								
2.								
3.								
4.								
5.								
Total								

Example 1.18 *Explain how you would tabulate statistics of death from principal diseases by sexes, in two different States in India for a period of five years.*

[I.C.W.A., July '70]

Solution

Table 1.2 Statistics of Death from Principal Diseases in Bihar and U.P., by Sex (1961-65)

Disease/Sex	Bihar 1961	Bihar 1962	Bihar 1963	Bihar 1964	Bihar 1965	U.P. 1961	U.P. 1962	U.P. 1963	U.P. 1964	U.P. 1965
1. Fever:										
Male										
Female										
Total										
2. Cholera:										
Male										
Female										
Total										
3. Dysentry:										
Male										
Female										
Total										
4. Small-pox:										
Male										
Female										
Total										
5. All Cases:										
Male										
Female										
Grand Total										

Example 1.19 *Prepare a blank table showing the particulars relating to students of the Bombay University classified according to their age, sex, faculty and three important religions.*

[C.A., Nov. '69]

Solution

Table 1.3 Number of Students in Bombay University (by Age, Sex, Faculty and Religion)

Faculty/ Age (years)	Hindu			Christian			Parsee			Total		
	Male	*Female*	*Total*	*Male*	*Female*	*Total*	*Male*	*Female*	*Total*	*Male*	*Female*	*Total*
1. Arts:												
Upto 15												
16 – 20												
Above 20												
Total												
2. Science:												
Upto 15												
16 – 20												
Above 20												
Total												
3. Medicine:												
Upto 15												
16 – 20												
Above 20												
Total												
4. Agriculture:												
Upto 15												
16 – 20												
Above 20												
Total												
All Faculties:												
Upto 15												
16 – 20												
Above 20												
Grand Total												

Example 1.20 *Represent the information contained in the following passage in a suitable tabular form:*

'The cropped area of vegetables (excluding potatoes) grown for human consumption in the United Kingdom rose in 1955–56 and was the highest since 1950–51. The cropped area increased to 509,000 acres, some 11,000 acres more than in 1954–55. The area of root vegetables increased by 8,100 acres to 62,400 acres, carrots alone increasing by 5,700 acres to 33,200 acres. The area of cabbage rose slightly, thus halting the steady decline since 1947–48; the cropped area was 75,700 acres with 74,800 acres in 1954–55. The cropped area of cauliflower and broccoli was 33,400 acres, 2,400 acres less than in 1954–55. Peas harvested dry decreased by about 9,800 acres to 121,800 acres,

but a larger area of beans, mainly broad beans and green peas, was grown. The area of broad beans increased by 2,600 acres to 7,300 acres and the area of green peas for canning and quick freezing rose by 7,000 acres to 50,400 acres'. [I.C.W.A., June '74]

Solution

Table 1.4 Cropped Area under Vegetables (excl. Potatoes) in U.K., 1954-56

(Hundred acres)

Type	*1954-55*	*1955-56*	*Increase (+) or Decrease (–)*
1. Root vegetables:	543	624	+ 81
Carrots	275	332	+ 57
Others	268	292	+ 24
2. Cabbage	748	757	+ 9
3. Cauliflower & Broccoli	358	334	– 24
4. Peas (harvested dry)	1,316	1,218	– 98
5. Beans:	2,015	2,157	+ 142
Broad beans	47	73	+ 26
Green Peas	434	504	+ 70
Miscellaneous	1,534	1,580	+ 46
Total	4,980	5,090*	+ 110

*Highest since 1950-51

Example 1.21 *The total number of accidents in Southern Railway in 1960 was 3500, and it decreased by 300 in 1961 and by 700 in 1962. The total number of accidents in Metre Gauge section showed a progressive increase from 1960 to 1962. It was 245 in 1960; 346 in 1961; and 428 in 1962. In the metre gauge section, "Not compensated" cases were 49 in 1960, 77 in 1961, and 108 in 1962. "Compensated" cases in the broad gauge section were 2867, 2587 and 2152 in these three years respectively.*

From the above report, you are required to prepare a neat table as per the rules of tabulation. [C.A., Nov. '71]

Solution

Table 1.5 Number of Accidents in Southern Railway During 1960-62

Section	*1960*	*1961*	*1962*
Metre Gauge:	245	346	428
Not compensated	49	77	108
Compensated	196	269	320
Broad Gauge:	3,255	2,854	2,372
Not compensated	388	267	220
Compensated	2,867	2,587	2,152
Total:	3,500	3,200	2,800
Not compensated	437	344	328
Compensated	3,063	2,856	2,472

Example 1.22 *In the White Paper on National Income and Expenditure in U.K. (1938–45) it was stated that the estimated number of individuals assessed to income tax in 1938 was 1745 thousand with annual gross income between £250 and £500, while in 1944 the corresponding number was 5200 thousand. In the higher range £500–£1000 the number rose from 500 thousand to 1400 thousand, between the same years, while in the £1000–£2000 group the increase was from 195 thousand to 520 thousand. The distribution of net income after deduction of tax gave for the first group 1820 thousand and 5050 thousand, for the second group 450 thousand and 830 thousand, and for the last group 155 thousand and 186 thousand respectively.*

Calculate appropriate statistical measures to bring out the relative difference in the numbers in each range as well as between the years. Calculate further percentages to bring out the effect of taxation in the three groups. Tabulate yours results together with the original data. [B.U., B.A.(Econ) '66; C.U., B.A. (Econ) '71]

Solution

Table 1.6 Number of Individuals Assessed to Income Tax(Estimated)—Years 1938 & 1944

Year		*Income Range (£)* 250–500	500–1000	1000–2000	*Total*
I. Number ('000) with Gross Income	1938	1745	500	195	2440
	1944	5200	1400	520	7120
Increase (%)		198	180	167	192
II. Number ('000) with Net Income after taxation	1938	1820	450	155	2425
	1944	5050	830	186	6066
Increase (%)		177	84	20	150
III. Increase (+) or Decrease (–) after taxation (%)	1938	(+) 4	(–) 10	(–) 21	(–) 1
	1944	(–) 3	(–) 41	(–) 64	(–) 15

Source: *White Paper on National Income and Erpenditure, U.K. (1938-45)*

1.10 MECHANICAL TABULATION

In large-scale surveys, the manual process of classification and tabulation of various types of data would involve huge time and expenditure, and therefore mechanical aids to tabulation are resorted to by what are known as '*Punched Card Systems.*' In this process, the information from the interviewers' schedules are transferred on to specially designed *Punched Cards*, and these are run through *Sorting* and *Tabulating Machines*, which give out the desired tables in printed form according to the desired classification.

The commontype 'Punched Cards' are thin cards approximately of size 3 inches by 7½ inches, having a number of columns, usually 80, with the column numbers printed thereon. Under each column there are printed figures 0, 1, 2, ... 9, one below the other at equidistant intervals. Some cards have provision for two more positions in each column at the top. When the completed schedules are available, the first task is to transfer the information on to the cards by means of rectangular holes punched at appropriate places thereon. For this purpose, *Code Numbers* are assigned to the various characters, like age, sex, income, profession, etc., and their various categories. One column or a group of columns is allotted to each character, and the different categories or classes of a character are assigned different positions in that column represented by the printed numbers 0 to 9 according to a predetermined arrangement. For example, we may allot column number 1 to the character age (code : 1 for '0 to 20 years', 2 for '20 to 30 years', 3 for '30 to 40 years', and so on; column number 2 to sex (coded : 1 for male, and 2 for female), etc.

A separate card is used for each member, and holes are cut thereon at appropriate positions by means of the *Punching Machine,* which looks like a typewriter keyboard and is operated manually by means of keys. The accuracy of the transcribed information is then verified by *Punched Card Verifier,* also operated manually, on which the punching operations are repeated. If a card is punched at a wrong position or not punched, the machine locks. The verified cards are then passed through a *Sorting Machine*, which automatically sorts out the cards into different groups and sub-groups. as desired, at speeds upto 40,000 cards per hour. The sorted cards are now put through an electric *Tabulating Machine*, which 'senses' the cards one by one by means of electrical contacts through the holes, and transforms the information into printed figures and words, as needed. The same cards may be used over and over again through the Sorting and Tabulating machines for obtaining different types of tables. The tabulating machine does not only tabulate the data at high speed, but is also capable of printing totals and sub-totals, and may be so set as to reproduce any part of the data contained in the cards at speeds ranging from 5,000 to 12,000 cards per hour.

EXERCISES

1. Examine whether the following variables are discrete or continuous—Age of a person, Size of land holding, Size of family, Temperature, Monthly sales in a shop. *[C.U., M.Com. '69]*
2. Explain the terms 'classification' and 'tabulation' and point out their importance in a statistical investigation. What precautions would you take in tabulating statistical data? *[I.C.W.A., Jan. '70]*
3. State the various reasons why errors are made in the interpretation of statistical data, and indicate how to guard against such errors. *[C.A., May '70 & '72]*
4. What do you mean by an error in statistics? Classify the different types of errors in statistics and discuss their effect on statistical inference. *[C.A., May '76]*
5. Draw up a table to show the number of wholly unemployed, temporarily stopped and the total unemployed, each class being divided into males and females for the following industries: Fishing, Coal mining, Iron-ore mining, Engineering and Ship building.

6. Draw up a blank table in which could be shown the number of persons employed in six industries, on two different dates, distinguishing males from females and among the latter, singles, married and widows. *[I.C.W.A., Jan. '73]*
7. Prepare a blank table showing the distribution of population according to sex and four religions in five age groups in seven different cities. *[C.U., B.A. (Econ) '72]*
8. Draw up a blank table showing the Exports and Imports during the years 1960, 1961, 1962, 1963 and 1964 relating to the ports Mumbai, Kolkata, Madras and other ports. The table should provide for the values and the balance of trade and the totals for each year. *[C.A., Nov. '68]*
9. You are given data on exports (both quantity and value) of Indian jute to U.K., U.S.A., U.S.S.R., Japan and Canada for 5 consecutive years. Suggest a suitable tabular representation by drawing a blank table. *[B.U., B.A. (Econ) '72]*
10. Draft a blank table to show the distribution of personnel in a manufacturing concern according to (a) Sex : male and female; (b) Three grades of salary: below Rs 300, Rs 300–Rs 500, Rs 500 and over; (c) Two periods : 1940 and 1950; (d) Three age groups : below 25, 25 and under 40, 40 and over. *[C.A., May '63]*
11. Tabulate the following information which appeared as part of a newspaper article on dishwashing machines—"An idea of the slow growth of the industry during a period of soaring sale in other appliances in competitive price ranges is given in a recent study by the trade journal *Electrical Merchandising*: In 1947, the survey showed, 120,000 motor-driven dishwashers were sold at an average price of $ 250. In 1948, sales rose to 225,000 at an average price of $ 275. In 1949, with the average price the same as the year before, the sales dropped to 160,000. In 1950, they jumped to 230,000 at an average price of $ 290 and in 1951 they continued to climb to 260,000 units even though the average price went up to $ 300. In 1952, sales dropped to 175,000 units and in 1953 they rose to only 180,000 despite the fact that average price remained at $ 300."
12. Present the following information in a tabular form and suggest a suitable title:

    'The production of 10.95 lakh tons of rice in Maharashtra in 1962–1963 was the lowest in the period since 1955–56. In 1963–64, however, it has shown a spectacular recovery and reached the level of 15.14 lakh tons. During 1963–64, wheat and bajri output decreased. The production of bajri which was 5.50 lakh tons in 1962–63 declined to 4.51 lakh tons in 1963–64. The production of wheat also decreased from 4.63 lakh tons in 1962–63 to 3.43 lakh tons in 1963–64. The area under pulses has shown a decreasing trend and the production was less by 22,000 tons in 1963–64 than the production of 8.89 lakh tons in 1962–63.' *[I.C.W.A., Jan. '70]*
13. An investigation conducted by the Education Department in a public library revealed the following facts. You are required to tabulate the information as neatly and clearly as you can:

"In 1950 the total number of the readers was 46,000 and they borrowed some 16,000 volumes. In 1960 the number of books borrowed increased by 4,000 and the borrowers by 50%. The classification was on the basis of three sections: Literature, Fiction and Illustrated News. There were 10,000 and 30,000 readers in the sections Literature and Fiction respectively in the year 1950. In the same year 2,000 and 10,000 books were lent in the sections Illustrated News and Fiction respectively. Marked changes were seen in 1960. There were 7,000 and 42,000 readers in the Literature and Fiction sections respectively. So also 4,000 and 13,000 books were lent in the sections Illustrated News and Fiction respectively." *[C.A., Nov. '66]*

14. Tabulate the following information which has been taken from *The Times Review of Industry,* 1958:

'In 1958 France exported 1,66,000 cars as against 1,07,000 in 1957 of which 64,000 went to Europe as compared with 46,000 in the same period in 1957. Some 34,000 went to oversea French territories as compared with 26,000 and 68,000 elsewhere as against 35,000. The increase to countries other than Europe is 40 per cent. The proportion of French production going to foreign markets was 27 per cent, last year. Renault's sales outside France amounted to 37 per cent of total production, while their total foreign sales of cars was 92,000.' *[C.U., B.A. (Econ) '69]*

15. The total population of Hong Kong aged 15 and over rose from 1853 thousand in 1961 to 2528 thousand in 1971. The male population rose from 945 thousand to 1280 thousand and the female population from 908 thousand to 1248 thousand. The distribution of population (above 15 years) by conjugal status was as follows in 1961: 504 thousand persons were never married, 1202 thousand were currently married, 135 thousand widowed and only 12 thousand divorced or separated. The corresponding figures (in thousand) were 914, 1481, 125 and 8 in 1971. Among males aged 15 or more in 1961, 334 thousand were never married, 588 thousand currently married, 17 thousand widowed and 6 thousand divorced or separated; the corresponding figures for 1971 were 546, 710, 20 and 4 (in thousand).
Present the data in a tabular form with appropriate headings and sub-headings. *[C.U., B.A.(Econ) '78]*

16. The following is an excerpt from the White Paper on the general budget for 1957-58, appearing in the publication "*Budget for 1957–58*" of the Ministry of Finance, Govt. of India:

"...The value of imports increased from Rs 418 crores in April-November, 1955, to Rs 535 crores in April-November, 1956. Of this increase of Rs 117 crores, Rs 96 crores was accounted for by the increase in the imports of machinery, iron and steel and other metals. Imports of machinery increased from Rs 73.5 crores in April-November, 1955 to Rs 105.8 crores in April-November, 1956; of iron and steel from Rs 34.3 crores to Rs 88 crores and of other metals from Rs 16.4 crores to Rs 26.2 crores.

The small decline in exports (from Rs 388.4 crores in April-November, 1955, to 378 crores in April-November, 1956) is mostly accounted for by the decline in exports of oil (from Rs 26.2 crores to Rs 13.4 crores) and of cotton (from Rs 24.6 crores to Rs 10.9 crores) which is largely explained by the poor crop of oil seeds and cotton in 1955–56. Exports of manganese ore also declined from Rs 10.6 crores in April-November, 1955 to Rs 6.8 crores in April-November, 1956; of cotton textiles from Rs 42.2 crores to Rs 40.1 crores and of jute manufactures from Rs 83.3 crores to Rs 79.5 crores."

Represent the information contained in the above excerpt in a suitable tabular form. *[B.U., B.A.(Econ) '68]*

17. Present the following information in a concise tabular form and indicate which type of lamp shows the greatest wastage during manufacture:

'Lamps are rejected at several manufacturing stages for different faults. 12,000 glass tubes are supplied to make 40-watt, 60-watt and 100-watt lamps in the ratio 1 : 2 : 3. At the stage I, 10 per cent of the 40-watt, 4 per cent of the 60-watt and 5 per cent of the 100-watt bulbs are broken. At the stage II, about 1 per cent of the remainder of the lamps have broken filaments. At the stage III, 100 100-watt lamps have badly soldered caps, and half as many have croooked caps; twice as many 40-watt and 60-watt lamps have these faults. At the stage IV, about 3 per cent are rejected for bad type-marking and 1 in every 100 are broken in the packing which follows.'

[I.C.W.A., July '71; C.U., B.A.(Econ) '67]

18. Mention different methods of tabulation applied to survey data. *[I.C.W.A., Dec. '73]*

ANSWERS

1. Continuous; continuous; discrete; continuous; discrete.

10.

Distribution of Personnel in a Manufacturing Concern

Age Group (Years)	Below 25		25–40		40 and over		Total		
Salary Grades (Rs)	*Male*	*Female*	*Male*	*Female*	*Male*	*Female*	*Male*	*Female*	*Total*
Year 1940: Below 300									
300–500									
500 and over									
Total									
Year 1950: Below 300									
300–500									
500 and over									
Total									

11.

Sale of Motor-driven Dishwashers (1947–1953)

Year	Number of Units sold ('000)	Average Price ($)
1947	120	250
1948	225	275
1949	160	275
1950	230	290
1951	260	300
1952	175	300
1953	180	300

Source: *Electrical Merchandising*

13.

Number of Readers in and Books Borrowed Library (Years from a Public 1950 & 1960)

Sections	Readers ('000)		Books Borrowed ('000)	
	1950	1960	1950	1960
1. Literature	10	7	4	3
2. Fiction	30	42	10	13
3. Illustrated News	6	20	2	4
Total	46	69	16	20

14.

Export of Motor Cars from France, 1957–1958

(Number '000)

Exported to	1957	1958
1. Europe	46	64
2. Oversea French Territories	26	34
3. Other Countries	35	68
Total	107	166

15.

Production and Export of Motor Cars (France, 1957)

	Company	Production ('000)	Export ('000)	Per cent Exported
(i)	Renault	250	92	37
(ii)	Other Companies	150	15	10
	Total	400	107	27

16.

Values of Exports and Imports, Govt. of India

(Rs crores)

Items	April–November 1955	April–November 1956
I. Imports:		
Machinery	73.5	105.8
Iron & Steel	34.3	88.0
Other metals	16.4	26.2
Miscellaneous	293.8	315.0
Total	418.0	535.0
II. Exports:		
Oil	26.2	13.4
Cotton	24.6	10.9
Manganese ore	10.6	6.8
Cotton textiles	42.2	40.1
Jute manufactures	83.3	79.5
Miscellaneous	201.5	227.3
Total	388.4	378.0

Source: "*Budget for 1957-58*", Ministry of Finance, Govt. of India.

17.

Wastage of Electric Lamps during Manufacture

Power of Lamp (watt)	Total No. of Glass Tubes Supplied	Number Rejected at Stages: I Broken bulbs	II Broken filament	III Bad cap	IV Bad type marking	Total	Rejections %
40	2,000	200	18	300	59	577	29
60	4,000	160	38	300	140	638	16
100	6,000	300	57	150	220	727	12
Total	12,000	600	113	750	419	1,942	16

40-watt lamps have the greatest wastage, viz. 29%.

2 CHARTS AND DIAGRAMS

2.1 OBJECTS OF DIAGRAMMATIC REPRESENTATION

Charts and diagrams are effective devices for vivid presentation of statistical data. The main object of diagrammatic representation is to emphasise the relative position of different subdivisions and not simply to record details.

Advantages—Diagrams are appealing to the eyes as well as to intellect, and are therefore helpful in assimilating the data readily and quickly. Moreover, a chart can clarify a complex problem and reveal hidden facts, which are not apparent from the tabular form. It is sometimes necessary in finding the trend in time series and also in finding relations between several sets of observations. In some cases, the graph may be used as a means of checking mistakes.

Disadvantages—Charts do not show details, which is possible in the tabular presentation. Graphical presentation can reveal only the approximate position. Also graphs and charts require much time to construct, whereas the desired information can be quickly conveyed by arranging the data in the form of a table.

2.2 TYPES OF CHARTS AND DIAGRAMS

The common types of charts and diagrams are:

1. Line diagram, or Graph
2. Bar diagram
3. Pie diagram
4. Pictogram
5. Histogram, Frequency polygon, and Ogive.

[**Note:** The words 'chart' and 'diagram' are used almost in the same sense. We also say *Line chart, Bar chart, Pie chart,* etc.]

(1) Line Diagram (or Graph)

This is the most common method of representing statistical data, especially used in business and commerce, where data are shown in accordance with the time of occurrence. The line diagram shows by means of a curve or a straight line, the relationship between two variables. Two straight lines, one horizontal and the other vertical (known as *X-axis* and *Y-axis* respectively), are drawn on a graph paper, which intersect at a point, called *origin.* The given data are represented as points on the graph paper. The position of any point is determined by its distances from the axes on the basis of the given data. The consecutive points thus obtained are joined by pieces of straight lines, giving the line diagram.

Two types of line diagrams are used—(i) *natural scale,* and (ii) *ratio scale*. Equal distances represent equal *amounts* of change in the natural scale, but equal *ratios* of change in the ratio scale (see Fig. 2.3).

Hints for Drawing Line Diagrams

1. Graph papers usually available in the market are not equal in length and breadth. The line diagram should by so planned that the horizontal axis is longer than the vertical axis. In Fig. 2.1 note that the diagram is longer horizontally.
2. The scales should be indicated below the horizontal axis and on the left of the vertical axis. It is not necessary to write such expressions as "1 small division = 5 units."
3. The variables represented by the two axes must be clearly shown. Note that in Fig. 2.1 the words 'Year and Month' appear below the horizontal axis, and the words 'Cheque Clearance' appear on the top of the vertical axis. The variable represented by the vertical axis is usually shown on the left of the axis (see Fig. 2.2).
4. The units of numbers indicated by the scales must be shown. Note the words 'Rs Crores' below 'Cheque Clearance.'
5. The horizontal scale may not start from zero; but the zero point of the vertical scale must be shown. In cases where the minimum observation shown by the vertical scale is well above the zero point, a break in the vertical axis must be shown (see Figs 2.1 and 2.8).
6. A brief but clear title must accompany the diagram.
7. If several line diagrams are shown on the same graph paper, they should preferably be drawn by distinctly different lines, e.g. thick, thin, solid, and broken lines (see Figs 2.2 and 2.8).

Example 2.1 *Draw a line diagram of the following data, relating to cheque clearance:*

Cheque Clearance (Crores of Rs)

Month / *Year*	*Jan.*	*Feb.*	*Mar.*	*Apr.*	*May*	*Jun.*	*Jul.*	*Aug.*	*Sep.*	*Oct.*	*Nov.*	*Dec.*
1958	*832*	*765*	*873*	*792*	*791*	*663*	*834*	*754*	*806*	*799*	*773*	*887*
1959	*894*	*828*	*946*	*923*	*849*	—	—	—	—	—	—	—

[C.U., M.Com. '61]

Solution

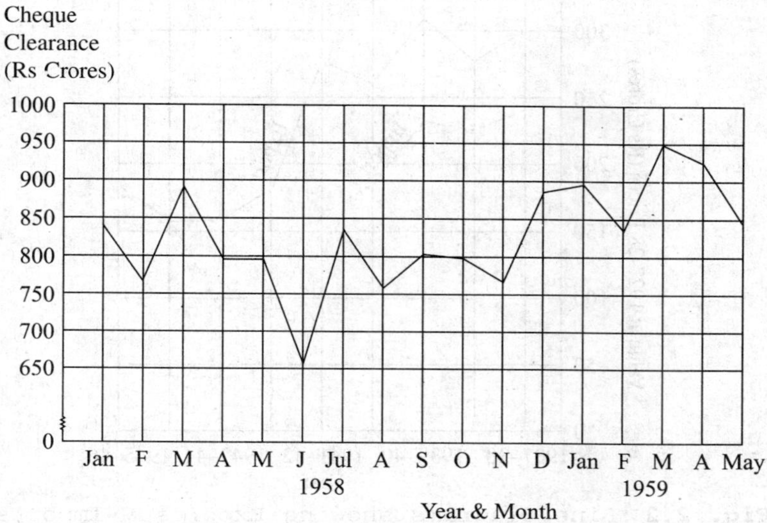

Fig. 2.1 Line Diagram showing Cheque Clearance

Example 2.2 *Draw line diagrams of the following statistical information with regard to exports and imports, and comment on their relationship:*

Year	*Value of Exports at 1927-28 Prices (Rs Crores)*	*Value of Imports at 1927–28 Prices (Rs Crores)*
1937–38	*301*	*243*
1938–39	*295*	*226*
1939–40	*309*	*230*
1940–41	*260*	*184*
1941–42	*276*	*168*
1942–43	*184*	*85*
1943–44	*158*	*89*
1944–45	*156*	*160*
1945–46	*182*	*177*

[C.U., B.A. (Econ) '64, '70]

Solution (See Fig. 2.2, next page).

Comments: (i) Values of exports generally exceed values of imports. (ii) Exports and imports closely follow the same pattern of change. (iii) While exports and imports showed a decreasing tendency since 1937–38 till 1943–44 an increasing tendency was noticed during the years that followed. This indicates the presence of cyclical fluctuations in both the series. Values of exports and imports are expected to rise further for some years after 1945–46.

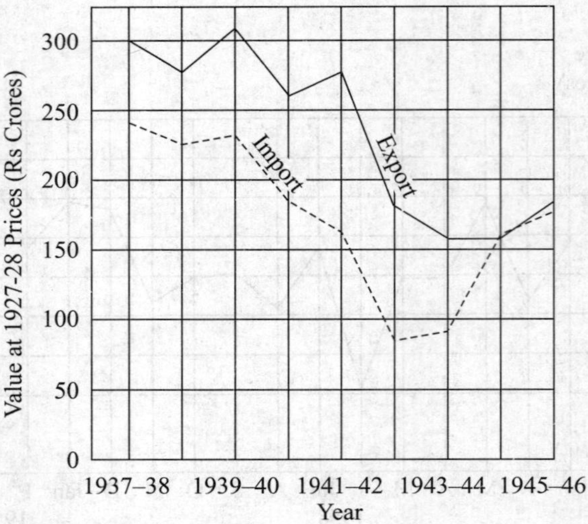

Fig. 2.2 Line Diagrams showing Exports & Imports (1937-1946)

Example 2.3 *Explain clearly the distinction between the natural scale and the logarithmic scale used in graphical presentation of data.*

[C.U., B.A.(Econ) '62; I.C.W.A., Jan. '71]

Solution In the usual type of graph papers, all rulings are shown equally apart, both horizontally and vertically. These are known as *natural scale* or *arithmetic scale* graph papers. There is a special type of graph paper (Fig. 2.5) in which the distances of rulings from the initial line are proportional to the logarithms of numbers, and hence the distances between consecutive rulings are not equal. Such a scale is known as *logarithmic scale* or *ratio scale.*

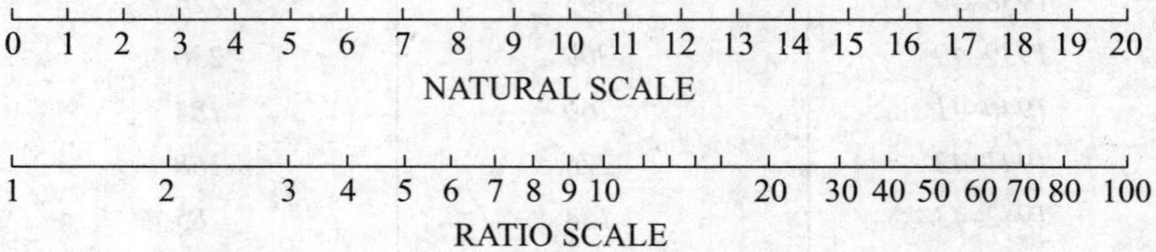

Fig. 2.3 Comparison of Natural Scale and Ratio Scale

The natural scale is used for showing absolute amount of change. In Fig. 2.1, an amount of 50 (Rs crores) increase either from 650 to 700, or from 900 to 950 is represented by the same distance in the vertical direction. However, in terms of percentage changes, these do not have the same significance. In the former, the increase is $\frac{50}{650} \times 100 = 7.7\%$, whereas in the latter case it is $\frac{50}{900} \times 100 = 5.6\%$. Such discrepancies in percentage changes can be readily detected when the data are shown on the ratio scale. Equal distances represent equal absolute changes in the natural scale, but equal percentage changes in the ratio scale. A series of observations having a common difference, i.e. in arithmetic progression, shows a straight line on the natural scale; but a straight line is obtained on the ratio scale only when the consecutive observations have a common ratio, i.e. are in geometric progression (Fig. 2.4).

In the natural scale graph, the base line of the vertical scale must show zero. But since the ratio scale shows proportionate changes, there is no zero point. The base line must show either 1, or 10, or 100, or 1000, etc.

Some graph papers have a ratio scale in the vertical direction, but natural scale in the horizontal direction. These are known as *semi-logarithmic* (or briefly *semi-log*) graph papers (Fig. 2.5). There are some others which show ratio scales in both directions and are known as *double-logarithmic* (or briefly *double-log)* graph papers.

Example 2.4 *Explain what is a semi-logarithmic graph. What purpose is served by such graphs and what are its uses?*

[I.C.W.A., Jan. '69; C.U., B.A.(Econ) '71]

Solution *Semi-logarithmic Graph* or *Ratio Chart* is a line diagram drawn on a special type of graph paper which shows the natural scale in the horizontal direction and the logarithmic or ratio scale in the vertical direction (Fig. 2.5). It shows the logarithmic scale in one direction only, and hence the name 'semi-logarithmic' ('semi' denotes 'half', e.g. semi-circle). In the semi-logarithmic graph paper, the vertical rulings are equispaced, but the horizontal rulings are not, their distances from the base line being proportional to the logarithms of the numbers represented. If a semi-logarithmic graph paper is not available, the ratio chart may be drawn on a natural scale graph paper by plotting the logarithms of values of the dependent variable y against the corresponding values of the independent variable x (Fig. 2.6).

Uses of Semi-logarithmic graph—(1) The semi-log graph is used where ratios of change are more important than absolute amounts of change. If the relationship between the variables follows the compound interest law, i.e. there is a constant ratio between the successive values of y corresponding to equidistant values of x, the equation will be of the form

$$y = ab^x$$

and the semi-log graph will give a perfect straight line (Fig. 2.4). In the analysis of time series, if therefore a given series of observations is suspected to show a constant rate of growth or decline, the simplest way to test this is to draw the semi-log graph and see whether the points lie on or nearabout a straight line.

(2) Semi-logarithmic graphs are useful in comparing fluctuations within the same series, checking established ratios, and reading percentage changes without numerical calculations. In the statistical analysis of business and economic activities, cases very often arise when we are more interested in the relative amounts of growth or decline than the absolute amounts of change. It may also become necessary to compare the rates of change in two or more series (Fig. 2.7). In such cases it is very easy to compare the rates from the slopes of semi-logarithmic charts. Two parallel lines on a semi-logarithmic graph paper will mean that the series represented by the lines are both changing at the same constant rate. A steeper slope of the semi-logarithmic chart at a point indicates a higher rate of growth than at another (Fig. 2.7).

(3) Semi-logarithmic graphs are also useful in representing a series with widely differing values (Example 2.6) or in comparing the rates of change of several series of observations in different units (Fig. 2.7).

Example 2.5 *From the following table, draw a ratio chart on a graph paper:*

Year	*1937*	*1938*	*1939*	*1940*	*1941*	*1942*	*1943*	*1944*	*1945*
Units produced	*2*	*4*	*8*	*16*	*32*	*64*	*128*	*256*	*512*

[C.U., M.Com. '56]

Solution Since the number of units produced each year is exactly double the number in the previous year, there is a common ratio between the successive observations. The ratio chart (Fig. 2.4) is therefore, a straight line.

[Note that the base line does not represent zero as is necessary in the natural scale. Also, equal distances in the vertical direction represent equal ratios of change, viz. 2 to 4, 4 to 8, etc.]

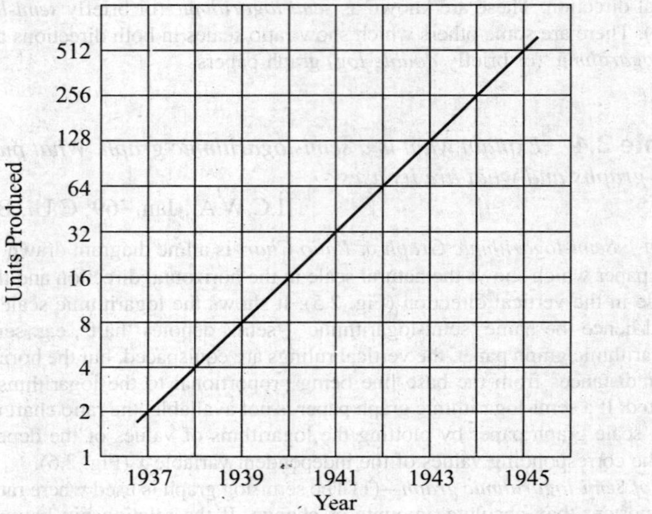

Fig. 2.4 Ratio Chart

Example 2.6 *The following table gives the total units produced at the beginning of the different years. Represent the data graphically and estimate the mid-year value for 1949 and 1953.*

Year	*1947*	*1948*	*1949*	*1950*	*1951*	*1952*	*1953*	*1954*	*1955*
Units Produced	*20*	*62*	*147*	*300*	*536*	*811*	*1104*	*1425*	*1755*

[I.C.W.A., Jan. '65]

Solution

Table 2.1 Calculation of Logarithms for Ratio Chart

Year	*Units Produced y*	*Log y (from log table)*	*log y (rounded) to 2 decimals)*
1947	20	1.3010	1.30
1948	62	1.7924	1.79
1949	147	2.1673	2.17
1950	300	2.4771	2.48
1951	536	2.7292	2.73
1952	811	2.9090	2.91
1953	1104	3.0430	3.04
1954	1425	3.1538	3.15
1955	1755	3.2442	3.24

It is found that the number of units produced differ widely—from only 20 in 1947 to 1755 in the year 1955. The ratio chart is the most appropriate graphical representation of such data.

(Drawn on Semi-Logarithmic Graph Paper)

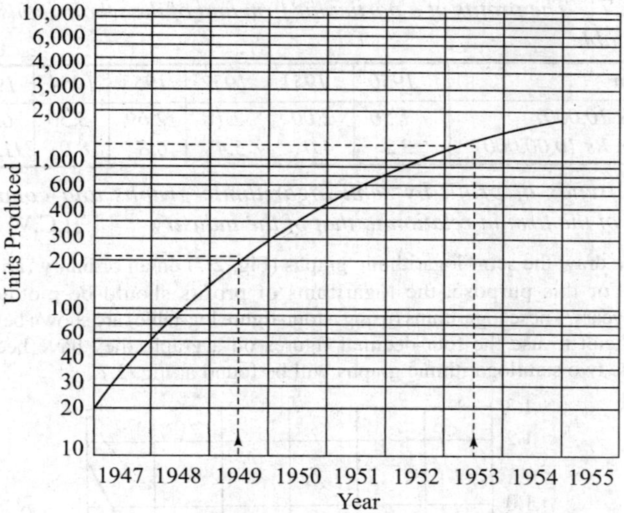

Fig. 2.5 Semi-Logarithmic Graph (or Ratio Chart)

(Drawn on Ordinary Graph Paper)

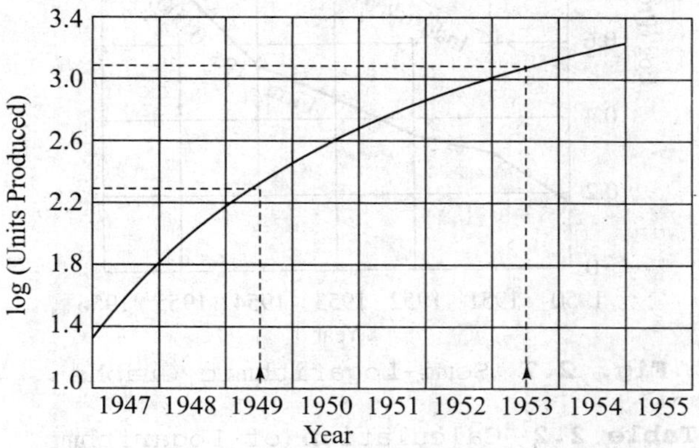

Fig. 2.6 Semi-Logarithmic Graph (or Ratio Chart)

If a semi-log graph paper is available, the ratio chart will be as shown in Fig. 2.5. However, the ratio chart may also be drawn on a natural scale graph paper by showing the logarithms of values along the vertical scale, as in Fig. 2.6. The values of log *y*, shown in the last column of Table 2.1, are plotted on an ordinary graph paper against the corresponding years, giving the ratio chart (Fig. 2.6).

In order to estimate the mid-year values for 1949 and 1953, two vertical lines are drawn starting from the mid-points of intervals representing such years on the horizontal axis, until they meet the ratio chart. From these points of intersection, two horizontal lines are now drawn to meet the vertical scales on the left. If the ratio chart is drawn on a semi-log graph paper (Fig. 2.5), the values can be read directly from the scale, and found to be approximately 210 and 1260. However, if the ratio chart is drawn on an ordinary graph paper (Fig. 2.6) the values of log y are read approximately to be 2.32 and 3.10. The antilogarithms of these values, viz. 209 and 1259, give the required estimates.

Example 2.7 *The profits of a particular firm and of the whole industry are given in following table:*

Year	1950	1951	1952	1953	1954	1955	1956
Firm (in Rs 10,000)	1.50	2.00	2.18	2.69	3.50	6.00	14.00
Industry (in Rs 10,00,000)	3.2	4.0	5.4	6.8	8.0	11.0	19.0

Compare trends of profit by semi-logarithmic graphs and comment on the performance of the firm in relation to that of the industry. [I.C.W.A., Jan. '69]

Solution We draw the semi-logarithmic graphs (Fig. 2.7) on an ordinary (arithmetic scale) graph paper. For this purpose, the logarithms of profits should be plotted against the corresponding years. These logarithms (using a four-figure log-table) are shown below. However, since it is difficult to use the four-decimal figures on a graph, they have been rounded to 2 decimals. The two semilogarithmic graphs will be found at Fig. 2.7.

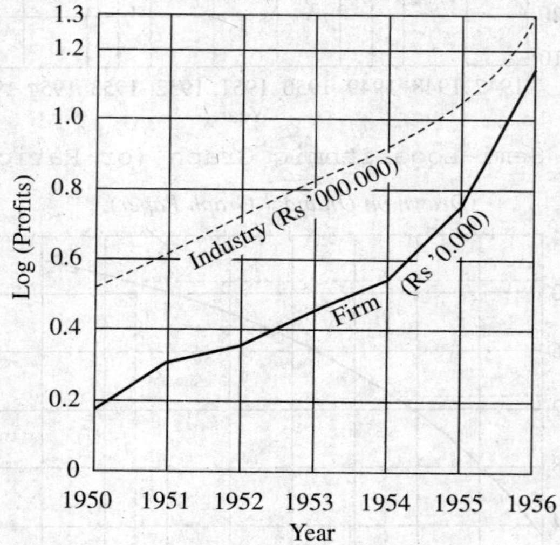

Fig. 2.7 Semi-Logarithmic Graphs

Table 2.2 Calculation of Logarithms

Year	Profits of Firm (Rs 10,000) y	Profits of Industry (Rs 10,00,00) Y	log y	log y
1950	1.50	3.2	0.1761 = 0.18	0.5051 = 0.51
1951	2.00	4.0	.3010 = .30	.6021 = .60
1952	2.18	5.4	.3385 = .34	.7324 = .73
1953	2.69	6.8	.4298 = .43	.8325 = .83
1954	3.50	8.0	.5441 = .54	.9031 = .90
1955	6.00	11.0	.7782 = .78	1.0414 = 1.04
1956	14.00	19.0	1.1461 = 1.15	1.2788 = 1.28

Comments—The slope of the semi logarithmic graph for the firm is found to be smaller than that for the industry at the beginning, indicating a smaller rate of growth in profits for the firm. However, from 1953 the graph for the firm rose more rapidly, and this shows a favourable trend of profits for the firm in relation to that for the industry as a whole.

Example 2.8 *Represent graphically from the following data the growth of the population of a State in India to show both relative growth and growth by absolute amount:*

Census year	*1871*	*1881*	*1891*	*1901*	*1911*	*1921*	*1931*	*1941*
Population (in lakhs)	*50.0*	*52.4*	*55.6*	*59.5*	*65.0*	*68.7*	*73.1*	*77.5*

[B.U., B.A. (Econ) '69]

Solution In Fig. 2.8, two line charts have been drawn, one showing the absolute growth (corresponding to the natural scale on the left) and another showing the relative growth (corresponding to the logarithmic scale on the right).

Table 2.3 Calculation of Logarithms

Year	*Population (in lakh)* *Absolute Value*	*Logarithm*	
1871	50.0	1.6990 = 1.70	(approx.)
1881	52.4	1.7193 = 1.72	"
1891	55.6	1.7451 = 1.75	"
1901	59.5	1.7745 = 1.77	"
1911	65.0	1.8129 = 1.81	"
1921	68.7	1.8370 = 1.84	"
1931	73.1	1.8639 = 1.86	"
1941	77.5	1.8893 = 1.89	"

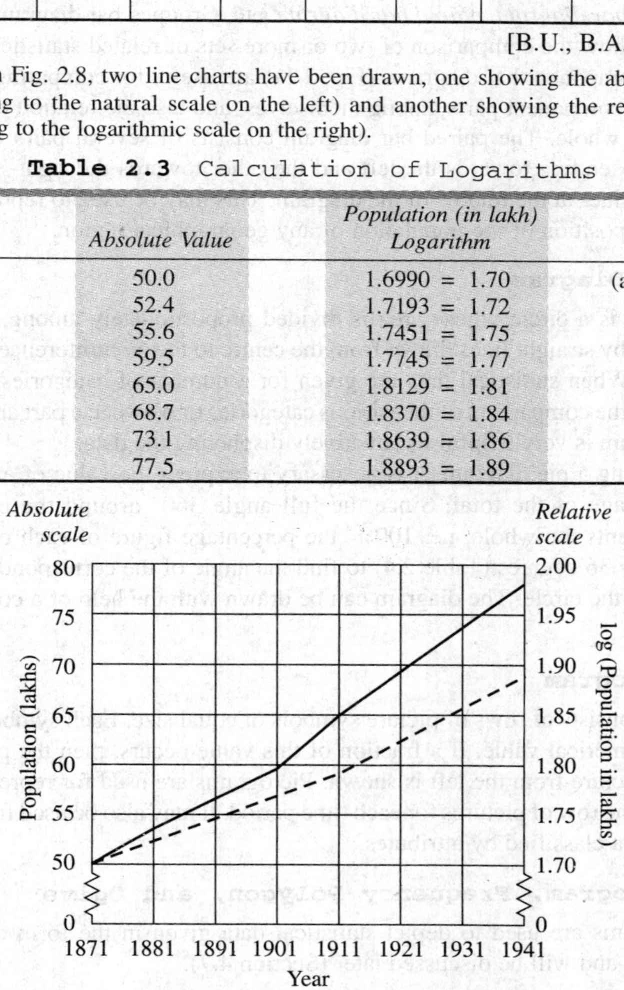

Solid Line—Absolute growth; *Broken Line*—Relative growth

Fig. 2.8 Graphs showing Population Growth (Absolute & Relative) of an Indian State

(2) Bar Diagram

Bar diagram consists of a group of equispaced rectangular bars, one for each category (or class) of given statistical data. The bars, starting from a common base line, must be of equal width and their lengths represent the values of statistical data. They are shaded or coloured suitably.

There are two types of bar diagrams–*vertical bar diagram* and *horizontal bar diagram.* Vertical bars (Fig. 2.10) are used to represent time series data or data classified by the values of a variable. Horizonal bars (Fig. 2.14) are used to depict data classified by attributes only.

For each of these types, we have again *grouped bar diagram, sub-divided* (or *component*) *bar diagram, paired bar diagram,* etc. Grouped bar diagrams (Fig. 2.12) are used to show the comparison of two or more sets of related statistical data. Sub-divided or component bar diagrams (Fig. 2.11) are useful for comparing the sizes of the different component parts among themselves, and also the relation between each part and the whole. The paired bar diagram consists of several pairs of horizontal bars—one extending towards the left and the other towards the right–starting from two vertical lines at the middle of the diagram. This may be used to represent the age and sex composition of the population of any geographical region.

(3) Pie Diagram

Pie diagram is a circle whose area is divided proportionately among the different components by straight lines drawn from the centre to the circumference of the circle (Fig. 2.15). When statistical data are given for a number of categories, and we are interested in the comparison of the various categories or between a part and the whole, such a diagram is very helpful in effectively displaying the data.

For drawing a pie diagram, it is necessary to express the value of each category as a percentage of the total. Since the full angle 360° around the centre of the circle represents the whole, i.e. 100%, the percentage figure of each component is multiplied by 36 degrees (Table 2.4) to find the angle of the corresponding sector at the centre of the circle. The diagram can be drawn with the help of a compass and a protractor.

(4) Pictogram

Pictogram consists of rows of picture symbols of equal size. Each symbol represents a definite numerical value. If a fraction of this value occurs, then the proportionate part of the picture from the left is shown. Pictograms are used for representing time series data, one row of pictures for each time period. It may also be used for displaying statistical data classified by attributes.

(5) Histogram, Frequency Polygon, and Ogive

These diagrams are used to depict statistical data given in the form of frequency distributions, and will be discussed later (Section 4.7).

Example 2.9 *The following table shows the numbers of agricultural and non-agricultural workers for the year 1860–1950. Graph the data using (a) line graphs, (b) bar charts, and (c) component bar chart. Add a small note on the diagrams.*

Year	1860	1870	1880	1890	1900	1910	1920	1930	1940	1950
Agricultural Workers (in millions)	6.2	6.9	8.6	9.9	10.9	11.6	11.4	10.5	8.8	6.8
Non-Agricultural Workers (in millions)	4.3	6.1	8.8	13.4	18.2	25.4	31.0	38.4	42.9	52.2

[C.U., M.Com. '67]

Solution In the component bar chart, the length of each bar must represent the total number of workers (agricultural and non-agricultural combined), in which separate lengths have to be shown for each component, viz. agricultural and non-agricultural. These totals are calculated below:

Year	1860	1870	1880	1890	1900	1910	1920	1930	1940	1950
Agricultural	6.2	6.9	8.6	9.9	10.9	11.6	11.4	10.5	8.8	6.8
Non-agricultural	4.3	6.1	8.8	13.4	18.2	25.4	31.0	38.4	42.9	52.2
Total	10.5	13.0	17.4	23.3	29.1	37.0	42.4	48.9	51.7	59.0

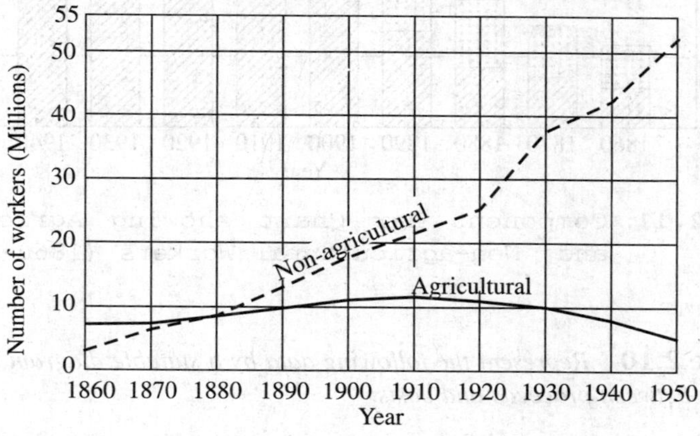

Fig. 2.9 Line Graphs showing Agricultural and Non-agricultural Workers (1860-1950)

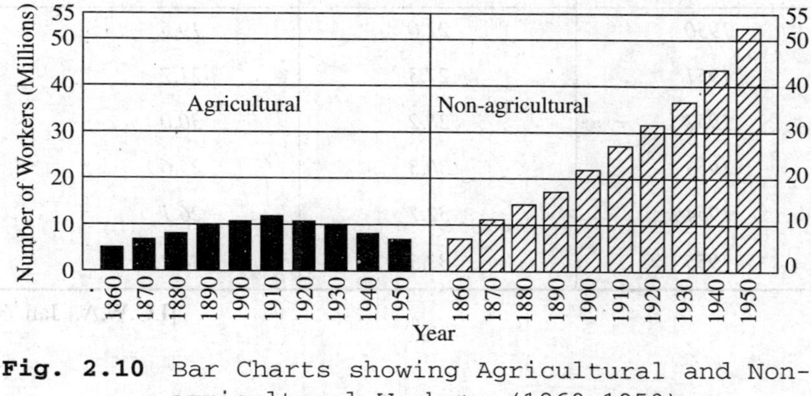

Fig. 2.10 Bar Charts showing Agricultural and Non-agricultural Workers (1860-1950)

It is observed from these diagrams that the number of agricultural workers systematically increased since the year 1860 to about double its figure in the year 1910, and then again it came down gradually. The number of non-agricultural workers, however, increased consistently, nearly 12 times in 90 years since 1860. The trend appears to be approximately a straight line since 1880, indicating a constant annual increase in the number of non-agricultural workers. A comparison of the diagrams for Agricultural and Non-agricultural shows that people are gradually leaving agriculture and engaging themselves more and more to other means of livelihood.

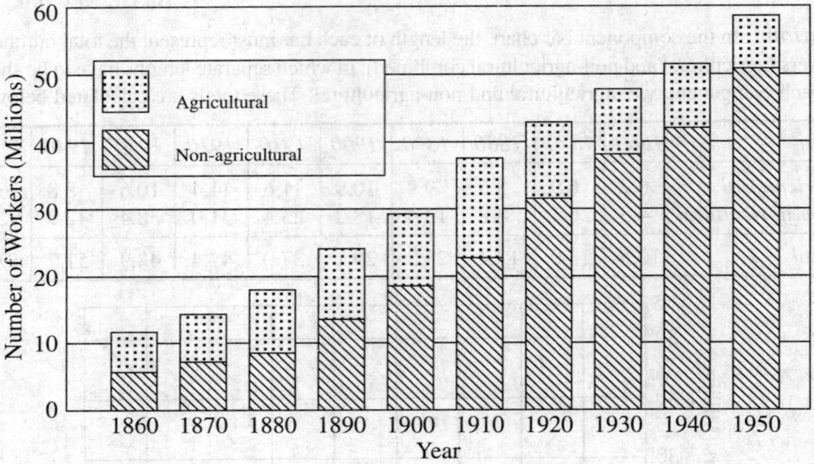

Fig. 2.11 Component Bar Chart showing Agricultural and Non-agricultural Workers (1860-1950)

Example 2.10 *Represent the following data by a suitable diagram showing the difference between proceeds and costs:*

Proceeds and Costs of a firm (in thousands of rupees)

Year	*Total Proceeds*	*Total Costs*
1950	*22.0*	*19.5*
1951	*27.3*	*21.7*
1952	*28.2*	*30.0*
1953	*30.3*	*25.6*
1954	*32.7*	*26.1*
1955	*33.3*	*34.2*

[I.C.W.A., Jan. '66]

Solution We draw a grouped bar diagram for the data.

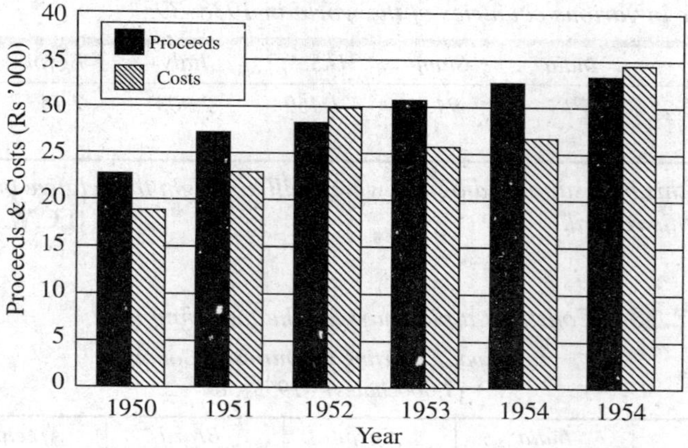

Fig. 2.12 Grouped Bar Diagram

Example 2.11 *The following figures show the actual outlay for public sector in India's first two Five-year Plans under different heads of development expenditure:*

Head of Development		*Percentage of Plan I*	*Total Outlay Plan II*
Agriculture & Community Development	...	*14.8*	*11.7*
Irrigation & Power	...	*29.7*	*18.9*
Industry & Mining	...	*5.0*	*24.1*
Transport & Communications	...	*26.4*	*27.0*
Social Services	...	*21.0*	*16.5*
Miscellaneous	...	*3.1*	*1.8*
Total		*100.0*	*100.0*

Represent the data by means of a sub-divided bar chart.

Solution

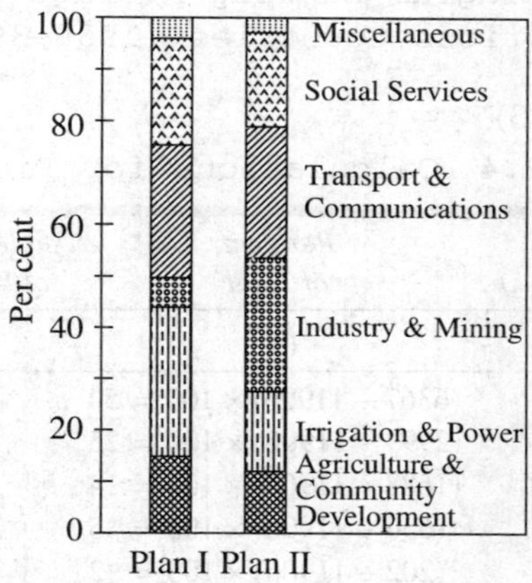

Fig. 2.13 Sub-Divided Bar Chart showing Actual Outlay in India's Five Year Plans I & II

Example 2.12 *The following table gives the average approximate yield of rice in lbs. per acre in various countries of the world in 1938–39:*

Country	*India*	*Siam*	*U.S.A.*	*Italy*	*Egypt*	*Japan*
Yield in lbs. per acre	*728*	*943*	*1,469*	*2,903*	*2,153*	*2,276*

Indicate this by a suitable diagram which will highlight the relative backwardness of India in this regard. [I.C.W.A., Jan. '64]

Example 2.13 *Construct a pie chart for the following data:*

Principal Exporting Countries of Cotton
(1,000 bales)—1955–56

U.S.A.	*India*	*Egypt*	*Brazil*	*Argentina*
6,367	*2,999*	*1,688*	*650*	*202*

[C.U., M.Com. '74]

Solution (Example 2.12)

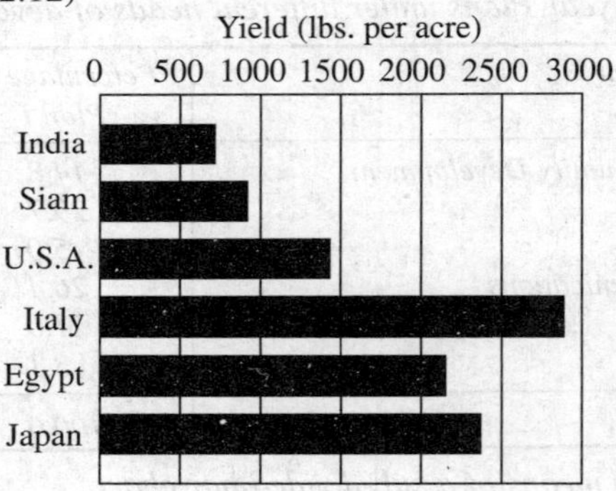

Fig. 2.14 Bar Diagram showing Average Yield of Rice in Various Countries (1938-39)

Solution (Example 2.13)

Table 2.4 Calculations for Pie Chart

Country	*Exports ('000 bales)*	*Per cent of Total*	*Angle (degrees) at the Centre of Pie Chart col. (3) × 3.6*
(1)	(2)	(3)	(4)
U.S.A.	6,367	(6367 ÷ 11906) × 100 = 54	194.4
India	2,999	(2999 ÷ 11906) × 100 = 25	90.0
Egypt	1;688	(1688 ÷ 11906) × 100 = 14	50.4
Brazil	650	(650 ÷ 11906) × 100 = 5	18.0
Argentina	202	(202 ÷ 11906) × 100 = 2	7.2
Total	11,906	100	360.0

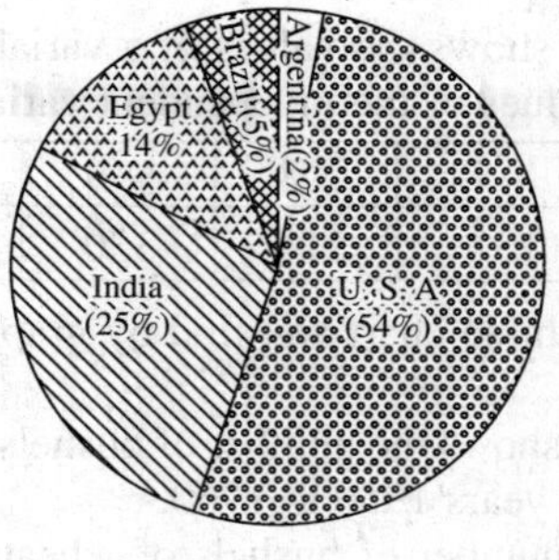

Fig. 2.15 Pie Chart showing Principal Exporting Countries of Cotton (1955-56)

EXERCISES

1. Briefly describe one diagrammatic method for representing (i) a time series, (ii) distribution of a total among its components, (iii) a frequency distribution of individuals classified according to an attribute. *[I.C.W.A., Dec. '73]*
2. Indicate the diagrams you would consider most appropriate to represent the following statistical data, stating briefly the reasons for your choice: (i) Marks scored by two selected candidates in each of six subjects tested at an examination; (ii) Total value of Indian exports and imports during the years 1956–1972; (iii) Middleclass cost of living index numbers in Bombay and Calcutta during the years 1961–1972; (iv) Distribution of civil conditions of persons enumerated in 1971-census. *[B.U., B.A. (Econ) '73]*
3. Represent the following statistical information graphically:

Year	1924	1925	1926	1927	1928	1929	1930
Monthly Average Production	609	522	205	608	551	632	516

[C.U., B.Com. (Hons) '65]

4. Plot the following data relating to population of India so as to indicate the proportionate increase in population from one period to another:

Year	1872	1881	1891	1901	1911	1921	1931	1941
Population (in millions)	210	250	290	295	315	320	350	390

[C.U., B.A. (Econ) '62]

5. Represent the information contained in the following table in a component part chart:

Commodity pattern of India's Exports (Percentage)

	1956-'57	1957-'58	1958-'59
Capital Goods	0.29	0.31	0.30
Intermediate Goods	45.82	46.87	44.19
Consumer Goods	50.50	47.32	48.19
Unclassified	3.39	5.50	7.32
Total	100.00	100.00	100.00

[C.U., B.Com. (Hons.) '67]

6. The following table shows the values of a variable y corresponding to some given equidistant values of the independent variable x:

x ...	7	8	9	10	11	12
y ...	132	214	330	486	688	942

Draw a semi-logarithmic chart and find by graphical interpolation the value of y, when $x = 10.5$ *[I.C.W.A., Jan. '71]*

7. The following table shows the number of bushels of wheat and corn produced in a farm during the years 1950 to 1960.
Express the yearly number of bushels of wheat and corn as percentages of total annual production. Graph the percentages by component bar charts.

Year	1950	1951	1952	1953	1954	1955	1956	1957	1958	1959	1960
No. of Bushels of Wheat	200	185	225	250	240	195	210	225	250	230	235
No. of Bushels of Corn	75	90	100	85	80	100	110	105	95	110	100

[Dip. Soc. Welfare, May '68]

8. Of the life insurance policy dividends paid in the United States 21% were taken in cash, 31% were used to pay premiums, 18% were used to purchase additional paid-up life insurance, 30% were left with life insurance companies to earn interest. Construct a pie diagram showing these different uses of policy dividends. *[C.U., M.Com. '62]*

9. A summary of the estimated receipts and expenditures of Government of India for a particular year, is given below:

Receipts	*Millions of Rupees*	*Expenditure*	*Millions of Rupees*
Direct taxes on income	2076.0	Interest on public debt	1143.3
Customs duties	1680.0	Education and health	272.6
Other taxes	2955.6	National defence	2774.5
Revenue from public undertakings	1660.7	Transfers to States	3755.3
Other receipts	1189.1	Other current expenditures	1754.4
		Capital expenditures	5702.2
		Other loans and advances	759.8
Total	9561.4	Total	16162.1

Draw suitable diagrams to show the relative importance of different heads of income and expenditure. *[C.U., B.A.(Econ) '73]*

ANSWERS

1. (i) Line diagram, or Vertical bar diagram; (ii) Component bar diagram, or Pie diagram; (iii) Horizontal bar diagram.
2. (i) Grouped horizontal bar diagram; (ii) Grouped vertical bar diagram; (iii) Two line diagrams drawn on the same graph paper; (iv) Pie diagram.
3. Line diagram.
4. Semi-logarithmic chart.
5. Component bar diagram.
6. Plot *log y* against *x*, and join the points by free-hand curve.
7. Component bar diagram using percentages.
8. Pie Diagram.
9. Two component bar diagrams, one for receipts and another for expenditure.

3 USEFUL MATHEMATICAL DEVICES

3.1 ROUNDING OF NUMBERS

It is known that the number 54.38 rounded to the nearest unit is 54, because the digit 3 next to the unit's place is less than 5; when rounded to one decimal, it is 54.4, because the digit 8 next to the first decimal place is more than 5. Difficulties arise when the next digit to be dropped is exactly 5. The rules are:

(a) When this 5 is followed by digits all of which are not zeros, then the preceding digit is increased by one. 72.5002 rounded to the nearest unit becomes 73 and 12.7537 rounded to one decimal is 12.8.

(b) When this 5 is the last digit or is followed by zeros only, it has become conventional to round the preceding digit to an *even* number. 73.5 rounded to the nearest unit is 74; 12.85 rounded to one decimal is 12.8; 41500 rounded to the nearest thousand is 42 thousand.

Example 3.1

Quantity	*Rounded to*	*Rounded Number*
(a) 3.245002	*nearest hundredth*	*3.25*
(b) 3.25501	" "	*3.26*
(c) 3.245	" "	*3.24*
(d) 3.255	" "	*3.26*
(e) 274.95	*nearest tenth*	*275.0*
(f) 5.99735	*nearest hundredth*	*6.00*
(g) 289,523	*nearest thousand*	*290 thousand*
(h) 8,500,000	*nearest million*	*8 million*
(i) 33.5	*nearest unit*	*34*

3.2 ABSOLUTE, RELATIVE AND PERCENTAGE ERRORS

Most of us are accustomed to think of 5.2 as meaning the same thing as 5.20 and 5.200000...to an unlimited number of decimal places. But 5.2 cm actually means a measurement between 5.15 and 5.25 cm. Similarly, 5.20 would mean a number between 5.195 and 5.205, i.e.,

5.2 means a number between 5.2 ± .05

5.20 means a number between 5.20 ± .005

Unless otherwise stated, *an observation of a continuous variate should be interpreted as extending half a unit of the last place below to half a unit above the recorded value.*

Absolute error of a measurement is its difference from the true value. *Relative error* is the ratio of absolute error to the true value. It is usually expressed as a percentage, and then called *percentage error*.

Absolute Error = Difference between measurement and true value.

$$\text{Relative Error} = \frac{\text{Absolute Error}}{\text{True value}}$$

Percentage Error = Relative Error × 100

The maximum *Absolute error* in the sum or difference of two measurements is equal to the sum of the separate absolute errors. The maximum *relative error* in the product or quotient of two measurements is a approximately equal to the sum of the separate relative errors. Thus, if the measurements x_1 and x_2 have absolute errors e_1 and e_2 respectively, then the maximum absolute error either in $(x_1 + x_2)$ or in $(x_1 - x_2)$ is $(e_1 + e_2)$. Again, the maximum relative error either in x_1x_2 or in x_1/x_2 or in x_2/x_1 is $\left(\frac{e_1}{x_1} + \frac{e_2}{x_2}\right)$.

Example 3.2 *The true value of π upto 5 decimals is 3.14159. Find the absolute error, relative error, and percentage error, if π is approximated by 22/7.*

Solution 22/7 = 3.14286 (upto 5 decimals)

Hence, Absolute error = 3.14286 – 3.14159 = .00127

$$\text{Relative error} = \left(\frac{.00127}{3.14159}\right) = .0004$$

Percentage error = .0004 × 100 = .04

Example 3.3 *The sides of a rectangular field are measured to be 127 ± 0.2 ft. and 237 ± 0.4 ft. Find the possible error in the sum of the two sides and also the limits of the true value of the sum of the two sides.* [I.C.W.A., Jan. '65]

Solution The absolute errors are $e_1 = 0.2$ and $e_2 = 0.4$. Hence the possible error, i.e. the maximum absolute error, in the sum is given by

$$e_1 + e_2 = 0.2 + 0.4 = 0.6 \text{ ft.}$$

The limits of the true value of the sum are, therefore,

$$(127 + 237) - 0.6 \quad \text{and} \quad (127 + 237) + 0.6$$

i.e. 363.4 ft. and 364.6 ft. *Ans.* 0.6 ft; 363.4 to 364.6 ft.

3.3 SIGNIFICANT FIGURES

The significant figures of a quantity are those starting from the first non-zero digit on the left to the end of the last digit accurately specified on the right.

Example 3.4

Quantity	Number of Significant Figures
(a) 58	two
(b) 604	three
(c) 1008	four
(d) 95080	five
(e) 7.0	two
(f) 0.80	two
(g) 18.00	four
(h) 0.003	one
(i) 0.0030	two
(j) 3.20	three

3.4 SOME SHORT PROCESSES OF CALCULATION

(1) *Multiplication by 5, 25, or 125*—Place 0, 00 or 000 after the multiplicand and divide by 2, 4, or 8.

Example 3.5 (a) $849 \times 5 = 8490 \div 2 = 4245$
(b) $436 \times 25 = 43600 \div 4 = 10900$
(c) $6874 \times 125 = 6874000 \div 8 = 859250$

Multiplication by 25 becomes easier in the following way: For 17×25 let us think of (17×25 paise), i.e. 17 twentyfive-paise coins which make Rs 4.25 P. = 425 P. Hence, $17 \times 25 = 425$. Similarly, 434×25 reminds us of 434 twentyfive-paise coins = Rs 108.50 P. leading to the result $434 \times 25 = 10850$.

(2) *Division by 5, 25, or 125*—Multiply by 2, 4, or 8 and shift the decimal point one, two, or three places to the left.

Example 3.6 (a) $849 \div 5$ = (849×2) and shift the decimal point one place = 169.8
(b) $436 \div 25$ = (436×4) and shift the decimal point two places = 17.44
(c) $6845 \div 125$ = (6845×8) and shift the decimal point three places = 54.760.

(3) *Multiplication (division) by 5, .5, .05, .005 etc.*—Divide (multiply) the given number by 2 and adjust the decimal point.

Example 3.7 *(a) To find 8.49×5, divide 849 by 2 mentally, getting 4245 (forget the decimal point completely for the present). Our answer will be one amongst 4245, 424.5, 42.45, 4.245, .4245, etc., i.e. the non-zero digits from the left must be 4245; and it is now a matter of fixing the decimal point suitably. Can we not guess the result? Yes, we can; the result is nearly $8 \times 5 = 40$. The correct answer must be 42.45.*
(b) $.068 \times .05 \rightarrow 68 \div 2 = 34 \rightarrow .0034$ (because $.07 \times .05 = .0035$)
(c) $6.7 \times .005 \rightarrow 67 \div 2 \rightarrow 335 \rightarrow .0335$ (because $7 \times .005 = .035$)
(d) $84.6 \times 50 \rightarrow 846 \div 2 = 423 \rightarrow 4230$ (since $80 \times 50 = 4000$)

(e) $8.49 \div 5 \rightarrow 849 \times 2 = 1698 \rightarrow 1.698$ (since $8 \div 5 = 1$ ...)

(f) $84.6 \div 50 = \dfrac{84.6 \times 2}{50 \times 2} = \dfrac{169.2}{100} = 1.692$

(g) $.846 \div .05 = \dfrac{.846 \times 2}{.05 \times 2} = \dfrac{1.692}{.1} = \dfrac{16.92}{1} = 16.92$

(4) *Multiplication (division) by 25, 2.5, .25, .025 etc.*—Divide (multiply) by 4 and adjust the decimal point.

Example 3.8 (a) $48 \times 2.5 \rightarrow 48 \div 4 = 12 \rightarrow 120$ (since $48 \times 2 = 96$)

(b) $14.2 \times .25 \rightarrow 142 \div 4 \rightarrow 355 \rightarrow 3.55$ (since $14 \times .2 = 2.8$)

(c) $483.2 \times .025 \rightarrow 4832 \div 4 = 1208 \rightarrow 12.08$ (since $500 \times .025 = 12.5$)

(d) $48 \div 2.5 = \dfrac{48 \times 4}{2.5 \times 4} = \dfrac{192}{10} = 19.2$

(e) $483.2 \div 250 = \dfrac{483.2 \times 4}{250 \times 4} = \dfrac{1932.8}{1000} = 1.9328$

(f) $1.42 \div .025 = \dfrac{1.42 \times 4}{.025 \times 4} = \dfrac{5.68}{.1} = \dfrac{56.8}{1} = 56.8$

(5) *Division by a number ending in 5*—If the divisor is a multiple of 5, multiply both the numerator and denominator by 2; in case the denominator is a multiple of 25, use 4 instead of 2.

Example 3.9 (a) $\dfrac{44}{3.5} = \dfrac{44 \times 2}{3.5 \times 2} = \dfrac{88}{7} = 12.57$

(b) $\dfrac{3.8}{75} = \dfrac{3.8 \times 4}{.75 \times 4} = \dfrac{15.2}{3} = 5.07$

(c) $\dfrac{382}{150} = \dfrac{382 \times 2}{150 \times 2} = \dfrac{764}{300} = 2.55$

(d) $\dfrac{246}{27.5} = \dfrac{246 \times 4}{27.5 \times 4} = \dfrac{984}{110} = \dfrac{98.4}{11} = 8.95$

(6) *Multiplication by 9 (or 11)*—Place a zero to the right of the multiplicand and subtract (add) the multiplicand.

Example 3.10 (a) $234 \times 9 \rightarrow$

$$\begin{array}{r} 2340 \\ -234 \\ \hline 2106 \end{array}$$

(b) $234 \times 11 \rightarrow$

$$\begin{array}{r} 2340 \\ +234 \\ \hline 2574 \\ \hline \end{array}$$

For multiplication by 99, place two zeros to the right and then subtract the multiplicand.

7. *Subtracting decimal fraction from whole number*—The integral part of the result will be one less than the difference between the whole number and the integral part of

the number to be subtracted. For the decimal part of the result, subtract from 9 each of the digits following the decimal point, except the last digit, which is subtracted from 10.

Example 3.11 *8 – 5.498063 = 2.501937*

Because, 8 – 5 – 1 = 2; 9 – 4 = 5; 9 – 9 = 0; 9 – 8 = 1; 9 – 0 = 9; 9 – 6 = 3; and finally 10 – 3 = 7.

(8) *Fixing the position of decimal point in division*—It is a common practice that in divisions involving decimal fractions, the decimals of the numerator and the denominator are shifted to the right by the same number of places, so that both of them become whole numbers. If now the numerator is less than the denominator, many students find it difficult to decide how many zeros will follow the decimal point in the quotient before the first significant figure appears. The following simple rule may be found useful:

In the denominator take as many digits from the left, such that the number so formed just exceeds the numerator. The number of digits in the denominator in excess of this will give the exact number of zeros after the decimal point preceding the first significant figure of the quotient.

Example 3.12 *(a) Find the number of zeros after the decimal point in the quotient* $\frac{243}{51807}$.

(b) $23 \div 3798 = .00$

(c) $19 \div 1265 = .0$

(d) $173 \div 5297 = .0$

(e) $3.84 \div 423 = 384 \div 42300 = .00$

(f) $0.663 \div 0.83 = 663 \div 830 = .$

(g) $.07 \div 664 = 7 \div 66400 = .000$

(h) $.0002 \div .039 = 2 \div 390 = .00$

(i) $0.24 \div 72.113 = 240 \div 72113 = .00$

Solution (a) The first digit from the left in the denominator 51807, viz. 5, is less than the numerator 243. Next consider the number 51 formed by the first two digits. This is also less than the numerator 243. Again consider the first 3 digits of denominator, giving 518. This is greater than 243 and hence here we stop. Since there are 2 more digits 0 and 7 remaining in the denominator, there will be exactly 2 zeros in the result after the decimal point.

$$\frac{243}{51807} = .0046 \ldots$$

For obtaining the significant figures in the result, ignore all decimals both in the numerator and in the denominator; put sufficient number of zeros to the right of the numerator and then divide.

```
51907) 243000 ( 46....
       207228
       ------
        357720
        310842
        ------
         46878
```

3.5 ROOTS AND RECIPROCALS EXPRESSED AS POWER

Example 3.13

(a) $\sqrt{x} = x^{\frac{1}{2}}$ (g) $\frac{1}{x} = x^{-1}$

(b) $\sqrt[3]{x} = x^{\frac{1}{3}}$ (h) $\frac{1}{x^2} = x^{-2}$

(c) $\sqrt[n]{x} = x^{\frac{1}{n}}$ (i) $\frac{1}{x^n} = x^{-n}$

(d) $x\sqrt{x} = x.\, x^{\frac{1}{2}} = x^{\frac{3}{2}}$ (j) $\frac{1}{\sqrt{x}} = x^{-\frac{1}{2}}$

(e) $\sqrt[3]{x^2} = (x^2)^{\frac{1}{3}} = x^{\frac{2}{3}}$ (k) $\frac{1}{x\sqrt{x}} = x^{-\frac{3}{2}}$

(f) $\sqrt[n]{x^m} = (x^m)^{\frac{1}{n}} = x^{\frac{m}{n}}$ (I) $\frac{1}{\sqrt[3]{x^2}} = x^{-\frac{2}{3}}$

3.6 LOGARITHM

Every positive number can be expressed as a power of 10. This power is called *logarithm to the base 10* or simply *logarithm* of the number, briefly written as log. For example, since $100 = 10^2$, log 100 = 2; $.001 = 10^{-3}$, therefore log .001 = – 3. The number 69 when expressed as a power of 10 will be

$$69 = 10^{1.8388}$$

This means that

$$\log 69 = 1.8388$$

The result 1.8388 is not exact, but correct only upto 4 decimal places. Logarithms of numbers correct to certain decimals are obtained from log-tables. For example, a four-figure log table is used to obtain logarithms correct to 4 decimals only.

(i) How to find the logarithm?

The logarithm of any number consists of an integral part, called *characteristic*, and a decimal part, called *mantissa*.

The characteristic of any number greater than 1 is positive, and is equal to the number of digits before the decimal point minus one. The characteristic of a number less than 1 is negative, and is equal to the number of consecutive zeros immediately following the decimal point plus one. A negative characteristic, say – 3, is written as 3 (read *bar* 3).

Example 3.14

Number	*Characteristic*
(a) 176	2
(b) 69.58	1
(c) 3.47064	0
(d) 0.754	$\bar{1}$ (i.e. – 1)
(e) .075403	$\bar{2}$
(f) .0000327	$\bar{5}$
(g) 1000	3

The mantissa of any number is obtained from log-tables. In a four-figure log-table, series of four-figure numbers are shown in rows against 10, 11, 12, …99 at the extreme left of the table. These four-figure numbers are also arranged in columns with headings 0, 1, 2, …9. At the extreme right of the table, the mean-differences are shown with column headings 1, 2, …9. Such a log-table can be used to find the logarithm of a number not exceeding 4 significant figures.

The mantissa is independent of the position of decimal point in the number whose logarithm is required. Thus, the mantissa of 695.8 is the same as that of 6958 or of 0.6958 or even of .0006958.

In order to obtain the mantissa of 6958, we look up the number 69 (by the first two digits) at the extreme left column of the table. Then moving horizontally towards the right we come under the column-heading 5 (the third digit) to the figure 8420, which is the mantissa of 695. Moving further to the right under the mean-difference column with heading 8 (the fourth digit), we get 5, which is added to 8420, giving 8425. Thus the mantissa of 6958 is .8425.

Example 3.15

Number		Mantissa
(a)	*176*	*.2455*
(b)	*3470*	*.5403*
(c)	*1254*	*.0983*
(d)	*7506*	*.8754*
(e)	*32*	*.5051*
(f)	*3*	*.4771*
(g)	*300,000*	*.4771*

An important point to note is that the characteristic may be positive of negative, but the *mantissa is always positive*.

Example 3.16

Number	Logarithm	
(a) 6958	*3.8425*	
(b) 695.8	*2.8425*	
(c) 69.58	*1.8425*	
(d) 6.958	*0.8425*	
(e) .6958	$\bar{1}$*.8425*	*(i.e. – 1 + .8425 = – .1575)*
(f) .06958	$\bar{2}$*.8425*	*(i.e. – 2 + .8425 = – 1.1575)*
(g) .006958	$\bar{3}$*.8425*	
(h) 176	*2.2455*	
(i) 0.75	$\bar{1}$*.8751*	

(ii) Rules for logarithms

(1) The logarithm of the product of two or more numbers is the sum of their logarithms.

$$\log (a \times b) = (\log a) + (\log b) \tag{3.6.1}$$

Example 3.17 *(a)* $\log (47 \times 8) = (\log 47) + (\log 8)$
$= 1.6721 + 0.9031$
$= 2.5752.$

This may be verified directly. $\log .376 = 2.5752$

(b) $\log (439.4 \times .0347) = (\log 439.4) + (\log .0347)$
$= 2.6429 + \bar{2}.5403 = 2.6429 - 2 + .5403 = 1.1832$

(c) $\log (x_1, x_2, x_3) = \log x_1 + \log x_2 + \log x_3$

(2) The logarithm of the ratio of two numbers is equal to the logarithm of numerator minus the logarithm of denominator.

$$\log \left(\frac{a}{b}\right) = (\log a) - (\log b) \tag{3.6.2}$$

Example 3.18 *(a)* $\log \left(\frac{758.3}{44.78}\right) = (\log 758.3) - (\log 44.78)$
$= 2.8799 = 1.6511 = 1.2288$

(b) $\log \left(\frac{.035}{44.78}\right) = \log .035 - \log 44.78$
$= \bar{2}.5441 - 1.6511 = -2 + .5441 - 1.6511$
$= -3.1070$

(3) The logarithm of any number raised to a certain power is equal to the product of the power and the logarithm of the number.

$$\log (a^b) = b(\log a) \tag{3.6.3}$$

Example 3.19 *(a)* $\log (4.8)^3 = 3 \times (\log 4.8)$
$= 3 \times 0.6812 = 2.0436$

(b) $\log (4.8)^{\frac{1}{3}} = \frac{1}{3} \times (\log 4.8) = \frac{1}{3} \times 0.6812 = 0.2271$

(c) $\log \left(1 + \frac{r}{100}\right)^n = n \log \left(1 + \frac{r}{100}\right)$

(d) $\log P \left(1 + \frac{r}{100}\right)^n = \log P + \log \left(1 + \frac{r}{100}\right)^n$
$= \log P + n \log \left(1 + \frac{r}{100}\right)$

(e) $\log \left(x_1^{f_1} \cdot x_2^{f_2} \ldots x_n^{f_n}\right)^{\frac{1}{N}} = \frac{1}{N} \log \left(x_1^{f_1} \cdot x_2^{f_2} \ldots x_n^{f_n}\right)$
$= \frac{1}{N}\left[\log x_1^{f_1} + \log x_2^{f_2} + \ldots + \log x_n^{f_n}\right]$

$$= \frac{1}{N}\left[f_1 (\log x_1) + f_2 (\log x_2) + \ldots + f_n (\log x_n) \right]$$

(iii) How to find the antilogarithm?

The antilogarithm of a number is that quantity whose logarithm is the given number. Since log 695.8 = 2.8425, therefore

antilog 2.8425 = 695.8.

In order to find antilog 2.8425 by using the tables, we first consisder only the decimal part, viz, .8425. The process of reading the antilog-table is exactly similar to that of the log-table. The first two digits (including zeros) immediately after the decimal point of the number are found from the extreme left column of the table, the third digit from the column-headings, and the fourth digit from the mean-difference column, thus obtaining 6950 + 8 = 6958. The decimal point has to be inserted at a place depending on the value of the characteristic. Since the integral part of the number 2.8425 is 2, hence we must have three digits in the integral part of the antilog, giving 695.8.

Example 3.20

Number	*Antilogarithm*
(a) 2.034	108.1
(b) – 0.8734 (= $\bar{1}$.1266)	0.1339
(c) – 3.8734 (= $\bar{4}$ –.1266)	.0001339
(d) .0357	1.086
(e) 3.2678	1852

(iv) Uses of logarithms

Logarithms are used to find the product, quotient, powers and roots of numbers.

Example 3.21 *Find the values of*

(a) $439.4 \times .0347 \times 6.8$ (b) $\dfrac{0.6745 \times 8.351}{.00737 \times 4905}$

(c) $35(.09)^4 (.91)^3$ (d) $\dfrac{3.845}{(2.6408)^3}$

(e) $\sqrt[3]{4.8}$ (f) $\dfrac{129}{\sqrt{108 \times 176}}$

(g) $\dfrac{114.72}{\sqrt{548.76 \times 171.20}}$

Solution (a) log (439.4 × .0347 × 6.8)

= log 439.4 + log .0347 + log 6.8

= 2.6429 + $\bar{2}$.5403 + 0.8325
= 2.6429 + (–2 + .5403) + 0.8325
= 2.0157

∴ 439.4 × .0347 × 6.8 = antilog 2.0157 = 103.7

(b) Let $x = \dfrac{0.6745 \times 8.351}{.00737 \times 4905}$

$$\begin{aligned}
\log x &= \log(0.6745 \times 8.351) - \log(.00737 \times 4905)\\
&= (\log 0.6745 + \log 8.351) - (\log .00737 + \log 4905)\\
&= (\bar{1}.8290) + 0.9218) - (\bar{3}.8675 + 3.6906)\\
&= (-1 + .8290 + .9218) - (-3 + .8675 + 3.6906)\\
&= -1 + 1.7508 - 1.5581\\
&= -1 + .1927 = \bar{1}.1927
\end{aligned}$$

$\therefore \quad x = \text{antilog } \bar{1}.1927 = 0.1559$

(c) Let $x = 35(.09)^4 (.91)^3$

$$\begin{aligned}
\log x &= \log 35 + \log(.09)^4 + \log(.91)^3\\
&= \log 35 + 4(\log .09) + 3(\log .91)\\
&= 1.5441 + 4(\bar{2}.9542) + 3(\bar{1}.9590)\\
&= 1.5441 + 4(-2 + .9542) + 3(-1 + .9590)\\
&= 8.2379 - 11 = .2379 - 3 = \bar{3}.2379
\end{aligned}$$

$x = \text{antilog } \bar{3}.2379 = .001730$

(d) Let $x = \dfrac{3.854}{(2.6408)^3}$

$$\begin{aligned}
\log x &= \log 3.854 - \log(2.6408)^3\\
&= \log 3.854 - 3(\log 2.6408)\\
&= 0.5860 - 3(0.4218);\ \text{using } \log 2.641 \text{ for } \log 2.6408\\
&= 0.5860 - 1.2654 = (0.5860 - 0.2654) - 1\\
&= .3206 - 1 = \bar{1}.3206
\end{aligned}$$

$\therefore \quad x = \text{antilog} - \bar{1}.3206 = .2092$

(e) $\sqrt[3]{4.8} = (4.8)^{\frac{1}{3}}$, see Sec. 3.5

$$\log \sqrt[3]{4.8} = \log (4.8)^{\frac{1}{3}} = \frac{1}{3}(\log 4.8) = \frac{1}{3}(.6812) = 0.2271$$

$\therefore \quad \sqrt[3]{4.8} = \text{antilog } .2271 = 1.687$

(f) Let $x = \dfrac{129}{\sqrt{108 \times 176}} = \dfrac{129}{(108 \times 176)^{\frac{1}{2}}}$

$$\begin{aligned}
\log x &= \log 129 - \frac{1}{2}\log(108 \times 176)\\
&= \log 129 - \frac{1}{2}(\log 108 + \log 176)\\
&= 2.1106 - \frac{1}{2}(2.0334 + 2.2455)\\
&= 2.1106 - 2.1394 = \bar{1}.9712
\end{aligned}$$

$\therefore \quad x = \text{antilog } \bar{1}.9712 = 0.9358$

(g) For using the four-figure log table, we round off the numbers to 4 significant digits.

Let $\quad x = \dfrac{114.72}{\sqrt{548.76 \times 171.20}} = \dfrac{114.7}{\sqrt{548.8 \times 171.2}}$ (approx.)

$$\log x = \log 114.7 - \frac{1}{2}(\log 548.8 + \log 171.2)$$
$$= 2.0596 - \frac{1}{2}(2.7394 + 2.2335)$$
$$= 2.0596 - 2.4864 = 1.5732$$
$$\therefore \quad x = \text{antilog } \bar{1}.5732 = .3743$$

3.7 A.P. SERIES AND G.P. SERIES

A series of observations is said to be in Arithmetic Progression (A.P.), if the consecutive values have a common difference.

Example 3.22 *(i) 1, 2, 3, 4, …*
(ii) 5, 8, 11, 14, …
(iii) 7, 5, 3, 1, –1, –3, –5, …

The A.P. series is algebraically represented by

$$a, (a + d), (a + 2d), (a + 3d), \ldots \tag{3.7.1}$$

where a = initial term, and d = common difference. The nth term (t_n) and the sum (S_n) upto n terms of the A.P. series are given by

$$t_n = a + (n - 1)d, \; S_n = [2a + (n - 1)d]\frac{n}{2} \tag{3.7.2}$$

A series of observations is said to be in Geometric Progression (G.P.), if the consecutive values have a common ratio.

Example 3.23 *(i) 5, 10, 20, 40, 80, 160, …*

(ii) $4, 1, \frac{1}{4}, \frac{1}{16}, \frac{1}{64}, \ldots$

The G.P. series is algebraically represented by

$$a, ar, ar^2, ar^3 \ldots \tag{3.7.3}$$

where a = initial term, and r = common ratio. The nth term (t_n') and the sum (S_n') upto n terms of the G.P. series are given by

$$t_n' = ar^{n-1} \qquad S_n' = \frac{a\left(1 - r^n\right)}{\left(1 - r\right)}$$

Example 3.24 *Two men X and Y started working for a certain company at similar iobs on Jan. 1, 1950. X asked for an initial salary of Rs 300 with an annual increment of Rs 30. Y asked for an initial salary of Rs 200 with a rise of Rs 15 every six months. What are their respective salaries on Dec. 31, 1959, assuming that the arrangements remained unaltered? Find the total amount paid to each of them as salary during the period.* [C.U., B.A. (Econ) '68]

Solution The initial monthly salary of *X* is Rs 300 and increases every year by Rs 30. Hence his salary figures in successive *years* will be in A.P. with a common difference of Rs 30. Similarly, the salary figures of *Y* in successive *half-years* will be in A.P. with a common difference of Rs 15. For *X*, Dec. 31, 1959 falls in the 10th *year* of serivice, and for *Y* it falls in the 20th *half-year*. Thus, for the two A.P. series, we have

Men	Initial Term (a)	Common Difference (d)	No. of Terms (n)
X	300	30	10
Y	200	15	20

Using the first formula of (3.7.2), the salaries as on Dec. 31, 1959 are

$$t_{10} = 300 + (10 - 1) \times 30 = 570 \text{ Rs (for } X)$$
$$t_{20} = 200 + (20 - 1) \times 15 = 485 \text{ Rs (for } Y)$$

For determining the total amount paid to each of them, we apply the second formula of (3.7.2). Taking one month's salary from each year, the total amount paid to X in each of the 12 months during the 10 year period is

$$S_{10} = \{2 \times 300 + (10 - 1) \times 30\} \frac{10}{2} = 4{,}350 \text{ Rs}$$

Therefore, the total amount paid to X throughout the period is

$$= 4{,}350 \times 12 = 52{,}200 \text{ Rs}$$

Similarly, taking one month's salary from each half-year, the total amount paid to Y in each of the 6 months during the 20 half-years is

$$S_{20} = \{2 \times 200 + (20 - 1) \times 15\} \frac{20}{2} = 6{,}850 \text{ Rs}$$

Therefore, the total amount paid to Y throughout the period is

$$= 6{,}850 \times 6 = 41{,}100 \text{ Rs}$$

Ans. (a) Rs 570, Rs 485; (b) Rs 52,200, Rs 41,100.

3.8 SUM AND SUM OF THE SQUARES OF NUMBERS

Sum of the first n numbers $= 1 + 2 + 3 + \ldots + n = \dfrac{n(n+1)}{2}$

Sum of the squares of the first n numbers

$$= 1^2 + 2^2 + 3^2 + \ldots + n^2 = \frac{n(n+1)(2n+1)}{6}$$

3.9 SIMPLE INTEREST LAW AND COMPOUND INTEREST LAW

If P denotes the principal invested, i the interest rate (i.e. interest on Re 1 for 1 period), n the number of periods, and A denotes the amount (i.e. principal + interest) after n periods, then

$$A = P(1 + ni) \quad \text{at Simple interest} \qquad (3.9.1)$$
$$A = P(1 + i)^n \quad \text{at Compound interest} \qquad (3.9.2)$$

If the rates of interst vary in different periods, say i_1 for n_1 periods, i_2 for the next n_2 periods, i_3 for the next n_3 periods then

$$A = P(1 + n_1 i_1 + n_2 i_2 + n_3 i_3) \text{ at Simple interest} \qquad (3.9.3)$$
$$A = P(1 + i_1)^{n_1} (1 + i_2)^{n_2} (1 + i_3)^{n_3} \text{ at Compound interest} \qquad (3.9.4)$$

3.10 PERMUTATION AND COMBINATION

The *permutation* of *n* things taken *r* at a time (denoted by ^np_r) is the total number of possible orders of a arrangement by considering *r* things at a time out of *n* things.

$$^np_r = n(n-1)(n-2)(n-3)\ldots(n-r+1)$$

When $r = n$, we have $^nP_n = n(n-1)(n-2)\ldots$ 3.2.1 = Product of all numbers from 1 to *n*. This is written as *n*! (read 'factorial *n*')

$$n! = 1 \times 2 \times 3 \times 4 \times \ldots \times n$$

It may be verified that

$$1! = 1;\ 2! = 2;\ 3! = 6;\ 4! = 24;\ 5! = 120;$$

Customarily, we write 0! = 1.

Examle 3.25 *(i)* $^5P_2 = 5 \times 4 = 20$; *(ii)* $^7P_3 = 7 \times 6 \times 5 = 210$.

The combination of n things taken r at a time (denoted by ^nC_r) is the total number of possible groups that can be formed by considering r things at a time out of n, without any regard to the order of arrangement.

$$^nC_r = \frac{n(n-1)(n-2)(n-3)\ldots(n-r+1)}{r!} = \frac{n!}{r!(n-r)!}$$

Putting r = 0 and n successively, we have

$$^nC_0 = \frac{n!}{0!n!} = 1; \quad ^nC_n = \frac{n!}{0!n!} = 1$$

A useful relation is $^nC_r = {}^nC_{n-r}$

Example 3.26 *(i)* $^6C_2 = \dfrac{6 \times 5}{1 \times 2} = 15$

(ii) $^8C_3 = \dfrac{8 \times 7 \times 6}{1 \times 2 \times 3} = 56$

(iii) $^8C_6 = {}^8C_2 = \dfrac{8 \times 7}{1 \times 2} = 28$

(iv) $^{20}C_{16} = {}^{20}C_4 = \dfrac{20 \times 19 \times 18 \times 17}{1 \times 2 \times 3 \times 4} = 4845$

3.11 BINOMIAL SERIES AND BINOMIAL COEFFICIENTS

We know that

$$(a+b)^2 = a^2 + 2ab + b2$$
$$(a+b)^3 = a^3 + 3a^2b + 3ab^2 + b^3$$

In general, when *n* is a positive integer,

$$(a+b)^n = (^nC_0)a^n + (^nC_1)a^{n-1}b + (^nC_2)a^{n-2}b^2 + \ldots + (^nC_n)b^n$$
$$(a-b)^n = (^nC_0)a^n - (^nC_1)a^{n-1}b + (^nC_2)a^{n-2}b^2 - \ldots + (-)^n(^nC_n)b^n \quad (3.11.1)$$

The series expansion on the right is known as *binomial series* and the coefficients

$$^nC_0,\ ^nC_1,\ ^nC_2,\ \ldots,\ ^nC_n$$

are called *binomial coefficients*.

Table 3.1 Binomial Coefficients

n	*Binomial Coefficients*										
1	1,	1									
2	1,	2,	1								
3	1,	3,	3,	1							
4	1,	4,	6,	4,	1						
5	1,	5,	10,	10,	5,	1					
6	1,	6,	15,	20,	15,	6,	1				
7	1,	7,	21,	35,	35,	21,	7,	1			
8	1,	8,	28,	56,	70,	56,	28,	8,	1		
9	1,	9,	36,	84,	126,	126,	84,	36,	9,	1	
10	1,	10,	45,	120,	210,	252,	210,	120,	45,	10,	1

Note that the binomial coefficients are symmetrical, i.e. the coefficients equidistant from either end are equal. Using the binomial coefficients in Table 3.1, we can easily write down the binomial series for any power upto 10.

Example 3.27 (i) $(1 + x)^7 = 1 + 7x + 21x^2 + 35x^3 + 35x^4 + 21x^5 + 7x^6 + x^7$

(ii) $(x_i - c)^4 = x_i^4 - 4x_i^3 c + 6x_i^2 c^2 - 4x_i c^3 + c^4$

(iii) $(E - 1)^5 = E^5 - 5E^4 + 10E^3 - 10E^2 + 5E - 1$

3.12 INEQUALITIES

The following symbols have meanings shown against each:

$<$ less than

$\le$ less than, or equal to

$>$ greater than

$\ge$ greater than, or equal to

(i) If $a > b$, then $a + c > b + c$; $a - c > b - c$

but $-a < -b$; $\dfrac{1}{a} < \dfrac{1}{b}$

(ii) Cauchy-Schwarz inequality:

$$(a_1^2 + a_2^2 + \dots + a_n^2)(b_1^2 + b_2^2 + \dots + b_n^2) \ge (a_1 b_1 + a_2 b_2 + \dots + a_n b_n)^2$$

The two sides are equal only when $\dfrac{a_1}{b_1} = \dfrac{a_2}{b_2} = \dfrac{a_3}{b_3} = \dots = \dfrac{a_n}{b_n}$

3.13 CONCEPT OF 'FUNCTION'

If two variables x and y are so related that the value of y depends on the value of x, then y is said to be a *function of* x. Sometimes, x is called the *independent variable* (to which we can assign values at pleasure), and y is called the *dependent variable* (whose values depend on those assigned to the independent variable).

Example 3.28 (a) $y = 7 + 4x$
(b) $y = 9x^2 - 5x + 6$
(c) $y = x^3 - x - 5$
(d) $y = 5 \times 2x$
(e) $y = 4(\log x)^2 + (\log x) + 4$

In each of the above cases, y will be called a *function of* x, the functional form being given by the expression on the right. In general, any mathematical expression involving x is a function of x, and this is often denoted by writing $f(x)$, $g(x)$ etc. For example, if

$$f(x) = x^3 - x - 5,$$

then the value of the function when $x = 4$ is

$$f(4) = 4^3 - 4 - 5 = 55.$$

3.14 POLYNOMIAL

Polynomial in x is a special function in which the individual terms contain only positive integral powers of x, e.g. x, x^2, x^3, etc. The *degree* of the polynomial is the highest power of x contained in the polynomial.

Example 3.29 (a) $4x + 7$ is a *1st degree polynomial or linear function.*
(b) $9x^2 - 5x + 6$ *is a 2nd degree polynomial or parabolic function.*
(c) $x^3 + 5$ *is a 3rd degree polynomial or cubic function.*
(d) $2x^4 + 7x^3 - 4x^2 + 5$ *is a 4th degree polynomial.*
(e) $x(x - 1)(x - 2) + 8x(x - 1) + 4x + 3$ *is a cubic function.*
(f) $a_0 + a_1x + a_2x^2 + a_3x^3 + \ldots + a_nx^n$
is an n-th degree polynomial, provided a_n *is not zero.*

If $f(x)$ represents a polynomial in x of degree n, then $y = f(x)$ geometrically represents a curve, called the *n-th degree curve.*

Example 3.30 (a) $y = 4x + 7$ *represents a 1st degree curve or straight line.*
(b) $y = 9x^2 - 5x + 6$ *represents a 2nd degree curve or parabola.*
(c) $y = a_0 + a_1x + a_2x^2 + \ldots + a_nx^n$ *represents an n-th degree curve, provided* a_n *is not zero.*

3.15 SIGMA (Σ) NOTATION

The Greek capital letter Σ, when placed before a mathematical expression, signifies the *sum* of a series. For instance, $\sum_{r=1}^{20}(r^2 + 3)$ denotes the sum of a series of 20 terms obtained by putting $r = 1, 2, 3, \ldots, 20$ successively in the expression $(r^2 + 3)$, i.e.

$$\sum_{r=1}^{20}(r^2 + 3) = (1^2 + 3) + (2^2 + 3) + (3^2 + 3) \ldots + (20^2 + 3).$$

The expression $(r^2 + 3)$, which follows Σ, may therefore be taken as a general term of the series.

Similarly, if $x_1, x_2, \ldots x_n$ represent a series of n values, their sum may be written as

$$\sum_{i=1}^{n} x_i = x_1 + x_2 + \ldots + x_n \qquad (3.15.1)$$

This is read as *summation of x sub i, for i equal to* 1 *to n.* Very often it happens that there is no necessity of writing the *limits of summation,* it being implied that the lower and the upper limits are 1 and n respectively. Thus,

$$\sum_{1}^{n} x_i, \quad \sum_{i}^{n} x_i, \quad \sum x_1, \quad \sum x$$

are all used to denote the sum (3.15.1)

Example 3.31 *Write the expanded form:*

(i) $\sum_{i=2}^{7} (x_1 - 3)^2$ (ii) $\sum_{i=5}^{9} x_i^3$ (iii) $\sum^{40} (x_i + 4)$

(iv) $\sum_{i=1}^{n} f_i x_i$ (v) $\sum_{i} ax_i$ (vi) $\sum_{1}^{n} \log x_i$

(vii) $\sum_{x=1}^{n} x$ (viii) $\sum_{x=1}^{n} x^2$ (ix) $\sum_{x=1}^{n} (2x - 1)$

(x) $\sum_{x=0}^{n} {}^nC_x$ (xi) $\sum_{x=0}^{n} {}^nC_x p^x q^{n-x}$

Solution (i) The limits of summation are from 2 to 7. Putting $i = 2, 3, 4, 5, 6, 7$ successively in the algebraic expression $(x_1 - 3)^2$, we get $(x_2 - 3)^2, (x_3 - 3)^2, (x_4 - 3)^2, (x_5 - 3)^2, (x_6 - 3)^2, (x_7 - 3)^2$. The sum of these terms gives the expanded form.

$$(x_1 - 3)^2 = (x_2 - 3)^2 + (x_3 - 3)^2 + (x_4 - 3)^2 + (x_5 - 3)^2 + (x_6 - 3)^2 + (x_7 - 3)^2$$

(ii) $x_5^3 + x_6^3 + x_7^3 + x_8^3 + x_9^3$

(iii) $(x_1 + 4) + (x_2 + 4) + (x_3 + 4) + \ldots + (x_{50} + 4)$

(iv) $f_1x_1 + f_2x_2 + \ldots + f_nx_n$

(v) $ax_1 + ax_2 + \ldots + ax_n$

(vi) $\log x_1 + \log x_2 + \ldots + \log x_n$

(vii) $1 + 2 + 3 + \ldots + n$

(viii) $1^2 + 2^2 + 3^2 + \ldots + n^2$

(ix) $1 + 3 + 5 + \ldots + (2n - 1)$

(x) ${}^nC_0 + {}^nC_1 + {}^nC_2 + \ldots + {}^nC_n$

(xi) ${}^nC_0 q^n + {}^nC_1 pq^{n-1} + {}^nC_2 p^2 q^{n-2} + \ldots + {}^nC_n p^n$

Example 3.32 Write the contracted form with Σ notation:

(i) $x_1 + x_2 + \ldots + x_n$

(ii) $(x_1 + y_1) + (x_2 + y_2) + \ldots + (x_n + y_n)$

(iii) $f_1x_1^2 + f_2x_2^2 + \ldots + f_nx_n^2$

(iv) $(2x_1 + 7) + (2x_2 + 7) + \ldots + (2x_9 + 7)$

(v) $x_1(x_1 - 1) + x_2(x_2 - 1) + \ldots + x_n(x_n - 1)$

(vi) $f_1(\log x_1) + f_2(\log x_2) + \ldots + f_n(\log x_n)$

Solution

(i) $\sum_{i=1}^{n} x_i$ (ii) $\sum_{i=1}^{n} (x_i + y_i)$ (iii) $\sum_{i=1}^{n} f_i x_i^2$

(iv) $\sum_{i=1}^{n} (2x_i + 7)$ (v) $\sum_{i=1}^{n} x_i(x_i - 1)$ (vi) $\sum_{i=1}^{n} f_i(\log x_i)$

If a series of values $x_1, x_2, \ldots x_n$ and the corresponding values of another series $f_1, f_2, \ldots, f_n$ are shown vertically in the columns of a table, then

Σx denotes the total of *x-column*
Σfx " " " " *fx-column*
Σfx^2 " " " " *fx^2-column*

Table 3.2 Meaning of Σx, Σfx, Σfx^2

	x	f	fx	fx^2
	x_1	f_1	f_1x_1	$f_1x_1^2$
	x_2	f_2	f_2x_2	$f_2x_2^2$
	…	…	…	…
	x_n	f_n	f_nx_n	$f_nx_n^2$
Total	Σx	Σf	Σfx	Σfx^2

Example 3.33 *Given* $x_1 = 6$, $x_2 = -4$, $x_3 = 5$, $x_4 = 8$,
$f_1 = 3$, $f_2 = 1$, $f_3 = 2$, $f_4 = 4$,

find the values of

(i) Σx *(ii)* Σfx *(iii)* Σx^2

(iv) $\Sigma f(x + 4)$ *(v)* $\Sigma\left(\frac{f}{x}\right)$ *(vi)* $\left(\Sigma x\right)^2$

(vii) $\frac{\Sigma fx}{\Sigma f}$ *(viii)* Σfx^2 *(ix)* $\Sigma fx(x - 1)$

Solution

Table 3.3 Calculations for Σx, Σfx, etc.

	x	f	fx	x^2	$(x + 4)$	$f(x + 4)$	$\frac{f}{x}$	fx^2	$fx(x - 1)$
	(1)	(2)	(3)	(4)	(5)	(6)	(7)	(8)	(9)
	6	3	18	36	10	30	0.5	108	90
	– 4	1	– 4	16	0	0	–.25	16	20
	5	2	10	25	9	18	.4	50	40
	8	4	32	64	12	48	.5	256	224
Total	15	10	56	141	31	96	1.15	430	374

Hints on calculation (Table 3.3):

col. 3 = (col. 1) (col. 2) col. 4 = (col. 1)^2

col. 5 = (col. 1) + 4 col. 6 = (col. 2) (col. 5)

col. 7 = $\dfrac{\text{col.2}}{\text{col.1}}$ col. 8 = (col. 1) (col. 3), or (col. 2) (col. 4)

col. 9 = (col. 3) (col. 1 minus 1)

(i) Σx = Total of x-column = 15

(ii) Σfx = Total of fx-column = 56

(iii) Σx^2 = Total of x^2-column = 141

(iv) $\Sigma f(x+4)$ = Total of $f(x+4)$ column = 96

(v) $\Sigma\left(\dfrac{f}{x}\right)$ = Total of column with heading $\dfrac{f}{x}$ = 1.15

(vi) $(\Sigma x)^2$ = (Total of x-column)^2 = 15^2 = 225

(vii) $\dfrac{\Sigma fx}{\Sigma f} = \dfrac{\text{Total of } fx\text{-column}}{\text{Total of } f\text{-column}} = \dfrac{56}{10} = 5.6$

(viii) Σfx^2 = Total of fx^2-column = 430

(ix) $\Sigma fx(x-1)$ = Total of column with heading $fx(x-1)$ = 374

3 Rules for Σ notation

(Rule I) If Σ appears before an expression containing two or more terms, then Σ may be placed before each term. (3.15.2)

Example 3.34

(i) $\Sigma(x_i + y_i) = \Sigma x_i + \Sigma y_i$

(ii) $\Sigma(x_i - y_i) = \Sigma x_i + \Sigma y_i$

(iii) $\Sigma(2f_ix_i - 3y_i + 8) = \Sigma 2f_ix_i - \Sigma 3y_i + \Sigma 8$

(iv) $\Sigma(ax_i^2 + bx_i + c) = \Sigma ax_i^2 + \Sigma bx_i + \Sigma c$

(v) $\Sigma f_i(x_i - 5) = \Sigma(f_ix_i - 5y_i) = \Sigma f_ix_i - \Sigma 5f_i$

(vi) $\Sigma f_i(x_i - 5)^2 = \Sigma f_ix_i^2 - \Sigma 10f_ix_i + \Sigma 25f_i$

Solution (i) $\Sigma(x_i + y_i) = (x_1 + y_1) + (x_2 + y_2) + \ldots + (x_n + y_n)$

$= x_1 + y_1 + x_2 + y_2 + \ldots + x_n + y_n$

$= (x_1 + x_2 + \ldots + x_n) + (y_1 + y_2 + \ldots + y_n)$

$= \Sigma x_i + \Sigma y_i$

(ii) $\Sigma(x_i - y_i) = (x_1 - y_1) + (x_2 - y_2) + \ldots + (x_n - y_n)$

$= x_1 - y_1 + x_2 - y_2 + \ldots + x_n - y_n$

$= x_1 + x_2 + \ldots + x_n - y_1 - y_2 - \ldots - y_n$

$= (x_1 + x_2 + \ldots + x_n) - (y_1 + y_2 + \ldots + y_n)$

$= \Sigma x_i - \Sigma y_i$

(Rule II) If Σ appears before a term containing a constant factor (i.e. which is independent of i), then the constant factor may come in front of Σ. (3.15.3)

Example 3.35 (*i*) $\Sigma\, cx_i = c\, \Sigma y_i$, *Where c is a constant.*

(*ii*) $\Sigma\, 8(x_i + 3)^2 = 8\, \Sigma\, (x_i + 3)^2$.

Solution (i) $\Sigma\, cx_i = cx_1 + cx_2 + \ldots + cx_n$

$$= c(x_1 + x_2 + \ldots + x_n) = c\Sigma x_i$$

(Rule III) If Σ appears before a constant, then the constant is multiplied by the total number of terms included in the sum. (3.15.4)

Example 3.36 (*i*) $\sum_{i=1}^{n} c = cn.$

(*ii*) $\sum_{8}^{8} 13 \times 6 = 78$

Solution (i) $\sum_{i=1}^{n} c = c + c + \ldots + c$ (one c for each value of i)

$$= cn.$$

Example 3.37 *Show that*

$$\Sigma f_i(x_i + 1)^4 = \Sigma f_i x_i^4 + 4\Sigma f_i x_i^3 + 6\Sigma f_i x_i^2 + 4\Sigma f_i x_i + N,$$

where $N = \Sigma f_i$

Solution $\Sigma f_i (x_i + 1)^4 = \Sigma f_i (x_i^4 + 4x_i^3 + 6x_i^2 + 4x_i + 1)$

$$= \Sigma f_i x_i^4 + 4f_i x_i^3 + 6f_i x_i^2 + 4f_i x_i + f_i)$$

$$= \Sigma f_i x_i^4 + \Sigma 4f_i x_i^3 + \Sigma\, 6f_i x_i^2 + \Sigma\, 4f_i x_i + \Sigma f_i \quad \text{(by Rule 1)}$$

$$= \Sigma f_i x_i^4 + 4\Sigma f_i x_i^3 + 6\Sigma f_i x_i^2 + 4\Sigma f_i x_i + N \quad \text{(by Rule I1)}$$

Example 3.38

If $m_1 = \frac{1}{n}\sum_{1}^{n} x_i$, $m_2 = \frac{1}{n}\sum_{2}^{n+1} x_i$ *and* $m_3 = \frac{1}{n}\sum_{3}^{n+2} x_i$, *show that*

$$(i)\ m_2 = m_1 + \frac{x_{n+1} - x_1}{n}$$

$$(ii)\ m_3 = m_2 + \frac{x_{n+2} - x_2}{n}$$

(C.U., M. Com. '63)

Solution

(i) Right hand side $= m_1 + \dfrac{x_{n+1} - x_1}{n}$

$$= \frac{1}{n}\sum_1^n x_i + \frac{1}{n}(x_{n+1} - x_1)$$

$$= \frac{1}{n}\left(\sum_1^n x_i + x_{n+1} - x_1\right)$$

$$= \frac{1}{n}(x_1 + x_2 + \dots + x_n + x_{n+1} - x_1)$$

$$= \frac{1}{n}(x_2 + \dots + x_n + x_{n+1})$$

$$= \frac{1}{n}\sum_2^{n+1} x_i = m_2 = \text{Left hand side.} \qquad \text{(shown)}$$

(ii) Right hand side $= m_2 + \dfrac{x_{n+2} - x_2}{n}$

$$= \frac{1}{n}\sum_2^{n+1} x_i + \frac{1}{n}(x_{n+2} - x_2)$$

$$= \frac{1}{n}\left(\sum_2^{n+1} x_i + x_{n+2} - x_2\right)$$

$$= \frac{1}{n}(x_2 + x_3 + \dots + x_{n+1} + x_{n+2} - x_2)$$

$$= \frac{1}{n}(x_3 + \dots + x_{n+1} + x_{n+2})$$

$$= \frac{1}{n}\sum_3^{n+2} x_i = m_3 = \text{Left hand side.} \qquad \text{(shown)}$$

3.16 SIMPLE INTERPOLATION

Interpolation denotes the process of finding the intermediate value of a function from a set of given values of that function (Chapter 10). For example, suppose the following pairs of values of x and y are known:

x	y
4	17
7	32

What is the value of y when x has any specified value lying between the given values, say $x = 5.2$? This is a problem of interpolation. In *simple interpolation,* we assume that the values of y change uniformly with those of x. When x has increased by 3(from 4 to 7) y has increased by 15 (from 17 to 32), i.e. y increases by 5, as x increases by unity. When $x = 5.2$, the increase in x is $5.2 - 4 = 1.2$, for which the increase in y should be $1.2 \times 5 = 6$. The required value of y is, therefore, $17 + 6 = 23$.

The following alternative method will be found very convenient. Arrange the problem in the tabular form shown below:

	x	y	
D_1 [d_1 {	4	17	} d_2] D_2
	5.2 →	← Y	
	7	32	

The partial differences in the values of x and y are

$$d_1 = 5.2 - 4, \quad d_2 = Y - 17$$

and the total differences are

$$D_1 = 7 - 4, \quad D_2 = 32 - 17$$

Since we assume a uniform rate of change in the given interval, the partial differences must be proportional to the corresponding total differences.

$$\frac{\text{Partial difference in } x}{\text{Total difference in } x} = \frac{\text{Partial difference in } y}{\text{Total difference in } y}$$

i.e.

$$\frac{d_1}{D_1} = \frac{d_2}{D_2} \tag{3.16.1}$$

Substituting the values,

$$\frac{5.2 - 4}{7 - 4} = \frac{Y - 17}{32 - 17}$$

or,

$$\frac{1.2}{3} = \frac{Y - 17}{15}$$

or,

$$Y - 17 = \frac{1.2}{3} \times 15 = 1.2 \times 5 = 6$$

∴

$$Y = 17 + 6 = 23.$$

The relation (3.16.1) may also be used to find the unknown value of x corresponding to any specified y, intermediate between the given values.

Example 3.39 *From the following table, find (i) the value of y, when x = 19.3, (ii) the value of x, when y = 12.8, using simple interpolation:*

x	12	17	22	27	32
y	2.8	5.1	8.8	12.0	15.2

Solution (i) $x = 19.3$ lies between the tabulated values $x = 17$ and $x = 22$; hence, the corresponding value of y, say Y, lies between $y = 5.1$ and $y = 8.8$, as indicated by arrow marks. (ii) Similarly, $y = 12.8$ lies between the given values $y = 12.0$ and $y = 15.2$, and hence the corresponding value of x, say X, lies between $x = 27$ and $x = 32$, as indicated. Now using (3.16.1),

x	y
12	2.8
17	5.1
19.3→	←Y
22	8.8
27	12.0
X→	←12.8
32	15.2

(i) $$\frac{19.3 - 17}{22 - 17} = \frac{Y - 5.1}{8.8 - 5.1}$$

or, $$\frac{2.3}{5} = \frac{Y - 5.1}{3.7}$$

$$\therefore\ Y - 5.1 = \frac{2.3}{5} \times 3.7 = 1.702$$

$$\therefore\ Y = 5.1 + 1.702 = 6.802$$

(ii) $$\frac{X - 27}{32 - 27} = \frac{12.8 - 12.0}{15.2 - 12.0}$$

Solving this we get $X = 28.25$

Ans. (i) 6.802, (ii) 28.25.

4 FREQUENCY DISTRIBUTION

4.1 OBSERVATION, FREQUENCY

Suppose that the daily number of car accidents in a certain city recorded over a period of 30 days are as shown below:

Table 4.1 Daily Number of Car Accidents

4	6	5	4	4	6
3	5	4	5	3	5
4	4	5	6	4	3
5	5	3	3	5	4
7	6	5	5	4	5

The variable observed here is "the daily number of car accidents", and the data obtained are known as *observations* or *observed values* or *measurements*. If n observations on a variable x are available, they are usually denoted by

$$x_1, x_2, \ldots, x_n$$

x_1 denotes the first observation on x, x_2 denotes the second observation, etc. and finally x_n denotes the nth observation on x. In general, x_i denotes the ith observation on $x (i = 1, 2, \ldots, n)$. For example, in Table 4.1, $x_1 = 4$, $x_2 = 3$, $x_3 = 4$, …, $x_{30} = 5$. The 30 observations are not all different; some of them are repeated. The distinct observations are known as *values of the variable.*

Frequency of a value of the variable is the number of times it occurs in a given series of observations. In Table 4.1, 3 occurs 5 times, 4 occurs 9 times, 5 occurs 11 times, 6 occurs 4 times, and 7 occurs once. Hence, the frequencies of the values 3, 4, 5, 6, 7 are respectively 5, 9, 11, 4, 1.

4.2 SIMPLE SERIES (OR UNGROUPED DATA) AND FREQUENCY DISTRIBUTION

'*Simple series*' or '*ungrouped data*' are a series of observations recorded without any definite systematic arrangement. e.g. Table 4.1. If however, a large number of observations are available in the form of a simple series, the mind cannot properly grasp their significance. Some method of condensation of data is therefore necessary. If the order in which the observations have been recorded is immaterial, one way of doing this is to write down the values of the variable in order of magnitude and show the corresponding frequencies side by side in a tabular form. Such a table is called *Frequency Table*, or more usually, *Frequency Distribution* of the variable.

'Frequency Distribution' is a statistical table which shows the values of the variable arranged in order of magnitude, either individually or in groups, and also the corresponding frequencies side by side. There are two types of frequency distributions:

(a) Simple frequency distribution,
(b) Grouped frequency distribution.

Simple frequency distribution shows the values of the variable individually (Table 4.2), whereas *grouped frequency distribution* (Table 4.3) shows the values of the variable in groups of intervals.

Table 4.2 Simple Frequency Distribution

Daily Number of Car Accidents	*Frequency (No. of days)*
3	5
4	9
5	11
6	4
7	1
Total	30

Consider the following data showing the ages (in years) of 200 unemployed persons who registered their names in an Employment Exchange on a certain day.

Ages (in years) of 200 Unemployed Persons									
33	26	33	30	24	25	28	22	17	30
26	17	29	22	18	40	20	33	21	25
25	20	21	29	27	26	19	37	32	21
20	40	20	20	27	19	20	20	29	20
21	49	34	26	29	31	25	33	23	16
21	20	23	29	23	17	23	27	43	22
24	26	29	29	20	17	32	23	24	25
36	19	28	26	31	52	25	57	16	26
27	19	24	19	39	23	24	25	17	19
31	18	26	23	19	30	20	19	50	40
22	27	17	22	43	33	22	17	21	25
24	46	27	20	26	20	21	24	22	21
25	22	24	32	24	20	22	15	34	22
22	20	37	19	24	21	28	27	19	31
22	24	17	29	24	17	23	21	21	31
31	18	22	19	33	25	21	25	23	21
24	18	15	18	24	19	22	22	16	21
17	18	25	31	26	29	24	31	23	18
31	24	23	21	24	18	21	19	30	24
24	22	30	19	25	21	21	25	45	24

The first step towards summarisation of data is to arrange the figures in the form of a simple frequency distribution as follows:

Age (yrs.)	*Frequency*	*Age (yrs.)*	*Frequency*	*Age (yrs.)*	*Frequency*
15	2	26	10	39	1
16	3	27	7	40	3
17	10	28	3	43	2
18	8	29	9	45	1
19	14	30	5	46	1
20	15	31	8	49	1
21	19	32	3	50	1
22	16	33	6	52	1
23	11	34	2	57	1
24	20	36	1		
25	14	37	2	Total	200

Although this table gives a somewhat clear picture, there are still too many figures to assimilate, and the age distribution of unemployed persons in general is not very prominent. The usual procedure in such cases is to group the observations into a number of intervals and show the frequency of occurrence of all values included therein, separately for each interval (Table 4.3). Such a frequency table is known as *Grouped Frequency Distribution* (Note that the frequency of each individual observation cannot be obtained from a grouped frequency distribution).

Table 4.3 Grouped Frequency Distribution

Age in Years	*Frequency (No. of Persons)*
15–19	37
20–24	81
25–29	43
30–34	24
35–44	9
45–59	6
Total	200

An inspection of the frequency distribution gives a quick idea about the central value (average) in the series, and shows how the observations vary ('dispersion') around the average. The histogram and frequency polygon drawn from a frequency distribution also throw light on the shape ('skewness' and 'kurtosis') of the frequency curve. Again, all recorded data may be thought of as a sample coming from a certain 'population'. The frequency distribution may be used to appproximate the 'probability distribution' (Chapter 12) in the population, which is fundamental in solving statistical problems. Thus, the important uses of a frequency distribution are: (1) Condensation and summarization of a large mass of data; (2) Visualizing the important characteristics of data, like average, dispersion, skewness and kurtosis; (3) Approximating the nature of probability distribution in the population.

4.3 USEFUL TERMS ASSOCIATED WITH GROUPED FREQUENCY DISTRIBUTIONS

(a) *Class interval,* or *Class;*
(b) *Class frequency,* and *Total frequency;*
(c) *Class limits—Lower class limit, Upper class limit;*
(d) *Class boundaries—Lower class boundary, Upper class boundary;*
(e) *Class mark,* or *Mid-value,* or *Mid-point of class interval;*
(f) *Width,* or *Size of class interval;*
(g) *Frequency density.*

We shall explain these terms with reference to Table 4.3, as illustrated in Table 4.4.

(1) Class interval (or Class)

When a large number of observations varying in a wide range are available, these are usually classified in several groups according to the size of values. Each of these groups, defined by an interval, is called *Class interval,* or *simply Class*. In Table 4.4, column (1), the class intervals of ages (in years) are 15–19, 20–24, etc. There are 6 classes in the frequency distribution the last class being 45–59.

When one end of a class is not specified, the class is called an *open-end class*. A frequency distribution may have either one or two open-end classes (see Tables 4.13 and 4.14). The necessity of open-end classes arises when there are relatively few observations which are far apart from the rest. In such a case, it is not considered worth-while to show several classes with zero frequencies (called *empty class*) before reaching a class with a very small frequency.

(2) Class frequency, Total frequency

The number of observations falling within a class is called its *Class frequency*, or simply *Frequency*. The sum of all the class frequencies is called *Total frequency*. In Table 4.4, column (2), the class frequencies are 37, 81, etc. and the total frequency is 200. Total frequency shows the total number of observations considered in the frequency distribution.

Table 4.4 Class Limits, Class Boundaries, etc. Illustrated (Data: Table 4.3)

Class Interval	*Class Frequency*	*Class Limits*		*Class Boundaries*		*Class Mark*	*Width of Class*	*Frequency Density*	*Relative Frequency*
		Lower	*Upper*	*Lower*	*Upper*				
(1)	*(2)*	*(3)*	*(4)*	*(5)*	*(6)*	*(7)*	*(8)*	*(9)*	*(10)*
15–19	37	15	19	14.5	19.5	17	5	7.4	0.185
20–24	81	20	24	19.5	24.5	22	5	16.2	.405
25–29	43	25	29	24.5	29.5	27	5	8.6	.215
30–34	24	30	34	29.5	34.5	32	5	4.8	.120
35–44	9	35	44	34.5	44.5	39.5	10	0.9	.045
45–59	6	45	59	44.5	59.5	52	15	0.4	.030
Total	200	–	–	–	–	–	–	–	1.000

(3) Class limits

All recorded data or observations are discrete in character. For a discrete variable, the values themselves are isolated, e.g. records regarding the number of workers employed in a factory will show data such as 226, 105, 873, 66, 484, etc. (no fractional numbers are possible). Although a continuous variable can theoretically take any value, all observations are rounded to a certain unit for convenience. For example, if it is decided to record the height of persons to the nearest inch, we get observations like 58, 67, 63, etc.; if the decision is to record them to the nearest tenths of an inch we may get 57.8, 67.5, 63.0, etc. In any case, all observations will be uniformly rounded, either all whole numbers, or all showing upto one place of decimal, and so on.

When a grouped frequency distribution is constructed (Section 4.4) from these collected data, the values of the variable are shown in several classes. For determining the class frequencies it is necessary that these classes are mutually exclusive, i.e. be such that any observation is contained in only one class; for example 54–56, 57–59, 60–62, etc. (not like 54–56, 56–58, 58–60, etc.), or 54.1–56.0, 56.1–58.0, 58.1–60.0. etc. (not like 54.0–56.0, 56.0–58.0, 58.0–60.0 etc.). Note that classes will be written to the same degree of precision as the original data.

In the construction of a grouped frequency distribution, the class intervals must therefore be defined by pairs of numbers such that the upper end of one class does not coincide with the lower end of the immediately-following class. *The two numbers used to specify the limits of a class interval for the purpose of tallying the original observations into the various classes, are called 'Class Limits'*. The smaller of the pair is known as *Lower Class Limit* and the larger as *Upper Class Limit* with reference to the particular class. See Table 4.4, cols. (3) and (4). Class limits are used solely for the construction of grouped frequency distribution from ungrouped data.

(4) Class boundaries

When measurements are taken on a continuous variable, all data are recorded nearest to a certain unit. Thus, if ages are recorded to the nearest whole number of years, any age between 14.5 years and 15.5 years is recorded as 15 years. Similarly, '19 years' denotes an age between 18.5 years and 19.5 years. Hence, the class interval 15–19 actually includes all ages between 14.5 and 19.5 years. *These most extreme values which would ever be included in a class interval are called 'Class Boundaries'*. Class boundaries are, in fact, the real limits of a class interval. The lower extreme point is called *Lower Class Boundary*, and the upper extreme point is called *Upper Class Boundary* with reference to any particular class. See Table 4.4, columns (5) and (6).

Class boundaries may be calculated from class limits by applying the following rule:

$$\textit{Lower class boundary} = \textit{Lower class limit} - \tfrac{1}{2}d$$
$$\textit{Upper class boundary} = \textit{Upper class limit} + \tfrac{1}{2}d$$

where d is the common difference between the upper class limit of any class interval and the lower class limit of the next class interval. In fact, if observations are recorded to the nearest unit, $d = 1$; if to the nearest tenth of a unit, $d = 0.1$, etc. In Table 4.4, $d = 1$. Excepting the first and the last, all other class boundaries lie midway between the upper limit of a class and the lower limit of the next higher class.

Note that the upper boundary of any class coincides with the lower boundary of the next class; but the upper limit of any class is different from the lower limit of the next class. This fact may be utilized in deciding whether the pairs of numbers defining the class intervals in a given frequency distribution are really class limits or class boundaries. Class limits are used only for the construction of a grouped frequency distribution, but in all statistical calculations and diagrams involving end-points of classes (e.g. median, mode, quartiles, and histogram, ogive etc.), class boundaries are used.

(5) Class mark (or Mid-value or Mid-point)

The value exactly at the middle of a class interval is called *Class mark or Mid-value.* If lies half-way between the class limits or between the class boundaries. See Table 4.4, column (7).

Class mark = (Lower class limit + Upper class limit) ÷ 2
= (Lower class boundary + Upper class boundary) ÷ 2

Class mark is used as a representative value of the class interval, for the calculation of mean, standard deviation, mean deviation, etc.

(6) Width (or Size) of class

Width of *Size* of a class is the difference between the lower and the upper class boundaries (*not* class limits). See Table 4.4, column (8).

Width of class = Upper class boundary – Lower class boundary.

In the construction of a frequency distribution, it is generally preferable to have classes of equal width. This simplifies the calculation of some statistical measures (e.g. mean, standard deviation) by short-cut methdos. Unequal width of classes is resorted to when some of the observations are few and far away from the rest. The use of equal width in such cases may result in some '*empty classes*', i.e. classes with zero frequency.

If all the classes are of equal width, this common width must be equal to the difference between (i) the successive lower class limits/boundaries, (ii) the successive upper class limits/boundaries, and (iii) the successive class marks. Therefore, in order to see, whether all the classes have a common width, the following test may be applied: Calculate mentally the differences between the successive lower class limits/ boundaries, and also the successive upper class limits/boundaries. If these differences are all equal, the classes have a common width, which is equal to this common difference. If not, all classes do not have the same width.

If no confusion arises, sometimes 'class interval' is itself used to denote its width, e.g. 'equal class intervals', or 'class interval 10 paise' (see Example 4 : 6).

(7) Frequency density

Frequency density of a class is its frequency per unit width. It shows the concentration of frequency in a class and is given by the formula

$$\textit{Frequency density} = \frac{\textit{Class frequency}}{\textit{Width of the class}}$$

See Table 4.4, column (9). Frequency density is used in drawing histogram, when the classes are of unequal width.

Example 4.1 *(a) Simple Frequency Distribution:*

Table 4.5 and 4.6 illustrate two simple frequency distributions, the former relating to a discrete variable and the latter to a continuous variable.

Table 4.5 Showing Number of Telephone Calls Received in 245 One-minute Intervals

No. of Calls	*Frequency*
0	*14*
1	*21*
2	*25*
3	*43*
4	*51*
5	*40*
6	*39*
7	*12*
Total	*245*

(*Source: I.C.W.A., June '74*)

Table 4.6 Showing Heights of School Children

Height (inch)	*No. of Children*
60	*2*
61	*0*
62	*15*
63	*29*
64	*25*
65	*12*
66	*10*
67	*4*
68	*3*
Total	*100*

(*Source: I.C.W.A., Jan. '73*)

(b) Grouped Frequency Distribution:

(i) Tables 4.7 and 4.8 relate to discrete variables and the classes have been defined by using class limits.

Table 4.7 Showing Wages of 65 Employees

Wages (Rs)	*No. of Employees*
50.00 – 59.99	*8*
60.00 – 69.99	*10*
70.00 – 79.99	*16*
80.00 – 89.99	*14*
90.00 – 99.99	*10*
100.00 – 109.99	*5*
110.00 – 119.99	*2*
Total	*65*

(*Source: B.U., B.A. (Econ) '72*)

Table 4.8 Showing Marks Obtained in Statistics

Mark	*No. of Candidates*
30–39	2
40–49	3
50–59	11
60–69	20
70–79	32
80–89	25
90–99	7
Total	100

(Source: C.U., M.Com. '72)

(ii) Tables 4.9 and 4.10 relate to discrete variables, but the classes are defined by using class boundaries. Here, the number of values (if any) which are exactly equal to the class boundaries, is supposed to be small compared to the total frequency as to influence the pattern of the distribution to any appreciable extent. Table 4.9 also illustrates a frequency distribution with unequal width for class intervals.

Table 4.9 Showing Annual Sales of 213 Firms

Annual Sales (Rs '000)	*No. of Firms*
0 – 20	20
20 – 50	50
50 – 100	69
100 – 250	30
250 – 500	25
500 – 1000	19
Total	213

(Source: C.U., M.Com. '72)

Table 4.10 Showing Distribution of Daily Wages

Wages (Rs)	*No. of Workers*
2 – 3	15
3 – 4	30
4 – 5	44
5 – 6	60
6 – 7	30
7 – 8	14
8 – 9	7

(Source: C.A., Nov. '72)

(iii) Tables 4.11 and 4.12 relate to continuous variables, the classes having been defined by using class limits in the former and class boundaries in the latter.

Table 4.11 Showing Weights of 100 School-children

Weight (lbs.)	*Frequency*
65 – 69	*8*
70 – 74	*15*
75 – 79	*18*
80 – 84	*25*
85 – 89	*14*
90 – 94	*9*
95 – 99	*6*
100 – 104	*5*
Total	*100*

(Source: I.C.W.A., June '74)

Table 4.12 Showing Hospitalisation Time

Treatment Duration (days)	*Number of Patients*
0 – 10	*460*
10 – 20	*730*
20 – 30	*260*
30 – 40	*150*
40 – 50	*90*
Total	*1690*

(Source: C.U., M.Com. '71)

(iv) Tables 4.13 and 4.14 illustrate grouped frequency distributions with open-end classes–either end or both ends may be open.

Table 4.13 Showing Income of Employees in an Industry

Income (Rs)	*Frequency*
0 – 50	*90*
50 – 100	*150*
100 – 150	*100*
150 – 200	*80*
200 – 250	*70*
250 – over	*10*
Total	*500*

(Source: I.C.W.A., June '73)

Table 4.14 Showing Income Distribution

Monthly Income (Rs)	*Frequency*
Upto 100	*9*
101 – 150	*51*
151 – 200	*120*
201 – 300	*240*
301 – 500	*136*
501 – 750	*33*
751 – 1000	*9*
Above 1000	*2*
Total	*600*

(Source: I.C.W.A., Jan. '72)

(v) Tables 4.15, 4.16 and 4.17 show class intervals in which the upper ends have not been stated explicitly (they should not be confused with open-end classes). In Tables 4.15 and 4.16, the class intervals are respectively 10–19, 20–29, …, 70–79 and 25–26, 26–27, …, 35–36. In Table 4.17, the class intervals should be taken to imply '57.5 and above, but below 60.0', '60.0 and above, but below 62.5' and so on.

Table 4.15

Age (years)	*Frequency*
10 – 19	*3*
20 –	*61*
30 –	*223*
40 –	*137*
50 –	*53*
60 –	*19*
70 –	*4*
	500

(Source: I.C.W.A., July '67)

Table 4.16

Weekly Earnings (Rs)	*No. of Persons*
25–	*25*
26–	*70*
27–	*210*
28–	*275*
29–	*430*
30–	*550*
31–	*340*
32–	*130*
33–	*90*
34–	*55*
35–36	*25*
	2200

(Source: I.C.W.A., Dec. '73)

Table 4.17

Heights (inch)	*Number*
57.5 –	*6*
60.0 –	*26*
62.5 –	*190*
65.0 –	*281*
67.5 –	*412*
70.0 –	*127*
72.5 –	*38*
	1080

(Source: I.C.W.A., July '65)

(vi) Table 4.18 shows 4 different forms of class intervals, in each of which the class boundaries are same, viz. as in case B. Note that an observation 25, if any, is included in the 2nd class interval of C, but included in the 1st class interval of D. In cases A and B, the true significance of class intervals is generally understood from the text of the collected data. However, no difficulty arises in statistical calculations, since class boundaries only will be used for the purpose.

Table 4.18 Different Forms of Class Intervals

A	*B*	*C*						*D*				
20–	*20–25*	*20*	*and*	*above*	*but*	*below*	*25*	*Above*	*20*	*but*	*upto*	*25*
25–	*25–30*	*25*	*"*	*"*	*"*	*"*	*30*	*"*	*25*	*"*	*"*	*30*
30–	*30–35*	*30*	*"*	*"*	*"*	*"*	*35*	*"*	*30*	*"*	*"*	*35*
35–	*35–40*	*35*	*"*	*"*	*"*	*"*	*40*	*"*	*35*	*"*	*"*	*40*

Example 4.2 *Find the class boundaries, if the class limits are 9–10$\frac{3}{4}$; 11–12$\frac{3}{4}$; 13–14$\frac{3}{4}$; … the measurements being made nearest to $\frac{1}{8}$ of an unit. Explain why you should consider the class boundaries as the real limits for data of continuous type.*

[C.U., B.A. (Econ) '68]

Solution (There is a mistake in the question). The measurements have been made nearest to $\frac{1}{4}$ of a unit, and not $\frac{1}{8}$, as stated. Hence, $d = \frac{1}{4}$. This may also be seen from the following:

$$d = 11 - 10\frac{3}{4} = 13 - 12\frac{3}{4} = \ldots = \frac{1}{4} \text{ in each case.}$$

So $d/2 = \frac{1}{8}$. Subtracting $d/2$ from each of the lower class limits and adding $d/2$ to each of the upper class limits, the class boundaries are $\left(9 - \frac{1}{8}\right)$ and $\left(10\frac{3}{4} + \frac{1}{8}\right)$; $\left(11 - \frac{1}{8}\right)$ and

$\left(12\frac{3}{4}+\frac{1}{8}\right)$; $\left(13-\frac{1}{8}\right)$ and $\left(14\frac{3}{4}+\frac{1}{8}\right)$ etc. The required class boundaries are, therefore.

$$8\frac{7}{8}-10\frac{7}{8};\ 10\frac{7}{8}-12\frac{7}{8};\ 12\frac{7}{8}-14\frac{7}{8};\ ...$$

When observations are taken in respect of a continuous variable, all measurements are made nearest to a certain unit, e.g. heights measured to the nearest inch. Hence, any height between 61.5 and 62.5 inches is recorded as 62 (to the nearest inch); similarly any height between 63.5 and 64.5 is recorded as 64. If 62–64 be a class interval of the distribution of heights (inches), then the class interval actually includes all heights between 61.5 and 64.5 inches, i.e. the class boundaries. As such, the class boundaries should be considered as the real limits for the class interval.

Example 4.3 *Find the class boundaries in each of the following cases, when the class intervals given are:*

(a) 20–25; 25–30; 30–35; ...
(b) 20–24; 25–29; 30–34; ...
(c) 20–24.9; 25–29.9; 30–34.9; ...
(d) 20.00–24.99; 25.00–29.99; 30.00–34.99; ...
(e) 20.01–25.00; 25.01–30.00; 30.01–35.00; ...
(f) 20– ; 25– ; 30– ; ... ; 55–60.
(g) 20– ; 25– ; 30– ; ... ; 55–59.
(h) 20– ; 25– ; 30– ; ... ; 55–.
(i) 20– ; 25– ; 30– ; ... ; 55–. (Measurements are given to the nearest unit).
(j) 20– ; 25– ; 30– ; ... ; 55–. (Measurements are given to the nearest 1/8 of a unit).
(k) Above 20 but not exceeding 25;
" 25; " " " 30;
" 30 " " " 35; etc.
(l) 20 and above but below 25;
25 " " " " 30;
30 " " " " 35; etc.
(m) 20–24; 25–29; 30–34; ... (Age last birthday)

Solution (a), (f), (h), (k), (l), (m):–
20–25; 25–30; 30–35; ...

(b), (g), (i):
19.5–24.5; 24.5–29.5; 29.5–34.5; ...

(c) 19.95–24.95; 24.95–29.95; 29.95–34.95; ...
(d) 19.995–24.995; 24.995–29.995; 29.995–34.995; ...
(e) 20.005–25.005; 25.005–30.005; 30.005–35.005; ...
(j) $19\frac{15}{16}-24\frac{15}{16};\ 24\frac{15}{16}-29\frac{15}{16};\ 29\frac{15}{16}-34\frac{15}{16};...$

4.4 CONSTRUCTION OF FREQUENCY DISTRIBUTION

Example 4.4 *Describe the various steps in the construction of a frequency distribution from unclassified data.* [C.U., M.Com. '69, '74; B.A.(Econ) '72, '74]

Solution The steps in the construction of a frequency distribution from ungrouped data are as follows:–

1. Find the largest and the smallest observations in the given data, and then calculate the *range,* i.e. the difference between them.
2. Divide the range into a suitable number of class intervals, by means of class limits. The number of class intervals should not ordinarily be less than 5 nor more than 15, depending on the number of observations available. The more the number of observations, the larger will be the number of class intervals. Class limits should be so chosen that most of the observations lie near the class marks. The class intervals should preferably be of the same width. In special circumstances, class intervals of unequal width may also be used.
3. The number of observations failling in each class interval, i.e. class frequency, is determined by *tally marks* (see Table 4.19).
4. A table is now prepared showing the class intervals in the first column and the corresponding class frequencies in the second column. This is the required frequency distribution.

Example 4.5 *Discuss the problems in the construction of a frequency distribution from raw data, with particular reference to the choice of number of classes and class limits.* [I.C.W.A., Jan '72; C.A., May '72; C.U., M.Com. '71; B.A. (Econ)'64]

Solution In constructing a grouped frequency distribution, the two main problems involved are: (1) choice of the number of classes, and (b) choice of the class limits. Although there are no hard and fast rules in this regard, it is generally agreed that the number of classes should neither be very large (because, in that case the frequency distribution will be very lengthy), nor very small (because, in that case the true pattern of the distribution of observations will be obscured). As a working rule, this number should lie between 5 and 15, depending on the number of observations available. When the number of observations is small, some authors suggest the use of Sturges' formula

$$n = 1 + 3.3 (\log N)$$

where n denotes the number of classes and N the total frequency. In practice, the number of classes is fixed from other considerations depending on the particular situation.

The second problem, viz. the choice of class limits is to some extent dependent on the number of classes. For this purpose, we first find the maximum and the minimum of the observations and calculate the 'range', i.e. the difference between these two. If we like to have classes of equal width, then the approximate width of the classes may be obtained on dividing the range by the number of such classes. The lower limit of the first class may not coincide with the minimum observation. In most cases we prefer to have classes of width 5 or its multiple, e.g. 15–19, 20–24 etc., or 60.01–80.00, 80.01–100.00, 100.01–120.00 etc., one of the class limits being preferably a multiple of 5. Again, sometimes it is found that some particular values of the variable occur more frequently than others. In such cases, the class limits should be so chosen that the highly frequent values lie near the class marks. This will minimise the errors of grouping (see Section 7.6) in the calculation of various statistical measures like standard deviation, moments, etc. when class intervals are represented by their class marks.

Example 4.6 *You are given below the wages paid to some workers in a small factory. Form a frequency distribution with class interval 10 paise:*

Wages in Rs								
1.10	*1.13*	*1.44*	*1.44*	*1.27*	*1.17*	*1.98*	*1.36*	*1.30*
1.27	*1.24*	*1.73*	*1.51*	*1.12*	*1.42*	*1.03*	*1.58*	*1.46*
1.40	*1.21*	*1.62*	*1.31*	*1.55*	*1.33*	*1.04*	*1.48*	*1.20*
1.60	*1.70*	*1.09*	*1.49*	*1.86*	*1.95*	*1.51*	*1.82*	*1.42*
1.29	*1.54*	*1.38*	*1.87*	*1.41*	*1.77*	*1.15*	*1.57*	*1.07*
1.65	*1.36*	*1.67*	*1.41*	*1.55*	*1.22*	*1.69*	*1.67*	*1.34*
1.45	*1.39*	*1.25*	*1.26*	*1.75*	*1.57*	*1.53*	*1.37*	*1.59*
1.19	*1.52*	*1.56*	*1.32*	*1.81*	*1.40*	*1.47*	*1.38*	*1.62*
1.76	*1.28*	*1.92*	*1.46*	*1.46*	*1.35*	*1.16*	*1.42*	*1.78*
1.68	*1.47*	*1.37*	*1.35*	*1.47*	*1.43*	*1.66*	*1.56*	*1.48*

[C.A., May '67]

Solution From the given observations, we find that

Maximum value = 1.98
Minimum value = 1.03

The width of the classes is to be 10 paise (Note that in the question 'class interval 10 paise' implies that the *width* of the classes is to be 10 paise). Let us take the class limits 1.01–1.10; 1.11–1.20; 1.21–1.30; ...; 1.91–2.00. These are shown in a *tally sheet* (Table 4.19) and the given observations are represented by *tally marks* one by one against the appropriate classes. The tally marks are shown is groups of five to facilitate counting, every fifth tally mark in a class interval being placed across the preceding four.

Table 4.19 Tally Sheet

Class Limits	*Tally Marks*	*Frequency*
1.01–1.10	𝍸	5
1.11–120	𝍸 II	7
1.21–130	𝍸 𝍸	10
1.31–1.40	𝍸 𝍸 𝍸	15
1.41–1.50	𝍸 𝍸 𝍸 III	18
1.51–160	𝍸 𝍸 IIII	14
1.61–1.70	𝍸 IIII	9
1.71–1.80	𝍸	5
1.81–190	IIII	4
1.91–2.00	III	3
Total		90

The required frequency distribution is shown below:

Table 4.20 Frequency Distribution of Wages

Wages (Rs)	*Frequency*
1.01 – 1.10	5
1.11 – 1.20	7
1.21 – 1.30	10
1.31 – 1.40	15
1.41 – 1.50	18
1.51 – 1.60	14
1.61 – 1.70	9
1.71 – 1.80	5
1.81 – 1.90	4
1.91 – 2.00	3
Total	90

Example 4.7 *In a test run of 50 cars of the same model, the average mileage covered per gallon of fuel consumed by each car was as follows:*

18.6	33.4	25.3	27.8	30.6	31.9	33.0	26.3	24.9	29.4
20.0	26.2	28.1	33.1	37.5	22.5	39.1	32.9	33.8	52.6
32.5	34.6	32.7	9.5	38.5	29.6	25.3	49.5	30.1	27.9
26.9	23.8	36.0	38.0	27.5	32.3	34.2	23.1	34.7	29.0
34.1	38.6	25.9	40.6	53.8	29.3	36.8	27.1	34.9	31.6

Arrange the data in a frequency table with six classes. Clearly indicate the class boundaries and the class marks. [B.U., B.A.(Econ) '73]

Solution [**Note:** (i) Since the given data are rounded to one decimal place, the *class limits* must also be shown in one decimal (see Table 4.21). Class limits in whole numbers like 9–16, 17–24, …, or in two decimals like 9.01–16.50, 16.51–24.00, … will not do here.

(ii) Unlike Example 4.6, the *width of classes* is not specified here. So the common width has to be determined from the data, keeping in view the total number of classes in the proposed frequency distribution.]

In the given data, *Maximum value* = 53.8 and *Minimum value* = 9.5, so that *Range* = 53.8–9.5 = 44.3. If we are to use 6 classes of equal width, the common width will be approximately 44.3 ÷ 6 = 7.38 = 7.5 (suppose). Using the class limits 9.1–16.5, 16.6–24.0, … 46.6–54.0, and by tally marking, the frequency table has been obtained as shown below:

Table 4.21 Frequency Table showing Average Mileage Covered per Gallon of Fuel by 50 Cars

Class Limits	*Frequency*	*Class Boundaries*	*Class Mark*
9.1–16.5	1	9.05–16.55	12.8
16.6–24.0	5	16.55–24.05	20.3
24.1–31.5	18	24.05–31.55	27.8
31.6–39.0	21	31.55–39.05	35.3
39.1–46.5	2	39.05–46.55	42.8
46.6–54.0	3	46.55–54.05	50.3
Total	50	–	–

4.5 CUMULATIVE FREQUENCY DISTRIBUTION

In statistical investigations, sometimes we are interested in the number of observations smaller than (or greater than) a given value. In such cases, or chief concern is the accumulated frequency upto (or above) some value of the variable. This accumulated frequency is known as 'cumulative frequency'.

Cumulative Frequency corresponding to a specified value of the variable may be defined as the number of observations smaller than (or greater than) that value.

The number of observations 'upto' a given value is called *less-than cumulative frequency*; and the number of observations 'greater than' a value is called the *more-than cumulative frequency.* 'Cumulative frequency' only refers to the less-than type. When a frequency distribution is given, the cumulative totals of frequencies give the cumulative frequencies. When a grouped frequency distribution relates to a variable of the continuous type, the cumulative frequencies calculated therefrom must be shown against the class boundary points (i.e. end-points of classess). Cumulative frequency expressed as a percentage of total frequency, is known as *Cumulative Percentage.*

A table showing the cumulative frequencies against values of the variable systematically arranged in increasing (or decreasing) order is known as Cumulative Frequency Distribution. If cumulative percentages are shown, instead of cumulative frequencies, the table is called *Cumulative Percentage Distribution.*

Example 4.8 *What do you mean by a cumulative frequency distribution? Point out its special advantages and uses.* [C.A., May '67; I.C.W.A., Jan. '71]

Solution *Cumulative Frequency Distribution* is a statistical table which shows the values of the variable and the corresponding cumulative frequencies. It can be derived from a grouped frequency distribution by writing down the consecutive class boundary points and noting the number of observations less than (or more than) each class boundary point. A cumulative frequency distribution is of advantage in such studies as the number or percentage of observations less than (or more than) a specified value of the variable.

Uses—Cumulative frequency distribution is used (i) to find the number of observations less than (or more than) any given value; (ii) to find the number of observations falling between any two specified values of the variable; (iii) to find median, quartiles, deciles and percentiles (Section 5.18) graphically; i.e. in general, to find the value of the variable below (or above) which a specified number or percentage of the total frequency lies.

Example 4.9 *Construct (a) the cumulative frequency distribution, and (b) the grouped frequency distribution, from the following data:*

Value	*Frequency*
Less than 10	*4*
" " 20	*16*
" " 30	*40*
" " 40	*76*
" " 50	*96*
" " 60	*112*
" " 70	*120*
" " 80	*125*

Solution The cumulative frequency distribution shows the values of the variable (class boundaries) and the corresponding cumulative frequencies (less-than type):

Table 4.22 Cumulative Frequency Distribution

Value	*Cumulative Frequency (less-than)*
10	4
20	16
30	40
40	76
50	96
60	112
70	120
80	125

The grouped frequency distribution shows the values of the variable in class intervals and the corresponding class frequencies:

Table 4.23 Grouped Frequency Distribution

Value	*Frequency*
0 – 10	4
10 – 20	12
20 – 30	24
30 – 40	36
40 – 50	20
50 – 60	16
60 – 70	8
70 – 80	5
Total	125

Note: (i) Unless otherwise stated, or implied from the given data, the variable should be treated as continuous.

(ii) For a continuous variable, expressions like "less than 10", "upto 10", and "less than or equal to 10" are considered equivalent. However for a discrete variable, whereas the last two convey the same meaning, the first expression, which excludes the value 'exactly 10' should be treated as different.

(iii) For the value "less than 40" the 'frequency' is 76. However, if the words "less than" are omitted, we must write 'cumulative frequency' in place of 'frequency', e.g. for the value 40 the cumulative frequency is 76 (Table 4.22).

(iv) In the absence of any information to the contrary, the lower boundary of the first class interval in Table 4.23 is assumed to be 0; this makes all the classes of equal width, as is usual in practice.

Example 4.10 *From the following data, calculate*

(a) "less-than" and "more than" cumulative frequencies,
(b) cumulative frequency distribution, and
(c) cumulative percentage distribution:

Age (years)	*Frequency*
15 – 19	*37*
20 – 24	*81*
25 – 29	*43*
30 – 34	*24*
35 – 44	*9*
45 – 59	*6*
Total	*200*

Solution [Working note: The class boundaries are 14.5–19.5, 19.5–24.5, etc. (Table 4.4). The *class boundary points* are therefore, 14.5, 19.5, 24.5, 29.5, 34.5, 44.5, 59.5. There is no observation below 14.5, and hence its cumulative frequency is 0; the frequency below 19.5 is 37; the frequency below 24.5 is (37 + 81), i.e. 118; the frequency below 29.5 is (37 + 81 + 43), i.e. 161; and so on. The "less-than" cumulative frequencies corresponding to the above class boundary points are therefore 0, 37, (37 + 81), (37 + 81 + 43), (37 + 81 + 43 + 24), (37 + 81 + 43 + 24 + 9), (37 + 81 + 43 + 24 + 9 + 6), i.e. 0, 37, 118, 161, 185, 194, 200.

The "more-than", cumulative frequencies are calculated by cumulating the class frequencies from the higher end of the values. Corresponding to the class boundary points in reverse order 59.5, 44.5, 34.5, 29.5, 24.5, 19.5, 14.5 these are respectively 0, 6, (6 + 9), (6 + 9 + 24), (6 + 9 + 24 + 43), (6 + 9 + 24 + 43 + 81), (6 + 9 + 24 + 43 + 81 + 37), i.e. 0, 6, 15, 39, 82, 163, 200.

It should be noted that the "less-than" cumulative frequency corresponding to the highest class boundary point and the "more-than" cumulative frequency corresponding to the lowest class boundary point must be equal to the total frequency N. Also, the sum of the "less-than" and the "more-than" cumulative frequencies corresponding to any value is equal to the total frequency]

Table 4.24 Cumulative Frequencies

Age (years)	*Cumulative Frequency*	
	"less than"	*"more-than"*
14.5	0	200 = N
19.5	37	163
24.5	118	82
29.5	161	39
34.5	185	15
44.5	194	6
39.5	200 = N	0

The cumulative frequency distribution shows the values of the variable (class boundary points) and the corresponding cumulative frequencies (less-than type) side by side:

Table 4.25 Cumulative Frequency Distribution

Age (years)	*Cumulative Frequency ("less-than")*
14.5	0
19.5	37
24.5	118
29.5	161
34.5	185
44.5	194
59.5	200 = N

The cumulative percentage distribution shows the values of the variable (class boundary points) and the corresponding cumulative percentages.

Table 4.26 Cumulative Percentage Distribution

Age (years)	*Cumulative Percentage ("less-than")*
14.5	0
19.5	18.5
24.5	59.0
29.5	80.5
34.5	92.5
44.5	97.0
59.5	100.0

Example 4.11 *Estimate the number of candidates securing not more than 48 marks and number of candidates whose marks lie between 48 and 62:*

Marks Obtained	*Candidates*
Not more than 45	*447*
Not more than 50	*484*
Not more than 55	*505*
Not more than 60	*511*
Not more than 65	*514*

[I.C.W.A., July '66]

Solution [Working note: The number of candidates securing not more than 48 marks is actually the cumulative frequency corresponding to 48. This cumulative frequency can be found by simple interpolation from a cumulative frequency distribution. Again, if we find the number of candidates securing not more than 62 marks, i.e. cumulative frequency corresponding to 62, then the difference between the cumulative frequencies corresponding to 48 and 62 would give the number of candidates whose marks lie between 48 and 62].

Table 4.27 Cumulative Frequency Distribution

Marks	*Cumulative Frequency* ("*less-than*")
45	447
48→	←y
50	484
55	505
60	511
62→	←z
65	514

Now, 48 lies between the marks 45 and 50 in the first column, and hence its cumulative frequency (y) must lie between the corresponding cumulative frequencies 447 and 484 in the second column, as shown by the arrow mark. Applying formula (3.16.1) for simple interpolation (p. 77)

$$d_1/D_1 = d_2/D_2$$

i.e. $$\frac{48-45}{50-45} = \frac{y-447}{484-447}$$

or, $$\frac{3}{5} = \frac{y-447}{37}$$

or, $$y - 447 = \frac{3}{5} \times 37 = 22.2$$

or, $$y = 447 + 22.2 = 469.2 = 469 \text{ (approx.)}$$

That is, the number of candidates securing not more than 48 marks is 469.

Again, 62 lies between 60 and 65 in the first column, and its cumulative frequency (z) must lie between the corresponding cumulative frequencies 511 and 514 in the second column. Applying simple interpolation, as before,

$$\frac{62-60}{65-60} = \frac{z-511}{514-511}$$

or, $$\frac{2}{5} = \frac{z-511}{3}$$

or, $$z - 511 = \frac{2}{5} \times 3 = 1.2$$

∴ $$z = 511 + 1.2 = 512.2 = 512 \text{ (approx)}.$$

Therefore, the number of candidates securing marks between 48 and 62 is

$$z - y = 512 - 469 = 43.$$

Ans. 469; 43.

Example 4.12 *Calculate*

(a) *the number of cases between 112 and 134,*

(b) *Number less than 112,*

(c) *Number greater than 134, from the following:*

Class limit	*90–100*	*100–110*	*110–120*	*120–130*	*130–140*	*140–150*	*150–160*
Frequency	*16*	*22*	*45*	*60*	*50*	*24*	*10*

[C.A., May '69]

Solution The required numbers can be obtained by applying simple interpolation in the cumulative frequency distribution.

(a) Number of cases between 112 and 134
= Number of cases less than 134 – Number of cases less than 112.
= Cumulative frequency corresponding to 134 – Cumulative frequency corresponding to 112.

(b) Number less than 112 = Cumulative frequency (less-than) corresponding to 112.

(c) Number greater than 134 = Cumulative frequency (more-than) corresponding to 134.
= Total frequency – Cumulative frequency (less-than) corresponding to 134.

Table 4.28 Cumulative Frequency Distribution

Class Boundary	*Cumulative Frequency (less-than)*
90	0
100	16
110	38
112→	←y
120	83
130	143
134→	←z
140	193
150	217
160	227 = N

To find the cumulative frequency y, we have

$$\frac{112-110}{120-110} = \frac{y-38}{83-38}$$

or, $$\frac{2}{10} = \frac{y-38}{45}$$

or, $$y-38 = \frac{2}{10} \times 45 = 9$$

$\therefore$ $$y = 38 + 9 = 47$$

Again, to find the cumulative frequency z,

$$\frac{134-130}{140-130} = \frac{z-143}{193-143}$$

or, $$\frac{4}{10} = \frac{z-143}{50}$$

or, $$z-143 = \frac{4}{10} \times 50 = 20$$

$\therefore$ $$z = 143 + 20 = 163$$

Therefore, (a) Number of cases between 112 and 134 = 163–47 = 116,

(b) Number less than 112 = 47,

(c) Number greater than 134 = 227 – 163 = 64. *Ans.* (a) 116; (b) 47; (c) 64.

4.6 RELATIVE FREQUENCY DISTRIBUTION

Relative frequency denotes the class frequency expressed as a fraction of the total frequency.

$$\text{Relative frequency} = \frac{\text{Class frequency}}{\text{Total frequency}}$$

The sum of all the relative frequencies is equal to unity [see Table 4.4, column (10)]. When relative frequencies are shown against the corresponding classes, the table is known as *Relative Frequency Distribution.*

Table 4.29 Relative Frequency Distribution

(Data: Table 4.3)

Age (years)	*Relative Frequency*
15–19	0.185
20–24	.405
25–29	.215
30–34	.120
35–44	.045
45–59	.030
Total	1.000

4.7 DIAGRAMMATIC REPRESENTATION OF FREQUENCY DISTRIBUTIONS

The diagrams commonly used to depict statistical data, given in the form of a frequency distribution, are:

1. Histogram,
2. Frequency polygon,
3. Ogive (or Cumulative frequency polygon)

(1) Histogram

Histogram is the most common form of diagrammatic representation of a grouped frequency distribution. It consists of a set of adjoining rectangles drawn on a horizontal base line, with *areas proportional to the class frequencies*. The width of rectangles, one for each class, extends over the class boundaries (*not* class limits) shown on the horizontal scale. When all classes have equal width, the heights of rectangles will be proportional to the class frequencies and it is then customary to take the heights numerically equal to the class frequencies (Fig. 4.1). If, however, the classes are of unequal width the rectangles will also be unequal width, and therefore the heights must be proportional to the frequency densities (Fig. 4.5). Because, then

$$\begin{aligned}\text{Area of each rectangle} &= \text{width} \times \text{height}\\ &= (\text{width of class}) \times (\text{frequency density})\end{aligned}$$

$$= (\text{width of class}) \times \frac{\text{class frequency}}{\text{width of class}}$$
$$= \text{class frequency.}$$

Although vertical bar chart and histogram may appear some-what alike, the main point of distinction between them is that the consecutive rectangles in a histogram have no space in between, but the bar diagram must have equal spaces left between the consecutive rectangular bars. Also, the rectangles in a bar diagram must be of equal width, but those in a histogram are proportional to the widths of classes, and hence may also be unequal. The histogram is, after all, an *area diagram,* which emphasises the widths of rectangles between the class boundaries. But in a bar diagram only the height is all-important, the spacing and the width of bars being arbitrary.

Uses—(1) The series of rectangles is a histogram give a visual representation of the relative sizes of the various groups, and the entire distribution of total frequency among the different classes becomes at once visible. (2) The surface of the tops of rectangles also gives an idea of the nature of frequency curve (Section 4.8) for the population, from which the given data may be assumed to be a sample only. (3) The histogram may be used to find the mode graphically (Fig. 5.4).

(2) Frequency Polygon

Frequency Polygon is the graphical representation alternative to histogram and may be looked upon as derived from histogram by joining the mid-points of the tops of consecutive rectangles. It is generally used in cases when all the classes have a common width. In actual construction, therefore, the frequency polygon is obtained by joining the successive points whose abscissae represent the mid-values and ordinates represent the corresponding class frequencies. The two end-points are joined to the base line at the mid-values of the empty classes at each end of the frequency distribution. Thus the frequency polygon has the same area as the histogram, provided the width of all classes is the same. The frequency polygon is particularly useful in representing simple frequency distributions of a discrete variable (see Fig. 4.6).

Use—The frequency polygon gives us an approximate idea of the shape of frequency curve.

(3) Ogive (or Cumulative Frequency Polygon)

Ogive is the graphical representation of a cumulative frequency distribution, and hence is also called *Cumulative Frequency Polygon.* When cumulative frequencies are plotted against the corresponding class boundaries and the successive points are joined by straight lines, the line diagram obtained is known as *Ogive* or *Cumulative Frequency Polygon.* The ogive is of "less-than" or "more-than" type according as the cumulative frequencies used are of "less-than" or "more-than" type. The "less-than ogive" starts from the lowest class boundary on the horizontal axis and gradually rising upward ends at the highest class boundary corresponding to the cumulative frequency N, i.e. the total frequency. It looks like an elongated letter S (Fig. 4.2). The "more-than ogive" has the appearance of an elongated S, turned upside down (Fig. 5.3). Unequal widths of classes in the frequency distribution do not cause any difficulty in the construction of an ogive.

Uses—Since the ogive is only a diagrammatic representation of the cumulative frequency distribution, it has the same uses as those of the latter, but by graphical means. (1) The ogive is used to find the median, quartiles, deciles and percentiles, or the value of the variable such that its cumulative frequency is a specified number. (2) It is also useful in finding the cumulative frequency corresponding to a given value of the variable; and (3) to find the number of observations which are expected to lie between two given values.

Example 4.13 *Draw histogram, frequency polygon and ogives (both "less-than" and "more-than" types) for the following frequency distribution:*

Wages (Rs)	*50–59*	*60–69*	*70–79*	*80–89*	*90–99*	*100–109*	*110–119*
No. of employees	*8*	*10*	*16*	*14*	*10*	*5*	*2*

Solution [**Note:** Here the class intervals are defined by class limits and so we have to find the class boundaries for drawing the histogram. All the classes have the same width and therefore when drawing the histogram, heights of the rectangles may be represented by class frequencies. Also for drawing the ogives, we have to calculate both 'less-than' and 'more-than' cumulative frequencies.]

Table 4.30 Calculations for Drawing Histogram

Class Limits	*Class Boundaries*	*Frequency*
50–59	49.5–59.5	8
60–69	59.5–69.5	10
70–79	69.5–79.5	16
80–89	79.5–89.5	14
90–99	89.5–99.5	10
100–109	99.5–109.5	5
110–119	109.5–119.5	2

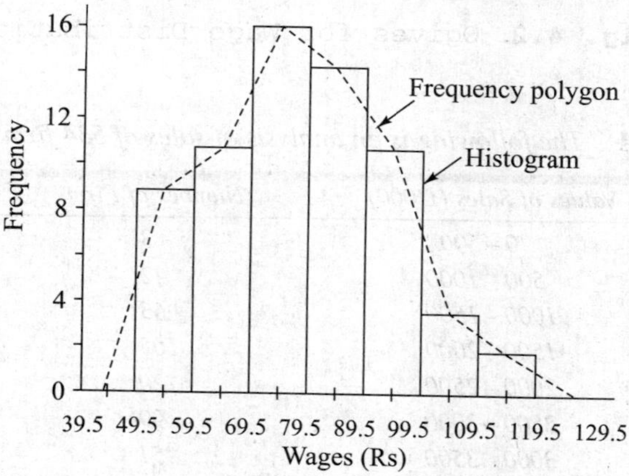

Fig. 4.1 Histogram and Frequency Polygon for Wage Distribution

Table 4.31 Calculations for Drawing Ogives

Class Boundary	*Cumulative Frequency*	
	'less-than'	*'more-than'*
49.5	0	65 = *N*
59.5	8	57
69.5	18	47
79.5	34	31
89.5	48	17
99.5	58	7
109.5	63	2
119.5	65 = *N*	0

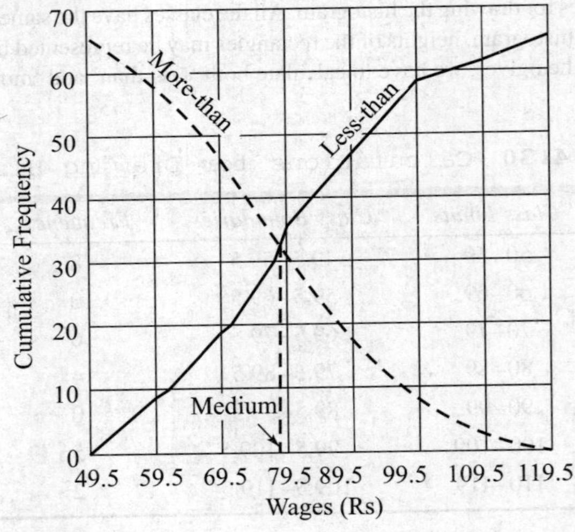

Fig. 4.2 Ogives for Wage Distribution

Example 4.14 *The following is an analysis of sales of 534 firms in an industry:*

Values of Sales (£'000)	*Number of Firms*
0 – 500	*3*
500 – 1000	*42*
1000 – 1500	*63*
1500 – 2000	*105*
2000 – 2500	*120*
2500 – 3000	*99*
3000 – 3500	*51*
3500 – 4000	*47*
4000 – 4500	*4*

Draw the histogram and the less than cumulative frequency polygon of the above distribution. Find, graphically, the number of firms whose sales lie between £ 12,00,000 and £ 26,00,000. [C.U., B.A. (Econ.) '63]

Solution [**Note:** Here the class intervals have been defined by class boundaries, as evident from the fact that the upper end of any class coincides with the lower end of the next class. Therefore, it is not necessary to construct a table for calculation of class boundaries like Table Table 4.30].

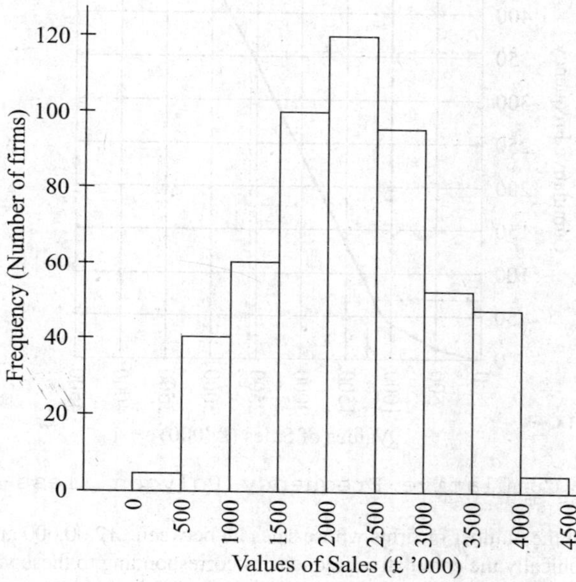

Fig. 4.3 Histogram

In order to draw the cumulative frequency polygon, we have to construct a cumulative frequency distribution.

Table 4.32 Cumulative Frequency Distribution

Class Boundary	*Cumulative Frequency (less-than)*
0	0
500	3
1000	45
1500	108
2000	213
2500	333
3000	432
3500	483
4000	530
4500	534 = N

Taking the class boundaries along the horizontal axis and the less-than cumulative frequencies along the vertical axis, the data in Table 4.32 are plotted as points on a graph paper and the consecutive points are joined by strainght lines. The line diagram obtained is the required less-than cumulative frequency polygon (Fig. 4.4).

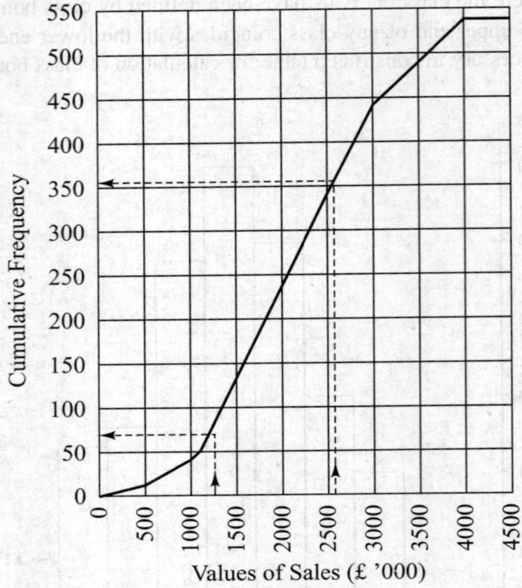

Fig. 4.4 Cumulative Frequency Polygon (less-than)

In order to obtain the number of firms whose sales lie between £12,00,000 and £26,00,000 we have to find graphically the cumulative frequencies corresponding to these values. For this purpose, two perpendiculars (broken lines in Fig. 4.4) are erected at the points 1200 and 2600 on the horizontal scale (since the units are £ '000). From the points of intersection of these perpendiculars with the cumulative frequency polygon, horizontal lines are then drawn meeting the vertical scale at two other points. Their values are read from the vertical scale, and found to be 70 and 353 approximately. This means that the cumulative frequencies corresponding to 1200 and 2600 units on the horizontal axis are respectively 70 and 353. The difference, viz. 353 – 70 = 283, gives the frequency between 1200 and 2600 units. That is, the number of firms whose sales lie between £12,00,000 and £26,00,000 is 283. *Ans.* 283

Example 4.15 *Draw the histogram of the following frequency distribution and find the proportion of firms with annual sales greater than Rs 70,000:*

Annual Sales (Rs '000) :	*0–20*	*20–50*	*50–100*	*100–250*	*250–500*	*500–1000*
No. of firms :	*20*	*50*	*69*	*30*	*25*	*19*

[C.U.,M.Com. '72]

Solution [**Note:** Here, class boundaries are given. The class intervals are also of unequal width, so that the heights of rectangles in the histogram must be proportional to *frequency densities.*]

Table 4.33 Calculations for Drawing Histogram

Class Interval (Rs '000)	*Frequency (f)*	*Width of Class (w)*	*Frequency Density (f ÷ w)*
0–20	20	20	1.00
20–50	50	30	1.67
50–100	69	50	1.38
100–250	30	150	0.20
250–500	25	250	0.10
500–1000	19	500	0.04

The histogram is shown at Fig. 4.5.

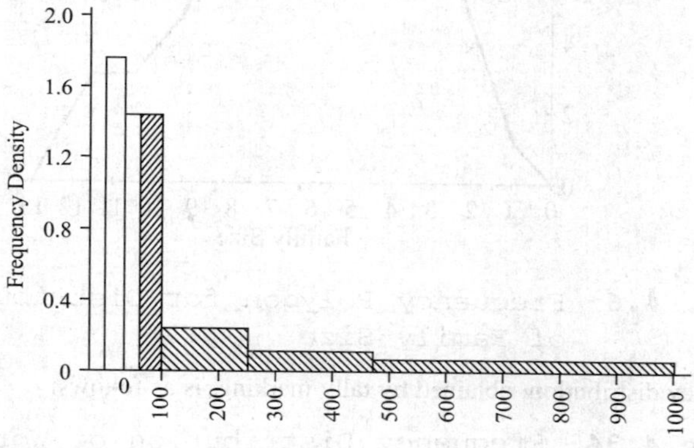

Fig. 4.5 Histogram for the Distribution of Sales

The proportion of firms with annual sales greater than Rs 70,000 is given by the proportion of area under the histogram which lies to the right of the vertical line at 70 units (Rs '000) on the horizontal axis. Assuming that frequency 69 in the class interval 50–100 is uniformly distributed in the whole interval, the proportional part of the area (i.e. frequency) between 70 and 100 units is

$$(100 \times 70) \times 1.38 = 41.4.$$

Therefore, the frequency above 70 units is

$$41.4 + 30 + 25 + 19 = 115.4.$$

Since the total frequency is 213,

$\therefore$ Required proportion of firms = $115.4 \div 213 = 0.54$

Example 4.16 *In a sample survey of 60 workers' families living in a factory area, the following data were obtained, as regards the number of members in the families. Form a frequency distribution.*

5	11	4	6	3	10	5	7	9	6
6	2	3	7	8	6	4	3	6	5
9	5	6	4	7	1	5	8	6	2
6	4	8	7	5	12	4	7	10	6
7	8	3	6	7	5	5	8	6	4
6	11	5	2	6	9	7	3	7	5

Draw a frequency polygon for the above distribution. [I.C.W.A., July '69]

Solution

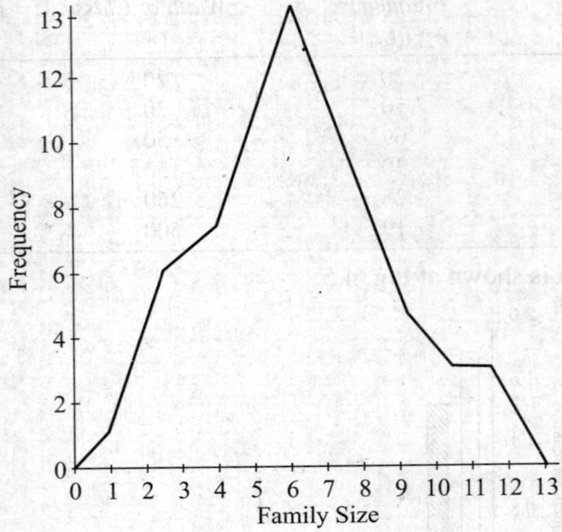

Fig. 4.6 Frequency Polygon for Distribution of Family Size

The frequency distribution, obtained by tally marking, is as follows:

Table 4.34 Frequency Distribution of Workers' Family Size

Family Size	1	2	3	4	5	6	7	8	9	10	11	12	Total
Frequency	1	3	5	6	10	13	9	5	3	2	2	1	60

The frequency polygon for the above distribution is shown at Fig. 4.6.

Example 4.17 *Draw a suitable diagram to represent the frequency distribution of the daily number of car accidents shown in Table 4.2 (page 67). Also construct a cumulative frequency distribution and draw the cumulative frequency diagram.*

Solution A simple frequency distribution relating to a discrete variable is represented by *Frequency Bar Diagram.* (In all other cases, a histogram is used.) This diagram consists essentially of a group of vertical thick lines (or bars) drawn at such points which represent values of the variable on the horizontal axis. The heights of bars are proportional to frequencies.

Table 4.35 Cumulative Frequency Distribution

No. of Car Accidents	*Cumulative Frequency*
3	5
4	14
5	25
6	29
7	30

(Data: Table 4.2)

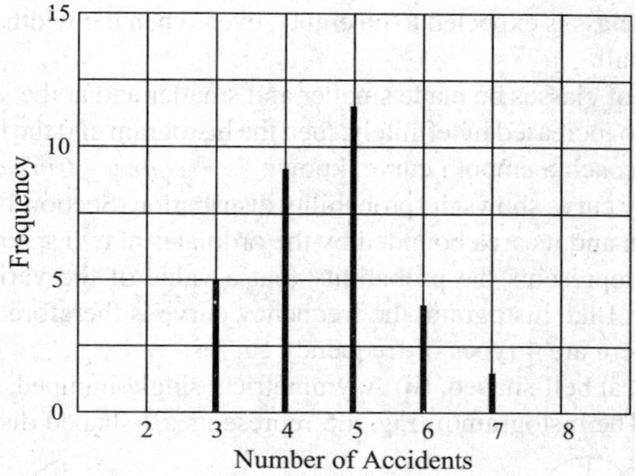

Fig. 4.7 Frequency Bar Diagram

It will be noticed that the cumulative frequency diagram (Fig. 4.8) for a discrete variable from a simple frequency distribution is a *Step Diagram*, the cumulative frequency remaining constant between two successive values of the variable.

(Data: Table 4.35)

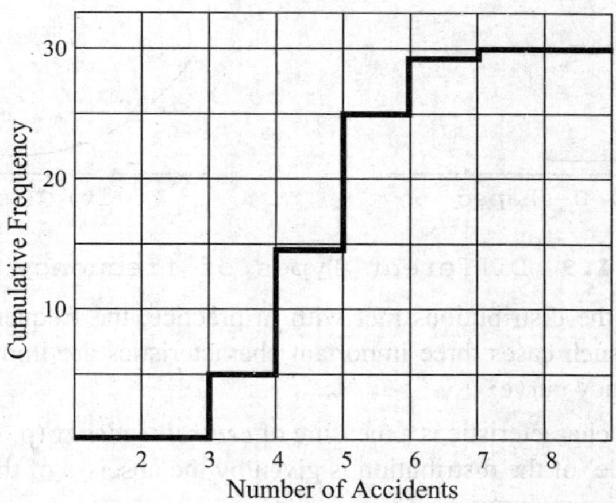

Fig. 4.8 Cumulative Frequency Diagram

4.8 FREQUENCY CURVE

The histogram gives us only an approximate idea of the nature of distribution with a limited number of observations. By drawing rectangles in a histogram, we assume that the observations in any class are uniformly distributed throughout the range of values within the class boundaries, although this may not be really true. The widths of classes could be made smaller; but, in that case some of the classes may not have any class frequency, and the true pattern of the distribution in the population will not be known. If, however, the number of observations is very large, the position will improve and no class is expected to be empty, even when the widths of the classes are considerably small.

If the widths of classes be made smaller and smaller and at the same time the total frequency be also increased indefinitely, then the histogram and the frequency polygon will closely approach a smooth curve, known as *Frequency Curve.*

The frequency curve shows the probability distribution (Section 12.1) of the variable in the population and its area bounded by the ordinates at two specified points on the horizontal axis represents the probability that a value of the variable lies between those two limits. Like histogram, the frequency curve is therefore an area diagram.

Generally, there are 4 types of frequency curves:

(i) Symmetrical bell-shaped, (ii) Asymmetrical single-humped, (iii) J-shaped, and (iv) U-shaped. The histogram in Fig. 4.5 represents a J-shaped distribution.

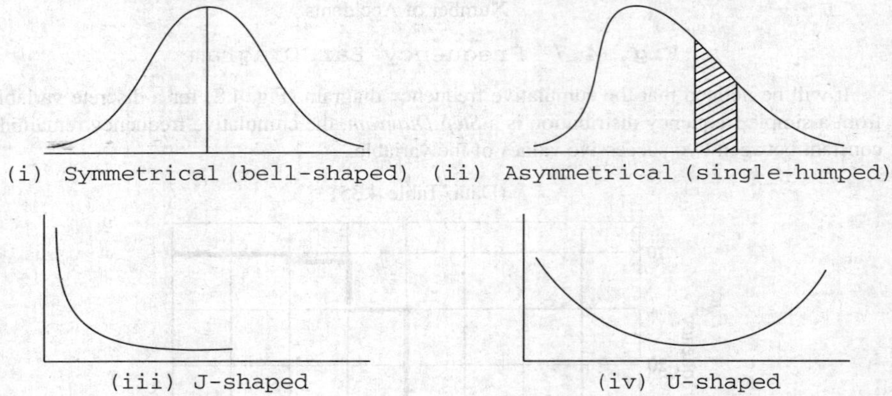

Fig. 4.9 Different Types of frequency Curves

For most of the distributions met with in practice, the frequency curve is bell-shaped, and in such cases three important characteristics are immediately apparent from the frequency curve:

1. The first characteristic is a measure of *central tendency* (p. 119). In particular, the 'mode' of the distribution is given by the abscissa of the highest point of the frequency curve.
2. The second characteristic is a measure of *dispersion* (p. 204). In particular, the 'range', i.e. the maximum possible discrepancy between any two values, is given by the distance between the two points at which the frequency curve meets the hoizontal axis.
3. The third characteristic is the *shape* of the frequency curve, i.e. whether the curve is symmetrical or not; and if not, a measure of the degree of "skewness" (p. 271). A symmetrical curve indicates that mean, median and mode are equal. For asymmetrical curves this is not true, the mean being greater or less than the mode according as the longer tail of the curve lies to the right for to the left (Fig. 7.1).

A very important symmetrical (bell-shaped) frequency curve used in statistical theory is the so-called "*Normal curve*"; the corresponding probability distribution is called "Normal distribution" (Chapter 12). However, this curve does not adequately represent most of the data occurring in business and commerce.

EXERCISES

1. What is a frequency distribution? What considerations should be given in selecting the class-intervals while preparing a frequency distribution? *[I.C.W.A., June '75 (old)]*
2. Define the terms: frequency, relative frequency, cumulative frequency (of less-than and more-than types), and frequency density. *[B.U., B.A.(Econ) '73]*
3. Below is given the distribution of heights of a group of 60 students:

Height (in cm)	145.0–149.9	150.0–154.9	155.0–159.9	160.0–164.9
No. of students	2	5	9	15
	165.0–169.9	170.0–174.9	175.0–179.9	180.0–184.9
	16	7	5	1

Explain the terms 'class limits' and 'class boundaries' with reference to this distribution. *[I.C.W.A., Dec. '75]*

4. Discuss in detail how you will construct a frequency table from raw data relating to a continuous variable. *[C.U., M.Com. '74; W.B.H.S. '78]*
5. Form a frequency distribution with the following:

7, 4, 3, 5, 6; 3, 3, 2, 4, 3; 4, 3, 3, 4, 4;
3, 2, 2, 4, 3; 5, 4, 3, 4, 3; 4, 3, 1, 2, 3;

[C.U., B.Com. '71]

6. The following are the monthly salaries of 20 employees:

(Rs) 130, 62, 145, 118, 125; 76, 151, 142, 110, 98;
95, 116, 100, 103, 71; 85, 80, 122, 132, 95;

Form a frequency distribution with class intervals Rs 61–80, 81–100, 101–120, 121–140 and 141–160. *[C.U., B.Com. '74]*

7. The data below give the marks secured by 70 candidates in a certain examination:

21	31	35	52	64	74	89	53	42	7
22	35	43	67	76	35	46	26	32	40
72	43	38	41	63	71	28	32	45	54
15	18	52	73	86	50	39	55	47	12
44	58	67	85	39	40	50	65	72	69
57	63	5	56	79	37	24	54	82	49
51	54	68	29	34	44	58	62	59	65

Construct a frequency distribution of the marks, taking classes of uniform width of 10 marks and 0 as the lower limit of the lower-most class. *[I.C.W.A., June '74]*

8. Ages at death (years) of 50 persons of a town are given below:

36	48	50	45	49	31	50	48	43	42
37	32	40	39	41	47	45	39	43	47
38	39	37	40	32	52	56	31	54	36
51	46	41	55	58	31	42	53	32	44
53	36	60	59	41	53	58	36	38	60

Arrange the data in a frequency distribution in 10 class intervals, and obtain the percentage frequency in each class interval. *[C.U., B.Com. '72]*

9. Form an ordinary frequency table from the following cumulative frequency distribution of marks obtained by 22 students:

Marks	*Number of students*
Below 10	3
" 20	8
" 30	17
" 40	20
" 50	22

[I.C.W.A., June '77]

10. From the following data, calculate the "percentage" of workers getting wages: (a) more than Rs 44, (b) between Rs 22 and Rs 58.

Wages (Rs) :	0–10	10–20	20–30	30–40	40–50	50–60	60–70	70–80	Total
No. of workers :	20	45	85	160	70	55	35	30	500

[C.A., May '76]

11. What do you mean by a histogram? Explain with the help of an example and illustration how it is constructed in the case of a continuous variable with unequal class intervals. *[C.A., Nov. '73]*
12. What is a Histogram? How is it constructed?
13. Draw the histogram of the following frequency distribution of heights of 100 college students:

Height (cm)	141–150	151–160	161–170	171–180	181–190	Total
Frequency	5	16	56	19	4	100

[W.B.H.S. '78]

14. Draw histogram and frequency polygon to present the following data:

Income (Rs)	100–149	150–199	200–249	250–299	300–349	350–399
No. of individuals	21	32	52	105	62	43
	400–449	450–499	Total			
	18	9	342			

[I.C.W.A., June '78]

15. Draw the histogram of the distribution given below and obtain the number of firms whose sales lie between Rs 12,00,000 and Rs 26,00,000:

Values of Sales (Rs '000)	No. of Firms
0 – 500	3
500 – 1000	42
1000 – 2500	288
2500 – 3500	150
3500 – 4500	51

Draw also the cumulative frequency polygon. *[C.U., B.A. (Econ) '71]*

16. What are cumulative frequencies? How do you represent them diagrammatically for discrete and continuous distributions? *[I.C.W.A., June '74]*
17. What is a cumulative frequency graph? Explain its uses. Draw a cumulative frequency graph and estimate the number of persons between the ages 30–32 in the following table:

Age	20–25	25–30	30–35	35–40	40–45	45–50	50–55	55–60
No. of persons	50	70	100	180	150	120	70	59

[C.U., M.Com. '68]

18. Describe the two types of 'ogive' and discuss their uses. *[I.C.W.A., Dec. '75 (old)]*
19. Draw an ogive from the following data and find graphically the number of observations lying between 360 and 440:

Value	*Number of observations*
More than 200	400
More than 250	370
More than 300	315
More than 350	220
More than 400	115
More than 500	45
More than 600	15
More than 700	0

[*I.C.W.A., Jan.* '72]

20. Draw less-than Ogive on the basis of the data given below:

Mid point	18	25	32	39	46	53	60	
Frequency	10	15	32	42	26	12	9	$N = 146$

[C.A., May '74]

ANSWERS

3. *Class limits* : 145.0 and 149.9, 150.0 and 154.9, etc.
   Class boundaries : 144.95 and 149.95, 149.95 and 154.95, etc.

5.

Value:	1	2	3	4	5	6	7	Total
Frequency:	1	4	12	9	2	1	1	30

6.

Salary (Rs):	61–80	81–100	101–120	121–140	141–160	Total
Frequency:	4	5	4	4	3	20

7.

Marks:	0–9	10–19	20–29	30–39	40–49	50–59	60–69	70–79	80–89	Total
Frequency:	2	3	6	11	12	15	10	7	4	70

8.

Age (years):	31–33	34–36	37–39	40–42	43–45	46–48	49–51	52–54	55–57	58–60	Total
Frequency:	6	4	7	7	5	5	4	5	2	5	50
Percentage frequency:	12	8	14	14	10	10	8	10	4	10	100

9.

Marks:	0–9	10–19	20–29	30–39	40–49	Total
Frequency:	3	5	9	3	2	22

10. (a) 32.4, (b) 68.4

15. 265 17. 40 19. 112

20. **Hint:** Class boundary points are 14.5, 21.5, 28.5; etc.

5 MEASURES OF CENTRAL TENDENCY

5.1 AVERAGES OR MEASURES OF CENTRAL TENDENCY

The word 'average' denotes a 'representative' or 'typical value' of a whole set of observations. It is a single figure which describes the entire series of observations with their varying sizes. Since a typical value usually occupies a central position, so that some observations are larger and some others are smaller than it, averages are also known as *measures of central tendency.*

There are 3 measures of central tendency—Mean, Median, and Mode. Again, Mean is of 3 types—Arithmetic Mean (A.M.), Geometric Mean (G.M.) and Harmonic Mean (H.M.). The words 'mean' and 'average' only refer to Arithmetic Mean.

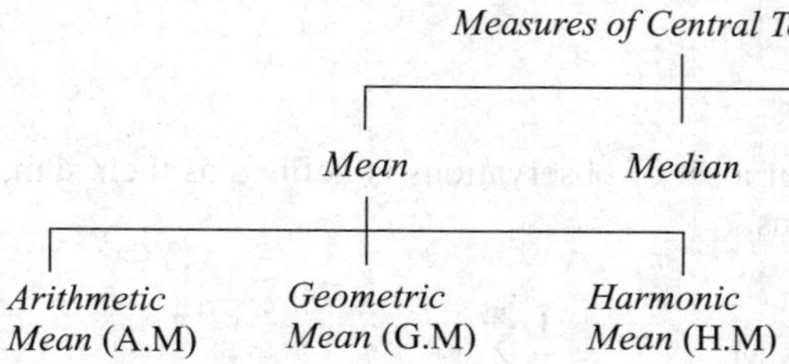

Example 5.1 *What are the desirable properties of a statistical average? Examine how far the arithmetic mean, the geometric mean, the median and the mode possess these properties.*

[C.U., B.A.(Econ) '62; M.Com. '72; B.U., B.A.(Econ) '73; C.A., Nov. '67, May '74, May '76]

Solution According to G.U. Yule, the desirable properties of a satisfactory average are as follows:

(i) It should be rigidly defined;
(ii) It should be based on all observations;
(iii) It should be readily comprehensible;
(iv) It should be easily calculated;
(v) It should be affected as little as possible by sampling fluctuations; and
(vi) It should readily lend itself to algebraic treatment.

Arithmetic mean is the simplest and easiest to understand, and fulfils all the criteria of a satisfactory average. It is rigidly defined, based on all observations, and also easy to calculate. If the arithmetic mean and the number of observations in each of several groups are given, they can be algebraically combined to find the arithmetic mean of the composite group. Arithmetic mean is also the most stable of all the measures of central tendency, so far as sampling fluctuations are concerned. If many samples are drawn from the same statistical population, arithmetic mean will be found to fluctuate less than any other measure of central tendency.

Geometric mean is rigidly defined, based on all observations, and is suited for algebraic treatment. However, it is difficult to calculate and is not readily understood by common people.

Median is rigidly defined and based on all observations. But it is not so well-understood as arithmetic mean or mode. Median is easy to calculate. However, it shows greater fluctuation from sample to sample, and hence is less reliable than arithmetic mean. It can not be treated algebraically.

Although the technical term 'mode' may not be known to all, its concept is well-understood and easily comprehensible. A dealer of shoes will be more interested in the mode of different sizes of shoes, he sells, than arithmetic mean or median. Mode is that size of shoes which is sold most. It is rigidly defined, and based on all observations. However, it is very difficult to calculate mode accurately. Mode does not lend itself to algebraic treatment as arithmetic mean or geometric mean. It is less reliable and less stable as regards sampling fluctuations.

Example 5.2 *Give a critical review of the different measures of central tendency, with examples.*

[C.U., M.Com. '64, '71; B.Com(Hons) '66; B.A.(Econ) '67, '73, '75]

Solution There are three measures of central tendency: (1) Mean, (2) Median, (3) Mode. Again, Mean is of three types—(i) Arithmetic Mean, (ii) Geometric Mean, and (iii) Harmonic Mean.

(1) Mean

Arithmetic Mean (A.M.) of a set of observations is defined as their sum, divided by the number of observations.

Formula: (i) Simple A.M. $\bar{x} = \frac{1}{n}\sum_{i=1}^{n} x_i$

(ii) Weighted A.M.

$$\bar{x} = \frac{1}{N}\sum_{i=1}^{n} f_i x_i, \quad \text{where } N = \sum_{i=1}^{n} f_i$$

Illustration 1. A.M. of 8, 1, 6 is

$$\bar{x} = \frac{1}{3}(8+1+6) = \frac{1}{3}\times 15 = 5$$

Illustration 2. A.M. of 8, 1, 6 with frequencies (or weights) 3, 2, 5 respectively is (here $N = 3 + 2 + 5 = 10$)

$$\bar{x} = \frac{1}{10}(3\times 8 + 2\times 1 + 5\times 6) = 56/10 = 5.6$$

A.M. is simple to calculate and easy to understand. It has also certain algebraic properties, for which it is widely used. For instance, given the mean ($\bar{x}_i$), and the number of observations (n_i) in each of several groups, the mean ($\bar{x}$) of the composite group can be calculated by the formula

$$N\bar{x} = \Sigma n_i \bar{x}_i \qquad \text{where } N = \Sigma n_i$$

Illustration 3. If 7, 4, 9 be the A.M. of three separate groups containing 6, 2, 5 observations respectively, then the A.M. of all the 13 observations is given by

$$13\,\bar{x} = 6 \times 7 + 2 \times 4 + 5 \times 9 = 95$$

$$\therefore \quad \bar{x} = 95/13 = 7.31$$

An important property of A.M. is that the algebraic sum of deviations of observations from their A.M. is zero.

$$\Sigma(x_i - \bar{x}) = 0, \qquad \text{where } \bar{x} = \text{simple A.M.}$$

$$\Sigma f_i(x_i - \bar{x}) = 0, \qquad \text{where } \bar{x} = \text{weighted A.M.}$$

Illustration 4. The algebraic sum of deviations of 8, 1, 6 from their A.M., viz 5 (see *Illustration 1*), is

$$(8 - 5) + (1 - 5) + (6 - 5) = 3 + (-4) + 1 = 0.$$

Illustration 5. When the observations 8, 1, 6 are weighted by 3, 2, 5 respectively, A.M. is 5.6 (see *Illustration 2*), and the algebraic sum of the deviations is

$$3(8 - 5.6) + 2\,(1 - 5.6) + 5(6 - 5.6)$$
$$= 3 \times 2.4 + 2 \times (-4.6) + 5 \times 0.4$$
$$= 7.2 - 9.2 + 2.0 = 0$$

However, A.M. suffers from the disadvantage that it is highly affected by the presence of extreme values, i.e. extremely large or small observations.

Illustration 6. A.M. of 2, 5, 8, 4, 6, 9 is 34/6 = 5.67; but A.M. of 2, 5, 8, 4, 6, 9, 71 is 105/7 = 15, which is nearly 3 times the previous A.M. 5.67, because of the presence of a very large observation 71.

Geometric Mean (G.M.) of a set of n observations is the n-th root of their product.

Formula: (i) Simple G.M. $\quad g = \sqrt[n]{x_1 x_2 \ldots x_n}$

(ii) Weighted G.M. $\quad G = \sqrt[N]{x_1^{f_1}\, x_2^{f_2} \ldots x_n^{f_n}}$

Illustration 7. (a) G.M. of 8, 1, 6 is

$$g = \sqrt[3]{8 \times 1 \times 6} = \sqrt[3]{48}$$

(b) G.M. of 8, 1, 6 with weights 3, 2, 5 respectively is

$$G = \sqrt[10]{8^3 \times 1^2 \times 6^5} = \sqrt[10]{3981312}$$

However, for numerical calculations, the formulae are written with logarithms:

(i) $\log g = \dfrac{1}{n} \sum_i \log x_i$

(ii) $\log G = \dfrac{1}{N} \sum_i f_i \left(\log x_i\right)$

Illustration 8. (a) G.M. of 8, 1, 6 is calculated by

$$\log g = \frac{1}{3}(\log 8 + \log 1 + \log 6)$$

$$= \frac{1}{3}(0.9031 + 0 + 0.7782) = 0.5604.$$

$$\therefore \quad g = \text{antilog } 0.5604 = 3.63.$$

(b) G.M. of 8, 1, 6 weighted by 3, 2, 5 respectively is given by

$$\log G = \frac{1}{10}(3 \times \log 8 + 2 \times \log 1 + 5 \times \log 6)$$

$$= \frac{1}{10}(3 \times 0.9031 + 2 \times 0 + 5 \times 0.7782)$$

$$= 0.6600$$

$$\therefore \quad G = \text{antilog } 0.6600 = 4.57$$

G.M. is more difficult to calculate than A.M. But, since G.M. is less affected by the presence of extreme values, it is sometimes used in the construction of index numbers.

Harmonic Mean (H.M.) is the reciprocal of the arithmetic mean of reciprocals of observations.

Formula: (i) Simple H.M. $h = \dfrac{n}{\sum_i \left(\dfrac{1}{x_i}\right)}$

(ii) Weighted H.M. $H = \dfrac{N}{\sum_i \left(\dfrac{f_i}{x_i}\right)}$

Illustration 9. H.M. of 8,1, 6 is

$$h = \frac{3}{\frac{1}{8}+\frac{1}{1}+\frac{1}{6}} = \frac{3}{\frac{31}{24}} = \frac{72}{31} = 2.32$$

Illustration 10. H.M. of 8, 1, 6 weighted by 3, 2, 5 respectively is

$$H = \frac{10}{\frac{3}{8}+\frac{2}{1}+\frac{5}{6}} = \frac{10}{\frac{77}{24}} = \frac{240}{77} = 3.12$$

H.M. has a very limited use.

For any given set of observations, A.M. is never less than G.M., and G.M. is never less than H.M.

$$A.M. \geq G.M. \geq H.M.$$

Illustration 11. From *Illustrations 1, 8(a)* and *9* we find that for the 3 observations 8, 1, 6:

$$A.M. = 5; \quad G.M. = 3.63; \quad H.M. = 2.32;$$

which support the above theory.

Illustration 12. From *Illustrations* 2, 8(b) and 10, we find that for the observations 8, 1, 6, weighted by 3, 2, 5 respectively:

$$A.M. = 5.6; \quad G.M. = 4.57; \quad H.M. = 3.12$$

This verifies the theory.

(2) Median

Median of a set of observations is the value of the middle-most item when they are arranged in order of magnitude. It can be calculated from a grouped frequency distribution either (i) by using simple interpolation in a cumulative frequency distribution, or (ii) by using the formula:

$$Median = l_1 + \frac{(N/2) - F}{f_m} \times c$$

where

l_1 = lower boundary of the median class;
N = total frequency;
F = cumulative frequency corresponding to l_1;
f_m = frequency of the median class;
c = width of the median class.

Median is, in a certain sense, the real measure of central tendency, as it gives the value of the most central observation. It is unaffected by extreme values, and can be easily calculated from frequency distributions with open-end classes.

Illustration 13. To find the median of 2, 5, 8, 4, 9, 6, 71, we arrange the observations in order of magnitude

2, 4, 5, (6), 8, 9, 71.

Here, the middle-most item is 6 : i.e. Median = 6. If the large observation 71 is replaced by a still large value, median remains unaltered.

(3) Mode

Mode of a set of observations is that value which occurs with the maximum frequency. It is the most typical or fashionable value and at times represents the true characteristic of the frequency distribution as a measure of central tendency. In the case of a simple frequency distribution, mode can be found by inspection only. However, in the case of a grouped frequency distribution, it is difficult to find the mode accurately. It is generally calculated by the formula:

$$Mode = l_1 + \frac{d_1}{d_1 + d_2} \times c$$

where

l_1 = lower boundary of the modal class;
d_1 = difference of frequencies in the modal class and the preceding class;
d_2 = difference of frequencies in the modal class and the following class;
c = common width of classes.

The formula is applicable only when all classes have the same width. Mode has certain peculiarities too. When all observations occur with equal frequency, mode does not exist. Again, there is more than one mode, if two or more values occur with the maximum frequency.

Illustration 14. (i) Mode of the observations 2, 5, 8, 4, 3, 4, 5, 2, 4 is 4, because 4 occurs the maximum number of times.

(ii) For the observations 5, 3, 6, 3, 5, 6, there is no mode.

(iii) For the observations 5, 3, 6, 3, 5, 10, 7, 2, there are two modes, viz. 3 and 5.

5.2 ARITHMETIC MEAN (A.M.)

Arithmetic Mean of a set of observations is defined to be their sum, divided by the number of observations.

Given n observations $x_1, x_2, ..., x_n$, their A.M., denoted by the symbol $\overline{x}$ (read *x-bar*), is

$$\overline{x} = \frac{x_1 + x_2 + \ldots + x_n}{n} = \frac{1}{n}\Sigma x \quad (5.2.1)$$

If $x_1, x_2, ..., x_n$ have frequencies $f_1, f_2, ..., f_n$ respectively, i.e. x_1 occurs f_1 times, x_2 occurs f_2 times and so on, then the sum of all the $f_1 + f_2 + ... + f_n$ observations is

$$\underbrace{x_1 + x_1 + \ldots + x_1}_{f_1 \text{ terms}} + \underbrace{x_2 + x_2 + \ldots + x_2}_{f_2 \text{ terms}} \ldots + \underbrace{x_n + x_n + \ldots + x_n}_{f_n \text{ terms}}$$

$$= f_1x_1 + f_2x_2 + \ldots + f_nx_n$$

Hence, the arithmetic mean is

$$\overline{x} = \frac{f_1x_1 + f_2x_2 + \ldots + f_nx_n}{f_1 + f_2 + \ldots + f_n} = \frac{\Sigma fx}{N} \quad (5.2.2)$$

where $N = \Sigma f$ is the total frequency. This is sometimes referred to as *Weighted arithmetic mean*, as distinct from *Simple arithmetic mean*.

Example 5.3 *Distinguish between simple and weighted average, and state the circumstances under which the latter should be employed.*

[I.C.W.A., June. '73; C.A., Nov. '66 & '73, May '70, '76]

Solution *Simple Average* (sometimes called *Unweighted Average*) of a set of observations is the average calculated by considering each different value as equally important. If however, in the context of the problem, all values are not considered as equally important, the *relative importance* of the different values is indicated by assigning certain numbers, called *weights.* Let $x_1, x_2, ..., x_n$ denote the values and $w_1, w_2, ..., w_n$ be the corresponding weights. Then

$$\text{Simple Average} = \frac{x_1 + x_2 + \ldots + x_n}{n} = \frac{\Sigma x}{n}$$

$$\text{Weighted Average} = \frac{w_1x_1 + w_2x_2 + \ldots + w_nx_n}{w_1 + w_2 + \ldots + w_n} = \frac{\Sigma wx}{\Sigma w}$$

When a frequency distribution is given, the frequencies of values are themselves treated as weights, and the formulae for averages (i.e. arithmetic means) are

$$\text{Simple A.M.} = \frac{\Sigma x}{n}\,;\ \text{Weighted A.M.} = \frac{\Sigma fx}{N} \quad (5.2.3)$$

where $N = \Sigma f$. Thus for the simple average each different value is considered only once, while for weighted average each value is considered as many times as it occurs. Simple average is obtained on dividing the total of a set of observations by their number. Weighted average is obtained on multiplying the values of the variable by the corresponding weights, and then dividing the sum of products by the sum of weights.

Simple and weighted averages are equal only when all the weights are equal, i.e. $w_1 = w_2 = ... = w_n$ or $f_1 = f_2 = ... = f_n$. Even if the weight of one observation be different from the rest, the two averages will be unequal. Again, if each weight is multiplied (or divided) by a constant, the weighted average will remain the same.

Simple average is used when there is nothing to suspect that the values to be averaged are comparable. However, in cases where the values differ in accuracy in varying degrees or are conceptually such as would influence the final result differently, weighted average is used. If the weights are known, weighted average is more appropriate, and must be used.

[**Note:** (i) The word 'average' used in 'simple average' and 'weighted average' generally refers to A.M., G.M. or H.M., e.g. Simple A.M., Weighted G.M., etc. Expressions like 'simple median' and 'weighted mode' are not used.

(ii) Although the words 'frequency' and 'weight' are used almost synonymously, there is a slight difference between the two. Frequency of an observation, being the number of times it occurs, must always be a whole number. Weight of an observation is its '*relative importance*' as compared with the other values, and may therefore be expressed either in whole numbers or percentages or fractions. Weights and frequencies are proportional. Frequencies are generally used as weights.

(iii) Refer to *Illustrations 1* and *2* in Example 5.2.

$$\text{Simple A.M.} = \frac{1}{3}(8 + 1 + 6) = \frac{1}{3} \times 8 + \frac{1}{3} \times \frac{1}{3} \times 6$$

Simple A.M. of 3 observations is thus obtained by adding one-third part of each observation.

$$\text{Weighted A.M.} = \frac{1}{10}(3 \times 8 + 2 \times 1 + 5 \times 6) = \frac{2}{10} \times 8 + \frac{3}{10} \times 1 + \frac{5}{10} \times 6$$

This shows that in the weighted A.M., we do not take equal parts of the observations but $\frac{2}{10}$ part of 8, $\frac{2}{10}$ part of 1, $\frac{5}{10}$ part of 6. The individual parts must of course add to 1 in both the cases, viz. $\frac{1}{3} + \frac{1}{3} + \frac{1}{3} = 1$, and also $\frac{2}{10} + \frac{2}{10} + \frac{5}{10} = 1$.]

Example 5.4 *Calculate the weighted mean price of a table from the following data, assuming that weights are proportional to the number of tables sold:*

Price per Table (Rs)	*36*	*40*	*44*	*48*
Number of Tables Sold	*14*	*11*	*9*	*6*

[C.A., Nov. '62]

Solution

Table 5.1 Calculations for Weighted Mean

Price (Rs) per Table (x)	*No. of Tables Sold (f)*	*fx*
36	14	504
40	11	440
44	9	396
48	6	288
Total	40	1628

$$\text{Weighted Mean} = \frac{\sum fx}{N} = \frac{1628}{40} = 40.70 \text{ Rs}$$

[**Note:** Simple Mean = $\frac{30 + 40 + 44 + 48}{4} = \frac{168}{4} = 4200$ Rs] *Ans.* Rs 40.70 P

Example 5.5 *Calculate the simple and the weighted arithmetic average price per ton of coal purchased by an industry for the half year. Account for the difference between the two.*

Month	*Price per Ton*	*Tons Purchased*
Jan.	*Rs 42.50*	*25*
Feb.	*Rs 51.25*	*30*
Mar.	*Rs 50.00*	*40*
Apr.	*Rs 52.00*	*50*
May	*Rs 44.25*	*10*
June	*Rs 54.00*	*45*

[C.A., Nov. '66]

Solution In order to find the weighted average price per ton, we have to use the 'tons purchased' as weights (Note the words 'average *price*'; so 'price' must be the variable x).

Table 5.2 Calculations for Simple and Weighted Arithmetic Average

Month	*Price (Rs) per Ton (x)*	*Tons Purchased (f)*	*fx*
Jan.	42.50	25	1,062.50
Feb.	51.25	30	1,537.50
Mar.	50.00	40	2,000.00
Apr.	52.00	50	2,600.00
May	44.25	10	442.50
June	54.00	45	2,430.00
Total	294.00	200 = N	10,072.50

$$\text{Simple arithmetic average} = \frac{\Sigma x}{n} = \frac{294.00}{6} = 49.00 \text{ Rs}$$

$$\text{Weighted arithmetic average} = \frac{\Sigma fx}{N} = \frac{10072.50}{200} = 50.36 \text{ Rs}$$

Simple average and weighted average are equal, only when all weights are equal. In the present case, the difference between simple and weighted arithmetic averages has arisen, because the weights are not all equal.

Ans. Rs 49.00; Rs 50.36 P

Example 5.6 *The relative importance of the following eight groups of family expenditure was found to be: food 348, rent 88, clothing 97, fuel 65, household durable goods 71, miscellaneous goods 35, services 79, drink and tobacco 217. The corresponding increases in price for October 1950 based on 1947 gave the following eight percentages 25, 1, 22, 18, 14, 13, 11 and 4. Calculate the average percentage increase in price for the whole group.*

[B.U., B.A.(Econ) '67]

Solution

[**Note:** Since the relative importance (i.e. weight) is given, we must calculate the weighted average].

Table 5.3 Calculations for Weighted Average

Group	*Percentage Increase in Price (x)*	*Weight (f)*	*fx*
1. *Food*	25	348	8,700
2. *Rent*	1	88	88
3. *Clothing*	22	97	2,134
4. *Fuel*	18	65	1,170
5. *Household durable goods*	14	71	994
6. *Miscellaneous goods*	13	35	455
7. *Services*	11	79	869
8. *Drink & Tobacco*	4	217	868
Total	–	1,000	15,278

$$\text{Average percentage increase} = \frac{\Sigma fx}{\Sigma f} = \frac{15,278}{1,000} = 15.278$$

Ans. 15.278

Example 5.7 *The following table gives the marks of two candidates:*

Subject	*Weight*	*Marks of*	
		Ist Candidate	*2nd Candidate*
A	*1*	*70*	*80*
B	*2*	*65*	*64*
C	*3*	*58*	*56*
D	*4*	*63*	*60*

Find the weighted average mark of each candidate. By what figure would the 2nd candidate have had to increase his marks in subject B, all other marks remaining same, in order that both the candidates have the same place?

[C.U., M.Com. '63]

Solution

Table 5.4 Calculations for Weighted Arithmetic Mean

Subject	Marks of 1st Candi. (x)	Marks of 2nd Candi. (y)	Weight (f)	fx	fy
A	70	80	1	70	80
B	65	64	2	130	128
C	58	56	3	174	168
D	63	60	4	252	240
Total	–	–	10	626	616

Weighted averages (arithmetic mean) of marks are:

$$\text{(a) 1st candidate} = \frac{\Sigma fx}{\Sigma f} = \frac{260}{10} = 62.6$$

$$\text{(b) 2nd candidate} = \frac{\Sigma fx}{\Sigma f} = \frac{616}{10} = 61.6$$

(*Second Part*) Let the required increase in marks be K. Then the marks of 2nd candidate in subject B would be $(64 + K)$. Hence, in Table 5.4, the number 64 against subject B in the 3rd column would be $64 + K$; and consequently in the last column 128 should be replaced by $2 \times (64 + K) = 128 + 2K$. The total of the last column would then be $616 + 2K$. In order that both candidates have the same place, their total marks should be equal. Hence

$$616 + 2K = \text{Total marks of 1st candidate} = 626$$

or, $\quad 2K = 626 - 616 = 10$

$\therefore \quad K = 10/2 = 5.$ *Ans.* (i) 62.6, 61.6, (ii) 5

5.3 IMPORTANT PROPERTIES OF A.M.

(a) The total of a set of observations is equal to the product of their number and the A.M.

(i) $\Sigma x_i = n\,\overline{x}$; (ii) $\Sigma f_i x_i = N\overline{x}$ (5.3.1)

If we assume that observation x_i are all different, then the first relation in (5.3.1) implies that the simple sum is equal to the product of their number and the simple A.M. The second relation implies that the weighted sum is equal to the product of the sum of weights and the weighted A.M.

(b) The sum of the deviations of a set of observations from their A.M. is always zero.

(i) $\Sigma(x_i - \overline{x}) = 0$, where $\overline{x} = \Sigma x_i / n$

(ii) $\Sigma f_i(x_i - \overline{x}) = 0$, where $\overline{x} = \Sigma f_i x_i / N$ (5.3.2)

[**Note:** '*Deviation of A from B*' is defined to be the quantity A–B (not B–A). Deviation may be positive, or negative, or zero, according as A is greater than, or less than, or equal to B. For example, deviation of 8 from 2 is $8 - 2 = 6$; deviation of 4 from 7 is $4 - 7 = -3$; deviation of x_i from A is $x_i - A$; deviation of x_i from $\overline{x}$ is $x_i - \overline{x}$.]

Relations (5.3.2) imply that the simple (weighted) sum of the deviations of observations from the simple (weighted) A.M. is always zero (Example 5:2, *Illustrations 4 & 5*).

(c) If two variables x and z are so related that $z = ax + b$ for each $x = x_i$, where a and b are constants, then

$$\bar{z} = a\bar{x} + b \tag{5.3.3}$$

In particular, if $y = \dfrac{x-c}{d}$, where c and d are constants, then

$$\bar{x} = c + d\bar{y} \tag{5.3.4}$$

Relation (5.3.4) implies that if each of the observations x_i is increased, decreased, multilied of divided by a constant, then the mean $\bar{x}$ also will be similarly affected.

(d) If a group of n_1 observations has A.M. $\bar{x}_1$, and another group of n_2 observations has A.M. $\bar{x}_2$, then the A.M. ($\bar{x}$) of the composite group of $n_1 + n_2$ (= N, say) observations is given by

$$N\bar{x} = n_1\bar{x}_1 + n_2\bar{x}_2 \tag{5.3.5}$$

This can be generalised to any number of groups:

$$N\bar{x} = \Sigma n_i\bar{x}_i \quad \text{where } N = \Sigma n_i \tag{5.3.6}$$

(e) The sum of the squares of deviations of a set of observations has the smallest value, when deviations are taken from their A.M.

(i) $\Sigma(x_i - A)^2$ is minimum, when A = simple A.M.

(ii) $\Sigma f_i(x_i - A)^2$ is minimum, when A = weighted A.M. (5.3.7)

Example 5.8 *Show that if $\bar{x}$ is the arithmetic mean of the quantities $x_1, x_2, \ldots, x_n$ then $\sum_{1}^{n}(x_i - \bar{x}) = 0$.* [C.U., M.Com. '63, '65]

Or,

Prove that the sum of the deviations of $x_1, x_2, \ldots, x_n$ from their mean $\bar{x}$ is equal to zero. [I.C.W.A., Dec. '76 (old)]

Solution (*First method*) Let $x_1, x_2, \ldots, x_n$ be a series of numbers, and $\bar{x}$ their arithmetic mean, defined by

$$\bar{x} = \frac{x_1 + x_2 + \ldots + x_n}{n}$$

The deviations of the numbers from their A.M. ($\bar{x}$) are

$$(x_1 - \bar{x}), (x_2 - \bar{x}), \ldots, (x_n - \bar{x})$$

The algebraic sum of these deviations is

$$\begin{aligned}
&= (x_1 - \bar{x}), + (x_2 - \bar{x}), + \ldots + (x_n - \bar{x}) \\
&= x_1 + x_2 + \ldots + x_n - \bar{x} - \bar{x} - \ldots - \bar{x} \ (n \text{ times}) \\
&= (x_1 + x_2 + \ldots + x_n) - n\bar{x} \\
&= n\bar{x} - n\bar{x} \qquad \text{by (5.3.1)} \\
&= 0.
\end{aligned}$$

(*Second method*)

By definition $\bar{x} = \dfrac{1}{n}\Sigma x_i$

The deviation of x_i from the mean $\bar{x}$ is $(x_i - \bar{x})$, $i = 1, 2, \ldots, n$.

Hence, the algebraic sum of all the deviations is

$$\sum_{1}^{n}(x_i - \bar{x})$$

$$= \sum_{1}^{n} x_i - \sum_{1}^{n} \bar{x} \qquad \text{by (3.15.2)}$$

$$= n\bar{x} - \sum_{1}^{n} \bar{x} \qquad \text{by (5.3.1)}$$

$$= n\bar{x} - n\bar{x} \qquad \text{by (3.15.4)}$$

$$= 0.$$

Example 5.9 *Show that if $\bar{x}$ be the arithmetic mean of the values x_i weighted by f_i $(i = 1, 2, ..., n)$, then*

$$\sum_{1}^{n} f_i(x_i - \bar{x}) = 0.$$

Solution (*First method*) The arithmetic mean of the values x_i weighted by f_i $(i = 1, 2, ..., n)$ is, by definition,

$$\bar{x} = \frac{f_1x_1 + f_2x_2 + ... + f_nx_n}{N}$$

where $N = f_1 + f_2 + ... + f_n$. Therefore,

$$\sum_{1}^{n} f_i(x_i - \bar{x}) = f_1(x_1 - \bar{x}) + f_2(x_2 - \bar{x}) + ... + f_n(x_n - \bar{x})$$

$$= f_1x_1 - f_1\bar{x} + f_2x_2 - f_2\bar{x} + ... + f_nx_n - f_n\bar{x}$$

$$= f_1x_1 + f_2x_2 + ... + f_nx_n - f_1\bar{x} - f_2\bar{x} - ... - f_n\bar{x}$$

$$= (f_1x_1 + f_2x_2 + ... + f_nx_n) - \bar{x}(f_1 + f_2 + ... + f_n)$$

$$= N\bar{x} - \bar{x}.N = N\bar{x} - N\bar{x} = 0.$$

(*Second method*) $\bar{x} = \dfrac{1}{N}\Sigma f_i x_i$, where $N = \Sigma f_i$

$$\therefore \quad \Sigma f_i(x_i - \bar{x}) = \Sigma(f_ix_i - f_i\bar{x})$$

$$= \Sigma f_ix_i - \Sigma f_i\bar{x}, \qquad \text{by (3.15.2)}$$

$$= \Sigma f_ix_i - \bar{x}\Sigma f_i, \qquad \text{by (3.15.3)}$$

$$= \Sigma f_ix_i - \left(\frac{1}{N}\Sigma f_ix_i\right)\cdot N,$$

$$= \Sigma f_ix_i - \Sigma f_ix_i$$

$$= 0.$$

Example 5.10 *If $y_i = x_i - c$, $(i = 1, 2, ..., n)$ where c is a constant, prove that $\bar{x} = c + y$.*

Solution [Note that y_i is the deviation of x_i from c.]

Since $y_i = x_i - c$, therefore $x_i = c + y_i$

Multiplying both sides by f_i and then summing over all values of $i = 1, 2, ..., n$ we have

$$\sum_{1}^{n} f_ix_i = \sum_{1}^{n} f_i(c + y_i)$$

$$= \Sigma(f_i c + f_i y_i)$$
$$= \Sigma f_i c + \Sigma f_i y_i \quad \text{by (3.15.2)}$$
$$= c\Sigma f_i + \Sigma f_i y_i \quad \text{by (3.15.3)}$$
$$= cN + \Sigma f_i y_i \quad \text{since } \Sigma f_i = N$$

Hence,
$$\bar{x} = \frac{1}{N}\Sigma f_i x_i = \frac{1}{N}(cN + \Sigma f_i y_i)$$
$$= c + \frac{1}{N}\Sigma f_i y_i = c + y$$

The significance of this result is that if $y_1, y_2 \ldots, y_n$ are the deviations of $x_1, x_2, \ldots, x_n$ from an arbitary constant c, then

Mean of x = c + Mean of y.

Example 5.11 *If* $y_i = \dfrac{x_i - c}{d}$ *(i = 1, 2, ..., n) where c and d are constants, prove that* $\bar{x} = c + d\bar{y}$. [B.U., B.A. (Econ) '67; I.C.W.A., July '72]

Solution **[Note:** y_1 may be called *deviation of x_i from c in units of d*. The constants c and d are known as *origin* and *scale* respectively of the values y_i]

Since
$$y_i = \frac{x_i - c}{d}, \text{ we have } dy_i = x_i - c$$
or,
$$x_i = c + dy_i$$

Multiplying both sides by f_i, we get $f_i x_i = f_i(c + dy_i)$

Now, summing over all values of $i = 1, 2, \ldots, n$

$$\Sigma f_i x_i = \Sigma f_i (c + dy_i) = \Sigma(f_i c + d f_i y_i)$$
$$= \Sigma f_i c + \Sigma d f_i y_i \quad \text{by (3.15.2)}$$
$$= c\Sigma f_i + d\,\Sigma f_i y_i \quad \text{by (3.15.3)}$$
$$= cN + d\,\Sigma f_i y_i \quad \text{writing } N = \Sigma f_i$$

Hence,
$$\bar{x} = \frac{1}{N}\Sigma f_i x_i = \frac{1}{N}(cN + d\Sigma f_i y_i)$$
$$= c + d\left(\frac{1}{N}\Sigma f_i y_i\right)$$
$$= c + d\bar{y}$$

This means that if $y_1, y_2, \ldots, y_n$ be the deviations of $x_1, x_2, \ldots, x_n$ from an arbitrary constant c, in units of another constant d, then

Mean of x = c + d × (Mean of y)

It may be noted that the result of Example 5.10 may be obtained from this, on putting $d = 1$.

Example 5.12 *Prove that* $\sum_{i=1}^{n}(x_i - A)^{2/n}$ *is the least when* $A = \bar{x}$, *where* $x_1, x_2, \ldots, x_n$ *are the observations, A is any arbitrary constant and* $\bar{x}$ *the arithmetic mean.*

[C.U., B.A. (Econ) '65]

Or,

Prove that for a given set of observations the sum of the squares of deviations is the minimum, when deviations are taken from the arithmetic mean.

[I.C.W.A., Jan. '71]

Solution $\sum_{i=1}^{n}(x_i - A)^2 / n$ will be the least, when $\sum_{i=1}^{n}(x_i - A)^2$ is so. Now, we can write

$$x_i - A = (x_i - \bar{x}) + (\bar{x} - A)$$

Therefore,

$$\Sigma(x_i - A)^2 = \Sigma\{(x_i - \bar{x}) + (\bar{x} - A)\}^2$$
$$= \Sigma(x_i - \bar{x})^2 + 2(x_i - \bar{x})(\bar{x} - A) + (\bar{x} - A)^2\}$$
$$= \Sigma(x_i - \bar{x})^2 + \Sigma 2(x_i - \bar{x})(\bar{x} - A) + \Sigma(\bar{x} - A)^2, \text{ by (3.15.2)}$$
$$= \Sigma(x_i - \bar{x})^2 + 2(\bar{x} - A)\,\Sigma(x_i - \bar{x}) + n(\bar{x} - A)^2, \text{ by (3.15.3) \& (3.15.4)}$$
$$= \Sigma(x_i - \bar{x})^2 + 2(\bar{x} - A).\,0 + n(\bar{x} - A)^2, \text{ by (5.3.2)}$$

i.e.

$$\Sigma(x_i - A)^2 = \Sigma(x_i - \bar{x})^2 + n(\bar{x} - A)^2.$$

Both terms on the right are positive; because, the first is the sum of n squares $(x_i - \bar{x})^2$, and the second is the product of n (a positive integer) and a square. But we have only to choose the value of A which makes $\Sigma(x_i - A)^2$ the minimum possible, and this will be achieved when the second term on the right has the minimum possible value, viz. 0, i.e.

$$n(\bar{x} - A)^2 = 0$$

or,

$$(\bar{x} - A)^2 = 0$$

or,

$$(\bar{x} - A) = 0$$

$\therefore$

$$A = \bar{x}.$$

Example 5.13 *Show that if* $S = f_k + (f_k + f_{k-1}) + (f_k + f_{k-1} + f_{k-2}) + \ldots + (f_k + f_{k-1} + \ldots + f_2)$ *where* $f_1, f_2, \ldots, f_k$ *are the class frequencies, then A.M.* $= x_1 + \dfrac{hS}{N}$; x_1 *= mid-point of the first class, h = class interval, and N = total frequency.* [C.U., B.A. (Econ) '70]

Solution Let $x_1, x_2, \ldots, x_k$ be the mid-points of the successive classes, and $f_1, f_2, \ldots f_k$ be the corresponding class frequencies. Since the classes have the same width h here 'class interval' has been used in the sense 'width of class'), the successive mid-points $x_1, x_2, \ldots, x_k$ must also have the same difference h, i.e. $x_2 = x_1 + h$, $x_3 = x_1 + 2h$, ..., $x_k = x_1 + (k - 1)h$.

Table 5.5 Calculations for Mean

Mid-point *(x)*	*Class Frequency (f)*	*Cumulative Frequency "more-than"*	$y = \dfrac{x - x_1}{h}$	*fy*
x_1	f_1	–	0	0
x_2	f_2	$F_2 = f_k + f_{k-1} + \ldots + f_2$	1	f_2
x_3	f_3	$F_3 = f_k + f_{k-1} + \ldots + f_3$	2	$2f_3$
...	...	... ...	...	
x_{k-2}	f_{k-2}	$F_{k-2} = f_k + f_{k-1} + f_{k-2}$	$k - 3$	$(k-3)f_{k-2}$
x_{k-1}	f_{k-1}	$F_{k-1} = f_k + f_{k-1}$	$k - 2$	$(k-2)f_{k-1}$
x_k	f_k	$F_k = f_k$	$k - 1$	$(k-1)f_k$
Total	N	S	–	Σfy

Let $F_1, F_2, \ldots F_1$ denote the "more-than" cumulative frequencies. It will be noticed from Table 5.5, that the given expression is

$$S = F_k + F_{k-1} + \ldots + F_2$$

i.e. the sum of "more-than" cumulative frequencies excepting the first (viz. F_1). Also $f_2 = F_2 - F_3, f_3 = F_3 - F_4, \ldots, f_{k-2} = F_{k-2} - F_{k-1}$,

$$f_{k-1} = f_{k-1} - F_k, \text{ and } f_k = F_k \quad \text{...(i)}$$

If we write $y_i = \dfrac{x_i - x_1}{h}$, $(i = 1, 2, \ldots k)$, then using (5.4.2, page 116),

$$\bar{x} = x_1 + h\bar{y} = x_1 + h\left(\frac{\Sigma fy}{N}\right) \quad \text{...(ii)}$$

But, $y_1 = 0, y_2 = 1, y_3 = 2, \ldots, y_k = (k-1)$. Therefore, from the last column of Table 5.5, and using (i) we find

$$\begin{aligned}
\Sigma fy &= 0 + f_2 + 2f_3 + \ldots + (k-3)f_{k-2} + (k-2)f_{k-1} + (k-1)f_k \\
&= 0 + (F_2 - F_3) + 2(F_3 - F_4) + \ldots + (k-3)(F_{k-2} - F_{k-1}) \\
&\qquad + (k-2)(F_{k-1} - F_k) + (k-1)F_k \\
&= F_2 - F_3 + 2F_3 - 2F_4 + \ldots + (k-3)F_{k-2} - (k-3)F_{k-1} \\
&\qquad + (k-2)F_{k-1} - (k-2)F_k + (k-1)F_k \\
&= F_2 + F_3 + F_4 + \ldots + F_{k-1} + F_k \\
&= S
\end{aligned}$$

Hence, from *(ii)* we have $\bar{x} = x_1 + \dfrac{hS}{N}$

[**Note:** This relation gives us another method of calculating the A.M. from a grouped frequency distribution with a common width.]

$$\text{A.M.} = (\text{1st mid-point}) + \frac{(\text{Common width}) \times (\text{Total of more-than cumulative frequency, except the Ist})}{\text{Total frequency}}$$

Illustration: (Data of Example 5.15, page 117)

Mid-point	*Class Frequency*	*Cumulative Frequency ("more-than")*
$55 = x_1$	8	–
65	10	57
75	16	47
85	14	31
95	10	17
105	5	7
115	2	2
Total	$65 = N$	$161 = S$

Here, $x_1 = 55$, $h = 10$, $N = 65$, $S = 161$. Using the above relation

$$\bar{x} = 55 + 10 \times \frac{161}{65} = 79.77$$

This agrees with the result obtained in Example 5.15, using the direct method.

5.4 SIMPLIFIED CALCULATION FOR A.M.

Although A.M. can be calculated by applying (5.2.3) directly, the calculations can be considerably simplified based on the following theory (see Examples 5.10 and 5.11)–

(I) If $y_1, y_2, ..., y_n$ represent the deviations of $x_1, x_2, ..., x_n$ from an arbitrary constant c, then

$$\textit{Mean of } x = c + \textit{Mean of } y$$

Stated in symbols,

if $$y = x - c, \text{ then } \bar{x} = c + \bar{y} \quad (5.4.1)$$

Thus, in order to find the A.M., we can always reduce the given observations x_i by subtracting a convenient number c and obtain deviations $y_i = x_i - c$. The mean of these deviatiation y_i, which can be calculated easily, is added to the constant c, giving the required mean of x (see Example 5.14).

(II) If $y_1, y_2, ..., y_n$ represent the deviations of $x_1, x_2, ..., x_n$ from an arbitrary constant c, in units of another constant d, then

$$\textit{Mean of } x = c + d \cdot (\textit{Mean of } y)$$

Stated in symbols.

if $$y = \frac{x-c}{d}, \text{ then } \bar{x} = c + d\bar{y} \quad (5.4.2)$$

In some cases, the values of x_i are such that after subtracting a constant c, the deviations can further be reduced on division by a constant factor. If this is possible, the deviations $x_i - c$ are divided by the constant factor d, obtaining step-deviations $y_i = (x_i - c)/d$. These will be small numbers compared to the original values x_i. The A.M. of step-deviations is multiplied by the divisor d, and the product is added to the constant c, giving the required mean of x. This method is particularly useful when the consecutive values of x_i have a common difference (see Example 5.15). Then the step deviations can be reduced to 0, 1, 2, 3, ... and –1, –2, –3, ... whatever be the values of x_i.

For the calculation of mean from a grouped frequency distribution, see Example 5.16.

Example 5.14 *The index numbers of 4 commodities were 92, 125, 180 and 80, and the weights 12, 7, 6 and 9 are assigned to these commodities. Find the combined arithmetic average index number.* [C.A., May '65]

Solution We have to find the A.M. of 92, 125, 180, 80 weighted by 12, 7, 6, 9 respectively. By direct calculation

$$\bar{x} = \frac{\Sigma fx}{N} = \frac{12 \times 92 + 7 \times 125 + 6 \times 180 + 9 \times 80}{12 + 7 + 6 + 9}$$

$$= \frac{1104 + 875 + 1080 + 720}{34} = 3779/34 = 111.1.$$

The calculations can be very much simplified, when we apply the method of deviations shown below:

Table 5.6 Calculations for Arithmetic Average

Commodity	*Index Number* x	*Weight* f	$y = x - 100$	fy
A	92	12	–8	–96
B	125	7	25	175
C	180	6	80	480
D	80	9	–20	180
Total	–	34	–	379

We know that, if $y = x - c$, then $\bar{x} = c + \bar{y}$. Here $c = 100$, and

$$\bar{y} = \Sigma fy/N = 379/34 = 11.1$$

Therefore, $\bar{x} = c + \bar{y} = 100 + 11.1 = 111.1$ *Ans.* 111.1

Example 5.15 *Compute the mean weekly wages of the 65 employees working in a factory from the frequency table, using the coding method (i.e. transforming x to a new variate)–*

x	55.00	65.00	75.00	85.00	95.00	105.00	115.00
f	8	10	16	14	10	5	2

[Dip. Soc. Welfare, May '68]

Solution [**Note:** When the successive values of x have a common difference, the calculation of mean will be very much simplified if we transform x to a new variable y, defined by

$$y = \frac{x - (\text{one of the given values of } x, \text{ preferably near the middle})}{\text{common difference}}$$

The values of y can be written down without calculation: Put 0 against the value of x which has been taken as the origin; –1, –2, –3, ... successively as we move away towards the smaller values of x; and 1, 2, 3, ... as we move away towards the higher values of x]

We introduce the new variable $y = \dfrac{x - 85.00}{10}$; so that

Table 5.7 Calculations for Mean

x	f	$y = \dfrac{x - 85.00}{10}$	fy
55.00	8	–3	–24
65.00	10	–2	–20
75.00	16	–1	–16
85.00	14	0	0
95.00	10	1	10
105.00	5	2	10
115.00	2	3	6
Total	65	–	– 34

$$\bar{x} = c + d\bar{y} = 85.00 + 10\bar{y}$$

$$\bar{y} = \Sigma fy/N = -34/65$$

$$\bar{x} = 85.00 + 10\left(\frac{-34}{65}\right)$$

$$= 85.00 - \frac{340}{65}$$

$$= 85.00 - 5.23 = 79.77$$ *Ans.* 79.77

Example 5.16 *How do you calculate the mean of a grouped frequency distribution?*

Solution In a grouped frequency distribution, let $x_1, x_2, \ldots x_n$ represent the mid-values of classes with class frequencies $f_1, f_2, \ldots, f_n$ respectively. Let $y_1, y_2, \ldots, y_n$ represent the deviations of these mid-values from an arbitrary origin c, in units of another constant d; i.e. $y_i = (x_i - c)/d$. Then

$$\text{Mean} = c + d\bar{y} = c + d\left(\frac{\Sigma fy}{N}\right) \qquad \text{where } N = \Sigma f.$$

Here, the mid-value is taken as representative of all observations falling within each class. The computation is done in the following tabular form:

Table 5.8 Calculations for Mean

Class Interval	*Frequency* f	*Mid-value* x	*Deviation* $= x - c$	$y = \frac{x-c}{d}$	fy
(1)	(2)	(3)	(4)	(5)	(6)
	f_1	x_1		y_1	
	f_2	x_2		y_2	
	:	:		:	
	:	:		:	
	f_n	x_n		y_n	
Total	$N = \Sigma f$	–	–	–	Σfy

The steps in the calculation are as follows:

1. From the given data, the mid-values—col. (3)—are found for each class interval. This is done by averaging the two class limits or class boundaries, as given in the frequency distribution.
2. One of the mid-values is taken as the arbitrary origin c, preferably the one corresponding to the maximum frequency, if it lies near the middle of the distribution. The deviation of each mid-value from the arbitrary origin c is calculated, as shown in col. (4).
3. The deviations are now divided by their highest common factor d and the result, called '*step deviation*', is shown in column (5). If the class intervals have a common width, d will be equal to this width. [See Table 5.9].
4. The step deviations are multipled by the corresponding frequencies, and the total of these products is obtained; see col. (6).
5. This total is now multiplied by the divisor d, used in col. (5), and the product divided by the total frequency N.
6. The result is added to the arbitrary origin c, to give the arithmetic mean of the frequency distribution.

$$\text{Mean} = \text{Arbitrary origin} + \frac{\text{total of col. (6)}}{\text{total of col. (2)}} \times (\text{divisor})$$

Example 5.17 *Given the following frequency distribution, calculate the mean:*

Monthly Wages (Rs)	*12.5–17.5*	*17.5–22.5*	*22.5–27.5*	*27.5–32.5*	*32.5–37.5*
Number of Workers	*2*	*22*	*10*	*14*	*3*
	37.5–42.5	*42.5–47.5*	*47.5–52.5*	*52.5–57.5*	*Total*
	4	*6*	*1*	*1*	*63*

[I.C.W.A., Dec. '77]

Solution

Table 5.9 Calculations for Mean

Class Interval	Frequency (f)	Mid-value (x)	$y = \dfrac{x - 30.0}{5}$	fy
12.5–17.5	2	15.0	–3	–6
17.5–22.5	22	20.0	–2	–44
22.5–27.5	10	25.0	–1	–10
27.5–32.5	14	30.0	0	0
32.5–37.5	3	35.0	1	3
37.5–42.5	4	40.0	2	8
42.5–47.5	6	45.0	3	18
47.5–52.5	1	50.0	4	4
52.5–57.5	1	55.0	5	5
Total	63	–	–	–22

[**Note:** If the successive mid-values are found to have a common difference, it is unnecessary to calculate col. (4) of Table 5.8. The step deviation in col. (5) can be obtained easily by the following rule: Take *any one of the mid-values* (preferably one near the middle of the table) as origin c, and the *common differences of mid-values* as the value of d. Put 0 against the value of x which has been taken as origin c, put –1, –2, –3, ... successively as we move away towards the smaller values of x, and 1, 2, 3, ... towards the higher values).

We know that if $y = (x - c)/d$, then $\bar{x} = c + d\bar{y}$ (Example 5.11)

Here $y = \dfrac{x - 30.0}{5}$ (i.e. $c = 30.0$, $d = 5$), and $\bar{y} = \dfrac{\Sigma fy}{N} = \dfrac{-22}{63}$

Therefore, $$\bar{x} = c + d\bar{y} = 30.0 + 5 \times \left(\frac{-22}{63}\right) = 30.0 - \frac{110}{63}$$

$$= 30.0 - 1.75 = 28.25 \text{ Rs}$$

Ans. Rs 28.25

Example 5.18 *Compute the mean from the following frequency distribution of earners by monthly income:*

Income (Rs)	*Below 200*	*200–399*	*400–599*	*600–799*	*800–999*	*1000–1199*
No. of earners	*25*	*72*	*47*	*22*	*13*	*7*

[C.U., B.A.(Econ) '78]

Solution

Note: (i) Here 'class limits' have been used to define the class intervals, but in the Example 5.17 'class boundaries' were used.

(ii) It is generally assumed that the open-end class has the same width as the adjacent class.

Table 5.10 Calculations for Mean

Class Interval	Frequency (f)	Mid-value (x)	$y = \frac{x - 499.5}{200}$	fy
0–199	25	99.5	–2	–50
200–399	72	299.5	–1	–72
400–599	47	499.5	0	0
600–799	22	699.5	1	22
800–999	13	899.5	2	26
1000–1199	7	1099.5	3	21
Total	186	–	–	–53

If $y = (x - c)/d$, then $\bar{x} = c + d\bar{y}$. Here $c = 499.5$, $d = 200$

$$\therefore \quad \bar{x} = 499.5 + 200 \times \left(\frac{-53}{186}\right) = 499.5 - 56.99 = 442.51 \text{ Rs}$$

Example 5.19

Find the mean from the following distribution:

Age in years	*15–19*	*20–24*	*25–29*	*30–34*	*35–44*	*45–59*	*Total*
No. of persons	*37*	*81*	*43*	*24*	*9*	*6*	*200*

Solution

Table 5.11 Calculations for Arithmetic Mean

Class Interval	Frequency (f)	Mid-value x	Deviation $(x - 27)$	$y = \frac{x - 27}{5}$	fy
15–19	37	17	–10	–2	–74
20–24	81	22	–5	–1	–81
25–29	43	27	0	0	0
30–34	24	32	5	1	24
35–44	9	39.5	12.5	2.5	22.5
45–59	6	52	25	5	30
Total	200	–	–	–	–78.5

If $\quad y = \frac{x - c}{d}$ then $\bar{x} = c + d\bar{y}$.

Here $\quad c = 27, d = 5, \bar{y} = \frac{\Sigma fy}{N} = \frac{-78.5}{200}$

$$\therefore \quad \bar{x} = 27 + 5\left(\frac{-78.5}{200}\right) = 27 - 1.96 = 25.04 \text{ years.}$$

[**Note:** (i) Here the classes are of unequal width.

(ii) The deviations in Table 5.11 have no common factor; but since most of them are divisible by 5, we have taken $d = 5$ in order to reduce the calculations. This example shows that the values of y may not be whole numbers.

(iii) The A.M., or in fact any measure of central tendency, must be stated in the same unit in which the observations are given.]

Example 5.20 *Find the missing frequencies in the following frequency distribution, when it is known that A.M. = 11.09*

Class Limits	*9.3–9.7*	*9.8–10.2*	*10.3–10.7*	*10.8–11.2*	*11.3–11.7*
Frequency	*2*	*5*	f_3	f_4	*14*
	11.8–12.2	*12.3–12.7*	*12.8–13.2*	*Total*	
	6	*3*	*1*	*60*	

[C.U., M.Com. '74]

Solution We proceed to calculate the mean in the usual way.

Table 5.12 Calculations for Arithmetic Mean

Class Limits	*Frequency (f)*	*Mid-value (x)*	$y = \dfrac{x - 11.0}{0.5}$	*fy*
9.3–9.7	2	9.5	–3	–6
9.8–10.2	5	10.0	–2	–10
10.3–10.7	f_2	10.5	–1	$-f_3$
10.8–11.2	f_4	11.0	0	0
11.3–11.7	14	11.5	1	14
11.8–12.2	6	12.0	2	12
12.3–12.7	3	12.5	3	9
12.8–13.2	1	13.0	4	4
Total	60	–	–	$23 - f_3$

By the formula $\bar{x} = c + d\bar{y}$, and since $\bar{x} = 11.09$ is given, we have

$$11.09 = 11.0 + 0.5 \times \frac{23 - f_3}{60}$$

or, $$11.09 - 11.0 = \frac{0.5}{60} \times (23 - f_3)$$

i.e. $$0.09 = \frac{1}{120}\,(23 - f_3)$$

or, $$10.8 = 23 - f_3$$

$\therefore$ $$f_3 = 23 - 10.8 = 12.2$$

Since the frequency f_3 must be a whole number, $f_3 = 12$. Again, total frequency is

$$2 + 5 + f_3 + f_4 + 14 + 6 + 3 + 1 = 60;$$

or, $$31 + f_3 + f_4 = 60$$

or, $$f_3 + f_4 = 60 - 31 = 29$$

Since $f_3 = 12$, we have $f_4 = 17$. *Ans.* $f_3 = 12, f_4 = 17$

Example 5.21 *The following table gives the monthly income of 45 employees in a factory. The total monthly income of 5 employees in the class 'Rs 400 and above' is Rs 6000. Find the mean.*

Income (Rs)	220–249	250–279	280–309	310–339	340–369	370–399
Frequency	6	8	12	5	5	4
	400 & above	Total				
	5	45				

[I.C.W.A., Dec. '75-old]

Solution [**Note:** Since the total income of the 5 employees in the open-end class 'Rs 400 and above' is given, it is unnecessary to make any assumption, as in Example 5.18]

We first calculate the mean income of the 40 employees, excluding those in the open-end class.

Putting $y = \dfrac{x - 294.5}{30}$, we find that $\Sigma fy = 7$ (The calculations are not shown here). Hence

$$\bar{x} = 294.5 + 30 \times \left(\frac{7}{40}\right) = 294.5 + 5.25 = \text{Rs } 299.75$$

The total income of the 40 employees is then Rs 299.75 × 40 = Rs 11,990. To this is added the total income Rs 6,000 (given) of the 5 employees in the open-end class, so that for all the 45 employees, Total income = Rs 11,990 + Rs 6,000 = Rs 17,990. Therefore,

$$\text{Mean income} = \frac{17,990}{45} = \text{Rs } 399.78$$

Example 5.22 *Put the following information into a frequency distribution and obtain the arithmetic mean (assuming the range of salary is Rs 0–500):*

"For a group of wage earners, 20%, 40%, 70% and 80% of the wage earners receive less than Rs 50, 120, 300 and 350 respectively; and 5% are receiving Rs 400 and over." [C.U., B.A. (Econ) '65]

Solution We first arrange the data in a cumulative percentage distribution (Table 5.13), from which the grouped frequency distribution (Table 5.14) has been obtained, by taking differences of successive cumulative percentages (see Example 4.9, page 95-96).

Table 5.13
Cumulative Percentage Distribution

Salary (Rs)	Cumulative Percentage
0	0
50	20
120	40
300	70
350	80
400	95
500	100

Table 5.14
Frequency Distribution of Wages

Salary (Rs)	Percentage of Wage Earners
0–50	20
50–120	20
120–300	30
300–350	10
350–400	15
400–500	5
Total	100

Now, we calculate the arithmetic mean from Table 5.14 using the percentage of wage earners as weights (Note that actual frequencies are not known; but this will not affect the calculation of mean).

Table 5.15 Calculations for Arithmetic Mean

Salary (Rs)	Percentage of Wage Earners (f)	Mid-value (x)	Deviation x − 210	$y = \frac{x-210}{5}$	fy
0–50	20	25	−185	−37	−740
50–120	20	85	−125	−25	−500
120–300	30	210	0	0	0
300–350	10	325	115	23	230
350–400	15	375	165	33	495
400–500	5	450	240	48	240
Total	100	–	–	–	−275

If $y = \frac{x-c}{d}$, then $\bar{x} = c + d\bar{y}$. Here $y = \frac{x-210}{5}$, so that $c = 210$ and $d = 5$. Therefore,

$$\bar{x} = 210 + 5\left(\frac{-275}{100}\right) = 210 - 13.75 = 196.25 \text{ Rs}$$

Ans. Rs 196.25

Example 5.23 *(a) Show that the weighted arithmetic mean is unaffected, if all the weights are multiplied by some common factor.* [C.U., B.A.(Econ) '77]

(b) Prove that simple A.M. and weighted A.M. are equal, when all the weights are equal.

Solution (a) Let the observations $x_1, x_2, ..., x_n$ have weights $w_1, w_2, ..., w_n$ respectively. Then

$$\text{Weighted A.M.} = \frac{\Sigma w_i x_i}{\Sigma w_i}$$

It each of the weights is multiplied by a constant k, then using the new weights $kw_1, kw_2, ..., kw_n$, we have

$$\text{Weighted A.M.} = \frac{\Sigma(kw_i)x_i}{\Sigma(kw_i)} = \frac{k\Sigma w_i x_i}{k\Sigma w_i} = \frac{\Sigma w_i x_i}{\Sigma w_i}$$

which is the same as obtained earlier with the original weights. This shows that the weighted A.M. is unaffected, if all the weights are multiplied by a constant.

(b) Simple A.M. $= \frac{\Sigma x_i}{n}$ Weighted A.M. $= \frac{\Sigma w_i x_i}{\Sigma w_i}$

If all the weights are equal, i.e. $w_1 = w_2 = \cdots = w_n = w$ (suppose), then

$$\text{Weighted A.M.} = \frac{\Sigma w x_i}{\Sigma w} = \frac{w\Sigma x_i}{wn} = \frac{\Sigma x_i}{n} = \text{Simple A.M.}$$

This proves the statement.

Example 5.24 *The table given below has been constructed from data obtained from a factory showing the distribution of the number of processed articles per day per person and the rate of payment:*

Table 5.16

Daily No. of Articles Processed per Person	No. of Persons Processing	Rate of Payment per Article Processed (nP)
80–99	12	3.1
100–119	63	3.2
120–139	87	3.3
140–159	56	3.4
160–169	3	3.5

Calculate the rate of payment per person per article processed.

[I.C.W.A., July '64; C.U., B.A.(Econ.) '65, '69]

Solution Let us denote by

x : Rate of payment per article processed,

n : Number of persons processing

z : Daily number of articles processed per person.

Then, the number of articles processed at the rate x_i is

$$n_i z_i = f_i \text{ (suppose)}$$

(Here, z is the midvalue of the class interval). We have to find the weighted A.M. of x, the weights being f.

Since the successive values of x have a common difference, we apply the formula $\bar{x} = c + d\bar{y}$.

Table 5.17 Calculations for Weighted Arithmetic Mean

Rate of Payment per Article (nP) x	No. of Persons Processing n	Daily No. of Articles Processed z	$nz = f$	$y = \dfrac{x - 3.3}{0.1}$	fy
3.1	12	89.5	1,074.0	–2	–2,148.0
3.2	63	109.5	6,898.5	–1	–6,898.5
3.3	87	129.5	11,266.5	0	0
3.4	56	149.5	8,372.0	1	8,372.0
3.5	3	164.5	493.5	2	987.0
Total	–	–	28,104.5	–	312.5

$$\bar{x} = c + d\bar{y} = 3.3 + 0.1\ (312.5/28104.5)$$

$$= 3.3 + 0.001$$

$$= 3.301 \text{ (approx.)}$$

Ans. 3.301 nP.

5.5 MEAN OF COMPOSITE GROUP

If two groups contain n_1 and n_2 observations with means $\bar{x}_1$ and $\bar{x}_2$ respectively, then the mean $(\bar{x})$ of the composite group of $n_1 + n_2$ observations is given by the relation

$$N\bar{x} = n_1\bar{x}_1 + n_2\bar{x}_2 \qquad (5.5.1)$$

where $N = n_1 + n_2$.

It will be found convenient to arrange the data in the following tabular form:

Table 5.18 Data for Calculation of Composite Mean

Characteristics	Groups I	Groups II	Composite Group
No. of Observations	n_1	n_2	N
Mean	$\bar{x}_1$	$\bar{x}_2$	$\bar{x}$

[**Note:** The product of the two quantities in Composite Group is equal to the sum of the products of quantities for all the Groups.]

In general, if there are several groups, the i-th group containing n_i observations with mean $\bar{x}_i$, then the mean ($\bar{x}$) of the composite group is given by the relation

$$N\bar{x} = \Sigma n_i \bar{x}_i \tag{5.5.2}$$

where $N = \Sigma n_i$. In this case also the data may be laid out in the form as shown in Table 5.18, all the groups being shown one after the other, followed by the composite group (see Table 5.21).

Example 5.25 *There are two branches of an establishment employing 100 and 80 persons respectively. If the arithmetic means of the monthly salaries paid by the two branches are Rs 275 and Rs 225 respectively, find the arithmetic mean of the salaries of the employees of the establishment as a whole.* [C.A., Nov. '63]

Solution

Table 5.19 Mean of Composite Group

Characteristics	Groups I	Groups II	Composite Group
No. of Observations	$n_1 = 100$	$n_2 = 80$	$N = 180$
Mean salary (Rs)	$\bar{x}_1 = 275$	$\bar{x}_2 = 225$	$\bar{x} = ?$

Applying (5.5.1), $N\bar{x} = n_1\bar{x}_1 + n_2\bar{x}_2$

or, $180\bar{x} = 100 \times 275 + 80 \times 225 = 45{,}500$

$\therefore$ $\bar{x} = 45500/180 = 252.78$ Rs

Example 5.26 *The mean monthly salary paid to all employees in a certain company was rupees 500. The mean monthly salaries paid to male and female employees were 520 and 420 rupees respectively. Obtain the percentage of male to female employees in the company.* [C.U., M.Com. '71]

Solution We apply the formula for mean of composite group:

$$N\bar{x} = n_1\bar{x}_1 + n_2\bar{x}_2$$

The given information is shown in the following table:

Table 5.20 Mean of Composite Group

Characteristics	*Groups* I (Males)	II (Females)	*Composite Group (All employees)*
No. of Employees	n_1	n_2	$N = n_1 + n_2$
Mean Salary (Rs)	$\bar{x}_1 = 520$	$\bar{x}_2 = 420$	$\bar{x} = 500$

Substituting the values in the formula,

$$(n_1 + n_2) \times 500 = n_1 \times 520 + n_2 \times 420$$

or, $$500n_1 + 500n_2 = 520n_1 + 420n_2$$

or, $$500n_1 - 520n_1 = 420n_2 - 500n_2$$

or, $$-20n_1 = -80n_2$$

$\therefore$ $$\frac{n_1}{n_2} = \frac{-80}{-20} = \frac{4}{1}$$

i.e. $$n_1 : n_2 = 4 : 1$$

Hence, percentage of male employees = $\frac{4}{4+1} \times 100 = 80$

Percentage of female employees = 100–80 = 20 *Ans.* Males 80%, Females 20%.

Example 5.27 *The mean age of a group of 100 children was 9.35 years. The mean age of 25 of them was 8.75 years and that of another 65 was 10.51 years. What was the mean age of the remainder?* [C.U., M.Com. '65]

Solution We use the formula for mean of the composite group of 100, observations, the ages of 25 children with mean 8.75 years forming Group I, those of another 65 with mean 10.51 years forming Group II, and the remaining 10 (because in all there are 100) forming Group III.

Table 5.21 Mean of Composite Group

Characteristics	*Groups* I	II	III	*Composite Group*
No. of Observations	$n_1 = 25$	$n_2 = 65$	$n_3 = 10$	$N = 100$
Mean (years)	$\bar{x}_1 = 8.75$	$\bar{x}_2 = 10.51$	$\bar{x}_3 = ?$	$\bar{x} = 9.35$

The formula is $$N\bar{x} = n_1\bar{x}_1 + n_2\bar{x}_2 + n_3\bar{x}_3$$

or, $$100 \times 9.35 = 25 \times 8.75 + 65 \times 10.51 + 10\bar{x}_3$$

or, $$935 = 218.75 + 683.15 + 10\bar{x}_3$$

or, $$10\bar{x}_3 = 935 - 218.75 - 683.15 = 33.10$$

$\therefore$ $$\bar{x}_3 = 33.10/10 = 3.31 \text{ years.}$$

Ans. 3.31 years

5.6 GEOMETRIC MEAN (G.M.)

Geometric Mean of a group of *n* observations is the *n*-th root of their product. It is defined only when all observations have the same sign, and none of them is zero.

Given n observations $x_1, x_2, ..., x_n$,

$$\text{G.M.} = \sqrt[n]{x_1 \times x_2 \times \ldots \times x_n} \tag{5.6.1}$$

This may also be written as $(x_1x_2...x_n)^{\frac{1}{n}}$, see Example 3.13 (c), p. 49.

If however, $x_1, x_2, ..., x_n$ have frequencies $f_1, f_2, ..., f_n$ respectively, the product of all the $N(= f_1 + f_2 + ... + f_n)$ observations is

$$\underbrace{x_1.x_1 \ldots x_1}_{f_1 \text{ terms}} . \underbrace{x_2.x_2 \ldots x_2}_{f_2 \text{ terms}} \ldots \underbrace{x_n.x_n \ldots x_n}_{f_n \text{ terms}} = x_1^{f_1} x_2^{f_2} \ldots x_n^{f_n}$$

so that

$$\text{G.M.} = \sqrt[N]{x_1^{f_1} x_2^{f_2} \ldots x_n^{f_n}} \tag{5.6.2}$$

This may also be written in the form $\left(x_1^{f_1} x_2^{f_2} \ldots x_n^{f_n}\right)^{\frac{1}{N}}$ where $N = \Sigma f_i$ is the total frequency.

As in the case of A.M., we have *Simple geometric mean* and *Weighted geometric mean*, given by the formulae

$$\text{Simple G.M. } (g) = (x_1x_2 \ldots x_n)^{\frac{1}{n}}$$

$$\text{Weighted G.M. } (G) = \left(x_1^{f_1} x_2^{f_2} \ldots x_n^{f_n}\right)^{\frac{1}{N}} \tag{5.6.3}$$

They are equal, only when all weights are equal. For practical calculations, these formulae cannot be applied directly. Taking logarithms of both sides [see Examples 3.17(c), 3.19(e), 3.31(vi), 3.32(vi)] we have

$$\log g = \frac{1}{n}(\log x_1 + \log x_2 + \ldots + \log x_n)$$

$$= \frac{1}{n}\Sigma \log x$$

$$\log G = \frac{1}{N}[f_1(\log x_1) + f_2(\log x_2) + \ldots + f_n(\log x_n)]$$

$$= \frac{1}{N}\Sigma f_i (\log x_i) \tag{5.6.4}$$

Uses—Geometric mean is useful in averaging ratios, rates and percentages. It is, therefore, considered as the most appropriate type in finding the average rate under compound interest, depreciation of machines and, sometimes, in the growth of living organism, when the compound interest law (p. 55) holds. G.M. is particularly useful in economics and business in the construction of index numbers. It makes the index numbers time reversible and gives equal weight to equal ratios of change. If a series of numbers are in geometric progression, either exactly or approximately, G.M. is the appropriate average to use.

5.7 PROPERTIES OF G.M.

(a) The product of a group of n observations is equal to the n-th power of their G.M.

$$x_1x_2 \ldots x_n = g^n \tag{5.7.1}$$

This follows from (5.6.3)

(b) The logarithm of G.M. of a set of observations is equal to the A.M. of their logarithms (see 5.6.4)

$$\log g = \frac{1}{N} \Sigma \log x_i; \quad \log G = \frac{1}{N} \Sigma f_i (\log x_i) \qquad (5.7.2)$$

(c) If G_1, G_2, ... be the geometric means of several groups having n_1, n_2, ... observations respectively, then G.M. (G) of the composite group is given by their weighted geometric mean.

$$G = \sqrt[N]{G_1^{n_1} G_2^{n_2} \ldots}$$

i.e.
$$\text{Log } G = \frac{1}{N} \Sigma n_i (\log G_i) \qquad (5.7.3)$$

where $N = n_1 + n_2 + \ldots$

Example 5.28 *(a) If you have n distinct observations on a variable x given by x_1, x_2, ..., x_n with frequencies f_1, f_2, ..., f_n respectively, write the formulae for (i) the simple geometric mean, and (ii) the weighted geometric mean. Are these formulae directly applied for numerical computation? If not, show how this is done. State the usefulness of geometric mean in the construction of index numbers.*

(b) Apply the geometric mean to find the general index from the following group indices by assigning the given weights:

Group	A	B	C	D	E	F
Group index	118	120	97	107	111	93
Weight	4	1	2	6	5	2

[I.C.W.A., July '67]

Solution (a) The formulae are

$$\text{Simple Geometric Mean } (g) = \sqrt[n]{x_1 x_2 \ldots x_n}$$

$$\text{Weighted Geometric Mean } (G) = \sqrt[N]{x_1^{f_1} x_2^{f_2} \ldots x_n^{f_n}}$$

where $N = f_1 + f_2 + \ldots + f_n$. These formulae cannot be directly applied for numerical calculations. Taking logarithms, we have

$$\log g = \frac{1}{n} \Sigma \log x_i; \qquad \log G = \frac{1}{N} \Sigma f_i (\log x_i)$$

That is, the logarithm of simple (weighted) G.M. is equal to simple (weighted) A.M. of the logarithms of observations. The antilog of A.M., so found, gives G.M.

G.M. is sometimes preferred to A.M. in the construction of index numbers, although it involves more laborious numerical calculations. This is so, because G.M. is less affected than A.M. by the presence of extremely large or small values. Since in the construction of index numbers, ratios of change are more important than absolute magnitudes of change, G.M. is likely to be more typical of the changing prices than A.M. or any other average.

(b)

Table 5.22 Calculations for Geometric Mean

Group	*Group Index* (x)	*Weight* (f)	*log x*	*f (log x)*
A	118	4	2.0719	8.2876
B	120	1	2.0792	2.0792
C	97	2	1.9868	3.9736
D	107	6	2.0294	12.1764
E	111	5	2.0453	10.2265
F	93	2	1.9685	3.9370
Total	–	20	–	40.6803

$$\log G = \frac{1}{N}\Sigma f_i(\log x_i) = \frac{40.6803}{20} = 2.0340$$

$$G = \text{antilog } 2.0340 = 108.1$$ [see Example 3.20(a)]

Ans. 108.1

Example 5.29 *The geometric mean of 10 observations on a certain variable was calculated to be 16.2. It was later discovered that one of the observations was wrongly recorded as 12.9, when in fact it was 21.9. Apply appropriate correction and calculate the correct geometric mean.* [I.C.W.A. (Inter), June '74]

Solution Applying (5.7.1), the product of 10 observations (including the incorrect value 12.9) is $(16.2)^{10}$. When the incorrect 12.9 is replaced by the correct 21.9, the product would be

$$g^{10} = \frac{(16.2)^{10}}{12.9} \times 21.9$$

where g denotes the new geometric mean. Taking logarithms

$$10(\log g) = 10(\log 16.2) + \log 21.9 - \log 12.9$$
$$= 10(1.2095) + 1.3404 - 1.1106 = 12.3248$$

$\therefore$ $\log g = 1.23248 = 1.2325$ (approx.), so that $g = 17.08$

Ans. 17.08

Example 5.30 *The weighted geometric mean of the four numbers 8, 25, 17 and 30 is 15.3. If the weights of the first three numbers are 5, 3 and 4 respectively, find the weight of the fourth number.* [I.C.W.A., Jan '71]

Solution We use the formula

$$\log G = \frac{1}{N}\Sigma f(\log x)$$

Denoting by w the weight of the fourth number 30, the calculations are shown below:

Table 5.23 Calculations for Geometric Mean

x	f	$\log x$	$f(\log x)$
8	5	0.9031	4.5155
25	3	1.3979	4.1937
17	4	1.2304	4.9216
30	w	1.4771	1.4771w
Total	12 + w	–	13.6308 + 1.4771w

Since G.M. is given to be $G = 15.3$ and log 15.3 = 1.1847, we have

$$1.1847 = \frac{13.6308 + 1.4771w}{12 + w}$$

By cross multiplication,

$$14.2164 + 1.1847w = 13.6308 + 1.4771w$$

or, $$1.1847w - 1.4771w = 13.6308 - 14.2164$$

or, $$-0.2924w = -0.5856$$

$\therefore$ $$w = 0.5856 \div 0.2924 = 2.$$

Ans. 2

Example 5.31 *The production of portland cement increased from 99 million barrels is 1951 to 176 million in 1958. Assuming that the production increased at a constant annual rate, find the average annual rate of increase.*

Solution Since the annual *rate* of increase was a constant, here the Compound Interest Law (3.9.2) holds: $A = P\,(1+i)^n$, where i denotes the rate per unit.

$$A = 176 \text{ million} = 176 \times 10^6$$
$$P = 99 \text{ million} = 99 \times 10^6$$
$$n = 7 \text{ (from 1951 to 1958).}$$

Substituting these values

$$176 \times 10^6 = 99 \times 10^6\,(1 + i)^7$$

or, $$176 = 99\,(1 + i)^7$$

Taking logarithms of both sides,

$$\log 176 = \log 99 + 7 \log (1 + i)$$ [see Example 3.19(d)]

or, $$2.2455 = 1.9956 + 7 \log (1 + i),$$ [see Example 3.16(h)]

or, $$7 \log (1 + i) = 2.2455 - 1.9956 = 0.2499$$

$\therefore$ $$\log (1 + i) = 0.2499 \div 7 = 0.0357$$

or, $$(1 + i) = \text{antilog } 0.0357 = 1.086$$ [see Example 3.20(d)]

or, $$i = 1.086 - 1 = 0.086 = 8.6\%$$

Ans. 8.6%.

Example 5.32 *Compared to the previous year, the overhead expenses went up by 32% in 1961; it increased by 40% in the next year and by 50% in the following year. Calculate the average rate of increase in the overhead expenses over the three years. Explain clearly the reason for the choice of the average.* [C.A., Nov. '67]

Solution Let i be the average rate of increase per year over the 3 year period. Here, since the increase in any year is based on the expenses of the previous year and *not* on any fixed expense, hence the Compound Interest Law (3.9.4, p 55) will hold.

$$P(1+i)^3 = p\left(1+\frac{32}{100}\right)\left(1+\frac{40}{100}\right)\left(1+\frac{50}{100}\right)$$
$$= P \times 1.32 \times 1.40 \times 1.50$$

where P denotes the overhead expenses in 1960.

or, $(1+i)^3 = 1.32 \times 1.40 \times 1.50$

Taking logarithms of both sides

$$3 \log (1+i) = \log 1.32 + \log 1.40 + \log 1.50$$
$$= 0.1206 + 0.1461 + 0.1761$$
$$= 0.4428$$

or, $\log (1+i) = 0.4428 \div 3 = 0.1476$

$\therefore$ $1+i = \text{antilog } 0.1476 = 1.405$

or, $i = 1.405 - 1 = 0.405,$
$= 40.5\%.$

Ans. 40.5%

Example 5.33 *A sum of money was invested for 4 years at 2% per annum and then the accumulated sum was invested for 3 years at 3% per annum. What is the average compound interest rate?* [B.U., B.A.(Econ) '70]

Solution Let P be the sum invested and A the amount after the whole period of 7 years. Using (3.9.4)

$$A = P(1+i_1)^{n_1}(1+i_2)^{n_2} = p\left(1+\frac{2}{100}\right)^4\left(1+\frac{3}{100}\right)^3$$
$$= P(1.02)^4 (1.03)^3.$$

Had the compound interest rate remained a constant i throughout the 7 year period yielding the same amount as before, this rate i would be considered as the average compound interest rate.

$$A = P(1+i)^n = P(1+i)^7$$

Therefore, $P(1+i)^7 = P(1.02)^4(1.03)^3$

Cancelling P from both sides and then taking logarithms

$$7 \log (1+i) = 4 (\log 1.02) + 3 (\log 1.03)$$
$$= 4(.0086) + 3(.0128) = .0728$$

or, $\log (1+i) = .0728 \div 7 = .0104$

or, $1+i = \text{antilog } .0104 = 1.024$

$i = 1.024 - 1 = .024 = 2.4\%.$

Ans. 2.4% per annum.

Example 5.34 *A machine depreciates 25% of its value during the first two years, 10% for the next 3 years and 2% in value for the next five years, depreciation being calculated on the diminishing value. If the value of the machine is Rs 5000, find the depreciated value after 10 years.* [C.U., B.A.(Econ) '64]

Solution In the calculation of depreciated values, the Compound Interest Law (3.9.4) should be applied, with i negative.

$$A = P(1-i_1)^{n_1}(1-i_2)^{n_2}(1-i_3)^{n_3}$$
$$= 5000\left(1-\frac{25}{100}\right)^2\left(1-\frac{10}{100}\right)^3\left(1-\frac{2}{100}\right)^5$$
$$= 5000 \times (0.75)^2 \times (0.90)^3 \times (0.98)^5$$

$$
\begin{aligned}
\therefore \quad \log A &= \log 5000 + 2(\log 0.75) + 3(\log 0.90) + 5(\log 0.98)\\
&= 3.6990 + 2\times \bar{1}.8751 + 3\times \bar{1}.9542 + 5\times \bar{1}.9912 \quad \text{[see Example 3.16(i)]}\\
&= 3.6990 + 2(-1 + .8751) + 3(-1 + .9542) + 5(-1 + .9912)\\
&= 3.6990 + (-2 + 1.7502) + (-3 + 2.8626) + (-5 + 4.9560)\\
&= 13.2678 - 10 = 3.2678\\
\therefore \quad A &= \text{antilog } 3.2678 = 1852 \text{ Rs} \quad \text{[see Example 3.20 (e)]}
\end{aligned}
$$

Ans. Rs 1852

Example 5.35 *A firm buys Rs 10,000 worth of plant on January 1, 1939, and arranges for depreciation according to the following schedule:*

Category	*Cost (Rs)*	*Depreciations rate*	
General Plant	*4,000*	*5%*	*on reducing balance*
Electrical Plant	*1,000*	*7.5%*	*" " "*
Process Plant	*3,000*	*10%*	*" " "*
Lorries	*2,000*	*20%*	*" " "*

Find the average depreciation rates for 1939 and 1940. Give reasons for the difference between the two averages. [C.U., B.A.(Econ) '66 ; I.C.W.A., June '73]

Solution Let P be the cost and i the rate of depreciation; then the depreciated value after 1 year would be $P(1 - i)$, and after 2 years would be $P(1 - i)^2$.

Table 5.24 Calculations for Depreciated Values

Category	*Cost (Rs)* *(P)*	*Depreciation rate* *(i)*	*(1–i)*	*$(1-i)^2$*	*Depreciated* *1-1-1940* *p(1 – i)*	*Values on* *1-1-1941* *$p(1-i)^2$*
General Plant	4,000	.05	.95	.9025	3,800	3,610
Electrical Plant	1,000	.075	.925	.856	925	856
Process Plant	3,000	.10	.90	.81	2,700	2,430
Lorries	2,000	.20	.80	.64	1,600	1,280
Total	10,000	—	—	—	9,025	8,176

Amount of depreciation during 1939 is 10,000 – 9,025 = 975 Rs

$$\therefore \text{ Average rate of depreciation} = \frac{975}{10,000} = .0975 = 9.75\%$$

Amount of depreciation during 1940 is 9025 – 8176 = 849 Rs

$$\therefore \text{ Average rate of depreciation} = \frac{849}{9025} = .0941 = 9.41\%$$

Average depreciation rates during the years 1939 and 1940 are different, because the rates for various categories are not equal. The two average rates will be the same, only when the rates of depreciation for different categories are equal.

Ans. (i) 9.75%, (ii) 9.41%

Example 5.36 *Three groups of observations contain 8, 7 and 5 observations. Their geometric means are 8.52, 10.12 and 7.75 respectively. Find the geometric mean of the 20 observations in the single group formed by pooling the three groups.* [I.C.W.A., Dec. '76]

Solution We apply the formula (5.7.3) for the G.M. of composite group.

Here $n_1 = 8$ $n_2 = 7$ $n_3 = 5$ $N = 20$

$G_1 = 8.52$ $G_2 = 10.12$ $G_3 = 7.75$ $G = ?$

Substituting the values in the formula,

$$\log G = \frac{1}{20}\left[8(\log 8.52) + 7(\log 10.12) + 5(\log 7.75)\right]$$

$$= \frac{1}{20}[8 \times 0.9304 + 7 \times 1.0052 + 5 \times 0.8893]$$

$$= 18.9261 \div 20 = 0.9463$$

Therefore, G = antilog 0.9463 = 8.837

5.8 HARMONIC MEAN (H.M.)

Harmonic Mean of a set of observations is the reciprocal of the arithmetic mean of their reciprocals. Like G.M., H.M. is defined only when no observation is zero.

[**Note:** 'Reciprocal of k' denotes $1/k$]

$$\text{Simple H.M.} = \frac{n}{\dfrac{1}{x_2} + \dfrac{1}{x_2} + \ldots + \dfrac{1}{x_n}} = \frac{n}{\sum\left(\dfrac{1}{x_i}\right)} \quad (5.8.1)$$

$$\text{Weighted H.M.} = \frac{N}{\dfrac{f_1}{x_1} + \dfrac{f_2}{x_2} + \ldots + \dfrac{f_n}{x_n}} = \frac{N}{\sum\left(\dfrac{f_i}{x_i}\right)} \quad (5.8.2)$$

They are equal only when all weights are equal (like A.M. and G.M.)

Uses—The use of H.M. is very limited. H.M. gives the largest weight to the smallest item and the smallest weight to the largest item. Hence, when there are a few extremely large or small values, H.M. is preferable to A.M. as an average. H.M. is also useful in averages involving time, rate and price.

Example 5.37 *Point out the mistake or ambiguity in the following statement: 'A person goes from X to Y on cycle at 20 m.p.h. and returns at 24 m.p.h. His average speed was 22 m.p.h.'* [C.A., May '70]

Solution Average speed will not be the A.M., viz (20 + 24)/2 = 22 m.p.h. Since the distance travelled in each case was the same, the appropriate average here will be the H.M.

$$\text{Average speed} = \frac{2}{\dfrac{1}{20} + \dfrac{1}{24}} = \frac{240}{11} = 21\frac{9}{11} \text{ m.p.h.} = 21.82 \text{ m.p.h.}$$

The A.M. will be used only if the duration (i.e. time) of journey at 20 m.p.h. and also at 24 m.p.h. were the same. *Ans.* 21.82 m.p.h.

Example 5.38 *If the interest paid on each of three different sums of money yielding 5 per cent, 6 per cent and 8 per cent simple interest per annum respectively is the same, what is the average yield per cent on the total sum invested?* [I.C.W.A., Jan. '73]

Solution H.M. is the appropriate average, here.

$$\text{Average} = \text{H.M. of 5, 6 and 8} = \frac{3}{\frac{1}{5}+\frac{1}{6}+\frac{1}{8}} = \frac{360}{59}$$

$$= 6.1 \text{ per cent}$$

Example 5.39 *A motor car covered a distance of 50 miles four times. The first time at 50 m.p.h., the second at 20 m.p.h., the third at 40 m.p.h. and the fourth at 25 m.p.h. Calculate the average speed and explain the choice of the average.* [C.A., Nov. '67]

Solution The average speed will be given by the harmonic mean of the four speeds 50, 20, 40, 25, each weighted by the distance covered, viz. 50. However, since the weights are equal, simple H.M. and weighted H.M. will be the same; i.e. simple H.M. of the four speeds will give the required average speed.

$$\text{Average speed} = \frac{4}{\frac{1}{50}+\frac{1}{20}+\frac{1}{40}+\frac{1}{25}} = \frac{800}{27} = 30 \text{ m.p.h.}$$

[**Important Note:** Which average to use–A.M. or H.M.?

In averaging rates and ratios involving speed, time and distance; or price, quantity and amount of money spent, sometimes it becomes difficult to decide whether A.M. or H.M. is the appropriate average. In such cases, the following rule will be found helpful in coming to the right decision:

If the given ratios are stated as *x units per y*, then for finding the average ratio.

use (i) H.M. when *x*'s are given

(ii) A.M. when *y*'s are given

For example, in order to find the average speed shown in *miles per hour* (m.p.h.) use H.M. when *miles* (i.e. distances covered) are given; but use A.M. when *hours* (i.e. time of journey) are given. The given miles or hours, as the case may be, should be used as weights when finding the average.]

Example 5.40 *(a) A man travelled 12 miles at 4 m.p.h. and again 10 miles at 5 m.p.h. What was the average speed?*

(b) A man travelled 12 hours at 4 m.p.h. and again 10 hours at 5 m.p.h. What was the average speed?

Solution (a) Here, the speeds are shown in *miles per hour* (m.p.h.); we are also given the *miles* travelled. Hence, H.M. of the speeds 4 and 5, weighted by miles 12 and 10, will be the appropriate average.

x	f	(f/x)
4	12	3
5	10	2
Total	22	5

$$\text{Average speed} = \frac{N}{\sum\left(\frac{f}{x}\right)}$$

$$= \frac{22}{5} = 4.40 \text{ m.p.h.}$$

(b) Here, the speeds are shown in m.p.h., as before. But we are given the *hours* travelled. Hence, A.M. of the speeds 4 and 5, weighted by hours 12 and 10, will be the appropriate average.

x	f	fx
4	12	48
5	10	50
Total	22	98

$$\text{Average speed} = \frac{\Sigma fx}{N}$$

$$= \frac{98}{22} = 4.45 \text{ m.p.h.}$$

Ans. (a) 4.40 m.p.h.; (b) 4.45 m.p.h.

Example 5.41 *A person bought 6 rupees worth of orange from five markets at 15 p., 20 p., 25 p., 30 p. and 50 p. per orange respectively. What is the average price of an orange?*

What would be the average price, if he had purchased 20 oranges from each market?

[B.U., B.A.(Econ) '73]

Solution [**Note:** Here, the variable is *price per orange* (in paise per number). In the first case, each value of the variable has weight 600 p, and hence H.M. is the appropriate average. In the second case, each value has weight 20 oranges, and so A.M. is to be used. Again, weights being equal for all values, simple and weighted averages are the same.]

Case I. Average price = H.M. of 15, 20, 25, 30 and 50

$$= \frac{5}{\frac{1}{15}+\frac{1}{20}+\frac{1}{25}+\frac{1}{30}+\frac{1}{50}} = \frac{1500}{63} = 24 \text{ p.}$$

Case II. Average price = A.M. of 15, 20, 25, 30 and 50

$$= (15 + 20 + 25 + 30 + 50) \div 5 = 28 \text{ p.}$$

5.9 ADVANTAGES AND DISADVANTAGES OF A.M., G.M., H.M.

(1) Arithmetic Mean

Advantages— (1) A.M. is readily understood, and hence needs no explanation, when used.

(2) The computation of A.M. is easy and does not involve any laborious numerical calculations. Even if all observations are not known individually, A.M. can be found, provided their sum and the number of observations are known.

(3) It can be calculated without arranging the data in order of magnitude (as required for median), or in the form of a frequency distribution (as required for mode).

(4) A.M. can be treated algebraically. Given the A.M. and the number of observations in each of several groups, A.M. of the composite group can be easily determined by using an algebraic formula.

(5) It is a very stable and reliable average as regards sampling fluctuations. If many samples are drawn from the same population and each time several measures of central tendency calculated, it will be found that A.M. fluctuates less from sample to sample than any other measures.

Disadvantages—(1) A.M. cannot be obtained by inspection, as in the case of median or mode.

(2) It cannot be calculated unless the exact magnitude of all observations and their number is known accurately. If some of the extreme values are missing, the accuracy of A.M. is greatly affected. A.M. cannot therefore be calculated from a grouped frequency distribution with open-end classes, unless some assumptions are made regarding the sizes of these classes.

(3) The greatest disadvantage of A.M. is that it is highly affected by the presence of even a few extremely large or small observations.

(4) A.M. may not be an actual value of the variable. (For example, A.M. of the number of children per family may come to 2.54. However, although there may be 2 children or 3 children in a family, 2.54 children is meaningless.)

(2) Geometric Mean

Advantages— (1) Given the product and the number of observations, G.M. can be calculated, even if the magnitude of each observation is not known individually.

(2) G.M. is particularly suitable for averaging rates, ratios and percentages. It is therefore used in the construction of index numbers.

(3) Unlike A.M., G.M. is not affected very much by the presence of extremely large or small observations.

(4) It can be treated algebraically. G.M. of the composite group can be determined, if the G.M. and the number of observations in each of the constituent groups are known.

Disadvantages— (1) The significance of G.M. is not easy to understand.

(2) It cannot be calculated, if any of the observations is zero. Also, if all the observations do not have the same sign, G.M. will be meaningless.

(3) It is difficult to calculate. The calculation of G.M. needs knowledge of logarithms.

(4) Like A.M., G.M. may not be an actual value of the variable.

(3) Harmonic Mean

Advantages— (1) H.M. is advantageous when it is desired to give greater weight to smaller observations and less weight to the larger.

(2) It is used in averages involving time, rate and price.

Disadvantages— (1) H.M has an abstract character, and its meaning is not generally understood by common people. As such, it has only a limited use.

(2) Like G.M., it cannot be calculated, if any of the observations is zero.

(3) It may not be an actual value of the variable.

5.10 RELATIONS BETWEEN A.M., G.M., H.M.

(1) For any given set of observations, A.M. is greater than or equal to G.M., and G.M. is greater than or equal to H.M.

$$\text{A.M.} \geq \text{G.M.} \geq \text{H.M.} \tag{5.10.1}$$

They are equal, only when all observations are equal.

(2) For two observations only,

$$\frac{\text{A.M.}}{\text{G.M.}} = \frac{\text{G.M.}}{\text{H.M.}} \tag{5.10.2}$$

This means that G.M. not only lies between A.M. and H.M., but $(G.M.)^2 = A.M. \times H.M.$, provided there are only two observations.

Example 5.42 *Prove that*

$$A.M. \geq G.M. \geq H.M.$$

where A.M., G.M., and H.M. represent arithmetic, geometric and harmonic means respectively. [C.U., B.Com. (Hons) '64, '68]

Or,

Prove that the arithmetic mean is greater than or equal to the geometric mean, and state the case when the two averages are equal. [C.U., B.A.(Econ) '66, '68]

Solution Let $x_1, x_2, ..., x_n$ be a set of n observations (all positive). Their *A.M.*, *G.M.*, and *H.M.* (denoted by *A*, *G* and *H* respectively) are

$$A = \frac{x_1 + x_2 + \ldots + x_n}{n} \; ; \quad G = \left(x_1 x_2 \ldots x_n\right)^{\frac{1}{n}}$$

$$H = \frac{n}{\frac{1}{x_1} + \frac{1}{x_2} + \ldots + \frac{1}{x_n}}$$

Considering only two observations x_1 and x_2, we see that

$$\left(\sqrt{x_1} - \sqrt{x_2}\right)^2 \geq 0,$$

because the left side is a square quantity.

or, $$x_1 + x_2 - 2\sqrt{x_1 x_2} \geq 0$$

or, $$x_1 + x_2 \geq 2\sqrt{x_1 x_2}$$

or, $$\frac{x_1 + x_2}{2} \geq \sqrt{x_1 x_2} \qquad \ldots \text{(i)}$$

i.e. $$A.M. \geq G.M., \quad \text{when } n = 2. \qquad \ldots \text{(ii)}$$

Similarly, considering only the observations x_3 and x_4, we have

$$\frac{x_3 + x_4}{2} \geq \sqrt{x_3 x_4} \qquad \ldots \text{(iii)}$$

If we now consider the two quantities $\frac{x_1 + x_2}{2}$ and $\frac{x_3 + x_4}{2}$, we must have, by (ii),

$$\frac{\frac{x_1 + x_2}{2} + \frac{x_3 + x_4}{2}}{2} \geq \sqrt{\frac{x_1 + x_2}{2} \cdot \frac{x_3 + x_4}{2}}$$

or, $$\frac{x_1 + x_2 + x_3 + x_4}{4} \geq \sqrt{\frac{x_1 + x_2}{2} \cdot \frac{x_3 + x_4}{2}} \quad \text{... (iv)}$$

But, $$\frac{x_1 + x_2}{2} \cdot \frac{x_3 + x_4}{2} \geq \sqrt{x_1 x_2} \cdot \sqrt{x_3 x_4} \quad \text{... (v)}$$

because each of the two terms on the left is greater than or equal to the corresponding term on the right, by (i) and (iii).

Substituting from (v) in (iv),

$$\frac{x_1 + x_2 + x_3 + x_4}{4} \geq \sqrt{\sqrt{x_1 x_2} \cdot \sqrt{x_3 x_4}}$$

or, $$\frac{x_1 + x_2 + x_3 + x_4}{4} \geq \sqrt[4]{x_1 x_2 x_3 x_4} \quad \text{(vi)}$$

i.e., $A.M. \geq G.M.$, when $n = 4$.

Proceeding this way, it can be shown that $A.M. \geq G.M.$, whenever $n = 2$ or 4 or 8 or 16 or 32 etc. i.e., of the form $n = 2^m$, where m is a positive integer. But we have to prove the result for any value of n. For this purpose, let us suppose that the given value of n lies between two such values 2^{m-1} and 2^m, i.e. $2^{m-1} < n < 2^m$.

We now consider 2^m (= N, say) values, consisting of the n given observations $x_1, x_2, \ldots, x_n$, and $(N - n)$ further values each equal to A, i.e. $(x_1 + x_2 + \ldots + x_n)/n$.

$$\underbrace{x_1,\ x_2,\ \ldots\ x_n}_{n \text{ terms}}\ \underbrace{A,\ A,\ \ldots\ A}_{(N-n) \text{ terms}}$$

A.M. of these N values is

$$\frac{x_1 + x_2 + \ldots + x_n + A + A \ldots + A}{N} = \frac{nA + (N - n)A}{N} = A$$

Also, G.M. of these N values is

$$(x_1 x_2 \ldots x_n . A . A \ldots A)^{\frac{1}{N}} = (G^n \cdot A^{N-n})^{\frac{1}{N}}$$

Since $A.M. \geq G.M.$ for $N = 2^m$ values, therefore, in the present case

$$A \geq (G^n \cdot A^{N-n})^{\frac{1}{N}}$$

Raising both sides to the power N, we have

$$A^N \geq G^n \cdot A^{N-n}$$

Simplifying, we get

$$A^n \geq G^n$$

or, $$A \geq G$$

This completes the proof that $A.M. \geq G.M.$ for any number of observations.

Using this result, we shall now prove that $G \geq H$. For this purpose, let us consider the n values , $\frac{1}{x_1}, \frac{1}{x_1}, \ldots, \frac{1}{x_n}$

The A.M. of these values is

$$\frac{\frac{1}{x_1} + \frac{1}{x_2} + \ldots + \frac{1}{x_n}}{n} = \frac{1}{H}$$

and their G.M. is

$$\left(\frac{1}{x_1}\cdot\frac{1}{x_2}\cdots\frac{1}{x_n}\right)^{\frac{1}{n}} = \left(\frac{1}{x_1x_2\ldots x_n}\right)^{\frac{1}{n}} = \left(\frac{1}{G^n}\right)^{\frac{1}{n}} = \frac{1}{G}$$

Since, we have proved that $A.M. \geq G.M.$, in the present case

$$\frac{1}{H} \geq \frac{1}{G}$$

i.e. $$G \geq H$$ [Section 3.12(i)]

Combining the two results $A \geq G$ and $G \geq H$, we have $A \geq G \geq H$; i.e. in general,

$$A.M. \geq G.M. \geq H.M.$$

We shall now prove that $A.M. = G.M. = H.M.$, only when all the observations have the same value, i.e. $x_1 = k$, $x_2 = k$, …, $x_n = k$. In such a situation

$$A.M. = \frac{k + k + \cdots + k}{n} = \frac{nk}{n} = k$$

$$G.M. = (k.k.\cdots k)^{\frac{1}{n}} = (k^n)^{\frac{1}{n}} = k$$

$$H.M. = \frac{n}{\frac{1}{k}+\frac{1}{k}+\cdots+\frac{1}{k}} = \frac{n}{(n/k)} = k$$

and hence $A.M. = G.M. = H.M.$, when all the observations are equal.

Example 5.43 *If x_1 and x_2 are two positive values of a variate, prove that their geometric mean is equal to the geometric mean of their arithmetic and harmonic means.* [I.C.W.A., Jan. '66]

Solution Let *A, G* and *H* represent respectively *A.M., G.M.* and *H.M.* of the two values x_1 and x_2. Then, by definition,

$$A = \frac{x_1 + x_2}{2}\ ;\ G\ \sqrt{x_1x_2}$$

$$H = \frac{2}{\frac{1}{x_1}+\frac{1}{x_2}} = \frac{2}{\frac{x_1+x_2}{x_1x_2}} = \frac{2x_1x_2}{x_1+x_2}$$

∴ $$A \times H = \frac{x_1+x_2}{2} \times \frac{2x_1x_2}{x_1+x_2} = x_1x_2 = \left(\sqrt{x_1x_2}\right)^2 = G^2$$

Thus, $$A \times H = G^2$$

or, $$G^2 = A \times H$$

or, $$G = \sqrt{A \times H}$$

= Geometric mean of *A* and *H*. *(proved)*

Example 5.44 *The arithmetic mean of two observations is 25 and their geometric mean is 15. Find (i) their harmonic mean and (ii) the two observations.*

[I.C.W.A., Jan. '72]

Solution For two observations $A.M. \times H.M. = (G.M.)^2$

Substituting the values of $A.M.$ and $G.M.$,

$$25 \times H.M. = (15)^2,$$

so that

$$H.M. = (15)^2/25 = 225/25 = 9.$$

Again, if x and y be the two observations

$$A.M. = \frac{x+y}{2} = 25 \text{ and } G.M. = \sqrt{xy} = 15$$

i.e. $x + y = 50$ and $xy = 225$. Substituting the value of y from the first equation into the second, we have

$$x(50 - x) = 225 \quad \text{or,} \quad 50x - x^2 = 225$$

$$\text{or,} \quad x^2 - 50x + 225 = 0; \quad \text{or,} \quad (x - 45)(x - 5) = 0$$

$$\therefore \quad x = 45, 5. \quad \text{Hence, } y = 5, 45.$$

In any case, the two observations are 45 and 5.

[**Note:** Alternatively, first calculate the two observations by the above method and then calculate the H.M. directly]. *Ans.* (i) 9, (ii) 45, 5.

5.11 MEDIAN

Median of a set of observations is the middle-most value when the observations are arranged in order of magnitude. The number of observations smaller than Median is the same as the number greater than it. Thus, Median divides the observations into two equal parts. Median is, in a certain sense, the real measure of central tendency, as it gives the value of the most central observation. It is unaffected by the presence of extremely large or small observations and can be calculated from frequency distributions with open-end classes. Median finds the largest application in psychological and achievement tests, e.g. to find the boy of average intelligence, the candidates may be ranked in order of intelligence and the median employed.

An important property of Median is that for any given set of observations the sum of absolute deviations from median is the least.

5.12 CALCULATION OF MEDIAN

The median is calculated as follows:

(a) *From simple series*—The given data are arranged in order of magnitude. If the number of observations be odd, the value of the middle-most item is the median. However, if the number be even, the arithmetic mean of the two middle-most items is taken as median (Example 5.45).

(b) *From simple frequency distribution*—The cumulative frequency ("less than" type) corresponding to each distinct value of the variable is calculated. If the total frequency be N, the value of the variable corresponding to cumulative frequency $(N + 1)/2$ gives the median.

(c) *From grouped frequency distribution*—Median from a grouped frequency distribution is that value which corresponds to cumulative frequency $N/2$ [Sometimes $(N + 1)/2$ is used as in (b) above; but this procedure is not correct]. Median from a grouped frequency distribution can be calculated by any of the following methods:

(i) *By the application of formula for median:* The cumulative frequencies are calculated. The class in which cumulative frequency $N/2$ lies, is called the *median class.* Now we apply the formula:

$$Median = l_1 + \frac{N/2 - F}{f_m} \times c \qquad (5.12.1)$$

where l_1 = lower boundary of median class;
N = total frequency;
F = cumulative frequency below l_1;
f_m = frequency of median class;
c = width of median class.

(ii) *By the application of simple interpolation in a cumulative frequency distribution:* If F_1 and F_2 be the cumulative frequencies shown in the table which are just smaller than and just larger than $N/2$, and they correspond to the class boundaries l_1 and l_2 respectively, then

$$\frac{Median - l_1}{l_2 - l_1} = \frac{(N/2) - F_1}{F_2 - F_1} \qquad (5.12.2)$$

[**Note:** l_1 and l_2 are the lower and the upper boundaries of the median class, and $N/2$ lies between F_1 and F_2].

(iii) *Graphical method*—An approximate value of median can be obtained graphically from ogive, or cumulative frequency polygon (page 104). Draw a horizontal line from the point $N/2$ on the vertical scale showing the cumulative frequencies, until it meets the ogive (either less-than or more-than type). From the point of intersection, a perpendicular is now drawn on the horizontal axis. The position of the foot of the perpendicular is read from the horizontal scale showing values of the variable, and this gives the median. If both ogives (less-than and more-than) are available on the same graph paper, the position of the foot of the perpendicular drawn from te point of intersection of the two ogives, gives the median (Figs 5.2, 5.3 and 4.2 at pages 148, 149, 90).

5.13 ADVANTAGES AND DISADVANTAGES OF MEDIAN

Advantages—(1) Median is not difficult to understand, although it is not so popular as the arithmetic mean.

(2) It is easy to calculate. Even when all the observations are not known, median can be calculated, provided the general location of all observations and values near the middle are available. Median can also be calculated without difficulty from grouped frequency distributions with classes of unequal width or with open-end classes (Example 5.50).

(3) Median is applicable to qualitative data in psychological and social studies, where numerical measurements may not be available, but it is possible to rank the objects in some order.

Disadvantages—(1) For the calculation of median, the data must be arranged.

(2) Unlike A.M. or G.M., it cannot be treated algebraically. Given the medians of several groups of observations, median of the composite group can not be determined.

(3) If is desired to give greater importance to large or small values, median is unsuitable.

(4) The calculation of median from a grouped frequency distribution is based on simple interpolation, which assumes that the observations in the median class are uniformly distributed. But in reality, this may not be true.

(5) Median is affected more by sampling fluctuations than the arithmetic mean, and is therefore, less reliable.

Example 5.45 *Find the median of (i) 32, 22, 29, 17, 40, 26, 21; (ii) 3.1, 2.6, 5.0, 4.7, 2.4, 3.9, 5.1, 3.6.*

Solution (i) The data when arranged are: 17, 21, 22, (26), 29, 32, 40. Since the middlemost item is 26, we have *Median* = 26.

(ii) The data are arranged: 2.4, 2.6, 3.1, (3.6), (3.9), 4.7, 5.0, 5.1. There are two middlemost items, viz. 3.6 and 3.9. So,

$$Median = \frac{3.6 + 3.9}{2} = 3.75$$

Example 5.46 *Find the median from the following simple frequency distribution:*

x	*0*	*1*	*2*	*3*	*4*	*5*	*6*	*Total*
f	*7*	*44*	*35*	*16*	*9*	*4*	*1*	*116*

Solution

Table 5.25 Calculation for Median

x	*f*	*Cumulative Frequency*
0	7	7
1	44	51
2	←35	←86
3	16	102
4	9	111
5	4	115
6	1	116 = *N*

Here, $(N + 1)/2 = 117/2 = 58.5$. From the last column of Table 5.25 it is found that $(N + 1)/2 = 58.5$ is greater than cumulative frequency 51, but is smaller than the next cumulative frequency 86, corresponding to $x = 2$. Hence, median is 2. (Also see Example 5.57).

[**Note:** Median is the value of x corresponding to cumulative frequency 58.5. This implies that when all the 116 observations are arranged in order of magnitude, median will be just midway between the 58th and 59th observations. Last column of Table 5.25 shows that the smallest 7 observations are each 0; 8th, 9th, … 51st observations are each 1; 52nd, 53rd, 86th observations are each 2; and so on. Thus, 58th and 59th observations are each 2, so that 58.5th observation, viz. median, is $(2 + 2)/2 = 2$].

Ans. 2.

Example 5.47 *Show that the expression for median is*

$$Median = l_1 + \frac{\left(\frac{N}{2} - \Sigma f_1\right)}{f_{med}} \times C$$

where l_1 is the lower limit of the class containing the median, Σf_1 is the summation of the frequencies in all classes below l_1, f_{med} is the frequency of the class in which the median falls and c is the class interval, and N is the total number of observations.

[C.U., M.Com. '65]

Solution If l_1 and l_2 represent the lower and the upper boundaries (note that l_1 is the *lower boundary,* and NOT the "lower limit", as stated in the question) of the class containing median, then cumulative frequency below l_1, viz. Σf_1, is smaller than $N/2$; and cumulative frequency below l_2 is greater than $N/2$. Also $l_2 - l_1 = c$. Therefore, the frequency between l_1 and the median is $\left(\frac{N}{2} - \Sigma f_1\right)$. Since the frequency in the median class of width c is f_{med}, the width per unit frequency is $\frac{c}{f_{med}}$, and the difference of median beyond l_1 for a further increase of $\left(\frac{N}{2} - \Sigma f_1\right)$ in frequency is proportionately $\frac{c}{f_{med}} \times \left(\frac{N}{2} - \Sigma f_1\right)$

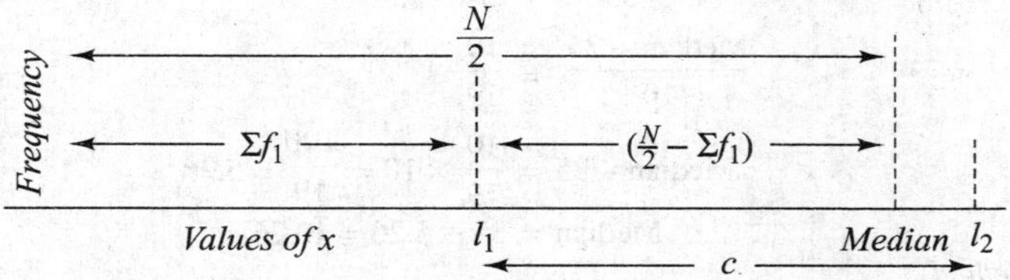

Fig. 5.1 Diagram Explaining Formula for Median

Hence,

$$\text{Median} = l_1 + \frac{c}{f_{med}} \times \left(\frac{N}{2} - \Sigma f_1\right)$$

$$= l_1 + \frac{\left(\frac{N}{2} - \Sigma f_1\right)}{f_{med}} \times c$$

Example 5.48 *Find the median and the median class of the data given below:*

Class Boundaries	*15–25*	*25–35*	*35–45*	*45–55*	*55–65*	*65–75*
Frequency	*4*	*11*	*19*	*14*	*0*	*2*

[I.C.W.A., Jan. '65]

Solution (*First method*) For a grouped frequency distribution

Median = Value of the variable corresponding to cumulative frequency N/2.

This can be found from a cumulative frequency distribution by using simple interpolation (page 63) as shown below:

Table 5.26 Cumulative Frequency Distribution

Class Boundary	*Cumulative Frequency (less-than)*
15	0
25	4
35	15
Median →	←$N/2 = 25$
45	34
55	48
65	48
75	$50 = N$

Since $N/2 = 25$ lies between the cumulative frequencies 15 and 34, the corresponding value of the variable, viz. median, must lie in the interval between 35 and 45. The *median class* is, therefore, (35 – 45). Now applying simple interpolation

$$\frac{\text{Median} - 35}{45 - 35} = \frac{25 - 15}{34 - 15}$$

or,
$$\frac{\text{Median} - 35}{10} = \frac{10}{19}$$

or,
$$\text{Median} - 35 = \frac{10}{19} \times 10 = \frac{100}{19} = 5.26$$

or,
$$\text{Median} = 35 + 5.26 = 40.26$$

(*Second method*)

We first calculate the cumulative frequencies against each class interval as shown below:

Table 5.27 Calculation of Cumulative Frequency

Class Boundaries	*Frequency*	*Cumulative Frequency*
15–25	4	4
25–35	11	$15 = F$
(35–45)	$19 = f_m$	34
45–55	14	48
55–65	0	48
65–75	2	$50 = N$

From the last col. of the table it will be seen that $N/2 = 25$ is more than cumulative frequency 15, but is less than the next cumulative frequency 34; hence the *median class* is 35–45. Now, we can apply formula (5.12.1), page 141; here $l_1 = 35$, $N = 50$, $F = 15$, $f_m = 19$, $c = 45–35 = 10$.

$$\therefore \quad \text{Median} = 35 + \frac{25 - 15}{19} \times 10 = 35 + 5.26 = 40.26$$

Ans. (i) 40.26; (ii) 35–45

Example 5.49 *The following is the table which gives you the distribution of marks secured by some students in an examination:*

Marks	*0–20*	*21–30*	*31–40*	*41–50*	*51–60*	*61–70*	*71–80*
No. of Students	*42*	*38*	*120*	*84*	*48*	*36*	*31*

Find: (i) Median marks. (ii) The percentage of failure if the minimum for a pass is 35 marks. [C.A., Nov. '69]

Solution [**Note:** Here class limits are given, and therefore we have to find class boundaries for the cumulative frequency distribution.]

Table 5.28 Cumulative Frequency Distribution

Marks (class boundaries)	*Cumulative Frequency*
20.5	42
30.5	80
34.5→	←F
Median→	←$N/2$ = 199.5
40.5	200
50.5	284
60.5	332
70.5	368
80.5	399 = N

(i) Median is the value of the variable corresponding to cumulative frequency $N/2$, i.e. 199.5. Using simple interpolation in the above cumulative frequency distribution,

$$\frac{\text{Median} - 30.5}{40.5 - 30.5} = \frac{199.5 - 80}{200 - 80}$$

Solving this, we get

$$\text{Median} = 30.5 + \frac{119.5 - 10}{120} = 40.46.$$

(ii) The minimum for a pass is 35. Assuming that 'marks' is a continuous variable and that marks are rounded to the nearest whole number, the mark 35 actually represents the interval (34.5–35.5). The minimum for a pass on the continuous scale is therefore 34.5. The number (F) of students obtaining less than 34.5 is the cumulative frequency corresponding to 34.5 marks. Using simple interpolation in Table 5.28 again, we have

$$\frac{34.5 - 30.5}{40.5 - 30.5} = \frac{F - 80}{200 - 80}$$

Solving, $$F = 80 + \frac{4}{10} \times 120 = 128$$

The percentage of failures is therefore $\frac{128}{399} \times 100 = 32.08$

Ans. (i) 40.46, (ii) 32.08

Example 5.50 *You are given below a certain statistical distribution:*

Value Less than	*100*	*100–200*	*200–300*	*300–400*	*400 & above*	*Total*
Frequency	*40*	*89*	*148*	*64*	*39*	*380*

Calculate the most suitable average giving reasons for your choice, [C.A., Nov. '73]

Solution The most suitable average from a frequency distribution with open-end classes is Median. Here, the total frequency $N = 380$; so $N/2 = 190$.

Table 5.29 Cumulative Frequency Distribution

Class Boundary	*Cumulative Frequency (less-than)*
100	40
200	129
Median→	←$N/2$ = 190
300	277
400	341
…	380 = N

Using simple interpolation (page 63, formula 3.16.1),

$$\frac{\text{Median} - 200}{300 - 200} = \frac{190 - 129}{277 - 129}$$

Solving this, we get

$$\text{Median} = 200 + \frac{61}{148} \times 100 = 241.2$$

Example 5.51 *The table below gives the diastolic blood pressure of 250 men. The readings were made to the nearest millimetre and the central value of each group is given:*

Blood Pressure (mm):	*60*	*65*	*70*	*75*	*80*	*85*	*90*	*95*
Number of Men:	*4*	*5*	*31*	*39*	*114*	*30*	*25*	*2*

Calculate from the data the mean and the median. [I.C.W.A., July '70]

Solution [**Note:** For the calculation of mean, we have to use the given midvalues. However, for the calculation of median, the class boundaries are required. Since the given midvalues are found to have a common difference, the classes must be of equal width and the class boundaries will lie exactly midway between the successive midvalues.]

Table 5.30 Calculations for Mean and Median

Mid-values x	*Class Boundaries*	*Frequency* f	$y = \frac{x-75}{5}$	fy	*Cumulative Frequency*
60	57.5–62.5	4	–3	–12	4
65	62.5–67.5	5	–2	–10	9
70	67.5–72.5	31	–1	–31	40
75	72.5–77.5	39	0	0	79
80	(77.5–82.5)	114	1	114	193
85	82.5–87.5	30	2	60	223
90	87.5–92.5	25	3	75	248
95	92.5–97.5	2	4	8	250 = N
Total	—	250 = N	—	204	—

Using (5.4.2)

$$\text{Mean} = 75 + 5\left(\frac{204}{250}\right) = 79.08 \text{ mm.}$$

Again, it is found that $N/2 = 125$ exceeds the cumulative frequency 79, but is less than the next higher cumulative frequency 193. Hence the median class is (77.5–82.5). Now using formula (5.12.1)

$$\text{Median} = 77.5 + \frac{125-79}{114}(82.5 - 77.5) = 79.52 \text{ mm.}$$

Ans. Mean = 9.08 mm., Median = 79.52 mm.

Exmaple 5.52 *Draw one of the ogives for the following data and find the median wage:*

Weekly Wages in Rs	*0–20*	*20–40*	*40–60*	*60–80*	*80–100*
Number of Workers	*40*	*51*	*64*	*38*	*7*

[C.A.Nov.'63]

Solution Let us draw the ogive of 'less-than' type. For this purpose, we have to construct a cumulative frequency distribution (Note that in the data class boundaries are given) as shown below.

Class Boundary	0	20	40	60	80	100
Cumulative Frequecny	0	40	91	155	193	200 = *N*

The cumulative frequecies are plotted on a graph paper against class boundaries, and the ogive (Fig. 5.2) is drawn. From the point $N/2 = 200/2 = 100$ on the vertical scale, a horizontal line is drawn meeting the ogive. From the point of intersection, a perpendicular (dotted line in the figure) is now drawn on the horizontal axis. The point showing the foot of the perpendicular is now read from the scale, and is found to be approximately 43. Thus, median = 43.

Ans. 43.

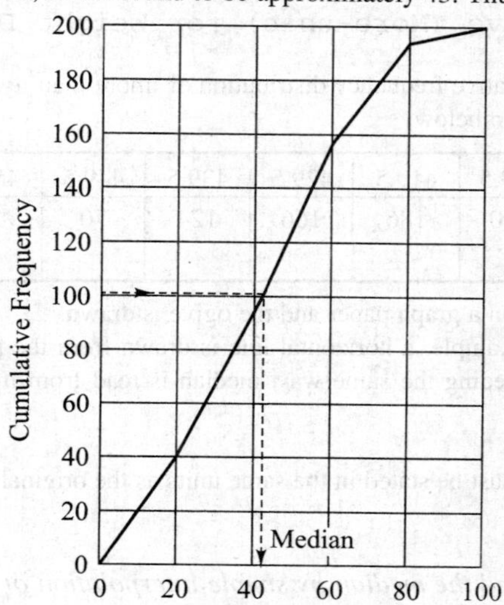

Fig. 5.2 Ogive (Less-than) for Wage Distribution

Example 5.53 *Draw an ogive of 'more than' type on the data given below. Deduce from it the median of the distribution.*

Wt. in gms.	*410–419*	*420–429*	*430–439*	*440–449*	*450–459*	*460–469*	*470–479*
Frequency	*14*	*20*	*42*	*54*	*45*	*18*	*7*

[I.C.W.A., July '68]

Solution

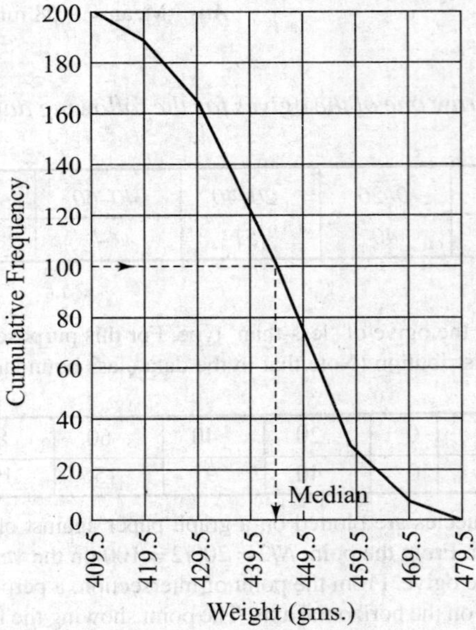

Fig. 5.3 Ogive (More-than) for Weight Distribution

We construct a cumulative frequency distribution of 'more-than' type (Note that here class limits are given) as shown below:

Class Boundary	409.5	419.5	429.5	439.5	449.5	459.5	469.5	479.5
Cumulative Frequency	200	186	166	124	70	25	7	0

The data are plotted on a graph paper and the ogive is drawn.

As in the previous example, a horizontal line is drawn from the point $N/2 = 100$ on the vertical scale, and proceeding the same way, median is read from the horizontal scale, and found to be 444 gms.

Ans. 444 gms.

[**Note:** The median must be stated in the same units as the original data].

Exmaple 5.54 *Find the median by simple interpolation or graphically from the following distribution:*

X :	1	2	3	4–6	7–9	10–12	13–20	21–28	29–36
Freq. :	10	5	3	9	6	2	1	10	15

Should interpolation and graphic method give the same value of median? If so, why?

[C.U., B.A. (Econ) '68]

Solution We have to construct a cumulative frequency distribution from the given data:

Class Boundary:	1.5	2.5	3.5	6.5	Median ↓	9.5	12.5	20.5	28.5	36.5
Cumulative Frequency:	10	15	18	27	↑ $N/2 = 30.5$	33	35	36	46	61 = N

Now, $N/2 = 30.5$ lies between the consecutive cumulative frequencies 27 and 33 in the cumulative frequency distribution. Therefore, Median (i.e. the value of the variable corresponding to cumulative frequency 30.5) must lie between the class boundaries 6.5 and 9.5. Applying simple interpolation, we have

$$\frac{\text{Median} - 6.5}{9.5 - 6.5} = \frac{30.5 - 27}{33 - 27}$$

or,
$$\frac{\text{Median}-6.5}{3} = \frac{3.5}{6}$$

or,
$$\text{Median} - 6.5 = \frac{3.5}{6} \times 3 = 1.75$$

$\therefore$
$$\text{Median} = 6.5 + 1.75 = 8.25.$$

[**Note:** If we apply formula (5.12.2), $l_1 = 6.5$, $l_2 = 9.5$, $F_1 = 27$, $F_2 = 33$].

Simple interpolation and graphic method should both give the same value of median; because they are based on the same principles, viz. uniform distribution of frequencies in the median class. Simple interpolation involves arithmetic calculation and the graphic method uses the ogive. In practical calculations, however, the graphic method gives only an approximate result.

Example 5.55 *You are given the following incomplete frequency distribution. It is known that the total frequency is 1,000 and that the median is 413.11. Estimate by calculation the missing frequencies.*

Values	300–325	325–350	350–375	375–400	400–425	425–450
Frequency	5	17	80	?	326	?
					450–475	475–500
					88	9

[C.A., May '77]

Solution Let a and b denote the missing frequencies of classes 375–400 and 425–450 respectively. Then the total frequency is

$$5 + 17 + 80 + a + 326 + b + 88 + 9 = 1000 \text{ (given)}$$

or,
$$525 + a + b = 1000$$

$\therefore \quad a + b = 475 \quad \ldots \quad \text{(i)}$

Again, since the given value of median, *viz.* 413.11, lies in the class interval 400–425, this must be the median class. So, in the formula for median (5.12.1) at p. 141 we have $l_1 = 400$, $f_m = 326$, $F = 5 + 17 + 80 + a = 102 + a$, $c = 25$ and $N = 1000$.

$$\text{Median} = l_1 + \frac{(N/2) - F}{f_m} \times c$$

or, $$413.11 = 400 + \frac{500 - (102 + a)}{326} \times 25$$

or, $$413.11 = 400 + \frac{398 - a}{326} \times 25$$

or, $$(413.11 - 400) \times \frac{326}{25} = (398 - a)$$

or, $$171 = 398 - a,$$

$$\therefore \quad a = 227 \quad \ldots \text{(ii)}$$

Using (i) and (ii), we get $b = 248$. *Ans.* 227, 248

Example 5.56 *Show that* $\sum_{i=1}^{n} (x_i - A)^2$ *is least if* $A = \bar{x}$, *but* $\sum_{i=1}^{n} | x_i - A |$ *is least if* A = *median.* [C.U., B.Com. (Hons.)' 63; B.A. (Econ)' 71]

Solution *(First Part)* See Example 5.12, page 113

(Second Part) Let us arrange the given observations x_1. x_2, ..., x_n in increasing order of magnitude and let the ordered series be denoted by

$$X_1, X_2, \ldots X_n \quad \ldots \quad (1)$$

X_1 is the smallest and X_n the largest among the given observation.

Since X_1, X_2, ..., X_n is only a different arrangement of the same observation x_1 x_2, ..., x_n we have

$$\sum_{i=1}^{n} | x_i - A | = \sum_{i=1}^{n} | x_i - A |$$

The number of observations may be either odd or even, and we shall prove the result separartely for each case:

Case I. When n is odd, say $n = 2m + 1$.

$$\sum_{i=1}^{n} | x_i - A | = \sum_{i=1}^{2m+1} | x_i - A |$$

$$= | X_1 - A | + | X_2 - A | + \ldots + | X_m - A |$$
$$+ | X_{m+1} - A | + | X_{m+2} - A | + \ldots + | X_{2m} - A |$$
$$+ | X_{m+1} - A | \quad \ldots (2)$$

There are $2m + 1$ terms on the right, and we group them in pairs, the terms equidistant from either end being paired together. Now the sum of the first and the last terms, viz.

$$| X_1 - A | + | X_{2m+1} - A |$$

has the minimum value, only when A lies between X_1 and X_{2m+1}, i.e $X_1 \le A \le X_{2m+1}$.
The sum of the 2nd terms from either end, viz.

$$|X_2 - A| + |X_{2m} - A|$$

has the minimum value, only when A lies between X_2 and X_{2m}, i.e. $X_2 \le A \le X_{2m}$.
Continuing this way, the sum of the last pair, viz.

$$|X_m - A| + |X_{m+2} - A|$$

has the minimum value, only when A lies between X_m and X_{m+2}, i.e. $X_m \le A \le X_{m+2}$.
There is still another term, viz.

$$|X_{m+1} - A|$$

which has the minimum value zero, only when $A = X_{m+1}$.

It will now be noticed that *(i)* X_{m+1} is the median, because it occupies the middle-most position of the ordered series $X_1, X_2, \ldots X_m, X_{m+1}, X_{m+2}, \ldots, X_{2m+1}$; and (ii) if $A = X_{m+1}$, it satisfies all the conditions for minimising the sums of the terms in pairs, and also the remaining term. Thus for odd n,

$$\sum_{i=1}^{n} |x_i - A| \text{ is minimum, when } A = \text{median.}$$

Case II. When n is even, say $n = 2m$

$$\sum_{i=1}^{n} |x_i - A| = \sum_{i=1}^{2m} |X_i - A|$$
$$= |X_1 - A| + |X_2 - A| + \ldots + |X_m - A| + |X_{m+1} - A| + |X_{m+2} - A| + \ldots + |X_{2m} - A| \quad \ldots (3)$$

As in Case I, we group the terms in pairs, there being exactly m pairs now. Proceeding the same way as before, we find that the sum on the right of (3) is minimum, only when A lies between the two middle-most values X_m and X_{m+1} of the orered series $X_1, X_2, \ldots X_m, X_{m+1}, \ldots X_{2m}$. In particular, if we take $A = \text{median} = \frac{1}{2}(X_m + X_{m+1})$, we see that it lies beteen X_m and X_{m+1}, and as such minimises (3). Thus, for even n,

$$\sum_{i=1}^{n} |x_i - A| \text{ is minimum, when } A = \text{median.}$$

Combining the results of Case I and Case. II, we conclude that in general

$$\sum_{i=1}^{n} |x_i - A| \text{ is the least, when } A = \text{median.}$$

(Alternative Proof for the Second Part)

Let $\Delta = \sum_{i=1}^{n} |x_i - A|$ = Sum of the distances from A,

r = Number of observations smaller than A.

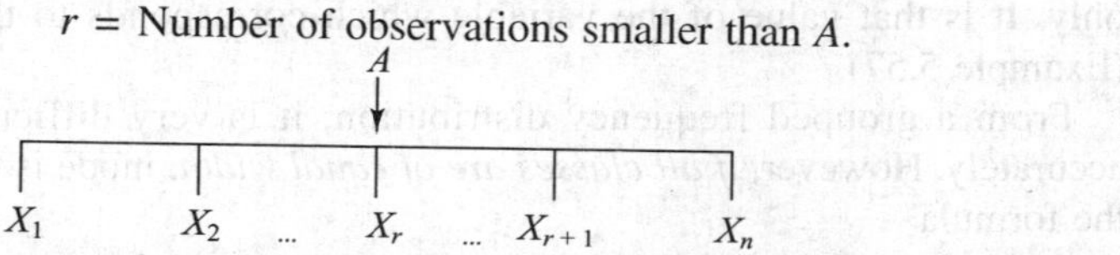

If Δ' represents the sum of the distances from the $(r + 1)$th value X_{r+1}, then the distance of each of the r values $X_1, X_2, \ldots, X_r$, will increase by $(X_{r+1} - A) = c$ (say) and

the distance of each of the $(n-r)$ values $X_{r+1} \ldots, X_n$ will diminish by c, so that ultimately

$$\Delta' = \Delta + rc - (n-r)c = \Delta - (n-2r)c$$

Since c is positive,

$$\Delta' < \Delta, \text{ when } r < n/2$$
$$\Delta' = \Delta, \text{ when } r = n/2$$
$$\Delta' > \Delta, \text{ when } r > n/2$$

These relations imply that, if the origin is shifted to the next higher value, the sum of the distances from A.

(i) will reduce further, so long as less than half the total number of observations is smaller than A.

(ii) will remain constant, when $r = n/2$, i.e. exactly half the total number of observations is smaller than A (and naturally the same number greater than A).

(iii) will increase, if more than half the total number of observations is smaller than A.

By (i) and (iii) we find that the minimum value of Δ will be attained when A lies between the $\frac{n}{2}$ th and the $\left(\frac{n}{2}+1\right)$ th values. If n is odd, say $n = 2m + 1$, Δ will therefore be minimum, when A coincides with the $(m + 1)$th value, i.e A = Median. Again if n is even, say $n = 2m$, any value of A lying between the mth and $(m + 1)$th given values (in order of magnitude) will minimise Δ, and by (ii), this Δ will remain a constant. Conventionally, we take Median = A.M. of the mth and the $(m + 1)$th values. Hence, in this case also, Δ will be minimum when A = Median.

This proves that $\Sigma|x_i - A|$ is the least, when A = Median.

5.14 MODE

Mode of a given set of observations is that value which occurs with the maximum frequency. It is the most typical or prevalent value, and at times represents the true characteristic of the distribution as a measure of central tendency. The modal wage of a group of workers is that wage which the largest number of workers receive, and as such this wage may be considered as the representative wage of the group. Mode is used in business, because it is most likely to occur. Meteorological forecasts are, in fact, based on mode.

5.15 CALCULATION OF MODE

From a simple series, mode can be determined by locating that value, which occurs the maximum number of times [Example 5.2, *Illustration 14(i)*].

From a simple frequency distribution, mode can be determined by inspection only. It is that value of the variable which corresponds to the largest frequency (Example 5.57).

From a grouped frequency distribution, it is very difficult to find the mode accurately. However, *if all classes are of equal width*, mode is usually calculated by the formula

$$Mode = l_1 + \frac{d_1}{d_1 + d_2} \times c \qquad (5.15.1)$$

where

l_1 = lower boundary of the *modal class* (i.e the class containing the largest frequency);

d_1 = difference of the largest frequency and the frequency of class just preceding the modal class;

d_2 = difference of the largest frequency and the frequency of class just following the modal class;

c = common width of classes.

If f_0, f_{-1}, f_1 represent the frequencies of the modal class, the class just preceding and the class just following it, then

$$d_1 = f_0 - f_{-1} \quad \text{and} \quad d_2 = f_0 - f_1,$$

so that (5.15.1) may also be written in the form (see Example 5.6l)

$$Mode = l_1 + \frac{f_0 - f_{-1}}{2f_0 - f_{-1} - f_1} \times c \qquad (5.15.1a)$$

Formula (5.15.1) or (5.15.la) is based on the assumption that the distances of mode from the two boundaries of modal class are proportional to the differences in frequencies of the modal class and its two adjoining classes, viz. d_1 and d_2 (Fig. 5.4, Page 154).

If however the frequency distribution has classes of unequal width formula (5.15.1) of (5.15.1a) cannot be applied. In this case, an approximate value of mode is usually obtained by the relation (5.17.1), when the mean and median are known.

5.16 ADVANTAGES AND DISADVANTAGES OF MODE

Advantages

1. From a simple frequency distribution, mode can be obtained only by inspection. Also, for a simple series with a small number of observations, mode can often be determined without any calculation.
2. It is unaffected by the presence of extreme values.
3. Unlike A. M., it can be calculated from frequency distributions, with open-end classes. In fact, if the modal class and its two adjoining classes together with their class frequencies are available, mode can be determined, provided it is known that all classes are of equal width.

Disadvantages

1. Mode has no significance unless a large number of observations is available.
2. It is a peculiar measure of central tendency. For any given set of observations, it is always possible to find the values of A.M., G.M. H.M., or median. But mode may not exist. When all values occur with equal frequency, there is no mode. On the other hand if two or more values have the same maximum frequency, there is more than one mode (A distribution is called *unimodal, bimodal* or *multimodal*, when it has one, two or more than two modes). See *Illustration 14*, Example 5.2 (p. 106)
3. From a grouped frequency distribution, it is difficult to locate the mode accurately. An approximate value of mode is obtained by the formula (5.15.1)

or (5.15.1a), only when all classes are of equal width, and the class with the largest frequency is preceded and followed by two other classes. Even then, if class frequencies do not show any gradual tendency to maximum concentration in a class, or the largest frequency occurs in two or more classes, the formula is inapplicable.

4. For the calculation of mode, the data must be arranged in the form of a frequency distribution.
5. Mode cannot be treated algebraically.

Example 5.57 *The numbers of telephone calls received in 245 successive one minute intervals at an exchange are shown in the following frequency distribution:*

No. of Calls	*0*	*1*	*2*	*3*	*4*	*5*	*6*	*7*	*Total*
Frequency	*14*	*21*	*25*	*43*	*51*	*40*	*39*	*12*	*245*

Evaluate the mean, median and mode. [I.C.W.A., June '74]

Solution [**Hint:** (i) $\Sigma fx = 922$, $N = 245$, $\bar{x} = \Sigma fx/N = 3.8$

(ii) Proceeding as in Example 5.46 (page, 166)

Median = Value corresponding to cumulative frequency $(N + 1)/2$, i.e. 123. = 4

(iii) Mode is the value of the variable corresponding to the highest frequency 51; i.e. Mode = 4]

Ans. Mean = 3.8. Median–4, Mode = 4.

Example 5.58 *Find the modal wage from the following:*

Wages (Rs)	*50.00–59.99*	*60.00–69.99*	*70.00–79.99*	*80.00–89.99*
No. of Employees	*8*	*10*	*16*	*14*
	90.00–99.99	*100.00–109.99*	*110.00–119.99*	
	10	*5*	*2*	

[B.U.,B.A.(Econ) '72]

Solution The class 70.00–79.99, having the maximum frequency 16, is the '*modal class*'; its lower boundary is 69.995. Since the classes are of equal width, formula (5.15.1a) can be applied: $l_1 = 69.995$, $f_0 = 16$, $f_{-1} = 10$, $f_1 = 14$, $c = 10$.

$$\text{Mode} = 69.995 + \frac{16 - 10}{2 \times 16 - 10 - 14} \times 10 = 77.50 \text{ Rs.}$$

Example 5.59 *The monthly profits in rupees of 100 shops are distributed as follows:*

Profits per Shop	*0–100*	*100–200*	*200–300*	*300–400*	*400–500*	*500–600*
No. of Shops	*12*	*18*	*27*	*20*	*17*	*6*

Draw the histogram to the data and hence find the modal value. Check this value by direct calculation. [I.C.W.A., Jan.'64]

Solution The histogram has been drawn (Fig. 5.4)

In the histogram, the top right corner of the highest rectangle is joined by a straight line to the top right corner of the preceding rectangle. Similarly, the top left corner of the highest rectangle is joined to the top left corner of the following rectangle. From the point of intersection of these two lines (dotted lines in Fig. 5.4) a perpendicular is drawn on the horizontal axis. The foot of the perpendicular indicates the Mode. This is read from the horizontal scale and the modal value is found to be 256 (in Rs) approximately.

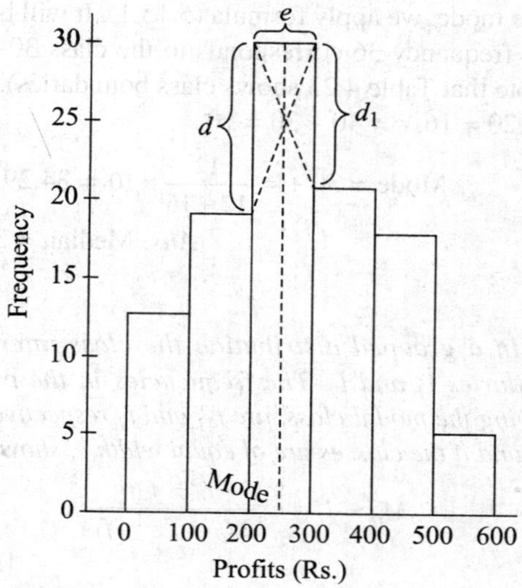

Fig. 5.4 Histogram for Distribution of Profits

In order to calculate the Mode directly, we find that the largest class frequency, viz, 27, lies in the class 200–300, and hance this is the *modal class*. Therefore, in formula (5.15.1), $l_1 = 200$, $d_1 = 27 - 18 = 9$, $d_2 = 27 - 20 = 7$, $c = 300 - 200 = 100$.

$$\text{Mode} = 200 + \frac{9}{9+7} \times 100 = 256.25 \text{ Rs.}$$

Ans. Rs 256.25 P.

Example 5.60 *Calculate the Median and Mode from the following data:*

Value	*Frequency*
Less than 10	4
Less than 20	16
Less than 30	40
Less than 40	76
Less than 50	96
Less than 60	112
Less than 70	120
Less than 80	125

[C.A., Nov. '67]

Solution We have to construct a cumulative frequency distribution for the calculation of Median and a grouped frequency distribution for the calculation of Mode (see Example 4.9, page 82). Using simple interpolation in. Table 4.22, we can now find the value corresponding to cumulative frequency $N/2 = 125/2 = 62.5$, giving the median.

$$\frac{\text{Median} - 30}{40 - 30} = \frac{62.5 - 40}{76 - 40}$$

Solving this, we get Median = 36.25.

In order to find the mode, we apply formula (5.15.1). It will be found in Table 4.23 that the largest class frequency 36 corresponds to the class 30-40, and hence this is the 'modal class' (note that Table 4.23 shows class boundaries). Here $l_1 = 30$, $d_1 = 36 - 24 = 12$, $d_2 = 36 - 20 = 16$, $c = 40 - 30 = 10$.

$$\therefore \quad \text{Mode} = 30 + = \frac{12}{12 + 16} \times 10 = 34.29$$

Ans. Median = 36.25, Mode = 34.29.

Example 5.61 *In a grouped distribution the class interval with the highest frequency has boundaries l_1 and l_2. The frequencies in the two adjoining classes, preceding and following the modal class, are f_{-1} and f_1 respectively. If the frequency of the modal class is f_0 and if the classes are of equal width c, show that mode is given by*

$$M_0 = l_1 + c \cdot \frac{f_0 - f_{-1}}{2f_0 - f_{-1} - f_1}$$

[B.U.B.A, (Econ) '71]

Solution (Refer to the histogram at Fig. 5.4, p, 155), Since all classes are of equal width, the frequencies f_{-1}, f_0, f_1, represent the heights of the 3 adjacent rectangles (starting from the left) with the tallest at the middle. The vertical lines forming boundaries of the tallest rectangle meet the x-axis at l_1 and l_2, so that the Mode (M_0) lies in between them, i.e. $l_1 < M_0 < l_2$ (in the Fig. $l_1 = 200$, $l_2 = 300$) and $c = l_2 - l_1$.

Let d_1 and d_2 represent the differences in heights of the tallest rectangle from the two adjacent ones (see figure), i.e. $d_1 = f_0 - f_{-1}$, and $d_2 = f_0 - f_1$. If d_1 and d_2 were equal, the mode would lie exactly midway between l_1 and l_2. In other words, distances $M_0 - l_1$, and $l_2 - M_0$ would be equal. The distance of M_0 from l_1, is expected to be larger (smaller) than its distance from l_2 as d_1 is more (less) than d_2. If we assume that the distance of the mode from the lower (upper) boundary is proportional to the corresponding difference in the heights of rectangles to the left (right), we have

$$\frac{M_0 - l_1}{l_2 - M_0} = \frac{d_1}{d_2}$$

or, $$d_2 (M_0 - l_1) = d_1 (l_2 - M_0)$$

or, $$M_0(d_1 + d_2) = d_1 l_2 + d_2 l_1$$

$$= d_1(l_1 + c) + d_2 l_1, \quad \text{since } l_2 - l_1 = c$$

$$= l_1 (d_1 + d_2) + d_1 c$$

Dividing both sides by $(d_1 + d_2)$, we get (5.15. 1), viz.

$$M_0 = l_1 + \frac{d_1}{d_1 + d_2} \cdot c$$

Now substituting the values $d_1, = f_0 - f_{-1}$ and $d_2 = f_0 - f_1$, the required result follows.

5.17 RELATION BETWEEN MEAN, MEDIAN, MODE

For unimodal distributions of moderate skewness (Section 7.7) the following approximate relation has been found to hold:

$$\text{Mean} - \text{Mode} = 3(\text{Mean}-\text{Median}) \qquad (5.17.1)$$

Sometimes this relation may be utilised for the calculation of mode. When the distribution is symmetrical, mean, median and mode coincide. In particular, for the "normal distribution" (Section 12.6) mean, median and mode are all equal.

In most frequency distributions, it has been observed that the three measures of central tendency, viz. Mean, Median and Mode, obey the approximate relation (5.17.1), provided the distribution is not very skew. This relation is therefore applied to estimate one of them when the values of the other two are known.

Example 5.62 *Arithmetic Mean = 26.8, Median = 27.9. What is the value of Mode ? Explain the principle involved.* [C.A. Nov. '72]

Solution Using (5.17.1), $26.8 - \text{Mode} = 3(26.8 - 27.9)$

or, $26.8 - \text{Mode} = -3.3$; $\therefore$ $\text{Mode} = 30.1$

5.18 PARTITION VALUES—QUARTILES, DECILES, PERCENTILES

Just as median divides the total number of observations into two equal parts, there are similar other measures which are used to divide, or partition, the observations into a fixed number of parts, say 4, 10 or 100. These are collectively known as *partition values* or *quantiles* or *fractiles*. Some of the important types of partition values are

(a) Median, (b) Quartiles, (c) Deciles, (d) Percentiles.

We know that Median is the middle-most value of a set of observations, i.e. it divides the total number of observations into 2 equal parts. The number of observations smaller than median is the same as the number larger than it. For data of continuous type, exactly one-half of the observations are smaller than median, i.e. median is the value of the variable corresponding to cumulative frequency *N*/2. These ideas are extended to Quartiles, Deciles and Percentiles,

Quartiles are such values which divide the total number of observations into 4 equal parts. Obviously, there are 3 quartiles—

(i) *First quartile* (or *Lower quartile*) : Q_1

(ii) *Second quartile*, (or *Middle quartile*) : Q_2

(iii) *Third quartile* (or *Upper quartile*) : Q_3

The number of observations smaller than Q_1, is the same as the number lying between Q_1 and Q_2, or between Q_2 and Q_3, or larger than Q_3. For data of continuous type, one-quarter of the observations is smaller than Q_1, two-quarters are smaller than Q_2. and three-quarters are smaller than Q_3. This means that Q_1, Q_2, Q_3 are values of the variable corresponding to 'less-than' cumulative frequencies $N/4$, $2N/4$, $3N/4$ respectively. Since, $2N/4 = N/2$, it is evident that the second quartile Q_2 is the same as median.

$$Q_1 < Q_2 < Q_3; \quad Q_3 = \text{Median}. \quad (5.18.1)$$

Quartiles are used for measuring central tendency, dispersion and skewness. For instance, the second quartile Q_2 is itself taken as a measure of central tendency, where it is known as *Median*. The lower and the upper quartiles, viz. Q_1 and Q_3, are used to define *Quartile Deviation* (6.3.1), which is a measure of dispersion. In *Bowley's formula* (7.7.3) for skewness all the three quartiles are used.

$$\text{Median} = Q_2$$

$$\textit{Quartile Deviation} = \frac{Q_3 - Q_1}{2}$$

$$\textit{Skewness} = \frac{Q_3 - 2Q_2 + Q_1}{Q_3 - Q_1}$$

[**Note:** (i) Students must not carry a false impression that there is a common difference between the successive quartiles, or that $Q_3 - Q_2 = Q_2 - Q_1$. Only the number of observations' (i e. frequency) between them is the same. In fact,

Less than Q_1, the frequency = $N/4$
Between Q_1 and Q_2, the frequency = $N/4$
Between Q_2 and Q_3, the frequency = $N/4$
Above Q_3, the frequency = $N/4$

(ii) Again, when asked to define quartiles, many students often write $Q_2 = N/4$, $Q_2 = 2N/4$, $Q_2 = 3N/4$. This is incorrect. The correct position is

Corresponding to Q_1, the cumulative frequency = N/4
Corresponding to Q_2, the cumulative frequency = 2N/4
Corresponding to Q_3, the cumulative frequency = 3N/4]

Deciles are such values which divide the total number of observations into 10 equal parts. There are 9 deciles $D_1, D_2, .., D_9$ called the *first decile*, the *second decile*, etc. The number of observations smaller than D_1, or between two successive deciles, or larger than D_9 is the same. For data of continuous type, $D_1, D_2, .., D_9$ correspond to cumulative frequencies $N/10$, $2N/10$, .. , $9N/10$ respectively.

$$D_1 < D_2 < \ldots < D_9; \quad D_5 = Q_2 = \textit{Median}. \quad (5.18.2)$$

Percentiles are such values which divide the total number of observations into 100 equal parts. There are 99 percentiles $P_1, P_2,.. P_{99}$, called the *first percentile*, the *second percentile*, and so on. The *k*-th percentile (P_k) is, therefore, that value of the variable upto which lie exactly $k\%$ of the total number of observations. Hence, P_k corresponds to 'less-than' cumulative frequency $kN/100$. In particular,

$$P_{10} = D_1, \quad P_{20} = D_2, .., P_{90} = D_9$$
$$P_{25} = Q_1, \quad P_{50} = D_5 = Q_2 = \text{Median}, \quad P_{75} = Q_3$$
$$P_1 < P_2 < \ldots < P_{99} \quad (5.18.3)$$

5.19 CALCULATION OF PARTITION VALUES

The method of calculation of quartiles, deciles, percentiles is exactly the same as that employed for median, using simple interpolation (page 77) in a cumulative frequency distribution. These may also be obtained graphically from ogive. For example, the 7th decile D_7 is the abscissa of that point on the ogive whose ordinate is $7N/10$.

(a) *From simple series*—The given data are arranged in increasing order of magnitude and a number showing the rank is attached to each observation. The smallest value is given rank 1, the next higher value rank 2, etc. and the largest value is given rank n. The ranks of partition values are as follows:

$$\text{Rank of Median} = \frac{1}{2}(n+1)$$

$$\text{Rank of } Q_1 = \frac{1}{4}(n+1)$$

$$\text{Rank of } Q_3 = \frac{3}{4}(n+1) \tag{5.19.1}$$

$$\text{Rank of } D_k = \frac{k}{10}(n+1)$$

$$\text{Rank of } P_k = \frac{k}{100}(n+1)$$

Using simple interpolation, the value of the varible corresponding to the appropriate rank is determined, giving the partition value (see Example 5.63)

$$\textit{Median} = \text{Value corresponding to rank } \frac{1}{2}(n+1)$$

$$Q_1 = \text{Value corresponding to rank } \frac{1}{4}(n+1)$$

$$Q_3 = \text{Value corresponding to rank } \frac{3}{4}(n+1) \tag{5.19.2}$$

$$D_k = \text{Value corresponding to rank } \frac{k}{10}(n+1)$$

$$P_k = \text{Value corresponding to rank } \frac{k}{100}(n+1)$$

(b) *From simple frequency distribution*—The cumulative frequency corresponding to each distinct value of the variable is calculated (see Table 5.25, p. 142). If the total frequency be N.

$$\textit{Median} = \text{Value corresponding to cumulative frequency } \frac{1}{2}(N+1)$$

$$Q_1 = \text{Value corresponding to cumulative frequency } \frac{1}{4}(N+1)$$

$$Q_3 = \text{Value corresponding to cumulative frequency } \frac{3}{4}(N+1) \tag{5.19.3}$$

$$D_k = \text{Value corresponding to cumulative frequency } \frac{k}{10}(N+1)$$

$$P_k = \text{Value corresponding to cumulative frequency } \frac{k}{100}(N+1)$$

(c) *From grouped frequency distribution*—(i) *By appliction of simple interpolation* —A cumulative frequency distribution is constructed showing the class boundaries and the corresponding cumulative frequencies ('less-than'type). Using simple interpolation, we now find (see Examples 5.64, 5.65, 5.66)

$$\begin{aligned}
\textit{Median} &= \text{Value corresponding to cumulative frequency } \frac{1}{2}N \\
Q_1 &= \text{Value corresponding to cumulative frequency } \frac{1}{4}N \\
Q_3 &= \text{Value corresponding to cumulative frequency } \frac{3}{4}N \qquad (5.19.4)\\
D_k &= \text{Value corresponding to cumulative frequency } \frac{k}{10}N \\
P_k &= \text{Value corresponding to cumulative frequency } \frac{k}{100}N
\end{aligned}$$

If the cumulative frequencies are viewed as ranks, some authors define median as the value corresponding to rank $N/2$, $Q1$ corresponding to rank $N/4$, etc. (see Example 5.68).

(ii) *Graphical method*—An ogive ('less-than' type) is drawn. From this ogive (see Example 5.66)

$$\begin{aligned}
\textit{Median} &= \text{Abscissa corresponding to ordinate } \frac{1}{2}N \\
Q_1 &= \text{Abscissa corresponding to ordinate } \frac{1}{4}N \\
Q_3 &= \text{Abscissa corresponding to ordinate } \frac{3}{4}N \qquad (5.19.5)\\
D_k &= \text{Abscissa corresponding to ordinate } \frac{k}{10}N \\
P_k &= \text{Abscissa corresponding to ordinate } \frac{k}{100}N
\end{aligned}$$

Example 5.63 *Obtain the values of Median and the two Quartiles: 391 384 591 407 672 522 777 773 2488 1490*

[C.A. May '69]

Solution The given values are arranged in increasing order of magnitude and ranked (Table 5.31). Here, $n = 10$; so using (5.19.1)

$$\begin{aligned}
\text{Rank of Median} &= \frac{1}{2}(10+1) = 5.5 \\
\text{Rank of } Q_1 &= \frac{1}{4}(10+1) = 2.75 \\
\text{Rank of } Q_3 &= \frac{3}{4}(10+1) = 8.25
\end{aligned}$$

Table 5.31 Ordered Values and Their Ranks

Rank	*Value*
1	384
2	391
2.75 →	← Q_1
3	407
4	522
5	591
5.5 →	←Median
6	672
7	733
8	777
8.25 →	← Q_3
9	1490
10	2488

Using simple interpolation, we have

$$\frac{5.5-5}{6-5} = \frac{\text{Median} - 591}{672-591}; \quad \frac{2.75-2}{3-2} = \frac{Q_1 - 391}{407-391}$$

$$\frac{8.25-8}{9-8} = \frac{Q_3 - 777}{1490-777}$$

Solving the three equations,

$$\text{Median} = 631.5; \; Q_1 = 403; \; Q_3 = 955.25$$

[**Note:** Since the second quartile Q_3 is the same as Median, for practical purposes we speak of two quartiles only, viz. the lower quartile Q_1, and the upper quartile Q_3].

Example 5.64 *Determine Q_1, Q_3 and the median (Q_2) from the following frequency table:*

Marks in English	*10–19*	*20–29*	*30–39*	*40–49*	*50–59*	*60–69*	*Total*
Frequency	*8*	*11*	*15*	*17*	*12*	*7*	*70*

[B.U.,B.A. (Econ) '73]

Solution In a grouped frequency distribution, Q_1 Q_2 and Q_3 are values of the variable corresponding to cumulative frequencies $N/4$, $2N/4$ and $3N/4$ respectively.

Table 5.32 Cumulative Frequency Distribution

Marks (Class boundary)	*Cumulative Frequency (less-than)*
9.5	0
19.5	8
Q_1→	←$N/4 = 17.5$
29.5	19
39.5	34
Q_2→	←$2N/4 = 35$
49.5	51
Q_3→	← $3N/4 = 52.5$
59.5	63
69.5	$70 = N$

[**Note:** The values of $N/4$, $2N/4$, $3N/4$ are first obtained numerically, and then their positions are indicated by arrow marks in the column of cumulative, frequencies. Q_1, Q_2, Q_3 are then shown in the corresponding spaces under the first column. Here, since $N/4 = 70/4 = 17.5$ exceeds cumulative frequency 8, but is smaller than the next cumulative frequency 19, . $N/4$ is shown in the space between 8 and 19. Q_1 therefore must lie in between the corresponding values of the variable, viz. 19.5 and 29.5 Similarly, $2N/4 = 35$ lies between the cumulative frequencies 34 and 51, and hence the position of Q_2, is shown between the corresponding values 39.5 and 49.5. Similarly for Q_3]

Using simple interpolation (3.16.1), we have

$$\frac{Q_1 - 19.5}{29.5 - 19.5} = \frac{17.5 - 8}{19 - 8}; \quad \frac{Q_2 - 39.5}{49.5 - 39.5} = \frac{35 - 34}{51 - 34}$$

$$\frac{Q_3 - 49.5}{59.5 - 49.5} = \frac{52.5 - 51}{63 - 51}$$

Solving these equations, we get $Q_1 = 28$, $Q_2 = 40$, $Q_3 = 51$

Example 5.65 *From the following data calculate the first quartile and 4th decile :*

X	*0–5*	*5–10*	*10–15*	*15–25*	*25–35*	*35–60*	*60–80*
Freq.	*12*	*30*	*51*	*84*	*66*	*50*	*7*

[C.U., B.Com. (Hons) '69]

Solution (Note that here all classes do not have the same width ; but this does not cause any difficulty in the calculation of partition values). The total frequency is $N = 300$. Hence, using (5.19.4), the first quartile Q_1 and the fourth decile D_4 are those values of X which correspond to cumulative frequencies $\frac{1}{4} \times 300 = 75$ and $\frac{4}{10} \times 300 = 120$ respectively.

Table 5.33 Cumulative Frequency Distribution

Class Boundary	*Cumulative Frequency (less-than)*
0	0
5	12
10	42
$Q_1\rightarrow$	$\leftarrow N/4 = 75$
15	93
$D_4\rightarrow$	$\leftarrow 4/N\ 10 = 120$
25	177
35	243
60	293
80	$300 = N$

Using simple interpolation,

$$\frac{Q_1 - 10}{15 - 10} = \frac{75 - 42}{93 - 42}, \text{ hence we get } Q_1 = 13.24$$

$$\frac{D_4 - 15}{25 - 15} = \frac{120 - 93}{177 - 93}, \text{ hence we get } Q_4 = 18.21$$

Ans. $Q_1 = 13.24$, $D_4 = 18.21$

Example 5.66

Height in inches	57.5–	60.0–	62.5–	65.0–	67.5–	70.0–	72.5
Number ...	6	26	190	281	412	7	38

Compute the third decile and the quartiles of the distribution given above. Check your results by graphical means. [I.C.W.A., July '65]

Solution By (5.19.4), D_3, Q_1, Q_2, and Q_3 are values of the variable corresponding to cumulative frequencies $\frac{3}{10} \times 1080 = 324$, $\frac{1}{4} \times 1080 = 270$, $\frac{2}{4} \times 1080 = 540$, and $\frac{3}{4} \times 1080 = 810$ resperectively. Their positions are shown by arrow marks in the table below.
(Note that here the upper ends of classes have not been stated explicity. But the class boundaries can be obtained; see Table 4.17 at p. 76 and Example 4.3 (*h*) at p. 77)

Table 5.34 Cumulative Frequency Distribution

Class Boundary	*Cumulative Frequency (less-than)*
57.5	0
60.0	6
62.5	32
65.0	222
$Q_1 \rightarrow$	$\leftarrow N/4 = 270$
$D_2 \rightarrow$	$\leftarrow 3N/10 = 324$
67.5	503
$Q_2 \rightarrow$	$\leftarrow 2\,N/4 = 540$
$Q_3 \rightarrow$	$\leftarrow 3\,N/4 = 810$
70.0	915
72.5	1042
75.0	$1080 = N$

See Example 4.3(h), page 77

Now, using simple interpolation

(a) $\frac{Q_1 - 65.0}{67.5 - 65.0} = \frac{270 - 222}{503 - 222}$; or, $Q_1 = 65.0 + \frac{48}{281} \times 2.5$
$= 65.43$ inches

(b) $\frac{D_3 - 65.0}{67.5 - 65.0} = \frac{324 - 222}{503 - 222}$; or, $D_3 = 65.0 + \frac{102}{281} \times 2.5$
$= 65.91$ inches

(c) $\frac{Q_2 - 67.5}{70.0 - 67.5} = \frac{540 - 503}{915 - 503}$; or, $Q_2 = 67.5 + \frac{37}{412} \times 2.5$
$= 67.72$ inches

(d) $\frac{Q_3 - 67.5}{70.0 - 67.5} = \frac{810 - 503}{915 - 503}$; or, $Q3 = 67.5 + \frac{307}{412} \times 2.5$
$= 69.36$ inches

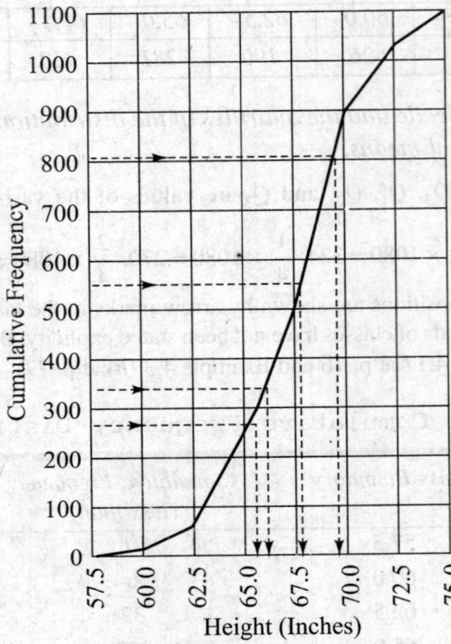

Fig. 5.5 Ogive and Location of Partition Values

We draw the ogive (less-than type) from the data in Table 5.34 (Fig. 5.5). Horizontal lines are drawn from cumulative frequencies 270, 324, 540, 810 on the vertical scale meeting the ogive. The abscissae of the points of intersection are read approximately from the horizontal scale, giving the values of Q_1, D_3, Q_2, Q_3 respectively. (Students are advised to draw the ogive and check how far the graphical values agree with those obtained by direct calculation. The more accurate the drawing, the better will be the agreement. Graphical values are only approximate, and are hardly expected to agree exactly with those obtained by arthmetic interpolation).

Example 5.67 *In a group of 500 wage earners the weekly wages of 4% were under Rs 30 and those of 15% were under Rs 32.50. The median and quartile wages were Rs 52.25, Rs 42.75 and Rs 60.50; the fourth and sixth decile wages were Rs 48.75 and Rs 55.25 respectively.*

Put the above information in the form of a frequency distribution and calculate the second and third deciles. [C.U., B.A. (Econ) '74]

Solution Here the variable is 'weekly wage (Rs)'. Let us first compile the given information:

(i) Total frequency $N = 500$,

(ii) 4% of total frequency lie under 30,

(iii) 15% of total frequency lie under 32.50,

(iv) Median = 52.25, $Q_1 = 42.75$, $Q_3 = 60.50$ (note that of the two quartiles Q_1 is the smaller); i.e. 50%, 25% and 75% of total frequency lie under 52.25, 42.75 and 60.50 respectively.

(v) $D_4 = 48.75$, $D_6 = 55.25$; i.e. 40% and 60% of total frequency lie under 48.75 and 55.25 respectively.

This information can be re-written as follows:

(a) Under 30.00 the frequency is 4% of 500, i.e. 20
(b) Under 32.50 the frequency is 15% of 500, i.e. 75
(c) Under 52.25 the frequency is 50% of 500, i.e. 250
(d) Under 42.75 the frequency is 25% of 500, i.e. 125
(e) Under 60.50 the frequency is 75% of 500, i.e. 375
(f) Under 48.75 the frequency is 40% of 500, i.e. 200
(g) Under 55.25 the frequency is 60% of 500, i.e. 300

Now arranging the results at (a) to (g) above in increasing order of values of the variable, we have the following cumulative frequency distribution. The second and third deciles, viz. D_2 and D_3, are values of the variable corresponding to cumulative frequencies $2N/10 = 100$ and $3N/10 = 150$ respectively. These may be obtained from Table 5.35, using simple interpolation.

Table 5.35 Cumulative Frequency Distribution

Weekly Wage (Rs)	*Cumulative Frequency*
30.00	20
32.50	75
$D_2 \rightarrow$	$\leftarrow 2N/10 = 100$
42.75	125
$D_3 \rightarrow$	$\leftarrow 3N/10 = 150$
48.75	200
52.25	250
55.25	300
60.50	375

$$\frac{D_2 - 32.50}{42.75 - 32.50} = \frac{100 - 75}{125 - 75}; \quad \frac{D_3 - 42.75}{48.75 - 42.75} = \frac{150 - 125}{200 - 125}$$

Solving these equations, we get $D_2 = 37.625$ Rs and $D_3 = 44.75$ Rs

The frequency distribution (grouped) is now constructed from Table 5.35 (as in Example 4.9).

Tabel 5.36 Frequency Distribution of Wages

Weekly Wages (Rs)	*Frequency*
Under 30.00	20
30.00–32.50	55
32.50–42.75	50
42.75–48.75	75
48.75–52.25	50
52.25–55.25	50
55.25–60.50	75
60.50 & above	125
Total	500

Ans. D_2 = Rs 37.62 P, D_3 = Rs 44.75 P

EXERCISES

1. Find the mean and the median of: 88, 72, 33, 29, 70, 54, 86, 91, 57, 61.
[C.U., B.Com. '73]
2. Find the mean, median and mode of the following numbers: 7, 4, 3, 5, 6, 3, 3, 2, 4, 3, 4, 3, 3, 4, 4, 3, 2, 2, 4, 3, 5, 4, 3, 4, 3, 4, 3, 1, 2, 3. *[C.U., B.Com. '71]*
3. Evaluate the arithmetic mean, median, and mode for the following distribution of 'number of telephone calls received per one-minute interval':

No. of Calls	0	1	2	3	4	5	6	7	8
Frequency	5	22	31	43	51	40	35	15	3

[B.U., B.A.(Econ) '72]

4. Calculate the simple and weighted average from the following and account for the difference between the two:

Price per ton (Rs P.)	45.60	50.70	42.45
Tons Purchased	135	40	25

[C.A., Nov. '72]

5. The numbers 3.2, 5.8, 7.9 and 4.5 have frequencies x, $(x + 2)$, $(x - 3)$ and $(x + 6)$ respectively. If the arithmetic mean is 4.876, find the value of x.
[C.U., M.Com. '73]
6. Calculate the arthmetic mean from the following data:

(i)

Class Interval	50–59	60–69	70–79	80–89	90–99	100–109	110–119
Frequency	14	38	44	54	45	30	25

(ii)

Height in Inches	57.5	–60.0	–62.5	–65.0	–67.5	–70.0	–72.5–
Number of Men	6	26	190	281	412	127	38

(iii)

Weight in lbs.	137.5–147.5	147.5–157.5	157.5–167.5	167.5–177.5	177.5–187.5	187.5–197.5	197.5–217.5	217.5–247.5
Number of Men	2	5	4	5	7	5	3	1

(iv)

X	20–30	30–50	50–100	100–200	200–350	350–550
Freq.	2	9	11	52	10	3

7. From the data in the following table calculate the average marks of the M.Com. examinees in Statistics at a class test:

Marks	30–39	40–49	50–59	60–69	70–79	80–89	90–99
No. of Examinees	2	3	11	20	32	25	7

[C.U., M.Com. '72]

8. The following table gives the rise in price of 300 commodities between two dates. Calculate the mean rise in price:

% Increase	0–	5–	10–	15–	25–	35–	45–	60–80
Frequency	12	30	51	84	66	35	15	7

[Dip. Soc. Welfare, '71]

9. The following are the monthly salaries (in Rs) of 30 employees in a firm:

140	139	126	114	100	88	62	77	99	103
108	129	144	148	134	63	69	148	132	118
142	116	123	104	95	80	85	106	123	133

The firm gave bonus of Rs 10, 15, 20, 25, 30, 35 for individuals in the respective salary groups: 'exceeding Rs 60 but not exceeding Rs 75'; 'exceeding Rs 75 but not exceeding Rs 90'; and so on upto 'exceeding Rs 135 but not exceeding Rs 150'. Find the average bonus paid per worker.

[I.C.W.A., Dec. '76-old]

10. For the variable x, taking the values 0, 1, 2, ..., k, the cumulative frequencies of more-than type are $F_0, F_1, F_2, \ldots, F_k$. Show that $\bar{x} = \sum_{i=1}^{k} F_i / n$, where n is the total frequency.

[B.U., B.A. (Econ) '73]

11. (a) The arthmetic mean calculated from the following frequency distribution is known to be 67.45 inches. Find the value of f_3.

Height (inches)	60–62	63–65	66–68	69–71	72–74
Frequency	15	54	f_3	81	24

[I.C.W.A., July '71]

(b) The expenditure of 1000 families is given below:

Expenditure (Rs):	40–59	60–79	80–99	100–119	120–139
No. of Families:	50	?	500	?	50

The median and mean for the distribution are both Rs 87.50 P. Calculate the missing frequencies. *[I.C.W.A., June '78]*

12. Find out the missing frequencies of the following data, given that A.M. is 67.45 inches.

Height (inches)	60–62	63–65	66–68	69–71	72–74	Total
No. of Students	5	18	f_3	f_4	8	100

[Dip. Management, '72]

13. (i) The average marks obtained in an examination by two groups of students was found to be 75 and 85 respectively. Determine the ratio of students in the two groups, if the average mark for all students was 80.

[B.U., B.A. (Econ] '69]

(ii) On a certain examination, the average grade of all students in class A is 68.4 and all students in class B is 71.2. If the average of both classes combined is 70.0, what is the ratio of the number of students in class A to the number in class B? *[C.U., M.Com. '62]*

(iii) The mean age of a combined group of men and women is 30 years. If the mean age of the group of men is 32 and that of the group of women is 27, find out the percentage of men and women in the group.

[C.A., Nov. '65]

14. Out of the total population in a certain town in South Africa, 60% belonged to the Black Race and the rest belonged to the White Race. It was estimated that their mean incomes were respectively 2,000 and 5,000 pounds. Find the average income of the entire town. *[C.A., May '68]*

15. (i) A factory has five sections employing 105, 184, 130, 93 & 125 workers. The mean earnings in a certain week per worker are Rs 13.80, Rs 15.00, Rs 15.20, Rs 18.20 and Rs 14.20 for the 5 sections. Determine the mean earning per worker of the whole factory. *[Dip. Management, '70]*

(ii) In a survey of locality the following figures regarding the income of the people in different occupations were received. Find out the average per capita income:

Occupation	*Average Income (in Rs)*	*Number of People*
Business	500	700
Labour	300	300
Craftmanship	200	200
Other	400	100

[D.S.W., Nov. '70]

16. The following shows some data collected for three regions of a country:

Region	*No. of Inhabitants (million)*	*Percentage of Literates*	*Average Annual Income per Person (Rs)*
A	10	52	850
B	5	68	620
C	18	39	730

Obtain the over-all figures for the three regions taken together.

[C.U., B.A. (Econ) '77]

17. The population of India in 1951 and in 1961 were 361 and 439 million respectively. (i) What was the average percentage increase per year during the period? (ii) If the average rate of increase from 1961 to 1971 remains the same, what would be the population in 1971? *[I.C.W.A., July '71]*

18. A man gets three successive annual increments in salary of 20%, 30% and 25%, each percentage being reckoned on his salary at the end of the previous year. How much better or worse off would he have been if he had been given 3 annual increments of 25% each, reckoned in the same way?

[I.C.W.A., Dec. '74-Old]

19. A machine is assumed to depreciate 40% in value in the first year, 25% in the second year and 10% per annum for the next 3 years, each percentage being calculated on the diminishing value. What is the average percentage depreciation, reckoned on the diminishing value, for the 5 years?

[B.U., B.A. (Econ) '66; C.U., M.Com. '73]

20. The G.M. of 4 observations is 47, and the G.M. of 6 others is 40. Find the G.M. of all the 10 observations.

21. The geometric mean of six numbers is 75. If the geometric mean of four of them is 67, what is the geometric mean of the other two? *[B.U., B.A. (Econ) '71]*

22. You fly to a place X in a Boeing at a speed of 500 miles per hour and come back from X, following the same route, at a speed of 160 m.p.h. What is your average speed for the to-and-fro journey? *[C.U., M.Com. '72]*
23. An aeroplane flies around a square the sides of which measure 100 Kms. each. The aeroplane covers at a speed of 100 Kms. per hour the first side, at 200 Kms per hour the second side, at 300 Kms. per hour the third side, and at 400 Kms per hour the fourth side. Use the correct mean to find the average speed round the square. *[I.C.W.A., June '78]*
24. If two grades of oranges sell at 10 for Re. 1 and 20 for Re. 1 respectively, calculate the average price per orange, stating your assumptions explicitly. *[B.U., B.A. (Econ) '69]*
25. (a) The weights (in lb.) of 8 persons are: 138, 143, 141, 139, 152, 148, 160 and 267. Find the average weight using a suitable form of average. Give reasons for your choice.
    (b) What is the suitable form of average in each of the following cases? (i) The length of a rod is measured by a tape 10 times. You are to estimate the length of the rod by averaging these 10 determinations. (ii) A person purchases 5 rupees worth of eggs from 10 different markets. You are to find the average number of eggs per rupee for all the markets taken together. (iii) You are given the population of India for the censuses of 1961 and 1971. You are to find the population of India at the middle of the period by averaging these population figures, assuming a constant rate of increase of population. *[B.U., B.A. (Econ) '72]*
26. Find the mean and the median for the following data, and comment on the shape of the distribution:

Weight in Kg	36–40	41–45	46–50	51–55	56–60	61–65	66–70
No. of Persons	14	26	40	53	50	37	25

[I.C.W.A., June '75-old]

27. The G.M., H.M. and A.M. of three observations are 3.63, 3.27 and 4 respectively. Find the observations. *[C.U., M.Com. '75]*
28. Using a suitable formula calculate the median value from the following data:

Midvalue	115	125	135	145	155	165	175	185	195	Total
Frequency	6	25	48	72	116	60	38	22	3	390

[C.A., Nov. '66]

29. In a group of 1000 wage earners the monthly wages of 4% are below Rs 60 and those of 15% are under Rs 62.50. 15% earned Rs 95 and over, and 5% got Rs 100 and over. Find the median wage. *[B.U., B.A. (Econ) '70]*
30. The table below gives the frequency distribution of weights of 80 applies:

Weight (gms)	110–119	120–129	130–139	140–149	150–159	160–169	170–179	180–189
Frequency	5	7	12	20	16	10	7	3

Draw the cumulative frequency diagram and hence determine the median weight of an apple.

31. Draw the less-than Ogive and estimate the value of median on the basis of the data given below:

Mid-Point	18	25	32	39	46	53	60	
Frequency	10	15	32	42	26	12	9	$N = 146$

Verify the above result by calculation. *[C.A., May '74]*

32. An incomplete frequency distribution is given below:

Height (inches)	5.1–6.0	6.1–7.0	7.1–8.0	8.1–9.0	9.1–10.0	10.1–11.0	11.1–12.0
No. of Plants	3	8	27	?	17	11	9

It is known that the median height of the plant is 8.53 inches. Calculate the missing frequency. *[I.C.W.A., Jan. '72]*

33. In the following data two class frequencies are missing:

C.I.	100–110	110–120	120–130	130–140	140–150	150–160
Frequency	4	7	15	?	40	?
	160–170	170–180	180–190	190–200		
	16	10	6	3		

However it was possible to ascertain that the total number of frequencies was 150 and that the median has been correctly found out as 146.25. Find the two missing frequencies. *[C.A., May '73]*

34. In each of the following cases, explain whether the description applies to the mean, median or both:
    (i) can be calculated from a frequency distribution with open end intervals.
    (ii) the values of all items are taken into consideration in the calculation,
    (iii) the values of extreme items do not influence the average,
    (iv) in a distribution with a single peak and moderate skewness to the right, it is closer to the concentration of the distribution. *[C.A., Nov. '65]*

35. Calculate the value of the mode by the usual formula (after grouping if necessary):

x	10–20	20–30	30–40	40–50	50–60	60–70	70–80	80–90	90–100	100–110
f	4	6	5	10	20	22	24	6	2	1

[C.A., May '74]

36. From the following distribution of weekly earnings, calculate (i) the most usual wage, and (ii) the percentage earning more than Rs. 31.50.

Weekly Earnings (Rs)	25–	26–	27–	28–	29–	30–	31–	32–	
No. of Persons	25	70	210	275	430	550	340	130	
					33–	34–	35–	36	Total
					90	55	25		2200

[I.C.W.A., Dec. '73]

37. Find the mean and mode for the following:

Year under	10	20	30	40	50	60
Number of Persons	15	32	51	78	97	109

[B.U., B.A. (Econ) '73]

38. From the following cumulative frequency distribution of marks obtained by 22 students, calculate (a) Arithmetic mean, (b) Median, and (c) Mode.

Marks	*No. of Students*
Below 10	3
Below 20	8
Below 30	17
Below 40	20
Below 50	22

[I.C.W.A., June '77]

39. The table below gives the numbers (f) of candidates obtaining marks x or higher in a certain examination (all marks are given in whole numbers):

x	10	20	30	40	50	60	70	80	90	100
f	140	133	118	100	75	45	25	9	2	0

Calculate the mean and the median marks obtained by the candidates.

[I.C.W.A. June '75]

40. Calculate the values of (i) mean, (ii) median, and (iii) the two quartiles:

Income

(Rs '000): Under	1	1–2	2–3	3–5	5–10	10–25	25–50	50–100	100–1000
No. of Persons	13	90	81	117	66	27	6	2	2

[C.U., M.Com. '74]

41. In a moderately asymmetrical distribution, the mean and the median are respectively 25.6 and 26.1 inches. What is the mode of the distribution?

[I.C.W.A., Jan. '71]

42. In a moderately skewed distribution, A. Mean = 24.6, and the mode = 26.1. Find the value of the Median and explain the reason for the method employed.

[C.A., Nov. '67]

43. Calculate arithmetic mean, and median of the frequency distribution given below. Hence calculate the mode using the empirical relation between the three.

Class Limits	130–134	135–139	140–144	145–149	150–154	155–159	160–164
Frequency	5	15	28	24	17	10	1

[I.C.W.A., Dec. '74]

44. Compute the median and the upper quartile of the following:

Intelligence Quotient (IQ)	55–64	65–74	75–84	85–94	95–104	105–114	115–124	125–134	135–144
No. of Students	2	20	79	184	302	207	82	24	4

[C.U., B.A. (Econ) '76]

45. The weekly wages earned by the hundred workers of a factory are set out in the following table:

Weekly Wages (Rs)	12.5–17.5	17.5–22.5	22.5–27.5	27.5–32.5		
No. of Workers	12	16	25	14		
	32.5–37.5	37.5–42.5	42.5–47.5	47.5–52.5	52.5–57.5	
	13	10	6	3	1	

Calculate the three quantiles of the above distribution, taking $n/4$, $2n/4$, and $3n/4$, as their respective ranks. [C.A., Nov. '63]

46. The following table shows the age distribution of heads of families in a certain country during the year 1957. Find the median, the third quartile and the second decile of the distribution. Check your results by the graphical method:

Age of Head of Family (Yrs)	Under 25	25–29	30–34	35–44	45–54
Number (million)	2.3	4.1	5.3	10.6	9.7
	55–64	65–74	Above 74	Total	
	6.8	4.4	1.8	45.0	

[I.C.W.A., Jan. '73]

47. For an income distribution of a group of men 20 per cent of men have income below Rs 30, 35 per cent below Rs 70, 60 per cent below Rs 150 and 80 per cent below Rs 250. The first and third quartiles are Rs 50 and Rs 170. Put the above information in a cumulative frequency distribution and find the median. [C.U., M.Com. '66]

48. "For a group of 5000 workers the weekly wages vary from Rs 20 to Rs 80. The wages of 4 per cent of the workers are under Rs 25 and those of 10 per cent are under Rs 30; 15 per cent of the workers earn Rs 60 and over, and 5 per cent of them get Rs 70 and over. The quartile wages are Rs 40 and Rs 54, and the sixth decile is Rs 50". Put the above information in the form of a frequency distribution and find the mean wage therefrom. [I.C.W.A., Jan. '71]

49. For a certain group of 'Saree' weavers of Varanasi, the median and quartile of earnings per week are Rs 44.30, Rs 43.00 and Rs 45.90 respectively. Ten per cent of the group earn under Rs 42 per week and 13% earn Rs 47 and over, and 6% Rs 48 and over. The range of earnings per week is Rs 40–Rs 50. Put the data into a frequency distribution. [C.U., B.A. (Econ) '70]

50. Comment on the following statements:
 (a) "The median of a distribution is $N/2$, the lower quartile is $N/4$ and the upper quartile is $3N/4$". (Here N denotes the total frequency). [I.C.W.A. Dec. '74]
 (b) "If Q_1, Q_2, Q_3 be respectively the lower quartile, median and the upper quartile of a distribution, then $Q_2 - Q_1 = Q_3 - Q_2$. [I.C.W.A. June '75]

51. Given below is the frequency distribution of carbon content (present) in 150 determinations of a certain mixed powder.

Per cent Carbon	4.0–4.1	4.2–4.3	4.4–4.5	4.6–4.7	4.8–4.9	5.0–5.1
Frequency	1	2	7	20	25	30
				5.2–5.3	5.4–5.5	5.6–5.7
				10	25	30

Compute the arithmetic mean and median. [I.C.W.A., Dec. '78]

52. *Write True or False:* (i) The mean, median and mode are located at the same point in a frequency distribution.
(ii) Extreme values in the series affect the utility of the mean but not the median or mode. *[D.S.W., '78]*

53. Compute the arithmetic mean, median and mode of the following distribution and explain their relationships:
Monthly income

(Rs)	0–75	75–150	150–225	225–300	300–375	375–450
Frequency	15	200	250	225	10	5

[C.U., M.Com. '76]

54. (a) *If* $z_1 = x_1 + y_1,\ z_2 = x_2 + y_2,\ \ldots\ z_n = x_n + y_n$, then prove that $\overline{z} = \overline{x} + \overline{y}$, where the symbols have their usual meaning.
(b) Prove that the logarithm of geometric mean of observations is the arithmetic mean of logarithms of the observations. *[D.M., '73]*

55. The following are the population figures (in thousands) of 10 cities. Find the median: 2488, 1490, 777, 733, 522, 672, 591, 407, 387 and 391. *[D.M., '78]*

ANSWERS

1. 64.1, 65.5
2. 3.47, 3, 3
3. 3.90, 4, 4
4. Rs 46.25, Rs 46.23
5. 5
6. (i) 85.22, (ii) 67.45 inches, (iii) 176.72 lbs. (iv) 150.98
7. 72.5
8. 24.1%
9. Rs 24.50
10. (see Example 5.13)
11. (a) 126, (b) 250, 150
12. $f_3 = 42, f_4 = 27$
13. (i) 1 : 1, (ii) 3 : 4 (iii) Men 60%, Women 40%
14. £3200
15. (i) Rs 15.15, (ii) Rs 400
16. 33 million, 47.3%, Rs 749.70
17. 2%, 533.9 million
18. He would have received $R/320$ more (R = starting salary)
19. 20%
20. 42.67
21. 94.0
22. 242.4 m.p.h.
23. 192 k m.p.h.
24. 7.5 Paise (assuming equal numbers of the two grades are sold)
25. (a) Median = 145.5 lbs. (Since there is one extremely large value, A.M. will be unsuitable. Mode does not exist; morover it will be inappropriate with only 8 observations). (b) (i) A.M., (ii) H.M., (iii) G.M.

26. Mean 54.3 Kg., Median 54.5 Kg. Since Mean is less than Median, distribution has negative skewness; i.e the longer tail of the frequency curve lies to the left (see pages 114 and 272)
27. 2, 4, 6
28. 153.8
29. Rs 78.75
30. 147.5 gms.
31. 38.2
32. 25
33. 24, 25
34. (i) Median, (ii) Mean, (iii) Median, (iv) Median.
35. Since the frequency of the class immediately following the modal class has declined abruptly, re-group the given classes in two's giving 10–30, 30–50, 50–70, 70–90, 90–110 with frequencies 10 15, 42, 30, 3. Mode = 63.8
36. (i) Mode Rs 30.36, (ii) 21.4%
37. 29.95, 35.00 (years)
38. 22.7, 22.8, 23.5
39. 50.7, 51.2
40. (i) 8.06, (ii) 3.31, (iii) 1.98, 5.15 (all in Rs '000)
41. 27.1 inches
42. 25.1
43. 145.35, 144.92, 144.06
44. 100.03, 108.90
45. Quratiles 21.56, 26.90, 35.38 (Rs)
46. 44.7, 57.1, 32.0 years
47. Rs 118
48.

Weekly Wage (Rs)	20–25	25–30	30–40	40–50	50–54	54–60
Number of Workers	200	300	750	1750	750	500
	60–70	70–80	Total			
	500	250	5000	Mean = Rs 4730		

49.

Weekly Earnings (Rs.)	40.00–	42.00–	43.00–	44.30–
Percentage of Weavers	10	15	25	25
	45.90–	47.00–	48.00-50.00	Total
	12	7	6	100

50. (a) The statement is not correct. Median is the value of the $N/2$-th item when the observations are arranged in order of magnitude, and not $N/2$ itself. In fact, $N/2$ is the *rank* of median. Similarly for the lower and the upper quartiles. (b) The statement is not generally correct. The number of observations (i.e. frequency) lying between Q_1 and Q_2 is the same as the number lying between Q_2 and Q_3. The differnces $(Q_2–Q_1)$ and $(Q_3–Q_2)$ may not be equal.
51. 5.118%, 5.083%
52. (i) False, (ii) True
53. (Rs) 190.69, 191.25, 200.00.
55. 631.5 thousand.

6 MEASURES OF DISPERSION

6.1 MEANING AND NECESSITY OF 'MEASURES OF DISPERSION'

The word *dispersion* is used to denote the 'degree of heterogeneity' in the data. It is an important characteristic indicating the extent to which observations vary among themselves. The dispersion of a given set of observations will be zero, only when all of them are equal. The wider the discrepancy from one observation to another, the larger will be the dispersion.

It may be noted that a measure of central tendency exhibits quite a different characteristic of data, viz. the size of a typical value representing a whole set of observations. Two series of observations may have the same average, but unequal dispersion. For example, consider the following:

Series A—	30,	32,	36,	38,	39, Total	175
Series B—	4,	19,	36,	53,	63, Total	175

These series with equal number of observations have the same mean and also the same mediam, so that when we consider their averages the two series are not different. However, it may be noticed that observations in Series A are close to one another, while those in Series B are widely dissimilar. We say that the *dispersion* (or *variability*) of Series B is more than that of Series A. Thus averages alone are not sufficient to reveal all the characteristics of data and hence the necessity of measures of dispersion.

A measure of dispersion is designed to state numerically the extent to which individual observations vary on the average. There are several measures of dispersion.

Measures of Dispersion

- *Absolute Measures*
  - *Range*
  - *Quartile Deviation*
  - *Mean Deviation*
  - *Standard Deviation*
- *Relative Measures*
  - *Coefficient of Variation*
  - *Coefficient of Quartile Deviation*
  - *Coefficient of Mean Deviation*

Example 6.1 *State the requisites of a satisfactory measure of dispersion and examine in their light any two common measures of dispersion.*

[I.C.W.A., Dec.'74]

Solution According to G.U. Yule, a good measure of dispersion should obey conditions similar to those (Example 5.1) for a satisfactory average:

(i) It should be rigidly defined.
(ii) It should be based on all observations.
(iii) It should be readily comprehensible.
(iv) It should be fairly easily calculated.
(v) It should be affected as little as possible by fluctuations of sampling; and
(vi) It should readily lend itself to algebraic treatment.

Standard Deviation (S.D.) obeys the majority of these requisites. It is rigidly defined, based on all observations, calculated fairly easily, and can be treated algebraically. Given the mean, the standard deviation and the number of observations in each of several groups, S.D. of the composite group can be determined. No other measure of dispersion has this property which makes S.D. the most suitable among the different measures. S.D. is also the least affected by sampling fluctuations. However, the significance of S.D. is not easy to understand. It is a round-about measure obtained by squaring deviations from mean, finding the average of these squares, and then taking the square-root (called 'Root-Mean-Square Deviation from mean).

Quartile Deviation is also rigidly defined and based on all observations. It has two definite advantages over S.D., viz. that it can be calculated with great ease and has a very simple meaning. It is half the difference between the two quartiles Q_1 and Q_3, which include the central 50% of observations. It can also be calculated from frequency distributions with open-end classes. However, it has no simple algebraic properties and its behaviour as regards sampling fluctuations is difficult to decide.

Example 6.2 *What are the different measures of variability of observations?*

[C.U., B.A.(Econ) '72; M.Com. '71]

Solution The measures of variability (or dispersion) in common use are:

(1) Range,
(2) Quartile Deviation (or Semi-interquartile Range),
(3) Mean Deviation, (or Mean Absolute Deviation),
(4) Standard Deviation,
(5) Coefficient of Variation.

Two other measures of dispersion, rarely used, are

(6) Coefficient of Quartile Deviation,
(7) Coefficient of Mean Deviation.

The measures (1) to (4) are known as 'absolute measures', and the remaining three, viz. the coefficients of dispersion, are called 'relative measures.'

Range is the difference between the largest and the smallest observations. It is dimple to understand and easy to calculate, but is not generally used because of its many disadvantages.

$$\text{Range} = \text{Maximum value} - \text{Minimum value}$$

Illustration 1. For the observations (in Rs) 6, 4, 1, 6, 5, 10, 3, the maximun and the minimum values are 10 and 1. Therefore, Range = 10 – 1 = 9 Rs.

Quartile Deviation (also called *Semi-interquartile Range*) is defined as half the difference between the lower and the upper quartiles. This is also not very much used, because it does not take into account the variability of all observations.

$$\text{Quartile Deviation} = \frac{Q_3 - Q_1}{2}$$

Illustration 2. The data in Illustration 1, arranged in order of magnitude, are 1, 3, 4, 5, 6, 6, 10, so that $Q_1 = 3$, $Q_3 = 6$. Therefore, Quartile Deviation = $\frac{1}{2}(6 - 3) = 1.5$ Rs.

Mean Deviation (also called *Mean Absolute Deviation*) is the arithmetic mean of absolute deviations (i.e. differences) from mean or any other specified value. Although it defends on all observations, the procedure of neglecting the signs and taking absolute deviations only makes algebraic treatment difficult.

$$\text{Mean Deviation} = \frac{1}{n}\Sigma\left|x_1 - \bar{x}\right|$$

Illustration 3. For the data in Illustration 1, mean = 35/7 = 5 Rs and deviations from mean are 1, –1,– 4, 1, 0, 5, –2, so that absolute deviations (i.e. deviations without regard to signs) are 1, 1, 4, 1, 0, 5, 2. The arithmetic mean of these absolute deviations is Mean Deviation = (1+ 1+ 4 + 1+ 0 + 5 + 2)/7 = 14/7 = 2 Rs.

Standard Deviation is the square-root of the arithmetic mean of squares of deviations from arithmetic mean. It is the most useful among all measures of dispersion because of many advantage.

$$\text{Standard Deviation} = \sqrt{\frac{1}{n}\Sigma(x_i - \bar{x})^2}$$

Illustration 4. In Illustration 3, the deviations from mean have been found. Therefore, Standard Deviation

$$\text{S.D.} = \sqrt{\{1^2 + (-1)^2 + (-4)^2 + 1^2 + 0^2 + 5^2 + (-2)^2\}/7}$$

$$= \sqrt{48/7} = \sqrt{6\,86} = 2.6 \text{ Rs}$$

Absolute measures of dispersion are expressed in the same unit as the observations, and therefore, cannot be used for comparing the variability of two or more series given in different units. The relative measures, defined below, are pure numbers and are employed for the purpose.

$$\text{Coefficient of Variation} = \frac{\text{Standard Deviation}}{\text{Mean}} \times 100$$

$$\text{Coefficient of Quartile Deviation} = \frac{\text{Quartile Deviation}}{\text{Median}} \times 100$$

Coifficient of Mean Deviation

$$= \frac{\text{Mean Deviation about mean (or median)}}{\text{Mean (or median)}} \times 100$$

Illustration 5. For the data in illustration 1, Mean = 5 Rs, Standard Deviation = 2.6 Rs Therefore,

$$\text{Coefficient of Variation} = \frac{2.6 \text{ Rs}}{5 \text{ Rs}} \times 100 = 52.$$

Median = 5 Rs, and Q.D. = 1.5 Rs. Therefore,

$$\text{Coefficient of Quartile Deviation} = \frac{1.5 \text{ Rs}}{5 \text{ Rs}} \times 100 = 30$$

Mean = 5 Rs and Mean Deviation = 2 Rs. Therefore,

$$\text{Coefficient of Mean Deviation} = \frac{2 \text{ Rs}}{5 \text{ Rs}} \times 100 = 40.$$

[**Note:** All the relative measures of dispersion are defined as the ratio of an absolute measure of dispersion to the corresponding measure of central tendency, and the ratio is expressed as a percentage.]

Example 6.3 *Explain the advantages and disadvantages of the different measures of dispersion.* [C.U., M.Com. '72 ; B.A. (Econ) '72, '74]

Solution The advantages of Range are that it is very easy to understand and simple to calculate. However, Range has many disadvantage, which make it virtually useless as a reliable measure of dispersion. It is highly affected by the presence of even one extremely high or low value (Example 6.6). It does not depend on all observations. Lastly, Range cannot be calculated from frequency distributions with open-end classes.

Unlike Range, Quartile Deviation is dependent only on the smallest and the largest of the central 50% observations, and does not take into account the variability of the highest 25% and the lowest 25% observations. It can be calculated from frequency distributions with open-end classes (Example 6.8). However, it does not depend on the actual magnitudes of all observations.

Mean Absolute Deviation is based on all observations and hence takes into account the variability of each of them. However, the practice of neglecting signs and taking absolute deviations makes it difficult to be treated algebraically.

Standard Deviation is comparatively more difficult to calculate and not easy to understand. But, it has many advantages, for which it is most extensively used as a measure of dispersion. It is based on all observations, and is suitable for algebraic treatment. Given the mean, standard deviation and number of observations in each of several groups, standard deviation of the composite group can be calculated (Examples 6.39, 6.40). It is also the least affected by sampling fluctuations (see (4) of Example 6.12).

All absolute measures of dispersion suffer from the disadvantage that they are expressed in the same unit as original observations, and hence are unsuitable for comparing the dispersion of two or more sets of observations given in different units, e.g. heights in inches, and weights in lbs. However, relative measures are pure numbers and therefore may be used for this purpose (Example 6.54). Again, there are cases where it is found that although the unit of measurement is the same, the averages differ widely and in such cases a relative measure, e.g. Coefficient of Variation, provides a more useful measure of variability than standard deviation itself (Examples 6.48, 6.55, 6.56). Coefficient of Quartile Deviation and Coefficient of Mean Deviation are not generally used.

Example 6.4 *Distinguish between absolute and relative measures of dispersion.* [C.A., Nov. '71; I.C.W.A., Jan. '73]

Solution Measures of dispersion are of two types—Absolute and Relative. (1) Absolute measures of dispersion are: Range, Quartile deviation, Mean deviation, and Standard deviation. Relative measures are: Coefficient of Variation, Coefficient of Quartile deviation, and Coefficient of Mean deviation.

(2) Absolute measures are expressed in the same unit in which the observations are given. Relative measures are obtained by expressing an absolute measure as percentage of a measure of central tendency and hence relative measures are independent of the units of measurement.

(3) Usually the absolute measures are employed for measuring dispersion. But for purposes of comparing the dispersion in different series, relative measures are used. Again when two sets of data are given in dissimilar units, there is no otber alternative but to use a relative measure for comparison (Example 6.54).

(4) Relative measures may also be used to compare the relative accuracy of data. But absolute measures cannot be so used (Example 6.46).

6.2 RANGE

Range of a set of observations is the difference between the maximum and the minimum values.

$$Range = Maximum\ value - Minimum\ value \quad (6.2.1)$$

Example 6.5 *Discuss the advantages and limitations of 'range' as a measure of variability of observations.* [B.U., B.A.(Econ) '77]

Solution Range is the simplest of all measures of dispersion.

Advantages—(i) It is easy to understand. Range is the difference between the two extremes, viz. the largest and the smallest values, and as such represents the maximum possible difference between any two observations.

(ii) Range is also simple to calculate. This simplicity in the calculation of Range has led to its extensive application in Statistical Quality Control (Chapter 17). Here, 'Control Chart for Range' and estimates of variability based on Range are preferred.

However, Range has many limitations.

Limitations—(i) It does not depend on all observations, and is based on only the largest and the smallest among them. The values of intermediate observations, are not at all necessary for its calculation.

(ii) It is highly affected by extreme values. The presence of even one very high or low value is sufficient to increase Range many times.

(iii) It does not take into account the form of the distribution. The same value of Range might be obtained from a symmetrical or a J-shaped distribution, although the nature of dispersion in the two are completely different.

(iv) The greatest disadvantage of Range is that it cannot be calculated from frequency distributions with open-end classes (Example 6.8).

Example 6.6 *The weight of 11 forty-year old men were 148, 154, 158, 160, 161, 162, 166, 170, 182, 195 and 236 pounds. If the heaviest man is omitted, what is the percentage change in the range?* [C.U., M.Com '68]

Solutoin In the given data, maximum value = 236, minimum value = 148. Hence,

Range = 236 – 148 = 88 pounds.

If the heaviest man is omitted, the maximum value 236 is cancelled, and among the remaining 10 observations, maximum = 195, minimum = 148.

Range (revised) = 195–148 = 47 pounds

Change in range = 88 – 47 = 41, out of the original 88 pounds.

$\therefore$ Precentage change in range $= \frac{41}{88} \times 100 = 47$ *Ans.* 47%

6.3 QUARTILE DEVIATION (OR SEMI-INTERQUARTILE RANGE)

Quartile Deviation is defined as half the difference between the upper and the lower quartiles.

$$Quartile\ Deviation = \frac{Q_3 - Q_1}{2} \tag{6.3.1}$$

The difference $Q_3 - Q_1$ being the distance between the two quartiles, may be called *interquartile range*, and half of this is *Semi-interquartile Range* ('Semi' denotes 'half', e.g. semi-circle). Thus, the name 'Semi-interquartile Range' itself gives the definition of Quartile Deviation.

Quartile Deviation (Q.D.) is dependent on the two quartiles and does not take into account the variability of the largest 25% or the smallest 25% of observations. It is therefore unaffected by extreme values. Since in most cases the central 50% of observations tend to be fairly typical, Q.D. affords a convenient measure of dispersion. It can be calculated from frequency distributions with open-end classes (Example 6.8). Q.D. is thus superior to Range in many ways. Its unpopularity lies in the fact that Q.D. does not depend on the magnitudes of all observations. The calculation of Q.D. only depends on that of the two quartiles, Q_1 and Q_3, which can be found from a cumulative frequency distribution using simple interpolation (Example 6.7).

Example 6.7 *Calculate the quartile deviation from the following:*

Class Interval	*10–15*	*15–20*	*20–25*	*25–30*	*30–40*	*40–50*	*50–60*	*60–70*	*Total*
Frequency	*4*	*12*	*16*	*22*	*10*	*8*	*6*	*4*	*82*

[C.A., Nov. '76]

Solution In order to compute Quartile Deviation, we have to find Q_1 and Q_3, i.e. values of the variable corresponding to cumulative frequencies $N/4$ and $3N/4$. Here, total frequency $N = 82$. Therefore, $N/4 = 20.5$ and $3N/4 = 61.5$.

TABLE 6.1 Cumulative Frequency Distribution

Class Boundary	*Cumulative Frequency (less-than)*
10	0
15	4
20	16
$Q_1 \rightarrow$	$\leftarrow N/4 = 20.5$
25	32
30	54
$Q_3 \rightarrow$	$\leftarrow 3N/4 = 61.5$
40	64
50	72
60	78
70	$82 = N$

Applying simple interpolation (page 77),

$$\frac{Q_1 - 20}{25 - 20} = \frac{20.5 - 16}{32 - 16}$$

or, $$\frac{Q_1 - 20}{5} = \frac{4.5}{16}$$

or, $$Q_1 - 20 = \frac{4.5}{16} \times 5 = 1.4$$

∴ $$Q_1 = 20 + 1.4 = 21.4$$

Similary, $$\frac{Q_3 - 30}{40 - 30} = \frac{61.5 - 54}{64 - 54}$$

or, $$\frac{Q_3 - 30}{10} = \frac{7.5}{10}$$

or, $$Q_3 - 30 = \frac{7.5}{10} \times 10 = 7.5$$

∴ $$Q_3 = 30 + 7.5 = 37.5$$

$$\text{Quartile Deviation} = \frac{Q_3 - Q_1}{2} = \frac{37.5 - 21.4}{2} = 8.0$$

Example 6.8 *Calculate the appropriate measure of dispersion from the following data:*

Wages in Rupees per Week	*No. of Wage Earners*
less than 35	*14*
35–37	*62*
38–40	*99*
41–43	*18*
over 43	*7*

[I.C.W.A., Jan.'64]

Solution Since there are open-end classes in the frequency distribution. Quartile Deviation would be the most appropriate measure of dispersion for the data. So, we have to determine the quartiles Q_1 and Q_3, which can be done from a cumulative frequency distribution using simple interpolation, as in Example 6.7.

TABLE 6.2 Cumulative Frequency Distribution

Wages (Rs) per Week	*Cumulative Frequency*
34.5	14
$Q_1 \rightarrow$	$\leftarrow N/4 = 50$
37.5	76
$Q_3 \rightarrow$	$\leftarrow 3N/4 = 150$
40.5	175
43.5	193
...	200 = N

Applying simple interpolation,

$$\frac{Q_1 - 34.5}{37.5 - 34.5} = \frac{50 - 14}{76 - 14}, \frac{Q_3 - 37.5}{40.5 - 37.5} = \frac{150 - 76}{175 - 76}$$

Solving $Q_1 = 36.24$ Rs., $Q_3 = 39.74$ Rs. Therefore,

Quartile Deviation = (39.74 – 36.24) /2 = 1.75 Rs.

[**Note:** (i) Because of the presence of open-end classes in the frequency distribution, Range cannot be determined. Mean Deviation and Standard Deviation also cannot be calculated, as mid-values for the open end classes cannot be obtained.

(2) Quartile deviation, and in fact all absolute measures of dispersion, must be expressed in the same unit in which observations are given].

6.4 MEAN DEVIATION (OR MEAN ABSOLUTE DEVIATION)

Mean Deviation (also called *Mean Absolute Deviation*) of a set of observations is the arithmetic mean of absolute deviations from mean or any other specified value (The term 'absolute deviation' used here refers to deviation without regard to sign, and is commonly known as 'difference').

Given the observations $x_1, x_2, ..., x_n$, in order to 'find 'Mean Deviation about A', we first obtain the deviations $(x_1 - A), (x_2 - A), ..., (x_n - A)$. Some of these deviations may be positive and some negetive. If we write $|x_i - A|$ to denote the positive value of $(x_i - A)$, whatever be the actual sign, the sum of these 'absolute deviations' is

$$|x_1 - A| + |x_2 - A| + \ldots + |x_n - A| = \Sigma |x_i - A|$$

and A.M. of the absolute deviations is

$$\textit{Mean Deviation about } A = \frac{1}{n} \Sigma |x_i - A| \quad (6.4.1)$$

Mean Deviation (M.D.) is usually calculated about arithmetic mean ($\bar{x}$), and hence 'Mean Deviation' only refers to M.D. about mean.

For simple series,

$$\textit{Mean Deviation} = \frac{1}{n} \Sigma |x_i - \bar{x}| \quad (6.4.2)$$

For frequency distribution,

$$\textit{Mean Deviation} = \frac{1}{N} \Sigma f_i |x_i - \bar{x}| \quad (6.4.3)$$

An important property of M.D. is that it has the minimum value when deviations are taken from median, i.e. M.D. about median is the least (see Examples 6.17, 6.56).

Example 6.9 *Calculate the Mean Deviation of the following values about the medean: 8, 15, 53, 49, 19, 62, 7, 15, 95, 77* [C.U., B.Com. '73]

Solution Since there are an even number of observations, viz. 10, the medean is the average of the two middlemost observations, when arranged in order of magnitude: 7, 8, 15, 15, (19, 49), 53, 62, 77, 95.

∴ Median = (19 + 49)/2 = 34

TABLE 6.3 Calculations for Mean Deviation

x	| x—median | i.e. difference from median
8	26
15	19
53	19
49	15
19	15
62	28
7	27
15	19
95	61
77	43
Total	272

Applying (6.4.1)

Mean Deviation about median

$$= \frac{1}{n} \Sigma |x - \text{median}|$$

$$= \frac{1}{10} \times 272$$

$$= 27.2$$

Example 6.10 *Find the mean deviaton of the following series:*

x	*10*	*11*	*12*	*13*	*14*	*Total*
Frequency	*3*	*12*	*18*	*12*	*3*	*48*

[C.A., May '63]

Solution

TABLE 6.4 Calculations for Mean Deviation

x	f	fx	$\|x - \bar{x}\|$	$f\|x - \bar{x}\|$
10	3	30	2	6
11	12	132	1	12
12	18	216	0	0
13	12	156	1	12
14	3	42	2	6
Total	48	576	–	36

$\bar{x} = \Sigma fx/N = 576/48 = 12$

Applying (6.4.3), Mean Deviation = 36/48 = 0.75 *Ans.* 0.75

Example 6.11 *Measurements of the lengths in feet of 50 iron rods are distributed as follows:*

Class Boundary	*Frequency*
2.35–2.45	*1*
2.45–2.55	*4*
2.55–2.65	*7*
2.65–2.75	*15*
2.75–2.85	*11*
2.85–2.95	*10*
2.95–3.05	*2*

Find, to two decimal places, the value of the mean deviation.

[C.U., M.Com. '65]

Solution Let $x_1, x_2, ..., x_n$ represent mid-values of classes with frequencies $f_1, f_2, ..., f_n$ respectively and $\bar{x}$ be their arithmetic mean. Then

$$\text{Mean Deviation} = \frac{1}{N}\Sigma f\,|x - \bar{x}|$$

where $N = \Sigma f$

TABLE 6.5 Calculations for Mean Deviation

Class Boundaries (ft.)	*Frequency f*	*Mid-value x*	$y = \frac{x - 2.70}{0.10}$	*fy*	$x - \bar{x}$	$f\,\|x - \bar{x}\|$
(1)	(2)	(3)	(4)	(5)	(6)	(7)
2.35–2.45	1	2.40	–3	–3	–0.338	0.338
2.45–2.55	4	2.50	–2	–8	–0.238	0.952
2.55–2.65	7	2.60	–1	–7	–0.138	0.966
2.65–2.75	15	2.70	0	0	–0.038	0.570
2.75–2.85	11	2.80	1	11	0.062	0.682
2.85–2.95	10	2.90	2	20	0.162	1.620
2.95–3.05	2	3.00	3	6	0.262	0.524
Total	50	–	–	19	–	5.652

By (5.3.4), $\bar{x} = c + d\bar{y} = 2.70 + 0.10 \times \frac{19}{50} = 2.738$ ft. Using this, cols. (6) and (7) are then calculated in the above Table.

Mean Deviation = 5.652/50 = 0.11 ft. (upto 2 decimals).

6.5 STANDARD DEVIATION (S.D.)

Standard Deviation of a set of observations is the square-root of the arithmetic mean of squares of deviations from arithmetic mean. In short, S.D. may be defined as "Root-Mean-Square-Deviation from mean". It is usually denoted by the Greek small letter σ (sigma). If $x_1, x_2, ... x_n$ be a set of observations and $\bar{x}$ their A.M. then,

Deviations from mean: $(x_1 - \bar{x}), (x_2 - \bar{x}), \ldots, (x_n - \bar{x})$
Square-Deviations from mean: $(x_1 - \bar{x})^2, (x_2 - \bar{x})^2, \ldots, (x_n - \bar{x})^2$
Mean-Square-Deviation from mean:

$$\frac{1}{n}[(x_1 - \bar{x})^2 + (x_2 - \bar{x})^2 + \ldots + (x_n - \bar{x})^2] = \frac{1}{n}\Sigma(x_i - \bar{x})^2$$

Root-Mean Square-Deviation from mean, i.e.

$$\textit{Standard Deviation } (\sigma) = \sqrt{\frac{1}{n}\Sigma(x_i - \bar{x})^2} \qquad (6.5.1)$$

The square of standard deviatin, i.e. σ^2, is known as *variance.*
Variance = $(\text{S.D})^2$
For simple series,

$$\sigma^2 = \frac{1}{n}\Sigma(x_i - \bar{x})^2 \qquad (6.5.2)$$

For frequency distribution,

$$\sigma^2 = \frac{1}{N}\Sigma f_i (x_i - \bar{x})^2 \qquad (6.5.3)$$

S.D. is always considered as positive. Thus, S.D. is the positive square-root of Variance.

6.6 IMPORTANT PROPERTIES OF S.D.

(a) S.D. is independent of the change of origin ; i.e. if $y = x - c$, where c is a constant, then

$$S.D.\ of\ x = S.D.\ of\ y$$

In symbols,

$$\sigma_x = \sigma_y \qquad (6.6.1)$$

This implies that the same S.D. will be obtained if each of the observations is increased or decreased by a constant.

(b) If two variables x and z are so related that $z = ax + b$ for each $x = x_i$, where a and b are constants, then

$$\sigma_x = |a| \,.\, \sigma_x \qquad (6.6.2)$$

where $|a|$ denotes the positive value of a.

In particular, if $y = (x - c)/d$, where c and d are constants (d positive), then

$$\sigma_x = d.\ \sigma_y \qquad (6.6.3)$$

This implies that S.D. does not depend on origin, but depends on scale of measurement. If each observation is multiplied or divided by a constant, S.D. will also be similarly affected.

(c) If a group of n_1 observations has mean $\bar{x}_1$ and S.D. σ_1, and another group of n_2 observations has mean $\bar{x}_2$ and S.D. σ_2, then S.D. (σ) of the composite group of $n_1 + n_2$ ($= N$, say) observations can be obtained by the formula

$$N\sigma^2 = (n_1\sigma_1^2 + n_2\sigma_2^2) + (n_1 d_1^2 + n_2 d_2^2) \qquad (6.6.4)$$

where $d_1 = \bar{x}_1 - \bar{x}$, $d_2 = \bar{x}_2 - \bar{x}$, and $N\bar{x} = n_1\bar{x}_1 + n_2\bar{x}_2$.

Relation (6 .6.4) may be extended to any number of groups:

$$N\sigma^2 = \Sigma n_i \sigma_i^2 + \Sigma n_i d_i^2 \qquad (6.6.5)$$

where $d_i = \bar{x}_i - \bar{x}$, $N = \Sigma n_i$, and $\bar{x}$ is the mean of composite group, given by $N\bar{x} = \Sigma n_i \bar{x}_i$ (see 5.3.6).

(d) S.D. is the minimum root-mean-square-deviation, i.e.

$$\sqrt{\frac{1}{n}\Sigma(x_i - \bar{x})^2} \le \sqrt{\frac{1}{n}\Sigma(x_i - A)^2} \qquad (6.6.6)$$

whatever be the value of A (see Example 6.16).

Example 6.12 *Explain why the standard deviation is regarded as superior to other measures of dispersion. What is its chief defect?*

[C.U., B.A.(Econ) '77; I.C.W.A, Dec. '75, June '76, B.U., B.A. '72, '74; '76]

Solution There are four absolute measures of dispersion—Range, Quartile Deviation, Mean Deviation, and Standard Deviation. Among these, S.D. is considered superior, because it possesses almost all the requisites of a good measure of dispersion (see Example 6.1, page 176).

(1) It is rigidly defined. S.D. of a set of observations is the square-root of the mean of squared deviations from mean.

(2) It is based on all observations. Even if one of the observations is changed, S.D. changes. However, Range and Quartile Deviation do not possess this property. The whole set of observations intermediate between the largest and the smallest values may be changed without affecting Range. Similarly, Quartile Deviation is based only on the two extremes of the central 50% of observations, and is independent of the magnitudes of other observations, so long as the two quartiles do not change.

(3) Although not so simple as Range or Quartile Deviation, the calculation of S.D. is not very difficult, and does not necessitate any special technique for computation. A change of origin does not affect S.D. As such, observations may be reduced conveniently by subtracting a constant, and then the formula applied to the reduced values (Examples 6.29, 6.28). If possible, the scale may also be changed to simplify the calculations further; but in that case, the S.D. obtained must be multiplied later (Example 6.30).

(4) S.D. is the least affected by fluctuations of sampling. If several independent samples are drawn from the same statistical population and each time all the four absolute measures of dispersion calculated, it will be found that S.D. fluctuates less from sample to sample than any other measure of disperion.

(5) The unique property which makes S.D. superior to other measures of dispersion is that it is amenable to algebraic treatment. Given the number of observations, mean and S.D. for each of several groups, S.D. of the composite group can be calculated (Section 6.8).

Because of these advantages, S.D. is almost exclusively used unless there are definite reasons for using any other measure.

The chief defect of S.D. is that it cannot be used for comparing the dispersion of two or more series of observations given in different units. S.D. is also difficult to understand, and the process of squaring deviations from mean, and then taking the square-root of the mean of these squared deviations seems to be a complicated affair.

Example 6.13 *Prove that the standard deviation is independent of change of origin.* [IC.W.A June '76-old]

Solution Let $x_1, x_2, ..., x_n$ be a set of n observations and let $y_1, y_2, ..., y_n$ be the quantities derived from them by shifting the origin to an arbitrary constant c, i.e. $y_i = x_i - c$ ($i = 1, 2, ... n$). We have to prove that $\sigma_x = \sigma_y$. By definition (6.5.2),

$$\sigma_x^2 = \Sigma(x_i - \bar{x})^2/n, \quad \text{where } \bar{x} = \Sigma x_i/n.$$

Now, $y_i = x_i - c$, so that $\bar{x} = c + \bar{y}$ (see Example 5.10, p. 112)

Therefore, $x_i - \bar{x} = (c + y_i) - (c + \bar{y}) = y_i - \bar{y}$

Substituing this $\sigma_x^2 = \Sigma(x_i - \bar{x})^2/n = \Sigma(y_i - \bar{y})^2/n = \sigma_y^2$

or, $$\sigma_x = \sigma_y$$

This shows that σ_x is exactly equal to σ_y, proving thereby that S.D. has not changed by the choice of new origin c.

Example 6.14 *Prove that the standard deviation is independent of any change of origin, but is dependent on the change of scale.*

[I.C.W.A., July '71, Dec. '76; Dip. Management, May '72]

Solution Let $x_1, x_2, \ldots, x_n$ be a set of observations. If we change the origin of x to c and the scale to d (positive), i.e. write

$$y_i = \frac{x_i - c}{d}, \quad \text{then } x_i = c + dy_i$$

Hence, $\bar{x} = c + d\bar{y}$ (Example 5.11, page 113)

or, $x_i - \bar{x} = (c + dy_i) - (c + d\bar{y}) = d(y_i - \bar{y})$

∴ $\sigma_x^2 = \Sigma(x_i - \bar{x})^2/n = \Sigma\{d(y_i - \bar{y})\}^2/n = d^2\Sigma(y_i - \bar{y})^2/n$

or, $\sigma_x^2 = d^2.\sigma_y^2$, so that $\sigma_x = d.\sigma_y$.

It is observed from the result that on the right hand side, the new origin c is absent, but the scale d is present. This proves that S.D. is unaffected by any change of origin, but depends on scale.

Example 6.15 *Prove that the standard deviation calculated from two values x_1 and x_2 of a variable x is equal to half their difference.* [I.C.W.A., Jan. '68]

Solution By definition (6.5.2),

$$\sigma^2 = \frac{1}{n}\Sigma(x_i - \bar{x})^2 = \frac{1}{2}\{(x_1 - \bar{x})^2 + (x_2 - \bar{x})^2\}.$$

Since there are only two values x_1 and x_2, $\bar{x} = \frac{1}{2}(x_1 + x_2)$.

$$\therefore \quad x_1 - \bar{x} = x_1 - \frac{1}{2}(x_1 + x_2) = \frac{x_1 - x_2}{2}$$

$$x_2 - \bar{x} = x_2 - \frac{1}{2}(x_1 + x_2) = \frac{x_2 - x_1}{2}$$

$$\therefore \quad \sigma^2 = \frac{1}{2}\left\{\left(\frac{x_1 - x_2}{2}\right)^2 + \left(\frac{x_2 - x_1}{2}\right)^2\right\}$$

$$= \frac{1}{2}\left\{\left(\frac{x_1 - x_2}{2}\right)^2 + \left(\frac{x_1 - x_2}{2}\right)^2\right\}$$

$$= \frac{1}{4}(x_1 - x_2)^2$$

$$\therefore \quad \sigma = \frac{1}{2}|x_1 - x_2|$$

[**Note:** It is necessary to take the positive value of $(x_1 - x_2)$, i.e. the difference of x_1 and x_2, since S.D. is always positive.]

Example 6.16 *Show that the standard deviation is the minimum root-mean-square deviation.* [B.U., B.A. (Econ) '72]

Solution Let $x_1, x_2, \ldots, x_n$ be a given set of observations. By definition (see page 184),

S.D. = Root-Mean-Square deviation from mean

$$= \sqrt{\Sigma(x_i - \bar{x})^2 / n} \qquad \ldots \text{(i)}$$

If instead of taking deviations from mean, we consider deviations from an arbitrary constant A, then

$$\text{Root-Mean-Square deviation from } A = \sqrt{\Sigma(x_i - A)^2 / n} \qquad \ldots \text{(ii)}$$

We have to show that (ii) cannot be smaller than (i) whatever be the value of A.

Now, it has been shown that (see Example 5.12, page 113)

$$\Sigma(x_i - A)^2 = \Sigma(x_i - \bar{x})^2 + n(\bar{x} - A)^2$$

Since the second term on the right can never be negative

$$\Sigma(x_i - A)^2 \geq \Sigma(x_i - \bar{x})^2$$

Dividing both sides by n and then taking the square-root, we have

$$\sqrt{\Sigma(x_i - A)^2/n} \geq \sqrt{\Sigma(x_i - \bar{x})^2/n} \qquad \textit{(shown)}$$

Example 6.17 *Show that the mean absolute deviation is least when the deviations are taken from median, but $\Sigma(x_i - A)^2/n$ is least when $A = \bar{x}$, the arithmetic mean of $x_1, x_2, \ldots, x_n$.* [C.U., B.A. (Econ) '67, '73, '75; B.U., B.A. (Econ) '67; D.M. '77]

Solution (*First Part*) In Example 5.56 (page 150), it has been proved that $\Sigma|x_i - A|$ is minimum when A = median.

Hence, Mean Absolute Deviation about A, i.e. $\Sigma|x_i - A|/n$, is the least, when A = median.

(*Second Part*) see Example 5.12 (page 113).

Example 6.18 *Show that $\bar{x}$ if is the arithmetic mean of the quantities $x_1, x_2, \ldots, x_n$, then*

$$\sum_1^n (x_i - \bar{x})^2 = \sum_1^n x_i^2 - n\bar{x}^2$$

[C.U., M.Com. '63; I.C.W.A., Jan. '69; D.M. '76]

Solution

(*First method*) By definition, $\bar{x} = \dfrac{1}{n}\sum_1^n x_i$

so that

$$n\bar{x} = \sum_1^n x_i \qquad \ldots \text{(i)}$$

Therefore,

$$\sum_1^n (x_i - \bar{x})^2 = \sum_1^n (x_i^2 - 2x_i\bar{x} + \bar{x}^2)$$

$$= \sum_1^n x_i^2 - \sum_1^n 2x_i\bar{x} + \sum_1^n \bar{x}^2 \qquad \text{by (3.15.2)}$$

$$= \sum_1^n x_i^2 - 2\bar{x}\sum_1^n x_i + n\bar{x}^2, \qquad \text{by (3.15.3) \& (3.15.4)}$$

$$= \sum_1^n x_i^2 - 2\bar{x}.n\bar{x} + n\bar{x}^2, \qquad \text{by (i) above}$$

$$= \sum_1^n x_i^2 - 2n\bar{x}^2 + n\bar{x}^2$$

$$= \sum_1^n x_i^2 - n\bar{x}^2.$$

(*Second method*)

$$\sum_1^n (x_1 - \bar{x})^2 = (x_1 - \bar{x})^2 + (x^2 - \bar{x})^2 + \cdots + (x_n - \bar{x})^2$$
$$= (x_1^2 - 2x_1\bar{x} + \bar{x}^2) + (x_2^2 - 2x_2\bar{x} + \bar{x}^2) + \ldots + (x_n^2 - 2x_n\bar{x} + \bar{x}^2).$$
$$= (x_1^2 + x_2^2 + \ldots + x_n^2) - 2x_1\bar{x} - 2x_2\bar{x} - \ldots - 2x_n\bar{x} + (\bar{x}^2 + \bar{x}^2 + \ldots + \bar{x}^2).$$
$$= \sum_1^n x_i^2 - 2\bar{x}(x_1 + x_2 + \ldots + x_n) + n\bar{x}^2$$
$$= \sum_1^n x_i^2 - 2\bar{x}.n\bar{x} + n\bar{x}^2$$
$$= \sum_1^n x_i^2 - 2n\bar{x}^2 + n\bar{x}^2$$
$$= \sum_1^n x_i^2 - n\bar{x}^2. \qquad \text{(Shown)}$$

Example 6.19 *Show that if $\bar{x}$ is the arithmetic mean of the values x_i (i = 1, 2, ..., n), then*

$$\sum_{i=1}^n f_i(x_i - \bar{x})^2 = \sum_{i=1}^n f_i x_i^2 - N\bar{x}^2, \quad \textit{where } N = \sum_{i=1}^n f_i.$$

[C.U., M.Com. '64]

Solution (*First method*) If f_i represents the frequency of the value $x_i (i = 1, 2, ..., n)$, then by definition, the arithmetic mean

$$\bar{x} = \frac{1}{N}\sum_{i=1}^n f_i x_i \quad \text{where } N = \sum_{i=1}^n f_i$$

By cross multiplication, $N\bar{x} = \sum_{i=1}^n f_i x_i$ ... (i)

Therefore, $\sum_{i=1}^n f_i(x_i - \bar{x})^2$

$$= \sum_{i=1}^n f_i(x_i^2 - 2x_i\bar{x} + \bar{x}^2)$$

$$= \sum_{i=1}^{n} (f_i x_i^2 - 2 f_i x_i \overline{x} + f_i \overline{x}^2)$$

$$= \sum_{i=1}^{n} f_i x_i^2 - \sum_{i=1}^{n} 2 f_i x_i \overline{x} + \sum_{i=1}^{n} f_i \overline{x}^2 \qquad \text{by (3.15.2)}$$

$$= \sum_{i=1}^{n} f_i x_i^2 - 2\overline{x} \sum_{i=1}^{n} f_i x_i + \overline{x}^2 \sum_{i=1}^{n} f_i \qquad \text{by (3.15.3)}$$

$$= \sum_{i=1}^{n} f_i x_i^2 - 2\overline{x}.N\overline{x} + \overline{x}^2.N \qquad \text{by (i) above}$$

$$= \sum_{i=1}^{n} f_i x_i^2 - 2N\overline{x}^2 + N\overline{x}^2$$

$$= \sum_{i=1}^{n} f_i x_i^2 - N\overline{x}^2 \qquad \text{(Shown)}$$

(*Second method*) The arthmetic mean $\overline{x}$ of the values $x_1, x_2, ..., x_n$ (having frequencies $f_1, f_2, ..., f_n$ respectively) is given by

$$\overline{x} = \frac{f_1 x_1 + f_2 x_2 + \ldots + f_n x_n}{N}, \qquad \text{where } N = \sum_{i=1}^{n} f_i$$

By cross multiplication, $N\overline{x} = f_1x_1 + f_2x_2 + ... + f_nx_n$. ... (ii)

Hence, $\sum_{i=1}^{n} f_i (x_i - \overline{x})^2$

$$= f_1(x_1 - \overline{x})^2 + f_2(x_2 - \overline{x})^2 + ... + f_n(x_n - \overline{x})^2$$
$$= f_1(x_1^2 - 2x_1\overline{x} + \overline{x}^2) + f_2(x_2^2 - 2x_2\overline{x} + \overline{x}^2) + ... + f_n(x_n^2 - 2x_n\overline{x} + \overline{x}^2)$$
$$= (f_1x_1^2 - 2f_1x_1\overline{x} + f_1\overline{x}^2) + (f_2x_2^2 - 2f_2x_2\overline{x} + f_2\overline{x}^2) + ... + (f_nx_n^2 - 2f_nx_n\overline{x} + f_n\overline{x}^2)$$
$$= f_1x_1^2 + f_2x_2^2 + ... + f_nx_n^2) - 2f_1x_1\overline{x} - 2f_2x_2\overline{x} - ... - 2f_nx_n\overline{x} + (f_1\overline{x}^2 + f_2\overline{x}^2 + ... + f_n\overline{x}^2)$$
$$= \sum_{i=1}^{n} f_i x_i^2 - 2\overline{x}(f_1 x_1 + f_2 x_2 + \ldots + f_n x_n) + \overline{x}^2(f_1 + f_2 + ... + f_n)$$
$$= \sum_{i=1}^{n} f_i x_i^2 - 2\overline{x}\ N\overline{x} + \overline{x}^2 + \sum_{i=1}^{n} f_i \qquad \text{by (ii)}$$
$$= \sum_{i=1}^{n} f_i x_i^2 - 2N\overline{x}^2 + \overline{x}^2 N$$
$$= \sum_{i=1}^{n} f_i x_i^2 - 2N\overline{x}^2 + N\overline{x}^2$$
$$= \sum_{i=1}^{n} f_i x_i^2 - N\overline{x}^2 \qquad \text{(Shown)}$$

Example 6.20 *Show that the formula for standard deviation (σ) of $x_1, x_2, ..., x_n$ can be expressed as*

$$\sigma^2 = \frac{\Sigma x^2}{n} - \left(\frac{\Sigma x}{n}\right)^2 \qquad \textit{for simple series}$$

$$\sigma^2 = \frac{\Sigma f x^2}{N} - \left(\frac{\Sigma f x}{N}\right)^2 \qquad \textit{for frequency distributions}$$

Solution For simple series

$$\sigma^2 = \frac{1}{n}\Sigma(x_i - \bar{x})^2, \qquad \text{by definition (6.5.2)}$$

$$= \frac{1}{n}(\Sigma x_i^2 \; n\bar{x}^2), \qquad \text{by Example 6.18}$$

$$= \frac{\Sigma x_i^2}{n} - \bar{x}^2$$

$$= \frac{\Sigma x^2}{n} - \left(\frac{\Sigma x}{n}\right)^2, \text{ since } \bar{x} = \frac{\Sigma x}{n}.$$

For frequency distributions,

$$\sigma^2 = \frac{1}{N}\Sigma f_i(x_i - \bar{x})^2, \qquad \text{by definition (6.5.3)}$$

$$= \frac{1}{N}(\Sigma f_i x_i^2 - N\bar{x}^2), \qquad \text{by Example 6.19}$$

$$= \frac{\Sigma f_i x_i^2}{N} - \bar{x}^2$$

$$= \frac{\Sigma fx^2}{N} - \left(\frac{\Sigma fx}{N}\right)^2, \text{ since } \bar{x} = \frac{\Sigma fx}{N}.$$

Example 6.21 *The standard deviation calculated from a set of 32 observations is 5. If the sum of the observations is 80, what is the sum of the squares of these observations?* [I.C.W.A., Jan. '71]

Solution We are given $n = 32$, $\sigma = 5$, $\Sigma x = 80$. It is required to find the value of Σx^2. Now from Example 6.20)

$$\sigma^2 = \frac{\Sigma x^2}{n} - \left(\frac{\Sigma x}{n}\right)^2$$

Substituting the values, $25 = \dfrac{\Sigma x^2}{32} - \left(\dfrac{80}{32}\right)^2$

Solving this, we get $\Sigma x^2 = 1000$. *Ans.* 1000.

Example 6.22 *Find the mean and the standard deviation of the first n natural numbers.* [Dip. Management, May '67]

Solution The first n natural numbers are 1, 2, 3, ... n. The sum and the sum of squares of these n numbers are (see page 55)

$$\Sigma x = 1 + 2 + 3 + \ldots + n = \frac{n(n+1)}{2}$$

$$\Sigma x^2 = 1^2 + 2^2 + 3^2 + \ldots + n^2 = \frac{n(n+1)(2n+1)}{6}$$

$$\therefore \quad \text{Mean } (\bar{x}) \; \frac{\Sigma x}{n} = \frac{n(n+1)}{2n} = \frac{n+1}{2}$$

$$\sigma^2 = \frac{\Sigma x^2}{n} - \left(\frac{\Sigma x}{n}\right)^2, \quad \text{by Example 6.20}$$

$$= \frac{n(n+1)(2n+1)}{6} - \left\{\frac{n(n+1)}{2n}\right\}^2$$

$$= \frac{(n+1)(2n+1)}{6} - \frac{(n+1)^2}{4}$$

$$= \frac{2(n+1)(2n+1) - 3(n+1)^2}{12}$$

$$= \frac{(n+1)(4n+2-3n-3)}{12} = \frac{(n+1)(n-1)}{12} = \frac{n^2-1}{12}$$

Therefore, Standard deviation (σ) = $\sqrt{(n^2-1)/12}$

Ans. Mean = $(n+1)/2$, S.D. = $\sqrt{(n^2-1)/12}$.

Example 6.23 *The arithmetic mean and variance of n values $x_1, x_2, ..., x_n$ are 0 and σ^2 respectively, and y is a variable defined by $y = x^2$. Prove that $\bar{y} = \sigma^2$.*

[C.U., B.A. (Econ) '72; D.M. '78]

Solution We are given $\bar{x} = 0$, i.e. $\Sigma x/n = 0$, so that $\Sigma x = 0$.

Again, $\sigma_x^2 = \sigma^2$; or, $\Sigma x^2/n - (\Sigma x/n)^2 = \sigma^2$

or, $\Sigma x^2/n - (0/n)^2 = \sigma^2$

or, $\Sigma x^2/n = \sigma^2$

Since, $y = x^2$, we have $\bar{y} = \Sigma y/n = \Sigma x^2/n = \sigma^2$ (Proved)

Example 6.24 *The mean and S.D. calculated from 20 observations are 15 and 10 respectively. If an additional observation 5, left out through oversight, be included in the calculations, find the corrected mean and S.D.* [I.C.W.A., Jan. '69]

Solution The sum of the 20 observations is $\Sigma x = 20 \times 15 = 300$. The sum of the squares of these 20 observations can be obtained from

$$\sigma^2 = \frac{\Sigma x^2}{n} - \left(\frac{\Sigma x}{n}\right)^2; \quad \text{or,} \quad 10^2 = \frac{\Sigma x^2}{20} - \left(\frac{300}{20}\right)^2$$

so that $\Sigma x^2 = 6500$. When the additional observation 5 is included, for the 21 observations, the corrected values are $\Sigma x = 300 + 5 = 305$; $\Sigma x^2 = 6500 + 5^2 = 6525$. Therefore, Mean = 305/21 = 14.52;

$$(\text{S.D.})^2 = \frac{6525}{21} - \left(\frac{305}{21}\right)^2 = \frac{44000}{441};$$

$$\therefore \quad \text{S.D.} = \sqrt{44000/441} = 9.99$$

Ans. Mean = 14.52; S.D. = 9.99

Example 6.25 *The mean income per month of a friendly society of 25 members is Rs 350 and the standard deviation is Rs 50. Five more members are admitted to the society and their incomes in Rs per month are 260, 300, 320, 490 and 590. Find the mean and standard deviation of income for the new group of 30 members.* [I.C.W.A., July '69]

Solution Let us suppose that the income figures of 25 members form Group I and those of 5 new members form Group II. We have to find the mean and S.D. of the composite group of 30 observations.

For group I, (as in Example 6.24) $\Sigma x = 25 \times 350 = 8{,}750$

and $50^2 = \Sigma x^2/25 - (350)^2$.

Solving we get $\Sigma x^2 = 3{,}125{,}000$.

For group II, $\Sigma x = 260 + 300 + 320 + 490 + 590 = 1{,}960$

$$\Sigma x^2 = 260^2 + 300^2 + 320^2 + 490^2 + 590^2 = 848{,}200$$

Hence, for the composite group of $n = 30$ observations,

$$\Sigma x = 8{,}750 + 1{,}960 = 10{,}710$$

$$\Sigma x^2 = 3{,}125{,}000 + 848{,}200 = 3{,}973{,}200$$

Therefore, $\bar{x} = 10710/30 = 357$ Rs

$$\sigma^2 = \frac{3973200}{30} - \left(\frac{10710}{30}\right)^2 = 4991; \sigma = \sqrt{4991} = 70.65 \text{ Rs}$$

Ans. Mean = 357 Rs; S.D. = 70.65 Rs

Example 6.26 *The mean and standard deviation of 20 items is found to be 10 and 2 respectively. At the time of checking it was found that one item 8 was incorrect. Calculate the mean and standard deviation, if (i) the wrong item is omitted, and (ii) it is replaced by 12.* [I.C.W.A., Dec. '77]

Solution (As in Example 6.24) We have for the 20 items (including the incorrect item 8)

$$\Sigma x = 20 \times 10 = 200 \text{ and } 2^2 = \Sigma x^2/20 - (10)^2, \text{ giving}$$

$$\Sigma x = 200 \text{ and } \Sigma x^2 = 2080$$

(i) When the incorrect item 8 is omitted, we have for the remaining 19 items, $\Sigma x = 200 - 8 = 192$, $\Sigma x^2 = 2080 - 8^2 = 2016$. Now applying the formulae for mean and variance,

$$\text{Mean} = 192/19 = 10.11$$

$$(\text{S.D.})^2 = 2016/19 - 192/19)^2 = 3.9889$$

$$\therefore \quad \text{S.D.} = \sqrt{3.9889} - 1.997$$

(ii) When the incorrect item 8 is replaced by 12, then for the 20 correct items $\Sigma x = 192 + 12 = 204$, $\Sigma x^2 = 2016 + 12^2 = 2160$. Hence, Mean = 204/20 = 10.2

$$(\text{S.D.})^2 = 2160/20 - (204/20)^2 = 3.96$$

$$\therefore \quad \text{S.D.} = \sqrt{3.96} = 1.990$$

Ans. (i) 10.11, 1,997; (ii) 10.2, 1.990

Example 6.27 *For a distribution of 280 observations, mean = 54 and standard deviation = 3. On checking it was discovered that observations which should correctly read as 62 and 82 had been wrongly recorded as 64 and 80 respectively. Calculate the correct values of mean and standard deviation.* [C.U., B.A. (Econ) '73, '75]

Solution For the 280 observations (which include the incorrect values 64 and 80), using the formulae for mean and variance, viz.

$$\bar{x} = \Sigma x/n, \quad \sigma^2 = \Sigma x^2/n - (\Sigma x/n)^2$$

we have $54 = \Sigma x/280$, so that $\Sigma x = 280 \times 54$

$3^2 = \Sigma x^2/280 - (54)^2$. Hence, $\Sigma x^2 = 280 \times 2925$

When 64 and 80 are replaced by 62 and 82, the corrected values are

$$\Sigma x = 280 \times 54 - (64 + 80) + (62 + 82) = 280 \times 54$$

$$\Sigma x^2 = 280 \times 2925 - (64^2 + 80^2) + (62^2 + 82^2) = 819072$$

Using these in the formulae for mean and variance,

$$\text{Mean} = (280 \times 54)/280 = 54$$

$$(\text{S.D.})^2 = 819072/280 - (54)^2 = 9.257$$

$$\therefore \quad \text{S.D.} = \sqrt{9.257} = 3.04$$

Ans. Mean = 54, S.D. = 3.04

6.7 CALCULATION OF STANDARD DEVIATION (σ)

If the observations are small, S.D. can be calculated by using the following relations (see Example 6.20).

For simple series

$$\sigma^2 = \frac{\Sigma x^2}{n} - \left(\frac{\Sigma x}{n}\right)^2 \qquad (6.7.1)$$

For frequency distributions

$$\sigma^2 = \frac{\Sigma f x^2}{N} - \left(\frac{\Sigma f x}{N}\right)^2 \qquad (6.7.2)$$

The calculations can, however, be simplified based on the following results (see Examples 6.13, 6.14, at p. 186, 187)

(I) if $y_1, y_2, ..., y_n$ represent the deviations of $x_1, x_2, ..., x_n$ from an arbitrary constant c, then

S.D. of x = S.D. of y

In symbols, if $y = x - c$, then $\sigma_x = \sigma_y$. (6.7.3)

This result implies that in order to find the S.D. we can always reduce the given observations by subtracting a convenient number. S.D. of the reduced values gives the S.D. of the original values of x.

(II) If $y_1, y_2, ..., y_n$ represent the deviations of $x_1, x_2, ..., x_n$ from an arbitrary constant c, in units of another constant d, then

S.D. of $x = d$ (S.D. of y).

In symbols, if $y = \dfrac{x - c}{d}$, then $\sigma_x = d.\sigma_y$ (6.7.4)

This implies that, if after taking deviations of x from a constant c, it is possible to reduce these deviations on division by a constant d, then the S.D. of these reduced values when multiplied by d, will give the S.D. of the original values of x. The result is particularly suitable when the consecutive values of x have a common difference. In such cases the values of y can be reduced to 0, 1, 2, 3, ... and –1, –2, –3, ..., whatever be the values of x.

σ_y used in (6.7.3) or (6.7.4) is calculated by using formulae similar to (6.7.1) or (6.7.2), replacing x by y, i.e.

$$\sigma_y^2 = \frac{\Sigma y^2}{n} - \left(\frac{\Sigma y}{n}\right)^2 \quad \text{for simple series}$$

$$\sigma_y^2 = \frac{\Sigma f y^2}{N} - \left(\frac{\Sigma f y}{N}\right)^2 \quad \text{for frequency distributions}$$

For the calculation of S.D. from a grouped frequency distribution, see Example 6.32.

Example 6.28 *Find the standard deviation from the following data: 49, 63, 46, 59, 65, 52, 60, 54.* [C.U., B.Com. '72]

Solution Since the standard deviation remains unaffected by change of origin, we may reduce each observation by 50 (suppose) and write $y = x - 50$. By (6.7.3),

$$\sigma_x = \sigma_y \quad \text{where } \sigma_y^2 = \frac{\Sigma y^2}{n} - \left(\frac{\Sigma y}{n}\right)^2$$

Table 6.6 Calculations for S.D.

x	$y = x - 50$	y^2
49	–1	1
63	13	169
46	–4	16
59	9	81
65	15	225
52	2	4
60	10	100
54	4	16
Total	48	612

Here, $n = 8$, $\Sigma y = 48$, $\Sigma y^2 = 612$

$$\therefore \quad \sigma_y^2 = \frac{612}{8} - \left(\frac{48}{8}\right)^2$$

$$= 76.5 - (6)^2 = 40.5$$

$$\sigma_y = \sqrt{40.5} = 6.36$$

Hence, $\sigma_x = \sigma_y = 636$

Ans. 6.36

[**Note:** Students are advised to try this problem by taking different origins. For instance, if the origin is taken at 55, i.e. $y = x - 55$, we have the following table.

Table 6.7 Calculations for S.D.

x	$y = x - 55$	y^2
49	−6	36
63	8	64
46	−9	81
59	4	16
65	10	100
52	−3	9
60	5	25
54	−1	1
Total	8	332

$$\sigma_y^2 = \frac{332}{8} - \left(\frac{8}{8}\right)^2$$

$$= 41.5 - (1)^2 = 40.5$$

$\therefore \quad \sigma_y = \sqrt{40.5} = 6.36$

Hence, $\quad \sigma_x = 6.36$

We get the same result as before. This verifies that S.D. does not depend on the origin. However, numerical computations will be small, when a value close to the mean is taken as origin.

Example 6.29 *Compute the standard deviation of household size from the following frequency distribution of 500 households:*

Household Size	*1*	*2*	*3*	*4*	*5*	*6*	*7*	*8*	*9*
No. of Households	*92*	*49*	*52*	*82*	*102*	*60*	*35*	*24*	*4*

[C.U., B.A. (Econ) '77]

Solution Since the S.D. is unaffected by change of origin, we write $y = x - 4$; then $\sigma_x = \sigma_y$.

Table 6.8 Calculations for S.D.

x	f	$y = x - 4$	fy	fy^2
1	92	−3	−276	828
2	49	−2	−98	196
3	52	−1	−52	52
4	82	0	0	0
5	102	1	102	102
6	60	2	120	240
7	35	3	105	315
8	24	4	96	384
9	4	5	20	100
Total	500	–	17	2217

$$\sigma_y^2 = \frac{\Sigma f y^2}{N} - \left(\frac{\Sigma f y}{N}\right)^2 = \frac{2217}{500} - \left(\frac{17}{500}\right)^2$$

$$= 4.434 - .0012 = 4.4328$$

$$\sigma y = \sqrt{4.4328} = 2.11;$$

$$\therefore \quad \sigma_x = \sigma_y^2 = 2.11$$

Ans. 2.11

Example 6.30 *Find out the S.D. for the following table:*

x	*5*	*15*	*25*	*35*	*45*	*55*	*65*	*75*
f	*3*	*7*	*9*	*23*	*15*	*8*	*6*	*4*

[D.S.W., Nov. '70]

Solution Since the successive values of *x* have a common difference, we use formula (6.7.4).

Table 6.9 Calculations for S.D.

x	*f*	$y = \dfrac{x-35}{10}$	*fy*	fy^2
5	3	−3	− 9	27
15	7	−2	− 14	28
25	9	−1	− 9	9
35	23	0	0	0
45	15	1	15	15
55	8	2	16	32
65	6	3	18	54
75	4	4	16	64
Total	75	–	33	229

$$\sigma_y^2 = \frac{\Sigma fy^2}{N} - \left(\frac{\Sigma fy}{N}\right)^2 = \frac{229}{75} - \left(\frac{33}{75}\right)^2 = 3.053 - (0.44)^2 = 2.859$$

$$\therefore \quad \sigma_y = \sqrt{2.859} = 1.69. \quad \sigma_x = d.\ \sigma_y = 10 \times 1.69 = 16.9$$

[**Note:** For the calculation of S.D. only an extra column fy^2 is needed beyond what is necessary for the calculation of A.M.]

Example 6.31 *How do you calculate the Standard Deviation of a grouped variate?*

[C.U., M.Com.'61]

Solution For finding the standard deviation from a grouped frequency distribution, the mid-values are taken as representatives of all observations falling in the respective classes (see Example 5.16). If $x_1, x_2, ..., x_n$ be the mid-values of classes with frequencies $f_1, f_2, ..., f_n$ respectively, and $y_1, y_2, ..., y_n$ represent the deviations of these mid-values from an arbitrary origin *c*, in units of another constant *d*, then $\sigma_x = d.\ \sigma_y$, where

$$\sigma_y^2 = \frac{\Sigma fy^2}{N} - \left(\frac{\Sigma fy}{N}\right)^2 \text{ and } N = \Sigma f.$$

Calculations are done in the following tabular form:

Table 6.10 Calculations for S.D.

Class Interval	Frequency f	Mid-value x	Deviation $= x - c$	$y = \dfrac{x-c}{d}$	fy	fy^2
(1)	(2)	(3)	(4)	(5)	(6)	(7)
	f_1	x_1		y_1		
	f_2	x_2		y_2		
	$\vdots$	$\vdots$		$\vdots$		
	$\vdots$	$\vdots$		$\vdots$		
	f_n	x_n		y_n		
Total	$N = \Sigma f$	–	–	–	Σfy	Σfy^2

The steps in the calculation are as follows:

(i) From the given class intervals, we find the mid-values in col. (3). This is done by taking the arithmetic mean of class limits or class boundaries, as given in the frequency distribution.

(ii) One of the mid-values is chosen as the arbitrary origin c, preferably the mid-value corresponding to the maximum frequency, if it lies near the middle of the distribution. Deviation of each mid-value from c is calculated, as shown in col. (4).

(iii) These deviations are divided by their highest common factor (d) and the results, called *step deviations* (y), are shown in col. (5). If all class intervals have the same width, d will be equal to this width. (In such a case, it will be found that we get 0 against the mid-value which has been taken as origin, and –1, –2, –3, ... successively as we move away towards the smaller mid-values, and +1, +2, +3, ... towards the higher mid-values).

(iv) Cols. (2) and (5) are multiplied to get col. (6).

(v) Cols. (5) and (6) are multiplied to complete col. (7).

(vi) Total of cols. (2), (6) and (7) are obtained, and these values are substituted in the formula to obtain the required S.D.

Example 6.32 *Compute the standard deviation from the following distribution of marks obtained by 90 students:*

Marks	*20–29*	*30–39*	*40–49*	*50–59*	*60–69*	*70–79*	*80–89*	*90–99*
No. of Students	*5*	*12*	*15*	*20*	*18*	*10*	*6*	*4*

[I.C.W.A., June '76]

Solution

Table 6.11 Calculations for S.D.

Class Interval	Frequency f	Mid-value x	$y = \dfrac{x - 54.5}{10}$	fy	fy^2
(1)	(2)	(3)	(4)	(5)	(6)
20–29	5	24.5	–3	–15	45
30–39	12	34.5	–2	–24	48
40–49	15	44.5	–1	–15	15
50–59	20	54.5	0	0	0

(Contd.)

Class Interval	Frequency f	Mid-value x	$y = \frac{x-54.5}{10}$	fy	fy^2
60–69	18	64.5	1	18	18
70–79	10	74.5	2	20	40
80–89	6	84.5	3	18	54
90–99	4	94.5	4	16	64
Total	90	–	–	18	284

If $y = \frac{x-c}{d}$, then $\sigma_x = d.\sigma_y$. Here in col. (4), $y = \frac{x-54.5}{10}$; hence $c = 54.5$, $d = 10$.

$$\sigma_y^2 = \frac{\Sigma fy^2}{N} - \left(\frac{\Sigma fy}{N}\right)^2 = \frac{284}{90} - \left(\frac{18}{90}\right)^2$$

$$= 3.156 - .04 = 3.116$$

$$\therefore \quad \sigma_y = \sqrt{3.116} = 1.77;$$

$$\sigma_x = 10 \times 1.77 = 17.7$$

Ans. 17.7

Example 6.33 *Find the standard deviation from the following frequency distribution:*

Height in Inches	*No. of Students*
Over 60 but not more than 62	*35*
Over 62 but not more than 64	*27*
Over 64 but not more than 66	*20*
Over 66 but not more than 68	*13*
Over 68 but not more than 70	*5*
	100

[C.U., B.Com. '76]

Solution

Table 6.12 Calculations for S.D.

Height* (inch)	Frequency f	Mid-value x	$y = \frac{x-65}{2}$	fy	fy^2
(1)	(2)	(3)	(4)	(5)	(6)
60–62	35	61	–2	–70	140
62–64	27	63	–1	–27	27
64–66	20	65	0	0	0
66–68	13	67	1	13	13
68–70	5	69	2	10	20
Total	100	–	–	–74	200

* See Table 4.18(D) and Example 4.3(1) at p, 76–77.

If $y = (x - c)/d$, then $\sigma_x = d.\sigma_y$

$$\sigma_y^2 = \frac{\Sigma fy^2}{N} - \left(\frac{\Sigma fy}{N}\right)^2 = \frac{200}{100} - \left(\frac{-74}{100}\right)^2$$

$$= 2.00 - (-0.74)^2 = 2.00 - 0.5476 = 1.4524$$

$$\sigma_y = \sqrt{1.4524} = 1.21;\ \sigma_x = 2 \times 1.21 = 24 \text{ inches.}$$

Example 6.34 *Calculate the standard deviation for the following frequency distribution. Find out the numbers which will lie between $\bar{x} \pm s$ and $\bar{x} \pm 2s$.*

Values	*0–5*	*5–10*	*10–15*	*15–20*	*20–25*	*25–30*	*30–35*	*35–40*
Frequency	2	5	7	13	21	16	8	3

[D.S.W., Nov. '78]

Solution (Note that here *s* denotes the standard deviation).

Table 6.13 Calculations for Mean and S.D.

Class Interval	*Frequency (f)*	*Mid-value (x)*	$y = \frac{x-22.5}{5}$	*fy*	*fy*²
(1)	(2)	(3)	(4)	(5)	(6)
0–5	2	2.5	–4	–8	32
5–10	5	7.5	–3	–15	45
10–15	7	12.5	–2	–14	28
15–20	13	17.5	–1	–13	13
20–25	21	22.5	0	0	0
25–30	16	27.5	1	16	16
30–35	8	32.5	2	16	32
35–40	3	37.5	3	9	27
Total	75	–	–	–9	193

If $y = (x - c)/d$, then (i) $\bar{x} = c + d\bar{y}$, and (ii) $s_x = d.s_y$

Here in col. (4), $y = (x - 22.5)/5$; so, $c = 22.5$, $d = 5$.

$$\bar{y} = \Sigma fy/N = -9/75 = -0.12$$

$$\bar{x} = 22.5 + 5(-0.12) = 22.5 - 0.6 = 21.9$$

Again, $s_y{}^2 = \Sigma fy^2/N - (\Sigma fy/N)^2 = 193/75 - (-9/75)^2$

$$= 2.5733 - .0144 = 2.5589$$

$\therefore$ $s_y = \sqrt{2.5589} = 1.60$; $s_x = 5 \times 1.60 = 8.0$. Thus, for the given frequency distribution, we have $\bar{x} = 21.9$, $s = 8.0$.

In order to find the number of observations lying between $\bar{x} \pm s$, we use simple interpolation in a cumulative frequency distribution (see Example 4.12).

$\bar{x} \pm s = 21.9 \pm 8.0 = 21.9 + 8.0$ and $21.9 - 8.0 = 29.9$ and 13.9

Table 6.14 Comulative Frequency Distribution

Class Boundary	*Cumulative Frequency*
0	0
5	2
5.9→	←*C*
10	7
13.9→	←*A*
15	14
20	27
25	48
29.9→	←*B*
30	64
35	72
37.9→	←*D*
40	75

If *A* and *B* denote cumulative frequencies corresponding to 13.9 and 29.9 respectively, then

$$\frac{13.9-10}{15-10}=\frac{A-7}{14-7},\ \frac{29.9-25}{30-25}=\frac{B-48}{64-48}$$

Solving, $A = 12.46$, $B = 63.68$

Hence, the number of observations lying between $\bar{x} \pm s$, i.e. between 13.9 and 29.9, is $B - A = 63.48 - 12.46 = 51$

Similarly, $\bar{x} \pm 2s = 21.9 \pm 2 \times 8.0 = 21.9 \pm 16.0 = 37.9$ and 5.9

If *C* and *D* be the cumulative frequencies of 5.9 and 37.9 respectively, then

$$\frac{5.9-5}{10-5}=\frac{C-2}{7-2},\ \frac{37.9-35}{40-35}=\frac{D-72}{75-72}$$

whence $C = 2.9$ and $D = 73.74$. Hence the frequency lying between $\bar{x} \pm 2s$ is $D - C = 73.74 - 2.9 = 71$.

Ans. 8.0; 51 and 71.

Example 6.35 *Find the standard deviation of the following distribution:*

X	*0–500*	*500–1000*	*1000–1500*	*1500–2000*	*2000–3000*
Frequency	*90*	*218*	*86*	*41*	*15*

[C.U., B.A.(Econ) '74]

Solution

Table 6.15 Calculations for S.D.

X	*Frequency* (*f*.)	*Mid-value* (*x*)	*x* – 750	$y=\frac{x-750}{250}$	*fy*	fy^2
0–500	90	250	–500	–2	–180	360
500–1000	218	750	0	0	0	0
1000–1500	86	1250	500	2	172	344
1500–2000	41	1750	1000	4	164	656
2000–3000	15	2500	1750	7	105	735
Total	450	–	–	–	261	2095

$$\sigma_y^2 = \frac{\Sigma fy^2}{n} - \left(\frac{\Sigma fy}{N}\right)^2 = \frac{2095}{450} - \left(\frac{261}{450}\right)^2 = 4.6556 - (0.58)^2 = 4.3192$$

$$\sigma_x = d.\sigma_y = 250 \times \sqrt{4.3192} = 250 \times 2.0783 = 519.6$$

Example 6.36 *Calculate the mean and the S.D. from the data in Example 6.34 by the **Method of Summation.***

Solution *Method of Summation* (similar to that shown in the Table at p. 115) may be used to obtain the mean and the S.D. from a frequency distribution, if the class intervals are of equal width.

Table 6.16 Method of Summation for Mean and S.D.

Class Interval	*Frequency (f)*	*1st Cumulation*	*2nd Cumulation*
(1)	(2)	(3)	(4)
0–5	2	2	2
5–10	5	7	9
10–15	7	14	23
15–20	13	27	50
20–25	21	48	98
25–30	16	64	162
30–35	8	72	234
35–40	3	–	–
Total	75	234	578

Note: Col. (3) shows cumulative totals of col. (2), i.e. cumulation frequencies.
Col. (4) shows cumulative totals of col. (3).

Let x_n denote the mid-value of the last class interval, and h the common width of classes (here $x_n = 37.5$ and $h = 5$).

$$F_1 = \frac{\text{Total of 1st cumulation}}{N} = \frac{234}{75} = 3.12$$

$$F_2 = \frac{\text{Total of 2nd cumulation}}{N} = \frac{578}{75} = 7.7067$$

Then, Mean $= x_n - h.F_1 = 37.5 - 5(3.12) = 21.9$

$$\text{S.D.} = h\sqrt{2F_2 - F_1 - F_1^2} = 5\sqrt{15.4134 - 3.12 - 9.7344}$$

$$= 5\sqrt{2.5590} = 8.0$$

The results agree with those obtained in Example 6.34.

Example 6.37 *A variable X assumes two distinct values, 0 and 1. Of the total number of observations N, the fraction p of N are ones and the fraction q of N are zeros. Find the mean and standard deviation of the N observations.*

[C.U., B.Com.(Hons) '64]

Solution The variable X has two distinct values 0 and 1 with frequencies qN and pN respectively, the total frequency being N. Hence, $qN + pN = N$; or, $N(q + p) = N$; $\therefore \quad q + p = 1$.

Table 6.17 Calculations for Mean and S.D.

X	*Frequency* f	fX	fX^2
0	qN	0	0
1	pN	pN	pN
Total	N	pN	pN

Therefore, $$\bar{x} = \frac{\Sigma fX}{\Sigma f} = \frac{pN}{N} = p$$

$$\sigma^2 = \frac{\Sigma fX^2}{N} - \left(\frac{\Sigma fX}{N}\right)^2 = \frac{pN}{N} - \left(\frac{pN}{N}\right)^2$$
$$= p - p^2 = p(1 - p) = pq \text{ (since } q + p = 1)$$

$$\therefore \quad \sigma = \sqrt{pq}$$ *Ans.* Mean = p, S.D. = $\sqrt{pq}$.

6.8 S.D. OF COMPOSITE GROUP

If two groups contain n_1 and n_2 observations with means $\bar{x}_1$ and $\bar{x}_2$, and standard deviations σ_1 and σ_2 respectively, then the S.D. (σ) of the composite group is given by

$$N\sigma^2 = (n_1\sigma_1^2 + n_2\sigma_2^2) + (n_1 d_1^2 + n_2 d_2^2) \quad (6.8.1)$$

where $d_1 = \bar{x}_1 - \bar{x}$, $d_2 = \bar{x}_2 - \bar{x}$, and $\bar{x}$ is the mean of composite group, given by (5.5.1), viz. $N\bar{x} = n_1\bar{x}_1 + n_2\bar{x}_2$. (see Example 6.43)

It will usually be found convenient to arrange the data (see Table 5.18, page 125) in the following tabular form:

Table 6.18 Data for S.D. of Composite Group

Characteristics	*Groups* I	II	*Composite Group*
No. of Observations	n_1	n_2	N
Mean	$\bar{x}_1$	$\bar{x}_2$	$\bar{x}$
Standard Deviation	σ_1	σ_2	σ

In general, if there are several group, the i-th group containing n_i observations with mean $\bar{x}_i$ and S.D. σ_i, then the S.D. (σ) of the composite group is given by

$$N\sigma^2 = \Sigma n_i\sigma_i^2 + \Sigma n_i d_i^2 \quad (6.8.2)$$

where $N\bar{x} = \Sigma n_i\bar{x}_i$, $d_i = \bar{x}_i - \bar{x}$, and $N = \Sigma n_i$.

Example 6.38 *For a group of 50 boys the mean score and the standard deviation of scores on a test are 59.5 and 8.38. For a group of 40 girls the same results are 54.0 and 8.23. Find the mean and S.D. of the combined group of 90 children.*

[C.U., M.Com. '72]

Solution

Table 6.19 Data for Mean and S.D. of Combined Group

Characteristics	Group I (Boys)	Group II (Girls)	Combined Group
No. of Observations	$n_1 = 50$	$n_2 = 40$	$N = 90$
Mean	$\bar{x}_1 = 59.5$	$\bar{x}_2 = 54.0$	$\bar{x} = ?$
Standard Deviation	$\sigma_1 = 8.38$	$\sigma_2 = 8.23$	$\sigma = ?$

The formulae for mean and S.D. of combined group are

(1) $N\bar{x} = n_1\bar{x}_1 + n_2\bar{x}_2$

(2) $N\sigma^2 = (n_1\sigma_1^2 + n_2\sigma_2^2) + (n_1d_1^2 + n_2d_2^2)$

Using (1), $90\bar{x} = 50 \times 59.5 + 40 \times 54.0 = 5135.0$

$\therefore$ $\bar{x} = 5135.0/90 = 57.06$

Hence, $d_1 = 59.5 - 57.06 = 2.44$, $d_2 = 54.0 - 57.06 = -3.06$

Using (2), $90\sigma^2 = \{50 \times (8.38)^2 + 40 \times (8.23)^2\} + \{50 \times (2.44)^2 + 40 \times (-3.06)^2\}$

$= (3511.22 + 2709.32) + (297.68 + 374.54) = 6892.76$

$\sigma^2 = 6892.76/90 = 76.59, \qquad \sigma = \sqrt{76.59} = 8.75$

Ans. 57.06, 8.75

Example 6.39 *Find the standard deviation of the composite set from the means and standard deviations of its component sets as given in the following table:*

	First Set	Second Set	Third Set
Number of Observations	200	250	300
Arithmetic Mean	25	10	15
Standard Deviation	3	4	5

[B.U., B.A.(Econ) '70]

Solution

Table 6.20 Mean and S.D. of Component Sets

Characteristics	Set I	Set II	Set III	Composite Set
No. of Observations	$n_1 = 200$	$n_2 = 250$	$n_3 = 300$	$N = 750$
Mean	$\bar{x}_1 = 25$	$\bar{x}_2 = 10$	$\bar{x}_3 = 15$	$\bar{x} = ?$
Standard Deviation	$\sigma_1 = 3$	$\sigma_2 = 4$	$\sigma_2 = 5$	$\sigma = ?$

The required formulae are

(1) $N\bar{x} = n_1\bar{x}_1 + n_2\bar{x}_2 + n_2\bar{x}_3$ (where $N = n_1 + n_2 + n_2$)

(2) $N\sigma^2 = (n_1\sigma_1^2 + n_2\sigma_2^2 + n_3\sigma_3^2) + (n_1{d_1}^2 + n_2d_2^2 + n_3d_3^2)$

Putting the values in (1),

$750\bar{x} = 200 \times 25 + 250 \times 10 + 300 \times 15 = 12{,}000$

$\bar{x} = 12000/750 = 16$

So, $d_1 = \bar{x}_1 - \bar{x} = 9, \quad d_2 = \bar{x}_2 - \bar{x} = -6, \quad d_3 = \bar{x}_3 - \bar{x} = -1.$

Now using (2),

$$750\bar{\sigma}^2 = (200 \times 3^2 + 250 \times 4^2 + 300 \times 5^2) + \{200 \times 9^2 + 250\,(-6)^2 + 300\,(-1)^2\}$$
$$= (1800 + 4000 + 7500) + (16200 + 9000 + 300) = 38800$$
$$\sigma^2 = 38800/750 = 51.73; \qquad \sigma = \sqrt{51.73} = 7.19$$

Example 6.40 *The mean and the variance of a group of 100 observations are 6.5 and 3.0 respectively. 55 of these observations have mean 6.6 and standard deviation 1.5. Find the mean and the S.D. of the remaining 45 observations.*

[I.C.W.A., Dec. '75-old]

Solution We assume that the 100 observations have been split up into two groups–Group I containing 55 observations with mean ($\bar{x}_1$) = 6.6 and standard deviation (σ_1) = 1.5, and Group II containing 45 observations. The composite group of 100 observations has mean ($\bar{x}$) = 6.5 and variance (σ^2) = 3.0. It is required to find the mean ($\bar{x}_2$) and s.d. (σ_2) of Group II.

Table 6.21 Data for Mean and S.D. of Composite Group

Characteristics	*Groups* I	*Groups* II	*Composite Group*
No. of Observations	$n_1 = 55$	$n_2 = 45$	$N = 100$
Mean	$\bar{x}_1 = 6.6$	$\bar{x}_2 = ?$	$\bar{x} = 6.5$
Standard Deviation	$\sigma_1 = 1.5$	$\sigma_2 = ?$	$\sigma = \sqrt{3.0}$

Formulae for Mean and S.D. of composite group are

(i) $N\bar{x} = n_1\bar{x}_1 + n_2\bar{x}_2$

(ii) $N\sigma^2 = (n_1\sigma_1^2 + n_2\sigma_2^2) + (n_1 d_1^2 + n_2 d_2^2)$

Substituting the values in (i), $100 \times 6.5 = 55 \times 6.6 + 45\bar{x}_2$; hence,

$$\bar{x}_2 = 287/45 = 6.38$$

Therefore, $d_1 = \bar{x}_1 - \bar{x} = 6.6 - 6.5 = 0.1$

$d_2 = \bar{x}_2 - \bar{x} = 6.38 - 6.5 = -0.12$

Now, applying (ii),

$$100 \times \left(\sqrt{3.0}\right)^2 = 55 \times (1.5)^2 + 45\sigma_2^2 + 55 \times (0.1)^2 + 45\,(-0.12)^2$$

or, $300 = 123.75 + 45\sigma_2^2 + 0.55 + 0.648$

or, $45\sigma_2^2 = 300 - 123.75 - 0.55 - 0.648 = 175.052$

$\therefore$ $\sigma_2^2 = 175.052 \div 45 = 3.89$

$\sigma_2 = \sqrt{3.89} = 1.97$ *Ans.* 6.38, 1.97

Example 6.41 *In a batch of 10 children, the I.Q. of a dull boy is 36 below the average I.Q. of the other children. Show that the standard deviation of I.Q. for all the children cannot be less than 10.8. If this standard deviation is actually 11.4, determine what the s.d. will be, when the dull boy is left out.* [C.U., M.Com. '75]

Solution Let a denote the average I.Q. of the 9 children (excluding the dull boy). Then, I.Q. of the dull boy is $a - 36$. We separate the 10 children into two groups—Group I consisting of the 9 children and Group II including only the dull boy.

Table 6.22 Mean and S.D. of Composite Group

Characteristics	*Group I*	*Group II*	*Composite Group*
No. of Observations	$n_1 = 9$	$n_2 = 1$	$N = 10$
Mean	$\bar{x}_1 = a$	$\bar{x}_2 = a - 36$	$\bar{x}$
Standard Deviation	σ_1	$\sigma_2 = 0$	σ

Since the s.d. of one observation is zero, therefore $\sigma_2 = 0$.

(1) $N\bar{x} = n_1\bar{x}_1 + n_2\bar{x}_2$

or, $10\bar{x} = 9a + 1(a - 36) = 10a - 36$

$\therefore$ $\bar{x} = a - 3.6$

$d_1 = \bar{x}_1 - \bar{x} = a - (a - 3.6) = 3.6$

$d_2 = \bar{x}_2 - \bar{x} = (a - 36) - (a - 3.6) = -32.4$

(2) $N\sigma^2 = (n_1\sigma_1^2 + n_2\sigma_2^2) + (n_1d_1^2 + n_2d_2^2)$

or, $10\sigma^2 = (9\sigma_1^2 + 1\times 0) + \{9(3.6)^2 + 1\times(-32.4)^2\}$

or, $10\sigma^2 = 9\sigma_1^2 + 1166.4$ ... (i)

or, $\sigma^2 = 0.9\sigma_1^2 + 116.64$

$\therefore$ $\sigma^2 \geq 116.64$ (since σ_1^2 cannot be negative)

or, $\sigma \geq \sqrt{116.64}$, i.e. 10.8

If $\sigma = 11.4$, then from (i), $10 \times (11.4)^2 = 9\sigma_1^2 + 1166.4$

Solving, $\sigma = \sqrt{14.8} = 3.85$

Example 6.42 *Particulars regarding income of two villages are given below:*

	Village X	*Village Y*
Number of People	600	500
Average Income in Rs	175	186
Variance of Income in Rs	100	81

(a) In which village is the variation in income greater? (b) What is the total income of both villages put together? (c) What is the average income of the people in X and Y villages put together? (d) What is the combined Standard Deviation of the villages X and Y put together? [C.A., May '74]

Solution (a) Standard Deviation (S.D.) provides a measure of the variation or discrepancy among observations. Since S.D. = $\sqrt{\text{Variance}}$, the S.D.s of villages *X* and *Y* are respectively $\sigma_1 = \sqrt{100} = 10$, $\sigma_2 = \sqrt{81} = 9$. Since σ_1 is larger than σ_2, we conclude that the variation in income is greater in village *X*.

(b) Total incomes of villages *X* and *Y* are respectively 600 × 175 = 105,000 Rs and 500 × 186 = 93,000 Rs. Hence the combined total income of both villages put together is 105,000 + 93,000 = 198,000 Rs.

(c) The total number of people $N = n_1 + n_2 = 600 + 500 = 1100$, whose total income is Rs 198,000. Average income = 198,000 ÷ 1100 = 180 Rs (We could also apply the formula $N\bar{x} = n_1\bar{x}_1 + n_2\bar{x}_2$).

(d) We use the formula $N\sigma^2 = (n_1\sigma_1^2 + n_2\sigma_2^2) + (n_1d_1^2 + n_2d_2^2)$.

Here $d_1 = \bar{x}_1 - \bar{x} = 175 - 180 = -5$, $d_2 = \bar{x}_2 - \bar{x} = 186 - 180 = 6$.

$\therefore$ $1100\sigma^2 = (600 \times 100 + 500 \times 81) + (600 \times 25 + 500 \times 36) = 133{,}500$

$\sigma^2 = 133500/1100 = 121.36$; $\sigma = \sqrt{121.36} = 11.02$ Rs

Ans. (a) Village *X*, (b) Rs 198,000, (c) Rs 180, (d) Rs 11.02

Example 6.43 *If n_1, σ_1 and $\bar{x}_1$ be the number of observations, standard deviation and mean of a set of observations, and n_2, σ_2, and $\bar{x}_2$ be those for a second set of observations, prove that the standard deviation σ of the combined set of $(n_1 + n_2)$ observations is given by*

$$(n_1 + n_2)\sigma^2 = n_1\sigma_1^2 + n_2\sigma_2^2 + n_1d_1^2 + n_2d_2^2$$

where $d_1 = \bar{x}_1 - m$, $d_2 = \bar{x}_2 - m$, *and* $m = (n_1\bar{x}_1 + n_2\bar{x}_2)/(n_1 + n_2)$.

[C.U., B.A.(Econ) '68, '75]

Solution Let x_i $(i = 1, 2, ..., n_1)$ and x_j' $(j = 1, 2, ..., n_2)$ be the observations in the 1st set and 2nd set respectively. By definition

$$\bar{x}_1 = \sum_i x_i / n_1 \qquad \sigma_1^2 = \sum_i (x_i - \bar{x}_1)^2/n_1$$

$$\bar{x}_2 = \sum_j x_j' \Big/ n_2 \qquad \sigma_2^2 = \sum_j (x_j' - \bar{x}_2)^2 \Big/ n_2$$

By cross-multiplication, we get

$$\left.\begin{aligned} \sum_i x_i &= n_1\bar{x}_1 & \sum_i (x_i - \bar{x}_1)^2 &= n_1\sigma_1^2 \\ \sum_j x_j &= n_2\bar{x}_2 & \sum_j (x_j' - \bar{x}_2)^2 &= n_2\sigma_2^2 \end{aligned}\right\} \quad \dots (1)$$

Now considering the combined set of $(n_1 + n_2)$ observations x_i and x_i', the mean (m) and standard deviation (σ) are given by

$$m = \left(\sum_i x_i + \sum_j x_j'\right) \Bigg/ (n_1 + n_2)$$

$$= (n_1\bar{x}_1 + n_2\bar{x}_2)/(n_1 + n_2), \text{ by (1)}$$

and

$$\sigma^2 = \left\{\sum_i (x_i - m)^2 + \sum_j \left(x_j' - m\right)^2\right\} \Bigg/ (n_1 + n_2) \quad \dots (2)$$

But, $\quad x_i - m = (x_i - \bar{x}_1) + (\bar{x}_1 - m) = (x_i - \bar{x}_1) + d_1$, where $d_1 = \bar{x}_1 - m$

$$\therefore \quad \sum_i (x_i - m)^2 = \sum_i \{(x_i - \bar{x}_1) + d_1\}^2,$$

$$= \sum_i \{(x_i - \bar{x}_1)^2 + d_1^2 + 2(x_i - \bar{x}_1)d_1\}$$

$$= \sum_i (x_i - \bar{x}_1)^2 + \sum_i d_1^2 + \sum_i 2(x_i - \bar{x}_1)\, d_1; \quad \text{by (3.15.2)}$$

$$= \sum_i (x_i - \bar{x}_1)^2 + n_1d_1^2 + 2d_1 \sum_i (x_i - \bar{x}_1),$$

by (3.15.4) and (3.15.3)

$$= n_1\sigma_1^2 + n_1d_1^2 + 2d_1 \times 0$$

$$= n_1\sigma_2^2 + n_1d_1^2$$

because, $\quad \sum_i (x_i - \bar{x}_1) = \sum_i x_i - \sum_i \bar{x}_1 = n_1\bar{x}_1 - n_1\bar{x}_1 = 0.$

Similarly, $\quad \sum_j (x_j' - m)^2 = \sum_j \{(x_j' - \bar{x}_2) + d_2\}^2 = n_2\sigma_2^2 + n_2d_2^2.$

Substituting in (2),

$$\sigma^2 = \{(n_1\sigma_1^2 + n_1d_1^2) + (n_2\sigma_2^2 + n_2d_2^2)\}/(n_1 + n_2)$$

i.e. $\quad (n_1 + n_2)\, \sigma^2 = n_1\sigma_1^2 + n_2\sigma_2^2 + n_1d_1^2 + n_2d_2^2$

Example 6.44 *Show that the combined standard deviation of two distributions pooled together is given by the expression*

$$NS^2 = n_1 s_1^2 + n_2 s_2^2 + \frac{n_1 n_2}{N}(\bar{x}_1 - \bar{x}_2)^2$$

Where the symbols have their usual meaning.

[C.U., B.Com.(Hons) '67; B.A.(Econ) '73]

Solution Here n_1, s_1 and $\bar{x}_1$ represent the number of observations, S.D. and mean of the first distribution; n_2, s_2, $\bar{x}_2$ those of the second distribution; and N, S represent the total number of observations and S.D. of the combined distribution respectively. Obviously, $N = n_1 + n_2$.

From Example 6.43, we have (changing the notations)

$$NS^2 = n_1 s_1^2 + n_2 s_2^2 + n_1 d_1^2 + n_2 d_2^2$$

But, $$d_1 = \bar{x}_1 - m = \bar{x}_1 - \frac{n_1\bar{x}_1 + n_2\bar{x}_2}{n_1 + n_2} = \frac{n_2(\bar{x}_2 - \bar{x}_2)}{n_1 + n_2}$$

and similarly, $$d_2 = \frac{n_1(\bar{x}_2 + \bar{x}_1)}{n_1 + n_2}$$

$$\therefore \quad n_1 d_1^2 + n_2 d_2^2 = n_1 \frac{n_2^2(\bar{x}_1 - \bar{x}_2)^2}{(n_1 + n_2)^2} + n_2 \frac{n_1^2(\bar{x}_2 - \bar{x}_1)^2}{(n_1 + n_2)^2}$$

$$= \frac{n_1 n_2^2(\bar{x}_1 - \bar{x}_2)^2 + n_1^2 n_2(\bar{x}_1 - \bar{x}_2)^2}{(n_1 + n_2)^2}$$

$$= \frac{n_1 n_2(\bar{x}_1 - \bar{x}_2)^2}{(n_1 + n_2)^2}(n_2 + n_1)$$

$$= \frac{n_1 n_2(\bar{x}_1 - \bar{x}_2)^2}{n_1 + n_2} = \frac{n_1 n_2}{N}(\bar{x}_1 - \bar{x}_2)^2$$

Hence, the result.

6.9 RELATION BETWEEN S.D. AND OTHER MEASURES

Many of the distributions occuring in practice have frequency curves of the symmetrical bell shaped type. Such frequency curves may be regarded as approximations to an important curve, known as "Normal" curve (Chapter 12). Some of the properties of this curve therefore hold approximately for observed frequency distributions. For instance, about 99.7% of the area under normal curve lies between *mean* ± 3σ, so that the effective range is 6σ. Hence, for any distribution the S.D. may be taken to be about one-sixth of the range.

Again, approximately 68% and 95% of area under normal curve are included between (*mean* ± σ) and (*mean* ± 2σ) respectively. For moderately asymmetrical distributions, therefore, about two-thirds of the total number of observations are expected to lie between *mean* ± *s.d.* (see Example 6.34).

It can be proved (Example 6.45) that for any given set of observations S.D. is never less than M.D.

$$S.D. \geq M.D. \qquad (6.9.1)$$

For the normal distribution, Q.D. = 0.67σ and M.D. = 0.80σ (appr.). These results form the very basis of the following approximate relations which hold for any distribution, not far from normal:

$$\text{Range} = 6\sigma, \qquad \text{Q.D.} = \frac{2}{3}\sigma, \qquad \text{M.D.} = \frac{4}{5}\sigma \tag{6.9.2}$$

Example 6.45 *Show that* $\sum_{i=1}^{n} x_i^2 \geq \left(\sum_{i=1}^{n} x_i\right)^2 \Big/ n$, *where* $x_1, x_2, \ldots x_n$ *are any given values of the variable x. Hence, or otherwise, prove that the mean deviation about the mean cannot exceed the standard deviation. When will the two be equal?*

[B.U., B.A.(Econ) '74]

Solution We have shown (Example 6.18) that $\Sigma(x_i - \bar{x})^2 = \Sigma x_i^2 - n\bar{x}^2$. But the L.H.S. being the sum of the squares cannot be negative; i.e. $\Sigma(x_i - \bar{x})^2 \geq 0$. Hence

$$\Sigma x_i^2 \geq n\bar{x}^2, \qquad \text{or,} \qquad \Sigma x_i^2 \geq (\Sigma x_i)^2/n$$

The two sides are equal, only when all x_i's are equal.

The above inequality holds for any given set of real numbers $x_1, x_2, \ldots x_n$. In particular, if we replace x_i by $|x_i - \bar{x}|$, then

$$\Sigma\{|x_i - \bar{x}|\}^2 \geq \{\Sigma|x_i - \bar{x}|\}^2/n$$

or, $$\Sigma(x_i - \bar{x})^2 \geq \{\Sigma|x_i - \bar{x}|\}^2/n$$

or, $$\Sigma(x_i - \bar{x})^2/n \geq \{\Sigma|x_i - \bar{x}|/n\}^2$$

or, $$(\text{S.D.})^2 \geq (\text{M.D.})^2$$

or, $$\text{S.D.} \geq \text{M.D.}$$

The equality sign holds only when all $|x_i - \bar{x}|$ are equal, i.e. all the observations x_i are equal (then S.D. = M.D. = 0)

6.10 RELATIVE MEASURES OF DISPERSION

The four measures—Range, Quartile Deviation, Mean Deviation and Standard Deviation, are expressed in the same units as the original observations, and are called *Absolute measures* of variability. So, they cannot be used for comparing the variability of two or more distributions given in different units. In order to meet such situations, the *Relative measures* of variability have been introduced which are independent of the units of measurement. There are 3 such measures:

(i) *Coefficient of Variation*

$$= 100 \times \frac{\text{Standard Deviation}}{\text{Mean}}$$

(ii) *Coefficient of Quartile Deviation*

$$= 100 \times \frac{\text{Quartile Deviation}}{\text{Median}}$$

(iii) *Coefficient of Mean Deviation*

$$= 100 \times \frac{\text{Mean Deviation}}{\text{Mean or Median}}$$

Among these, coefficient of variation is the most important and is used in almost all cases.

The relative measures of variability may also be used to measure the precision of observations, although given in the same unit.

Example 6.46 *Define 'coefficient of variation'. What are the special uses of this measure?* [B.U., B.A. (Econ) '76; D.M. '73; D.S.W. '77; M.B.A. '79]

Solution *Coefficient of Variation* is defined by the formula:

$$\text{Coefficient of Variation} = \frac{\text{Standard Deviation}}{\text{Mean}} \times 100$$

It is a relative measure of dispersion and is independent of the unit of measurement.

Uses—(i) The absolute measures cannot be used for comparing the dispersion of two or more distributions given in different units. The coefficient of variation is a pure number and as such may be employed for the purpose (Example 6.54).

(ii) Again, where the means of distributions are widely different, although given in the same unit, a relative measure will be more logical than an absolute measure of dispersion (Examples 6.48, 6.55, 6.56).

(iii) It may also be used to compare the precision of two or more sets of measurements. If repeated measurements of the length of a 10 ft. rod and also of a 100 ft. long road (the exact lengths not given) are taken, it will be found that the observations will vary among themselves. If both sets of measurements are taken with the same care, we expect that S.D. of the former will be less than that of the latter. The same S.D. in both sets will imply that the first set of measurements were less precise than the second. The coefficient of variation will however reveal the correct position.

Example 6.47 *If the record of amounts spent by you on various evenings gives a coefficient of variation of 50 per cent and a standard deviation of Rs 4.00, what was your average expenditure?* [D.S.W., Nov '73]

Solution By definition, $\text{Coefficient of Variation} = \frac{\text{S.D.}}{\text{Mean}} \times 100$

or, $50 = \frac{4.00 \text{ Rs}}{\text{Mean}} \times 100$; Solving, Mean = 8.00 Rs

Example 6.48 *From some financial statistics, it is found that the monthly average Electricity Charges was Rs 2,460 and S.D. Rs 120. The monthly average Direct Wages was Rs 42,000 and S.D. Rs 1,200. State which is the more variable with proper reasons.* [C.A., Nov. '67]

Solution The given data are (in Rs)

	Electricity Charges	*Direct Wages*
Mean	2,460	42,000
S.D.	120	1,200

Since the means in the two cases differ widely, although given in the same unit, viz. Rs, it will be inappropriate to employ S.D. for comparing the variability. The coefficient of variation (C.V.) will be more logical here.

$$\text{C.V. (for Electricity Charges)} = \frac{\text{Rs } 120}{\text{Rs } 2460} \times 100 = 4.9$$

$$\text{C.V. (for Direct Wages)} = \frac{\text{Rs } 1200}{\text{Rs } 42000} \times 100 = 2.9$$

Since the coefficient of variation for Electricity Charges is the larger, Electricity Charges are more variable than Direct Wages.

Example 6.49 *In an industrial establishment, the coefficients of variation of wages of male and female workers were 55% and 70% respectively. The standard deviations were Rs 22.00 and Rs 15.40 respectively. Calculate the combined average wages for all the workers, if 80% of the workers were males.* [C.A., Nov. '75]

Solution Let $\bar{x}_1$ and $\bar{x}_2$ denote the mean wages of male and female workers respectively. Then, as in Example 6.47,

$$55 = \frac{\text{Rs } 22.00}{\bar{x}_1} \times 100 \quad \text{and} \quad 70 = \frac{\text{Rs } 15.40}{\bar{x}_2} \times 100.$$

Solving we get $\bar{x}_1$ = Rs 40.00, $\bar{x}_2$ = Rs 22.00.

If the total number of workers be 100, the numbers of male and female workers are $n_1 = 80$ and $n_2 = 20$. Now applying formula (5.5.1) for mean of a composite group (p. 146), $100\bar{x} - 80 \times 40.00 + 20 \times 22.00$, from which $\bar{x}$ = 36.40 Rs *Ans.* Rs 36.40

Example 6.50 *Find the coefficient of variation from the following:*

Weight (gm)	*110–119*	*120–129*	*130–139*	*140–149*	*150–159*
Frequency	*5*	*7*	*12*	*20*	*16*
			160–169	*170–179*	*180–189*
			10	*7*	*3*

[I.C.W.A., Dec. '76]

Solution In order to find the coefficient of variation, we have to calculate both the mean and the standard deviation.

Table 6.23 Calculations for Mean and S.D.

Weight (gm)	*Frequency (f)*	*Mid-value (x)*	$y = \frac{x - 144.5}{10}$	*fy*	*fy*^{2}
110–119	5	114.5	–3	–15	45
120–129	7	124.5	–2	–14	28
130–139	12	134.5	–1	–12	12
140–149	20	144.5	0	0	0
150–159	16	154.5	1	16	16
160–169	10	164.5	2	20	40
170–179	7	174.5	3	21	63
180–189	3	184.5	4	12	48
Total	80	–	–	28	252

Proceeding as in Example 6.34,

$$\text{Mean } (\bar{x}) = 144.5 + 10\ (28/80) = 148.0 \text{ gm.}$$

$$\sigma_y^2 = 252/80 - (28/80)^2 = 3.15 - (0.35)^2 = 3.0275$$

$$\therefore \quad \sigma_y = \sqrt{3.0275} = 1.74; \qquad \sigma_x = 10 \times 1.74 = 17.4 \text{ gm.}$$

Coefficient of variation = (17.4 + 148.0) × 100 = 12

Example 6.51 *The scores of two batsmen, A and B, in ten innings during a certain season, are as under:*

A	*32*	*28*	*47*	*63*	*71*	*39*	*10*	*60*	*96*	*14*
B	*19*	*31*	*48*	*53*	*67*	*90*	*10*	*62*	*40*	*80*

Find which of the batsmen is more consistent in scoring. [I.C.W.A., Jan. '70]

Solution We use Coefficient of Variation for measuring dispersion. The batsman with the smaller dispersion is more consistent. The mean and S.D. for simple series are obtained by using formulae (5.4.1) and (6.7.3): If $y = x - c$, then (i) $\bar{x} = c + y$ –, (ii) $\sigma_x = \sigma_y$

Table 6.24 Calculations for Mean and S.D.

Cricketer A			Cricketer B		
x	$y = x^{-50}$	y^2	x	$y = x^{-50}$	y^2
32	–18	324	19	–31	961
28	–22	484	31	–19	361
47	–3	9	48	–2	4
63	13	169	53	3	9
71	21	441	67	17	289
39	–11	121	90	40	1600
10	–40	1600	10	–40	1600
60	10	100	62	12	144
96	46	2116	40	–10	100
14	–36	1296	80	30	900
Total	–40	6660	Total	0	5968

For Cricketer A:

$$\text{Mean} = 50 + (-40/10) = 46$$

$$\text{S.D.} = \sqrt{\frac{6660}{10} - \left(\frac{-40}{10}\right)^2} = \sqrt{650} = 25.5$$

$$\therefore \quad \text{C.V.} = \frac{25.5}{46} \times 100 = 55$$

For Cricketer B:

$$\text{Mean} = 50 + (0/10) = 50$$

$$\text{S.D.} = \sqrt{\frac{5968}{10} - \left(\frac{0}{10}\right)^2} = \sqrt{596.8} = 24.4$$

$$\therefore \quad \text{C.V.} = \frac{24.4}{50} \times 100 = 49$$

Since for Cricketer B, the coefficient of variation is smaller, he is more consistent.

Ans. B.

Example 6.52 *From the data given below, state which series is more variable (Use standard deviation):*

Variable	*Series A*	*Series B*
10–20	*10*	*18*
20–30	*18*	*22*
30–40	*32*	*40*
40–50	*40*	*32*
50–60	*22*	*18*
60–70	*18*	*10*

[C.A., Nov. '71]

Solution Means and standard deviations for the two series are (calculations have not been shown here) as follows:

Series A: Mean = 42.14, S.D. = 14.06

Series B: Mean = 37.86, S.D. = 14.06

If we use standard deviation directly, it is not possible to decide which series is more variable; because the standard deviations are equal. However, since the means of the two Series are different, a relative measure of dispersion (using standard deviation), viz. Coefficient of Variation, would be appropriate.

$$\text{C.V. (for Series A)} = \frac{14.06}{42.14} \times 100 = 33$$

$$\text{C.V. (for Series B)} = \frac{14.06}{37.86} \times 100 = 37$$

Since the Coefficient of Variation for Series B is larger than that for Series A, the former is more variable.

Ans. Series B

Example 6.53 *Given the following results relating to two groups containing 20 and 30 observations, calculate the coefficient of variation of all the 50 observations by combining both the groups:*

	Groups	
	I	*II*
Σx	*45*	*55*
Σx^2	*118*	*132*

[I.C.W.A., Jan. '68]

Solution For the combined group,

$n = 20 + 30 = 50;$ $\Sigma x = 45 + 55 = 100;$

$\Sigma x^2 = 118 + 132 = 250;$ $\therefore$ Mean $= \Sigma x/n = 100/50 = 2.$

$$\sigma^2 = \frac{\Sigma x^2}{n} - \left(\frac{\Sigma x}{n}\right)^2 = \frac{250}{50} - \left(\frac{100}{50}\right)^2 = 5 - 4 = 1. \quad \therefore \quad \sigma = \sqrt{1} = 1.$$

$$\therefore \text{ Coefficient of Variation} = \frac{\text{S.D.}}{\text{Mean}} \times 100 = \frac{1}{2} \times 100 = 50.$$

Ans. 50.

Example 6.54 *The following frequency distributions have been constructed from measurements of heights (in inches) and weights (in lbs.) of the same group of adult persons. Which is the more variable, height or weight?*

Mid-point (height)	*Frequency*	*Mid-point (weight)*	*Frequency*
60	10	84	10
62	10	94	10
64	30	104	15
66	30	114	20
68	15	124	15
70	5	134	10
		144	10
	100	154	10
			100

[I.C.W.A., July '68]

Solution The variability of the two distributions, given in different units, can only be compared by means of a relative measure. We calculate their coefficients of variation, and therefore have to find their means and S.Ds.

Table 6.25 Calculations for Mean and S.D. (Height)

Mid-point (X)	f	$x = \frac{X-64}{2}$	fx	fx^2
60	10	–2	–20	40
62	10	–1	–10	10
64	30	0	0	0
66	30	1	30	30
68	15	2	30	60
70	5	3	15	45
Total	100	–	45	185

$$\bar{X} = c + d\bar{x} = 64 + 2 \times \frac{45}{100} = 64.9 \text{ inches}$$

$$\sigma_X = d\sqrt{\frac{\Sigma fx^2}{N} - \left(\frac{\Sigma fx}{N}\right)^2} = 2 \times \sqrt{\frac{185}{100} - \left(\frac{45}{100}\right)^2} = 2.57 \text{ inches.}$$

Table 6.26 Calculations for Mean and S.D. (Weight)

Mid-point (*Y*)	*f*	$y = \frac{Y - 114}{10}$	*fy*	fy^2
80	10	–3	–30	90
94	10	–2	–20	40
104	15	–1	–15	15
114	20	0	0	0
124	15	1	15	15
134	10	2	20	40
144	10	3	30	90
154	10	4	40	160
Total	100	–	40	450

$$\bar{Y} = c + d\bar{y} = 114 + \frac{40}{100} \times 10 = 118 \text{ lbs.}$$

$$\sigma_y = d\sqrt{\frac{\Sigma fy^2}{N} - \left(\frac{\Sigma fy}{N}\right)^2} = 10 \times \sqrt{\frac{450}{100} - \left(\frac{40}{100}\right)^2} = 20.83 \text{ lbs.}$$

$$\text{C.V. (for height)} = \frac{2.57 \text{ inches}}{64.9 \text{ inches}} \times 100 = 4$$

$$\text{C.V. (for weight)} = \frac{20.83 \text{ lbs.}}{118 \text{ lbs.}} \times 100 = 18$$

Since, the coefficient of variation for weight is greater than that for height, weight is the more variable.

Example 6.55 *You are given two variables A and B. Using Quartile Deviation, state which is more variable.*

A.	*Mid point*	*15*	*20*	*25*	*30*	*35*	*40*	*45*	
	Frequency	*15*	*33*	*56*	*103*	*40*	*32*	*10*	
B.	*Mid point*	*100*	*150*	*200*	*250*	*300*	*350*	*400*	*450*
	Frequency	*340*	*492*	*890*	*1420*	*620*	*360*	*187*	*140*

[C.A., Nov. '70]

Solution It is seen that averages of the two distributions differ widely (Averages being measures of central tendency will lie near the middle of the distributions, viz. near about 30 and 250 respectively). It will, therefore, be inappropriate to compare the variability of distributions based on an absolute measure of dispersion. A relative measure will be more logical here. Using Quartile Deviation, the relative measure is

$$\text{Coefficient of Quartile Deviation} = 100 \times \frac{\text{Quartile Deviation}}{\text{Median}}$$

$$= 100 \times \frac{Q_3 - Q_1}{2\,(\text{Median})}$$

Medians and quartiles (Q_1 and Q_3) for the two distributions are (Calculations have not been shown here. Note that here mid-points are given, and therefore it is necessary to obtain the class boundaries for the calculation):

For A: $Q_1 = 24.67$ Median = 29.47 $Q_3 = 33.72$

For B: $Q_1 = 190.74$ Median = 242.69 $Q_3 = 290.71$

Coefficient of Quartile Deviation (for A)

$$= 100 \times \frac{33.72 - 24.67}{2(29.47)} = 15$$

Coefficient of Quartile Deviation (for B)

$$= 100 \times \frac{290.71 - 190.74}{2(242.69)} = 21$$

Since the Coefficient of Quartile Deviation for B is larger than that for A, B is more variable (i.e. has larger variability) than A.

Example 6.56 *With median as the base, calculate mean deviation and compare the variability of the two series A and B.*

Series A	*3484*	*4572*	*4124*	*3682*	*5624*	*4388*	*3680*	*4308*
Series B	*487*	*508*	*620*	*382*	*408*	*266*	*186*	*218*

[C.A., May '73]

Solution (As in the previous Example, the averages of the two series differ widely, as is evident from the sizes of items. Hence, for comparing variability, we have to employ a relative measure). Using *Mean Deviation from median*, the relative measure of dispersion is

$$\text{Coefficient of Mean Deviation} = \frac{\text{M.D. about median}}{\text{Median}} \times 100$$

For the determination of Median, we arrange the values of each series in order of magnitude. Since there are 8 observations in each series, the average of the 4th and the 5th items gives the median. The ordered values are 3484, 3680, 3682, 4124, 4308, 4388, 4572, 5624 for series A, and 186, 218, 266, 382, 408, 487, 508, 620 for series B.

Median for series A = (4124 + 4308)/2 = 4216

Median for series B = (382 + 408)/2 = 395

Table 6.27 Calculations for M.D. from Median

Series A		*Series B*	
x	$\lvert x - 4216 \rvert$	*x*	$\lvert x - 395 \rvert$
3484	732	487	92
4572	356	508	113
4124	92	620	225
3682	534	382	13
5624	1408	408	13
4388	172	266	129
3680	536	186	209
4308	92	218	177
Total	3922	Total	971

Using formula (6.4.1), Mean Deviations about median are 3922/8 = 490.25 for Series A, and 971/8 = 121.375 for Series B.

The Coefficients of Mean Deviation are

$$\frac{490.25}{4216} \times 100 = 12 \text{ (for Series A)}$$

$$\frac{121.375}{395} \times 100 = 31 \text{ (for Series B)}$$

Since the coefficient of M.D. for Series B, viz. 31, is the larger, we conclude that Scries B has greater variability.

6.11 LORENZ CURVE

Lorenz Curve is a diagram for showing the dispersion of a group. It is, in effect, a cumulative percentage curve, combining the percentage of items under review with the percentage of the factor (say, wealth distribution) among the items. If wealth were equally distributed among the people, the curve would be the straight line *ACB,* connecting the two extremes of the scales. In practice, however, curves like *ADB* are obtained. The less the area between the Lorenz curve *ADB* and the diagonal straight line *ACB*, the greater is the homogeneity in the distribution of wealth, i.e. less is the dispersion. On the other hand, the larger the area, the larger is the percentage of poor people and greater is the concentration of wealth in the hands of a few, Lorenz curve does not yield a numerical measure. It is, in this respect, inferior to the familiar measures of dispersion, e.g. Range, Standard deviation, etc. But the advantage is that it affords a picture of the dispersion at a glance. Lorenz curve is useful in such studies as the distribution of land, wages and income among the population of a country or the distribution of profits over different groups in business.

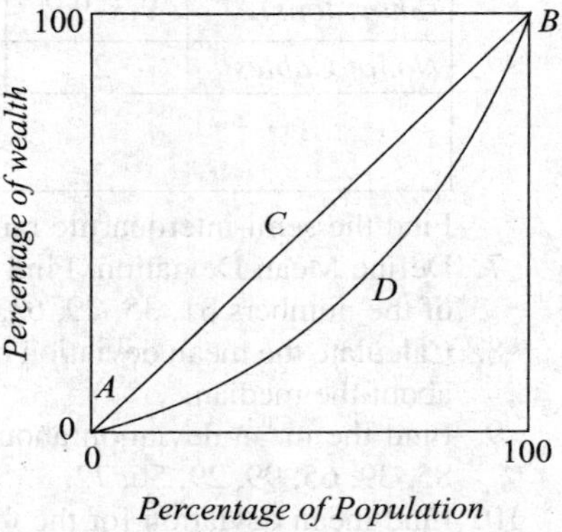

Percentage of Population

Fig. 6.1 Lorenz Curve

EXERCISES

1. If each item is reduced by 10, what effect would this have on (i) the arithmetic mean, (ii) the range, and (iii) the standard deviation? *[C.A., May '64]*
2. If the variables are increased or decreased (i) by the same amount, (ii) by the same proportion, what will be the effect on standard deviation? *[B.U., B.A.(Econ) '68; I.C.W.A., Dec. '76–old]*
3. (i) If the first quartile is 142 and the semi-interquartile range is 18, what is the third quartile?
   (ii) The coefficient of variation is 40 and the mean is 30; find the standard deviation. *[C.U., M.Com. '69]*
4. Find out the range of the following data:

Height (inches)	60–62	63–65	66–68	69–71	72–74
No. of Students	8	27	42	18	5

[D.S.W., Nov.'73]

5. Calculate the quartile deviation and its coefficient from the following:

Cl. Interval	10–15	15–20	20–25	25–30	30–40	40–50	50–60	60–70	Total
Frequency	4	12	16	22	10	8	6	4	82

[C.A., Nov.'76]

6. The following table shows the distribution of the maximum loads supported by certain cables produced by a company:

Maximum load (Short tons)	9.3–9.7	9.8–10.2	10.3–10.7	10.8–11.2	11.3–11.7
No. of Cables	2	5	12	17	14
			11.8–12.2	12.3–12.7	12.8–13.2
			6	3	1

Find the semi-interquartile range. *[D.S.W., '68]*

7. Define Mean Deviation. Find the mean deviation about the arithmetic mean of the numbers 31, 35, 29, 63, 55, 72, 37. *[B.U., B.Com.'76]*
8. Calculate the mean deviation of the following: 13, 84, 68, 24, 96, 139, 84, 27, about the median. *[B.U., B.Com.'77]*
9. Find the mean deviation about median from the following data: 46, 79, 26, 85, 39, 65, 99, 29, 56, 72. *[C.U., B.Com. '77]*
10. Find mean deviation for the following frequency distribution:

Variable	3	5	7	9	11	13
Frequency	2	7	10	9	5	1

[D.M.(Suppl.) '77]

11. Calculate the mean deviation from the following data, relating to heights (to the nearest inch) of 100 children:

Height (inches)	60	61	62	63	64	65	66	67	68
No. of Children	2	0	15	29	25	12	10	4	3

[I.C.W.A., Jan.'73]

12. Calculate mean deviation from median from the following:

Class Intervals	2–4	4–6	6–8	8–10
Frequencies	3	4	2	1

[I.C.W.A., Dec. '77]

13. In a certain distribution of $N = 25$ measurements it was found that $\bar{x} = 56$ inches and S.D. = 2 inches. After these results were computed it was discovered that a mistake had been made in one of the measurements which was recorded as 64 inches. Find the mean and standard deviation, if the incorrect measurement is omitted. *[C.U., M.Com. '62]*
14. The mean and S.D. of a group of 25 observations were found to be 30 and 3 respectively. After the calculations were made, it was found that two of the observations were incorrect, which were recorded as 29 and 31. Find the mean and S.D. if the incorrect observations are excluded. *[C.U., B.Com.(Hons) '68]*
15. The mean and standard deviation of a group of 100 observations were found to be 20 and 3 respectively. After the calculations were made it was found that

three of the observations were incorrect which were recorded as 21, 21 and 18. Find the mean and s.d. if the incorrect observations are omitted. *[C.U., B.A.(Econ) '65]*

16. The mean and the standard deviation of a sample of size 10 were found to be 9.5 and 2.5 respectively. Later on, an additional observation became available. This was 15.0 and was included in the original sample. Find the mean and the standard deviation of the 11 observations. *[I.C.W.A.. June '75]*
17. The mean and the s.d. of a sample of 100 observations were calculated as 40 and 5.1 respectively, by a student who by mistake took one observation as 50 instead of 40. Calculate the correct s.d. *[I.C.W.A., Dec. '76-old]*
18. For a distribution of 280 observations mean and standard deviation were found to be 54 and 3 respectively. On checking it was discovered that two observations, which should correctly read as 62 and 82, had been wrongly recorded as 64 and 80 respectively. Calculate the correct values of mean and s.d. *[C.U., B.A.(Econ) '75]*
19. $\bar{X}$ is the mean of X_1, X_2 and X_3. If x_1, x_2, x_3, are the deviations of X_1, X_2, X_3, from $\bar{X}$ respectively, prove that
$$x_1^2 + x_2^2 + x_3^2 = X_1^2 + X_2^2 + X_3^2 - 3\bar{X}^2.$$
[C.U., B.A (Econ) '69]
20. Let x_1, x_2, ..., xn be a set of observations. Suppose we compute $y_i = a + bx_i$ (i = 1, 2, ... n), where a and b are constants. Express the s.d. of the y's in terms of the s.d. of the x's and comment on the relation between the two. *[C.U., B.A.(Econ) '78]*
21. 'If the mean and the standard deviation of n observations $x_1, x_2, ..., x_n$ be $\bar{x}$ and σ respectively, then the mean and the standard deviation of $-x_1, -x_2, ..., -x_n$ will be $-\bar{x}$ and $-\sigma$ respectively'—Comment. *[I.C.W.A., June '75]*
22. If d^2 = mean square deviation about A, σ = standard deviation, and $A - \bar{x} = a$, then show that $d^2 = \sigma^2 + a^2$. *[D.S.W., '71]*
23. Calculate the standard deviation from the following series: 20, 85, 120, 60, 40. *[B.U.. B.Com. '71]*
24. Find the standard deviation of weights (to the nearest pound) of 15 students given below: 138, 156, 147, 115, 145, 132, 163, 158, 130, 123, 103, 109, 100, 105, 106. *[B.U., B.A.(Econ) '72]*
25. Calculate the s.d. of the following observations:

    240.12, 240.13, 240.15, 240.12, 240.17,
    240.15, 240.17, 240.16, 240.22, 240.21. *[I.C.W.A., June '76]*
26. Find the standard deviation for the distribution given below:

x	1	2	3	4	5	6	7
Frequency	10	20	30	35	14	10	2

[Dip. Management, '67]

27. Find the s.d. from the following table giving the age distribution of 540 members of a Parliament:

Age in Years	30	40	50	60	70
No. of Members	64	132	153	140	51

[C.U., B.Com. '78]

28. Find the s.d. from the following frequency distribution:

Wt.(lbs.)	120–124	125–129	130–134	135–139	140–144	145–149	Total
No. of Boys	12	25	28	15	12	8	100

[B.U., B.Com, '74]

29. The weights of a certain product produced in a factory are given below from a sample of 121 articles :

Weight (oz.)	3.0–3.1	3.1–3.2	3.2–3.3	3.3–3.4	3.4–3.5	3.5–3.6
Frequency	5	10	12	20	25	18
	3.6–3.7	3.7–3.8	3.8–3.9	3.9–4.0		
	10	8	8	5		

Calculate the arithmetic mean and standard deviation. *[C.A., May '72]*

30. Compute the standard deviation of the following data:

Weekly Wages in Rs	*Number of Men*
30 and under 40	8
40 and under 50	12
50 and under 60	6
60 and under 70	4
70 and under 80	10

[B.U., B,Com. '73]

31. Find the standard deviation of the following distribution:

Turnover (Rs '000 p.a.)	50–100	100–150	150–200	200–250	250–300	300–350	350–400
No. of Firms	5	8	9	12	18	23	17

[C.U., B.Com. '74]

32. Compute the arithmetic mean, standard deviation and mean deviation about the mean for the following data:

Scores	4–5	6–7	8–9	10–11	12–13	14–15	Total
f	4	10	20	15	8	3	60

[I.C.W.A., Dec.'78]

33. Compute the S.D. of income from the following:

Income (Rs)	Below 200	200–399	400–599	600–799	800–999	1000–1199
No. of Earners	25	72	47	22	13	7

[C.U., B.A.(Econ) '78]

34. Find the mean and the S.D. from the following frequency distribution:

Weight (lb.)	131–140	141–150	151–160	161–170	171–180
No. of Persons	2	5	4	9	7
	181–190	191–210	211–240		
	5	3	1		

[I.C.W.A.. Jan'71]

35. Calculate the proportion of farms in which costs of production are within the range A.M. ± S D. in the following distribution:

Costs of Production (Rs per 5 litres)	4–6	6–8	8–10	10–12	12–14	14–16	Total
No. of Dairy Farms	13	111	182	105	19	7	437

[I.C.W.A., Dec. '73]

36. Out of 400 observations, 100 observatioas have the value one and the rest of the observations are zero. Find the mean and s.d. of 400 observations together. *[B.U., B.A.(Econ) '66]*

37. Two samples of sizes 60 and 90 have 52 and 48 as the respective arithmetic means, and 9 and 12 as the respective standard deviations. Find the arithmetic mean and the standard deviation of the combined sample of size 150. *[I C.W.A., July '70]*

38. The means of two samples of sizes 50 and 100 respectively are 54.4 and 50.3 and the standard deviations are 8 and 7. Obtain the mean and standard deviation of the sample of size 150 obtained by combining the two simples. (Give answers correct to one decimal place.) *[D.M. '72; I.C.W.A., Dec. '77 & June '78]*

39. An analysis of monthly wages paid to workers in two firms *A* and *B*, belonging to the same industry, gives the following results:

	Firm A	*Firm B*
Number of Wage Earners	550	650
Average Monthiy Wages	Rs 50	Rs 45
S.D. of the Distribution of Wages	Rs ($\sqrt{90}$)	Rs ($\sqrt{120}$)

Answer the following questions with proper justifications:

(a) Which firm *A* or *B* pays out larger amount as monthly wages?

(b) In which firm *A* or *B* is there greater variability in individual wages?

(c) What are the measures of (i) average monthly wages, and (ii) standard deviation in the distribution of individual wages of all workers in the two firms taken together? *[I.C.W.A., June '77]*

40. A company has three establishments E_1, E_2 and E_3 in three cities. Analysis of the monthly salaries paid to the employees in the three establishments is given below:

	E_1	E_2	E_3
Number of Employees	20	25	40
Average Monthly Salary (Rs)	305	300	340
S.D. of Monthly Salaries (Rs)	50	40	45

Find the average and the standard deviation of the monthly salaries of all the 85 employees in the company. *[I.C.W.A., Dec. '76]*

41. Three sets of values of the variable x have means 26.3, 27.0 and 28.5 and standard deviations 4.5, 3.9 and 4.8. If the three sets have respectively 50. 60 and 55 values, what would be the mean and variance of x, if the three sets are taken together? *[B.U., B.A.(Econ) '77]*

42. The mean and the variance calculated from a group of 80 observations are 63.2 and 25.93 respectively. If 60 of these observations have mean 64.8 and S.D. 4. find the mean and the s.d. of the remaining 20 observations. *[I.CW.A., July '71]*

43. A group has the following measurements:

$$\bar{x} = 10, \quad \sigma^2 = 4 \text{ and } n = 60.$$

A subgroup of the above has $\bar{x}_1 = 11$, $\sigma_1^2 = 2.25$ and $n_1 = 40$. Find the mean and standard deviation of the other subgroup.

[I.C.W.A., June '73; M.B.A. '78]

44. The following data refer to the dividend (%) paid by two companies *A* and *B* over the last 7 years.

A	4	8	4	15	10	11	9
B	12	8	3	15	6	4	10

Calculate the coefficients of variation and comment.

[I.C.W.A., June'75-old]

45. From the prices of shares *x* and *y* below find out which is more stable in value:

x	35	54	52	53	56	58	52	50	51	49
y	108	107	105	105	106	107	104	103	104	101

[I.C.W.A., Dec. '76-old]

46. Calculate the coefficient of variation from the following data, showing Grades of 100 students in M.A. Mathematics:

Grades	30–39	40–49	50–59	60–69	70–79	80–89	90–99
Frequency	2	3	11	20	32	25	7

[C.U., M.Com. '73]

47. The mean life in days and standard deviation for two types of electric bulbs are given below:

	Mean Life in days	*Standard Deviation in days*
Type I	310	9
Type II	260	14

Compare the relative variability of life of the two types of bulbs.

[B.U., B.A.(Econ) '65]

48. You are given the distribution of wages ia two factories *X* and *Y*.

Wages (Rs)	50–100	100–150	150–200	200–250	250–300	300–350
No. of Workers X	2	9	29	54	11	5
No. of Workers Y	6	11	18	32	27	11

State in which factory the wages are more variable (Use Standard Deviation and Mean.)*[C.A., May '75]*

49. Calculate a suitable measure of dispersion for the following distribution:

Cotton Consumed in Thousand Candles	0–2	2–4	4–6	6–8	8–10	10–12	12–14	14–16	16–18	18–20	20–22
No. of Mills	5	13	12	11	8	4	1	3	1	1	2

How does this dispersion compare with a S.D. of 3 lbs for the weight of yarns per spindle) among mills producing yarns of average weight 20 lbs. per spindle? *[I.C.W.A., June '74]*

50. In a small town, a survey was conducted in respect of profit made by retail shops. The following results were obtained:

Profit or Loss (Rs '000)	−4 to −3	−3 to −2	−2 to −1	−1 to 0	0 to 1
No. of Shops	4	10	22	28	38
	1 to 2	2 to 3	3 to 4	4 to 5	5 to 6
	56	40	24	18	10

Calculate: (i) the average profit made by a retail shop;
(ii) total profit made by all the shops;
(iii) the coefficient of variation of earnings.,

[C.A., Nov. '77]

ANSWERS

1. (i) A.M. reduces by 10; (ii) & (iii) Range and S.D. Unchanged.
2. (i) Unchanged, (ii) Changes by the same proportion.
3. (i) 178, (ii) 12
4. 15 inches (= 74.5 – 59.5)
5. 8.05, 30
6. 0.49 short ton.
7. 14.9
8. 33.9
9. 20.4
10. 2.02
11. 1.24 inches
12. 1.48
13. 55.67, 1.18 (inches)
14. 30, 3.1
15. 20, 3.04
16. 10, 2.86
17. 5.0
18. 54, 3.04
19. (see Example 6.18, for $n = 3$)
20. $\sigma_y = |b|.\sigma_x$
21. Mean = $-\bar{x}$, S.D. = σ[see (5.3.3) & (6.6.2); $z = -x$, when a = -1 & $b = 0$]
22. (see Example 5.12)
23. 34.9
24. 21 lbs.
25. 0.033
26. 1.4
28. 11.64 years
28. 7.18 lbs.
29. 3.47, 0.23 (oz.)
40. Rs. 14.80
31. 88.4 (Rs '000)
32. 9.23, 2.48, 2.03
33. Rs. 251.40
34. 169.25, 19.5 (lbs.)
35. 0.65
36. 1/4, $\sqrt{3/4}$
37. 49.6, 11.1
38. 51.7, 7.6
39. B, B; 47.29 & 10.60 (Rs)
40. 320.00, 48.69 (Rs)
41. 27.29 Rs, 20.16 $(\text{Rs})^2$
42. 58.4, 5.0
43. 8, 1.22
44. 41.5, 48.6; Dividend paid by B less stable.
45. Share *y*; (C.V. 11.6 & 1.9)
46. 18.1
47. Type II more variable (C.V. 2.9 & 5.4)
48. *Y*(C.V. 23 & 30)
49. $\sigma = 4.78$ units; (C.V. 67 & 45). Dispersion for cotton consumption more.
50. 1.348 & 337 (Rs '000); 153.

7 MOMENTS, SKEWNESS AND KURTOSIS

7.1 MOMENTS

Given n observations $x_1, x_2, ..., x_n$, and an arbitrary constant A,

$\frac{1}{n}\ \Sigma(x - A)$ is called the *1st moment about A,*

$\frac{1}{n}\ \Sigma(x - A)^2$ is called the *2nd moment about A,* (7.1.1)

$\frac{1}{n}\ \Sigma(x - A)^3$ is called the *3rd moment about A,*

and so on. Let us denote these moments successively by m_1', m_2', m_3', etc. (Sometimes, these are also represented by μ_1', μ_2', μ_3', etc.)

Then $\quad m_1' = \Sigma(x - A)/n = (\Sigma x - \Sigma A)/n = (\Sigma x - nA)/n$

$$= \bar{x} - A \tag{7.1.2}$$

i.e. the 1st moment about A equals $(\bar{x} - A)$.

Moments about zero (i.e. when $A = 0$), and moments about mean (i.e. when $A = \bar{x}$) are particularly important.

(1) Moments about zero (or Raw moments)

1st moment about zero $= \frac{1}{n}\ \Sigma x = \bar{x}$

2nd moment about zero $= \frac{1}{n}\ \Sigma x^2$ (7.1.3)

3rd moment about zero $= \frac{1}{n}\ \Sigma x^3$

and so on. Note that the 1st moment about zero is the mean $\bar{x}$.

$$m_1' = \bar{x} \tag{7.1.4}$$

(2) Moments about mean (or Central moments)

1st moment about mean $= \frac{1}{n}\ \Sigma(x - \bar{x}) = 0$

$$\text{2nd moment about mean} = \frac{1}{n}\ \Sigma(x-\bar{x})^2 = \sigma^2$$

$$\text{3rd moment about mean} = \frac{1}{n}\ \Sigma(x-\bar{x})^3$$

$$\text{4th moment about mean} = \frac{1}{n}\ \Sigma(x-\bar{x})^4 \tag{7.1.5}$$

and so on. These are usually denoted by m_1, m_2, m_3, m_4, etc. (Sometimes, these are represented by μ_1, μ_2, μ_3, μ_4, etc.) Note that the 1st central moment is always zero, and the 2nd central moment is the variance σ^2. By (5.3.2) and (6.5.2),

$$m_1 = 0, \qquad m_2 = \sigma^2 \tag{7.1.6}$$

From the second relation, we find that the standard deviation is the square-root of the second central moment m_2.

The 3rd central moment m_3 is used to measure skewness (Section 7.7) and the 4th central moment m_4 to measure kurtosis (Section 7.8). Higher order moments m_5, m_6 etc. are seldom used.

In general, given n observations $x_1, x_2, \ldots, x_n$, the r-th order, moments ($r = 0, 1, 2, \ldots$) are defined as follows:

$$r\text{-th moment about } A: \quad m'_r = \frac{1}{n}\ \Sigma(x-A)^r$$

$$r\text{-th raw moment:} \quad m'_r = \frac{1}{n}\ \Sigma x^r$$

$$r\text{-th central moment:} \quad m_r = \frac{1}{n}\ \Sigma(x-\bar{x})^r \tag{7.1.7}$$

For a frequency distribution,

$$r\text{-th moment about } A: \quad m'_r = \frac{1}{N}\ \Sigma f(x-A)^r$$

$$r\text{-th raw moment:} \quad m'_r = \frac{1}{N}\ \Sigma f x^r$$

$$r\text{-th central moment:} \quad m_r = \frac{1}{N}\ \Sigma f(x-\bar{x})^r \tag{7.1.8}$$

where $N = \Sigma f$. [Note that moments about mean are written without dashes (′), but moments about any other origin, i.e. non-central moments, with dashes.]

There are important relations between central and non-central moments. For example, if the non-central moments (m'_1, m'_2, m'_3 etc.) about any arbitrary origin A are known, the central moments can be obtained by using the relations (7.2.2), viz.

$$\begin{aligned} m_2 &= m'_2 - m'^2_1 \\ m_3 &= m'_3 - 3m'_2 m'_1 + 2m'^3_1 \\ m_4 &= m'_4 - 4m'_3 m'_1 + 6m'_2 m'^2_1 - 3m'^4_1 \end{aligned} \tag{7.1.9}$$

In particular, using the first two moments m_1' and m_2', abut an arbitrary origin A, the mean and variance may be obtained:

$$\bar{x} = m'_1 + A, \qquad \sigma^2 = m'_2 - m'^2_1 \tag{7.1.10}$$

Example 7.1 *The first two moments of a distribution about the value 5 of the variable are 2 and 20. Find the mean and the variance.* [I.C.W.A. June '77]

Solution Using formulae (7.1.10), we have here

$$m_1' = 2,\ m_2' = 20, \text{ and } A = 5.$$

$$\therefore \quad \bar{x} = 2 + 5 = 7,$$

$$\sigma^2 = 20 - 2^2 = 16.$$

Ans. 7, 16.

Example 7.2 *The first four moments of a distribution about the value 3 are 2, 10, 40 and 218. Find the moments about the origin and mean.*

Solution Given

$$\Sigma(x-3)/n = 2, \qquad \Sigma(x-3)^2/n = 10,$$
$$\Sigma(x-3)^3/n = 40, \qquad \Sigma(x-3)^4/n = 218,$$

we have to find (a) $\Sigma x/n$, $\Sigma x^2/n$, $\Sigma x^3/n$, $\Sigma x^4/n$; and (b) m_1, m_2, m_3, m_4.

(a) $\Sigma(x-3)/n = 2;$

or, $(\Sigma x - 3n)/n = 2$

or, $\Sigma x/n - 3 = 2$

$\therefore$ $\Sigma x/n = 2 + 3 = 5$ ... (i)

$\Sigma(x-3)^2/n = 10$

or, $\Sigma(x^2 - 6x + 9)/n = 10$

or, $\Sigma x^2/n - 6\,\Sigma x/n + 9 = 10$

or, $\Sigma x^2/n - 6 \times 5 + 9 = 10$; putting $\Sigma x/n = 5$ from (i)

$\therefore$ $\Sigma x^2/n = 10 + 30 - 9 = 31$ ... (ii)

Again, $\Sigma(x-3)^3/n = 40$

Using the binomial $(a-b)^3 = a^3 - 3a^2b + 3ab^2 - b^3$ we have

$\Sigma(x^3 - 3.x^2.3 + 3.x.3^2 - 3^3)/n = 40$

or, $\Sigma x^3/n - 9\,\Sigma x^2/n + 27\,\Sigma x/n - 27 = 40$

or, $\Sigma x^3/n - 9 \times 31 + 27 \times 5 - 27 = 40$; from (i) and (ii).

$\therefore$ $\Sigma x^3/n = 40 + 279 - 135 + 27 = 211$ ... (iii)

Also, $\Sigma(x-3)4/n = 218$

Using the binomial expansion $(a-b)^4 = a^4 - 4a^3b + 6a^2b^2 - 4ab^3 + b^1$,

$\Sigma(x^4 - 4.x^3.3 + 6.x^2.3^2 - 4.x.3^3 + 3^4)/n = 218$

or, $\Sigma x^1/n - 12\,\Sigma x^3/n + 54\,\Sigma x^2/n - 108\Sigma x/n + 81 = 218$

or, $\Sigma x^1/n - 12 \times 211 + 54 \times 31 - 108 \times 5 + 81 = 218$;

from (i), (ii) & (iii)

$\therefore$ $\Sigma x^4/n = 218 + 2532 - 1674 + 540 - 81 = 1535$ ... (iv)

(*b*) The moments about 3 are given as

$$m_1' = 2,\ m_2' = 10,\ m_3' = 40,\ m_4' = 218$$

Hence, using relations (7.1.9), the moments about mean are

$$m_1 = 0 \text{ (in all cases)}$$
$$m_2 = m_2' - m_1'^2 = 10 - 2^2 = 6$$
$$m_3 = m_3' - 3m_2'm_1' + 2m_1'^3$$
$$= 40 - 3 \times 10 \times 2 + 2 \times 2^3 = -4$$
$$m_4 = m_4' - 4m_3'm_1' + 6m_2'm_1'^2 - 3m_1'^4$$
$$= 218 - 4 \times 40 \times 2 + 6 \times 10 \times 2^2 - 3 \times 2^4$$
$$= 218 - 320 + 240 - 48 = 90$$

[Note that in relations (7.1.9.), the moments about any arbitrary origin may be used to obtain the central moments. Hence, we may also use the moments about origin, obtained earlier, viz. 5, 31, 211, 1535, to find the central moments.

$$m_2 = 31 - 5^2 = 6$$
$$m_3 = 211 - 3 \times 31 \times 5 + 2 \times 5^3 = 211 - 465 + 250 = -4$$

$$m_4 = 1535 - 4 \times 211 \times 5 + 6 \times 31 \times 5^2 - 3 \times 5^1$$
$$= 1535 - 4220 + 4650 - 1875 = 90]$$

Ans. 5, 31, 211, 1535; 0, 6, – 4, 90

Example 7.3 *The arithmetic mean of a certain distribution is 5. The second and the third moments about the mean are 20 and 140 respectively. Find the third moment of the distribution about 10.* [I.C.W.A., June '74]

Solution Given $\Sigma(x - 5)^2/n = 20$ and $\Sigma(x - 5)^3/n = 140$, we have to find the value of $\Sigma(x - 10)^3/n$. Applying the binomial expansion $(a - b)^3 = a^3 - 3a^2b + 3ab^2 - b^3$, we have

$$(x - 10)^3 = \{(x - 5) - 5\}^3$$
$$= (x - 5)^3 - 15(x - 5)^2 + 75(x - 5) - 125$$

Summing over all values of x and then dividing both sides by n,

$$\Sigma(x - 10)^3/n = \Sigma(x - 5)^3/n - 15\Sigma(x - 5)^2/n + 75\Sigma(x - 5)/n - 125$$

But $\Sigma(x - 5)/n = (\Sigma x - \Sigma 5)/n = (\bar{x} - 5) = 0$ (since A.M. = 5). Substituting the values,

$$\Sigma(x - 10)^3/n = 140 - 15 \times 20 + 75 \times 0 - 125$$
$$= 140 - 300 + 0 - 125 = -285$$

Ans. –285

Example 7.4 *If m_2' and m_2 are respectively the second moment about an arbitrary origin a and that about $\bar{x}$, then show that $m_2' = m_2 + d^2$, where $d = \bar{x} - a$.* [C.U., M.Com. '76]

Solution By definition

$$m_2' = \Sigma(x - a)^2/n, \quad m_2 = \Sigma(x - \bar{x})^2/n$$

We may write $(x - a)^2 = \{(x - \bar{x}) + (\bar{x} - a)\}^2$

$$= (x - \bar{x})^2 + (\bar{x} - a)^2 + 2(x - \bar{x})(\bar{x} - a)$$

$\therefore$ $$\Sigma(x - a)^2/n = \Sigma(x - \bar{x})^2/n + \Sigma(\bar{x} - a)^2/n + 2(\bar{x} - a)\,\Sigma(x - \bar{x})/n$$

or, $$m_2' = m_2 + n(\bar{x} - a)^2/n + 2(\bar{x} - a).0/n$$

or, $$m_2' = m_2 + (\bar{x} - a)^2$$

or, $$m_2' = m_2 + d^2$$ (Shown)

Example 7.5 *Show how moments are used to describe the characteristics of a distribution, viz. central tendency, dispersion, skewness, and kurtosis.* [I.C.W.A., Dec. '73, June '75]

Solution All the important characteristics of a frequency distribution can be described by moments. A measure of central tendency is given by the first moment about zero, which is the arithmetic mean. By (7.1.4)

$$\bar{x} = m_1'$$

A measure of dispersion is given by the second central moment, which is the square of standard deviation. By (7.1.6)

$$\sigma = \sqrt{m_2}$$

Measures of skewness and kurtosis are provided by the 3rd and the 4th central moments. By (7.7.4) and (7.8.1)

$$\text{Skewness } (\gamma_1) = \frac{m_3}{\left(\sqrt{m_2}\right)^3}; \quad \text{Kurtosis } (\gamma_2) = \frac{m_4}{m_2^2} - 3.$$

7.2 RELATION BETWEEN CENTRAL AND NON-CENTRAL MOMENTS

(I) *Formula for m_r in terms of m_r' and moments of lower order:*

$$m_r = \Sigma(x_i - \bar{x})^r/n \qquad m_r' = \Sigma(x_i - A)^r/n$$

Let us write

$$x_i - \bar{x} = (x_i - A) - (\bar{x} - A) = \{(x_i - A) - d\}, \text{ (suppose)}$$

Using the binomial expansion

$$(x_i - \bar{x})^r = (x_i - A)^r - {}^rC_1(x_i - A)^{r-1}d + {}^rC_2(x_i - A)^{r-2}d^2 - \ldots + (-1)^r d^r$$

Summing over all values of $i = 1, 2, \ldots, n$

$$\Sigma(x_i - \bar{x})^r = \Sigma(x_i - A)^r - {}^rC_1 d\Sigma(x_i - A)^{r-1} + {}^rC_2 d^2\Sigma(x_i - A)^{r-2} - \ldots + (-1)^r nd^r.$$

Now dividing both sides by n,

$$m_r = m_r' - {}^rC_1 m_{r-1}' d + {}^rC_2 m_{r-2}' d^2 - \ldots + (-1)^r d^r \qquad (7.2.1)$$

where $d = \bar{x} - A = m_1'$ (see 7.1.2).

In particular, putting $r = 1, 2, 3, 4$, we get

$$m_1 = m_1' - d$$
$$m_2 = m_2' - 2m_1'd + d^2$$
$$m_3 = m_3' - 3m_2'd + 3m_1'd^2 - d^3$$
$$m_4 = m_4' - 4m_3'd + 6m_2'd^2 - 4m_1'd^3 + d^4$$

Writing $d = m_1'$ and simplifying, the central moments (m_r) when expressed in terms of the moments (m_r') about any origin are

$$m_1 = 0, \qquad \text{as it should be, by (7.1.6)}$$
$$m_2 = m_2' - m_1'^2$$
$$m_3 = m_3' - 3m_2'm_1' + 2m_1'^3 \qquad (7.2.2)$$
$$m_4 = m_4' - 4m_3'm_1' + 6m_2'm_1'^2 - 3m_1'^4$$

(II) *Formula for m_r' in terms of m_r and moments of lower order*

We write $x_i - A = (x_i - \bar{x}) + (\bar{x} - A) = (x_i - \bar{x}) + d$, (suppose)

Therefore, using the binomial expansion

$$(x_i - A)^r = (x_i - \bar{x})^r + {}^rC_1(x_i = \bar{x})^{r-1}d + {}^rC_2(x_i - \bar{x})^{r-2}d^2 + \ldots + d^r$$

Summing over all values of $i = 1, 2, \ldots, n$

$$\Sigma(x_i - A)^r = \Sigma(x_i - \bar{x})^r + {}^rC_1 d\Sigma(x_i - \bar{x})^{r-1} + {}^rC_2 d^2\Sigma(x_i - \bar{x})^{r-2} + \ldots + nd^r.$$

Dividing both sides by n,

$$m_r' = m_r + {}^rC_1 m_{r-1} d + {}^rC_2 m_{r-2} d^2 + \ldots + d^r \qquad (7.2.3)$$

where $d = \bar{x} - A = m_1'$ is the first moment about A.

In particular, putting $r = 1, 2, 3, 4$ we get

$$m_1' = m_1 + d$$
$$m_2' = m_2 + 2m_1d + d^2$$
$$m_3' = m_3 + 3m_2d + 3m_1d^2 + d^3$$
$$m_4' = m_4 + 4m_3d + 6m_2d^2 + 4m_1d^3 + d^4$$

Since $m_1 = 0$ and $d = \bar{x} - A = m_1'$, we have

$$m_1' = m_1' \text{ (as expected)}$$
$$m_2' = m_2 + m_1'^2$$
$$m_3' = m_3 + 3m_2m_1' + m_1'^3 \qquad (7.2.4)$$
$$m_4' = m_4 + 4m_3m_1' + 6m_2m_1'^2 + m_1'^4$$

Example 7.6 *Let $a_r = \mu_r/\sigma^r$ where μ_r is the rth moment about mean and σ is the standard deviation. Show that $\alpha_1 = 0$ and $\alpha_2 = 1$.* [C.U., B.Com. (Hons) '67]

Solution The rth moment about mean, is defined as

$$\mu_r = \Sigma(x_i - \overline{x})^r/n$$

Putting $r = 1$ and 2 successively, we have

$$\mu_1 = \Sigma(x_i - \overline{x})/n = 0/n = 0, \qquad \text{by (5.3.2)}$$

$$\mu_2 = \Sigma(x_i - \overline{x})^2/n = \sigma^2. \qquad \text{by (6.5.2)}$$

Also, we are given $a_r = \mu_r/\sigma^2$. Therefore,

$$\alpha_1 = \mu_1/\sigma = 0/\sigma = 0, \quad \alpha_2 = \mu_2/\sigma^2 = \sigma^2/\sigma^2 = 1$$

7.3 BETA-COEFFICIENTS AND GAMMA-COEFFICIENTS

'*Beta-coefficients*' are defined as follows:

$$\beta_1 = \frac{m_3^2}{m_2^3}; \qquad \beta_2 = \frac{m_4}{m_2^2} \qquad (7.3.1)$$

(Note that β is the Greek small letter 'beta').

The beta-coefficients can never be negative, and are pure numbers, independent of origin and scale of observations. Since for a symmetrical distribution all odd order moments are zero, $m_3 = 0$ (see Example 7.17); consequently $\beta_1 = 0$, when the distribution is symmetrical.

Frequency distributions are classified as *leptokurtic, platykurtic,* or *mesokurtic,* according as the value of β_2 is greater than, less than, or equal to 3. Beta-coefficients are used for measuring skewness and kurtosis. (Sections 7.7 and 7.8)

'*Gamma-coefficients*' are defined as follows:

$$\gamma_1 = \sqrt{\beta_1}; \qquad \gamma_2 = \beta_2 - 3 \qquad (7.3.2)$$

(Note that γ is the Greek small letter 'gamma'). γ_1 must have the same sign as m_3. The gamma-coefficients may be positive, negative, or zero; but are pure numbers like the beta-coefficients. These are used as measures of 'skewness' and 'kurtosis'.

$$\text{Skewness } (\gamma_1) = \sqrt{\beta_1} = \frac{m_3}{\sigma^3} \qquad (7.3.3)$$

$$\text{Kurtosis } (\gamma_2) = \beta_2 - 3 = \frac{m_4}{\sigma^4} - 3$$

Distributions are said to be 'positively skew', 'negatively skew', or 'symmetrical', according as γ_1 is positive, negative, or zero. Similarly positive, negative, or zero values of γ_2 are associated with 'leptokurtic', 'platykurtic', or 'mesokurtic' distributions.

From the results of Example 7.2, we have

$$\beta_1 = \frac{(-4)^2}{6^3} = .074; \quad \beta_2 = \frac{90}{6^2} = 2.5$$

Hence, $\gamma_1 = \sqrt{.074} = -0.27$ (since $m_3 = -4$ is negative), and $\gamma_2 = 2.5 - 3 = -0.5$. The distribution is therefore negatively skew and platykurtic.

For the 'Normal distribution' (Chapter 12), $\beta_1 = 0$ and $\beta_2 = 3$; i.e. $\gamma_1 = 0$ and $\gamma_2 = 0$. If, therefore, for any given distribution γ_1 and γ_2 do not significantly differ from 0, the data may be assumed to belong to a 'normal' distribution.

7.4 STANDARDIZED VARIABLE

In statistical theores, it is often found convenient to express deviations from mean ($\bar{x}$) in terms of standard deviation (σ) as unit.

$$z = \frac{x - \bar{x}}{\sigma} = \frac{\text{Variable} - \text{Mean}}{\text{S.D.}} \tag{7.4.1}$$

is called the *Standardized variable* or *Standardized form of x*. The standardized variable z is independent of origin and unit of measurement. If the variable x represents length in ft., its mean $\bar{x}$ and S.D. σ will both be expressed in ft., so that the numerator and the denominator of z will be in ft., and hence z is a pure number.

Mean of the standardized variable z is zero, and its S.D. is 1.

$$\bar{z} = 0, \quad \sigma_z = 1 \tag{7.4.2}$$

Since the mean is zero, the rth central moment of z is the same as the rth raw moment of z, and is usually denoted by α_r

$$\alpha_r = m_r/\sigma^r \tag{7.4.3}$$

Note that $\alpha_1 = 0$, $\alpha_2 = 1$, $\alpha_3 = \sqrt{\beta_1} = \gamma_1$, $\alpha_4 = \beta_2 = \gamma_2 + 3$.

Example 7.7 *If m_r and α_r denote the rth central moments of x and of the standardized variable $z = (x - \bar{x})/\sigma$ respectively, where σ is the S.D. of x, then show that $\alpha_r = m_r/\sigma^r$.*

Solution By definition, $m_r = \Sigma(x_i - \bar{x})^r/n$, $\alpha_r = \Sigma(z_i - \bar{x})^r/n$

But $z_i = (x_i - \bar{x})/\sigma$; hence by (5.3.2)

$$\bar{z} = \frac{1}{n}\Sigma z_i = \frac{1}{n}\Sigma\left(\frac{x_i - \bar{x}}{\sigma}\right) = \frac{1}{n\sigma}\Sigma(x_i - \bar{x}) = 0$$

$$\therefore \quad \alpha_r = \Sigma(z_i - 0)^r/n = \Sigma z_i^r/n$$

$$= \frac{1}{n}\Sigma\left(\frac{x_i - \bar{x}}{\sigma}\right)^r = \frac{1}{n}\Sigma\frac{(x_i - \bar{x})^r}{\sigma^r} = \frac{1}{\sigma^r}\Sigma\frac{(x_i - \bar{x})^r}{n} = \frac{m_r}{\sigma^r}$$

7.5 MOMENTS OF FREQUENCY DISTRIBUTIONS

If $x_1, x_2, \ldots, x_n$ have frequencies $f_1, f_2, \ldots, f_n$ respectively, the *r-th moment about A* is defined as

$$m_r' = \frac{1}{N}\Sigma f(x - A)^r \tag{7.5.1}$$

where $N = \Sigma f$.

The *r-th central moment* is similarly defined as

$$m_r = \frac{1}{N}\Sigma f(x - \bar{x})^r \tag{7.5.2}$$

where $\bar{x} = \Sigma fx/N$.

In the case of grouped frequency distributions, the midvalues are taken as representatives of the respective classes and $x_1, x_2, \ldots, x_n$ denote these midvalues. If the successive mid values have a common difference, the calculation of central moments can be simplified.

If $y = (x - c)/d$, where c and d are constants, the r-th central moment of x is equal to d^r times the r-th central moment of y.

$$m_{r(x)} = d^r.m_{r(y)} \tag{7.5.3}$$

Since the values of y are usually small, $m_{r(y)}$ can be calculated easily from the raw moments of y. Finally, $m_{r(y)}$ is multiplied by d^r to give the required central moments of x. In particular, for the standardized variable $z = (x - \bar{x})/\sigma$, we have $c = \bar{x}$ and $d = \sigma$, so that $m_{r(x)} = \sigma^r.m_{r(z)}$; or, $m_{r(z)} = m_{r(x)}/\sigma^r$; i.e. $\alpha_r = m_r/\sigma^r$, the result derived in Example 7.7.

Example 7.8 *Show that the central moments are invariant under change of origin, but not under change of scale.* [I.C.W.A. Dec. '75]

Solution Let $y = (x - c)/d$. Then, by (5.3.4), $\bar{x} = c + d\bar{y}$.

$\therefore$ $x - \bar{x} = (c + dy) - (c + d\bar{y}) = d(y - \bar{y})$

Hence, $m_{r(x)} = \Sigma(x - \bar{x})^r/N = \Sigma f d^r(y - \bar{y})^r/N$

$= d^r \Sigma f(y - \bar{y})^r/N = d^r.m_{r(y)}$

Since the expression on the right hand side is independent of c, but contains d, the result follows.

Example 7.9 *Find the second, the third and the fourth central moments of the frequency distribution given below. Hence, find a measure of skewness (γ_1) and a measure of kurtosis (γ_2).*

Class Limits	*110.0–114.9*	*115.0–119.9*	*120.0–124.9*	*125.0–129.9*
Frequency	5	15	20	35
	130.0–134.9	*135.0–139.9*	*140.0–144.9*	*Total*
	10	10	5	100

[I.C.W.A., June '76]

Solution

Table 7.1 Calculations for Moments

Mid-value x	f	$y = \dfrac{x - 127.45}{5}$	fy	fy^2	fy^3	fy^4	$f(y+1)^4$
112.45	5	–3	–15	45	–135	405	80
117.45	15	–2	–30	60	–120	240	15
122.45	20	–1	–20	20	–20	20	0
127.45	35	0	0	0	0	0	35
132.45	10	1	10	10	10	10	160
137.45	10	2	20	40	80	160	810
142.45	5	3	15	45	135	405	1280
Totals	100	–	–20	220	–50	1240	2380

Charlier's check:

C.V.L. Charlier gave a simple and effective check on the calculations for moments from a grouped frequency distribution. Expanding $(y + 1)^4$ by binomial theorem, multiplying by f, and then summing, we have the identity

$$\Sigma f(y + 1)^4 = \Sigma fy^4 + 4\Sigma fy^3 + 6\Sigma fy^2 + 4\Sigma fy + N \quad (7.5.4)$$

Along with the columns for fy, fy^2, fy^3, fy^4 needed for the calculation of moments upto the 4th order, an extra column for $f(y + 1)^4$ is also calculated. If it is found that the numerical values obtained from the table satisfy the identity (7.5.4), then we are sure that the calculations are correct.

Applying Charlier's check on the calculations in Table 7.1,

$$\begin{aligned} 2380 &= 1240 + 4(-50) + 6(220) + 4(-20) + 100 \\ &= 1240 - 200 + 1320 - 80 + 100 \\ &= 2660 - 280 \\ &= 2380 \end{aligned}$$

Hence, the calculations in Table 7.1 are correct.

Raw moments of y:

$$\begin{aligned} m_1' &= \Sigma fy/N = -20/100 = -0.2 \\ m_2' &= \Sigma fy^2/N = 220/100 = 2.2 \\ m_3' &= \Sigma fy^3/N \; - 50/100 = -0.5 \\ m_4' &= \Sigma fy^4/N = 1240/100 = 12.4 \end{aligned}$$

Central moments of y: Applying (7.2.2), we have

$$\begin{aligned} m_2 &= m_2' - (m_1')^2 = 2.2 - (-0.2)^2 = 2.16 \\ m_3 &= m_3' - 3m_2'm_1' + 2(m_1')^3 \\ &= -0.5 - 3(2.2)(-0.2) + 2(-0.2)^3 \\ &= -0.5 + 1.32 - .016 \\ &= 0.804 \\ m_4 &= m_4' - 4m_3'm_1' + 6m_2'(m_1')^2 - 3(m_1')^4 \\ &= 12.4 - 4(-0.5)(-0.2) + 6(2.2)(-0.2)^2 - 3(-0.2)^4 \\ &= 12.4 - 0.4 + 0.528 - .0048 \\ &= 12.5232 \end{aligned}$$

Central moments of x: Using (7.5.3), we have

$$\begin{aligned} m_{2(x)} &= d^2.m_{2(y)} = 5^2 \times 2.16 = 54.0 \\ m_{3(x)} &= d^3.m_{3(y)} = 5^3 \times 0.804 = 100.5 \\ m_{4(x)} &= d^4.m_{4(y)} = 5^4 \times 12.5232 = 7827.0 \end{aligned}$$

Also, $\bar{x} = c + d\bar{y} = 127.45 + 5(-0.2) = 126.45$

Since the beta-coefficients (7.3.1) and gamma-coefficients (7.3.2) are independent of origin and scale, we may use the moments of y:

$$\beta_1 = \frac{(0.804)^2}{(2.16)^3} = 0.064; \qquad \beta_2 = \frac{12.5232}{(2.16)^2} = 2.68$$

$$\therefore \quad \gamma_1 = \sqrt{0.064} = 0.25; \qquad \gamma_2 = 2.68 - 3 = -0.32$$

The distribution is positively skew and platykurtic.

Ans. 54.0, 100.5, 7827.0; 0.25, – 0.32.

7.6 SHEPPARD'S CORRECTION FOR ERRORS DUE TO GROUPING

For the calculation of mean, variance and other moments from a grouped frequency distribution, the mid-value of a class interval is taken as representative of all observations falling within that class interval. This, however, introduces some error in the calculated values, known as *error due to grouping*. The presence of error will be evident from the fact that moments calculated from ungrouped data will, in general, be different from those calculated from the grouped frequency distribution. To minimize such errors, the class intervals may be made smaller, or the class limits so chosen that the mid-values coincide with the majority of observations falling within a class interval. Even then, the differences can not be removed altogether. However, under certain conditions it is possible to make corrections for these discrepancies by using certain formulae due to W.F. Sheppard; which are known as *Sheppard's corrections for errors due to grouping*. So far as the first four moments are concerned:

(i) No correction is necessary for the mean m_1' and the third central moment m_3.

(ii) Corrections for the 2nd and the 4th central moments are m_2 (corrected) = m_2 (uncorrected) $- \dfrac{c^2}{12}$

$$m_4(\text{corrected}) = m_4\ (\text{uncorrected}) - \frac{c^2}{2} m_2\ (\text{uncorrected}) + \frac{7c^4}{240}$$

where c is the common width of class intervals.

Sheppard's corrections should be applied only when:

(a) the distribution relates to a continuous variable, and is of moderate asymmetry;

(b) the frequency becomes smaller and smaller approaching zero at each end of the distribution.

These corrections are not applicable to J- or U-shaped distributions or to very skew distributions. Moreover, unless the total frequency is fairly large, the corrections will be of little practical importance.

Example 7.10 *Apply Sheppard's corrections to the central moments calculated in Example 7.9.*

Solution The moments calculated from the grouped data are m_2 = 54.0; m_3 = 100.5; m_4 = 7827.0; and c = 5

$\therefore$ corrected m_2 = 54.0 – 5^2/12 = 54.0 – 2.1 = 51.9

corrected m_3 = 100.5

corrected m_4 = 7827.0 – $\frac{1}{2}$ (5^2)(54.0) + 7(5^4)/240

= 7827.0 – 675.0 + 18.2 = 7170.2

7.7 SKEWNESS

A frequency distribution is said to be 'symmetrical', if the frequencies are symmetrically distributed about mean, i.e. when values of the variable equidistant from mean have equal frequencies.

Illustration 1. Symmetrical distributions:

(i)	x :	10	15	20	25	30
	f :	3	7	16	7	3

(ii)	x	10	15	20	25	30	35
	f	3	7	16	16	7	3
(iii)	x :	5–8	9–12	13–16	17–20	21–24	
	f :	7	18	23	18	7	
(iv)	x :	5–8	9–12	13–16	17–20	21–24	25–28
	f :	7	18	23	23	18	7

Note that in the above distributions, the means are respectively 20, 22.5, 14.5 and 16.5. The median and the mode for each also have the same value. In fact, for any symmetrical distribution mean, median and mode are equal.

In general, however, frequency distributions are not symmetrical; some are slightly asymmetrical and some others may be highly asymmetrical.

Illustration 2. Asymmetrical (or Skew) distributions:

(i)	x :	5–8	9–12	13–16	17–20	21–24
	f :	7	18	23	16	7
(ii)	x :	5–8	9–12	13–16	17–20	21–24
	f :	7	18	23	10	3

Here the frequencies are not symmetrically distributed; in distribution (i) the extent of asymmetry is small, while in (ii) it is comparatively larger.

The word "*skewness*" is used to denote the 'extent of asymmetry' in the data. When the frequency distribution is not symmetrical, it is said to be '*skew*'. The word 'skewness' literally denotes 'asymmetry, or 'lack of symmetry', and 'skew' denotes 'asymmetrical'. A symmetrical distribution has therefore zero skewness. Skewness may also be positive or negative.

Skewness is measured by the following formulae:

(1) *Pearsons' first measure–*

$$\text{Skewness} = \frac{\text{Mean} - \text{Mode}}{\text{Standard Deviation}} \tag{7.7.1}$$

(2) *Pearson's second measure–*

$$\text{Skewness} = \frac{3\,(\text{Mean} - \text{Median})}{\text{Standard Deviation}} \tag{7.7.2}$$

(3) *Bowley's measure–*

$$\text{Skewness} = \frac{(Q_3 - Q_2) - (Q_2 - Q_1)}{(Q_3 - Q_2) + (Q_2 - Q_1)}$$

$$= \frac{Q_3 - 2Q_2 + Q_1}{Q_3 - Q_1} \tag{7.7.3}$$

where Q_1, Q_2, Q_3 denote the first, second and third quartiles of the distribution.

(4) *Moment measure–*

$$\text{Skewness } (\gamma_1) = \frac{m_3}{\sigma^3} = \frac{m_3}{\left(\sqrt{m_2}\right)^3} \tag{7.7.4}$$

where m_2 and m_3 are the second and third central moments, and σ denotes the S.D. [Note that this is the third moment α_3 of the standardized variable z, i.e. $\alpha_3 = m_3/\sigma^3 = \sqrt{\beta_1} = \gamma_1$; see (7.3.3) and (7.4.3)].

It should be noted that all the measures of skewness are pure numbers (which do not depend upon the units of measurement), and have the value zero when the distribution is symmetrical.

(**Note:** The measures of skewness (7.7.1) to (7.7.4) are also sometimes known as 'coefficients of skewness')

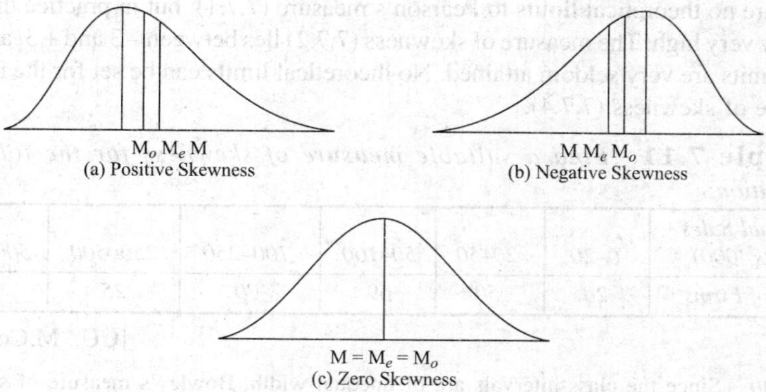

Fig. 7.1 Positions of Mean (M), Median (M_e), Mode(M_o) for Different Types of Skewness

Discussion on different measures—For a symmetrical distribution Mean, Median and Mode are equal. The more the asymmetry in the data, the larger is the discrepancy between them. So the difference between Mean and Mode may be taken as a measure of skewness. This difference, when judged relative to S.D., gives Pearson's measure (7.7.1). If Mode is not known accurately, the approximate relation *Mean – Mode* = 3 (*Mean – Median*) is utilised in (7.7.2). Note that skewness measured by (7.7.1) and (7.7.2) is positive when Mean is larger than Median and Mode, and negative when Mean is less.

Bowley's measure of skewness (7.7.3) has been developed from the following viewpoint. For a symmetrical distribution, Q_2 (i.e. Median) lies exactly midway between Q_1 and Q_3; for a positively skew distribution (i.e. when the longer "tail" of the frequency curve lies towards the right, Fig. 7.1a) Q_3 will be wider away from Q_2 than Q_1, and for a negatively skew distribution (Fig. 7.1b) the reverse will be the case. The difference of the midpoint of Q_1 and Q_3 from the Median Q_2, taken relative to Quartile Deviation, gives (7.7.3).

$$\text{Skewness} = \frac{\frac{1}{2}(Q_1 + Q_3) - Q_2}{\frac{1}{2}(Q_3 - Q_1)} = \frac{Q_3 - 2Q_2 + Q_1}{Q_3 - Q_1}$$

As regards the moment measure of skewness (7.7.4), it may be noted that in a symmetrical distribution, for each positive value of $(x_i - \overline{x})$ there is a corresponding negative value. When these deviation $(x_i - \overline{x})$ are cubed, positive values retain their positive sign and negative values the negative sign, so that $m_3 = \Sigma f_i(x_i - \bar{x})^3/N$ will be zero. For positively skew distributions (Fig. 7.1a), large positive values of $(x_i - \bar{x})$ are magnified considerably when cubed, and ultimately the sum of positive cubed deviations outweigh the negative cubes, making m_3 positive. In a similar manner, for negatively skew distributions (Fig. 7.1b), m_3 becomes negative.

Limits of different measures—Bowley's measure (7.7.3) lies between – 1 and + 1. There are no theoretical limits to Pearson's measure (7.7.1); but in practice the value is rarely very high. The measure of skewness (7.7.2) lies between – 3 and + 3, although these limits are very seldom attained. No theoretical limits can be set for the moment measure of skewness (7.7.4).

Example 7.11 *Find a suitable measure of skewness for the following distribution:*

Annual Sales (Rs '000)	*0–20*	*20–50*	*50–100*	*100–250*	*250–500*	*500–1000*
No. of Firms	*20*	*50*	*69*	*30*	*25*	*19*

[CU., M.Com. '72]

Solution Since the class-intervals are of unequal width, Bowley's measure of skewness (7.7.3), which is based on quartiles, will be appropriate here (Note that unequal widths of class intervals do not cause any difficulty in the calculation of partition values; however, other measures of skewness will involve laborious calculations for mean, s.d., moments, etc.). Using simple interpolation in a cumulative frequency distribution (as in Example 5.64), we find that (in Rs '000) $Q_1 = 39.95$, $Q_2 = 76.45$, $Q_3 = 203.75$ (Calculations are not shown here). Putting these values in formula (7.7.3):

$$\text{Skewness} = \frac{203.75 - 2 \times 76.45 + 39.95}{203.75 - 39.95} = \frac{90.80}{163.80} = +0.55$$

Example 7.12 *Calculate the coefficient of skewness based on quartiles from the following:*

More than	0	5474	More than	60	2718
More than	10	5426	More than	70	1406
More than	20	5259	More than	80	764
More than	30	5023	More than	90	370
More than	40	4475	More than	100	160
More than	50	3712	More than	110	39

[C.A., Nov. '74]

Solution The measure of skewness based on quartiles is given by Bowley's formula (7.7.3). We have to find the quartiles Q_1, Q_2, Q_3, by using simple interpolation (page 64) in the following table showing the cumulative frequencies of more-than type.

Table 7.2 Cumulative Frequencies

x	*Cum. Freq. (more-than)*
0	5474 = N
10	5426
20	5259
30	5023
40	4475
$Q_1\rightarrow$	$\leftarrow 3N/4 = 4105.5$
50	3712
$Q_2\rightarrow$	$\leftarrow 2N/4 = 2737$
60	2718
70	1406
$Q_3\rightarrow$	$\leftarrow N/4 = 1368.5$
80	764
90	370
100	160
110	39

Since Q_1, Q_2, Q_3 are values of x corresponding to (less-than) cumulative frequencies $N/4$, $2N/4$, $3N/4$ respectively, they also correspond to more-than cumulative frequencies $3N/4$, $2N/4$, $N/4$.

Using simple interpolation,

$$\frac{Q_1 - 40}{50 - 40} = \frac{4105.5 - 4475}{3712 - 4475}$$

$$\frac{Q_2 - 50}{60 - 50} = \frac{2737 - 3712}{2718 - 3712}$$

$$\frac{Q_3 - 70}{80 - 70} = \frac{1368.5 - 1406}{764 - 1406}$$

Solving, we get $Q_1 = 44.84$, $Q_2 = 59.81$, $Q_3 = 70.58$

Substituting these values in the formula

$$\text{Skewness} = \frac{70.58 - 2\times 59.81 + 44.84}{70.58 - 44.84} = \frac{-4.20}{25.74} = -0.16$$

Example 7.13 *A frequency distribution gives the following results: (i) Coefficient of variation = 5, (ii) Variance = 4, (iii) Karl Pearson's coefficient of skewness = 0.5. Find the mean and the mode of the distribution.* [I.C.W.A., Dec. '75 & '76-old]

Solution From (ii), S.D. = $\sqrt{(\text{Variance})} = \sqrt{4} = 2$.

Substituting the values of S.D. and C.V. in the formula

$$\text{C.V.} = \frac{\text{S.D.}}{\text{Mean}} \times 100, \quad \text{we get } 5 = \frac{2}{\text{Mean}} \times 100$$

Solving this, Mean = 40. Now, putting the values of Mean and Skewness in formula (7.7.1), viz.

$$\text{Skewness} = \frac{\text{Mean–Mode}}{\text{S.D.}}, \quad 0.5 = \frac{40 - \text{Mode}}{2}$$

we get Mode = 39. *Ans.* 40, 39

Example 7.14 *From the data given below, calculate the coefficient of variation: Pearson's measure of skewness = 0.42; Arithmetic mean = 86; Median = 80.* [C.A., Nov. '75-old]

Solution Using A.M. and Median, Pearson's measure of skewness is

$$\text{Skewness} = \frac{3(\text{Mean} - \text{Median})}{\text{S.D.}}; \quad \text{or,} \quad 0.42 = \frac{3(86-80)}{\text{S.D.}}$$

Solving this, S.D. = 18/0.42 = 43. Hence

$$\text{C.V.} = \frac{\text{S.D.}}{\text{Mean}} \times 100 = \frac{43}{86} \times 100 = 50.$$ *Ans.* 50

Example 7.15 *If the first quartile is 142 and the semi-interquartile range is 18, find the median (assuming the distribution to be symmetrical about mean or median).* [I.C.W.A., Dec. '78]

Solution Semi-interquartile range is given by the formula $(Q_3 - Q_1)/2$. Hence, 18 = $(Q_3 - 142)/2$. Solving, $Q_3 = 178$.

For a symmetrical distribution, Skewness = 0. Using quartiles,

$$\text{Skewness} = \frac{Q_3 - 2.Q_2 + Q_1}{Q_3 - Q_1} = 0;$$

i.e. $Q_3 - 2.Q_2 + Q_1 = 0$ or, $178 - 2.Q_2 + 142 = 0$.

Solving this, $Q_2 = 160$. *Ans.* Median = 160

Example 7.16 *The following facts were gathered from a firm before and after an industrial dispute:*

			Before Dispute	*After Dispute*
Mean Wages	...	*(Rs)*	*185*	*190*
Median Wages	...	*(Rs)*	*182*	*180*
Modal Wages	...	*(Rs)*	*176*	*160*
Quartiles	...	*(Rs)*	*175 and 192*	*175 and 195*
S.D.	...	*(Rs)*	*13*	*19*
Number Employed	...		*600*	*550*

Compare the position of the firm before and after the dispute by making use of the above data as fully as possible. [C.A., May '76-old]

Solution (a) Number of workers has decreased by 50, from 600 to 550 as a result of the dispute.

(b) Although the mean wage has slightly increased, the firm saves Rs 6,500 (after dispute) in respect of the monthly salary bill:

Total wages (before dispute) = 600 × 185 = 1,11,000 Rs
Total wages (after dispute) = 550 × 190 = 1,04,500 Rs
Difference = 6,500 Rs

(c) The median and modal wages have decreased. Before the dispute, 50% of the workers used to get Rs 182 and above, but after the dispute, workers in this category are less than 50%. Similarly, most of the workers are being paid around Rs 160 (after dispute) as against Rs 180 (before dispute).

(d) The first quartile Q_1 has not changed. The second quartile Q_2 (i.e. Median) has decreased slightly, but the third quartile Q_3 has increased. The significance of this information is as shown below:

Category of Workers	*Wages (Rs)* *Before Dispute*	*After Dispute*
(A) Lowest-paid 25%	Upto 175	Upto 175
(B) Next higher group of 25%	175–182	175–180
(C) Next higher group of 25%	182–192	180–195
(D) Highest-paid 25%	Above 192	Above 195

Category (A) workers are not affected. The next higher category (B) workers are now confined to a narrower range of salary. But the highest-paid categories (C) and (D) are now generally paid more after the dispute.

(e) S.D. has increased from Rs 13 to Rs 19, implying thereby that the variability in individual wages has increased after dispute.

C.V. (before dispute) = (13 ÷ 185) × 100 = 7
C.V. (after dispute) = (19 ÷ 190) × 100 = 10

The variability relative to mean has also increased.

(f) Measures of skewness are:

	Before dispute	*After dispute*
Pearson's measure	$\frac{185-176}{13} = 0.69$	$\frac{190-160}{19} = 1.58$
Bowley's measure	$\frac{192-2(182)+175}{192-175} = 0.18$	$\frac{195-2(180)+175}{195-175} = 0.50$

Skewness, i.e. the asymmetry in the distribution of wages has increased considerably after dispute.

Example 7.17 *Show that the measure of skewness based on third central moment is a pure number and is zero for symmetrical distribution.*

[C.U., B.Com. (Hons) '65]

Solution The measure of skewness based on the third central moment is

$$\gamma_1 = \frac{m_3}{\sigma^3} = \frac{\Sigma f(x-\bar{x})^3}{N\sigma^3}$$

where m_3 denotes the third central moment of the distribution, and σ is the standard deviation.

Let us suppose that the variable x represents length in ft. Then $\bar{x}$ and σ will be in ft., so that $(x - \bar{x})$ will be in ft. and $(x - \bar{x})^3$ will have "cubic ft." as unit. So, m_3 will be in cubic ft. and $\gamma_1 = m_3/\sigma^3$ is a ratio whose numerator is in cubic ft. and denominator also in cubic ft. Hence, the ratio is a pure number, independent of the unit of measurement.

If the values of x are symmetrically distributed about the mean, then for every positive value of $(x - \bar{x})$ there is a corresponding negative value. When these deviations $(x - \bar{x})$ are cubed, the negative values retain the negative sign and positive values the positive sign. Hence in $\Sigma f(x-\overline{x})^3$, positive and negative values cancel each other and become zero. Thus, γ_1 is zero for a symmetrical distribution. If the frequency curve has a longer tail towards the right, positive values will outweigh the negative values and m_3 will be positive, so that γ_1 will be positive. The case for a negatively skew distribution is just the opposite.

Example 7.18 *The frequency distribution of a large number of balls when classified according to their radius is symmetrical. Show that if we classify the balls according to their volume, the resulting frequency distribution will be skewed.*

[C.U., B.A. (Econ) '66]

Solution Let q_1, q_2, and q_3 represent the first, second and third quartiles of the frequency distribution of radius R. Then 25% of the balls have radius upto q_1, 50% have radius upto q_2, and 75% upto q_3. It is given that the frequency distribution of radius is symmetrical; hence the first quartile q_1 and the third quartile q_3 are equidistant from the second quartile q_2, on either side. Let

$$q_1 = q_2 - h, \qquad q_3 = q_2 + h, \qquad (h \text{ positive})$$

If now Q_1, Q_2 and Q_3 represent the corresponding quartiles of the frequency distribution of volume V, then 25% of the same balls have volume upto Q_1, 50% have volume upto Q_2, and 75% upto Q_3. Again, since the volume V of a ball of radius R is given by

$$V = \frac{4}{3}\pi R^3$$

and 25% of the balls have radius upto q_1, the same balls have volume upto Q_1. Substituting Q_1 and q_1 for V and R respectively

$$Q_1 = \frac{4}{3}\pi\, q_1^3 = \frac{4}{3}\pi\,(q_2 - h)^3$$

Similarly,

$$Q_2 = \frac{4}{3}\pi\, q_2^3$$

$$Q_3 = \frac{4}{3}\pi\, q_3^3 = \frac{4}{3}\pi\,(q_2 + h)^3$$

Using Bowley's formula (7.7.3) for the frequency distribution of volume V

$$\text{Skewness} = \frac{Q_3 - 2Q_2 + Q_1}{Q_3 - Q_1} = \frac{\frac{4}{3}\pi\{(q_2+h)^3 - 2q_2^3 + (q_2-h)^3\}}{\frac{4}{3}\pi\{(q_2+h)^3 - (q_2-h)^3\}}$$

$$= \frac{6q_2h^2}{2(3q_2^2h + h^3)} = \frac{3q_2h}{3q_2^2 + h^2}$$

This is positive (and *not zero*), because both q_2 and h are positive. Thus, the frequency distribution of volume is positively skew.

7.8 KURTOSIS

Kurtosis refers to the degree of "peakedness" of the frequency curve. Two distributions may have the same average, dispersion and skewness; yet, in one there may be high concentration of values near the mode, showing a sharper peak in the frequency curve than in the other. This characteristic of the frequency distribution is known as "kurtosis". The only measure of kurtosis is based on moments, viz.

$$\text{Kurtosis } (\gamma_2) = \frac{m_4}{\sigma^4} - 3 = \beta_2 - 3 \tag{7.8.1}$$

where m_4 and σ denote the fourth central moment and S.D. respectively. [Note that m_4/σ^4 is the fourth moment α_4 of the standardized variable z, i.e. $\alpha_4 = m_4/\sigma^4 = \beta_2$; see (7.3.3) and (7.4.3)].

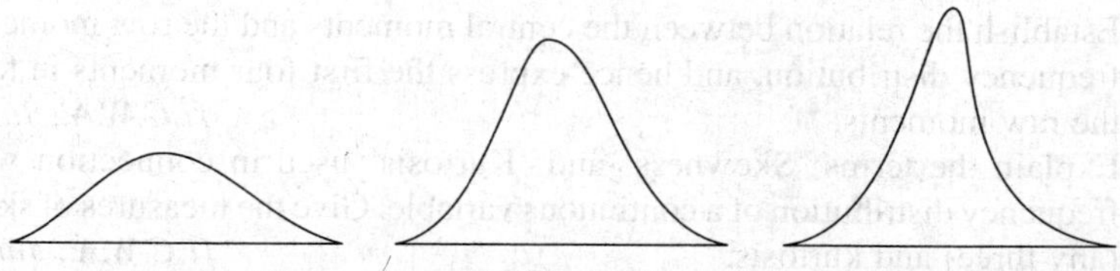

(a) Platykurtic (b) Mesokurtic (c) Leptokurtic

Fig, 7.2 Different Types of Kurtosis

$\beta_2 < 3$, for Platykurtic distribution
$\beta_2 = 3$, for Mesokurtic distribution
$\beta_2 > 3$, for Leptokurtic distribution

For the 'Normal Distribution' (Chapter 12), which is neither very peaked nor flat-topped, $\beta_2 = 3$. This important distribution is taken as a standard for measuring kurtosis, and it has become customary to use $\gamma_2 = \beta_2 - 3$ as a measure of kurtosis.

A distribution is said to be "*platykurtic*", when γ_2 is negative; it is said to be "*mesokurtic*" when $\gamma_2 = 0$, and "*leptokurtic*" when γ_2 is positive.

The frequency curve for a platykurtic distribution is relatively flat-topped, and for a leptokurtic distribution it has a relatively high peak. A mesokurtic distribution (e.g. Normal distribution) is of moderate peakedness.

EXERCISES

1. What is meant by 'moment' of a distribution? What are the 'raw' and the 'central' moments? *[I.C.W.A., June & Dec. '75]*
2. The first two moments of a distribution about the value 4 are –1.5 and 2.7. (a) Find the moments about zero. (b) Also calculate the mean and S.D.
3. The first three moments of a distribution about the value 3 of the variable are 2, 10 and 30 respectively. Obtain the first three moments about zero. Show also that the variance of the distribution is 6. *[I.C.W.A., Jan. '64]*
4. The first four moments about the value 1 are 2.6, 10.2, 43.4 and 192.6 respectively. Find the A.M. and the first four moments about 4.
5. The first three moments of a distribution about the value 7, calculated from a set of 9 observations, are 0.2, 19.4 and –41.0. Find the measures of central tendency, dispersion, and also the third moment about the origin. *[I.C.W.A., Dec. '75]*
6. Find the first, the second and the third central moments of the frequency distribution of expenditure (Rs per month) given below:

Expenditure	3–6	6–9	9–12	12–15	15–18	18–21	21–24	Total
No. of Families	28	292	389	212	59	18	2	1000

[I.C.W.A., June '78]

7. Find the first four moments and the values of β_1 and β_2 from the following frequency distribution:

x	21–24	25–28	29–32	33–36	37–40	41–44
f	40	90	190	110	50	20

Also, find the measures of skewness and kurtosis.

8. Establish the relation between the central moments and the row moments of a frequency distribution, and hence express the first four moments in terms of the raw moments. *[I.C.W.A., June '77]*

9. Explain the terms 'Skewness' and 'Kurtosis' used in connection with the frequency distribution of a continuous variable. Give the measures of skewness (any three) and kurtosis. *[I.C.W.A., June '77]*

10. Find the coefficient of skewness = (Mean – Mode)/S.D.:

Marks	55–58	58–61	61–64	64–67	67–70
Frequency	12	17	23	18	11

[D.M., '78]

11. Calculate Pearson's measure of skewness on the basis of Mean, Mode and Standard deviation–

x	14.5	15.5	16.5	17.5	18.5	19.5	20.5	21.5
f	35	40	48	100	125	87	43	22

[C.A., May '75]

12. Calculate from the undernoted table the measure of skewness based on Mean, Median and Standard Deviation:

X	100–200	200–300	300–400	400–500	500–600	600–700	700–800	800–900
F	45	88	146	206	79	52	30	14

[C.A., Nov. '73]

13. Calculate the measure of skewness based on quartiles and median from the following data:

Variable	10–20	20–30	30–40	40–50	50–60	60–70	70–80
Frequency	358	2417	976	129	62	18	10

[C.A., May '72]

14. Calculate the coefficient of skewness, based on quartiles:

Class Limits	10–19	20–29	30–39	40–49	50–59	60–69	70–79	80–89
Frequency	5	9	14	20	25	15	8	4

[I.C.W.A., Dec. '76]

15. Calculate Bowley's measure of skewness from the following:

x	10–14	15–19	20–29	30–39	40–49	50–59
f	786	924	320	172	96	32

[C.A., Nov. '75-old]

16. Compute Bowley's and Pearson's measures of skewness:

Monthly Income (Rs)	0–75	75–150	150–225	225–300	300–375	375–450
Frequency	15	200	250	225	10	5

[C.U., M.Com. '76]

17. Calculate the quartile measure of skewness for the distribution of time taken by 100 workers to complete a job:

Time (seconds)	–12	13–15	16–18	19–21	22–24	25–27	28–
No. of Workers	4	16	22	28	15	9	6

[I.C.W.A., Dec. '74-old & D.M., '77]

18. Find the appropriate measure of skewness from the following:

Age (years)	Below 20	20–25	25–30	30–35	35–40
No. of Employees	13	29	46	60	112
			40–45	45–55	55 and above
			94	45	21

[I.C.W.A., Dec. '75]

19. Compute an appropriate measure of skewness from:

Marks (per cent)	Under 30	30–40	40–50	50–60	60–70	70–80	80–90
No. of Students	45	40	24	12	9	3	2

[C.U., M.Com. '75]

20. Calculate with the use of quartiles the coefficient of skewness for the following frequency distribution:

Under Years	10	20	30	40	50	60
No. of Persons	15	32	51	78	97	109

[C.U., M.Com. '61]

21. Coefficient of skewness = – 0.375, Mean = 62, Median = 65. Find the value of standard deviation. *[C.A., Nov. '72]*

22. The measure of skewness for a certain distribution is – 0.8. If the lower and the upper quartiles are 44.1 and 56.6 respectively, find the median. *[I.C.W.A., Jan. '71]*

23. The median, mode and coefficient of skewness for a certain distribution are respectively 17.4, 15.3 and 0.35. Calculate the coefficient of variation. *[I.C.W.A., Jan. '73]*

24. The mean, median and the coefficient of variation of the weekly wages of a group of workers are respectively Rs 45, Rs 42 and 40. Find the (i) mode, (ii) variance, and (iii) coefficient of skewness, for the distribution of wages. *[I.C.W.A., Jan. '72]*

25. You are given: Mean = 50, C.V. = 40%, Sk. = – 0.4. Find the S.D., Mode and Median. *[C.A., Nov. '73]*
26. The first two moments of a distribution about the value 4 of the variable are – 1.5 and 2.7. It is also known that the median of the distribution is 2.1. Comment on the shape of the distribution. *[I.C.W.A., June '74-old]*
27. Define moments and quartiles. Show how they provide measures of dispersion and skewness. *[I.C.W.A., June '73]*
28. The following facts were gathered before and after an industrial dispute:

	Before Dispute	*After Dispute*
No. of Workers Employed	516	508
Mean Wages (Rs)	49.50	51.75
Median Wages (Rs)	52.70	50.00
Variance of Wages (Rs^2)	100.00	121.00

Compare the position before and after the dispute in respect of (a) total wages, (b) modal wages, (c) standard deviation, (d) coefficient of variation, (e) skewness. *[C.A., Nov. '70]*

ANSWERS

2. (a) 2.5, 6.7; (b) 2.5, 0.671
3. 5, 31, 201
4. 3.6; –0.4, 3.6, –5.2, 22.8
5. Mean = 72, S.D. = 4.4; 738.8
6. 0, 9.4 Rs^2, 17.4 Rs^3
7. mean = 31.3 Rs, $m_2 = 23.04$ Rs^2, $m_3 = 26.11$ Rs^3, $m_4 = 1496.68$ Rs^4; $\beta_1 = .056$, $\beta_2 = 2.82$; Skewness (γ_1) = + 0.24, Kurtosis (γ_2) = – 0.18
10. (62.46 – 62.64)/3.76 – 0.048
11. (18.07 – 18.40)/1.77 = – 0.19
12. 3(430.6 – 424.8/158.6 = + 0.11
13. + 0.13 (quartiles 22.63, 26.73, 32.07)
14. – 0.103 (quartiles 37.36, 50.30, 60.83)
15. + 0.105 (quartiles 13.21, 16.55, 20.67)
16. – .0070 (quartiles 135.47, 191.25, 246.25 Rs) and – 0.14 (mean = 190.69, mode = 200, S.D. = 67.36 Rs).
17. – .0063 (quartiles 16.18, 19.36, 22.50 sec.)
18. Bowley's measure = – 0.103 (quartiles 31.42, 37.77, 42.93 yr.)
19. Bowley's measure = – .081 (quartiles 22.50, 35.62, 46.77 p.c.).
20. – 0.14 (quartiles 17.2, 31.3, 42.0 years)
21. 24 22. 55.35 23. 49
24. (i) Rs 36, (ii) 324 Rs^2, (iii) 0.5. 25. 20, 58, 52.67
26. Sk. = + 1.8 (Frequency curve is asymmetrical and has the longer tail on the right hand side.
28. (a) 25,542 and 26,289 Rs, i.e. 3% increase; (b) 59.10 and 46.50 Rs; (c) 10 and 11 Rs; (d) 20.2 and 21.3; (e) – 0.96 and + 0.48.

8 CURVE FITTING AND METHOD OF LEAST SQUARES

8.1 CURVE FITTING

When observations in respect of two variables are available, very often a relation is found to exist between them. For example, height and weight of persons are interdependent, expenditure depends on income, yield of a crop depends on the amount of rainfall, production depends on price, etc. Frequency, it is found desirable to express this relationship between variables by means of some mathematical equation, representing a certain geometrical curve. The process of finding such a curve or its equation on the basis of a given set of observations is called *curve fitting.*

We list below the equations of some common types of curves:

1. $y = a + bx$	*Straight line*
2. $y = a + bx + cx^2$	*Parabola*
3. $y = a + bx + cx^2 + dx^3$	*Cubic curve*
4. $y = a + bx + cx^2 + ... + px^n$	*Polynomial of n-th degree*
5. $y = ab^x$	*Exponential curve*
6. $y = a + bc^x$	*Modified exponential curve*
7. $\frac{1}{y} = a + bc^x$	*Logistic curve*
8. $\log y = a + bc^x$	*Gompertz curve*
9. $y = ax^b$	*Geometric curve*

The variables x and y are often referred to as *independent* variable and *dependent* variable respectively. (All letters $a, b, c, d,.., p$, except x and y, appearing in the above equations represent constants.)

The choice of the appropriate equation to be fitted to a given set of observations is often facilitated by the following rules:

(i) Plot the corresponding observations of x and y as points on a graph paper. If the pattern of points shows approximately a linear path, use the straight line. (or, if the successive differences in y corresponding to equidistant values of x are approximately equal, use the straight line).

(ii) If *log y* when plotted against x shows a linear path, use the exponential curve.

(iii) If *log y* when plotted against $\log x$ show a linear path, use the geometric curve.

8.2 STRAIGHT LINE AND PARABOLA

Straight Line

Straight line is the geometrical representation of an equation of the form

$$Ax + By + C = 0 \tag{8.2.1}$$

where A, B, C are constants. When B is not zero, i.e. the equation contains a term in y, the equation of the straight line can be solved for y, giving $y = (-A/B)x + (-C/B)$, which is of the form

$$y = a + bx \tag{8.2.2}$$

where a and b are new constants. This is the form we shall generally use to represent a straight line. Sometimes, the form $y = mx + c$ is also used. [Note that the equations $y = a + bx$ and $y = mx + c$ are of the same form. The left hand side contains only y and the right hand side contains a constant and another term involving x.]

Illustration 1. The following equations represent straight lines:

(i) $y = 3 + 2x$
(ii) $y = 2x$
(iii) $y = 5$
(iv) $5x + 2y - 6 = 0$
(v) $ax + by + c = 0$
(vi) $4y + 7 = 0$
(vii) $3x - 8 = 0$
(viii) $y = 1600 + 200t$
(ix) $y - 3 = 2.8(x - 5)$
(x) $x - 4 = -5(y + 8)$

[Note that the equation of a straight line contains terms involving only single power of one or both the variables and a constant term. There must not be any term involving (say) x^2, xy, $\sqrt{x}$, $1/x$, etc. Here, equations (iv) to (vii) are of the form $Ax + By + C = 0$, and (i), (ii), (iii), (viii) belong to the type $y = a + bx$. Equations (ix) and (x) may also be brought to any of these two forms.]

Slope of Straight Line

In the equation $y = a + bx$, the coefficient of x on the right, viz. b, is called the '*slope*' of the straight line. The slope may be positive, negative, or zero. If the equation is given in this form, the slope can be obtained simply by inspection. If however the equation is given in the form $Ax + By + C = 0$, it should be reduced to the form $y = a + bx$ and then the slope easily obtained. If the equation of the straight line does not contain a term in y, the slope cannot be determined.

Illustration 2. (i) The slopes of equations (i), (ii), (iii) and (viii) in Illustration 1 above, are respectively 2, 2, 0 and 200.

(ii) To find the slope of the straight line $5x + 2y - 6 = 0$, let us write the equation as $2y = 6 - 5x$. Dividing both sides by 2, we get $y = 3 - \frac{5}{2}x$ which is of the form $y = a + bx$. The slope is seen to be $-\frac{5}{2}$.

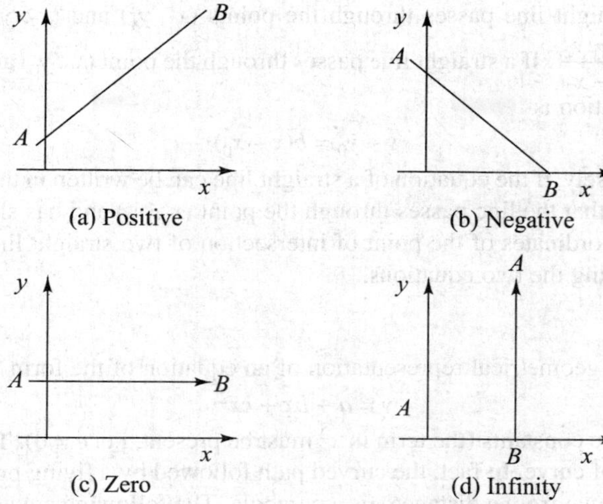

Fig. 8.1 Slopes of Straight Lines

(iii) The equation $y = 5$, can be written as $y = 5 + 0.x$. Thus, the slope of the straight line is zero.

(iv) The slope of the straight line $3x - 8 = 0$ is indeterminate.

When the slope is positive, y increases as x increases; when the slope is negative, y decreases as x increases; when the slope is zero, y remains a constant whatever be the value of x. Slope represents the amount of change (increase or decrease) in the value of y for a unit increase in the value of x. Conversely, if it is found that the change in y per unit increase in x is always the same, then the relation between x and y can be given by an equation of the form $y = a + bx$.

Illustration 3. (i) In the equation $y = 3 + 2x$, for $x = 0, 1, 2, 3$, the corresponding values are $y = 3, 5, 7, 9$. It is seen that the values of *y increase*, as x increases; because the slope is positive. Also the increase in y for successive values of x is always 2, which is the slope of the straight line.

(ii) In the equation $y = 6 - 2x$, for $x = 0, 1, 2, 3$, the values of y are 6, 4, 2, 0. That is, the values of *y decrease*, as x increases, the successive changes being –2, which is the slope.

Geometrically, the slope depends on the inclination of the straight line with the x-axis. Two parallel straight lines have the same slope. When the slope is zero, the straight line is parallel to the x-axis. As the slope increases, the inclination also increases. When the slope is positive, the straight line is inclined towards the right; when the slope is negative, the line is inclined towards the left (Fig. 8.1).

Some useful information about straight lines are:

(a) The constant term a on the right of the equation $y = a + bx$ is called the *y-intercept*, i.e. the value of y at the point where the line crosses the vertical axis ($x = 0$). The quantity b is called the *slope*.

(b) If a straight line $y = a + bx$ passes through the point (x_0, y_0), the co-ordinates must satisfy the equation, i.e. $y_0 = a + bx_0$.

(c) If a straight line passes through the points (x_1, y_1) and (x_2, y_2), its slope is $b = \dfrac{y_1 - y_2}{x_1 - x_2}$. If a straight line passes through the point (x_0, y_0) and has slope b, its equation is

$$y - y_0 = b(x - x_0) \tag{8.2.3}$$

Conversely, if the equation of a straight line can be written in this form we are certain that the line passes through the point (x_0, y_0) and has slope b.

(d) The co-ordinates of the point of intersection of two straight lines is obtained by solving the two equations.

Parabola

Parabola is the geometrical representation of an equation of the form

$$y = a + bx + cx^2 \tag{8.2.4}$$

where a, b, c are constants (the term in x^2 must be present, i.e. $c \neq 0$). The parabola is a special type of curve. In fact, the curved path followed by a flying projectile (e.g. a cricket ball thrown from a distance), is a parabola. The following equations represent parabolas:

(i) $y = 3 - 2x + 7x^2$ (ii) $y = 6 - 4x^2$

(iii) $y = 0.5x^2$ (iv) $y = a + bx + cx(x - 1)$

Example 8.1 *(a) Find the slopes of the following straight lines:*

(i) $y - 3 = 2.8(x - 5)$; (ii) $x - 4 = -5(y + 8)$

(b) Show that the straight line $3x - 2y + 7 = 0$ passes through the point (5, 11).

(c) Find the slopes of the straight lines passing through the points (i) (3, 8) and (5, 11), (ii) (– 2, 8) and (4, 12), (iii) (8, 5) and (12, 2).

(d) Find the equation of the straight line passing through the points (3, 8) and (5, 11).

(e) Show that the straight line $y - 3 = 2.8(x - 5)$ passes through the point (5, 3) and has slope 2.8.

(f) Find the slopes of the straight lines (i) $y - \bar{y} = b(x - \bar{x})$ and (ii) $x - \bar{x} = b'(y - \bar{y})$, where $\bar{x}$, $\bar{y}$, b and b' are constants.

(g) Show that the lines $2x - 3y - 1 = 0$ and $3x + 4y - 10 = 0$ intersect at the point (2, 1).

Solution (a) (i) The equation can be written as $y = 3 + 2.8\,(x - 5)$. It is seen that the left-hand side (L.H.S.) contains y only, and on the right-hand side (R.H.S.) the coefficient of x is clearly 2.8, which is the slope.

(ii) Dividing both sides by – 5, we get $\dfrac{x-4}{-5} = y + 8$, which is the same as $y + 8 = -\dfrac{1}{5}(x - 4)$. As discussed in (i), the slope is $-\dfrac{1}{5}$.

(b) Put $x^s = 5$ and $y = 11$, and we see that L.H.S. $= 3x - 2y + 7 = 3 \times 5 - 2 \times 11 + 7 = 15 - 22 + 7 = 0 =$ R.H.S. Thus the equation $3x - 2y + 7 = 0$ is satisfied. Hence the result.

(c) (i) Slope $= \dfrac{8-11}{3-5} = \dfrac{-3}{-2} = 3/2$. (ii) 2/3, (iii) – 3/4.

(d) As in (c), the slope is 3/2. Since the straight line passes through the point (3, 8) and has slope 3/2, by (8.2.3), its equation is $y - 8 = (3/2)\,(x - 3)$; simplifying, we get $3x - 2y + 7 = 0$.

(e) Put $x = 5$, $y = 8$, and we find that each side is zero; i.e. L.H.S. = R.H.S. Hence the straight line passes through (5, 8). See solution (a) (i) above.

(f) (i) This is of the form (8.2.3). Hence the straight line passes through the point $(\bar{x}, \bar{y})$ and has slope b. See solution (e) above.

(ii) Dividing both sides by b', we get $\dfrac{x-\bar{x}}{b'} = y - \bar{y}$, which is the same as $y - \bar{y}(1/b')$ $(x - \bar{x})$. This is of the form (8.2.3), indicating that the straight line passes through the point $(\bar{x}, \bar{y})$ and has slope $1/b'$.

[**Note:** It is interesting to observe that since both the lines $y - \bar{y} = b(x - \bar{x})$ and $x - \bar{x} = b'(y - \bar{y})$ pass through the same point $(\bar{x}, \bar{y})$, they intersect at the point].

(g) Solving the two equations, we get $x = 2$, $y = 1$. Hence the point of intersection is (2, 1).

8.3 FREE-HAND METHOD OF CURVE FITTING

When the given data are plotted as points on a graph paper, it is often possible to draw a smooth curve through the cluster of points, which appears best to represent their pattern. The smooth curve so drawn is called a *free-hand curve*. It may be noted that the free-hand curve depends entirely on individual judgement, and may either be a straight line or a curved line. If the pattern of points is linear, the equation of a straight line of the form $y = a + bx$ is obtained by choosing two points on the line. Let (1, 5) and (4, 11) be two such points. Substituting for x and y in the equation $y = a + bx$, we find the relation $5 = a + b$ and $11 = a + 4b$, solving which we get $a = 3$ and $b = 2$. Hence $y = 3 + 2x$ is the equation of the fitted free-hand curve, which may be used to estimate the value of y for any given value of x. Similarly, for other types of curves, it is necessary to choose as many points on the smooth curve as there are constants in the equation. The free-hand method has the disadvantage that different individuals will get different curves and equations.

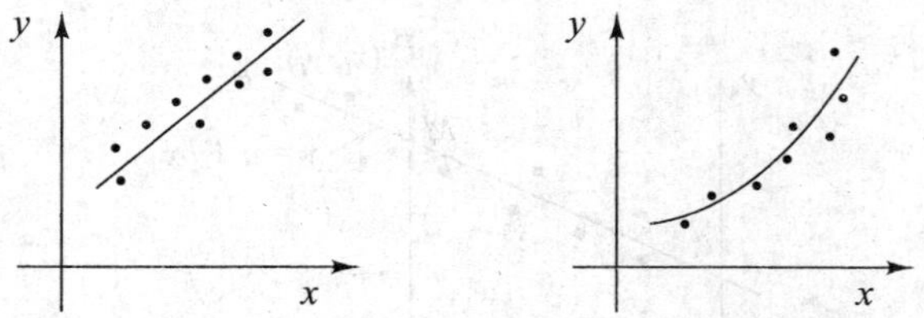

(a) Linear (b) Curvilinear

Fig. 8.2 Free-hand Curves

8.4 METHOD OF LEAST SQUARES

Method of Least Squares is a device for finding the equation of a specified type of curve, which best fits a given set of observations. The method depends upon the *Principle of Least Squares*, which suggests that for the "best-fitting" curve, the *sum of the squares of differences* between the observed and the corresponding estimated values should be the minimum possible.

Suppose, we are given n pairs of observations $(x_1, y_1), (x_2, y_2), \ldots (x_n, y_n)$, and it is required to fit a straight line to these data. The general equation of a straight line $y = a + bx$ is taken, where a and b are constants. Any values for a and b would give a straight line, and once these values are obtained, an estimate of y can be had by substituting the values of x. That is to say, the estimated values of y when $x = x_1, x_2, \ldots, x_n$ would be more $a + bx_1, a + bx_2, \ldots, a + bx_n$ respectively. In order that the equation $y = a + bx$ gives a good representation of the relationship between x and y, it is desirable that the estimated values $a + bx_1, a + bx_2, \ldots, a + bx_n$ are, on the whole, close enough to the corresponding observed values $y_1, y_2, \ldots, y_n$. For the best fitting straight line, therefore, our problem is only to choose such values of a and b for the equation $y = a + bx$ which will provide estimates of y as close as possible to the observed values. This can be done in different ways. However, according to the Principle of Least Squares, the "best-fitting" equation is interpreted as that which minimises the *sum of the squares* of differences $\Sigma(y_i - a - bx_i)^2$, i.e.

$$(y_1 - a - bx_1)^2 + (y_2 - a - bx_2)^2 + \ldots + (y_n - a - bx_n)^2$$

Principle of Fitting Straight Line by Least Squares Method

x (1)	*Observed y* (2)	*Estimated y = a + bx* (3)	*Difference = (2) – (3)* (4)	*(Difference)²* (5)
x_1	y_1	$a + bx_1$	$y_1 - a - bx_1$	$(y_1 - a - bx_1)^2$
x_2	y_2	$a + bx_2$	$y_2 - a - bx_2$	$(y_2 - a - bx_2)^2$
⋮	⋮	⋮	⋮	⋮
x_n	y_n	$a + bx_n$	$y_n - a - bx_n$	$(y_n - a - bx_n)^2$
Total		–	–	$\Sigma(y_i - a - bx_i)^2$

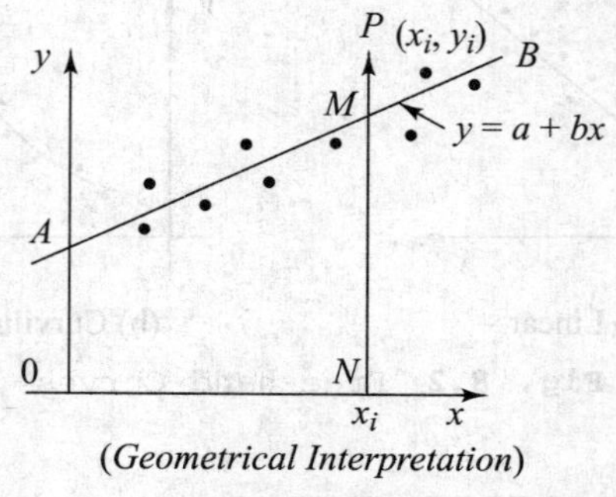

(*Geometrical Interpretation*)

Fig. 8.3 Method of Least Squares

In the geometric sense, the problem of finding the best-fitting straight line is as follows: If the pairs of observations $(x_1, y_1), (x_2, y_2), \ldots, (x_n, y_n)$ are plotted as points on a graph paper, and all possible straight lines are drawn on it (Fig. 8.3), that straight line will be considered to be the "best-fitting" for which the *sum of the squares of vertical distances* PM between the plotted points P and the line AB is the least.

Fortunately, we do not have to find the values of a and b (or geometrically select the straight line) by trial. The problem is tackled by mathematical methods. This leads to a set of equations, called *normal equations*, solving which we get the desired values of a and b. [See Mathematical Note at the end of this Chapter]

Similarly, the method of least squares can be used to fit other types of curves, e.g. parabola, exponential curve, geometric curve, etc. Method of Least Squares is applied to find regression lines (Section 9.12) and also in the determination of trend in time series (Section 14.4).

8.5 FITTING STRAIGHT LINE

Let

$$y = a + bx \qquad (8.5.1)$$

be the equation of the straight line to be fitted to a given set of n pairs of observations $(x_1, y_1), (x_2, y_2), \ldots, (x_n, y_n)$. Applying the method of least squares, the values of a and b are so determined as to minimise $\sum_{i=1}^{n}(y_i - a - bx_i)^2$. This leads to the *normal equations*

$$\Sigma y = an + b\Sigma x$$

$$\Sigma xy = a\Sigma x + b\Sigma x^2 \qquad (8.5.2)$$

Here, the values of n Σx, Σy, Σx^2 and Σxy are substituted on the basis of the given data. We have then two equations involving a and b, solving which the values of a and b are obtained.

[**Note:** The normal equation (8.5.2) can be easily remembered in the following way: Multiply both sides of (8.5.1) by the coefficients of a and b, viz. 1 and x, respectively; and then sum, using Σ notation. Thus

$$\Sigma y = \Sigma(a + bx) = \Sigma a + \Sigma bx = an + b\Sigma x$$

giving the first equation, and

$$\Sigma xy = \Sigma x\,(a + bx) = \Sigma(ax + bx^2) = \Sigma ax + \Sigma bx^2 = a\Sigma x + b\Sigma x^2$$

giving the second equation. However, it should be borne in mind that this is not a derivation of the normal equations, but only a means of remembering them].

8.6 SIMPLIFIED CALCULATIONS

The labour in computation can often be minimised by a transformation of one or both the variables, either by change of origin only or by changes of origin and scale both.

1. If we change the origin of x only, i.e. write $X = x - c$, where c is an arbitrary constant, then x is replaced by X in the equations (8.5.2), giving

$$\Sigma y = an + b\Sigma X$$

$$\Sigma Xy = a\Sigma X + b\Sigma X^2 \qquad (8.6.1)$$

2. If we change the origins of both x and y. i.e. write $X = x - c$ and $Y = y - c'$, where c and c are arbitrary constants, then x and y are replaced by X and Y in the equation (8.5.2), so that

$$\Sigma Y = an + b\Sigma X$$

$$\Sigma XY = a\Sigma X + b\Sigma X^2 \tag{8.6.2}$$

3. In case the successive values of the independent variable x are found to have a common difference, two special transformations are available for the cases, when (i) n is odd, and (ii) n is even.

(i) when n is odd, write

$$u = \frac{x - (\text{central value of } x)}{\text{common difference}} \tag{8.6.3}$$

The values of u will be 0 at the middle; 1, 2, 3, 4, ... for the successive values; and – 1, – 2, – 3, – 4, ... for the preceding values (see Tables 8.5, 8.8, 8.11, 8.12)

(ii) when n is even, write

$$u = \frac{x - (\text{mean of two central values of } x)}{\frac{1}{2}(\text{common difference})} \tag{8.6.4}$$

The values of u will be successively 1, 3, 5, 7, ... for the lower half, and –1, –3, –5, –7, ... for the upper half, starting form the middle (see Tables 8.6, 8.7).

If we write the equation of straight line in the form $y = a + bu$, the normal equations will be

$$\Sigma y = an + b\Sigma u$$

$$\Sigma uy = a\Sigma u + b\,\Sigma u^2 \tag{8.6.5}$$

But since by the above transformation $\Sigma u = 0$, these equations take very simple forms

$$\Sigma y = an$$

$$\Sigma uy = b\Sigma u^2 \tag{8.6.6}$$

from which the values of a and b can easily determined.

However, in all cases, finally we have to rewrite the transformed variables X, Y, u in terms of the original variables x and y, and then simplify the result, to get the equation of the fitted straight line.

Example 8.2 *Determine the equation of a straight line which best fits the following data:*

x	10	12	13	16	17	20	25
y	19	22	24	27	29	33	37

[C.U., M.Com,' 62; D.S.W., '69]

Solution (*First method*) Let $y = a + bx$ be the equation of the best fitting straight line by method of least squares. The constants a and b are obtained by solving the normal equations (8.5.2),

$$\Sigma y = an + b\Sigma x, \qquad \Sigma xy = a\Sigma x + b\Sigma x^2$$

where n is the number of pairs of observations.

Table 8.1 Calculations for Fitting Straight Line

x	y	x^2	xy
10	19	100	190
12	22	144	264
13	24	169	312
16	27	256	432
17	29	289	493
20	33	400	660
25	37	625	925
Total 113	191	1983	3276

Putting $\Sigma y = 191$, $n = 7$, $\Sigma x = 113$, $\Sigma xy = 3276$, $\Sigma x^2 = 1983$ in the normal equations, we have

$$191 = 7a + 113b \quad \text{...(i)}$$

$$3276 = 113a + 1983b \quad \text{...(ii)}$$

Multiplying (i) by 113 and (ii) by 7, and subtracting

$$\begin{aligned} 21{,}583 &= 791a + 12{,}769b \\ 22{,}932 &= 791a + 13{,}881b \\ \hline -1{,}349 &= \quad -1{,}112b \end{aligned} \qquad \therefore \quad b = \frac{-1349}{-1112} = 1.21$$

Putting the value of b in (i) we have

$$7a = 191 - 113 \times 1.21 = 54.27, \text{ so that } a = 54.27/7 = 7.75$$

Now, substituting the values of a and b in $y = a + bx$, the equation of the fitted straight line is

$$y = 7.75 + 1.21x \quad \text{...(iii)}$$

[**Note:** The accuracy of equation (iii) may be verified by observing that its both sides will be equal, when x and y are replaced by their means $\bar{x} = 113/7 = 16.14$ and $\bar{y} = 191/7 = 27.29$. Here $\bar{y} = 27.29$ and $7.75 + 1.21 \times 16.14 = 27.28$. The discrepancy of 1 in the last place is due to rounding error].

Ans. $y = 7.75 + 1.21x$

(*Second method*) Let $y = a + bX$ be the equation of the best fitting straight line, where we write $X = x - 16$. The normal equations for determining the values of a and b are (8.6.1), viz.

$$\Sigma y = an + b\Sigma X, \quad \Sigma Xy = a\Sigma X + b\Sigma X^2$$

Table 8.2 Calculations for Fitting Straight Line

x	y	$X = x - 16$	X^2	X_y
10	19	–6	36	–114
12	22	–4	16	–88
13	24	–3	9	–72
16	27	0	0	0
17	29	1	1	29
20	33	4	16	132
25	37	9	81	333
Total	191	1	159	220

Substituting the values $\Sigma y = 191$, $\Sigma X = 1$, $\Sigma X^2 = 159$, $\Sigma Xy = 220$ and $n = 7$ in the normal equations, we get

$$191 = 7a + b \quad \text{...(iv)}$$

$$220 = a + 159b \quad \text{...(v)}$$

Multiplying (v) by 7 and then subtracting from (iv), we have

$$\begin{aligned} 191 &= 7a + b \\ 1540 &= 7a + 1113b \\ \hline -1349 &= \quad -1112b \end{aligned} \qquad \therefore \quad b = \frac{-1349}{-1112} = 1.21$$

Substituting this value in (iv) we get

$$7a = 191 - 1.21 = 189.79, \qquad \therefore \quad a = 189.79/7 = 27.11$$

Hence, the equation of the best-fitting straight line is

$$y = a + bX$$

or,
$$\begin{aligned} y &= 27.11 + 1.21\,(x - 16) \\ &= 27.11 + 1.21x - 19.36 \\ &= 7.75 + 1.21x \end{aligned}$$

(*Third method*) Let $Y = a + bX$ be the equation of the best fitting straight line, where $X = x - 16$ and $Y = y - 26$ (Deviations should be taken from values close to the respective means; here $\bar{x} = 113/7 = 16.1$ and $\bar{y} = 191/7 = 27.3$). The normal equations are (8.6.2):

$$\Sigma Y = an + b\Sigma X, \qquad \Sigma XY = a\Sigma X + b\Sigma X^2$$

Table 8.3 Calculations for Fitting Straight Line

x	y	$X = x - 16$	$Y = y - 26$	X^2	XY
10	19	−6	−7	36	42
12	22	−4	−4	16	16
13	24	−3	−2	9	6
16	27	0	1	0	0
17	29	1	3	1	3
20	33	4	7	16	28
25	37	9	11	81	99
Total	–	1	9	159	194

Substituting the values in the normal equations, we get

$$9 = 7a + b \quad \text{...(vi)}$$

$$194 = a + 159b \quad \text{...(vii)}$$

Multiplying (vii) by 7 and substracting from (vi),

$$\begin{aligned} 9 &= 7a + b \\ 1358 &= 7a + 1113b \\ \hline -1349 &= \quad -1112b \end{aligned}\ ; \qquad \therefore \quad b = \frac{-1349}{-1112} = 1.21$$

Putting the value in (vi), $7a = 7.79$; $a = 7.79/7 = 1.11$

Therefore, $Y = a + bX$

or, $y - 26 = 1.11 + 1.21(x - 16)$

or, $y = 26 + 1.11 + 1.21x - 19.36$

or, $y = 7.75 + 1.21x$

[**Note:** The *First method* should be employed when the values of x and y are small, say x not exceeding 10 and y not exceeding 100. The *Second method* will be found convenient when the values of x exceed 10 and those of y are not very large. When the values of x and y are both large, the *Third method* should be employed. In case, the values of the independent variable x are found to have a common difference, the special methods shown in Examples 8.4 and 8.5 are the most convenient, requiring the minimum possible calculations].

Example 8.3 *Fit a straight line by the method of least squares to the following data:*

Age	*21*	*42*	*38*	*64*	*53*	*61*	*47*
Absence (no. of days)	*4*	*14*	*10*	*38*	*19*	*34*	*17*

Estimate the probable number of days absent when age is 40 years.

[C.U., B.A. (Econ)' 70]

Solution Let x denote age (in years) and y denote the number of days absent, and $y = a + bX$ be the equation of the fitted straight line, where $X = x - 47$ (**Note:** The origin of X should preferably be taken near the mean of the given values of x). Using the method of the squares, the normal equations for determining the values of a and b are

$$\Sigma y = an + b\Sigma X \qquad \Sigma Xy = a\Sigma X + b\Sigma X^2$$

Table 8.4 Calculations for Fitting Straight Line

Age (x)	*Absence (y)*	$X = x - 47$	X^2	*Xy*
21	4	– 26	676	– 104
42	14	– 5	25	– 70
38	10	– 9	81	– 90
64	38	17	289	646
53	19	6	36	114
61	34	14	196	476
47	17	0	0	0
Total 326	136	–3	1303	972

Putting the values from the table in the normal equations, we have

$$136 = 7a - 3b, \qquad 972 = -3a + 1303b$$

Solving these two equations we get $a = 19.77$ and $b = 0.791$. The equation of the fitted straight line is therefore

$$y = a + bX$$

or, $$y = 19.77 + 0.791\,(x - 47)$$

i.e. $$y = 0.791x - 17.41$$

When $x = 40$, $y = 0.791 \times 40 - 17.41 = 14$ days (approx.)

Ans. $y = 0.791x - 17.41$; 14 days

Example 8.4 *Fit a straight line to the following data by the method of least squares:*

x	*15*	*20*	*25*	*30*	*35*
y	*12*	*14*	*18*	*25*	*31*

Solution [**Note:** Here, the successive values of x have a common difference and n is *odd*; so, the transformation (8.6.3) will be the most suitable.]

Let us write the equation of the straight line in the form

$$y = a + bu \qquad \text{...(i)}$$

where $u = \dfrac{(x-25)}{5}$. The normal equation are

$$\Sigma y = an + b\Sigma u, \qquad \Sigma uy = a\Sigma u + b\Sigma u^2$$

Table 8.5 Calculations for Fitting Straight Line

x	y	$u = \dfrac{(x-25)}{5}$	u^2	uy
15	12	– 2	4	– 24
20	14	– 1	1	– 14
25	18	0	0	0
30	25	1	1	25
35	31	2	4	62
Total	100	0	10	49

Putting the values from Table 8.5 in the normal equations,

$$\begin{cases} 100 = a(5) + b(0) \\ 49 = a(0) + b(10) \end{cases}$$

or,

$$\begin{cases} 100 = 5a \\ 49 = 10b \end{cases}$$

Hence $a = 100/5 = 20$ and $b = 49/10 = 4.9$. Substituting these values in (i), and rewriting u is terms of x,

$$y = 20 + 4.9\left(\frac{x-25}{5}\right)$$

or

$$y = 20 + (0.98x - 24.5) = -4.5 + 0.98x$$

Thus, the required equation of the straight line is

$$y = -4.5 + 0.98x \qquad \text{...(ii)}$$

Ans. $y = -4.5 + 0.98x$

Example 8.5 *Fit a straight line to the following data by the method of least squares:*

x	15	20	25	30	35	40
y	12	14	18	25	31	44

Solution [**Note:** As in the previous Example, here also successive values of x have a common difference; but n is *even*. So the transformation (8.6.4) will be used.]

Let the equation of the straight line be $y = a + bu$, where $u = \dfrac{(x-27.5)}{2.5}$. The normal equations are the same as (8.6.5)

Table 8.6 Calculations for Fitting Straight Line

x	y	$u = \frac{(x-27.5)}{2.5}$	u^2	uy
15	12	– 5	25	– 60
20	14	– 3	9	– 42
25	18	– 1	1	– 18
30	25	1	1	25
35	31	3	9	93
40	44	5	25	220
Total	114	0	70	218

Substituting the values from Table 8.6 in the normal equations,

$$\begin{cases} 144 = a(6) + b(0) \\ 218 = a(0) + b(70) \end{cases}$$

or

$$\begin{cases} 6a = 144 \\ 70b = 218 \end{cases}$$

Hence $a = 144/6 = 24$, and $b = 218/70 = 3.114$.

Putting the values of a and b and rewriting u in terms of x,

$$y = 24 + 3.114\left(\frac{x - 27.5}{2.5}\right)$$

i.e.

$$y = -10.25 + 1.246x$$

Example 8.6 *The data below show the lengths (y) in cm. attained by a coiled spring corresponding to various weights (x) in gm. Fit a straight line of the form y = a + bx.*

x (gm.)	*100*	*200*	*300*	*400*	*500*	*600*
y (cm.)	*90.2*	*92.3*	*94.2*	*96.3*	*98.2*	*100.3*

[I.C.W. A., July '71]

Solution Let us write the equation of the straight line as $Y = A + BX$, where $X = (x - 350)/50$ and $Y = y - 90$. Using the method of least squares, the normal equations for determining the values of A and B are $\Sigma Y = An + B\Sigma X$, $\Sigma XY = A\Sigma X + B\Sigma X^2$

Table 8.7 Calculations for Fitting Straight Line

x	y	$X = \frac{x-350}{50}$	$Y = y - 90$	X^2	XY
100	90.2	– 5	0.2	25	– 1.0
200	92.3	– 3	2.3	9	– 6.9
300	94.2	– 1	4.2	1	– 4.2
400	96.3	1	6.3	1	6.3
500	98.2	3	8.2	9	24.6
600	100.3	5	10.3	25	51.5
Total	–	0	31.5	70	70.3

Substituting the values from the table in the normal equations,

$$\begin{cases} 31.5 = A \times 6 + B \times 0 \\ 70.3 = A \times 0 + B \times 70 \end{cases} \qquad \begin{cases} 6A = 31.5 \\ 70B = 70.3 \end{cases}$$

Solving, $A = 5.25$, $B = 1.004$. The equation of the straight line is therefore,

$$Y = A + BX$$

or,
$$y - 90 = 5.25 + 1.004\left(\frac{x - 350}{50}\right)$$

Simplifying we get $y = 88.22 + 0201x$ as the equation of the fitted straight line.

Ans. $y = 88.22 + .0201x$.

Example 8.7 *Fit straight line $y = mx + c$ to the following data and plot the observed and fitted values on the same graph paper:*

x	1	3	5	7	9	11	13	15	17
y	10	15	20	27	31	35	30	35	40

[C.U., B.A. (Econ) '74]

Solution (*First method*) [**Note:** There are $n = 9$ (odd) pairs of observations, and the values of x have a common difference 2. So we may use the special transformation as in Example 8.4]

Let us write the equation $y = mx + c$ in the form $y = a + bu$, where $u = (x - 9)/2$. The normal equations are,

$$\Sigma y = an + b\Sigma u, \qquad \Sigma uy = a\Sigma u + b\Sigma u^2.$$

Table 8.8 Fitting Straight Line

x	Observed Values (y)	$u = \frac{x-9}{2}$	u^2	uy	bu	Fitted Values $a + bu$
1	2	3	4	5	6	7
1	10	–4	16	–40	–13.88	13.12 = 13.1
3	15	–3	9	–45	–10.41	16.59 = 16.6
5	20	–2	4	–40	–6.94	20.06 = 20.1
7	27	–1	1	–27	–3.47	23.53 = 23.5
9	31	0	0	0	0	27.00 = 27.0
11	35	1	1	35	3.47	30.47 = 30.5
13	30	2	4	60	6.94	33.94 = 33.9
15	35	3	9	105	10.41	37.41 = 37.4
17	40	4	16	160	13.88	40.88 = 40.9
Total	243	0	60	208	–	243.0

Substituting the values in the normal equations, we get (since $\Sigma u = 0$), $243 = 9a$, $208 = 60b$. Therefore, $a = 243/9 = 27$, $b = 208/60 = 3.467$. The equation of the fitted line is therefore $y = a + bu$; i.e.

$$y = 27 + 3.467u$$

or,
$$y = 27 + 3.467\left(\frac{x - 9}{2}\right)$$

On simplification, we get $y = 11.40 + 1.73x$

The fitted values of y, corresponding to $x = 1, 3, 5, \ldots$, i.e. $u = -4, -3, -2, \ldots$, may be obtained from the equation $y = 27 + 3.467u$, and are shown in col. (7), Table 8.8.

The observed values are plotted on a graph paper as points with co-ordinates (1, 10), (3, 15), (5, 20), etc. The fitted values are also plotted with co-ordinates (1, 13.1), (3, 16.6), etc., and they lie on a straight line.

[**Note:** (*i*) The total of the observed values in col. (2) will be the same as the total of the fitted values in col. (7).

(ii) Since the successive values of u in col. (3) increase by 1, the values in col. (7) will increase by $b = 3.47$. When $x = 1$, i.e. $x = -4$, the fitted value is $27 + 3.47(-4) = 27 - 13.88 = 13.12$. The successive values may then be obtained by repeatedly adding 3.47; e.g. $13.12 + 3.47 = 16.59$; $16.59 + 3.47 = 20.06$; etc.]

(*Second method*) If we use the direct method with the equation as $y = mx + c$, then the normal equations will be

$$\Sigma y = m\,\Sigma x + cn, \qquad \Sigma xy = m\Sigma x^2 + c\Sigma x$$

Table 8.9 Fitting Straight Line

	Observed Values				Fitted Values
x	y	x^2	xy	mx	$mx + c$
1	10	1	10	1.73	13.13 = 13.1
3	15	9	45	5.19	16.59 = 16.6
5	20	25	100	8.65	20.05 = 20.0
7	27	49	189	12.11	23.51 = 23.5
9	31	81	279	15.57	26.97 = 27.0
11	35	121	385	19.03	30.43 = 30.4
13	30	169	390	22.49	33.89 = 33.9
15	35	225	525	25.95	37.35 = 37.4
17	40	289	680	29.41	40.81 = 40.8
Total 81	243	969	2603	–	–

Putting the values, the normal equations are $243 = 81m + 9c$ and $2603 = 969m + 81c$. Solving, $m = 1.73$, $c = 11.40$. Hence, the equation of the fitted line is $y = 1.73x + 11.40$.

Putting $x = 1, 3, 5$, etc. successively, we get the fitted values, as shown in the last column (Table 8.9). Note that some fitted values in Tables 8.8 and 8.9 differ slightly in the last place due to rounding error. Those in Table 8.8 are correct to one decimal.

8.7 FITTING PARABOLA

Let
$$y = a + bx + cx^2 \tag{8.7.1}$$
be the equation of the parabola to be fitted to a given set of n pairs of observations $(x_1, y_1), (x_2, y_2), \ldots, (x_n, y_n)$. Using the method of least squares, the constants a, b, c of the best fitting parabola are obtained by solving the normal equations

$$\left.\begin{aligned} \Sigma n &= an + b\Sigma x + c\Sigma x^2 \\ \Sigma xy &= a\Sigma x + b\Sigma x^2 + c\Sigma x^3 \\ \Sigma x^2 y &= a\Sigma x^2 + b\Sigma x^3 + c\Sigma x^4 \end{aligned}\right\} \tag{8.7.2}$$

[**Note:** (1) we have 3 equations in order to find the 3 constants a, b, c.

(2) As stated in the Note, p. 290, it will be found convenient to remember the normal equations in the following way: Multiply both sides of (8.7.1) by the coefficients of a, b, c, viz. 1, x, x^2 respectively, and then sum, using Σ notation.

Thus

$$\Sigma y = \Sigma(a + bx + cx^2) = \Sigma a + \Sigma bx + \Sigma cx^2 = an + b\Sigma x + c\Sigma x^2$$
$$\Sigma xy = \Sigma x(a + bx + cx^2) = \Sigma ax + \Sigma bx^2 + \Sigma cx^3 = a\Sigma x + b\Sigma x^2 + c\Sigma x^3$$
$$\Sigma x^2 y = \Sigma x^2(a + bx + cx^2) = \Sigma ax^2 + \Sigma bx^3 + \Sigma cx^4 = a\Sigma x^2 + b\Sigma x^3 + c\Sigma x^4]$$

In practical problems, simplications in calculation are often introduced either by change of origin of the independent variable x, or of both the variables x and y, as stated in (1) and (2) of Section 8.6 (p. 251 – 252). When the successive values of x have a common difference, the transformations suggested at (3) of Section 8.6 would, of course, be the most convenient.

Example 8.8 *The profits y (Rs lakh) of a certain company in the xth year of its life are given by*

x	*1*	*2*	*3*	*4*	*5*
y	*2.18*	*2.44*	*2.78*	*3.25*	*3.83*

Fit a second degree parabola $y = a + bx + cx^2$ to the data.

[I.C.W.A., Jan '68]

Solution (*First method*) The constants a, b, c appearing in the equation

$$y = a + bx + cx^2 \quad \text{...(i)}$$

are obtained by solving the normal equations (8.7.2).

Table 8.10 Calculations for Fitting Parabola

x	y	x^2	x^3	x^4	xy	x^2y
1	2.18	1	1	1	2.18	2.18
2	2.44	4	8	16	4.88	9.76
3	2.78	9	27	81	8.34	25.02
4	3.25	16	64	256	13.00	52.00
5	3.83	25	125	625	19.15	95.75
Total 15	14.48	55	225	979	47.55	184.71

Substituting the values from Table 8.10 (here $n = 5$)

$$14.48 = 5a + 15b + 55c \quad \text{...(ii)}$$
$$47.55 = 15a + 55b + 225c \quad \text{...(iii)}$$
$$184.71 = 55a + 225b + 979c \quad \text{...(iv)}$$

Multiplying (ii) by 3 and subtracting from (iii),

$$47.55 = 15a + 55b + 225c$$
$$43.44 = 15a + 45b + 165c$$
$$4.11 = 10b + 60c$$

i.e. $10b + 60c = 4.11$ ...(v)

Again multiplying (ii) by 11 and subtracting from (iv),

$$184.71 = 55a + 225b + 979c$$
$$159.28 = 55a + 165b + 605c$$
$$25.43 = 60b + 374c$$

i.e. $$60b + 374c = 25.43 \quad \text{...(vi)}$$

Solving (v) and (vi), we get $b = .081$ and $c = .055$. Putting these values in (ii), we have $a = 2.048$. Now putting the values of a, b and c in (i), the required equation of the parabola is

$$y = 2.048 + .081x + .005x^2 \quad \text{...(vii)}$$

(*Second method*)

[**Note:** Since the successive values of x have a common difference, and n is odd, we use the transformed variable u, defined by (8.6.3)].

Let $$y = a + bu + cu^2 \quad \text{...(viii)}$$

be the equation of the parabola, where $u = x - 3$. Applying the method of least squares, this normal equations for determining the constants a, b, c are

$$\begin{aligned} \Sigma y &= an + b\Sigma u + c\Sigma u^2 \\ \Sigma uy &= a\Sigma u + b\Sigma u^2 + c\Sigma u^3 \\ \Sigma u^2 y &= a\Sigma u^2 + b\Sigma u^3 + c\Sigma u^4 \end{aligned} \quad \text{...(ix)}$$

where n is the number of pairs of observations given.

Table 8.11 Calculations for the Fitting Parabola

x	y	$u = x - 3$	u^2	u^3	u^4	uy	u^2y
1	2.18	–2	4	–8	16	–4.36	8.72
2	2.44	–1	1	–1	1	–2.44	2.44
3	2.78	0	0	0	0	0	0
4	3.25	1	1	1	1	3.25	3.25
5	3.83	2	4	8	16	7.66	15.32
15	14.48	0	10	0	34	4.11	29.73

Putting the values from Table 8.11 in the normal equations

$$\begin{cases} 14.48 = a(5) + b\,(0) + c(10) \\ 4.11 = a(0) + b(10) + c(0) \\ 29.73 = a(10) + b(0) + c(34) \end{cases}$$

i.e.
$$\begin{cases} 14.48 = 5a + 10c \\ 4.11 = 10b \\ 29.73 = 10a + 34c \end{cases}$$

From the second equation, $b = 0.411$. Also solving the other two equations, $a = 2.786$ and $c = .055$. Now putting the values of a, b, c in (viii), and rewriting u in terms of x, i.e. $u = x - 3$,

$$\begin{aligned} y &= 2.786 + 0.411\,(x - 3) + .055(x - 3)^2 \\ &= 2.786 + 0.411\,(x - 3) + .055\,(x^2 - 6x + 9) \end{aligned}$$

Simplifying, we get

$$y = 2.048 + .081x + .055x^2$$

[**Check:** The accuracy of the fitted equation (vii) may be verified by observing that both sides of the equation will have the same value when y, x, x^2 are replaced by their means. In the present case,

$$\begin{aligned} \text{Mean of } y &= \Sigma y/n = 14.48/5 = 2.896 \\ \text{Mean of } x &= \Sigma x/n = 15/5 = 3 \\ \text{Mean of } x^2 &= \Sigma x^2/n = 55/5 = 11 \end{aligned}$$

Putting these values in (vii).

$$2.896 = 2.048 + .081 \times 3 + 0.55 \times 11 = 2.048 + 0.243 + 0.605 = 2.896].$$

Example 8.9 *Fit a second degree parabola to the given data, using the method of least squares:*

Length (in.)	*7.5*	*10.0*	*12.5*	*15.0*	*17.5*	*20.0*	*22.5*
Weight (oz.)	*1.9*	*4.5*	*10.1*	*17.6*	*27.8*	*40.8*	*56.9*

Solution Let x and y represent the length and the weight respectively. The equation of the parabola may be written as

$$y = a + bu + cu^2 \qquad \text{...(i)}$$

where
$$u = \frac{x - 15.0}{2.5}$$

[**Note:** If we use the direct method, the calculations would be large. However, since the values of x have a common difference, viz. 2.5, and n is odd, we use the transformed variable (8.6.3)].

Using the method of least squares, the normal equations for determining the constants a, b, c are

$$\Sigma y = an + b\Sigma u + c\Sigma u^2$$
$$\Sigma uy = a\Sigma v + b\Sigma u^2 + c\Sigma u^3$$
$$\Sigma u^2 y = a\Sigma u^2 + b\Sigma u^3 + c\Sigma u^4$$

Table 8.12 Calculations for Fitting Parabola

x	y	u	u^2	u^3	u^4	uy	u^2y
7.5	1.9	– 3	9	– 27	81	– 5.7	17.1
10.0	4.5	– 2	4	– 8	16	– 9.0	18.0
12.5	10.1	– 1	1	– 1	1	– 10.1	10.1
15.0	17.6	0	0	0	0	0	0
17.5	27.8	1	1	1	1	27.8	27.8
20.0	40.8	2	4	8	16	81.6	163.2
22.5	56.9	3	9	27	81	170.7	512.1
Total	159.6	0	28	0	196	255.3	748.3

Substituting for Σy, Σu, Σu^2 etc. in the normal equations (here $n = 7$), we get

$$159.6 = 7a + 28c,\ 255.3 = 28b,\ 748.3 = 28a + 196c$$

Solving $\qquad a = 17.6,\ b = 9.1 \text{ and } c = 1.3$

Putting these values in (i), the equation of the parabola is

$$y = 17.6 + 9.1u + 1.3u^2, \text{ where } u = (x - 15)/2.5 = (0.4x - 6).$$

Substituting for u, we have

$$y = 17.6 + 9.1(0.4x - 6) + 1.3(0.4x - 6)^2$$
$$= 17.6 + 3.64x - 54.6 + 1.3(0.16x^2 - 4.8x + 36)$$
$$= -37.0 + 3.64x + 0.208x^2 - 6.24x + 46.8$$

or, $$y = 9.8 - 2.60x + 0.208x^2 \qquad \text{...(ii)}$$

This is the required equation of the parabola.

Ans. $y = 9.8 - 2.60x + 0.208x^2$

8.8 FITTING EXPONENTIAL AND GEOMETRIC CURVES

In order to fit curves with equations of the form $y = ab^x$ and $y = ax^b$, the procedure is to take logarithms of both sides and then form normal equations. For example, to fit the *Exponential Curve* $y = ab^x$, we take logarithms of both sides, obtaining

$$\log y = (\log a) + x(\log b).$$

This can be written as

$$Y = A + Bx$$

where $Y = \log y$, $A = \log a$, $B = \log b$. The normal equations are then

$$\Sigma Y = An + B\Sigma x, \qquad \Sigma xY = A\Sigma x + B\Sigma x^2$$

These equations are solved for A and B, and then taking antilog, we find the values of a and b.

Similarly, the *Geometric Curve* $y = ax^b$ may be written as

$$\log y = (\log a) + b(\log x)$$

i.e.

$$Y = A + bX$$

where $Y = \log y$, $A = \log a$, $X = \log x$. The normal equations are

$$\Sigma Y = An + b\Sigma X, \qquad \Sigma XY = A\Sigma X + b\Sigma X^2$$

These are solved for A and b, and then a = antilog A is obtained.

Example 8.10 *Fit an equation of the form $y = ab^x$ to the following data:*

x	2	3	4	5	6
y	144	172.8	207.4	248.8	298.5

[B.U., B.A. (Econ) '69]

Solution Taking logarithms of both sides in the equation $y = ab^x$, we have $\log y = \log a + x(\log b)$, *i.e.*

$$Y = A + Bx$$

where $Y = \log y$, $A = \log a$ and $B = \log b$. Applying the method of least squares, the normal equations are

$$\Sigma Y = An + B\Sigma x, \qquad \Sigma xY = A\Sigma x + B\Sigma x^2$$

Table 8.13 Fitting Exponential Curve

	x	y	$Y = \log y$	x^2	xY
	2	144	2.1584	4	4.3168
	3	172.8	2.2375	9	6.7125
	4	207.4	2.3168	16	9.2672
	5	248.8	2.3959	25	11.9795
	6	298.5	2.4749	36	14.8494
Total	20	–	11.5835	90	47.1254

Substituting the values in the normal equations.

$$11.5835 = 5A + 20B$$
$$47.1254 = 20A + 90B$$

Solving these equations we get $A = 2.00014$ and $B = 0.07914$; i.e. upto 4 decimals $A = 2.0001$, $B = 0.0791$. Thus

$$\begin{cases} \log a = 2.0001 \\ \log b = 0.0791 \end{cases} \quad \text{or,} \quad \begin{cases} a = \text{antilog } 2.0001 = 100.0 = 100 \text{ approx.} \\ b = \text{antilog } 0.0791 = 1.200 = 1.2 \text{ approx.} \end{cases}$$

Therefore, the equation of the fitted curve is $y = 100(1.2)^x$.

Example 8.11 *Determine the constants of the curve $y = ax^n$ which best fits the data given below:*

x:	4	5	6	7	8
y:	8	12.5	18	24.5	32

Solution Taking logarithms of both sides in the equation $y = ax^n$, we have $\log y = (\log a) + n(\log x)$;

i.e. $$Y = A + nX \qquad \text{...(i)}$$

where $Y = \log y$, $A = \log a$, $X = \log x$. The normal equations for determining the constants A and n in (i) are

$$\Sigma Y = Ar + n\Sigma X, \qquad \Sigma XY = A\Sigma X + n\Sigma X^2$$

where r is the number of pairs of observations given (here, $r = 5$).

Table 8.14 Fitting Geometric Curve

x	y	x = log x from log-table	Rounded	Y = log y from log-table	rounded	X^2	XY
(1)	(2)	(3)	(4)	(5)	(6)	(7)	(8)
4	8	0.6021	0.60	0.9031	0.90	0.3600	0.5400
5	12.5	0.6990	0.70	1.0969	1.10	0.4900	0.7700
6	18	0.7782	0.78	1.2553	1.26	0.6084	0.9828
7	24.5	0.8451	0.85	1.3892	1.39	0.7225	1.1815
8	32	0.9031	0.90	1.5051	1.51	0.8100	1.3590
Total	–	–	3.83	–	6.16	2.9909	4.8333

***Note:** Values of log x and log y, obtained from 4-figure log tables, are rounded to two demicals for simplifying the calculations.

Putting the values from Table 8.14, we have

$$6.16 = 5A + 3.83n \qquad \text{...(ii)}$$
$$4.8333 = 3.83A + 2.9909n \qquad \text{...(iii)}$$

Dividing both sides of (ii) by 5 and both sides of (iii) 3.83,

$$1.232 = A + 0.766n \qquad \text{...(iv)}$$
$$1.262 = A + 0.781n \qquad \text{...(v)}$$

Subtracting $-0.030 = -0.015n$; $\therefore\ n = 2$

Putting $n = 2$ in (iv), $A = 1.232 - 2(0.766) = -0.3$

i.e. $\log a = -0.3 = \bar{1}.7$; or, $a = \text{antilog } (\bar{1}.7) = 0.5$

Thus, the constants of the curve $y = ax^n$ are $a = 0.5$ and $n = 2$, and the equation of the curve is $y = 0.5x^2$ (note that this is the equation of a parabola).

Ans. $a = 0.5$, $n = 2$

EXERCISES

1. Explain the method of least squares for fitting a straight line $y = mx + c$ to a given set of n pairs of observations $(x_1, y_1), (x_2, y_2), ...(x_n, y_n)$. Indicate some uses of this method in business statistics. *[I.C.W.A., Dec '74 (old)]*
2. Fit a straight line of the form $y = a + bx$ to each of the following sets of data:

(i)

x	2	5	6	8	9
y	8	14	19	20	31

(ii)

x	1	3	5	7	9	11
y	16	12	10	7	5	4

(iii)

x	4	6	8	12	15	17	22
y	46	42	40	36	30	25	19

(iv)

x	1.0	1.5	2.0	2.5	3.0	3.5	4.0
y	5.3	5.7	6.3	7.2	8.2	8.7	8.4

(v)

x	70	80	90	100	110	120
y	553	547	539	533	527	520

(vi)

x	5	7	9	11	13	15
y	1.7	2.4	2.8	3.4	3.7	4.4

3. Given the following data on quantities exchanged and prices, fit a linear demand curve $y = a + bx$.

Price (x)	16	10	8	9	5	4	3
Quantity (y)	70	85	100	115	120	124	130

[C.U., M.Com. (Business Econ) '70]

4. Apply the principle of least squares to fit a straight line $y = a + bx$ to the following data:

x:	2	4	6	8	10	12	14
y:	10	14	15	16	15	17	18

[C.U., B.Sc. (Math. Hons) '68]

5. Fit a straight line to the following data and estimate the most probable yield of rice for 40 inches of water.

Water x (inches):	12	18	24	30	36	42	48
Yield y (tons):	5.27	5.68	6.25	7.21	8.02	8.71	8.42

[C.U., M.Com. '64]

6. Calculate the values of m and k for the equation $y = mx + k$ to show the regression of profit per unit of output on output.

Output x ('000)	5	7	9	11	13	15
Profit per unit of Output y (Rs)	1.7	2.4	2.8	3.4	3.7	4.4

Estimate the profit per unit of output when there is an output of 10,500.

[*I.C.W.A., June '73*]

7. The following data relate to results of a fertiliser experiment on crop yields:

Units of Fertiliser used (X)	0	2	4	6	8	10
Units of Yield (Y)	110	113	118	119	120	118

Fit a straight line to the above data and estimate the amounts of yield when units of fertiliser used are 3 and 7 respectively. [*C.U., M.Com. (Old) '69*]

8. The weights (in 1bs.) of a calf taken at weekly intervals are given below. Fit a straight line, and calculate the average rate of growth per week.

Age (x)	1	2	3	4	5	6	7	8	9	10
Weight (y)	52.5	58.7	65.0	70.2	75.4	81.1	87.2	95.5	102.2	106.4

[**Hint:** Since n even, take origin of x at 5.5 and 1/2 as unit. i.e. $y = a + by$, where $u = (x - 5.5)/(1/2) = 2x - 11$. The values of u will be as usual –9, –7, –5, –3, –1, 1, 3, 5, 7, 9].

9. In the following table, S is the weight of potassium bromide which will dissolve in 100 gms of water at $T°C$. Fit an equation of the form $S = mT + b$ by the method of least squares. Use this relation of estimate S, when $T = 50°$

T	0	20	40	60	80
S	54	65	75	85	96

[*I.C.W.A., Jan. '65*]

10. Fit a curve of the form $y = ab^x$ to the following data:

x	2	3	4	5	6	7
y	640	512	410	328	262	210

11. Fit a curve of the form $y = ax^b$ to the following data:

x	1	2	3	4	5
y	5.0	6.3	7.2	7.9	8.5

12. Estimate the constants of the Pareto Curve $n = Ax^{-a}$ which fits the data below: 1945 – 46: Number of Net Incomes more than Rs x after Tax

Income (Rs x)	*Number* (n)
150	14,000,000
500	825,000
1000	173,000
2000	35,500

[*I.C.W.A., Dec. '73*]

[**Hint:** Use logarithm and reduce the curve to straight line form; i.e. $\log n = (\log A) + (-a)(\log x)$; or, $Y = a' + b'X$]

13. Fit a second degree parabola ($y = a + bx + cx^2$), to the following data:

x	0	1	2	3	4
y	1	5	10	22	38

[*C.U., M.Com. '66*]

14. Fit a parabola of the second degree to the following data taking x as the independent variable ($y = a + bx + cx^2$), by the method of least squares.

x	0	1	2	3	4
y	1	1.8	1.3	2.5	6.3

Find out the difference between the actual value of y and the value of y obtained from the fitted curve when $x = 2$. [*I.C.W.A. July '65*]

15. The profits (in '000 Rs) of a company in the x-th year of its life are given below. Fit a parabola and estimate its profits in the sixth year.

x	1	2	3	4	5
y	1250	1400	1650	1950	2300

[**Hint:** Write the equation in the form $v = a + bu + cu^2$, where $u = x - 3$ and $v = (y - 1650)/50$].

ANSWERS

2. (i) $y = 1.0 + 2.9x$ (ii) $y = 16.2 - 1.2x$
(iii) $y = 52 - 1.5x$ (iv) $y = 4.04 + 1.23x$
(v) $y = 599.2 - 0.66x$ (vi) $y = 0.50 + 0.257x$
3. $y = 143.0 - 4.667x$
4. $y = 10.72 + 0.536x$
5. $y = 3.99 + 0.103x$; 8.11 tons
6. $y = 0.50 + 0.257x$; 3.20 Rs
7. $Y = 111.88 + 0.886X$; 114.5 and 118.1 units.
8. $y = 46.13 + 6.053x$; 6.053 lbs.
9. $S = 0.52T + 54.2$; 80.2 units
10. $y = 1000(0.8)^x$
11. $y = 5.0x^{0.33}$
12. $\log A = 12.22$, $a = -2.33$
13. $y = 1.42 + 0.26x + 2.21x^2$
14. $y = 1.42 - 1.07x + 0.55x^2$; 0.18
15. $y = 1140 + 72.1x + 32.2x^2$; Rs 2732

MATHEMATICAL NOTE

How the Normal Equations are Derived

(1) Fitting straight line $y = a + bx$ (page 290):

Applying the Method of Least Squares, the constants a and b are so chosen as to minimise $\sum_{i=1}^{n}(y_i - a - bx_i)^2$. This is a problem of Differential Calculus. Taking partial derivatives with respect to a and b, and equating them to zero, we get

$$\frac{\partial}{\partial a}\Sigma(y - a - bx)^2 = 0, \qquad \frac{\partial}{\partial b}\Sigma(y - a - bx)^2 = 0$$

i.e. $\begin{cases} \Sigma(y - a - bx)\,(-2) = 0 \\ \Sigma(y - a - bx)\,(-2x) = 0 \end{cases}$ or, $\begin{cases} \Sigma(y - a - bx) = 0 \\ \Sigma x(y - a - bx) = 0 \end{cases}$

These give the normal equations (8.5.2).

(2) Fitting parabola $y = a + bx + cx^2$ (page 299):

Here, the constants a, b, c are so chosen as to minimise $\sum_{i=1}^{n} (y_i - a - bx_i - cx_i^2)^2$.

Taking partial derivatives with respect to a, b, c, and equating them to zero,

$$\frac{\partial}{\partial a}\Sigma(y - a - bx - cx^2)^2 = 0; \qquad \frac{\partial}{\partial b}\Sigma(y - a - bx - cx^2)^2 = 0;$$

$$\frac{\partial}{\partial c}\Sigma(y - a - bx - cx^2)^2 = 0$$

These lead to the normal equation (8.7.2).

9 CORRELATION AND REGRESSION

9.1 CONCEPTS OF 'CORRELATION' AND 'REGRESSION'

So far we have studied some characteristics of one variable only, e.g. mean of the distribution of height, standard deviation of weight, skewness of the distribution of income. But, many situations arise in which we may have to study two variables simultaneously, say x and y. For example, the variables may be (i) the amount of rainfall and yield of a certain crop, (ii) the height and weight of a group of children, (iii) income and expenditure of several families, (iv) ages of husband and wife; etc. There are two main problems involved in such studies :

Firstly, the data may reveal some association between x and y, and we may be interested to measure numerically the strength of this association between the variables. Such a measure will determine how well a linear or other equation explains the relationship between the variables. This is the problem of *Correlation.*

Secondly, there may be one variable of particular interest, and the other variable, regarded as an auxiliary variable, may be studied for its possible aid in throwing some light on the former. In such a case, one is then interested in using a mathematical equation for making estimates or predictions regarding the principal variable. This equation is known as a *Regression Equation*, and the problem of making predictions on the basis of the equation is called the problem of *Regression.*

In short, correlation is concerned with the measurement of the 'strength of association' between variable; while regression is concerned with the 'prediction' of the most likely value of one variable when the value of the other variable is known.

In *simple correlation* and *simple regression* (also called *linear correlation* and *linear regression*), we consider the simplest kind of relationship, viz. a linear relationship, as the regression equation. Simple correlation is, therefore, concerned with the strength of linear trpe of relationship between the variables.

9.2 BIVARIATE DATA

The word '*bivariate*' is used to describe situations in which two characters are measured on each individual or item, the characters being represented by two variables. Statistical data relating to the simultaneous measurement of two variables are called *bivariate data.* (Data relating to one variable only are called *univariate data*). The observations on each individual are then paired, one for each variable—(x_1, y_1), (x_2, y_2), ..., (x_n, y_n), suppose.

Table 9.1 Bivariate Data Showing Height and Weight of 15 Individuals

Height (inches)	*Weight (lbs.)*	*Height (Inches)*	*Weight (lbs.)*
68	135	66	131
64	127	70	149
69	149	70	130
62	133	59	118
67	145	72	139
63	114	71	136
67	152	67	136
68	157		

The above data show that height (in inches) and weight (in lbs.) for the 1st individual are (68, 135), for the 2nd individual (64, 127), ..., for the last individual (67, 136).

9.3 BIVARIATE FREQUENCY DISTRIBUTION

Example 9.1 *Explain clearly the method of constructing a bivariate frequency distribution and explain the meaning of marginal distribution, conditional distribution and conditional mean values with reference to such a table.*

[C.U. B.Com.(Hons.)' 65]

Solution In bivariate data, when a large number of pairs of observations is available, it becomes necessary to condense the data in the form of a two-way frequency table, called *Bivariate Frequency Table,* or *Bivariate Frequency Distribution,* or *Correlation Table.*

The method of constructing a bivariate frequency distribution is similar to that of a univariate frequency distribution (p. 91). Suppose, we are given N pairs of observations (x_1, y_1), (x_2, y_2), ..., (x_N, y_N) relating to two variables x and y. At first we find the maximum and the minimum values in each of the x- and y-series, based on which the number of class intervals and the class limits are fixed. If there are m class intervals for the x-series and n for the y-series, we construct a table with m rows and n columns, thus giving $m \times n$ rectangular spaces, called *cells*. The class limits of the x-series and y-series are shown as row-headings and column-headings respectively. Each pair of observations is now represented by a tally mark (/) placed side by side, in the appropriate cell. Every fifth tally mark, in any cell is however placed across the preceding four to form a group of 5 tally marks (page 93). When all the N pairs of observations have been shown in the two-way table by tally marks, these marks are counted and the number of observations falling in each cell, i.e. the *cell frequencies*, are obtained. A two-way table like Table 9.2 is finally prepared showing the class intervals for x and y and the corresponding frequencies shown in the cells, instead of the tally marks. This is the Bivariate Frequency Distribution.

Table 9.2 Bivariate Frequency Distribution

Weight (lbs.)

Height (inches) x \\ y	*111–120*	*121–130*	*131–140*	*141–150*	*151–160*	*Total*
58–61	1					1
62–65	1	1	1			3
66–69			3	2	2	7
70–73		1	2	1		4
Total	2	2	6	3	2	15 = N

Source: *Data of Table 9.1*

A univariate frequency distribution, derived from the bivariate frequency distribution, *irrespective of the values assumed by the other variable*, is called a *Marginal Distribution.* Totals of call frequencies shown in the last row and the last column and called *marginal totals* or *marginal frequencies*. The marginal totals in the last column represent class frequencies of the distribution of height, called the 'marginal distribution of height' whose class intervals are shown in the first column. Similarly, the marginal totals in the last row represent class frequencies of the 'marginal distribution of weight', whose class intervals are shown in the first row.

Table 9.3 Marginal Distribution of Height

Height (inches)	*Frequency*
58–61	1
62–65	3
66–69	7
70–73	4
Total	15

Table 9.4 Marginal Distribution of Weight

Height (lbs.)	*Frequency*
111–120	2
121–130	2
131–140	6
141–150	3
151—160	2
Total	15

A univariate frequency distribution, derived from the bivariate frequency distribution, *for a specified value (or class interval) of the other variable*, is called a *Conditional Distribution.*

Table 9.5 Conditional Distribution of Height
(when weight 131–140 lbs.)

Height (inches)	*Frequency*
58–61	0
62–65	1
66–69	3
70–73	2
Total	6

Table 9.6 Conditional Distribution of Weight
(when height 70–73 inches)

Weight (lbs.)	*Frequency*
111–120	0
121–130	1
131–140	2
141–150	1
151–160	0
Total	4

There is only one marginal distribution of x and one marginal distribution of y. But there is one conditional distribution of x corresponding to each class interval of y; and similarly, corresponding to each class interval of x, there is one conditional distribution of y.

The *Conditional Mean Value* is the arithmetic mean calculated from the conditional distribution.

Example 9.2 *The ages of twenty husbands and wives are given below : From a two-way frequency table showing the relationship between the ages of husbands and wives with class intervals 20–25, 25–30 etc.*

S. No.	*1*	*2*	*3*	*4*	*5*	*6*	*7*	*8*	*9*	*10*	*11*	*12*	*13*	*14*	*15*	*16*	*17*	*18*	*19*	*20*
Age of Husband	*28*	*37*	*42*	*25*	*29*	*47*	*37*	*35*	*23*	*41*	*27*	*39*	*23*	*33*	*36*	*32*	*22*	*29*	*38*	*48*
Age of Wife	*23*	*30*	*40*	*26*	*25*	*41*	*35*	*25*	*21*	*38*	*24*	*34*	*20*	*31*	*29*	*35*	*23*	*27*	*34*	*47*

[C.A., May '70]

Solution

[**Note:** For the construction of a frequency distribution, we must use *class limits*. But in the question, the class intervals have been specified by means of class boundaries, viz. 20–25, 25–30, etc. Our class limits should therefore be interpreted either as "20 and above but below 25; 25 and above but below 30; etc.", or as "Above 20 but not exceeding 25; Above 25 but not exceeding 30; etc." (see Example 4.3 (*k*, 1), p. 77). The latter type is not applicable here, as the age 20 of wife at Sl. No. 13 cannot be accommodated. The given class intervals 20–25, 25–30, etc. must therefore be interpreted in the sense "20 and above but below 25, 25 and above but below 30, etc."]

Table 9.7 Two-way Frequency Table showing Ages of Husbands and Wives

Age of Wife (years)

Age of Husband (years)	*20 – 25**	*25–30*	*30–35*	*35–40*	*40–45*	*45–50*	*Total*
20–25*	3						3
25–30	2	3					5
30–35			1	1			2
35–40		2	3	1			6
40–45				1	1		2
45–50					1	1	2
Total	5	5	4	3	2	1	20

*Class intervals 20–25, 25–30 etc. represent "20 and above but below 25, 25 and above but below 30, etc."

Example 9.3 *Find the conditional mean values of y when x = 1, 3 and 5:*

x \ *y*	*0*	*3*	*6*	*9*	*12*
1	*1*	*2*	*1*	*0*	*1*
3	*2*	*3*	*4*	*1*	*0*
5	*3*	*1*	*3*	*1*	*6*

[C.U., B.Com. (Hons.)' 68]

Solution

Table 9.8 Calculations for Conditional Mean Values of y

(a) when $x = 1$

y	*f*	*fy*
0	1	0
3	2	6
6	1	6
9	0	0
12	1	12
Total	5	24

(b) when $x = 3$

y	*f*	*fy*
0	2	0
3	3	9
6	4	24
9	1	9
12	0	0
Total	10	42

(c) when $x = 5$

y	*f*	*fy*
0	3	0
3	1	3
6	3	18
9	1	9
12	6	72
Total	14	102

$$\bar{y} = \frac{\Sigma fy}{N} = \frac{24}{5} = 4.8 \qquad \bar{y} = \frac{42}{10} = 4.2 \qquad \bar{y} = \frac{102}{14} = 7.3$$

The conditional mean values of y when $x = 1, 3, 5$ are respectively 4.8, 4.2, 7.3.

9.4 SCATTER DIAGRAM

Example 9.4 *What is a Scatter Diagram ? Explain how this can be used to indicate the degree and type of association between two variables.*

[D.S.W., Nov., '70; I.C.W.A., Jan. '67, July '71, Dec. '73, '74; C.U., M.Com. '63, '67; B.A. (Econ) '71, '73, '75; B.U., B.A. '77]

Solution When statistical data relating to the simultaneous measurement on two variables are available, each pair of observations can be geometrically represented by a point on the graph paper—the values of one variable being shown along the *X*-axis and those of the other variable along *Y*-axis. If there are *n* pairs of observations, finally the graph paper will contain *n* points. This diagrammatic representation of bivariate data is known as *Scatter Diagram* (Figs 9.1 and 9.2).

A scatter diagram indicates the nature of association between the two variables, i.e., the type of correlation between them. If the pattern of points (or dots) on the scatter diagram shows a linear path diagonally across the graph paper from the the bottom left-hand corner to the top right, correlation will be positive (Fig. 9.1a). In other words, association between the variables is direct, indicating thereby that high values of one variable are in general, associated with high values of the other variable, and low values are associated with low values.

On the other hand, if the pattern of dots be such as to indicate a straight line path from the upper left-hand corner to the bottom right, correlation is negative, i.e. the association is indirect, high values of one variable being associated with low values of the other (Fig. 9.1b).

When the dots do not indicate any straight line tendency, but a swarm (Fig. 9.1c), or concentration around a curved line, correlation is small (Fig. 9.1d). In fact, if no straight line tendency is noticed, correlation will be zero.

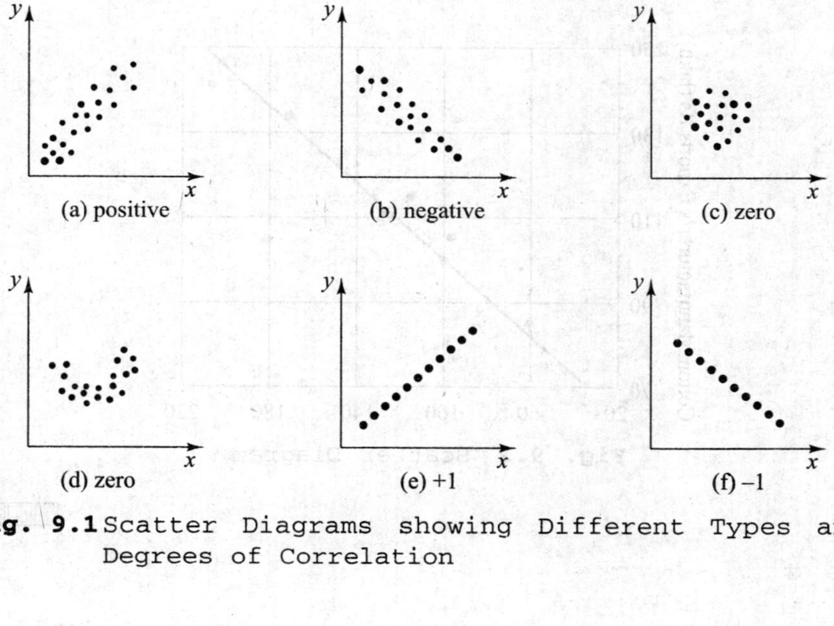

Fig. 9.1 Scatter Diagrams showing Different Types and Degrees of Correlation

When the dots lie exactly on a straight line, correlation is perfect—the correlation coefficient being + 1 or – 1, according as the slope of the straight line is positive or negative (Figs 9.1e, 9.1f).

The scatter diagram also gives an indication of the degree of linear correlation between the variables, i.e. whether correlation is high or low. If the plotted points on the scatter diagram lie approximately on, or near about, a straight line (Figs 9.1e, 9.1f, 9.2), correlation coefficient will be nearly one, numerically. The more scattered the points are around a straight line, the less is the correlation coefficient (Figs 9.1a, 9.1b). A scatter diagram like Fig. 9.1c will indicate almost zero correlation.

Example 9.5 *Construct the scatter diagram of the data given below. Draw a free hand straight line through the group of points and from your diagram discuss the probable amount of correlation:*

	Average Values in Lakhs							
	1946	*1947*	*1948*	*1949*	*1950*	*1951*	*1952*	*1953*
Raw Cotton Imports	*47*	*64*	*100*	*97*	*126*	*203*	*171*	*115*
Cotton Manufacturers' Exports	*70*	*85*	*100*	*103*	*111*	*139*	*133*	*115*

[C.U., M.Com. '69, 72]

Solution The variables here are Raw Cotton Imports (x, say) and Cotton Manufactures' Exports (y). The 8 pairs of observations (47, 70), (64, 85), ..., (115, 115) are plotted as points on a graph paper, giving the scatter diagram (Fig. 9.2). A free hand straight line has been drawn. It is found that out of 8 points, 4 lie almost on the straight line and the remaining 4 are also very close to it. This indicates that the correlatioin coefficient between the variables x and y is numerically very high, say about 0.95. Again since the slope of the straight line is positive, the correlation coefficient is also positive. The amount of correlation is thus estimated to be about + 0.95 (The calculated value of the correlation coefficient for the data is + 0.98).

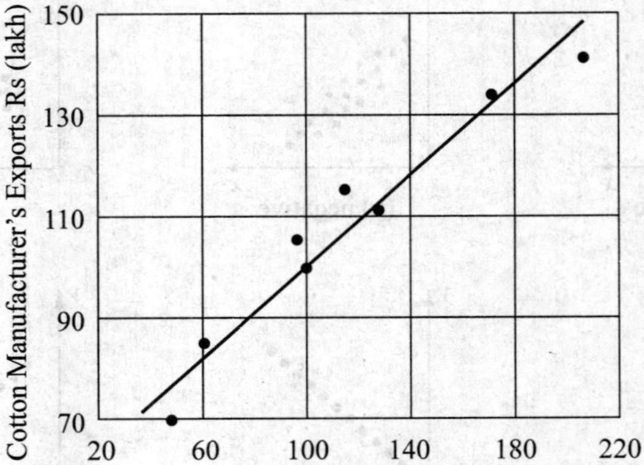

Fig. 9.2 Scatter Diagram

9.5 CORRELATION

The word "correlation" is used to denote the *degree of association between variables.* If two variables x and y are so related that variations in the magnitude of one variable tend to be accompanied by variations in the magnitude of the other variable, they are said to be *correlated.* If y tends to increase as x increases, the variables are said to be *positively correlated.* If y tends to decreases as x increases, the variables are *negatively correlated.* If the values of y are not affected by changes in the values of x, the variables are said to be *uncorrelated* (see Example 9.6).

Correlation may also be *linear* or *non-linear.* If the amount of change in one variable tends to bear a constant ratio to the amount of change in the other variable, then correlation is said to be 'linear'; because the scatter diagram would show a linear path. Here, we shall be concerned with *linear correlation* or *simple correlation* only. This is measured by *'Correlation Coefficient'* (Section 9.7).

9.6 COVARIANCE

Given a set of n pairs of observations $(x_1, y_1), (x_2, y_2), \ldots, (x_n, y_n)$ relating to two variables x and y, the *Covariance of x and y*, usually represented by *cov* (x, y), is defined as

$$\text{cov}(x, y) = \frac{1}{n}\Sigma(x-\bar{x})(y-\bar{y}) \tag{9.6.1}$$

Expanding the expanssion on the right, it can be shown that

$$\text{cov}(x, y) = \frac{\Sigma xy}{n} - \left(\frac{\Sigma x}{n}\right)\left(\frac{\Sigma y}{n}\right) \tag{9.6.2}$$

This form is generally used for calculations. Covariance has properties similar to those of variance, i.e. the square of S.D.

(i) If $X = x - c$ and $Y = y - c'$, where c, c' are constants, then

$$\text{cov}(x, y) = \text{cov}(X, Y) \tag{9.6.3}$$

(ii) If $u = (x - c)/d$ and $v = (y - c')/d'$, where $c, c', d\ d'$ are constants, then

$$\text{cov}(x, y) = dd'\ .\ \text{cov}(u, v) \tag{9.6.4}$$

However, while variance must be always positive, covariance may be positive, negative or zero.

[**Note:** By definition,

$$\text{Variance of } x = \sigma_x^2 = \frac{1}{n}\Sigma(x-\bar{x})^2 = \frac{1}{n}\Sigma(x-\bar{x})(x-\bar{x})$$

$$\text{Variance of } y = \sigma_y^2 = \frac{1}{n}\Sigma(y-\bar{y})^2 = \frac{1}{n}\Sigma(y-\bar{y})(y-\bar{y})$$

$$\text{Covariance of } x \text{ and } y = \frac{1}{n}\Sigma(x-\bar{x})(y-\bar{y})$$

For the variances, both factors in the expression on the right are same, either both $(x-\bar{x})$, or both $(y-\bar{y})$; both for the covariance one factor is $(x-\bar{x})$ and another $(y-\bar{y})$. Again, the working formulae are

$$\text{Variance of } x = \frac{\Sigma x^2}{n} - \left(\frac{\Sigma x}{n}\right)^2 = \frac{\Sigma x.x}{n} - \left(\frac{\Sigma x}{n}\right)\left(\frac{\Sigma x}{n}\right)$$

$$\text{Variance of } y = \frac{\Sigma y^2}{n} - \left(\frac{\Sigma y}{n}\right)^2 = \frac{\Sigma y.y}{n} - \left(\frac{\Sigma y}{n}\right)\left(\frac{\Sigma y}{n}\right)$$

$$\text{Covariance of } x \text{ and } y \qquad = \frac{\Sigma x.y}{n} - \left(\frac{\Sigma x}{n}\right)\left(\frac{\Sigma y}{n}\right)$$

For the variances, the first term on the right contains $x.x$ or $y.\ y$ and for the covariance it is $x.y$, i.e. one is x and the other y; in the second term similarly instead of both factors same, $\Sigma x/n$ or $\Sigma y/n$, we have one from x and another from y, i.e. $(\Sigma x/n)\ (\Sigma y/n)$.

It is thus seen that covarience is a variance like quantity, but obtained by the combination of two variables. Covariance may, therefore, be looked upon as a 'conjoint variance'.

It is also interesting to note that covariance of a variable and the same variable is the variance itself. Replacing y by x in (9.6.1) or (9.6.2), we see that cov. $(x, x) = \sigma_x^2$; similarly, cov $(y.\ y) = \sigma_y^2$]

9.7 CORRELATION COEFFICIENT (r)

Let $(x_1, y_1), (x_2, y_2), \ldots, (x_n, y_n)$ be a given set of n pairs of observations on two variables x and y. The *Correlation Coefficient*, or *Coefficient of Correlation,* between x and y (denoted by the symbol r) is then defined as

$$r = \frac{\text{cov}(x, y)}{\sigma_x \sigma_y} \qquad (9.7.1)$$

where σ_x and σ_y are the standard derivations of x and y respectively, and cov (x, y) denotes the coveriance of x and y (Section 9.6). This expression is known as *Pearson's product-moment formula,* and is used as a measure of linear correlation between x and y.

The formula for r may be written in various other forms. Putting the explicit expressions for cov(x, y), σ_x and σ_y in (9.7.1), and multiplying both the numerator and the denominator by n, we have

$$r = \frac{\Sigma\,(x - \bar{x})\,(y - \bar{y})}{\sqrt{[\Sigma(x - \bar{x})^2\ .\ 2(y - \bar{y})^2]}} \qquad (9.7.2)$$

Now, expanding the expressions

$$r = \frac{\Sigma\ xy - n\bar{x}\bar{y}}{\sqrt{[(\Sigma x^2 - n\bar{x}^2)\ .\ (\Sigma y^2 - n\bar{y}^2)]}} \qquad (9.7.3)$$

Multiplying the numerator and the denominator by n again, and since $n\bar{x} = \Sigma x$ and $n\bar{y} = \Sigma y$, we may write

$$r = \frac{n\Sigma xy - (\Sigma\, x)\ (\Sigma\, y)}{\sqrt{[\{n\ \Sigma x^2 - (\Sigma\ x)^2\} \cdot \{n\ \Sigma y^2 - (\Sigma y)^2\}]}} \qquad (9.7.4)$$

[**Note:** In all the forms shown above, the denominators contains two factors under the are-root. They may be obtained on replacing y by x, and x by y, in the numerator.]

9.8 PROPERTIES OF CORRELATION COEFFICIENT

(i) The correlation coefficient r is independent of the choice of both origin and scale of observations. This means that if

$$u = \frac{x - c}{d} \quad \text{and} \quad v = \frac{y - c'}{d'}$$

where c, c', d, d' are arbitrary constants (d and d' positive), then

$$r_{xy} = r_{uv} \tag{9.8.1}$$

i.e. *Correlation coefficient between x and y*
= Correlation coefficient between u and v

In genreal, if $X = a + bx$, $Y = a' + b'y$, then

$$r_{XY} = \pm\, r_{xy} \tag{9.8.2}$$

according as b and b' have the same sign, or opposite signs.

(ii) The correlation coefficient r is a pure number and is independent of the units of measurement. This means that if, for example, x represents height in inches and y weight in lbs., then the correlation coefficient between x and y will neigther be in inches nor in lbs. or any other unit, but only a number.

(iii) The correlation coefficient r lies between -1 and $+1$; i.e. r cannot exceed 1 numerically.

$$-1 \le r \le +1 \tag{9.8.3}$$

9.9 CALCULATION OF r

Correlation coefficient (r) is unaffected by the choice of origin and scale of one or both the variables (Property (i), Section 9.8). Therefore, it can be calculated from a given set of a n pairs of observations $(x_1, y_1), (x_2, y_2), ..., (x_n, y_n)$ as follows:

(I) If $X = x - c$ and $Y = y - c'$, (here c, c' are constants), then

$$r_{xy} = r_{XY} = \frac{\text{cov}(X, Y)}{\sigma_X\, \sigma_Y} \tag{9.9.1}$$

$$\text{where } \sigma_X^2 = \frac{\Sigma X^2}{n} - \left(\frac{\Sigma X}{n}\right)^2, \quad \sigma_Y^2 = \frac{\Sigma Y^2}{n} - \left(\frac{\Sigma Y}{n}\right)^2$$

$$\text{cov}(X, Y) = \frac{\Sigma XY}{n} - \left(\frac{\Sigma X}{n}\right)\left(\frac{\Sigma Y}{n}\right) \tag{9.9.2}$$

Thus, we can always reduce the given values of x and y on subtracting convenient numbers c and c', and obtain deviations $X = x - c$, $Y = y - c'$. From these reduced values X and Y, the two standard deviations and the covariance, viz. σ_X, σ_Y and cov (X, Y), are now calculated, and finally the correlation coefficient r_{XY} between them.

This will be exactly equal to the correlation coefficient r_{xy} between the original values of x and y (see Example 9.9).

(II) If $u = \dfrac{x - c}{d}$ and $v = \dfrac{y - c'}{d'}$, (here c, c', d, d' are constants and d, d' are positive), then

$$r_{xy} = r_{uv} = \frac{\text{cov}(u, v)}{\sigma_u \sigma_v} \tag{9.9.3}$$

$$\text{where} \quad \sigma_u^2 = \frac{\Sigma u^2}{n} - \left(\frac{\Sigma u}{n}\right)^2, \quad \sigma_v^2 = \frac{\Sigma v^2}{n} - \left(\frac{\Sigma v}{n}\right)^2$$

$$\text{cov}(u, v) = \frac{\Sigma uv}{n} - \left(\frac{\Sigma u}{n}\right)\left(\frac{\Sigma v}{n}\right) \tag{9.9.4}$$

In some cases, the given values of x and y may be such that it is further possible to reduce the deviations $x - c$ and $y - c'$ on division by constant factors d and d', i.e. $u = (x - c)/d$, $v = (y - c')/d'$. From these values of u and v, the two standard deviations σ_u, σ_v and the covariance cov(u, v) are calculated, and finally the correlation coefficient r_{uv} between u and v is obtained. This will be exactly equal to the correlation coefficient r_{xy} between the original values of x and y (see Example 9.11).

Example 9.6 *State in each case whether you would expect to find a positive correlation, a negative correlation or no correlation:*

(i) The ages of husband and wives,
(ii) Shoe size and intelligence,
(iii) Insurance companies' profit and the number of claims they have to pay,
(iv) Years of education and income,
(v) Amount of rainfall and yield of crop.

Solution (i) Positive, (ii) No correlation, (iii) Negative, (iv) Positive, (v) Positive.

Example 9.7 *Calculate the correlation coefficient and its probable error from the following results:*

$$\sum_{i=1}^{100} x_i = 280, \quad \sum_{i=1}^{100} y_i = 60, \quad \sum_{i=1}^{100} x_i^2 = 2384, \quad \sum_{i=1}^{100} y_i^2 = 117, \quad \sum_{i=1}^{100} x_i y_i = 438$$

[I.C.W.A., July '71]

Solution

$$\text{Correlation coefficient} \quad r = \frac{\text{cov}(x, y)}{\sigma_x \sigma_y}$$

$$\text{Here} \quad \sigma_x^2 = \frac{\Sigma x^2}{n} - \left(\frac{\Sigma x}{n}\right)^2 = \frac{2384}{100} - \left(\frac{280}{100}\right)^2 = 16$$

$$\sigma_y^2 = \frac{\Sigma y^2}{n} - \left(\frac{\Sigma y}{n}\right)^2 = \frac{117}{100} - \left(\frac{60}{100}\right)^2 = 0.81$$

$$\text{cov}(x, y) = \frac{\Sigma xy}{n} - \left(\frac{\Sigma x}{n}\right)\left(\frac{\Sigma y}{n}\right) = \frac{438}{100} - \frac{280}{100} \times \frac{60}{100} = 2.7$$

Substituting the values,

$$r = \frac{2.7}{\sqrt{16}\,\sqrt{0.81}} = \frac{2.7}{4 \times 0.9} = 0.75$$

The Probable Error (P. E.) of correlation coefficient is

$$\text{P.E.} = 0.6745\frac{1-r^2}{\sqrt{n}}$$

$$= 0.6745 \times \frac{1-(0.75)^2}{\sqrt{100}} = 0.0295.$$

Ans. $r = 0.75$, P.E. = 0.0295

Example 9.8 *While calculating the coefficient of correlation between two variables x and y, the following results were obtained : n = 25, Σx = 125, Σy = 100, Σx² = 650, Σy² = 460, Σxy = 508. It was however later discovered at the time of checking that two pairs of observations (x, y) were copied (6, 14) and (8, 6), while the correct values were (8, 12) and (6, 8) respectively. Determine the correct value of the coefficient of correlation.* [I.C.W.A., June '77, '78]

Solution For the two incorrect pairs of observations

x : 6, 8; $\Sigma x = 6 + 8 = 14; \Sigma x^2 = 6^2 + 8^2 = 100;$

y : 14, 6; $\Sigma y = 14 + 6 = 20; \Sigma y^2 = 14^2 + 6^2 = 232;$

$\Sigma xy = 6 \times 14 + 8 \times 6 = 132.$

For the two correct pairs

x : 8, 6; $\Sigma x = 8 + 6 = 14; \Sigma x^2 = 8^2 + 6^2 = 100;$

y : 12, 8; $\Sigma y = 12 + 8 = 20; \Sigma y^2 = 12^2 + 8^2 = 208;$

$\Sigma xy = 8 \times 12 + 6 \times 8 = 144.$

When the incorrect pairs of observations are replaced by the correct pairs, the revised results for the 25 pairs of observations are:

$\Sigma x = 125 - 14 + 14 = 125;$ $\Sigma y = 100 - 20 + 20 = 100;$

$\Sigma x^2 = 650 - 100 + 100 = 650;$ $\Sigma y^2 = 460 - 232 + 280 = 436;$

$\Sigma xy = 508 - 132 + 144 = 520$

Using these results,

$\sigma_x^2 = 650/25 - (125/25)^2 = 26 - 25 = 1;$ $\sigma_x = 1$

$\sigma_y^2 = 436/25 - (100/25)^2 = 17.44 - 16 = 1.44;$ $\sigma_y = 1.2$

$\text{cov}(x, y) = 520/25 - (125/25)(100/25) = 20.8 - 20 = 0.8$

$$r = \frac{\text{cov}(x, y)}{\sigma_x\sigma_y} = \frac{0.8}{1 \times 1.2} = \frac{2}{3}$$

Ans. 2/3.

Example 9.9 *Find the coefficient of correlation from the following data:*

X	65	63	67	64	68	62	70	66
Y	68	66	68	65	69	66	68	65

[C.U., B.A. (Econ) '69]

Solution Since the correlation coefficient is unaffected by change of origin (and also scale), let us change the origins of X and Y to 65 and 67 respectively, i.e. write $x = X - 65$, $y = Y - 67$.

Table 9.9 Calculations for Correlation Coefficient

X	Y	x = X – 65	y = Y – 67	x^2	y^2	xy
(1)	(2)	(3)	(4)	(5)	(6)	(7)
65	68	0	1	0	1	0
63	66	– 2	– 1	4	1	2
67	68	2	1	4	1	2
64	65	– 1	– 2	1	4	2
68	69	3	2	9	4	6
62	66	– 3	– 1	9	1	3
70	68	5	1	25	1	5
66	65	1	– 2	1	4	– 2
525	535	5	– 1	53	17	18

$$\sigma_x^2 = \frac{53}{8} - \left(\frac{5}{8}\right)^2 = \frac{399}{64}$$

$$\sigma_y^2 = \frac{17}{8} - \left(\frac{-1}{8}\right)^2 = \frac{135}{64}$$

$$\text{cov}(x, y) = \frac{18}{8} - \left(\frac{5}{8}\right)\left(\frac{-1}{8}\right) = \frac{149}{64}$$

$$r_{xy} = \frac{\text{cov}(x,\ y)}{\sigma_x\ \sigma_y} = \frac{\frac{149}{64}}{\sqrt{\frac{399}{64}}\sqrt{\frac{135}{64}}} = \frac{149}{\sqrt{(399 \times 135)}} = 0.64$$

Therefore, the required correlation coefficienty is 0.64.

[**Note:** (i) The origins should be changed to values near the means. Here, the means of X and Y are 525/8 = 65.625 and 535/8 = 66.875 respectively.

(*ii*) The following check may be introduced for the accuracy of figures in cols. (3) and (4), on which the calculations in the subsequent columns depend. (3) must be 525 – 8 × 65 = 525 – 520 = 5. Similarly the total of col. (2) may be used to check the total of col. (4), viz. 535 – 8 × 67 = 535 – 536 = – 1.]

Example 9.10 *Marks of 10 students in Mathematics and Statistics are given below:*

Mathematics (X)	*32*	*38*	*48*	*43*	*40*	*22*	*41*	*69*	*35*	*64*
Statistics (Y)	*30*	*31*	*38*	*43*	*33*	*11*	*27*	*76*	*40*	*59*

Calculate (i) product-moment correlation coefficient, and (ii) its standard error.

[I.C.W.A., June'74]

Solution Since the correlation coefficient is unaffected by changes of origin, we write $x = X - 43$, $y = Y - 38$, and construct a table like Table 9.9. It will be seen that

$$n = 10,\ \Sigma x = 2,\ \Sigma y = 8,\ \Sigma x^2 = 1806,\ \Sigma y^2 = 2902,\ \Sigma xy = 2140.$$

Using these results, $\sigma_x^2 = 180.56$, $\sigma_y^2 = 289.56$, $\text{cov}(x, y) = 213.84$.

$$r = \frac{213.84}{\sqrt{(180.56 \times 289.56)}} = 0.94$$

Standard Error (*S.E.*) of correlation coefficient is

$$S.E. = \frac{1 - r^2}{\sqrt{n}} = \frac{1 - (0.94)^2}{\sqrt{10}} = 0.04$$ *Ans.* (i) 0.94, (ii) .04

Example 9.11 *Calculate the coefficient of correlation from the following data :*

x	*2.52*	*2.49*	*2.49*	*2.45*	*2.43*	*2.42*	*2.41*	*2.40*
y	*730*	*710*	*770*	*890*	*970*	*1020*	*970*	*1040*

[C.U., B.A. (Econ) '73 : M.Com. '76]

Solution Since the coefficient of correlation is unaffected by changes of origin and scale, let us write

$$u = \frac{x - 2.45}{.01}, \quad v = \frac{y - 900}{10}$$

Table 9.10 Calculations for Correlation Coefficient

x	y	$u = \frac{x - 2.45}{.01}$	$v = \frac{y - 900}{10}$	u^2	v^2	uv
2.52	730	7	– 17	49	289	– 119
2.49	710	4	– 19	16	361	– 76
2.49	770	4	– 13	16	169	– 52
2.45	890	0	– 1	0	1	0
2.43	970	– 2	7	4	49	– 14
2.42	1020	– 3	12	9	144	– 36
2.41	970	– 4	7	16	49	– 28
2.40	1040	– 5	14	25	196	– 70
Total 19.61	7100	1	– 10	135	1258	– 395

[**Note:** (i) For the calculation of correlation coefficient, a change of origin is always possible, and must be made; a change of scale, wherever possible, should also be made to simplify the calculations. In the present case, the given values of y are divisible by 10, indicating that changes of both origin and scale and possible. Also, since the values of x are given in two decimals, division by .01 (i.e. multiplication by 100) will change them to whole numbers, and consequently worries about fixing the decimal point can be avoided.

(ii) Another important point to note here is how to effect changes of origin and scale both in one step. When data are given in two decimals the origin should be changed to a conveninet two-decimal number near the mean, and the scale shall be changed to 01. Here, $u = (x - 2.45)/.01 = 100_x - 245$. This shows that for obtaining the values of u, each x is to be multiplied by 100 (which can be achieved simply by ignoring the decimal point) and then 245 subtracted. Thus, 2.52 changes to 252, and 252 – 245 = 7; 2.49 changes to 2.49 and 249 – 245 = 4; etc. (For data given in one-decimal figures, the origin should be changed to a one-decimal number and scale to 0.1).

Again, $v = (y - 900)/10 = y/10 - 90$. This shows that for obtaining the values of v, each y is to be divided by 10 (i.e. ignore the last zero) and then subtract 90. For example, 730 changes to 73 and 73 – 90 = – 17; 710 changes to 71 and 71 – 90 = – 19; etc.]

Now using formula (9.9.3),

$$\sigma_u^2 = \Sigma u^2/n - (\Sigma u/n)^2 = 135/8 - (1/8)^2 = 1079/64$$

$$\sigma_v^2 = \Sigma v^2/n - (\Sigma v/n)^2 = 1258/8 - (-10/8)^2 = 9964/64$$

$$\text{cov}(u, v) = \Sigma uv/n - (\Sigma u/n)(\Sigma v/n) = -395/8 - (1/8)(-10/8) = -3150/64$$

$$r = \frac{\frac{-3150}{64}}{\sqrt{\frac{1079}{64}}\sqrt{\frac{9964}{64}}} = \frac{-3150}{\sqrt{1079 \times 9964}} = -0.96$$

Example 9.12 *The following table gives the index numbers of industrial production in a country and the number of registered unemployed persons in the same country during the eight consecutive years. Calculate the coefficient of correlation and comment on the result.*

Year	*1954*	*1955*	*1956*	*1957*	*1958*	*1959*	*1960*	*1961*
Index of industrial production	*100*	*102*	*103*	*105*	*106*	*104*	*103*	*98*
No. of registered unemployed (in thousands)	*10.5*	*11.4*	*13.0*	*11.5*	*12.0*	*12.5*	*15.6*	*20.8*

[I.C.W.A., July '64; C.U., B.A. (Econ)' 65]

Solution Let x denote the index of industrial production the y the number of registered unemployed (in thousands). Since the correlation coefficient remains unaffected by changes of origin and scale, we write $u = x - 100$ and $v = (y - 13.0)/0.1$

$$\sigma_u^2 = \Sigma u^2/n - (\Sigma u/n)^2 = 103/8 - (21/8)^2 = 6.0$$

$$\sigma_v^2 = \Sigma v^2/n - (\Sigma v/n)^2 = 7991/8 - (33/8)^2 = 982$$

$$\text{cov}(u, v) = \Sigma uv/n - (\Sigma u/n)(\Sigma v/n) = -265/8 - (21/8)(33/8) = -44$$

Table 9.11 Calculations for Correlation Coefficient

x (1)	*y* (2)	*u = x – 100* (3)	*v = (y – 13.0)/0.1* (4)	u^2 (5)	v^2 (6)	*uv* (7)
100	10.5	0	– 25	0	625	0
102	11.4	2	– 16	4	256	– 32
103	13.0	3	0	9	0	0
105	11.5	5	– 15	25	225	– 75
106	12.0	6	– 10	36	100	– 60
104	12.5	4	– 5	16	25	– 20
103	15.6	3	26	9	676	78
98	20.8	– 2	78	4	6084	– 156
Total 821	107.3	21	33	103	7991	– 265

$$\therefore \quad r_{xy} = r_{uv} = \frac{\text{cov}(u, v)}{\sigma_u \sigma_v} = \frac{-44}{\sqrt{(6.0 \times 982)}} = -0.57$$

Comments—The scatter diagram (not show here) does not indicate any straight line path and hence $r = -0.57$ does not really serve as a useful measure of association between the variables. In fact, if the data for 1961 are excluded, we find $r = +0.21$, a result which differs much from the former. The observed correlation coefficient $r = -0.57$ may be called *non-sense correlation* (Page 286).

Example 9.13 *Prove that the correlation coefficient does not depend on the origin or scale of the observations.*

[C.U., B.A. (Econ) '66, M. Com. '76; B.U., B.A. (Econ)'68; I.C.W.A., July '70, Jan. '72, June '75]

Solution Let x and y be the variables, and (x_1, y_1), (x_2, y_2), ..., (x_n, y_n) denote n pairs of observations. Let us change the origins of x and y to a and c, and the units of measurement (i.e. scales) to b and d respectively, and write

$$u_i = \frac{x_i - a}{b}, \quad v_i = \frac{y_i - c}{d} \qquad \text{(i)}$$

where a, b, c, d are arbitrary constants (b and d positive). We have to prove that the correlation coefficient between x and y is the same as that between u and v; i.e. $r_{xy} = r_{uv}$.

From (i), $x_i = a + bu_i, \quad y_i = c + dv_i$

Therefore, $\bar{x} = \Sigma x_i/n = \Sigma(a + bu_i)/n = (na + b\,\Sigma u_i)/n = a + b\,\bar{u}$

Similarly, $\bar{y} = c + d\,\bar{v}$

Hence, $\sigma_x^2 = \Sigma(x_i - \bar{x})^2/n = \Sigma(a + bu_i - a - b\,\bar{u})^2/n$

$= b^2\,\Sigma(u_i - \bar{u})^2/n = b^2 \,.\, \sigma_u^2$

$\therefore \quad \sigma_x = b \,.\, \sigma_u$ (since b is positive)

Similarly, $\sigma_y = d.\ \sigma_v$ (since d is positive) (ii)

Again, $\text{cov}(x, y) = \Sigma(x_i - \bar{x})(y_i - \bar{y})/n$

$= \Sigma\{(a + bu_i - a - b\,\bar{u})(c + dv_i - c - d\,\bar{v})\}/n$

$= bd\,\Sigma(u_i - \bar{u})(v_i - \bar{v})/n$

$= bd.\ \text{cov}(u, v)$ (iii)

Substituting the values from (ii) and (iii),

$$r_{xy} = \frac{\text{cov}(x, y)}{\sigma_x \sigma_y} = \frac{bd.\,\text{cov}(u, v)}{(b.\,\sigma_u)\,(d.\,\sigma_v)} = \frac{\text{cov}(u, v)}{\sigma_u \sigma_v} = r_{uv}$$

i.e. Correlation coefficient between x and y

= Correlation coefficient between u and v.

Example 9.14 *If $X_i = (x_i - a)/b$, $Y_i = (y_i - c)/d$, $(i = 1, 2, ..., n)$ where a, b, c, d are arbitrary constants, prove that $r_{xy} = r_{XY}$, if b and d are of the same sign, and $r_{xy} = -r_{XY}$ if they are of opposite signs.*

[C.U., B.A. (Econ) '68, ' 70; M. Com.' 74; B.U., B.A., (Econ)'72]

Solution As in Example 9.13, $\sigma_x^2 = b^2.\sigma_X^2$, $\sigma_y^2 = d^2 . \sigma_Y^2$, and cov$(x, y) = bd.$ cov(X, Y). These results are true whatever be the signs of b and d. But since the standard deviation is always positive, we must have $\sigma_x = |b| . \sigma_X$, $\sigma_y = |d| . \sigma_y$ (see 6.6.2, page 185)

$$r_{xy} = \frac{\text{cov}(x, y)}{\sigma_x \ \sigma_y} = \frac{bd.\ \text{cov}(X, Y)}{|b| . \sigma_X \ |d| . \sigma_Y}$$

$$= \frac{bd}{|b| . |d|} . \frac{\text{cov}(X, Y)}{\sigma_X \ \sigma_Y} = \frac{bd}{|b| . |d|} r_{XY}$$

If b and d are of the same sign (i.e. either both positive or both negative), the product bd will be positive, and exactly equal to $|b| . |d|$, which is always positive, Therefore,

$$\frac{bd}{|b| . |d|} = +1, \text{ so that } \quad r_{xy} = r_{XY}.$$

If b and d are of opposite signs (i.e. one of them positive and the other negative), the product bd will be negative, but numerically equal to $|b| . |d|$. Therefore,

$$\frac{bd}{|b| . |d|} = -1, \text{ so that } r_{xy} = -r_{XY}. \quad \textit{(Proved)}$$

Example 9.15 *If the correlation coefficient between x and y is 0.5, what would be the correlation coefficient between 5x and – 3y?*

Solution Let us write $X = 5x$ and $Y = -3y$. Then by formula (9.8.2), we have (here $a = 0$, $b = 5$, $a' = 0$, $b' = -3$),

$$r_{XY} = -r_{xy} = -0.5$$

since the coefficients of x and y (viz. 5 and –3) have opposite signs.

Example 9.16 *Prove that the correlation coefficient r lies between – 1 and + 1.*
[C.U., B.A.(Econ) '67, '74; M.Com. '74; B.U., B.A. (Econ) '65, '68, '72, '73]

Solution Let x and y be the variables, and (x_1, y_1), (x_2, y_2), ..., (x_n, y_n) denote n pairs of observations, with means $\bar{x}$, $\bar{y}$ and standard deviations σ_x, σ_y respectively. If we write

$$u_i = \frac{x_i - \bar{x}}{\sigma_x} \quad \text{and} \quad v_i = \frac{y_i - \bar{y}}{\sigma_y}$$

then $$\Sigma u_i^2 = \sum \left(\frac{x_i - \bar{x}}{\sigma_x} \right)^2 = \frac{\Sigma (x_i - \bar{x})^2}{\sigma_x^{\ 2}} = \frac{n\ \sigma_x^2}{\sigma_x^2} = n$$

Similarly, $$\Sigma v_i^2 = n$$

Again, $$\Sigma u_i v_i = \sum \left(\frac{x_i - \bar{x}}{\sigma_x} \right) \left(\frac{y_i - \bar{y}}{\sigma_y} \right) = \frac{\Sigma (x_i - \bar{x})\ (y_i - \bar{y})}{\sigma_x \ \sigma_y}$$

$$= \frac{n\ .\ \text{cov}(x, y)}{\sigma_x \sigma_y} = nr$$

where r denotes the correlation coefficient between x and y.

Now, $(u_i + v_i)^2$ can never be negative, because it is a perfect square. Hence, the sum of all such squares for $i = 1, 2, ..., n$, cannot be negative; i.e.

$$\Sigma (u_i + v_i)^2 \geq 0$$

or, $$\Sigma u_i^2 + \Sigma v_i^2 + 2\Sigma u_i v_i \geq 0$$

or, $$n + n + 2nr \geq 0$$

or, $$2n + 2nr \geq 0$$

or, $$2n(1 + r) \geq 0$$

or, $$1 + r \geq 0$$

or, $$r \geq -1; \quad \text{i.e.} -1 \leq r$$

Similarly, since $(u_i - v_i)^2$ cannot be negative,

$$\Sigma(u_i - v_i)^2 \geq 0$$

or, $$\Sigma u_i^2 + \Sigma v_i^2 - 2\Sigma u_i v_i \geq 0$$

or, $$n + n - 2nr > 0$$

or, $$2n(1 - r) \geq 0$$

or, $$1 - r \geq 0$$

or, $$1 \geq r; \quad \text{i.e. } r \leq 1$$

Combining the results $-1 \leq r$ and $r \leq 1$, we have $-1 \leq r \leq +1$. This proves that the correlation coefficient lies between -1 and $+1$.

9.10 INTERPRETATION AND USE OF r

Example 9.17 *Discuss the various uses and limitations of coefficient of correlation, indicating the difficulties involved in the interpretation of the coefficient of correlation.*

[C.U., B.A.(Econ) '66; M.Com '68]

Solution *Uses*—

(i) The coefficient of correlation (r) is a measure of the degree of association between two variables. For comparing two series of observations, it is sometimes necessary to determine whether they are associated or not, and to establish relations of cause and effect. The coefficient of correlation furnishes a method of determining numerically the existence of such causal connection between them, e.g. whether weights of school boys are connected with their parents salaries. When the pattern of dots in the scatter diagram is linear, the correlation coefficient (r) can be considered as a useful measure of such relationship. A positive value of r indicates that high values of one variable are in general associated with high values of the other, and low values with low values. When r is negative, high values of one variable are in general associated with low values of the other.

(*ii*) Again, the proportion of variation explained by regression is equal to the square of correlation coefficient (Section 9.14), i.e.

$$r^2 = \textit{Proportion of variation explained by regression}$$

The value of r^2, therefore, enables us to state the relative amount of variation in the dependent variable which can be explained by the regression equation.

(*iii*) The coefficient of correlation also helps in estimating the value of the dependent variable, when the value of the independent variable is known. Thus, for a given x, the estimated value of y is obtained from the regression equation of y on x (page 291)

$$y - \bar{y} = r\frac{\sigma_y}{\sigma_x}(x - \bar{x})$$

When $r = 0$, neither y nor x can be estimated by a linear function of the other variable. Again, when $r = \pm 1$, the two regression lines coincide, and the value of one variable can be accurately predicted by the linear regression equation.

(iv) In educational and psychological measurements, coefficients of correlation is used in problems of *reliability and validity* of tests.

Limitations—

(i) In linear correlation, it is assumed that there is a straight line relationship between the variables. A small value of r therefore indicates only a poor *linear type* of relationship between the variables. This however, does not rule out the possibility that the association is very cloose, but the relationship is non-linear (Example 9.21). Before using r as a measure of the degree of association between the variables, it is, therefore, advisable to draw a scatter diagram and see whether the pattern of points is linear or not.

(ii) Again, a high value of r also does not imply that there is a direct cause-and-effect relationship between the variables. The high value of r may be generated solely due to the influence of a third variable affecting both. In this cause, the effect of the third variable should be eliminated from the first two and then the 'partial' correlation coefficient between them found out (Section 9.17).

(iii) Sometimes, it may happen that two series of observations show a high correlation coefficient even though there is no logical basis for any relationship between them. For example, a renowned statistician observed a high correlation between the stork population is Oslo, Norway, over a period of several years and the number of babies born there each year. Yet it is hard to develop a theory as to why this should be. Such correlation is said to be *Spurious correlation* or *Non-sense correlation.* One should apply common-sense is deciding whether the association indicated by the value of r is real or spurious (Example 9.12).

(iv) If the data are not reasonably homogeneous the coefficient of correlation may give a misleading picture of the extent of association. For example, if the scatter diagram shows the points in separate clusters or groups, the correlation coefficient based on all the groups taken together may be very high; yet if seperate values of r are computed for each group, they may be close to zero. If some reasonable basis can be found for separating the data into groups, it is desirable to compute values of r for each group.

9.11 VARIANCE OF THE SUM (DIFFERENCE) OF TWO SERIES

Let $(x_1, y_1), (x_2, y_2), \ldots, (x_n, y_n)$ be a set of n pairs of observations on the variables x and y, and a new series $(x + y)$ is formed by combining the corresponding values of the two series, giving $x_1 + y_1, x_2 + y_2, \ldots, x_n + y_n$. If we denote the variances of x-series, y-series and $(x + y)$-series by var(x), var(y) and var($x + y$) respectively, and cov(x, y) denotes the covariance between x and y, then

$$\text{var}(x + y) = \text{var}(x) + \text{var}(y) + 2 \cdot \text{cov}(x, y) \quad (9.11.1)$$

Similarly, if another series $(x - y)$ is formed from the given data, then

$$\text{var}(x - y) = \text{var}(x) + \text{var}(y) - 2.\ \text{cov}(x, y) \quad (9.11.2)$$

In the usual notations, if we write $\sigma_x, \sigma_y, \sigma_{x+y}, \sigma_{x-y}$ to denote the standard deviations of $x, y, x+y, x-y$ respectively, the above formulae may be written as

$$\sigma_{x+y}^2 = \sigma_x^2 + \sigma_y^2 + 2r\sigma_x\sigma_y \quad (9.11.3)$$

$$\sigma_{x-y}^2 = \sigma_x^2 + \sigma_y^2 - 2r\sigma_x\sigma_y \quad (9.11.4)$$

When x and y are uncorrelated i.e. $r = 0$, then cov $(x, y) = 0$, and the last terms on the right of the above formulae vanish. Hence

$$\text{var}(x+y) = \text{var}(x-y) = \text{var}(x) + \text{var}(y)$$

i.e.
$$\sigma_{x+y}^2 = \sigma_{x-y}^2 = \sigma_x^2 + \sigma_y^2 \quad (9.11.5)$$

This implies that the variance of the sum (or difference) of uncorrelated variables is equal to the *sum* of their variances.

In general, if $z = ax + by$, where a and b are constants, then

$$\text{var}(ax+by) = a^2.\text{var}(x) + b^2.\text{var}(y) + 2ab.\text{cov}(x, y)$$

i.e.
$$\sigma_z^2 = a^2\,\sigma_x^2 + b^2\,\sigma_y^2 + 2ab\,r\,\sigma_x\,\sigma_y \quad (9.11.6)$$

Note that formulae (9.11.3) and (9.11.4) may be obtained from this by putting $a = 1, b = 1$, and $a = 1, b = -1$ respectively.

Again, solving (9.11.3) and (9.11.4) for r, we see that

$$r = \frac{\sigma_{x+y}^2 - \sigma_x^2 - \sigma_y^2}{2\sigma_x\sigma_y}, \quad r = \frac{\sigma_x^2 + \sigma_y^2 - \sigma_{x-y}^2}{2\sigma_x\sigma_y} \quad (9.11.7)$$

Thus, the standard deviations of x, y and $x + y$ (or $x - y$) may be used to find the correlation coefficient between x and y.

Example 9.18 *n pairs of values of two variables x and y are given. The variances of x, y and (x – y) are σ_x^2, σ_y^2 and σ_{x-y}^2 respectively. Show that the correlation coefficient r_{xy} between x and y is given by*

$$\frac{\sigma_x^2 + \sigma_y^2 - \sigma_{x-y}^2}{2\sigma_x\sigma_y}$$

[I.C.W.A., July '66, '72; C.U., B.A..(Econ) '72]

Solution Let (x_i, y_i), $i = 1, 2, ..., n$, denote n pairs of observations with means $\bar{x}$, $\bar{y}$ and standard deviations σ_x, σ_y for the x-series and y-series respectively. If we write $u_i = x_i - y_i$, then $\bar{u} = \bar{x} - \bar{y}$.

$$u_i - \bar{u} = (x_i - y_i) - (\bar{x} - \bar{y}) = (x_i - \bar{x}) - (y_i - \bar{y})$$

$$\begin{aligned}\therefore \quad \sigma_u^2 &= \Sigma(u_i - \bar{u})^2/n \\ &= \Sigma\{(x_i - \bar{x}) - (y_i - \bar{y})\}^2/n \\ &= \Sigma(x_i - \bar{x})^2/n - 2\Sigma(x_i - \bar{x})(y_i - \bar{y})/n + \Sigma(y_i - \bar{y})^2/n \\ &= \sigma_x^2 - 2\,.\,\text{cov}(x, y) + \sigma_y^2 \\ &= \sigma_x^2 - 2r_{xy}\,\sigma_x\,\sigma_y + \sigma_y^2,\end{aligned}$$

because by definition $r_{xy} = \dfrac{\text{cov}(x, y)}{\sigma_x\ \sigma_y}$ and hence $\text{cov}(x, y) = r_{xy}\,\sigma_x\,\sigma_y$

i.e. $\sigma_{x-y}^2 = \sigma_x^2 - 2r_{xy}\,\sigma_x\,\sigma_y + \sigma_y^2$

or, $2r_{xy}\,\sigma_x\,\sigma_y = \sigma_x^2 + \sigma_y^2 - \sigma_{x-y}^2$

or, $r_{xy} = (\sigma_x^2 + \sigma_y^2 - \sigma_{x-y}^2)/(2\sigma_x\sigma_y)$ (Shown)

Example 9. 19 *The differences $(y_i - x_i)$, $i = 1, 2, \ldots, 15$, are given for a set of 15 pairs of observations below :*

15, 15, 13, 14, 11, 13, 12, 12, 15, 15, 15, 19, 19, 19, 14.

If $\bar{x} = 49.34$, $\bar{y} = 64.07$, $\sigma_x = 3.53$, $\sigma_y = 4.30$, find the coefficient of correlation between x and y.

[C.U., B.Sc. '64; B.A. (Econ) '67]

Solution We apply the result of the previous Example. If $u_i = y_i - x_i$, then $\bar{u} = \bar{y} - \bar{x}$. In the present case, the mean of the given differences is (15 + 15 + 13 + 14 + ... + 19 + 14)/15 = 221/15 = 14.73, which agrees with $\bar{y} - \bar{x}$ = 64.07 – 49.34 = 14.73. For the calculation of σ_{y-x}^2, we use deviations from 15 (Table 9.12).

Table 9.12 Calculations for S.D.

Deviation from 15

$y - x$	(z)	z^2
15	0	0
15	0	0
13	– 2	4
14	– 1	1
11	– 4	16
13	– 2	4
12	– 3	9
12	– 3	9
15	0	0
15	0	0
15	0	0
19	4	16
19	4	16
19	4	16
14	– 1	1
Total	– 4	92

$$\sigma_{y-x}^2 = \Sigma z^2/n - (\Sigma z/n)^2$$
$$= 92/15 - (-4/15)^2 = 6.06$$

$$\therefore \quad \sigma_{x-y}^2 = \sigma_{y-x}^2 = 6.06$$

Now, putting the values

$$r_{xy} = \frac{\sigma_x^2 + \sigma_y^2 - \sigma_{x-y}^2}{2\sigma_x\sigma_y}$$

$$= \frac{(3.53)^2 + (4.30)^2 - 6.06}{2 \times 3.53 \times 4.30}$$

$$= \frac{24.8909}{30.3580} = 0.82.$$ *Ans.* 0.82.

Example 9.20 *If three uncorrelated variables x_1, x_2 and x_3 have the same standard deviation, find the coefficient of correlation between $x_1 + x_2$ and $x_2 + x_3$.*

[B.U., B.A. (Econ)' 71]

Solution Let us write $u = x_1 + x_2$, $v = x_2 + x_3$. It is required to find r_{uv}. Let σ_1, σ_2, σ_3, denote the standard deviations of the variables x_1, x_2, x_3 respectively; and cov(x_1, x_2), cov(x_2, x_3), cov(x_1, x_3) denote the covariances between pairs of variables. Since x_1, x_2, x_3 are uncorrelated.

$$\text{cov}(x_1, x_2) = 0, \quad \text{cov}(x_2, x_3) = 0, \quad \text{cov}(x_1, x_3) = 0$$

Also since the variables are assumed to have the same s.d.,

$$\sigma_1 = \sigma_2 = \sigma_3 = \sigma \text{ (suppose)}$$

Using the formula (9.11.5) for uncorrelated variables

$$\sigma_u^2 = \sigma_1^2 + \sigma_2^2 = 2\sigma^2; \quad \sigma_v^2 = \sigma_2^2 + \sigma_3^2 = 2\sigma^2$$

If $\bar{x}_1$, $\bar{x}_2$, $\bar{x}_3$ denote the means, then $\bar{u} = \bar{x}_1 + \bar{x}_2$ and $\bar{v} = \bar{x}_2 + \bar{x}_3$.

$$\begin{aligned}
\text{cov}(u, v) &= \Sigma(u - \bar{u})(v - \bar{v})/n \\
&= \Sigma[\{(x_1 - \bar{x}) + (x_2 - \bar{x}_2)\}\{(x_2 - \bar{x}_2) + (x_3 - \bar{x}_3)\}]/n \\
&= \Sigma(x_1 - \bar{x}_1)(x_2 - \bar{x}_2)/n + \Sigma(x_1 - \bar{x}_1)(x_3 - \bar{x}_3)/n + \Sigma(x_2 - \bar{x}_2)^2/n \\
&\quad + \Sigma(x_2 - \bar{x}_2)(x_3 - \bar{x}_3)/n \\
&= \text{cov}(x_1, x_2) + \text{cov}(x_1, x_3) + \sigma_2^2 + \text{cov}(x_2, x_3) \\
&= 0 + 0 + \sigma^2 + 0 = \sigma^2
\end{aligned}$$

Therefore, $$r_{uv} = \frac{\text{cov}(u, v)}{\sigma_u \sigma_v} = \frac{\sigma^2}{2\sigma^2} = 0.5.$$

Example 9.21 *If two variates are independent, their correlation coefficient is zero. Is the converse true? Explain by means of an example.*

[I.C.W.A., June '64, '77]

Solution If two variates are independent, their correlation coefficient is zero. But the converse is *not* true. A zero correlation coefficient does not necessarily signify that the variables are independent. It only implies that there is no linear relationship between the variables. However, the possible existence of a non-linear relationship cannot be ruled out altogether. For example, let us consider the following data:

x	–3	–2	–1	0	1	2	3
y	9	4	1	0	1	4	9

Here $\Sigma x = 0$, $\Sigma y = 28$, $\Sigma xy = 0$, $n = 7$. Therefore, using (9.6.2.),

$$\text{cov}(x, y) = \frac{\Sigma xy}{n} - \left(\frac{\Sigma x}{n}\right)\left(\frac{\Sigma y}{n}\right) = \frac{0}{7} - \left(\frac{0}{7}\right)\left(\frac{28}{7}\right) = 0.$$

Hence, $$r_{xy} = \text{cov}(x, y)/(\sigma_x\sigma_y) = 0/(\sigma_x\sigma_y) = 0$$

i.e. the correlation coefficient between x and y is zero. But it may be noticed that x and y are bound by the relation $y = x^2$. So, x and y are not independent. Thus the correlation coefficient may be zero, even when the variables are not independent.

9.12 REGRESSION

The word "regression" is used to denote *estimation* or *prediction* of the average value of one variable for a specified value of the other variable. The estimation is done by means of suitable equations, derived on the basis of available bivariate data. Such an equation is known as a *Regression equation* an its geometrical representation is called a *Regression curve*.

In *linear regression* (or *simple regression*) the relationship between the variables is assumed to be linear. The estimate of y (say, y') is obtained from an equation of the form

$$y' - \bar{y} = b_{yx}(x - \bar{x}) \qquad (i)$$

and the estimate of x (say, x') from another equation (usually different from the former) of the form

$$x' - \bar{x} = b_{xy}(y - \bar{y}) \qquad (ii)$$

Equation (i) is known as *Regression equation of y on x,* and equation (ii) as *Regression equation of x on y.* The coefficient b_{yx} appearing in the regression equation of y on x is known as the *Regression coefficient of y on x.* Similarly, b_{xy} is called the *Regression coefficient of x on y.* The geometrical representation of linear regression equations (i) and (ii) are known as *Regression lines.* These lines are "best fitting" straight lines obtained by the Method of Least Squares (page 249).

Example 9.22 *Derive the regression equations of y on x and x on y.*

[C.U., B.A. (Econ) '71, '73, '78]

Or,

Obtain the equations of the two lines of regression for a bivariate distribution.

[I.C.W.A., July '70]

Solution (1) *Regression Equation of y on x*

The regression equation of y on x is the equation of the best-fitting straight line in the form $y = a + bx$, obtained by the method of least squares (Section 8.5, page 251).

Let $(x_1, y_1), (x_2, y_2), ..., (x_n, y_n)$ be a set of n pairs of observations, and let us fit a straight line of the form

$$y = a + bx \qquad \text{(i)}$$

to these data (here, x is considered to be the independent variable and y and dependent variable, i.e. a value of y is obtained when a value of x is given). Applying the method of least squares, the constants a and b are obtained by solving the normoal equations (8.5.2), viz.

$$\Sigma y = an + b\,\Sigma x \qquad \text{(ii)}$$

$$\Sigma xy = a\,\Sigma x + b\,\Sigma x^2 \qquad \text{(iii)}$$

Dividing both sides of (ii) by n, we get $\bar{y} = a + b\bar{x}$ so that $a = \bar{y} - b\bar{x}$.

Substituting this in equation (*i*),

$$y - \bar{y} = b(x - \bar{x}) \qquad \text{(iv)}$$

Again, multiplying (ii) by Σx and (iii) by n, we have

$$(\Sigma x)(\Sigma y) = na(\Sigma x) + b(\Sigma x)^2$$

$$n(\Sigma xy) = na(\Sigma x) + nb(\Sigma x^2)$$

Subtracting the first from the second,

$$n\Sigma xy - (\Sigma x)(\Sigma y) = b\{n\,\Sigma x^2 - (\Sigma x)^2\}$$

$$\therefore \qquad b = \frac{n\,\Sigma xy - (\Sigma x)\,(\Sigma y)}{n\,\Sigma x^2 - (\Sigma x)^2}$$

Dividing both numerator and denominator by n^2,

$$b = \frac{\Sigma xy/n - (\Sigma x/n)\,(\Sigma y/n)}{n\,\Sigma x^2/n - (\Sigma x/n)^2} = \frac{\text{cov}(x,\ y)}{\sigma_x^{\ 2}} \qquad \text{(v)}$$

Writing b with the usual subscripts, we have from (iv),

$$y - \bar{y} = b_{yx}\,(x - \bar{x}) \qquad \text{(vi)}$$

where $b_{yx} = \text{cov}\,(x, y)/\sigma_x^{\ 2}$. This is the required regression equation of y on x. Since $r = \text{cov}(x, y)/(\sigma_x \sigma_y)$, we see that $\text{cov}(x, y) = r\sigma_x\ \sigma_y$.

Substituting this

$$b_{yx} = \frac{\text{cov}(x,\ y)}{\sigma_x^{\ 2}} = r\,\frac{\sigma_y}{\sigma_x} \qquad \text{(vii)}$$

(2) *Regression Equation of x on y*

If we fit an equation of the form $x = a' + b'\,y$ (assuming x to be the dependent variable and y the independent variable), we obtain the regression equation of x on y. Applying the method of least squares, the normal equations for determined a' and b' are

$$\Sigma x = a'\,n + b'\,\Sigma y; \quad \Sigma xy = a'\,\Sigma y + b'\,\Sigma y^2 \qquad \text{(viii)}$$

Proceeding the same way as before, the regression equation of x on y is found to be

$$x - \bar{x} = b_{xy}\,(y - \bar{y}) \qquad \text{(ix)}$$

where $$b_{xy} = \frac{\text{cov}(x,\ y)}{\sigma_y^{\ 2}} = r\,\frac{\sigma_x}{\sigma_y} \qquad \text{(x)}$$

9.13 PROPERTIES OF LINEAR REGRESSION

The regression equation of y on x is

$$y - \bar{y} = b_{yx}\,(x - \bar{x}) \qquad (9.13.1)$$

wehre $b_{yx} = \dfrac{\text{cov}(x,\ y)}{\sigma_x^{\ 2}} = r\,\dfrac{\sigma_y}{\sigma_x}$. This equation is used to estimate y, when the value of x is known.

The regression equation of x on y is

$$x - \bar{x} = b_{xy}\,(y - \bar{y}) \qquad (9.13.2)$$

where $b_{xy} = \dfrac{\text{cov}(x, y)}{\sigma_y^{\ 2}} = r\,\dfrac{\sigma_x}{\sigma_y}$. This equation is used to estimate x, when the value of y is known.

Substituting the values of b_{yx} and b_{xy}, the regression equations (9.13.1) and (9.13.2) may also be written as

$$\frac{y - \bar{y}}{\sigma_y} = r\left(\frac{x - \bar{x}}{\sigma_x}\right) \quad \text{and} \quad \frac{x - \bar{x}}{\sigma_x} = r\left(\frac{y - \bar{y}}{\sigma_y}\right) \qquad (9.13.3)$$

respectively.

It will be noticed that both regression equations are satisfied (i.e. the value of the right-hand side is equal to the value of the left hand side), when we put $x = \bar{x}$ and $y = \bar{y}$. Geometrically, this implies that both the regression lines pass through, and hence intersect at , the point $(\bar{x}, \bar{y})$. The slopes of the regression lines are respectively b_{yx} and $1/b_{xy}$ [Example 8.1 (*f*) page 248].

The two regression lines are usually different. However, since they must always intersect at the point $(\bar{x}, \bar{y})$, the line will coincide when their slopes are equal, i.e. $b_{yx} = 1/b_{xy}$; or $b_{yx} \cdot b_{xy} = 1$; i.e. $r^2 = 1$; or, $r = \pm 1$.

When $r = +1$, we find from (9.13.3) that both the regression equation reduce to $\dfrac{y - \bar{y}}{\sigma_y} = \dfrac{\bar{x} - \bar{x}}{\sigma_x}$. It is evident that the common regression line represented by this equation passes through the point $(\bar{x}, \bar{y})$ with a positive slope σ_y/σ_x.

When $r = -1$, as in the previous case, the regression equations reduce to $\dfrac{y - \bar{y}}{\sigma_y} = -\left(\dfrac{x - \bar{x}}{\sigma_x}\right)$. This common regression line passes through the point $(\bar{x}, \bar{y})$ with a negative slope $(-\sigma_y/\sigma_x)$.

When $r = 0$, from (9.13.3) we find that regression equations reduce to $\dfrac{y - \bar{y}}{\sigma_y} = 0$ and $\dfrac{x - \bar{x}}{\sigma_x} = 0$ respectively. That is, the regression equations become $y = \bar{y}$ and $x = \bar{x}$. The regression line of y on x, viz. $y = \bar{y}$, is now a straight line parallel to the x-axis; and the regression line of x on y, viz. $x = \bar{x}$, is a straight line parallel to the y-axis. The regression lines are thus perpendicular to each other, nevertheless intersecting at $(\bar{x}, \bar{y})$ as usual. Since now the regression equation of y on x does not contain x, it cannot be used to estimate y as a linear function of x. Similarly, the regression equation of x on y cannot be used to estimate x on the basis of y.

Properties of Linear Regression

(1) There are two linear regression equations—

(i) Regression equation of y on x:

$$y = \bar{y} = b_{yx}(x - \bar{x}) \tag{9.13.4}$$

(ii) Regression equation of x on y:

$$x = \bar{x} = b_{xy}(y - \bar{y}) \tag{9.13.5}$$

where b_{yx} and b_{xy} are respectively the regression coefficient of y on x and the regression coefficient of x on y.

$$b_{yx} = \frac{\operatorname{cov}(x, y)}{\sigma_x^2} = r\frac{\sigma_y}{\sigma_x} \tag{9.13.6}$$

$$b_{xy} = \frac{\operatorname{cov}(x, y)}{\sigma_y^2} = r\frac{\sigma_x}{\sigma_y} \tag{9.13.7}$$

(2) The product of the two regression coefficients is equal to the square of correlation coefficient.

$$b_{yx} \cdot b_{xy} = r^2 \tag{9.13.8}$$

(3) r, b_{yx} and b_{xy}, all have the same sign. If the correlation coefficient r is zero, the regression coefficients b_{yx} and b_{xy} are also zero.
(4) The regression lines always intersect at the point $(\bar{x}, \bar{y})$. The slopes of the regression line of y on x and the regressioin line of x on y are respectively b_{yx} and $1/b_{xy}$.
(5) The angle between the two regression lines depends on the correlation coefficient r. When $r = 0$, the two lines are perpendicualr to each other; when $r = +1$, or $r = -1$, they coincide. As r increases numerically from 0 to 1, the angle between the regression lines diminishes from 90° to 0°.
(6) The two regression equations are usually different. However, when $r = \pm 1$, they become identical; and in this case, there is an exact linear relationship between the variables. When $r = 0$, the regression equations reduce to $y = \bar{y}$ and $x = \bar{x}$, and neither y nor x can be estimated from linear regression equations.

Example 9.23 *Find the regression of x on y from the following data :*

$$\Sigma x = 24 \qquad \Sigma y = 44 \qquad \Sigma xy = 306$$
$$\Sigma x^2 = 164 \qquad \Sigma y^2 = 574 \qquad n = 4$$

Find the estimated value of x, when y = 6. [C.U., B.A.(Econ) '65]

Solution The regression equation of x on y is

$$x - \bar{x} = b_{xy}(y - \bar{y}), \text{ where } b_{xy} = \text{cov}(x, y)/\sigma_y^2.$$

Here, $\bar{x} = \Sigma x/n = 24/4 = 6$, $\bar{y} = \Sigma y/n = 44/4 = 11$.

By (9.6.2), $\text{cov}(x, y) = 306/4 - (24/4)(44/4) = 10.5$

Also, $\sigma_y^2 = 574/4 - (44/4)^2 = 22.5$; $\therefore$ $b_{xy} = 10.5/22.5 = 0.467$

The regression equation of x on y is then

$$x - 6 = 0.467(y - 11); \quad \text{or,} \quad x = 0.467y + 0.86$$

When $y = 6$, we have $x = 0.467 \times 6 + 0.86 = 3.7$

Ans. $x = 0.467y + 0.86$; 3.7

Example 9.24 *The following data show the maximum and the minimum temperatures (degrees centigrade) on a certain day at ten important centres located throughout India :*

Max. temp. (x)	*29*	*23*	*25*	*15*	*27*	*29*	*24*	*31*	*22*	*28*
Min. temp. (y)	*8*	*3*	*7*	*5*	*8*	*19*	*10*	*7*	*5*	*11*

It is known that for hte given data $\Sigma x^2 = 6595$, $\Sigma y^2 = 867$, $\Sigma xy = 2193$. Draw a scatter diagram and show regression line of y on x thereon. [I.C.W.A., July '72]

Solution With the given values of (x, y), viz. (29, 8), (23, 3), ..., points are plotted on a graph paper treating the pairs of values as co-ordiantes. This will be the scatter diagram.

From the given data, find $n = 10$, $\Sigma x = 253$, $\Sigma y = 83$. Also given $\Sigma x^2 = 6595$, $\Sigma y^2 = 867$, $\Sigma xy = 2193$. With these results it is possible to find the regression equation of y on x, viz. $y - \bar{y} = b_{yx}(x - \bar{x})$. Proceeding as in the previous example, $\bar{x} = 25.3$, $\bar{y} = 8.3$, $\text{cov}(x, y) = 9.31$, $\sigma_z^2 = 19.41$. Hence $b_{yx} = 9.31/19.41 = 0.48$, and the regression equation is $y - 8.3 = 0.48(x - 25.3)$; or, $y = 0.48x - 3.84$.

In order to show this line on the scatter diagram, any two points satisfying this equation are taken. For example, when $x = 20$, $y = 5.76$; when $x = 30$, $y = 10.56$. The straight line joining the points (20, 5.76) and (30, 10.56) will be the required regression line.

Example 9.25 *From the following results, obtain the two regression equations and estimate the yield of crops when the rainfall is 22 cms, and the rainfall when the yield is 600 kg:*

	y (Yield in kg.)	*x (Rainfall in cm.)*
Mean	*508.4*	*26.7*
S.D.	*36.8*	*4.6*

Coefficient of correlation between yield and rainfall = 0.52

[C.A., May '76]

Solution For estimating yield (y) we have to use the regression equation of y on x, and for rainfall (x) the regression equation of x on y. Given $\bar{x} = 26.6$, $\bar{y} = 508.4$, $\sigma_x = 4.6$, $\sigma_y = 36.8$, $r = 0.52$,

$$b_{yx} = 0.52 \times 36.8/4.6 = 4.16,$$
$$b_{xy} = 0.52 \times 4.6/36.8 = .065$$

The regression equations are therefore $y - 508.4 = 4.16\ (x - 26.7)$ and $x - 26.7 = .065(y - 508.4)$ respectively, which simplify to

$$y = 4.16x + 397.33, \text{ and } x = .065y - 6.346$$

When $x = 22$, $y = 4.16 \times 22 + 397.33 = 488.8$ kg.

When $y = 600$, $x = .065 \times 600 - 6.346 = 32.7$ cm.

Example 9.26 *The following results were obtained from records of age (x) and systolic blood pressure (y) of a group of 10 women:*

	x	*y*
Mean ...	*53*	*142*
Variance ...	*130*	*165*

$$\Sigma(x - \bar{x})(y - \bar{y}) = 1220$$

Find the appropriate regression equation and use it to estimate the blood pressure of a women whose age is 45. [I.C.W.A., Jan. '73]

Solution The appropriate equation is the regression equation of y on x, viz. $y - \bar{y} = b_{yx}(x - \bar{x})$, where $b_{yx} = \text{cov}(x, y)/\sigma_x^2$.

But $\text{cov}(x, y) = \Sigma(x - \bar{x})(y - \bar{y})/n = 1220/10 = 122$,

so that $b_{yx} = 122/130 = 0.94$.

The regression equation is therefore $y - 142 = 0.94\ (x - 53)$; or, $y = 0.94x + 92.18$. When $x = 45$, $y = 0.94 \times 92.18 = 134.5$. The estimated blood pressure is 134.5.

Ans. $y = 0.94x + 92.18$; 134.5

Example 9.27 *Marks obtained by 12 students in the college test (x) and the university test (y) are as follows:*

x	*41*	*45*	*50*	*68*	*47*	*77*	*90*	*100*	*80*	*100*	*40*	*43*
y	*60*	*63*	*60*	*48*	*85*	*56*	*53*	*91*	*74*	*98*	*65*	*43*

What is your estimate of the marks a student could have obtained in the university test if he obtained 60 in the college test but was ill at the time of the university test?

[B.U., B.A. (Econ) '70]

Solution For estimating y, we use the regression equation of y on x, *viz.* $y - \bar{y} = b_{yx}(x - \bar{x})$, where $b_{yx} = \text{cov}(x, y)/\sigma_x^2$. Therefore, we have to find $\bar{x}$, $\bar{y}$, cov(x, y) and σ_x^2.

Table 9.13 Calculations for Regression

x	y	$X = x - 60$	$Y = y - 60$	X^2	XY
41	60	– 19	0	361	0
45	63	– 15	3	225	– 45
50	60	– 10	0	100	0
68	48	8	– 12	64	– 96
47	85	– 13	25	169	– 325
77	56	17	– 4	289	– 68
90	53	30	– 7	900	– 210
100	91	40	31	1,600	1,240
80	74	20	14	400	280
100	98	40	38	1,600	1,520
40	65	– 20	5	400	– 100
43	43	– 17	– 17	289	289
Total 781	796	61	76	6,397	2,485

$\bar{x} = 781/12 = 65.08$, $\bar{y} = 796/12 = 66.33$

Since the variance and covariance are unaffected by changes of origin,

$$\text{cov}(x, y) = \text{cov}(X, Y) = \Sigma XY/n - (\Sigma X/\text{n})(\Sigma Y/n)$$
$$= 2485/12 - (61/12)(76/12) = 25184/144$$
$$\sigma_x^2 = \Sigma X^2/n - (\Sigma X/n)^2 = 6397/12 - (61/12)^2 = 73043/144$$

Therefore, $$b_{yx} = \frac{25184/144}{73043/144} = \frac{25184}{73043} = 0.345$$

The regression equation is $y - 66.33 = 0.345(x - 65.08)$;

i.e. $$y = 43.88 + 0.345x$$

When $x = 60$, $y = 43.88 + 0.345 \times 60 = 64.58 = 65$ (approx.)

Ans. $y = 43.88 + 0.345x$; 65

[**Note:** In Examples 9.26 and 9.27, the use of formula $b_{yx} = r\sigma_y / \sigma_x$ would involve laborious calculations for r, and must be avoided.]

Example 9.28 *Derive the regression line which you consider more important from the following series of observations :*

Output (thousand)	*5*	*7*	*9*	*11*	*13*	*15*
Profit per unit of Output (Rs)	*1.70*	*2.40*	*2.80*	*3.40*	*3.70*	*4.40*

[B.U., B.A.(Econ) '66; C.U., B.A. (Econ) '71; I.C.W.A., June '73]

Solution Let x represent 'Output (thousand)' and y represent 'Profit per unit of output (Rs.)'. It is considered that the regression line of y on x is the more important; because, using the equation it will be possible to find the expected profit based on output of the firm.

[**Note:** Here the successive values of x are equidistant. Also, since the number of pairs of observations is even, viz. 6, we apply the transformation (8.6.4) at p. 252. for simplifying the calculations]

Table 9.14 Calculations for Regression

x	y	$u = x - 10$	u^2	uy
5	1.70	– 5	25	– 8.50
7	2.40	– 3	9	– 7.20
9	2.80	– 1	1	– 2.80
11	3.40	1	1	3.40
13	3.70	3	9	11.10
15	4.40	5	25	22.00
Total 60	18.40	0	70	18.00

Proceeding as in Example 9.27, $\bar{x} = 60/6 = 10$, $\bar{y} = 18.406 = 3.07$

$$\text{cov}(x, y) = \text{cov}(u, y) = \Sigma uy/n - (\Sigma u/n)(\Sigma y/n)$$
$$= 18.00/6 - (0/6)\,(18.40/6) = 3.00$$
$$\sigma_x^2 = \sigma_u^2 = \Sigma u^2/n - (\Sigma u/n)^2 = 70/6 - (0/6)^2 = 70/6$$
$$b_{yx} = 18.00/70 = 0.257$$

The regression equation of y on x is $y - \bar{y} = b_{yx}\,(x - \bar{x})$;

i.e., $y - 3.07 = 0.257(x - 10)$; or, $y = 0.257x + 0.50$

Example 9.29 *Find the equation of the line of regression of x on y for the following data:*

x	1.0	1.5	2.0	2.5	3.0	3.5	4.0
y	5.3	5.7	6.3	7.2	8.2	8.7	8.4

[C.U., B.A. (Econ) '72]

Solution The regression equation of x on y is $x - \bar{x} = b_{xy}\,(y - \bar{y})$, where $b_{xy} = \text{cov}(x, y)/\sigma_y^2$. For simplifying the calculations, let us make a change of origin and scale for both the variables

$$u = \frac{x - 2.5}{0.5}, \quad v = \frac{y - 7.0}{0.1}$$

[**Note:** Here the values of x are equidistant and n is odd. So, we use the transformation (8.6.3), page 252].

$$\bar{x} = 17.5/7 = 2.5, \ \bar{y} = 49.8/7 = 7.11$$
$$\text{cov}(u, v) = \Sigma uv/n - (\Sigma u/n)\,(\Sigma v/n) = 172/7 - (0/7)(8/7) = 172/7$$
$$\sigma_v^2 = \Sigma v^2/n - (\Sigma v/n)^2 = 1140/7 - (8/7)^2 = 7916/49$$

Using (9.6.4), $\text{cov}(x, y) = d.d.'\ \text{cov}(u, v) = (0.5)\,(0.1)(172/7)$

Table 9.15 Calculations for Regression

x	y	u	v	v^2	uv
1.0	5.3	– 3	– 17	289	51
1.5	5.7	– 2	– 13	169	26
2.0	6.3	– 1	– 7	49	7
2.5	7.2	0	2	4	0
3.0	8.2	1	12	144	12
3.5	8.7	2	17	289	34
4.0	8.4	3	14	196	42
Total 17.5	49.8	0	8	1,140	172

Using (6.7.4), $\sigma_y^2 = d^2 . \sigma_v^2 = (0.1)^2 (7916/49)$

Hence, $$b_{xy} = \frac{\text{cov}(x, y)}{\sigma_y^2} = \frac{0.5 \times 0.1 \times (172/2)}{(0.1)^2 \times (7916/49)} = 0.76$$

Therefore, the regression equation of x on y is

$$x - 2.5 = 0.76(y - 7.11); \quad \text{i.e.} \quad x = 0.76y - 2.90.$$

Example 9.30 *From the following bivariate frequency distribution, calculate (i) the coefficient of correlation and (ii) the regression equation of marks (y) on age (x)–*

Age (years)

Marks	*15–20*	*20–25*	*25–30*	*30–35*	*35–40*	*Total*
0–19	*4*	*2*				*6*
20–39	*6*	*5*	*3*	*1*		*15*
40–59		*9*	*4*	*2*	*1*	*16*
60–79		*7*	*4*	*1*		*12*
80–99			*1*			*1*
Total	*10*	*23*	*12*	*4*	*1*	*50*

Solution Using the results from Table 9.16, we have

$$\sigma_u^2 = \frac{\Sigma f_1 u^2}{N} - \left(\frac{\Sigma f_1 u}{N}\right)^2 = \frac{71}{50} - \left(\frac{-37}{50}\right)^2 = 0.8724$$

Table 9.16 Calculations for Correlation and Regression

x_u / y_v	*17.5* / *– 2*	*22.5* / *– 1*	*27.5* / *0*	*32.5* / *1*	*37.5* / *2*	*Totals* / f_2	f_2v	f_2v^2	*fuv*
9.5–2	4 (16)	2 (4)				6	– 12	24	20
29.5–1	6 (12)	5 (5)	3 (0)	1 (– 1)		15	– 15	15	16
49.5–0		9 (0)	4 (0)	2 (0)	1 (0)	16	0	0	0
69.5 – 1		7 (– 7)	4 (0)	1 (1)		12	12	12	– 6
89.5 – 2			1 (0)			1	2	4	0
Totals									
f_1	10	23	12	4	1	50 = N	– 13	55	30
f_1u	– 20	–23	0	4	2	– 37			
f_1u^2	40	23	0	4	4	71			
fuv	28	2	0	0	0	30			

Explanation of symbols used in Table 9.16:

(i) x and y represent mid-values of the class intervals of age and marks respectively.

(ii) f_1 and f_2 represent marginal frequencies of the distribution of x and y respectively.

(iii) $u = \dfrac{x - 27.5}{5}$ and $v = \dfrac{y - 49.5}{20}$.

(iv) Number within brackets in each cell is the product of that cell frequency and the corresponding values of u and v.

(v) fuv represent totals of the numbers within brackets, mentioned at (iv) above, in any row or column.

$$\sigma_v^2 = \frac{\Sigma f_2 v^2}{N} - \left(\frac{\Sigma f_2 v}{N}\right)^2 = \frac{55}{50} - \left(\frac{-13}{50}\right)^2 = 1.0324$$

$$\text{cov}(u, v) = \frac{\Sigma fuv}{N} - \left(\frac{\Sigma f_1 u}{N}\right)\left(\frac{\Sigma f_2 v}{N}\right) = \frac{30}{50} - \left(\frac{-37}{50}\right)\left(\frac{-13}{50}\right) = 0.4076$$

$$\sigma_u = \sqrt{0.8724} = 0.93, \sigma_v = \sqrt{1.0324} = 1.02$$

$$\therefore \quad r_{xy} = r_{uv} = \frac{\text{cov}(u, v)}{\sigma_u \sigma_v} = \frac{0.4076}{0.93 \times 1.02} = 0.43$$

Again, since $u = (x - 27.5)/5$ and $v = (y - 49.5)/20$,

$$\bar{x} = 27.5 + 5(-37/50) = 23.8$$

$$\bar{y} = 49.5 + 20(-13/50) = 44.3$$

Applying (9.6.4) and (6.6.3),

$$\text{cov}(x, y) = 5 \times 20 \times \text{cov}(u, v) = 100 \times 0.4076$$

$$\sigma_x^2 = 5^2 \times \sigma_u^2 = .25 \times 0.8724$$

$$\therefore \quad b_{yx} = \frac{\text{cov}(x, y)}{\sigma_x^2} = \frac{100 \times 0.4076}{25 \times 0.8724} = 1.87$$

The regression equation of y on x is $y - \bar{y} = b_{yx}(x - \bar{x})$

i.e. $\quad y - 44.3 = 1.87(x - 23.8)$

or, $\quad y = 1.87x - 0.21$

Ans. $r = +0.43$, $y = 1.87x - 0.21$

Example 9.31 *Let the lines of regression concerning two variables x and y be given by y = 32 – x and x = 13 – 0.25y. Obtain the values of the means and the correlation coefficient.* [C.U., B.Sc. '71]

Solution Since the regression lines intersect at ($\bar{x}$, $\bar{y}$), the means will be obtained by solving the two equations. Solving $y = 32 - x$ and $x = 13 - 0.25y$, we get $x = 6.7$ and $y = 25.3$. So, $\bar{x} = 6.7$, $\bar{y} = 25.3$.

$y = 32 - x$ is the regression equation of y on x; therefore the 'regression coefficient of y on x' is the coefficient of x on the right; i.e. $b_{yx} = -1$. Similarly, $x = 13 - 0.25y$ being the regression equation of x on y, the 'regression coefficient of x on y' is the coefficient of y on the right; i.e. $b_{xy} = -0.25$.

$$r^2 = b_{yx} \times b_{xy} = (-1)(-0.25) = 0.25;$$

$$r = \sqrt{0.25} = \pm 0.5$$

But since the regression coefficients are negative (note that both must have the same sign), the correlation coefficient also must be negative, i.e. $r = -0.5$.

Ans. $\bar{x} = 6.7$, $\bar{y} = 25.3$, $r = -0.5$.

Example 9.32 *If x and y satisfy the relationship y = –5 + 6x, what is the product-moment correlation?* [B.U., B.A. (Econ) '73]

Solution Since the relationship between x and y is exact and *linear*, there is perfect correlation between the variables; i.e. $r = \pm 1$. But the slope of the straight line (viz. + 6) is positive. Hence r must be positive, and $r = +1$.

[**Note:** We speak of two types of relationship between the variables—(i) 'exact' or 'functional' relationship, and (ii) 'statistical' relationship. If to any specified value of one variable there corresponds some definite value or values of another variable, the relationship is exact. However, in many practical situations, no such relationship exists, e.g. between Height and Weight, or Income and Expenditure, etc. We then try to find the best possible estimates of one variable by means of an equation connecting them. This equation, known as 'Regression Equation', shows the statistical relationship between the variables.]

Example 9.33 *Regression of savings(s) of a family on income (y) may be expressed as* $s = a + \frac{y}{m}$, *where a and m are constants. In a random sample of 100 families the variance of savings in one-quarter of the variance of incomes and the correlation is found to be 0.4. Obtain the estimate of m.*

[I.C.W.A., June '74-old]

Solution The given regression equation of s on y can be written as $s = a + (1/m)y$, which is the equation of a straight line. Therefore the regression coefficient of s on y is the coefficent of y on the right;

i.e. $b_{sy} = 1/m$ (i)

But $b_{sy} = r\dfrac{\sigma_s}{\sigma_y}$ (ii)

However, we are given that $\sigma_s^2 = (1/4)\,\sigma_y^2$ and $r = 0.4$, so that $\dfrac{\sigma_s}{\sigma_y} = \dfrac{1}{2}$

Hence, using (ii), $b_{sy} = (0.4) \times \dfrac{1}{2} = 0.2$

Now, from (i), $0.2 = 1/m; \quad \therefore \quad m = 1/0.2 = 5$

Example 9.34 *You are given that the variance of x is 9. The regression equations are 8x – 10y + 66 = 0 and 40x – 18y = 214. Find (i) Average values of x and y, (ii) Correlation coefficient between the two variables, (iii) Standard deviation of y.*

[C.A., Nov. '77; I.C.W.A., June '77]

Solution

[**Note:** The regression equations of y on x and of x on y are usually shown in the forms $y = a + bx$ and $x = a' + b'y$ respectively, and in such cases it is not difficult to find the regression coefficients (Example 9.31). If however the equations are shown otherwise, they should be brought to the usual forms and then the regression coefficients can be easily determined. In the present problem, it is not known which one is the regression equation of y on x and which one the regression equation of x on y. A method has been shown for identifying the appropriate equations.]

Since the regression lines always intersect at the point ($\bar{x}$, $\bar{y}$), the averages of x and y will be obtained by solving the two equations [see Example 8.1 (g), p. 248]. Multiplying the first equation by 5 and then subtracting the result from the second equation, we have $32y = 544$, so that $y = 17$. Now putting this value in any of the given equations and solving we get $x = 13$. Therefore, the averages of x and y are 13 and 17 respectively.

Let us assume that $8x - 10y + 66 = 0$ represents the regression equation of x on y, and $40x - 18y = 214$ the regression equation of y on x. These equations can be re-written as $x = (5/4)y - (33/4)$ and $y = (20/9)\,x - (107/9)$ respectively. The regression coefficients should then be $b_{xy} = 5/4$ and $b_{yx} = 20/9$.

$$\therefore \qquad r^2 = b_{xy} \,.\, b_{yx} = (5/4)\,(20/9) = 25/9$$

This result is impossible, because r cannot exceed 1 numberically. So, our assumptions must be wrong.

The correct position must be the other way round; namely. $8x - 10y + 66 = 0$ must be the regression equation of y on x, and $40x - 18y = 214$ the regression equation of x on y. The two equations when expressed in the usual forms give $y = (4/5)x + (33/5)$ and $x = (9/20)y + (107/20)$. The correct values of the regression coefficients are $b_{yx} = 4/5$, $b_{xy} = 9/20$.

i.e., $$r = \sqrt{(4/5)(9/20)} = +0.6$$

(The positive value of the square-root is taken, since the regression coefficients are positive).

Again, given $\sigma_x^2 = 9$, i.e. $\sigma_x = 3$, using the relation $b_{yx} = 4/5$,

we have $$r\frac{\sigma_y}{\sigma_x} = 0.8; \quad \text{or} \quad 0.6 \times \frac{\sigma_y}{3} = 0.8$$

This gives $\sigma_y = 4$. *Ans.* 13, 17; + 06; 4

Example 9.35 *Prove that the coefficient of correlation is the geometric mean of the coefficients of regression.*

[I.C.W.A., Jan. '65, '69, June '74, Dec. '76; B.U., B.A., (Econ) '66; C.U., M.Com. '64, '68, '71]

Solution The regression coefficients b_{yx} and b_{xy} may be expressed in terms of the correlation coefficient (r) and the standard deviations of x and y (viz. σ_x and σ_y) by the relations

$$b_{yx} = r\frac{\sigma_y}{\sigma_x}, \quad b_{xy} = r\frac{\sigma_x}{\sigma_y}$$

$$\therefore \qquad b_{yx} \times b_{xy} = \left(r\frac{\sigma_y}{\sigma_x}\right)\left(r\frac{\sigma_x}{\sigma_y}\right) = r^2$$

or, $$r^2 = b_{yx} \times b_{xy}; \quad \therefore \quad r = \sqrt{b_{yx} \times b_{xy}}$$

Thus, the correlation coefficient r is the geometric mean of the two regression coefficients b_{yx} and b_{xy}.

Example 9.36 *What are regression lines? Explain why we have two regression lines and why these two lines are identical if r the correlation coefficient is + 1 or – 1.*

[C.U., B.A. (Econ) '63; M.Com. '66; I.C.W.A., Jan. '73, Dec. '77]

Solution If bivariate data are plotted as points on a graph papaer, it will be found that the concentration of points follows a certain pattern showing the relationship between the variables. When the trend of points is found to be linear, we determine the best-fitting straight line by the Method of Least Squares. Such straight lines which are used to obtain best estimates of one variable for given values of the other, are called *regression lines.* If x is considered as the independent variable and y the dependent variable, a linear equation of the form $y = a + bx$ is fitted to the pairs of observations (x_1, y_1), (x_2, y_2), ... (x_n, y_n) leading to the equation

$$y - \bar{y} = r\frac{\sigma_y}{\sigma_x}(x - \bar{x}) \qquad \text{(i)}$$

This is known as the *regression line of y on x*. If however, the variable x is considered to be dependent on y, a linear equation of the form $x = a' + b'y$ is fitted, yielding

$$x - \bar{x} = r\frac{\sigma_x}{\sigma_y}(y - \bar{y}) \qquad \text{(ii)}$$

This is known as the *regression line of x on y*. Equation (i) is used to obtain best estimates of y for given values of x, and equation (ii) to obtain best estimates of x for given values of y.

There are two regression lines, which are separately used to obtain best estimates of each variable. The same equation can not serve both the purposes; because the two equations (i) and (ii) are derived under two different assumptions. In deriving equation (i) it is assumed that the values of x are known exactly and those of y are subject to error. In deriving equation (ii) the assumption is just the reverse, i.e. the values of y are known exactly and those of x only are subject to error.

Since the two regression lines (i) and (ii) both pass through $(\bar{x}, \bar{y})$, they will be identical if their slopes are equal, i.e. if $b_{yx} = 1/b_{xy}$

or, $r\frac{\sigma_y}{\sigma_x} = \frac{1}{r}\frac{\sigma_y}{\sigma_x}$; simplifying, we get $r = \pm 1$.

9.14 EXPLAINED VARIATION AND UNEXPLAINED VARIATION

If y_i' represents the *estimated* value of y from the regression equation of y on x (note that y_i denotes the *observed* value) when $x = x_i$, i.e., $y_i' - \bar{y} = b_{yx}(x_i - \bar{x})$, then it can be shown that

$$\Sigma(y_i - \bar{y})^2 = \Sigma(y_i - y_i') + \Sigma(y_i' - \bar{y})^2$$

$\Sigma(y_i - \bar{y})^2$ is called *Total variation* of the observed values of y. $\Sigma(y_i - y_i')^2$ is the sum of the squares of vertical distances of the points on the scatter diagram around the regression line of y on x, and as such is called *Variation around the regression line* or *Unexplained variation* or *Residual variation*. $\Sigma(y_i' - \bar{y})^2$ is called *Variation due to regression* or *Explained variation.*

It can be shown that

$$\frac{\textit{Explained variation}}{\textit{Total variation}} = r^2$$

i.e. *Proportion of Total variation explained by regression* = r^2

We have, thus, a new interpretation of the correlation coefficient, viz. the square of the correlation coefficient is equal to the proportion of total variation explained by regression.

When $r^2 = 1$, the whole of the total variation is explained by regression, so that the unexplained variation is zero and each $y_i = y_i'$. All the points on the scatter diagram lie on the regression line of y on x (the two regression lines now coincide), and there is *perfect linear dependence* between the variables. For a given value of x, we have a fixed value of y.

When $r^2 = 0$, the explained variation is zero and each $y_i' = \bar{y}$. That is, the estimated value of y is a constant $\bar{y}$, whatever be the value of x. In this case there is *no linear dependence* between the variables.

r^2 may, therefore, be used as a measure of the degree of linear dependence between the variables. It may be noted that the sign of the correlation coefficient r only indicates whether x and y move in the same direction or opposite directions; but r^2 is always positive and gives a better measure (than r) of the degree of linear dependence.

Example 9.37 *For a set of pairs of observations $(x_1, y_1), (x_2, y_2), \ldots (x_n, y_n)$, prove that the total variation $\Sigma(y_i - \bar{y})^2$ can be written as*

$$\Sigma(y_i - \bar{y})^2 = \Sigma(y_i - y_i')^2 + \Sigma(y_i' - \bar{y})^2$$

where y_i' represents the values of y for given values of $x = x_1, x_2, \ldots, x_n$ obtained from the regression equation of y on x.

Explain the significance of the identity given above with reference to regression equations.

[C.U., B.A. (Econ) '67 '69; I.C.W.A., Dec. '76]

Solution The regression equation of y on x is $y - \bar{y} = b_{yx}(x - \bar{x})$ where $b_{yx} = \text{cov}(x, y)/\sigma_x^2$. If y_i' represents the estimated values of y when $x = x_i$, then $y_i' - \bar{y} = b_{yx}(x_i - \bar{x})$;

i.e.,
$$y_i' = \bar{y} + b_{yx}(x_i - \bar{x}) \quad \text{(i)}$$

Now,
$$(y_i - \bar{y})^2 = \{(y_i - y_i)\} + (y_i - \bar{y})\}^2$$
$$= (y_i - y_i')^2 + (y_i' - \bar{y})^2 + 2(y_i - y_i')(y_i' - \bar{y})$$

Summing both sides over all values of $i = 1, 2, \ldots n$.

$$\Sigma(y_i - \bar{y})^2 = \Sigma(y_i - y_i')^2 + \Sigma(y_i' - \bar{y})^2 + 2\Sigma(y_i - y_i')(y_i' - \bar{y}) \quad \text{(ii)}$$

Considering the last term on the right and substituting for y_i' from (i),

$$\Sigma(y_i - y_i')(y_i' - \bar{y}) = \Sigma[\{(y_i - \bar{y}) - b_{yx}(x_i - \bar{x})\}\{b_{yx}(x_i - \bar{x})\}]$$
$$= \Sigma[b_{yx}(x_i - \bar{x})(y_i - \bar{y}) - b_{yx}^{\,2}(x_i - \bar{x})^2]$$
$$= b_{yx}\Sigma(x_i - \bar{x})(y_i - \bar{y}) - b_{yx}^{\,2}\Sigma(x_i - \bar{x})^2$$
$$= b_{yx}[n.\text{cov}(x, y) - b_{yx}\,.\,n\sigma_x^2], \text{ because}$$
$$\text{cov}(x, y) = \Sigma(x_i - \bar{x})(y_i - \bar{y})/n \quad \text{and} \quad \sigma_x^2 = \Sigma(x_i - \bar{x})^2/n$$
$$= b_{yx}\left[n.\text{ cov}(x, y) - \frac{\text{cov}(x, y)}{\sigma_x^2}\,.\,n\sigma_x^2\right]$$
$$= b_{yx}[n \text{ cov}(x, y) - n.\text{cov}(x, y)] = b_{yx} \times 0 = 0.$$

Hence, from (ii)

$$\Sigma(y_i - \bar{y})^2 = \Sigma(y_i - y_i')^2 + \Sigma(y_i' - \bar{y})^2 \quad \text{(iii)}$$

Let us suppose that x represents the height and y the weight of several persons. Then the estimated weight y_i' of the person with height x_i is given by (i). It can not be expected that the actual weight y_i will be equal to its estimate y_i' obtained on the basis of his height x_i, i.e. the regression value. But y_i' provides *some* estimate of the weight of the person, i.e. a part of the actual weight y_i can be estimated or *explained* by the regression value y_i'. The difference $(y_i - y_i')$, called the *residual*, is the *unexplained* part of the weight.

The significance of the identity (iii) is that it is possible to split up the total variation of the observed values of y into two parts—a part (the second term on the right) explained by regression, i.e. which is entirely due to variation in the values of x, and another part which is due to other factors, unexplained by regression. This signifies that

Total variation = Unexplained variation + Explained variation.

9.15 REGRESSION CURVE IN BIVARIATE FREQUENCY DISTRIBUTION

With reference to a bivariate frequency distribution, the *Regression Curve of y on x* may be defined as that curve which shows the relationship between the conditional mean values of y and the corresponding specified values (or class marks) of x. If this curve happens to be a straight line, we say that the 'regression of y on x is linear', if it is a parabola, we come to the case of 'parabolic regression'.

Similarly, the *Regression Curve of x on y* may be defined as that curve which shows the relationship between the conditional mean values of x and the corresponding values of y; if this curve happens to be a straight line, we come to the case of linear regression, and say that the 'regression of x on y is linear'. It is possible for either regression to be linear or both.

Example 9.38 *In the table below verify that the means of the x-arrays are collinear and also those of the y-arrays and hence show that the correlation coefficient is – 0.535.*

y \ x	0	1	2	3	4
0	–	–	–	4	3
1	–	–	18	36	9
2	–	12	54	36	3
3	1	12	18	4	–

[I.C.W.A., June '73]

Solution (Note that the "means of the x-arrays" refers to the 'conditional mean values of x, and "means of the y-arrays" denotes the 'conditional mean values of y'; see Example 9.3, page 272). We may proceed as follows–

x	*Total of y-array*	*Marginal Frequency*	*Mean of y-array*
0	$1 \times 3 = 3$	1	$3/1 = 3$
1	$12 \times 2 + 12 \times 3 = 60$	24	$60/24 = 2.5$
2	$18 \times 1 + 54 \times 2 + 18 \times 3 = 180$	90	$180/90 = 2$
3	$4 \times 0 + 36 \times 1 + 36 \times 2 + 4 \times 3 = 120$	80	$120/80 = 1.5$
4	$3 \times 0 + 9 \times 1 + 3 \times 2 = 15$	15	$15/15 = 1$

The means of y-arrays corresponding to $x = 0, 1, 2, 3, 4$ are 3, 2.5, 2, 1.5, 1 respectively. Since, for unit increments in x these means change by equal amounts, namely – 0.5, they are collinear. This implies that if the means of y-arrays are plotted as points on a graph paper against the corresponding values of x, these points will lie on a straight line, which is the regression line of y on x. The regression coefficient of y on x, which measures the change in the value of y for unit increment in x, is thus found to be $b_{yx} = -0.5$.

y	*Total of x-array*	*Marginal Frequency*	*Mean of x-array*
0	$4 \times 3 + 3 \times 4 = 24$	7	24/7
1	$18 \times 2 + 36 \times 3 + 9 \times 4 = 180$	63	180/63 = 20/7
2	$12 \times 1 + 54 \times 2 + 36 \times 3 + 3 \times 4 = 240$	105	240/105 = 16/7
3	$1 \times 0 + 12 \times 1 + 18 \times 2 + 4 \times 3 = 60$	35	60/35 = 12/7

Here again it is seen that the means of x-arrays change by equal amounts, namely $-4/7$, for unit increments in y, and so these means are collinear. The regression of x on y is linear, and the regression coefficient of x on y is $b_{xy} = -4/7$.

Since the correlation coefficient can be obtained from the relation $r = \sqrt{b_{yx} \cdot b_{xy}}$, with a sign common to both b_{yx} and b_{xy},

$$r = \sqrt{(-0.5)(-4/7)} = \sqrt{2/7} = -0.535$$

9.16 RANK CORRELATION

The product-moment correlation coefficient (r) is calculated by using 'values' of the variables. But many situations arise in which either precise measurements are not available, or the characters cannot be measured at all. For example, in order to find the extent of association between 'intelligence' and 'effciency in salesmanship' for a group of salesmen, we could attempt to measure the two qualities by allotting marks to each salesman. But this method is open to many objections, and an exact measurement of the two qualities is not at all possible.

Difficulties of this type however disappear if we arrange the individuals in order of merit or proficiency in the possession of the qualities, using numbers 1, 2, 3, ... The individuals are then said to be *ranked* and the number allotted to a particular candidate is called his *rank*. For each individual, we then have a pair of ranks, one in each character. The correlation coefficient between the two series of ranks is called '*Rank Correlation Coefficient*'. It is given by the formula

$$R = 1 - \frac{6\Sigma d^2}{n^3 - n} \tag{9.16.1}$$

where d represents the difference of the ranks of an individual in the two characters, and n is the number of individuals. This formula is also known as *Spearman's formula* for rank correlation coefficient.

The rank correlation coeffcient lies between -1 and $+1$.

$$-1 \le R \le +1$$

It has the maximum value $+1$, when the ranks in the two characters are equal, i.e. individuals with ranks 1, 2, 3, ... n in any character have exactly the same ranks 1, 2, 3, ...n respectively in the other. Again, R has the minimum value -1, when the ranks are just the opposite, i.e. individuals with ranks 1, 2, 3, ... n in one character have ranks $n, n-1, \ldots 3, 2, 1$ in the other.

In the calculation of rank correlation coefficient from given scores, if several individuals have the same score in any character, they must be allotted the same ranks and we are then concerned with what are known as '*tied ranks*'. In dealing with such cases, the usual way is to allot the average rank to each of these individuals, and then calculate the product-moment correlation coefficient from these ranks (see Example 9.42). However in such cases one way of correcting formula (9.16.1) is to increase Σd^2 by $(t^3 - t)/12$ in respect of each tie, where t denotes the number of individuals involved in a tie wheither in the first or second series. The modified formula for rank correlation coefficient, when there are ties, is then

$$R' = 1 - \frac{6\{\Sigma d^2 + \Sigma(t^3 - t)/12\}}{n^2 - n} \qquad (9.16.2)$$

Uses—

(1) The rank correlation coefficient R is used as a measure of the 'degree of association between two attributes', where measurements on the characters are not available, but it is possible to rank the individuals in some order without difficulty (Note that the product-moment correlation coefficient r is a measure of the degree of association between two variables). It is therefore applied in vocational and applied psychology for finding the degree of correspondence between abilities.

(2) It is also used in cases where exact or reliable measurements are not available. In educational tests it is known to all that the usual method of allotting marks to the candidates is open to the objection that different examiners would probably give different marks to the same candidate. The product-moment correlation coefficient r between scores would not therefore accurately reflect the relation ship between the abilities. However, if the candidates are ranked, the discrepancies in marks given by examiners would not seriously affect the value of rank correlation coefficient.

(3) Even when exact measurements are available, the rank correlation coefficient R provides a quick estiamte of the degree of association between the variables. The laborious calculations for the product-moment correlation coefficient r are replaced by ranking the individuals and then using formula (9.16.1) or (9.16.2).

Example 9.39 *In a contest, two judges ranked seven candidates in order of their preference as in the following table:*

Candidates	*A*	*B*	*C*	*D*	*E*	*F*	*G*
Ranks by Judge I	2	1	4	5	3	7	6
Ranks by Judge II	3	4	2	5	1	6	7

Calculate the rank correlation coefficient.

Solution

Table 9.17 Calculations for Rank Correlation Coefficient

Candidates	*Ranks by* Judge I x	Judge II y	$d = x - y$	d^2
A	2	3	– 1	1
B	1	4	– 3	9

(*Contd.*)

Candidates	Ranks by Judge I x	Judge II y	$d = x - y$	d^2
C	4	2	2	4
D	5	5	0	0
E	3	1	2	4
F	7	6	1	1
G	6	7	– 1	1
Total	—	—	0	20

Here, $n = 7, \Sigma d^2 = 20,$

$$\therefore \quad R = 1 - \frac{6\ \Sigma d^2}{(n^3 - n)} = 1 - \frac{6 \times 20}{(7^3 - 7)} = 0.64. \qquad \textit{Ans. } R = 0.64$$

Example 9.40 *In a certain examination 10 students obtained the following marks in Mathematics and Physics. Find Spearman's rank correlation coefficient.*

Student (Roll No.)	*1*	*2*	*3*	*4*	*5*	*6*	*7*	*8*	*9*	*10*
Marks in Mathematics	*90*	*30*	*82*	*45*	*32*	*65*	*40*	*88*	*73*	*66*
Marks in Physics	*85*	*42*	*75*	*68*	*45*	*63*	*60*	*90*	*62*	*58*

[I.C.W.A., June '76-old]

Solution We have to otbain the ranks of students in the two subjects and calculate the rank correlation coefficient. For example, in Mathematics the student with Roll No. 1 gets the highest marks 90 and is given rank 1. Roll No. 8 with 88 marks has rank 2, and so on (see col. (3), Table 9.18.)

Table 9.18 Calculations for Rank Correlation Coefficient

Roll No. (1)	*Mathematics Marks (X) (2)*	*Rank (x) (3)*	*Physics Marks (Y) (4)*	*Rank (y) (5)*	$d = x - y$ *(6)*	d^2 *(7)*
1	90	1	85	2	– 1	1
2	30	10	42	10	0	0
3	82	3	75	3	0	0
4	45	7	68	4	3	9
5	32	9	45	9	0	0
6	65	6	63	5	1	1
7	40	8	60	7	1	1
8	88	2	90	1	1	1
9	73	4	62	6	– 2	4
10	66	5	58	8	– 3	9
Total	—	—	—	—	0	26

Applying formula (9.6.1),

$$R = 1 - \frac{6 \times 26}{10^3 - 10} = 1 - \frac{26}{165} = +0.84$$

[**Note:** It is interesting to note that Spearman's rank correlation coefficient $R = 0.84$, calculated above, is in good agreement with Pearson's product-moment correlation coefficient $r = 0.86$ calculated from actual marks.]

Example 9.41 *Deduce Spearman's formula for rank correlation coefficient.*

[B.U., B.A. (Econ) '73]

Solution Let $(x_1, y_1), (x_2, y_2), ..., (x_n, y_n)$ denote the ranks of n individuals in the two characters. Spearman's rank correlation coefficient is the product-mement correlation coefficient between these ranks, treating them as values of variables.

It may be noted that $x_1, x_2, ..., x_n$ being ranks (and not the actual values) are only the numbers 1, 2, ..., n arranged in some order. Similarly $y_1, y_2, ..., y_n$ are also the same numbers, but possibly in a different order [see Table 9.18, cols (3) and (5)].

$$\Sigma x = 1 + 2 + 3 + ... + n = n(n + 1)/2 \qquad \text{(see p 67)}$$

$$\Sigma x^2 = 1^2 + 2^2 + 3^2 + ... n^2 = n(n + 1)(2n + 1)/6$$

$$\therefore \quad \bar{x} = \Sigma x/n = (n + 1)/2$$

$$\sigma_x^2 = \frac{\Sigma x^2}{n} - \left(\frac{\Sigma x}{n}\right)^2 = \frac{n(n+1)(2n+1)}{6n} - \frac{(n+1)^2}{4} = \frac{n^2 - 1}{12}$$

Since the x-series and the y-series comprise the same numbers, their means and variances will be equal. Therefore

$$\bar{x} = \bar{y} = \frac{n+1}{2}; \quad \sigma_x^2 = \sigma_y^2 = \frac{n^2 - 1}{12}$$

In order to find cov(x, y), we proceed as follows. Let us write

$$d_i = x_i - y_i$$

Since $\bar{x}$ and $\bar{y}$ are equal, we may write $d_i = (x_i - \bar{x}) - (y_i - \bar{y})$.

$$\therefore \quad \Sigma d_i^2/n = \Sigma(x_i - \bar{x})^2/n + \Sigma(y_i - \bar{y})^2/n - 2\Sigma(x_i - \bar{x})(y_i - \bar{y})/n$$

$$= \sigma_x^2 + \sigma_y^2 - 2\,\text{cov}(x, y) = \frac{n^2 - 1}{6} - 2\,\text{cov}(x, y)$$

Transposing and dividing by 2, we find that

$$\text{cov}(x, y) = \frac{n^2 - 1}{12} - \frac{\Sigma d^2}{2n}$$

$$\therefore \quad R = \frac{\text{cov}(x, y)}{\sigma_x \sigma_y} = \frac{\dfrac{n^2 - 1}{12} - \dfrac{\Sigma d^2}{2n}}{\sqrt{\dfrac{n^2 - 1}{12}}\sqrt{\dfrac{n^2 - 1}{2n}}} = 1 - \frac{6\Sigma d^2}{n(n^2 - 1)}$$

Example 9.42 *In the following table are recorded data showing the test scores made by 10 salesmen on an intalligence test and their weekly sales:*

Salesmen	*1*	*2*	*3*	*4*	*5*	*6*	*7*	*8*	*9*	*10*
Test Scores	*50*	*70*	*50*	*60*	*80*	*50*	*90*	*50*	*60*	*60*
Sales ('000 Rs)	*25*	*60*	*45*	*50*	*45*	*20*	*55*	*30*	*45*	*30*

Calculate the rank correlation coefficient between intelligence and efficiency in salesmanship.

Note: *It will be noticed that in Test Scores, 3 salesmen (viz. 4, 9 and 10) with score 60 each tie for the fourth place, and hence are given the average rank (4 + 5 + 5)/3*

= 5; again, 4 salesmen (viz. 1, 3, 6 and 8) tie for the seventh place with score 50 each and are given the average rank (7 + 8 + 9 + 10)/4 = 8.5. Similarly in Sales, 3 salesmen tie for the fourth place and 2 salesmen tie for the seventh place, where average ranks are given].

Solution

Table 9.19 Calculations for Rank Correlation

Salesmen	Intelligence: Test score	Intelligence: Rank (x)	Sales: Amount ('000 Rs)	Sales: Rank (y)	d = x − y	d^2
1	50	8.5	25	9	− 0.5	0.25
2	70	3	60	1	2	4
3	50	8.5	45	5	3.5	12.25
4	60	5	50	3	2	4
5	80	2	45	5	− 3	9
6	50	8.5	20	10	− 1.5	2.25
7	90	1	55	2	− 1	1
8	50	8.5	30	7.5	1	1
9	60	5	45	5	0	0
10	60	5	30	7.5	− 2.5	6.25
Total	–	–	–	–	0	40

In the two series together, there are in all 4 ties, viz. one tie involving 4 individuals, 2 ties involving 3 individuals each, and 1 tie with 2 individuals. Hence,

$$\frac{\Sigma(t^3 - t)}{12} = \frac{4^3 - 4}{12} + \frac{3^3 - 3}{12} + \frac{3^3 - 3}{12} + \frac{23 - 2}{12}$$

$$= 5 + 2 + 2 + 0.5 = 9.5$$

Substituting the values in formula (9.16.2) we get

$$R' = 1 - \frac{6(40 + 9.5)}{10^3 - 10} = 0.70$$

9.17 MULTIPLE CORRELATION AND PARTIAL CORRELATION

In bivariate distributions, we know that when the values of one variable are influenced by those of another, the simple correlation coefficient r provides a measure of the degree of association between these two variables, and estimates of the values of a variable are obtained from the appropriate linear regression equation. But, very often it happens that the values of a variable are influenced not only by another single variable, but by several others. For example, (i) the yield of crop depends on rainfall, temperature, amount of fertiliser applied, etc.; (ii) the weight of a person depends on his height as well as his chest measurement; (iii) the yield of dry fibre from jute plants depends on the height of plant, the diameter of stem, etc. In each of these cases are concerned with the joint distribution of three or more variables, called *Multivariate Distribution*.

The problems in multivariate distribution are of three types:

(a) To estimate the value of one variable from those of several others; this is the problem of *Multiple Regression*.

(b) To measure the extent of *combined influence* of a group of variables upon another varaible; this is the problem of *Multiple Correlation*.

(c) To examine the influence of one variable upon another, *after eliminating the effects* of all other variables; this is the problem of *Partial Correlation*.

Let us consider the cases of three variables, denoted by x_1, x_2, x_3. The simplest kind of relationship which can be used for estimating x_1 in terms of x_2 and x_3 is an equation of the form

$$x_1 = a + bx_2 + cx_3,$$

where a, b, c are constants. Applying the method of least squares, the constants are determined. The equation is usually written as

$$x_1 - \bar{x}_1 = b_{12.3}(x_2 - \bar{x}_2) + b_{13.2}(x_3 - \bar{x}_3) \tag{9.17.1}$$

where $\bar{x}_1, \bar{x}_2, \bar{x}_3$ are the means of variables x_1, x_2, x_3 respectively, and $b_{12.3}$, $b_{13.2}$ are certain constants. This is known as *regression equation of x_1 on x_2 and x_3* (someimes called *multile regression equation*). The constants $b_{12.3}$ and $b_{13.2}$ are called *partial regression coefficients*, and are given by

$$b_{12.3} = \frac{\sigma_1}{\sigma_2}\left(\frac{r_{12} - r_{13}r_{23}}{1 - r_{23}^2}\right); \; b_{13.2} = \frac{\sigma_1}{\sigma_3}\left(\frac{r_{13} - r_{12}r_{23}}{1 - r_{23}^2}\right) \tag{9.17.2}$$

where $\sigma_1, \sigma_2, \sigma_3$ are the standard deviation of the variables x_1, x_2, x_3, and r_{12}, r_{13}, r_{23} denote the product-moment correlation coefficients between x_1 and x_{21}, between x_1 and x_3, and between x_2 and x_3 respectively.

The product-moment correlation coefficient between the observed values of x_1 and the estimated values from the regression equation (9.17.1) is called the *Multiple Correlation Coefficient* of x_1 *on x_2 and x_3*, and is given by

$$R_{1\cdot 23} = \sqrt{\frac{r_{12}^2 + r_{13}^2 - 2r_{12}r_{23}r_{13}}{1 - r_{23}^2}} \tag{9.17.3}$$

It is always considered as positive.

$$0 \le R_{1.23} \le 1. \tag{9.17.4}$$

In order to understand the significance of *Partial Correlation Coefficient*, let us consider the regressions of x_1 on x_3 and of x_2 on x_3. viz. $x_1 = a + b_{13}x_3$ and $x_2 = a' + b_{22}x_3$. We do not expect that $a + b_{13}x_3$ gives a very good estimate of x_1, but it shows only that part of x_1 which can be estimated as a linear function of x_3 alone.

The difference

$$x_1 - (a + b_{13}x_3) = x_{1.3} \text{ (suppose)}$$

is the part of x_1 that remains, after the effect of x_3 has been eliminated. Similarly,

$$x_2 - (a' + b_{23}\,x_3) = x_{2.3} \text{ (suppose)}$$

is the part of x_2 that remains, after the effect of x_3 has been eliminated.

The product-moment correlation coefficient between the residuals $x_{1.3}$ and $x_{2.3}$ is called *Partial Correlation Coefficient between x_1 and x_2*, and is given by

$$r_{12.3} = \frac{r_{12} - r_{13}r_{23}}{\sqrt{(1 - r_{13}^2)(1 - r_{23}^2)}} \tag{9.17.5}$$

The partial correlation coefficient lies between – 1 and + 1.

$$-1 \le r_{12.3} \le 1 \tag{9.17.6}$$

[**Note:**

(1) r_{12} is the correlation coefficient between x_1 and x_2, *ignoring* altogether any influences of x_3. But $r_{12,3}$ is the correlation coefficient between x_1 and x_2, after eliminating the influence of x_3 from both. r_{12} is sometimes called *Total Correlation Coefficient*, in order to distinguish it from *Partial* Correlation Coefficient $r_{12.3}$.

(2) Just as r_{12} is the same as r_{21}, similarly $r_{12.3}$ is the same as $r_{21\cdot3}$. The multiple correlation coefficient $R_{1.23}$ is the same $R_{1.32}$. But just as the *Total Regression Coefficients* b_{12} and b_{21} are different, similarly the partial regression coefficients $b_{12.3}$ and $b_{21.3}$ are also different. $b_{12.3}$ is the coefficient of x_2 in the multiple regression equation for estimating x_1 as a linear function of x_2 and x_3; but $b_{21.3}$ is the coefficient of x_1 in the multiple regression equation for estimating x_2 as a linear function of x_1 and x_3.

(3) The regression equation of x_2 and x_1 and x_3 may be obtained from (9.17.1) by interchanging the subscripts 1 and 2. The multiple correlation coefficient $R_{2.13}$ may also be obtained from (9.17.3) similarly. The partial correlation $r_{23.1} = r_{32.1}$ may be obtained from (9.17.5) by interchanging the subcripts 1 and 3.]

Important Results

(i) $1 - R_{1.23}{}^2 = (1 - r_{12}^2)(1 - r_{13.2}^2)$

$$= (1 - r_{13}^2)(1 - r_{13.1}^2) \tag{9.17.7}$$

(ii) $R_{1.23} \ge | r_{12} |$ and $R_{1.23} \ge | r_{13} |$ (9.17.8)

(iii) If $R_{1.23} = 0$, then $r_{12} = r_{13} = 0$ (9.17.9)

(iv) Just as the product of total regression coefficients b_{12} and b_{21} is the square of total correlation coefficient r_{12}, i.e. $b_{12} \times b_{21} = r_{12}{}^2$, similarly the product of partial regression coefficients $b_{12.3}$ and $b_{21.3}$ is the square of partial correlation coefficient $r_{12.3}$, i.e.

$$b_{12.3} \times b_{21.3} = r_{12.3}^2 \tag{9.17.10}$$

This is, probably, the reason why the regression coefficients $b_{12.3}$ etc. are called 'partial' regression coefficients, and not 'multiple' regression coefficients.

(v) $$b_{12.3} = \frac{b_{12} - b_{13}b_{32}}{1 - b_{23}b_{32}} \tag{9.17.11}$$

Example 9.43 *Given r_{12} = + 0.65, r_{13} = + 0.60 and r_{23} = + 0.90, calculate the values of $r_{12.3}$ and $R_{1.23}$ (the symbols have their usual significance).*

Solution Using formula (9.17.5) and (9.17.3), we have

$$r_{12.3} = \frac{0.65 - 0.6 \times 0.9}{\sqrt{(1 - 0.6^2)(1 - 0.9^2)}} = \frac{0.11}{0.349} = 0.32$$

$$R_{1.32} = \sqrt{\frac{0.65^2 + 0.6^2 - 2 \times 0.65 \times 0.6 \times 0.9}{1 - 0.9^2}} = \sqrt{\frac{0.0805}{0.19}} = 0.651.$$

EXERCISES

1. The data given below relate to the heights and weights of 20 persons. You are rquired to form a two-way frequency table with class-intervals 62″ to 64″, 64″ to 66″ and so on, and 115 to 125 lbs., 125 to 135 lbs., and so on.

S. No.	1	2	3	4	5	6	7	8	9	10	11	12	13
Height	70	65	65	64	69	63	65	70	71	62	70	67	63
Weight	170	135	136	137	148	124	117	128	143	129	163	139	122
							14	15	16	17	18	19	20
							68	67	69	66	68	67	67
							134	140	132	120	148	129	152

[*C.A., May '66*]

2. Calculate r from the following given results: $n = 10$; $\Sigma x = 125$; $\Sigma x^2 = 1585$; $\Sigma y = 80$; $\Sigma y^2 = 650$; $\Sigma xy = 1007$.

3. Find the coefficient of correlation from the following results:

$$\sum_{i=1}^{8} X = 42.2, \sum_{i=1}^{8} Y = 46.4, \sum_{i=1}^{8} X^2 = 291.20,$$

$$\sum_{i=1}^{8} Y^2 = 290.52, \sum_{i=1}^{8} XY = 230.42$$

4. Obtain the correlation coefficient from the following:

x –	6	2	10	4	8
y –	9	11	5	8	7

[D.S.W., '77]

5. Calculate the coefficient of correlation for the ages of husband and wife :

Age of Husband	23	27	28	29	30	31	33	35	36	39
Age of Wife	18	22	23	24	25	26	28	29	30	32

[I.C.W.A., Jan. '70]

6. Calculate the correlation coefficient r_{xy} from the following:

x	65	66	67	67	68	69	70	72
y	67	68	65	68	72	72	69	71

[D.M., '78]

7. Calculate the coefficient of correlation between x and y:

x	155	157	153	151	159	162	158
y	118	129	125	124	129	133	127

[C.U., B.A. (Econ) '75]

8. Calculate Pearson's coefficient of correlation from the following data using 44 and 26 as the origins of X and Y respectively :

X	43	44	46	40	44	42	45	42	38	40	42	57
Y	29	31	19	18	19	27	27	29	41	30	26	10

[*C.A., May '78*]

9. Specimens of similarly treated alloy steel containing various percentages of nickel are tested for toughness with the following results:

Toughness (arbitrary units)	47	50	52	52	54	56	58	59
Percentage of Nickel	2.7	2.7	2.8	2.8	2.9	3.2	3.2	3.3
			60	60	62	64	65	66
			3.4	3.5	3.5	3.6	3.7	3.8

Find the coefficient of correlation between 'toughness' as measured by the test, and 'percentage content of Nickel' in the alloy steel.

[C.U., M. Com. '65; D.S.W., May '71]

10. Calculate the coefficient of correlation from the following data:

Export of Raw Cotton (Rs crores)	42	44	58	55	89	98	66
Import of Manufactured Goods (Rs crores)	56	49	53	58	65	76	58

Calculate also the standard error of the coefficient of correlation.

[I.C.W.A., July '65]

11. Determine the correlation coefficient between x and y:

x	5	7	9	11	13	15
y	1.7	2.4	2.8	3.4	3.7	4.4

[Dip. Management, '67]

12. From the following figures, calculate the coefficient of correlation between the income and the general level of prices:

Income (X)	360	420	500	550	600	640	680	720	750
General Level of Prices (Y)	100	104	115	160	180	290	300	320	330

[C.U., M. Com. '68]

13. The following data give the hardness (x) and tensile strength (y) for some specimens of a material, in certain units. Find the correlation coefficient and calculate its probable error:

x	23.3	17.5	17.8	20.7	18.1	20.9	22.9	20.8
y	4.2	3.8	4.6	3.2	5.2	4.7	4.4	5.6

[I.C.W. A., Jan. '72]

14. The following table gives the saving-bank deposits in billions of dollars and strikes and lock-outs in thousands over a number of years. Compute the correlation coefficient and comment on the results.

Saving Deposits	5.1	5.4	5.5	5.9	6.5	6.0	7.2
Strickes and Lock-outs	3.8	4.4	3.3	3.6	3.3	2.3	1.0

[I.C.W.A., Jan. '64]

15. Two positively correlated variables x_1 and x_2 have variances σ_1^2 and σ_2^2 respectively. Determine the value of the constant a such that $x_1 + ax_2$ and $x_1 + \frac{\sigma_1}{\sigma_2} x_2$ are uncorrelated.

[B.U., B.A.(Econ) '72]

16. State whether the correlation will be positive or negative in the following cases:
    (i) Age and income;
    (ii) Speed of an automobile and the distance required to stop the car after applying brakes;
    (iii) Sale of woollen garments and day-temperature;
    (iv) Sale of cold-drinks and day-temperature;
    (v) Production and price per unit.

17. Given $\Sigma x = 56$, $\Sigma y = 40$, $\Sigma x^2 = 524$
    $\Sigma y^2 = 256$, $\Sigma xy = 364$, $n = 8$,
    find (i) the correlation coefficient and (ii) the regression equation of x on y.
    [I.C.W.A., July '67]

18. The following sums have been obtained from 100 observations-pairs:
    $\Sigma x = 12{,}500$, $\Sigma y = 8{,}000$, $\Sigma x^2 = 1{,}585{,}000$,
    $\Sigma y^2 = 648{,}100$, $\Sigma xy = 1{,}007{,}425$.
    (i) Find the regression of y on x, and estimate the value of y when $x = 130$.
    (ii) Compute the correlation coefficient (r) between x and y and state what you learn from the value of r obtained by you.
    [C.U., B.A. (Econ) '76]

19. Given the following totals for 10 pairs of observtions on two characters x and y, obtain the two regression equations and hence calculate the correlation coefficient:
    $\Sigma x = 12$, $\Sigma y = 4$, $\Sigma x^2 = 16.20$, $\Sigma y^2 = 1.96$, $\Sigma xy = 5.2$
    [M.B.A. '79, D.M. '77]

20. Estimate from the information given below, the probable crop yield, when rainfall is 29 inches:

	Mean	*S.D.*
Rainfall in inches	25	3
Yield in units per acre	40	6

    Coefficient of correlation between the variables: 0.65
    [C.U., B.SC. '73]

21. The correlation coefficient between two variates x and y is $r = 0.60$. If $\sigma_x = 1.50$, $\sigma_y = 2.00, \bar{x} = 10, \bar{y} = 20$, find the equations of the regression lines of (i) y on x, (ii) x on y.
    [I.C.W.A., Dec. '77]

22. The following data pertain to the marks in two subjects, say A and B.
    Mean marks in $A = 39.5$, Mean marks in $B = 47.5$
    S.D. of marks in $A = 10.8$, S.D. of marks in $B = 16.8$
    Coefficient of correlation between marks in A and $B = 0.42$
    Obtain the equations of the two regression lines and then estiamte the marks in B for candidates who secured 50 marks in A.
    [I.C.W.A., June '78]

23. Given the following results of the height and weight of 1000 men students:
    $\bar{x} = 68$ inches, $\bar{y} = 150$ lbs., $r = 0.60$, $\sigma_x = 2.50$ inches, $\sigma_y = 20.000$ lbs. John

Doe weights 200 lbs., Richard Roe is five feet tall. Estimate the height of Doe from his weight, and weight of Roe from his height.

[C.U., M. Com. '72, '76]

24. From the following data find the coefficient of linear correlation between X and Y. Determine also the regression line of Y on X, and then make an estimate of the value of Y when $X = 12$.

X	1	3	4	6	8	9	11	14
Y	1	2	4	4	5	7	8	9

[I.C.W.A., June '75]

25. Obtain the lines of regression for the following data:

(X)	1	2	3	4	5	6	7	8	9
(Y)	9	8	10	12	11	13	14	16	15

[C.U., M.Com. '68; D.S.W. '78; I.C.W.A., Dec. '78]

26. Find the two lines of regression from the following data:

Age of Husband (x)	25	22	28	26	35	20	22	40	20	18
Age of Wife (y)	18	15	20	17	22	14	16	21	15	14

Hence, estimate (i) the age of hushand when the age of wife is 19, (ii) the age of wife when the age of hushand is 30.

[C.U., M.Com. '70]

27. From the following data, obtain the two regression equations:

Sales	91	97	108	121	67	124	51	73	111	57
Purchases	71	75	69	97	70	91	39	61	80	47

[C.A., May '77]

28. Obtain the equation of the line of regression of yield of rice (y) on water (x) from the data given in the following table:

Water in Inches (x)	12	18	24	30	36	42	48
Yield in Tons (y)	5.27	5.68	6.25	7.21	8.02	8.71	8.42

Estimate the most probable yield of rice for 40 inches of water.

[C.U., M.Com. '64; I.C.W. A., Dec. '76]

29. If the regression equation of y on x be $Y = 0.57x + 6.93$ and the regression equation of x on y be $X = 1.12y - 2.46$, find the correlation coefficient between x and y.

[B.U., B.A. (Econ) '72]

30. For some bivariate data the following results were obtained. The mean value of $X = 53.2$, the mean value of $Y = 27.9$, the regression coefficient of Y on $X = -1.5$, and the regression coefficient of X on $Y = -0.2$. Find the (i) most probable value of Y when $X = 60$, (ii) r the coefficient of correlation between X and Y.

[C.U., M.Com '74]

31. The regression equation calculated from a given set of observations are $x = -0.2y + 4.2$, $y = -0.8x + 8.4$. Calculate (i) x and y, (ii) r, (iii) the estimated value of y when $x = 4$. *[I.C.W.A., Jan. '68; C.U., B Com. (Hons) '69]*

32. The two regression lines involving the two variables x and y are $Y = 5.6 + 1.2x$ and $X = 12.5 + 0.6y$. Find the means of x and y and their correlation coefficient.

[W.B.H.S., '78]

33. Two vairates have the least squares regression lines $x + 4y + 3 = 0$ and $4x + 9y + 5 = 0$. Find their mean values and the correlation coefficient.

[I.C.W.A., July '70; M.B.A. '78]

34. Two lines of regression are given by $x = 2y = 5$ and $2x + 3y = 8$, and $\sigma_x^2 = 12$. Calculate the values of $\bar{x}$, $\bar{y}$, σ_y and r.

[I.C.W.A., Dec '76-old]

35. In order to find the correlation coefficient between two variates x and y from 12 pairs of observations, the following calculations were made:
$\Sigma x = 30$, $\Sigma y = 5$, $\Sigma x^2 = 670$, $\Sigma y^2 = 285$, $\Sigma xy = 334$.
On subsequent varification it was found that the pair ($x = 11$, $y = 4$) was copied wrongly, the correct values being ($x = 10$, $y = 14$). Find the correct value of correlation coefficient.

[I.C.W.A., June '75-old]

36. Obtain the linear regression equation that you consider more relevant for the following set of paired observations and given reasons why you consider it to be so:

Age	56	42	72	36	63	47	55	49	38	42	68	60
Blood Pressure	147	125	160	118	149	128	150	145	115	140	152	155

Also estimate the blood pressure of a person whose age is 45.

[C.U., M.Com., '73]

37. For the variables x and y the equations of two regression linear are $4x - 5y + 33 = 0$ and $20x - 9y = 107$. Identify the regression line of y on x and that of x on y. What is the estimated value of y, when $x = 10$? If this estimate is denoted by y_0, find the estimated value of x when $y = y_0$

[C.U., B.A. (Econ) '75]

38. State the meaing of the terms *explained variation* and *unexplained variation*, used in the theory of regression. If the coefficient of correlation between two variables X and Y be 0.83, what percentage of total variation remains unexplained by the regresion equation?

[I.C.W.A., Dec. '75]

39. Given: Unexplained variation = 19.70, and Explained variation = 19.22, determine the coefficient of correlation.

[I.C.W.A., Dec. '76]

40. In a contest, two judeges ranked eight candidates *A, B, C, D, E, F, G* and *H* in order of their preference, as shown in the following table. Find the rank correlation coefficient.

	A	*B*	*C*	*D*	*E*	*F*	*G*	*H*
First Judge	5	2	8	1	4	6	3	7
Second Judge	4	5	7	3	2	8	1	6

[I.C.W.A., June '75]

41. Compute the correlation coefficient of the following ranks of a group of students in two examinations. What conclusion do you draw from the result?

Roll Nos.	1	2	3	4	5	6	7	8	9	10
Ranks in B.Com. Exam.	1	5	8	6	7	4	2	3	9	10
Ranks in M.Com. Exam.	2	1	5	7	6	3	4	8	10	9

[C.U., M.Com. '75]

42. Ten competitors in a musical contest were ranked by 3 judges *A, B, C* in the following order :

Ranks by A	1	6	5	10	3	2	4	9	7	8
Ranks by B	3	5	8	4	7	10	2	1	6	9
Ranks by C	6	4	9	8	1	2	3	10	5	7

Using Rank Correlation method, discuss which pair of judges has the nearest approach to common likings in music.

[I.C.W.A., Dec. '78]

43. Ten students obtained the following marks in Mathematics and Statistics. Calculate the rank correlation coefficient.

Student (Roll No.)	1	2	3	4	5	6	7	8	9	10
Marks in Mathematics	78	36	98	25	75	82	90	62	65	39
Marks in Statistics	84	51	91	60	68	62	86	58	53	47

44. Given the following coefficients : $r_{12} = 0.41$, $r_{13} = 0.71$, $r_{23} = 0.5$ Find $r_{12.3}$, $r_{13.2}$ and $R_{1.23}$, where the symbols have their usual significance.

[C.U., M.Com. '74]

45. In a three-variate multiple correlation analysis, the following results were found:

$$\bar{x}_1 = 60 \qquad \bar{x}_2 = 70 \qquad \bar{x}_3 = 100$$
$$s_1 = 3 \qquad s_2 = 4 \qquad s_3 = 5$$
$$r_{12} = 0.7 \qquad r_{13} = 0.6 \qquad r_{23} = 0.4$$

the symbols having their usual singificance. Find the regression of x_1 on x_2 and x_3, and the multiple correlation coefficient $R_{1.23}$.

[B.U., M.A.(Econ) '68]

ANSWERS

1. (See Example 9.2) 2. + 0.47 3. – 0.37
4. – 0.92 5. + 0.996 6. 0.60
7. + 0.71 8. – 0.73 9. + 0.98
10. + 0.90, 0.072 11. + 0.995 12. + 0.94
13. – 0.072, 0.237 14. – 0.82 15. $-\sigma_1/\sigma_2$
16. (i) positive, (ii) positive, (iii) negative, (iv) positive, (v) negative
17. + 0.98; $x = 1.5y - 0.5$ 18. $y = 0.33x + 38.75$; 81.65; 0.55
19. $y = 0.222x + 0.133$; $x = 1.11y + 0.756$; 0.50

20. 45.2 units per acre
21. $y = 0.8x + 12$; $x = 0.45y + 1$
22. $y = 0.65x + 21.82$; $x = 0.27y + 26.68$; 54
23. 71.75 inches; 111.60 lbs.
24. + 0.98; $y = 0.64x + 0.52$; 8.2
25. $X = 0.95Y - 6.4$; $Y = 0.95X + 7.25$
26. $x = 2.23y - 12.76$; $y = 0.39x + 7.22$; (i) 30, (ii) 19.
27. $y = 0.613x + 14.83$; $x = 1.360y - 5.2$
28. $y = 3.99 + 0.103x$; 8.11 tons.
29. $r = \sqrt{0.57 \times 1.12} = +0.80$
30. (i) 17.7, (ii) –0.55
31. (i) 3 and 6, (ii) –0.4, (iii) 5.2
32. 56.64, 73.57; + 0.85
33. $\overline{x} = 1, \overline{y} = -1, r = -0.75$
34. $\overline{x} = 1, \overline{y} = 2, \sigma_y = 2, r = -\sqrt{3/2}$
35. + 0.77
36. Regression equation of blood pressure (y) on age (x) is $y = 1.14x + 80.67$; 132
37. First equation regression line of y on x; $y_0 = 14.6$; 11.92
38. 31%
39. $|r| = 0.70$
40. 2/3
41. +0.64
42. Rank corr. coeffs. between *A* & *B*, *A* & *C* and *B* & *C* respectively are – 0.21, + 0.64, – 0.30. Since + 0.64 is the largest, Judges *A* and *C* hae the nearest approach to common likings in music.
43. +0.82
44. +0.09, +0.64, +0.71
45. $x_1 = 8.3 + 0.41x_2 + 0.23x_3$; $R_{1.23} = 0.78$

10 INTERPOLATION

10.1 INTRODUCTION

'*Interpolation*' has been defined as the 'art of reading between the lines of a table', and the term usually denotes the process of finding the intermediate value of a function from a set of given values of that function. For example, using the following given values of x and y

x:	1	3	5	7	9	11
y:	1	15	65	175	369	671

we may be required to find, i.e. *interpolate,* the value of y when $x = 4$. Since the mathematical relationship between x and y is not known, we have necessarily to base our calculations on the given data.

In *Simple Interpolation* (page 63) we use only two pairs of observations on either side of the unknown value, and assume a constant rate of change (i.e. linear relationship) in this interval.

x	y
3	15
4 →	← ?
5	65

Using (3.16.1),

$$\frac{4-3}{5-3} = \frac{y-15}{65-15}.$$ Solving, we get $y = 40$.

However, the mathematical relationship between the values of x and y given above is $y = \frac{1}{2}(x^3 + x)$, so that when $x = 4$, the value of y is 34. The discrepancy of the interpolated value $y = 40$ from the true value $y = 34$ is due to the fact that we have not utilised the whole of the given data, but only two pairs of observations. Applying formulae for interpolation, stated in the following pages, it is possible (see Example 10.12) in some cases, to obtain the value of y exactly, and if not to obtain a close approximation to the true value.

10.2 FINITE DIFFERENCES: Δ AND *E* OPERATORS

In problems of interpolation, the independent variable x is often known as '*argument*', and the dependent variable or the function $y = f(x)$ is known as '*entry*'. Let x_0, x_1, x_2, ..., x_n denote a set of equidistant values of the argument, i.e.

$$x_1 - x_0 = x_2 - x_1 = ... = x_n - x_{n-1} = h$$

where h is a constant; and $y_0, y_1, y_2, ..., y_n$ denote the corresponding values of the entry. Differences of the successive values of y, viz.

$$(y_1 - y_0), (y_2 - y_1), (y_3 - y_2), ..., (y_n - y_{n-1})$$

are called *finite differences of the first order* (or simply *first differences*), and are denoted by

$$\Delta y_0, \Delta y_1, \Delta y_2, ..., \Delta y_{n-1}$$

respectively. The differences of the successive first order differences Δy, namely

$$(\Delta y_1 - \Delta y_0), (\Delta y_2 - \Delta y_1), ..., (\Delta y_{n-1} - \Delta y_{n-2})$$

are known as *finite differences of the second order* (or simply *second differences*) and are denoted by

$$\Delta^2 y_0, \Delta^2 y_1, ..., \Delta^2 y_{n-2}$$

respectively. Similarly, the *third differences* $\Delta^3 y$, the *fourth differences* $\Delta^4 y$, and differences of higher order may be defined.

Argument *(x)*	*Entry* *(y)*	*First* *Differences* (Δy)	*Second* *Differences* $(\Delta^2 y)$
x_0	y_0		
		$y_1 - y_0 = \Delta y_0$	
x_1	y_1		$\Delta y_1 - \Delta y_0 = \Delta^2 y_0$
		$y_2 - y_1 = \Delta y_1$	
x_2	y_2		$\Delta y_2 - \Delta y_1 = \Delta^2 y_1$
		$y_3 - y_2 = \Delta y_2$	
x_3	y_3		$\Delta y_3 - \Delta y_2 = \Delta^2 y_2$
		$y_4 - y_3 = \Delta y_3$	
x_4	y_4		

A table which shows the finite differences is known as *Difference Table*.

Table 10.1 Difference Table

Argument x	*Entry* y	*First* *Differences* Δy	*second* *Differences* $\Delta^2 y$	*Third* *Differences* $\Delta^2 y$	*Fourth* *Differences* $\Delta^4 y$
$x_0 = 1$	$y_0 = 1$				
		14			
$x_1 = 3$	$y_1 = 15$		36		
		50		24	
$x_2 = 5$	$y_2 = 65$		60		0
		110		24	
$x_3 = 7$	$y_3 = 175$		84		0
		194		24	
$x_4 = 9$	$y_4 = 369$		108		
		302			
$x_5 = 11$	$y_5 = 671$				

In symbols, the difference table may be shown as follows:

Table 10.2 Difference Table

x	y	Δy	$\Delta^2 y$	$\Delta^3 y$	$\Delta^4 y$	$\Delta^5 y$
x_0	y_0					
		Δy_0				
x_1	y_1		$\Delta^2 y_0$			
		Δy_1		$\Delta^3 y_0$		
x_2	y_2		$\Delta^2 y_1$		$\Delta^4 y_0$	
		Δy_2		$\Delta^3 y_1$		$\Delta^5 y_0$
x_3	y_3		$\Delta^2 y_2$		$\Delta^4 y_1$	
		Δy_3		$\Delta^3 y_3$		
x_4	y_4		$\Delta^2 y_3$			
		Δy_4				
x_5	y_5					

The arrangement of differences should be noted. Each difference is written between the two quantities which are used to calculate it.

The initial term y_0 of the entry is called the *leading term* and the initial terms in the difference columns, viz. Δy_0, $\Delta^2 y_0$, $\Delta^3 y_0$ etc. are called *leading differences*.

[**Note:** Δ (*called delta*) is not a quantity multiplied to $y^2 s$, but an 'operator' like the root-sign $\sqrt{}$ or log. These operators separately have no meaning, but when operating on a number, e.g. $\sqrt{9}$ or log 7, have a real significance. Again, Δ^2 does not imply the square of Δ, but the repetition of operation by Δ twice].

Another symbolic operator E is defined as follows:

$$Ey_0 = y_1, \quad Ey_1 = y_2, \quad Ey_2 = y_3, \quad \dots\dots$$

Both the operators Δ and E can be applied repeatedly, the repeated operations being indicated by Δ^2, Δ^3, ... and E^2, E^3, etc. Thus

$$\begin{aligned}\Delta^2 y_0 &= \Delta(\Delta y_0) = \Delta y_1 - \Delta y_0\\ &= (y_2 - y_1) - (y_1 - y_0) = y_2 - 2y_1 + y_0\\ \Delta^3 y_0 &= \Delta(\Delta^2 y_0) = \Delta^2 y_1 - \Delta^2 y_0\\ &= (y_3 - 2y_2 + y_1) - (y_2 - 2y_1 + y_0)\\ &= y_3 - 3y_2 + 3y_1 - y_0\end{aligned}$$

Similarly,
$$\begin{aligned}E^2 y_0 &= E(Ey_0) = E(y_1) = y_2\\ E^3 y_0 &= E(E^2 y_0) = E(y_2) = y_3; \quad \text{etc.}\end{aligned}$$

From the definitions, we may in general write

$$\Delta y_r = y_{r+1} - y_r \tag{10.2.1}$$

$$Ey_r = y_{r+1} \tag{10.2.2}$$

These operators may thus be interpolated in the following manner:

(a) Δ when prefixed to y_r implies that y_r is to be subtracted from the next value of the entry y_{r+1}.

(b) E when prefixed to y_r denotes the next value of the entry y_{r+1}.

From (10.2.1) and (10.2.2) we find that

$$Ey_r = y_r + \Delta y_r$$

or, $$Ey_r = (1 + \Delta)\, y_r \quad \text{(suppose)} \tag{10.2.3}$$

Omitting y_r from both sides, we find that the operators E and Δ are connected by the symbolic relation

$$E = 1 + \Delta \tag{10.2.4}$$

This does not mean that 1 when added to Δ gives E, but that the operation by E is equivalent to the operation by $(1 + \Delta)$. It may be shown that the above relation follows certain algebraic rules.

As shown earlier, we have

$$\begin{aligned} \Delta\ y_0 &= y_1 - y_0 \\ \Delta^2\ y_0 &= y_2 - 2y_1 + y_0 \\ \Delta^3\ y_0 &= y_3 - 3y_2 + 3y_1 - y_0 \\ \Delta^4\ y_0 &= y_4 - 4y_3 + 6y_2 - 4y_1 + y_0 \end{aligned} \tag{10.2.5}$$

Alternatively, we may write

$$\begin{aligned} \Delta\ y_0 &= Ey_0 - y_0 \\ \Delta^2\ y_0 &= E^2y_0 - 2Ey_0 + y_0 \\ \Delta^3\ y_0 &= E^3y_0 - 3E^2y_0 + 3Ey_0 - y_0 \\ \Delta^4\ y_0 &= E^4y_0 - 4E^3y_0 + 6E^2y_0 - 4Ey_0 + y_0 \end{aligned}$$

With the operators only (removing y_0's from both sides)

$$\begin{aligned} \Delta &= E - 1 \\ \Delta^2 &= E^2 - 2E + 1 = (E + 1)^2, \\ \Delta^3 &= E^3 - 3E^2 + 3E - 1 = (E - 1)^3 \\ \Delta^4 &= E^4 - 4E^3 + 6E^2 - 4E + 1 = (E - 1)^4 \end{aligned}$$

Thus, we have developed a convenient method of expressing the finite difference of any order in terms of the entries.

(see Example 10.1)

Example 10.1 *Express $\Delta^5 y_0$ in terms of $y_0, y_1, y_2, \ldots$*

Solution Using (10.2.4), we replace Δ by $E - 1$ and expand $(E - 1)^5$ by the binomial series (see pages 68-69).

$$\begin{aligned} \Delta^5 y_0 &= (E - 1)^5\ y_0 \\ &= (E^5 - 5E^4 + 10E^3 - 10E^2 + 5E - 1)\ y_0 \qquad \text{[Example 3.27 (iii)]} \\ &= E^5y_0 - 5E^4y_0 + 10E^3y_0 - 10E^2y_0 + 5Ey_0 - y_0 \\ &= y_5 - 5y_4 + 10y_3 - 10y_2 + 5y_1 - y_0. \end{aligned}$$

Example 10.2 *If $y = c_0 + c_1x + \frac{c_2}{2!}\, x(x - 1)$ passes through the points $(0, y_0)$, $(1, y_1)$, $(2, y_2)$ calculate c_0, c_1, c_2 in terms of differences of y^2s.*

[I.C.W.A., July '68]

Solution The statement that the curve passes through the points $(0, y_0)$, $(1, y_1)$ and $(2, y_2)$ implies that when $x = 0, 1, 2$, the values of y are respectively y_0, y_1, y_2.

Putting $x = 0$,

$$y_0 = c_0 + c_1 \times 0 + \frac{c_2}{2!} \times 0(0 - 1)$$

i.e. $$y_0 = c_0 \qquad \ldots\text{(i)}$$

Putting $x = 1$,

$$y_1 = c_0 + c_1 \times 1 + \frac{c_2}{2!} \times 1(1-1)$$

or, $y_1 = y_0 + c_1 + 0$ (since $c_0 = y_0$)

i.e. $c_1 = y_1 - y_0 = \Delta y_0$ ...(ii)

Putting $x = 2$,

$$y_2 = c_0 + c_1 \times 2 + \frac{c_2}{2!} \times 2(2-1)$$

or, $y_2 = y_0 + 2(y_1 - y_0) + c_2$, from (i) and (ii)

i.e.

$$\begin{aligned} c_2 &= y_2 - y_0 - 2(y_1 - y_0) \\ &= y_2 - 2y_1 + y_0 \qquad \text{...(iii)} \\ &= E^2 y_0 - 2Ey_0 + y_0 \\ &= (E^2 - 2E + 1)\, y_0 = (E-1)^2\, y_0 \\ &= \Delta^2 y_0 \end{aligned}$$

Thus, $c_0 = y_0$, $c_1 = \Delta y_0$, $c_2 = \Delta^2 y_0$.

10.3 DIFFERENCES OF A POLYNOMIAL FUNCTION

Theorem: If y is a polynomial of the n-th degree [Example 3.29 (f), page 58], then the n-th differences are constants, and differences of higher order are all zero.

The third differences in Table 10.1 are found to be all equal and fourth differences are zero; because $y = \frac{1}{2}(x^3 + x)$ is a polynomial of the third degree.

Example 10.3 *Given $u_1 = (12 - x)(4 + x)$, $u_2 = (5 - x)(4 - x)$, $u_3 = (x + 18)(x + 18)(x + 6)$, $u_4 = 94$, obtain a value of x such that second differences of u are constant.*

[B.U., B.A. (Econ) '72]

Solution If the 2nd differences are constant, the 3rd difference (and all higher order differences) will be zero. In particular,

$$\Delta^3 u_1 = 0; \quad \text{i.e.} \quad (E-1)^3 u_1 = 0$$

Expanding as in the previous example, and substituting the values

$$u_4 - 3 \,.\, u_3 + 3 \,.\, u_2 - u_2 = 0$$

or, $94 - 3(x + 18)(x + 6) + 3(5 - x)(4 - x) - (12 - x)(4 + x) = 0$.

Simplifying, $x^2 - 107x - 218 = 0$; or, $(x + 2)(x - 109) = 0$.

$\therefore$ $x = -2$, or 109.

Example 10.4 *Estimate U_2 from the following table:*

x:	1	2	3	4	5
U_x:	7	*	13	21	37

State the necessary assumptions made. [I.C.W.A., June '77]

Solution Since only 4 values of the function are given, we assume that U_x is a 3rd degree polynomial in x. So the 4th differences may be regarded as zero; i.e. $\Delta^4 U_x = 0$. In particular, $\Delta^4 U_1 = 0$; or, $(E-1)^4 U_1 = 0$;

or, $(E^4 - 4E^3 + 6E^2 - 4E + 1)\, U_1 = 0$

or, $U_5 - 4U_4 + 6U_3 - 4U_2 + U_1 = 0$
or, $37 - 4(21) + 6(13) - 4U_2 + 7 = 0$
or, $38 - 4U_2 = 0. \quad \therefore \quad U_2 = 9.5$

[**Note:** It is always possible to pass a straight line, i.e. first degree polynomial, through any 2 given points; a parabola, i.e. second degree polynomial through any 3 given points; a third degree polynomial through any 4 given points; etc.]

Example 10.5 *The monthly average number of deaths of infants (under age 1 year) in Calcutta are as shown below. Find estimates for the missing terms.*

Year ...	*1950*	*1951*	*1952*	*1953*	*1954*	*1955*
Number of deaths (Monthly average)	*939*	*?*	*907*	*841*	*?*	*773*

Solution Since only 4 values are given, we assume a third degree polynomial, so that the fourth differences are zero. Denoting the entries corresponding to the years 1950, 1951, 1952 etc. by $y_0, y_1, y_2, \ldots$, we have then

$$\Delta^4 y_0 = 0, \qquad \Delta^4 y_1 = 0$$

or, $$(E-1)^4 y_0 = 0, \qquad (E-1)^4 y_1 = 0$$

Expanding, we get $$y_4 - 4y_3 + 6y_2 - 4y_1 + y_0 = 0$$
$$y_5 - 4y_4 + 6y_3 - 4y_2 + y_1 = 0$$

Putting $y_0 = 939, y_2 = 907, y_3 = 841, y_5 = 773$, we have

$$y_4 - 4(841) + 6(907) - 4y_1 + 939 = 0$$
$$773 - 4y_4 + 6(841) - 4(907) + y_1 = 0$$

or, $y_4 - 4y_1 = -3017$ and $y_1 - 4y_4 = -2191$

Solving these two equations, we get $y_1 = 951, y_4 = 785$

Thus, the missing terms are 951 for 1951 and 785 for the year 1954.

Example 10.6 *The values of a function f(x) are given for certain values of x:*

x	*4*	*5*	*6*	*7*	*8*
f(x)	*3.11*	*2.96*	*2.85*	*—*	*2.70*

Obtain the best approximation of f(7). [I.C.W.A, June '74-old]

Solution Since only 4 values are known, we assume a third degree polynomial for $f(x)$, so that 4th and higher order differences are zero. In particular $\Delta^4 f(4) = 0$;

or, $$(E-1)^4 f(4) = 0$$

i.e. $$f(8) - 4.f(7) + 6.f(6) - 4.f(5) + f(4) = 0$$

or, $$2.70 - 4.f(7) + 6 \times 2.85 - 4 \times 2.96 + 3.11 = 0$$
$$f(7) = 2.7675.$$

[**Note:** This method can be applied only when, the values of the argument, including those for which the function is to be determined (e.g. 4, 5, 6, 7, 8) are found to have a common difference. The method will not work if, for example, it is required to find $f(5.5)$ or $f(7.3)$ from the above data. Lagrange's formula is however applicable in all cases; see Example 10.15]

Example 10.7 *Show that third differences of a cubic function are constant.*

[I.C.W.A., June '74; C.U., B.A. (Econ) '76]

Solution Let us write the cubic function as

$$y = f(x) = a_0 + a_1x + a_2x^2 + a_3x^3$$

where a_0, a_1, a_2, a_3 are arbitrary constants. When the successive values of the argument x increase by a constant h, then the next higher value of the function y will be obtained on replacing x by $x + h$, i.e.

$$f(x + h) = a_0 + a_1(x + h) + a_2(x + h)^2 + a_3(x + h)^3$$

Subtracting $f(x)$ from the next value of the entry, viz. $f(x + h)$,

$$\begin{aligned}\Delta y &= \Delta f(x) = f(x + h) - f(x),\\ &= a_1\{(x + h) - x\} + a_2\{(x + h)^2 - x^2\} + a_3\{(x + h)^3 - x^3\}\\ &= a_2(x + h - x) + a_2(x^2 + 2xh + h^2 - x^2) + a_3(x^3 + 3x^2h + 3xh^2 + h^3 - x^3)\\ &= (a_1h + a_2h^2 + a_3h^3) + (2a_2h + 3a_3h^2)\, x + (3a_3h)\, x^2\\ &= b_0 + b_1x + b_2x^2 \quad \text{(suppose)}\end{aligned}$$

where b_0, b_1, b_2 are also constants. Applying Δ operator again,

$$\begin{aligned}\Delta^2 y &= \Delta^2 f(x) = \Delta[\Delta f(x)]\\ &= \Delta f(x + h) - \Delta f(x)\\ &= b_0 + b_1(x + h) + b_2(x + h)^2 - b_0 - b_1x - b_2x^2\\ &= b_1(x + h - x) + b_2(x^2 + 2xh + h^2 - x^2)\\ &= (b_1h + b_2h^2) + (2b_2h)\, x\\ &= c_0 + c_1x \qquad \text{(suppose)}\end{aligned}$$

Repeating the operation by Δ,

$$\begin{aligned}\Delta^3 y &= \Delta^3 f(x) = \Delta[\Delta^2 f(x)] = \Delta^2 f(x + h) - \Delta^2 f(x)\\ &= c_0 + c_1(x + h) - c_0 - c_1x = c_1h\end{aligned}$$

which is independent of x and hence a constant. This shows that the third differences of a cubic function are constants.

Example 10.8 *Find the polynomial function f(x) for which it is known that f(2) = f(3) = 27, f(4) = 77 and f(5) = 165.*

Solution (*First method*) Since only 4 values of the function are known, we may assume that the polynomial is of 3rd degree.

$$f(x) = a + bx + cx^2 + dx^3$$

where a, b, c, d are certain constants to be determined. Putting $x = 2, 3, 4, 5$ successively,

$$\begin{aligned}f(2) &= a + b \times 2 + c \times 2^2 + d \times 2^3 \quad \text{i.e.} \quad 27 = a + 2b + 4c + 8d\\ f(3) &= a + b \times 3 + c \times 3^2 + d \times 3^3 \qquad 27 = a + 3b + 9c + 27d\\ f(4) &= a + b \times 4 + c \times 4^2 + d \times 4^3 \qquad 77 = a + 4b + 16c + 64d\\ f(5) &= a + b \times 5 + c \times 5^2 + d \times 5^3 \qquad 165 = a + 5b + 25c + 125d\end{aligned}$$

Solving these 4 equations, we get the four constants, viz. $a = 225$, $b = -177$, $c = 43$, $d = -2$. Hence the required polynomial is

$$f(x) = 225 - 177x + 43x^2 - 2x^3$$

(*Second method*) The difficulty in solving 5 equations in the previous method can be avoided if we write the 3rd degree polynomial in the form

$$f(x) = A + B(x - 2) + C(x - 2)(x - 3) + D(x - 2)(x - 3)(x - 4)$$

where A, B, C, D are constants to be determined.

Putting $x = 2$, $f(2) = A + 0 + 0 + 0$

or, $27 = A$, i.e. $A = 27$ ...(i)

Putting $x = 3$, $f(3) = A + B + 0 + 0$

or, $27 = 27 + B$ (since $A = 27$)

i.e. $B = 0$ ...(ii)

Putting $x = 4$, $f(4) = A + 2B + 2C + 0$

or, $77 = 27 + 0 + 2C$ (since $A = 27$, $B = 0$)

i.e. $C = 25$ ...(*iii*)

Putting $x = 5$, $f(5) = A + 3B + 6C + 6D$

or, $165 = 27 + 0 + 150 + 6D$

i.e. $D = -2$ ...(*iv*)

Substituting the values of A, B, C, D, we get

$$f(x) = 27 + 25(x-2)(x-3) - 2(x-2)(x-3)(x-4)$$

or, $$f(x) = 225 - 117x + 43x^2 - 2x^3$$

(*Third method*) Since the function is given for equidistant values of x(viz. 2, 3, 4, 5), we can construct a difference table:

x	$f(x)$	$\Delta f(x)$	$\Delta^2 f(x)$	$\Delta^3 f(x)$
2	27			
		0		
3	27		50	
		50		-12
4	77		38	
		88		
5	165			

$$f(x) = E^{x-2} f(2) = (1+\Delta)^{x-2} f(2)$$

$$= [1 + (x-2)\Delta + \frac{(x-2)(x-3)}{1 \times 2}\Delta^2 + \frac{(x-2)(x-3)(x-4)}{1 \times 2 \times 3}\Delta^3] f(2)$$

$$= f(2) + (x-2) . \Delta f(2) + \frac{(x-2)(x-3)}{2}\Delta^2 f(2) + \frac{(x-2)(x-3)(x-4)}{6}\Delta^3 f(2).$$

$$= 27 + 0 + 25(x-2)(x-3) - 2(x-2)(x-3)(x-4)$$

This is of the same form as obtained in the 2nd method [Note that it is unnecessary to continue the expansion beyond 3rd differences, since higher order differences are zero.]

Example 10.9 *Given for a function f(x),*

$f(0) = 1$; $f(1) + f(2) = 10$; $f(3) + f(4) + f(5) = 65$; find $f(4)$. [I.C.W.A., Dec. '73]

Solution Since only 3 values are known, we may assume a 2nd degree polynomial for the function, and write $f(x) = a + bx + cx^2$.

Putting $x = 0, f(0) = a + b \times 0 + c \times 0^2 = a$

Putting $x = 1$ and 2, and adding

$$f(1) + f(2) = (a + b + c) + (a + 2b + 4c) = 2a + 3b + 5c$$

Putting $x = 3, 4, 5$, and adding

$$f(3) + f(4) + f(5) = (a + 3b + 9c) + (a + 4b + 16c) + (a + 5b + 25c)$$
$$= 3a + 12b + 50c$$

Thus, from the given results, we have

$$a = 1;\ 2a + 3b + 5c = 10;\ 3a + 12b + 50c = 65.$$

Solving these equations we get $a = 1$, $b = 1$, $c = 1$. Hence $f(x) = 1 + x + x^2$, so that $f(4) = 1 + 4 + 4^2 = 21$.

Example 10.10 *If u is any function of x, and given that*

$u_0 + u_8 = 56$ $\quad$ $u_2 + u_6 = 32$

$u_1 + u_7 = 42$ $\quad$ $u_3 + u_5 = 26$

find the value of u_4.

Solution Since only 4 values are available, we assume a 3rd degree polynomial for u, so that the 4th and higher order differences are zero. In particular, $\Delta^8 u_0 = 0$; or, $(E-1)^8 u_0 = 0$. Expanding,

$$u_8 - 8u_7 + 28u_6 - 56u_5 + 70u_4 - 56u_3 + 28u_2 - 8u_1 + u_0 = 0$$

or, $$(u_8 + u_0) - 8(u_7 + u_1) + 28(u_6 + u_2) - 56(u_5 + u_3) + 70u_4 = 0$$

Substituting the values and solving, we get $u_4 = 12$.

10.4 NEWTON'S FORWARD INTERPOLATION FORMULA

Let $y_0, y_1, y_2, \ldots y_n$ be some tabulated values of a function $y = f(x)$ corresponding to the equidistant values $x = x_0, x_1, x_2, \ldots x_n$.

$$x_1 - x_0 = x_2 - x_1 = x_3 - x_2 = \ldots = x_n - x_{n-1} = h \text{ (say)}.$$

It is required to find the value of y corresponding to an intermediate value of x lying near the beginning of the tabulated values. This is given by *Newton's Forward Interpolation Formula*:

$$y = y_0 + u\Delta y_0 + \frac{u(u-1)}{1\times 2}\Delta^2 y_0 + \frac{u(u-1)(u-2)}{1\times 2\times 3}\Delta^3 y_0 + \ldots$$

$$\ldots + \frac{u(u-1)(u-2)\ldots(u-n+1)}{1\times 2\times 3\times \ldots \times n}\Delta^n y_0. \quad (10.4.1)$$

where $$u = \frac{x - x_0}{h}$$

Derivation of Newton's Forward Formula

All interpolation formulae are based on the following principle. Within the range of $n+1$ given values, the tabulated function $y = f(x)$ can be replaced by an n-th degree polynomial $\phi(x)$, which coincides with $f(x)$ when $x = x_0, x_1, x_2, \ldots, x_n$; i.e.

$$\phi(x_0) = y_0,\ \phi(x_1) = y_1,\ \phi(x_2) = y_2,\ \ldots,\ \phi(x_n) = y_n \quad \ldots(1)$$

Let us write the n-th degree polynomial $\phi(x)$ in the form

$$\phi(x) = a_0 + a_1(x - x_0) + a_2(x - x_0)(x - x_1) + a_3(x - x_0)(x - x_1)(x - x_2)$$
$$+ \ldots + a_n(x - x_0)(x - x_1)\ldots(x - x_{n-1}). \quad \ldots(2)$$

The $n+1$ constants $a_0, a_1, a_2, \ldots, a_n$ are so determined as to satisfy the $n+1$ relations (1).

Putting $x = x_0$ in (2), we have

$$\phi(x_0) = a_0$$

because all other terms will contain $(x_0 - x_0)$ and therefore vanish.

or, $y_0 = a_0$; i.e. $a_0 = y_0$

Now putting $x = x_1$ in (2)

$$\phi(x_1) = a_0 + a_1(x_1 - x_0) \text{ because the other terms vanish.}$$

or, $$y_1 = y_0 + a_1 h, \quad \text{since } x_1 - x_0 = h \text{ (given)}$$

or, $$a_1 h = y_1 - y_0$$

$$\therefore \quad a_1 = \frac{y_1 - y_0}{h} = \frac{\Delta y_0}{h}$$

Again, putting $x = x_2$ in (2),

$$\phi(x_2) = a_0 + a_1(x_3 - x_0) + a_2(x_2 - x_0)(x_3 - x_1)$$

or, $$y_2 = y_0 + \frac{y_1 - y_0}{h}(2h) + a_2(2h^2)(h)$$

or, $$y_2 = y_0 + 2(y_1 - y_0) + a_2(2h^2)$$

Solving, we get

$$a_2 = \frac{y_2 - 2y_1 + y_0}{2h^2} = \frac{\Delta^2 y_0}{2h^2}$$

Now putting $x = x_3$ in (2),

$$\phi(x_3) = a_0 + a_1(x_3 - x_0) + a_2(x_3 - x_0)(x_3 - x_1) + a_3(x_3 - x_0)(x_3 - x_1)(x_3 - x_2)$$

or, $$y_3 = y_0 + \frac{y_1 - y_0}{h}(3h) + \frac{y_2 - 2y_1 + y_0}{2h^2}(3h)(2h) + a_3(3h)(2h)(h)$$

or, $$y_3 = y_0 + 3(y_1 - y_0) + 3(y_2 - 2y_1 + y_0) + a_3(1 \times 2 \times 3)\, h^3$$

Solving, $$a_3 = \frac{y_3 - 3y_2 + 3y_1 - y_0}{(1 \times 2 \times 3)h^3} = \frac{\Delta^3 y_0}{(1 \times 2 \times 3)h^3}$$

Similarly, putting $x = x_4, \ldots, x_n$ successively in (2), we find that

$$a_4 = \frac{\Delta^4 y_0}{(1 \times 2 \times 3 \times 4)h^4}, \ldots\ a_n = \frac{\Delta^n y_0}{(1 \times 2 \times \ldots \times n)\, h^n}$$

Substituting the values of $a_0, a_1, a_2, \ldots, a_n$ in (2),

$$y = \phi(x) = y_0 + \frac{\Delta y_0}{h}(x - x_0) + \frac{\Delta^2 y_0}{(1 \times 2)h^2}(x - x_0)(x - x_1)$$

$$+ \frac{\Delta^3 y_0}{(1 \times 2 \times 3)h^3}(x - x_0)(x - x_1)(x - x_2) + \ldots$$

$$+ \frac{\Delta^n y_0}{(1 \times 2 \times \ldots \times n)h^n}(x - x_0)(x - x_1) \ldots (x - x_{n-1}) \qquad \ldots(3)$$

If we write $\frac{x - x_0}{h} = u$, i.e. $x = x_0 + hu$, then

$$x - x_0 = hu,$$
$$x - x_1 = (x_0 + hu) - (x_0 + h) = h(u - 1),$$
$$x - x_2 = (x_0 + hu) - (x_0 + 2h) = h(u - 2),$$
$$\ldots \qquad \ldots \qquad \ldots$$
$$x - x_{n-1} = (x_0 + hu) - \{x_0 + (n - 1)h\} = h(u - n + 1)$$

Substituting these values in (3), we have

$$y = y_0 + \Delta y_0 . u + \frac{\Delta^2 y_0}{1 \times 2} u(u - 1) + \frac{\Delta^3 y_0}{1 \times 2 \times 3} u(u - 1)(u - 2)$$

$$+ \ldots \ldots + \frac{\Delta^n y_0}{1 \times 2 \times \ldots \times n} u(u - 1)(u - 2) \ldots (u - n + 1).$$

Rearranging the terms on the right, we get Newton's forward interpolation formula (10.4.1).

Uses—Newton's forward interpolation formula can be used only when the values of the argument are equidistant. Since the formula contains only the leading term y_0 and the leading differences Δy_0, $\Delta^2 y_0$, etc., it is most suitable for interpolation near the beginning of the tabulated values. It may also be used to find the value of the function corresponding to values slightly before the first tabulated value of argument (The process of finding the value of the function outside the range of tabulated values is known as *Extrapolation*).

Example 10.11 *From the following table, find by interpolation the value of f(x) for x = 24:*

x	20	25	30	35	40
$f(x)$	30.5	34.5	40	47.75	59.25

Solution Since the given values of x are equidistant and the value $x = 24$ lies near the beginning of these values, we use Newton's forward interpolation formula (10.4.1).

Table 10.3 Difference Table

x	$y = f(x)$	Δy	$\Delta^2 y$	$\Delta^3 y$	$\Delta^4 y$
20	30.5				
		4.0			
25	34.5		1.5		
		5.5		0.75	
30	40		2.25		0.75
		7.75		1.50	
35	47.75		3.75		
		11.50			
40	59.25				

Hence, $u = \dfrac{x - x_0}{h} = \dfrac{24 - 20}{5} = 0.8$. Applying (10.4.1),

$$y = 30.5 + 0.8(4.0) + \frac{0.8(0.8-1)}{2}(1.5) + \frac{0.8(0.8-1)(0.8-2)}{6}(0.75) + \frac{0.8(0.8-1)\ (0.8-2)\ (0.8) - 3)}{24}(0.75)$$

$$= 30.5 + 3.2 - 0.12 + 0.024 - 0.0132$$

$$= 33.5908 = 33.59$$

Ans. 33.59

Example 10.12 *Given the following table, find* $(3.24)^3$:

x	31	32	33	34	35
x^3	29791	32768	35937	39304	42875

Solution Since the successive values of x have a common difference, we shall use Newton's forward interpolation formula and find $y = (32.4)^3$, from which the value of $(3.24)^3$ can be easily obtained by shifting the decimal point three places to the left.

Table 10.4 Difference Table

x	$y = x^3$	Δy	$\Delta^2 y$	$\Delta^3 y$
31	29,791			
		2,977		
32	32,768		192	
		3,169		6
33	35,937		198	
		3,367		6
34	39,304		204	
		3,571		
35	42,875			

Here $u = (x - x_0)/h = (32.4 - 31)/1 = 1.4$. Using (10.4.1),

$$y = (32.4)^3 = 29791 + 1.4(2977) + \frac{1.4(1.4-1)}{2}(192) + \frac{1.4(1.4-1)(1.4-2)}{6} \quad (6)$$
$$= 29791 + 4167.8 + 53.76 - 0.336 = 34012.224$$

Hence, $(3.24)^3 = 34.012,224$. This result is exact.

10.5 NEWTON'S BACKWARD INTERPOLATION FORMULA

As in Section 10.4, let $x_0, x_1, x_2, ..., x_n$ be some equidistant values of the argument and $y_0, y_1, y_2, ..., y_n$ the corresponding entries. It is required to find the value of y corresponding to a specified value of x lying near the end of the tabulated values. This is given by *Newton's Background Interpolation Formula*:

$$y = y_n + v\Delta y_{n-1} + \frac{v(v+1)}{1\times 2}\Delta^2 y_{n-2} + \frac{v(v+1)(v+2)}{1\times 2\times 3}\Delta^3 y_{n-3} + \ldots$$
$$\ldots + \frac{v(v+1)(v+2)\ldots(v+n-1)}{1\times 2\times 3\times\ldots\times n}\Delta^n y_0. \quad (10.5.1)$$

where $v = \dfrac{x - x_n}{h}$.

Derivation of Newton's backward formula

As before, within the range of the given values, the tabulated function $y = f(x)$ is replaced by an n-th degree polynomial $\phi(x)$, which coincides with $f(x)$ when $x = x_0$, $x_1, ..., x_n$; i.e.

$$\phi(x_0) = y_0, \phi(x_1) = y_1, \phi(x_2) = y_2, ..., \phi(x_n) = y_n \quad ...(1)$$

We write the n-th degree polynomial $\phi(x)$ in the form

$$\phi(x) = a_0 + a_1(x - x_n) + a_2(x - x_n)(x - x_{n-1}) + a_3(x - x_n)(x - x_{n-1})(x - x_{n-2})$$
$$+ \ldots + a_n(x - x_n)(x - x_{n-1}) \ldots (x - x_1). \quad ...(2)$$

The $n + 1$ constants $a_0, a_1, a_2, ... a_n$ are so determined as to satisfy the $n + 1$ relations (1).

Putting $x = x_n$ in (2), we have

$$\phi(x_n) = a_0$$

because all other terms contain the factor $(x_n - x_n)$ and therefore vanish.

or, $\quad y_n = a_0$; i.e. $\quad a_0 = y_n$

Now putting $x = x_{n-1}$ in (2),

$$\phi(x_{n-1}) = a_0 + a_1(x_{n-1} - x_n), \text{ because the other terms vanish;}$$

or, $$y_{n-1} = y_n + a_1(-h), \quad \text{since} \quad x_n - x_{n-1} = h$$

or, $$a_1 h = y_n - y_{n-1} = \Delta y_{n-1}$$

$$\therefore \qquad a_1 = \frac{y_n - y_{n-1}}{h} = \frac{\Delta y_{n-1}}{h}$$

Again, putting $x = x_{n-2}$ in (2),

$$\phi(x_{n-2}) = a_0 + a_1(x_{n-2} - x_n) + a_2(x_{n-2} - x_n)(x_{n-2} - x_{n-1})$$

or, $$y_{n-2} = y_n + \frac{y_n - y_{n-1}}{h}(-2h) + a_2(-2h)(-h)$$

or, $$y_{n-2} = y_n - 2(y_n - y_{n-1}) + a_2(2h^2)$$

Solving, $$a_2 = \frac{y_n - 2y_{n-1} + y_{n-2}}{2h^2} = \frac{\Delta^2 y_{n-2}}{2h^2}$$

Similarly, putting $x = x_{n-3}, x_{n-4}, \ldots, x_0$ successively in (2), we get

$$a_3 = \frac{\Delta^3 y_{n-3}}{(1 \times 2 \times 3)h^3},$$

$$a_4 = \frac{\Delta^4 y_{n-4}}{(1 \times 2 \times 3 \times 4)h^4}, \ldots a_n = \frac{\Delta^n y_0}{(1 \times 2 \times \ldots \times n)h^n}$$

Substituting the values of $a_0, a_1, a_2, \ldots, a_n$ in (2),

$$y = y_n + \frac{\Delta y_{n-1}}{h}(x - x_n) + \frac{\Delta^2 y_{n-2}}{(1 \times 2)h^2}(x - x_n)(x - x_{n-1})$$

$$+ \frac{\Delta^3 y_{n-3}}{(1 \times 2 \times 3)h^3}(x - x_n)(x - x_{n-1})(x - x_{n-2}) + \ldots$$

$$\ldots + \frac{\Delta^n y_0}{(1 \times 2 \times \ldots \times n)h^n}(x - x_n)(x - x_{n-1}) \ldots (x - x_1) \qquad \ldots(3)$$

If we write $\frac{x - x_n}{h} = v$, i.e. $x = x_n + hv$, then

$$x - x_n = hv$$
$$x - x_{n-1} = (x_n + hv) - (x_n - h) = h(v + 1)$$
$$x - x_{n-2} = (x_n + hv) - (x_n - 2h) = h(v + 2)$$
$$\ldots \qquad \ldots$$
$$x - x_1 = (x_n + hv) - \{x_n - (n - 1)h\} = h(v + n - 1)$$

When these values are substituted in (3), we have

$$y = y_n + \Delta y_{n-1} v + \frac{\Delta^2 y_{n-2}}{1 \times 2} v(v + 1) + \frac{\Delta^2 y_{n-3}}{1 \times 2 \times 3} v(v + 1)(v + 2)$$

$$+ \ldots \ldots + \frac{\Delta^n y_0}{1 \times 2 \times \ldots \times n} v(v + 1)(v + 2) \ldots (v + n - 1).$$

Rearranging the terms on the right, we get Newton's backward interpolation formula (10.5.1).

Uses—Newton's backward interpolation formula can be used only when the values of the argument are equidistant. Since the formula contains only the terms at the bottom of the difference table (Table 10.2), it is most suitable for interpolation near the end of the tabulated values. It may also be used to extrapolate the value of the function corresponding to an argument slightly beyound the last tabulated valued.

[**Note:** (i) Usually, each term in formulae (10.4.1) and (10.5.1) is less than the preceding. Hence, in numerical problems it is not necessary to calculate all the terms given in the formula. Only the first few terms are sufficient to obtain the required estimate of y to the same degree of accuracy as the given entries.

(ii) If all the terms in formulae (10.4.1) and (10.5.1) are used, Newton's forward formula and backward formula would give exactly the same result.]

Example 10.13 *Given the following table, estimate y when x = 0.35, by using Newton's backward interpolation formula.*

x:	*0*	*0.1*	*0.2*	*0.3*	*0.4*
y:	*1*	*1.095*	*1.179*	*1.251*	*1.310*

(Give answer correct to 3 decimal places). [I.C.W.A., June '78]

Solution (For simplicity, we shall ignore the decimal point while computing the differences).

Table 10.5 Difference Table

x	y	Δy	$\Delta^2 y$	$\Delta^3 y$	$\Delta^4 y$
0	1				
		95			
0.1	1.095		– 11		
		84		– 1	
0.2	1.179		– 12		0
		72		– 1	
0.3	1.251		– 13		
		59			
0.4	1.310				

Here $v = (x - x_n)/h = (0.35 - 0.4)/0.1 = -0.5$

Using 3 decimals,

$$y = 1310 + (-0.5) \times 59 + \frac{(-0.5)(-0.5+1)}{2} \times (-13) + \frac{(-0.5)(-0.5+1)(-0.5+2)}{6} \times (-1)$$

$$= 1310 - 29.5 + 1.6 + 0.1 = 1282.2$$

$$= 1282$$

Adjusting the decimal point, the required value is 1.282.

10.6 CENTRAL DIFFERENCE FORMULAE-STIRLING'S AND BESSEL'S

When the successive values of the argument are equidistant, Newton's forward and backward formulae are generally applied in all cases of interpolation. However, for interpolation near the middle of the set of tabulated values, *Central Difference* formulae, which utilize differences near the central part of the difference table, are found more useful; because the successive terms converge more rapidly than in Newton's forward and backward formulae.

Let the function $y = f(x)$ be tabulated for some equidistant values of the argument

$$(x_0 - nh), \ldots (x_0 - 2h), (x_0 - h), x_0, (x_0 + h), (x_0 + 2h), \ldots (x_0 + nh)$$

the common difference being h; and let the corresponding entries be denoted by

$$y_{-n}, \quad \ldots \quad y_{-2}, \quad y_{-1}, \quad y_0, \quad y_1, \quad y_2, \quad \ldots \quad y_n.$$

It is required to interolate the value of the function $y = f(x)$ for a value of x in the interval $(x_0 - h)$ to $(x_0 + h)$.

1. Stirling's Interpolation Formula

$$y = y_0 + u \cdot \frac{(\Delta y_0 + \Delta y_{-1})}{2} + \frac{u^2}{2!} \cdot \Delta^2 y_{-1} + \frac{u(u^2 - 1^2)}{3!} \cdot \frac{(\Delta^3 y_{-1} + \Delta^3 y_{-2})}{2}$$
$$+ \frac{u^2(u^2 - 1^2)}{4!} \cdot \Delta^4 y_{-2} + \frac{u(u^2 - 1^2)(u^2 - 2^2)(\Delta^5 y_{-2} + \Delta^5 y_{-3})}{5!} + \ldots \quad (10.6.1)$$

where $u = (x - x_0)/h$.

Stirling's formula is appropriate when the value of u lies in the interval – 0.25 to + 0.25. It uses y_0 and even order differences on the horizontal line through y_0, but the average of odd order differences above and below that line.

2. Bessel's Interpolation Formula

$$y = \frac{(y_0 + y_1)}{2} + v \cdot \Delta y_0 + \frac{\left(v^2 - \frac{1}{4}\right)}{2!} \cdot \frac{(\Delta^2 y_0 + \Delta^2 y_{-1})}{2} + \frac{v\left(v^2 - \frac{1}{4}\right)}{3!} \cdot \Delta^3 y_{-1}.$$
$$+ \frac{\left(v^2 - \frac{1}{4}\right)\left(v^2 - \frac{9}{4}\right)}{4!} \cdot \frac{(\Delta^4 y_{-1}\ \Delta^4 y_{-2})}{2} + \frac{v\left(v^2 - \frac{1}{4}\right)\left(v^2 - \frac{9}{4}\right)}{5!} . \Delta^5 y_{-2} + \ldots \quad (10.6.2)$$

where $u = (x - x_0)/h$ and $v = u - \frac{1}{2}$.

Bessel's formula is appropriate for values of u lying in the interval + 0.25 to + 0.75, i.e. for interpolation in the middle-half of two tabulated values. It uses the average of y_0 and y_1, and odd order differences on the horizontal line mid-way between them, but the average of even order differences above and below that line.

Bessel's formula is especially found useful for interpolating the value of the function *exactly* at the middle of two given values, i.e. when $u = \frac{1}{2}$ so that $v = 0$. Every alternate term in Bessel's formula (10.6.2) then vanish. This special case of Bessel's formula is known as "*Formula for Interpolating to Halves*".

Example 10.14 *Mantissae of 4 numbers are given below:*

Numbers	*4,200*	*4,210*	*4,220*	*4,230*
Mantissae	*6,232,493*	*6,242,821,*	*6,253,125,*	*6,263,404*

Find the matnissa of the logarithm of 4,213.

Solution Since the values of the argument are equidistant, and 4213 lies near the middle of the tabulated values, we use a *central difference formula*. Let us take $x_0 = 4210$; then

$$u = (x - x_0)/h = (4213 - 4210)/10 = 0.3$$

Bessel's interpolation formula will be appropriate here.

Table 10.6 Difference Table

x	y	Δy	$\Delta^2 y$	$\Delta^3 y$
$x_{-1} = 4200$	$y_{-1} = 6{,}232{,}493$			
		$\Delta y_{-1} = 10{,}328$		
$x_0 = 4210$	$y_0 = 6{,}242{,}821$		$\Delta^2 y_{-1} = -24$	
		$\Delta y_0 = 10{,}304$		$\Delta^2 y_{-1} = -1$
$x_1 = 4220$	$y_1 = 6{,}253{,}125$		$\Delta^2 y_0 = -25$	
		$\Delta y_1 = 10{,}279$		
$x_2 = 4230$	$y_2 = 6{,}263{,}404$			

Here, $v = u - \frac{1}{2} = 0.3 - 0.5 = -0.2$. Substituting the values from Table 10.7 in formula (10.6.2),

$$\begin{aligned} y &= \frac{(6,242,821 + 6,253,125)}{2} + (-0.2)(10{,}304) \\ &\quad + \frac{(0.04 - 0.25)}{2} \cdot \frac{(-24 - 25)}{2} + \frac{(-0.2)(0.04 - 0.25)}{6}(-1). \\ &= 6{,}247{,}973 - 2{,}060.8 + \frac{(-0.21)(-49)}{4} + \frac{(-0.2)(-0.21)(-1)}{6} \\ &= 6{,}247{,}973 - 2{,}060.8 + 2.6 - 0.0 \\ &= 6{,}245{,}914.8 \\ &= 6{,}245{,}915. \end{aligned}$$

This result is correct to the last place.

10.7 LAGRANGE'S INTERPOLATION FORMULA

Let $y_0, y_1, y_2, \dots, y_n$ denote the tabulated values of a function $y = f(x)$ corresponding to the values of the argument $x_0, x_1, x_2, \dots, x_n$ (which may *not* be equidistant). It is required to find the value of y corresponding to a specified value of x lying in between the given values. This is obtained by using Lagrange's Interpolation Formula:

$$\begin{aligned} y &= \frac{(x - x_1)(x - x_2)\dots(x - x_n)}{(x_0 - x_1)(x_0 - x_2)\dots(x_0 - x_n)} y_0 + \frac{(x - x_0)(x - x_2)\dots(x - x_n)}{(x_1 - x_0)(x_1 - x_2)\dots(x_1 - x_n)} y_1 \\ &\quad + \dots \quad \dots \quad \dots \quad \dots \\ &\quad + \frac{(x - x_0)(x - x_1)\dots(x - x_{n-1})}{(x_n - x_0)(x_n - x_1)\dots(x_n - x_{n-1})} y_n \end{aligned} \tag{10.7.1}$$

Derivation of Lagrange's formula

As before, the tabulated function $y = f(x)$ is replaced by an n-th degree polynomial $\phi(x)$, which coincides with $f(x)$ when $x = x_0, x_1, ..., x_n$; i.e.

$$\phi(x_0) = y_0,\ \phi(x_1) = y_1,\ \phi(x_2) = y_2,\ ....,\ \phi(x_n) = y_n \qquad ...(1)$$

We write the n-th degree polynomial $\phi(x)$ in the form

$$\begin{aligned}\phi(x) = {} & a_0(x - x_1)(x - x_2)(x - x_3) \ldots (x - x_n)\\ & + a_1(x - x_0)(x - x_2)(x - x_3) \ldots (x - x_n)\\ & + a_2(x - x_0)(x - x_1)(x - x_3) \ldots (x - x_n)\\ & + \ldots \quad \ldots \quad \ldots\\ & + a_n(x - x_0)(x - x_1)(x - x_2) \ldots (x - x_{n-1}).\end{aligned} \qquad ...(2)$$

[Note that each term on the right is a polynomial of degree n; $(x - x_0)$ is absent from the 1st term, $(x - x_1)$ from the 2nd term, and so on, finally $(x - x_n)$ is absent from the last term]. Then $n + 1$ constants $a_0, a_1, a_2, ..., a_n$ are so determined as to satisfy the $n + 1$ relations (1).

Putting $x = x_0$ in (2), we have

$$y_0 = a_0(x_0 - x_1)(x_0 - x_2)(x_0 - x_3) \ldots (x_0 - x_n),$$

because all other terms contain the factor $(x_0 - x_0)$ and vanish indentically. Hence

$$a_0 = \frac{y_0}{(x_0 - x_1)(x_0 - x_2)(x_0 - x_3) \ldots (x_0 - x_n)}$$

Similarly, putting $x = x_1, x_2, ... x_n$ successively in (2), we find that

$$a_1 = \frac{y_1}{(x_1 - x_0)(x_1 - x_2)(x_0 - x_3) \ldots (x_0 - x_n)}$$

$$a_2 = \frac{y_2}{(x_2 - x_0)(x_2 - x_1)(x_2 - x_3) \ldots (x_2 - x_n)}$$

$$\ldots \qquad \ldots \qquad \ldots$$

$$a_n = \frac{y_n}{(x_n - x_0)(x_n - x_1)(x_n - x_2) \ldots (x_n - x_{n-1})}$$

Substituting in the values of $a_0, a_1, a_2, ..., a_n$ in (2), we get Lagrange's interpolation formula (10.7.1).

Uses—Lagrange's interpolation formula is generally used when the values of the argument are not equidistant, though it can be used for equidistant arguments also. It is applicable in any part of the tabulated values, and does not require the construction of a difference table. Lagrange's formula may also be used to find the value of the argument for a given value of the function, i.e. to find the value of x for a given value of y (This process is known as '*Inverse Interpolation*; page 335).

[**Note:** (i) Lagrange's formula may be used in all cases, whether the values of x differ by a constant or not; or whether the value of x, corresponding to which the value of y is to be found, lies near the beginning or end or middle of the tabulated values. However, if the given values of x are equidistant, it is preferable to use Newton's Forward or Backward Formula, because the application of Lagrange's formula would involve more laborious computations.

(ii) When the values of the argument are equidistant, Lagrange's formula may be reduced to Newton's Forward or Backward formula.]

Example 10.15 *The values of a function f(x) are given for certain values of x:*

x	4	5	6	8
$f(x)$	3.11	2.96	2.85	2.70

Obtain the best approximation of f(7). [I.C.W.A., June '74-old]

Solution Since the successive values of x for the tabulated function $f(x)$ are not equidistant, Newton's forward or backward formula cannot be applied. Applying Lagrange's formula,

$$y = f(x) = \frac{(7-5)(7-6)(7-8)}{(4-5)(4-6)(4-8)} \times 3.11 + \frac{(7-4)(7-6)(7-8)}{(5-4)(5-6)(5-8)} \times 2.96$$

$$+ \frac{(7-4)(7-5)(7-8)}{(6-4)(6-5)(6-8)} \times 2.85 + \frac{(7-4)(7-5)(7-6)}{(8-4)(8-5)(8-6)} \times 2.70$$

$$= \frac{2 \times 1 \times (-1)}{(-1)\,(-2)\,(-4)} \times 3.11 + \frac{3 \times 1 \times (-1)}{1 \times (-1)(-3)} \times 2.96$$

$$+ \frac{3 \times 3 \times (-1)}{2 \times 1 \times (-2)} \times 2.85 + \frac{3 \times 2 \times 1}{4 \times 3 \times 2} \times 2.70$$

$$= \frac{1}{4} \times (3.11) + (-1) \times (2.96) + \frac{3}{2} \times (2.85) + \frac{1}{4} \times (2.70)$$

$$= 0.7775 - 2.9600 + 4.2750 + 0.6750$$

$= 2.7675$ (see Example 10.6)

[**Note:** In numerical calculations with Lagrange's formula, the sum of the coefficients of the values of y is always 1. Here

$$\frac{1}{4} + (-1) + \frac{3}{2} + \frac{1}{4} = 1.$$

This gives a check on the accuracy of calculations]

10.8 INVERSE INTERPOLATION

Given a set of tabulated values of a function $y = f(x)$ corresponding to some values of the argument x, the process of finding the value of the argument for an intermediate value of the function is called '*Inverse Interpolation*'.

Let $y_0, y_1, ..., y_n$ be a set of tabulated values of a function $y = f(x)$ corresponding to some given values (usually equidistant) of the argument $x = x_0, x_1, ..., x_n$. In some cases it may be required to determine the value of x for another value of y lying in between $y_0, y_1, ..., y_n$. The process of finding such a value of x is known as 'inverse interpolation'. If y is a function of x, say $y = f(x)$, in many cases it is possible to treat x also as a function of y, say $x = g(y)$. Since Lagrange's interpolation formula is applicable for unequal intervals, we can use this formula for inverse interpolation, interchanging the role of the argument x and the function y.

$$x = \frac{(y-y_1)(y-y_2)(y-y_3)\dots(y-y_n)}{(y_0-y_1)(y_0-y_2)(y_0-y_3)\dots(y_0-y_n)}\, x_0$$

$$+ \frac{(y-y_0)(y-y_2)(y-y_3)\dots(y-y_n)}{(y_1-y_0)(y_1-y_2)(y_1-y_3)\dots(y_1-y_n)}\, x_1$$

$$+\frac{(y-y_0)(y-y_1)(y-y_3)\ldots(y-y_n)}{(y_2-y_0)(y_2-y_1)(y_2-y_3)\ldots(y_2-y_n)}x_2$$
$$+\ \ldots\quad \ldots\quad \ldots$$
$$+\frac{(y-y_0)(y-y_1)(y-y_2)\ldots(y-y_{n-1})}{(y_n-y_0)(y_n-y_1)(y_n-y_2)\ldots(y_n-y_{n-1})}x_n \qquad (10.8.1)$$

Example 10.16 *The values of a function f(x) are given for certain values of x:*

x	*4*	*5*	*6*	*8*
f(x)	*3.11*	*2.96*	*2.85*	*2.70*

Find the value of x for which f(x) = 3.00.

Solution Applying the formula (10.8.1),

$$x=\frac{(3.00-2.96)\,(3.00-2.85)\,(3.00-2.70)}{(3.11-2.96)\,(3.11-2.85)\,(3.11-2.70)}\times 4$$
$$+\frac{(3.00-3.11)\,(3.00-2.85)\,(3.00-2.70)}{(2.96-3.11)\,(2.96-2.85)\,(2.96-2.70)}\times 5$$
$$+\frac{(3.00-3.11)\,(3.00-2.96)\,(3.00-2.70)}{(2.85-3.11)\,(2.85-2.96)\,(2.85-2.70)}\times 6$$
$$+\frac{(3.00-3.11)\,(3.00-2.96)\,(3.00-2.85)}{(2.70-3.11)\,(2.70-2.96)\,(2.70-2.85)}\times 8$$
$$=\frac{.04\times .15\times .30}{.15\times .26\times .41}\times 4+\frac{(-.11)\times .15\times .30}{(-.15)\times .11\times .26}\times 5$$
$$+\frac{(-.11)\times .04\times .30}{(-.26)\,(-.11)\,(.15)}\times 6+\frac{(-.11)\,(.04)\,(.15)}{(-.41)\,(-.26)\,(-.15)}\times 8$$
$$=\frac{60}{533}\times(4)+\frac{15}{13}\times(5)+\frac{-4}{13}\times(6)+\frac{22}{533}\times(8)$$
$$\left[\text{Check: } \frac{60}{533}+\frac{15}{13}+\frac{(-4)}{13}+\frac{22}{533}=1\right]$$
$$=0.450+5.769-1.846+0.330=4.70$$

Example 10.17 *The mode of a certain frequency curve y = f(x) is attained at x – 9.1 and the value of the frequency function f(x) for x = 8.9, 9.0 and 9.3 are respectively equal to 0.30, 0.35 and 0.25. Calculate the approximate value of f(x) at the mode.*

[I.C.W.A., Dec. '78]

Solution We are given the following information:

x	8.9	9.0	9.3
$y = f(x)$	0.30	0.35	0.25

It is required to find the value of y when $x = 9.1$. Using Lagrange's interpolation formula (10.7.1),

$$y = \frac{(9.1-9.0)\,(9.1-9.3)}{(8.9-9.0)\,(8.9-9.3)} \times 0.30 + \frac{(9.1-8.9)(9.1-9.3)}{(9.0-8.9)(9.0-9.3)} \times 0.35$$

$$+ \frac{(9.1-8.9)(9.1-9.0)}{(9.3-8.9)(9.3-9.0)} \times 0.25$$

$$= \frac{0.1\,(-0.2)}{(-0.1)\,(-0.4)} \times 0.30 + \frac{0.2(-0.2)}{0.1(-0.3)} \times 0.35 + \frac{0.2 \times 0.1}{0.4 \times 0.3} \times 0.25$$

$$= \left(\frac{-1}{2}\right) \times 0.30 + \left(\frac{4}{3}\right) \times 0.35 + \left(\frac{1}{6}\right) \times 0.25.$$

$$= \frac{1}{6} \times (-0.90 + 2.80 + 0.25) = 0.358$$

Note that this result is larger than any of the given values of y, because $y = f(x)$ is the maximum at the Mode.

Example 10.18 *Using three pairs of observations (x_0, y_0), (x_1, y_1), (x_2, y_2), show that Lagrange's interpolation formula reduces to Newton's forward formula, when x_0, x_1, x_2 are equidistant.*

Solution Using Lagrange's formula (10.7.1), we have

$$y = \frac{(x-x_1)(x-x_2)}{(x_0-x_1)(x_0-x_2)}\, y_0 + \frac{(x-x_0)(x-x_2)}{(x_1-x_0)(x_1-x_2)}\, y_1 + \frac{(x-x_0)(x-x_1)}{(x_2-x_0)(x_2-x_1)}\, y_2$$

Here, since $x_1 - x_0 = x_2 - x_1 = h$, therefore

$$x_1 = x_0 + h; \qquad x_2 = x_0 + 2h$$

Also, writing $u = (x - x_0)/h$, we have $x = x_0 + hu$

$$\therefore \quad x - x_0 = (x_0 + hu) - x_0 = hu$$

$$x - x_1 = (x_0 + hu) - (x_0 + h) = h(u-1)$$

$$x - x_2 = (x_0 + hu) - (x_0 + 2h) = h(u-2)$$

Substituting these values,

$$y = \frac{h(u-1).h(u-2)}{(-h)(-2h)}\, y_0 + \frac{hu.h(u-2)}{(h)(-h)}\, y_1 + \frac{hu\,.\,h(u-1)}{(2h)(h)}\, y_2$$

$$= \frac{(u-1)(u-2)}{2}\, y_0 - \frac{u(u-2)}{1}\, y_1 + \frac{u(u-2)}{1}\, y_2$$

$$= \frac{(u-1)\,(u-2)}{2}\, y_0 - \frac{u(u-2)}{1}\,(1+\Delta)y_0 + \frac{u(u-1)}{2}\,(1+\Delta)^2 y_0$$

$$= \frac{(u-1)(u-2)}{2}\, y_0 - \frac{u(u-2)}{1}(y_0 + \Delta\, y_0) + \frac{u(u-1)}{2}\,(y_0 + 2\Delta y_0 + \Delta^2 y_0)$$

$$= y_0 \left\{\frac{(u-1)\,(u-2)}{2} - \frac{u(u-2)}{1} + \frac{u(u-1)}{2}\right\}$$

$$+ \Delta\, y_0 \left\{-\frac{u(u-2)}{1} + \frac{u(u-2)}{2}(2)\right\} + \Delta^2 y_0 \left\{\frac{u(u-1)}{2}\right\}$$

i.e. $\quad y = y_0 + y\Delta y_0 + \dfrac{u(u-1)}{2}\,\Delta^2 y_0$

Thus, Lagrange's interpolation formula reduces to Newton's forward interpolation formula, for equidistant values of the argument.

EXERCISES

1. The following data show the monthly average number of deaths under one year in a certain large city. Find the missing term:

Year	1960	1961	1962	1963	1964
Number of Deaths (monthly average)	940	?	907	843	798

[I.C.W.A., Jan. '72]

2. The following gives the amount y of cement in thousants of tons manufactured in India in the year x. Find the missing term.

x	1946	1948	1950	1952	1954	1956
y	39	85	?	151	264	388

[I.C.W.A., July '69]

3. The growth of population in India, according to the decemial census, is shown below:

Year	1901	1911	1921	1931	1941	1951
Population (lakh)	2384	2552	2514	2791	...	3613

The census figure for 1941 is not given here. Give an estimate of the actual population for 1941.

[C.U., B.Sc.'71]

4. Below are given the values of a function U_x for certain values of x:

x	0	1	2	3	4
U_x	1	0	5	22	57

Construct the table of differences. What does the table suggest? Use this table to find U_5.

[I.C.W.A., Dec.'76]

5. Form a difference table and find the values of y_3 and y_9 from the following: $y_4 = 135$, $y_5 = 122$, $y_6 = 1015$, $y_7 = 2016$, $y_8 = 3591$.
6. (a) Given $y_0 = 8$, $y_1 = 6$, $y_2 = 4$, $y_4 = 24$, find the value of y_3.
   (b) Given $y_2 = 5$, $y_5 = 122$, $y_6 = 193$, find the values of y_3 and y_4.
   (c) Given $u_5 = 3$, $u_{10} = 0$, $u_{20} = -24$, $u_{25} = -33$, estimate u_{15} under suitable assumption.
7. Find the missing term:

x	0	1	2	3	4
$f(x)$	1	3	9	?	81

8. Find y for $x = 2$, from the following table:

x	...	0	1	3	4	5
y	...	39	85	151	264	388

[I.C.W.A., Jan.'69]

9. Find $f(5)$ from the following data:
   $f(3) = 4, \quad f(4) = 13, \quad f(6) = 43$

10. Find the polynomial function $f(x)$ from the following values:
$f(3) = -1, \quad f(4) = 5, \quad f(5) = 15$

11. Given the following table, find the function $f(x)$, assuming it to be a polynomial of the 3rd degree in x:

x	0	1	2	3
$f(x)$	1	2	11	34

[I.C.W.A., June '75]

12. u_x is a polynomial in x. Given the following table, find u_x.

x	0	1	2	5
u_x	2	3	12	147

13. Below are given the values of a function $f(x)$ for certain values of x. Find $f(2)$, stating your assumption.

x	0	1	3	4
$f(x)$	5	6	50	105

[I.C.W.A., Dec. '75]

14. Find $f(x)$, given that
$$f(0) = -3, \quad f(1) = 6, \quad f(2) = 8, \quad f(3) = 12.$$
(State your assumptions, if any). Hence find $f(6)$. [I.C.W.A., June '76]

15. Using any algebraic method, find the value of y when $x = 6$,

x	3	7	9	10
y	16.8	12.0	7.2	6.3

[I.C.W.A., Dec. '75-old]

16. For a certain polynomial function y_x it is known that
$$y_1 = -1, \quad y_2 + y_3 = -1, \quad y_4 + y_5 + y_6 = 61.$$
Find y_x, and hence the value of y_3.

17. Given $u_0 + u_6 = -107$, $u_1 + u_5 = -36$, $u_2 + u_4 = -3$, find the value of u_3.

18. From the following table estimate by interpolation the number of units of a commodity supplied when the price is Rs 4:

Price in Rs	1	3	5	7	9
No. of units Supplied	256	625	935	1201	1433

19. The following table gives the expectation of life e_x° at age x. Calculate the expectation of life at age 12 by using Newton's forward interpolation formula.

x	10	15	20	25	30	35
e_x°	35.4	32.2	29.1	26.0	23.1	20.4

[I.C.W.A., Dec. '77]

20. The following shows the values of a function $y = f(x)$ for a number of values of x:

x:	0.5	0.6	0.7	0.8	0.9
y:	0.35207	0.33322	0.31225	0.28969	0.26609

Obtain the value of y when $x = 0.58$, using a suitable interpolation formula.
[C.U., B.A. (Econ) '76]

21. The table below gives the average number of years of life remaining to persons who survive to exact age x, for male African population of Belgian Congo:

x	0	5	10	15	20
$°e_x$	37.64	44.04	41.40	37.78	34.41

Obtain $°e_2$ approximately. *[I.C.W.A., Dec. '73]*

22. State Newton's Forward Interpolation formula, and use it to find $\sqrt{5.5}$, given that $\sqrt{5} = 2.236, \sqrt{6} = 2.449, \sqrt{7} = 2.646$ and $\sqrt{8} = 2.828$. *[I.C.W.A., June '74]*

23. The following table shows the number of earners earning incomes exceeding different amounts during a certain period:

Income (Rs)	50,000	75,000	100,00	125,000	150,000
No. of Earners	412	304	225	147	88

Find the number of earners earning more than Rs 60,000 by linear interpolation and also by using Newton's forward formula.

[C.U., B.A. (Econ) '77]

24. Using suitable interpolation formulae, calculate the values of y, when (i) $x = 10$, and (ii) $x = 25$.

x	7	11	15	19	23	27
y	20,256	20,625	21,2)6,	22,407,	24,098	26,511

25. Find the value of sin 48° from the following table:

x (degrees)	30	35	40	45	50
sin x	.5000	.5736	.6428	.7071	.7660

26. Using Newton's interpolation formula, find the number of factories earning less than Rs 65,000 as profits, from the following data:

Profits (Rs '000)	30–40	40–50	50–60	60–70	70–80
No. of Factories	34	43	56	39	29

[I.C.W.A., Dec. '75]

27. Apply the appropriate interpolation formula to find log 3.146, given

log 3.141 = 0.497, 0679   log 3.144 = 0.497, 4825
log 3.142 = 0.497, 2062   log 3.145 = 0.497, 62.05
log 3.143 = 0.497, 3444

(Find correct upto 7 decimal places). *[I.C.W.A., Dec. '78]*

28. Use Stirling's interpolation formula to find the value of the probability integral (P) when $X = 1.52$:

X...	1.3	1.4	1.5	1.6	1.7
P = prob. integral	.90320	.91924	.93319	.94520	.95543

29. Given the following cube-roots, find by Bessel's interpolation formula, the cube root of 102.5:

Number	101	102	103	104
Cube-root	4.657,0095	4.672,3287	4.687,5481	4.702,6694

[B.U., B.A. (Econ) '65]

30. Find by using Bessel's interpolation formula, the expectation of life at age 22 from the following data:

Age (x)	10	15	20	25	30	35
Exp. of Life (y)	35.4	32.2	29.1	26.0	23.1	20.4

31. The following table gives the normal weight of a baby during the first six months of life:

Age in Months	0	2	3	5	6
Weight in lbs.	5	7	8	10	12

Estimate the weight of a baby at the age of 4 months.

[I.C.W.A., Jan. '70]

32. State Lagrange's interpolation formula. Use it to find $f(x)$ when $x = 0$, given

x	–1	–2	2	4
$f(x)$	–1	–9	11	69

[I.C.W.A., Dec. '74]

33. State Lagrange's interpolation formula. Use it to find the value of U_4 of a function U_x, given that

$U_1 = 10, \quad U_2 = 15, \quad U_5 = 42.$ *[I.C.W.A., Dec. '76]*

34. Using Lagrange's formula or otherwise, obtain the value of log 95 approximately from the following table:

x	95	97	98	99
log x	1.977,7236	1.986,7717	1.991,2261	1.995,6352

[C.U., B.Com. (Hons) '66, '69]

35. Given $\log_{10} 654 = 2.8156$, $\log_{10} 658 = 2.8182$
$\log_{10} 659 = 2.8189$, $\log_{10} 661 = 2.8202$,
find by Lagrange's interpolation formula $\log_{10} 656$ (Retain 4 decimal places in your answer). *[I.C.W.A., June '78]*

36. Find the value of x for which $y = 40$:

x	10	12	15	20
y	25	32	35	45

ANSWERS

1. 952
2. 96.4
3. 3322 lakh
4. 3rd differences constant; hence, U_x may be given by a 3rd degree polynomial. $U_5 = 116$.
5. 16; 5920
6. (a) 8; (b) 28, 67; (c) –11
7. 31
8. 96.4
9. 26
10. $2x^2 - 8x + 5$
11. $x^3 + x^2 - x + 1$
12. $x^3 + x^2 - x + 2$
13. 19; assuming $f(x)$ to be a 3rd degree polynomial.

14. 126 : assuming $f(x)$ to be a 3rd degree polynomial.

$$f(x) = \frac{1}{2}(3x^3 - 16x^2 + 31x - 6)$$

15. 14.7; (use *Second method*, Example 10.8).

16. $2x^2 - 7x + 4$; 1
17. 3.2
18. 786
19. 34.1
20. 0.33718
21. 42.08
22. 2.345
23. 369; 362
24. 20,510; 25,205
25. 0.7431
26. 156 (use Cumulative frequency distribution)
27. 0.497, 7584
28. 0.93574
29. 4.679, 9508
30. 27.85
31. $8\frac{8}{9}$ lbs.
32. 1
33. 31
34. 1.9822, 7115 (see Examples 10.6 and 10.15)
35. 2.8168
36. 19.56

11 THEORY OF PROBABILITY

11.1 INTRODUCTION

The word *probability* literally denotes 'chance', and the *theory of probability* deals with laws governing the chances of occurrence of phenomena which are unpredictable in nature.

Although historically the probability theory originated from the games of chance played by tossing coins, throwing dice, drawing cards etc., its importance has enormously increased in recent years. Today, the notions of probability find important applications in almost all disciplines—physics, chemistry, biology, psychology, education, economics, business, industry, engineering etc. The concept of probability plays a vital role in statistics.

11.2 RANDOM EXPERIMENT, OUTCOME, EVENT

(1) **Random Experiment:** When a coin is tossed (as is done before the start of a cricket match), either Head or Tail appears. But the result of any toss cannot be predicted in advance, and is said to 'depend on chance'. Similarly, when a die is tossed and the number coming on the uppermost face is observed, one of the numbers 1, 2, 3, 4, 5, 6 definitely appears, but nobody can assert with certainty which result will materialise in any particular toss.

The word *experiment* is used to describe an act which can be repeated under some given conditions. *Random experiments* are those experiments whose results depend on chance (The word 'random' may be taken as equivalent to 'depending on chance').

Example 11.1 *Each of the following may be called a random experiment:*

(a) Tossing a coin (or several coins).
(b) Throwing a die (or several dice).
(c) Drawing cards from a pack.
(d) Drawing balls from a box containing given numbers of white and black balls.
(e) Studying the distribution of boys and girls in families having three children.

(2) **Outcome:** The result of a random experiment will be called an outcome. The possible outcomes of a random experiment may often be described in several ways.

Example 11.2

(a) In the random experiment of tossing a coin, there are 2 possible outcomes—'*Head*' and '*Tail*'; or in symbols

$$H, \quad T$$

(b) In the random experiment of throwing a six-faced die an observing the number of points that appear, the possible outcomes at

$$1, 2, 3, 4, 5, 6$$

In the same experiment, the possible outcomes could also be stated as

'*Odd number of points*'; '*Even number of points*'

(c) In the random experiment of two tosses of a coin (or, a single toss of two coins simultaneously), we may list the possible outcome as

$$HH, \quad HT, \quad TH, \quad TT$$

where the two letters indicate the results of the first and the second tosses (or coins) respectively. If our interest lies in the number of heads turned up, the possible outcomes may also be stated as

2 *heads*, 1 *head*, 0 *head*

(d) There are 36 possible outcomes in the random experiment of a single throw of 2 dice (or, two throws of one die), as follows:

(1, 1)	(2, 1)	(3, 1)	(4, 1)	(5, 1)	(6, 1)
(1, 2)	(2, 2)	(3, 2)	(4, 2)	(5, 2)	(6, 2)
(1, 3)	(2, 3)	(3, 3)	(4, 3)	(5, 3)	(6, 3)
(1, 4)	(2, 4)	(3, 4)	(4, 4)	(5, 4)	(6, 4)
(1, 5)	(2, 5)	(3, 5)	(4, 5)	(5, 5)	(6, 5)
(1, 6)	(2, 6)	(3, 6)	(4, 6)	(5, 6)	(6, 6)

each pair of numbers indicating the results of the first and the second dice (or throws) respectively. If we are interested in the sum of the points on the two dice, the possible outcomes may again be listed as

$$2, \quad 3, \quad 4, \quad 5, \quad 6, \quad 7, \quad 8, \quad 9, \quad 10, \quad 11, \quad 12,$$

because the sum of points on the two dice may vary from 2 to 12.

(e) In the random experiment of studying the distribution of boys (B) and girls (G in families having 3 children, there are 8 possible outcomes

$$BBB, \quad BBG, \quad BGB, \quad BGG, \quad GBB, \quad GBG, \quad GGB, \quad GGG$$

(the symbol *BGG*, for instance, denoting the outcome 'eldest child boy, 2nd and 3rd girls'); or, in terms of the number of boys, 4 possible outcomes.

3 *boys*, 2 *boys*, 1 *boy*, 0 *boy*

or, in terms of the number of girls, 4 possible outcomes

0 *girl*, 1 *girl*, 2 *girls*, 3 *girls*

(3) **Event:** In the theory of probability, the term event is used to denote any phenomenon which occurs in a random experiment. In effect, one or more outcomes are said to constitute an "event". Events may be 'elementary' or 'composite'. An event is said to be *elementary*, if it cannot be decomposed into simpler events. A *composite* event is an aggregate of several elementary events.

[**Note:** Readers will note that the words "outcome" and "event" may appear to have been used almost in the same sense. In the axiomatic theory of probability. (Section 11.13), 'outcome' is an undefined concept connected with a random experiment, just in the same way as 'point' is undefined to geometry. However, a slight distinction is made between the two words. The 'outcomes' are generally thought of as the ultimate results of a random experiment which cannot be split up further. Hence these are sometimes referred to as 'elementary outcomes.' Any "event" (i.e., phenomenon) connected with a random experiment is then defined as an aggregate (or 'set' in the mathematical sense) of certain specified outcomes. In particular, a set containing only a single outcome is called an *elementary event.*

For example, when a die is thrown, there are six possible outcomes 1, 2, 3, 4, 5, 6. The aggregate of three outcomes 1, 3, 5 may be said to form the event 'odd number', the aggregate of two outcomes 5, 6 may be called the event 'at least five', and the aggregate (if we use the word in a general sense) of just one outcome 6 may be called the event 'six']

Example 11.3

(a) When a coin is tossed, we may speak of the events '*Head*' and '*Tail*', each of which is an elementary event.
(b) When 2 coins are tossed, the event 'both heads, is an elementary event (*HH*). But 'one head and one tail' is a composite event consisting of the elementary events *HT* and *TH*.
(c) When a die is tossed, the event 'odd number of points' is composite, because it may be split up into elementary events 1, 3, 5.
(d) When two dice are thrown, the event 'total 8 points' is composite, consisting of the elementary events (2, 6), (3, 5), (4, 4), (5, 3) and (6, 2). But the event 'total 12' is elementary, consisting of the only outcome (6, 6), which cannot be decomposed.

Tree Diagram provides a systematic method of enumerating the elementary events of a random experiment. This diagram shows the outcomes at each stage of the experiment in an ordered form through the different 'branches' of the 'tree'.

Example 11.4 *For the possible distribution of boys and girls in families having 3 children, the tree diagram is shown below:*

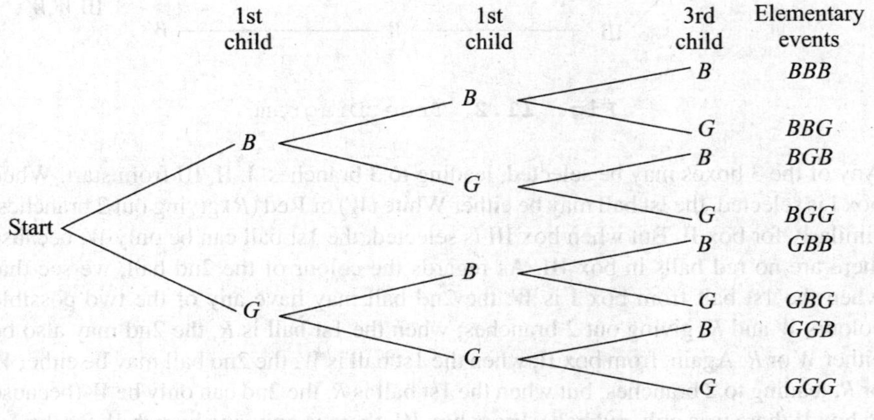

Fig. 11.1 Tree Diagram

The 1st child may have two possible outcomes, viz. Boy or Girl. Thus, two branches emerge from "Start", giving out *B* and *G*. Corresponding to *B* here, there are two possible outcomes *B* or *G* for the 2nd child, represented by the two branches emerging from the branch *B* of the 1st child. Similarly, from the branch *G* of the 1st child, there are two branches for the 2nd child, showing *B* and *G*. From each *B* and *G* for the 2nd child, 2 branches again spread out for the 3rd child *B* or *G*. When all different branches of the tree are followed from Start, the ordered results yield all possible elementary events.

Example 11.5 *There are 3 identical boxes:*

Box I contains 2 white and 5 red balls

Box II contains 4 white and 1 red balls

Box III contains 8 white balls only.

One box is selected at random; 2 balls are then successively taken out of it, and their colour noted. The elementary events arising out of the experiment, regard being had as to the number of the selected box and the colours of the balls drawn, can be represented by the following tree diagram:

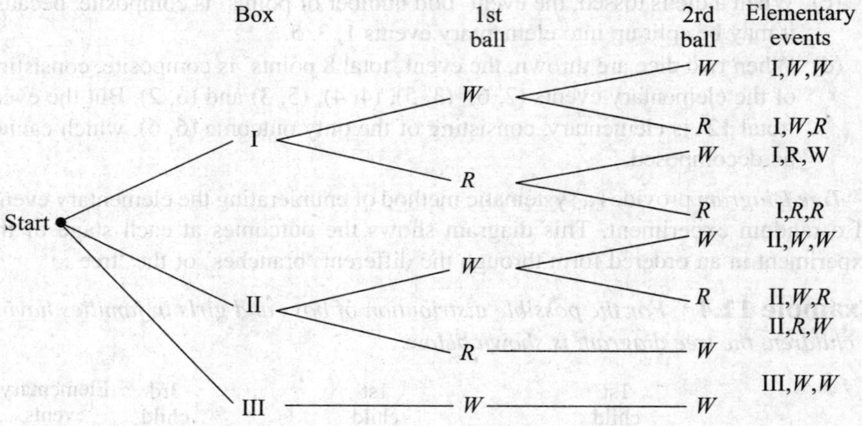

Fig. 11.2 Tree Diagram

Any of the 3 boxes may be selected, leading to 3 branches, I, II, III from start. When box I is selected, the Ist ball may be either White (*W*) or Red (*R*) giving out 2 branches; similarly for box II. But when box III is selected, the 1st ball can be only *W*, because there are no red balls in box III. As regards the colour of the 2nd ball, we see that when the 1st ball from box I is *W*, the 2nd ball may have any of the two possible colours *W* and *R*, giving out 2 branches; when the 1st ball is *R*, the 2nd may also be either *W* or *R*. Again, from box II, when the 1st ball is *W*, the 2nd ball may be either *W* or *R*, leading to 2 branches; but when the 1st ball is *R*, the 2nd can only be *W* (because in box II there was only red ball). From box III, there is only one branch *W* for the 1st ball, and the 2nd ball must also be *W*, leading to only one branch. The last column shows the complete set of elementary events.

11.3 IMPORTANT TERMINOLOGY

(1) **Mutually Exclusive:** Events are said to be 'mutually exclusive', when two or more of them can not occur simultaneously. This means that mutually exclusive events can occur only one at a time, and the occurrence of any event signifies impossibility of the remaining events in any particular performance of the random experiment.

[**Note:** Mutually exclusive events can always be connected by the words "either ... or". Events *A, B, C, D* are mutually exclusive, only if *either A or B or C or D* can occur.]

Example 11.6

(a) In tossing a coin, the elementary events 'Head' and 'Tail' are mutually exclusive. Because in any toss either Head occurs, or Tail occurs; and the occurrence of Head as well as Tail in any toss is impossible. The occurrence of any of them signifies non-occurrence of the other.

(b) When a die is thrown, let the symbols $E_1, E_2, E_3, E_4, E_5, E_6$ denote respectively the appearance of points 1, 2, 3, 4, 5, 6, on the face coming uppermost. These events are mutually exclusive, because the appearance of a particular face implies non-occurrence of any of the remaining faces. Only one of the faces can appear in any particular throw of the die. Again, let

A denote the event 'Odd number of points'

B denote the event 'Even number of points'

These events are mutually exclusive because the result of any throw shows either an odd number or an even number, but not both odd and even numbers simultaneously. (Note that $E_1, E_2, ..., E_6$ are elementary events, while *A* and *B* are composite events).

(c) When 2 coins are tossed, let

A_1 denote the event *HH*

A_2 denote the event *HT*

A_3 denote the event *TH*

A_4 denote the event *TT*

These 4 events are mutually exclusive because the result of any toss yields either A_1 or A_2 or A_3 or A_4. The result cannot be, for instance, *HH* as well as *TH*.

Again, in this experiment, let

B_1 denote the event 'Two heads'

B_2 denote the event 'One head'

B_3 denote the event 'No head'

Events B_1, B_2 and B_3 are mutually exclusive; because when 2 coins are tossed they show either both heads, or only one head (and one tail), or no head (i.e., both tails). Note that here B_1 and B_2 are elementary events, while B_2 is a composite event, composed of the elementary events *HT* and *TH*.

In the same experiment, let

C_1 denote the event 'Same result on the 2 coins'

C_2 denote the event 'Different results on the 2 coins'

The two events are mutually exclusive (Note that both C_1 and C_2 are composite events).

(2) **Exhaustive:** Several events are said to form an *exhaustive set*, if at least one of them must necessarily occur. The complete group of all possible elementary events of a random experiment gives an exhaustive set of events.

Example 11.7

(a) In Example 11.6(a), the two events 'Head' and 'Tail' form an exhaustive set, because one of these two must necessarily occur.

(b) In Example 11.6(b), the six events, E_1, E_2, E_3 ... E_6 form an exhaustive set. Similarly, the events A and B are also exhaustive.

(c) In Example 11.6(c), each of the groups of events (A_1, A_2, A_3, A_4), (B_1, B_2, B_3) and (C_1, C_2) forms an exhaustive set.

(d) Let us consider only 4 of the events E_2, E_3, E_4, E_6, as defined in Example 11.6 (b). These evens are mutually exclusive, because if one of them occurs, another cannot. However, the four events are not exhaustive, because the complete group of elementary events of the experiment includes some other events, viz. E_1 and E_5.

(e) When a die is thrown, let

A denote the event 'Odd number of points'

B denote the event 'At least 2 points'

These two events together form an exhaustive set; because the possible outcomes of this experiment can be described by one or the other of the events A and B. However, they are *not* mutually exclusive, because the occurrence of event A does not necessarily imply non-occurrence of B.

(3) **Trial:** Any particular performance of the random experiment is usually called a *trial*.

(4) **Cases Favourable to an Event:** Among all the possible outcomes of a random experiment, those cases which entail occurrence of an event A are called '*cases favourable to A*'.

Example 11.8

(a) When a die is thrown, these are six possible outcomes, viz 1, 2, 3, 4, 5, 6. Among these, 3 cases (viz. 1, 3, 5) are favourable to the event 'odd number of points', and 3 cases (viz. 2, 4, 6) are favourable to the event 'even number of points'.

(b) When 2 coins are tossed, out of the 4 possible outcomes *HH*, *HT*, *TH*, *TT*, there are 3 cases favourable to the event 'at least one head', (viz. *HH*, *HT*, *TH*), and 2 cases favourable to the event 'one head' (viz. *HT*, *TH*).

(c) When a card is drawn from a full pack, there are 52 possible outcomes; because any of the 52 cards may be drawn. Out of these only 4 cases are favourable to the event 'ace', viz. the cases when the card drawn is either an ace of spade, or an ace of heart, or of diamond, or of club. There are 13 cases favourable to the event 'spade' because the drawing of any of the 13 spade cards will entail drawing a spade. Similarly, there are 26 cases favourable to the event 'red card'.

(5) **Equally Likely:** The outcomes of a random experiment are said to be '*equally likely*', if after taking into consideration all relevant evidence, none of them can be expected in preference to another.

Example 11.9

(a) If a coin is unbiased (i.e., is perfectly homogenous and is uniform in both faces), there is no reason to expect that, for instance, heads will appear more often than tails, or vice versa. The two outcomes. Head and Tail are therefore equally likely.

(b) If a die is unbiased (i.e., is a perfect geometrical cube in shape and made of homogeneous material) there is no reason to suspect that in any throw of the die some particular face will come up more frequently than another face. The six outcomes showing points 1, 2, 3, 4,, 5, 6 are therefore considered 'equally likely.'

(c) If two unbiased coins are thrown, the elementary events *HH*, *HT*, *TH* and *TT* are equally likely. But the outcomes '2 heads' '1 head' and 'no head' are not equally likely.

(d) Out of a full pack of 52 cards, two cards may be drawn in $^{52}C_2 = \frac{52 \times 51}{1 \times 2}$ = 1326 ways, and the drawing of a pair of cards gives rise to an outcome. Since the drawing of 2 cards may reveal any of these 1326 possible combinations, the outcomes are equally likely.

11.4 TECHNIQUES OF COUNTING

Some mathematical methods are shown below, which are often helpful for determining without direct enumeration the number of outcomes of a random experiment or the number of cases favourable to an event. These are referred to as "*Combinatorial Methods*"

1. *Fundamental Principle of Counting*: If several processes can be performed in the following manner: the first process in *p* ways, the second in *q* ways, the third in *r* ways, and so on, then the total number of ways in which the whole process can be performed in the order indicated is given by the product.

$$p \,.\, q \,.\, r \ldots \qquad (11.4.1)$$

2. *Permutation*: The total number of ways of arranging (called *permutation*) *n* distinct objects taken *r* at a time is given by

$$^nP_r = n(n-1)\,(n-2)\ldots(n-r+1) \qquad (11.4.2)$$

3. *Arrangement in a Line or Circle*: The total number of ways in which *n* distinct objects can be arranged among themselves is

   (i) in a line $n\,! = 1\,.\,2\,.\,3 \ldots n$ (11.4.3)

   (ii) in a circle $(n-1)\,!$ (11.4.4)

4. *Permutation with Repetition*: The number of ways of arranging *n* objects, among which *p* are alike, *q* are alike, *r* are alike, etc. is

$$\frac{n\,!}{p\,!\,q\,!\,r\,!\ldots} \qquad (11.4.5)$$

5. *Combination*: The total number of possible groups (called *combination*) that can be formed by taking r objects out of n distinct objects is given by

$$^{n}C_{r} = \frac{n(n-1)(n-2)\ ...\ (n-r+1)}{r\,!}$$

$$= \frac{n\,!}{r\,!\,(n-r)\,!} \qquad (11.4.6)$$

6. *Combination (any number at a time)*: The total number of ways of forming groups by taking any number from n distinct objects is

$$^{n}C_{1} + {}^{n}C_{2} + {}^{n}C_{3} + \ldots + {}^{n}C_{n} = 2^{n} - 1 \qquad (11.4.7)$$

7. *Choosing Balls from an Urn*: The total number of ways of choosing a white balls and b black balls from an urn containing A white and B black balls is

$$^{A}C_{a}\,.\,{}^{B}C_{b} \qquad (11.4.8)$$

This may be extended to more than two categories of balls.

8. *Ordered Partitions (Distinct objects)*: The total number of ways of distributing n distinct objects into r compartments marked 1, 2, ..., r is

$$r^{n} \qquad (11.4.9)$$

The number of ways in which the n objects can be distributed so that the compartments contain respectively $n_1, n_2, \ldots, n_r$ object is

$$\frac{n\,!}{n_{1}\,!\,n_{2}\,!\ldots n_{r}\,!} \qquad (11.4.10)$$

9. *Ordered Partitions (Identical objects)*: The total number of ways of distributing n identical objects into r compartments marked 1, 2, ..., r is

$$^{n+r-1}C_{r-1} \qquad (11.4.11)$$

If none of the compartments should remain empty, the total number of ways of distributing the balls is

$$^{n-1}C_{r-1} \qquad (11.4.12)$$

10. *Sum of Points on the Dice*: When n dice are thrown, the number of ways of getting a total of r points is given by the

$$\text{Coefficient of } x^{r} \text{ in } (x + x^{2} + x^{3} + x^{4} + x^{5} + x^{6})^{n} \qquad (11.4.13)$$

11. *Dearrangements and Matches*: If n objects numbered 1, 2, 3, ..., n are distributed at random in n places also numbered 1, 2, 3, ..., n, a "*match*" is said to occur if an object occupies the place corresponding to its number. The number of permutations in which no match occurs is

$$t_{n} = n\,!\left\{1 - \frac{1}{1\,!} + \frac{1}{2\,!} - \frac{1}{3\,!} + \ldots + (-1)^{n}\frac{1}{n\,!}\right\} \qquad (11.4.14)$$

This is also known as "*derangement*".

The number of permutations of n objects in which exactly r matches occur is

$$^{n}C_{r}\; t_{n-r} = \frac{n\,!}{r\,!}\left\{1 - \frac{1}{1\,!} + \frac{1}{2\,!} - \frac{1}{3\,!} + \ldots + \frac{(-1)^{n-r}}{(n-r)\,!}\right\} \qquad (11.4.15)$$

11.5 CLASSICAL (OR 'A PRIORI') DEFINITION OF PROBABILITY

If a random experiment has n possible outcomes, which are *mutually exclusive, exhaustive* and *equally likely,* and m of these are favourable to an event A, then the 'probability' of the event is defined as the ratio m/n. In symbols

$$P(A) = \frac{m}{n} \tag{11.5.1}$$

$$\textit{Probability of an event} = \frac{\textit{Number of outcomes favourable to the event}}{\textit{Total number of mutually exclusive, exhaustive and equally likely outcomes of the random experiment}}$$

It may be noted that 'probability', as defined above, is only a ratio of two numbers, in which the numerator (m) is the number of favourable cases and the denominator (n) is the total number of possible outcomes satisfying certain conditions. Therefore, for the calculation of probability the undernoted steps should be followed:

(*Step 1*): Enumerate all the possible outcomes of the experiment, such that they satisfy the 3 criteria of being 'mutually exclusive', 'exhaustive', and 'equally likely'. Count the number (n) of such outcomes.

(*Step 2*): Check how many of these cases are favourable to the event for which the probability is desired. Let this number be m.

(*Step 3*): Divide m by n, and the result gives the probability of the event.

Probability, as defined by (11.5.1), always lies between 0 and 1.

$$0 \le p \le 1 \tag{11.5.2}$$

The minimum value of p, viz. 0, is attained when none of the outcomes is favourable to the event, i.e., $m = 0$. The event is then said to be '*impossible*'. The maximum value of p, viz. 1, is attained when all the possible outcomes are favourable to the event, i.e, $m = n$. The event is then said to be '*certain*'.

Defects of Classical Definition

(i) It is based on the feasibility of subdividing the possible outcomes of the experiment into 'mutually exclusive', 'exhaustive' and 'equally likely' cases. Unless this can be done, the formula is inapplicable.

(ii) The phrase 'equally likely' appearing in the classical definition, is synonymous with 'equally probable', which means that we are trying to define probability in terms of equal probabilities. How do you know whether the probabilities are equal, before you can measure them ? The definition is thus circular in nature.

(iii) The definition has only limited applications in coin-tossing, die-throwing and similar games of chance. Using this definition, we cannot, for example, find the probability that an Indian aged 25 will die before reaching the age 50 (such probabilities are required to be calculated for fixing the premium rates in life insurance). Thus, it may not be practically possible to enumerate all the outcomes of a random experiment.

(iv) The definition fails, when the number of possible outcomes is infinitely large.

Example 11.10 *What is the probability of obtaining 'Head' in a single toss of an unbiased coin ?*

Solution When a coin is tossed there are two possible outcomes viz. Head and Tail. These two outcomes are mutually exclusive and exhaustive. Moreover, since the coin is unbiased, the outcomes are also equally likely. Out of these two mutually exclusive, exhaustive and equally likely outcomes, only one case is favourable to the event 'Head'. Thus, using the classical definition we have

$$\text{Probability of obtaining 'Head'} = \frac{m}{n} = \frac{1}{2}$$

Example 11.11 *Two unbiased coins are tossed. What is the probability of obtaining (a) both heads, (b) one head and one tail, (c) both tails, (d) at least one head?*

Solution The experiment has 4 possible outcomes, viz. *HH, HT, TH, TT* (the two letters in each case denoting the results on the 1st and 2nd coins respectively). These are mutually exclusive, exhaustive and equally likely (Examples 11.6c, 11.7c, 11.9c). Thus $n = 4$. The cases favourable to the events are as follows:

Event	*Favourable cases*	*Number of favourable cases*
(a) *Both heads*	*HH*	1
(b) *One head and one tail*	*HT, TH*	2
(c) *Both tails*	*TT*	1
(d) *At least one head*	*HH, HT, TH*	3

Applying the classical definition (11.5.1)

(a) *P (both heads)* = 1/4

(b) *P (one head and one tail)* = 2/4 = 1/2

(c) *P (both tails)* = 1/4

(d) *P (both least one head)* = 3/4

Example 11.12 *A die is tossed and the number of points appearing on the uppermost face is observed. What is the probability of obtaining (a) an even number, (b) an odd number, (c) less than 3, (d) a "six" ?*

Solution In this experiment there are 6 possible outcomes, viz. 1, 2, 3, 4, 5, 6. These are mutually exclusive and exhaustive. If the die is assumed to be unbiased, then the 6 outcomes are also equally likely, Thus $n = 6$. Out of them.

3 cases (viz., 2, 4, 6) are favourable to 'even number'
3 cases (viz., 2, 3, 5) are favourable to 'odd number'
2 cases (viz., 1, 2) are favourable to 'less than 3'
1 case (viz., 6) is favourable to 'six'

Thus,

P(even number) = 3/6 = 1/2
P(odd number) = 3/6 = 1/2
P(less than 3) = 2/6 = 1/3
P(six) = 1/6

Example 11.13 *When two unbiased coins are tossed. what is the probability of obtaining (a) 3 heads, (b) not more than 3 heads?*

Solution There are 4 possible outcomes, viz, *HH, HT, TH, TT*. These are mutually exclusive, exhaustive and equally likely. Hence $n = 4$. Since none of these cases is favourable to the event '3 heads', $m = 0$. Therefore

$$P(3\ heads) = .0/4 = 0$$

(Note that in a toss of two coins, 3 heads are impossible).

Again, we see that all the four outcomes are favourable to the event 'not more than 3 heads' ($m = 4$). So

$$P(not\ more\ than\ 3\ heads) = 4/4 = 1$$

(Note that in a toss of 2 coins it is certain that we get not more than 3 heads; in fact, at most 2 heads may be obtained.)

Example 11.14 *Two coins are tossed. Find the probability of getting both heads or both tails.*

Solution Assuming that the coins are unbiased, there are 4 outcomes, viz, *HH, HT, TH, TT*, which are mutually exclusive, exhaustive and equally likely. Out of them, 2 cases, viz. *HH* and *TT*, are favourable to the event "both heads or both tails". Hence by the classical definition of probability.

$$p = 2/4 = 1/2$$

Example 11.15 *Two dice with faces marked 1, 2, 3, 4, 5, 6 are thrown simultaneously and the points on the dice are multiplied together. Find the probability that the product is 12.*

Solution If two dice are thrown, there are 36 possible outcomes (see Example 11.2(d), page 2), shown as pairs of numbers corresponding to the points on the 1st and the 2nd dice. These outcomes are mutually exclusive and exhaustive, and also equally likely on the assumption that the dice are unbiased. Of them only the following 4 cases are favourable to the event "product 12"—(2, 6), (3, 4), (4, 3), (6, 2). So, using the classical definition,

$$p = \frac{4}{36} = \frac{1}{9}.$$

Example 11.16 *A bag contains 6 white and 4 black balls. One ball is drawn. What is the probability that it is white ?*

Solution Let us put a serial number on each ball as follows. White balls: 1, 2, 3, 4, 5, 6 and Black balls: 7, 8, 9, 10. There are 10 outcomes as regards the number on the selected ball, because any of the 10 balls could be drawn. Since the balls are assumed to be identical except in colour, any of the balls is as likely to appear as any other ball. Out of these 10 possible outcomes, which are mutually exclusive, exhaustive and equally likely, only 6 cases (viz. when the number of the selected ball is 1, or 2, ... or 6) are favourable to the event "white ball". Hence,

$$p = \frac{6}{10} = \frac{3}{5}.$$

Note that here

$$\text{Probability} = \frac{\text{Number of white balls}}{\text{Total number of balls}}$$

Example 11.17 *If 2 balls are drawn one after another from a bag containing 3 white and 5 black balls, what is the probability that*

(i) *the fist ball is white and the 2nd is black;*

(ii) *one ball is white and the other is black ?* [D.M. '78]

Solution Let us put serial numbers on the balls as follows: Whites – 1, 2, 3; Black – 4, 5, 6, 7, 8. In order to find the total number of possible outcomes, we see that the 1st ball may be selected in 8 ways, because any of the 8 balls may be drawn. Corresponding to each way of drawing the 1st ball the 2nd ball may be drawn in 7 ways (because there remain only 7 balls in the bag after the 1st ball has been drawn). Hence the 2 balls may be drawn in $8 \times 7 = 56$ ways. Since the balls are identical in all respects except in colour, these 56 cases are mutually exclusive, exhaustive and equally likely.

(i) As regards the number of favourable cases, the 1st ball will be white only if any of the balls numbered 1, 2, 3 is drawn i.e., in 3 ways. When the 1st ball has been obtained in any of these 3 ways, the 2nd ball will be black only if any of the balls numbered 4, 5, 6, 7, 8 is drawn, i.e., in 5 ways. Hence the number of cases favourable to the event is $3 \times 5 = 15$. So,

$$p = \frac{15}{56}.$$

(ii) We have seen that the number of ways of drawing a white ball and a black ball in the order (white, black) is 15. Similarly, it can be shown that the number of ways of drawing a white ball and a black ball in the order (black, white) is $5 \times 3 = 15$. Hence the number of cases favourable to the event "one ball is white and the other black", irrespective of the order in which they are drawn, is $(15 + 15) = 30$. Therefore,

$$p = \frac{30}{56} = \frac{15}{28}$$

Example 11.18 *Two cards are drawn from a full pack of 52 cards. Find the probability that (i) both are red cards, (ii) one is a heart and the other is a diamond.*

Solution *(First method) (Ignoring the order of drawing).* Two cards may be drawn out of the pack of 52 cards in ${}^{52}C_2 = 1326$ ways. These outcomes are mutually exclusive, exhaustive and equally likely.

(i) The number of cases favourable to the event "both red cards" in ${}^{26}C_2 = 325$, because the pack contains only 26 red cards.

$$p = \frac{325}{1326} = \frac{25}{102}$$

(ii) In order to find the number of cases favourable to the event "1 heart and 1 diamond", we see that 1 heart can be drawn in 13 ways and similarly 1 diamond can be drawn in 13 ways (because each pack contains 13 cards of each suit: Spades, Hearts, Diamonds, Clubs). Since any of the heart card can be combined with any of diamond card to give a group of 1 heart and 1 diamond, the number of cases favourable to the event is $13 \times 13 = 169$.

$$p = \frac{169}{1326} = \frac{13}{102}$$

(Second method) (Considering the order of drawing) The 1st card may be drawn in 52 ways, and corresponding to each way of drawing the 1st card the 2nd card may be drawn in 51 ways. Hence the total number of cases, considering the order, is $52 \times 51 = 2652$.

(i) Since the 1st red card can be drawn in 26 ways, and thereafter the 2nd red card in 25 ways, the number of cases favourable to "two red cards" is 26 × 25 = 650.

$$p = \frac{650}{2652} = \frac{25}{102}$$

(ii) Here the number of cases favourable to the order (heart diamond) is 13 × 13 = 169; similarly the number of cases favourable to the order (diamond, heart) is also 169. Hence, the number of cases favourable to "one heart and one diamond" is (169 + 169) = 338

$$p = \frac{328}{2652} = \frac{13}{102} \text{ (as before)}$$

Example 11.19 *What is the probability that all 3 children in a family have different birthdays? (Assume, 1 year = 365 days).*

Solution The 1st child may be born on any of 365 days of the year; the 2nd also on any of the 365 days, and similarly the 3rd child. Hence, the total number of possible ways in which the 3 children have birthdays is 365 × 365 × 365. These cases are mutually exclusive, exhaustive and equally likely. As regards the number of favourable cases out of these, we note that the 1st child may have any of the 365 days of the year as its birthday. In order that the 2nd child has a birthday different from that of the 1st, it should have been born on any of the 364 remaining days of the year; similarly the 3rd should be born on any of the remaining 363 days. So, the number of cases favourable to the event "different birthdays" is 365 × 364 × 363.

$$p = \frac{365 \times 364 \times 363}{365 \times 365 \times 365} = 0.992$$

Example 11.20 *Five persons A, B, C, D, E occupy seats in a row at random. What is the probability that A and B sit next to each other?*

Solution Five persons can arrange themselves in a row, without restriction, in 5 ! = 1 × 2 × 3 × 4 × 5 = 120 ways. Considering *A* and *B* together, they can arrange themselves in 4 ! × 2 = (1 × 2 × 3 × 4) × 2 = 48 ways; because *A* may be to the left or to the right of *B*.

$$p = \frac{48}{120} = \frac{2}{5}$$

[**Note:** The phrase 'at random' denotes 'with equal probability'.]

Example 11.21 *A batch contains 10 articles of which 4 are defective. If 3 articles are chosen at random, what is the probability that none of them is defective?*

Solution Total number of ways of selecting 3 articles (without restriction) out of 10 is $^{10}C_3 = 120$. If none of the selected articles is defective, they must form a group out of the 6 non-defective articles. Hence the number of favourable cases is $^6C_3 = 20$.

$$p = \frac{20}{120} = \frac{1}{6}$$

Example 11.22 *10 distinguishable balls are distributed at random into 4 boxes. What is the probability that a specified box contains exactly 2 balls?*

Solution The first ball may go to any of the four boxes and hence may be distributed in 4 ways. The 2nd ball may also be distributed in 4 ways, the 3rd ball in 4 ways, ..., the 10th ball in 4 ways. Hence the total number of ways in which the 10 balls may be distributed into the 4 boxes is $4 \times 4 \times 4$ ... (10 times) = 4^{10} (see 11.4.9). In order to find the number of favourable cases, we see that the specified box may receive any of the ${}^{10}C_2$ groups of 2 balls out of 10. When two such balls have gone into that box the remaining 8 balls may be distributed in any manner into the remaining 3 boxes in 3^8 ways. Hence the total number of favourable cases is ${}^{10}C_2 \times 3^8 = 45 \times 3^8$. Therefore,

$$p = 45 \times \frac{3^8}{4^{10}}$$

Example 11.23 *If 10 persons are arranged at random (i) in a line (ii) in a ring, find the probability that 2 particular persons will be next to each other.*

Solution

(i) Ten persons can be arranged in a line in 10 ! ways (see formula 11.4.3), which are mutually exclusive, exhaustive and equally likely. To find the number of favourable cases, i.e., when the two particular persons remain together, we consider the two as one (supporting they are fastened together), so that the number of permutations is 9 !. But the two persons can arrange themselves in 2 ! ways without affecting the position of any of the remaining persons. Hence the total number of favourable cases is 9 ! 2 !. Using the classical definition,

$$p_1 = \frac{9\,!\,2\,!}{10\,!} = \frac{1}{5}$$

(ii) Ten persons can arrange themselves in a *ring* in 9 ! ways (see formula 11.4.4). Arguing as before, the number of favourable cases is 8 ! 2 !. Hence

$$p_2 = \frac{8\,!\,2\,!}{9\,!} = \frac{2}{9}$$

Example 11.24 *X and Y stand in a line at random with 10 other people. What is the probability that there are 3 people between X and Y?*

(C.U., B.A., (Econ) '66; B.Sc. (Math.) 70; B.U., B.A., (Econ) '67

Solution *(First method)* There are 12 people including *X* and *Y* and they can arrange themselves in n = 12 ! ways. These n arrangements are mutually exclusive, exhaustive and equally likely. To find the number of arrangements (m) favourable to the event (viz. 3 people standing between *X* and *Y*), let us first consider the case when *X* occupies place 1 and *Y* occupies place 5 (there are 3 people in between), so that the remaining 10 places can be filled up in 10 ! ways.

Place	1	2	3	4	5	6	7	8	9	10	11	12
	X				Y							

If *X* and *Y* interchange their positions, we have another set of 10! arrangements. Thus, with *X* and *Y* in places (1, 5) the total number of arrangements favourable to the event is (2×10 !).

In fact, with *X* and *Y* occupying any two specified places, the number of arrangement is (2×10)!. Among the total number of n arrangements, only those cases will be favourable to the event when *X* and *Y* occupy the following 8 pairs of places:

(1, 5), (2, 6), (3, 7), (4, 8), (5, 9), (6, 10), (7, 11), (8, 12)

Hence, the total number of arrangements favourable to the event is $m = 8(2 \times 10\,!)$. By the classical definition of probability (11.5.1),

$$p = \frac{m}{n} = \frac{8(2\times 10\,!)}{12\,!} = \frac{8\times 2}{12\times 11} = \frac{4}{33}$$

(*Second method*) Of the 12 places, *X* and *Y* could occupy any 2 places in ${}^{12}C_2 = 66$ ways, which are mutually exclusive, exhaustive and equally likely. There will be 3 persons between *X* and *Y*, if the latter occupy places (1, 5), (2, 6), (3, 7), (4, 8), (5, 9), (6, 10), (7, 11), (8, 12). Hence the number of favourable cases is 8. So, the required probability = 8/66 = 4/33.

Example 11.25 *Twelve persons, amongst whom are X and Y, are seated at random at a round table, What is the probability that there are 3 persons between X and Y?*

Solution *X* can occupy any of the 12 seats, and correspondingly *Y* any of the remaining 11 seats. Hence *X* and *Y* can occupy seats in $12 \times 11 = 132$ ways. There will be 3 persons in between if *X* and *Y* occupy seats (1, 5), (2, 6), (3, 7), (4, 8), (5, 9), (6, 10), (7, 11), (8, 12), (9, 1), (10, 2), (11, 3), (12, 4). Since *X* and *Y* may interchange their seats, the total number of favourable cases is $12 \times 2 = 24$.

$$\text{Required probability} = \frac{24}{132} = \frac{2}{11}$$

(Note that the permutation of *n* persons in a *ring* or *circle* is $(n-1)\,!$ but around a '*round table*' is $n\,!$; because in the former the positions are relative to their neighbours, but in the latter relative to their seat numbers.

Example 11.26 *A lady declares that by tasting a cup of tea with milk, she can discriminate whether milk or tea infusion was first added to the cup. In order to test these assertion, 10 cups of tea are prepared—5 in one way and 5 in the other, and presented to the lady for judgement in a random order. Assuming that the lady has no discrimination power, calculate the probability that she would judge correctly all the cups, it being known to her that 5 are of each kind.*

What is the probability, if the tea cups were presented to the lady in 5 pairs—each pair consisting of cups of each kind—in a random order?

Solution Since the 5 cups of each kind, prepared with milk (*M*) or tea (*T*) infusion first added, are identical, the total number of permutations of the 10 cups is (see formula 11.4.5).

$$\frac{10\,!}{5\,!\,5\,!} = 252.$$

Thus, there are 252 different possible ways of presenting the cups to the lady, and these are mutually exclusive, exhaustive and equally likely. Only one of these agrees with the lady's assertion, suppose *M T T T M M M T M T M*. So the required probability is 1/252.

In the second case, when the cups are presented in 5 pairs, the total number of permutations is

$$2 \times 2 \times 2 \times 2 \times 2 = 32$$

because each pair can be permuted in 2 ways—either (*M*, *T*) or (*T*, *M*). So there are now 32 possible ways of presenting the cups to the lady, As before, only one of these agrees with the lady's statement, say (*T M*), (*M T*), (*M T*), (*T M*), and (*M T*). The required probability is therefore 1/32. *Ans.* 1/252, 1/32.

Example 11.27 *A box contains twenty tickets of identical appearance, the tickets being numbered 1, 2, 3, ..., 20. If 3 tickets are chosen at random, find the probability that the numbers on the drawn tickets are in arithmetic progression.*

[C.U., B.Sc. '71]

Solution Three tickets can be drawn out of 20 in ${}^{20}C_3$ ways. The total number of possible outcomes, which are mutually exclusive, exhaustive and equally likely, is therefore

$$n = {}^{20}C_3 \ \frac{20 \times 19 \times 18}{1 \times 2 \times 3} = 1140$$

The 3 numbers on the drawn tickets will be in A.P., if they have a common difference of either 1, or 2, or 3, ... or at most 9.

With a common difference 1, there are 18 sets, viz.

(1, 2, 3), (2, 3, 4), (3, 4, 5), ... (18, 19, 20),

With a common difference 2, there are 16 sets, viz.

(1, 3, 5), (2, 4, 6), (3, 5, 7), ... (16, 18, 20),

With a common difference 3, there are 14 sets, viz.

(1, 4, 7), (2, 5, 8), (3, 6, 9), ... (14, 17, 20),

and so on. Proceeding this way, finally

With a common difference 9, there are only 2 sets, viz.

(1, 10, 19), (2, 11, 20).

Therefore, the total number of sets of 3 numbers in A.P., whatever be the common difference, i.e., number of favourable cases, is

$$m = 18 + 16 + 14 + ... + 2 = 90$$

Therefore, the required probability is

$$p = \frac{m}{n} = \frac{90}{1140} = \frac{3}{38}$$

Example 11.28 *Four cards are drawn at random from a full pack. What is the probability that they belong to (i) 4 different suits, (ii) different suits and denominations ?*

Solution The 4 cards may be assumed to have been drawn one by one without replacement. The first card can be drawn in 52 ways, the second card in 51 ways, the third in 50 and the fourth in 49 ways. So, the total number of ways in which the cards can be drawn (attention being paid to the order) is 52 × 51 × 50 × 49. (formula 11.4.1)

(i) In order to find the number of cases favourable to the event 'one card drawn from each suit', we see that the first drawn card may belong to any suit, and hence may be chosen in 52 ways. There now remain 51 cards of which 12 belong to the same suit as the first card and 39 belong to other suits. Since the 4 cards are to belong to different suits, the second card should come from the 39. Thus the second card may be drawn in 39 ways; similarly the third card in 26 ways and the fourth card in 13 ways. The number of favourable cases is therefore 52 × 39 × 26 × 13. The required probability is

$$\frac{52 \times 39 \times 26 \times 13}{52 \times 51 \times 50 \times 49} = \frac{2197}{20825}$$ (see Example 11.50)

(ii) In order to calculate the number of favourable cases, the first card can, as before, be drawn in 52 ways. The second card must not belong to the same suit and denomination as the first and can be drawn in 36 ways (because in the remaining 51 cards, 39 belong to other suits of which 3 are of the same denominating as the first and should be excluded). Similarly, the third card may be drawn in 22 ways and the fourth in 10 ways. The number of favourable cases is then 52 × 36 × 22 × 10. The required probability is

$$\frac{52 \times 36 \times 22 \times 10}{52 \times 51 \times 50 \times 49} = \frac{264}{4165}$$

Example 11.29 *A group of 2n boys and 2n girls is divided at random into two equal batches. Find the probability that each batch will be equally divided into boys and girls.*

Solution The group of $4n$ boys and girls will be divided into two equal batches, if $2n$ out of them are selected to form one batch (the remaining $2n$ will form another batch). This selection can be done in ${}^{6n}C_{2n}$ ways. In order that each batch consists of equal numbers of boys and girls, the first batch of $2n$ selected persons should contain n boys and n girls. So, the number of favourable cases is ${}^{2n}C_n \cdot {}^{2n}C_n = ({}^{2n}C_n)^2$. Hence, the required probability is

$$({}^{2n}C_n)^2/({}^{4n}C_{2n}).$$

Example 11.30 *From a pack of 52 cards, an even number of cards is drawn. Show that the probability that these consist half of red and half black is*

$$\frac{\left\{\dfrac{52\,!}{(26\,!)^2} - 1\right\}}{(2^{51} - 1)}$$

Solution An even number of cards drawn may either be 2. or 4, or 6, ... or 52. From the full pack of 52 cards, 2 cards can be drawn in ${}^{52}C_2$ ways, 4 cards in ${}^{52}C_4$ ways, and so on. So, the total number of ways of drawing as even number of cards is

$${}^{52}C_2 + {}^{52}C_4 + {}^{52}C_6 + \ldots + {}^{52}C_{52} = (2^{51} - 1)$$

Assuming that all the different ways of drawing an even number of cards are equally probable irrespective of their number, these cases are mutually exclusive, exhaustive and equally likely.

When 2 cards are drawn, the number of ways of getting 1 red and 1 black cards is (see formula 11.4.8) ${}^{26}C_1 \times {}^{26}C_1 = ({}^{26}C_1)^2$. When 4 cards are drawn, the number of ways of getting 2 red and 2 black cards is similarly ${}^{26}C_2 \times {}^{26}C_2 = ({}^{26}C_2)^2$; and so on. Hence the number of favourable cases is

$$({}^{26}C_1)^2 + ({}^{26}C_2)^2 + ({}^{26}C_3)^2 + \ldots + ({}^{26}C_{26})^2$$

$$= {}^{52}C_{26} - 1 = \frac{52\,!}{26\,!\,26\,!} - 1$$

Using the classical definition of probability, the result follows.

[**Note:** Binomial coefficients satisfy the following relations:

(a) ${}^nC_0 + {}^nC_1 + {}^nC_2 + \ldots + {}^nC_n = 2^n$

(b) ${}^nC_0 + {}^nC_2 + {}^nC_4 + \ldots = {}^nC_1 + {}^nC_3 + {}^nC_5 + \ldots = 2^{n-1}$

(c) $({}^nC_0)^2 + ({}^nC_1)^2 + ({}^nC_2)^2 + \ldots + ({}^nC_n)^2 = {}^{2n}C_n$]

Example 11.31 *10 dissimilar balls are distributed at random into 4 boxes marked A, B, C, D. Find the probability that these boxes contain respectively 2, 4, 4, 0 balls.*

Solution The total number of possible ways of distribution is 4^{10} The number of ways of distributing the balls so that Boxes A, B, C, D contain respectively 2, 4, 4, 0 balls is (see formulae 11.4.9 and 11.4.10)

$$\frac{10\,!}{2\,!\,4\,!\,4\,!\,0\,!} = 3150$$

Hence, the required probability is $3150/4^{10}$.

Example 11.32 *10 identical balls are distributed at random into 4 boxes marked A, B, C, D. Find the probability that these boxes contain respectively 2, 4, 4, 0 balls.*

Solution The total number of possible ways of distribution is (see formula 11.4.11)

$$^{10+4-1}C_{4-1} = {}^{13}C_3 = 286$$

Since the given distribution of balls (namely 2, 4, 4, 0 in that order) is only a particular case of these 286 possible cases, which are mutually exclusive, exhaustive and equally likely, the number of favourable cases is 1. Hence the required probability is

$$p = \frac{1}{286}$$

Example 11.32A *15 identical objects are distributed at random into 4 boxes numbered 1, 2, 3, 4. Find the probability that (i) each box contains at least 2 objects, (ii) no box is empty.*

Solution 15 identical objects may be distributed in 4 boxes in (see formula 11.4.11)

$$^{15+4-1}C_{4-1} = {}^{18}C_3 = 816$$

ways. These $n = 816$ cases are mutually exclusive, exhaustive and equally likely.

(i) If each box is to contain at least 2 objects, we first place 2 objects in each box. The remaining $15 - 8 = 7$ objects may then be distributed in any manner among the 4 boxes in

$$^{7+4-1}C_{4-1} = {}^{10}C_3 = 120$$

ways. Thus, the number of cases favourable to the event "at least two objects in each box" is $m = 120$. Hence, by the classical definition, the required probability is

$$p_1 = \frac{120}{816} = \frac{5}{34}$$

(ii) No box is empty, if each box contains at least 1 object. As before, the number of ways of distributing the 15 objects so that each box contains at least 1 object is obtained by first placing 1 object in each box, and then considering the number of ways of distributing the remaining $15 - 4 = 11$ objects in any manner. The number of favourable cases is therefore, by formula (11.4.11), given by

$$^{11+4-1}C_{4-1} = {}^{14}C_3 = 364.$$

i.e., $m = 364$. Note that we may also apply formula (11.4.12) and obtain

$$^{15-1}C_{4-1} = {}^{14}C_3 = 364$$

Hence, the required probability is

$$p_2 = \frac{384}{816} = \frac{91}{204}$$

Ans. (i) 5/34, (ii) 91/204.

Example 11.33 *What is the probability of obtaining a sum of 10 in a single throw with 5 dice?*

Solution When n dice are thrown, there are 6^n possible outcomes, among which the number of outcomes giving a total of r spots, is given be (see 11.4.13) the coefficient of x^r in the expansion of

$$(x + x^2 + x^3 + x^4 + x^5 + x^6)^n = \frac{x^n(1 - x^6)^n}{(1 - x)^n} = x^n (1 - x^6)^n (1 - x)^{-n}$$

i.e. the coefficient of x^{r-n} in the expansion of $(1 - x^6)^n (1 - x)^{-n}$.

In the present case, $r = 10$, $n = 5$. Therefore, the number of cases favourable to a total of 10 spots on the 5 dice is the

Coefficient of x^{10-5} in the expansion of $(1 - x^6)^5 (1 - x)^{-5}$

$$= \text{Coefficient of } x^5 \text{ in } (1 - 5x^6 + ...)(1 + 5x + ... + {}^{4+t}C_4 x^t + ...)$$

$$= {}^9C_4 = \frac{9 \times 8 \times 7 \times 6}{1 \times 2 \times 3 \times 4} = 126$$

Therefore, the required probability $= \dfrac{126}{6^5} = \dfrac{126}{7776} = \dfrac{7}{432}$

Example 11.34 *Five different objects numbered 1, 2, 3, 4, 5 are placed at random into 5 places also marked 1, 2, 3, 4, 5. What is the probability that (i) no object occupies the place corresponding to its number, (ii) exactly 2 objects are in their correct places ?*

Solution 5 objects can be distributed in 5 places in 5 ! = 120 ways (see formula 11.4.3). These are mutually exclusive, exhaustive and equally likely.

(i) The number of arrangements in which no match occurs is (formula 11.4.14)

$$t_5 = 5\,!\left\{1 - \frac{1}{1!} + \frac{1}{2\,!} + \frac{1}{3\,!} + \frac{1}{4\,!} - \frac{1}{5\,!}\right\}$$

$$= 120\left(1 - 1 + \frac{1}{2} - \frac{1}{6} + \frac{1}{24} - \frac{1}{120}\right)$$

$$= 120 \times \frac{11}{30} = 44$$

Hence, the probability of no match is 44/120 = 11/30.

(ii) The two objects occupying the correct places may be any one of ${}^5C_2 = 10$ groups. Corresponding to any of these, the number of arrangements in which the remaining 3 objects are not in their proper places is

$$t_3 = 3\,!\left\{1 - \frac{1}{1!} + \frac{1}{2\,!} - \frac{1}{3\,!}\right\}$$

$$= 6\left(\frac{1}{2} - \frac{1}{6}\right) = 2$$

Hence, the number of favourable cases is $10 \times 2 = 20$. The required probability of exactly 2 matches is therefore 20/120 = 1/6.

$P(A)$	denotes '*probability* of occurrence of *event A*'.
$\bar{A}$	denotes '*non-occurrence of event A*' (or, 'event *complementary to A*', i.e. *opposite A*)
$P(\bar{A})$	denotes '*probability* that event *A fails to occur*'.
$P(A + B)$ or $P(A \cup B)$	denotes 'probability of occurrence of *at least one* of the events *A* and *B*', (i.e., either *A*, or *B*, or both). In particular, if *A* and *B* are mutually exclusive, both *A* and *B* cannot occur simultaneouly, and then *P(A + B)* signifies the probability of occurrence of *either A or B*'.

P(AB) or $P(A \cap B)$ denotes 'probability of occurrence of *A as well as B*', or 'probability of the *compound event AB*'. This is also called *Compound Probability.*

P(A/B) denotes 'probability of *occurrence of A, assuming that B has actually occurred'*.

P(A/B) is called *Conditional Probability*. In contrast, P(A) may be called *Unconditional Probability*, which means the probability of occurrence of *A*, irrespective of whether *B* has occurred or not.

11.6 THEOREMS OF PROBABILITY

(I) Theorem of Total Probability

If two events *A* and *B* are mutually exclusive, then the probability of occurrence of *either A* or *B* is given by the sum of their probabilities, i.e.,

Probability of (*A* or *B*) = Probability of *A* + Probability of *B*.

In symbols,

$$P(A + B) = P(A) + P(B) \tag{11.6.1}$$

This is also known as *Addition Theorem.*

Proof

Let us suppose that a random experiment has *n* possible outcomes, which are mutually exclusive, exhaustive and equally likely. If m_1 of these cases are favourable to the event. *A*, and m_2 cases are favourable to the event *B*, then the probabilities of these events are, by the classical definition,

$$P(A) = \frac{m_1}{n}, \quad P(B) = \frac{m_2}{n}$$

However, since the events *A* and *B* are mutually exclusive (i.e., both of them cannot occur simultaneously), the m_1 cases favourale to *A* are completely distinct from the m_2 cases favourable to *B*. The number of cases favourbale to 'either *A* or *B*' is therefore $(m_1 + m_2)$

$$\therefore \quad P(A + B) = \frac{m_1 + m_2}{n}$$

But, $$\frac{m_1 + m_2}{n} = \frac{m_1}{n} + \frac{m_2}{n} = P(A) + P(B)$$

Hence, $$P(A + B) = P(A) + P(B) \quad \text{(Proved)}$$

Deductions from Theorem of Total Probability:

1. Theorem of Total Probability can be extended to any number of mutually exclusive events. If events $A_1, A_2, ..., A_k$ are mutually exclusive, then the probability of occurrence of any of them is given by the sum of the probabilities of the events.

   $$P(A_1 + A_2 + ... + A_k) = P(A_1) + P(A_2) + ... + P(A_k) \tag{11.6.2}$$

   In particular, for three mutualy exclusive events *A, B, C*, the probability of occurrence of either *A* or *B* or *C* is

   $$P(A + B + C) = P(A) + P(B) + P(C) \tag{11.6.3}$$

2. The probability of the event complementary to *A* is given by

   $$P(\bar{A}) = 1 - P(A) \tag{11.6.4}$$

or $$P(A) = 1 - P(\bar{A}) \tag{11.6.5}$$

Because, the event A and the complementary event $\bar{A}$ are mutually exclusive. So, using (11.6.1),

$$P(A + \bar{A}) = P(A) + P(\bar{A})$$

Also, the events A and $\bar{A}$ are exhaustive, so that one of them must necessarily occur; i.e., the event that any of them occurs is a certainty.

$$\therefore \quad P(A + \bar{A}) = 1$$

Eqating the right hand sides of the above two relations,

$$P(A) + P(\bar{A}) = 1$$

Hence the results.

Formula (11.6.4) gives us a method of finding the probability of the complementary event. Again, in many problems it is found comparatively easier to calculate the probability of complementary event by direct methods, and then the probability of the event is given by (11.6.5); see Examples 11.41, 11. 48 (ii), 11.53 (iii), 11.54.

3. The probability of occurrence of at least one of the 2 events A and B (which may not be mutually exclusive), is given by

$$P(A + B) = P(A) + P(B) - P(AB) \tag{11.6.6}$$

Because, the event $(A + B)$ means the occurrence of one of the following mutually exclusive events: AB, $A\bar{B}$, $\bar{A}B$. Therefore, by the extended Theorem of Total Probability (11.6.2)

$$P(A + B) = P(AB + A\bar{B} + \bar{A}B)$$

$$= P(AB) + (A\bar{B}) + P(\bar{A}B)$$

Again, coccurrence of either AB or $A\bar{B}$ signifies occurrence of A. Hence by (11.6.1),

$$P(A) = P(AB) + P(A\bar{B}); \text{ i.e.,}$$

$$P(A\bar{B}) = P(A) - P(AB)$$

Similarly, $$P(\bar{A}B) = P(B) - P(AB)$$

Using these relations,

$$P(A + B) = P(AB) + \{P(A) - P(AB)\} + \{P(B) - P(AB)\}$$

$$= P(A) + P(B) - P(AB)$$

The probability of occurrence of at least one of the 3 events A, B, C (which may not be mutually exclusive), is

$$P(A + B + C) = P(A) + P(B) + P(C) - P(AB) - P(AC) - P(BC) + P(ABC) \tag{11.6.7}$$

4. Boole's inequality

$$P(A + B) \le P(A) + P(B) \tag{11.6.8}$$

This follows from (11.6.6). Since the probability of an event can never be negative, the minimum value of $P(AB)$ is zero. So, the R.H.S. of (11.6.6) may have the maximum value $P(A) + P(B)$. Hence the result. The equality sign holds, when $P(AB) = 0$ i.e., A and B are mutually exclusive events.

5. Boneferroni's inequality:

$$P(A\,B) \ge P(A) + P(B) - 1 \tag{11.6.9}$$

This also follows from (11.6.6). Since the probability of an event can never exceed 1, the maximum value of $P(A + B)$ is 1; i.e.,

$$1 \geq P(A) + P(B) - P(AB)$$

Transposing, the result folows.

(II) Theorem of Compound Probability

The probability of occurrence of the event *A as well as B* is given by the product of (unconditional) probability of *A* and conditional probability of *B*, assuming that *A* has actualy occurred. i.e.,

Probability of (*A and B*) = Probability of $A \times$ Conditional probability of *B*, assuming *A*.

In symbols, $$P(AB) = P(A) \cdot P(B/A) \qquad (11.6.10)$$

This is also known as *Multiplication Theorem.*

Proof

Suppose that a random experiment has *n* mutually exclusive, exhaustive and equally likely outcomes, among which *m* cases are favourbale to an event *A*. So, the unconditinal probability of *A* is

$$P(A) = \frac{m}{n}$$

Out of these *m* cases, let m_1 cases be favourable to anotehr event *B* also, i.e., the number of cases favourable to *A* as well as *B* is m_1. Hence, by the 'classical definition of probability.

$$P(AB) = \frac{m_1}{n}$$

It may be noticed that once event *A* is known to have actually occurred, the occurrence of *B* as well is limited to only m_1 cases out of *m* (in which *A* occurs). So, the conditional probability of *B*, assuming that *A* has already occurred, is

$$P(B/A) = \frac{m_1}{m}$$

We find that $$\frac{m_1}{n} = \frac{m}{n} \cdot \frac{m_1}{m}$$

i.e., $$P(AB) = P(A) \,.\, P(B/A)$$

Theorem of Compund Probability can be extended to several events. For example, the probability of occurrence of the event *A as well as B as well as C* is given by

$$P(ABC) = P(A) \,.\, P(B/A) \,.\, P(C/AB) \qquad (11.6.10a)$$

where $P(C/AB)$ denotes the conditional probability of event *C*, assuming that both *A* and *B* have actually occurred. Similarly for four events *A, B, C, D,*

$$P(ABCD) = P(A).P(B/A).P(C/AB).P(D/ABC) \qquad (11.6.10b)$$

Independent Events

Several events are considered to be "*independent*" in the probability sense, or *statistically independent,* if the probability of occurrence of any of them remains unaffected by the supplementary knowledge regarding occurrence or non-occurrence of any number of the remaining events.

If 2 dice are tossed, the event (*A*) 'six' on Die I and (*B*) 'six' on Die II, are independent; because the probability of obtaining a six suppose, on Die II, which is $\frac{1}{6}$ does not change, if it be known that Die I has resulted in a six or not. This means that the unconditional probability of *B* is the same as conditional probabilities, given *A* or $\bar{A}$.

$$P(B) = P(B/A) = P(B/\bar{A}) \quad (11.6.11)$$

Similarly, if 2 balls are drawn *with replacement* from a box containing 4 white and 6 black balls, the events (*A*) 'white ball is 1st draw' and (*B*) 'white ball in 2nd draw' are independent. However, if the drawing is *without replacement, A* and *B* will be *dependent events;* because $P(B/A) = \frac{1}{3}$ $P(B/\bar{A}) = \frac{4}{9}$ and they are not equal.

Deductions from Theroem of Compound Probability

(1) If events *A* and *B* are *independent,* the probability of occurrence of *A* as well as *B* is given by the product of their probabilities.

$$P(AB) = P(A).P(B) \quad (11.6.12)$$

Because, in this case *P*(*B*/*A*) is unaffected by the supplementary knowledge regarding the occurrence or non-occurrence of *A*, so that *P*(*B*/*A*) = *P*(*B*). Hence, from (11.6.10) the result follows.

Relation (11.6.12) is also taken as a suficient condition for independence of two events. If *A* and *B* are independent, (11.6.12) holds. Conversely, for any two events *A* and *B*, if (11.6.12) holds, the events are said to be 'independent'.

Two events A and B are said to be "independent", if the probability of occurrence of A as well as B is given by the product of the probability of A and the probability of B. In symbols,

$$P(AB) = P(A).P(B)$$

For three independent events, *A, B, C*

$$P(ABC) = P(A).P(B).P(C) \quad (11.6.13)$$

The result may be extended to any number of events. If events $A_1, A_2, ..., A_k$ are independent, then the probability that all of them occur simultaneusly, is

$$P(A_1A_2 ... A_k) = P(A_1) . P(A_2) ... P(A_k) \quad (11.6.14)$$

(2) The occurrence of an event *B* may be associated with the occurrence or with the non-occurrence of another event *A*. This means that event *B* may be assumed to be composed of two mutually exclusive compound events *AB* and $\bar{A}B$. (note that in either case *B* occurs). Hence, by the theorem of total probability

$$P(B) = P(AB) + P(\bar{A}B)$$

Again by the theorem of compound probability,

$$P(AB) = P(A).P(B/A)$$

and $$P(\bar{A}B) = P(\bar{A}).P(B/\bar{A}).$$ Hence

$$P(B) = P(A).P(B/A) + P(\bar{A}).P(B/\bar{A}) \quad (11.6.15)$$

(3) The conditional probability of an event *B*, on the assumption that another event *A* has actually occurred, is given by

$$P(B/A) = \frac{P(AB)}{P(A)} \quad (11.6.16)$$

This follows from the theorem of compound probability. Similarly,

$$P(A/B) = \frac{P(AB)}{P(B)} \tag{11.6.17}$$

Thus the Theorem of Compound Probability enables us to find a formula for the calculation of conditional probability.

Important Formulae and Results

1. *Addition Theorem of Probability:* If events A and B are 'mutually exclusive', then
$$P(A + B) = P(A) + P(B)$$
2. In general, for any two events A and B,
$$P(A + B) = P(A) + P(B) - P(AB)$$
3. For three mutually exclusive events A, B, C
$$P(A + B + C) = P(A) + P(B) + P(C)$$
This can be extended to any number of mutually exclusive events.
4. For any three events A, B, C
$$P(A + B + C) = P(A) + P(B) + P(C) - P(AB) - P(AC) - P(BC) + P(ABC)$$
5. Probability of the complementary event:
$$P(\bar{A}) = 1 - P(A)$$
6. Boole's inequality:
$$P(A + B) \le P(A) + P(B)$$
The sign of equality holds, when A and B are mutually exclusive.
7. *Multiplication Theorem of Probability:* For any two events A and B,
$$P(AB) = P(A) \,.\, P(B/A) = P(B).P(A/B)$$
8. If events A and B are 'independent', then
$$P(AB) = P(A) \,.\, P(B)$$
Conversely, if this relation holds, then the events A and B are said to be "independent".
9. For any three events A, B, C
$$P(ABC) = P(A).P(B/A).P(C/AB)$$
10. For three *independent* events A, B, C,
$$P(ABC) = P(A).P(B).P(C)$$
This can be extended to any number of independent events.
11. If evetns A and B are independent, then
$$P(A) = P(A/B) = P(A/\bar{B})$$
$$P(B) = P(B/A) = P(B/\bar{A})$$
This is the conditional and unconditional probability are equal.
12. $$P(A + B) = 1 - P(\bar{A}\,\bar{B}) = 1 - P(\bar{A}).P(\bar{B}),$$
if A and B are independent.

13. $$P(B) = P(AB) + P(\bar{A}B) = P(A).P(B/A) + P(\bar{A}).P(B/\bar{A})$$

14. *Conditional Probability:*

$$P(B/A) = \frac{P(AB)}{P(A)}; \text{ provided } P(A) \neq 0.$$

$$P(A/B) = \frac{P(AB)}{P(B)}; \text{ provided } P(B) \neq 0.$$

Example 11.35 *Given* $P(A) = \frac{1}{2}$, $P(B) = \frac{1}{3}$, $P(AB) = \frac{1}{4}$,

(a) *find the values of the following probabilities:*

$P(\bar{A})$, $P(A + B)$, $P(A/B)$, $P(\bar{A}B)$, $P(\bar{A}\,\bar{B})$, $P(\bar{A} + B)$

(b) *State whether the events A and B are (i) mutually exclusive, (ii) exhaustive, (iii) equally likely, (iv) independent.*

Solution

(a) $$P(\bar{A}) = 1 - P(A) = 1 - \frac{1}{2} = \frac{1}{2}$$

$$P(A + B) = P(A) + P(B) - P(AB) = \frac{1}{2} + \frac{1}{3} - \frac{1}{4} = \frac{7}{12}$$

$$P(A/B) = \frac{P(AB)}{P(B)} = \frac{\frac{1}{4}}{\frac{1}{3}} = \frac{3}{4}$$

$$P(\bar{A}B) = P(B) - P(AB) = \frac{1}{3} - \frac{1}{4} = \frac{1}{12}$$

$$P(\bar{A}\,\bar{B}) = 1 - P(A + B) = 1 - \frac{7}{12} = \frac{5}{12}$$

$$P(\bar{A} + B) = P(\bar{A}) + P(B) - P(\bar{A}B) = \frac{1}{2} + \frac{1}{3} - \frac{1}{12} = \frac{3}{4}$$

(b) (i) No; because $P(AB) \neq 0$. Here, $P(AB) = \frac{1}{4}$

(ii) No; because $P(A + B) \neq 1$. Here, $P(A + B) = \frac{7}{12}$

(iii) No; because $P(A)$ and $P(B)$ are not equal. Here $P(A) = \frac{1}{2}$, and $P(B) = \frac{1}{3}$

(iv) No; because $P(AB) \neq P(A).P(B)$. Here $P(AB) = \frac{1}{4}$, but $P(A) . P(B) = \frac{1}{6}$

Example 11.36 *Given that* $P(A) = \frac{3}{8}$, $P(B) = \frac{5}{3}$ *and* $P(A + B) = \frac{3}{4}$, *find* $P(A/B)$ *and* $P(B/A)$. *Are A and B independet?* [W.B.H.S., '78]

Solution Using the relation

$$P(A+B) = P(A) + P(B) - P(AB),$$

$$\frac{3}{4} = \frac{3}{8} + \frac{5}{8} - P(AB)$$

Solving, we get $P(AB) = \frac{1}{4}$. Hence,

$$P(A/B) = \frac{P(AB)}{P(B)} = \frac{\frac{1}{4}}{\frac{5}{8}} = \frac{2}{5};$$

$$P(B/A) = \frac{P(AB)}{P(A)} = \frac{\frac{1}{4}}{\frac{3}{8}} = \frac{2}{3}$$

We have, $P(AB) = \frac{1}{4}$ and $P(A)\,P(B) = \frac{3}{8} \times \frac{5}{8} = \frac{15}{64}$.

Since $P(AB) \neq P(A).P(B)$, so events A and B are not independent.

Also, note that $P(A) \neq P(A/B)$ and $P(B) \neq P(B/A)$.

Example 11.37 *For any two events A and B, prove that*

$$P(AB) \leq P(A) \leq P(A+B) \leq P(A) + P(B)$$

Solution From the Multiplication Theorem, we have

$$P(AB) = P(A)P(B/A)$$

Since the factor $P(B/A)$ on the R.H.S. is a 'probability', it is neither negative, nor greater than 1. Hence, $P(AB)$ can never exceed $P(A)$;

i.e., $$P(AB) \leq P(A) \qquad \text{...(i)}$$

Again,

$$\begin{aligned} P(A+B) &= P(A) + P(B) - P(AB) \\ &= P(A) + P(AB + \bar{A}B) - P(AB) \\ &= P(A) + P(AB) + P(\bar{A}B) - P(AB) \\ &= P(A) + P(\bar{A}B) \end{aligned}$$

(because AB and $\bar{A}B$ are two mutually exclusive and exhaustive forms of the event B). But $P(\bar{A}B)$ cannot be negative; hence

$$P(A) \leq P(A+B) \qquad \text{...(ii)}$$

Using the relation $P(A+B) = P(A) + P(B) - P(AB)$, we see that since $P(AB)$ is non-negative, $P(A+B)$ cannot exceed $P(A) + P(B)$;

i.e. $$P(A+B) \leq P(A) + P(B) \qquad \text{...(iii)}$$

Combining the inequalities (i), (ii) & (iii), the results follow.

Example 11.38 *If events A and B are independent, prove that* $\bar{A}$ *and* $\bar{B}$ *are also independent.*

Solution From the definition of independence (11.6.12),

$$P(AB) = P(A).P(B).$$

Therefore,

$$\begin{aligned} P(A+B) &= P(A) + P(B) - P(AB) \\ &= P(A) + P(B) - P(A).P(B) \end{aligned}$$

Now, since the event $\overline{A}\overline{B}$ is complementary to the event $A + B$, (i.e. non-occurrence of both A and B, is complementary to occurrence of at least one of A and B), hence

$$\begin{aligned} P(\overline{A}\overline{B}) &= 1 - P(A + B) \\ &= 1 - \{P(A) + P(B) - P(A).P(B)\} \\ &= \{1 - P(A)\}\{1 - P(B)\} \\ &= P(\overline{A}).P(\overline{B}) \end{aligned}$$

We have shown that

$$P(\overline{A}\overline{B}) = P(\overline{A}).P(\overline{B})$$

So, events $\overline{A}$ and $\overline{B}$ are independent.

Example 11.39 *The odds in favour of an event A are 3:4. The odds against another independent event B are 7:4. What is the probability that at least one of the events will happen?*

Solution

"Odds *in favour of* A are $a : b$" signifies $\quad P(A) = \dfrac{a}{a+b}$

"Odds *against* A are $a : b$" signifies $\quad P(A) = \dfrac{b}{a+b}$

Here, the probabilities of occurrence of A and B are $P(A) = \dfrac{3}{7}$ and $P(B) = \dfrac{4}{11}$. Also, since A and B are independent, we have $P(AB) = P(A).\ P(B) = \dfrac{3}{7} \times \dfrac{4}{11} = \dfrac{12}{77}$. The probability of occurrence of at least one of the events A and B is given by

$$\begin{aligned} P(A + B) &= P(A) + P(B) - P(AB) \\ &= \frac{3}{7} + \frac{4}{11} - \frac{12}{77} = \frac{7}{11} \end{aligned}$$

Example 11.40 *A card is drawn from each of two well-shuffled packs of cards. Find the probability that at least one of them is an ace.* [C.U., B.Sc. (Math) '73]

Solution Let us denote by

A = event that the card from Pack I is an ace,

B = event that the card from Pack II is an ace,

It is required to find $P(A + B)$.

(First method) $\quad P(A + B) = P(A) + P(B) - P(AB)$

Since there are 4 aces in a pack of 52 cards, $P(A) = 4/52 = 1/13$. Similarly, $P(B) = 1/13$. The events A and B are independent, because the drawing of an ace or otherwise from one pack does not in any way affect the probability of drawing an acc from another pack. So,

$$P(AB) = P(A).P(B) = \frac{1}{13} \times \frac{1}{13} = \frac{1}{169}$$

Substituting the values,

$$P(A + B) = \frac{1}{13} + \frac{1}{13} - \frac{1}{169} = \frac{25}{169}$$

(Second method) The event complementary to 'drawing at least one ace' is that 'none of the drawn cards is an ace', i.e. the card drawn from Pack I is a non-ace as well as the card from Pack II is

a non-ace (in symbols $\bar{A}\bar{B}$). So,

$$P(A + B) = 1 - P(\bar{A}\bar{B})$$

Since A and B are independent, so also are $\bar{A}$ and $\bar{B}$. Hence,

$$P(\bar{A}\bar{B}) = P(\bar{A}).P(\bar{B})$$
$$= \{1 - P(A)\}.\{1 - P(B)\}$$
$$= \left(1 - \frac{1}{13}\right)\left(1 - \frac{1}{13}\right) = \frac{144}{169}$$

$$\therefore \quad P(A + B) = 1 - \frac{144}{169} = \frac{25}{169}.$$

Example 11.41 *An article manufactured by a company consists of two parts I and II. In the process of manufacture of part I, 9 out of 100 are likely to be defective. Similarly, 5 out of 100 are likely to be defective in the manufacture of part II. Calculate the probability that the assembled article will not be defective.* [C.A., Nov. 76]

Solution Let

A = event that part I is not defective,
B = event that part II is not defective.

We have to find the compound probability $P(AB)$ that part I is not defective, as well as part II is not defective. But

$$P(A) = \text{Probability that part I is not defective}$$
$$= 1 - \text{Probability that part I is defective}$$
$$= 1 - \frac{9}{100} = 0.91$$

Similarly, $$P(B) = 1 - \frac{5}{100} = 0.95.$$

Assuming that event A and B are independent, the required probability is

$$P(AB) = P(A).P(B) = 0.91 \times 0.95 = 0.8645$$

Example 11.42 *One urn contains 2 white and 2 black balls; a second urn contains 2 white and 4 black balls. (i) If one ball is chosen from each urn, what is the probability that they will be of the same colour? (ii) If an urn is selected at random and one ball is drawn from it, what is the probability that it will be a white ball?*

[C.U., M.Com. '75]

Solution

(i) The event (E) 'both the drawn balls are of the same colour' has two mutually exclusive forms, either E_1 (both white) or E_2 (both black). So, by the theorem of total probability.

$$P(E) = P(E_1) + P(E_2)$$

But E_1 is a compound event formed by two *independent* events of drawing a white ball from each urn. Hence $P(E_1) = \frac{2}{4} \times \frac{2}{6} = \frac{1}{6}$. Similarly, E_2 is also a compound event, and $P(E_2) = \frac{2}{4} \times \frac{4}{6} = \frac{1}{3}$. Hence, the required probability is $\frac{1}{6} + \frac{1}{3} = \frac{1}{2}$.

(ii) A white ball can be selected in two mutually exclusive ways:
(A) when urn I is selected and a white ball is drawn from it;
(B) when urn II is selected and a white ball is drawn from it.

To find the probability of event A, we note that the selection of urn I itself depends on chance and has probability $\frac{1}{2}$. Once urn I has been selected, the selection of a white ball from it again depends on chance and has probability $\frac{2}{4}$. Therefore, by the theorem of compound probability, $P(4) = \frac{1}{2} \times \frac{2}{4} = \frac{1}{4}$. In the same way, $P(B) = \frac{1}{2} \times \frac{2}{6} = \frac{1}{6}$. Now, using the theorem of total probability, the required probability of drawing a white ball (irrespective of the choice of urn) is $\frac{1}{4} + \frac{1}{6} = \frac{5}{12}$.

Example 11.43 *A salesman has a 80% chance of making a sale to each customer. The behaviour of successive customers is assumed to be independent. If two customers X and Y enter the shop, what is the probability that the salesman will make a sale?*

Solution Let A and B denote the events 'sale to customer X' and 'sale to customer Y' respectively. We have to find the probability that a sale is made to at least one of the two customers X and Y, i.e. $P(A + B)$. Since the complementary event is that no sale is made, i.e. no sale to X as well as no sale to Y, hence

$$P(A + B) = 1 - P(\overline{A}\overline{B}) = 1 - P(\overline{A}).P(\overline{B})$$

because the events A and B are independent. But $P(\overline{A}) = 1 - P(A) = 1 - \frac{50}{100} = \frac{1}{5}$. Similarly, $P(\overline{B}) = \frac{1}{5}$. Hence, the required probability is $1 - \frac{1}{5} \times \frac{1}{5} = \frac{24}{25}$.

Example 11.44 *There is a 50–50 chance that a contractor's firm A will bid for the construction of a multi-storeyed building. Another firm B submits a bid and the probability is 3/4 that it will get the job, provided firm A does not bid. If firm A submits a bid, the probability that firm B will get the job is only 1/3. What is the probability that firm B will get the job?*

Solution Let

A = event that firm A submits the bid;
B = event that firm B gets the job.

We are given $P(A) = 1/2$, $P(B/\overline{A}) = 3/4$ and $P(B/A) = 1/3$. Using (11.6.15),

$$\begin{aligned} P(B) &= P(AB) + P(\overline{A}B) \\ &= P(A).P(B/A) + P(\overline{A}).P(B/\overline{A}) \\ &= \left(\frac{1}{2} \times \frac{1}{3}\right) + \left(\frac{1}{2} \times \frac{3}{4}\right) = \frac{13}{24} \end{aligned}$$

Example 11.45 *Two players A and B toss a die alternately. He who first throws a "six" wins the game. If A begins, what is the probability that he wins? What is the probability of B winning the game?*

Solution A wins the game if any of the following events happens—

(E_1) A gets a six in the first toss;

(E_2) A fails, B fails, and then A gets a six.

(E_3) A fails, B fails, then again A fails, B fails, and then A gets a six.

This may continue indefinitely.

Since the probability of getting six is $\frac{1}{6}$ in each trial, and the probability of failing to get a six is $\frac{5}{6}$ and also the elementary events in successive trials are independent,

$$P(E_1) = \frac{1}{6}$$

$$P(E_2) = \frac{5}{6}.\frac{5}{6}.\frac{1}{6} = \frac{1}{6}.\left(\frac{5}{6}\right)^2$$

$$P(E_3) = \frac{5}{6}.\frac{5}{6}.\frac{5}{6}.\frac{5}{6}.\frac{1}{6} = \frac{1}{6}\left(\frac{5}{6}\right)^4$$

and so on. Using the theorem of total probability,

Probability that A wins

$$= P(E_1) + P(E_2) + P(E_3) + \dots \dots$$

$$= \frac{1}{6} + \frac{1}{6}\left(\frac{5}{6}\right)^2 + \frac{1}{6}\left(\frac{5}{6}\right)^4 + \dots \dots$$

$$= \frac{\frac{1}{6}}{1-\left(\frac{5}{6}\right)^2} = \frac{6}{11}$$

[**Note :** Sum of infinite G.P. series ($|r| < 1$) $a + ar + ar^2 + \dots = \frac{a}{1-r}$]

B wins the game, if either

(F_1) A fails to get a six in the first trial, but in the next trial B gets a six.

(F_2) A fails, B fails, and then A fails, but in the next trial B gets a six.

(F_3) A fails B fails, A fails, B fails, A fails, and then B gets a six.

and so on. Proceeding as before,

Probability that B wins

$$= P(F_1) + P(F_2) + P(F_3) + \dots \dots$$

$$= \frac{5}{6}.\frac{1}{6} + \frac{5}{6}.\frac{5}{6}.\frac{5}{6}.\frac{1}{6} + \frac{5}{6}.\frac{5}{6}.\frac{5}{6}.\frac{5}{6}.\frac{5}{6}.\frac{1}{6} + \dots \dots$$

$$= \frac{5}{36} + \frac{5}{36}\left(\frac{5}{6}\right)^2 + \frac{5}{36}\left(\frac{5}{6}\right)^4 + \dots \dots$$

$$= \frac{\frac{5}{36}}{1-\left(\frac{5}{6}\right)^2} = \frac{5}{11}$$

Ans. $\frac{6}{11}, \frac{5}{11}$

Example 11.46 *If $P(X) = i) = P_4$ and $P(Y = j) = q_i$, $(i, j = 1, 2, ..., n)$ where X and Y are two mutually independent random variables, prove that $P(X + Y = n) = \sum_{i=1}^{n-1} p_i q_{n-i}$*

[C.U., B.A. (Econ) '72, '75]

Solution The event $X + Y = n$ can occur in the following mutually exclusive ways:

$$(X = 1, Y = n - 1), (X = 2, Y = n - 2), ... (X = n - 1, Y = 1)$$

Therefore, by the Theorem of Total Probability,

$$P(X + Y = n) = P(X = 1, Y = n - 1) + P(X = 2, Y = n - 2) + ... + P(X = n - 1, Y = 1)$$

$$= \sum_{i=1}^{n-1} P(X = i, Y = n - i)$$

However, since X and Y are independent, by (11.6.12, page 23),

$$P(X = i, Y = n - i = P(X = i) \times P(Y = n - i) = P_i q_{n-}$$

Hence the result.

11.7 DRAWING WITHOUT REPLACEMENT

A box contains A white and B black balls. If $(a + b)$ balls are drawn at random, the probability that among them exactly a are white and b are black is [see Fig. 11.3(i)]

$$\frac{{}^{A}C_{a} \cdot {}^{B}C_{b}}{{}^{A+B}C_{a+b}} \tag{11.7.1}$$

There are ${}^{A+B}C_{a+b}$ possible ways of forming groups of $(a + b)$ balls out of a total of $(A + B)$ balls, and these groups are mutually exclusive, exhaustive and equally likely.

(i) Two categories (ii) Three categories

Fig. 11.3 Drawing with Replacement (Urn Models)

However, a group of a white balls can be obtained in ${}^{A}C_{a}$ ways because there are A white balls in the box. Similarly, a group of b black balls can be obtained in ${}^{B}C_{b}$ ways. Since any of the ${}^{A}C_{a}$ groups of a white balls can be combined with any of the ${}^{B}C_{b}$ groups of b black balls, the number of groups of $(a + b)$ balls favourable to the event is ${}^{A}C_{a} \cdot {}^{B}C_{b}$. Hence, by the classical definition, the required probability is given by (11.7.1).

[Note: (i) The phrase, 'at random signifies that all possible drawings are 'equally likely'.

(ii) If instead of drawing the $(a + b)$ balls all at a time, the balls are drawn one by one, $(a + b)$ times in succession (a ball drawn once *not having been returned* to the box), then the same probability (11.7.1) is obtained. Hence, such drawing is also call '*drawing without replacement*', as distinct from '*drawing with replacement*' (Section 11.8).]

If the box contains balls of 3 different categories A of them white. B black and C red, and $(a + b + c)$ balls are drawn, then the probability that among them exactly a are white, b black and c red, is [see Fig. 11.3(ii)]

$$\frac{{}^{A}C_{a}.{}^{B}C_{b}.{}^{C}C_{c}}{{}^{A+B+C}C_{a+b+c}} \tag{11.7.2}$$

Example 11.47 *A bag contains 8 white and 6 black balls. If 5 balls are drawn at random, what is the probability that 3 are white and 2 black?*

Solution 5 balls can be drawn out of 14 in ${}^{14}C_5$ ways, and these cases are mutually exclusive, exhaustive and equally likely. However, a group of 3 white balls can be drawn out of 8 in ${}^{8}C_3$ ways, and 2 black balls out of 6 in ${}^{6}C_2$ ways. So the number of favourable cases is ${}^{8}C_3.{}^{6}C_2$. By the classical definition

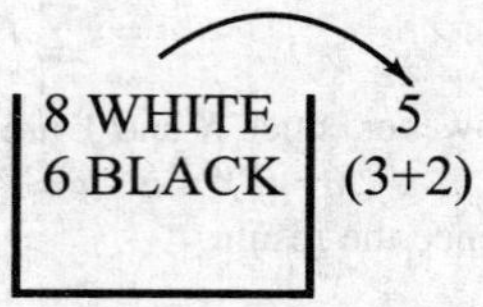

Fig. 1.4 Drawing Balls

$$P = \frac{{}^{8}C_3.{}^{6}C_2}{{}^{14}C_5}$$

But $\quad {}^{8}C_3 = \dfrac{8\times 7\times 6}{1\times 2\times 3} = 56;\ {}^{6}C_2 = \dfrac{6\times 5}{1\times 2} = 15;$ (see p. 56)

$${}^{14}C_5 = \frac{14\times 13\times 12\times 11\times 10}{1\times 2\times 3\times 4\times 5} = 14\times 13\times 11$$

$$\therefore \quad P = \frac{56\times 15}{14\times 13\times 11} = \frac{60}{143}$$

(*Second method*) We may assume that the balls have been drawn one by one without replacement.

Of the 5 balls drawn, the first one may be any of the 14 in the bag, the 2nd may be one of the remaining 13, and so on. So, the total number of ways in which 5 balls may be drawn, considering the order in which they appear, is $14 \times 13 \times 12 \times 11 \times 10$.

Now, let us consider the number of ways in which 3 white and 2 black balls may be obtained in a particular order; for example, the first 3 are white and the last 2 are black. This number is $(8 \times 7 \times 6) \times (6 \times 5)$. Hence the probability of drawing only white balls in the first 3 draws and only black balls in the last 2 draws is

$$\frac{(8\times 7\times 6)\times(6\times 5)}{14\times 13\times 12\times 11\times 10} = \frac{6}{143}$$

But, obviously this is the probability of having 3 white and 2 black balls in any other order. In our problem, the order is immaterial, and hence the required probability of having 3 white and 2 black balls is, by the Theorem of Total Probability,

$$\frac{6}{143} + \frac{6}{143} + \frac{6}{143} + \dots (k \text{ times}) = \frac{6}{143}\times k$$

where k is the number of arrangements (i.e., permutations) in which 3 white and 2 black balls may appear. Since this is given by

$k = \dfrac{5!}{3!\,2!} = 10$, the required probability is $\dfrac{6}{143}\times 10 = \dfrac{60}{143}$

Example 11.48 *Five men in a company of 20 are graduates. If 3 men are picked out of the 20 at random, what is the probability that they are all graduates? What is the probability of at least one graduate?* [C.U., B.A. (Econ) '73]

Solution (i) Applying (11.7.1), let

A = Number of graduates in the company = 5
a = Number of graduates in the sample = 3
B = Number of non-graduates in the company = 20 – 5 = 15
b = Number of non-graduates in the sample = 0

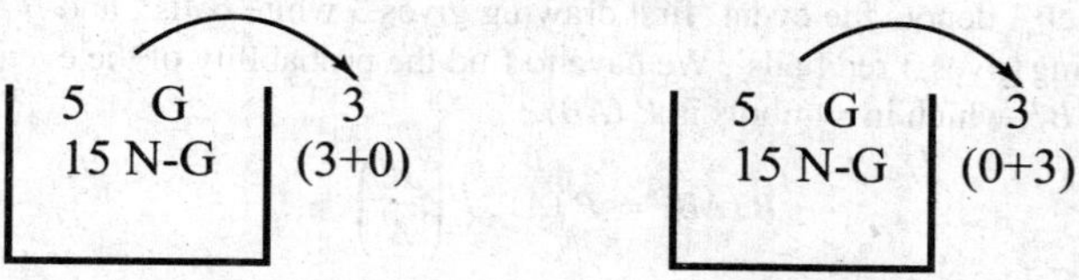

(i) All graduates (ii) None graduate

Fig. 11.5 Drawing from Two Categories

The probability that all are graduates is

$$p_1 = \frac{{}^5C_3 . {}^{15}C_0}{{}^{20}C_3}$$

But,

$${}^5C_3 = \frac{5 \times 4 \times 3}{1 \times 2 \times 3} = 10$$

$${}^{15}C_0 = 1; \; {}^{20}C_3 = \frac{20 \times 19 \times 18}{1 \times 2 \times 3} = 1440,$$

$$\therefore \quad p_1 = \frac{10 \times 1}{1140} = \frac{1}{114}$$

(ii) In order to find the probability of at least one graduate, it will be easier to find the probability of the complementary event, viz. that 'none is a graduate' (i.e. all 3 are non-graduates), so that $A = 5$, $a = 0$, $B = 15$, $b = 3$. Applying (11.7.1), the complementary probability is

$$\frac{{}^5C_0 . {}^{15}C_3}{{}^{20}C_3} = \frac{1 \times 455}{1140} = \frac{91}{228}$$

Hence, by (11.6.5), the required probability is

$$p_2 = 1 - \frac{91}{228} = \frac{137}{228}$$

(*Alternative method*)

(iii) The event 'at least one graduate' can be split up into three mutually exclusive events:

1. exactly 1 graduate and 2 non-graduates.
2. exactly 2 graduates and 1 non-graduate.
3. exactly 3 graduates and 0 non-graduate.

The probabilities of these cases are, rspectively

$$\frac{{}^5C_1 \times {}^{15}C_2}{{}^{20}C_3} = \frac{525}{1140}; \quad \frac{{}^5C_2 \times {}^{15}C_1}{{}^{20}C_3} = \frac{150}{1140};$$

$$\frac{{}^5C_3 \times {}^{15}C_0}{{}^{20}C_3} = \frac{10}{1140}$$

By the Theorem of Total Probability (11.6.2), the required probability p_2 is given by the sum of these probabilities. Hence,

$$p_2 = \frac{525}{1140} + \frac{150}{1140} + \frac{10}{1140} = \frac{137}{228} \qquad \textit{Ans. } \frac{1}{114}, \frac{137}{228}$$

Example 11.49 *A bag contains 8 red and 5 white balls. Two successive draws of 3 balls are made without replacement. Find the probability that the first drawing will give 3 white balls and the second 3 red balls.* [CA., May '78]

Solution Let A denote the event 'first drawing gives 3 white balls', and B denote the event 'second drawing gives 3 red balls'. We have to find the probability of the event "A and B", i.e. "A as well as B", which in symbols is $P(AB)$.

$$P(AB) = P(A) \,.\, P\left(\frac{B}{A}\right)$$

Using (11.7.1)

$$P(A) = \frac{{}^8C_0 \cdot {}^5C_3}{{}^{18}C_3} = \frac{1 \times 10}{286} = \frac{5}{143}$$

In order to find the conditional probability $P\left(\frac{B}{A}\right)$, we assume that event A has actually happened, i.e. white balls have been taken out, so that there remain 8 red and 2 white balls in the bag after the first drawing. The probability of getting 3 red balls now is

$$P\left(\frac{B}{A}\right) = \frac{{}^8C_8 \cdot {}^2C_0}{{}^{10}C_8} = \frac{56 \times 1}{120} = \frac{7}{15}$$

Hence, the required probability is $P(AB) = \frac{5}{143} \times \frac{7}{15} = \frac{7}{429}$

Example 11.50 *Four cards are drawn at random from a full pack. What is the probability that they belong to different suits?*

Solution In the pack of 52 cards, there are 13 cards of each suit—13 spades, 13 hearts, 13 diamonds and 13 clubs. If the 4 drawn cards are to belong to four different suits, we should have 1 card from each suit. Hence the required probability is (extension of formula 11.7.2).

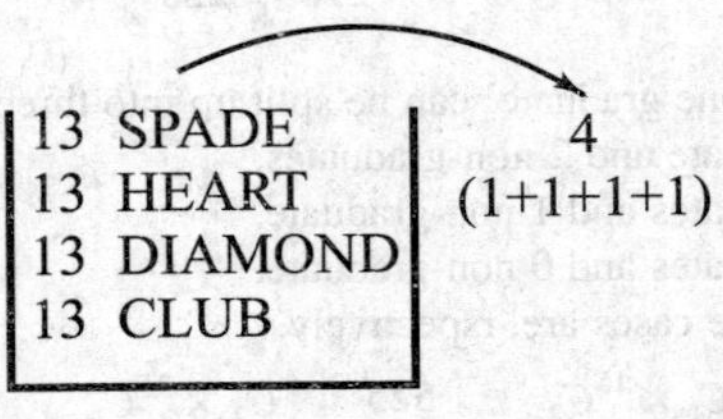

Fig. 11.6 Drawing from Four Categories

$$\frac{{}^{13}C_1 \cdot {}^{13}C_1 \cdot {}^{13}C_1 \cdot {}^{13}C_1}{{}^{52}C_4} = \frac{2197}{20825} \quad \text{(see Example 11.28)}$$

11.8 REPEATED TRIALS—DRAWING WITH REPLACEMENT

In a certain experiment, the probability of occurrence of an event is p and consequently the probability of its non-occurrence is $1 - p = q$, suppose. In n repeated trials of the experiment, if p remains a constant in each trial, then the probability that the event occurs i time is

$$^nC_r\, p^r\, q^{n-r} \tag{11.8.1}$$

where $p + q = 1$.

The performance of an experiment is usually called a "trial"; the occurrence of the event is called a "success" and its non-occurrence a "failure". Thus, the probability of r successes (and mutually $n - r$ failures) in n *independent* trials is given by (11.8.1), where p is the probability of success *in each trial.*

Example 11.51 *Find the probability that there will be exactly successes in n independent trials (n ≥ r), where p is the probability of success in a single trial.*

Solution Let us, at the beginning, consider the event in which the first r trials result in success in each trial, and the remaining $n - r$ trials result in failures only. Now, the probability of success is p in each trial, so that the probability of failure is $1 - p = q$, say. Since the n trials are *independent*, the probability of r consecutive success, followed by $n - r$ consecutive failures is, by the theorem of compound probability, given by

$$\underbrace{p \times p \times \ldots p}_{r \text{ times}} \times \underbrace{q \times q \times \ldots \times q}_{(n-r) \text{ times}} = p^r\, q^{n-r}$$

The probability of obtaining r successes (and hence $n - r$ failures) in any other specified order is similarly $p^r\, q^{n-r}$. However, we are not interested in any particular arrangement of the r successes and the $n - r$ failures, but overall number of r successes in n trials, whatever be the arrangement in which they appear. But, r successes and $n - r$ failures in n trails may arise in $\dfrac{n!}{r!\,(n-r)!} = {}^nC_r$ mutually exclusive ways, in each of which the probability is the same, viz. $p^r\, q^{n-r}$. Hence, by the theorem of total probability, the probability that there will be r successes, irrespective of the order in which they appear, is

$$p^r\, q^{n-r} + p^r q^{n-r} + \ldots\ ({}^nC_r \text{ times}) = {}^nC_r\, p^r\, q^{n-r}.$$

[**Note:** This problem may be identified with drawing n balls from a box in which the proportions of white and black balls are p and q. If balls are drawn one by one, n times in succession, the ball drawn each time *being returned* to the box before the next drawing, then the probability of obtaining r white balls is given by (11.8.1). Such experiments are, therefore, called "*drawing with replacement*". [Also see Note (ii) Section 11.7, page 31].

Example 11.52 *A coin is tossed 10 times. Find the probability of getting (i) exactly 6 heads, and (ii) 9 heads and 1 tail.*

Solution Let us describe the appearance of 'head' in one toss of the coin as "success". Then, assuming that the coin is unbiased,

$$p = \text{probability of success in each trial} = \frac{1}{2}$$

$$q = 1 - p = 1 - \frac{1}{2} = \frac{1}{2}$$

Also, since the probability of occurrence of a head or a tail in any tossing is not affected by the results of any other tossing, the trials are *independent*. Therefore, we apply (11.8.1), with $n = 10, p = \frac{1}{2}, q = \frac{1}{2}$. (i) Here, $r = 6$. Therefore, the probability of 6 heads is

$$^{10}C_6 \left(\frac{1}{2}\right)^6 \left(\frac{1}{2}\right)^{10-6} = \frac{210}{2^{10}} = \frac{105}{512}$$

(ii) We have to find the probability of 9 successes (and obviously the remaining 1 is a failure). Putting $r = 9$, the required probability is

$$^{10}C_9 \left(\frac{1}{2}\right)^9 \left(\frac{1}{2}\right)^{10-9} = \frac{10}{2^{10}} = \frac{5}{512}$$

Example 11.53 *The probability that an entering college student will be a graduate is 0.4. Determine the probability that out of 5 entering students, (i) none, (ii) one, (iii) at least one, will be a graduate.*

Solution Let the event 'an entering college student will be a graduate' be called a "success". Then p = probability of success in each trial = 0.4. Assuming that the success or failure of one student does not affect the result of any other student, the trials may be considered as *independent*.

Hence, putting $n = 5, p = 0.4, q = 1 - p = 0.6$ in (11.8.1)

(i) Probability that none is a graduate

$= $ probability of 0 success $= {}^3C_0\ (0.4)^0\ (0.6)^{5-0}$

$= (1)\ (1)\ (0.6)^5 = 0.07776$

(ii) Probability that one is a graduate

$= $ probability of one success $= {}^5C_1\ (0.4)^1\ (0.6)^{5-1}$

$= 5\ (0.4)\ (0.6)^4 = 0.2592$

(iii) The complementary event is 'none is a graduate'. Hence, applying (11.6.5).

Probability that at least one is graduate

= 1– (probability that none is a graduate)

= 1 – .07776, from (i) above.

= .92224.

Example 11.54 *A machine produces on the average 2 per cent defectives. If 4 articles are chosen randomly, find the probability that there are at least 2 defective articles.*

Solution Denoting the occurrence of a defective article as "success", we find that

$$p = \text{probability of a defective article} = 20\% = \frac{2}{100} = 02$$

$$q = 1 - 0.2 = .98$$

Since the occurrence of a defective article does not affect the probability of another article being defective or not, hence the trials are independent. Therefore, applying (11.8.1), we have to find the probability of at least 2 successes in 4 trials. The complementary event consists of two mutually exclusive cases, viz. (i) exactly 0 success in 4 trials, and (ii) exactly 1 success in 4 trials. The probabilities of these cases are respectively

$$^4C_0\ (.02)^0\ (.98)^{40} = (1)\ (1)\ (.98)^4 = .922 \text{ (approx.)}$$

$$^4C_1\ (.02)^1\ (.98)^{-1} = (4)\ (.02)\ (.98)^3 = .075 \text{ (approx.)}$$

By the Theorem of Total Probability, the probability of the complementary event is, therefore, .922 + .075 = .997. The required probability is 1 – .997 = .003. *Ans.* 0.003

11.9 BAYES' THEOREM

An event A can occur only if one of the mutually exclusive and exhaustive set of events $B_1, B_2, \ldots B_n$ occurs. Suppose that the unconditional probabilities

$$P(B_1), P(B_2), \ldots P(B_n)$$

and the conditional probabilities

$$P, \left(\frac{A}{B_1}\right), P\left(\frac{A}{B_2}\right), \ldots P\left(\frac{A}{B_n}\right)$$

are known. Then the conditional probability $P\left(\frac{B_i}{A}\right)$ of a specified event B_i, when A is stated to have actually occurred, is given by

$$P\left(\frac{B_i}{A}\right) = \frac{P(B_i).P\left(\frac{A}{B_i}\right)}{\sum_{i=1}^{n} P(B_i).P\left(\frac{A}{B_i}\right)} \tag{11.9.1}$$

This is known as *Bayes' Theorem.*

Proof

The event A can happens in n mutually exclusive ways $B_1 A, B_2 A, \ldots B_n A$, i.e. either when B_1 has occurred, or $B_2, \ldots$ or B_n. So by the theorem of total probability (11.6.2)

$$\begin{aligned} P(A) &= P(B_1A + B_2A + \ldots + B_nA) \\ &= P(B_1A) + P(B_2A) + \ldots + P(B_nA), \quad \because \text{ mutually exclusive} \\ &= P(B_1) \cdot P\left(\frac{A}{B_1}\right) + P(B_2) \cdot \left(\frac{A}{B_2}\right) + \ldots + P(B_n) \cdot P\left(\frac{A}{B_n}\right), \end{aligned}$$

using the Multiplication Theorem (11.6.10)

$$= \sum_{i=1}^{n} P(B_i) \cdot P\left(\frac{A}{B_i}\right)$$

Again, $P(AB_i) = P(A) \cdot P\left(\frac{B_i}{A}\right)$

$$P(B_i A) = P(B_i) \cdot P\left(\frac{A}{B_i}\right)$$

Since the events AB_i and $B_i A$ are equivalent, their probabilities are equal. Hence

$P(A) \cdot P\left(\frac{B_i}{A}\right) = P(B_i) \cdot P\left(\frac{A}{B_i}\right)$, so that

$$P\left(\frac{B_i}{A}\right) = \frac{P(B_i) \cdot P\left(\frac{A}{B_i}\right)}{P(A)}$$

Substituting for $P(A)$ from above, the theorem is proved.

Formula (11.9.1) is also known as "Bayes" formula for probabilities of hypothesis", because $B_1, B_2, \ldots B_n$ may be considered as hypothesis which account for the occurrence of A. The probabilities $P(B_1), P(B_2), \ldots P(B_n)$ are called *'a priori'* probabilities of the hypothesis (i.e. probabilities prior to any knowledge as to the occurrence or non-occurrence of A); while $P\left(\frac{B_1}{A}\right), P\left(\frac{B_2}{A}\right), \ldots P\left(\frac{B_n}{A}\right)$ are known as a *'a posteriori'* probabilities of the same hypothesis (i.e. probabilities after the occurrence of A is definitely known).

Example 11.55 *Two boxes contain respectively 4 white and 2 black, and 1 white and 3 black balls. One ball is transferred from the first box into the second, and then one ball is drawn from the latter. It turns out to be black. What is the probability that the transferred ball was white?*

Solution There are two hypotheses:

B_1 = the transferred ball was white;

B_2 = the transferred bass was black.

The event A which is stated to have actually happened after the occurrence of B_1 or B_2, is

A = the ball drawn from the 2nd box is black.

We have to find the probability $P\left(\frac{B_i}{A}\right)$.

Now, $P(B_1) = \frac{4}{6} = \frac{2}{3}, P(B_2) = \frac{2}{6} = \frac{1}{3}$.

Also, $P\left(\frac{A}{B_1}\right)$ = Probability that the ball drawn from the 2nd box is *black*, assuming that the transferred ball was *white*.

$= \frac{3}{5}$

since after transfer the 2nd box contains 5 balls (2 white and 3 black).

Similarly, $P\left(\frac{A}{B_2}\right)$ = Probability that the ball drawn from the 2nd box is *black*, assuming that the transferred ball was *black*

$= \frac{4}{5}$

Using Baye's formula (11.9.1)

$$P\left(\frac{B_1}{A}\right) = \frac{P(B_1) \cdot P\left(\frac{A}{B_1}\right)}{P(B_1) \cdot P\left(\frac{A}{B_1}\right) + P(B_2) \cdot P\left(\frac{A}{B_n}\right)} = \frac{\frac{2}{5} \times \frac{3}{5}}{\frac{2}{5} \times \frac{3}{5} \times \frac{1}{3} \times \frac{4}{5}} = \frac{3}{5} \qquad \textit{Ans. } \frac{3}{5}$$

Example 11.56 *Three identical boxes, I, II, III contain respectively 4 white and 3 red balls, 3 white and 7 red balls, and 2 white and 3 red balls. A box is chosen at random and a ball is drawn out of it. If the ball is found to be white, what is the probability that Box II was selected ?*

Solution Here A is the *observed* event that the 'drawn ball is white'. There are three hypotheses B_1, B_2, B_3, namely 'Box I was chosen', 'Box II was chosen', 'Box III was chosen' respectively. We have to find the probability of the hypothesis B_2, given that A has happened, i.e. $P\left(\frac{B_2}{A}\right)$.

Since the 3 boxes are identical in appearance.

$$P(B_1) = \frac{1}{3}, P(B_2) = \frac{1}{3}, P(B_n) = \frac{1}{3}$$

Also, $P\left(\frac{A}{B_1}\right)$ = Probability of getting a white ball, assuming that Box I was selected

$$= \frac{4}{7}$$

Similarly, $P\left(\frac{A}{B_2}\right) = \frac{3}{10}, P\left(\frac{A}{B_3}\right) = \frac{2}{5}$.

Hypothesis

(B_i)	$P(B_i)$	$P\left(\frac{A}{B_i}\right)$	$P(B_i) . P\left(\frac{A}{B_i}\right)$
B_1	$\frac{1}{3}$	$\frac{4}{7}$	$\frac{4}{21}$
B_2	$\frac{1}{3}$	$\frac{3}{10}$	$\frac{1}{10}$
B_3	$\frac{1}{3}$	$\frac{2}{5}$	$\frac{2}{15}$
Total	1	—	$\frac{89}{210}$

Note that $\Sigma P(B_i) = 1$, because B_i's are mutually exclusive and exhaustive.
Using Baye's formula (11.9.1).

$$P\left(\frac{B_2}{A}\right) = \frac{P(B_2) . P\left(\frac{A}{B_2}\right)}{\Sigma P(B_i) . P\left(\frac{A}{B_i}\right)} = \frac{\frac{1}{10}}{\frac{89}{210}} = \frac{21}{89}$$

[**Note:** (i) In problems involving Bayes' Theorem, some event A is known to have definitely occurred, through certain hypotheses B_1, B_2, ... B_n, and we are required to find the probability of one of these hypotheses say $P\left(\frac{B_i}{A}\right)$, after the occurrence of A is known.

(ii) In the numerator of Bayes' formula (11.9.1), start with the unconditional probability $P(B_i)$; the other factor is $P\left(\frac{A}{B_i}\right)$—not $P\left(\frac{B_i}{A}\right)$, which we are going to find, but A and B_i interchanging places. The denominator is the sum of all such products for all the hypotheses. Observe that the numerator and the denominator differ only in respect of a Σ sign.]

Example 11.57 *In a bolt factory, the machines M_1, M_2, M_3 manufacture respectively 25, 35 and 40 per cent of the total product. Of their output 5, 4 and 2 per cent respectively are defective bolts. One bolt is drawn at random from the product and is found to be defective. What is the probability that it was manufactured by machine M_3?*

Solution The observed event A is that the selected bolt is defective. The hypotheses B_1, B_2, B_3 are that the selected bolt comes from machines M_1, M_2, M_3 respectively. We have to find $P\left(\frac{B_3}{A}\right)$.

$$P(B_1) = 25\% = 0.25, P(B_2) = 35\% = 0.35, P(B_3) = 40\% = 0.40$$

$$P\left(\frac{A}{B_1}\right) = \text{Probability that a bolt is defective, if it comes from machine } M_1 = 5\% = .05$$

Similarly, $P\left(\frac{A}{B_2}\right) = 4\% = .04$, $P\left(\frac{A}{B_3}\right) = 2\% = .02$

B_i	$P(B_i)$	$P\left(\frac{A}{B_1}\right)$	$P(B_i)\,.\,P\left(\frac{A}{B_1}\right)$
B_1	0.25	.05	.0125
B_2	0.35	.04	.0140
B_3	0.40	.02	.0080
Total	1.00	—	.0345

Using Bayes' formula,

$$P\left(\frac{B_3}{A}\right) = \frac{P(B_3)\,.\,P\left(\frac{A}{B_3}\right)}{\Sigma P(B_i)\,.\,P\left(\frac{A}{B_i}\right)} = \frac{.0080}{.0345} = \frac{16}{69}$$

11.10 MATHEMATICAL EXPECTATION

Suppose that a random experiment has n mutually exclusive and exhaustive outcomes corresponding to which a variable x takes the values

$$x_1, \quad x_2, \quad \ldots \quad x_n$$

with probabilities

$$p_1, \quad p_2, \quad \ldots \quad p_n.$$

respectively. Then the *mathematical expectation* (or *Expected Value*) of the variable x, denoted by $E(x)$, is defined as

$$E(x) = p_1x_1 + p_2x_2 + \dots + p_nx_n$$
$$= \Sigma p_i\, x_i \qquad (11.10.1)$$

where $\Sigma p_i = 1$. That is, the mathematical expectation is the sum of products of the different possible values of the variable and the corresponding probabilities.

If $E(x) = m$, the mathematical expectation of $(x - m)^2$ is called the Variance of x, and usually written as Var (x) or σ^2.

$$\text{Var}(x) = E(x - m)^2$$
$$= p_1(x_1 - m)^2 + p_2(x_2 - m)^2 + \dots + p_n(x_n - m)^2$$
$$= \Sigma p_i(x_i - m)^2 \qquad (11.10.2)$$

It can be shown that

$$\sigma^2 = \Sigma p_i x_i^2 - m^2 \qquad (11.10.3)$$

where $m = E(x)$. The square-root of variance is called *Standard Deviation*. It is always considered positive and usually denoted by the symbol σ.

Theorems on Mathematical Expectation

Theorem I

The mathematical expectation of the sum of two variables x and y is the sum of their mathematical expectations.

$$E(x + y) = E(x) + E(y) \qquad (11.10.4)$$

Theorem II

The mathematical expectation of the product of two *independent* variables x and y is the product of their mathematical expectations.

$$E(x \cdot y) = E(x) \cdot E(y) \qquad (11.10.5)$$

provided x and y are independent.

The theorems may be extended to more than two variables.

Example 11.58 *(i) An unbiased die is thrown. What is the mathematical expectation of the number of points?*

(ii) If 4 dice are thrown, find the expected sum of points.

Solution

(i) The variable x is the 'number of points', which takes the values 1, 2, 3, 4, 5, 6 with probabilities $\frac{1}{6}$ for each.

$$\therefore E(x) = \left(\frac{1}{6}\times 1\right) + \left(\frac{1}{6}\times 2\right) + \left(\frac{1}{6}\times 3\right) + \left(\frac{1}{6}\times 4\right) + \left(\frac{1}{6}\times 5\right) + \left(\frac{1}{6}\times 6\right)$$
$$= \frac{1}{6}(1 + 2 + 3 + 4 + 5 + 6) = 3.5$$

(ii) Let x, y, z, w represent the points obtained on the 4 dice. The sum of the points on the dice is

$$s = x + y + z + w$$

The mathematical expectation of the sum is the sum of mathematical expectations.

$$E(s) = E(x) + E(y) + E(z) + E(w)$$

Since x, y, z, w represent individual points on the 4 dice, we have from (i) above,

$$E(x) = 3.5, E(y) = 3.5, E(z) = 3.5, E(w) = 3.5$$

Hence, the mathematical expectation of the sum of points on the 4 dice is

$$E(s) = 3.5 + 3.5 + 3.5 + 3.5 = 14.$$

Ans. 3.5, 14.

Example 11.59 *A box contains 4 white and 6 black balls. If 3 balls are drawn at random, find the mathematical expectation of the number of white balls.*

Solution Here the variable x is the number of white balls obtained among the 3 balls drawn. The number of white balls may be 0, 1, 2, 3.

$$\text{Probability of 0 white ball} = \frac{{}^4C_0 \cdot {}^6C_3}{{}^{10}C_3} = \frac{1 \times 20}{120} = \frac{1}{6}$$

$$\text{Probability of 1 white ball} = \frac{{}^4C_1 \cdot {}^6C_2}{{}^{10}C_3} = \frac{4 \times 15}{120} = \frac{1}{2}$$

$$\text{Probability of 2 white balls} = \frac{{}^4C_2 \cdot {}^6C_1}{{}^{10}C_3} = \frac{6 \times 6}{120} = \frac{3}{10}$$

$$\text{Probability of 3 white balls} = \frac{{}^4C_3 \cdot {}^6C_0}{{}^{10}C_3} = \frac{4 \times 1}{120} = \frac{1}{30}$$

The possible values of x and the corresponding probabilities p are shown below:

x	p	px
0	$\frac{1}{6}$	0
1	$\frac{1}{2}$	$\frac{1}{2}$
2	$\frac{3}{10}$	$\frac{3}{5}$
3	$\frac{1}{30}$	$\frac{1}{10}$
Total	1	$\frac{6}{5}$

(Note that $\Sigma p_i = 1$)

$$\therefore \quad E(x) = \Sigma p_i x_i = \frac{6}{5} = 1.2$$

Example 11.60 *A man purchases a lottery ticket, in which he may win the first prize of Rs 10,000 with probability .0001 or the second prize of Rs 4,000 with probability .0004. Find this mathematical expectation.*

Solution The probability that the man does not get any prize is 1 – .0001 – .0004 = .9995. Applying (11.10.1).

$$E(x) = (.0001 \times 10.000 + .0004 \times 4.000 + .9995 \times 0) \text{ Rs}$$
$$= \text{Rs } 260$$

Example 11.61 *Two dice, with faces numbered 1 to 6, are thrown and their points are added. The thrower is given Rs 40 for a score of 12, but he has to pay Rs 2 if the score is less than 12. Find his expectation per throw.*

Solution With 2 dice, the probability of getting a total of 12, i.e. 6 on die I and 6 on die II, is $\frac{1}{36}$. Since the total score on the dice cannot exceed 12, the probability of getting a score less than 12 is $1 - \frac{35}{36} = \frac{35}{36}$. Considering his loss as a negative gain, we find that the thrower gets,

Rs 40 with probability $\frac{1}{36}$

Rs (– 2) with probability $\frac{35}{36}$

$$\therefore \quad \text{Expectation} = \frac{1}{36} \times 40 + \frac{35}{36} \times (-2) = -\frac{5}{6} \text{ Re}$$

i.e., he expects to lose Re $\frac{5}{6}$ per throw.

Example 11.62 *A man tosses a coin twice and is to receive Rs 2 if a head is shown, and lose Re 1 for a tail. Find this expected gain and the variance.*

Solution Assuming that the coin is unbiased, the possible outcomes, their probabilities and the corresponding values of gain (x) and are shown below:

Outcome	*Probability (p)*	*Gain (x)*	*px*	*px*^{2}
2 heads	$\frac{1}{4}$	$2 \times 2 = 4$	1	4
1 head and 1 tail	$\frac{1}{2}$	$1 \times 2 + 1\,(-1) = 1$	$\frac{1}{2}$	$\frac{1}{2}$
2 tails	$\frac{1}{4}$	$2\,(-1) = -2$	$-\frac{1}{2}$	1
Total	1	—	1	$\frac{11}{2}$

$$\text{Expected Gain } (m) = \Sigma\, p_i\, x_i = 1$$

$$\text{Variance } (\sigma^2) = \Sigma\, p_i\, x_i^2 - m^2 = \frac{11}{2} - 1^2 = \frac{9}{2} = 4.5$$

Ans. 1 Re, 4.5 Rs^2

Example 11.63 *A man has the choice of running either a hot-snack stall or an ice-cream stall at a seaside resort during the summer season. If it is a fairly cool summer he should make Rs 5000 by running the hot-snack stall, but if the summer is quite hot he can only expect to make Rs 1000. On the other hand, if he operates the ice-cream stall, his profit is estimated at Rs 6500 if the summer is hot, but only Rs 1000 if it is cool. There is a 40 per cent chance of the summer being hot. Should he opt for running the hot-snack stall or the ice-cream stall? Give mathematical arguments.* [I.C.W.A., Dec. '75]

Solution Probability of hot summer $= 40\% = \frac{40}{100} = 0.4$

Probability of cool summer $= 1 - 0.4 = 0.6$

The given data are tabulated below:

Condition of Summer Weather	*Probability (p)*	*Gain (x) from Hot-snack Stall*	*Gain (y) from Ice-cream Stall*
(i) Hot	0.4	Rs 1000	Rs 6500
(ii) Cool	0.6	Rs 5000	Rs 1000
Total	1.0	—	—

Expected gain from hot-snack stall = $E(x)$

$$= 0.4 \times 1000 + 0.6 \times 5000$$
$$= 400 + 3000 = 3400 \text{ Rs}$$

Expected gain from ice-cream stall = $E(y)$

$$= 0.4 \times 6500 + 0.6 \times 1000$$
$$= 2600 + 600 = 3200 \text{ Rs}$$

Since the expected gain from the hot-snack stall is larger, the man should opt for running the hot-snack stall.

Example 11.64 *If $y = ax + b$, where a and b are constants, prove that $E(y) = a.E(x) + b$.*

Solution Let the variable x take the values $x_1, x_2, \ldots x_n$ with probabilities $p_1, p_2, \ldots. p_n$ respectively, where $\Sigma p_i = 1$. The values assumed by the variable y are respectively

$$y_1 = ax_1 + b,\ y_2 = ax_2 + b,\ \ldots\ y_n = ax_n + b$$
$$E(y) = p_1(ax_1 + b) + p_2(ax_2 + b) + \ldots + p_n(ax_n + b)$$
$$= (p_1ax_1 + p_2ax_2 + \ldots + p_nax_n) + (p_1b + p_2b + \ldots + p_nb)$$
$$= a(p_1x_1 + p_2x_2 + \ldots + p_nx_n) + b(p_1 + p_2 + \ldots + p_n)$$
$$= a\,\Sigma p_i\,x_i + b\,\Sigma p_i$$
$$= a\,.\,E(x) + b \text{ (since } \Sigma p_i = 1)$$

(Proved)

Example 11.65 *A coin is tossed repeatedly until a head appears. Find the expected number of tosses required to obtain the first head.*

Solution Let p denote the probability of getting a head in a single toss, and $q = 1 - p$ be the probability of getting a tail. Consider the variable x = Number of tosses required to obtain the first head. x can take the values 1, 2, 3, ... ∞. In general, r tosses will be required, if the first $r - 1$ tosses result in tails only and the last toss gives a head. Since the successive trials are independent, the probability of such an occurrence, i.e.

$$P(x = r) = \underbrace{(q \cdot q \ldots. q)}_{r-1 \text{ times}} \cdot p = p \cdot q^{r-1}$$

Hence, $$E(x) = \sum_{r=1}^{\infty} (p.q^{r-1}).\, r = p \sum_{r=1}^{\infty} (r.q^{r-1})$$
$$= p(1 + 2q + 3q^2 + \ldots \ldots)$$
$$= p\,.\,(1 - q)^{-2}$$

Since $(1-x)^{-2} = 1 + 2x + 3x^2 + \dots$ for $|x| < 1$.

$$= \frac{p}{(1-q)^2} = \frac{p}{p^2} = \frac{1}{p} \qquad \textit{Ans. } \frac{1}{p}$$

11.11 OTHER APPROACHES TO PROBABILITY THEORY

In view of the inherent defects of classical definition (page 10), the probability theory has also been considered from other points of view.

Frequency (or 'Empirical') Theory: In N trials of a random experiment, if an event is found to occur f times, the relative frequency of occurrence of the event is $\frac{f}{N}$. If this relative frequency approaches a limiting value p, as N increases indefinitely, then p is called the "probability" of the event. In mathematical language.

$$p = \lim_{N \to \infty} \left(\frac{f}{N}\right) \tag{11.11.1}$$

The frequency definition of probability has the following has the following defects: (i) It is based on the stability of relative frequency $\frac{f}{N}$, as N becomes large. Hence, for its calculation a large number of experiments are to be performed under identical conditions. (ii) The limiting value is only a mathematical concept, and 'probability' cannot be measured exactly, even with a very large number of cases.

Example 11.66 1000 patients suffering from a disease, were treated with a certain medicine, and 880 of them were found to have been cured. The relative frequency of cures is therefore $\frac{880}{1000} = 0.880$. In another batch of 1500 patients, the number of cures with the medicine is 1300. So the relative frequency from the combined group is $\frac{(880+1300)}{2500} = 0.872$. A third batch of patients shows 450 cures, and therefore the relative frequency of cures in the total of 3000 patients is $\frac{(880+1300+450)}{3000} = 0.877$.

According to the frequency definition of probability, by continuing in this manner, the relative frequencies calculated from a larger group is expected to come closer and closer to a number, which is called the 'probability' of cure with the medicine. We may state that the estimate of this probability is about 0.88.

(b) Axiomatic Theory: This modern theory introduces 'probability' simply as a *number* associated with each event. It is based on certain axioms, which express the rules for operating with such numbers. This means that the probabilities of our events can be perfectly arbitrary, except that they must satisfy the axioms. The advantage of the axiomatic theory is that it embraces all situations irrespective of whether the possible outcomes of a random experiment are "equally likely" or not. The classical theory can be derived frm the axiomatic theory as a special case.

Some elementary concepts of the mathematical 'Set Theory' which is used in the Axiomatic Theory of Probability are given below. A more elaborate discussion will be found in "*Set Theory (with Problems and Exercises)*" by the present author.

11.12 SET THEORY

Set

Any well-defined collection of distinct objects is called a *set.* Each object of the set is called its *element* or *member.*

Illustration 1.

(a) The set of vowels in the English alphabet.
(b) The set of students who appeared at the B.Com. examination of Calcutta University in the year 1979.
(c) The set of numbers which are multiples of 4.
(d) The set of lines parallel to a given straight line.

A set is specified either by listing its members or by stating some property held by all members of the set and no non-member. The members of the set or the statement specifiying the members are written down within curly brackets "{ }". Sets are usually denoted by capital letters, such as *A*, *B*, *S*, *R*, etc. The elements of a set are denoted by small letters, such as *a*, *b*, *c*, *p*, etc. If *p* is an element of a set *A*, we write

$$p \in A$$

which is read "*p belongs to A*".

Illustration 2.

(a) $A = \{1, 3, 5, 7, 9)$ means that *A* is a set whose elements are 1, 3, 5, 7, 9. The arrangement of elements in the set is immaterial, e.g. {7, 1, 3, 9, 5}. We may also write the set in the form

$$A = \{x: x \text{ is a positive odd number}, x \leq 10\}$$

This is read "*A* is the set of elements *x*, such that *x* is a positive odd number and *x* less than or equal to 10." Note that the symbol ":" is used to denote "such that".

(b) $B = \{x: x$ is a river in Uttar Pradesh$\}$
(c) $N = \{1, 2, 3, 4, ...\}$ is the set of natural numbers.
(d) $R = \{x: x$ is real, $-2 \leq x \leq 1\}$ is the set of all real numbers lying between -2 and 1, both inclusive.

Subset

If each element of a set *A* is also an element of a set *B*, then *A* is called a *subset* of *B*, and we write

$$A \subseteq B \quad \text{or} \quad B \supseteq A$$

This is read as "*A* is *contained in B*" or "*B contains A*" respectively. Note that if $A \subseteq B$, then the elements of *B* may be either exactly the same as those of *A*, or the elements of *B* include all the elements of *A* and some more elements. The sign $\subseteq$ is similar to $\leq$ (i.e. less than or equal to) and has a some what similar interpretation in set theory.

If $A \subseteq B$ and $B \subseteq A$, i.e. *A* is a subset of *B* and *B* is a subset of *A*, then the sets *A* and *B* are said to be *equal* or *identical,* and we write

$$A = B$$

In such a case, sets A and B have exactly the same elements. If sets A and B are not equal, i.e. the elements of A are not the same as the elements of B, then we write $A \neq B$.

If $A \subseteq B$ but $A \neq B$, i.e. A is a subset of B but the sets are not equal, then A is said to be a *proper subset* of B. We write $A \subset B$.

Illustration 3.

(a) The set $A = \{e, i, a\}$ is a 'subset' of the set $B = \{a, e, i, o, u\}$, because all the elements of A are also elements of B. Thus $A \subseteq B$.

(b) The set $C = \{3, 7, 8\}$ is a subset of the set $D = \{8, 3, 7\}$, since each element of C is also an element of D. Thus $C \subseteq D$. Similarly, we see that $D \subseteq C$. Hence, $C = D$, i.e. the sets C and D are 'equal'.

(c) $E = \{2, 4, 6\}$ is a 'proper subset' of $S = \{1, 2, 3, 4, 5, 6\}$ because each element of E belongs to S, but $E \neq S$.

(d) The set $P = \{2, 8, 12\}$ is not a subset or $Q = \{1, 4, 5, 8, 12\}$ because the element 2 of P does not belong to Q.

Universal Set (or Space)

For many purposes we limit our discussion to subsets of some particular set, which is called the *universal set.* Thus any set is a subset of the universal set. Using geometrical terminology, the universal set is also called *Space,* and its elements are called *points.* The symbol S is generally used to denote the universal set.

Null Set

It is useful to consider a set which contains no elements at all. This is called the *Null Set* (or *Empty Set*), and is denoted by the symbol ϕ. It will be seen that the null set ϕ is somewhat equivalent to 0 (zero) in the number system. Note that the null set ϕ is distinct from the set $\{0\}$ which is a set containing only one element, viz. 0.

Illustration 4.

(a) In the study of income of Indian population, the set of incomes of all citizens of India is the universal set, and the incomes of different groups or classes of people are its subsets.

(b) THe set $\{x: x$ is a person who can jump to a height of 3 miles$\}$ is the null set, because there is no person who can jump to such a height.

Finite and Infinite sets

A set is said to be *finite,* if it is empty or contains a finite number of elements, (i.e. if the number of different elements of the set can be counted and the counting process can be completed). Otherwise, the set is said to be *infinite*.

Infinite sets are of two types—(i) *countably infinite,* and (ii) *uncountably infinite.* An infinite set is called *countable,* if its elements can be written down in some order. Otherwise an infinite set is called *uncountable*.

Illustration 5.

(a) The set of days of the week {*Sunday, Monday, Tuesday, Wednesday, Thursday, Friday, Saturday*} is finite, because the total number of elements of the set is a specified number 7.

(b) The set of even numbers $\{2, 4, 6, 8, \ldots \ldots\}$ is countably infinite, because the total number of elements is not finite and the elements can be arranged in an increasing order.

(c) The set $\{x: x \text{ is real}, 0 \leq x \leq 1\}$ is uncountably infinite, because the possible elemetns of the set is neither finite in number nor can be arranged in a sequence.

(d) Let $R = \{x: x \text{ is a mountain peaks in the world}\}$. Although it is difficult to count all the mountain peaks in the whole world, R is still a finite set.

Venn Diagram

A universal set S is often represented geometrically by the set of points inside a rectangle, and the subsets of S, such as A and B, are represented by sets of points inside closed areas and shaded. Such a diagrammatic representation of sets is known as *Venn Diagram* (Fig. 11.7).

Set Operations

Union

The *union* of sets A and B is the set of all elements which belong *either to A, or to B, or to both* (i.e. the set of elements which belong to *at least one* of the sets A and B). This is written as

$$A \cup B$$

and read "*A union B*". Some authors also write $A + B$ and call this the set-theoretic "*sum* of A and B", or simply "*A plus B*".

Illustration 6.

Let $A = \{1, 3, 4, 5, 6\}$ and $B = \{2, 4, 6, 8\}$. Then $A \cup B = \{1, 2, 3, 4, 5, 6, 8\}$.

Note: (i) By definition $A \cup B = B \cup A$ (*i.e.* the union of A and B is identical with the union of B and A).

(ii) A and B are subsets of $A \cup B$.

(iii) The union of several sets, viz. $A_1 \cup A_2 \cup \ldots \cup A_n$ is sometimes abbreviated as

$$\bigcup_{i=1}^{n} A_i$$

Intersection

The *intersection* of sets A and B is the set of all elements which belong to *both A and B* (i.e. the set of elements which are *common* to the sets A and B). This is written as

$$A \cap B$$

and read "*A intersection B*". Some authors write AB and call this the set-theoretic "*product* of A and B".

Illustration 7.

(a) Let $A = \{1, 3, 4, 5, 6\}$ and $B = \{2, 4, 6, 8\}$. Then $A \cap B = \{4, 6\}$.

(b) Let $C = \{1, 3, 5\}$ and $D = \{2, 4, 6\}$. Then $C \cap D = \phi$. (i.e. the intersection of C and D is the null set, since they have no common elements).

Note: (i) By definition $A \cap B = B \cap A$ (i.e. the intersection of sets A and B is identical with the intersection of B and A).

(ii) $A \cap B$ is a subset of A and also a subset of B.

(iii) The intersection of several sets, viz. $A_1 \cap A_2 \cap \ldots \cap A_n$ is sometimes abbreviated as

$$\bigcap_{i=1}^{n} A_i$$

Complement

The *complement* of a set A is the set of all elements of the universal set S, which *do not belong* to A. This is written as

$$A' \text{ or } \bar{A} \text{ or } A^\circ$$

and read "*A prime*" or "*A bar*" or "*A complement*" respectively.

Illustration 8.

Let the universal set be $S = \{1, 2, 3, 4, 5, 6\}$ and $A = \{2, 4\}$. Then $A' = \{1, 3, 5, 6\}$.

Note: (i) $A \cup A' = S$ and $A \cap A' = \phi$.

(i.e. the union of a set and its complement is the universal set; and the intersection of a set and its complement is the null set).

(ii) $S' = \phi$, and $\phi' = S$.

(i.e. universal set and null set are complementary to each other).

(iii) $(A')' = A$

(i.e. the complement of the complement of a set is the set itself).

Difference

The *difference* of two sets A and B is the set of all elements which *belong to A but do not belong to B*. This is written as

$$A - B$$

and read "*A difference B*", or simply "*A minus B*".

Illustration 9.

Let $A = \{1, 3, 4, 5, 6\}$ and $B = \{2, 4, 6, 8\}$. Then $A - B = \{1, 3, 5\}$ and $B - A = \{2, 8\}$.

Note: (i) $A - B$ is a subset of A; and $B - A$ is a subset of B.

(ii) $A - B \neq B - A$

(i.e. the difference of A and B is not identical with the difference of B and A).

(iii) $A - B = A \cap B'$

(i.e. the difference of A and B is identical with the intersection of A and B complement).

Disjoint Sets

Two sets A and B are said to be *disjoint*, if they have no common element (i.e. not a single element of B is contained in A, and none of A is contained in B). If A and B are disjoint sets then $A \cap B = \phi$.

Note: (i) Any set A and its complement A' are disjoint, because $A \cap A' = \phi$.

(ii) Any set A and the null set ϕ are disjoint, because $A \cap \phi = \phi$.

Illustration 10.

(a) Let $A = \{2, 6, 7, 9\}$ and $B = \{1, 3, 4, 5, 8\}$. Then A and B are disjoint sets, since no element is common to both A and B.

(b) The set of odd numbers and the set of even numbers are disjoint, since no number can be both odd and even.

(c) Let $A = \{a, b, c, f\}$ and $B = \{d, e, f, g, h\}$. Then A and B are not disjoint, since the element f is common to both.

Venn Diagrams for some important cases are shown below:

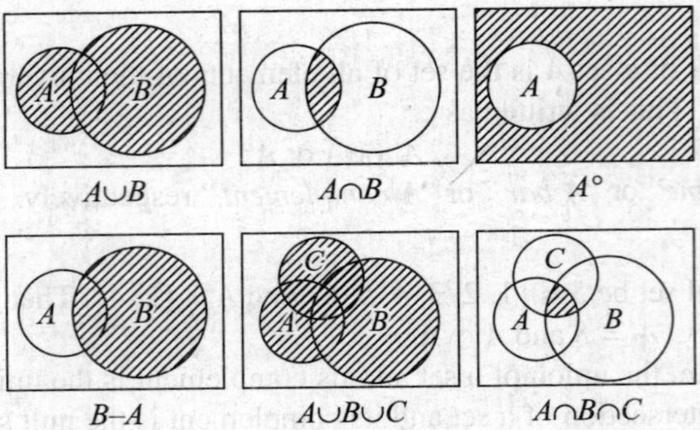

Fig. 11.7 Venn Diagrams

Algebra of Sets

Under the operations of union, intersection and complement, sets satisfy various laws, some of which have already been noted earlier. These are referred to as the algebra of sets:

1. Commutative Laws:
   $A \cup B = B \cup A$ $\qquad$ $A \cap B = B \cap A$
2. Associative Laws:
   $(A \cup B) \cup C = A \cup (B \cup C)$ $\qquad$ $(A \cap B) \cap C = A \cap (B \cap C)$
3. Distributive Laws:
   $A \cup (B \cap C) = (A \cup B) \cap (A \cup C)$ $\qquad$ $A \cap (B \cup C) = (A \cap B) \cup (A \cap C)$
4. De Morgan's Laws:
   $(A \cup B)' = A' \cap B'$ $\qquad$ $(A \cap B)' = A' \cup B'$

Besidees, we have:

5. $A \cup A = A$ $\qquad$ $A \cap A = A$
6. $A \cup \phi = A, \ A \cup S = S$ $\qquad$ $A \cap S = A, \ A \cap \phi = \phi$
7. $A \cup A' = S$ $\qquad$ $A \cap A' = \phi$
8. $(A')' = A$ $\qquad$ $S' = \phi, \ \phi = S$

11.13 SET AND PROBABILITY

Sample Space and Sample Point

In the axiomatic development of probability theory, we start with certain undefined objects, called '*outcomes*'. The 'set' of all possible outcomes of a given random experiment is called '*sample space*'. Any particular outcome (i.e. an element of the sampel space) is called a *Sample Point*. It may be noted that the sample space is not a geometrical space at all and has no dimension. It is the 'universal set' of outcomes. Often the sample space of a random experiment can be given in several ways, but there is usually one which provides the maximum information.

A sample space which is finite or countably infinite is called a '*discrete sample space*'. An uncountably infinite sample space is called a '*continuous sample space*'.

Illustration 1.

(a) In the random experiment of tossing a coin, the sample space S is the set

$$S = \{H, T\}$$

It has two sample points, viz. H and T.

(b) In the random experiment of tossing a die, the sample space is usually given as

$$S = \{1, 2, 3, 4, 5, 6\}.$$

We could also show the sampel space as S_1 = {even, odd}.

(c) In the experiment of tossing two coins

$$S = \{HH, HT, TH, TT\}$$

Note that here the sample points are ordered pairs of results (called 2-tuples) one from each coin. The sample space could also be given as $S_1 = \{0, 1, 2\}$ the sample points indicating the total number of heads obtained from the coins.

(d) In drawing a ball from an urn containing 2 white and 3 red balls, we could regard the balls as numbered, say white balls 1 and 2; red balls 3, 4 and 5. The sample space is then

$$S = \{1, 2, 3, 4, 5\}$$

(e) Suppose 2 balls are drawn from the above urn one at a time without replacement and the order in which they are drawn is taken into consideration. We shall write (4, 1) to denote the outcome when the first ball drawn bears the number 4 and the second ball bears the number 1. The sample space has now 20 points as follows:

$$S = \{(1, 2), (1, 3), (1, 4), (1, 5), (2, 1), (2, 3), 2, 4), (2, 5),$$
$$(3, 1), (3, 2), (3, 4), (3, 5), (4, 1), (4, 2), (4, 3), (4, 5),$$
$$(5, 1), (5, 2), (5, 3), (5, 4)\}$$

It will possibly be easier to understand if we number these 20 elements of S serially, and write

$$S = \{e_1, e_2, e_3, \dots \dots e_{20}\}$$

Note that here the sample points e_i are 2-tuples.

(f) When a coin is tossed 3 times (or 3 coins are tossed simultaneously), the sample space S contains 8 sample points.

$$S = \{HHH, HHT, HTH, HTT, THH, THT, TTH, TTT\}$$

The sample points are now 3-tuples.

(g) Toss a coin repeatedly until a head appears. Then

$$S = \{H, TH, TTH, TTTH, TTTTH, \dots \dots\}$$

If the number of times the coin need be tossed is counted, the sample space may be given as

$$S_1 = \{1, 2, 3, 4, 5, \dots \dots\}$$

(h) Suppose we observe the lifetime (in hours) of electric lamps of a particular brand, which are known to have a maximum life of 2000 hours. The sample space is

$$S = \{x: x \text{ is the lifetime in hours, } 0 \le x \le 2000\}$$

The sample spaces mentioned at (a) to (f) above are finite and at (g) are countably infinite. These are 'discrete' sampel spaces. The sample space at (h) is uncountably infinite and is therefore a 'continuous' sample space.

Event

An *event* is a 'set' of outcomes of a random experiment. In other words, event is a subset of the sample space S. Events may be 'elementary' or 'composite'. An *elementary event* is the set which contains a single sample points. A *composite event* contains more than one sample point.

In particular, the sample space S, which is a subset of itself, is also an event, called the *sure event* or *certain event*. The *impossible event* is the null set ϕ, i.e. the subset of S which contains no sample point at all.

Subevent

If $A \subseteq B$, where A and B are events of the sample space S, then A is called a *subevent* of B.

Illustration 2.

(a) In tossing a coin, the sample space is $S = \{H, T\}$. Let A = event that head appears; B = event that tail appears. Then,

$$A = \{H\}, B = \{T\}$$

Events A and B are elementary. Note the distinction between the sample point H, which is an element of S, and the event $\{H\}$, which is a subset of S.

(b) If the experiment of tossing a die is described by the sample space $S = \{1, 2, 3, 4, 5, 6\}$, then

A = event that an odd number appears $= \{1, 3, 5\}$

B = event that more than 4 appears $= \{5, 6\}$

C = event that "six" appears $= \{6\}$

Events A and B are composite, but event C is elementary.

(c) In drawing 2 balls from an urn containing 2 white and 3 red balls (*Illustration 1*-e),

E = event that the second ball is white

$= \{(1, 2), (2, 1), (3, 1), (3, 2), (4, 1), (4, 2), (5, 1), (5, 2)\}$

(d) In tossing two dice,

A = event that the sum of the points on the dice is 8

$= \{(2, 6), (3, 5), (4, 4), (5, 3), (6, 2)\}$

(e) In the random experiment of tossing 3 coins (*Illustration 1*-f)

A_1 = event that the 2nd coin shows Head

= set of sample points which show H in the second place

$= \{HHH, HHT, THH, THT\}$

A_2 = event that exactly 2 heads appear $= \{HHT, HTH, THH\}$

A_3 = event that all heads or all tails appear $= \{HHH, TTT\}$

(f) In the random experiment of tossing a coin until the first head (*Illustration 1*-g),

A = event that at most 3 tosses are required $= \{H, TH, TTH\}$

(g) In the random experiment of observing the lifetime of electric bulbs (*Illustration 1*-h}

B = Event that a lamp burns not less than 500 hours

$= \{x: 500 \leq x \leq 2000\}$

Remarks

If the sample space S is finite or countably infinite every subset of S is an event.

However, if the sample space is uncountable, then for technical reasons, certain subsets of S cannot be 'events'.

Events Derived by Set Operations

We can combine events to form new events using the various set operations. Thus if A and B are events, then

$A \cup B$ denotes the event "*at least one* on A and B", i.e. "either A or B or both". This is also written as $A + B$.

$A \cap B$ denotes the event "*A as well as B*", i.e. "both A and B". This is also written as AB, and called *the "compound event"*.

A' denotes the event "not-A", i.e. "opposite A". This is also written as $\overline{A}$, and called the *"complementary event"*.

$A - B$ denotes the event "A but not B", i.e. "both A and not-B". This may also be written as $A \cap B'$.

Mutually Exclusive Events

Two events A and B are said to be *mutually exclusive* or *disjoint*, if $A \cap B = \phi$ (*i.e.* A and B cannot occur simultaneously).

Note: (i) Events A and A' are mutually exclusive, because $A \cap A' = \phi$.
(ii) Any event A and the impossible evert ϕ are mutually exclusive, because $A \cap \phi = \phi$.

Illustration 3.

(a) In the experiment of tossing a die, we have $S = \{1, 2, 3, 4, 5, 6\}$. Let

$A = \{1, 3, 5\}$ = event that an odd number appears.
$B = \{2, 4, 6\}$ = event that an even number appears.
$C = \{5, 6\}$ = event that more than 4 appears.

Then

$A \cup C = \{1, 3, 5, 6\}$ = event that an odd number *or* more than 4 appears.
$B \cap C = \{6\}$ = event that a number which is even *as well as* more than 4 appears.
$A' = S - A = \{2, 4, 6\}$ = event that an odd number *does not* appear.
$A' \cup C = \{2, 4, 5, 6\}$ = event that an odd number does not appear *or* more than 4 appears.
$A \cap B = \phi$ = event that an odd number *as well as* an even number appears (this is impossible).

(b) Let the sample sapce $S = \{HH, HT, TH, TT\}$ describe the random experiment of tossing 2 coins, and

A = event that at least one head appears = $\{HH, HT, TH\}$.
B = event that the 2nd coin shows tail = $\{HT, TT\}$.

Then

$A \cup B$ = event that at least one head appears *or* the 2nd coin shows tail
$= \{HH, HT, TH, TT\} = S.$
$A \cap B$ = event that at least one head appears *and* the 2nd coin shows tail
$= \{HT\}.$
B = event that the 2nd coin *des not* show tail = $S - B$
$= \{HH, TH\}.$
$A - B$ = event that at least one head appears *but* the 2nd coin does *not* show tail
$= \{HH, TH\}.$

(c) In tossing 3 coins, $S = \{HHH, HHT, HTH, HTT, THH, THT, TTH, TTT\} = \{e_1, e_2, e_3, e_4, e_5, e_6, e_7, e_8\}$ suppose. Let

A_1 = event that the 2nd coin shows head = $\{e_1, e_2, e_5, e_6\}$

A_2 = event that exactly 2 heads appear = $\{e_2, e_3, e_5\}$

A_4 = event that 2 or more heads appear = $\{e_1, e_2, e_3, e_5\}$

$\therefore A_1 \cup A_2$ = event that *either* the 2nd coin shows head *or* exactly 2 heads appear = $\{e_1, e_2, e_3, e_5, e_6\}$.

A_2' = event that exactly 2 heads *do not* appear = $\{e_1, e_4, e_6, e_7, e_8\}$

Since, $A_2 \subseteq A_4$, we have $A_2 \cup A_4 = A_4$ and $A_2 \cap A_4 = A_2$. Hence

$A_1 \cap A_2 \cap A_4 = A_1 \cap (A_2 \cap A_4) = A_1 \cap A_2 = \{e_2, e_5\}$

= event that the 2nd coin shows head *as well as* 2 heads appear.

11.14 AXIOMS OF PROBABILITY

Let S be a sample space of a random experiment. If to each event A of the set of all possible events of S, we associate a real number $P(A)$, then $P(A)$ is called the "probability" of event A, if the following axioms hold:

Axiom 1. For every event A

$$P(A) \geq 0 \tag{11.14.1}$$

Axiom 2. For the sure event S

$$P(S) = 1 \tag{11.14.2}$$

Axiom 3. For any finite number or countably infinite number of mutually exclusive events $A_1, A_2, \ldots$ of S

$$P(A_1 \cup A_2 \cup \ldots) = P(A_1) + P(A_2) + \ldots \tag{11.14.3}$$

In particular, for two mutually exclusive events A and B

$$P(A \cup B) = P(A) + P(B) \tag{11.14.4}$$

It should be noted that we can speak of the 'probability' only if the event is a subset of a specified sample space S, and to each subset of S a real number, satisfying the axioms, can be assigned. A sample space on which 'probability' has been defined is called a *probability space*.

Deductions from the Axioms

Theorem I

The probability of the impossible event is zero.

$$P(\phi) = 0 \tag{11.14.5}$$

Proof

Any event A and the impossible event ϕ are mutually exclusive.

Also $A \cup \phi = A$. Hence, by Axiom 3,

$$P(A) = P(A \cup \phi) = P(A) + P(\phi)$$

$$\therefore \quad P(\phi) = 0.$$

Theorem II

The probability of the complementary event is

$$P(A') = 1 - P(A) \tag{11.14.6}$$

Proof

A and A' are mutually exclusive events, and

$$A \cup A' = S.$$

Hence, $P(S) = P(A \cup A') = P(A) + P(A')$, by Axiom 3.

i.e. $$1 = P(A) + P(A'), \quad \therefore \quad P(A') = 1 - P(A)$$

Theorem III

The probability of an event lies between 0 and 1.

$$0 \leq P(A) \leq 1 \tag{11.14.7}$$

Proof

By Axiom 1, $0 \leq P(A)$. Also, from (11.14.6), we have $P(A) = 1 - P(A')$. Since $P(A')$ is a probability, it cannot be negative (Axiom 1); therefore $P(A) \leq 1$. Combining both the inequalities, the result follows.

Theorem IV

If $A = A_1 \cup A_2 \cup \ldots \cup A_n$, where $A_1, A_2, \ldots A_n$ are mutually exclusive events, then

$$P(A) = P(A_1) + P(A_2) + \ldots + P(A_n) \tag{11.14.8}$$

In particular, if $A = S$, the sample space, then

$$P(A_1) + P(A_2) + \ldots + P(A_n) = 1 \tag{11.14.9}$$

Proof

Relation (11.14.8) follows from Axiom 3 for the case when the number of mutually exclusive events is finite, say n. In addition, using Axiom 2, the result (11.14.9) follows.

Theorem V

If $A \subseteq B$ (i.e. event A implies event B), then

$$P(A) \leq P(B) \tag{11.14.10}$$

Proof

If $A \subseteq B$, then events A and $A' \cap B$ are mutually exclusive, and their union $A \cup (A' \cap B) = B$. Hence, by Axiom 3,

$$P(B) = P(A) + P(A' \cap B)$$

Since by Axiom 1, $P(A' \cap B)$ cannot be negative, hence

$$P(B) \geq P(A).$$

Theorem VI

For any two events A and B,

$$P(A) = P(A \cap B) + P(A \cap B') \tag{11.14.11}$$

$$P(B) = P(A \cap B) + P(A' \cap B) \tag{11.14.12}$$

Proof

Events $A \cap B$ and $A \cap B'$ are mutually exclusive, and their union is the event A. Hence by Axiom 3, relation (11.14.11) can be proved. Similarly, event B is the union of mutually exclusive events $A \cap B$ and $A' \cap B$, and result (11.14.12) can be proved.

Theorem VII

For any two events A and B (which may or may not be mutually exclusive),

$$P(A \cup B) = P(A) + P(B) - P(A \cap B) \tag{11.14.13}$$

Proof

Events $A \cap B'$, $A \cap B$ and $A' \cap B$ are mutually exclusive, and their union is the event $A \cup B$. Hence, by Axiom 3,

$$P(A \cup B) = P(A \cap B') + P(A \cap B) + P(A' \cap B)$$

But from (11.14.11) and (11.14.12), we have

$$P(A \cap B') = P(A) - P(A \cap B) \text{ and } P(A' \cap B) = P(B) - P(A \cap B).$$

Substituting these values,

$$\begin{aligned} P(A \cap B) &= [P(A) - P(A \cap B)] + P(A \cap B) + [P(B) - P(A \cap B)] \\ &= P(A) + P(B) - P(A \cap B). \end{aligned}$$

Theorem VIII

For any three events A, B, C

$$\begin{aligned} P(A \cup B \cup C) = P(A) + P(B) + P(C) - P(A \cap B) \\ - P(A \cap C) - P(B \cap C) + P(A \cap B \cap C) \end{aligned} \quad (11.14.14)$$

Proof

In (11.14.13) let us replace the event B by $B \cup C$. Then

$$\begin{aligned} P[A \cup (B \cup C)] &= P(A) + P(B \cup C) - P[A \cap (B \cup C)] \\ &= P(A) + [P(B) + P(C) - P(B \cap C)] - P[(A \cap B) \cup (A \cap C)], \end{aligned}$$

by Distributive law.

Again using (11.14.13) for the union of events $A \cap B$ and $A \cap C$,

$$\begin{aligned} P[(A \cap B \cup (A \cap C)] &= P(A \cap B) + P(A \cap C) - P[(A \cap B) \cap (A \cap C)] \\ &= P(A \cap B) + P(A \cap C) - P(A \cap B \cap C) \end{aligned}$$

Hence the result follows.

11.15 FINITE PROBABILITY SPACE AND ASSIGNMENT OF PROBABILITIES

Suppose that a sample space is finite and consists of n elementary outcomes:

$$S = \{e_1, e_2, \dots e_n\}$$

Then a total number of 2^n possible events can be obtained from S. For example, in the random experiment of tossing 2 coins and observing the sequence of Heads and Tails, $S = \{HH, HT, TH, TT\} = \{e_1, e_2, e_3, e_4\}$, suppose. Here, the sample space S contains 4 outcomes and thereofore $2^4 = 16$ events

$$\phi, \{e_1\}, \{e_2\}, \{e_3\}, \{e_4\}, \{e_1, e_2\}, \{e_1, e_3\}, \{e_1, e_4\},$$
$$\{e_2, e_3\}, \{e_2, e_4\}, \{e_3, e_4\}, \{e_1, e_2, e_3\}, \{e_1, e_2, e_4\},$$
$$\{e_1, e_3, e_4\}, \{e_2, e_3, e_4\} \text{ and } \{e_1, e_2, e_3, e_4\} \text{ i.e. } S.$$

In order to define "probability" we need therefore specify 2^n values $P(A)$ for all 2^n events A, so that Axioms 1, 2, and 3 are satisfied.

In the finite probability space, it will however be sufficient if we assign only n real numbers $p_1, p_2, \dots p_n$ to the n elementary events

$$A_1 = \{e_1\}, A_2 = \{e_2\}, \dots A_n = \{e_n\}$$

respectively, such that

(i) $p_i \geq 0 \quad (i = 1, 2, \dots n)$ (11.15.1)

(ii) $p_1 + p_2 + \dots + p_n = 1$ (11.15.2)

This means that the 'numbers' p_i are arbitrary positive proper fractions, or some zeros, whose total is 1. These 'numbers' $p_1, p_2, \dots p_n$ are called "*probabilities*" associated with the elementary events $A_1, A_2, \dots A_n$ respectively.

The probability of an event A can then be computed as the sum of the probabilities of those elementary events whose union constitutes A. For example, if $A = \{e_1, e_3, e_4\}$, we can write

$$A = \{e_1, e_3, e_4\} = \{e_1\} \cup \{e_3\} \cup \{e_4\} = A_1 \cup A_3 \cup A_4$$

because the elementary events $\{e_i\}$ are mutually exclusive. Therefore by Axiom 3,

$$\begin{aligned} P(A) &= P(A_1) + P(A_3) + P(A_4) \\ &= p_1 + p_3 + p_4 \end{aligned}$$

11.16 FINITE EQUIPROBABLE SAMPLE SPACE AND CLASSICAL DEFINITION

A finite sample space $S = \{e_1, e_2, \ldots e_n\}$ is said to have "*equally likely*" outcomes, if the probabilities $p_1, p_2, \ldots, p_n$ assigned to all the n elementary events of S are equal; i.e. $p_1 = p_2 = \ldots = p_n$. The sample space S is then said to be "*equiprobable*".

By (11.15.1) and (11.15.2), since the probabilities $p_1, p_2, \ldots, p_n$ cannot be negative and their total must be 1, each p_i must have the same value $\frac{1}{n}$, i.e.

$$p_1 = p_2 = \ldots = p_n = \frac{1}{n}$$

Therefore, if an event A has m sample points, then the probability of A is

$$P(A) = \frac{1}{n} + \frac{1}{n} + \ldots \text{ (m times)}$$

or,

$$P(A) = \frac{m}{n} \tag{11.16.1}$$

Thus, if we have a finite equiprobable sample space of n elementary outcomes then the computation of probability of an event reduces to the computation of the number (m) of sample points which belong to the event A. The probability of A is then given by (11.16.1).

$$P(A) = \frac{\text{Number of outcomes favourable to } A}{\text{Total number of possible outcomes}} = \frac{m}{n}$$

This is equivalent to the 'classical' definition of probability (11.5.1).

The axiomatic definition of probability is thus more general, from which the classical definition can be obtained as a special case, i.e. when the sample space is finite and equiprobable.

Summary of Axiomatic Theory of Probability (Discrete Case)

I. (*Finite Case*) Let us suppose that a random experiment has a finite number (n) of possible 'elementary outcomes' $e_1, e_2, \ldots, e_n$. The set $S = \{e_1, e_2, \ldots e_n\}$ is called a 'Sample Space' connected with the random experiment and its elements e_i are called 'Sample Points'. Any set, say $\{e_1, e_2, e_5, e_7\}$, which is a subset of S, is called an "Evant". The sets $\{e_i\}$ consisting of single elements are called 'elementary events', while sets which consist of more than one element are called 'composite events'. In particular, the null set ϕ is called the 'impossible event' and the universal set S is called the 'sure event'.

Suppose that corresponding to the elementary events $\{e_1\}, \{e_2\}, \ldots \{e_n\}$ we are given real numbers $p_1, p_2 \ldots p_n$ respectively such that $p_i \geq 0$ and $\sum_1^n p_i = 1$. The numbers p_i are called "probabilities" assigned to the elementary events $A_i = \{e_i\}$. The probability of any event, say

$$A = \{e_1, e_2, \ldots p_m\}$$

is then given by the sum of the probabilities associated with those outcomes which belong to the event A (strictly speaking, with those elementary events whose union constitutes the given event A).

$$P(A) = p_1 + p_2 + \ldots + p_m$$

In particular, if the possible outcomes $e_1, e_2, \ldots e_n$ are "equally likely", the probabilities $p_1, p_2, \ldots p_n$ are equal and all have the same value $1/n$. The probability of the event A consisting of m sample points is then given by

$$P(A) = \frac{m}{n}$$

Thus the classical definition of probability is obtained from the axiomatic theory as a special case.

II. (*Countably Infinite Case*) The theories here are exactly the same as in Case I, with the following modifications:

The number of possible elementary outcomes $e_1, e_2, \ldots$ is countably infinite, and the sample space $S = \{e_1, e_2, \ldots\}$ is an infinite set. The probabilities $p_1, p_2, \ldots$ assigned to the elementary events are, as before, non-negative and their sum equals 1, i.e. $p_i \geq 0$ and $\sum_1^\infty p_i = 1$.

However, since the number of possible outcomes is infinite, the classical definition cannot be derived from this case.

Example 11.67 *There are 3 children in a family. What is the probability that they include (i) exactly 2 girls, (ii) not move than one girl?*

Solution The sample space is finite and contains 8 sample points

$$S = \{BBB, BBG, BGB, BGG, GBB, GBG, GGB. GGG\}$$

If we assume that the birth of boys and girls are equally likely, then S is equiprobable, so that the probability assigned to each elementary event is $\frac{1}{8}$.

Probability that there are exactly 2 girls

$$= P(\{BGG, GBG, GGB\}) = \frac{1}{8} \times 3 = \frac{3}{8}$$

Probability that not more than one is a girl

$$= P(\{BBB, BBG, BGB, GBB\}) = \frac{1}{8} \times 4 = \frac{1}{2}$$

Example 11.68 *Two urns marked I and II contain respectively 3 white, 2 black balls and 2 white, 4 black balls. If one ball is drawn from each urn, what is the probability that both are white balls?*

Solution Let us assume that the balls are numbered as follows:

Urn I: White balls 1, 2, 3; Black balls 4, 5.

Urn II: White balls 6,7; Black balls 8, 9, 10, 11.

The sample space S contains 30 sample points, each of which is an ordered pair of the form (2, 9), where the first number refers to the ball from Urn I and the second from Urn II. If the balls are identical in all respects except in colour, we may assume that the sample space is equiprobable, so that the probability assigned to each elementary event is $\frac{1}{30}$. The event A that both the balls are white is

$$A = \{(1, 6), (1, 7), (2, 6), (2, 7), (3, 6), (3, 7)\}$$

which contains 6 sample points. Thus $P(A) = \frac{1}{30} \times 6 = \frac{1}{5}$.

11.17 CONDITIONAL PROBABILITY

Let A and B be two events, such that $P(A) \neq 0$. Then the *Conditional probability* of B, *given that* A has occurred, is defined as

$$P\left(\frac{B}{A}\right) = \frac{P(A \cap B)}{P(A)} \qquad (11.17.1)$$

It follows that

$$P(A \cap B) = P(A) \, . \, P\left(\frac{B}{A}\right) \qquad (11.17.2)$$

This is known as Multiplication Theorem of probability.

Similarly, if $P(B) \neq 0$, the conditional probability of A, given that B has occurred, is

$$P\left(\frac{A}{B}\right) = \frac{P(A \cap B)}{P(B)} \qquad (11.17.3)$$

from which we get

$$P(A \cap B) = P(B) \, . \, P\left(\frac{A}{B}\right) \qquad (11.17.4)$$

Equating the right hand sides of (11.17.2) and (11.17.4),

$$P(A) \, . \, P\left(\frac{B}{A}\right) = P(B) \, . \, P\left(\frac{A}{B}\right) \qquad (11.17.5)$$

provided $P(A) \neq 0$, $P(B) \neq 0$. Thus we find that

$$P\left(\frac{A}{B}\right) = \frac{P(A)}{P(B)} P\left(\frac{B}{A}\right) \qquad (11.17.6)$$

Example 11.69 *There are 3 children in a family. Find the probability that all the children are boys, (i) if no prior information is available about the children, (ii) if it is known that the two eldest are boys, (iii) if it is known that at least two of them are boys.*

Solution The sample space is shown in Example 11.67. let A, C, D denote the events that 'all 3 are boys', 'eldest two are boys', and 'at least 2 of the children are boys' respectively.

Then

$$A = \{BBB\}, C = \{BBB, BBG\}, D = \{BBB, BBG, BGB, GBB\}$$

We are required to find (i) $P(A)$, (ii) $P\left(\frac{A}{C}\right)$, (iii) $P\left(\frac{A}{D}\right)$, By definition, $P(C) \neq 0$ and $P(D) \neq 0$,

$$P\left(\frac{A}{C}\right) = \frac{P(A \cap C)}{P(C)}, \quad P\left(\frac{A}{D}\right) = \frac{P(A \cap D)}{P(D)}$$

Since the sample space is finite (size 8) and equiprobable, and events A, C, D contain 1, 2 and 4 sample points respectively

$$P(A) = \frac{1}{8}, P(C) = \frac{2}{8}. P(D) = \frac{4}{8}.$$

Also, $P(A \cap C) = P(\{BBB\}) = \frac{1}{8}$, $P(A \cap D) = P(\{BBB\}) = \frac{1}{8}$, Thus,

Probability that all the children are boys $= P(A) = \frac{1}{8}$.

Probability that all the children are boys, given that the two eldest are boys

$$= P\left(\frac{A}{C}\right) = \frac{1}{8} \div \frac{2}{8} = \frac{1}{2}$$

Probability that all the children are boys, given that at least 2 of them are boys

$$= P\left(\frac{A}{D}\right) = \frac{1}{8} \div \frac{4}{8} = \frac{1}{4}$$

Example 11.70 *If A and B are mutually exclusive events and $P(A \cup B) \neq 0$, then prove that*

$$P\left(\frac{A}{A \cup B}\right) = \frac{P(A)}{P(A) + P(B)}$$ [C.U., B.Sc. '81]

Solution By the definition of conditional probability, and since $P(A \cup B) \neq 0$ (given),

$$P\left(\frac{A}{A \cup B}\right) = \frac{P[A \cap (A \cup B)]}{P(A \cup B)} \quad \text{(i)}$$

Also, since A and B are known to be mutually exclusive,

$$A \cap B = \phi \quad \text{and} \quad P(A \cup B) = P(A) + P(B) \quad \text{(ii)}$$

By the Algebra of sets (Distributive Law),

$$A \cap (A \cup B) = (A \cap A) \cup (A \cap B)$$
$$= A \cup \phi = A$$

$$\therefore \quad P[A \cap (A \cup B)] = P(A) \quad \text{(iii)}$$

Substituting from (ii) and (iii) on the R.H.S. of (i), the result follows.

11.18 INDEPENDENT EVENTS

An event B is considered to be *independent* of an event A, if the probability that B occurs is not influenced by the knowledge that A has occurred. In other words, the unconditional probability of B equals the conditional probability of B, given A, i.e.

$$P(B) = P\left(\frac{B}{A}\right) \tag{11.18.1}$$

Using (11.17.5), we see that in case $P(B) = P\left(\frac{B}{A}\right)$, then $P(A) = P\left(\frac{A}{B}\right)$. Thus, if B is independent of A, then A is also independent of B. We therefore say that events "*A and B are independent.*" From the Multiplication Theorem (11.17.2) it follows that $P(A \cap B) = P(A) \,.\, P(B)$. This equation is taken as a formal definition for independence of two events.

Events A and B are said to be "independent" (or stochastically independent), if and only if

$$P(A \cap B) = P(A) \,.\, P(B) \tag{11.18.2}$$

Otherwise, they are said to be "*dependent*" events,

If events A and B are independent, it can be shown (Example 11.72 that their complementary events A' and B' are also independent. Similarly, events A and B' and also events A' and B are independent. This means that if $P(A \cap B) = P(A) \,.\, P(B)$ then the following relations are also satisfied:

$$P(A' \cap B') = P(A') \,.\, P(B')$$
$$P(A \cap B') = P(A) \,.\, P(B')$$
$$P(A' \cap B) = P(A') \,.\, P(B).$$

Thus, we may replace any one or both the events by their complements A', B' on both sides of (11.18.2) and still get a true result.

Note: If A and B are *two* independent events then (11.18.2) holds. Conversely, if this relation holds, the events must be independent (in the sense that any combination of A or A' with B or B' will be independent). The relation (11.18.2) is therefore called a 'necessary and sufficient' condition for independence of two events.

Pairwise Independent Events

Several events $A_1, A_2, ..., A_n$ are said to be "*pairwise independent*", if every pair of these events are independent; i.e.

$$P(A_i \cap A_j) = P(A_i) \,.\, P(A_j) \tag{11.18.3}$$

for all values of $i, j = 1, 2, ..., n, (i \neq j)$.

For example, 3 events A, B, C will be said to be 'pairwise independent' if the following relations hold:

$$P(A \cap B) = P(A) \,.\, P(B)$$
$$P(A \cap C) = P(A) \,.\, P(C)$$
$$P(B \cap C) = P(B) \,.\, P(C)$$

Mutually Independent Events

Several events $A_1, A_2, ... A_n$ are said to be "*mutually independent*" (or simply *independent*), if the probability of the joint occurrence of any number of these events is equal to the product of their probabilities.

For example, 3 events A, B, C will be said to be 'independent' if every one of the following relations holds:

$$P(A \cap B) = P(A) \,.\, P(B)$$
$$P(A \cap C) = P(A) \,.\, P(C)$$
$$P(B \cap C) = P(B) \,.\, P(C) \quad (11.18.4)$$
and
$$P(A \cap B \cap C) = P(A) \,.\, P(B) \,.\, P(C)$$

In general, for the mutual independence of n events, $2^n - n - 1$ such conditions must be satisfied.

It may be noted here that if 3 or more events $A_1, A_2, ..., A_n$ are mutually independent, then the relation

$$P(A_1 \cap A_2 \cap ... \cap A_n) = P(A_1)\, P(A_2), ...\, P(A_n) \quad (11.18.5)$$

obviously holds (Besides some other relations must also hold):

If events $A_1, A_2, ..., A_n$ are independent, then their complements $A_1', A_2', ..., A_n'$ are also independent. In particular

$$P(A_1' \cap A_2' \cap ... \cap A_n') = P(A_1')\, P(A_2'), ...\, P(A_n') \quad (11.18.6)$$

Note:

1. If n events are mutually independent, they are necessarily pairwise independent. The converse is not true. Pairwise independent events may not be mutually independent.
2. If 3 or more events are independent, then the relation (11.18.5) must hold. But the converse is not true in general. In other words, if only (11.18.5) holds, the events $A_1, A_2, ... A_n$ $(n \geq 3)$ may not be independent. This relation is therefore a necessary, but not sufficient, condition for independence of *more than two* events.
3. If n events A_i are independent, then any combination of these events, taken any number at a time, is also independent; e.g. events A_2, A_4 and A_5 will be independent.
4. If n events A_i are independent and we replace some or all of them by their complements, then any combination of the new group of events, taken any number at a time, will also be independent; e.g. events A_1' A_3, A_5' and A_7' will be mutually independent. In particular, the n complementary events A_i' will also be independent.

Example 11.71 *A fair coin is tossed 3 times in succession. Show that the events 'first toss gives a head' and 'second toss gives a head' are independent.*

Solution The sample space $S = \{HHH, HHT, HTH, HTT, THH, THT, TTH, TTT\}$ is equiprobable, and contains 8 sample points.

Let A and B denote respectively the events 'first toss gives a head' and 'second toss gives a head'. Then

$$P(A) = P(\{HHH, HHT, HTH, HTT\}) = \frac{4}{8} = \frac{1}{2}$$

$$P(B) = P(\{HHH, HHT, THH, THT\}) = \frac{4}{8} = \frac{1}{2}$$

$$P(A \cap B) = P(\{HHH, HHT\}) = \frac{2}{8} = \frac{1}{4}$$

We find that $P(A) \,.\, P(B) = \frac{1}{2} \times \frac{1}{2} = \frac{1}{4} = P(A \cap B)$

So, A and B are independent events.

Example 11.72 *Let A and B be two independent events. Then show that (i) A and B^c, (ii) A^c and B^c are also independent.* [C.U., B.Com. (Hons) '81]

Solution Since A and B are independent events, we have

$$P(A \cap B) = P(A) \,.\, P(B) \tag{1}$$

Here A^c and B^c denote the complements of A and B respectively.

(i)
$$\begin{aligned} P(A \cap B^c) &= P(A) - P(A \cap B), && \text{by (11.14.11)} \\ &= P(A) - P(A) \,.\, P(B), && \text{by (1) above} \\ &= P(A) \,.\, \{1 - P(B)\} \\ &= P(A) \,.\, P(B^c), && \text{by (11.14.6)} \end{aligned}$$

Since $P(A \cap B^c) = P(A) \,.\, P(B^c)$, the events A and B^c are independent.

(ii)
$$\begin{aligned} P(A^c \cap B^c) &= P(A \cup B)^c, \text{ by De Morgan's Law} \\ &= 1 - P(A \cup B), \\ &= 1 - [P(A) + P(B) - P(A \cap B)], && \text{by (11.14.13)} \\ &= 1 - P(A) - P(B) + P(A) \,.\, P(B), && \text{by (1)} \\ &= [1 - P(A)]\,[1 - P(B)] \\ &= P(A^c) \,.\, P(B^c) \end{aligned}$$

Thus events A^c and B^c are independent.

Example 11.73 *(a) $A_1, A_2, \ldots A_n$ are independent events such that $P(A_i) = 1 - q_i$, $i = 1, 2, \ldots n$. Prove that*

$$P\left(\bigcup_{i=1}^{n} A_i\right) = 1 - q_1 q_2 \cdot\cdot\, q_n.$$

(b) In supplies of 3 components, viz. base, neck and switch, for an electric lamp, the percentages of defectives on a day were 5, 20 and 10 respectively. An assembled lamp is considered defective if at least one of the 3 components is defective. If components are selected randomly, what is the probability that an assembled lamp would be defective? [W.B.H.S. '79]

Solution

(a) Since $A_1, A_2, \ldots A_n$ are independent events, their complements $A_1', A_2', \ldots A_n'$ are also independent, having probabilities $P(A_i') = 1 - P(A_i) = q_i$ (given, $i = 1, 2 \ldots n$. So,

$$\begin{aligned} P(A_1' \cap A_2' \cap \ldots \cap A_n') &= P(A_1') \,.\, P(A_2'), \ldots P(A_n') \\ &= q_1 q_2 \ldots q_n \end{aligned}$$

$$\begin{aligned} \therefore \quad P\left(\bigcap_{i=1}^{n} A_i\right) &= P(A_1 \cup A_2 \cup \ldots \cup A_n) \\ &= 1 - P(A_1 \cup A_2 \cup \ldots \cup A_n)', \text{ since } P(A) = 1 - P(A') \\ &= 1 - P(A_1' \cap A_2' \cap \ldots \cap A_n'), \text{ by De Morgan's law} \\ &= 1 - q_1 q_2 \ldots q_n \end{aligned}$$

(b) Let A, B, C denote the events 'base is defective', 'neck is defective', 'switch is defective' respectively. Then (given)

$$P(A) = 5\% = \frac{1}{20}, \; P(B) = 20\% = \frac{1}{5}, \; P(C) = 10\% = \frac{1}{10}.$$

The probabilities of complementary events are

$$P(A') = \frac{19}{20}, \; P(B') = \frac{4}{5}, \; P(C') = \frac{9}{10}.$$

We assume that the events A, B, C are independent.

$\therefore$ Probability that the lamp is defective

$$\begin{aligned}&= \text{Probability that at least one component is defective}\\&= P(A \cup B \cup C) = P(A' \cap B' \cap C')'\\&= 1 - P(A' \cap B' \cap C')\\&= 1 - P(A') \cdot P(B') \cdot P(C'), \text{ since } A, B, C \text{ are assumed independent}\\&= 1 - \frac{19}{20} \times \frac{4}{5} \times \frac{9}{10}\\&= \frac{79}{250} = 0.316\end{aligned}$$

Comparison of Classical Theory and Axiomatic Theory

(i) In the classical theory, all theorems and results are obtianed by logical arguments. In the axiomatic theory, all results are derived from the axioms by using the mathematical properties of sets.

(ii) The classical theory is based upon the concept of "equally likely cases" when the number of possible outcomes is only finite. The axiomatic theory is quite general and embraces all cases whether equally likely or not and irrespective of whether the number of possible outcomes is finite or infinite.

(iii) The classical theory defines "event" simply as a phenomenon which may arise in the performance of the random experiment. The axiomatic theory defines the "event" strictly according to mathematical principles as a 'set', which is in effect a subset of the universal set of all possible outcomes, called the Sample Space.

(iv) In the classical theory, "probability" is defined as a ratio (i.e. fraction) of two positive whole numbers (the numerator may be zero) showing the number of cases favourable to the event and the total number of all possible outcomes which are equally likely. In the axiomatic theory, "probability" is simply a non-negative number associated with the event, i.e. probability is a set-function obeying the three axioms.

(v) The 'addition theorem' of classical theory is not derived in the axiomatic theory, but simply accepted as a rule by Axiom 3 (see 11.14.4). The 'multiplication theorem' is derived as a rule (see 11.17.2) which follows from the definition of conditional probability in axiomatic theory.

(vi) The concepts of "conditional probability" and "independent events" are introduced in the classical theory by logical arguments, whereas in the axiomatic theory these are defined by mathematical statements.

11.19 RANDOM VARIABLE

Let S be a sample space of some given random experiment. It has been observed that the outcomes (i.e. sample points of S) are not always numbers. We may however assign a real number to each sample point according to some definite rule. Such an assignment given us a "*function defined on the sample space S*". This function is called a *random variable* (or *stochastic variable*).

Random Variable X may be defined as a function which assigns a real number $X(e)$ to each sample point e of a given sample space S.

Illustration 1.

In the random experiment of tossing 2 coins, let the sample space be $S = \{HH, HT', TH, TT\}$. If X is the random variable denoting the 'number of heads', then we have assigned a number to each sample point as follows:

$$X(HH) = 2, \quad X(HT) = 1, \quad X(TH) = 1, \quad X(TT) = 0$$

(**Note:** Many other random variables could also be proposed on this sample space, e.g. the square of the number of tails, the difference of the numbers of heads and tails.)

Illustration 2.

Let $S = \{1, 2, 3, 4, 5, 6\}$ be a sample space of the random experiment of throwing a die. The sample points are 1, 2, 3, 4, 5, 6. We introduce the random variable X = number of points obtained when the die is thrown. Then

$$X(1) = 1, \; X(2) = 2, \quad X(3) = 3, \quad X(4) = 4, \quad X(5) = 5, \quad X(6) = 6$$

Suppose, we introduce another random variable Y on the same sample space, Y = (Number of points less 3)^2. Then

$$Y(1) = (1-3)^2 = 4 \qquad Y(4) = (4-3)^2 = 1$$
$$Y(2) = (2-3)^2 = 1 \qquad Y(5) = (5-3)^2 = 4$$
$$Y(3) = (3-3)^2 = 0 \qquad Y(6) = (6-3)^2 = 9$$

These are shown in the table below:

Sample point	1	2	3	4	5	6
Random variable Y	4	1	0	1	4	9

A random variable which assumes a finite number or a countably infinite number of values is called a *Discrete random variable.* If the random variable assumes an uncountably infinite number of values, it is called a *Continuous random variable.*

Illustration 3.

The random variable X in Illustration 1 takes only 3 values 0, 1, 2. In *Illustration 2*, X takes 6 values 1, 2, ... 6, and Y takes 4 values 0, 1, 4, 9. These are all discrete random variables with a finite number of values.

Discrete Probability Distribution

Let X be a discrete random variable which can assume the values $x_1, x_2, x_3, \ldots$ (arranged in an increasing order of magnitude) with probabilities $p_1, p_2, p_3 \ldots$ respectively ($\Sigma p_i = 1$). The specification of the set of values x_i together with their probabilities p_i $(i = 1, 2, 3, \ldots)$ defines the *discrete probability distribution* of X.

Let us write $f(x)$ to denote the probability that X takes a specified value x.

$$f(x) = P(X = x) \qquad (11.19.1)$$

The function $f(x)$ is called the *probability mass function* (p.m.f.), or simply *probability function,* of the discrete random variable X. It satisfies two conditions

(i) $f(x) \geq 0$, (ii) $\Sigma f(x) = 1$ (11.19.2)

the summation in (ii) being taken over all values of $x = x_1, x_2, x_3, \ldots$

$$f(x) = p_i \text{ when } x = x_i (i = 1, 2, \ldots)$$
$$= 0 \quad \text{otherwise} \qquad (11.19.3)$$

Example 11.74 *Find the probability distributions of the random variables in Illustrations 1 and 2.*

Solution

(i) Assuming that the coin is unbiased, the sample space is finite and equiprobable.

$$P(X = 0) = P(\{TT\}) = \frac{1}{4}$$

$$P(X = 1) = P(\{HT, TH\}) = \frac{2}{4} = \frac{1}{2}$$

$$P(X = 2) = P(\{HH\}) = \frac{1}{4}$$

The probability distribution of X is

x	0	1	2	Total
$f(x)$	$\frac{1}{4}$	$\frac{1}{2}$	$\frac{1}{4}$	1

(ii) In *Illustration 2*, assuming that the die is unbiased, the probability distribution of X is

x	1	2	3	4	5	6	Total
$f(x)$	$\frac{1}{6}$	$\frac{1}{6}$	$\frac{1}{6}$	$\frac{1}{6}$	$\frac{1}{6}$	$\frac{1}{6}$	1

Again, $P(Y = 0) = P(\{3\}) = \frac{1}{6}$

$$P(Y = 1) = P(\{2, 4\}) = \frac{2}{6} = \frac{1}{3}$$

$$P(Y = 4) = P(\{1, 5\}) = \frac{2}{6} = \frac{1}{3}$$

$$P(Y = 9) = P(\{6\}) = \frac{1}{6}$$

The probabiity distribution of Y is therefore

y	0	1	4	9	Total
$f(y)$	$\frac{1}{6}$	$\frac{1}{3}$	$\frac{1}{3}$	$\frac{1}{6}$	1

Note the distinction between capital letters X, Y which denote the random variables, and small letters x, y which denote any arbitrary values of these random variables.

Example 11.75 *Find the probability distribution of the number of tails when a coin is thrown repeatedly until the first head appears.*

Solution Here $S = \{H, TH, TTH, TTTH, \ldots\}$

Sample point	H	TH	TTH	$TTTH$	...
X	0	1	2	3	...

Assuming that the coin is unbiased,

$$P(X = 0) = P(\{H\}) = \frac{1}{2}$$

$$P(X = 1) = P(\{TH\}) = \left(\frac{1}{2}\right)^2$$

$$P(X = 2) = P(\{TTH\}) = \left(\frac{1}{2}\right)^3$$

$$P(X = 3) = P(\{TTTH\}) = \left(\frac{1}{2}\right)^4 \text{ etc.}$$

The probability distribution of X is

x	0	1	2	3	...	Total
$f(x)$	$\frac{1}{2}$	$\left(\frac{1}{2}\right)^2$	$\left(\frac{1}{2}\right)^3$	$\left(\frac{1}{2}\right)^4$	...	1

Continuous Probability Distribution

If X is a continuous random variable, the number of possible values which it can assume is uncountably infinite, and hence the probability function cannot be defined in the same manner as for a discrete random varaible. In this case we introduce a function $f(x)$ such that it satisfies two conditions.

$$\text{(i) } f(x) \geq 0 \quad \text{(ii) } \int_{-\infty}^{\infty} f(x)\, dx = 1. \tag{11.19.4}$$

The probability that X lies between two specified values c and d, is defined as

$$P(c \leq X \leq d) = \int_c^d f(x)\, dx \tag{11.19.5}$$

It may be shown that this definition of probability satisfies all the Axioms of probability (page 409). The function $f(x)$ is called the *probability density function* (p.d.f.), or simply *density function,* of the continuous random variable X. Any function satisfying both the conditions (11.19.4) may be accepted as a density function.

Example 11.76 *(a) Is the following a probability density function?*

$$f(x) = \begin{cases} 2x, & 0 < x \leq 1 \\ 4 - 2x, & 1 < x \leq 2 \\ 0, & \text{elsewhere} \end{cases}$$

(b) If the random variable X has the probability density function

$$f(x) = \begin{cases} \frac{1}{4}, & -2 \leq x \leq 2 \\ 0 & \text{elsewhere} \end{cases}$$

obtain P{(2x + 3) > 5}. (Here P denotes probability)

[C.U., B.Sc. (Econ) '81 (New)]

Solution

(a) If the function $f(x)$ is to be a p.d.f. it must satisfy the two conditions (i) and (ii) at (11.19.4). The function $f(x)$ defined here clearly satisfies (i). It is not negative at any value of x.

$$\int_{-\infty}^{\infty} f(x)dx = \int_{-\infty}^{0} f(x)dx + \int_0^1 f(x)dx + \int_1^2 f(x)dx + \int_2^{\infty} f(x)dx$$

$$= 0 + \int_0^1 (2x)dx + \int_1^2 (4-2x)dx + 0$$

(the first and the last integrals being zero, since $f(x)$ itself is defined to be zero in those two intervals)

$$= \left[x^2\right]_0^1 + \left[4x - x^2\right]_1^2$$
$$= 1 + 1 = 2$$

Since condition (ii) is not satisfied, the given function $f(x)$ is not a probability density function.

(b) The event $(2x + 3) > 5 \Rightarrow 2x > 2 \Rightarrow x > 1$. Therefore

$$P\{(2x + 3) > 5\} = P\{x > 1\}$$
$$= \int_1^2 f(x)\, dx \text{, since } f(x) = 0 \text{ for } x > 2.$$
$$= \int_1^2 \left(\frac{1}{4}\right) dx = \frac{1}{4}$$

[*Ans.* No.; $\frac{1}{4}$]

11.20 CUMULATIVE DISTRIBUTION FUNCTION (C.D.F.)

The probability that a random variable X takes a value less than or equal to a specified value x is a function of x, and is usually denoted by $F(x)$. This function is called *cumulative distribution function* (c.d.f.), or simply *distribution function,* of the random variable X.

$$F(x) = P(X \le x) \tag{11.20.1}$$

Since $F(x)$ represents probability, it must lie between 0 and 1.

$$0 \le F(x) \le 1 \tag{11.20.2}$$

For the discrete probability distribution of X, which assumes a finite number of values $x_1, x_2, \ldots x_n$ with probabilities $p_1, p_2 \cdots p_n$ respectively, the c.d.f. is given as follows:

$$\begin{aligned} F(x) &= 0 && \text{when } x < x_1 \\ &= p_1 && \text{when } x_1 \le x < x_2 \\ &= p_1 + p_2 && \text{when } x_2 \le x < x_3 \\ &= p_1 + p_2 + p_3 && \text{when } x_3 \le x < x_4 \\ &\ldots && \ldots \\ &= p_1 + p_2 + \ldots + p_n = 1 && \text{when } x_n \le x \end{aligned} \tag{11.20.3}$$

The probability that X lies in an interval, say between x_3 and x_7 (both inclusive), is then given by

$$\begin{aligned} P(x_3 \le X \le x_7) &= P(X \le x_7) - P(X \le x_2) \\ &= F(x_7) - F(x_2) \end{aligned} \tag{11.20.4}$$

For the continuous probability distribution of X, with probability density function $f(x)$ in the interval $(a \le x \le b)$, the c.d.f. is given as

$$F(x) = \int_a^x f(t)\, dt \tag{11.20.5}$$

If follows that the p.d.f. $f(x)$ can be obtained from the c.d.f. $F(x)$ dy differentiation with respect to x.

$$f(x) = \frac{d}{dx} F(x) \tag{11.20.6}$$

If the c.d.f. is known, the probabilities of intervals can be obtained from the relation

$$P(c \le X \le d) = F(d) - F(c) \tag{11.20.7}$$

In particular when c and d are equal

$$P(c \le X \le c) = F(c) - F(c) = 0$$

i.e.

$$P(X = c) = 0 \tag{11.20.8}$$

Thus, the probability that a continuous random variable takes any specified value is zero. As a consequence,

$$\begin{aligned} P(c \le X \le d) &= P(c \le X < d) \\ &= P(c < X \le d) = P(c < X < d). \end{aligned}$$

Example 11.77 *An unbiased coin is thrown three times. If the random variable X denotes the number of heads obtained, find the cumulative distribution function (c.d.f.) of X.*

Solution It can be easily shown that the probabilities of obtaining 0 head, 1 head, 2 heads, 3 heads are respectively $\frac{1}{8}, \frac{3}{8}, \frac{3}{8}, \frac{1}{8}$. The probability distribution of X is therefore

x	0	1	2	3	Total
$f(x)$	$\frac{1}{8}$	$\frac{3}{8}$	$\frac{3}{8}$	$\frac{1}{8}$	1

From this table, we can find the probabilities of obtaining 0 or less head, 1 or less head, 2 or less heads, 3 or less heads, as follows:

$$\begin{aligned} F(0) &= P(X \le 0) = P(X = 0) = \frac{1}{8} \\ F(1) &= P(X \le 1) \\ &= P(X = 0) + P(X = 1) = \frac{1}{8} + \frac{3}{8} = \frac{1}{2} \\ F(2) &= P(X \le 2) \\ &= P(X = 0) + P(X = 1) + P(X = 2) \\ &= \frac{1}{8} + \frac{3}{8} + \frac{3}{8} = \frac{7}{8} \\ F(3) &= P(X \le 3) \\ &= P(X = 0) + P(X = 1) + P(X = 2) + P(X = 3) \\ &= \frac{1}{8} + \frac{3}{8} + \frac{3}{8} + \frac{1}{8} = 1 \end{aligned}$$

Thus we have

x	0	1	2	3
$F(x)$	$\frac{1}{8}$	$\frac{1}{2}$	$\frac{7}{8}$	1

In general, if x denotes any arbitrary real number, the probability that the number of heads (X) is x or less can be given as

$$F(x) = 0, \qquad \text{when } x < 0$$

because we cannot have less than 0 head.

$$F(x) = \frac{1}{8}, \qquad \text{when } 0 \le x < 1$$

For example $F(0.7) = \frac{1}{8}$, because the probability of having 0.7 or less head is the same as that of having 0 head.

$$F(x) = \frac{1}{8} + \frac{3}{8} = \frac{1}{2}, \text{ when } 1 \le x < 2$$

For example, $F(11.34) = \frac{1}{2}$, because the probability of having 11.34 or less heads is the same as that of having 0 or 1 head. Similarly,

$$F(x) = \frac{1}{8}+\frac{3}{8}+\frac{3}{8}=\frac{7}{8}, \qquad \text{when } 2 \le x < 3$$

and $$F(x) = \frac{1}{8}+\frac{3}{8}+\frac{3}{8}+\frac{1}{8}=1, \qquad \text{when } 3 \le x$$

For example $F(8.75) = 1$, because the probability of having 8.75 heads or less is the same as that of having 0, 1, 2, 3 heads.

The c.d.f. of X is thus given as follows:

$$\begin{aligned} F(x) &= 0, && \text{when } x < 0 \\ &= \frac{1}{8}, && \text{when } 0 \le x < 1 \\ &= \frac{1}{2}, && \text{when } 1 \le x < 2 \\ &= \frac{7}{8}, && \text{when } 2 \le x < 3 \\ &= 1, && \text{when } 3 \le x \end{aligned}$$

Example 11.78 *The p.d.f. of a continuous random variable is*

$$y = k(x-1)(2-x); \qquad (1 \le x \le 2)$$

Determine

(i) the value of the constant k;

(ii) the cumulative distribution function;

(iii) the probability that X is less than $\frac{5}{4}$;

(iv) the probability that X is greater than $\frac{3}{2}$;

(v) the probability that X lies between $\frac{5}{4}$ *and* $\frac{3}{2}$.

Solution $y = k(x-1)(2-x) = k(3x - x^2 - 2)$

(i) Using the second relation of (11.19.4),

$$1 = \int_1^2 k(3x - x^2 - 2)\, dx$$

$$= k \cdot \left[\frac{3x^2}{2} - \frac{x^3}{3} - 2x\right]_1^2$$

$$= k\left[\left(6 - \frac{8}{3} - 4\right) - \left(\frac{3}{2} - \frac{1}{3} - 2\right)\right] = \frac{k}{6}$$

Thus, $\frac{k}{6} = 1$; i.e. $k = 6$

The complete p.d.f. is therefore $y = 6(3x - x^2 - 2)$

(ii) $F(x) = \int_1^x 6(3t - t^2 - 2)\, dt$, by (11.20.5)

$$= 6\left[\frac{3t^2}{2} - \frac{t^3}{3} - 2t\right]_1^x$$

$$= 6\left[\left(\frac{3x^2}{2} - \frac{x^3}{3} - 2x\right) - \left(\frac{3}{2} - \frac{1}{3} - 2\right)\right]$$

$$= 5 - 12x + 9x^2 - 2x^3$$

(iii) Probability that X is less than $\frac{5}{4}$

$$= P\left(X \le \frac{5}{4}\right) = F\left(\frac{5}{4}\right), \text{ by (11.20.1)}$$

$$= 5 - 12 \times \frac{5}{4} + 9 \times \frac{25}{16} - 2\,\frac{125}{64}$$

$$= \frac{5}{32}$$

(iv) Probability that X is less than $\frac{3}{2}$

$$= P\left(X < \frac{3}{2}\right) = P\left(X \le \frac{3}{2}\right) = F\left(\frac{3}{2}\right)$$

$$= 5 - 12 \times \frac{3}{2} + 9 \times \frac{9}{4} - 2 \times \frac{27}{8}$$

$$= \frac{1}{2}$$

$\therefore$ Probability that X is greater than $\frac{3}{2}$

$$= P\left(X > \frac{3}{2}\right) = 1 - P\left(X \le \frac{3}{2}\right)$$

$$= 1 - F\left(\frac{3}{2}\right) = 1 - \frac{1}{2} = \frac{1}{2}$$

(v) Probability that X lies between $\frac{5}{4}$ and $\frac{3}{2}$

$$= P\left(\frac{5}{4} \le X \le \frac{3}{2}\right) = F\left(\frac{3}{2}\right) - F\left(\frac{5}{4}\right)$$

$$= \frac{1}{2} - \frac{5}{32} = \frac{11}{32}$$

11.21 JOINT DISTRIBUTION OF TWO VARIABLES (DISCRETE)

Let S be a sample space of some given random experiment. We may assign two real numbers $X(e)$ and $Y(e)$ to each sample point e of S according to some given rules (Section 11.19). We then have two random variables X and Y defined on the sample space S.

The possible pairs of values (x, y) assumed by the two random variables X and Y together with the probabilities for all the pairs of values, gives the *Joint Distribution* of X and Y, or *Bivariate Distribution.*

Example 11.79 *Four unbiased coins are tossed. If X and Y denote respectively the 'number of heads' and the 'longest run of heads', construct the joint distribution of X and Y.*

Solution The sample space $S = \{e_1, e_2, \ldots e_{16}\}$ consisting of the following 16 sample points is equiprobable. Hence, the probability of each elementary event is $\frac{1}{16}$ (p. 404). The values (x, y) assumed by X and Y are shown against each sample point.

Sample point	x	y
e_1 = *HHHH*	4	4
e_2 = *HHHT*	3	3
e_3 = *HHTH*	3	2
e_4 = *HHTT*	2	2
e_5 = *HTHH*	3	2
e_6 = *HTHT*	2	1
e_7 = *HTTH*	2	1
e_8 = *HTTT*	1	1
e_9 = *THHH*	3	3
e_{10} = *THHT*	2	2
e_{11} = *THTH*	2	1
e_{12} = *THTT*	1	1
e_{13} = *TTHH*	2	2
e_{14} = *TTHT*	1	1
e_{15} = *TTTH*	1	1
e_{16} = *TTTT*	0	0

The distinct pairs of values (x, y) when systematically arranged, shown the following joint distribution of X and Y:

(x, y)	*Probability*
0, 0	$\frac{1}{16}$
1, 1	$\frac{4}{16}$
2, 1	$\frac{3}{16}$
2, 2	$\frac{3}{16}$
3, 2	$\frac{2}{16}$
3, 3	$\frac{2}{16}$
4, 4	$\frac{1}{16}$
Total	1

However, it is preferable to show the values x and y in a two-way table, with the probabilities written in the cells as in the table below:

Joint Distribution of X and Y

x \ y	*0*	*1*	*2*	*3*	*4*	*Total*
0	$\frac{1}{16}$					$\frac{1}{16}$
1		$\frac{4}{16}$				$\frac{4}{16}$
2		$\frac{3}{16}$	$\frac{3}{16}$			$\frac{6}{16}$
3			$\frac{2}{16}$	$\frac{2}{16}$		$\frac{4}{16}$
4					$\frac{1}{16}$	$\frac{1}{16}$
Total	$\frac{1}{16}$	$\frac{7}{16}$	$\frac{5}{16}$	$\frac{2}{16}$	$\frac{1}{16}$	1

EXERCISES

1. Define and illustrate the following terms—Mutually exclusive events, Exhaustive set of events, Independent events, and Complementary events. *[C.U., M. Com. '79; B.Com. (Hons) '80]*
2. Give the *classical definition* of probability. What are its limitations? *[C.U., B.Com. (Hons) '82; I.C.W.A., June '81; W.B.H.S. '82]*
3. State and prove the *Addition Theorem* of probability for two mutually exclusive events. What modification would you make if the events are *not* mutually exclusive?
4. With usual notations, prove that
$$P(A + B) = P(A) + P(B) - P(AB).$$
[W.B.H.S. '78; I.C.W.A., June '81]
5. (a) Establish the relation
$$P(A + B) \leq P(A) + P(B).$$
When does the sign of equality hold? *[W.B.H.S. '78]*
   (b) If events A and B are not mutually exclusive, show that
$$P(AB) \geq P(A) + P(B) - 1$$
6. What is meant by *Compound Event* in probability ? State and prove the Theorem of Compound Probability. *[C.U., B.Com. (Hons) '81; B.Sc. (Econ) '81]*
7. State and prove the *Multiplication Theorem* of probability. How is the result modified when the events are *independent*? *[C.U., M.Com. '80]*
8. Give the different definitions of probability and state their limitations if any. *[C.U., M.Com. '81]*
9. State whether each of the following statements is *True* or *False*:
   (a) The probability of an event cannot be zero.
   (b) The probability of a complementary event must not exceed one.
   (c) The probability of the simultaneous occurrence of two events can never exceed the sum of the probabilities of these events.
   (d) The probability of occurrence of at least one of two events is less than the probability of occurrence of each of these events.
   (e) If two events are mutually exclusive, their complements are also mutually exclusive.
   (f) Two events are said to be "independent", if the occurrence of one of them is unaffected by the occurrence of the other event.
   (g) The conditional probability of an event, given another event, can never be less than the probability of the joint occurrence of these events.
   (h) Two mutually exclusive events must be independent.
   (i) The mathematical expectation of the product of two random variables is given by he product of their mathematical expectations.
   (j) In a Bernoullian series of trials, the probability of success must remain a constant in each trial.
10. Examine the following relations and state which of them are *incorrect* (A and B are arbitrary events):

(a) $P(A + B) = P(A) + P(B)$
(b) $P(AB) \leq P(A) + P(B)$
(c) $P(A + B) \geq P(A)$
(d) $P(A \text{ and } B) \leq P(A \text{ or } B)$
(e) $P(B) \leq P(A \text{ and } B)$
(f) $P(A \cup B) + P(A \cap B) = P(A) + P(B)$
(g) $P(A \cap B) = P(A \cup B) + P(A \cup \bar{B}) + P(\bar{A} \cup B)$
(h) $P\left(\frac{A}{B}\right) . P\left(\frac{B}{A}\right) = 1$
(i) $P\left(\frac{A}{B}\right) \leq \frac{P(A)}{P(B)}$
(j) $P\left(\frac{\bar{A}}{B}\right) = 1 - P\left(\frac{A}{B}\right)$

11. State which of the following statements are *Correct* and which are *Wrong*:
   (a) For any event A, $P(A)$ is a non-negative real number.
   (b) If A and B are mutually exclusive events, then $P(A) + P(B) = 1$.
   (c) If event A imples event B, then $P(A) \leq P(B)$.
   (d) If $P\left(\frac{A}{B}\right) = P(A)$, then $P\left(\frac{B}{A}\right) = P(A)$.
   (e) If $P(A) \neq 0$ and $P(B) \neq 0$, then $\frac{P(A)}{P(B)} = \frac{P\left(\frac{A}{B}\right)}{P\left(\frac{B}{A}\right)}$
   (f) If A and B are independent events, then $P(AB) = P(A) . P(B)$.
   (g) If events A and B are independent, then A^c and B^c are dependent.
   (h) If A and B are independent events, then A and B are also independent.
   (i) If $P(ABC) = P(A) . P(B) . P(C)$, then events A, B, C are independent.
   (j) The probability mass function $f(x)$ of a discrete random variable X may exceed 1.

12. Let A be the event that a person is a male and B the event that a person is more than 6 feet tall. Give, in words, the events whose probabilities are represented by
   (i) $P(A \text{ or } B)$
   (ii) $P(A \text{ or } B) - P(A \text{ and } B)$,
   (iii) $1 - P(A)$. *[W.B.H.S. '80]*

13. In each of the following cases, pick out the correct alternative:
   (a) If an event cannot happen, the probability of the event will be: + 1, – 1, 0, none of these. *[C.A., Nov. '81]*
   (b) When two perfect coins are tossed simultaneously, the probability of getting at least one head is: $\frac{1}{2}, \frac{1}{4}, \frac{3}{4}, 1$. *[C.A., May '81]*

(c) A dice is thrown two times, and the sum of numbers on the faces up is noted. The probability of this sum being 11 is: $\frac{1}{6}$, $\frac{1}{36}$, $\frac{1}{18}$, none of these. [C.A., Nov. '82]

(d) Out of 120 tickets numbered consecutively from 1 to 120, one is drawn at random. What is the probability of getting a number which is a multiple of 5? $\frac{1}{24}, \frac{1}{8}, \frac{1}{5}, \frac{1}{16}$. [C.A., May '81]

(e) If two cards are drawn at random from a good deck, the probability that both the cards are of the same colour is: $\frac{1}{4}, \frac{1}{2}$, 1, none of these.

(f) If A_1, A_2, A_3 are equally likely, mutually exclusive and exhaustive, then $P(A_1)$ equals: 1, 0, $\frac{1}{2}, \frac{1}{3}$. [W.B.H.S. '82]

(g) If $P(A)$ = 0.2, $P(B)$ = 0.1, and $P(C)$ = 0.3, and A, B, C are mutually independent events, then the probabiilty that all the three events do not happen simultaneously is: .006, 0.504, 0.994, none of these.

(h) The probability of having at least one 'six' from 3 throws of a perfect dice is: $\left(\frac{5}{6}\right)^3, 1-\left(\frac{5}{6}\right)^3, \left(\frac{1}{6}\right)^3$, none of these. [W.B.H.S. '79]

(i) The probabilities of occurrence of events A and B are respectively 0.25 and 0.50. Both the events occur simultaneously with probability 0.14. The probability that neither A nor B occurs is: 0.39, 0.25, 0.11, none of these.

14. For two events A and B, let $P(A) = 0.4$, $P(A + B) = 0.7$ and $P(B) = p$. For what value of p are A and B (i) mutually exclusive, (ii) independent?

15. Three perfect coins are tossed together. What is the probability of getting at least one head? [C.A., Nov. '81]

16. If four unbiased coins are tossed, find the probability that there should be two tails. [C.U., B.Com. (Hons) '82]

17. A card is drawn at random from a well-shuffled pack of cards. What is the probability that it is a heart or a queen ? [C.A. May '82]

18. In a single cast with two dice find the chance of throwing 7 (i.e. for throwing two numbers whose sum is 7). [B.U., B.Com. '76]

19. An ordinary die is tossed twice and the difference between the numbers of spots turned up is noted. Find the probability of a difference of 3. [C.U., B.A. (Econ) '66]

20. Three balls are drawn at random from a bag containing 6 blue and 4 red balls. What is the chance that two balls are blue and one is red? [B.U., B.Com. '77]

21. An urn contains 8 white and 3 red balls. If 2 balls are drawn at random, find the probability that (i) both are white, (ii) both are red, (iii) one is of each colour. [C.U., B.A. (Econ) '73]

22. Five persons A, B, C, D, E speak at a meeting. What is the probability that A speaks immediately before B?

23. The nine digits 1, 2, 3, ..., 9 are arranged in random order to form a nine-digit number. Find the probability that 1, 2 and 3 appear as neighbours in the order mentioned. *[C.U., B.A. (Econ) '72]*
24. What is the chance that a leap year, selected at random, will contain 53 Sundays? *[C.U., B.Sc. '77]*
25. Three cards are drawn at random one after another from a full pack of Playing Cards. What is the probability that (i) the first two are spades and the third is a heart, (ii) two are spades and one is a heart?
26. A four-digit number is formed by the digits 1, 2, 3, 4 with no repetition. Find the probability that the number is (i) odd, (ii) divisible by 4.
27. Four dice are thrown. Find the probability that the sum ofthe numbers appearing will be 18.
28. There are 4 persons in a company. Find the probability that (i) all of them have different brithdays, (ii) at least 2 of them have the same birthday, (iii) exactly 2 of them have the same birthday (Assume 1 year = 365 days).
29. (i) If $P(A) = \frac{1}{4}$, $P(B) = \frac{2}{5}$, $P(A + B) = \frac{1}{2}$, find $P(AB)$ $P\left(\frac{A}{B}\right)$ and $P\left(\frac{B}{A}\right)$.

    (ii) If $P(A) = \frac{1}{2}$, $P(B) = \frac{3}{5}$, $P(AB) = \frac{1}{3}$, find $P(A + B)$, $P(\overline{A}\overline{B})$, $P(\overline{A} + \overline{B})$ and $P(A\overline{B})$.

    (iii) If A and B are independent events, and $P(A) = \frac{2}{3}$, $P(B) = \frac{3}{5}$, find $P(A + B)$, $P\left(\frac{\overline{A}}{B}\right)$ and $P(\overline{A}B)$.
30. Given $P(A\overline{B}) = \frac{1}{3}$, $P(A + B) = \frac{2}{3}$, find $P(B)$. What is $P(A)$, if $P(AB) = \frac{1}{6}$? *[C.U., B.Sc. '73]*
31. The probability that a contractor will get a plumbing contract is $\frac{2}{3}$, and the probability that he will not get an electric contract is $\frac{5}{9}$. If the probability of getting at least one contract is $\frac{4}{5}$, what is the probability that he will get both the contracts? *[C.A., May '79]*
32. From a set of 1000 cards serially numbered 1, 2, 3, ..., 1000, one card is drawn at random. Find the probability that the number found is a the multiple of (i) 12 or 18, (ii) 12 and 18.
33. The probability that Asok can solve a problem in Business Statistics is $\frac{4}{5}$, that Amal can solve it is $\frac{2}{3}$, and that Abdul can solve it is $\frac{3}{7}$. If all of them try independently, find the probability that the problem will be solved. *[C.U., B.Com. (Hons) '82]*

34. The odds against student X solving a Business Statistics problem are 8 to 6, and the odds in favour of student Y solving the same problem are 14 to 16. (i) What is the chance that the problem will be solved if they both try, independently of each other? (ii) What is the probability that neither solves the problem? *[C.A., Nov. '79]*
35. Two sets of candidates are competing for the position on the Board of Directors of a company. The probabilities that the first and second sets will win are 0.6 and 0.4 respectively. If the first set wins, the probability of introducing a new product is 0.8, and the corresponding probability if the second set wins is 0.3. What is the probability that the new product will be introduced? *[C.A., Nov. '78]*
36. Three groups of children contain respectively 3 girls and 1 boy, 2 girls and 2 boys, and 1 girl and 3 boys. One child is selected at random from each group. Find the chance that the selected group consists of 1 girl and 2 boys. *[C.U., B.Sc. '77]*
37. The probability that a teacher will give a surprise test during any class meeting is $\frac{1}{5}$. If a student is absent on two days, what is the probability that he will miss at least one test? *[C.U., B.Sc. '76]*
38. Three lots contain respectively 10%, 20% and 25% defective articles. One article is drawn at random from each lot. What is the probability that among them there is (i) exactly one defective, (ii) at least one defective?
39. A candidate is selected for interview for 3 posts. For the first post there are 3 candidates, for the second post there are 4, and for the third there are 2. What are the chances of getting at least one post? *[C.A., May '80]*
40. An urn contains 2 white and 2 black balls. Balls are drawn successively at random without replacement. What is the probability that a black ball appears (i) for the first time in the 3rd drawing, (ii) for the 2nd time in the 4th drawing ? *[C.U., B.Sc. '76]*
41. Five different letters are put inside 5 addressed envelopes by an illiterate servant. What is the probability that only 2 letters are placed in the correct envelopes ?
42. There are two identical boxes containing respectively 4 white and 3 red balls, and 3 white and 7 red balls. A box is chosen at random and a ball is drawn from it. Find the probability that the ball is white. *[C.U., B.Sc. (Math) '67]*
43. What is the chance of getting at least one defective item, if 3 items are drawn randomly from a lot containing 10 items, of which just 2 are defective ? *[W.B.H.S., '78]*
44. A lot contains 10 items of which 3 are defective. Three items are chosen from the lot at random one after another without replacement. Find the probability that all the three are defective. *[C.A., May '81]*
45. A packet of 10 electronic components is known to include 3 defectives. If 4 components are randomly chosen and tested, what is the probability of finding among them not more than one defective? *[C.U., B.Com. (Hons) '80; M.Com. '79]*
46. A bag contains 7 red and 5 white balls. 4 balls are drawn at random. What is the probability that (i) all of them are red; (ii) 2 of them are red and 2 white? *[C.U., M.Com. '72]*

47. A subcommittee of 6 members is to be formed out of a group consisting of 7 men and 4 ladies. Calculate the probability that the subcommittee will consist of (i) exactly 2 ladies, and (ii) at least 2 ladies. *[C.U., B.A. (Econ) '75]*
48. A bag contains 8 red balls and 5 white balls. Two successive draws of 3 balls are made without replacement. Find the probability that the first drawing will give 3 white balls and the second 3 red balls. *[C.A., May '78]*
49. A bag contains 8 white and 4 red balls. Two balls are first drawn at random and then without replacement another two balls are drawn at random. What is the probability of obtaining one white and one red ball in each drawing? *[C.U., M.Com. '80]*
50. In a bridge game, North and South have 9 spades between them. Find the probability that either East or West has no spades. (There are only 13 spades in a pack of 52 cards and each player has 13 cards. The players are designated by the positions they occupy, viz. North, South, East, West.) *[C.U., B.Sc. (Math) '68]*
51. A manufacturer supplies cheap clocks in lots of 50. A buyer, before taking a lot, tests a random sample of 5 clocks and if all are good, he accepts the lot. What is the probability that he accepts a lot containing 10 defective clocks? *[C.U., B.Com(Hons) '67]*
52. 5 cards are drawn from a pack of 52 well-shuffled cards. Find the probabiity that 4 are aces and 1 is a king. *[C.U., B.A. (Econ) '69]*
53. A bag contains 50 tickets numbered 1, 2, 3, ..., 50, of which 5 are drawn at random and arranged in ascending order of their numbers $x_1 < x_2 < ... < x_5$. What is the probability that $x_3 = 30$? *[C.U., B.Sc. (Math.) '66]*
54. A bag contains 5 red and 4 black balls. A ball is drawn at random from the bag and put into another bag which contains 3 red and 7 black balls. A ball is drawn randomly from the second bag. What is the probability that it is red? *[C.U., B.A.(Econ) '70]*
55. An experiment succeeds twice as often as it fails. What is the probability that is the next 6 trails there will be at least 4 successes?
56. If 20 dates are named at random, what is the probability that 3 of them will be Sundays?
57. A factory produces articles among which 20% are defective. If 5 articles are selected at random from a day's production, find the probability that there will be (i) exactly 2 defectives, (ii) not less than 2 defectives.
58. The incidence of occupational disease is such that on the average 20% of workers suffer from it. If 10 workers are selected at random, find the probability that (i) exactly 2 workers suffer from the disease, (ii) not more than 2 workers suffer from the disease. *[C.U., B.A. (Econ) '74]*
59. A marksman firing at a target hits the bulls-eye once in 3 shots on an average. If he fires 4 times, what is the probability of hitting the bulls-eye (i) twice, (ii) 3 times, (iii) not at all?
60. A bag contains 5 white and 2 black balls. The experiment of drawing a ball from the bag is performed 3 times. Calculate the probability of drawing 2 white and 1 black balls, if the ball is (i) replaced, (ii) not replaced, after each drawing.

61. A box contains 7 white and 5 black balls. If 3 balls are drawn simultaneously at random, what is the probability that they are not all of the same colour? Calculate the probability of the same event for the case where the balls are drawn in succession with replacement between drawings. *[C.U., M.Com. '81]*
62. Two dice are thrown *n* times in succession. What is the probability of obtaining doubel-six at least once? *[C.U., B.Sc. '75]*
63. (a) Define *mathematical expectation.* *[C.A., Nov. '82]*
    (b) An urn contains 7 white and 3 red balls. Two balls are drawn together at random from this urn. Find the mathematical expectation of the number of white balls drawn. *[C.U., B.A. (Econ) '78]*
64. A man draws 2 balls from a bag containing 3 white and 5 black balls. If he is to receive Rs 14 for every white ball which he draws and Rs 7 for every black ball, what is his expectation? *[C.U., B.Sc. '77]*
65. A bag contains 5 white and 7 black balls. Find the expectation of a man who is allowed to draw two balls from the bag and who is to receive one rupee for each black ball and two rupees for each white ball drawn. *[C.U., B.Com. (Hons) '82]*
66. If a person gains or loses an amount equal to the number appearing when a balanced die is rolled once, accordig to whether the number is even or odd, how much money can he expect per game in the long run? *[C.U., B.Sc. (Econ) '82]*
67. *A* and *B* play for a prize of Rs 99. The prize is to be won by a player who first throws a '3' with one die. *A* first throws, and if he fails *B* throws, and if he fails *A* again throws, and so on. Find their respective expectations. *[C.A., Nov. '81]*
68. If *x* is a discrete random variable and *k* an arbitrary constant, show that
    $E(kX) = k \cdot E(X).$
69. (a) For two jointly distributed random variables *X* and *Y*, prove that
    $E(X + Y) = E(X) + E(Y).$
    (b) Three coins whose faces are marked 1 and 2, are tossed. What is the expectation of the total value of numbers on their faces? *[C.A., Nov. '82]*
70. (a) If *X* and *Y* are *independent* random variables, show that
    $E(XY) = E(X) \cdot E(Y).$
    (b) A number is chosen at random from the set 1, 2, 3, ..., 100 and another number is chosen at random from the set 1, 2, 3, ..., 50. What is the expected value of the product ? *[C.U., B.Com. (Hons) '81]*
71. Three boxes of the same appearance have the following proportions of white and black balls—Box I: 1 white and 2 black; Box II: 2 white and 1 black; Box III: 2 white and 2 black. One of the boxes is selected at random and one ball is drawn randomly from it. It turns out to be white. What is the probability that the third box is chosen? *[C.U., B.Sc. '79]*
72. The contents of three urns are as follows: 1 white, 2 black, 3 red balls; 2 white, 1 black, 1 red; 4 white, 5 black, 3 red. One urn is chosen at random and 2 balls are drawn. They happen to be white and red. What is the probability that they came from the second urn? *[C.U., B.Sc. (Math) '67]*

73. Let A, B, C be the subsets of a sample space S defined by
$S = \{1, 2, 3, 4, 5, 6, 7, 8\}$, $A = \{1, 3, 5, 7\}$
$B = \{1, 2, 3, 4, 5\}$, $C = \{5, 6, 7, 8\}$.
Find the set (i) $A \cup B$, (ii) $A \cap B$, (iii) $A \cup B \cup C$, (iv) $(A \cup B)'$, (v) $(A \cap B)'$, (vi) $C'' \cap B'$, (vii) $A \cap B \cap C$.

74. Let $S = \{a, b, c, d, e\}$ be the Universal set and let $A = \{a, b, d\}$ and $B = \{b, d, e\}$ be two of its subsets. Find $(A \cap B)'$ and $(A \cup B)'$. *[C.U., B. Com. (Hons) '82]*

75. Explain the terms—Sample space; Sample point; Event; Elementary and Composite events, as used in the axiomatic theory of probability.

76. Suppose that an unbiased coin is tossed 3 times in succession, and that one is interested to know the number of *Heads* that fall. Using this illustration, explain the terms—Sample space, Event, Random variable. *[I.C.W.A., June '82]*

77. Suppose that a random experiment consists of throwing a coin 3 times in succession. Write down the sample space S. Mention two mutually exclusive *events* of this sample space. *[C.U., B.Com. (Hons) '82]*

78. A biased coin is tossed 3 times. Write down the sample space, and the event "the second toss shows a head". If the probability of getting a head with the coin is $\frac{1}{3}$, find the probability of this event.

79. Explain the axiomatic theory of probability and show how the classical definition can be derived from this as a special case.

80. Let S be a sample space and let A be any event in the field of events F. State three axioms to define a probability function $P(A)$ on the field F. Using these axioms prove that
(i) $P(A) + P(A') = 1$
(ii) $P(A \cup B) = P(A) + P(B) - P(A \cap B)$ for *any two* events A and B in F. *[C.U., B.Com. (Hons) '82]*

81. Assuming the result for two events and using set algebra, show that for 3 events A, B, C

$$P(A \cup B \cup C) = P(A) + P(B) + P(C) - P(A \cap B) - P(A \cap C) - P(B \cap C) + P(A \cap B \cap C).$$

82. A and B are two events of a sample space on which a probability function has been defined.
(i) If $P(A) = 1/2$, $P(B) = 3/8$ and $P(A \cap B) = 1/4$, find the values of $P(A \cup B)$, $P(B')$, $P(A' \cap B')$, $P(A' \cup B')$, $P(A' \cap B)$, $P(A/B)$, $P(B/A')$.
(ii) If $P(A \cup B) = 3/4$, $P(A) = 1/2$, $P(B') = 5/8$, find the values of $P(A \cap B)$, $P(A' \cup B)$, $P(A'/B)$.
(iii) If $P(A \cap B) = 1/4$, $P(A \cup B) = 3/4$, $P(B) = 1/3$, find $P(A \cup B')$, $P(A \cap B')$ $P(A/B')$
(iv) If $P(A) = 1/4$, $P(B/A) = 1/2$, $P(A/B) = 3/4$, find $P(B)$, $P(A \cup B)$, $P(A^c/B^c)$.

83. Let A_1 and A_2 be two events related to an experiment E. If it is given that $P(A_1) = 1/2$, $P(A_2) = 1/3$, and $P(A_1 \cap A_2) = 1/4$, then find $P([A_1 \cup A_2]^c)$ where c stands for the complement. *[C.U., B.Sc. '76]*

84. A box contains 40 envelops of which 25 are ordinary (not meant for airmail) and 16 are unstamped, while the number of unstamped ordinary envelopes is

10. What is the probability that an envelope chosen from the box is a stamped air-mail envelope? *[C.U., B.Sc. '76]*

85. Let A, B, C be three events related to an experiment. Under what conditions will the events be (i) exhaustive, (ii) mutually exclusive, (iii) independent ?

86. Let A, B, C be three arbitrary events. Find expressions for the following events, using the usual set-theory notations:
 (i) only A occurs,
 (ii) both A and B, but not C occur,
 (iii) all these events occur,
 (iv) at least one event occurs,
 (v) at least two events occur. *[C.U., B.Sc. (Econ) '81]*

87. One card is drawn at random from a well-shuffed pack of 52 playing cards. Let A denote the event that the card drawn is a Heart, and B denote the event that it is a face card (King, Queen or Jack). Assuming natural assignment of probabilities, evaluate (i) $P(A \cup B)$, (ii) $P(A \cap B')$.

88. The probability that a construction job will be finished in time is 17/20; the probability that there will be no strike is 3/4; and the probability that the construction job will be finished in time, assuming that there will be no strike, is 14/15. Find the probability that
 (i) The construction job will be finished in time and there will be no strike;
 (ii) there will be strike or the job will not be finished in time.

89. What is *conditional probability*? *[W.B.H.S. '78; I.C.W.A. June '82; C.U., M.Com. '76 & B.Sc. (Econ) '81]*

90. Given the information that family has 2 children and that at least one of these two children is a boy, find the probability that both are boys. *[C.U., B.Sc. '76]*

91. There are two events A and B. The probability that B happens is x when A has happened; and y when A has failed to happen. If the probability that A happens is p, find the probability that B happens

92. (a) Show that in general, $P(B/A) = 1 - P(\bar{B}/A)$.
 (b) If $P(A) > P(B)$, show that $P(A/B) > P(B/A)$.

93. When are *two* events said to be 'independent'? *[W.B.H.S. '78; C.U., B.Sc. (Econ) '81; I.C.W.A., June '82]*

94. It is 9 to 5 against a person who is 50 years living till he is 70 and 8 to 6 against a person who is 60 living till he is 80. Find the probability that at least one of them will be alive after 20 years. *[C.A., Nov. '81]*

95. Prove that if A and B are mutually exclusive events and $P(A) \neq 0$, $P(B) \neq 0$, then A and B are not stochastically independent. *[C.U., B.Sc. (Econ) '81]*

96. Distinguish between *pairwise independence and mutual independence* of events. *[C.U., B.Sc. (Econ) '82]*

97. (a) Define stochastic independence of three events.
 (b) Suppose that all the four possible outcomes e_1, e_2, e_3, e_4 of an experiment are equally likely. Define the events A, B, C as
 $A = \{e_1, e_4\}, B = \{e_2, e_4\}, C = \{e_3, e_4\}$.

What can you say about the dependence or independence of the events A, B, C? [C.U., B.Sc. '76]

98. One shot is fired form each of three guns. Let A, B, C denote the events that the target is hit by the first, second and the third gun respectively. Assuming that A, B, C are mutually independent events and the $P(A) = 0.5$, $P(B) = 0.6$, $P(C) = 0.8$, find the probability that at least on hit is registered.

99. There are three men aged 60, 65 and 70 years. The probability to live 5 years more is 0.8 for a 60 year old, 0.6 for a 65 year old, and 0.3 for a 70 year old person. Find the probability that at least two of the three persons will remain alive 5 years hence. [C.U., B.Com. (H) '81]

100. The probabilities of occurrence of 3 independent events are p, q, r. Find the probabilities of (i) occurrence of at least one of the events, (ii) occurrence of exactly one of the events (whichever occurs).

101. If $P(A \cap B \cap C) = 0$, show that

$$P(A \cup B/C) = P(A/C) + P(B/C)$$

102. If A, B, C are independent events, prove that A and $B \cup C$ are independent.

103. If events $A_1, A_2, \ldots A_n$ are independent, show that

$$P\left(\bigcup_{i=1}^{n} A_i\right) = 1 - P(A_1').P(A'_2). \ldots P(A_n')$$

104. A random experiment consists of 3 independent tosses of a fair coin. Let X be the random variable whose value for any outcome is the number of heads obtained. (i) Construct a table showing the probability distribution of X. Also show the cumulative distribution function (c.d.f.) of X.

105. A continuous random variable X follows uniform distribution with probability density function (p.d.f.)

$$f(x) = 1/2; \ (4 \le x \le 6)$$

(a) Find the probabilities (i) $P(4 \le X \le 5)$, (ii) $P(X < 4.2)$ (iii) $P(X < 3.2)$, (iv) $P(3 \le X \le 4.5)$, (v) $P(X > 5.5)$.

(b) Find the cumulative distribution function (c.d.f.) and hence determine that value of k such that $P(X < k) = 0.7$.

106. A random variable X has the probability density function

$$f(x) = cx^2, \ 0 \le x \le 1$$
$$= 0 \quad \text{otherwise}$$

Find the value of the constant c; and the probabilities $P(0 \le X \le \frac{1}{2})$, $P\left(X > \frac{3}{4}\right)$, $P\left(\frac{1}{4} < X < \frac{3}{4}\right)$. What is the cumulative distribution function?

107. (a) Explain the concept of *probability density function.*

(b) Show that $f(x)$ defined by

$$\begin{aligned} f(x) &= x; & 0 \le x \le 1 \\ &= k - x; & 1 \le x \le 2 \\ &= 0 & \text{elsewhere,} \end{aligned}$$

is a probability density function for a suitable value of the constant k. [C.U., B.Sc. (Math.) '73]

108. A continuous random variable X has the density function

$$f(x) = \frac{1}{2} - ax;\ (0 \le x \le 4)$$

where a is constant. (i) Determine the value of a and the probability that X lies between 2 and 3. (ii) Also find the cumulative distribution function (c.d.f.) and calcualte the probabilities $P(X \le 1)$, $P(X \ge 2.5)$, $P(|X - 2| < 0.5)$.

109. The length of life X (in hours) of a certain electronic components is assumed to follow a continuous distribution with density function $f(x) = k/x^2$; $(1000 \le x \le 1500)$.

Determine the constant k, and find the probability that a component selected at random will function for at least 1200 hours.

110. (a) Define the *distribution function* or a random variable X and state some of its properties. *[C.U., B.Sc. (Econ) '81]*

(b) Given the probability density function

$$f(x) = \frac{1}{2}e^{-x/2},\ x > 0$$

evaluate the distribution function. *[C.U., B.Sc. (Econ) '82]*

ANSWERS

9. True—(b), (c), (g), (j); False—(a), (d), (e), (f), (h), (i).
10. (a), (e), (g), (h)
11. Correct—(a), (c), (e), (f), (h); Wrong—(b), (d), (g), (i), (j).
12. (i) The person is either a male or more than 6 ft. tall;
    (ii) The person is either a male not more than 6 ft. tall, or a female but more than 6 ft. tall;
    (iii) The person is not a male.
13. (a) 0; (*b*) 3/4; (c) 1/18; (d) 1/5; (e) none of these; (f) 1/3; (g) 0.994; (h) $1 - (5/6)^3$; (i) 0.39
14. (i) 0.3; (ii) 0.5
15. 7/8
16. 3/8
17. 4/13
18. 1/6
19. 1/6
20. 1/2
21. 28/55, 3/55, 24/55
22. 1/5
23. 1/72
24. 2/7
25. 13/850, 39/850
26. 1/2, 1/4
27. 5/81
28. $(364 \times 363 \times 362)/365^3 = 0.984$; $1 - 0.984 = 0.016$; $(6 \times 364 \times 363)/365^3$
29. (i) 3/20, 3/8, 3/5; (ii) 23/30, 7/30, 2/3, 1/6; (iii) 13/15, 1/3, 1/5
30. 1/3, 1/2;
31. 14/45
32. 0.111, 0.027
33. 101/105
34. 73/105, 32/105
35. 0.6
36. 13/32
37. 9/25
38. 3/8, 23/50
39. 3/4
40. 1/6, 1/2
41. 1/6
42. 61/140
43. 8/15
44. 1/120
45. 2/3
46. 7/99, 14/33
47. 5/11, 53/66
48. 7/429
49. 112/495
50. 11/115
51. ${}^{40}C_5/{}^{50}C_5$
52. $4/{}^{52}C_5$

12 THEORETICAL DISTRIBUTIONS–BINOMIAL, POISSON, NORMAL

12.1 RANDOM VARIABLE AND PROBABILITY DISTRIBUTION

Random Variable has been defined as a 'function' which assumes real values on the outcomes of a random experiment (see Section 11.19). In repeated performances of the random experiment, the random variable will in general assume different values with a definite probability associated with each value or interval of values.

Probability Distribution (or simply *Distribution*) of a random variable is a statement specifying the set of its possible values together with their respective probabilities. When a random experiment is theoretically assumed to serve as a model, the probabilities can be given as a function of the random variable. The probability distribution concerned is then generally known as *theoretical distribution*. We shall discuss here mainly four important theoretical distributions, viz. Binomial, Poisson, Hypergeometric and Normal.

12.2 DISCRETE PROBABILITY DISTRIBUTION

Let a discrete random variable X assume the values $x_1, x_2, ..., x_n$ with probabilities $p_1, p_2, ..., p_n$ respectively, where $\Sigma p_i = 1$. The specification of the set of values x_i together with their probabilities p_i ($i = 1, 2, ..., n$) defines the probability distribution of the discrete random variable X, or in short, *discrete probability distribution* of X.

Table 12.1 Discrete Probability Distribution

Values (x)	x_1	x_2	x_3	...	...	x_n	Total
Probabilities (p)	p_1	p_2	p_3	...	...	p_n	1

Often, the probability (p) that the random variable X assumes a specified value (x) can be expressed in terms of a general mathematical expression $f(x)$.

$$f(x) = \text{Probability that } X \text{ assumes the value } x = P(X = x) \quad (12.2.1)$$

The function $f(x)$ is known as "*probability mass function*" (p.m.f), or sometimes '*probability function*' of the discrete random variable X.

If $f(x)$ is known, it is unnecessary to write the probabilities $p_1, p_2, ..., p_n$, because p_i may be obtained from $f(x)$ on putting $x = x_i$, i.e., $p_i = f(x_i)$. The discrete probability distribution is then defined simply by stating the mathematical expression for $f(x)$, along with the set of possible values $x_1, x_2, ..., x_n$.

Table 12.2 Theoretical Distribution (Discrete)

x	x_1	x_2	x_3	...	x_n	Total
$p = f(x)$	$f(x_1)$	$f(x_2)$	$f(x_3)$	...	$f(x_n)$	1

The discrete random variable X may also assume countably infinite number of possible values $x_1, x_2, ..., x_n, ...$ with p.m.f. $f(x)$.

The probability mass function $f(x)$ must satisfy the two conditions

(i) $f(x) \geq 0$, (ii) $\Sigma f(x) = 1$ (12.2.2)

where the summation is taken over all possible value x.

Illustration 1. The following table shows the discrete, probability distribution of a random variable X:

Value (x)	3	6	7	11	Total
Probability (p)	0.2	0.5	0.2	0.1	1

Illustration 2. If X denotes the 'number of points obtained when an unbiased die is thrown', the probability distribution of X is:

No. of Points (x)	1	2	3	4	5	6	Total
Probability (p)	1/6	1/6	1/6	1/6	1/6	1/6	1

This distribution may also be written as

$f(x) = 1/6; \quad (x = 1, 2, ..., 6)$

Illustration 3. If X denotes the 'number of heads obtained in 3 tosses of an unbiased coin', the probability distribution of X is:

No. of Heads (x)	0	1	2	3	Total
Probability (p)	1/8	3/8	3/8	1/8	1

These probabilities may also be given by the p.m.f. (Example 12.8)

$$f(x) = {}^3C_x \left(\frac{1}{2}\right)^3; \quad (x = 0, 1, 2, 3)$$

Illustration 4. An unbiased coin is thrown repeatedly until a head appears. If X denotes the 'number of tails preceding the first head' then its probability distribution is given below:

No. of Tails (x)	0	1	2	3	...	r	....	Total
Probability (p)	$\frac{1}{2}$	$\left(\frac{1}{2}\right)^2$	$\left(\frac{1}{2}\right)^3$	$\left(\frac{1}{2}\right)^4$	...	$\left(\frac{1}{2}\right)^{r+1}$	...	1

This distribution can be written by the p.m.f.

$$f(x) = \left(\frac{1}{2}\right)^{x+1} ; \quad (x = 0, 1, 2, ..., \infty)$$

[**Note:** (i) A discrete random variable (r.v.) can assume either a 'finite' number of possible values (see *Illustrations 1, 2, 3*) or a 'countably infinite' number of values (*Illustration 4*), with a definite probability (i.e. a non-negative real number) associated with each district value of the variable, such that the total probability is 1. The 'probability distribution' of a discrete r.v. shows how the total probability 1 is distributed over the different possible values of the variable. The probability of an interval of values is then given by the sum of the probability associated with those values which lie in that interval.

It will be shown later that a continuous random variable assumed an uncountably infinite' number of possible values. Probabilities are now associated with intervals of values, but not with each individual value of the variable. The probability is obtained by integrating the '*probability density function*' (p.d.f.) over the limits of the interval.

(ii) Generally, the capital letter X is used to denote the '*random variable*' and the small letter x to denote '*any specified value*' of the random variable. If however no confusion arises, the small letter x is often used for both.]

Example 12.1 *X is a discrete random variate having probability mass function:*

x	0	1	2	3	4	5	6	7
$P = (X = x)$	0	k	$2k$	$2k$	$3k$	k^2	$2k^2$	$7k^2 + k$

(i) Determine the constant k; (ii) Find $P(X < 6)$; (iii) What will be $P(X \geq 6)$?

[W.B.H.S. '81]

Solution (*i*) For a discrete random variable X, the probability mass function $f(x) = P(X = x)$ must satisfy the relations

$$\text{(I)}\ f(x) \geq 0, \qquad \text{(II)}\ \Sigma f(x) = 1$$

where the summation is taken over all possible values x. Using (II) here we have

$$0 + k + 2k + 2k + 3k + k^2 + 2k^2 + (7k^2 + k) = 1$$

or, $$10k^2 + 9k = 1$$

or, $$10k^2 + 9k - 1 = 0$$

or, $$(k + 1)(10k - 1) = 0;$$

$\therefore$ $$k = -1, \frac{1}{10}.$$

By virtue of (I), $k = -1$ is impossible. (Note that the probabilities corresponding to 1, 2, 3, 4 become negative). Hence $k = \frac{1}{10} = 0.1$

(ii) $$P(X < 6) = P(X = 0) + P(X = 1) + ... + P(X = 5)$$
$$= 0 + k + 2k + 2k + 3k + k^2$$
$$= 8k + k^2 = 8(0.1) + (0.1)^2 = 0.81$$

(iii) $P(X \geq 6) = 1 - P(X < 6)$, since the events $X < 6$ and $X \geq 6$ are complementary,
$= 1 - 0.81 = 0.19$

Ans. 0.1, 0.81, 0.19

12.3 EXPECTATIONS—MEAN, VARIANCE, MOMENTS (DISCRETE DISTRIBUTION)

Let a discrete random variable x assume the values $x_1, x_2, ..., x_n$ with probabilities $p_1, p_2, ..., p_n$ respectively. Then the *expectation,* or '*expected value*' of x—written $E(x)$—is defined as the sum of products of the different values of x and the corresponding probabilities.

$$E(x) = \Sigma p_i x_i \tag{12.3.1}$$

The expected value of x^2 is similarly defined as the sum of products of the squares of values and the corresponding probabilities.

$$E(x^2) = \Sigma p_i x_i^2 \tag{12.3.2}$$

In general, the expected value of any function $g(x)$ is defined as

$$E[g(x)] = \Sigma p_i g(x_i)$$

Hence, the expected value of a constant k is the constant k itself.

$$E(k) = k,$$

where k is a constant; because $E(k) = \Sigma p_i k = k\Sigma p_i = k$.

If the p.m.f. $f(x)$ is given, then the expectations are defined as follows:

$$E(x) = \Sigma x \,.\, f(x) \tag{12.3.3}$$

$$E(x^2) = \Sigma x^2 \,.\, f(x) \tag{12.3.4}$$

Mean of a probability distribution is the expected value of x.

$$\text{Mean } (\mu) = E(x) \tag{12.3.5}$$

Variance is the expected value of $(x - \mu)^2$, where μ is the mean.

$$\text{Variance } (\sigma^2) = E(x - \mu)^2$$

It may be shown that

$$\sigma^2 = E(x^2) - \mu^2 \tag{12.3.6}$$

The positive square-root of variance gives the *standard deviation* (σ).

Moments of a discrete distribution are defined as follows:

r-th moment about A: $\quad \mu_r' = X(x - A)^r = \Sigma(x - A)^r f(x)$

r-th raw moment: $\quad \mu_r' = E(x^r) = \Sigma x^r \,.\, f(x)$ (12.3.7)

r-th central moment: $\quad \mu_r = E(x - \mu)^r = \Sigma(x - \mu)^r \,.\, f(x)$

where $\mu = E(x)$ is the mean of the distribution. Note that

$$\mu_0' = \mu_0 = 1: \qquad \mu_1' = E(x) = \mu$$

$$\mu_1 = E(x - \mu) = \Sigma(x - \mu) \,.\, f(x) = \Sigma x \,.\, f(x) - \mu\Sigma f(x)$$

$$= \mu - \mu = 0$$

The central moments can be obtained from non-central moments using the relations (see 7.2.2)

$$\begin{aligned} \mu_2 &= \mu_2' - \mu_1'^2 \\ \mu_3 &= \mu_3' - 3\mu_2'\mu_1' + 2\mu_1'^3 \\ \mu_4 &= \mu_4' - 4\mu_3'\mu_1' + 6\mu_2'\mu_1'^2 - 3\mu_1'^4 \end{aligned} \tag{12.3.8}$$

Example 12.2 *A random variable has the following probability distribution:*

x	*4*	*5*	*6*	*8*
Probability	*0.1*	*0.3*	*0.4*	*0.2*

Find the expectation and the standard deviation of the random variable.

[C.U., B.Com. (Hons) '80]

Solution (Note that the sum of the probabilities is 1; $\Sigma p_i = 1$)

$$\text{Mean } (\mu) = E(x) = \Sigma p_i x_i$$
$$= (0.1 \times 4) + (0.3 \times 5) + (0.4 \times 6) + (0.2 \times 8)$$
$$= 0.4 + 1.5 + 2.4 + 1.6 = 5.9$$
$$E(x^2) = \Sigma p_i x_i^2 = (0.1 \times 4^2) + (0.3 \times 5^2) + (0.4 \times 6^2) + (0.2 \times 8^2)$$
$$= 1.6 + 7.5 + 14.4 + 12.8 = 36.3$$
$$\text{Variance } (\sigma^2) = E(x^2) - \mu^2, \text{ using (12.3.6)}$$
$$= 36.3 - (5.9)^2 = 36.3 - 34.81 = 1.49$$
$$\text{S.D. } (\sigma) = \sqrt{1.49} = 1.22$$

Ans. 5.9, 1.22

Example 12.3 *For what value of a will the function*

$f(x) = ax$; $x = 1, 2, 3, \ldots, n$ be the probability mass function of a discrete random variable x? Find the mean and variance of x.

[W.B.H.S. '78, '82]

Solution The conditions for any function $f(x)$ to be a p.m.f. are

(i) $f(x) \geq 0$, (ii) $\Sigma f(x) = 1$

From (i), we have $a > 0$, because none of the values of x is negative.

From (ii),
$$1 = \Sigma f(x) = \sum_{x=1}^{n} ax = a \sum_{1}^{n} x = a \cdot \frac{n(n+1)}{2}$$

$\therefore$
$$a = \frac{2}{n(n+1)}$$

$$\text{Mean} = E(x) = \Sigma x\, f(x) = \sum_{x=1}^{n} x \,.\, ax = a \sum_{1}^{n} x^2$$
$$= a(1^2 + 2^2 + \ldots + n^2)$$
$$= \frac{2}{n(n+1)} \cdot \frac{n(n+1)(2n+1)}{6} = \frac{n(n+1)}{3}$$

Again,
$$E(x^2) = \Sigma x^2 \,.\, f(x) = a \sum_{1}^{n} x^3$$
$$= a(1^3 + 2^3 + \ldots + n^3)$$
$$= \frac{2}{n(n+1)} \left\{\frac{n(n+1)}{2}\right\}^2 = \frac{n(n+1)}{2}$$

$\therefore$
$$\text{Variance} = E(x)^2 - \{E(x)\}^3 = \frac{n(n+1)}{2} - \left(\frac{2n+1}{3}\right)^2$$
$$= \frac{(n-1)(n+2)}{18}$$

Example 12.4 *A random variable X is defined as follows:*

$$Prob\ (X = 1) = p, \quad Prob\ (X = 0) = 1 - p$$

where $0 < p < 1$. Find the mean, variance and the central moments μ_2, μ_3, μ_4 of the distribution.

Solution The probability distribution of X is as follows:

Value (x)	1	0	Total
Probability	p	$1-p$	1

The raw moments are

$$\mu_1' = E(X) = p \times 1 + (1-p) \times 0 = p + 0 = p$$
$$\mu_2' = E(X^2) = p \times 1^2 + (1-p) \times 0^2 = p \times 1 + (1-p) \times 0 = p$$
$$\mu_3' = E(X^3) = p \times 1^3 + (1-p) \times 0^3 = p$$
$$\mu_4' = E(X^4) = p \times 1^4 + (1-p) \times 0^4 = p$$

Using (12.3.8)

$$\mu_2 = p - p^2 = p(1-p)$$
$$\mu_3 = p - 3p \,.\, p + 2p^3 = p - 3p^2 + 2p^3 = p(1-p)\,(1-2p)$$
$$\mu_4 = p - 4p \,.\, p + 6p \,.\, p^2 - 3p^4 = p(1-p)\,(1 - 3p + 3p^2)$$

Also, Mean $(\mu) = E(x) = \mu_1 = p$

Variance $(\sigma^2) = \mu_2 = p(1-p)$.

Example 12.5 *Obtain the expectation of the number of tails preceding the first head in an indefinite series of tosses of the same coin.*

[C.U., B.Sc.(Econ) '81 (New)]

Solution Let the random variable X denote the number of tails preceding the first head. Suppose that p denotes the probability of getting a head in one toss of the coin and $q\,(= 1 - p)$ denotes the probability of a tail. Then the probability that x tails precede the first head is

$P(X = x)$ = Probability of x successive tails, followed by a head,

$= q \,.\, q \,.\, q \,,\ldots,\, q \,.\, p$ (q's nothing x times)

$= q^x \,.\, p$

The p.m.f of X is thus given by

$$f(x) = pq^x; \quad (x = 0, 1, 2, 3, \ldots \infty)$$

[Check: (i) $f(x) \geq 0$; because none of p and q is negative

(ii) $$\sum_0^\infty f(x) = pq^0 + pq^1 + pq^2 + pq^3 + \ldots \ldots$$
$$= p(1 + q + q^2 + q^3 + \ldots \ldots$$
$$= p(1-q)^{-1} = p \,.\, p^{-1} = 1]$$

The expectation of the number of tails preceding the first head is

$$E(x) = \Sigma x \,.\, f(x) = \sum_0^\infty x \,.\, pq^x$$
$$= 0 + pq + 2pq^2 + 3pq^2 + \ldots \ldots$$
$$= pq(1 + 2q + 3q^2 + \ldots \ldots)$$
$$= pq(1-q)^{-2} = pq \,.\, p^{-2} = qp^{-1}$$
$$= q/p.$$

Ans. q/p

12.4 UNIFORM DISTRIBUTION (DISCRETE)

If a discrete random variable x assumes n possible values $x_1, x_2, ..., x_n$ with *equal probabilities*, then the probability that it takes any particular value is a constant $1/n$. The probability distribution defined by the probability mass function (p.m.f).

$$f(x) = 1/n; \quad (x = x_1, x_2, ..., x_n) \qquad (12.4.1)$$

is known as *Uniform Distribution.*

Table 12.3 Uniform Distribution (Discrete)

x	x_1	x_2	...	x_n	Total
$f(x)$	$1/n$	$1/n$	...	$1/n$	1

The number of points obtained in a single throw of an unbiased die follows uniform distribution with p.m.f

$$f(x) = 1/6, (x = 1, 2, 3, ..., 6).$$

No. of Points x	1	2	3	4	5	6	Total
Probability f(x)	1/6	1/6	1/6	1/6	1/6	1/6	1

[**Note:** (i) The p.m.f of a discrete uniform distribution is a constant (independent of x), and this constant is one divided by the number of possible values of x.

(ii) In a discrete uniform distribution, the possible values of the random variable may be any real numbers, and the successive values may not have a common difference. However, there must be a finite number of possible values.

Example 12.6 *If a discrete random variable x follows uniform distribution and assumes only the values 8, 9, 11, 15, 18, 20, find the probabilities (i) $P(x = 9)$, (ii) $P(x = 12)$, (iii) $P(x < 15)$, (iv) $P(x \leq 15)$, (v) $P(x > 15)$, (vi) $P(|x - 14| < 15)$.*

Solution Since x has a discrete uniform distribution with 6 possible values, the probability that it takes any particular value is a constant $\frac{1}{6}$ and the probability that it takes a value other than those given, is zero. Therefore,

(i) $$P(x = 9) = \frac{1}{6}$$

(ii) $$P(x = 12) = 0$$

(iii) Now, $x < 15$ implies that x takes only 3 values, viz. 8, 9, 11.

Hence, $$P(x < 15) = 3 \times \frac{1}{6} = \frac{1}{2}$$

(iv) Again, $x \leq 15$ implies that the variable takes 4 values (15 included), viz 8, 9, 11, 15. Hence

$$P(x \leq 15) = 4 \times \frac{1}{6} = \frac{2}{3}$$

(v) $x > 15$ implies only 2 possible values of x viz. 18 and 20.

$$P(x > 15) = 2 \times \frac{1}{6} = \frac{1}{3}$$

(vi) $|x - 14| < 5$ implies that the difference of x and 14 is to be less than 5; so x takes only the values 11, 15 and 18. Hence.

$$P(|x - 14| < 5) = 3 \times \frac{1}{6} = \frac{1}{2}$$

Ans. $\frac{1}{6}, 0, \frac{1}{2}, \frac{2}{3}, \frac{1}{3}, \frac{1}{2}$

Example 12.7 *Find the mean and the standard deviation of the uniform distribution* $f(x) = 1/n$; $(x = 1, 2, ..., n)$

Solution Mean $(\mu) = E(x) = \Sigma x \,.\, f(x)$

$$= \sum_{x=1}^{n} x \,.\, (1/n) = (1/n) \sum_{x=1}^{n} x$$

$$= (1/n)\,(1 + 2 + ... + r)$$

$$= \frac{n(n+1}{2n} = \frac{n+1}{2}$$

$$E(x^2) = \Sigma x^2 f(x) = \sum_{x=1}^{n} x^2 \,.\, (1/n)$$

$$= (1/n)\,(1^2 + 2^2 + ... + n^2)$$

$$= \frac{n(n+1)\,(2n+1)}{6n} = \frac{(n+1)\,(2n+1)}{6}$$

Using (12.3.6),

$$\sigma^2 = E(x^2) - \mu^2 = \frac{(n+1)\,(2n+1)}{6} - \left(\frac{n+1}{2}\right)^2$$

$$= \frac{n^2 - 1}{12}$$

$\therefore$ S.D. $(\sigma) = \sqrt{(n^2 - 1)/12}$ *Ans.* $(n + 1)/2$, $\sqrt{(n^2 - 1)/12}$

12.5 BINOMIAL DISTRIBUTION

Binomial distribution is a discrete probability distribution and is defined by the p.m.f.

$$f(x) = {}^nC_x\, p^x\, q^{n-x}; \quad (x = 0, 1, 2, ..., n) \qquad (12.5.1)$$

where p and q are positive fraction $(p + q = 1)$.

Suppose that we have a series of n independent trails in each of which the probability of occurrence of an event is fixed and constantly p. Then the probability that the event occurs exactly r times in n trials (Section 11.8) is ${}^xC_r\, p^r\, q^{n-r}$, where $q = 1 - p$ and r may assume any of the values 0, 1, 2, ..., n. In general, the occurrence of the event is called "*success*" and its non-occurrence is known as "*failure*". The generalised various of the theory is as follows:

In a series of *n independent trials*, if the probability of 'success' in each trial is a *constant p*, and the probability of 'failure' is q, then the probability of x successes (and obviously $n - x$ failures) is given by the Binomial distribution (12.5.1.)

Table 12.4 Binomial Distribution

x	0	1	2	...	n	Total
$f(x)$	q^n	${}^nC_1 pq^{n-1}$	${}^nC_2 p^2q^{n-2}$	...	p^n	1

Note: (i) In the expression for $f(x)$, besides the factor ${}^nC_x$, the power of p (i.e., probability of success) is the number of successes x , and the power of q (i.e., probability of failure) is the number of failures $n - x$.

(ii) The sum of the powers of p and q is always n, whatever be the number of successes.

(iii) These are $(n + 1)$ possible values of x, viz. 0, 1, 2, ..., n.

(iv) The total probability for all the $(n + 1)$ possible successes is 1.]

The distribution is known as 'binomial', because the probabilities are given by the binomial series.

$$(q + p)^n = q^n + {}^nC_1 pq^{n-1} + {}^nC_2 p^2q^{n-2} + ... + p^n. \tag{12.5.2}$$

A series of independent trials of a random experiment with a constant probability of success in each trial is called a "*Bernoullian series*", and this distribution is known as "Bernoulli's distribution", after the name of its discoverer James Bernoulli.

The two constants n and p appearing in the expression for $f(x)$, are known as '*parameters*' of the binomial distribution. If the values of parameters are known, the distribution is completely known $(q = 1 - p)$.

I should be noted that the binomial distribution holds under the following conditions:

(1) *The result of any trial can be classified only under two categories*; e.g., either head or tail in a throw of one coin; either a "six" or a "non-six" in one throw of a die; either a spade or a non-spade when one card is drawn; either a defective or a non-defective when an article is examined from a lot; i.e., in general, occurrence (called *success*) or non-occurrence (called *failure*) of a certain event.

(2) *The probability of success in each trial remains a constant,* and does not change from one trial to another. For example, the probability of obtaining a head in successive thrown of a coin is always $\frac{1}{2}$; the probability of obtaining a defective article from a batch does not change in successive drawings *with replacement*, and practically remains a constant even in drawings *without replacement,* when the batch is large.

(3) *The trials are independent,* so that the probability of success in any trial remains unaffected by the results of other trials. For example, in successive throws of a coin the occurrence of a head at any trial will in no way affect the probability of a head or a tail in any subsequent trial; or if several coins are thrown together, the occurrence of a head or a tail in any particular coin does not alter the probability of occurrence of a head in any other coin.

Important Properties

(1) Binomial distribution is a discrete probability distribution, where the random variable assumes a finite number of values 0, 1, 2, ..., n. The distribution is specified by two parameters n and p.

(2) Mean = np, variance = npq (12.5.3)

Standard deviation $(\sigma) = \sqrt{npq}$

(3) Binomial distribution may have either one or two modes. When $(n + 1)p$ is not an integer, Mode is the largest integer contained therein. However, when $(n + 1)p$ is itself an integer, there are two modes, viz. $(n + 1)p$ and $(n + 1)p - 1$.

(4) Skewness $(\gamma_1) = \dfrac{q-p}{\sqrt{npq}}$; Kurtosis $(\gamma_2) = \dfrac{1-6pq}{npq}$ (12.5.4)

When $p = q = \frac{1}{2}$, the distribution becomes symmetrical for all values of n.

i.e., $f(x) = f(n-x) = {}^nC_x \left(\frac{1}{2}\right)^n$.

(5) If x follows binomial distribution with parameters (n_1, p) and y follows binomial distribution with parameters $(n_2, -p)$, and x and y are statistically independent, then $(x + y)$ also follows binomial distribution with parameters $(n_1 + n_2, p)$. The result can be extended to several independent binomial variates with a common p.

(6) Binomial distribution may be obtained as a limiting case of Hypergeometric distribution.

Example 12.8 *Three coins are tossed. Find the probabilities of (i) 0 head, 1 head, 2 heads, 3 heads; (ii) more than one head; (iii) at least 1 head.*

Solution Here, the 'random experiment' consists in tossing 3 coins and observing the number of heads. The 'random variable' x is the number of heads obtained in a toss of 3 coins.

If we denote the occurrence of head as "success", and the coins are assumed to be unbiased, then

$$p = \text{Probability of "success" in a single trial}$$

$$= \text{Probability of head with a single coin} = \frac{1}{2}$$

$$q = 1 - p = 1 - \frac{1}{2} = \frac{1}{2}$$

$$n = \text{Number of independent trials} = 3$$

Since the value of p is constant for each coin and the trials are independent, the variable x follows binomial distribution with parameters $n = 3$, $p = \frac{1}{2}$. Therefore, the probability of x successes is

$$f(x) = {}^nC_x p^x q^{n-x} = {}^3C_x \left(\frac{1}{2}\right)^x \left(\frac{1}{2}\right)^{3-x} = {}^3C_x \left(\frac{1}{2}\right)^3$$

(i) Putting the values of $x = 0, 1, 2$, respectively,

$$f(0) = \text{Probability of 0 success}$$

$$= {}^3C_0 \left(\frac{1}{2}\right)^3 = 1 \times \frac{1}{8} = \frac{1}{8}$$

$$f(1) = \text{Probability of 1 success}$$

$$= {}^3C_1 \left(\frac{1}{2}\right)^3 = 3 \times \frac{1}{8} = \frac{3}{8}$$

$f(2)$ = Probability of 2 successes

$$= {}^3C_2 \left(\frac{1}{2}\right)^3 = 3 \times \frac{1}{8} = \frac{3}{8}$$

$f(3)$ = Probability of 3 successes

$$= {}^3C_3 \left(\frac{1}{2}\right)^3 = 1 \times \frac{1}{8} = \frac{1}{8}$$

(ii) Prob (more than one success) = $f(2) + f(3)$

$$= \frac{3}{8} + \frac{1}{8} = \frac{1}{2}$$

(iii) Prob (at least one success) = 1 – Prob (0 success)

$$= 1 - f(0) = 1 - \frac{1}{8} = \frac{7}{8}$$

Example 12.9 *Five coins are tossed 3200 times. Find the expected frequencies of the distribution of heads and tails, and tabulate the result. Calculate the mean number of heads and standard deviation.* [C.A., May '77]

Solution We assume that the coins are unbiased. The probability of getting a head (or tail) is $\frac{1}{2}$ for each coin at each trial. Also the successive trials are independent. So, the conditions of binomial distribution are fulfilled. The probabilities of 0 head, 1 head, ... 5 heads are then given by the successive terms in the expansion of the binomial expression

$$\left(\frac{1}{2}+\frac{1}{2}\right)^5 = \left(\frac{1}{2}\right)^5 + {}^5C_1\left(\frac{1}{2}\right)\left(\frac{1}{2}\right)^4 + {}^5C_2\left(\frac{1}{2}\right)^2\left(\frac{1}{2}\right)^3$$

$$+ {}^5C_3\left(\frac{1}{2}\right)^3\left(\frac{1}{2}\right)^2 + {}^5C_4\left(\frac{1}{2}\right)^4\left(\frac{1}{2}\right) + \left(\frac{1}{2}\right)^5$$

$$= \frac{1}{32} + \frac{5}{32} + \frac{10}{32} + \frac{10}{32} + \frac{5}{32} + \frac{1}{32}$$

The probabilities of the different numbers of heads and tails in a single toss of 5 coins, and the expected frequencies in 3200 tosses are shown below:

Number of Heads	Number of Tails	Probability	Expected Frequency = 3200 × Probability
0	5	1/32	100
1	4	5/32	500
2	3	10/32	1000
3	2	10/32	1000
4	1	5/32	500
5	0	1/32	100
Total		1	3200

The mean number of heads, and the standard deviation are now calculated from the following frequency distribution (see Examples 5.57 and 6.29):

Number of Heads (x)	0	1	2	3	4	5	Total
Frequency (f)	100	500	1000	1000	500	100	3200

Mean ($\bar{x}$) = 2.5, Standard deviation (σ) = $\sqrt{5/2}$ = 1.12

[**Note:** that the mean and s.d. calculated from the frequency distribution agree exactly with the formulae (12.5.3), viz.

$$np = 5 \times \frac{1}{2} = 2.5, \text{ s.d.} = \sqrt{npq} = \sqrt{5 \times \frac{1}{2} \times \frac{1}{2}} = \sqrt{\frac{5}{2}}$$

Example 12.10 *The overall percentage of failures in a certain examination is 40. What is the probability that out of a group of 6 candidates at least 4 passed the examinations?* [I.C.W.A. Dec.'74]

Solution Let us denote the event of a candidate passing the examination as "success". It is required to find the probability of at least 4 successes.

$$\begin{aligned} q &= \text{Probability of failure of a candidate} \\ &= 40\% = 40/100 = 2/5 \\ p &= \text{Probability of success in a single trial} \\ &= 1 - q = 1 - 2/5 = 3/5 \\ n &= \text{Number of candidates in the group} = 6 \end{aligned}$$

Using binomial distribution, the probability of x successes is

$$f(x) = {}^nC_x p^x q^{n-x} = {}^6C_x \left(\frac{3}{5}\right)^x \left(\frac{3}{5}\right)^{6-x}$$

Since 'at least 4' in a group of 6 implies 'either 4, or 5, or 6', the probability of at least 4 successes is given by the sum of probabilities $f(4) + f(5) + f(6)$. Putting $x = 4, 5, 6$ in the expression for $f(x)$, we have

$$f(4) = {}^6C_4 \left(\frac{3}{5}\right)^4 \left(\frac{3}{5}\right)^2 = 15 \times \frac{3^4 \times 2^2}{5^6} = \frac{4860}{15625}$$

$$f(5) = {}^6C_5 \left(\frac{3}{5}\right)^5 \left(\frac{3}{5}\right)^1 = 6 \times \frac{3^5 \times 2}{5^6} = \frac{2916}{15625}$$

$$f(6) = {}^6C_6 \left(\frac{3}{5}\right)^6 \left(\frac{3}{5}\right)^0 = 1 \times \frac{3^6 \times 1}{5^6} = \frac{729}{15625}$$

Thus, the required probability is

$$f(4) + f(5) + f(6) = \frac{4860 + 2916 + 729}{15625} = \frac{1701}{3125}$$

Example 12.11 *In 10 independent throws of a defective dice, the probability that an even number will appear 5 times is twice the probability that an even number will appear 4 times. Find the probability that an even number will not appear at all in 10 independent throws of the dice.*

Solution Let the occurrence of an even number be called a "success". We write

p = probability of getting an even number (i.e., success) in a single trial,

q = probability of not getting an even number (i.e., failure)

n = number of throws = 10

Using binomial distribution, the probability of x successes in 10 throws is

$$f(x) = {}^{10}C_x p^x q^{10-x}$$

It is given that

$$f(5) = 2 \times f(4)$$

or, $${}^{10}C_5 p^5 q^5 = 2 \times {}^{10}C_4 p^4 q^6$$

or, $$\frac{10!}{5!5!} p^5 q^5 = 2 \times \frac{10!}{4!6!} p^4 q^6$$

Simplifying, we get $3p = 5q$

or, $3p = 5(1-p)$

Solving we get $p = \frac{5}{8}$, so that $q = 1 - p = 1 - \frac{5}{8} = \frac{3}{8}$

$\therefore$ Probability that an even number will not appear at all (i.e. 0 success) is

$$f(0) = q^{10} = \left(\frac{3}{8}\right)^{10}$$

Ans. $\left(\frac{3}{8}\right)^{10}$

Example 12.12 *Find the mean and the standard deviation of binomial distribution with parameters n and p.*

Solution The p.m.f. of binomial distribution is

$$f(x) = {}^nC_x p^x q^{n-x}; \ (x = 0, 1, 2, ..., n)$$

Hence, by definition (12.3.3),

$$\text{Mean} = E(x) = \sum_x x f(x) = \sum_{x=0}^{n} x \,.\, {}^nC_x p^x q^{n-x}$$

$$= 0 \times ({}^nC_0 p^0 q^n) + 1 \times ({}^nC_1 p q^{n-1}) + 2 \times ({}^nC_2 p^2 q^{n-2}) + 3 \times ({}^nC_3 p^3 q^{n-3}) + ... + n \times ({}^nC_n p^n q^0)$$

$$= 0 + 1 \times npq^{n-1} + 2 \times \frac{n(n-1)}{1 \times 2} p^2 q^{n-2} + 3 \times \frac{n(n-1)(n-2)}{1 \times 2 \times 3} p^2 q^{n-3} + ... + n \times p^n$$

$$= npq^{n-1} + n(n-1) p^2 q^{n-2} + \frac{n(n-1)(n-2)}{1 \times 2} p^3 q^{n-3} + ... + n p^n$$

$$= np[q^{n-1} + (n-1) pq^{n-2} + \frac{n(n-1)(n-2)}{1 \times 2} p^2 q^{n-3} + ... + p^{n-1}]$$

$$= np\,[q^{n-1} + {}^{n-1}C_1 pq^{n-2} + {}^{n-1}C_2 pq^{n-3} + ... + {}^{n-1}C_{n-1} p^{n-1}]$$

$= np(q+p)^{n-1}$, replacing n by $n-1$ in (12.5.2)

$= np(1)^{n-1}$, because $p + q = 1$

$= np(1) = np$

Therefore, Mean $(\mu) = np$

Again, by (12.3.6) $\sigma^2 = E(x^2) - \mu^2$

But, $E(x^2) = Ex^2 \,.\, f(x) = \Sigma\{x(x-1) + x\} \,.\, f(x)$

$= \Sigma x(x-1) f(x) + \Sigma x \,.\, f(x)$

$= \Sigma x(x-1) \,.\, f(x) + \mu.$

Now, $\Sigma x(x-1).f(x) = \sum_{x=0}^{n} x(x-1).{}^nC_x p^x q^{n-x};$

$$= 0(-1)\{{}^nC_0 p^0 q^n\} + 1\times 0\{{}^nC_1 pq^{n-1}\} + 2\times 1\{{}^nC_2 p^2 q^{n-2}\} + 3\times 2\{{}^nC_3 p^3 q^{n-3}\} + \ldots + n(n-1)\{{}^nC_n p^n q^0\}$$

$$= 0 + 0 + 2\times 1\left\{\frac{n(n-1)}{1\times 2}p^2q^{n-2}\right\} + 3\times 2\left\{\frac{n(n-1)(n-2)}{1\times 2\times 3}p^2q^{n-2}\right\} + \ldots + n(n-1)\{1\cdot p^n\}$$

$$= n(n-1)p^2q^{n-2} + n(n-1)(n-2)p^3q^{n-3} + \ldots + n(n-1)p^n$$

$$= n(n-1)p^2[q^{n-2} + (n-2)pq^{n-3} + \ldots + p^{n-2}]$$

$$= n(n-1)p^2[q^{n-2} + {}^{n-2}C_1 pq^{n-3} + \ldots + {}^{n-2}C_{r-2}p^{n-2}]$$

$= n(n-1)p^2(q+p)^{n-2}$, replacing n by $n-2$ in (12.5.2)

$= n(n-1)p^2(1)^{n-2}$, since $p+q=1$

$= n(n-1)p^2$

$\therefore \quad E(x)^2 = n(n-1)p^2 + \mu$, and hence

$$\sigma^2 = E(x^2) - \mu^2$$

$$= n(n-1)p^2 + \mu - \mu^2 = n(n-1)p^2 + np - (np)^2$$

$$= n^2p^2 - np^2 + np - n^2p^2 = -np^2 + np$$

$= np(1-p) = npq, \qquad$ since $q = 1-p$.

$\therefore$ Standard Deviation $(\sigma) = \sqrt{npq}$

Ans. mean $= np$, s.d.$= \sqrt{npq}$

Example 12.13 *Arithmetic mean and standard deviation of a binomial distribution are respectively 4 and* $\sqrt{8/3}$. *Find the values of q and p.* [W.B.H.S. '81]

Solution Using the formulae for mean and s.d. of the binomial distribution

$$np = 4, \qquad \sqrt{npq} = \sqrt{8/3}$$

Squaring the second relation, $npq = 8/3$. Now putting $np = 4$ and solving we have $q = 2/3$. Hence, $p = 1 - q = 1/3$.

Putting the value of p in $np = 4$, we get $n = 12$. *Ans.* $n = 12, p = 1/3$.

Example 12.14 *Find the mode of binomial distribution with parameters n and p.*

Solution The 'mode' of a discrete distribution is the value of the variate which corresponds to the maximum probability.

In the binomial distribution, the random variable X assumes only the values 0, 1, 2, ..., n and the probabilities are given by

$$f(x) = P(X = x) = {}^nC_x p^x q^{n-x};\ (x = 0, 1, 2, \ldots, n)$$

The mode is that value (or values) among 0, 1, 2, ..., n corresponding to which $f(x)$ is the maximum.

$$\frac{f(x-1)}{f(x)} = \frac{{}^nC_{x-1}p^{x-1}q^{n-x+1}}{{}^nC_x p^x q^{n-x}} \quad (x \neq 0)$$

$$= \frac{n!}{(x-1)!(n-x+1)!} p^{x-1} \cdot q^{n-x+1} \bigg/ \frac{n!}{x!(n-x)!} p^x q^{n-x}$$

$$= \frac{xq}{(n-x+1)p} = \frac{xq}{(n+1)p - xp} \qquad \text{... (1)}$$

We find that:

(a) $f(x-1) < f(x)$

whenever $xq < (n+1)p - xp$

or, $x(p+q) < (n+1)p$; i.e. $x < (n+1)p$ ... (2)

(b) $f(x-1) = f(x)$

whenever $xq = (n+1)p - xp$; i.e. $x = (n+1)p$ ... (3)

(c) $f(x-1) > f(x)$

whenever $xq > (n+1)p - xp$; i.e. $x > (n+1)p$ ... (4)

We discuss the following cases:

Case I When (n + 1)p is not an integer.

Let r be the largest integer in $(n+1)p$. Since in a binomial distribution, x assumes only integral values, the result (3) is impossible.

From (2),

$$f(x-1) < f(x), \quad \text{when } x = 1, 2, 3, ..., r$$

or, $f(0) < f(1), \quad f(1) < f(2), \ldots f(r-1) < f(r)$

or, $f(0) < f(1) < f(2) < \ldots f(r-1) < f(r)$ ... (5)

From (4),

$$f(x-1) > f(x), \quad \text{when } x = r+1, r+2, \ldots, n$$

or, $f(r) < f(r+1), \quad f(r+1) > f(r+2), \ldots f(n-1) > f(n)$

or, $f(r) > f(r+1) > f(r+2) > \ldots > f(n)$ ... (6)

Combining the results at (5) and (6),

$$f(0) < f(1) < \ldots < f(r-1) < f(r) > f(r+1) > \ldots > f(n).$$

We find that the probability $f(r)$ corresponding to $x = r$ is the maximum, and therefore the mode is r.

Thus, when $(n+1)p$ is not an integer, the mode is the largest integer contained in $(n+1)p$.

Case II When (n + 1)p is an integer, say (n + 1)p = r.

From (2),

$$f(x-1) < f(x), \quad \text{when } x = 1, 2, 3, \ldots, r-1$$

or, $f(0) < f(1) < \ldots < f(r-1)$. ... (7)

From (3),

$$f(x-1) = f(x), \quad \text{when } x = r$$

or, $f(r-1) = f(r)$ ... (8)

From (4),

$$f(x-1) > f(x), \quad \text{when } x = r+1, r+2, \ldots, n$$

or, $f(r) > f(r+1) > \ldots > f(n)$ ... (9)

Combining the results at (7), (8) and (9),

$$f(0) < f(1) < \ldots < f(r-1) = f(r) > f(r+1) > \ldots > f(n).$$

We find that the probabilities $f(r)$ and $f(r-1)$ corresponding to $x = r$ and $x = r-1$ respectively are equal, and have the largest values. There are now two modes $r = r-1$, and the distribution is '*bimodal*' (i.e., have two modes).

Thus, when $(n + 1)p$ is not an integer, say r, the binomial distribution has two modes, viz. r and $r - 1$.

12.6 POISSON DISTRIBUTION

Poisson distribution is a discrete probability distribution and is defined by the probability mass function (p.m.f.)

$$f(x) = \frac{e^{-m} \cdot m^x}{x\,!}; \quad (x = 0, 1, 2, ..., \infty) \tag{12.6.1}$$

where m is positive. The quantity e is a mathematical constant and is given by the infinite series

$$e = 1 + \frac{1}{1\,!} + \frac{1}{2\,!} + \frac{1}{3\,!} + \dots = 2.718 \text{ approx} \tag{12.6.2}$$

The constant m appearing in (12.6.1) is known as 'parameter' of the Poisson distribution. It may be noted that the variable assumes an infinite number of values $x = 0, 1, 2, 3, ..., \infty$ with probabilities as shown below:

Table 12.5 Poisson Distribution

x	0	1	2	3	...	Total
$f(x)$	e^{-m}	$e^{-m} \cdot m$	$\frac{e^{-m} \cdot m^2}{2\,!}$	$\frac{e^{-m} \cdot m^3}{3\,!}$	...	1

This distribution is named after its discoverer S.D. Poisson.

Important Properties

(1) Poisson distribution is a discrete probability distribution where the random variable assumes a countably infinite number of values 0, 1, 2, ..., ∞. The distribution is completely specified, when the parameter m (positive) is known.

(2) Mean = m, Variance = m

Standard deviation (σ) = $\sqrt{m}$ (12.6.3)

(3) Poisson distribution may have either one or two modes (like the binomial distribution). When m is not an integer, mode is the largest integer contained in m. However, when m is itself an integer, there are two modes, viz m and $m - 1$.

(4) $$\text{Skewness } (\gamma_1) = \frac{1}{\sqrt{m}}, \quad \text{Kurtosis } (\gamma_2) = \frac{1}{m} \tag{12.6.4}$$

Thus, Poisson distribution is positively skew and leptokurtic.

(5) If x and y are independent Poisson variables with parameters m_1 and m_2 respectively then $(x + y)$ also follows Poisson distribution with parameter $(m_1 + m_2)$.

(6) Poisson distribution may be used as an approximation to binomial distribution, when p is small, and n is large, but np finite.

Some examples of Poisson variable are:

(i) Number of printing mistakes per page (or typographical errors per page by a skilled typist);

(ii) Number of goals scored in a foot ball match;

(iii) Number of telephone calls received in a telephone box per unit interval of time during a busy period;

(iv) Number of suicides (or deaths from a rare disease) per year in a given region;

(v) Number of bacteria present in a given liquid culture per unit area observed under the microscope;

(vi) Number of radio-active atoms decaying in a given interval of time;

(vii) Number of defects per unit area of sheet material (e.g. paper, cloth, or metal sheet);

(viii) Number of cars passing through a road crossing per unit time interval (e.g. 1 minute) during a busy period.

Example 12.15 *A random variable x follows Poisson distribution with parameter 3, Find the probabilities that the variable assumes the values (i) 0, 1, 2, 3; (ii) less than 3; (iii) at least 2, (Given that $e^{-3} = .0498$).*

Solution For the Poisson distribution with parameter $m = 3$, the probability of x successes is

$$f(x) = \frac{e^{-m} \cdot m^x}{x\,!} = \frac{e^{-3} \cdot 3^x}{x\,!}$$

(i) Putting $x = 0, 1, 2, 3$ successively, the required probabilities are

$$f(0) = \frac{e^{-3} \cdot 3^0}{0\,!} = \frac{e^{-3} \cdot 1}{1} = e^{-3} \text{ (since } 0\,! = 1 \text{ and } 3^0 = 1)$$
$$= .0498$$

$$f(1) = \frac{e^{-3} \cdot 3^1}{1!} = \frac{e^{-3} \cdot 3}{1} = 3 \cdot e^{-3}$$
$$= 3 \times .0498 = .1494$$

$$f(2) = \frac{e^{-3} \cdot 3^2}{2\,!} = \frac{e^{-3} \cdot 9}{2} = 4.5 \times e^{-3}$$
$$= 4.5 \times .0498 = .2241$$

$$f(3) = \frac{e^{-3} \cdot 3^3}{3\,!} = \frac{e^{-3} \cdot 27}{6} = 4.5 \times e^{-3}$$
$$= 4.5 \times .0498 = .2241$$

(ii) Since 'less than 3' implies 'either 0, or 1, or 2', the probability of less than 3 successes is given by the sum of the probabilities

$$f(0) + f(1) + f(2) = .0498 + .1494 + .2241 = 0.4233$$

(iii) Since 'at least 3' implies 'either 2, or 3, or 4, ..., ∞, the probability of at least 2 successes is given by the sum of the probabilities of the infinite number of terms. But since the sum of the probabilities for all values of x, viz. 0, 1, 2,, ..., ∞ is 1, i.e.,

$$f(0) + f(1) + f(2) + \ldots = 1$$

hence $$f(2) + f(3) + f(4) + \ldots = 1 - f(0) - f(1)$$
$$= 1 - .0498 - .1494 = .8008$$

Example 12.16 *A business firm receives on an average 2.5 telephone calls per day during the time period 10.00 – 10.05 a.m. Find the probability that on a certain day, the firm receives (i) no call; (ii) exactly 4 calls, during the same period. (Assume Poisson distribution; given $e^{-2.5} = .0821$).*

Solution The random variable x is the 'number of telephone calls received during the period 10.00 – 10.05 a.m.'. Since x is assumed to follow Poisson distribution, the parameter m is equal to the mean of the distribution; i.e., $m = 2.5$. Hence, the probability of x calls is

$$f(x) = \frac{e^{-2.5}(2.5)^x}{x\,!}$$

(i) Probability of no call during the period (i.e., $x = 0$) is

$$f(0) = \frac{e^{-2.5}(2.5)^0}{0\,!} = \frac{e^{-2.5}\times 1}{1} = e^{-2.5} = .0821$$

(ii) Probability of exactly 4 calls during the period (i.e., $x = 4$) is

$$f(4) = \frac{e^{-2.5}(2.5)^4}{4\,!} = \frac{.0821\times 39.0625}{24} = .1336$$

Ans. .0821, .1336

Example 12.17 *Let x be distributed in the Poisson form. If $P(x = 1) = P(x = 2)$, what is $P(x = 0 \text{ or } 1)$. Also find $E(x)$.*

Solution For Poisson distribution, the p.m.f. is

$$f(x) = \frac{e^{-m}m^x}{x\,!}$$

The probability that x assumes the value 1 is then given by $f(1)$, i.e., the value of $f(x)$ when $x = 1$.

$$P(x = 1) = f(1) = \frac{e^{-m}m^1}{1\,!} = e^{-m}.m$$

Similarly, the probability that x assumes the value 2 is $f(2)$.

$$P(x = 2) = f(2) = \frac{e^{-m}.m^2}{2\,!} = \frac{e^{-m}m^2}{2}$$

But it is given that

$$P(x = 1) = P(x = 2)$$

or,

$$e^{-m}.m = \frac{e^{-m}.m^2}{2}$$

Dividing both sides by $(e^{-m}\cdot m)$, we have

$$1 = \frac{m}{2}; \quad \text{i.e., } m = 2$$

Therefore, the probability that x assumes the value 0 or 1 is

$$\begin{aligned} P(x = 0 \text{ or } 1) &= P(x = 0) + P(x = 1) \\ &= f(0) + f(1) \\ &= e^{-m}(1 + m) = e^{-2}(1 + 2) = 3e^{-2} \end{aligned}$$

$$E(x) = m = 2.$$

Ans. $3e^{-2}$, 2.

Example 12.18 *Find the mean, and variance of Poisson distribution.*

[C.U., B.Sc.(Econ.)'81; M.Com. '75, '77]

Solution Poisson distribution is defined by the probability mass function (p.m.f.)

$$f(x) = e^{-m}\frac{m^x}{x\,!} \quad (x = 0, 1, 2, .., \infty)$$

$$\text{Mean} = E(x) = \sum_{x=0}^{\infty} x \cdot f(x)$$

$$= 0\,.\,f(0) + 1\,.\,f(1) + 2\,.\,f(2) + 3\,.\,f(3) + \ldots \text{ to infinity.}$$

$$= 0 + 1\,.\,e^{-m}\frac{m}{1\,!} + 2\,.\,e^{-m}\frac{m^2}{2\,!} + 3\,.\,e^{-m}\frac{m^3}{3\,!} + \ldots$$

$$= e^{-m}\,.\,m + e^{-m}\frac{m^2}{2\,!} + e^{-m}\,.\,\frac{m^2}{2\,!} + \ldots$$

$$= m\,.\,e^{-m}\left(1 + \frac{m}{1\,!} + \frac{m^2}{2\,!} + \ldots\right)$$

$$= m\,.\,e^{-m}\,.\,e^m$$

$$= m$$

$$\text{Variance} = E(x^2) - \{E(x)\}^2$$

$$= E(x^2) - m^2, \quad \text{since } E(x) = m$$

$$= E\{x(x-1) + x\} - m^2$$

$$= E\{x(x-1)\} + m - m^2$$

Now,

$$E\{x(x-1)\} = \sum_{x=0}^{\infty} x(x-1)\,.\,f(x)$$

$$= 0\,.\,f(0) + 1 \times 0\,.\,f(1) + 2 \times 1\,.\,f(2) + 3 \times 2\,.\,f(3) + \ldots$$

$$= 2\,.\,f(2) + 3 \times 2\,.\,f(3) + 4 \times 3\,.\,f(4) + \ldots$$

$$= 2\,.\,e^{-m}\frac{m^2}{2\,!} + 3 \times 2\,.\,e^{-m}\frac{m^3}{3\,!} + 4 \times 3\,.\,e^{-m}\frac{m^4}{4\,!} + \ldots$$

$$= e^{-m}\,.\,m^2 + e^{-m}\,.\,\frac{m^3}{1\,!} + e^{-m}\,.\,\frac{m^4}{2\,!} + \ldots$$

$$= m^2\,.\,e^{-m}\left(1 + \frac{m}{1\,!} + \frac{m^2}{2\,!} + \ldots\right)$$

$$= m^2\,e^{-m}\,.\,e^m = m^2$$

$$\therefore \quad \text{Variance} = m^2 + m - m^2 = m$$

Example 12.19 *Find the mode of Poisson distribution.*

[C.U., B.Sc. (Econ.). '81 (New)]

Solution In the Poisson distribution with parameter m, the random variable X assumes the values 0, 1, 2, ... ∞ and the probabilities are given by

$$f(x) = P(X = x) = e^{-m}\frac{m^x}{x\,!}\,; \quad (x = 0, 1, 2, .., \infty)$$

The mode is that value (or values) among 0, 1, 2, 3, ..., ∞ corresponding to which the probabilities $f(x)$ is the maximum.

$$\frac{f(x-1)}{f(x)} = \frac{e^{-m} \cdot \dfrac{m^{x-1}}{(x-1)!}}{e^{-m} \cdot \dfrac{m^x}{x!}} \qquad (x \neq 0$$

$$= \frac{x}{m} \qquad \text{... (1)}$$

We find that:

(a) $f(x-1) < f(x)$, whenever $x < m$ .... (2)

(b) $f(x-1) = f(x)$, whenever $x = m$ .... (3)

(c) $f(x-1) > f(x)$, whenever $x > m$ .... (4)

Case I When m is not an integer,

Let r be the integral part of m. Since in a Poisson distribution x assumes only non-negative integral values, the result (3) is impossible here.

From (2), $f(x-1) < f(x)$, when $x = 1, 2, 3, ..., r$

$\therefore$ $f(0) < f(1) < f(2) < ... < f(r)$ ... (5)

From (4), $f(x-1) > f(x)$, when $x = r+1, r+2, ...$

$\therefore$ $f(r) > f(r+1) > f(r+2) > ... ...$ ... (6)

Combining the results (5) and (6), we have

$$f(0) < f(1) < ... < f(r-1) < f(r) > f(r+1) > ...$$

Thus, the mode of Poisson distribution is r, i.e., the largest integer contained in the parameter m.

Case II When m is an integer, say $m = r$

From (2), $f(x-1) < f(x)$, when $x = 1, 2, 3, ..., (r-1)$

$\therefore$ $f(0) < f(1) < f(2) < ... < f(r-1)$ ... (7)

From (3), $f(x-1) = f(x)$, when $x = r$

$\therefore$ $f(r-1) = f(r)$ ... (8)

From (4), $f(x-1) > f(x)$, when $x = r+1, r+2, ...$

$\therefore$ $f(r) > f(r+1) > f(r+2) > ... ...$ ... (9)

Combining the results (7), (8) and (9)

$$f(0) < f(1) < ... < f(r-1) = f(r) > f(r+1) > ... ...$$

We find that two terms $f(r)$ and $f(r-1)$ are equal and the largest of all. Thus the Poisson distribution is bimodal when m is an integer, the modes being m and $m-1$.

Note: This method is identical with that for binomial distribution:

12.7 POISSON APPROXIMATION TO BINOMIAL DISTRIBUTION

Poisson distribution (12.6.1) may be obtained as a limiting case of Binomial distribution (12.5.1) under the following conditions:

(i) the number of trials n is infinitely large; i.e., $n \to \infty$;

(ii) the probability of success (p) is extremely small; i.e., $p \to 0$;

(iii) the mean $np = m$ is finite.

Under these conditions it can be shown that the probability of x successes in Binomial distribution can be closely approximated by the probability of x successes in Poisson distribution with parameter $m = np$. i.e.,

$$^nC_x p^x q^{n-x} \simeq \frac{e^{-m} m^x}{x!} \qquad (12.7.1)$$

Proof $$^nC_x p^x q^{n-x} = \frac{n(n-1)(n-2)\ldots(n-x+1)}{x!} p^x(1-p)^{n-x}$$

$$= \frac{1\left(1-\frac{1}{n}\right)\left(1-\frac{2}{n}\right)\ldots\left(1-\frac{x-1}{n}\right)}{x!}(np)^x\left(1-\frac{np}{n}\right)^{n-x}$$

$$= \frac{1}{x!}\left[1.\left(1-\frac{1}{n}\right)\left(1-\frac{2}{n}\right)\ldots\left(1-\frac{x-1}{n}\right)\right] m^x\left(1-\frac{m}{n}\right)^{n-x}$$

As $n \to \infty$, each of the factors $\left(1-\frac{1}{n}\right)$; $\left(1-\frac{2}{n}\right)$, ..., $\left(1-\frac{x-1}{n}\right)$ tends to 1 and $\left(1-\frac{m}{n}\right)^{n-x}$ tends to the limiting value e^{-m}. Thus, under the conditions stated above.

$$^nC_x p^x q^{n-x} \simeq \frac{1}{x!}[1, 1, 1, \ldots 1]\, m^x e^{-m}$$

$$= e^{-m}, \frac{m^x}{x!}$$

Since the probabilities of Poisson distribution are easier to compute than those of Binomial distribution, for practical purposes the Poisson approximation is used when p is less than 0.1 and np is not very large (say, less than 10).

Poisson distribution thus funds important application in such phenomenon where the probability of occurrence of an event (success) is extremely small, but the number of opportunities is infinitely large, so that the mean is something finite. Hence this distribution is sometimes known as the "distribution of rare events".

Example 12.20 *In turning out certain toys in a manufacturing process in a factory, the average number of defectives is 10%. What is the probability of getting exactly 3 defectives in a sample of 10 toys chosen at random, by using the Poisson approximation to the binomial distribution? (Take e = 2.72).* [I.C.W.A., Dec. '78]

Solution Let the occurrence of a defective toy be called a "success". The number of defectives (x) follows binomial distribution with parameters $n = 10$ and $p = 10\% = 0.1$. If this distribution is to be approximated by Poisson distribution

$$f(x) = e^{-m} \cdot \frac{m^x}{x!}$$

we have $m = np = 10 \times 0.1 = 1$. Hence the probability of 3 defective in the sample is

$$f(3) = e^{-1} \cdot \frac{1^3}{3!} = e^{-1} \cdot \frac{1}{6} = \frac{1}{6e}$$

$$= \frac{1}{6 \times 2.72} = 0.61$$

[**Note:** Using binomial distribution, the true probability is $^{10}C_3(0.1)^3 (0.9)^7$] = .057 (upto 3 decimals. Even for a small $n = 10$, this agrees fairly well with the Poisson approximation 061 (the difference being only .004).]

Example 12.21 *2% of the items made by a machine are defective. Find the probability that 3 or more items are defective in a sample of 100 items (Given e^{-1} = 0.368, e^{-2} = 0.135, e^{-3} = .0498).* [Dip. Management '77]

Solution The number of defectives (x) follows binomial distribution. Since $p = 2\% = 0.2$ is small, and $n = 100$ is fairly large, making $np = 100 \times .02 = 2$ a finite quantity, we use Poisson approximation to binomial distribution with $m = np = 2$.

$$f(x) = e^{-m} \cdot \frac{m^x}{x\,!} = e^{-2} \frac{2^x}{x\,!}$$

Probability of 3 or more defectives

$$= f(3) + f(4) + f(5) + \dots \dots$$

$$= 1 - [f(0) + f(1) + f(2)]$$

$$= 1 - e^{-2}\left[1 + 2 + \frac{2^2}{2\,!}\right]$$

$$= 1 - 0.135 \times 5$$

$$= 0.325$$

12.8 HYPERGEOMETRIC DISTRIBUTION

Suppose that a box contains N balls among which A balls are white and the remaining $N - A$ are black. If n balls are drawn at random *without replacement*, the probability of obtaining x white balls (and obviously $n - x$ black balls) among them is given by

$$f(x) = \frac{{}^AC_x \; {}^{N-A}C_{n-x}}{{}^NC_n}; \; (x = 0, 1, 2, \dots, m) \tag{12.8.1}$$

where m is the smaller of the positive integers n and A. The probability distribution defined by the probability mass function (12.8.1) is known as *Hypergeometric Distribution.*

It may be noted that (12.8.1) satisfies both the conditions for a p.m.f. viz. (i) $f(x) \geq 0$; because the numerator and denominator are positive integers; (ii) $\Sigma f(x) = 1$; because the L.H.S. represents the sum of probabilities for all mutually exclusively and exhaustive events $x = 0, 1, 2, \dots, m$.

[The relation (ii) may also be obtained by using the identity

$$\sum_{x=0}^{m} {}^AC_x \, {}^{N-A}C_{n-x} = {}^NC_n$$

i.e. $\quad {}^AC_0 \cdot {}^AC_n + {}^AC_1 \cdot {}^{N-A}C_{n-1} + {}^AC_2 \cdot {}^{N-A}C_{n-2} + \dots + {}^AC_m \cdot {}^{N-A}C_{n-m} = {}^NC_n$

This may be established by equating the coefficients of y^n on both sides of the identity $(1 + y)^A \cdot (1 + y)^{N-A} = (1 + y)^N$, where N and A are positive integers.]

The p.m.f. of Hypergeometric distribution may also be given in another form. Let us put $A = Np$ in (12.8.1), so that $p = A/N$ is the proportion of white balls in the box. Then $N - A = N(1 - p) = Nq$ (suppose), where $q = 1 - p$ is the proportion of black balls. Thus we can write

$$f(x) = \frac{{}^{Np}C_x \, {}^{Nq}C_{n-x}}{{}^{N}C_n}; \ (x = 0, 1, 2, ..., m) \tag{12.8.2}$$

where $p + q = 1$ (p and q being positive proper fractions), and m is the smaller of the numbers n and Np.

Hypergeometric distribution has three parameters, viz. N, n and A in (12.8.1), or N, n and p in (12.8.2).

[**Note:** The model for hypergeometric distribution (12.8.2) is somewhat analogous to that of binomial distribution (12.5.1). If n balls are drawn at random from a box containing Np white and Nq black balls ($p + q = 1$), then the probability of obtaining exactly x white balls is given by

$${}^{Np}C_x \, {}^{Nq}C_{n-x} / {}^{N}C_n$$

if the drawing is made *without replacement*, and

$${}^{n}C_x \, p_x \, q^{n-x}$$

if the drawing is made x *with replacement*.]

Important Properties:

1. Hypergeometric distribution is a discrete probability distribution where the random variable assumes a finite number of values 0, 1, 2, ..., m. It has three parameters—(N, n, A) in (12.8.1) and (N, n, p) in (12.8.2).
2. For the hypergeometric distribution (12.8.1)

$$\text{Mean} = \frac{nA}{N}; \quad \text{Variance} = \frac{nA(N-A)(N-n)}{N^2(N^2-1)} \tag{12.8.3}$$

These expressions take convenient forms on putting $A = Np$. Thus, for the hypergeometric distribution (12.8.2),

$$\text{Mean} = np, \quad \text{Variance} = npq\left(\frac{N-n}{N-1}\right) \tag{12.8.4}$$

[These results may be compared with those of binomial distribution at (2.5.2)]

3. The binomial distribution may be obtained as a limiting case of Hypergeometric distribution when $N \to \infty$

Proof The p.m.f. of hypergeometric distribution is

$$\frac{{}^{Np}C_x \, {}^{Nq}C_{n-x}}{{}^{N}C_n} = \frac{\dfrac{Np(Np-1)\ ...\ (Np-x+1)}{x\,!} \cdot \dfrac{Nq(Nq-1)\ ...\ (Nq-n+x+1)}{(n-x)\,!}}{\dfrac{N(N-1)\ ...\ (N-n+1)}{n\,!}}$$

$$= \frac{n\,!}{x\,!(n-x)\,!} \cdot \frac{\overbrace{Np.(Np-1)\ldots(Np-x+1)}^{x \text{ factors}} \cdot \overbrace{Nq.(Nq-1)\ldots(Nq-n+x+1)}^{n-x \text{ factors}}}{\underbrace{N(N-1)\ldots\ldots\ldots\ldots(N-n+1)}_{n \text{ factors}}}$$

$$= {}^nC_x \cdot \frac{p \cdot \left(p - \frac{1}{N}\right) \cdots \left(p - \frac{x-1}{N}\right) \cdot q \cdot \left(q - \frac{1}{N}\right) \cdots \left(q - \frac{n-x-1}{N}\right)}{1 \cdot \left(1 - \frac{1}{N}\right) \cdots \left(1 - \frac{n-1}{N}\right)}$$

(dividing each factor in the numerator and denominator by N)

$$\therefore \quad \underset{N \to \infty}{Lt} \frac{{}^{Np}C_x \cdot {}^{Nq}C_{n-x}}{{}^{N}C_n} = {}^nC_x \cdot \frac{(p \cdot p \ldots p)\,(q \cdot q \ldots q)}{1 \cdot 1 \ldots \ldots 1}$$

$$= {}^nC_x\, p^x\, q^{n-x}$$

this is the p.m.f. of binomial distribution. (Proved)

Example 12.22 *Two defective tube-lights are mixed with 8 non-defective tube-lights by mistake. A sample of 3 tubes is taken at random from the lot and tested for the number of defectives. Find the probability distribution of the 'number of defective tube-lights in the sample' and tabulate the probabilities. What are the mean and standard deviation of the number of defective obtained in the sample?*

Solution The 'number of defectives (x) in the sample' of 3 tubes may be $x = 0, 1, 2$ (because the whole lot contains only 2 defectives). Since the tubes may be assumed to have been drawn without replacement, x follows Hypergeometric distribution.

$$f(x) = \frac{{}^2C_x \cdot {}^3C_{3-x}}{{}^{10}C_3} = \frac{1}{120}\,({}^2C_x\, {}^3C_{3-x});\ (x = 0, 1, 2)$$

Putting $x = 0, 1, 2$, and simplifying.

$$f(0) = \frac{7}{15},\quad f(1) = \frac{7}{15},\quad f(2) = \frac{1}{15}.$$

The probability distribution of the number (x) of defective tube-lights is:

x	0	1	2	Total
Probability	7/15	7/15	1/15	1

Here $N = 10$, $n = 3$, and the proportion of defectives in the lot is $p = 2/10 = 0.2$. Using (12.8.4)

$$\text{Mean} = 3 \times 0.2 = 0.6$$

$$\text{Variance} = 3 \times 0.2 \times 0.8 \left(\frac{10-3}{10-1}\right) = \frac{3.36}{9}$$

$$\therefore \quad \text{Standard deviation} = \sqrt{3.36/9} = 1.83/3 = 0.61$$

12.9 MULTINOMIAL DISTRIBUTION

The binomial distribution is applicable only in cases when the possible outcomes are of 2 types—'success' and 'failure'. In general, if the possible outcomes of a random experiment are of k types, say $E_1, E_2, \ldots, E_k$ with constant probabilities $p_1, p_2, \ldots, p_k$ respectively, then the probability that in n independent trials E_1 occurs x_1 times, E_2 occurs x_2 times, ..., E_k occurs x_k times is given by

$$\frac{n!}{x_1!\,x_2!\,...\,x_k!}\,(p_1)^{x_1}\,(p_2)^{x_2}\,...\,(p_k)^{x_k} \qquad (12.9.1)$$

where $x_1 + x_2 + ... + x_k = n$ and $p_1 + p_2 + ... + p_k = 1$.

The discrete probability distribution defined by the probability function (12.9.1) is called *Multinomial Distribution*. The binomial distribution may be obtained as a special case of Multinomial distribution, when $k = 2$.

Example 12.23 *A die is rolled 6 times. Find the probability that each of the six faces appears exactly once.*

Solution In a single throw of the die, there are 6 possible outcomes, viz. 1, 2, ..., 6 with probability 1/6 for each (the die is assumed to be unbiased); and 6 independent trials are made because the die is rolled 6 times. Hence

$$p_1 = \frac{1}{6};\; p_2 = \frac{1}{6};\; p_3 = \frac{1}{6};\; p_4 = \frac{1}{6};\; p_5 = \frac{1}{6};\; p_6 = \frac{1}{6}$$

$k = 6$ and $n = 6$. We have to find the probability that

$$x_1 = 1,\quad x_2 = 1,\quad x_3 = 1,\quad x_4 = 1,\quad x_5 = 1,\quad x_6 = 1$$

Substituting the values in (12.9.1), the required probability is

$$\frac{6!}{1!1!1!1!1!1!}\left(\frac{1}{6}\right)^1\left(\frac{1}{6}\right)^1\left(\frac{1}{6}\right)^1\left(\frac{1}{6}\right)^1\left(\frac{1}{6}\right)^1\left(\frac{1}{6}\right)^1$$

$$= 720\left(\frac{1}{6}\right)^6 = \frac{720}{6^6} = \frac{5}{324}$$

Example 12.24 *A box contains 4 white, 3 black and 5 red balls. A ball is drawn from the box at random, its colour noted, and then the ball is replaced. If 6 balls are drawn in this manner (i.e., with replacement), find the probability that (i) 3 are white, 1 black and 2 red, (ii) 2 are white and 4 red.*

Solution Here the 6 trials are independent and the outcome of any trial may be one of the 3 types—white, black, red. Since there are 12 balls in the box (4 white, 3 black, 5 red), therefore, in a single trial.

$$p_1 = \text{probability of white ball} = \frac{4}{12} = \frac{1}{3}$$

$$p_2 = \text{probability of black ball} = \frac{3}{12} = \frac{1}{4}$$

$$p_3 = \text{probability of red ball} = \frac{5}{12}$$

(i) Probability that among $n = 6$ balls drawn, $x_1 = 3$, $x_2 = 1$ and $x_3 = 2$ is

$$\frac{6!}{6!1!2!}\left(\frac{1}{3}\right)^3\left(\frac{1}{6}\right)^1\left(\frac{5}{12}\right)^2$$

$$= \frac{720}{6\times 2}\left(\frac{1}{27}\right)\left(\frac{1}{4}\right)\left(\frac{25}{144}\right) = \frac{125}{1296}$$

(ii) Probability that 2 are white and 4 red ($x_1 = 2, x_2 = 0, x_3 = 4$) is

$$\frac{6!}{2!0!4!}\left(\frac{1}{3}\right)^2\left(\frac{1}{4}\right)^0\left(\frac{5}{12}\right)^4$$

$$= 15 \times \frac{1}{9} \times 1 \times \frac{625}{20736} = \frac{3125}{62208}$$

12.10 JOINT DISTRIBUTION OF TWO VARIABLES

Let us suppose that a random variable x assumes 3 possible values x_1, x_2, x_3 corresponding to each of which another random variable y assumes 4 possible values y_1, y_2, y_3, y_4. There are then 12 pairs of values (x_i, y_j), $i = 1, 2, 3; j = 1, 2, 3, 4$. Let p_{ij} denote the probability assigned to the pair (x_i, y_j).

$$p_{ij} = \text{Prob}\ (x = x_i, y = y_j) \qquad (12.10.1)$$

All the possible values of the pair (x, y) and the corresponding probabilities can be shown in a *Joint Distribution* (section 11.21) as follows:

Table 12.6 Joint Distribution of x and y

y/x	y_1	y_2	y_3	y_4	Total
x_1	p_{11}	p_{12}	p_{13}	p_{14}	$p_{1\cdot}$
x_2	p_{21}	p_{22}	p_{23}	p_{24}	$p_{2\cdot}$
x_3	p_{31}	p_{32}	p_{33}	p_{34}	$p_{3\cdot}$
Total	$p_{\cdot 1}$	$p_{\cdot 2}$	$p_{\cdot 3}$	$p_{\cdot 4}$	1

Obviously, the sum of the probabilities p_{ij} in all rows and columns is 1.

$$p_{11} + p_{12} + \ldots + p_{33} + p_{34} = 1$$

i.e.,

$$\sum_i \sum_j p_{ij} = 1 \qquad (12.10.2)$$

We use the symbols $p_{1\cdot}, p_{2\cdot}, p_{3\cdot}$ to denote the row-totals and $p_{\cdot 1}, p_{\cdot 2}, p_{\cdot 3}, p_{\cdot 4}$ to denote the column-totals. These are known as *marginal probabilities.*

[**Note:** that p_1 is the total of all p's with the first subscript 1, and $p_{\cdot 1}$ is the total of all p's whose second subscript is 1, etc.]

$$p_{i\cdot} = p_{i1} + p_{i2} + p_{i3} + p_{i4} = \sum_j p_{ij} \qquad (12.10.3)$$

$$p_{\cdot j} = p_{1j} + p_{2j} + p_{3j} + p_{4j} = \sum_i p_{ij}$$

The sum of the marginal probabilities for x (or for y) is therefore 1.

$$p_{1\cdot} + p_{2\cdot} + p_{3\cdot} = \sum_i p_{i\cdot} = 1$$

$$p_{\cdot 1} + p_{\cdot 2} + p_{\cdot 3} + p_{\cdot 4} = \sum_j p_{\cdot j} = 1 \qquad (12.10.4)$$

Marginal Distribution

Let us consider only the random variable x, which assumes the values x_1, x_2, x_3. The event $x = x_1$, for example, occurs in 4 mutually exclusive ways, in association with the values assumed by y, viz. as (x_1, y_1), (x_1, y_2) (x_1, y_3) (x_1, y_4). So, by the theorem of total probability,

$$P(x = x_1) = P(x = x_1, y = y_1) + P(x = x_1, y = y_2) + P(x = x_1, y = y_3) + P(x = x_1, y = y_4)$$

$$= p_{11} + p_{12} + p_{13} + p_{14}$$

$$= p_{1\cdot}$$

Similarly, $P(x = x_2) = p_{2\cdot}$ and $P(x = x_3) = p_{3\cdot}$

Thus the total of probabilities in any row (as shown in the margin) gives the probability of the corresponding value $x = x_i$.

The probability distribution of a random variable, obtained from the joint distribution *irrespective of the values assumed by the other variable,* is called the *Marginal Distribution*. It shows the possible values of the variable and their 'marginal probabilities'.

Table 12.7 Marginal Distribution of x

x	x_1	x_2	x_3	Total
Probability	$p_{1\cdot}$	$p_{2\cdot}$	$p_{3\cdot}$	1

Table 12.8 Marginal Distribution of y

y	y_1	y_2	y_3	y_4	Total
Probability	$p_{\cdot 1}$	$p_{\cdot 2}$	$p_{\cdot 3}$	$p_{\cdot 4}$	1

Conditional Distribution

By the definition of conditional probability (Section 11.17)

$$P\left(\frac{A}{B}\right) = \frac{P(A \cap B)}{P(B)}$$

Suppose that A and B represent the events $x = x_1$ and $y = y_3$ respectively. Then the conditional probability of the event $x = x_1$, given the $y = y_3$ has happened, is

$$P(x = x_1 / y = y_2) = \frac{P(x = x_1, y = y_3)}{P(y = y_3)} = \frac{p_{13}}{p_{\cdot 3}}$$

Similarly $P\,(x = x_2/y = y_3) = \dfrac{p_{23}}{p_{\cdot 3}}$, $P\,(x = x_3/y = y_3) = \dfrac{p_{33}}{p_{\cdot 3}}$

This means that once $y = y_3$ is known to have occurred, we fix our attention to the entries in the two-way table against y_3. Then the conditional probabilities of x_1, x_2, x_3 are obtained on dividing the probabilities p_{13}, p_{23}, p_{33} (shown against y_3) by their total ($p_{\cdot 3}$).

The probability distribution of a random variable, *given that the other variable has a specified value*, is called *Conditional Distribution*. It shows the possible values of the variable and their conditional probabilities.

Table 12.9 Conditional Distribution of x
(given $y = y_3$)

x	x_1	x_2	x_3	Total
Probability	$p_{13}/p_{\cdot 3}$	$p_{23}/p_{\cdot 3}$	$p_{33}/p_{\cdot 3}$	1

[**Note:** The probabilities shown here are those given in the two-way table against $y = y_3$, but adjusted to a total on division by their total $p_{\cdot 3}$]

Table 12.10 Conditional Distribution of y
(given $x = x_2$)

y	y_1	y_2	y_3	y_4	Total
Probability	$p_{21}/p_{2\cdot}$	$p_{22}/p_{2\cdot}$	$p_{23}/p_{2\cdot}$	$p_{24}/p_{2\cdot}$	1

Similarly, conditional distributions of x (or of y) can be stated for other given values.

Example 12.25 *The following tables gives the joint distribution of x and y.*

x \ y	2	3	7	Total
1	*.10*	*.25*	*.05*	*.40*
3	*.30*	*.15*	*.15*	*.60*
Total	*.40*	*.40*	*.20*	*1*

(i) How many pairs of values of x and y are possible? Write them down and show the probability of each pair separately.
(ii) Show the marginal distribution of x and y.
(iii) Show the conditional distribution of x, given y = 2.
(iv) Show the conditional distribution of y, given x = 3.
(v) Find the probabilities P (x < y), P (2x + y ≥ 9).

Solution

(i) There are 2 possible values of x, and 3 possible values of y, so that the number of possible pairs (x, y) is $2 \times 3 = 6$.

(x, y)	(1, 2)	(1, 3)	(1, 7)	(3, 2)	(3, 3)	(3, 7)	Total
Probability	.10	.25	.05	.30	.15	.15	1

(ii) *Marginal Distribution of x*

x	1	3	Total
Probability	.40	.60	1

Marginal Distribution of y

y	2	3	7	Total
Probability	.40	.40	.20	1

(iii) *Conditional Distribution of x (given y = 2)*

x	1	3	Total
Probability	$\frac{.10}{.40} = .25$	$\frac{.30}{.40} = .75$	1

(iv) *Conditional Distribution of y (given x = 3)*

y	2	3	7	Total
Probability	$\frac{.30}{.60} = .50$	$\frac{.15}{.60} = .25$	$\frac{.15}{.60} = .25$	1

(v) The event $x < y$ consists of the pairs of values (1, 2), (1, 3), (1, 7), (3, 7). Hence $P(x < y)$ is given by the sum of the probabilities of these pairs.

$$P(x < y) = .10 + .25 + .05 + .15 = 0.55$$

The event $2x + y \geq 9$ is satisfied by the pairs (1, 7), (3, 3), (3, 7).

$$\therefore \quad P(2x + y \geq 9) = .05 + .15 + .15 = .035$$

Independent Variables

The random variables x and y are said to be *"independent"*, if

$$p_{ij} = p_{i\cdot} \times p_{\cdot j} \tag{12.10.5}$$

i.e. $$P(x = x_i,\ y = y_i) = P(x = x_i) \times P(y = y_i)$$

for each and every pair of values $(x_i, y_i) : i = 1, 2, ... m ; j = 1, 2, ... n$. This means that if x and y are independent, then each entry p_{ij} in the joint distribution is the product of the corresponding marginal probabilities $p_{i\cdot}$ and $p_{\cdot f}$. If the condition (12.10.5) fails even for one pair of values (x_i, y_i), then the variables x and y will not be 'indepependent'.

Illustration In the following joint distribution, the variables x and y are 'independent'.

x \ *y*	0	2	4	Total
2	.06	.15	.09	.30
5	.14	.35	.21	.70
Total	.20	.50	.30	1

Note: That each cell probability is the product of the corresponding marginal probabilities of x and of y.

Bivariate Expectation and Covariance

If the joint distribution of x and y is given, the expectation of x (or of y) or of any function of a single variable can be calculated from the corresponding marginal distribution. Thus

$$E(x) = \sum_i p_i . x_i \qquad E(x^2) = \sum_j p_i . x_i^2$$

$$E(y) = \sum_j p_{.j}\, y_j \qquad E(y^2) = \sum_j p_j . y_j^2 \tag{12.10.6}$$

The variances of x and y are then obtained, as before

$$\text{Var}(x) = E(x^2) - \{E(x)\}^2$$
$$\text{Var}(y) = E(y^2) - \{E(y)\}^2 \tag{12.10.7}$$

In general, the bivariate expectation of any function of the random variables x and y is defined as the sum of products of all possible values of the function and the corresponding bivariate probabilities. For example

$$E(x + y) = \sum_i \sum_j p_{ij}\,(x_i + y_j)$$

$$E(xy) = \sum_i \sum_j p_{ij}\,(x_i\, y_j)$$

$$E(x - a) = \sum_i \sum_j p_{ij}\,(x_i - a) = \Sigma p_i (x_i - a)$$

$$E\{(x - a)\,(y - b)\} = \sum_i \sum_j p_{ij}\,(x_i - a)\,(y_j - b),$$

where a and b are constants.

If m_1, m_2 denote the means of x and y respectively, the expectation of $(x - m_2)$ $(y - m_2)$ is particularly important, and is called the *Covariance of x and y*, written as Covariance of x and y, written as Covariance (x, y).

$$\text{Cov}(x, y) = E\{(x - m_1)\,(y - m_2)\}$$
$$= E(xy) - E(x).E(y) \tag{12.10.8}$$

In case the variables x and y are independent

$$\text{Cov}(x, y) = 0 \tag{12.10.8a}$$

The following expectations are especially important:

(i) $E(x) = m_1$; $\quad E(y) = m_2$

(ii) $E(x - m_1) = 0$; $\quad E(y - m_2) = 0$

(iii) $E(x - m_1)^2 = \text{Var}(x)$; $\quad E(y - m_2)^2 = \text{Var}(y)$

(iv) $E\{(x - m_1)(y - m_2)\} = \text{Cov}(x, y)$ $\qquad$ (12.10.9)

Theorems of Expectation

Theorem I The expectation of the sum of two jointly distributed random variables x and y is the sum of the expectations.

$$E(x + y) = E(x) + E(y) \tag{12.10.10}$$

Proof Let x and y be jointly distributed discrete random variables, and suppose that x takes numerically different values $x_1, x_2, \ldots x_m$ while y takes numerically different values $y_1, y_2, \ldots y_3$. Let p_{ij} denote the probability that x assumes the value x_i and simultaneously y assumes the values y_j.

$$P(x = x_j, y = y_j) = p_{ij}$$

The marginal probabilities are then given by

$$P(x = x_i) = \sum_j p_{ij} = p_{i\cdot}$$

$$P(y = y_j) = \sum_i p_{ij} = p_{\cdot j}$$

Hence, $E(x) = \Sigma\, p_i \,.\, x_i \quad E(y) = \Sigma p_{\cdot j}\, y_j$

Since x and y can assume m and n different values, there are mn mutually exclusive cases for their sum $x + y$. In general, when x assumes the value x_i and y assumes the value y_j, the sum $x + y$ assumes the values $x_j + y_j$. Therefore,

$$P(x + y = x_i + y_j) = P(x = x_i, y = y_j) = p_{ij}$$

Hence, by definition of expectation

$$\begin{aligned} E(x + y) &= \sum_i \sum_j p_{ij}\,(x_i + y_j) \\ &= \sum_i \sum_j p_{ij}\, x_i + \sum_i \sum_j p_{ij}\, y_j \\ &= \sum_i x_i \sum_j p_{ij} + \sum_j y_j \sum_i p_{ij} \\ &= \sum_i x_i p_i + \sum_j y_j p_{\cdot j} \\ &= E(x) + E(y) \end{aligned}$$

This theorem can be extended to more than two variables.

Theorem II The expectation of the product of two *independent* random variables x and y is the product of their expectations.

$$E(x, y) = E(x)\,.\,E(y) \tag{12.10.11}$$

provided x and y are independent.

Proof Let x and y be independent discrete random variables and suppose that x takes m numerically different values $x_1, x_2, \ldots x_m$ and y takes n numerically different values $y_1, y_2, \ldots y_n$. Let us write

$$P(x = x_i, y = y_j) = p_{ij}$$

We have, as before,

$$E(x) = \sum_i p_i\,.\,x_i \text{ and } E(y) = \sum_j p_{\cdot j}\, y_j$$

Since x and y can assume m and n different values there are mn mutually exclusive cases for the product xy. In general, when x assumes the value x_i and y assumes the value y_j, the product xy assumes the value $x_i y_j$. Therefore,

$$P(xy = x_i y_j) = P(x = x_i, y = y_j) = p_{ij}$$

If x and y are *independent* random variables, we have

$$P(x = x_i, y = y_j) = P(x = x_i)\,.\,P(y = y_j)$$

i.e. $\quad p_{ij} = p_i\,.\,p_{\cdot j}$

Hence, by the definition of expectation

$$E(xy) = \sum_i \sum_j p_{ij}\,(x_i\, y_j)$$

$$=\sum_i \sum_j p_i . p_{.j} \, x_i \, y_j$$

$$=\sum_i p_i . x_i \sum_j p_{.j} \, y_j$$

$$=E(x) \, . \, E(y)$$

This theorem can be extended to several *mutually independent* random variables.

Correlation Coefficient

This *Correlation Coefficient* between x and y, which is denoted by the symbol ρ_{xy}, is defined as

$$\rho_{xy} = \frac{\text{Cov}\,(x, y)}{\sigma_x \sigma_y} \qquad (12.10.12)$$

where σ_x and σ_y are the standard deviations of x and y respectively. It has the following important properties :

(1) The correlation coefficient is independent of origin and scale of the variables. If $u = ax + b$ and $v = cy + d$, where a, b, c, d are constants, then

$$\rho_{uv} = \pm \, \rho_{xy}$$

according as a and c have the same sign or opposite signs.

(2) The correlation coefficient lies between – 1 and + 1.

$$-1 \le \sigma_{xy} \le +1$$

(3) If the variables x and y are independent, then $\rho_{xy} = 0$. But, if $\rho_{xy} = 0$, the variables may not be independent; they are then said to be *uncorrelated.*

Example 12.26 *The following table gives the joint distribution of x and y:*

x \ *y*	*0*	*1*	*2*
1	*0.3*	*0.2*	*0.1*
2	*0.1*	*0.0*	*0.3*

(i) Are x and y independent?

(ii) Determine the correlation coefficient between x and y. (W.B.H.S. '78)

Solution After completing the row and column-totals, the joint distribution of x and y is

x \ y	0	1	2	Total
1	0.3	0.2	0.1	0.6
2	0.1	0.0	0.3	0.4
Total	0.4	0.2	0.4	1

We find that $P(x = 1, y = 0) = 0.3$, $P(x = 1) = 0.6$ and $P(y = 0) = 0.4$. Since here $P(x = 1, y = 0) \ne P(x = 1) \, . \, P(y = 0)$ the variables are not independent.

From the marginal distribution of x,

$$E(x) = 0.6 \times 1 + 0.4 \times 2 = 1.4$$

$$E(x^2) = 0.6 \times 1^2 + 0.4 \times 2^2 = 2.2$$

$$\therefore \quad \text{Var}(x) = 2.2 - (1.4)^2 = 2.2 - 1.96 = 0.24\ ; \quad \sigma_x = \sqrt{0.24}$$

Similarly from the marginal distribution of y,

$$E(y) = 0.4 \times 0 + 0.2 \times 1 + 0.4 \times 2 = 1.0$$
$$E(y^2) = 0.4 \times 0^2 + 0.2 \times 1^2 + 0.4 \times 2^2 = 1.8$$

$$\therefore \quad \text{Var}(y) = 1.8 - (1.0)^2 = 0.8; \quad \sigma_y = \sqrt{0.8}$$

Again, from the joint distribution, we have

$$E(xy) = (1 \times 0) \times 0.3 + (1 \times 1) \times 0.2 + (1 \times 2) \times 0.1$$
$$+ (2 \times 0) \times 0.1 + (2 \times 1) \times 0.0 + (2 \times 2) \times 0.3$$
$$= 0 + 0.2 + 0.2 + 0 + 0 + 1.2 = 1.6$$
$$\text{Cov}(x, y) = E(xy) - E(x).E(y) = 1.6 - 1.4 \times 1.0 = 0.2$$

Now, using (12.10.9).

$$\rho_{xy} = \frac{0.2}{\sqrt{0.24 \times 0.8}} = \frac{0.2}{0.438} = 0.46$$

Example 12.27 *The marginal distributions of x and y are given in the following table*:

y \ x	5	7	Total
3			$\frac{1}{3}$
6			$\frac{2}{3}$
Total	$\frac{1}{2}$	$\frac{1}{2}$	1

If cov (x, y) = E(xy) – E(x). E(y) is known to be $-\frac{1}{2}$, *obtain the cell probabilities.*

Hence calculate $P(x > y)$ *and* $P(x = \frac{5}{y} = 6)$. [C.U., B.Sc. '74]

Solution We find that $E(x) = \frac{1}{2} \times 5 + \frac{1}{2} \times 7 = 6$

$$E(y) = \frac{1}{3} \times 3 + \frac{2}{3} \times 6 = 5$$

Let us write p for the unknown cell probability $P(x = 5, y = 3)$. Then the other cell probabilities can be obtained from the marginal totals.

y \ x	5	7	Total
3	p	$\frac{1}{3} - p$	$\frac{1}{3}$
6	$\frac{1}{2} - p$	$\frac{1}{6} + p$	$\frac{2}{3}$
Total	$\frac{1}{2}$	$\frac{1}{2}$	1

Using these cells probabilities,

$$E(xy) = p\,(5 \times 3) + (\frac{1}{3} - p)\,(7 \times 3) + (\frac{1}{2} - p)\,(5 \times 6) + (\frac{1}{6} + p)\,(7 \times 6)$$
$$= 15\,p + (7 - 21p) + (15 - 30p) + (7 + 42p)$$
$$= 6p + 29$$

$\therefore$ $\quad \text{Cov}\,(x, y) = E(xy) - E(x).E(y)$
$$= (6p + 29) - 6 \times 5 = 6p - 1$$

By the given condition,

$$6p - 1 = -\frac{1}{2}\ ; \quad \therefore p = \frac{1}{12}$$

The other cell probabilities are then

$$\frac{1}{2} - p, = \frac{1}{4},\ \frac{1}{2} - p = \frac{5}{12},\ \frac{1}{6} + p = \frac{1}{4}.$$

The completed joint distribution is

y \ x	5	7	Total
3	$\frac{1}{12}$	$\frac{1}{4}$	$\frac{1}{3}$
6	$\frac{5}{12}$	$\frac{1}{4}$	$\frac{2}{3}$
Total	$\frac{1}{2}$	$\frac{1}{2}$	1

$P\,(x > y)$ = Sum of the probabilities of the pairs (5, 3), (7, 3), (7, 6)

$$= \frac{1}{12} + \frac{1}{4} + \frac{1}{4} = \frac{7}{12}$$

$$P\,(x = 5/y = 6) = \frac{P\,(x = 5, y = 6)}{P\,(y = 6)} = \frac{\frac{5}{12}}{\frac{2}{3}} = \frac{5}{8}$$

Example 12.28 *Show that the correlation coefficient of x and y is zero if x and y are independence. Is the converse also true? Give reasons for your answer.*

[W.B.H.S., '80]

Solution Let the variables x and y be independent and have means m_1, m_2 and variances σ_x^2, σ_y^2 respectively. Then,

$$\text{Cov}\,(x, y) = E\{(x - m_1)\,(y - m_2)\}$$
$$= E(x - m_1)\,.\,E(y - m_2), \text{ since independent}$$
$$= 0.0 = 0$$

Therefore, the correlation coefficient is

$$\rho_{xy} = \frac{\text{Cov}(x, y)}{\sigma_x \sigma_y} = \frac{0}{\sigma_x \sigma_y} = 0$$

This shows that the correlation coefficient is zero, if x and y are independent.
However, if $\quad \rho_{xy} = 0$, we have

$$\text{Cov}(x, y) = 0$$

or, $$E(xy) - E(x) . E(y) = 0$$

or, $$E(xy) = E(x) . E(y)$$

or, $$\sum_i \sum_j p_{ij}\, x_j\, y_j = \sum_i p_i . x_i \sum_i p._j\, y_j$$

This does not ensure that $p_{ij} = p_j . p._j$, which is condition for independence of x and y. That is, the condition

$$P(x = x_j, y = y_j) = P(x = x_j) . P(y = y_j)$$

is not necessarily true for all pairs of values (x_i, y_j).

Therefore, if $\rho_{xy} = 0$, the variables x and y may not be independent. The converse is not true.

Example 12.29 *If $u = ax + b$ and $v = cy + d$, where a, b, c, d are constants, show that $\rho_{uv} = \pm \rho_{xy}$.*

Solution Let us suppose that the random variables x and y have means m_1, m_2 and variances σ_1^2, σ_2^2 respectively;

i.e. $$E(x) = m_1,\ E(y) = m_2$$

$$E(x - m_1)^2 = \sigma_1^2,\ E(y - m_2)^2 = \sigma_2^2$$

Also, $$\text{Cov}(x, y) = E\{(x - m_1)(y - m_2)\}$$

and $$\rho_{xy} = \frac{\text{Cov}(x, y)}{\sigma_1 \sigma_2}$$

If we write $$u = ax + b,\ v = cy + d$$

then $$m_u = E(u) = E(ax + b) = E(ax) + E(b)$$
$$= a.E(x) + b = am_1 + b$$

Similarly, $$m_v = E(v) = E(cy + d) = cm_2 + d$$

Hence, $$u - m_u = (ax + b) - (am_1 + b) = a(x - m_1)$$
$$v - m_v = (cy + d) - cm_2 + d) = c(y - m_2)$$

The variances of u and v are

$$\sigma_u^2 = E(u - m_u)^2 = E\{a^2(x - m_1)^2\}$$
$$= a^2.E(x - m_1)^2 = a^2 . \sigma_1^2$$

Similarly, $$\sigma_v^2 = c^2 . \sigma_2^2$$

Since by definition, the standard deviation is always positive

$$\sigma_u = |a| . \sigma_1, \quad \sigma_v = |c| . \sigma_2$$

Again, $$\text{Cov}(u, v) = E\{(u - m_u)(v - m_v)\}$$
$$= E\{ac(x - m_1)(y - m_2)\}$$
$$= ac.E\{(x - m_1)(y - m_2)\}$$
$$= ac. \text{Cov}(x, y)$$

The correlation coefficient between u and v is

$$\rho_{uv} = \frac{\text{Cov}(u, v)}{\sigma_u \sigma_v} = \frac{ac.\text{Cov}(x, y)}{|a|\sigma_1 |c|.\sigma_2}$$

$$= \frac{ac}{|a|.|c|} . \frac{\text{Cov}(x, y)}{\sigma_1 \sigma_2} = \frac{ac}{|a|.|c|} \rho_{xy}$$

When a and c have the same sign (i.e. both are positive, or both negative), the product ac is positive and is the same $|a| . |c|$; so that $\rho_{uv} = \rho_{xy}$.

When a and c have opposite signs (i.e. one is positive and the other is negative), the product ac is negative. But $|a| . |c|$ is always positive and has the same numerical values as ac; Hence $\rho_{uv} = -\rho_{xy}$.

Thus, $\rho_{uv} = \pm \rho_{xy}$

[*Illustration* If the correlation coefficient between x and y is -0.7, then the correlation coefficient between (i) $u = 2x + 3$ and $v = 5y - 8$ is also -0.7; (ii) $u = -2x + 3$ and $v = 5v$ is $+0.7$.]

Example 12.30 *Prove that the correlation coefficient p between two jointly distributed random variables x and y always like between – 1 and + 1.*

Solution Let the random variables x and y have means m_1, m_2 and variances σ_1^2, σ_2^2 respectively. For any real value of k, the square quantity $\{(x - m_1) + k(y - m_2)\}^2$ can never be negative, and hence its expected value

$$E\{(x - m_1) + k(y - m_2)\}^2 \geq 0$$

Expanding the L.H. and taking expectations

$$E(x - m_1)^2 + k^2.E(y - m_2)^2 + 2k.E\{(x - m_1)(y - m_2)\} \geq 0$$

or, $$\sigma_1{}^2 + k^2.\sigma_2{}^2 + 2k . \text{Cov}(x, y) \geq 0$$

or, $$\sigma_1{}^2 + k^2.\sigma_2{}^2 + 2k . \rho\sigma_1\sigma_2 \geq 0$$

or, $$(k^2\sigma_2{}^2 + 2k\,\sigma_2.\rho\sigma_1 + \rho^2\sigma_1{}^2) + \sigma_1{}^2(1 - \rho^2) \geq 0$$

i.e. $$(k\sigma_2 + \rho\sigma_2)^2 + \sigma_1{}^2(1 - \rho^2) \geq 0$$

Since this is true for any value of k, we have for $k = -\dfrac{\rho\sigma_1}{\sigma_2}$, the first term is zero and hence

$$\sigma_1^2(1 - \rho^2) \geq 0$$

or $$(1 - \rho^2) \geq 0$$

or, $$\rho^2 \leq 1$$

or, $$-1 \leq \rho \leq +1$$ (Proved)

Variance of the Sum (Difference)

If x and y are jointly distributed random variables, then

$$\text{Var}(x + y) = \text{Var}(x) + \text{Var}(y) + 2\,\text{Cov}(x, y)$$
$$\text{Var}(x - y) = \text{Var}(x) + \text{Var}(y) - 2\,\text{Cov}(x, y) \qquad (12.10.13)$$

Theorem $\text{Var}(ax + by) = a^2\,\text{Var}(x) + b^2\,\text{Var}(y) + 2ab\,\text{Cov}(x, y)$

where a and b are constants. (12.10.14)

Proof Suppose that x and y have means m_1, m_2 and variances σ_1^2, σ_2^2 respectively. If we write

$$z = ax + by$$

then $$E(z) = E(ax) + E(by) = a . E(x) + b.E(y)$$

or, $$m_z = am_1 + bm_2, \text{ where } m_z \text{ is the mean of } z$$

$\therefore$ $$z - m_z = (ax + by) - (am_1 + bm_2) = a(x - m_1) + b(y - m_2)$$

By definition,

$$\begin{aligned} \text{Var}(z) &= E(z - m_z)^2 \\ &= E[a(x - m_1) + b(y - m_2)]^2 \\ &= E[a^2(x - m_1)^2 + b^2 + (y - m_2)^2 + 2ab(x - m_1)(y - m_2)] \end{aligned}$$

$$= a^2 . E(x-m_1)^2 + b^2 (y-m_2)^2 + 2ab.E\{(x-m_1)(y-m_2)\}$$

i.e. $\quad \text{Var}(ax+by) = a^2 . \text{Var}(x) + b^2.\text{Var}(y) + 2ab.\text{Cov}(x, y) \quad$ (Proved)

It may be noted that the results at (12.10.13) may be obtained from (12.10.14) on putting $a = 1, b = 1$ and $a = 1, b = -1$.

When x and y are independent, we have $\text{Cov}(x, y) = 0$, and

hence
$$\text{Var}(x+y) = \text{Var}(x) + \text{Var}(y)$$
$$\text{Var}(x-y) = \text{Var}(x) + \text{Var}(y) \qquad (12.10.15)$$

Thus the variance of the sum or difference of two independent random variables is the sum of their variances.

It may be noted from (12.10.12) that

$$\text{Cov}(x, y) = \rho_{xy}\, \sigma_x\, \sigma_y$$

and so the results (12.10.13) may also be written in the form

$$\text{Var}(x+y) = \sigma_x^2 + \sigma_y^2 + 2\rho_{xy}\, \sigma_x\, \sigma_y$$
$$\text{Var}(x-y) = \sigma_x^2 + \sigma_y^2 - 2\rho_{xy}\, \sigma_x\, \sigma_y \qquad (12.10.16)$$

If x and y are uncorrelated, i.e. $\rho_{xy} = 0$ (they may not be independent) then

$$\text{Var}(x+y) = \sigma_x^2 + \sigma_y^2 = \text{Var}(x-y) \qquad (12.10.17)$$

Example 12.31 *If the random variables x and y have the same standard deviation, show that u = x + y and v = x – y are uncorrelated.*

Solution Let the variables x and y have means m_1, m_2 and a common standard deviation σ.

$$E(x) = m_1,\ E(y) = m_2,\ E(x-m_1)^2 = E(y-m_2)^2 = \sigma^2$$

The means of u and v are then

$$m_u = E(x) + E(y) = m_1 + m_2$$
$$m_v = E(x) - E(y) = m_1 - m_2$$

Therefore,

$$u - m_u = (x+y) - (m_1+m_2) = (x-m_1) + (y-m_2)$$
$$v - m_v = (x-y) - (m_1-m_2) = (x-m_1) - (y-m_2)$$

By definition, the covariance of u and v is

$$\begin{aligned}\text{Cov}(u, v) &= E\{(u-m_u)(v-m_v)\}\\ &= E[\{(x-m_1)+(y-m_2)\}\{(x-m_1)-(y-m_2)\}]\\ &= E[(x-m_1)^2 - (y-m_2)^2]\\ &= E(x-m_1)^2 - E(y-m_2)^2\\ &= \sigma^2 - \sigma^2\\ &= 0\end{aligned}$$

If σ_u and σ_v denote the standard deviations of u and v respectively, the correlation coefficient between u and v is

$$\rho = \frac{Cov(u, v)}{\sigma_u\, \sigma_v} = \frac{0}{\sigma_u \cdot \sigma_v} = 0$$

i.e. u and v are uncorrelated.

Example 12.32 *Let x and y be independent random variables with standard deviations σ_x and σ_y. Show that the correlation co-efficient between x and x + y is*

$$\frac{\sigma_x}{\sqrt{(\sigma_x^2 + \sigma_y^2)}}$$

Solution Let m_x and m_y denote the means of x and y respectively.

$$E(x) = m_x, \qquad E(y) = m_y$$
$$E(x - m_y)^2 = \sigma_x^2 \qquad E(y - m_y)^2 = \sigma_y^2$$

If we write $u = x + y$, we have to find the correlation coefficient between x and u.

$$\rho_{xu} = \frac{Cov\,(x,u)}{\sigma_x \sigma_u} \qquad ...(i)$$

The mean of u is

$$E\,(u) = E(x + y) = E(x) + E(y)$$

or, $$m_u = m_x + m_y$$

Hence, $$u - m_u = (x + y) - (m_x + m_y) = (x - m_x) + (y - m_y)$$

Now, $$\text{Var}\,(u) = \text{Var}\,(x + y) = \text{Var}\,(x) + \text{Var}\,(y) + 2.\text{Cov}\,(x, y)$$
$$= \sigma_x^2 + \sigma_y^2,$$

since x and y are independent, Cov $(x, y) = 0$

i.e. $$\sigma_u = \sqrt{(\sigma_x^2 + \sigma_y^2)} \qquad ...(ii)$$

Again, $$\text{Cov}\,(x, u) = E\{(x - m_x)\,(u - m_u)\}$$
$$= E\,[(x - m_x)\,\{(x - m_x) + (y - m_y)\}]$$
$$= E(x - m_x)^2 + E\{(x - m_x)\,(y - m_y)\}$$
$$= \sigma_x^2 + \text{Cov}\,(x,\,y)$$
$$= \sigma_x^2, \text{ since Cov}\,(x, y) = 0 \qquad ...(iii)$$

Substituting from (ii) and (iii) in (i), the required correlation coefficient is

$$\rho_{z,\,x+y} = \frac{\sigma_x^2}{\sigma_x \cdot \sqrt{(\sigma_x^2 + \sigma_y^2)}} = \frac{\sigma_x}{\sqrt{(\sigma_x^2 + \sigma_y^2)}}$$

Example 12.33 *If X and Y are independent binomial variables with parameters (m, p) and (n, p) respectively, show that the their sum X + Y has a binomial distribution with parameters (m + n, p).*

Solution The probability that X assumes an arbitrary value x is

$$P(X = x) = {}^mC_x\,p^z\,q^{m-x} \qquad ...(i)$$

for all values of $x = 0, 1, 2, \ldots m$, where $p + q = 1$. Similarly, the probability that Y assumes an arbitrary value y is

$$P(Y = y) = {}^nC_y\,p^y q^{n-y} \qquad ...(ii)$$

for all values of $y = 0, 1, 2, \ldots n$.

Let us write $Z = X + Y$. Since the random variable X can assume the values 0, 1, 2, ... m and Y can assume the values 0, 1, 2, ...n, their sum Z can assume the values 0, 1, 2, ..., $(m + n)$.

The event that $Z = X + Y$ can assume an arbitrary value $z\,(z = 0, 1, 2, \ldots.\ m + n)$ can materialise in the following mutually exclusive ways, viz. $(X = 0, Y = z)$, $(X = 1, Y = z - 1)$, ..., $(X = z, Y = 0)$. Therefore by the Addition Theorem of Probability.

$$P\,(Z = z) = P\,(X = 0, Y = z) + P\,(X = 1, Y = z - 1) + \ldots + P\,(X - z, Y = 0)$$

$$= \sum_{r \to 0}^{z} P\,(X = r, Y = z - r)$$

$$= \Sigma\, P(X = r)\,.\,P\,(Y = z - r),$$

since X and Y are independent.

Using (i) and (ii), we have

$$P(Z = z) = \sum_{r \to 0}^{z} {}^mC_r\, p^r\, q^{m-r}\, {}^nC_{z-r}\, p^{z-r}\, q^{n-(z-r)}$$

$$= p^z\, a^{m+n-z} \sum_{r \to 0}^{z} {}^mC_r \cdot {}^nC_{z-r}$$

$$= {}^{m+n}C_z\, p^z\, q^{m+n-z} \quad (z = 0, 1, 2, \ldots m + n)$$

This shows that the random variable $Z = X + Y$ has a binomial distribution with parameters $(m + n, p)$.

Example 12.34 *If X and Y are independent Poisson variates, find the conditional distribution of X, given X + Y. What are the mean and variance of this conditional distribution?*

Solution If X and Y are Poisson variates with parameters m and n respectively, then

$$P(X = x) = \frac{e^{-m} m^x}{x\,!}\;; \qquad P(Y = y) = \frac{e^{-n} n^y}{y\,!} \qquad \ldots\text{(i)}$$

$$x = 0, 1, 2, \ldots \infty; \qquad y = 0, 1, 2, \ldots \infty$$

We know that the sum of two independent Poisson variates also follows a Poisson distribution. Therefore, $Z = X + Z$ follows Poisson distribution with parameter $m + n$. This means that the probability of Z to assume a specified value z is given by

$$P(Z = z) = \frac{e^{-(m+n)}\,(m+n)^z}{z\,!} \qquad \ldots\text{(ii)}$$

$$z = 0, 1, 2, \ldots \infty$$

The conditional probability that X takes an arbitrary value x, when $Z = X + Y$ is known to have assume a fixed value z is, by definition,

$$P(X = x/Z = z) = \frac{P(X = x, Z = z)}{P(Z = z)} \qquad \ldots\text{(iii)}$$

$$\left[\text{Remember the formula } P(A/B) = \frac{P(A \cap B)}{P(B)}\right]$$

But the event ($X = x$ and $Z = z$) is the same as ($X = x$ and $X + Y = z$), which is equivalent to ($X = x$ and $Y = z - x$). Therefore,

$$P(X = x, Z = z) = P(X = x) \cdot P(Y = z - x)$$

since X and Y are independent

$$= \frac{e^{-m}\, m^x}{x\,!} \cdot \frac{e^{-n}\, n^{z-x}}{(z-x)\,!} \text{ using (i)}$$

$$= \frac{e^{-(m+n)} \cdot m^x n^{2-x}}{x\,!\,(z-x)\,!}\;; \qquad \ldots\text{(iv)}$$

$$x = 0, 1, 2, \ldots z$$

(**Note:** Since the values of a Poisson variable are only non-negative whole numbers, x and z are necessarily whose numbers. Also x cannot exceed z because the value attained by X can not be larger than that of $X + Y$).

Substituting from (ii) and (iv) in (iii), and simplifying

$$P(X = x/Z = z) = \frac{z!}{x!(z-x)!} \cdot \frac{m^x n^{z-x}}{(m+n)^z}$$

$$= {}^zC_x \left(\frac{m}{m+n}\right)^x \left(\frac{n}{m+n}\right)^{z-x}$$

$$(x = 0, 1, 2, \ldots z)$$

This is seen to be a binomial distribution with parameters z and $\frac{m}{(m+n)}$. Thus, the conditional distribution of X, given $X + Y$, is a binomial distribution. The mean and variance of this distribution are obtained by using the formulae for those of binomial distribution.

$$\text{Mean} = z \cdot \frac{m}{m+n}; \text{ Variance} = z \cdot \frac{m}{m+n} \cdot \frac{n}{m+n}$$

$$= \frac{zmn}{(m+n)^2}.$$

12.11 CONTINUOUS PROBABILITY DISTRIBUTION

Let x be a continuous random variable, which can assume any value in the interval (a, b), i.e. $a \le x \le b$. Since the number of possible values of x is uncountable infinite, we cannot assign a probability to each value of the variable, as is done in discrete probability distributions (Section 12.2). Therefore, in a continuous probability distribution we have to assign probabilities to intervals, and not to individual values.

As stated earlier (Section 11.19) for a continuous probability distribution let $f(x)$ be a non-negative function such that

$$P(c \le x \le d) = \int_c^d f(x)\, dx \qquad (12.11.1)$$

i.e. the probability of x lying in any given interval (c, d) can be obtained by integrating the function between the two limits of the interval. The function $f(x)$ is called the *probability density function* (p.d.f.) or simply *density function*, of the continuous random variable x. It satisfies the two conditions

$$\text{(i) } f(x) \ge 0 \qquad \text{(ii) } \int_c^b f(x)\, dx = 1 \qquad (12.11.2)$$

where $a \le x \le b$ is the range of x.

A continuous probability distribution is defined by stating the mathematical expression for the p.d.f. $f(x)$ and specifying the range (a, b). The curve represented by the equation

$$y = f(x)$$

is known as the '*probability curve*'. Geometrically, the integral of the p.d.f. represents the area under the probabilities curve, and hence

$P(c \le x \le d)$ = Area under probability curve between the vertical lines at c and d. (12.11.3)

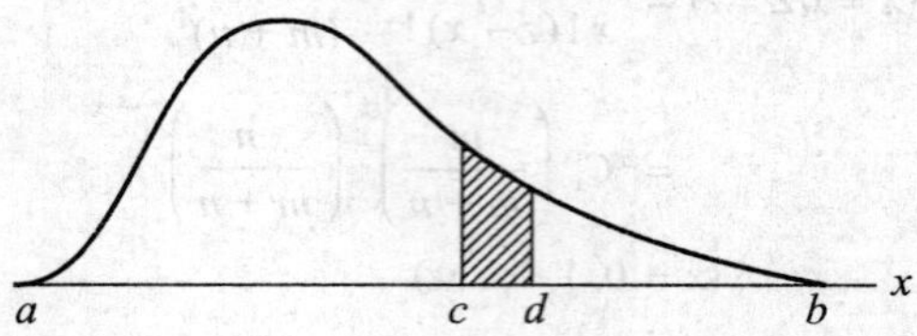

Fig. 12.1 Probability Curve

The cumulative distribution function (c.d.f.) $F(x)$, defined by (11.20.5) represents the probability that the random variable takes a value less than or equal to a specified value x. Geometrically, $F(x)$ shows the area under probability curve to the left of the ordinate at any specified value. For example,

$F(c) = P(x \le c)$
= Area under probability curve to the left of the vertical line at c. (12.11.4)

If the values of $F(x)$ are known for all values of x, then the probability distribution is completely given. Because then the probabilities of intervals can be obtained from the relation

$P(c \le x \le d)$ = Area between the ordinates at c and d
= (Area to the left of the ordinate at d) *minus* (Area to the left of the ordinate at c)
$= F(d) - F(c)$ (12.11.5)

Putting $c = d$, we see that

$P(c \le x \le c)$ = Area between the ordinates at c and c, which is zero.

i.e. $P(x = c) = 0$ (12.11.5a)

Thus, as shown in (11.20.8) earlier, for a continuous probability distribution, the probability that a random variable takes any specific value is zero.

Expectation—Mean, Variance (Continuous Distribution)

For a continuous probability distribution with p.d.f. $f(x)$ in the range (a, b), the expectation mean, variance, moments etc. are defined as follows :

$$\text{Mean } (\mu) = E(x) = \int_a^b x.f(x)\, dx \qquad (12.11.6)$$

$$\text{Variance } (\sigma^2) = E(x-\mu)^2 = \int_a^b (x-\mu)^2 . f(x)\, dx \qquad (12.11.7)$$

$$= E(x^2) - \mu^2 = \int_a^b x^2 f(x)\, dx - \mu^2 \qquad (12.11.8)$$

In general, the expectation of any function $g(x)$ is defined as

$$E\{g(x)\} = \int_a^b g(x) \cdot f(x)\, dx \qquad (12.11.9)$$

The moments are defined as follows:

r-th moment about A: $\quad \mu_r' = E(x - A)^r = \int_a^b (x - A^r) \cdot f(x)\, dx$

r-th raw moment: $\quad \mu_r' = E(x^r) = \int_a^b x^r \cdot f(x)\, dx$

r-th central moment: $\quad \mu_r = E(x - \mu)^r = \int_a^b (x - \mu^r)\, f(x)\, dx \qquad (12.11.10)$

The relations (12.3.8) should be used to find the central moments from non-central moments.

12.12 UNIFORM DISTRIBUTION (CONTINUOUS)

A continuous random variable x is said to follow *Uniform distribution*, if the probabilities associated with intervals of equal length are equal at any part of the range of values. The uniform distribution is defined by the density function

$$f(x) = \frac{1}{b-a}; \quad (a \le x \le b) \qquad (12.12.1)$$

Since the probability curve for this distribution looks like a rectangle of height $\frac{1}{(b-a)}$ over the range $(a \le x \le b)$, the distribution is also known as "*Rectangular Distribution*". The total area under the curve is 1 and the probability that x lies between any two specified values c and d (within the range) is

$$P(c \le x \le d) = \frac{d-c}{b-a} \qquad (12.12.2)$$

The cumulative distribution function (c.d.f.) is given by the part area of the rectangle to the left of the ordinate at x, i.e.

$$F(x) = \frac{x-a}{b-a}; \quad (a \le x \le b) \qquad (12.12.3)$$

Example 12.35 *If a continuous random variable x follows rectangular distribution in the range (2, 7), find the probabilities:*

(i) $P(2.5 \le x \le 4)$, *(ii)* $P(x \le 3.4)$

(iii) $P(x > 6)$ *(iv)* $P(x = 4.5)$.

Solution Using (12.12.2),

$$P(2.5 \le x \le 4) = \frac{4-2.5}{7-2} = \frac{1.5}{5} = 0.3$$

Using (12.12.3), $P(x \le 3.4) = \dfrac{3.4-2}{7-2} = \dfrac{1.4}{5} = 0.28$

$$P(x > 6) = 1 - P(x \le 6) = 1 - \frac{6-2}{7-2} = 0.2$$

Since x is a continuous random variable, by (12.11.5a),

$$P(x = 4.5) = 0$$

Ans. 0.3, 0.28, 0.2, 0

12.13 NORMAL DISTRIBUTION

Normal distribution, or *Gaussian distribution*, is a continuous probability distribution and is defined by the density function (p.d.f.)

$$f(x) = \frac{1}{\sigma\sqrt{2x}} e^{-\frac{(x-\mu)^2}{2\sigma^2}} \; ; (-\infty < x < \infty) \qquad (12.13.1)$$

where μ = mean, σ = standard deviation. [π and e are two mathematical constants having the approximate values $\frac{22}{7}$ and 2.718 (see 12.6.2). respectively]. μ and σ are known as '*parameters*' of this distribution

Although at first glance the expression (12.13.1) may look very complicated, this is the most useful distribution in theoretical statistics because of its many important characteristics. A random variable x is said to '*follow normal distribution*', or to be '*normally distributed*' if the probability density function is of the form (12.13.1). The probability curve of normal distribution is known as *Normal Curve*. The curve is symmetrical and bell-shaped (see Fig. 12.3) and the two tails extend to infinitely on either side.

If a random variable x is normally distributed with mean μ and standard deviation σ, then $z = \dfrac{(x-\mu)}{\sigma}$ is called the "*standard normal variable*" or "*normal deviate*". It has the density function

$$p(z) = \frac{1}{\sqrt{2x}} e^{-\frac{z^2}{2}} \; ; (-\infty < z < \infty) \qquad (12.13.2)$$

The continuous probability distribution defined by (12.13.2) is known as *Standard Normal Distribution*. This is in fact a special case of normal distribution with mean 0 and standard deviation 1.

Area under Normal Curve

As in all continuous probability distributions, the total area under the normal curve is 1; and the probability that x lies between c and d, denoted by $P(c \le x \le d)$, is given by the area under the curve between the vertical lines at c and d. This is also equal to the area under '*Standard normal curve*' between the vertical lines at the standardized values (page 230, Section 7.4) of c and d; i.e.

$P(c \le x \le d)$ = Area under '*standard normal curve*' between the vertical lines at c' and d'. (12.13.3) where $c' = \frac{(c - \mu)}{\sigma}$ and $d' = \frac{(d - \mu)}{\sigma}$.

Extensive tables showing the areas under standard normal curve are available for this purpose.

The cumulative distribution function (c.d.f.) of standard normal distribution, viz.

$\Phi(z)$ = Probability that the standard normal variable takes a value less than or equal to z.

= Area under '*standard normal curve*' to the *left* of the ordinate at z. (12.13.4)

Mathematically,

$$\Phi(z) = \int_{-\infty}^{z} \frac{1}{\sigma\sqrt{2\pi}} e^{-\frac{z^2}{2}} dz \qquad (12.13.4a)$$

and hence $\Phi(z)$ is also known as *Probability Integral* of standard normal distribution. The values of $\Phi(z)$ are given in statistical tables only for positive values of z. For negative values the relation

$$\Phi(-z) = 1 - \Phi(z) \qquad (12.13.5)$$

is used to find the area; because, due to the symmetry of standard normal curve about 0.

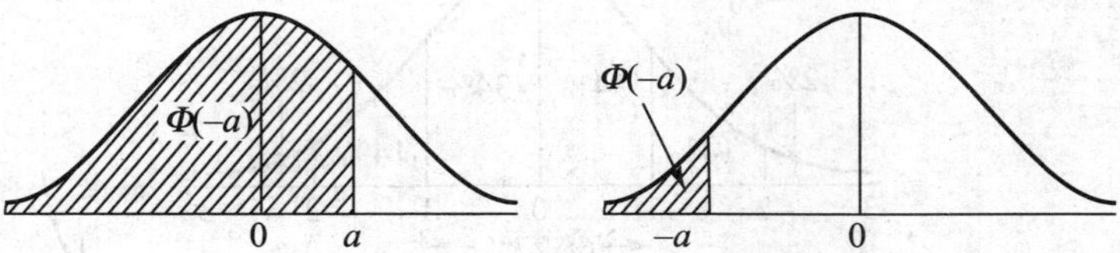

Fig. 12.2 Probability Integral of Standard Normal Distribution

$$(\textit{Area between } z = -a \text{ and } z = 0) \qquad (12.13.6)$$
$$= (\textit{Area between } z = 0 \text{ and } z = a)$$

Thus, probabilities of normal distribution can be calculated by using the relation

$$P(c \le x \le d) = \Phi(d') - \Phi(c') \qquad (12.13.7)$$

where c' and d' are the standardized values of c and d.

Properties of Normal Distribution

(i) The normal distribution has two parameters μ and σ.

(ii) Mean = μ; Standard Deviation = σ.

(iii) Mean, median and mode are equal, each being μ.

Mean = Median = Mode = μ

(iv) The quartiles are equidistant from mean. Approximately;

$$Q_1 = \mu - 0.67\,\sigma;\ Q_3 = \mu + 0.67\,\sigma$$

$\therefore$ Quartile deviation = $0.67\,\sigma$

Mean deviation = $0.80\,\sigma$

(v) All *odd* order central moments are zero.

$$\mu_1 = \mu_3 = 0$$

The second order and fourth order central moments are

$$\mu_2 = \sigma^2, \quad \mu_A = 3\sigma^4$$

In general, $\mu_{2r+1} = 0, \quad \mu_{2r} = 1.3.5... (2r-1)\,\sigma^{2r}$

(vi) $\beta_1 = 0, \quad \beta_2 = 3;$

Skewness $(\gamma_1) = 0$; Kurtosis $(\gamma_2) = 0$

(vii) The normal curve is bell-shaped and symmetrical about the line $x = \mu$. The two tails of the curve extend to infinity on both sides of the mean. The maximum ordinate is at $x = \mu$ and given by

$$y = \frac{1}{\sigma\sqrt{2x}}$$

[This is obtained from (12.13.1) by putting $x = \mu$]. As the distance from the mean increases, the curve comes closer and closer to the horizontal axis but never meets it.

(viii) The points of inflection of the normal curve are at $x = \mu \pm \sigma$. This means that at these points the normal curve changes its curvature from concave to convex and vice versa.

(ix) The percentage distribution of area under the 'standard normal curve' is broadly shown in Fig. 12.3.

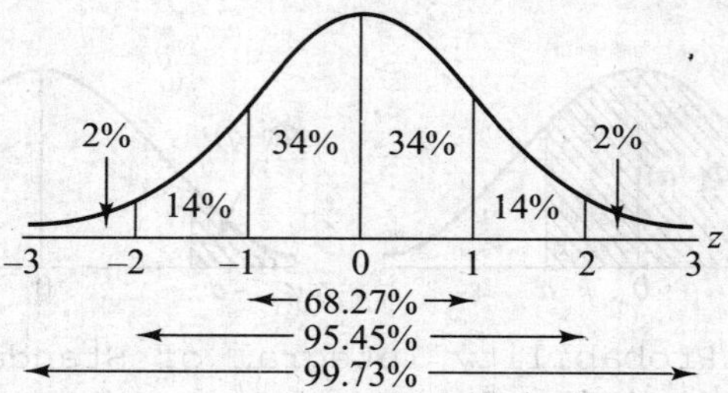

Area between $z = \pm 1$ is 68.27%
Area between $z = \pm 2$ is 95.45%
Area between $z = \pm 3$ is 99.73%

Fig. 12.3 Area under Standard Normal Curve

The last result is especially useful in the theory of Statistical Quality Control (S.Q.C.). It implies that if a random variable x is normally distributed with mean μ and s.d. σ, then almost all the values of x will lie between the limits $\mu - 3\sigma$ to $\mu + 3\sigma$, i.e. Mean ± 3 (S.D.).

(x) If x and y are independent normal variates with means μ_1 and μ_2, and variances σ_1^2 and σ_2^2 respectively, then $(x + y)$ is also a normal variate with mean $(\mu_1 + \mu_2)$ and variance $(\sigma_1^2 + \sigma_2^2)$.

Importance of Normal Distribution

The normal distribution has many important algebraic properties for which it is used widely in statistical theories. In most cases of physical, biological and psychological measurements, the data are found to follow normal distribution. *Statistical Quality*

Control (S.Q.C.) methods in the manufacturing industry and the *Theory of Errors* of observations in physical measurements are also based on normal distribution. The normal distribution is used (Section 12.14) to serve as approximations to the binomial and Poisson distributions. In the theory of sampling, it has been found that any statistic based on a large sample approximately follows normal distribution. The result considerably simplifies the work of testing statistics hypotheses and is also useful to find the confidence limits of parameters.

Example 12.36 *The height distribution of a group of 10,000 men is normal with man height 64.5″ and s.d. 4.5″. Find the number of men whose height is (a) less than 69″ but greater than 55.5″, (b) less than 55.5″, and (c) more than 73.5″.*

Solution The mean (μ) and standard deviation (σ) of the normal distribution are given to be $\mu = 64.5''$ and $\sigma = 4.5''$.

(a) Percentage of men whose height lies between 55.5″ and 69″

= Area under standard normal curve between the vertical lines at the corresponding standardized values, viz.

$$z = \frac{55.5 - 64.5}{4.5} = -2 \quad \text{and} \quad z = \frac{69 - 64.5}{4.5} = 1.$$

From the percentage distribution of area under the standard normal curve (Fig. 12.3), it is found that the area between $z = -2$ and $z = 1$ is (14% + 34% + 34%) = 82%. This means that 82% of the total number of 10,000 men are expected to have heights between 55.5″ and 69″. Hence, the required number of men is 82% of 10,000 i.e. $\frac{82}{100} \times 10{,}000 = 8.200$.

(b) Percentage of men whose height is less than 55.5″

= Area under standard normal curve to the *left* of the Standardized value

$$z = \frac{55.5 - 64.5}{4.5} = -2.$$

= 20% (see Fig. 12.3).

The number of men is therefore, 20% of 10,000 i.e. 200.

(c) Percentage of men whose height is *more* than 73.5″.

= Area under standard normal curve to the *right* of the Standardized value

$$z = \frac{73.5 - 64.5}{4.5} = 2.$$

= 20%

Hence, the number of men whose height is more than 73.5″ is 20% of 10,000. i.e. 200.

Ans. (a) 8,200; (b) 200; (c) 200.

Example 12.37 *The mean weight of 500 male students at a certain college is 151 lbs. and the standard deviation is 15 lbs. Assuming that the weights are normally distributed, find how many students weight (i) between 120 and 155 lbs., (ii) more than 155 lbs. [Given $\Phi(0.27) = 0.6064$ and $\Phi(2.07) = 0.9808$, where $\Phi(t)$ denotes the area under standard normal curve to the left of the ordinate at t.]*

Solution The mean (μ) and the standard deviation (σ) are:

$$\mu = 151 \text{ lbs.}, \quad \sigma = 15 \text{ lbs.}$$

(i) Proportion of students whose weights lie between 120 and 155 lbs. = Area under standard normal curve between the vertical lines at the standardized values, viz.

$$z = \frac{(120-151)}{15} = -2.07 \quad \text{and} \quad z = \frac{(155-151)}{15} = 0.27. \quad \text{Using (12.13.7).}$$

$$\begin{aligned} P(120 \le x \le 155) &= \Phi(0.27) - \Phi(-2.07) \\ &= \Phi(0.27) - \{1 - \Phi(2.07)\}, \quad \text{by} \quad (12.13.5) \\ &= 0.6064 - 1 + 0.9808 \\ &= 0.5872 \quad \text{[see Fig. 12.4 (a)]} \end{aligned}$$

The number of students whose weights lie between 120 and 155 lbs. is therefore $0.5872 \times 500 = 294$ (approx.)

(ii) Proportion of students who weigh more than 155 lbs. is

$$\begin{aligned} P(x > 155) &= 1 - P(x \le 155) \\ &= 1 - \Phi(0.27) = 1 - 0.6064 = 0.3964 \end{aligned}$$

The number of students who are expected to weigh more than 155 lbs is $0.3936 \times 500 = 197$ (approx.) *Ans.* 294, 197

Example 12.38 *A sample of 100 dry battery cells tested to find the length of life produced the following results : $\overline{x} = 12$ hours, $\sigma = 3$ hours. Assuming that the data are normally distributed, what percentage of battery cells are expected to have life (i) more than 15 hours, (ii) less than 6 hours, and (iii) between 10 and 14 hours?*

[I.C.W.A. June, '79]

Given	*z*	*2.5*	*2*	*1*	*0.67*
	Area	*.4938*	*.4772*	*.3413*	*.2487*

Solution [**Note:** The given table shows the area under standard normal curve between the ordinates at $z = 0$ and the specified values of z. This is clear from the fact that the area increases with larger values of z, and the area upto $z = 2.5$ is almost 0.5. Remember that the area under standard normal curve between $z = -3$ and $z = +3$ is almost 1, and hence the area from $z = 0$ upto $z = 3$ may be taken to be 0.5].

The percentage of battery cells with life between specified limits will be given by the proportionate are under the standard normal curve between their standardized values. Here, mean = 12, s.d. = 3.

$$\text{Standardized value of 15 hours} = \frac{(15-12)}{3} = +1$$

$$\text{Standardized value of 6 hours} = \frac{(6-12)}{3} = -2$$

$$\text{Standardized value of 10 hours} = \frac{(10-12)}{3} = -0.67$$

$$\text{Standardized value of 14 hours} = \frac{(14-12)}{3} = +0.67$$

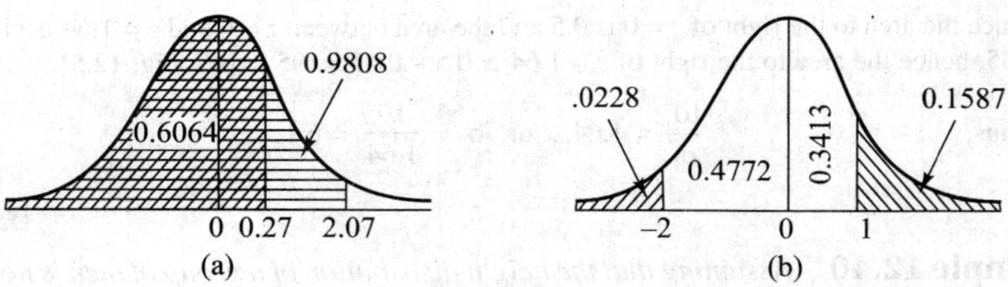

Fig. 12.4 Area under Standard Normal Curve

Proportion of battery cells with life *more* than 15 hours
= Area under standard normal curve to the *right* of $z = +1$
= (Total area to the right of $z = 0$) – (Area between $z = 0$ and $z = +1$)
= 0.5 – 0.3413 [see Fig. 12.4(b)]
= 0.1587 = 15.87%

Proportion with life *less* than 6 hours
= Area under the curve to the *left* side of $z = -2$
= Area to the *right* of $z = +2$, (since the curve is symmetrical about $z = 0$)
= 0.5 – 0.4772
= 0.0228 = 2.28%

Proportion with life between 10 and 14 hours
= Area under the curve between $z = -0.67$ and $z = +0.67$
= 2 × (Area between $z = 0$ and $z = +0.67$), since the curve is symmetrical
= 2 × 0.2487 [see Fig. 12.5 (a)]
= 0.4974 = 49.74%

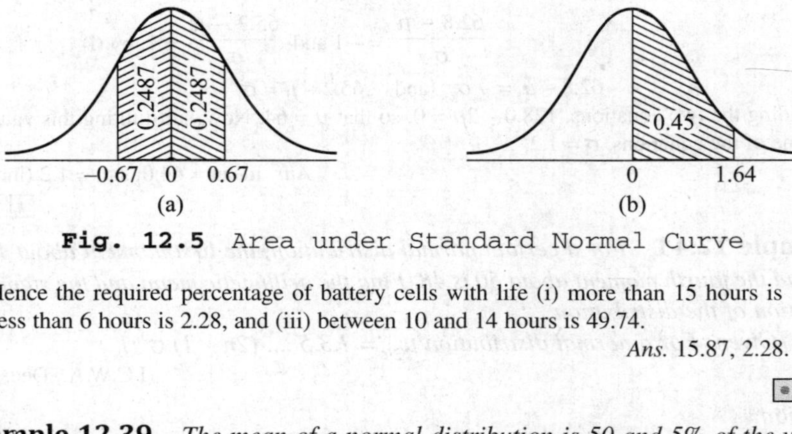

Fig. 12.5 Area under Standard Normal Curve

Hence the required percentage of battery cells with life (i) more than 15 hours is 15.87, (ii) less than 6 hours is 2.28, and (iii) between 10 and 14 hours is 49.74.

Ans. 15.87, 2.28. 49.74

Example 12.39 *The mean of a normal distribution is 50 and 5% of the values are greater than 60. Find the standard deviation of the distribution (Given that the area under standard normal curve between z = 0 and z = 1.64 is 0.45).*

(I.C.W.A., Dec. '79)

Solution The probability that x takes a value greater than 60 is 5%, i.e. .05. This must be the area under standard normal curve to the right of the ordinate at the standardized value

$$z = \frac{60-50}{\sigma} = \frac{10}{\sigma}$$

Since the area to the right of $z = 0$ is 0.5 and the area between $z = 0$ and $z = 1.64$ is given to be 0.45, hence the area to the right of $z = 1.64$ is $0.5 - 0.45 = .05$ [see Fig. 12.5].

Thus, $$\frac{10}{\sigma} = 1.64; \quad \text{or, } \sigma = \frac{10}{1.64} = 6.1.$$

Example 12.40 *Assuming that the height distribution of a group of men is normal, find the mean and standard deviation, if 84% of the men have heights less than 65.2 inches and 68% have height lying between 65.2 and 62.8 inches.*

(I.C.W.A. June '78)

Solution Let μ denote the mean and σ the standard deviation: and z_1 and z_2 represent the standardized values of 62.8″ and 65.2″,

i.e. $$z_1 = \frac{62.8 - \mu}{\sigma} \quad \text{and} \quad z_2 = \frac{65.2 - \mu}{\sigma} \qquad \text{...(i)}$$

(Note that since 65.2 is greater than 62.8, z_2 will be greater than z_1).

Since 84% of the men have height *below* 65.2″, this must be the area under the standard normal curve to the *left* of the standardized value z_2. But from the percentage distribution of area under the curve (Fig. 12.3) we find that the area to the left of $z = 1$ is 84%. Hence, z_2 must be identical with 1, i.e. $z_2 = 1$.

Again, it is given that 68% of the men have height between 65.2″ and 62.8″ ; i.e. the percentage of area under the standard normal curve between $z_2 = 1$ and a smaller value z_1 is 68%. From Fig. 12.3 it is found that the area between $z = 1$ and $z = -1$ is 68%. Hence, z_1 must be identical with –1, i.e. $z_1 = -1$.

Thus, we find that $z_1 = -1$ and $z_2 = 1$.

or, $$\frac{62.8 - \mu}{\sigma} = -1 \text{ and } \frac{65.2 - \mu}{\sigma} = 1, \text{ by (i)}$$

or, $$62.8 - \mu = -\sigma \quad \text{and} \quad 65.2 - \mu = \sigma$$

Adding the two equations, $128.0 - 2\mu = 0$, so that $\mu = 64$. Now, substituting this value of μ in one of the equations, $\sigma = 1.2$.

Ans. mean = 64.0; s.d. = 1.2 (inches)

Example 12.41 *For a certain normal distribution, the first moment about 10 is 40, and the fourth moment about 50 is 48. Find the arithmetic mean and the standard deviation of the distribution.*

[Given: For a normal distribution $\mu_{2n} = 1.3.5 \ldots. (2n - 1)\, \sigma^{2n}$]

(I.C.W.A., Dec. '78)

Solution

$$\text{Mean} = A + \text{first moment about } A, \text{ (see 7.1.10)}$$
$$= 10 + 40 = 50$$

Since the mean is 50, the fourth moment about 50 is the fourth central moment μ_4. Again, putting $n = 1, 2$ in the given formula,

$$\mu_2 = \sigma^2, \quad \mu_4 = (1 \times 3)\, \sigma^4 = 3\sigma^4$$

The first relation shows that σ^2 is the variance of normal distribution, i.e., the parameter σ is the s.d. From the second relation

$$3\sigma^4 = 48 \text{ (given)}; \quad \text{i.e. } \sigma = 2$$

Ans. Mean = 50, S.D. = 2.

12.14 NORMAL APPROXIMATION TO BINOMIAL (POISSON)

In the binomial distribution (12.5.1), if neither p or q is very small, but n is large, then the binomial distribution can be closely approximated by the normal distribution having the same mean and standard deviation as those of the binomial,

i.e. $\mu = np$ and $\sigma = \sqrt{npq}$

In order to find the probability $P(a \le x \le b)$ of the discrete binomial distribution by the continuous normal distribution, some corrections to the values of a and b are necessary. The number of successes is always a whole number, and hence the number of successes a corresponds to an interval $a - \frac{1}{2}$ to $a + \frac{1}{2}$ in the continuous scale (e.g. '47 successes' refers to any value between 46.5 and 47.5). Therefore, the number of successes a actually starts from $a - \frac{1}{2}$, and the number of successes b ends at $b + \frac{1}{2}$ in the normal distribution. The probability $P(a \le x \le b)$ that the number of successes lies between a and b (both inclusive) therefore, corresponds to $P(a - \frac{1}{2} \le x \le b + \frac{1}{2})$ of the normal distribution. Using (12.11.3), this probability is given by the area under standard normal curve between the vertical lines at the standardized values.

$P(a \le x \le b)$ of binomial distribution

$= P\left(a - \frac{1}{2} \le x \le b + \frac{1}{2}\right)$ of the approximating normal distribution.

= Area under standard normal curve between the vertical lines at a' and b'. (12.14.1)

where $a' = \dfrac{\left(a - \frac{1}{2}\right) - np}{\sqrt{npq}}$ and $b' = \dfrac{\left(b + \frac{1}{2}\right) - np}{\sqrt{npq}}$

The larger the value of n, the better is the approximation. But, for practical purposes, the approximation is generally used, if both p and q are larger than 0.1 and n is not less than 30.

When the parameter m is large. Poisson distribution can be approximated by the normal distribution with the same mean and standard deviation as those of Poisson,

i.e. mean = m and s.d. = $\sqrt{m}$

In practice, the approximation is used, when $m \ge 10$.

Example 12.42 *A fair coin is tossed 400 times. Using normal approximation, find the probability of obtaining (i) exactly 200 heads, (ii) less than 210 heads, (iii) between 190 and 210 heads, both inclusive. Given that the area under standard normal curve*

between $z = 0$ and $z = .05$ is .0199
between $z = 0$ and $z = .95$ is .3289
between $z = 0$ and $z = 1.05$ is .3531

Solution Let us denote the occurrence of a head as "success". The number of heads (x) follows binomial distribution with $n = 400$, and $p = \frac{1}{2}$. Since n is large, we use the normal approximation to the binomial with

$$\mu = np = 400 \times \frac{1}{2} = 200$$

$$\sigma = \sqrt{npq} = \sqrt{400 \times \frac{1}{2} \times \frac{1}{2}} = 10$$

(i) Probability of obtaining exactly 200 heads

$= P(x = 200)$ of binomial distribution

$= P(199.5 \le x \le 200.5)$ of normal distribution

= Area under standard normal curve between the standardized values z

$= \frac{(199.5 - 200)}{10} = -.05$ and $z = \frac{(200.5 - 200)}{10} = +.05$

$= 2 \times$ (Area between $z = 0$ and $z = +.05$)

$= 2 \times 0.199 = 0.4$ (approx.)

(ii) Probability of obtaining less than 210 heads

$= P(x < 210)$ of binomial distribution

$= P(x \le 209.5)$ of normal distribution

= Area under the standard normal curve to the *left* of

$$z = \frac{(209.5 - 200)}{10} = .95$$

= (Area to the left of $z = 0$) + (Area between $z = 0$ and $z = .95$)

$= 0.5 + 0.3289 = 0.83$ (approx.)

(iii) Probability of obtaining between 190 and 210 heads

$= P(190 \le x \le 210)$ of binomial distribution

$= P(189.5 \le x \le 210.5)$ of normal distribution

= Area under standard normal curve between $z = \frac{189.5 - 200}{10}$

$= -1.05$ and $z = \frac{210.5 - 200}{10} = +1.05$

$= 2 \times$ (Area between $x = 0$ and $z = 1.05$)

$= 2 \times .3531 = 0.71$ (approx.) *Ans.* 0.04, 0.83, 0.71

Example 12.43 *A variable x follows Poisson distribution with mean 16. Find the probability $P(x \ge 20)$. Given $F(0.875) = .8092$ where $F(x)$ is the standards normal distribution function.*

Solution Using normal approximation to Poisson distribution,

Mean (μ) = $m = 16$, S.D. (σ) = $\sqrt{m} = \sqrt{16} = 4$

$P(x \ge 20)$ of Poisson distribution

$= P(x \ge 19.5)$ of normal distribution

= Area under standard normal curve to the *right* of

$$z = \frac{(19.5 - 16)}{4} = 0.875$$

= 1 – (Area to the *left* of $z = 0.875$)

= 1 – F (0.875)

= 1 – 0.8092 = 0.19 (approx.)

12.15 CONTROL LIMIT THEOREM

The normal distribution is used as an approximation to the binomial and Poisson distributions, because the former can be obtained as a limiting case under certain conditions. The *Central Limit Theorem*, stated below, shows that a large class of distributions can in fact be approximated by the normal distribution.

Central Limit Theorem—Let $x_1, x_2, \ldots, x_n, \ldots$ be independent random variables which are identically distributed (i.e. all have the same probability distribution) with mean μ and variance σ^2. If $\bar{x}_n = \frac{(x_1 + x_2 + \ldots x_n)}{n}$ denotes the mean of the first n variables, then the distribution of

$$z_n = \frac{(\bar{x}_n - \mu)}{\frac{\sigma}{\sqrt{n}}}$$

approaches the Standard Normal Distribution, as n becomes larger and larger.

This shows that the mean of n independently and identically distributed random variables with mean μ and standard deviation σ can be approximated by the normal distribution with

$$\text{Mean} = \mu, \quad \text{S.D.} = \frac{\sigma}{\sqrt{n}}$$

whatever by the probability distribution of the variables, provided n is sufficiently large. (Note that the distribution of the random variables may be discrete or continuous and of any form). For practical purposes $n \geq 30$ gives a sufficiently good approximation, the accuracy increasing with larger values of n.

EXERCISES

1. Explain the concept of 'theoretical distribution' with reference to a discrete random variable.
2. What is meant by the probability distribution of a discrete random variable? *(W.B.H.S. '78; C.U., B.Sc., '79)*
3. What is meant by Probability Mean Function of a discrete random variable? Write down these functions for the Binomial as well as for Poisson distributions. *[I.C.W.A., June '79]*

4. What do you understand by 'expectation' of a random variable? State how you will find the 'mean' and 'standard deviation' of a discrete probability distribution with p.m.f. $f(x)$.
5. State whether each of the following is *True* or *False*:
   (a) The probability mass function of a random variable may be zero.
   (b) The expected number of points to be obtained from a throw of a perfect die is 3.
   (c) The expectation of a random variable cannot be negative.
   (d) The square of $E(x)$ is larger than $E(x^2)$.
   (e) If Var $(x) = 0$, then all values of x are equal.
6. A discrete r.v. assumes the values 1, 2, 3, ... 10 with probabilities proportional to 1, 2, 3, ... , 10 respectively. Calculate the mean and variance.
7. Let X be a random variable with p.m.f. $f(x) = \left(\frac{1}{32}\right){}^5C_x$ where $x = 0, 1, 2, \ldots 5$. Find the mean and standard deviation.
8. A discrete random variable x follows uniform distribution and takes only the values 6, 8, 11, 12, 17. Find the probabilities

   $P(x = 8)$, $P(x = 10)$, $P(x \le 12)$, $P(x < 12)$ $P(x > 10)$.
9. Define 'Binomial distribution' and state the conditions under which the distribution holds. *[I.C.W.A., June '74, Dec. '79]*
10. Four coins are tossed simultaneously. What is the probability of getting 2 heads and 2 tails? *[C.A., May '77]*
11. Eight coins are thrown simultaneously. Show that the probability of obtaining at least 6 heads is $\frac{37}{256}$. *[I.C.W.A. June, '74]*
12. Find the probability that in a family of 5 children there will be (i) at least one boy, (ii) at least one boy and one girl. (Assume that the probability of a female birth is 1/2). *[I.C.W.A., Dec. '76]*
13. If a sample of 5 items is drawn randomly form a lot containing 10% defective items, what is the probability of getting not more than one defective item? *[M.B.A. '77, Dip. Mgmt. '77]*
14. In a shooting competition, the probability of a man hitting a target is $\frac{1}{5}$. If the fires 5 times, what is the probability of hitting the target at least twice? *[C.U., M.Com., '75]*
15. Assume that on the average 30% of the candidates appearing in an examination from a certain college get First Division. What is the probability that out of a group of 4 such candidates not more than two will fail to get a First Division?
16. Let the probability of a patient recovering from a certain disease by 0.75. Find the distribution of the number of recoveries among 4 patients. Hence compute the mean and standard deviation. *[W.B.H.S. '80]*

17. The probability of a bomb hitting a target is $\frac{2}{5}$. Four direct hits are necessary to destroy a bridge completely. If 6 bombs are aimed at the bridge, what is the probability that the bridge will be destroyed?

18. Two dice are thrown n times in succession. What is the probability of obtaining double-six at least once? *[C.U., B.Sc., '75]*

19. (a) What is a Bernoullian series of trials? Give an example.
    (b) Find the probability of at least one success in a Bernoullian series of n trials with a probability of success p in each trial.

20. In an infinite series of Bernoulli trials with a probability p of success in a trial, find the probability of x failures preceding the first success, and hence find $E(x)$. *[C.U., B.Sc. '81]*

21. In 8 independent trials of a random experiment with a constant probability of success in each trial, it is known that the probability of 5 successes is exactly equal to the probability of 3 successes. If the experiment is repeated 6 times, what is the probability of obtaining 4 successes?

22. Suppose that half the population of a town are consumers of rice, 100 investigators are appointed to find out its truth. Each investigator interviews 10 individuals. How many investigators do you expect to report that three or less of the people interviewed are consumers of rice? *[I.C.W.A., June '79]*

23. (a) What is the probability of guessing correctly at least six of ten answers in True-False objective test?
    (b) The following statement cannot be true—why? "The mean of a Binomial distribution is 4 and the standard deviation is 3". *[I.C.W.A., Dec. '77]*

24. For a binomial distribution, the mean is 3 and the variance is 2. Find the values of n and p. Hence find the probability that X (the variable value) is 5. *[W.B.H.S. '78; M.B.A. '77]*

25. For a binomial distribution, the mean and S.D. are respectively 4 and $\sqrt{3}$. Calculate the probability of getting a non-zero value from this distribution. *[Dip. Mgmt. '78]*

26. Let x be a binomial distributed random variable with parameters n and p. For what value of p is Var (x) a maximum, if you assume that n is fixed? *[W.B.H.S. '80]*

27. If y is the proportion of success in a series of Bernoullian trials, then show that the standard deviation of y cannot be greater than $\frac{1}{(2\sqrt{n})}$.

28. Show that the binomial distribution is symmetric when $p = \frac{1}{2}$. *[W.B.H.S. '79]*

29. Find the mode of binomial distribution with parameters (i) $n = 11$, $p = 0.8$; (ii) $n = 8$, $p = \frac{1}{3}$.

30. Find the maximum term in the expansion of $\left(\frac{1}{3}+\frac{2}{3}\right)^6$. *[W.B.H.S. '80]*

31. (a) Find the most probable number of heads in $2n$ tosses of a fair coin.
    (b) If a fair die is tossed 100 times how many "sixes" are most likely to appear?

32. If the probability of a defective bolt is $\frac{1}{10}$, find (i) mean, (ii) variance, (iii) moment coefficient of skewness and (iv) kurtosis, for the distribution of defective bolts in a total of 400. *[I.C.W.A., June, '77, '80]*

33. Write down the expressions which define Binomial, Poisson and Normal probability distributions. Give 3 physical situations illustrating a Poisson random variable. *[I.C.W.A. June, '79]*

34. A random variable x follows Poisson distribution with parameter $m = 2$. Find the probabilities $P(x = 1)$, $P(x \leq 1)$, $P(x < 1)$, $P(x > 1)$, $P(1 \leq x \leq 3)$. Given $e^{-2} = 0.1353$.

35. The standard deviation of a Poisson distribution is 2. Find the probability that $x = 3$. (Given $e^{-4} = .0183$). *[I.C.W.A., Dec., '81]*

36. Is it possible that a Poisson distribution has the same mean and standard deviation? If so, what is the probability that the variable takes the value zero?

37. For a Poisson distribution, $P(x = 0) = Pr(x = 1)$. Find $Pr(x > 0)$. *[M.B.A., '79]*

38. A discrete random variable x follows Poisson distribution such that $P(x = 1) = P(x = 2)$. Find the mean and variance of the distribution. *[C.U., B.Com(Hons) '82]*

39. The probability that a Poisson variate X takes a positive value is $(1 - e^{-2})$. Find the (i) Mean, (ii) Mode, (iii) probability that X lies between -1 and 1.5.

40. If 3% of the bolts manufactured by a company are defective, what is the probability that in a sample of 200 bolts, 5 will be defective? (Given $\frac{1}{e^6} = 0.00248$). *[I.C.W.A., Dec., '75]*

41. The average number of misprints per page of a book is 2. Assuming Poisson distribution, what is the probability that a particular page is free from misprints? If the book contains 1000 pages, how many of the pages contain more than 2 misprints?

42. Suppose that the number of telephone calls an operator receives from 11.00 a.m. to 11.05 a.m. follows a Poisson distribution with $m = 3$. (i) Find the probabilities that the operator will receive no calls in that time interval tomorrow. (ii) Find the probability that in the next 3 days the operator will receive a total of 1 call in that time interval. ($e = 2.718$). *[C.U., M.Com. '78]*

43. The average number of defects per yard on a piece of cloth is 0.9. What is the probability that a one-yard piece chosen at random contains less than 2 defects? (Given $e^{0.9} = 2.46$).

44. A system contains 1000 components. Each component fails independently of the others and the probability of its failure in one months is $\frac{1}{1000}$. What is the probability that the system will function (i.e. no component fails) at the end of the month?

45. If 5% of the electrical bulbs manufactured by a company are defective, use Poisson distribution to find the probability that in a sample of 100 bulbs (i) none is defective, (ii) 5 bulbs will be defective. (Given $e^{-5} = .007$).
[I.C.W.A., Dec. '79]

46. Find the probability that at most 5 defective bolts will be found in a box of 200 bolts, if it is known that 2 per cent of such bolts are expected to be defective. You may take the distribution to be Poisson). (Given $e^{-4} = 0.0183$).
[I.C.W.A., June '80]

47. The manufacturer of a certain electronic component knows that 3% of his product is defective. He sells the components in boxes of 100 and guarantees that not more than 3 in any box will be defective. What is the probability that a box will fail to meet the guarantee? (Given $e^3 = 20.1$).

48. In a certain factory, blades are manufactured in packets of 10. There is a 0.2% probability for any blade to be defective. Using Poisson distribution calculate approximately the number of packets containing two defective blades in a consignment of 20,000 packets. (Given that $e^{-.02} = .9802$).
[I.C.W.A., June '81]

49. A car hire firm has 2 cars which it hires out day by day. The number of demands for a car on each day follows Poisson distribution with mean 1.5. Calculate the proportion of days on which (i) neither car is used, (ii) some demand is refused. (Given $e^{-1.5} = 0.2231$).

50. What is the probability of getting 3 white balls in a draw of 5 balls from a box containing 5 white and 4 black balls?

51. 10 cards are drawn at random one by one without replacement from a full pack of 52 playing cards. Find the mean and variance of the number of red cards obtained.

52. State the conditions under which a hypergeometric distribution can be approximated by a binomial distribution, which again can be approximated by a normal distribution.

53. An unbiased dice is thrown 7 times. Find the probability that the face 1 turns up twice, face 2 once, face 4 twice and face 6 twice.

54. A box contains 5 white, 2 blue and 3 red balls. If 6 balls are drawn with replacement, find the probability that there are (i) 2 white, 3 blue and 1 red balls; (ii) 4 white and 2 red balls; (iii) 2 balls of each colour.

55. Find the mean and variance of the rectangular distribution given by the p.d.f. $f(x) = 1, (0 \le x \le 1)$.

56. Find the mean and the standard deviation of the continuous probability distribution given by the p.d.f.

$$f(x) = y_0 x^2 (1 - x); \quad (0 \le x \le 1)$$

where y_0 is a constant.

57. A continuous variate X has the p.d.f.

$$f(x) = \frac{x}{2}, \quad 0 \le x \le 1$$
$$= \frac{1}{2} \quad 1 < x \le 2$$
$$= \frac{(3-x)}{2} \quad 2 < x \le 3.$$

Find the mean of the distribution.

58. Explain clearly what you understand by the following statement—"A random variable x follows Normal distribution". Write down the important properties of this distribution.

59. (a) Explain the concept of probability density function (p.d.f.). *[I.C.W.A., Dec. '75]*

(b) Write down the probability density function of a normal variate. What is the ordinate of the normal curve at its mean? State its important properties. *[C.U., M.Com. '78; B.Com. (Hons) 82; I.C.W.A., Dec. '80]*

60. Comment on the accuracy of the following results:

(i) For a binomial distribution, mean = 4 and variance = 3.

(ii) For a Poisson distribution, mean = 10 and s.d. = 5.

(iii) For a normal distribution, mean = 50, median = 52. *[I.C.W.A., Dec. '80]*

61. Comment on the following statements:

(a) For a binomial distribution, mean = 16, s.d. = 4.

(b) For a normal distribution, mean = 50, mode = 90.

(c) Any measure of skewness for Normal distribution is + 5.

(d) For Normal distribution, the value of the fourth central moment per unit of square of second central moment is + 3. *[C.U., M.Com. '78]*

62. State whether each of the following is *true* or *false*:

(a) The variance of a binomial distribution can never exceed the mean.

(b) The expected value of Poisson variate must be a positive integer.

(c) The mean of normal distribution cannot be negative.

(d) The Poisson distribution may have one or two modes.

63. The height distribution of a group of 2989 individuals is known to be normal distribution with mean 65″ and standard deviation 2.1″. Find the number of individuals whose heights lie between 60.8 and 67.1″. Find also the number of individuals whose height are above 67.1″

64. What is a 'Standard Normal distribution'? State some of its important properties. As a result of test on electric light bulbs, it was found that the lifetime of a particular make was distributed normally with an average life of 1000 burning hours and standard deviation of 200 hours. Out of 10,000 bulbs produced by the company how many bulbs are expected to fail (i) in the first 800 burning hours, (ii) between 800 and 1,200 burning hours. (Given $\Phi(1) = 0.84134$). *[C.U., M.Com. '76]*

65. Assume the mean height of soldiers to be 68.22 inches with a variance of 10.8 sq. inches. How many soldiers in a regiment of 1000 would you expect to be over 6 ft. tall? (Given that the area under the standard normal curve between $x = 0$ and $x = 0.35$ is 0.1368 and between $x = 0$ and $x = 1.15$ is 0.3749).

66. The mean I.Q. of a group of children is 90 with a standard deviation of 20. Assuming that I.Q. is normally distributed, find the percentage of children with I.Q. over 100. (Given $\Phi(0.5) = 0.6915$, where $\Phi(x)$ is the cumulative distribution function of standard normal distribution.)

67. 5000 candidates appeared at an examination, in which the minimum for a pass is 40 and the minimum for a distinction is 50. If it is known that the average mark obtained by the candidates is 43 and the S.D. is 7. Find how many of the candidates expect to get simply 'pass' and the number obtaining distinction. Assume normal distribution. (Given $\Phi(0.43) = 0.6664$, $\Phi(1) = 0.8413$).

68. In a sample of 120 workers in a factory, the mean and s.d. of wages were Rs 11.35 and Rs 3.03 respectively. Find the percentage of workers getting wages between Rs 9 and Rs 17 in the whole factor, assuming that the wages are normally distributed. (Given, Area under standard normal curve from $z = 0$ to $z = 0.78$ is 0.2823 and to $z = 1.86$ is 0.4686). *[C.A.. May '81]*

69. The mean of the inner diameters (in inch) of a sample of 200 tubes produced by a machine is 0.502 and the standard deviation is 0.005. The purpose for which these tubes are intended allows a maximum tolerance in the diameter of 0.496 to 0.508 (i.e. otherwise the tubes are considered defective). What percentage of the tubes produced by the machine is defective, if the diameters are found to be normally distributed? (Area under the standard normal curve between $z = 0$ and $z = 1.2$ is 0.3849). *[I.C.W.A., June '80]*

70. In a certain city, the daily supply of electric power (in mega watt) can be treated as a random variable having a normal distribution with mean 300 m.w. and s.d. 50 m.w. Since the supply is not a constant, the local authorities have imposed a system of rationing to deal with the problem. It is known that to ensure proper rationing a minimum of 250 m.w. is required, otherwise load-shedding has to be imposed. There is no need of rationing whenever supply exceed 350 m.w. On the other hand, the maximum consumption in the city can never exceed 425 m.w. Find the percentage of days in which the city experiences load-shedding, the percentage of days on which proper rationing is implemented, and the percentage of days when there is an excess of power. (Given, $\Phi(1) = .8413$, $\Phi(2.5) = .9983$). *[C.U. M.Com. '77]*

71. X follows a normal distribution whose mean is 12 and standard deviation is 4. Find $P(X \geq 20)$. Given $\int_{-\infty}^{2} \frac{1}{\sqrt{2\pi}} e^{-\frac{t^2}{2}} dt = 0.9772499.$ *[W.B.H.S. '81]*

72. In a normal distribution, 8% of the items are under 50 and 10% are over 60. Find the mean and s.d. Given $\int_{\alpha}^{\infty} \frac{1}{\sqrt{2x}} e^{-\frac{z^2}{2}} dz = 0.08$ or 0.10, according as $x = 1.4$ or 1.28.

73. A die is tossed 1200 times. Find the probability that the number of 'sixes' lies between 190 and 210. (Given that the area under the standard normal curve between $z = 0$ and $z = 0.78$ is .2823 and between $z = 0$ and $z = 0.81$ is .2910).

74. Show that the probability that the number of heads in 400 tosses of a fair coin lies between 180 and 220 is approximately 2, $\Phi(2) - 1$, where $\Phi(x)$ denotes the standard normal distribution functions.

75. A normal distribution has man 10.4 and s.d. 1.2. Find the limits within which an observations chosen at random almost certainly lies.

76. Player A tosses 5 coins and B tosses 8 coins. If the coins are unbiased, what is the probability of obtaining a total of 6 heads by the two players?

77. The number of deaths per day in a city due to road accidents and due to other causes independently follow Poisson distributions with parameters 2 and 6 respectively. Find the probability that the total number of deaths on a particular day is 2 or fewer. (Given $e^{-4} = .018$).

78. In an examination, marks obtained by the students in Mathematics, Physics and Chemistry are independently and normally distributed with means 50, 52, 48 and s.ds. 15, 12, 16 respectively. Find the probability that the total marks of a student are (i) 180 or more, (ii) 90 or less. [Given Φ (1.2) = 0.8849, Φ (2.4) = 0.9918].

79. When are two random variables said to 'independent'? Show that the covariance of two independent random variables is zero.

[C.U., B.A. (Econ) '81]

80. The joint probability distribution of the random variables x and y is shown below:

x \ y	0	1	2
2	.05	.10	.25
4	.15	.05	.15
6	.10	.10	.05

Find (a) the marginal distribution of x and y.

(b) the conditional distribution of x, given $y = 1$.

(c) the conditional distribution of y, given $x = 2$.

(d) the probability $P(x + y \geq 6)$.

Are x and y independent?

81. *Joint Distribution of x and y*

x \ y	2	3	4	Total
1	$\frac{1}{4}$	$\frac{1}{4}$	0	$\frac{1}{2}$
2	$\frac{1}{12}$	$\frac{1}{12}$	$\frac{1}{3}$	$\frac{1}{2}$
Total	$\frac{1}{3}$	$\frac{1}{3}$	$\frac{1}{3}$	1

Find the correlation coefficient between x and y.

82. Two random variables x and y are jointly distributed so that

$$P(x = 1) = \frac{1}{3}, \qquad P(x = 2) = \frac{2}{3}$$

$$P(y = 0) = \frac{2}{3}, \qquad P(y = 1) = \frac{1}{3}$$

$$P(x = 2, y = 1) = p$$

where $0 \leq p \leq \frac{1}{3}$. Find out the correlation coefficient between x and y. For what value of p are x and y uncorrelated? For this value of p, will x and y be independent, too? *[W.B.H.S. '79]*

83. Two discrete random variables x and y are jointly distributed with

$$P(x = 2) = P(y = 3) = P(y = 4) = \frac{1}{3};$$

$$P(x = 1, y = 2) = \frac{1}{4}; \qquad P(x = 2, y = 3) = \frac{1}{12}$$

Find (i) the conditional mean of y, given $x = 2$

(ii) the correlation coefficient between x and y.

(iii) the probability distribution of $z = x + y$.

84. Following is the joint distribution of two random variables x and y:

x \ y	1	2	3
0	0.1	0.3	0.1
2	0.2	0.1	0.2

(i) Compute $P(x = 2)$, $P(y = 3)$, $P(x = 2/y = 3)$, μ_x, μ_y, σ_x, σ_y and ρ_{xy}.

(ii) Are x and y independent? *[W.B.H.S. '80]*

85. If X and Y are jointly distributed random variables, and a, b, c, d are arbitrary constants, prove that

$$\text{Cov}(aX + b, cY + d) = ac\ \text{Cov}(X, Y).$$

86. A fair die is rolled twice independent. Let X and Y denote the number of points showing on the first roll and the second roll respectively. Define $U = X + Y$ and $V = X - Y$. Find the correlation coefficient between U and V. *[C.U., B.Sc. '80]*

87. State whether *True* or *False*:
   (a) If x and y are independent, then cov $(x, y) = 0$.
   (b) If cov $(x, y) = 0$, then x and y are independent.
   (c) If x and y are independent, then $E(xy) = E(x).E(y)$.
   (d) If $E(xy) = E(x).E(y)$, then x and y are independent.
   (e) If $\text{Var}(x + y) = \text{Var}(x) + \text{Var}(y)$, then $\rho_{xy} = 0$.
   (f) If the correlation coefficient between x and y is zero, then $\text{Var}(x + y) = \text{Var}(x) + \text{Var}(y)$.
   (g) If the random variables x and y are independent, then $\text{Var}(x - y) = \text{Var}(x) - \text{Var}(y)$.
   (h) If x and y are uncorrelated, then $(x + y)$ and $(x - y)$ are also uncorrelated.
   (i) Two independent variables must be uncorrelated.
   (j) Two uncorrelated variables must be independent.

88. If x and y are two independent Poisson variates with parameters m and n respectively, show that their sum $(x + y)$ follows Poisson distribution with parameter $(m + n)$.

89. If ρ denotes the correlation coefficient between x and y, show that the variance of $(x/\sigma_s + y/\sigma_y)$ is $2(1 + \rho)$.

90. x and y are jointly distributed random variables with standard deviations σ_x, σ_y and correlation coefficient ρ (positive). Find the value of k so that $x + ky$ and $x + (\sigma_x/\sigma_y)y$ are uncorrelated,

ANSWERS

5. True, False False, False, True
6. 7, 6
7. 5/2, √5/2
8. 1/5, 0, 4/5, 3/5, 3/5
10. 3/8
12. 31/32, 15/16
13. $1.4(0.9)^4$
14. 821/3125
15. 0.3483
16.

x	:	0	1	2	3	4	Total	Mean = 3
Prob	:	$\frac{1}{256}$	$\frac{3}{64}$	$\frac{27}{128}$	$\frac{27}{64}$	$\frac{81}{256}$	1	S. D. = √3/2

17. 112/625
18. $1 - (35/36)^n$
19. $1 - (1 - p)^n$
20. pq^x, p/q $(p + q = 1)$
21. 15/64
22. 17
23. (a) 193/512; (b) Mean of binomial cannot be less than $(\text{S.D.})^2$
24. 9, 1/3 ; 224/2187
25. $1 - (0.75)^{16}$
26. 1/2
29. 9; 2 and 3

30. 80/243
31. n; 16
32. 40, 36, 2/15, 23/1800
34. 0.2706, .4059, .1353, .5941, .7216
35. 0.1952
36. yes; e^{-1}
37. $1 - e^{-1}$
38. 2, 2
39. 2; 1 and 2; $3e^{-2}$
40. 0.16
41. e^{-2}; $1000(1 - 5e^{-2})$
42. $e^{-3} = .050$; $9e^{-9} = .0011$
43. 0.772
44. $e^{-1} = 0.37$
45. .007, 0.182
46. 0.78
47. $1 - 13\, e^{-3} = 0.35$
48. 4
49. 0.2231, 0.1913
50. 10/21
51. 5, 35/17
53. $105/6^6$
54. 9/250, 27/320, 81/1000
55. 1/2, 1/12
56. 3/5, 1/5
57. 3/2
59. $1/\sigma\sqrt{2\pi}$
60. (i) may be correct; for binomial $n = 16$, $p = 1/4$.
    (ii) Incorrect; for Poisson, mean = $(\text{S.D.})^2$.
    (iii) Incorrect; for normal, Mean = Median.

13 SAMPLING THEORY

13.1 MEANING AND OBJECTS OF 'SAMPLING'

'Sampling' denotes the selection of a part of the aggregate statistical material with a view to obtain information about the whole. This aggregate or totality of statistical information on a particular character of all the members covered by an investigation, is called *Population* or *Universe*. The selected part, which is used to ascertain the characteristics of population is called *sample* (Section 1.4, page 6). While taking a sample, the population is assumed to be composed of individual *units* or *members,* some of which are included in the sample. The total number of members of the population and the number included in the sample, called '*population size*' and '*sample size'* respectively.

A statistical population is *'finite'* or '*infinite'* according to its size. When the number of members of the population can be expressed as a definite quantity, the population is said to be 'finite'. Otherwise the population is 'infinite'. In particular, if a sample is known to have been drawn from a continuous probability distribution, then the population is infinite. A complete list of all the units in finite population, properly numbers for identification, is called a *Sampling Frame,* e.g., a list of all households in a given region as used in population census.

Again, the population may be '*existent*' or '*hypothetical*'. The population of incomes of all persons in a country is an example of existent population; because the members of the population really exist, though they may not be actually known because of practical difficulties. The population of points obtained in all possible throws of a die is an example of 'hypothetical' population.

If a sample of an adequate size is properly chosen and analysed, it is most likely to reveal the characteristics of the whole population, and the results obtained therefore can be fairly relied upon as if they were based on all members of the population. The possibility of reaching valid conclusions concerning a population by means of a properly chosen sample is based on two important laws: (i) Law of Statistical Regularity, and (ii) Law of Inertia of Large Numbers. The former states that a sample of reasonably large size when selected at random, is almost sure to represent the characteristics of the population (page, 11). The second law states that samples of large size show a high degree of stability, i.e., the results obtained theorem are expected to be very close to the population characteristics.

Some practical examples of sampling in Commerce and Industry are as follows:

(1) Sampling methods are used in Market Research for assessing customer behaviour, especially during launching of new products in the market, and also to gauge their opinion about the existing products.

(2) In industry sampling is done for Statistical Quality Control (Chapter 19). During manufacture, a few consecutive items are picked out from the production line at regular intervals of time, (say, half an hour) and these items are thoroughly tested. If the results are beyond certain permissible limits, the production process is shut down to find out the causes of variation.

(3) Sampling is also used by Stores people for incoming lots from suppliers. These are subjected to 'sampling inspection' lot by lot and the decision about the acceptance or otherwise of a lot is decided by examining only samples.

(4) Auditors use test check and test audit to cover only a small fraction of all entries in the books of accounts in order to verify the accuracy of all the entries, where either full verification is not possible or considered unnecessary.

There are 2 methods of collection of statistical data—(i) *Census method* or *Complete Enumeration,* and (ii) *Sample Survey method.* 'Census' denotes collection of data from each and every unit of the statistical population; while 'sample survey' refers to the study of population based only on a sample (Section 1.5).

Example 13.1 *Discuss the advantages of sampling method over census method of collection of statistical information.*
[C.U., M. Com, '73, '78, '80; B.A. (Econ) '75; I.C.W.A. June '74; W.B.H.S. '78, '82]

Solution (See Example 1.9)

Example 13.2 *What are the main objects of sampling?*

[I.C.W.A., June '75]

Solution The main objects of sampling are:

(i) to obtain the maximum information about the population with the minimum effort; and

(ii) to state the limits of accuracy of estimates based on samples.

In the study of a population by means of sample observations, we are generally interested in one or more variables. The first requirement towards the analysis of a sample is to form an idea of the probability distribution of the variable in the population. This probability distribution involves some constants (known as *parameters*), and consequently our next task is to form estimates of these unknown parameters on the basis of available data. For example if the probability distribution is assumed to be normal, there are two parameters μ and σ to be estimated. Usually there are several methods of estimating the same parameter, and naturally the question arises as to which among these estimates will be the most suitable under given circumstances. The study of various criteria for food estimators and the methods of estimation are treated in what is known as *Theory of Estimation.*

Again, on the basis of samples we may be interested to known whether a certain population parameter has a specified value, or to compare the parameters of two populations on the basis of two samples drawn from these populations. Such problems are treated in what is known as *Theory of Testing Statistical Hypothesis.*

The second object of sampling is to determine how precise the estimate would be, i.e., to state the degree of confidence that we may put on the estimates obtained. Since the sample is only a part of the population and is obtained by chance, any inference regarding population cannot be hundred per cent correct. Therefore a measure of precision of the estimate is given in terms of probability. This is done by using the probability distribution called *Sampling Distribution* of the estimate.

Example 13.3 *"Sampling is a necessity under certain conditions"—Explain this with illustrative examples.*

Solution For the collection of statistical data, sample survey method is usually preferred to census, because the former takes less time, money and labour. But many situations arise in which sampling is not an alternative to census, but is a must; because census is either impossible or useless.

Illustrations:

(i) In order to find the acreage under rice in India by the census method, each and every plot of land on which rice is grown within the geographical limits of Indian Union has to be visited and area under rise cultivation has to be measured. The method is quite impracticable, because of the enormous task of visiting the plots and the consequential time and money involved. Even if such a step is taken, it will take years before the data are completely available and the whole purpose of enquiry may be vitiated. However, estimates a of crop area are now obtained by the methods of sample survey within a short period of time.

(ii) Again, the rice merchant cannot afford to examine every single grain of rice he purchases. He has to depend on only a sample based on which he forms an idea about the quality of rice in the whole consignment. Thus when the population size is extremely large, e.g., in counting the number of 'sal' trees in a forest area sampling the only practical method.

(iii) A municipal corporation, which purchases thousands of electric bulbs for street lighting, in an attempt to know the average life of a lot supplied by a company, cannot go on testing each and every bulb until all of them burn out; because in that case nothing will be left for use. Instead, by testing only a few bulbs from the whole lot, an estimate of the average length of life is known. Similarly, in other destructive tests, e.g. measuring the resistance of glass sheets, or strength of bullets, there is no alternative but to resort to sampling.

(iv) When a new mine is discovered and it is necessary to know the quality of its contents, only a few ounces of ore, drilled out from here and there, are sufficient for chemical analysis. It is impossible and unnecessary to dig out the whole mine and then examine the quality.

(v) When the population is hypothetical, e.g. in order to test whether a coin is biased or not, it is tossed a number of times and decisions have to be arrived at on the basis of the results of the sample. Here sampling is the only method of studying the characteristics of the population.

13.2 SAMPLING ERROR AND BIAS

Example 13.4 *Explain what is meant by 'random error'.*

[C.U., M.Com., '73; I.C.W.A. June '74]

Solution Sampling means the selection of a part of the aggregate statistical material, called *population,* with a view to obtaining information about the whole aggregate.

However, the composition of sample is a matter of chance and hence even if a proper process of selection is employed, the results obtained from sample cannot be expected to represent the characteristics of population in toto. Some error, i.e., discrepancy, is inevitable. This is known as 'sampling error'. When random sampling is used for obtaining the sample, it is called *Random Sampling Error.* The magnitude of such errors depends on the size of the sample and the variability of the population. Random sampling error depends on chance, and if the proper process of selection is followed, the possible limits of the magnitude of such error can be theoretically calculated.

Example 13.5 *What is 'bias' and how does it arise in sampling? Discuss the various methods of reducing bias.* [C.A. May '78]

Solution In sampling, a properly selected part only of the universe is analysed and the results are made applicable to the whole. It cannot be expected that these results will agree in full with those of the universe, e.g. the mean of sample may differ from the mean of universe. Since a part can never represent the whole in toto, a certain amount of discrepency is involved in the sampling process. These discrepancies or errors are of two types—(i) *Sampling Error,* and (ii) *Non-sampling Error* or *Bias.*

Sampling error is inherent in the method of sampling and arises because the information collected relates only to a part. In actual practice, certain inaccuracies arise out of complex reaction between enumerators and informants, or in processing the collected data, which affect the conclusions derived therefrom. This kind of error is known as 'bias'. Bias arises solely due to human factor and is difficult to detect. It occurs even in census, where sampling error is completely absent. Bias may arise in several ways:

1. Bias may arise in selecting the sample itself. If the proper process of selection is not strictly adhered to, the investigator's desire or motive to obtain a certain result from the sample survey may influence his selection of the sample consciously or unconsciously. This type of bias is father serious in as much as its existence may not be immediately apparent. Such bias can be avoided only by following the scientific process of selection. For example, white drawing a random sample, the universally accepted random numbers should be used, instead of resorting to haphazard selection by the investigator.
2. Bias may arise because of substitution of a selected unit by another. For example, in a house-to-house survey, the next house is interviewed if no reply is forthcoming from the selected house. Bias introduced in such cases can only be avoided, if the selected sample is strictly adhered to.
3. Bias may arise from incomplete investigation, i.e. because of not investigating the whole of the selected sample. In a house-to-house survey, even if there be no substitution of the selected house by the next house, bias may arise if the selected house is not visited a second time, or if necessary, several times till the desired information is obtained. This type of bias arises particularly in postal questionnaires. Such bias can only be eliminated if the entire selected sample is investigated.
4. Bias may arise in the process of estimation of parameters. For instance, forecasts of crop-yield (say, rice production) in a State are computed on the basis of estimates of average yields per acre obtained for several regions and sub-regions. If the simple average of the recorded yields are used, instead of the weighted average, the ultimate result will be biased.

5. Again, bias may arise due to negligence or carelessness on the part of investigators in asking questions, or inaccuracies in recording, or false recording by dishonest investigators without actually visiting the sample. Arrangements should therefore be made to impart training to the staff in the collection of data. The importance of accurate information in the success of the investigation should also be impressed upon them. Sometimes, supervisory staff may be used to check the investigators in the actual collection of data. As soon as the collection of data begins; arrangements should be made to receive the completed forms in batches and scrutiny should begin immediately. If necessary, some of the returns may be referred back to enumerators for correcting the inconsistencies, or completing the defective returns.

13.3 TYPES OF SAMPLING

There are many different ways of selecting a sample. We describe below some of the important types of sampling— (1) Simple random sampling. (2) Purposive sampling. (3) Stratified sampling. (4) Systematic sampling. (5) Multi-stage sampling.

(1) Simple Random Sampling

Simple Random Sampling (also called *Random Sampling)* is the process of selection of a group of units in such a manner that every unit of the population has an "equal chance" of being included in the sample. The group of units thus obtained is called a *Simple Random Sample* (or *Random Sample* only). In practice, the members of the sample are drawn one by one.

There are two ways of drawing a simple random sample:

(a) **Simple Random Sampling With Replacement (SRSWR):** Simple random sampling is said to be "with replacement", when the sample members are drawn from the population one by one, and after each drawing, the selected population unit is noted and then returned to the population before the next one is drawn. This means that at each stage of the sampling process all the population units (including those obtained in earlier drawings) are considered for selection with equal probability. Thus the population remains the same before each drawing and any of the population units may appear more than once in the sample.

(b) **Simple Random Sampling Without Replacement (SRSWOR):** Simple random sampling is said to be "without replacement", when either the sample members are drawn all at a time, or drawn one by one in such a manner that after each drawing the selected unit is not returned to the population when the next one is drawn. This means that when drawing is made one by one, at each stage of the sampling process the population units already chosen are not considered for subsequent selections, but the drawing is made with equal probability only from those units not selected in any of the earlier drawings. It is evident that in simple random sampling without replacement from a finite population, the size of the population goes on diminishing as the sampling process continues. Consequently, no population unit can appear more than once in the sample.

Let a simple random sample of n members be drawn from a finite population consisting of N members (i.e. sample size = n, population size = N). Then

Total number of possible samples with distinct 'permutations' of members

(i) in SRSWR is N^n

(ii) in SRSWOR is ${}^N P_n = N(N-1)(N-2) \ldots (N-n+1)$

Among these, the number of cases favourable to each group (ignoring the order of drawing) or distinct 'combination' of members

(i) in SRSWR is not the same.

(ii) in SRSWOR is the same. viz. $n\,!$

For this reason, we generally consider

(i) in SRSWR, N^n possible samples with distinct 'permutations' each with a probability $\dfrac{1}{N^n}$.

(ii) in SRSWOR, ${}^N C_n = \dfrac{N\,!}{n\,!(N-n)\,!}$ possible samples with distinct 'combinations' each with a probability $\dfrac{1}{{}^N C_n}$.

Let $x_1, x_2, \ldots x_n$ represent a simple random sample from the population $X_1, X_2, \ldots$. X_N. In SRSWR, the probability of selection of any particular member of the population at any drawing remains a constant $1/N$; because before any drawing the population contains all the N members. It may be shown that this result is also true in SRSWOR, although the population size varies at each stage of selection. Thus the probability of obtaining the population member X_k (suppose) at the i-th drawing is a constant $1/N$ both in SRSWR and in SRSWOR; i.e.,

$$P(x_i = X_k) = \frac{1}{N}$$

for all $i = 1, 2, \ldots n$ and $k = 1, 2, \ldots N$.

Random sampling is the simplest and most important among the various sampling techniques. It is absolutely free from the influence of human bias, and hence also called *Lottery Sampling*. Chance alone determines whether one unit or the other is selected. The selection of a random sample is facilitated by what are known as *Random Numbers*.

Random sampling is the most appropriate in cases when the population is more or less homogeneous with respect to the characteristic under study. The theories of sampling distribution and test of significance are based on random sampling only.

Simple Sampling: This is a special case of simple random sampling in which the probability of selection of any particular member remains constant throughout the sampling process irrespective of whether the member had been selected earlier or not. Therefore SRSWR will always give a 'simple sample' from a finite or an infinite population. However SRSWOR gives a simple sample only when the population is infinite.

(2) Purposive Sampling

A sample which is selected on the basis of individual judgement of the sampler, is called a *Purposive Sample*. There is no special technique for selecting a purposive sample; but the sampler picks out a typical or representative sample according to his own judgement. It all depends on the personal factor and chance is not allowed to play at all. Consequently, there is much scope for bias and the degree of accuracy of the estimates is not known. Purposive sampling may be useful when the sample is small; but as the sample size increases the estimates become unreliable due to accumulation of bias. The advantage of purposive sampling is that whereas a random sample may vary widely from the average, a purposive sample will not.

(3) Stratified Sampling

In *Stratified Sampling*, the population is subdivided into several parts, called *strata*; and then a sub-sample is chosen from each of them. All the sub-samples combined together give the *Stratified Sample.* If the selection from strata is done by random sampling, the method is known as *Stratified Random Sampling.* The subdivision of the population into strata is done by purposive method, but the selection of sub-samples from within the strata depends purely on chance. Stratified random sampling may therefore be viewed as a mixture of both purposive and random sampling, and combines the advantages of both.

Stratified sampling is generally used when the population is heterogeneous. but can be subdivided into strata within each of which the heterogeneity is not so prominent. Some prior knowledge is, therefore necessary for subdivision into strata, called *stratification.* If a proper stratification can be made such that the strata differ from one another as much as possible, but there is much homogeneity within each of them, then a stratified sample will yield better estimates than a random sample of the same size. This is so, because in a stratified sample the different sections of the population are suitably represented through the sub-samples, while in random sampling some of these sections may be over-represented or under-represented or may even be omitted.

Usually the same fraction of members from each stratum is included in the sample, i.e., sub-sample sizes are made proportional to sub-population sizes. The principal purposes of stratification are: (i) to increase the overall estimates; (ii) to ensure that all sections of the population are adequately represented; (iii) to avoid a large size of the population; and (v) to avoid the heterogeneity of the population.

Uses—(i) Administrative convenience may dictate the use of stratified sampling. For example, in conducting a sample survey over West Bengal, the different districts may be taken as strata, so that the district authorities can supervise the survey and collect data more efficiently from their own region, (ii) Sometimes, different parts of the population involve different sampling problems, and special measures are necessary for dealing with these cases. Stratified sampling is extremely useful in such cases.

Advantages—(i) If data of a given precision are required for certain parts of the population, it is advisable to treat each part as a population in its own right, (ii) Stratification also brings about a gain in precision of the estimates obtained.

(4) Systematic Sampling

Systematic Sampling involves the selection of sample units at equal intervals, after all the units in the population have been arranged in some order. If the population size is finite, the units may be serially numbered and arranged. From the first k of these, a single unit is chosen at random. This unit and every k-th unit thereafter constitutes a *Systematic Sample.* In order to obtain a systematic sample of 500 villages out of

40,000 in West Bengal, i.e. one out of 80 on an average, all the villages have to be numbered serially. From the first 80 of these a village is selected at *random,* suppose with the serial number 27. Then the villages with serial numbers 27, 107, 187, 267, 347, ... constitute the systematic sample.

If the characteristic under study is independent of the order of arrangement of the units, then a systematic sample is practically equivalent to a random sample. The actual selection of the sample is easier and quicker. Systematic sampling is suitable when the units are described on serial numbered cards, e.g. workers listed on cards. Then the sample can be drawn easily by looking at the serial numbers. The sample may be biased it there are periodic features associated with the sampling interval.

(5) Multi-stage Sampling

Multi-stage Sampling refers to a sampling procedure which is carried out in several stages. The population is first divided into large groups, called *first-stage units.* These first-stage units are again divided into smaller units, called *second-stage units*—the second-stage units into *third-stage units,* and so on, until we reach the *ultimate units.* Initially some of the first-stage units are selected, from each of which some second-stage units are chosen; and the process is carried on from stage to stage until the selection of ultimate units. For example, in order to introduce a scheme on an experimental basis in the villages, we may have to select a few villages from the whole of the State. If we apply three-stage sampling. Sub-divisions may be used as first-stage units. Anchals forming the second-stage units and then Villages as the ultimate units.

Multi-stage sampling enables existing divisions and subdivisions of the population to be used as units at various stages and permits the field work to be concentrated, although a large area is covered. Another advantage is that the subdivision into second stage units need be carried out for only those first stage units which are included in the sample. It therefore helps in surveys of underdeveloped areas where no sampling frame is sufficiently detailed and accurate for subdivision of the natural units into reasonably small sampling units. Usually, considerable saving in cost is achieved through multi-stage sampling. However, this method is in general less accurate than any other sampling method using the same number of ultimate units by some single-stage process.

13.4 METHOD OF DRAWING RANDOM SAMPLE

Example 13.6 *What are random numbers ?*

[I.C.WA., June '74; W.B.H.S. '79; C.U., M.Com. '77, '80; M.B.A. '78]

Solution *Random numbers* (also called *Random sampling numbers)* refer to some well-known sequence of digits in which the successive figures appear in a perfectly 'random' order. This means that if a digit is blindly selected from the Random Number Table, any of the ten digits 0, 1, 2, ..., 9, is likely to occur with the same probability 1/10. Similarly, if two consecutive digits are taken, the number formed by them (in the order they appear) may be any of the hundred numbers 00, 01, 02, ... 99 with the same probability 1/100, and so on. The series of random numbers are usually available in groups of four digits to facilitate easy reading as follows:

4504	9523	3282	3756	0653
1014	7894	9307	0458	9983
1575	2458	9200	8566	7302

Of course, any haphazard arrangement of the digits 0 to 9 will not give random numbers. The random numbers have been prepared by special devices and tested for randomness, Some of the famous series of random number are those of Tippett (41,600 digits), Kendall and Babington Smith (100,000 digits). Fisher and Yates (15,000 digits), Rand Corporation (1,000,000 digits). Random numbers are used for drawing random samples.

Example 13.7 *The following table gives the grades of 100 students in mathematics. Draw a random sample of size ten from the group of students and estamte the mean grade from the sample:*

75, 58, 80, 86, 76, 65, 75, 76, 72, 86, 83, 82, 80, 88, 68,
66, 63, 60, 69, 80, 66, 87, 73, 58, 76, 74, 85, 96, 60, 72,
86, 79, 95, 84, 41, 76, 78, 74, 74, 56, 50, 80, 66, 96, 80,
68, 79, 73, 72, 73, 87, 77, 60, 87, 40, 82, 77, 87, 76, 82,
66, 81, 84, 72, 63, 59, 76, 52, 57, 78, 79, 92, 80, 65, 90,
75, 74, 98, 64, 45, 68, 57, 79, 79, 83, 35, 76, 88, 67, 75,
60, 52, 63, 80, 94, 34, 78, 64, 58, 56.

(Tables of Random Sampling Numbers to be supplied.)

Solution Here, the population size $N = 100$. Each of the population members is, therefore, allotted a two-digited identity number 01, 02, ..., 99 and 00 (for the 100th member), as shown below:

Table 13.1 Grades in Mathematics and Their Identity Numbers

Identity No.	01	02	03	04	05	06	07	08	09	10	11	12	13	14	15	16	17	18	19	20
Grade	75	58	80	86	76	65	75	76	72	86	83	82	80	88	68	66	63	60	69	80
Identity No.	21	22	23	24	25	26	27	28	29	30	31	32	33	34	35	36	37	38	39	40
Grade	66	87	73	58	76	74	85	96	60	72	86	79	95	84	41	76	78	74	74	56
Identity No.	41	42	43	44	45	46	47	48	49	50	51	52	53	54	55	56	57	58	59	60
Grade	50	80	66	96	80	68	79	73	72	73	87	77	60	87	40	82	77	87	76	82
Identity No.	61	62	63	64	65	66	67	68	69	70	71	72	73	74	75	76	77	78	79	80
Grade	66	81	84	72	63	59	76	52	57	78	79	92	80	65	90	75	74	98	64	45
Identity No.	81	82	83	84	85	86	87	88	89	90	91	92	93	94	95	96	97	98	99	00
Grade	68	57	79	79	83	35	76	88	67	75	60	52	63	80	94	34	78	64	58	56

We now take down random numbers in groups of two, starting from the digit 5 in the 2nd row and 15th column (this was done without looking at the table) and moving horizontally, using the random number table at page 511. The numbers obtained are

58, 99, 83, 15, 75, 24, 58, (rejected), 92, 00, 85, 66

The 7th number 58 is rejected, because it has already appeared once. The ten random numbers remaining and the grades having these as identity numbers (Table 13.1) are shown below:

Random Numbers	58	99	83	15	75	24	92	00	85	66
Grades	87	58	79	68	90	58	52	56	83	59

The random sample of size ten is, therefore, composed of grades

87, 58 79, 68, 90, 58, 52, 83, 59

The mean of the sample, which gives an estimate of the population mean, is

$$\bar{x} = (87 + 58 + ... + 83 + 59) \div 10 = \frac{690}{10} = 69$$

Incidentally, the population mean of the 100 grades is 72.66.

Example 13.8 *Draw a random sample of size 10 (without replacement) from the following data, stating clearly the procedure followed by you:*

45	*24*	*43*	*17*	*5*	*28*	*27*	*21*	*11*	*46*	
33	*26*	*24*	*14*	*34*	*21*	*25*	*48*	*35*	*38*	
26	*27*	*35*	*8*	*30*	*26*	*30*	*28*	*21*	*27*	
20	*13*	*23*	*36*	*38*	*20*	*25*	*31*	*24*	*18*	*12*

You may use the random sampling numbers given below:

5967	*8941*	*7989*	*3335*	*7577*	*9735*
3042	*8409*	*7053*	*5364*	*5872*	*1143*

[I.C.W.A., July '72]

Solution There are altogether $N = 41$ members and these are serially numbered horizontally from 1 to 41. Sicne the population size $N = 41$ is two-digited, the identity numbers must also be two-digited. But there are 100 two-digited numbers possible, viz., 00, 01, 02, ..., 99.

If we allot only one identity number to each member, many random numbers will be rejected before we get the required sample. Let us allot 2 identity numbers to each member as follows:

Serial Nos	1	2	3	...	...	39	40	41
Identity	01	02	03	....	...	39	40	41
Numbers	42	43	44	...	...	80	81	82

[**Note:** It will be noticed that identity numbers upto 41 are the same as the corresponding serial numbers, and the other identity numbers when divided by 41 leave remainders which are equal to the serial numbers. With identity number 82, however, the remainder is zero, and the last observation is referred to.]

Starting from the frist digit of the given random sampling numbers and proceeding row-wise, we now taken down groups of 2 consecutive digits.

59, 67, 89 (rejected), 41, 79, 89 (rejected), 33, 35, 75, 77

97 (rejected), 35, 30, 42, 84 (rejected), 09, 70, etc.

The numbers greater than 82 have been rejected, because the identity numbers are 01 to 82 only (If 00 apeared, it will be also have been rejected). The numbers greater than 41 are replaced by their remainders on division by 41, and we now have

18, 26, 41, 38, 33, 35, 34, 36, 35 (rejected), 30, 1, 9, 29 etc.

Since the drawing are *without* replacement, the second 35 occurring above, has been rejected. Of the remaining numbers, the first ten are

18, 26, 41, 38, 33, 35, 34, 36, 30 & 1

(Note that these numbers all lie between 1 and 41, and none is repeated). The members having these as serial numbers are respectively

48, 26, 12, 31, 23, 38, 36, 20, 27, 45

These observations comprise the required random sample of size 10.

(*Alternative method*) [This is a modification of the previous method. Note that 100 identity numbers 00, 01, 02, ..., 99 and 00 can be allotted equally among 50 (slightly higher than 41) members.]

Let the population members be serially numbered horizontally from 1 to 41, as before; and two-digited random numbers are taken down. These numbers are divided by 50 (not 41) and the remainers obtained. If a number is less than 50, the number itself is taken as the remainders. Any remainders between 42 and 49, and also 00, if any are rejected. Remainders which occur more than once are retained only at the first appearane: the repetitions are rejected. Thus the two-digited random numbers

59, 67, 89, 41, 79, 89, 33, 35, 75, 77, 97, 35, 30, 42, 84, 09 leave the remainders

9, 17, 39, 41, 29, 39 (rejected, because repeated), 33, 35, 25, 27, 47 (rejected because more than 41), 35 (rejected, because repeated), 30, etc.

From the numbers of now remaining the frist ten are only retained

9, 17, 39, 41, 29, 33, 35, 25, 27, 30

The members of the population which correspond to these as serial numbers, viz.

11, 25, 24, 12, 21, 23, 38, 30, 30, 27

constitute the required random sample.

Example 13.9 *Describe in detail ow you will select without replacement a random sample of 3 units from a population of 121 units using a procedure which does not involve rejection of a large number of random numbers.* [C.U., B.A. (E.con) '75]

Solution The method is described in the following steps:

1. All the 121 units of the population are serially numbered from 1 to 121.
2. Let us now take a page of random sampling numbers, and with closed eyes select a digit from the page. Starting from this and proceeding horizontally, consecutive digits are taken down in groups of three (because of largest serial number 121 is three-digited), giving several 3-digited numbers.
3. The random numbers are divided by 125 and the remainders obtained (Note that 1000 three-digited numbers are possible, viz. 001, 002, ..., 999 and 000; and 1000 is exactly divisible 125). If a number is less than 125, that number itself is taken as the remainder.
4. Any remainders between 122 and 124, and also 000, if any occurring, are rejected. If any remainder appears more than one, all subsequent repetitions are rejected.
5. From the numbers now remaining, only the first three numbers are taken and the population units corresponding to these as serial numbers give the random sapmle.

Example 13.9A *Draw a random sample of size 10 with replacement from the following frequency distribution and compare the sample mean with the population mean:*

Annual Sales (Rs '000)	*30–39*	*40–49*	*50–59*	*60–69*	*70–79*	*80–89*	*90–99*	*100–109*	Total
No. of Firms	*14*	*36*	*47*	*66*	*41*	*24*	*10*	*2*	*240*

Solution All the 240 firms considered in the frequency distribution are serially numbered—14 observations in the class interval 30–39 are numbered 1 to 14, 36 in the next interval are numbered 15 to 50, and so on, as shown in the table below.

Table 13.1A Calculation of Cumulative Frequencies

Class Interval	*Frequency*	*Mid value*	*Cumulative Frequency*	*Serial Nos. allotted*
30–39	14	34.5	14	1–14
40–49	36	44.5	50	15–50
50–59	47	54.5	97	51–97
60–69	66	64.5	163	98–163
70–79	41	74.5	204	164–204
80–89	24	84.5	228	205–228
90–99	10	94.5	238	229–238
100–109	2	104.5	240	239–240

Using the random sampling numbers shown in the Appendix (Table VI), we start from (Row 6, Col. 1) an take down 3-digited numbers proceeding horizontally. The numbers 250 and above are replaced by their remainders on division by 250. Any number between 241–249 or 000, if occurring, is now rejected.

Random Numbers	489	572	665	890	501	154	786	475	888	750	147
Remainders	239	72	165	140	1	154	36	225	138	000*	147

(*The remainder 000 is rejected). The first 10 of the numbers are now retained, and the firms corresponding to these as serial numbers constitute the sample of firms. (Note that there is no necessity of rejecting any number which is repeated, because the sample is drawn *with* replacement.)

The random sample is given by the mid-values of classes in which the selected firms are placed. This can be done by looking at the last column and then the 3rd col. of Table 13.1A.

Random Sample (with replacement)

Serial Nos	239	72	165	140	1	154	36	225	138	147	Total
Sample Values (Rs '000)	104.5	54.5	74.5	64.5	34.5	64.5	44.5	84.5	64.5	64.5	655.0

The sample mean is 655.0 ÷ 10 = 65.5 (Rs '000). The mean calculated from the the given frequency distribution (calculations not shown here) is 63.08 (Rs '000).

Sample mean ($\bar{x}$) = 65.5, Population Mean (μ) = 63.08.

13.5 SAMPLING DISTRIBUTION

Example 13.10 *Distinguish between Statistic and Parameter, giving examples.*
[W.B.H.S. '78, '82; I.C.W.A. June '73, '76, '79; Dec. '81; C.U., B.A. (Econ) '81]

Solution Any statistical measure calculated on the basis of sample observations is called a *Statistic;* e.g., sample mean, sample s.d., the proportion of defectives observed in the sample, etc. Any statistical measure based on all units in the population is called a *Parameter;* e.g., population mean, population s.d., proportion of defectives in the whole lot, etc.

In all sampling techniques (except Purposive Sampling), the composition of sample depends only on chance. Therefore, if many samples of a fixed size the drawn from a given population, the group of units constituting the sample also varies from one sample to another. Hence, the value of a statistic varies from sample to sample; but the parameter remains a constant. This variation in the value of a statistic is known as '*sampling fluctuation*'. The parameter has no sampling fluctuation. Usually parameters are unknown and statistics are used as estimates of parameters. The probability distribution of a statistic is called its '*sampling distribution*' and the standard deviation in the sampling distribution is called '*standard error*' of the statistic. However, since the parameter is constant it has neither a sampling distribution nor a standard error.

The following notations will be used to distinguish between statistic and parameter:

	Statistic (from Sample Values)	*Parameter (from all Population Values)*
Mean	$\bar{x}$	μ
Standard Deviation	S	σ
Proportion	p	P
rth Raw Moment	m_r'	μ_r'
rth Central Moment	m_r	μ_r

Example 13.11 *Explain clearly the concept of Sampling Distribution of a statistic.*

[W.B.H.S. '78, '82; C.U., B.A. (Econ) '74, '81; C.A., Nov. '78; I.C.W.A., June '73, '76, '78, '79, Dec. '74, '81]

Solution *Sampling Distribution* of a statistic may be defined as the probability law which the statistic follows, if repeated random samples of a fixed size are drawn from a specified population.

Let us consider a random sample $x_1, x_2, ..., x_n$ of size n drawn from a population containing N units. Let us further suppose that we are interested in the sampling distribution of the statistic $\bar{x}$ (i.e., sample mean), where $\bar{x} = (x_1 + x_2 + ... + x_n)/n$.

If the population size N is finite, there is a finite number (say K) of possible ways of drawing n units in the sample out of a total of N units in the population. For each of these K samples we calculate the value of $\bar{x}$ (see Tables 13.3 and 13.5. Here $N = 4$, $n = 2$, and $K = N^n = 16$ for SRSWR and $K = {}^NC_n = 6$ for SRSWOR). Although the K samples are distinct, the sample means may not be all different; but each of these occurs with equal probability. Thus, we can construct a table shwoing the set of possible values of the statistic $\bar{x}$ and also the probability that $\bar{x}$ will take each of these values. The probability distribution of the statistic $\bar{x}$, will be called '*sampling distribution*' of sample mean (Tables 13.4 and 13.6). The above method is quite general, and the sampling distribution of any other statistic, say, median, or standard deviation of the sample, may be obtained.

If, however, the number (N) of units in the population is large, the number (K) of possible distinct samples being even larger, the above method of finding the sampling distribution cannot be applied. In this case, the values of $\bar{x}$ obtained from a large number of samples may be arranged in the form of a relative frequency distribution. The limiting form of this relative frequency distribution, when the number of samples considered becomes infinitely large, will be called *sampling distribution* of the statistic. When the population is specified by a theoretical distribution (e.g., binomial, or normal), the sampling distribution can be theoretically obtained. The knowledge of sampling distribution is necessary in finding 'confidence limits' for parameters and in 'testing statistical hypotheses'.

13.6 TWO IMPORTANT SAMPLING DISTRIBUTION (LARGE SAMPLE)

(I) Sampling Distribution of Sample Mean

If $\overline{x}$ represents the mean of a random sample of size n, drawn from a population with mean μ and standard deviation (s.d.) σ, then the sampling distribution of $\overline{x}$ is approximately a *normal distribution* with

mean = μ and s.d. = standard error of $\overline{x}$ (13.6.1)

provided the sample size n is sufficiently large.

(II) Sampling Distribution of Sample Proportion

If p represents the proportion of defectives in a random sample of size n drawn from a lot with proportion of defectives P, then the sampling distribution of p is approximately a *normal distribution* with

mean = P and s.d. = standard error of p, (13.6.2)

provided the sample size n is sufficiently large.

[**Note:** Usually a sample of size 30 or more is considered as a 'large sample'. However, the larger the value of n the better is the approximation.]

13.7 STANDARD ERROR (S.E.)

Standard Error of a statistic is the standard deviation calcualted from the sampling distribution of the statistic. Some importnat formula for standard error are shown below:

1. For random samples of size n,

$$\text{S.E. of sample mean } (\overline{x}) = \frac{\sigma}{\sqrt{n}} \tag{13.7.1}$$

$$\text{S.E. of sample proportion } (p) = \sqrt{\frac{PQ}{n}} \tag{13.7.2}$$

where σ denotes the population standard deviation (s.d.) and P the population proportion ($P + Q = 1$).

If the random sample is drawn without replacement from a finite population of size N, then the above formulae are modified on multiplication by the correction factor $\sqrt{\frac{N-n}{N-1}}$ (see Example 13.14)

2. If $\overline{x}_1$ and $\overline{x}_2$ denote the means calculated from independent random samples of sizes n_1 and n_2 drawn from the populations with standard deviations σ_1 and σ_2 respectively, then

$$\text{S.E. of } (\overline{x}_1 - \overline{x}_2) = \sqrt{\frac{\sigma_1^2}{n_1} + \frac{\sigma_2^2}{n_2}} \tag{13.7.3}$$

In particular, if the poulations have the same s.d. σ, then

$$\text{S.E. of } (\overline{x}_1 - \overline{x}_2) = \sigma\sqrt{\frac{1}{n_1} + \frac{1}{n_2}} \tag{13.7.3a}$$

3. If p_1 and p_2 denote the proportions calcualted from independent random samples of sizes n_1 and n_2 and drawn from two populations with proportions P_1 and P_2 respectively, then

$$\text{S.E. (of } p_1 - p_2) = \sqrt{\frac{P_1Q_1}{n_1} + \frac{P_2Q_2}{n_2}} \tag{13.7.4}$$

In particular, if it is assume that the two population proportions P_1 and P_2 are equa, say $P_1 = P_2 = P$, then

$$\text{S.E. of } (p_1 - p_2) = \sqrt{PQ\left(\frac{1}{n_1} + \frac{1}{n_2}\right)} \tag{13.7.4a}$$

4. For random sample from a normal population with s.d. σ,

$$\text{S.E. of sample s.d. } (S) = \frac{\sigma}{\sqrt{2n}} \tag{13.7.5}$$

$$\text{S.E. of sample variance } (S^2) = \sigma^2\sqrt{\frac{2}{n}} \tag{13.7.6}$$

Formulae (13.7.5) and (13.7.6) are however approximate and used only when the sample size n is large (say greater than 50).

Example 13.12 *Discuss the concept of 'Standard error' of a statistic. What does the standard error of a statistic measure?*

[C.U. M.Com. '71 '77,; C.U. B.A. (Econ) '8; C.A., Nov. '78; I.C.W.A., June '73, '75, '76, '78; Dec. '81]

Solution Let $x_1, x_2, ..., x_n$ be a random sample of size n drawn from a specified population. On the basis of this sample, let us calculate the value of certain statistic, say, sample mean

$$\bar{x} = \frac{(x_1 + x_2 + ... x_n)}{n}$$

We repeat the process of drawing a random sample of the fixed size n a large number of times, and calcualte the value of the statistic (here, sample mean) for each sample. The relative frequency distribution of these sample means, when the number of samples considered is infinitely large, is called the *sampling distribution* of sample mean.

The sampling distribution of any statistic will have its own mean, standard deviation, moments, etc. *The standard deviation calculated from the sampling distribution of a statistic is called its 'Standard Error'*. The standard error gives a measure of dispersion of the concerned statistic. It depends on the sample size n and goes on diminishng as the sample size increases. It is used to set up 'confidence limits' for population parameters and in 'tests of significance'. Thus, the standard errors of sample mean ($\bar{x}$) and sample proportion (p) are used to find confidence limits for the population mean (μ) and the population proportion (P) respectively.

Example 13.13 *State the formulae for standard error of sample mean and sample proportion.* [M.B.A. '77; I.C.W.A., June '74]

Solution

(I) Standard Error of Sample Mean

If $\bar{x}$ represents the mean calcualted from a random sample of size n, and the standard deviation of the population be σ, then *Standard Error (S.E.) of sample mean* is given by

(i) $$\text{S.E.} = \frac{\sigma}{\sqrt{n}} \tag{13.7.7}$$

when either the population size is infinitely large, or the sample is drawn *with replacement* (i.e., in simple sampling).

(ii) $$\text{S.E.} = \frac{\sigma}{\sqrt{n}}\sqrt{\frac{N-n}{N-1}} \tag{13.7.8}$$

whent he population size N is finite, but the sample is drawn *without replacement.*

(II) Standard Error of Sample Proportion

If a random sample of size n is drawn from a population containing a proportion P of the units belonging to a certain category (e.g., proportion of defectives in a batch of articles), then *Standard Error (S.E.) of sample proportion* is given by

(i) $$\text{S.E.} = \sqrt{\frac{PQ}{n}} \tag{13.7.9}$$

when either the population size is infinitely large, or the sample is drawn *with replacement.* ($P + Q = 1$, i.e., $Q = 1 - P$).

(ii) $$\text{S.E.} = \sqrt{\frac{PQ}{n}}\sqrt{\frac{N-n}{N-1}} \tag{13.7.10}$$

when the population size N is finite, but the sample is drawn *without replacement.*

It may be noted that the formulae for S.E. in simple sampling are easier. In sampling without replacement from a finite population S.E. contains an exra factor $\sqrt{\frac{N-n}{N-1}}$, which is called *finite population correction* (f.p.c.)

Example 13.14 *A simple random sample of size 5 is drawn without replacement from a finite population consisting of 41 units. If the population standard deviation is 6.25, what is the standard error of sample mean? (Use finite population correction).* [B.U., B.A. (Econ) '73]

Solution Here, $n = 5$, $N = 41$, $\sigma = 6.25$. Using the formulae

$$\text{S.E. of mean} = \frac{\sigma}{\sqrt{n}} \cdot \sqrt{\frac{N-n}{N-1}}$$

$$= \frac{6.25}{\sqrt{5}}\sqrt{\frac{41-5}{41-1}}$$

$$= \frac{6.25 \times 6}{10\sqrt{2}} = 2.65$$

Example 13.15 *The safety limit of a crane is known to be 32 tons. The mean weight and the standard deviation of a large number of iron rods are 0.3 ton and 0.2 ton respectively. 100 rods are lifted at a time. Find the probability of an accident.*

Solution The accident will occur when the total weight of 100 rods exceeds 32 tons, i.e., the mean weight ($\bar{x}$) is greater than 0.32 ton. So, we have to find the probability that the mean of a random sample of size 100 exceeds 0.32.

The weight (x) of a single rod has population mean $\mu = 0.3$ and s.d. $\sigma = 0.2$. Since the sample size $n = 100$ is large, the sample mean ($\bar{x}$) follows normal distribution approximately (see Section 13.6) with

$$\text{mean} = \sigma = 0.3$$

$$\text{s.d.} = \frac{\sigma}{\sqrt{n}} = \frac{0.2}{\sqrt{100}} = 0.02$$

Therefore, the standardised value of the sample mean 0.32 is

$$z = \frac{0.32 - 0.3}{0.02} = 1$$

Probability that $\bar{x}$ is greater than 0.32

= Area under the standard normal curve to the right of the ordinate at $z = 1$

= 0.1587

(Also see Example 13.26 and 13.27, p. 173)

Example 13.16 *It has been found that 2% of the tools produced by a certain machine are defective. What is the probability that in a shipment of 400 such tools, 3% or more will be defective? (Probability that the normal deviate lies between 0 and 1.43 is 0.4236).* [B.U., B.A. (Econ) '72]

Solution Since the sample size $n = 400$ is lare, the sample proportion (p) is approximately normally distributed (see Section 13.6) with

$$\text{mean} = P = 2\% = .02$$

$$\text{s.d.} = \sqrt{\frac{PQ}{n}}\sqrt{\frac{.02 \times .98}{400}} = .007$$

Probability that the sample proportion p exceds .03

= Area under the standard normal curve to the right of the ordinate at the standardised value $z = \dfrac{.03 - .02}{.007} = 1.43$

= (Area to the right of 0) – (Area between 0 and 1.43)

= 0.5 – 0.4236 = .0764 *Ans.* .0764

Example 13.17 *Define Standard Error of the mean. How is it calculated ? Explain its uses.* [C.U. B.A. (Econ) '79; C.A., Nov. '73]

Solution *Standard Error (S.E.) of mean* may be defined as the standard deviation in the sampling distribution of mean. If several random samples, each having the same number of members, are drawn from a statistical population, it will be found that the means of these smaples are in general different from the mean of the population. A measure of the extent of this discrepancy, or 'error' in the sample mean, is provided by Standard Error of mean.

Applying the principles of probability, the formula for S.E. of mean can be theoretically derived. In fact, if σ denotes the standard deviation of the population and n represents the number of members included in the sample, then (see Examples 13.21 and 13.24)

$$\text{S.E. of mean} = \frac{\sigma}{\sqrt{n}}$$

If, however, the random sample is drawn without replacement from a finite population of size N, thenS.E. of mean

$$= \frac{\sigma}{\sqrt{n}} \cdot \sqrt{\frac{N-n}{N-1}}.$$

S.E. of mean is used to measure the extent of sampling error in the mean. If a sufficiently large number of members is included in the sample, the means obtained from many samples (all of equal size) will closly follow a pattern known as 'normal distribution' whatever be the nature of distribution in the population. S.E. of mean would then give a good account of the proportion of sample means different by given amounts from the population mean—about 2/3rd of them will differ by less than the S.E., about 95% by less than twice the S.E., and almost all sample means will differ from the population mean by less than 3 times the S.E.

Example 13.18 *How do you distinguish between 'standard error' and 'standard deviation'?*

[I.C.W.A., Dec. '78]

Solution *Standard Deviation* of a given set of values $x_1, x_2, \ldots, x_n$ is a measure of their variability or dispersion, and is defined as $\sigma = \sqrt{\frac{\Sigma(x-x)^2}{n}}$ · It is irrelevant to ask whether these values relate to a sample or the the population. We may calculate the standard deviation of sample observations or of all members in the population.

The term '*Standard Error*' is however, used in relation to a statistic, which is a measure based on the sample observations. As such, we may speak of the standard error only in connection with sampling. As the group of observations constituting a sample is likely to be different from that of another sample, the value of a statistic varies from sample to sample. A measure of this variabiity of the statistic is called *standard error.* Of course, the variability is measured by standard deviation. Thus, standard error is the standard deviation of all possible values of statistic in repeated samples of a fixed size from a given population. In other words, standard error is the standard deviation calculated from the sampling distribution.

Standard error depends on (i) the sampel size, (ii) the natue of the statistic e.g., mean variance, etc., (iii) the mathematical form of the sampling distribution, and (iv) the values of some of the parameters used in the sampling distribution. The reciprocal of standard error is sometimes used to measure the precision of the statistic as an estimate of a parameter. For example, since

S.E. of $\bar{x} = \frac{\sigma}{\sqrt{n}}$, the precision of $\bar{x}$ is $\frac{\sqrt{n}}{\sigma}$.

Hence, the precision of $\bar{x}$, which is used as an estimate of the population mean (μ) is directly proportional to the square-root of the sample size (n). This implies that to double the precision of the estimate we have to make the sample size four-fold.

Example 13.19 *For a population of six units, the values of a characteristic x are given below:*

$$3, \quad 9, \quad 6, \quad 5, \quad 7, \quad 10$$

Consider all possible samples of size two from the above population and show that the mean of the sample means is exactly equal to the population mean.

[C.U., M.Com. '73]

Solution The population mean is

$$\mu = \frac{(3+9+6+5+7+10)}{6} = \frac{20}{3}$$

The possible distinct samples (here random sampling *without replacement* is implied) and their means are shown below:

Table 13.2 Random Samples of Size Two (Without Replacement)

Serial No.	Sample Values	Total	Sample Mean	Serial No.	Sample Values	Total	Sample Mean
1	3, 9	12	6.0	9	9, 10	19	9.5
2	3, 6	9	4.5	10	6, 5	11	5.5
3	3, 5	8	4.0	11	6, 7	13	6.5
4	3, 7	10	5.0	12	6, 10	16	8.0
5	3, 10	13	6.5	13	5, 7	12	6.0
6	9, 6	15	7.5	14	5, 10	15	7.5
7	9, 5	14	7.0	15	7, 10	17	8.5
8	9, 7	16	8.0			Total	100.0

$$\therefore \quad \text{Mean of sample means} = \frac{100.0}{15} = \frac{20}{3}$$

This is exactly equal to the population mean $\mu = \frac{20}{3}$, show above.

[**Note:** If random sampling *with replacement* is applied, then also the mean of sample means will be exactly equal to the population mean]

Example 13.20 *A population consists of the four members 3, 7, 11, 15. Consider all possible samples of size two which can be drawn with replacement from this population. Find*

(i) the population mean,
(ii) the population standard deviation,
(iii) the mean of the sampling distribution of means,
(iv) the standard deviation of the sampling distribution of mean.

Verify (iii) and (iv) directly from (i) and (ii) by use of suitable formulae (which you are to members).

Solve this problem if sampling is without replacement. [I.C.W.A., Dec. '76]

Solution For the 4 members of the population

$$\Sigma x = 3 + 7 + 11 + 15 = 36$$
$$\Sigma x^2 = 3^2 + 7^2 + 11^2 + 15^2 = 404$$

(i) Population mean (μ) = $\dfrac{36}{4} = 9$

(ii) Population variance (σ^2) = $\dfrac{404}{4} - \left(\dfrac{36}{4}\right)^2 = 101 - 81 = 20$

$\therefore$ Population standard deviation (σ) = $\sqrt{20}$.

Case I

Simple Random Sampling With Replacement (SRSWR):

The possible random samples of size 2 (with replacement) and the sample means are shown below:

Table 13.3 Possible Samples (With Replacement)

Serial No.	*Sample Values*	*Total*	*Sample Mean*	*Serial No.*	*Sample Values*	*Total*	*Sample Mean*
1	3, 3	6	3	9	11, 3	14	7
2	3, 7	10	5	10	11, 7	18	9
3	3, 11	14	7	11	11, 11	22	11
4	3, 15	18	9	12	11, 15	26	13
5	7, 3	10	5	13	15, 3	18	9
6	7, 7	14	7	14	15, 7	22	11
7	7, 11	18	9	15	15, 11	26	13
8	7, 15	22	11	16	15, 15	30	15

$N^n = 16$ possible samples of size 2 (with replacement) are possible, and hence each of the 16 values of sample mean ($\bar{x}$), as shown in the last column, occurs with probability $\dfrac{1}{16}$. The distinct possible values of $\bar{x}$, are arranged in order of magnitude and the corresponding probabilities shown against each.

Table 13.4 Sampling Distribution of Mean (With Replacement)

Sample Mean ($\bar{x}$)	3	5	7	9	11	13	15	Total
Probability (*p*)	$\frac{1}{16}$	$\frac{1}{8}$	$\frac{3}{16}$	$\frac{1}{4}$	$\frac{3}{16}$	$\frac{1}{8}$	$\frac{1}{16}$	1

(iii) Mean of the sampling distributon of means:

$$E(\bar{x}) = \frac{1}{16}\times 3+\frac{1}{8}\times 5+\frac{3}{16}\times 7+\frac{1}{4}\times 9+\frac{3}{16}\times 11+\frac{1}{8}\times 13+\frac{1}{16}\times 15$$

$$= \frac{3}{16}+\frac{5}{8}+\frac{21}{16}+\frac{9}{4}+\frac{33}{16}+\frac{13}{8}+\frac{15}{16}$$

$$= 9 \qquad \text{...(A)}$$

(iv) Variance of the sampling distribution of means:

$$\text{Var}(\bar{x}) = E(\bar{x}^2) - \{E(\bar{x})\}^2$$

$$= \left(\frac{1}{16}\times 3^2 + \frac{1}{8}\times 5^2 + \frac{3}{16}\times 7^2 + \frac{1}{4}\times 9^2 + \frac{3}{16}\times 11^2 + \frac{1}{5}\times 13^2 + \frac{1}{16}\times 15^2\right) - 9^2$$

$$= \left(\frac{9}{16} + \frac{25}{8} + \frac{147}{16} + \frac{81}{4} + \frac{363}{16} + \frac{169}{8} + \frac{225}{16}\right) - 81$$

$$= 91 - 81 = 10$$

∴ Standard Deviation of the sampling distribution of means

$$\text{S.D.}(\bar{x}) = \sqrt{10} \qquad \text{...(B)}$$

Using formulae (13.6.1) and (13.7.7),

$$E(\bar{x}) = \mu = 9$$

$$\text{S.E.}(\bar{x}) = \frac{\sigma}{\sqrt{n}} = \frac{\sqrt{20}}{\sqrt{2}} = \sqrt{10}$$

These results agree exactly with those at (A) and (B) abve, obtained by direct calculation from the sampling distribution of mean.

Case II

Simple Random Sampling Without Replacement (SRSWOR):

Table 13.5 Possible Samples (Without Replacement)

Serial No.	*Sample Values*	*Total*	*Sample mean*	*Probability*
1	3, 7	10	5	1/6
2	3, 11	14	7	1/6
3	3, 15	18	9	1/6
4	7, 11	18	9	1/6
5	7, 15	22	11	1/6
6	11, 15	26	13	1/6

Table 13.6 Sampling Distribution of Mean (Without Replacement)

Sample Mean ($\bar{x}$)	5	7	9	11	13	Total
Probability (p)	$\frac{1}{6}$	$\frac{1}{6}$	$\frac{1}{3}$	$\frac{1}{6}$	$\frac{1}{6}$	1

Mean of the sampling distribution of means:

$$E(\bar{x}) = \frac{1}{6}\times 5 + \frac{1}{6}\times 7 + \frac{1}{3}\times 9 + \frac{1}{6}\times 11 + \frac{1}{6}\times 13$$

$$= \frac{5}{6} + \frac{7}{6} + 3 + \frac{11}{6} + \frac{13}{6} = 9$$

Variance of the sampling distribution of means:

$$\text{Var}(\bar{x}) = E(\bar{x}^2) - \{E(\bar{x})\}^2$$

$$= \left(\frac{1}{6}\times 5^2 + \frac{1}{6}\times 7^2 + \frac{1}{3}\times 9^2 + \frac{1}{6}\times 9^2 + \frac{1}{6}\times 11^2 + \frac{1}{6}\times 13^2\right) - 9^2$$

$$= \left(\frac{25}{6} + \frac{49}{6} + 27 + \frac{121}{6} + \frac{169}{6}\right) - 81$$

$$= \frac{263}{3} - 81 = \frac{20}{3}$$

Standard deviation of the sampling distribution of means:

$$\text{S.D.}(\bar{x}) = \sqrt{\frac{20}{3}}$$

Using formulae (13.6.1) and (13.7.8),

$$E(\bar{x}) = \mu = 9$$

$$\text{S.E.}(\bar{x}) = \frac{\sigma}{\sqrt{n}} \cdot \sqrt{\frac{N-n}{N-1}}$$

$$= \frac{\sqrt{20}}{\sqrt{2}} \sqrt{\frac{4-2}{4-1}} = \sqrt{\frac{20}{3}}$$

These results agree exactly with those obtained above by direct calculation from the sampling distribution of $\bar{x}$.

Example 13.21 *Show that the mean and standard error of sample mean ($\bar{x}$) from simple samples of size n are:*

$$E(\bar{x}) = \mu$$

$$S.E.(\bar{x}) = \frac{\sigma}{\sqrt{n}}$$

which μ and σ denote the mean and s.d. of the population.

Solution Let $x_1, x_2, ..., x_n$ denote the sample observations. The sample mean is

$$\bar{x} = \frac{1}{n}(x_1 + x_2 + ... + x_n) = \frac{1}{n}\sum x_i$$

where x_i is the i-th member of the sample (i.e., the member obtained at the i-th drawing).

In simple sampling, each of the sample members has the same probability distribution as the variable x in the population. Hence

$$E(x_i) = \mu = \text{Population mean}$$

$$\text{Var}(x_i) = E(x_i - \mu)^2 = \sigma^2 = \text{Population variance}$$

Therefore,

$$E(\bar{x}) = E\left[\frac{1}{n}(x_1 + x_2 + ... + x_n)\right]$$

$$= \frac{1}{n} E(x_1 + x_2 + ... + x_n)$$

$$= \frac{1}{n}[E(x_1) + (x_2) + ... + (x_n)]$$

i.e.,

$$E(\bar{x}) = \frac{1}{n}\sum E(x_i) = \frac{1}{n}\sum_{i=1}^{n} \mu$$

$$= \frac{1}{n} n\mu = \mu$$

Now,

$$\text{Var}(\bar{x}) = E(\bar{x} - \mu)^2$$

$$= E\left[\frac{1}{n}\sum (x_i - \mu)\right]^2 = \frac{1}{n^2} \cdot E\left[\sum (x_i - \mu)\right]^2$$

$$= \frac{1}{n^2} E\left[\sum (x_i - \mu)^2 + \sum_{i<j} 2(x_i - \mu)(x_j - \mu)\right]$$

$$= \frac{1}{n^2}\left[\sum E(x_i - \mu)^2 + \sum_{i<j} 2E(x_i - \mu)(x_j - \mu)\right]$$

[**Note:** $(a_1 + a_2 + a_3)^2 = (a_1^2 + a_2^2 + a_3^2) + (2a_1a_2 + 2a_1a_3 + 2a_2a_3)$

$$= \sum_i a_i^2 + \sum_{i<j} 2a_ia_j]$$

But
$$E(x_i - \mu)^2 = \sigma^2$$
$$E\{(x_i - \mu)(x_j - \mu)\} = E(x_i - \mu) . E(x_j - \mu)$$
since x_i and x_j are independent.
$$= 0 \times 0 = 0$$

Substituting these results,

$$\text{Var}(\bar{x}) = \frac{1}{n^2}\left[\sum_{i=1}^{n} \sigma^2 + \sum_{i<j} 2 \times 0\right]$$

$$= \frac{1}{n^2}\left[n\sigma^2 + 0\right] = \frac{\sigma^2}{n}$$

$$\therefore \quad \text{S.E.}(\bar{x}) = \sqrt{\text{Var}(\bar{x})} \quad \frac{\sigma}{\sqrt{n}}$$

[**Note:** In random sampling without replacement from a finite population, the result $E(\bar{x}) = \mu$ holds; because each of hte sample members x_i has individual the same probability distribution as x in the population.

But, S.E.$(\bar{x}) \neq \frac{\sigma}{\sqrt{n}}$, since x_i and x_j are *not independent* in this case.]

Example 13.22 *If $x_1, x_2, \ldots x_n$ is a simple random sample of size n from a finite population of N units, show that*

(i) in drawing "with" replacement (SRSWR).

$$P(x_i = X_l) = \frac{1}{N}$$

$$P(x_i = X_k, x_j = X_l) = \frac{1}{N^2}$$

(ii) in drawing "without" replacement (SRSWOR),

$$P(x_i = X_k) = \frac{1}{N}$$

$$P(x_i = X_k, x_j = X_l) = \frac{1}{N^2}(N-1)$$

where x_i and x_j denote the i-th and the j-th members of the sample, and X_k, X_l are two specified members of the poipulation.

Solution The simple random sample is drawn one by one and x_i, x_j denote sample members obtained at the *i*-th and *j*-th drawings respectively. Since the selection of a member in any drawing depends purely of chance, x_i and x_j may be considered as random variables.

(i) In SRSWR, the size and composition of the population remains exactly the same throughout the sampling process, and hence during the *i*-th drawing any of the *N* population members is likely to appear. These *N* cases are mutually exclusive, exhaustive and equally likely (because all the members are given 'equal chance'). The probability that a particular member X_k appears at the *i*-th drawing is therefore, by the classical definition, $\frac{1}{N}$. That is

$$P(x_i = X_k) = \frac{1}{N}$$

Since the sample is drawn "with" replacement, the selection is made from all the population members in any drawing, and hence the probability of selection of any member at the *j*-th drawing remains $\frac{1}{N}$, irrespective of what has actually been obtained in the *i*-th drawing. In other words, the random variables x_i and x_j are independent.

$$P(x_i = X_k, x_j = X_l) = P(x_i = X_k)\,(x_j = X_l)$$

$$= \left(\frac{1}{N}\right)\cdot\left(\frac{1}{N}\right) = \frac{1}{N^2}$$

(ii) In SRSWOR, the sample $x_1, x_2, \ldots x_n$ of size *n* may be drawn out of *N* units in the population in ${}^NP_n$ ways; because we may think of the *n* places of the sample units (corresponding to the *n* drawings) as being filled up b *n* things out of *N* in the population. These ${}^NP_n$ cases are mutually exclusive, exhaustive and equally likely. The number of cases favourable to the event that X_k is selected in the *i*-th drawing (i.e., $x_i = X_k$) may be obtained by considering X_k as having occupied the *i*-th place and filling up the remaining $(n-1)$ places by $(n-1)$ things out of the remaining $(N-1)$ things left in the population (excluding X_k), viz., ${}^{N-1}P_{n-1}$. Hence, by the classical definition of probability,

$$P(x_i = X_k) = \frac{{}^{N-1}P_{n-1}}{{}^NP_n} = \frac{(N-1)(N-2)\ldots(N-n+1)}{N(N-1)\ldots(N-n+1)} = \frac{1}{N}$$

Similarly, the number of cases favourable to the event that X_k is selected at the *i*-th drwing and X_l at the *j*-th drawing (i.e., $x_i = X_k, x_j = X_l$) is the by the number of permutations of $(n-2)$ things out of $(N-2)$, viz. ${}^{N-2}P_{n-2}$. Hence by the classical definition,

$$P(x_i = X_k, x_j = X_l) = \frac{{}^{N-2}P_{n-2}}{{}^NP_n} = \frac{1}{N(N-1)}$$

Example 13.23 *If $x_1, x_2, \ldots x_n$ is a simple random samle of size n from a finite population of N units with mean μ and variance σ^2, show that*

(a) $E(x_i) = \mu$

(b) $Var(x_i) = \sigma^2$

(c) $Cov(x_i, x_j) = 0$ *in SRSWR*

$= \dfrac{-\sigma^2}{N-1}$ *in SRSWOR*

where x_i and x_j denote the sample units obtained at the i-th and j-th drawings respectively.

Solution Let $X_1, X_2, \ldots, X_N$ denote the members of the population from which the sample is drawn. Then

$$\text{Population Mean } (\mu) = \frac{(X_1 + X_2 + \ldots + X_N)}{N} \qquad \ldots(1)$$

$$\text{Population variance } (\sigma^2) = \frac{\Sigma(X_k - \mu)^2}{N} \quad ...(2)$$

where the summation extends over all values of $k = 1, 2, ... N$.

(a) In SRSWR, any of the population members may appear at the i-th drawing; i.e., the random variable x_i take the possible values $X_1, X_2, ..., X_N$ with equal probability $\frac{1}{N}$ · (Example 13.22)

x_i :	X_1	X_2	...	...	X_N	Total
Prob. :	$\frac{1}{N}$	$\frac{1}{N}$	...	...	$\frac{1}{N}$	1

Therefore, by the definition of Expectation (page 385),

$$E(x_i) = \left(\frac{1}{N}\right)X_1 + \left(\frac{1}{N}\right)X_2 + \dots\dots + \left(\frac{1}{N}\right)X_N$$

$$= \frac{(X_1 + X_2 + \dots + X_n)}{N}$$

$$= \mu, \text{ by (1)}$$

(b) $\text{Var}(x_i) = E\{x_i - E(x_i)\}^2$, by definition

$$= E(x_i - \mu)^2, \quad \text{since } E(x_i) = \mu$$

$$= \sum_{k=1}^{N} (X_k - \mu)^2 \,.\, P(x_i = X_k)$$

$$= \sum_k (X_k - \mu)^2 \,.\left(\frac{1}{N}\right), \text{ from Example 13.22}$$

$$= \frac{\sum (X_k - \mu)^2}{N}$$

$$= \sigma^2, \text{ from (2)}$$

It may be noted that the results (a) and (b) hold both for SRSWR and SRSWOR.

(c) In SRSWR, the random variable x_i and x_j are independent. Hence,

$$\text{Cov}(x_i, x_j) = 0$$

In SRSWOR, the random variable x_i and x_j are not independent.

$\text{Cov}(x_i, x_j) = E[\{x_i - E(x_j)\}\ \{x_i - E(x_j)\}]$, by defintion

$$= E[(x_i - \mu)(x_j - \mu)]$$

$$= \sum_{k \neq l}\sum (X_k - \mu)(X_i - \mu).\ P(x_i = X_k, x_j = X_l)$$

$$= \sum_{k \neq l}\sum (X_k - \mu)(X_i - \mu).\ \frac{1}{N(N-1)}, \text{ by Example 13.22}$$

$$= \frac{1}{N(N-1)} \sum_{k=1}^{N} (X_k - \mu) \sum_{l \neq k=1}^{N} (X_l - \mu)$$

$$= \frac{1}{N(N-1)} \sum_k (X_k - \mu) \left\{ \sum_{l=1}^{N} (X_l - \mu) - (X_k - \mu) \right\}$$

$$= \frac{1}{N(N-1)} \sum_k (X_k - \mu)\{0 - (X_k - \mu)\}$$

$$= \frac{1}{N(N-1)} \sum \{-X_k - \mu)^2\}$$

$$= \frac{-1}{N-1} \cdot \frac{1}{N} \Sigma (X_k - \mu)^2$$

$$= \frac{-1}{N-1} \cdot \sigma^2$$

$$= \frac{-\sigma^2}{N-1}$$

Example 13.24 *Derive the formulae for standard error of sample mean in both simple random sampling with replacement (SRSWR) and simple random sampling without replacement (SRSWOR) from a finite population.* [W.B.H.S. '80]

Solution Let $x_1, x_2, ..., x_n$ denote a simple random sample of size n from a finite population of size N, with mean μ and standard deviation σ. The sampel mean is

$$\bar{x} = \frac{(x_1 + x_2 + ... + x_n)}{N} = \sum_{i=1}^{n} \frac{x_i}{n}$$

where x_i is the i-th member of the sample.

$$\therefore \quad E(\bar{x}) = E\left(\frac{1}{n} \Sigma x_i\right) = \frac{1}{n} \Sigma E(x_i)$$

But $E(x_i) = \mu$, in both SRSWR and SRSWOR (see Example 13.23)

$$\therefore \quad E(\bar{x}) = \sum_{i=1}^{n} \frac{\mu}{n} = \frac{n\mu}{n} = \mu \qquad ...(1)$$

The result holds for both SRSWR and SRSWOR. Hence

$$\text{Var}(\bar{x}) = E(\bar{x} - \mu)^2$$

$$= E\left[\sum \frac{(x_i - \mu)}{n}\right]^2$$

$$= \frac{1}{n^2} \cdot E\left[\sum (x_i - \mu)\right]^2$$

$$= \frac{1}{n^2} \cdot E\left[\sum_i (x_i - \mu)^2 + \sum\sum_{i \neq j} (x_i - \mu)(x_j - \mu)\right]$$

because $$\left(\sum a_i\right)^2 = \sum a_i^2 + \sum\sum_{i \neq j} a_i a_j$$

$$= \frac{1}{n^2} \cdot \left[\Sigma E(x_i - \mu)^2 + \Sigma\Sigma E\{(x_i - \mu)(x_j - \mu)\}\right]$$

But, $E(x_i - \mu)^2 = \text{Var}(x_i)$, $E\{(x_i - m)(x_j - m)\} = \text{Cov}(x_i, x_j)$

$$\therefore \quad \text{Var}(\bar{x}) = \frac{1}{n^2}\left[\sum_i \text{Var}(xi) + \sum\sum_{i \neq 1} \text{Cov}(x_i, x_j)\right] \qquad ...(2)$$

This result holds for both the case SRSWR and SRSWOR.

Case I

Simpl random sampling "with" replacement (SRSWR):

Here x_i and x_j are independent.

$$\text{Var}(x_i) = \sigma^2$$

$$\text{Cov}(x_i, x_j) = 0 \qquad \text{(see Example 13.23)}$$

Substituting these results in (2), we have

$$\text{Var}(\bar{x}) = \frac{1}{n^2}\left[\sum_i \sigma^2 + \sum_{i \neq j}\sum 0\right]$$

$$= \frac{1}{n^2}\left[n\sigma^2 + 0\right] = \frac{\sigma^2}{n}$$

Thus, in SRSWR, the standard error of sample mean is

$$\text{S.E.}(\bar{x}) = \frac{\sigma}{\sqrt{n}}$$

Case II

Simple random sampling "without" replacement (SRSWOR).

Here x_i and x_j are not independent

$$\text{Var}(x_i) = \sigma^2$$

$$\text{Cov}(x_i, x_j) = \frac{-\sigma^2}{N-1} \quad \text{(see Example 13.23)}$$

Substituting in (2),

$$\text{Var}(\bar{x}) = \frac{1}{n^2}\left[\sum_i \sigma^2 + \sum_{i\neq j}\sum\left(\frac{-\sigma^2}{N-1}\right)\right]$$

$$= \frac{1}{n^2}\left[n\sigma^2 - n(n-1)\frac{\sigma^2}{N-1}\right]$$

because there are $n(n-1)$ possible pairs of values $i \neq j$.

$$\text{Var}(\bar{x}) = \frac{n\sigma^2}{n^2}\left[1 - \frac{n-1}{N-1}\right] = \frac{\sigma^2}{n}\left(\frac{N-n}{N-1}\right)$$

Thus, in SRSWOR, the standard error of sample mean is

$$\text{S.E.}(\bar{x}) = \frac{\sigma}{\sqrt{n}}\sqrt{\frac{N-n}{N-1}}$$

Example 13.25 *Find the expectation and standard error of sample proportion in both simple random sampling with replacement and simple random sampling without replacement from a finite population.*

Solution Let a simple random simple of size n be drawn from a population in which the proportion of members belonging to a particular category (e.g., proportion of defective articles) be P. If x denotes the number of such members included in the sample, then

$$\text{Sample proportion } (p) = \frac{x}{n}$$

where $x = 0, 1, 2, ..., n$. Therefore

$$E(p) = \frac{E(x)}{n} \qquad ...(1)$$

$$\text{Var}(p) = \text{Var}\frac{(x)}{n^2} \qquad ...(2)$$

(Note that Sample proportion = p, Population proportion = P).

Case I

Simple random sampling "with" replacement (SRSWR).

Here the probability of "success" at each trial is a constant P, and the n trials are independent. Therefore, x follows *binomial* distribution (p. 433) with

$$E(x) = np$$
$$\text{Var}(x) = nPQ$$

where $Q = 1 - P$. Substituting these results in (1) and (2),

$$E(p) = \frac{nP}{n} = P$$
$$\text{Var}(p) = \frac{nPQ}{n^2} = \frac{PQ}{n}$$

Thus, in SRSWR,

$$E(p) = P$$
$$\text{S.E.}(p) = \sqrt{\frac{PQ}{n}}$$

Case II

Simple random sampling "without" replacement (SRSWOR).

Here the drawing is "without" replacement, and hence x follows *Hypergeomtric* distribution (page 447) with

$$E(x) = nP$$
$$\text{Var}(x) = nPQ\left(\frac{N-n}{N-1}\right)$$

Substituting in (1) and (2),

$$E(p) = P$$
$$\text{Var}(p) = \frac{PQ}{n}\cdot\left(\frac{N-n}{N-1}\right);$$

Thus, in SRSWOR,

$$E(p) = P$$
$$\text{S.E.}(p) = \sqrt{\frac{PQ}{n}\cdot\left(\frac{N-n}{N-1}\right)}$$

Example 13.26 *The mean of a certain normal distribution is equal to the standard error of the mean of samples of 25 from the distribution. Find the probability that the mean of a sample of 49 from the distribution will be negative. (Given: Area under standard normal curve to the left of the ordinate at 1.4 is 0.9192).*

Solution Let μ and σ denote the mean and s.d. of the normal distribution. Here, it is stated that

$$\text{Mean} = \text{S.E.}(\bar{x}) \text{ in samples of size 25}$$

i.e., $$\mu = \frac{\sigma}{\sqrt{25}}, \text{ since S.E.}(\bar{x}) = \frac{\sigma}{\sqrt{n}}$$

or, $$\sigma = 5\mu \qquad \text{... (i)}$$

We have to find the probability $P(\bar{x} < 0)$, when $n = 40$. Since the population is normal,

$$z = \frac{\bar{x} - \mu}{\sigma/\sqrt{n}} \text{ is a standard normal variate.}$$

$$\therefore \quad P(\bar{x} < 0) = P\left\{\frac{\bar{x}-\mu}{\sigma/\sqrt{n}} < \frac{0-\mu}{\sigma/\sqrt{n}}\right\}, \text{ when } n = 49,$$

$$= P\left\{z < \frac{-\mu\sqrt{49}}{5\mu}\right\}, \text{ since } \sigma = 5\mu, \text{ from (i)}$$

$$= P(z < -1.4)$$

$$= \phi(-1.4) = 1 - \phi(1.4)$$

$$= 1 - 0.9192 = 0.0808.$$

Ans. 0.0808

Example 13.27 *The guaranteed average life of a certain type of electric light bulbs is 1000 hours with a standard deviation of 125 hours. It is proposed to sample the output so as to assure that 90% of the bulbs do not fall short of the guaranteed average by more than 2.5 per cent. What should be the minimum size of the sample? (The area under standard normal curve from z = 0 to z = 1.28 is 0.4000).*

[I.C.W.A., Dec. 82]

Solution Let x denote the life (in hours) of the electric light bulbs. Assuming normal distribution, we have $\mu = 1000$, $\sigma = 125$. It is required to find the value of sample size (n) such that

$$P\{\bar{x} - \mu > -2.5\% \text{ of } \mu\} = 0.90$$

or, $$P\{\bar{x} - \mu > -25\} = 0.90, \qquad \text{since } \mu = 1000$$

or, $$P\left\{\frac{\bar{x}-\mu}{\sigma/\sqrt{n}} < \frac{-25}{\sigma/\sqrt{n}}\right\} = 0.90$$

or, $$P\{z > -\sqrt{n}/5\} = 0.90, \qquad \text{since } \sigma = 125$$

or, $$P\{z < -\sqrt{n}/5\} = 1 - 0.90 = 0.10$$

or, $$P\{z > \sqrt{n}/5\} = 0.10, \qquad \text{... (i)}$$

since the curve is symmetrical about $z = 0$

From the given area under standard normal curve, we find that the area to the *right* of $z = 1.28$ is $0.5 - 0.4000 = 0.10$. That is $P(z > 1.28) = 0.10$.

Comparing this with (i), we find that

$$\sqrt{n}/5 = 1.28, \quad \text{Hence } n = 40.96 = 41.$$

(Also see Example 14.38). *Ans.* 41

13.7A PROBABLE ERROR (P.E.)

The *probable error* of a statistic may be defined to be a quantity such that exactly half of the possible values of the statistic differ from its expected value by more than that quantity. If t is a statistic and θ its expected value, then the probability that the difference of t and θ exceeds the probable error (P.E.) is exactly $\frac{1}{2}$.

$$P\{|t - \theta| > \text{P.E.}\} = \frac{1}{2} \qquad (13.7\text{ A.1})$$

Consequently, $P\{|t-\theta|<\text{P.E.}\} = \frac{1}{2}$

Thus, there is a 50 : 50 chance that the statistic t differs from θ by more than or less than P.E. In this sense, P.E. measures the sampling fluctuation like S.E.

In random samples of size h from a normal population with mean μ and s.d σ, it can be shown that very nearly

$$\begin{aligned}\text{P.E. of mean} &= 0.6745\ \sigma/\sqrt{n}\\ &= 0.6745 \times (\text{S.E. of mean}) \qquad (13.7\text{ A.2})\end{aligned}$$

Even if the population is not normal, this relation helds approximately for large n. For this reason, *the probable error* (P.E.) is *sometimes defined as 0.6745 times the standard error* (S.E.). For example, in large samples, S.E. of correlation coefficient (r) is $(1-r^2)/\sqrt{n}$, so that

$$\text{P.E. of correlation coefficient} = 0.6745\ \frac{1-r^2}{\sqrt{n}} \qquad (13.7\text{ A.3})$$

P.E. is rarely used in statistical theory.

13.8 DISTRIBUTIONS USED IN SAMPLING THEORY

Four important probability distribution, which are derived from Normal distribution and used in sampling theory are:

(i) Standard Normal distribution.
(ii) Chi-square (x^2) distribution
(iii) Student's t distribution
(iv) Snedecor's F distribution

(A) Standard Normal Distribution

If a random variable x is normally distributed with mean μ and standard deviation σ, then

$$z = \frac{x-\mu}{\sigma} = \frac{\text{Normal variable} - \text{Mean}}{\text{S.D.}} \qquad (13.8.1)$$

is called a *Standard Normal Variate*. The probability distribution of z is called *Standard Normal Distribution*, and is defined by the p.d.f.

$$p(z) = \frac{1}{\sqrt{2\pi}} e^{-z^2/2};\ (-\infty < z < +\infty) \qquad (13.8.2)$$

Characteristics:

1. The standard normal distribution is a special case of normal distribution with the mean = 0, and s.d. = 1.
2. It has no parameters (unlike the normal distribution which has two parameters μ and σ).
3. The central moments are $\mu_2 = 1$, $\mu_3 = 0$, $\mu_4 = 3$.
   Also, $\beta_1 = 0$, $\beta_2 = 3$; Skewness (γ_1) = 0, Kurtosis (γ_2) = 0.
4. The standard normal curve is symmetrical about the mean 0, and the two tails of the curve extend to infinity on either side of the mean. The points of infection are at $z = \pm 1$.

(5) The following results regarding the area under the curve are especially useful:

Area between $z = \pm 3$ is 99.73%

Area between $z = \pm 2.58$ is 99%

Area between $z = \pm 1.96$ is 95% (13.8.3)

Area to the right of $z = 1.645$ is 5%

Area to the right of $z = 1.96$ is 2.5%

Area to the right of $z = 2.33$ is 1%

Area to the right of $z = 2.58$ is 0.5% (13.8.4)

Theorem I

If $\bar{x}$ denotes the mean of a random sample of size n, drawn from a normal population with mean μ and s.d. σ, then

$$z = \frac{\bar{x} - \mu}{\sigma / \sqrt{n}} \tag{13.8.5}$$

follows standard normal distribution.

Percentage Points:

In any probability distribution, the specified value which the random variable exceeds (or is exceeds by) with a defintie probability is known as *Percentage Point* of the distribution.

Let us denote by z_p the value of the standard normal variate z such that the probability of its being exceeded is p [Fig. 13.1(A)].

$$P(z > z_p) = p \tag{13.8.6}$$

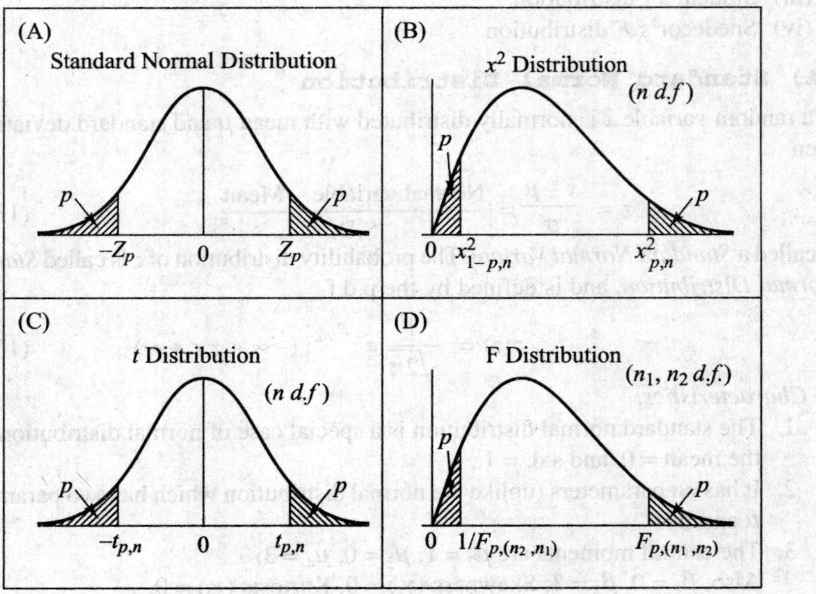

Fig. 13.1 Important Probability Distributions and Percentage Points

This means that the area under the curve to the right of the ordinate at z_p is p. For example, from the table of Area under Standard Normal Curve (Appendix, Table I) we find that the area to the *right* of the ordinate at $z = 1.645$ is .05; i.e.,

$$P(z > 1.645) = .05 = 5\%.$$

The *upper 5% point* of the standard normal distribution is thus 1.645, and we express this as $z_{.05} = 1.645$. The following percentage points are useful in sampling theory:

$$z_{.05} = 1.645, \quad z_{.025} = 1.96,$$
$$z_{.01} = 2.33, \quad z_{.005} = 2.58 \qquad (13.8.7)$$

Again because of symmetry of the standard normal curve, the area to the *left* of the ordinate at $z = -1.645$ is .05; i.e.,

$$P(z < -1.645) = .05.$$

The *lower 5% point* of the distribution is thus – 1.645. Since the area to the right of – 1.645 is 0.95, hence the lower 5% point is $z_{.95} = -1.645 = -z_{.05}$. In general, for the standard normal distribution, the lower percentage point is ust the negative of the upper percentage point; i.e.,

$$z_{1-p} = -z_p \qquad (13.8.8)$$

(B) Chi-square (χ^2) Distribution

A random variable x is said to follow *Chi-square* (χ^2) *distribution* if its p.d.f. is of the form

$$f(x) = K.e^{-x/2}\, x^{(n/2)-1};\ (0 < x < \infty) \qquad (13.8.9)$$

where K is a constant. The parameter n (positive integer) is called the *number of degrees of freedom,* (χ is a letter of the Greem alphabet and pronounced *ky*). A variable which follows chi-square distribution is called a *Chi-square variate,* and often denoted by the symbol χ^2.

In statistical work it is common to use the same symbol both for the random variable and a specified value of the variable. For example, in the χ^2 distribution with n degrees of freedom (d.f.) the percentage points are denoted by $\chi^2_{p,n}$ or breifly χ^2_p if the d.f. n is understood from the context. These are vlues of the variable such that $P(\chi^2 > \chi^2_p) = p$. Percentage points are given in Statistical Tables for various values of p (Appendix, Table II).

Characteristics:

1. Mean = n, S.D. = $\sqrt{2n}$ where n is the number of degrees of freedom (d.f.) of Chi-square distribution.
2. The chi-square curve is positively skew, and starting from 0 extends to infinity on the right [Fig. 13.1(B)].
3. If x and y are independent chi-square variates with d.f. n_2 and n_2 respectively, then their sum $(x + y)$ also follows chi-square distribution with d.f. $(n_1 + n_2)$.
4. When the d.f. n is large $\sqrt{2\chi^2} - \sqrt{2n-1}$ approximately follows the standard normal distribution.

Theorem II

If $z_1, z_2, \ldots z_n$ are n independent standard normal varieties, then

$$z_1^2 + z_2^2 + \ldots + z_n^2 = \sum_{i=1}^{n} z_i^{\,2} \qquad (13.8.10)$$

follows chi-square distribution with n degrees of freedom.

For this reason, often a χ^2 variate with n d.f. is defined to be the sum of the squares of n independent standard normal variates.

Theorem III

If $x_1, x_2, \ldots x_n$ is a random sample from a normal population with mean μ and s.d. σ, then

(i) $$\sum_{i=1}^{n} (x_i - \mu)^{2/\sigma^2} \qquad (13.8.11)$$

follows chi-square distribution with n degrees of freedom; and

(ii) $$\sum_{i=1}^{n} (x_i - \bar{x})^{2/\sigma^2} \qquad (13.8.12)$$

follows chi-square distribution with $(n - 1)$ degrees of fredom. (Here, $\bar{x}$ is the mean of the sample).

(C) Student's t Distribution

A random variable is said to folow *Student's t distribution,* or simply *t distribution,* if its p.d.f. is of the form

$$f(t) = K.\left(1 + \frac{t^2}{n}\right)^{-(n+1)/2} \qquad (13.8.13)$$

$$(-\infty < t < +\infty)$$

where K is a constant. The parameter n (positive integer) is called the *number of degrees of freedom* (d.f.). This distribution was discovered by W.S. Gossett who wrote under the pen-name "Student" and hence it is called *Student's distribution.* A variable which follows Student's distribution is denoted by the symbol t.

The percentage points of t distribution with n degrees of freedom are denoted by $t_{p,\,n}$ or briefly t_p if the d.f. n is understood. Since the t-curve is symmetrical about zero [Fig. 13.1(C)], $t_{1-p} = -t_p$; for example $t_{.95} = -t_{.03}$.

Characteristics:

1. Mean = 0, S.D. = $\sqrt{\dfrac{n}{n-2}}$ $(n > 2)$
2. The t-curve is symmetrical about 0, extending from $-\infty$ to $+\infty$ (like the standard normal curve). It has zero skewness and positive kurtosis (leptokurtic). i.e., $\beta_1 = 0$, $\beta_2 > 3$.
3. When the d.f. n is large, the t distribution can be approximated by the standard normal distribution.

Theorem IV

If z and y are independent random variables, where z follows standard normal distribution and y follows chi-square distribution with n degrees of freedom, then

$$t = \frac{z}{\sqrt{y/n}} \tag{13.8.14}$$

follows Student's t distribution with n degrees of freedom.

For this reason, often a t variate with n d.f. is defined as the ratio of a standard normal variate to the square root of an independent mean chi-square variate with n d.f. (Note that the d.f. of t is the same as that of chi-square used to define it).

If $x_1, x_2, \ldots x_n$ is a random sample from a normal population with mean μ and s.d. σ, then by Theorem 1,

$$z = \frac{\bar{x} - \mu}{\sigma/\sqrt{n}}$$

follows standard normal distribution, and by Theorem 3 (ii),

$$y = \frac{\Sigma(x_i - \bar{x})^2}{\sigma^2} = \frac{nS^2}{\sigma^2}$$

follows chi-square distribution with $(n - 1)$ d.f., where $S^2 = \Sigma(x_1 - \bar{x})^2/n$ is the sample variance. Also the variable z and v are independently distributed.

Theorem V

If a random sample of size n is drawn from a normal population with mean μ and s.d. σ, then

$$\frac{\bar{x} - \mu}{S/\sqrt{n-1}} \tag{13.8.15}$$

follows t distribution with $(n - 1)$ degrees of freedom, where $\bar{x}$ and S denote the mean and s.d. of the sample.

Theorem VI

If two independent random samples of sizes n_1 and n_2 are drawn from two normal populations with means μ_1 and μ_2 respectively and a common s.d. σ, then

$$\frac{(\bar{x}_1 - \bar{x}_2) - (\mu_1 - \mu_2)}{s\sqrt{1/n_1 + 1/n_2}} \tag{13.8.16}$$

follows t distribution with $(n_1 + n_2 - 2)$ degrees of freedom, where $\bar{x}_1$, $\bar{x}_2$ denote the means and S_1, S_2, the s.d.s. of the samples, and

$$s = \sqrt{\frac{n_1 S_1^2 + n_2 S_2^2}{n_1 + n_2 - 2}} \tag{13.8.17}$$

(D) Snedecor's *F* Distribution

A random variable is said to follow F distribution with degrees of freedom (n_1, n_2) if its p.d.f. is of the form

$$f(F) = K.\, F^{(n_1/2)-1}\,(n_2 + n_1 F)^{-(n_1+n_2)/2} \tag{13.8.18}$$

$$(0 < F < \infty)$$

where K is a constant. The distribution was discovered by G.W. Snedecor and named F in honour of the distinguished mathematical statistician Sir R.A. Fisher. The percentage points of F distribution with d.f. (n_1, n_2) are denoted by $F_{p,(n_1,n_2)}$ of briefly F_p if the d.f. are understood. The lower percentage points are given by

$$F_{1-p,(n_1,n_2)} = 1/F_{p,(n_2,n_1)} \tag{13.8.19}$$

i.e., the lower percentage point is the reciprocal of the upper percentage point with the order of d.f. reversed.

Characteristics:

1. Mean = $\dfrac{n_2}{n_2 - 2}$ Mode = $\dfrac{n_2(n_1 - 2)}{n_1(n_2 + 2)}$

   S.D. = $\left(\dfrac{n_2}{n_2 - 2}\right)\sqrt{\dfrac{2(n_1 + n_2 - 2)}{n_1(n_2 - 4)}}$

   provided they exist and are positive.
2. The F-curve is positively skew, and starting from 0 extends to infinity [Fig. 13.1(D)].

Theroem VII

If y_1 and y_2 are independent chi-square variates with degrees of freedom n_1 and n_2 respectively, then

$$F = \frac{y_1/n_1}{y_2/n_2} \tag{13.8.20}$$

follows F distribution with degrees of freedom (n_1, n_2).

In view of the results given in Theorem III, and as a consequence of the above theorem, we have

Theorem VIII

If $x_1, x_2, \ldots x_{n_1}$ and $y_1, y_2, \ldots y_{n_2}$, are independent random samples of sizes n_1 and n_2 respectively from two normal populations with means μ_1, μ_2 and common standard deviation σ, then

(i) $$\frac{\Sigma(x_i - \mu_1)^2/n_1}{\Sigma(y_i - \mu_2)^2/n_2} \tag{13.8.21}$$

follows F distribution with (n_1, n_2) d.f.

(ii) $$\frac{\Sigma(x_i - \bar{x})^2/(n_1 - 1)}{\Sigma(y_i - \bar{y})^2/(n_2 - 1)} \tag{13.8.22}$$

follows F distribution with $(n_1 - 1, n_2 - 1)$ d.f.

(iii) $$\frac{\Sigma(x_i - \mu_1)^2/n_1}{\Sigma(y_i - \bar{y})^2/(n_2 - 1)} \tag{13.8.23}$$

follows F distribution with $(n_1, n_2 - 1)$ d.f.

(iv) $$\frac{\Sigma(x_i - \bar{x})^2/(n_1 - 1)}{\Sigma(y_i - \mu_2)^2/n_2} \tag{13.8.24}$$

follows F distribution with $(n_1 - 1, n_2)$ d.f.

[**Note:** In F distribution, the first d.f. is always the divisor in the numerator and the second d.f. that in the denominator. The order of writing the two degrees of freedom is important].

Relations between Standard Normal, χ^2, t and F distribution

(a) A chi-square variate with n degrees of freedom (d.f.) is the sum of the squares of n independent standard normal variates.

$$\chi^2 = \sum_{i=1}^{n} z_i^2$$

(b) A t-variate with n d.f. is the ratio of a standard normal variate and the square-root of an independent chi-square variate when divided by its degree of freedom.

$$t = \frac{z}{\sqrt{y/n}}$$

(c) An F-variate with (n_1, n_2) d.f. is the ratio of two independent chi-square variates each divided by its own d.f. (chi-square with d.f. n_1 appears in the numerator).

$$F = \frac{y_1/n_1}{y_2/n_2}$$

(d) An F-variate with d.f. $(1, n)$ is equivalent to the square of a t-variate with n d.f. The realation between upper percentage points is

$$F_{p, (1, n)} = t^2_{p/2, n}$$

(e) When the the d.f. n is large, $\sqrt{2\chi^2} - \sqrt{2n-1}$ is approximately a standard normal variate.

(f) When the d.f. n is large, a t-variate is approximately a standard normal variate. The relation between percentage points is

$$z_p = t_{p,\infty}$$

(g) For large degrees of freedom, both the chi-square and the F distribution tend to nromal distributions.

EXERCISES

1. What are the main objects of sampling? Compare and contrast the merits and drawbacks of Sample and Census studies. *[C.U., M.Com. '80]*
2. Distinguish between a *population* and a *sample*. What is a random sample ? Describe some method of drawing such a sample from a finite population. *[C.U., M.Com '72]*
3. What are sampling and non-sampling errors? *[D.S.W. '78]*
4. Define simple random sampling. Distinguish between simple random sampling with replacement and simple random sampling without replacement from a finite population. *[W.B.H.S. '80, '82]*
5. Define 'simple random sampling' and 'stratified random sampling'. What are 'random numbers' and how can you use them? *[M.B.A. '78]*
6. Distinguish between simple random sampling and systematic sampling. *[I.C.W.A., June '81]*
7. Describe briefly any three types of sampling commonly used. *[C.A., Nov. '81]*

8. What are 'random number' and how can you use them in drawing a random sample from a finite population? Give the expression of the standard error of the sample mean. *[D.M.'78]*
9. (a) From a population with 20 members a random sample without replacement of 2 members is taken. Which of the following is the possible number of all such samples—400, 40, 380, 190? *[W.B.H.S. '78]*
   (b) A random sample of 2 individuals is to be drawn from a population of size 43. What is the possible number of distinct samples when sampling is (i) with replacement order of drawing to be taken into account), and (ii) without replacement (order of drawing to be ignored). *[W.B.H.S. '79]*
10. How can you draw a random sample of 10 patients from a hospital accommodating 310 patients? What do you mean by the sampling distribution of $\overline{x}$ and its standard error? *[D.S.W. '77]*
11. Explain how you will select without replacement a random sample of 3 units from a population of 124 units using a procedure which is fairly simple and does not involve the rejection of a large number of random number. *[(C.U., B.A., (Econ) '74]*
12. Describe in each of the following cases an appropriate method of drawing a random sample:
    (i) Drawing a random sample of 2 members with replacement from a family of 10.
    (ii) Drawing a random sample of 15 answer-scripts *without* replacement from 1243 answer-scripts.
    (iii) Drawing a random sample of 3 houses *without* replacement from a locality in which there are 4 streets with 10, 12, 4 and 7 houses respectively. *[W.B.H.S. '78]*
13. There are 220 students in a school. Their Roll Nos. are from 1 to 220. Explain the method of drawing a simple random sample of size 5 without replacement from the above population with the help of random sampling number series table. *[W.B.H.S. '81]*
14. Describe a procedure for drawing a random sample of size 8 from a population of size 24 (without replacement method). *[I.C.W.A., June '81]*
15. How can you draw a random sample of 100 workers from a population of 1000 workers? If in your sample the mean and S.D. of weakly wages come out as Rs 48 and Rs 11 respectively, how will you get the standard error of the sample mean ? *[D.S.W. '76]*
16. What is meant by 'stratified random sampling? Explain the procedure and advantages of stratification. *[I.C.W.A., July '70]*
17. Describe the techniques of 'systematic sampling' and 'stratified random sampling', and state the situations where each of them will be suitable. *[I.C.W.A., June '75]*
18. Define "statistic" and its "sampling distribution". *[C.U., M.Com. '79]*
19. Explain what is meant by 'sampling fluctuation' and 'sampling distribution' of a statistic. *[W.B.H.S. '82]*
20. What is meant by a 'statistic' and its 'standard error? Give expressions for the standard error of the sample mean and the sample proportion. *[M.B.A., '77]*

21. Explain briefly the two terms: Standard Deviation and Standard Error. Mention at least two utilities of Standard Error in the Sampling theory. *[I.C.W.A., Dec. '78]*
22. The values of a characteristic x of a population containing six units are given below: 2, 6, 5, 1, 7, 3. Take all possible samples of size two and verify that the mean of the population is exactly equal to the mean of the sample means. *[C.U., M.Com. '71]*
23. A population consists of 5 numbers (2, 3, 6, 8, 11). Consider all possible samples of size two which can be drawn with replacement from this population. Calculate the S.E. of sample means. *[I.C.W.A., June '78]*
24. Define simple random sampling with replacement (SRSWR). What do you know about the sampling distribution of the sample mean when sampling is SRSWR? What do you know about its standard error? *[C.U., B.A.(Econ) '76]*
25. Compute the standard error of the mean and construct the sampling distribution of the mean for simple random samples of two families each from a population of 5 families which is given below:

Family	A	B	C	D	E
Family Size	4	3	2	5	7

[C.U., M.Com. '74]

26. Ages of 5 persons have been recorded (in years) as 14, 19, 17, 20, 25. For random samples of size 3 drawn without replacement from this population, obtain the sampling distribution of $\overline{x}$. Show that the mean of $\overline{x}$ equals the population mean and obtain the S.E. of $\overline{x}$, directly from the sampling distribution and also by using the formula. *[W.B.H.S., '79]*
27. The mean μ of a certain population is equal to the standard error of mean of random samples of size 100 from that population. Find out the standard error of the mean of random samples of size 36 from the population, in terms, of μ. (Assume that the population size is very large.)
28. If it costs one rupee to draw one member of a sample, how much will it cost to take a random sample of such a size that the standard error of sample mean is 0.1% of population mean, when the population is very large-sized one with, its mean = 100 and s.d. = 1. How will you modify your answer, if the population size is 100, instead of very large?
29. Define standard error of a statistic and derive it when the statistic is sample proportion in a random sample of size n, drawn without replacement. *[W.B.H.S., '82]*
30. Explain each of the following statements, and state (giving reasons) whether it is true or false:
    (a) Random sampling is sampling in a haphazard manner.
    (b) The larger the sample (in random sampling), the smaller is the standard error of the sample mean.
    (c) The value of a statistic in a random sample must be near the value of the corresponding parameter. *[I.C.W.A., Dec. '82]*

31. Pick, out the correct alternative:
    (a) Sampling distribution is the distribution of : Parameter : Statistic : Sample : Population.
    (b) In random sample with replacement from a population with standard deviation σ, if the sample size is equal to the population size ($= N$), then the standard error of sample mean will be: 0 ; σ ; $\sigma/\sqrt{N}$.
32. If a random variable x is distributed normally around a mean 20 with a s.d. 3, describe the important characteristics of the probability distribution of $y = (x - 20)/3$. *[M.B.A. '78]*
33. A normal population has a mean 0.1 and a standard deviation 2.1. Find the probability that the mean of a sample of size 900 will be negative, it being given that the probability that the absolute value of a standard normal variate exceeds 1.43 is 0.153. *[C.U., B.Sc. (Math) '78]*
34. The variable x is normally distributed with mean 68 inches and s.d. 2.5 inches. What is the size of the sample whose mean shall not differ from the population mean by more than one inch, with probability 0.95? (Given that the area under standard normal curve to the right of the ordinate at 1.96 is 0.025).
35. A random sample of size 25 is taken from a $N(\mu, \sigma^2)$ with $\mu = 30$ and $\sigma^2 = 16$. Would the probability that the sample mean would lie between 25 and 35 be greater than 0.99? *[C.U., B.Sc. (Econ) '82]*
36. If $x_1, x_2, \ldots x_n$ be a random sample from a normal population with mean zero and variance σ^2, define "chi-square" as a function of these observations.
37. Describe the important characteristics of t and F distributions.
38. If X_1, X_2, X_3 be a random sample from $N(0, \sigma^2)$ population, what is the distribution of $(X_1^2 + X_2^2 + X_3^2)/\sigma^2$? State (without derivation) the sampling distributions of the statistics.

$$\frac{\sqrt{2}X_1}{\sqrt{(X_2^2 + X_3^2)}} \quad \text{and} \quad \frac{X_1^2}{X_2^2}$$

mentioning the appropriate degrees of freedom in each case.

ANSWERS

9. (a) 190, (b) 1849; 903.
15. 1.04 Rs
23. $\sqrt{5.4} = 2.32$
25. (i) Sampling Distribution of Sample Mean

Mean ($\bar{x}$)	2.5	3	3.5	4	4.5	5	5.5	6	Total
Probability	0.1	0.1	0.2	0.1	0.2	0.1	0.1	0.1	1.0

(ii) Standard Error of mean = $\sqrt{1.11}$

26. S.E. = $\sqrt{2.2}$ years
27. (5/3) μ

28. Rs 100; Rs 91
30. (a) False; it is sampling with equal probability.
    (b) True; as n increases, SE($\bar{x}$) decreases.
    (c) False; it depends on the sampling distribution of the statistic and its standard error.
31. (a) Statistic, (b) $\sigma / \sqrt{N}$
32. Standard normal
33. 0.0765
34. 24
35. Yes; $P(25 < \bar{x} < 35) = P(-6.25 < z < 6.25) > 0.99$
36. $\Sigma x_i^2/\sigma^2$
38. Chi-square distribution with 3 d.f.; t distribution with 2 d.f.; F distribution with (1, 1) d.f.

14 ESTIMATION AND TEST OF SIGNIFICANCE

14.1 INTRODUCTION

The object of sampling is to study the features of the population on the basis of sample observations. A carefully selection sample is expected to reveal these features, and hence we shall infer about the population from a statistical analysis of the sample. This process is known as *Statistical Inference.*

There are two types of problems. Firstly, we may have no information at all about some characteristics of the population, especially the values of the parameters involved in the distribution, and it is required to obtain estimates of these parameters. This is the problem of *Estimation.* Secondly, some information or hypothetical values of the parameters may be available, and it is required to test how far the hypothesis is tenable in the light of the information provided by the sample. This is the problem of *Test of Hypothesis* or *Test of Significance.*

14.2 THEORY OF ESTIMATION

Suppose we have a random sample $x_1, x_2, \ldots x_n$ on a variable x, whose distribution in the population involves an unknown parameter θ. It is required to find an estimate of θ on the basis of sample values. The estimation is done in two different ways: (i) *Point Estimation,* and (ii) *Interval Estimation.* In point estimation, the estimated value is given by a single quantity, which is a function of sample observations (i.e. statistic). This function is called the '*estimator*' of the parameter, and the value of the estimator in a particular sample is called an '*estimate*'. In interval estimation, an interval within which the parameter is expected to lie is given by using two quantities based on sample values. This is known as *Confidence Interval,* and the two quantities which are used to specify the interval, are known as *Confidence Limits.*

14.3 POINT ESTIMATION—CRITERIA FOR GOOD ESTIMATORS

Many functions of sample observations may be proposed as estimators of the same parameter. For example, either the mean or median or mode of the sample values may be used to estimate the parameter μ of the Normal distribution with p.d.f.

$$\frac{1}{\sigma\sqrt{2\pi}}\, e^{-(x-\mu)^2/2\sigma^2}$$

which we shall in future refer to as $N(\mu, \sigma^2)$. Naturally we have to choose one among these estimators on the basis of certain criteria.

According to R.A. Fisher, the criteria for a good estimator are

(i) Unbiasedness,
(ii) Consistency,
(iii) Efficiency,
(iv) Sufficiency.

Unbiasedness

A statistic t is said to be an *Unbiased Estimator* of a parameter θ, if the expected value of t is θ.

$$E(t) = \theta \tag{14.3.1}$$

Otherwise, the estimator is said to be '*biased*'. The bias of a statistic in estimating θ is given as

$$Bias = E(t) - \theta \tag{14.3.2}$$

Let $x_1, x_2, \ldots x_n$ be a random sample drawn from a population with mean μ and variance σ^2. Then

Sample mean $\qquad (\bar{x}) = \sum \frac{x_i}{n}$

Sample variance $\qquad (S^2) = \sum \frac{(x_i - \bar{x})^2}{n} \qquad (14.3.3)$

Theorem I The sample mean $\bar{x}$ is an *unbiased* estimator of the population mean μ; because

$$E(\bar{x}) = \mu \tag{14.3.4}$$

Theorem II The sample variance S^2 is a *biased* estimator of the population variance σ^2; because

$$E(S^2) = \frac{n-1}{n}\sigma^2 \neq \sigma^2 \tag{14.3.5}$$

Theorem III An unbiased estimator of the population variance σ^2 is given by

$$s^2 = \sum \frac{(x_i - \bar{x})^2}{(n-1)} \tag{14.3.6}$$

because $\qquad E(s^2) = \sigma^2 \qquad (14.3.7)$

Note the distinction between S^2 and s^2 in which only he denominators are different. S^2 is the variance of the sample observations, but s^2 is the 'unbiased estimator' of the variance (σ^2) in the population.

Example 14.1 *Show that the sample mean based on a simple random sample with replacement (srswr) is an unbiased estimator of the population mean.* [C.U., B.A(Econ) '78]

Solution Let $x_1, x_2, \ldots x_n$ be a simple random sample with replacement from a finite population of N members $X_1, X_2, \ldots X_N$ (suppose).

Sample mean $(\bar{x}) = (x_1 + x_2 + \ldots + x_n)/n$

Population mean $(\mu) = (X_1 + X_2 + \ldots + X_N)/N$

We have to show that $E(\bar{x}) = \mu$

Now, each of the sample members $x_1, x_2, \ldots x_n$ behaves like a random variable, because their values in any particular sample depend on chance. For instance, in srswr any of the population members $X_1, X_2, \ldots X_N$ may appear at the *i*-the drawing, i.e. x_i is a random variable with the following probability distribution:

x_i	X_1	X_2	...	X_N	Total
Prob	$\frac{1}{N}$	$\frac{1}{N}$	...	$\frac{1}{N}$	1

$$\text{Therefore,} \quad E(x_i) = \left(\frac{1}{N}\right) X_1 + (1 \,.\, N) \,.\, X_2 + \ldots + \left(\frac{1}{N}\right) . X_N$$

$$= (X_1 + X_2 + \ldots + X_N)/N$$

$$= \mu$$

$$\text{Hence,} \quad E(\bar{x}) = E\left[\frac{1}{n}(x_1 + x_2 + \ldots + x_n)\right]$$

$$= \frac{1}{n}[E(x_1) + E(x_2) + \ldots + E(x_n)]$$

$$= (\mu + \mu + \ldots + \mu)/n$$

$$= \frac{n\mu}{n} = \mu$$

This shows that $\bar{x}$ is an unbiased estimator of μ (also see Example 13.21 and 13.24).

[**Note:** This result holds in all cases of random sampling, irrespective of whether the sample is drawn 'with replacement' or 'without replacement' from a finite population or from an infinite population.]

Example 14.2 *If $x_1, x_2, \ldots x_n$ is a random sample from an infinite population with variance σ^2, and $\bar{x}$ is the sample mean, show that $\sum_{i=1}^{n} \frac{(x_i - \bar{x})^2}{n}$ is a biased estimator of σ^2, but the bias becomes negligible for large n. Give an unbiased estimator of σ^2 here.* [W.B.H.S. '79, '82]

Solution Let m and σ^2 be the mean and variance of the population. Then $E(x_i) = \mu$, Var $(x_i) = E(x_i - \mu)^2 = \sigma^2$ for each $i = 1, 2, \ldots n$. The variance of the sample is

$$S^2 = \frac{\Sigma(x_i - \bar{x})^2}{n}$$

We have to show that $E(S^2) \neq \sigma^2$

$$\text{Now,} \quad S^2 = \frac{\Sigma(x_i - \bar{x})^2}{n} = \frac{\Sigma x_i^2}{n - \bar{x}^2}$$

$$= \frac{\Sigma y_i^2}{n - \bar{y}^2}, \text{ where } y_i = x_i - \mu$$

($\because$ S.D. is unaffected by change of origin)

$$= \frac{\Sigma(x_i - \mu)^2}{n} - (\bar{x} - \mu)^2$$

$$\therefore \quad E(S^2) = \frac{\Sigma E(x_i - \mu)^2}{n} - E(\bar{x} - \mu)^2$$

$$= \frac{\Sigma\sigma^2}{n} - \text{Var}(\bar{x})$$

$$= \sigma^2 - \frac{\sigma^2}{n} \quad \text{(see 13.7.1)}$$

$$= \frac{n-1}{n}\sigma^2 \neq \sigma^2$$

This shows that S^2 is a *biased* estimator of σ^2.

$$\text{Bias} = E(S^2) - \sigma^2$$

$$= \frac{n-1}{n}\sigma^2 - \sigma^2 = \frac{-\sigma^2}{n}$$

Thus for large n, Bias will be negligibly small. If we write

$$s^2 = \frac{\Sigma(x_i - \bar{x})^2}{(n-1)}$$

(note that the divisor is $n - 1$ instead of n) we see that

$$s^2 = \frac{n}{n-1} S^2,$$

$$\therefore \quad E(s^2) = \frac{n}{n-1} E(S^2) = \frac{n}{n-1} \cdot \frac{n-1}{n}\sigma^2 = \sigma^2$$

Thus shows that s^2 is an *unbiased* estimator of σ^2.

Example 14.3 *The following observations constitute a random sample from an unknown population. Estimate the mean and standard deviation of the population. Also, find the estimate of standard error of sample mean.*

14, 19, 17, 20, 25. [M.B.A. '79]

Solution The unbiased estimators of the population mean (μ) and the population variance (σ^2) are $\bar{x} = \frac{\Sigma x_i}{n}$ and $s^2 = \frac{\Sigma(x_i - \bar{x})^2}{(n-1)}$ respectively. Here $n = 5$ and $\bar{x} = \frac{\Sigma x_i}{n} = \frac{95}{5} = 19$

$$\Sigma(x_i - \bar{x})^2 = (-5)^2 + 0^2 + (-2)^2 + 1^2 + 6^2 = 66$$

$$\therefore \quad s^2 = \frac{66}{4} = 16.5, \; s = \sqrt{16.5} = 4.06$$

The estimates of μ or σ are 19 and 4.06 respectively.

The standard error of sample mean is S.E. ($\bar{x}$) = $\frac{\sigma}{\sqrt{n}}$. But as σ is not known, it is estimated by s,

$$\text{Estimate of S.E. } (\bar{x}) = \frac{s}{\sqrt{n}} = \frac{\sqrt{\frac{66}{4}}}{\sqrt{5}} = \sqrt{\frac{66}{20}}$$
$$= \sqrt{3.3} = 1.82$$

Consistency

A desirable property of a good estimator is that its accuracy should increase when the sample size becomes larger. That is, the estimator is expected to come closer to the parameter as the size of the sample increases.

A statistic t_n computed from a sample of n observations is said to be a *Consistent Estimator* of a parameter θ, if it converges in probability to θ as n tends to infinity. This means that the larger the sample size (n), the less is the chance that the difference between t_n and θ will exceed any fixed value. In symbols, given any arbitrary small positive quantity $\in$,

$$\underset{n \to \infty}{\text{Lt}} = P\{|t_n - \theta| > \in\} = 0 \tag{14.3.8}$$

If $E(t_n) \to \theta$ and Var $(t_n) \to 0$ as $n \to \mu$, then the statistic t_n will be a 'consistent estimator' of θ.

Consistency is a limiting property. Moreover, several consistent estimators may exist for the same parameter. For example, in sampling from a Normal population $N(\mu, \sigma^2)$, both the sample mean and the sample median are consistent estimators of μ.

Efficiency and Minimum Variance

In order to make a choice amongst consistent estimators, we have to introduce the idea of 'efficiency'. Of two consistent estimators for the same parameter, the statistic with the smaller sampling variance is said to be "*more efficient*". Thus if t and t' are both consistent estimators of θ, and

$$\text{Var}(t) < \text{Var}(t')$$

then t is 'more efficient' than t' in estimating θ; because it is grouped more closely around θ and will on the average deviate less from θ.

If a consistent estimator exists whose sampling variance is less than that of any other consistent estimator, it is said to be "most efficient"; and it provides a standard for the measurement of '*efficiency*' of a statistic. If V_0 be the variance of the most efficient estimator and V be the variance of any other estimator, then the efficiency of the estimator is defined as

$$\text{Efficiency} = \frac{V_0}{V} \tag{14.3.9}$$

Obviously, the measure of efficiency cannot exceed 1. In sampling from a Normal population $N(\mu, \sigma^2)$, both the sample mean and the sample median are consistent estimators of μ, but

$$\text{Var}(\bar{x}) = \frac{\sigma^2}{n}, \quad \text{Var}(\textit{median}) = \frac{\pi\sigma^2}{2n}$$

(for large n). Since Var ($\bar{x}$) is smaller than Var (median), mean is more efficient than median in estimating the parameter μ. It can be shown that the sample mean is the most efficient estimator. Hence

$$\text{Efficiency of median} = \frac{\dfrac{\sigma^2}{n}}{\dfrac{\pi\sigma^2}{2n}} = \frac{2}{\pi} = 0.64 \text{ approx.}$$

A statistic t which has the minimum variance among all estimators of θ is called the *Minimum Variance* (MV) estimator.

A statistic which is unbiased and has also the minimum variance (i.e., most efficient) is said to be the *Minimum Variance Unbiased Estimator* (MVUE). The variance of MVUE is often given by the R.H.S. of (14.3.10).

Theorem IV In sampling from a Normal population $N(\mu, \sigma^2)$ the sample mean $\bar{x}$ is the Minimum Variance Unbiased Estimator for the parameter μ.

Let $x_1, x_2, ... x_n$ be a random sample and

$$T = a_1x_1 + a_2x_2 + \dots + a_nx_n$$

where $a_1, a_2, ... a_n$ are constants. If the linear function T is an unbiased estimator of a parameter and also has the minimum variance, it is said to be the *Best Linear Unbiased Estimator* (BLUE).

Theorem V The sample mean $\bar{x}$ is the Best Linear Unbiased Estimator of the population mean μ.

Example 14.4 *If T_1, T_2, T_3 are independent, unbiased estimates of θ and all have the same variance, which of the following unbiased estimates of θ would you prefer? $(T_1 + 2T_2 + T_3)/4$, $(2T_1 + T_2 + 2T_2)/5$, $(T_1 + T_2 + T_3)/3$.* [W.B.H.S. '78]

Solution The last expression, viz. $(T_1 + T_2 + T_3)/3$ would be the most preferable; because it is the *mean* of T_1, T_2, T_3 and hence the Best Linear Unbiased Estimator, i.e. has the minimum variance among all linear functions which may be proposed as unbiased estimators of θ.

Theorem VI (*Cramer-Rao inequality*) If t is an unbiased estimator of a parameter θ, based on a random sample from a continuous population with p.d.f. $f(x, \theta)$, then

$$\text{Var}(t) > \frac{1}{n\,E\left\{\dfrac{\partial}{\partial\theta}\log f(x,\theta)\right\}^2} \qquad (14.3.10)$$

An estimator whose sampling variance attains the lower bound given by (14.3.10) is called the *Minimum Variance Bound* (MVB) *estimator.* For example, the mean ($\bar{x}$) of a random sample from a normal population is a MVB estimator of μ.

Sufficiency

A statistic is said to be a '*sufficient estimator*' of a parameter θ, if it contains all information in the sample about θ. If a statistic t exists such that the joint distribution of the sample is expressible as the product of two factors, one of which is the sampling distribution of t and contains θ, but the other factor is independent of θ, then t will be a sufficient estimator of θ.

Thus, if $x_1, x_2, \dots x_n$ is a random sample from a population whose p.m.f. or p.d.f. is $f(x, \theta)$, and t is a sufficient estimator of θ then we can write

$$f(x_1, \theta), f(x_2, \theta). \dots f(x_n, \theta) = g(t, \theta) \,.\, h(x_1, x_2, \dots x_n)$$

where $g(t, \theta)$ is the sampling distribution of t and contains θ, but $h(x_1, x_2, \dots x_n)$ is independent of θ. Since the parameter θ occurring in the joint distribution of all the sample observations can be contained in the distribution of the statistic t, it is said that t alone can provide all 'information' about θ and is therefore "sufficient" for θ.

Sufficient estimators are the most desirable kind of estimators, but unfortunately they exist in only relatively few cases. If a sufficient estimator exists, it can be found by the Method of Maximum Likelihood.

Theorem VII In random sampling from a Normal population $N(\mu, \sigma^2)$, the sample mean $\bar{x}$ is a sufficient estimator of μ.

14.4 METHODS OF POINT ESTIMATION

(1) Method of Maximum Likelihood

This is a convenient method for finding an estimator which satisfies most of the criteria discussed earlier. Let $x_1, x_2, \dots x_n$ be a random sample from a population with p.m.f. (for discrete case) or p.d.f. (for continuous case) $f(x, \theta)$, where θ is the parameter. Then the joint distribution of the sample observations viz.

$$L = f(x_1, \theta), f(x_2, \theta), \dots f(x_n, \theta) \qquad (14.4.1)$$

is called the *Likelihood Function* of the sample.

The *Method of Maximum Likelihood* consists in choosing as an estimator of θ that statistic, which when substituted for θ, maximises the likelihood function L. Such a statistic is called a *maximum likelihood estimator* (*m.l.e.*). We shall denote the *m.l.e.* of θ by the symbol θ_0.

Since log L is maximum when L is maximum, in practice the m.l.e. of θ is obtained by maximising log L. This is achieved by differentiating log L partially with respect to θ, and using the two relations

$$\left[\frac{\partial}{\partial\theta}\log L\right]_{\theta=\theta_0} = 0, \quad \left[\frac{\partial^2}{\partial\theta^2}\log L\right]_{\theta=\theta_0} < 0 \qquad (14.4.2)$$

Properties of maximum likelihood estimator (m.l.e.)

(1) The m.l.e. is consistent, most efficient, and also sufficient, provided a sufficient estimator exists.

(2) The m.l.e. is not necessarily unbiased. But when the m.l.e. is biased, by a slight modification, it can be converted into an unbiased estimator.

(3) The m.l.e. tends to be distributed normally for large samples.

(4) The m.l.e. is invariant under functional transformations. This means that if T is an m.l.e. of θ, and $g(\theta)$ is a function of θ, then $g(T)$ is the m.l.e. of $g(\theta)$.

Example 14.5 *On the basis of a random sample find the maximum likelihood estimator of the parameter of a Poisson distribution.*

Solution The Poisson distribution with parameter m has p.m.f.

$$f(x, m) = \frac{e^{-m} \cdot m^x}{x!} \quad (x = 0, 1, 2, \dots \infty)$$

The likelihood function of the sample observations is

$$L = f(x_1, m) \cdot f(x_2, m). \dots f(x_n, m)$$

$$\therefore \quad \log L = \log f(x_1, m) + \log f(x_2, m) + \dots + \log f(x_n, m)$$

$$= \sum_{i=1}^{n} \log f(x_i, m)$$

$$= \sum [-m + x_i (\log m) - \log x_i !]$$

$$= -nm + (\log m)\, \Sigma x_i - \Sigma \log (x_i !)$$

Taking partial derivative of $\log L$ with respect to the parameter m,

$$\frac{\partial \log L}{\partial m} = -n + \frac{\Sigma x_i}{m} = -n + \frac{n\bar{x}}{m}$$

Now replacing m by m_0 and equating the result to zero,

$$\left[\frac{\partial \log L}{\partial m}\right]_{m = m_0} = -n + \frac{n\bar{x}}{m_0} = 0$$

Solving, we get $m_0 = \bar{x}$. Again,

$$\left[\frac{\partial^2 \log L}{\partial m^2}\right]_{m = m_0} = -\frac{n\bar{x}}{m_0^2} = -\frac{n\bar{x}}{\bar{x}^2} = -\frac{n}{\bar{x}} \text{ which is negative.}$$

This shows that $\log L$ is maximum at $m = m_0 = \bar{x}$. That is the m.l.e. of m is $m_0 = \bar{x}$, the sample mean.

Example 14.6 *Find the maximum likelihood estimator of the variance σ^2 of a Normal population $N(\mu, \sigma^2)$, when the parameter μ is known. Show that this estimator is unbiased.*

Solution The p.d.f. of Normal distribution is

$$f(x, \mu, \sigma^2) = \frac{1}{\sigma\sqrt{2\pi}}\, e^{-(x-\mu)^2/2\sigma^2} \quad ; (-\infty < x < +\infty)$$

The logarithm (to the base e) of the likelihood function L is

$$\log L = \sum_{i=1}^{n} \log f(x_i, \mu, \sigma^2)$$

$$= \sum \left[-\log \sigma - \frac{1}{2} \log (2x) - \frac{(x_i - \mu)^2}{2\sigma^2} \right]$$

$$= -\frac{n}{2} \log \sigma^2 - \frac{n}{2} \log (2x) - \frac{\Sigma (x_i - \mu)^2}{2\sigma^2}$$

Differentiating partially with respect to σ^2,

$$\frac{\partial}{\partial(\sigma^2)} \log L = -\frac{n}{2\sigma^2} + \frac{\Sigma (x_i - \mu)^2}{2(\sigma^2)^2}$$

The m.l.e. of σ^2 is obtained by solving

$$-\frac{n}{2\sigma^2}+\frac{\Sigma(x_i-\mu)^2}{2\sigma_0^{\ 4}}=0$$

$$\therefore \qquad \sigma_0^{\ 2}=\sum_{i=1}^{n}\frac{(x_i-\mu)^2}{n}$$

It can be shown that $\left[\dfrac{\partial^2 \log L}{\partial(\sigma^2)^2}\right]_{\sigma^2=\sigma_0^2}=\dfrac{-n}{2\sigma_0^4}$

which is negative. Thus the maximum likelihood estimator of σ^2 is

$$\sigma_0^{\ 2}=\sum\frac{(x_i-\mu)^2}{n},\ (\mu \text{ known})$$

Again, since $x_1, x_2, \ldots x_n$ is a random sample and μ is the population mean, we have $E(x_i-\mu)^2=\sigma^2$. Therefore,

$$E(\sigma_0^2)=\sum_{i=1}^{n}\frac{E(x_i-\mu)^2}{n}=\frac{\Sigma\sigma^2}{n}=\sigma^2$$

Thus, $\sigma_0^{\ 2}$ is an unbiased estimator of σ^2.

Example 14.7 *Find the m.l.e. of the parameters μ and σ^2 in random samples from a $N(\mu, \sigma^2)$ population, when both the parameters are unknown.*

Solution As in the preceding example,

$$\log L=-\frac{n}{2}\log\sigma^2-n\log\sqrt{2\pi}-\frac{\Sigma(x_i-\mu)^2}{2\sigma^2}$$

$$\therefore \qquad \left[\frac{\partial \log L}{\partial\mu}\right]_{\mu=\mu_0}=\frac{-1}{2\sigma^2}\Sigma 2(x_i-\mu_0)(-1)=0$$

This gives $\Sigma(x_i-\mu_0)=0$; i.e., $\mu_0=\bar{x}$, the sample mean. The m.l.e. of the parameter μ is the sample mean $\bar{x}$. (Note that this estimator is *unbiased*).

Proceeding as in Example 14.6 we have $\sigma_0^2=\dfrac{\Sigma(x_i-\mu)^2}{n}$. But since the parameter μ is not known, it is replaced by the m.l.e. $\mu_2=\bar{x}$. The m.l.e. of σ^2 is now

$$\sigma_0^{\ 2}=\frac{\Sigma(x_i-\bar{x})^2}{n}=S^2$$

which is the sample variance. (Note that this estimator is *biased*).

Example 14.8 *A tossed a biased coin 50 times and got head 20 times, while B tossed it 90 times and got 40 heads. Find the maximum likelihood estimate of the probability of getting head when the coin is tossed.*

Solution Let P be the unknown probability of obtaining a head. Using binomial distribution,

Probability of 20 heads in 50 tosses $= {}^{50}C_{20}\,P^{20}\,(1-P)^{30}$

Probability of 40 heads in 90 tosses $= {}^{90}C_{40}\,P^{40}\,(1-P)^{50}$

The likelihood function is given by the product of these probabilities:

$$L = {}^{50}C_{20} \cdot {}^{90}C_{40}\, P^{30}\, (1 - P)^{30}$$

$$\therefore \quad \log L = \log({}^{50}C_{20}\, {}^{90}C_{40}) + 60 \log P + 80 \log (1 - P)$$

Hence,
$$\frac{\partial \log L}{\partial P} = \frac{60}{P} - \frac{80}{1 - P}$$

The maximum likelihood estimate P_0 is therefore obtained by solving

$$\frac{60}{P_0} - \frac{80}{1 - P_0} = 0.$$

This gives
$$P_0 = \frac{60}{140} = \frac{3}{7}.$$
Ans. 3/7

(2) Method of Moments

The *Method of Moments* consists in equating the first few moments of the population with the corresponding moments of the sample, i.e. setting

$$\mu_r' = m_r' \tag{14.4.3}$$

where $\mu_r' = E(x^r)$ and $m_r' = \dfrac{\Sigma x_i^r}{n}$. Since the parameters enter into the population moments, these relations when solved for the parameters give the estimates by the method of moments. Of course, this method is applicable only when the population moments exist. The method is generally applied for fitting theoretical distributions to observed data.

Example 14.9 *Estimate the parameter p of the binomial distribution by the method of moments (when n is known).*

Solution For the binomial distribution $\mu_1' = E(x) = np$. Also $m_1' = \bar{x}$. Setting $\mu_1' = m_1'$, we have $np = \bar{x}$. Thus

$$p = \frac{\bar{x}}{n}$$

i.e. the estimated value of p is given by the sample mean divided by the parameter n (known).

Example 14.10 *Find the estimates of μ and σ in the Normal population $N(\mu, \sigma^2)$ by the method of moments.*

Solution Equate the first two moments of the population and the sample, $\mu_1' = m_1'$ and $\mu_2' = m_2'$, i.e. $\mu_2 = m_2$. Thus

$$\mu = \bar{x} \text{ and } \sigma^2 = S^2, \text{ the sample variance.}$$

The parameters μ and σ are estimated by the sample mean $\bar{x}$ and the sample standard deviation S respectively.

14.5 INTERVAL ESTIMATION

In the theory of point estimation, developed earlier, any unknown parameter is estimated by a single quantity. Thus the sample mean ($\bar{x}$) is used to estimate the population mean (μ), and the sample proportion (p) is taken as an estimator of the population proportion (P). A single estimator of this kind, however good it may be, cannot be expected to coincide with the true value of the parameter, and may in some cases differ widely from it. In the theory of interval estimation, it is desired to find an interval which is expected to include the unknown parameter with a specified probability.

Let $x_1, x_2, \ldots x_n$ be a random sample from a population of a known mathematical form which involves an unknown parameter θ. We would try to find two functions t_1 and t_2 based on sample observations such that the probability of θ being included in the interval (t_1, t_2) has a given value, say c.

$$P(t_1 \leq \theta \leq t_2) = c \tag{14.5.1}$$

Such an interval, when it exists, is called a *Confidence Interval* for θ. The two quantities t_1 and t_2 which serve as the lower and upper limits of the interval are known as *Confidence Limits*. The probability (c) with which the confidence interval will include the true value of the parameter is known as *Confidence Coefficient* of the interval.

The significance of confidence limits is that if many independent random samples are drawn from the same population and the confidence interval is calculated from each sample, then the parameter will actually be included in the intervals in c proportion of cases in the long run. Thus the estimate of the parameter is stated as an interval with a specified degree of confidence.

The calculation of confidence limits is based on the knowledge of sampling distribution of an appropriate statistic. Suppose, we have a random sample of size n from a Normal population $N(\mu, \sigma^2)$, where the variance σ^2 is known. It is required to find 95% confidence limits for the unknown parameter μ. We know that the sample mean ($\bar{x}$) follows normal distribution with mean μ and variance σ^2/n, and so

$$z = \frac{\bar{x} - \mu}{\dfrac{\sigma}{\sqrt{n}}}$$

has a standard normal distribution (see 13.8.5, page 518). Since 95% of the area under the standard normal curve lies between the ordinates at $z = \pm\, 1.96$ (see 13.8.3), we have

$$P\left[-1.96 \leq \frac{\bar{x} - \mu}{\dfrac{\sigma}{\sqrt{n}}} \leq 1.96\right] = 0.95$$

i.e. in 95% of cases the following inequalities hold

$$-\,1.96 \leq \frac{\bar{x} - \mu}{\dfrac{\sigma}{\sqrt{n}}} \leq 1.96$$

Separating out μ we get

$$\bar{x} - 1.96\frac{\sigma}{\sqrt{n}} \leq \mu \leq \bar{x} + 1.96\frac{\sigma}{\sqrt{n}}$$

The interval $\left(\bar{x} - 1.96\frac{\sigma}{\sqrt{n}},\ \bar{x} + 1.96\frac{\sigma}{\sqrt{n}}\right)$ is known as the 95% confidence interval for μ, and the 95% confidence limits are

$$\bar{x} \pm 1.96\frac{\sigma}{\sqrt{n}}$$

Again, 99% of area under the standard normal curve lies between the ordinates at $z = \pm 2.58$, and 99.73% (i.e. almost whole) of the area lies between $z = \pm 3$. Hence proceeding exactly in the same manner, the 99% confidence limits for μ are

$$\bar{x} \pm 2.58\frac{\sigma}{\sqrt{n}}$$

and almost sure limits for μ are

$$\bar{x} \pm 3\frac{\sigma}{\sqrt{n}}$$

In fact, using values from the normal probability integral table (showing areas under standard normal curve), confidence limits corresponding to any specified percentage can be obtained. These are *exact* confidence limits.

In some cases, the population may not be truely a normal distribution, but the sampling distributions of statistics based on large samples are approximately normal. For example, the sample mean ($\bar{x}$) based on a large random sample drawn (with or without replacement) from any population is approximately normally distributed. Similarly, the sample proportion (p) calculated from a large random sample has approximately a normal distribution (see Section 13.6). It is therefore possible to utilise the properties relating to the percentage of area under the standard normal curve (page 518) to find approximate confidence limits for the population mean μ and the population proportion P, provided the sample size n is large.

Approximate Confidence Limits (large samples) any Distribution

(1) for Mean μ:

95% confidence limits $= \bar{x} \pm 1.96$ (S.E. of $\bar{x}$)

99% confidence limits $= \bar{x} \pm 2.58$ (S.E. of $\bar{x}$)

Almost sure limits $= \bar{x} \pm 3$ (S.E. of $\bar{x}$) (14.5.2)

(see Example 14.12)

(2) for Proportion P:

95% confidence limits $= p \pm 1.96$ (S.E. of p)

99% confidence limits $= p \pm 2.58$ (S.E. of p)

Almost sure limits $= p \pm 3$ (S.E. of p) (14.5.3)

(see Example 14.11 and 14.13)

(3) for Difference of Means $(\mu_1 - \mu_2)$:

95% confidence limits $= (\bar{x}_1 - \bar{x}_2) \pm 1.96$ (S.E. of $\bar{x}_1 - \bar{x}_2$)

99% confidence limits $= (\bar{x}_1 - \bar{x}_2) \pm 2.58$ (S.E. of $\bar{x}_1 - \bar{x}_2$)

Almost sure limits = $(\bar{x}_1 - \bar{x}_2 \pm 3$ (S.E. of $\bar{x}_1 - \bar{x}_2)$ (14.5.4)
(see Example 14.29)

(4) for Difference of Proportions $(P_1 - P_2)$:

95% confidence limits $= (p_1 - p_2) \pm 1.96$ (S.E. of $p_1 - p_2$)

99% confidence limits $= (p_1 - p_2) \pm 2.58$ (S.E. of $p_1 - p_2$)

Almost sure limits $= (p_1 - p_2) \pm 3$ (S.E. of $p_1 - p_2$) (14.5.5)

(see Example 14.23)

Note: (i) The *'probable limits'* (without any reference to the degree of confidence) may be taken to be *'almost sure limits'* in all the abvoe cases.

(ii) The formulae for S.E. involve population parameters. If these parameters are not known, an approximate value of S.E. may be obtained by substituting the statistic for the corresponding parameter.]

Example 14.11 *A sample of 6500 screws is taken form a large consignment and 75 are found to be defective. Estimate the percentage of defectives in the consignment and assign limits within which the percentage lies.* [I.C.W.A. (old), June '75]

Solution There are 75 defectives in a sample of size $n = 600$. Therefore, the sample proportion of defectives is

$$p = \frac{75}{600} = \frac{1}{8} = 12.5\%$$

This may be taken as an estimate of the percentage of defectives (P) in the whole consignment ('Point estimation').

The 'limits' to the percentage of defectives refer to the confidence limits, which may be given as $p \pm 3$(S.E. of p).

$$\text{S.E. of } p = \sqrt{\frac{PQ}{n}}$$

$$= \sqrt{\frac{pq}{n}} \text{ approximately;}$$

(since the population proportion P is not known).

$$= \sqrt{\frac{\frac{1}{8}\left(1 - \frac{1}{8}\right)}{600}} = \frac{1}{80}\sqrt{\frac{7}{6}} = .0135$$

Thus, the limits for P are

$$\frac{1}{8} \pm 3 \times .0135 = .125 \pm .0405$$

$$= .1655 \text{ and } .0845 = 16.55\% \text{ and } 8.45\%$$

The limits to the percentage of defectives in the consignment are 8.45% to 16.55% ('Interval estimation').

Ans. 12.5%; 8.45% to 16.55%.

Example 14.12 *A random sample of 100 ball bearings selected from a shipment of 2000 ball bearings has an average diameter of 0.354 inch with a S.D. = .048 inch. Find 95% confidence interval for the average diameter of these 2000 ball bearings.* [C.U., M.Com. '72; I.C.W.A., Dec '79]

Solution Theory: If a random sample of large size n is drawn *without replacement* from a *finite population* of size N, hen the 95% confidence limits for the population mean μ are $\overline{x} \pm$ 1.96 (S.E. of x), where $\overline{x}$ denotes the sample mean and

$$\text{S.E. of } \overline{x} = \frac{\sigma}{\sqrt{n}}\sqrt{\frac{N-n}{N-1}},$$

σ denoting the Standard Deviation (S.D.) of the population. Here,

Sample size (n) = 100, Population size (N) = 2000

Sample mean ($\overline{x}$) = 0.354, Sample S.D. (S) = .048

Since σ is not known, an approximate value of S.E. is obtained on replacing the population S.D. (σ) by the sample S.D. (S).

$$\text{S.E. of } \overline{x} = \frac{S}{\sqrt{n}}\sqrt{\frac{N-n}{N-1}} \text{ approximately}$$

$$= \frac{.048}{\sqrt{100}}\sqrt{\frac{2000-100}{2000-1}} = .0047$$

The 95% confidence limits for the population mean μ are

$$\overline{x} \pm 1.96 \text{ (S.E. of } \overline{x}) = 0.354 \pm 1.96 \times .0047$$
$$= 0.354 \pm .0092 = 0.3632 \text{ and } 0.3448$$

Thus, the 95% confidence interval is (0.3448 to 0.3632) inch.

Example 14.13 *A random sample of 100 articles taken from an large batch of articles contains 5 defective articles. (a) Set up 96 per cent confidence limits for the proportion of defective articles in the batch. (b) If the batch contains 2696 articles set up 95% confidence interval for the proportion of defective articles.*

[I.C.W.A., June '74]

Solution (a) The 96% confidence limits for the population proportion (P) are given by $p \pm$ 2.05 (S.E. of p), where p is the sample proportion.

$$\text{S.E. of } p = \sqrt{\frac{PQ}{n}}$$

Since the formula involves the unknown population proportion P, an approximate value of S.E. is obtained on replacing the population proportion (P) by the sample proportion (p). Putting $n = 100$ and $p = 5/100 = .05$, ($q = 1 - p = .95$)

$$\text{S.E. of } p = \sqrt{\frac{pq}{n}} = \sqrt{\frac{.05 \times .95}{100}} = .022$$

Hence, the 96% confidence limits for P are

$$p \pm 2.05 \text{ (S.E. of } p) = .05 \pm 2.05 \times .022 = .05 \pm .045$$
$$= .05 + .045 \text{ and } .05 - .045$$
$$= .095 \text{ and } .005$$

(b) The 95% confidence limits for proportion (P) are given by $p \pm 1.96$ (S.E. of p). But, when the population is of a finite size N,

$$\text{S.E. of } p = \sqrt{\frac{pq}{n}}\sqrt{\frac{N-n}{N-1}} \text{ (approximately)}$$

Here, $n = 100$, $N = 2696$, $p = .05$. Putting these values

$$\text{S.E. of } p = \sqrt{\frac{05 \times .95}{100}}\sqrt{\frac{2696-100}{2696-1}} = .022\sqrt{\frac{2596}{2695}}$$
$$= .022 \times .963 = .022 \times .98 = .0216 \text{ (approx.)}$$

Hence, the required 95% confidence limits for P are

$$p \pm 1.96(\text{S.E. of } p) = .05 \pm 1.96 \times .0216 = .092 \text{ and } .008$$

The 95% *confidence interval* for the proportion of defective articles is 008 to .092.

Ans. .005 and .095; .008 to .092

Example 14.14 *10 Life Insurance Polices in a sample of 200 taken out of 50,000 were found to be insured for less than Rs 5,000. How many policies can be reasonably expected to be insured for less than Rs 5,000 in the whole lot at 95%. Confidence level?* [C.A., Nov. '80]

Solution Let us first find the confidence limits for the 'proportion' of life insurance policies insured for less than Rs 5,000 in the whole lot. Here,

Population size (N) = 50,000,

Sample size (n) = 200

$$\text{Sample proportion } (p) = \frac{10}{200} = .05$$

Using (13.7.10),

$$\text{S.E. of } p = \sqrt{\frac{.05 \times .95}{200}}\sqrt{\frac{50{,}000-200}{50{,}000-1}}$$
$$= \sqrt{\frac{.05 \times .95 \times 49800}{200 \times 50{,}000}} \text{ (approx.)}$$
$$= .0154$$

95% confidence limits for the population proportion (P) are

$$p \pm 1.96 \text{ (S.E. of } p) = .05 \pm 1.96 \times .0154$$
$$= .05 \pm .030 = .080 \text{ and } .020$$

The means that out of the lot of 50,000, the 'proportion' of policies insured for less than Rs 5000 lies between .020 and .080, with probability 95%. Thus the 'number' of such policies lies between 50,000 × .020 = 1000 and 50,000 × .080 = 4,000.

Ans. Between 1,000 and 4,000.

Exact Confidence Limits (any sample size) Normal Distribution

In the following cases, it is assumed that samples are drawn at random from Normal populations.

(5) for Mean μ: (s.d. known)

$$\text{95\% confidence limits less} = \bar{x} \pm 1.96\left(\frac{\sigma}{\sqrt{n}}\right)$$

$$\text{99\% confidence limits} = \bar{x} \pm 2.58\left(\frac{\sigma}{\sqrt{n}}\right)$$

(see Examples 14.15 and 14.37) (14.5.6)

(6) for Difference of Means ($\mu_1 - \mu_2$): (s.d.s known)

$$95\% \text{ confidence limits} = (\bar{x}_1 - \bar{x}_2) \pm 1.96\sqrt{\frac{\sigma_1^2}{n_1} + \frac{\sigma_2^2}{n_2}}$$

$$99\% \text{ confidence limits} = (\bar{x}_1 - \bar{x}_2) \pm 2.58\sqrt{\frac{\sigma_1^2}{n_1} + \frac{\sigma_2^2}{n_2}} \qquad (14.5.7)$$

(see Example 14.39)

Example 14.15 *A random sample of size 10 was drawn from a normal population with an unknown mean and a variance of 44.1 (inch)^2. If the observations are (in inches): 65, 71, 80, 76, 78, 82, 68, 72, 65 and 81, obtain the 95% confidence interval for the population mean.*

Solution We are given $n = 10$, $\sigma^2 = 44.1$ and $\Sigma x_i = 738$.

$$\therefore \qquad \bar{x} = \frac{738}{10} = 73.8.$$

Since the population s.d. σ is known, using formula (14.5.6), 95% confidence limits for μ are given by

$$73.8 \pm 1.96\,\frac{\sqrt{44.1}}{\sqrt{10}} = 73.8 \pm 1.96 \times \sqrt{4.41}$$

$$= 73.8 \pm 1.96 \times 2.1$$

$$= 73.8 \pm 4.1 = 77.9 \text{ and } 69.7.$$

The 95% confidence interval for μ is therefore 69.7 to 77.9 inches.

(7) for Mean *m*: (s.d. unknown)

In random samples from a Normal population $N(\mu, \sigma^2)$

$$t = \frac{\bar{x} - \mu}{S/\sqrt{n-1}}$$

follows t distribution with $(n - 1)$ degree of freedom, where S is the sample s.d. (see 13.8.15). If $t_{.025}$ denotes the upper 2.5% point of t distribution with $(n - 1)$ d.f., then the 95% confidence interval for μ is obtained from

$$-t_{.025} \le \frac{\bar{x} - \mu}{S/\sqrt{n-1}} \le t_{.025}$$

Hence, for the population mean μ

$$95\% \text{ confidence limits} = \bar{x} \pm t_{.025}\left(\frac{S}{\sqrt{n-1}}\right) \qquad (14.5.8a)$$

$$99\% \text{ confidence limits} = \bar{x} \pm t_{.005}\left(\frac{S}{\sqrt{n-1}}\right) \qquad (14.5.8b)$$

(see Examples 14.16 and 14.40)

(8) for Difference of Means ($\mu_1 - \mu_2$): (common s.d. unknown)

Assuming that two independent samples are drawn from two Normal populations with means μ_1, μ_2 but a common *unknown* s.d. σ.

$$t = \frac{(\bar{x}_1 - \bar{x}_2) - (\mu_1 - \mu_2)}{s\sqrt{1}\dfrac{}{n_1} + \dfrac{1}{n_1}}$$

follows t distribution with $(n_1 + n_2 - 2)$ d.f. (see 13.8.16), where

$$s^2 = \frac{(n_1 S_1^2 + n_2 S_2^2}{(n_1 + n_2 - 2)}$$

Hence, with 95% probability the following inequalities hold

$$-t_{.025} \le \frac{(\bar{x}_1 - \bar{x}_2) - (\mu_1 - \mu_2)}{s\sqrt{\left(\dfrac{1}{n_1} + \dfrac{1}{n_2}\right)}} \le t_{.025}$$

from which the 95% confidence limits for $(\mu_1 - \mu_2)$ are

$$(\bar{x}_1 - \bar{x}_2) \pm t_{.025} \cdot s\sqrt{\left(\frac{1}{n_1} + \frac{1}{n_2}\right)} \qquad (14.5.9a)$$

Similarly the 99% confidence limits for $(\mu_1 - \mu_2)$ are

$$(\bar{x}_1 - \bar{x}_2) \pm t_{.005} \cdot s\sqrt{\left(\frac{1}{n_1} + \frac{1}{n_2}\right)} \qquad (14.5.9b)$$

(see Example 14.43)

Example 14.16 *A random sample of 10 students of class II was selected from schools in a certain region, and their weights recorded are shown below (in. lb.): 38, 46, 45, 40, 35, 39, 44, 45, 33, 37. Find 95% confidence limits within which the mean weight of all such students in the region is expected to lie. (Given $t_{.025} = 2.262$ for 9 d.f. and 2.228 for 10 d.f.).*

Solution From the given data, $\bar{x} = \dfrac{402}{10} = 40.2$. To calculate the s.d. (S), we take deviations from 40, i.e. $d = x - 40$.

$-2, 6, 5, 0, -5, \quad -1, 4, 5, -7, -3$

$\Sigma d = 2, \quad \Sigma d^2 = 190$

$$\therefore \qquad S^2 = \frac{\Sigma d^2}{n} - \left(\frac{\Sigma d}{n}\right)^2 = \frac{190}{10} - \left(\frac{2}{10}\right)^2 = 18.96$$

$$S = \sqrt{18.96} = 4.35.$$

Since the population s.d. σ is *unknown* the 95% confidence limits for μ are (see formula 14.5.8a)

$$40.2 \pm 2.262 \times 4.35/\sqrt{9} \quad \text{(degrees of freedom = 9)}$$

$$= 40.2 \pm 3.28 = 36.92 \text{ and } 43.48$$

The 95% confidence limits for the mean weight are (in lb.) 36.9 and 43.5.

(9) for Variance σ^2:

Case I

(*mean known*)—In random samples from $N(\mu, \sigma^2)$ population, $\Sigma(x_i - \mu)^2/\sigma^2$ follows chi-square distribution with n degrees of freedom (see 13.8.11, page 520). If $\chi^2_{.975}$ and $\chi^2_{.025}$ denote the lower and the upper 2.5% points of chi-square distribution with n d.f., then with probability 95% we have

$$\chi^2_{.975} \leq \sum \frac{(x_i - \mu)^2}{\sigma^2} \leq \chi^2_{.025}$$

From which the 95% confidence interval for σ^2 is

$$\frac{\Sigma(x_i - \mu)^2}{\chi^2_{.025}} \leq \sigma^2 \leq \frac{\Sigma(x_i - \mu)^2}{\chi^2_{.975}} \tag{14.5.10}$$

Case II

(*mean unknown*)—In this case (see 13.8.12) $\frac{nS^2}{\sigma^2} = \frac{\Sigma(x_i - \bar{x})^2}{\sigma^2}$ follows chi-square distribution with $(n - 1)$ degrees of freedom. Using the lower and the upper 2.5% points of chi-square distribution with $(n - 1)$ d.f., we have with probability 95% the following inequalities

$$\chi^2_{.975} \leq \frac{nS^2}{\sigma^2} \leq \chi^2_{.025}$$

from which the 95% confidence interval for σ^2 can be given as

$$\frac{nS^2}{\chi^2_{.025}} \leq \sigma^2 \leq \frac{nS^2}{\chi^2_{.975}} \tag{14.5.11}$$

Example 14.17 *The standard deviation of a random sample of size 12 drawn from a normal population is 5.5. Calculate the 95% confidence limits for the standard deviation (σ) in the population (Given $\chi^2_{.975} = 3.82$ and $\chi^2_{.025} = 21.92$ for 11 degrees of freedom).*

Solution Here $n = 12$ and the sample s.d. $(S) = 5.5$. Substituting the values in formula (14.5.11), the 95% confidence interval for σ^2 is

$$\frac{12 \times (5.5)^2}{21.92} \leq \sigma^2 \leq \frac{12 \times (5.5)^2}{3.82}$$

or, $$16.56 \leq \sigma^2 \leq 95.03$$

i.e., $$4.1 \leq \sigma \leq 9.7$$

The 95% confidence limits for the population s.d. (σ) are 4.1 and 9.7.

Example 14.18 *A sample of size 8 from a normal population yields as the ubiased estimate of population variance the value 4.4. Obtain the 99% confidence limits for the population variance σ^2 (Given $\chi^2_{.975} = 0.99$ and $\chi^2_{.005} = 20.3$ for 7 d.f.).*

Solution Here $n = 8$, and the unbiased estimate $s^2 = 4.4$. So, $nS^2 = (n - 1)s^2 = 7 \times 4.4 = 30.8$. Hence the 99% confidence limits for σ^2 are obtained from

$$\frac{nS^2}{\chi^2_{.005}} \le \sigma^2 \le \frac{nS^2}{\chi^2_{.995}}$$

or, $$\frac{30.8}{20.3} \le \sigma^2 \le \frac{30.8}{0.99}; \quad \text{i.e.,} \quad 1.52 \le \sigma^2 \le 31.1$$

(10) for Variance-Ratio $\frac{\sigma_1^2}{\sigma_2^2}$: (means unknown)

If two independent random samples of sizes n_1 and n_2 are drawn from two Normal populations with unknown means μ_1, μ_2 and variances σ_1^2, σ_2^2 respectively, then $\frac{s_1^2 / \sigma_1^2}{s_2^2 / \sigma_2^2}$ follows F distribution with degrees of freedom $(n_1 - 1, n_2 - 1)$. If $F_{.975}$ and $F_{.025}$ denote the lower and the upper 2.5% points of F distribution, we have with probability 95% the following inequalities

$$F_{.975} \le \frac{s_1^2 / \sigma_1^2}{s_2^2 / \sigma_2^2} \le F_{.025} \qquad (14.5.12)$$

The 95% confidence interval for σ_1^2/σ_2^2 can be obtained from this as

$$\frac{1}{F_{.025}} \cdot \frac{s_1^2}{s_2^2} \le \frac{\sigma_1^2}{\sigma_2^2} \le \frac{1}{F_{.975}} \cdot \frac{s_1^2}{s_2^2} \qquad (14.5.13)$$

where s_1^2 and s_2^2 denote unbiased estimators of σ_1^2, σ_2^2 respectively from the two samples. (see Example 14.51)

14.6 THEORY OF TEST OF SIGNIFICANCE

Statistical Hypothesis

In many practical problems, statisticians are called upon to make decisions about a statistical population on the basis of sample observations. For example, given a random sample, it may be required to decide whether the population, from which the sample has been obtained, is a normal distribution with mean = 40 and s.d. = 3. In attempting to reach such decisions, it is necessary to make certain assumptions or guesses about the characteristics of population, particularly about the probability distribution or the values of its parameters. Such an assumption or statement about the population is called a *Statistical Hypothesis.* The validity of a hypothesis will be tested by analysing the sample. The procedure which enables us to decide whether a certain hypothesis is true or not, is called *Test of Significance* or *Test of Hypothesis.*

Null Hypothesis and Alternative Hypothesis

In tests of significance, we start with a certain hypothesis about the population characteristics. This is called *Null Hypothesis,* and denoted by the symbol H_0. For example, the null hypothesis may be that the population mean is 40. We write

$$H_0 \ (\mu = 40)$$

Any hypothesis which differes from the null hypothesis is called *Alternative Hypothesis,* and is denoted by the symbol H_1. The null hypothesis is tested against an alternative hypothesis which in the above case, may be either that the population mean is not 40, or that it is greater than 40, or tat it is less than 40; i.e. any one of

$$H_1(\mu \neq 40), \quad H_1(\mu > 40), \quad H_1(\mu < 40)$$

The sample is then analysed to decide whether to 'reject' or not to reject the null hypothesis H_0. For this purpose, we choose a suitable statistic, called *"Test Statistic"* and find its sampling distribution, assuming that H_0 is really true. The 'observed value' of the statistic in the sample will in general be different from the 'expected value' because of sampling fluctuations. If the difference between them is large, the null hypothesis H_0 is rejected, and we doubt the validity of our assumption. If the difference is not large, H_0 is not rejected, and the difference may be considered to have arisen solely due to fluctuations of sampling. It is therefore necessary to decide how much of difference is tolerable before we are able to conclude that the null hypothesis is acceptable.

Level of Significance and Critical Region

The decision about rejection or otherwise of the null hypothesis is based on probability considerations. Assuming the null hypothesis to be true, we calculate the *probability of obtaining a differnce equal to or greater than the observed difference.* If this probability is found to be small, say less than .05, the conclusion is that the observed value of the statistic is rather unusual and has arisen because the underlying assumption (i.e. null hypothesis) is not true. We say that the observed difference is *significant at 5 per cent level,* and hence the 'null hypothesis is rejected' at 5 per cent *level of significance.* If, however, this probability is not very small, say more than .05, the observed difference cannot be considered to be unusual and is attributed to sampling fluctuations only. The difference is, now, said to be *not significant* at 5 per cent level, and we conclude that there is no reason to reject the null hypothesis' at 5 per cent level of significance. It has become customary to use 5% and 1% levels of significance, although othe levels, such as 2% or 0.5% may also be used.

Without actually going to calculate this probability, the test of significance may be simplified as follows. From the sampling distribution of the statistic, we find the maximum difference which is exceeded in (say (5) per cent of cases. If the observed difference is larger than this value, the null hypothesis is rejected. If it is less, there is no reason to reject the null hypothesis.

Suppose, the smapling distribution of the statistic is a normal distribution. Since the area under normal curve outside the ordinates at *mean ± 1.96 (s.d.)* is only 5%, the probability that the observed value of the statistic differs from the expected value of 1.96 times the S.E. or more is .05; and the probability of a larger difference will be stillo smaller. If, therefore.

$$z = \frac{(\text{Observed value}) - (\text{Expected value})}{\text{S.E.}} \tag{14.6.1}$$

is either greater than 1.96 or less than –1.96 (i.e. numerically greater than 1.96, the null hypothesis H_0 is rejected at 5% level of significance. The set of values $z \geq 1.96$ or ≤ -1.96, i.e.

$$|z| \geq 1.96$$

constitutes what is called the *Critical Region* for the test. Similarly since the area outside *mean* ± 2.58 (*s.d.*) is only 1%. H_0 is rejected at 1% level of significance, if z numerically exceeds 2.58, i.e. the critical region is $|z| \geq 2.58$ at 1% level.

Using the sampling distribution of an appropriate test statistic we are thus able to establish the maximum difference at a specified level between the observed and expected values that is consistent with the null hypothesis H_0. The set of values of the test statistic corresponding to this difference which lead to the acceptance of H_0 is called *Region of acceptance.* Conversely, the set of values of the test statistic leading to the rejection of H_0 is referred to as *Region of Rejection* or "*Critical Region*" of the test. The value of the statistic which lies at the boundary of the regions of acceptance and rejection is called *Critical Value.* When the null hypothesis is true, the probability of observed value ofthe test statistic falling in the critical region is often called the "*Size of Critical Region*".

Size of Critical Region $\leq$ *Level of Significance*

However, for a continuous population, the critical region is so determined that its Size equals the Level of Significance (α).

Two-tailed and One-tailed Tests

Our discussions above were centred round testing the significance of 'difference' between the observed and expected values, i.e. whether the observed value is *significantly different* from (i.e. either larger or smaller than) the expected value, as could arise due to fluctuations of random sampling. In the illustration, the null hypothesis is tested against "*both-sided alternatives*" ($\mu > 40$ or $\mu < 40$), i.e.

$$H_0\ (\mu = 40) \quad \text{against } H_1(\mu \neq 40)$$

Thus assuming H_0 to be true, we would be looking for large differences on both sides of the expected value, i.e. in "both tails" of the distribution. Such tests are, therefore, called "*two-tailed tests*".

Sometimes we are interested in tests for large differences on one side only i.e. in one 'one tail' of the distribution. For example, whether a new process of manufacture produces bricks with a 'higher' breaking strength, or whether a change in the production technique yields 'lower' percentage of defectives. These are known as "*one-tailed tests*".

For testing the null hypothesis against "*one-sided alternatives (right side)*" ($\mu > 40$), i.e.,

$$H_0\ (\mu = 40) \quad \text{against } H_1\ (\mu > 40)$$

the calculated value of the statistic z is compared with 1.645, since 5% of the area under the standard normal curve lies to the right of 1.645. If the observed value of z exceeds 1.645, the null hypothesis H_0 is rejected at 5% level of significance. If a 1% level were used, we would replace 1.645 by 2.33. Thus the critical regions for test at 5% and 1% levels are $z \geq 1.645$ and $z \geq 2.33$ respectively.

For testing the null hypothesis against "*one-sided alternatives (left side)*" ($\mu < 40$) i.e.

$$H_0\ (\mu = 40) \quad \text{against } H_1\ (\mu < 40)$$

the value of z is compared with -1.645 for significance at 5% level, and with -2.33 for significance at 1% level. The critical regions are now $z \leq -1.645$ and $z \leq -2.33$ for 5% and 1% levels respectively.

In fact, the sampling distributions of many of the commonly-used statistics can be approximated by normal distributions as the sample size increases, so that these rules are applicable in most cases when the sample size is 'large', say, more than 30.

It is evident that the same null hypothesis may be tested against alternative hypothesis of different types depending on the nature of the problem (see Table 14.1). Correspondingly, the type of test and the critical region associated with each test will also be different (Table 14.2).

Table 14.1 Formulation of Null Hypothesis and ALternative Hypothesis

Problem	*Null Hypothesis*	*Alternative Hypothesis*
Test whether $\mu = 40$, or		
(1) μ is *different* from 40	$H_0\ (\mu = 40)$	$H_1(\mu \neq 40)$
(2) μ is *more* than 40	$H_0\ (\mu = 40)$	$H_1(\mu > 40)$
(3) μ is *less* than 40	$H_0\ (\mu = 40)$	$H_1(\mu < 40)$

Table 14.2 Type of Test and Critical Region

ALternative Hypothesis	*Type of Alternative*	*Type of Test*	*Critical Region*
$H_1\ (\mu \neq 40)$	Both-sided	Two-tailed	Both Tails
$H_1\ (\mu > 40)$	One-sided	One-tailed	Right Tail
$H_1\ (\mu < 40)$	One-sided	One-tailed	Left Tail

Using z-statistic, the critical regions at α level of significance are shown in Fig. 14.1.

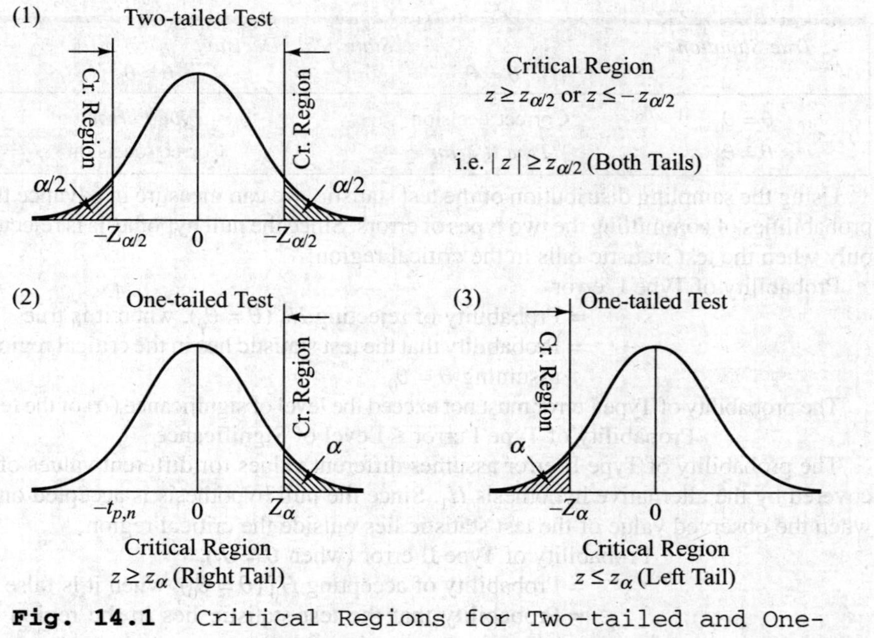

Fig. 14.1 Critical Regions for Two-tailed and One-tailed Tests

Type I and Type II Errors

The procedure of testing statistical hypothesis does not guarantee that all decisions are perfectly accurate. At times, the tests may lead to erroneous conclusions. This is so, because the decision is taken on the basis of sample values, which are themselves fluctuating and depend purely on chance.

The errors in statistical decisions are of two types:

(i) **Type I error** —This is the error committed by the test in.

rejecting a true null hypothersis.

(ii) **Type II error** —This is the error committed by the test in

accepting a false null hypothesis.

Considering the previous illustration for testing whether the population mean in 40, i.e. H_0 ($\mu = 40$), let us imagine that we have a random sample from a population whose mean is really 40. If we apply the test for H_0 ($\mu = 40$), we might find that the value of test statistic lies in the critical region, thereby leading to the conclusion that the population mean is not 40; i.e. the test rejects the null hypothesis although it is true. We have thus committed what is known as "*Type I error*" or " *Error of First Kind*".

On the other hand, suppose that we have a random sample from a population whose mean is known to be different from 40, say 43. If we apply the test for H_0 ($\mu = 40$), the value of the test statistic may, by chance, lie in the acceptance region, leading to the conclusion that the mean may be 40; i.e. the test does not reject the null hypothesis H_0 ($\mu = 40$), although it is false. This is again another form of incorrect decision, and the error thus committed is known as "*Type II error*" or "*Error of Second Kind*".

Table 14.3 Errors in Test of Significance H_0 ($\theta = \theta_0$)

True Situation	*Statistical Decision*	
	$\theta = \theta_0$	$\theta \neq \theta_0$
$\theta = \theta_0$	Correct decision	*Type I error*
$\theta \neq \theta_0$	*Type II error*	Correct decision

Using the sampling distribution of the test statistic, we can measure in advance the probabilities of committing the two types of errors. Since the null hypothesis is rejected only when the test statistic falls in the critical region,

Probability of Type I error

= Probability of rejecting H_0 ($\theta = \theta_0$), when it is true

= Probability that the test statistic lies in the critical region, assuming $\theta = \theta_0$.

The probability of Type I error must not exceed the level of significance (α) of the test.

Probability of Type I error ≤ Level of Significance

The probability of Type II error assumes different values for different values of θ covered by the alternative hypothesis H_1. Since the null hypothesis is accepted only when the observed value of the test statistic lies outside the critical region.

Probability of Type II error (when $\theta = \theta_1$),

= Probability of accepting H_0 ($\theta = \theta_0$), when it is false

= Probability that the test statistic lies in the region of acceptance, assuming $\theta = \theta_1$.

The probability of Type I error is necessary for constructing a test of significance. It is in fact the 'Size of the Critical Region'. The probability of Type II error is used to measure the "power" of the test in detecting falsity of the null hypothesis.

It is desirable that the test procedure be so framed which minimises both the types of error. But this is not possible, because for a given sample size an attempt to reduce one type of error is generally accompanied by an increase in the other type of error. The tests of significance are designed so as to limit the probability of Type I error to a specified value (usually 5% or 1%) and at the same time to minimise the probability of Type II error. Note that when the population has a continuous distribution,

Probability of Type I error

= Level of significance

= Size of critical region

Power of a Test

The null hypothesis H_0 ($\theta = \theta_0$) is accepted when the observed value of test statistic lies the critical region, as determined by the test procedure. Suppose that the true value of θ is not θ_0, but another value θ_1, i.e. a specified alternative hypothesis $H_1(\theta = \theta_1)$ is true. Type II error is committed if H_0 is not rejected, i.e. the test statistic lies outside the critical region. Hence the probability of Type II error is a function of θ_1, because now $\theta = \theta_1$ is assumed to be true.

If $\beta(\theta_1)$ denotes the probability of Type II error, when $\theta = \theta_1$ is true, the complementary probability $1-\beta(\theta_1)$ is called *Power* of the test against the specified alternative H_1 ($\theta = \theta_1$).

Power = 1 – Probability of Type II error

= Probability of rejecting H_0 when H_1 is true

Obviously, we would like a test to be as 'powerful' as possible for all critical regions of the same size. Treated as a function of θ, the expression $P(\theta) = 1 - \beta(\theta)$ is called *Power Function* of the test for θ_0 against θ. The curve obtained by plotting $P(\theta)$ against all possible values of θ, is known as *Power Curve.*

Example 14.19 *The fraction of defective items in a large lot is P. To test the null hypothesis H_0: $P = 0.2$, one considers the number f of defectives in a sample of 8 items and accepts the hypothesis if $f \leq 6$, and rejects the hypothesis otherwise. What is the probability of type I error of this test? What is the probability of type II error corresponding to $P = 0.1$?* [W.B.H.S., '79]

Solution We are going to test whether the fraction (P) of defectives in the lot is 0.2 or not.

Null hypothesis H_0 ($P = 0.2$). Alternative hypothesis $H_1(P \neq 0.2)$. The test procedure is as follows:

(1) Take a random sample of 8 items from the lot.

(2) Count the number of defectives in the sample.

(3) *Accept* H_0, if the number of defectives found in the sample is 6 or less ($f \leq 6$), and conclude that the fraction of defectives in the lot may be 0.2.

Reject H_0, if the number of defectives actually obtained in the sample is 7 or 8. Conclusion: The fraction of defectives in the lot is not 0.2; i.e. H_0 is not tenable.

It may be seen that the number of defectives (f) in the sample is a random variable, which follows Binomial distribution with parameters $n = 8$ and P. The probability of r defectives is ${}^8C_r P^r (1 - P)^{8-r}$.

Probability of Type I error

$= $ Probability of rejecting H_0, when H_0 is true.

$= $ Probability of 7 or 8 defectives, when $P = 0.2$

$= {}^8C_7\ (0.2)^7(0.8)^1 + (0.2)^8$

$= 0.0000,8448$

Probability of Type II error

$= $ Probability of accepting H_0, when a specified H_1 is true.

$= $ Probability of 6 or less defectives, when $P = 0.1$.

$= 1 - $ Probability of 7 or 8 defectives, when $P = 0.1$.

$= 1 - [{}^8C_7(0.1)^7(0.9)^1 + (0.1)^8]$

$= 0.9999,9927$

Definition of Useful Terms

(1) **Statistical Hypothesis.** Any statement or assertion about a statistical population or the values of its parameters is called a *Statistical Hypothesis.* There are two types of hypothesis—Simple and Composite.

(2) **Simple Hypothesis.** A statistical hypothesis which specifies the population completely (i.e. the probability distribution and all parameters are known) is called a *Simple Hypothesis.*

(3) **Composite Hypothesis.** A statistical hypothesis which does not specify the population completely (i.e. either the form of probability distribution or some parameters remain unknown) is called a *Composite Hypothesis.*

(4) **Test of Hypothesis (or Test of Significance).** A *Test of Hypothesis* is a procedure which specifies a set of "rules for decision" whether to 'accept' or 'reject' the hypothesis under consideration (i.e. null hypothesis).

(5) **Null Hypothesis.** A statistical hypothesis whichis set up (i.e. assumed) and whose validity is tested for possible rejection on the basis of sample observations is called a *Null Hypothesis.* It is denoted by H_0 and tested against alternatives. Tests of hypothesis deal with rejection or acceptance of null hypothesis only.

(6) **Alternative Hypothesis.** A statistical hypothesis which differs from the null hypothesis is called an *Alternative Hypothesis,* and is denoted by H_1. The alternative hypothesis is not tested, but its acceptance (rejection) depends on the rejection (acceptance) of the null hypothesis. Alternative hypothesis contradicts the null hypothesis. The choice of an appropriate critical region depends on the type of alternative hypothesis, viz. whether both-sided, one-sided (right/left) or specified alternative.

(7) **Test Statistic.** A function of sample observations (i.e. statistic) whose computed value determines the final decision regarding acceptance or rejection of H_0, is called a *Test Statistic.* The appropriate test statistic has to be chosen very carefully and a knowledge of its sampling distribution under H_0 (i.e. when the null hypothesis is true) is essential in framing the decision rules. If the value of the test statistic falls in the critical region, the null hypothesis is rejected.

(8) **Critical Region.** The set of values of the test statistic which lead to rejection of the null hypothesis is called *Critical Region* of the test. The probability with which a true null hypothesis is rejected by the test is often referred to as *"Size" of the Critical Region.* Geometrically, a sample $x_1, x_2, \dots x_n$, of size n is looked upon as just a point x, called *Sample point,* within the region of all possible samples, called the *Sample Space* (W). The critical region is then defined as a subset (w) of those sample points which lead to the rejection of the null hypothesis.

(9) **Level of Significance.** The maximum probability with which a true null hypothesis is rejected is known as *Level of Significance* of the test, and is denoted by α. In framing decision rules, the level of significance is arbitrarily chosen in advance depending on the consequences of a statistical decision. Customarily, 5% or 1% level of significance is taken, although other levels such as 2% or ½% is also used. The level of significance α is used to indicate the upper limit of the probability of committing Type I error, i.e. the size of the critical region.

(10) **Type I Error (or Error of First Kind).** This is the error committed in rejecting a null hypothesis by the test when it is really true. The critical region is so determined that the probability of Type I error does not exceed the level of significance of the test.

(11) **Type II Error (or Error of Second Kind).** This is the error committed in accepting a null hypothesis by the test when it is really false. The probability of Type II error depends on the specified value of the alternative hypothesis, and is used in evaluating the efficiency of a test.

(12) **Power.** The probability of rejecting a false null hypothesis is called *Power of the test.* Therefore, Power is the probability of drawing a correct conclusion by the test, when the null hypothesis is false. For a specified value of the parameter consistent with the alternative hypothesis,

Power = 1 – *Probability of Type II error.*

If the power is graphically plotted against all specified alternatives, the curve obtained is known as *Power Curve.* The power of a test determines the extent to which a test can discriminate between true and false null hypothesis.

Steps in Test of Significance

(1) *Set up the "Null Hypothesis" H_0 and the "Alternative Hypothesis" H_1* on the basis of the given problem. The null hypothesis usually specifies the values of some parameters involved in the population: H_0 ($\theta = \theta_0$). The alternative hypothesis may be any one of the following types: $H_1(\theta \neq \theta_1)$ $H_1(\theta > \theta_0)$, $H_1(\theta < \theta_0)$. The type of alternative hypothesis determines whether to use a two-tailed or one-tailed test (right or left tail).

(2) *State the appropriate "test statistic" T and also its sampling distribution, when the null hypothesis is true.* In large sample tests the statistic $z = (T - \theta_0)$/S.E. (T), which approximately follows Standard Normal distribution, is often used. In small sample tests, the population is assumed to be Normal and various test statistics are used which follow Standard Normal, Chi-square, t or F distribution exactly.

(3) *Select the "level of significance" α of the test,* if it is not specified in the given problem. This represents the maximum probability of committing a Type I error, i.e. of making a wrong decision by the test procedure when in fact the null hypothesis is true. Usually, a 5% or 1% level of significance is used (If nothing is mentioned, use 5% level).

(4) *Find the "critical region" of the test atthe chosen level of significance.* This represents the set of values of the test statistic which lead to rejection of the null hypothesis. The critical region always appears in one or both tails of the distribution, depending on whether the alternative hypothesis is one-sided or both-sided. The area in the tails (called 'size of the critical region') must be equal to the level of significance α. For a one-tailed test, α appears in one tail and for a two-tailed test $\alpha/2$ appears in each tail of the distribution. The critical region is

$$T \geq T_{\alpha/2} \text{ or } T \leq T_{1-\alpha/2} \quad \text{when } H_1 \ (\theta \neq \theta_0)$$
$$T \geq T_{\alpha} \quad \text{when } H_1 \ (\theta > \theta_0)$$
$$T \leq T_{1-\alpha} \quad \text{when } H_1 \ (\theta < \theta_0)$$

where T_α is the value of T such that the area to its right is α.

(5) *Compute the value of the test statistic T on the basis of sample data the null hypothesis.* In large sample tests, if some parameters remain unknown they should be estimated from the sample.

(6) *If the computed value of test statistic T lies in the critical region, "reject H_0"; otherwise "do not reject H_0".* The decision regarding rejection or otherwise of H_0 is made after a comparison of the computed value of T with the critical value (i.e. boundary value of the appropriate critical region).

(7) *Write the conclusion in plain non-technical language.* If H_0 is rejected, the interpretation is: "*the data are not consistent with the assumption that the null hypothesis is true and hence H_0 is not tenable*". If H_0 is not rejected, "*the data cannot provide any evidence against the null hypothesis and hence H_0 may be accepted to the true*". The conclusion should preferably be given in the words stated in the problem.

14.7 LARGE SAMPLE TESTS

The following test of significance are valid only for large sample sizes. These tests are also called "approximate tests", because the sampling distributions used are only approximately true, when the number of observations in the sample is large (say, more than 30). The accuracy however inreases with larger sample sizes.

(A) Using Normal Distribution

(1) Test for a specified proportion A random sample of size n shows that the proportion of members possessing a certain attribute (e.g. defectives) is p. It is required to test the hypothesis that the proportion P in the population has a specified value P_0.

$$H_0 \ (P = P_0)$$

If the sample size n is large, we use the test statistic

$$z = \frac{p - P_0}{(\text{S.E. of } p)} \tag{14.7.1}$$

which approximately follows standard normal distribution.

For testing significance at 5% level, the rules are as follows:

(i) If the alternative hypothesis is that the population proportion P is 'different' from P_0, reject H_0 when the value of z lies outside the range–1.96 to 1.96.

H_1 $(P \neq P_0)$; Critical Region $|z| \geq 1.96$

(ii) If the alternative hypothesis is that the population proportion P is 'greater' than P_0, reject H_0 when the value of z is greater than 1.645.

$H_1(P > P_0)$; Critical Region $z \geq 1.645$

(iii) If the alternative hypothesis is that the population proportion P is 'less' than P_0, reject H_0 when the value of z is less than – 1.645.

$H_1(P < P_0)$; Critical Region $z \leq -1.645$

Otherwise, do not reject the null hypothesis H_0.

For testing at 1% level, the values 1.96, 1.645 and –1.645, given above, should be replaced by 2.58, 2.33 and –2.33 respectively.

Table 14.4 Rejection Rules for $H_0(P = P_0)$

Alternative Hypothesis	*Critical Region*	
H_1	*5% level*	*1% level*
$P \neq P_0$	$\lvert z \rvert \geq 1.96$	$\lvert z \rvert \geq 2.58$
$P > P_0$	$z \geq 1.645$	$z \geq 2.33$
$P < P_0$	$z \leq 1.645$	$z \leq -2.33$

Confidence limits for the population proportion P are (see 14.5.2)

95% confidence limits = $p \pm 1.96$ (S.E. of p)

99% confidence limits = $p \pm 2.58$ (S.E. of p) (14.7.2)

where S.E. of $p = \sqrt{\frac{pq}{n}}$, if H_0 is rejected.

Example 14.20 *A dice was thrown 400 times and 'six' resulted 80 times. Do the data justify the hypothesis of an unbiased dice?* [I.C.W.A. June '77]

Solution Let us assume that the dice is unbiased, i.e. the null hypothesis is that the probability of obtaining a 'six' with the dice is $\frac{1}{6}$,

$$H_0\left(p = \frac{1}{6}\right)$$

Alternatively the dice is not ubiased, *i.e.* the alternative hypothesis

$$H_1\left(p \neq \frac{1}{6}\right)$$

Since 'six' occurred 80 times out of 400, the *observed value* of the proportion (p) of 'six' is

$$p = \frac{80}{400} = 0.2$$

On the assumption that H_0 is true (i.e. the dice is ubiased), the expected value of the proportion (P_0) of 'six' is

$$P_0 = \frac{1}{6} = 0.167$$

$$(\text{S.E. of } p) = \sqrt{\frac{\frac{1}{6} \times \frac{5}{6}}{400}} = \frac{\sqrt{5}}{120} = .0186$$

Using the proportion of 'six' in the sample, our test statistic is

$$z = \frac{\text{Observed value} - \text{Expected value}}{\text{Standard Error}}$$

$$= \frac{0.2 - 0.167}{.0186} = 1.77$$

When H_0 is true, the statistic z follows Standard Normal distribution. Since the value of z does not fall in the Critical Region (critical region at 5% level $|z| \geq 1.96$), it is not significant at 5% level. We have, therefore, no reject the null hypothesis, and conclude that the dice may be unbiased.

Example 14.21 *In a big city 325 men out of 600 were found to be smokers. Does this information support the conclusion that the majority of men in this city are smokers? (State the hypothesis clearly).* [I.C.W.A., June '82]

Solution Null Hypothesis is that the proportion of smokers in the whole city is 50%, i.e. $\frac{50}{100} = 0.5$.

$$H_0\ (P = 0.5)$$

We are interested to see if the proportion of smokers is *more than* 50%; i.e. Alternative Hypothesis is

$$H_1(P > 0.5)$$

The proportion of smokers in the sample is 325 out of $n = 600$. Using sample proportion,

$$\text{Observed Value } (p) = \frac{325}{600} = 0.542$$

If the null hypothesis H_0 is true,

$$\text{Expected Value } (P_0) = 0.5$$

and

$$\text{S.E. of } p = \sqrt{\frac{0.5(1-0.5)}{600}} = 0.0204$$

The test statistic is

$$z = \frac{\text{Observed value} - \text{Expected value}}{\text{S.E.}} = \frac{0.542 - 0.5}{0.0204} = 2.1$$

Since the alternative hypothesis $H_1(P > 0.5)$ is one-sided, the Critical Region of the test is one-tailed. At 5% level of significance

Critical Region is $z \geq 1.645$

(Note that the tail-area of Standard Normal turve for $z \geq 1.645$ is 5%).

The value of test statistic z, viz. 2.1, lies in the critical region, and hence is "significant". We therefore *reject* the null hypothesis at 5% level of significance and conclude that the data support the hypothesis that majority of men in the city are smokers.

Example 14.22 *A manufacturer claimed that at least 90% of the components which he supplied, conformed to specifications. A random sample of 200 components showed that only 164 were upto the standard. Test his claim at 1% level of significance.*

Solution Null hypothesis is that the proportion of components conforming to specifications is 90%, i.e. 0.90

$$H_0\ (P = 0.9)$$

The alternative hypothesis isthat it is less than 90%, $H_1\ (P < 0.9)$. The proportion of articles conformingto specifications, i.e.

$$\text{Observed value} = \frac{164}{200} = 0.82$$

Assuming that H_0 is true,

$$\text{Expected value} = 0.9$$

$$\text{Standard Error} = \sqrt{\frac{0.9 \times 0.1}{200}} = .0212$$

The value of the test statistic is

$$z = \frac{0.82 - 0.9}{.0212} = -3.77$$

Since the value of z, viz. –3.77, is less than –2.33 (critical region at 1% level is $z \le -2.33$), the null hypothesis is rejected. The test therefore reveals that the manufacturer's claim is not justified.

[**Note:** (1) Examples 14.20 to 14.22 all relate to "Test for a specified proportion", but the tests are different depending on the type of alternative hypothesis. Example 14.20 shows a "two tailed" test, Example 14.21 a "one-tailed" (right tail) test and Example 14.22 a "one-tailed" (left tail) test.

(2) As far as possible, the Examples below illustrate various types of tests—two tailed tests, followed by the two kinds of one-tailed tests.]

(2) Test for equality of two proportions Let p_1 and p_2 be the proportions in two large random samples of sizes n_1 and n_2 drawn respectively from two populations. We are interested to test whether the proportions in the two populations are equal or not, i.e. whether the difference $(p_1 - p_2)$ as observed in the samples has arisen only due to fluctuations of sampling.

$$H_0(P_1 = P_2) \quad \text{against } H_1(P_1 \neq P_2)$$

If H_0 is true, i.e. the two population proportions are equal, say P, the estimate of P is

$$p = \frac{n_1 p_1 + n_2 p_2}{n_1 + n_2} \tag{14.7.3}$$

The expected value of the difference $(p_1 - p_2)$ is then 0, and

$$\text{S.E. of } p_1 - p_2 = \sqrt{pq\left(\frac{1}{n_1} + \frac{1}{n_2}\right)}, \quad \text{where } p + q = 1$$

If the sample sizes n_1 and n_2 are large the test statistic

$$z = \frac{p_1 - p_2}{\text{S.E.}} \quad (14.7.4)$$

approximately follows standard normal distribution. The critical regions are

$$|z| \geq 1.96 \text{ at 5\% level of significance}$$
$$|z| \geq 2.58 \text{ at 1\% level of significance}$$

Confidence limits for $P_1 - P_2$ are (see 14.5.5)

$$95\% \text{ confidence limits} = (p_1 - p_2) \pm 1.96 \text{ (S.E.)}$$
$$99\% \text{ confidence limits} = (p_1 - p_2) \pm 2.58 \text{ (S.E.)} \quad (14.7.5)$$

where $\text{S.E.} = \sqrt{\frac{p_1q_1}{n_1} + \frac{p_2q_2}{n_2}}$, if H_0 is rejected.

Example 14.23 *In a large city A, 20 per cent of a random sample of 900 school children had defective eye-sight. In another large city B, 5 per cent of a random sample of 1600 children had the same defect. Is this difference between the two proportions significant? Obtain 95% confidence limits for the differnece in the population proportions.*

Solution Null hypothesis is that the population proportions are equal.

$$H_0\ (P_1 = P_2)$$

Alternative hypothesis is that they are not equal, $H_1(P_1 \neq P_2)$

$$n_1 = 900 \quad p_1 = 20\% = 0.20$$
$$n_2 = 1600,\ p_2 = 15\% = 0.15$$

Under H_0, the estimate of the equal but unknown proportion P is

$$p = \frac{900 \times 0.20 + 1600 \times 0.15}{900 + 1600} = 0.168$$

$$\text{S.E. of } (p_1 - p_2) = \sqrt{0.168 \times 0.832\left(\frac{1}{900} + \frac{1}{1600}\right)}$$

$$= \sqrt{.0002427} = .0156$$

$$\therefore \quad z = \frac{0.20 - 0.15}{.0156} = 3.21$$

Since the value of z falls in the critical region $|z| \geq 2.58$, it is significant at 1% level. We therefore reject the null hypothesis H_0 and conclude that the difference between the two proportions is significant; i.e. P_1 and P_2 are not equal.

[**Note:** This is a *two-tailed* test.]

The 95% confidence limits for the difference $P_1 - P_2$ are

$$(p_1 - p_2) \pm 1.96 \text{ (S.E. of } p_1 - p_2)$$

where S.E. of $(p_1 - p_2)$ is given by (13.7.4) at page 160. But since P_1 and P_2 are not known, and the significance test has shwon that $P_1 \neq P_2$, they are estimated by the sample proportions $p_1 = 0.20$ and $p_2 = 0.15$ respectively.

$$\text{S.E. of } p_1 - p_2 = \sqrt{\frac{P_1Q_1}{n_1} + \frac{P_2Q_2}{n_2}}$$

$$= \sqrt{\frac{p_1q_1}{n_1} + \frac{p_2q_2}{n_2}} \text{ approximately}$$

$$= \sqrt{\frac{0.20 \times 0.80}{900} + \frac{0.15 \times 0.85}{1600}} = .016$$

Hence, the 95% confidence limits for $P_1 - P_2$ are
(0.20 – 0.156) ± 1.96 (.016) = .05 ± .031 = .019 and .081

Example 14.24 *A machine produced 20 defective articles in a batch of 400. After overhauling it produced 10 defectives in a batch of 300. Has the machine imporved?* [C.A., Nov. '81]

Solution Null Hypothesis is that the proportions of defectives before and after overhauling are equal. Alternative Hypothesis is that the proportion of defectives has decreased after overhauling.

$$H_0\ (P_1 = P_2) \quad \text{against} \quad H_1\ (P_1 > P_2)$$

We have

Sample size	*Sample proportion*
Sample 1: $n_1 = 400$	$p_1 = \frac{20}{400} = \frac{1}{20}$
Sample 2: $n_2 = 300$	$p_2 = \frac{10}{300} = \frac{1}{30}$

Assuming that the null hypothesis is true, the estimate of the common proportion in the population is

$$p = \frac{20+10}{400+300} = \frac{3}{70}$$

Taking the difference of proportions $(p_1 - p_2)$,

$$\text{Observed value} = \frac{1}{20} - \frac{1}{30} = .0167$$

Expected value = 0 (the two proportions assumed equal)

$$\text{S.E. of } (p_1 - p_2) = \sqrt{\frac{3}{70} \times \frac{67}{70}\left(\frac{1}{400} + \frac{1}{900}\right)} = .0155$$

Using the test statistic (14.7.4),

$$z = \frac{\text{Observed value} - \text{Expected value}}{\text{S.E.}} = \frac{.0167}{.0155} = 1.08$$

Here the alternative hypothesis H_1 $(P_1 > P_2)$ is one-sided, and the critical region is therefore given by one tail of the Standard Normal curve. At 5% level, the Critical Region is $z > 1.645$.

Since the observed value of z, viz. 1.08 does not fall in the critical region (i.e. is not greater than 1.645), it is "not significant". We have therefore *no reason to reject* the null hypothesis and conclude that the two population proportrions may be equal; i.e. the machine has not imporved after overhauling.

[**Note:** This is a *one-tailed* (right tail) test. The next one shows a *one-tailed* (left tail) test.]

Example 14.25 *In a certain city 100 men in a sample of 400 were found to be smokers. In another city, the number of smokers was 300 in a random sample of 800. Does this indicate that there is a greater proportion of smokers in the second city than in the first?*

Solution Null hypothesis $H_0\ (P_1 = P_2)$

Alternative hypothesis $H_1(P_1 < P_2)$

$$n_1 = 400, \quad p_1 = \frac{100}{400} = 0.25$$

$$n_2 = 800, \quad p_2 = \frac{300}{800} = 0.375$$

Under H_0 the estimate of the equal but unknown proportion is

$$p = \frac{100+300}{400+800} = \frac{1}{3}$$

Expected value of $(p_1 - p_2)$ is 0, and

$$\text{S.E. of } p_1 - p_2 = \sqrt{\frac{1}{3} \times \frac{2}{3}\left(\frac{1}{400} + \frac{1}{900}\right)} = .029$$

$$\therefore \quad z = \frac{0.25 - 0.375}{.029} = -4.31$$

Here the alternative hypothesis $H_1(P_1 < P_2)$ is one sided, and for this one-tailed test the critical regions are

$$z \leq -1.645 \text{ at 5\% level of significance}$$

$$z \leq -2.33 \text{ at 1\% level of significance}$$

Since the observed value of z is less than -2.33, it is significant at 1% level. We reject the null hypothesis and conclude that the proportions of smokers is greater in the second city than in the first.

[**Note:** Which Level of Significance to use—5% or 1%?

(1) The critical region at 1% level is always included in the critical region at 5% level (i.e. the former is a *subset* of the latter). Therefore, if an observed value of the test statistic is found to be 'significcant' at 1% level, it is no doubt significant at 5% level also. On the other hand, if the observed value is 'not significant' at 5% level, it is obviously not significant at 1% level.

(2) The following general rules may be followed in the choice of level of significance of a test: Consider the critical regions both at 5% and 1% levels. If the observed value of the test statistic is

(a) *not significant at 5% level,* mention that only. It is unnecessary to consider significance at 1% level. See Examples 14.20, 24, 26, 40, etc.

(b) *significant at 5% level, but not significant of 1% level,* mention only that it is significant at 5% level. Do not mention that the observed value is not significant at 1% level. See Examples 14.21 and 42.

(c) *significant at 1% level,* mention that only. It is useless to mention significance at 5% level. See Examples 14.23, 25, 27 to 31, etc.

If the level of significance is mentioned in the problem, there is no choice, but to use it. See Examples 14.22, 32, 33, 36, 39, 41, 43 to 45, etc.

(3) An observed value of the test statistic that is significant at 5% level is said to be simply "significant"; if significant at 1% level, it is said to be "highly significant".]

(3) Test for a specified mean (large sample) A random sample of large size n gives a sample mean $\bar{x}$. It is required to test the hypothesis that the population mean μ has a specified value μ_0.

$$H_0\ (\mu = \mu_0)$$

For large n, the sampling distribution of sample mean ($\bar{x}$) is approximately normal (see 13.6.1). Therefore, when H_0 is true, the test statistic

$$z = \frac{\bar{x} - \mu_0}{\text{S.E. of } \bar{x}} \tag{14.7.6}$$

approximately follows standard normal distribution. The critical region of the test, which depends on the nature of the alternative hypothesis, is given in the table below.

Table 14.5 Rejection Rules for H_0 $(\mu = \mu_0)$

Alternative Hypothesis	*Critical Region*	
H_1	*5% level*	*1% level*
$\mu \neq \mu_0$	$\lvert z \rvert \geq 1.96$	$\lvert z \rvert \geq 2.58$
$\mu > \mu_0$	$z \geq 1.645$	$z \geq 2.33$
$\mu < \mu_0$	$z \leq -1.645$	$z \leq -2.33$

Confidence limits for the population proportion μ are (see 14.5.2)

$$95\% \text{ confidence limits} = \bar{x} \pm 1.96 \text{ (S.E. of } \bar{x})$$

$$99\% \text{ confidence limits} = \bar{x} \pm 2.58 \text{ (S.E. of } \bar{x}) \tag{14.7.7.}$$

where S.E. of $\bar{x} = \dfrac{\sigma}{\sqrt{n}}$. If the population s.d. σ is not known, it is replaced by the sample s.d. S.

(4) Test for equality of two means (large samples) Suppose we have two *independent* random samples of large sizes n_1 and n_2 from two populations, and the sample means are $\bar{x}_1$ and $\bar{x}_2$ respectively. It is required to test whether the two populations have the same means.

The standard error of the difference of means ($\bar{x}_1 - \bar{x}_2$) is

$$\text{S.E.} = \sqrt{\frac{\sigma_1^2}{n_1} + \frac{\sigma_2^2}{n_2}} \tag{14.7.8}$$

where σ_1 and σ_2 are the standard deviations in the two populations. In order to test the null hypothesis that the means are equal

$$H_0(\mu_1 = \mu_2)$$

we use the test statistic

$$\text{S.E.} = \frac{\bar{x}_1 - \bar{x}_2}{\text{S.E.}} \tag{14.7.9}$$

which approximately follows standard normal distibution. For testing equality of means against H_1 ($\mu_1 \neq \mu_2$) the critical regions are:

$$|z| \geq 1.96 \text{ at 5\% level of significance}$$
$$|z| \geq 2.58 \text{ at 1\% level of significance}$$

Confidence limits for $\mu_1 - \mu_2$ are (see 4.5.4)

$$95\% \text{ confidence limits} = (\bar{x}_1 - \bar{x}_2) \pm 1.96 \text{ (S.E.)}$$
$$99\% \text{ confidence limits} = (\bar{x}_1 - \bar{x}_2) \pm 2.58 \text{ (S.E.)} \qquad (14.7.10.)$$

where S.E. is given by (14.7.8).

Example 14.26 *A sample of 400 male students is found to have a mean height of 171.38 cms. Can it be reasonably regarded as a sample from a large population with mean height (17.17 cms and s.d. 3.30 cms).* [I.C.W.A., June '80]

Solution We assume that the sample really comes from a large population with mean 171.17 and s.d. 3.30 (cms). Null Hypothesis is

$$H_0 \ (\mu = 171.17, \sigma = 3.30)$$

Alternative Hypothesis is that the sample does not come from such a population.

$$H_1(\mu \neq 171.17)$$

Since the sample size $n = 400$ is large, the sample mean ($\bar{x}$) is approximately normally distributed.

$$\text{Observed value } (\bar{x}) = 171.38$$
$$\text{Expected value } (\mu_0) = 171.17$$
$$\text{S.E. of } \bar{x} = \frac{\sigma}{\sqrt{n}} = \frac{3.30}{\sqrt{400}} = 0.165$$

When H_0 is true, the z-statistic (14.7.6) follows N (0, 1).

$$z = \frac{\text{Observed value} - \text{Expected value}}{\text{S.E.}} = \frac{171.38 - 171.17}{0.165} = 1.27$$

Since the alternative hypothesis is both sided (i.e. μ is either more than or less than 171.17), trhe critical region of the test is also two tailed (i.e. given by two tails of Standard Normal curve). At 5% level

Critical Region is $|z| \geq 1.96$

The value of test statistic z(viz. 1.27) does not fall in the critical region and hence is "not significant". We have therefore *no reason to reject* the null hypothesis at 5% level of significance and conclude that the sample may be regarded as having arisen from the given population.

Example 14.27 *An automatic machine was designed to pack exactly 2.0 kg of Vanaspati. A sample of 100 tins was examined to test the machine. The average weight was found to be 1.94 kg with Standard Deviation 0.10 kg. Is the machine working properly?* [C.A., May '68]

Solution Given Sampel size (n) = 100

Sample mean ($\bar{x}$) = 1.94 kg

Sample s.d. (S) = 0.10 kg

it is required to test the hypothesis that the population mean is 2.0 kg

Null hypothesis: $H_0(\mu = 2.0 \text{ kg})$

Alternative hypothesis: $H_1(\mu \neq 2.0 \text{ kg})$

Since the sample size (n) is large, the sample mean (x) is approximately normally distributed with mean μ and S.E. = $\sigma/\sqrt{n}$. However, since the population s.d. σ is not known, an aproximate value of S.E. is

$$\text{S.E.} = \frac{S}{\sqrt{n}} = \frac{0.10}{\sqrt{100}} = 0.01$$

Therefore, $$z = \frac{1.94 - 2.0}{0.01} = -6$$

Since | z | = 6 exceeds 2.58., we reject the null hypothesis at 1% level of significance and conclude that the machine is not functioning properly.

[**Note:** Examples 14.26 and 27 are *two-tailed tests* for specified mean, the former with σ *known,* and the latter with σ *unknown*. Example 14.28 is a *one-tailed test* (σ *unknown).*]

Example 14.28 *A machine part was designed to withstand an average pressure of 120 units. A random sample of size 100 from a large batch was tested and it was found that the average pressure which these parts can withstand is 105 units with a standard deviation of 20 units. Test whether the batch meets the specifications.* [C.U., M.Com. '72, '74]

Solution Null hypothesis: Mean pressure in the population which a machine part can withstand is 120 units. Alternnative hypothesis: Mean pressure is smaller than 120 units. $H_0(\mu = 120)$ against $H_1(\mu < 120)$.

$$\text{S.E. of } x = \frac{S}{\sqrt{n}} \text{ approximately, (since } \sigma \text{ is not known)}$$

$$= \frac{20}{100} = 2.$$

Here, sample mean is $\bar{x}$ =105. Therefore,

$$z = \frac{105 - 120}{2} = -7.5$$

Since the valuje of $z = -7.5$ is smaller than -2.33 (critical region at 1% level for the one-tailed test is $z \leq -2.33$), we reject the null hypothesis at 1% level of significance and conclude that the average breaking strength in the population is less than 120 units, i.e. that the batch of machine parts does not meet the specification. (If the value is found to be significant at 1% level, we have stronger grounds for rejection of H_0 than at 5% level).

Example 14.29 *In a certain factory there are two different processes of manufacturing the same item. The average weight in a sample of 250 items produced from one process is found to be 120 grammes with a S.D. of 12 grammes; the corresponding figures in a sample of 400 items from the other process are 124 and 4. Compute the Standard Error of difference between the two sample means. Is this difference significant? Also find 99% confidence limits for the differenced in the average weights of items produced by the two processes.* [I.C.W.A., Dec. '78]

Solution Null hypothesis is that the average weights of items produced from the two processes are equal. Alternative hypothesis is that the averages are not equal. H_0 ($\mu = \mu_2$) against $H_1(\mu_1 \neq \mu_2)$.

Sample sizes	$n_1 = 250$	$n_2 = 400$
Sample means	$\bar{x}_1 = 120$ gms	$\bar{x}_2 = 124$ gms
Sample S.D.s	$S_1 = 12$ gms	$S_2 = 14$ gms

Since the population S.D.s are not known, we use the sample S.D.s to calculate the Standard Error of the difference ($\bar{x}_1 - \bar{x}_2$) between the two sample means (see 14.7.8)

$$\text{S.E. of } \bar{x}_1 - \bar{x}_2 = \sqrt{\frac{S_1^2}{n_1} + \frac{S_2^2}{n_2}}$$

$$= \sqrt{\frac{12^2}{250} + \frac{14^2}{400}} = 1.03$$

The observed value of the test statistic is

$$z = \frac{\bar{x}_1 - \bar{x}_2}{\text{S.E.}} = \frac{120 - 124}{1.03} = -3.9$$

Critical region for both sided alternatives $H_1(\mu_1 \neq \mu_2)$ is $|z| \geq 1.96$ at 5% level and $|z| \geq 2.58$ at 1% level of significance. Since $|z| = 3.9$ exceeds 2.58, we reject H_0 at 1% level of significance, and conclude that the difference between two means is significant; i.e. $\mu_1 \neq \mu_2$.

99% confidence limits for $\mu_1 - \mu_2$ are(see 14.7.10).

$$(\bar{x}_1 - \bar{x}_2) \pm 2.58 \text{ (S.E.)} = (120 - 124) \pm 2.58(1.03)$$

$$= -4 \pm 2.66 = -1.34 \text{ and } -6.66$$

This means that confidence limits for the difference $\mu_2 - \mu_1$ are 1.34 and 6.66 gms.

Example 14.30 *The means of two large samples of sizes 1000 and 2000 are 67.5 and 68.0 respectively. Test the equality of means of the two populations each with s.d. 2.5. (No credit if the null and alternative hypothesis are not stated. Assumptions should be stated clearly).* [I.C.W.A., June '82]

Solution Null Hypothesis is that the means of the two populations are equal, and Alternative Hypothesis is that they are not equal.

$$H_0(\mu_1 = \mu_2) \text{ against } H_1(\mu_1 \neq \mu_2)$$

Since the population sizes are large, the appropriate test statistic is

$$z = \frac{\bar{x}_1 - \bar{x}_2}{\text{S.E. of } (\bar{x}_1 - \bar{x}_2)}$$

which follows Standard Normal distribution, when H_0 is true. Given

$$n_1 = 1000 \qquad n_2 = 2000$$

$$\bar{x}_1 = 67.5 \qquad \bar{x}_2 = 68.0$$

Population s.d. (σ) = 2.5 (common for both)

Using (13.7.3a),

$$\text{S.E. of } (\bar{x}_1 - \bar{x}_2 = 2.5\sqrt{\frac{1}{1000} + \frac{1}{2000}}$$

$$= 2.5\sqrt{.0015} = 0.097$$

$$\text{Observed value of } z = \frac{(67.5 - 68.0)}{0.097} = -5.2$$

The alternative hypothesis is both-sided, and so the Critical Region of the test is given by both tails of Standard Normal curve. At 1% level

Critical Region : $|z| \geq 2.58$

In the present case, $|z| = |-5.2| = 5.2$ is actually greaterthan 2.58. So, we *reject* the null hypothesis H_0 at 1% level of significance and conclude that the means of the two populations are *not equal*.

[**Note:** Examples 14.29 to 31 are all *two-tailed tests* for equality of two means. The two population s.d.s are (i) *unknown* in Example 14.29; (ii) *known* and *equal*, but sample s.d.s are not given in Example 14.30; and (iii) *known* and *equal,* but the sample s.d.s are also given in Example 14.31.]

Example 14.31 *The mean yield of wheat from a district A was 210 lbs. with S.D. = 10 lbs. per acre from a sample of 100 plots. In another district B, the mean yield was 220 lbs. with S.D. = 12 lbs. from a sample of 150 plots. Assuming that the standard deviation of yield in the entire state was 11 lbs., test whether there is any significant difference between the mean yield of crops in the two districts.* [C.A., May '76]

Solution Let μ_1 and μ_2 denote the mean yields of crops in districts A and B respectively. H_0 $(\mu_1 = \mu_2)$ against $H_1(\mu_1 \neq \mu_2)$

Given	$n_1 = 100$	$n_2 = 150$
	$\bar{x}_1 = 210$ lbs.	$\bar{x}_2 = 220$ lbs.
	$S_1 = 10$ lbs.	$S_2 = 12$ lbs.

Population S.D. $(\sigma) = 11$ lbs.

We may assume that the S.D. of yield in the whole state is the S.D. of yields in the two districts. That is, the two populations have the same S.D. $(\sigma) = 11$ lbs. Using (13.7.3a)

$$\text{S.E. of } \bar{x}_1 - \bar{x}_2 = \sigma\sqrt{\frac{1}{n_1} + \frac{1}{n_2}}$$

$$= 11\sqrt{\frac{1}{100} + \frac{1}{150}} = 1.42$$

The observed value of the test statistic is

$$z = \frac{\bar{x}_1 - \bar{x}_2}{\text{S.E.}} = \frac{210 - 220}{1.42} = -7.04$$

Since $|z| = 7.04$ exceeds 2.58, we reject H_0 at 1% level and conclude that there is a significant difference in the mean yields of crops in the two districts.

(B) Using Chi-square (χ^2) Distribution

The chi-square distribution (Page 519) is used in both large sample and small sample tests. It is mainly used in

(1) Test for goodness of fit.
(2) Test for independence of attributes.
(3) Test for a specified standard deviation (Small Sample test).

(1) Test for goodness of fit (Pearsonian χ^2) The test, devised by Karl Pearson, is used to decide whether the observations are in good agreement with a hypothetical distribution, i.e. whether the sample may be spposed to have arisen from a specified population. The observed frequencies (f_0) of different classes are compared with the expected frequencies (f_e) by he test statistic

$$\chi^2 = \sum\left\{\frac{(f_0 - f_e)^2}{f_e}\right\} \tag{14.7.11}$$

This is called "*Pearsonian Chi-square*" or "*goodness-of-fit chi-square*". When the null hypothesis, viz.

H_0 (Date are in agreement with the hypothetical population)

is true, the statistic approximately follows chi-square distribution with $(k-1)$ degrees of freedom, where k is the number of classes. If the observed value of the statistic exceeds the tabulated value of χ^2 at a given level, the null hypothesis is rejected.

Example 14.32 *In his experiments on pea-breeding, Mendel obtained the following frequencies of seeds : Round and yellow—315; Wrinkled and yellow—101; Round and green—108; Wrinkled and green—32; Total—556. Theory predicts that the frequencies should be in the proportions 9 : 3 : 3 : 1. Examine the correspondence between theory and observations. (Given that 5% value of χ^2 for 3 d.f. is 7.815).*

Solution On the assumption that the data agree with theory, and that the divergences have arisen only due to sampling fluctuations, the probabilities for the classes should be $\frac{9}{16}, \frac{3}{16}, \frac{3}{16}, \frac{1}{16}$ respectively. The expected frequencies are calculated below:

Class	*Observed Frequency*	*Expected Frequency*
Round and yellow	315	$\frac{9}{16} \times 556 = 313$
Wrinkled and yellow	101	$\frac{3}{16} \times 556 = 104$
Round and green	108	$\frac{3}{16} \times 556 = 104$
Wrinkled ande green	32	$\frac{3}{16} \times 556 = 35$
	556	556

Note: The totals for Observed and Expected frequencies must be equal.

$$\therefore \quad \chi^2 = \frac{(315-313)^2}{313} + \frac{(101-104)^2}{104} + \frac{(108-104)^2}{104} + \frac{(32-35)^2}{35}$$

$$= \frac{4}{313} + \frac{9}{104} + \frac{16}{104} + \frac{9}{35}$$

$$= .01 + .09 + .15 + .26 = 0.51$$

Since there are 4 classes. Degrees of freedom = (4 – 1) = 3

We are given that the 5% value of χ^2 for 3 d.f. is 7.815. Since the observed value 0.51 is less than the tabulated value, the null hypothesis cannot be rejected. We conclude that the observations support the theory.

Example 14.33 *A dice was thrown 60 times with the following results:*

Face	*1*	*2*	*3*	*4*	*5*	*6*	*Total*
Frequency	*6*	*10*	*8*	*13*	*11*	*12*	*60*

Are the data consistent with the hypothesis that the dice is unbiased?

(Given $\chi^2_{.01} = 15.09$ for 5 degrees of freedom).

Solution Null hypothesis is that the dice is unbiased. Then the probability of each face is $\frac{1}{6}$, and the expected frequency is $60 \times \frac{1}{6} = 10$ for each.

Observed Frequency (f_0)	6	10	8	13	11	12
Expected Frequency (f_e)	10	10	10	10	10	10
$(f_0 - f_e)^2$	16	0	4	9	1	4

$$\therefore \quad \chi^2 = \frac{16}{10} + \frac{0}{10} + \frac{4}{10} + \frac{9}{10} + \frac{1}{10} + \frac{4}{10} = 3.4$$

There are 6 classes, Degrees of freedom = (6 – 1) = 5

Since the observed value of χ^2 (viz. 3.4) is less than the tabulated value 15.09 at 1% for 5 degrees of freedom, we cannot reject the null hypothesis at 1% level of significance. The conclusion is that the data are in agreement with the hypothesis of an unbiased dice.

Example 14.34 *5 identical coins are tossed 320 times, and the number of heads appearing each time is recorded. The results are:*

Number of Heads	*0*	*1*	*2*	*3*	*4*	*5*	*Total*
Frequency	*14*	*45*	*80*	*112*	*61*	*8*	*320*

Would you conclude that the coins are biased ? (Given $\chi^2_{.05} = 11.07$ and $\chi^2_{.01} = 15.09$ for 5 degrees of freedom).

Solution This is a problem for testing "goodness of fit" of a binomial distribution. The null hypothesis is that the coins are unbiased, i.e. the probability of obtaining head is 1/2 for each coin.

H_0 (data support Binomial distribution with $P = 1/2$)

The test statistic is $\chi^2 = \sum \left\{ \frac{(f_0 - f_e)^2}{f_e} \right\}$.

If H_0 is true, i.e. the coins are assumed to tbe unbiased, the probability of obtaining r heads in one throw of the set of 5 coins is given by the binomial distribution

$$P(r) = {}^5C_r \left(\frac{1}{2}\right)^r \left(\frac{1}{2}\right)^{5-r} = \frac{{}^5C_r}{32}$$

So, the expected frequency (f_e) of r heads in 320 tosses is

$$f_e = 320 \,.\, P(r) = 10 \times {}^5C_r$$

These are shown below for different values of r :

Number of Heads (r)	0	1	2	3	4	5	Total
Observed Frequency (f_0)	14	45	80	112	61	8	320
Expected Frequency (f_e)	10	50	100	100	50	10	320

$$\therefore \qquad \chi^2 = \frac{4^2}{10} + \frac{5^2}{50} + \frac{20^2}{100} + \frac{12^2}{100} + \frac{11^2}{50} + \frac{2^2}{10}$$

$$= 1.6 + 0.5 + 4.0 + 1.44 + 2.42 + 0.4 = 10.36.$$

There are 6 classes; so Degrees of freedom = 6 – 1 = 5

Since the observed value of χ^2 (viz. 10.36) is less than the tabulated value at 5% level (viz. 11.07), we cannot reject H_0 at 5% level of significance. We therefore conclude that the coins may not be biased.

(2) Test for independence of attributes [**Note: (1) Contingency Table**—In Statistics, sometimes we have to deal with "attributes" (see Section 1.2) or qualitative characters of members which cannot be measured accurately, although the members can be divided into two or more categories with respect to the attributes. Let us consider two attributes A and B, where A is shown in m categories $A_1, A_2, ..., A_m$, and B in n categories $B_1, B_2, ..., B_n$. The data can be shown in the form of a two-way table with m rows and n columns, as in a 'bivariate' frequency distribution (Tables 9.2 and 9.7). This two-way frequency table for attributes is known as $(m \times n)$ *Contingency Table* (see Example 14.35). The frequency of members belonging to both the categories A_i and B_j simultaneously is shown in the cell at the i-th row and j-th column, and denoted by (A_iB_j).

Similarly, (A_i) and (B_j) denote the frequency of members belonging to categories A_i and B_j respectively, and N the total frequency, as in the table below.

(3 × 4) Contingency Table

		Attribute B				
		B_1	B_2	B_3	B_4	Total
Attribute A	A_1	(A_1B_1)	(A_1B_2)	(A_1B_3)	(A_1B_4)	(A_1)
	A_2	(A_2B_1)	(A_2B_2)	(A_2B_3)	(A_2B_4)	(A_2)
	A_3	(A_3B_1)	(A_3B_2)	(A_3B_3)	(A_3B_4)	(A_3)
Total		(B_1)	(B_2)	(B_3)	(B_4)	N

Often, the members are divided into two categories only in respect of each attribute, according to the presence or absence of attribute A (denoted by A and α) and presence or absence of attribute B (denoted by B and β). See Example 14.36.

(2 × 2) Contingency Table

		Attribute A		
		A	α	Total
Attribute B	B	$(AB) = a$	$(\alpha B) = b$	$(B) = a + b$
	β	$(A\beta) = c$	$(\alpha\beta) = d$	$(\beta) = c + d$
Total		$(A) = a + c$	$(\alpha) = b + d$	N

(2) Independence of Attributes—Two attributes are said to be "indepentent", if they are unrelated to each other; i.e., the presence or absence of any attribute among the members of the population does in no way influence whether the other attribute will be present. In general, if there are m categories A_i for attribute A and n categories B_j for attribute B, then A and B will be independent if

$$(A_iB_j) = \frac{(A_i)\,(B_j)}{N}$$

for all values of i and j. This means that in the population each cell frequency will be equal to the product of the corresponding row and column-totals divided by the total number of members in the population. In particular, for a (2×2) contingency table,

$$(AB) = \frac{(A)\,(B)}{N}$$

If the attributes are not independent, they are said to be "associated".

(3) Association of Attributes—The word "association" is used to indicate the degree of relationship between attributes (the corresponding word for variable is "correlation"). If two attributes are not "independent", they are said to be "associated"; i.e. there is "association" between the two attributes. For a (2×2) table, a measure of association is given by Yule's "Coefficient of Association".

$$Q = \frac{(AB)\,(\alpha\beta) - (A\beta)\,(\alpha B)}{(AB)\,(\alpha\beta) + (A\beta)\,(\alpha B)} = \frac{ad - bc}{ad + bc}$$

Suppose that the observations are classified simultanceously according to two attributes, and the frequencies (f_0) in the difference categories are shown in a two-way table (known as *Contingency Table*). On the basis of cell frequencies it is required to test whether the two attributes are associated or not. Under the null hypothesis

H_0 (Attributes are independent)

the expected frequency (f_e) of any cell is given by

$$f_e = \frac{(\text{Row total}) \times (\text{Column total})}{\text{Total frequency}} = \frac{(A_i)\,(B_j)}{N}$$

Then we calculate the statistic

$$\chi^2 = \sum \left\{ \frac{(f_o - f_e)^2}{f_e} \right\} \qquad (14.7.12)$$

which approximately follows chi-square distribution with

d.f. = (number of rows – 1) × (number of columns – 1)

If the calculated value of χ^2 exceeds the tabulated value for the given d.f. and at a specified level, the null hypothesis is rejected, and we conclude that the attributes are not independent, but associated.

Simplified formula for (2 × 2) table

Suppose, the contingency table has only 2 rows and 2 columns with the four cell frequencies a, b, c, d as shown below:

			Total
	a	b	R_1
	c	d	R_2
Total	C_1	C_2	N

[**Note:** It has become customary to use a for the frequency in the *upper left cell* and d for that in the *lower right cell; b* and c are frequencies in the other diagonal positions.]

For simplicity we write R_1, R_2 to denote the row totals, and C_1, C_2 to denote the column totals; i.e.

$$R_1 = a + b, \quad R_2 = c + d, \quad C_1 = a + c, \quad C_2 = b + d$$
$$N = a + b + c + d = R_1 + R_2 = C_1 + C_2$$

In this case, formula (14.7.12) reduces to

$$\chi^2 = \frac{N(ad - bc)^2}{R_1 R_2\ C_1 C_2} \tag{14.7.13}$$

with degrees of freedom = (2 – 1) × (2 – 1) = 1.

Yates' correction

For a (2 × 2) table, there is only one degree of freedom; i.e. only one of the four cell frequencies can be arbitrarily given, if the row and column totals should remain fixed. It is therefore necessary to make a correction to formula (14.7.13), so that its approximation to the continuous chi-square distribution can be improved.

If $ad > bc$, reduce a and d by $\frac{1}{2}$, and increase b and c by $\frac{1}{2}$;

If $ad < bc$, increase a and d by $\frac{1}{2}$, and reduce b and c by $\frac{1}{2}$.

This is known as *Yates' Correction for Continuity.*

$$\chi^2 \text{ (corrected)} = \frac{N\{|\ ad - bc\ | - N/2\}^2}{R_1\ R_2\ C_1\ C_2} \tag{14.7.14}$$

It should be noted that Yates' correction can be applied only for 1 d.f.

Example 14.35 *A random sample of 500 students were classified according to economic condition of their family and also according to merit, as shown below:*

Merit	*Economic Condition*			*Total*
	Rich	*Middle class*	*Poor*	
Meritorious	42	137	61	240
Not Meritorious	58	113	89	260
Total	100	250	150	500

Test whether the two attributes Merit and Economic Condition are associated or not, (Given, $\chi^2_{.05} = 5.99$ *and* $\chi^2_{.01} = 9.21$ *for 2 d.f.)*

Solution Null hypothesis is that the attributes are independent. The expected frequencies are calculated as follows:

Table 14.6 Calculation of Expected Frequencies

	Rich	*Middle Class*	*Poor*	*Total*
Meritorious	$\frac{240\times100}{500}=48$	$\frac{240\times250}{500}=120$	$\frac{240\times150}{500}=72$	240
Not Meritorious	$\frac{260\times100}{500}=52$	$\frac{260\times250}{500}=130$	$\frac{260\times150}{500}=78$	260
Total	100	250	150	500

$$\chi^2=\frac{(42-48)^2}{48}+\frac{(137-120)^2}{120}+\frac{(61-72)^2}{72}$$

$$+\frac{(58-52)^2}{52}+\frac{(113-130)^2}{130}+\frac{(89-78)^2}{78}$$

$$=0.75+2.41+1.68+0.69+2.22+1.55=9.30$$

Degrees of freedom = $(2-1)\times(3-1)=1\times2=2$

For 2 d.f. the tabulated value of χ^2 at 5% level is 5.99 and at 1% level is 9.21. Since the observed value of χ^2 exceeds even the 1% tablulated value, it is highly significant. We reject the null hypothesis and conclude tha Merit and Economic Condition are not independent; i.e. they are associated.

Example 14.36 *In a survey of 200 boys, of which 75 were intelligent, 40 had skilled fathers; while 85 of the unintelligent boys had unskilled fathers. Do these figures support the hypothesis that skilled fathers have intelligent boys? Use χ^2 test. Value of χ^2 for 1 d.f. at 5% level is 3.84.* [C.A., Nov. '82]

Solution The data are shown in the following (2 × 2) table:

Intelligence of son	*Skill of father* *Skilled*	*Unskilled*	*Total*
Intelligent	**40**	35	75
Unintelligent	40	**85**	125
Total	80	120	**200**

[**Note:** Figures in bold type are only given.]

Null Hypothesis is that the two attributes "Skill of father" and "Intelligence of son" are *independent,* and the Alternative Hypothesis is that they are not independent. On the hypothesis of independence, the test statistic (14.7.14) follows χ^2 distribution with 1 d.f.

$$\chi^2=\frac{200\,\{|\,40\times85-35\times40\,|-100\}^2}{80\times120\times75\times125}$$

$$= \frac{200 \times 1900 \times 1900}{80 \times 120 \times 75 \times 125} = \frac{361}{45} = 8.02$$

Since the observed value of the statistic 8.02 is greater than the tabulated value 3.84 (given), it is significant. We therefore reject the null hypothesis at 5% level of significance and conclude that the attributes are *not independent*; i.e. the data support the alternative hypothesis that 'skilled fathers have intelligent boys'.

14.8 SMALL SAMPLE TESTS

These are "exact tests" and are applicable to samples of any size, although they are generally used for small samples; say $n < 30$.

(A) Using Normal Distribution

(1) Test for a specified population mean (s.d. known)

Consider a random sample of size n from a *normal* population, whose standard deviation σ is known. It is required to test the null hypothesis that the mean of population has a specified value.

$$H_0\ (\mu = \mu_0)$$

The sampling distribution of sample mean ($\bar{x}$) is exactly normal (see theorem I, page 518) with standard error S.E. $= \dfrac{\sigma}{\sqrt{n}}$. When the null hypothesis is true, the test statistic

$$z = \frac{\bar{x} - \mu_0}{\text{S.E.}} = \frac{\bar{x} - \mu_0}{\dfrac{\sigma}{\sqrt{n}}} \tag{14.8.1}$$

follows standard normal distribution exactly. If the value of z falls in the critical region, shown below, the null hypothesis is rejected.

Table 14.7 Rejection Rules for $H_0\ (\mu = \mu_0)$

Alternative Hypothesis H_1	*Critical Region* 5% level	*Critical Region* 1% level
$\mu \neq \mu_0$	$\lvert z \rvert \geq 1.96$	$\lvert z \rvert \geq 2.58$
$\mu > \mu_0$	$z \geq 1.645$	$z \geq 2.33$
$\mu < \mu_0$	$z \leq -1.645$	$z \leq -2.33$

Confidence limits for the population proportion μ are (see 14.5.6):

$$95\% \text{ confidence limits} = \bar{x} \pm 1.96\left(\frac{\sigma}{\sqrt{n}}\right)$$

$$99\% \text{ confidence limits} = \bar{x} \pm 2.58\left(\frac{\sigma}{\sqrt{n}}\right) \tag{14.8.2}$$

(2) Test for equality of two means (s.d.s known) Consider two *independent* sampels of sizes n_1 and n_2 from two *normal* populations with unknown means, but *known* standard deviations σ_1 and σ_2 respectively. It is required to test the hypothesis that the population means are equal.

$$H_0(\mu_1 = \mu_2)$$

The sampling distribution of the difference of sample means $(\bar{x}_1 - \bar{x}_2)$ is exactly normal with standard error

$$\text{S.E.} = \sqrt{\frac{\sigma_1^2}{n_1} + \frac{\sigma_2^2}{n_2}} \tag{14.8.3}$$

The test statistic

$$z = \frac{\bar{x}_1 - \bar{x}_2}{\sqrt{\frac{\sigma_1^2}{n_1} + \frac{\sigma_2^2}{n_2}}} \tag{14.8.4}$$

follows standard normal distribution exactly. If the value of z falls in the critical region, the null hypothesis is rejected.

Table 14.8 Rejection Rules for H_0 $(\mu_1 = \mu_2)$

Alternative Hypothesis	*Critical Region*	
H_1	*5% level*	*1% level*
$\mu \neq \mu_2$	$\lvert z \rvert \geq 1.96$	$\lvert z \rvert \geq 2.58$
$\mu > \mu_2$	$z \geq 1.645$	$z \geq 2.33$
$\mu < \mu_2$	$z \leq -1.645$	$z \leq -2.33$

Confidence limits for the difference of population means $(\mu_1 - \mu_2)$ are (see 14.5.7):

$$95\% \text{ confidence limits} = (\bar{x}_1 - \bar{x}_2) \pm 1.96 \text{ (S.E.)}$$

$$99\% \text{ confidence limits} = (\bar{x}_1 - \bar{x}_2) \pm 2.58 \text{ (S.E.)} \tag{14.8.5}$$

where S.E. is given by (14.8.3).

Example 14.37 *A random sample of size 10 was taken from a normal poulation, whose variance is known to be 7.056 sq. inches. If the observations are (in inches) 65, 71, 64, 71, 70, 69, 64, 63, 67 and 68, test the hypothesis that the population mean is 69 inches. Also obtain 99% confidence limits for the population mean.*

Solution Null hypothesis is that the population mean is 69 inches, and the alternative hypothesis is that it is not so; i.e. H_0 $(\mu = 69)$ against $H_1(\mu \neq 69)$. Population s.d. $(\sigma) = \sqrt{7.056}$, and $n = 10$.

For the given data sample mean $\bar{x} = 67.2$. The value of test statistic is

$$z = \frac{67.2 - 69}{\frac{\sqrt{7.056}}{\sqrt{10}}} = \frac{-1.8}{\sqrt{.7056}} = -2.14$$

Since $|z| = 2.14$ exceeds 1.96 (the critical region at 5% level is $|z| \geq 1.96$), the null hypothesis is rejected. We conclude that the population mean is not 69 inches.

99% confidence limits for μ are $67.2 \pm 2.58\left(\frac{\sqrt{7.056}}{\sqrt{10}}\right)$

$$= 67.2 \pm 2.17 = 65.03 \text{ and } 69.37 \text{ inches.}$$

Example 14.38 *It is required to estimate the mean of a normal populations using a sample sufficiently large so that the probability will be 0.95 that the sample mean will not differ from the population mean by more than 25% of the population s.d. How large should the sample be?*

Solution We have to find the sample size (n) so that

$$P\left(|\bar{x} - \mu| \leq \frac{\sigma}{4}\right) = 0.95$$

i.e.
$$P\left(\left|\frac{\bar{x} - \mu}{\frac{\sigma}{\sqrt{n}}}\right| \leq \frac{\frac{\sigma}{4}}{\frac{\sigma}{\sqrt{n}}}\right) = 0.95$$

i.e.
$$P\left(|z| \leq \sqrt{\frac{n}{4}}\right) = 0.95 \qquad \text{(i)}$$

where z is a standard normal variate (see 13.8.5). But we know that (see 13.8.3)

$$P(|z| \leq 1.96) = 0.95 \qquad \text{(ii)}$$

Comparing (i) and (ii),

$$\sqrt{\frac{n}{4}} = 1.96; \quad \therefore \quad n = (4 \times 1.96)^2 = 61.47$$

Since the sample size works out to be larger than 61, the least value of n is 62.

Example 14.39 *Two independent random samples were taken from two normal populations and the following information is given:*

	Population I	*Population II*
Sample Size	$n_1 = 10$	$n_2 = 12$
Sample Mean	$\bar{x}_1 = 20$	$\bar{x}_2 = 27$
Population s.d.	$\sigma_1 = 8$	$\sigma_2 = 6$

Is it likely that the mean of Population I is smaller than that of Population II? (Use 1% level of significance). Also find 99% confidence limits for the difference of population means.

Solution Null hypothesis $H_0(\mu_1 = \mu_2)$; i.e. we assume that the means are equal. Alternative hypothesis $H_1(\mu_1 < \mu_2)$. The observed value of z is

$$z = \frac{20 - 27}{\sqrt{\frac{8^2}{10} + \frac{6^2}{12}}} = \frac{-7}{3.07} = -2.28$$

Since the value of z is not less than –2.33 (critical region at 1% is $z \le -2.33$), it is not significant at 1% level. We conclude that the mean of Population I may not be smaller than that of Population II.

99% confidence limits for $(\mu_1 - \mu_2)$ are

$$(\bar{x}_1 - \bar{x}_2) \pm 2.58 \text{ (S.E.)}$$
$$= (20 - 27) \pm 2.58 \times 3.07$$
$$= -7 \pm 7.92 = -14.92 \text{ and } 0.92$$

(B) Using t distribution

(1) Test for a specified mean (s.d. unknown) For testing the null hypothesis H_0 $(\mu = \mu_0)$, if the population standard deviation (σ) is not known, formula (14.8.1) cannot be used. However the "unbiased" estimator of the population variance (σ^2) is

$$s^2 = \frac{n}{n-1} S^2$$

where S^2 is the variance of a sample of size n. Thus the estimate of σ is $s = \sqrt{\frac{n}{n-1}}\, S$. If this is substituted for σ in formula (14.8.1), then the statistic (known as **Student's *t***)

$$t = \frac{\bar{x} - \mu_0}{\frac{s}{\sqrt{n}}} = \frac{\bar{x} - \mu_0}{\frac{S}{\sqrt{n-1}}} \qquad (14.8.6)$$

follows t distribution (13.8.15) with degrees of freedom $(n - 1)$. For tests of significance, the tables of t must therefore be used. If $t_{.025}$ denotes the value of t distribution corresponding to the upper tail area 025, then the null hypothesis is rejected, if $|t| \ge t_{.025}$.

95% confidence limits for the population mean μ are (see 14.5.8a)

$$\bar{x} \pm t_{.025} \frac{s}{\sqrt{n}} \quad \text{or,} \quad \bar{x} \pm t_{.025} \frac{S}{\sqrt{n-1}} \qquad (14.8.7)$$

99% confidence limits for μ are

$$\bar{x} \pm t_{.005} \frac{s}{\sqrt{n}} \quad \text{or,} \quad \bar{x} \pm t_{.005} \frac{S}{\sqrt{n-1}} \qquad (14.8.7a)$$

(2) Test for equality of two means (s.d.s unknown) Consider two *independent* random samples of sizes n_1 and n_2 from two normal populations with means μ_1 and μ_2. It is required to test the hypothesis that themeans are equal.

$$H_0(\mu_1 = \mu_2)$$

If the two population *standard deviations are assumed to be equal,* an "unbiased" estimator of the common variance is given by

$$s^2 = \frac{n_1 S_1^2 + n_2 S_2^2}{n_1 + n_2 - 2} \tag{14.8.8}$$

where S_1, S_2 are are the sample standard deviations.

If this is substituted for σ_1^2 and σ_2^2 in formula (14.8.4) the statistic

$$t = \frac{\bar{x}_1 - \bar{x}_2}{s\sqrt{\frac{1}{n_1} + \frac{1}{n_2}}} \tag{14.8.9}$$

(known as **Fisher's** t) follows t distribution (13.8.16) with degrees of freedom $(n_1 + n_2 - 2)$.

Confidence limits for $(\mu_1 - \mu_2)$ are (see 4.5.9a):

$$95\% \text{ confidence limits} = (\bar{x}_1 - \bar{x}_2) \pm t_{.025} \times s\sqrt{\frac{1}{n_1} + \frac{1}{n_2}}$$

$$99\% \text{ confidence limits} = (\bar{x}_1 - \bar{x}_2) \pm t_{.005} \times s\sqrt{\frac{1}{n_1} + \frac{1}{n_2}} \tag{14.8.10}$$

Example 14.40 *A random sample of size 20 from a normal population gives a sample mean of 42 and sample standard deviation of 6. Test the hypothesis that the population mean is 44. State clearly the alternative hypothesis you allow for and the level of significance adopted.* [C.A., Nov. '74]

Solution Null hypothesis is that the population mean is 44. $H_0(\mu = 44)$. Alternative hypothesis is that the population mean is different from 44. $H_2(\mu \neq 44)$, i.e. both-sided alternatives.

Since the population s.d. (σ) is not known, we use t test. Here, n = 20, $\bar{x}$ = 42 and S = 6. Therefore using (14.8.6)

$$t = \frac{42 - 44}{\frac{6}{\sqrt{19}}} = -1.45$$

$$\text{Degrees of freedom} = (20 - 1) = 19$$

From the tables of t distribution, we find for 19 d.f. the percentage points of t are respectively $t_{.025} = 2.09$ and $t_{.005} = 2.86$. We have to use a two-tailed test, because the alternatives are both-sided. Since $|t| = 1.45$ is less than the 5% tabulated value corresponding to the two tails (viz. 2.09) there is no reason to reject the null hypothesis at 5% level of significance, and we conclude that the population mean may be 44.

[99% confidence limits for μ are $42 \pm 2.86 \times \left(\frac{6}{\sqrt{19}}\right)$

$= 42 \pm 3.94 = 38.06$ and 45.94]

Example 4.41 *A fertilizer mixing machine is set to give 12 kg of nitrate for every quintal bag of fertiliser. Ten 100 kg bags are examined. The percentages of nitrate are : 11, 14, 13, 12, 13, 12, 13, 14, 11, 12. Is there reason to believe that the machine is defective ? Value of t for 9 d.f. is 2.262.* [C.A., Nov. '82]

Solution *Assumption*—The 'weight (in kg) of nitratre in one bag of the fertiliser' follows normal distribution with mean μ and s.d. σ.

Null Hypothesis $H_0(\mu = 12)$; Alternative Hypothesis $H_1(\mu \neq 12)$. Since σ is not known the appropriate test statistic is *Student's t,* given by (14.8.6). When H_0 is true, this follows t distribution with $(n - 1)$ d.f.

From the given data, we calculate the sample mean ($\bar{x}$) and the sample s.d. (S).

$$\bar{x} = \frac{\Sigma x}{n} = \frac{125}{10} = 12.5$$

For the s.d. take deviations from 12, i.e. $d = x - 12$.

$-1, 2, 1, 0, 1, 0, 1, 2, -1, 0$; $\Sigma d = 5$, $\Sigma d^2 = 12$

$$\therefore \qquad S^2 = \frac{\Sigma d^2}{n} - \left(\frac{\Sigma d}{n}\right)^2 = \frac{13}{10} - \left(\frac{5}{10}\right)^2 = 1.05$$

Thus, $\qquad \bar{x} = 12.5, \quad S = \sqrt{1.05} = 1.02$

Using the second form in (14.8.6),

$$t = \frac{12.5 - 12}{\dfrac{1.02}{\sqrt{(10-1)}}} = \frac{0.5 \times 3}{1.02} = 1.47$$

Since the alternative hypothesis is both-sided, the critical region is given by both tails of the t distribution with 9 d.f. At 5%level,

Critical Region : $|t| \geq 2.262$

The observed value of $t = 1.47$ does not fall in the critical region (because it is not more than 2.262) and is therefore *not significant.* So we cannot reject the null hypothesis and conclude that the population mean is 12; i.e. the data do not indicate that the machine is defective.

[**Note:** The previous two examples show the use of *Student's t* statistic in two-tailed tests. In Example 14.40, the sample mean and s.d. are given, while in Example 14.41, these are obtained from the sample data. Example 14.42 shows a one-tailed test (right tail), using *Student's t.*]

Example 4.42 *A soap manufacturing company was distributing a particular brand of soap through a large number of retail shops. Before a heavy advertisemant campaign, the mean sales per week per shop was 140 dozens. After the campaign, a sample of 26 shops was taken and the mean sales was found to be 147 dozens with standard deviation 16. Can you consider the advertisement effective?* [C.A., Nov. '78]

Solution We assume that the sales per week follow normal distribution with mean $\mu = 140$ (σ unknown); and it is required to test whether the average sales after the campaign is larger than 140. $H_0(\mu = 140)$ against one sided alternative $H_1(\mu > 140)$.

$$t = \frac{\bar{x} - \mu_0}{\dfrac{S}{\sqrt{n-1}}} = \frac{147 - 140}{\dfrac{16}{\sqrt{25}}} = 2.19$$

From Statistical tables, we find that for 25 degrees of freedom, the value of t corresponding to upper 5% and 1% "tail areas" are $t_{.05} = 1.708$ and $t_{.01} = 2.485$. Since the observed value of t, namely 2.19, is larger than $t_{.05,}$ we reject H_0 and conclude that μ is greater than 140; i.e. the advertisement has been effective in increasing the sales.

Example 14.43 *Two types of batteries are tested for their length of life and the following data are obtained :*

	No. of Sample	*Mean life in Hours*	*Variance*
Type A	9	600	121
Type B	8	640	144

Is there a significant difference in the two means? Value of t for 15 degrees of freedom at 5% level is 2.131. [C.A., Nov. '82]

Solution *Assumption:* (1) The two populations are normal distrubutions with means $\mu_1 \cdot \mu_2$ and a common s.d. σ. (2) The two samples are randomly drawn and independent.

$$H_0(\mu_1 = \mu_2) \text{ against } H_1(\mu_1 \neq \mu_2)$$

Given

	Population A	*Population B*
Sample Size	$n_1 = 9$	$n_2 = 8$
Sample Mean	$\bar{x}_1 = 600$	$\bar{x}_2 = 640$
Sample Variance	$S_1^2 = 121$	$S_2^2 = 144$

(Compare with Example 14.39)

The appropriate test statistic is *Fisher's* t (14.8.9) which, under H_0, follows t distribution with $(n_1 + n_2 - 2)$ d.f. An estimate of the common but unknown s.d. (σ) is obtained from (14.8.8)

$$s = \sqrt{\frac{9 \times 121 + 8 \times 144}{9 + 8 - 2}} = \sqrt{149.4} = 12.2$$

The observed value of the statistic is

$$t = \frac{600 - 640}{12.2\sqrt{\frac{1}{9} + \frac{1}{8}}} = \frac{-40}{12.2 \times 0.486} = -6.7$$

Degrees of freedom = 9 + 8 – 2 = 15

Since the alternative hypothesis is both-sided, th test is two-tailed. The critical region is given by both the tails of t distribution. At 5% level

Critical Region: $|t| \geq 2.131$

Since the observed value falls in the critical region (because $|t| = |-6.7| = 6.7$ is larger than 2.131), we reject the null hypothesis at 5% level of significance and conclude that there is 'significant difference in the two means'.

[Confidence interval for the difference $(\mu_1 - \mu_2)$ is given by (14.5.9a)

$$95\% \text{ confidence limits} = (600 - 640) \pm 2.131 \times 12.2\sqrt{\frac{1}{9} + \frac{1}{8}}$$

$$= -40 \pm 12.63$$

$$= -27.37 \text{ and } -52.63$$

Thus, 95% confidence interval for $(\mu_2 - \mu_1)$ is 27.37 to 52.63.]

[**Note:** Examples 14.43 and 44 illustrate the use of *Fisher's t* statistic, the former showing a two-tailed test and the latter a one-tailed test (left tail).]

Example 4.44 *A group of 5 patients treated with medicine A weigh 42, 39, 48, 60,41 kg; a secnd group of 7 patients from the same Hospital treated with medicine B weigh 38, 42, 56, 64, 68, 69, 62 kg. Do you agree with the claim that medicine B increases the weight significantly ? (The value of't' at 5% level of significance for 10 degrees freedom is 2.2281).* [I.C.W.A., June '79]

Solution This is a problem for testing equality of two population means based on independent samples. We assume that the samples are drawn from normal populations having the same standard deviation σ (unknown) and means μ_1 and μ_2 respectively. The null hypothesis is that patients treated with medicines A and B have the same average weight (in the population) and the alternative hypothesis is that medicine B increases the average weight. $H_0(\mu_1 = \mu_2)$ against the alternative $H_1(\mu_1 < \mu_2)$.

The appropriate statistic is Fisher's t (14.8.9). Since the alternative hypothesis is one-sided H_1 $(\mu_1 < \mu_2)$, the null hypothesis H_0 will be rejected, if $t \leq - t_{.05}$.

From the given data we obtain the following results:

	Population A	*Population B*
Sample Size	$n_1 = 5$	$n_2 = 7$
Sample Mean	$\bar{x}_1 = 46$	$\bar{x}_2 = 57$
Sample Variance	$S_1^2 = 58$	$S_2^2 = \dfrac{926}{7}$

As in the previous example, the estimate of σ is

$$s = \sqrt{\frac{5 \times 58 + 7 \times \left(\dfrac{926}{7}\right)}{5+7-2}} = \sqrt{121.6} = 11.03$$

$$\therefore \quad t = \frac{46-57}{11.03\sqrt{\left(\dfrac{1}{5}+\dfrac{1}{7}\right)}} = \frac{-11}{6.46} = -1.70$$

Since the observed value of t, viz. -1.70, is not less than -2.2281, we cannot reject H_0 at 5% level of significance. Thus the claim is not justified.

[*Comment*: The given value 2.2281 corresponds to "both tail areas" equal to 5%, and is not appropriate for this test. The correct value 1.812 corresponding to one tail area 5% for 10 d.f. should have been given and used for comparison].

(3) Test for equality of two means—Paired t Suppose, we have a random sample of n pairs of observations $(x_1, y_1), (x_2, y_2), \ldots, (x_n, y_n)$ from a *bivariate normal* population. The values of x and y are now *correlated* and no longer independent. It is required to test whether the means of x and y in the population are equal, i.e. H_0 $(\mu_x = \mu_y)$.

Let $d_i = x_i - y_i$ denote the difference (with proper sign) in the values of x and y for the i-th pair ($i = 1, 2, \ldots n$).

The test statistic is

$$t = \frac{\bar{d}}{S / \sqrt{n-1}} \tag{14.8.10a}$$

where $\bar{d}$ and S denote respectively the mean and standard deviation of the differences d_i; i.e.

$$\bar{d} = \frac{\Sigma d_i}{n} \tag{14.8.10b}$$

$$S^2 = \frac{\Sigma(d_i - \bar{d})^2}{n} = \frac{\Sigma d_i^2}{n} - \left(\frac{\Sigma d_i}{n}\right)^2$$

When H_0 is true, the statistic (14.8.10a), known as *Paired t*, follows t distribution with $(n-1)$ degrees of freedom. The critical region is given by two tails or one tail (right or left tail) of t distribution depending on the alternative hypothesis H_1 $(\mu_x \neq \mu_y)$, H_1 $(\mu_x > \mu_y)$ or $H_1(\mu_x < \mu_y)$.

Remarks: Paired t-test for $H_0(\mu_x = \mu_y)$ may be viewed as Student's t-test ofr H_0 $(\mu_d = 0)$, where $d = x - y$; because $\mu_d = \mu_x - \mu_y = 0$, when $\mu_x = \mu_y$. Thus, for testing equality of means from correlated pairs (x, y), we take deviations $d = x - y$ and apply Student's t-test for a specified mean 0, using the values of d.

Example 14.45 *An I.O. test was administered to 5 persons before and after they were trained. The results are given below:*

	I	*II*	*III*	*IV*	*V*
I.Q. before Training	*110*	*120*	*123*	*132*	*125*
I.Q. after Training	*120*	*118*	*125*	*136*	*121*

Test whether there is any change in I.Q. after the training programme. (Given $t_{0.01(4)} = 4.6$) [C.A., May '80]

Solution This is a problem of *Paired t test.* Because, the scores before training (x) and after training (y) are not independent, but the latter are likely to be affected by the former; i.e. x and y are correlated.

Null Hypothesis: $H_0(\mu_x = \mu_y)$; i.e. there is no change in the mean scores before and after the training, *Alternative Hypothesis:* H_i $(\mu_x \neq \mu_y)$. The test statistic is (14.8.10a).

Calculations for Paired t

						Total
x	110	120	123	132	125	
y	120	118	125	136	121	—
$d = x - y$	− 10	2	− 2	− 4	4	− 10
d^2	100	4	4	16	16	140

By (14.8.10b), $\bar{d} = -\frac{10}{5} = -2$

$$S^2 = \frac{140}{5} - \left(\frac{-10}{5}\right)^2 = 28 - 4 = 24$$

The observed value of the test statistic is

$$t = \frac{-2}{\frac{\sqrt{24}}{\sqrt{5-1}}} = \frac{-2}{\sqrt{6}}$$

$$= \frac{-2\sqrt{6}}{6} = \frac{-2.45}{3} = -0.82$$

Critical Region at 1% level of significance is $|t| \geq 4.6$. Since the observed value $|t| = 0.82$ is less than 4.6, H_0 cannot be rejected and we conclude that there is no significant change in the mean I.Q. after the training programme.

[**Note:** (i) In this book, we have used the notation $t_{.01(4)}$ to denote the value of t for 4 d.f. corresponding to *one tail* area (right tail) .01, i.e. 1%. Some authors use the same notation $t_{.01(4)}$ corresponding to the *two tail* areas 1%, i.e. right tail area 1/2%, as has been done in the above problem. However, whatever be the notation followed, the value 4.6 shown here is the appropriate critical value at 1% level for the two-tailed t-test with 4 d.f.

(ii) Example 14.45 shows a *two-tailed* test, and Example 14.45A *a one-tailed* (right tail) test.]

Example 14.45A *A certain stimulus administered to each of 12 patients resulted in the following changes in blood pressure:*

5,	2,	8,	– 1,	3,	0,	– 2,	1,	5,	0,	4,	6.

Can it be concluded that the stimulus will in general be accompanied by an increase in blood pressure ? (Given for 11 d.f., $t_{.05} = 2.2$) [I.C.W.A., Dec. '80]

Solution Here, the 'changes' $d = x - y$ in blood pressure are given, i.e. x is the (final) blood pressure after administering the stimulus and y is the initial blood pressure. We are required to test whether the mean blood pressure has increased, i.e. μ_x is greater than μ_y.

Null Hypothesis: H_0 ($\mu_x = \mu_y$); i.e. mean blood pressure has not changed. *Alternative Hypothesis:* $H_1(\mu_x > \mu_y)$; i.e. mean blood pressure *increases* after the stimulus.

When H_0 is true, the statistic (14.8.10a) follows t distribution with $(n-1)$ dgrees of freedom. For the given data,

$$n = 12, \quad \Sigma d_i = 31, \quad \Sigma d_i^2 = 185$$

$$\therefore \quad \bar{d} = \frac{31}{12} = 2.58, \text{ and}$$

$$S^2 = \frac{185}{12} - \left(\frac{31}{12}\right)^2 = \frac{1259}{144} \text{ Hence.}$$

$$t = \frac{\bar{d}}{\sqrt{\frac{S^2}{(n-1)}}} = \frac{2.58}{\sqrt{\frac{1259}{144 \times 11}}} = \frac{2.58 \times 12}{\sqrt{114.5}}$$

$$= \frac{30.96}{10.7} = 2.89$$

Degrees of freedom = 12 – 1 = 11

Since the alternative hypothesis is one-sided, this is a one-tailed test, and the critical region is given by the right tail.

Critical Region: $t \geq t_{.05}$ at 5% level of significance.

Since the calculated value 2.89 is larger than the tabulated value 2.2 (given), it is 'significant'. We therefore reject H_0 and conclude that "the stimulus will in general be accompanied by an increase in blood pressure".

[**Note:** The value of t given in the problem is not correct. This should be $t_{.05} = 1.8$ corresponding to the right tail area of t distribution for 11 d.f. The given value actually corresponds to the two tail areas and is appropriate only for a two-tailed test, i.e. $t_{.025} = 2.2$. However, the 'conclusion' remains unaltered.]

(4) Test for 'significance' of observed correlation coefficient [see (14.9.3)]

(C) Using Chi-square (χ^2) Distribution

(1) Test for a specified value of s.d. Suppose, we have a random sample $x_1, x_2, \ldots, x_n$ from a normal population with mean μ and s.d. σ (both unknown). It is required to test the hypothesis that the population s.d. σ has a specified value σ_0. Null hypothesis is $H_0(\sigma = \sigma_0)$ against alternative hypothesis $H_1(\sigma > \sigma_0)$. We use the test statistic

$$\chi^2 = \frac{\Sigma(x_i - \overline{x})^2}{\sigma_0^2} = \frac{nS^2}{\sigma_0^2} \tag{14.8.11}$$

where S is the s.d. of the sample. When the null hypothesis H_0 is true, the statistic follows chi-square distribution with $(n - 1)$ d.f. If the observed value of χ^2 exceeds the tabulated 5% value, we conclude that the population s.d. is not σ_0.

Example 14.46 *A random sample of size 20 from a normal population gives a sample mean of 42 and a sample standard deviation of 6. Test the hypothesis that the population s.d. is 9. Clearly state the alternative hypothesis you allow for and the level of significance adopted.* [C.A., Nov. '74]

Solution Given that Sample size (n) = 20. Sample s.d. (S) = 6, it is required to test the null hypothesis H_0 ($\sigma = 9$) against the alternative hypothesis $H_1(\sigma > 9)$.

$$\chi^2 = \frac{nS^2}{\sigma_0^2} = \frac{20 \times 6^2}{9^2} = 8.89$$

From the tables we find that for 19 d.f., 5% value of $\chi^2 = 30.14$.

Since the observed value of χ^2, viz. 8.89, is less than the 5% tabulated value, it is not significant at 5% level. Therefore we cannot reject the null hypothesis, and conclude that the population s.d. may be 9.

[**Note:** In testing for a specified s.d. (σ), the sample mean ($\overline{x}$) is not required; but in testing for a specified mean (μ) the sample s.d. (S) is also necessary (see 14.8.6).]

Example 14.47 *Weights (in kg) of 10 students are given below: 38, 40, 45, 53, 47, 43, 55, 48, 52, 49. Can we say that variance of the distribution of weights of all students from which the above sample of 10 students was drawn, is equal to 20 square kg?* [C.A., Nov.'79]

Degrees of Freedom	$\chi^2_{.05}$	$\chi^2_{.01}$
9	16.92	21.67
10	18.31	23.21

Solution We assume that the sample comes from a normal population with variance 20 sq kg (mean unknown).

$$H_0\ (\sigma^2 = 20) \text{ against } H_1(\sigma^2 > 20)$$

The appropriate test statistic is χ^2 defined by (14.8.11). For the calculation of $\Sigma(x - \bar{x})^2$, we take deviations from an arbitrary origin 45: – 7, – 5, 0, 8, 2, – 2, 10, 3, 7, 4; $\Sigma d = 20$ and $\Sigma d^2 = 320$.

$$\Sigma(x_i - \bar{x})^2 = \Sigma d^2 - \frac{(\Sigma d)^2}{n} = 320 - \frac{20^2}{10} = 280$$

$$\chi^2 = \frac{\Sigma(x_i - \bar{x})^2}{\sigma_0^2} = \frac{280}{20} = 14$$

$$\text{d.f.} = 10 - 1 = 9$$

Critical region at 5% level is $\chi^2 \geq \chi^2_{.05}$ for 9 d.f. Since the observed value of $\chi 2 = 14$ is less than 16.92, we have no reason to reject H_0 and thus conclude that the population variance may be 20 sq. kg.

[**Note:** 95% confidence limits for σ may be obtained from (see 14.5.11)

$$\chi^2_{.975} \leq \frac{280}{\sigma^2} \leq \chi^2_{.025}$$

or, $$270 \leq \frac{280}{\sigma^2} \leq 19.02$$

or, $$\frac{1}{19.02} \leq \frac{\sigma^2}{280} \leq \frac{1}{2.70}$$

or, $$\frac{280}{19.02} \leq \frac{\sigma^2}{280} \leq \frac{280}{2.70}$$

or, $$14.7 \leq \sigma^2 \leq 103.7$$

$\therefore$ $$3.8 \leq \sigma \leq 10.2]$$

(D) Using *F* distribution

(1) Test for equality of two s.d.s (means unknown) Given two *independent* random samples of sizes n_1 and n_2 from two normal populations with unknown means, we may be required to test the hypothesis that the population s.d.s are equal, H_0 ($\sigma_1 = \sigma_2$). The estimates of variances σ_1^2 and σ_2^2 are obtained from the samples (see p. 549)

$$s_1^2 = \left(\frac{n_1}{n_1 - 1}\right) S_1^2; \; s_2^2 = \left(\frac{n_2}{n_2 - 1}\right) S_2^2$$

where S_1 and S_2 are the standard deviations in the samples. If the null hypothesis is true, the statistic (see 13.8.22)

$$F = \frac{s_1^2}{s_2^2} \tag{14.8.12}$$

follows F distribution with d.f. $(n_1 - 1, n_2 - 1)$. If the observed value of F is larger than the 5% tabulated value, we reject the null hypothesis H_0 and conclude that the standard deviations in the two populations are not equal.

Example 14.48 *The standard deviations calculated from two random samples of sizes 9 and 13 are 2.1 and 1.8 respectively. May the samples be regarded as drawn from normal populations with the same s.d.? (The 5% value of F from tables with d.f. 8 and 12 is $F_{.05} = 2.85$)*

Solution Here, $n_1 = 9$, $n_2 = 13$, $S_1 = 2.1$ and $S_2 = 1.8$. The "unbiased" estimates of the variances σ_1^2 and σ_2^2 in the two populations are respectively

$$s_1^2 = \frac{9 \times (2.1)^2}{8} = 4.96, \quad s_2^2 = \frac{13 \times (1.8)^2}{12} = 3.51.$$

We have to test $H_0(\sigma_1 = \sigma_2)$ against $H_1\ (\sigma_1 > \sigma_2)$. The value of the statistic is

$$F = \frac{4.96}{3.51} = 1.41$$

Degrees of freedom are (8, 12).

Since the observed value of F, viz. 1.41, is less than the 5% tabulated value (viz. 2.85) corresponding to d.f. (8, 12), we cannot reject the null hypothesis at 5% level of significance. The conclusion is that the population s.d.s may be equal.

[**Note :** F statistic is the ratio of two independent "unbiased estimators" of the population variance, and not simply the ratio of sample variances.]

Example 14.49 *Two random samples are drawn from two populations and the following results were obtained :*

Sample I	*16*	*17*	*18*	*19*	*20*	*21*	*22*	*24*	*26*	*27*		
Sample II	*19*	*22*	*23*	*25*	*26*	*28*	*29*	*30*	*31*	*32*	*35*	*36*

Find the variances of the two samples and test whether the two populations have the same variance. [C.A., Nov. '81]

Solution For Samples I and II, the means are

$$\bar{x}_1 = \frac{210}{10} = 21, \; \bar{x}_2 = \frac{336}{12} = 28$$

Table 14.9 Calculation of Sample Variances

Sample I (x)	16	17	18	19	20	21	22	24	26	27		Total
$x-\bar{x}_1$	−5	−4	−3	−2	−1	0	1	3	5	6		0
$(x-\bar{x}_1)^2$	25	16	9	4	1	0	1	9	25	36		126
Sample II (x)	19	22	23	25	26	28	29	30	31	32	35	36	Total
$x-\bar{x}_2$	−9	−6	−5	−3	−2	0	1	2	3	4	7	8	0
$(x-\bar{x}_2)^2$	81	36	25	9	4	0	1	4	9	16	49	64	298

$$S_1^2 = \text{Variance of Sample I} = \frac{\Sigma(x-\bar{x}_1)^2}{n_1} = \frac{126}{10} = 12.6$$

$$S_2^2 = \text{Variance of Sample II} = \frac{\Sigma(x-\bar{x}_2)^2}{n_2} = \frac{298}{12} = 24.83$$

For testing the equality of population variances, $H_0(\sigma_1^2 = \sigma_2^2)$, we use F statistic (14.8.12).

$$n_1 = 10, \quad n_2 = 12, \quad S_1^2 = 12.6, \quad S_2^2 = 24.83$$

The unbiased estimates of population variances are

$$s_1^2 = \frac{10\times 12.6}{9} = 14, \quad s_2^2 = \frac{12\times 24.83}{11} = 27.1$$

Since $s_2^2 > s_1^2$, the test statistic is

$$F = \frac{s_2^2}{s_1^2} = \frac{27.1}{14} = 1.94; \text{ d.f. } (11, 9)$$

Alternative Hypothesis $H_1(\sigma_2^2 > \sigma_1^2)$. The tabulated value at 5% level is $F_{.05} = 3.10$ (approx.).

Critical Region: $F \geq 3.10$

Since the calculated value $F = 1.94$ is less than 3.10, it is not significant, and we conclude that the two populations may have the same variance.

Example 14.50 *In a sample of 8 observations, the sum of the squared deviations of items from the mean was 94.5. In another sample of 10 observations, the value was found to be 101.7. Test whether the difference is significant at 5% level. (You are given that at 5% level, critical value of F for $v_1 = 7$ and $v_2 = 9$ degrees of freedom is 3.29 and for $v_1 = 8$ and $v_2 = 10$ d.f. its value is 3.07.* [C.A., May '79]

Solution This is a problem of testing for equality of two population variances, based on small samples.

$$H_0(\sigma_1^2 = \sigma_2^2) \text{ against } H_1(\sigma_1^2 > \sigma_2^2)$$

The test statistic is F defined by (14.8.12). If H_0 is true, the estimates of the common variance from two samples are (see 14.3.6):

$$s_1^2 = \frac{94.5}{(8-1)} = 13.5, \quad s_2^2 = \frac{101.7}{(10-1)} = 11.3$$

$$F = \frac{13.5}{11.3} = 1.19 \text{ (Degrees of freedom 7 and 9)}$$

Critical region at 5% level is $F \geq F_{.05}$. Since the observed value of $F = 1.19$ is less than the critical value 3.29 (for $v_1 = 7$, $v_2 = 9$), we have no reason to reject H_0, and thus conclude that there is no significant difference between the two variances.

Example 4.51 *With the data of Example 14.50, obtain the 95% confidence limits for the ratio of variances* $\dfrac{\sigma_1^2}{\sigma_2^2}$. *(Given, the upper 2.5% value of F with d.f. (7, 9) is 4.20 and d.f. (9, 7) is 4.82).*

Solution From (14.5.12) we have with probability 95%

$$\text{Lower 2.5\% value of } F \text{ with d.f. } (7, 9) \leq \frac{\dfrac{s_1^2}{\sigma_1^2}}{\dfrac{s_2^2}{\sigma_2^2}} \leq \text{Upper 2.5\% value of with d.f. } (7, 9)$$

But, using (13.8.19),

Lower 2.5% value of F with d.f. (7, 9)

$$= 1/[\text{Upper 2.5\% value of } F \text{ with d.f. } (9, 7)]$$

$$= \frac{1}{4.82}$$

Putting the values,

$$\frac{1}{4.82} \leq \frac{13.5}{11.3} \cdot \frac{\sigma_2^2}{\sigma_1^2} \leq 4.20$$

or,
$$\frac{1}{4.82} \leq 1.19 \frac{\sigma_2^2}{\sigma_1^2} \leq 4.20$$

or,
$$4.82 \geq \frac{1}{1.19} \cdot \frac{\sigma_1^2}{\sigma_2^2} \geq \frac{1}{4.20}$$

or,
$$\frac{1.19}{4.20} \leq \frac{\sigma_1^2}{\sigma_2^2} \leq 1.19 \times 4.82$$

i.e.
$$0.28 \leq \frac{\sigma_1^2}{\sigma_2^2} \leq 5.74$$

The 95% confidence limits for σ_1^2/σ_2^2 are 0.28 and 5.74, and the confidence interval is 0.28 and 5.74.

Example 14.52 *The following results were obtained from two independent random samples:*

	Sample Size	*Mean*	*Standard Deviation*
Sample I	5	25	2.1
Sample II	6	29	4.0

Test whether the two samples may be regarded as drawn from the same Normal population.

Given $t_{.025} = 1.82$ *for 9 d.f. and* $f_{.05} = 6.26$ *for (5, 4) d.f.*

Solution A normal distribution has two parameters, namely mean μ and variance σ^2. In order to see if the samples come from the same normal population, we have to test for

(i) equality of two means $H_0\ (\mu_1 = \mu_2)$, by t-test,

(ii) equality of two variances $H_0\ (\sigma_1^2 = \sigma_2^2)$, by F-test.

For (i) we must assume that the population variances are equal, and hence we first apply F-test for (ii) using the test statistic (14.8.12).

$$s_1^2 = \frac{n_1 S_1^2}{(n_1 - 1)} = \frac{5 \times (2.1)^2}{4} = 5.51$$

$$s_2^2 = \frac{n_2 S_2^2}{(n_2 - 1)} = \frac{6 \times (4.0)^2}{5} = 19.20$$

Since $s_2^2 > s_1^2$, we set up $H_1(\sigma_2^2 > \sigma_1^2)$.

$$F = \frac{s_2^2}{s_1^2} = \frac{19.20}{5.51} = 3.48;\ \text{d.f. } (5, 4)$$

[**Note:** In the computation of F, the larger variance appears in the numerator]

Critical Region: $F \geq 6.26$

Since the observed value $F = 3.48$ does not fall in the critical region, it is not significant. We therefore cannot reject the null hypothesis at 5% level and conclude that the population *variances may be equal.*

We can now apply t-test for (i), using *Fisher's t* statistic (14.8.9). $H_1(\mu_1 \neq \mu_2)$. An estimate of the unknown common variance in the population is

$$s^2 = \frac{n_1 S_1^2 + n_2 S_2^2}{n_1 + n_2 - 2} = \frac{5(2.1)^2 + 6(4.0)^2}{5 + 6 - 2} = \frac{118.05}{9} = 13.12$$

$$\therefore \quad t = \frac{25 - 29}{\sqrt{13.12}\sqrt{\frac{1}{5} + \frac{1}{6}}} = \frac{-4}{2.19} = -1.83;\ \text{d.f. } 9$$

Since the alternative hypothesis is two-sided, the critical region is at the two tails of the distribution. At 5% level, the critical region is $|t| \geq t_{.025}$. From Tables $t_{.025} = 2.26$ for 9 d.f.

Critical Region: $|t| \geq 2.26$

Since $|t| = 1.82$ is not greater than 2.26, the observed value is not significant. We cannot reject H_0 at 5% level and conclude that the population *means may be equal*.

Both the hypotheses $H_0(\mu_1 = \mu_2)$ and $H_0(\sigma_1^2 = \sigma_2^2)$ are found to be tenable. We conclude that the two samples may be regarded as drawn from the same Normal population.

14.9 TESTS FOR CORRELATION COEFFICIENT

In sampling from a *bivariate normal* population, the sampling distribution of sample correlation coefficient (r) is far from normal even for large sample size (n). Therefore, tests for correlation coefficient have to be treated separately depending on the largeness of sample size n and whether the population correlation coefficient is $\rho = 0$ or $\rho \neq 0$, i.e. whether the variables are *uncorrelated* or *correlated* in the population.

(1) Test for a specified correlation coefficient (very large sample)

When a sample size (n) is considerably large (say more than 500), the sample correlation coefficient (r) may be assumed to follow roughly a normal distribution with mean ρ (*i.e.*, population correlation coefficient) and standard error (S.E. of r) $= \dfrac{(1-\rho^2)}{\sqrt{n}}$. Hence,

$$z = \frac{r-\rho}{(\text{S.E. of } r)} \tag{14.9.1}$$

has approximately a Standard Normal distribution. This may therefore be used to test the hypothesis H_0 ($\rho = \rho_0$) i.e., whether the sample may be assumed to have arisen from a population with correlation coefficient $\rho = \rho_0$, and also for finding approximate confidence limits for ρ.

95% confidence limits = $r \pm 1.96$ (S.E. of r)

99% confidence limits = $r \pm 2.58$ (S.E. of r) (4.9.2)

where (S.E. of r) = $\dfrac{(1-r^2)}{\sqrt{n}}$ approx.

Example 14.53 *A random sample of 900 pairs of observations from a bivariate normal population shows a correlation coefficient of 0.74. Is it likely that the population correlation coefficient is 0.8? Also find the 99% confidence limits for the correlation coefficient in the population.*

Solution Null Hypothesis is that the population correlation coefficient is 0.8 and Alternative Hypothesis is that it is not 0.8. $H_0(\rho = 0.8)$ against $H_1(\rho = 0.8)$.

Sample size $n = 900$ is quite large. Hence, assuming approximate normal distribution for the sample correlation coefficient and under H_0,

Observed value (r) = 0.74

Expected value (ρ) = 0.8

$$\text{S.E. of } r = \frac{(1-\rho^2)}{\sqrt{n}} = \frac{1-0.64}{\sqrt{900}} = .012$$

The test statistic (14.9.1) is

$$z = \frac{r-\rho}{\text{S.E. of } r} = \frac{0.74-0.8}{.012} = -5$$

The alternative hypothesis is both-sided, and hence this is a two-tailed test. Critical Region is $|z| \geq 1.96$ at 5% level, and $|z| \geq 2.58$ at 1% level.

Since, here $|z| = 5$ is larger than 2.58, z is "highly significant". We therefore reject H_0 and conclude that the population correlation coefficient is not 0.8.

In order to find the confidence limits, we have

$$\text{S.E. of } r = \frac{1-r^2}{\sqrt{n}} \text{ approx. (since } \rho \text{ is now unknown).}$$

$$= \frac{1-0.5476}{\sqrt{900}} = .015$$

99% confidence limits for ρ are

$$r \pm 2.58 \text{ (S.E. of } r) = 0.74 \pm 2.58\ (.015)$$

$$= .74 \pm .04 = 0.78 \text{ and } 0.70$$

(2) Test for 'significance' of observed correlation coefficient

Suppose, we have a random sample of n pairs of observations (x_1, y_1), (x_2, y_2), ..., (x_n, y_n) from a bivariate normal population and let r be the observed correlation coefficient in the sample. It is required to test if this sample correlation coefficient is 'significant' of any correlation in the population (i.e., whether the value of the population correlation coefficient ρ is zero and the observed value of r has arisen due to fluctuation of sampling).

R.A. Fisher has shown that when the Null Hypothesis $H_0(\rho = 0)$ is true, the test statistic

$$t = \frac{r\sqrt{n-2}}{\sqrt{1-r^2}} \tag{14.9.3}$$

follows Student's t distribution with $(n-2)$ degrees of freedom. This statistic is used for testing the significance of r, as in the following examples.

Example 14.54 *A random sample of 18 pairs of observations from a bivariate normal population gives a correlation coefficient of 0.3. It is likely that the variables are uncorrelated in the population ? (At 5% level of significance the value of t for 16 d.f. is 2.12).*

Solution Here $n = 18$, $r = 0.3$.

Null Hypothesis $H_0(\rho = 0)$; Alternative Hypothesis $H_1(\rho \neq 0)$. When H_0 is true, the appropriate test statistic (14.9.3) follows Student's t distribution with $(n-2) = 16$ d.f.

$$t = \frac{0.3\sqrt{18-2}}{\sqrt{1-.09}} = \frac{0.3 \times 4}{0.95} = 1.26$$

Since the alternative hypothesis is both-sided (either p is less than or more than zero), the test is two-tailed.

Critical Region: $|t| \geq 2.12$

The observed value of t does not fall in the critical region (because $|t| = 1.26$ is not more than 2.12) the null hypothesis cannot be rejected at 5% level of significance and we coclude that the variables may be uncorrelated in the population.

Example 14.55 *Find the least value of r in samples of 18 pairs of observation from a bivariate normal population, which is significant at 5% level. (Value of t at 5% level for 16 d.f. is 2.12).*

Solution The value of r for $n = 18$ will be significant at 5% level, if

$$|t| = \left| \frac{r\sqrt{n-2}}{\sqrt{1-r^2}} \right| \geq 2.12$$

Squaring both sides and putting $n = 18$,

$$\frac{r^2(18-2)}{1-r^2} \geq 4.5$$

or, $16r^2 \geq 4.5 - 4.5r^2$

or, $20.5r^2 \geq 4.5$

or, $r^2 \geq 0.22$

or, $|r| \geq 0.47$

The least value of r is 0.47 (numerically).

Fisher's Z-transformation of Correlation Coefficient

It has been stated that when $H_0(\rho = 0)$ is true, the statistic (14.9.3) follows t distribution, and t-test is used for the case $\rho = 0$. This is an exact test. If $\rho \neq 0$, the statistic (14.9.3) does not follow either normal or any other commonly-used distribution, even for large samples. In this case, R.A. Fisher has suggested the following transforation of r to a new variable

$$Z = \frac{1}{2} \log_e \frac{1+r}{1-r} \tag{14.9.4}$$

This is known as **Fisher's Z-transformation** of correlation coefficient. Fisher has shown that even for small samples sizes, Z approximately follows normal distribution with mean μ and s.d. σ, given by

$$\mu = \frac{1}{2} \log_e \frac{1+\rho}{1-\rho}, \qquad \sigma = \frac{1}{\sqrt{n-3}} \tag{14.9.5}$$

i.e., $$U = (Z - \mu)\sqrt{n-3} \tag{14.9.6}$$

is approximately a standard normal variate.

Fisher's Z transformation is used to

(i) test for a specified value of $\rho \neq 0$, and find confidence limits for ρ;

(ii) test for equality of two correlation coefficients.

(3) Test for a specified correlation coefficient (any sample size)

A random sample of n pairs of observations from a bivariate normal population shows a correlation coefficient r. It is required to test the hypothesis that the population correlation coefficient ρ has a specified value. $H_0(\rho = \rho_0)$.

Using Fisher's Z-transformation (14.9.4) and when H_0 is true, the test statistic

$$U = (Z - \mu_0)\sqrt{n-3} \qquad (14.9.7)$$

follows Standard Normal distribution approximately, where μ_0 is the value of μ from (14.9.5) when $\rho = \rho_0$.

[**Note:** (i) Since Fisher's Z-transformation uses logarithm to the base e, for practical calculations we may change the base to 10.

$$Z = \frac{1}{2}\log_e \frac{1+r}{1-r} = \frac{1}{2}\log_{10}\frac{1+r}{1-r} \times \log_e 10$$

$$= \frac{1}{2}\log_{10}\frac{1+r}{1-r} \times 2.30, \text{ since } \log_e 10 = 2.30 \text{ approx.}$$

$$= 115 \log 10 \frac{1+r}{1-r} \text{ approximately} \qquad (14.9.8)$$

(ii) For testing $H_0(\rho = 0)$, the t-test (14.9.3) is preferable, because it is an *exact* test. Fisher's Z-transformation may also be used, but this is an *approximate* test.]

Example 14.56 *A random sample of 28 pairs of observations shows a correlation coefficient of 0.74. Is it reasonable to believe that the sample comes from a bivariate normal population with correlation coefficient 0.6?*

Solution Null Hypothesis $H_0(\rho = 0.6)$

Alternative Hypothesis $H_1(\rho \neq 0.6)$

Using Fisher's Z-transformation (14.9.4) and H_0 is true (here $r = 0.74$, $\rho_0 = 0.6$ and $n = 28$):

$$Z = 1.15\log_{10}\frac{1+r}{1-r} = 1.15\log_{10}\left(\frac{1.74}{0.26}\right)$$

$$= 1.15\log_{10}\frac{174}{26}$$

$$= 1.15(\log 174 - \log 26)$$

$$= 1.15(2.2405 - 1.4150)$$

$$= 1.15(0.8255) = 0.949$$

$$\mu_0 = 1.15\log_{10}\frac{1+\rho_0}{1-\rho_0} = 1.15\log_{10}\frac{1.6}{0.4}$$

$$= 1.15\log 4 = 1.15(0.6021) = 0.692$$

The test statistic is (14.9.7), which follows standard normal distribution under H_0.

$$U = (0.949 - 0.692)\sqrt{28-3}$$

$$= 1.285$$

Since here $|U| < 1.96$, the observed value of the test statistic is not significant at 5% level. We therefore, cannot reject H_0 and conclude that the sample might come from a bivariate normal population with correlation coefficient 0.6.

Example 14.57 *A correlation coefficient of 0.2 is discovered in a sample of 28 pairs. Use Z test to find out if this is significantly different from zero.*

[C.A., Nov. '82]

Solution Our null hypothesis is that the correlation coefficient in the population is zero, and the alternative hypothesis is that it is not zero. $H_0(\rho = 0)$ against $H_1(\rho \neq 0)$. We are given $r = 0.2$, $n = 28$, $\rho_0 = 0$.

Using Fisher's Z-transformation (14.9.4) and when H_0 is true,

$$\begin{aligned} Z &= 1.15 \log \frac{1+0.2}{1-0.2} \\ &= 1.15 \log \frac{1.2}{0.8} \\ &= 1.15 \log 1.5 \\ &= 1.15\,(0.1761) = 0.203 \end{aligned}$$

$$\begin{aligned} \mu_0 &= 1.15 \log \frac{1+0}{1-0} = 1.15 \log 1 \\ &= 1.15 \times 0 = 0 \end{aligned}$$

The test statistic (14.9.7) has the observed value

$$U = (0.203 - 0)\,\sqrt{28-3} = 0.203 \times 5 = 1.015$$

Since $|U| < 1.96$, we cannot reject H_0 at 5% level of significance, and conclude that the population correlation coefficient may be zero; the observed correlation coefficient of 0.2 is not significantly different from zero.

[**Note:** The *t*-test (14.9.3) is more appropriate here, because it is an exact test; whereas Z-test is only an approximate test.]

(4) Test for equality of two correlation coefficients

Suppose that two *independent* random samples of n_1 and n_2 pairs of observations, drawn from bivariate normal populations, show correlation coefficient r_1 and r_2 respectively. It is required to test the hypothesis that the correlation coefficients in the two populations are equal. $H_0(\rho_1 = \rho_2)$.

Using Fisher's Z-transformation (14.9.4), and when H_0 is true,

$$Z_1 = \frac{1}{2} \log_e \frac{1+r_1}{1-r_1}, \qquad Z_2 = \frac{1}{2} \log_e \frac{1+r_2}{1-r_2}$$

are approximately normally distributed and independent with the same mean, but variances

$$\sigma_1^2 = \frac{1}{(n_1-3)}, \qquad \sigma_2^2 = \frac{1}{(n_2-3)}$$

respectively. Consequently, $(Z_1 - Z_2)$ is normally distributed with mean 0 and variance

$\dfrac{1}{n_1-3} + \dfrac{1}{n_2-3}$; i.e.,

$$U = \frac{Z_1 - Z_2}{\sqrt{\left(\frac{1}{n_1 - 3} + \frac{1}{n_2 - 3}\right)}} \quad (14.9.9)$$

is approxiamtely a standard normal variate. If the Alternative Hypothesis is $H_1(\rho_1 \neq \rho_2)$, the observed value of $|U|$ is compared with 1.96 or 2.58 for significance at 5% or 1% level. In case the observed value of U is significant, we conclude that the two population correlation coefficients are not equal.

Example 14.58 *The correlation coefficients 0.45 and 0.70 were obtained from two independent random samples of 19 and 28 pairs of observations respectively drawn from bivariate normal populations. Do these results support the hypothesis that the correlation coefficients in the two populations are equal?*

Solution The null hypothesis is that the correlation coefficients in the two populations are equal, and the alternative hypothesis is that they are not equal. $H_0(\rho_1 = \rho_2)$ against $H_1(\rho_1 \neq \rho_2)$.

Using Fisher's Z-transformation of correlation coefficient, the test statistic is (14.9.9). Here, $r_1 = 0.45$, $r_2 = 0.70$, $n_1 = 19$, $n_2 = 28$.

$$\begin{aligned} Z_1 &= 1.15 \log_{10}\left(\frac{1.45}{0.55}\right) \\ &= 1.15 \log_{10}\left(\frac{145}{55}\right) \\ &= 1.15 (\log 145 - \log 55) \\ &= 1.15 (2.1614 - 1.7404) \\ &= 1.15 (0.4210) = 0.484 \end{aligned}$$

$$\begin{aligned} Z_2 &= 1.15 \log_{10}\left(\frac{1.70}{0.30}\right) \\ &= 1.15 \log_{10}\left(\frac{17}{3}\right) \\ &= 1.15 (\log 17 - \log 3) \\ &= 1.15 (1.2304 - 0.4771) \\ &= 1.15 (0.7533) = 0.866 \end{aligned}$$

$$\therefore \quad U = \frac{0.484 - 0.866}{\sqrt{\frac{1}{16} + \frac{1}{25}}} = \frac{-0.382 \times 20}{\sqrt{41}} = -1.19$$

Since here $|U| < 1.96$, the observed value of the test statistic is not significant at 5% level. The results therefore do not provide any evidence against H_0, and we conclude that the samples might come from populations with equal correlation coefficients.

Table 14.10 Formulae for Confidence Limits

Confidence Limits for (Parameter) (1)	*Confidence Coefficient* (2)	*Confidence Limits* (3)	*Reference to Formula* (4)
(Large Samples)—Approximate Limits			
(1) Population Mean (μ):			
	95%	$\bar{x} \pm 1.96$ (S.E.)	
	99%	$\bar{x} \pm 2.58$ (S.E.)	(14.5.2)
	Almost sure	$\bar{x} \pm 3$ (S.E.)	
where S.E. $= \dfrac{\sigma}{\sqrt{n}}$ or $\dfrac{S}{\sqrt{n}}$ approx.			
(2) Population Proportion (P):			
	95%	$p \pm 1.96$ (S.E.)	
	99%	$p \pm 2.58$ (S.E.)	(14.5.3)
	Almost sure	$p \pm 3$ (S.E.)	
where S.E. $= \sqrt{\dfrac{pq}{n}}$ (approx.), $p + q = 1$ [**Note:** In sampling without replacemnt from a finite population, the formulae for S.E. above should be multiplied by the correction factor $\sqrt{\dfrac{(N-n)}{(N-1)}}$			
(3) Difference of Population Means ($\mu_1 - \mu_2$):			
	95%	$(\bar{x}_1 - \bar{x}_2) \pm 1.96$ (S.E.)	
	99%	$(\bar{x}_1 - \bar{x}_2) \pm 2.58$ (S.E.)	(14.5.4)
	Almost sure	$(\bar{x}_1 - \bar{x}_2) \pm 3$ (S.E.)	
where S.E. $= \sqrt{\dfrac{\sigma_1^2}{n_1} + \dfrac{\sigma_2^2}{n_2}}$ or $\sqrt{\dfrac{S_1^2}{n_1} + \dfrac{S_2^2}{n_2}}$ approx. [**Note:** If $\sigma_1^2 = \sigma_2^2$ (but unknown), they are estimated by $S^2 = \dfrac{(n_1 S_1^2 + n_2 S_2^2)}{(n_1 + n_2)}$			
(4) Difference of Population Proportions ($P_1 - P_2$):			
	95%	$(p_1 - p_2) \pm 1.96$ (S.E.)	
	99%	$(p_1 - p_2) \pm 2.58$ (S.E.)	(14.5.5)
	Almost sure	$(p_1 - p_2) \pm 3$ (S.E.)	
where S.E. $= \sqrt{\dfrac{p_1 q_1}{n_1} + \dfrac{p_2 q_2}{n_2}}$ approx.			

(Contd)

(1)	(2)	(3)	(4)
(Large Samples)—Approximate Limits			
(5) Population Correlation Coefficient (ρ) :			
	95%	$r \pm 1.96$ (S.E.)	
	99%	$r \pm 2.58$ (S.E.)	(14.9.2)
	Almost sure	$r \pm 3$ (S.E.)	
where S.E. $= \dfrac{(1-r^2)}{\sqrt{n}}$ approx.			
(Small Samples)—Exact Limits *Assumption: Normal Population*			
(6) Population Mean (μ) :			
(a) σ known			
	95%	$\bar{x} \pm 1.96$ (S.E.)	
	99%	$\bar{x} \pm 2.58$ (S.E.)	(14.5.6)
where S.E. $= \dfrac{\sigma}{\sqrt{n}}$			
(b) σ unknown			
	95%	$\bar{x} \pm t_{.025} \dfrac{S}{\sqrt{n-1}}$	
	99%	$\bar{x} \pm t_{.005} \dfrac{S}{\sqrt{n-1}}$	(14.5.8a,b)
where S = sample s.d.; t has *d.f.* $(n-1)$			
(7) Difference of Population Means ($\mu_1 - \mu_2$):			
(a) σ_1, σ_2 known			
	95%	$(\bar{x}_1 - \bar{x}_2) \pm 1.96$ (S.E.)	
	99%	$(\bar{x}_1 - \bar{x}_2) \pm 2.58$ (S.E.)	(14.5.7)
where S.E. $= \sqrt{\dfrac{\sigma_1^2}{n_1} + \dfrac{\sigma_2^2}{n_2}}$			
(b) $\sigma_1 = \sigma_2$ unknown			
	95%	$(\bar{x}_1 - \bar{x}_2) \pm t_{.025}\, s\sqrt{\dfrac{1}{n_1} + \dfrac{1}{n_2}}$	
	99%	$(\bar{x}_1 - \bar{x}_2) \pm t_{.005}\, s\sqrt{\dfrac{1}{n_1} + \dfrac{1}{n_2}}$	(14.5.9a, b)
where $s^2 = \dfrac{(n_1 S_1^2 + n_2 S_2^2)}{(n_1 + n_2 - 2)}$; t has d.f. $(n_1 + n_2 - 2)$			

(Contd)

(1)	(2)	(3)	(4)

(Small Samples)—Exact Limits

Assumption: Normal Population

(8) Population Variance (σ^2) :

(a) μ known

95% $$\frac{\Sigma(x_i - \mu)^2}{\chi^2_{.025}} \le \sigma^2 \le \frac{\Sigma(x_i - \mu)^2}{\chi^2_{.975}}$$

99% $$\frac{\Sigma(x_i - \mu)^2}{\chi^2_{.005}} \le \sigma^2 \le \frac{\Sigma(x_i - \mu)^2}{\chi^2_{,995}}$$

(χ^2 has d.f. n). (14.5.10)

(b) μ unknown

95% $$\frac{nS^2}{\chi^2_{.005}} \le \sigma^2 \le \frac{nS^2}{\chi^2_{.975}}$$

99% $$\frac{nS^2}{\chi^2_{.005}} \le \sigma^2 \le \frac{nS^2}{\chi^2_{.995}}$$ (14.5.11)

(χ^2 has d.f. $n - 1$)

(9) Ratio of Population Variances $\left(\frac{\sigma_1^2}{\sigma_2^2}\right)$: μ_1, μ_2 unknown

95% $$\frac{s_1^2}{s_2^2 \, . \, F_{.025}} \le \frac{\sigma_1^2}{\sigma_2^2} \le \frac{s_1^2}{s_2^2 \, . \, F_{.975}}$$

99% $$\frac{s_1^2}{s_2^2 \cdot F_{.005}} \le \frac{\sigma_1^2}{\sigma_2^2} \le \frac{s_1^2}{s_2^2 \, . \, F_{.995}}$$ (14.5.13)

where $s_1^2 = \dfrac{n_1 S_1^2}{(n_1 - 1)}$, $s_2^2 = \dfrac{n_2 S_2^2}{(n_2 - 1)}$; S_1, S_2 being sample standard deviations; F has d.f. $(n_1 - 1, n_2 - 1)$; $s_1^2 > s_2^2$

Table 14.11 Test of Significance (Summary)

Test for Parameter & Null Hypothesis H_0	*Test Statistic*	*Sampling Distribution under H_0*	*Alternative Hypothesis H_1*	*Level of Significance*	*Reject H_0 if (i.e. Critical Region)*
(1)	*(2)*	*(3)*	*(4)*	*(5)*	*(6)*
Large Sample Tests (Approximate)					
(1) Test for a Specified Proportion: $H_0(P = P_0)$					
	$z = \dfrac{P - P_0}{\sqrt{\dfrac{P_0 Q_0}{n}}}$	Standard Normal	$P \ne P_0$	.05 .01	$\|z\| \ge 1.96$ $\|z\| \ge 2.58$

(Contd)

(1)	(2)	(3)	(4)	(5)	(6)
Large Sample Tests (Approximate)					
		(see 14.7.1)		$P > P_0$	.05 $z \geq 1.645$ .01 $z \geq 2.33$
				$P > P_0$	.05 $z \leq -1.645$ .01 $z \leq -2.33$
(2) Test for Equality of Two Proportions: $H_0(P_1 = P_2)$					
	$z = \dfrac{p_1 - p_2}{\sqrt{pq\left(\dfrac{1}{n_1} + \dfrac{1}{n_2}\right)}}$	Standard Normal		$P_1 \neq P_2$	.05 $\lvert z \rvert \geq 1.96$ .01 $\lvert z \rvert \geq 2.58$
		(see 14.7.1)		$P_1 > P_2$	.05 $z \geq 1.645$ .01 $z \geq 2.33$
	where $p = \dfrac{n_1 p_1 + n_2 p_2}{n_1 + n_2}$			$P_1 > P_2$	.05 $z \leq -1.645$ .01 $z \leq -2.33$
(3) Test for a Specified Mean : $H_0(\mu = \mu_0)$					
	$z = \dfrac{\bar{x} - \mu_0}{\dfrac{\sigma}{\sqrt{n}}}$	Standard Normal		$\mu \neq \mu_0$	.05 $\lvert z \rvert \geq 1.96$ .01 $\lvert z \rvert \geq 2.58$
	(σ known)	(see 14.7.6)		$\mu > \mu_0$	.05 $z \geq 1.645$ .01 $z \geq 2.33$
	or $= \dfrac{\bar{x} - \mu_0}{\dfrac{S}{\sqrt{n}}}$ (σ unknown)			$\mu > \mu_0$	.05 $z \leq -1.645$ .01 $z \leq -2.33$
(4) Test for a Specified Mean: $H_0(\mu_1 = \mu_2)$					
	$z = \dfrac{\bar{x}_1 - \bar{x}_2}{\sqrt{\dfrac{\sigma_1^2}{n_1} + \dfrac{\sigma_2^2}{n_2}}}$	Standard Normal		$\mu_1 \neq \mu_2$	.05 $\lvert z \rvert \geq 1.96$ .01 $\lvert z \rvert \geq 2.58$
	(σ_1, σ_2 known)	(see 14.7.9)		$\mu_1 > \mu_2$	.05 $z \geq 1.645$ .01 $z \geq 2.33$
	or $= \dfrac{\bar{x}_1 - \bar{x}_2}{\sqrt{\dfrac{S_1^2}{n_1} + \dfrac{S_2^2}{n_2}}}$ (σ_1, σ_2 unknown)			$\mu_1 > \mu_2$	.05 $z \leq -1.645$ .01 $z \leq -2.33$

(Contd)

(1)	(2)	(3)	(4)	(5)	(6)

Large Sample Tests (Approximate)

(5) Test for Goodness fo Fit : H_0*(Data support theory)*
(Pearsonian χ^2)

Test statistic	Distribution	Alternative	Critical region
$\chi^2 = \sum \frac{(f_0 - f_e)^2}{f_e}$	χ^2 d.f. = $k - 1$ (see 14.7.11)	Data do not support theory	.05 $\chi^2 \geq \chi^2_{.05}$; .01 $\chi^2 \geq \chi^2_{.01}$

(6) Test for Independence of Attributes: H_0 *(Attributes are Independent)*
(Contingency χ^2)

Test statistic	Distribution	Alternative	Critical region
$\chi^2 = \sum \frac{(f_0 - f_e)^2}{f_e}$	χ^2 d.f. = $(m-1)(n-1)$ (see 14.7.12)	Attributes are associated	.05 $\chi^2 \geq \chi^2_{.05}$; .01 $\chi^2 \geq \chi^2_{.01}$

where

$$f_e = \frac{(A_i)\ (B_j)}{N}$$

For 2 × 2 table, apply Yates' correction:

$$\chi^2 = \frac{N(|ad - bc| - N/2)^2}{R_1\ R_2\ C_1\ C_8}\text{; d.f.} = 1 \text{ (see 14.7.14)}$$

(7) Test for a Specified Correlation Coefficient: H_0 $(\rho = \rho_0)$

(a) for very large sample

Test statistic	Distribution	Alternative	Critical region
$z = \frac{r - \rho_0}{\frac{1 - \rho_0}{\sqrt{n}}}$	Standard Normal (see 14.9.1)	$\rho \neq \rho_0$	.05 $\|z\| \geq 1.96$; .01 $\|z\| \geq 2.58$

(b) using Fisher's Z-transformation

Test statistic	Distribution	Alternative	Critical region
$U = (Z - \mu_0)\ \sqrt{n-3}$	Standard Normal	$\rho \neq \rho_0$	.05 $\|z\| \geq 1.96$; .01 $\|z\| \geq 2.58$

where

$$Z = \frac{1}{2}\log_e \frac{1+r}{1-r} \quad \text{(see 14.9.7)}$$

$$\mu_0 = \frac{1}{2}\log_e \frac{1+\rho_0}{1-\rho_0}$$

(8) Test for Equality of Two Correlation Coefficients : H_0 $(\rho_1 = \rho_2)$

Test statistic	Distribution	Alternative	Critical region
$U = \frac{Z_1 - Z_2}{\sqrt{\frac{1}{n_1 - 3} + \frac{1}{n_2 - 3}}}$	Standard Normal (see 14.9.9)	$\rho_1 \neq \rho_2$	.05 $\|z\| \geq 1.96$; .01 $\|z\| \geq 2.58$

where

$$Z_1 = \frac{1}{2}\log_e \frac{1+r}{1-r}; \qquad Z_2 = \frac{1}{2}\log_e \frac{1+r_2}{1-r_2}$$

(Contd)

(1)	(2)	(3)	(4)	(5)	(6)
		Small Sample Tests (Exact)			
		(Assumption: Normal population)			
(9) Test for a Specified Mean: $H_0(\mu = \mu_0)$					
(a)	**σ known**				
	$z = \dfrac{\bar{x} - \mu_0}{\sigma / \sqrt{n}}$	Standard Normal	$\mu \neq \mu_0$	.05	$\lvert z \rvert \geq 1.96$
				.01	$\lvert z \rvert \geq 2.58$
		(see 14.8.1)	$\mu > \mu_0$	.05	$z \geq 1.645$
				.01	$z \geq 2.33$
			$\mu > \mu_0$	.05	$z \leq -1.645$
				.01	$z \leq -2.33$
(b)	**σ unknown**				
	(Student's t)	t	$\mu \neq \mu_0$	.05	$\lvert t \rvert \geq t_{.025}$
				.01	$\lvert t \rvert \geq t_{.005}$
	$t = \dfrac{\bar{x} - \mu_0}{S / \sqrt{n-1}}$	d.f. = $n - 1$ (see 14.8.6)	$\mu > \mu_0$	.05	$t \geq t_{.05}$
				.01	$t \geq t_{.01}$
			$\mu < \mu_0$	.05	$t \leq -t_{.05}$
				.01	$t \leq -t_{.005}$
(10) Test for Equality of Two Mean: $H_0(\mu_1 = \mu_2)$					
(a) σ_1, σ_2 known (two independent samples)					
	$z = \dfrac{\bar{x}_1 - \bar{x}_2}{\sqrt{\dfrac{\sigma_1^2}{n_1} + \dfrac{\sigma_2^2}{n_2}}}$	Standard Normal	$\mu_1 \neq \mu_2$	.05	$\lvert z \rvert \geq 1.96$
				.01	$\lvert z \rvert \geq 2.58$
		(see 14.8.7)	$\mu_1 > \mu_2$	.05	$z \geq 1.645$
				.01	$z \geq 2.33$
			$\mu_1 < \mu_2$	.05	$z \leq -1.645$
				.01	$z \leq -2.33$
(b) $\sigma_1 = \sigma_2$ unknown (two independent samples)					
	(Fisher S t)				
	$t = \dfrac{\bar{x}_1 - \bar{x}_2}{s\sqrt{\dfrac{1}{n_1} + \dfrac{1}{n_2}}}$	t d.f.= $n_1 + n_2 - 2$	$\mu_1 \neq \mu_2$	.05	$\lvert t \rvert \geq t_{.025}$
				.01	$\lvert t \rvert \geq t_{.005}$
	where	(see 14.8.9)	$\mu_1 > \mu_2$	.05	$t \geq t_{.05}$
				.01	$t \geq t_{.01}$
	$s^2 = \dfrac{n_1 S_1^2 + n_2 S_2^2}{n_1 + n_2 - 2}$		$\mu_1 < \mu_2$	.05	$t \leq -t_{.05}$
				.01	$t \leq -t_{.005}$

(Contd)

(1)	(2)	(3)	(4)	(5)	(6)
Small Sample Tests (Exact) *(Assumption: Normal population)*					
(c) From Correlated Pairs : $H_0\ (\mu_x = \mu_y)$ (Paired t)					
$t = \dfrac{\bar{d}}{S/\sqrt{n-1}}$		t d.f. $= n-1$	$\mu_1 < \mu_2$	.05 .01	$\lvert t \rvert \geq -t_{.025}$ $\lvert t \rvert \geq -t_{.005}$
where $d = x - y$		(see 14.8.10 a)	$\mu_x \neq \mu_y$	.05 .01	$t \geq t_{.05}$ $t \geq t_{.01}$
			$\mu_x < \mu_y$	.05 .01	$t \leq -t_{.05}$ $t \leq -t_{.01}$
(11) Test for a 'significance' of observed Correlation Coefficient : $H_0\ (\rho = 0)$					
$t = \dfrac{r\sqrt{n-2}}{\sqrt{1-r^2}}$		t d.f. $= n-2$ (see 14.9.3)	$\rho \neq 0$	.05 .01	$\lvert t \rvert \geq t_{.025}$ $\lvert t \rvert \geq t_{.005}$
(12) Test for a Specified s.d. : $H_0\ (\sigma = \sigma_0)$					
(a) μ known					
$\chi^2 = \dfrac{\Sigma(x_i - \mu)^2}{\sigma_0^2}$		χ^2 d.f. $= n$	$\sigma > \sigma_0$	.05 .01	$\chi^2 \geq \chi^2_{.05}$ $\chi^2 \geq \chi^2_{.01}$
(b) μ unknown					
$\chi_2 = \dfrac{\Sigma(x_i - \bar{x})^2}{\sigma_0^2} = \dfrac{nS^2}{\sigma_0^2}$		χ^2 d.f. $= n-1$ (see 14.8.11)	$\sigma > \sigma_0$	.05 .01	$\chi^2 \geq \chi^2_{.05}$ $\chi^2 \geq \chi^2_{.01}$
(13) Test for Equality of Two s.d. (μ_1, μ_2) unknown): $H_0\ (\sigma_1 = \sigma_2)$					
$F = \dfrac{s_1^2}{s_2^2};\ (s_1 > s_2)$ $= \dfrac{n_1 S_1^2/(n_1-1)}{n_2 S_2/(n_2-1)}$ where S_1, S_2 are sample s.d.;		F d.f. $= (n_1 - 1, n_2 - 1)$ (see 14.8.12)	$\sigma_1 > \sigma_2$	.05 .01	$F \geq F_{.05}$ $F \geq F_{.01}$

EXERCISES

1. State the criteria of a good estimator. Give illustrations.
2. Distinguish between *point estimation* and *interval estimation*. *[C.U., B.Sc.(Econ) '81]*
3. On the basis of the following random sample, obtain unbaised estimates of the mean μ and variance σ^2 in the population: 38, 32, 39, 37, 27, 35, 28, 32, 30, 35.
4. Define a *consistent estimator* and give an example.
5. Show that in random samples from a normal population with a known variance, the sample mean $\bar{x}$ is unbiased and consistent for stimation the population mean μ.
6. What is a '*sufficient statistic*'? Why is it so called? Give an example. *[C.U., B.Sc. '77]*
7. Explain the terms '*unbiased estimator*' and '*minimum variance estimator*'. *[W.B.H.S. '78, '83]*
8. (a) Let T_1 and T_2 be statistics with expectations $E(T_1) = \theta_1 + \theta_2$ and $E(T_2) = \theta_1 - \theta_2$. Find unbiased estimators of θ_1 and θ^2. *[W.B.H.S. '78]*

   (b) If T_1 and T_2 be statistics with expectations $E(T_1) = 2\theta_1 + 3\theta_2$ and $E(T_2) = \theta_1 + \theta_2$, find unbiased estimators of parameters θ_1 and θ_2. *[W.B.H.S. '83]*
9. What do you understand by Best Linear Unbiased Estimator (*BLUE*)? Give an example.
10. Describe the '*method of maximum likelihood*' for the estimation of unknown parameters. State the important properties of maximum likelihood estimators.
11. If $x_1, x_2, \ldots\ldots x_n$ is a random sample from a normal population with zero mean, obtain the maximum likelihood estimator of the variance, and show that it is unbaised.
12. On the basis of a random sample from a binomial population with parameters (n, P), find the maximum likelihood estimator of P. Show that this estimator is unbiased.
13. Suppose $x_1, x_2, \ldots\ldots x_n$ are independent random observations from a univariate normal population with mean μ and variance one. Find the maximum likelihood estimator for μ. *[C.U., B.Sc. '78]*
14. Show that the estimates of the Poisson parameter obtained by the method of moments and by the method of maximum likelihood are identical.
15. Show that both the Method of Moments and the Method of Maximum Likelihood give identical estimates of the parameters of a normal population. *[C.U., B.Sc. (Econ) '82]*
16. Explain the concept of *confidence interval, confidence limits* and *confidence coefficient*.
17. The mean height obtained from a sample of size 100 taken randomly from a population is 64 inches. If the s.d. of the height distribution of the population is 3 inches, set up probable limits to the mean height of the population.

18. A random sample of 400 is taken from a large number of coins. The mean weight of the coins in the sample is 28.57 gms and the s.d. is 1.25 gms. What are the limits which have a 95% chance of including the mean weight of all the coins.
19. A random sample of 100 days shows an average daily sale of Rs 50 with a s.d. of Rs 10 in a particular shop. Assuming a normal distribution, construct a 95% confidence interval for the expected sale per day. *[C.U., B.Sc. (Econ) '82]*
20. In a random sample of size 100 taken from apopulation of size 1,000, the mean and S.D. of a sample characteristic are found to be 4.8 and 1.1 respectively. Find the 95% confidence interval for population mean.
21. A random sample of size 1000 selected from a large bulk of mass produced machine parts contains 6% defectives. What information can be inferred about the precentage of defective in the bulk? *[C.U., M .Com, '73]*
22. A random sample of 500 pineapples was taken from a large consignment and 65 of them wre found to be bad. Show that the Standard Error of the proportion of bad ones in a sample of this size is 0.015 and deduce that the percentage of bad pineapples in the consignment certainly lies between 8.5 and 17.5. *[I.C.W.A., Dec. '78; June '82]*
23. In a sample of 400 oranges from a large consignment, 40 fruits were considered defective. Estimate the percentage of defective oranges in the whole consignment and assign limits within which the percentage will probably lie. *[C.A., Nov. '75]*
24. A factory is producing 50,000 pairs of shoes daily. From a sample of 500 pairs, 2% were found to be of sub-standard quality. Estimate the number of pairs that can be reasonably expected to be spolied in the daily production and assign limits at 95% level of confidence. *[C.A., May '79]*
25. Out of 20,000 customers' ledger accounts, a sample of 600 accounts was taken to test the accuracy of posting and balancing, and 45 mistakes were found. Assign limits within which the number of defective cases can be expected at 5% level. (z = 2.00). *[C.A., May '76]*
26. (a) Under what circumstances can the normal distribution be used to find confidence limits of the populations mean?
    (b) What is Student's t distribution? When is it used to construct a confidence interval estimate of the population mean?
27. On the basis of a random sample from a normal population with a known variance σ^2, obtain 99% confidence limits for the population mean μ. What will be the confidence limits, if the variance is unknown?
28. A sample of 16 items is randomly drawn from a normal population with unknown mean and s.d. = 12. If the mean of the sample is 40, find 95% interval estimate of the population mean.
29. Suppose that a random sample of size 10, drawn from a normal population, has mean 40 and s.d. 12. Find 99% confidence limits for the population mean. (Given $t_{.005}$ = 3.25 for 9 d.f.).

30. The marks obtained by 17 candidates in an examination have a mean 57 and variance 64. Find 99% confidence limits for the mean of the population of marks, assuming it to be normal. For 16 degrees of freedom $P(|t| > 2.9) = 0.01$. *[C.U., B.Sc.(Math) '77]*

31. Two independent random samples of sizes 8 and 6 are drawn from normal populations with unknown means μ_1, μ_2 and variances 32 and 30 respectively. If the sample means are 68.1 and 60.4 respectively, find 95% confidence limits for the difference of population means.

32. Two independent random sample of size 8 and 6 are drawn from two normal populations whose means and variances are unknown. If the samples have means 28.3 and 20.8, and standard deviations 6 and 5 respectively, find 95% confidence limits for the difference of population means. State the necessary assumption. (Value of t for 12 d.f. is $t_{.025} = 2.18$).

33. The following random sample was obtained from a normal population: 13, 10, 11, 20, 8, 16. Find the 95% confidence limits for the population standard deviation, when the population mean is (i) known to be 14, (ii) unknown. Given

d.f.	$X^2_{.975}$	$X^2_{.025}$
5	0.83	12.8
6	1.24	14.4

34. For tow independent random samples of sizes $n_1 = 8$ and $n_2 = 10$, the standard deviations are respectively $S_1 = 7$, $S_2 = 6$. Assuming that the populations are normal, find the 95% confidence limits for the variance-ratio σ_1^2/σ_2^2. Given $F_{.995} = 0.117$ and $F_{.005} = 6.88$ for (7, 9) d.f.

35. (a) What is meant by a test of a null hypothesis? Wha tare Type I and Type II errors? *[W.B.H.S. '79]*

    (b) Explain the terms: Tests of Significance, Null hypothesis, Critical region, Level of Significance. The explanations must be as brief as possible. *[I.C.W.A., Dec.'78]*

36. Explain clearly the following terms in connection with testing of hypothesis: (i) Region of Acceptance, (ii) First kind of error, (iii) Second kind of error. *[W.B.H.S., '83]*

37. In order to test whether a coin is perfect. I shall toss it six times. I shall reject the null hypothesis of perfectness if and only if I get no head or 6 heads. What is then the probability of Type I error for my test? *[W.B.H.S., '79]*

38. A professor is concerned with the effectiveness of a given teaching technique. What null hypothesis is he testing if he is committing a Type I error when he erroneously concludes that the particular teaching technique is effective? *[W.B.H.S., '80]*

39. In order to test whether a coin is perfect, the coin is tossed 5 times. The null hypothesis of perfectness is rejected if and only if more than 4 heads are obtained. What is the probability of Type I error? Find the probability of Type II error when the corresponding probability of head is 0.2. *[W.B.H.S., '82]*

40. The proportion of defective items in a large lot of items is p. To test the hypothesis $p = 0.2$, we take a random sample of 8 items and accept the hypothesis if the number of defectives in the sample is 6 or less. Find the probability of type I error of the test. What is the type II error if $p = 0.3$? *[W.B.H.S., '78]*
41. A coin is tossed 900 times and heads appear 490 times. Does this result support the hypothesis that the coin is unbaised? *[I.C.W.A., Dec. '74]*
42. A sample of size 600 persons selected at random from a large city shows that the percentage of male in the sample is 53%. It is believed that male to total population ratio in the city is 1/2. Test whether this belief is confirmed by the observation. Also, find 95% confidence limits for the percentage of males in the whole city.
43. A person threw 10 dice 500 times and obtained 2560 times 4, 5 or 6. Can this be attributed to fluctuations of sampling? *[C.A., May '75]*
44. In a sample of 500 people in Kerala, 280 are tea drinkers and the rest are coffee drinkers. Can we assume that both coffee and tea are equally popular in this State at 1% level of significance? *[C.A., May '78]*
45. Over a long period, a travelling salesman's records show a proportion of successful calls of 1/3. He tries a new technique and achieves 19 successes in 40 calls. Does this signify that the new technique is effective?
46. In a sample of 400 parts manufactured by a factory, the number of defective parts was found to be 30. The compay however claims that only 5% of their product is defective. Is the claim tenable? *[I.C.W.A., Dec. '75]*
47. A manufacturer claimed that at least 95% of the equipment which he supplied to a factory conformed to specifications. An examination of a sample of 200 pieces of equipment revealed that 18 were faulty. Test his claim at a significance level of (i) .05, (ii) .01. *[I.C.W.A., Dec. '77; C.A., Nov. '79]*
48. A manufacturer claimed that at least 90% of the components which hw supplied, conformed to specifications. A random sample of 200 components showed that only 164 were upto the standard. Test this claim at 5% level of significance. *[C.U., B.Sc.(Econ) '81]*
49. A Life Insurance Company has 1500 policies averaging Rs 2,000 on lives at age 30. From the experience table, it is found that of 100,000 alive at age 30, only 99,000 are alive at age 31. Find tha lower and the upper values of the amount that Company will have to pay out in insurance during the year. *[C.A., May '82]*
50. In a certain district A, 450 persons were considered regular consumers of Tea our of a sample of 1000 persons. In another district B, 400 were regular consumers of Tea out of a sample of 800 persons. Do these facts reveal a significant difference between the two district as far as tea-drinking habit is concerned? (Use 5% level) *[C.A., Nov. '75]*
51. In a sample of 600 students of a certain college, 400 are found to use dot pens. In another college from a sample of 900 students 450 were found to use dot pens. Test whether the two colleges are significantly different with respect to the habit of using dot pens. (Null and alternativge hypotheses should be stated clearly). *[I.C.W.A., Dec. '82]*

52. Before an increase in excise duty on tea, 400 people out of a sample of 500 persons were found to be tea drinkers. After an increase in duty, 400 people were tea drinkers in a sample of 600 people. Using standard error of proportion, state whether there is a significant decrease in the consumption of tea. *[C.A., May '77]*
53. In a large city, 20% of a random sample of 1100 school boys had a certain physical defect. In anther large city, 200 out of a random sample of 900 school boys had the same defect. Do you think that the percentage is less in the former city? Calculate the 95% confidence limits for the difference.
54. Is it likely that a sample of 300 items whose mean is 16.0 is a random sample from a large population whose mean is 16.8 and standard deviation 5.2? *[I.C.W.A., June '73; M.B.A. '78]*
55. An automatic machine fills tea in sealed tins with mean weight of tea 1 kg. and Standard Deviation = 1 gm. A random sample of 50 tins was examined, and it was found that their mean weight was 999.50 gms. Using standard error of the mean, state whether the machine is working properly or not. *[C.A., Nov. '73]*
56. The mean lifetime of a sample of 100 fluorescent-light bulbs produced by a company is computed to be 1570 hours with a standard deviation of 120 hours. The company claims that the average life of the bulbs is 1600 hours. Using a level of significance of .05 is the claim acceptable? *[I.C.W.A., June, '76]*
57. A sample of 900 members has a mean 3.4 cm. and s.d. 2.61cm. Can the sample be regarded as drawn from a population with mean 3.25 cm? Find the 95% confidence limits for the population mean. *[C.U., B.Sc. '78]*
58. A new variety of potato grown on 250 plots gave rise to a mean yield of 82.7 mds. per acre with a s.d. of 14.6 mds. per acre. Is it reasonable to assert that the new variety is superior in yield to the standard variety with an established yield of 80.2 mds. per acre?
59. The mean breaking strength of the cables supplied by a manufacturer is 1800 with a stadard deviation of 100. By a new technique in the manufacturing process it is claimed that the breaking strength of the cables has increased. In order to test this claim a sample of 50 cables is tested. It is found that the mean breaking strength is 1850. Can we support the claim at a 0.01 level of significance? *[I.C.W.A., Dec. '80]*
60. The average number of defective articles per day in a certain factory is claimed to be less than the average for all the factories. The average for all the factories is 30.5. A random sample of 100 days showed the following distribution:

Class Limits	16–20	21–25	26–30	31–35	36–40	Total
No. of Days	12	22	20	30	16	100

Calculate the mean and standard deviation of the sample and use it to claim that the average is less than the figure for all factories, at 5% level of significance. Given $Z(1.645) = 0.95$. *[C.A., May '79]*

61. Random samples of size 500 and 400 have means 11.5 and 10.9 respectively. Can the samples be regarded as drawn from the same population of s.d. 5? Find 99% confidence limits for the difference of means.

62. Two salesmen *A* and *B* are working in a certain district. From a sample survey conducted by the Head Office, the following results were obatined. State whether there is any significant difference in the average sales between the two salesmen:

	A	*B*
No. of Sales	20	18
Average Sales in Rs	170	205
S.D. in Rs	20	25

[C.A., Nov. '77]

63. Intelligence tests on two groups—one group consisting of 121 girls and the other group consisting of 81 boys—give the following results:

Group of Girls	Mean 84	Standard Deviation 10
Group of Boys	Mean 81	Standard Deviation 12

Examine if the difference is significant. [I.C.W.A., Dec. '81]

64. A simple sample of heights of 6400 Englishmen has a mean of 67.85 inches and a s.d. of 2.56 inches, while a simple sample of heights of 1600 Australians has a mean of 68.55 and a s.d. of 2.52 inches. Do the data indicate that Australians are on the average taller than the Englishmen?

[C.U., B.Sc.(Math) '67]

65. Define Pearsonian χ^2 and discuss its uses in testing of hypothesis.

[W.B.H.S. '82]

66. Of 160 offsprings of a certain cross between guinea pigs, 102 were red, 24 were black and 34 were white. According to a genetic model the probabilities of red, black and white are respectively 9/16, 3/16 and 1/4. Test at 5% significance level, If the data are consistent with the model. For 2 degrees of freedom $P(x^2 > 5.99) = 0.05$. [C.U., B.Sc.(Math) '77, '79]

67. 200 digits from 0 to 9 are taken at random from a page of a certain random from number table. The frequency distribution of the digits is given:

Digit	0	1	2	3	4	5	6	7	8	9
Frequency	18	19	13	21	16	25	22	20	21	25

Can this table be regarded as random? (Given $\chi^2_{\cdot 05,\, 9} = 16.92$).

[W.B.H.S., '81]

68. In 60 throws of a die, face one turned up 6 times, face two or three 18 times, face four or five 24 times, and face six 12 times. Test at 10% significance level if the die is honest, it being given that $P(\chi^2 > 6.25) = 0.1$ for 3 degrees of freedom.

[C.U., B.Sc.(Math) '75]

69. Discuss the importanceof χ^2 test. How is it used to test the association between attributes? [C.A.,Nov. '82]

70. A random sample of 200 students in a Calcutta college were asked the question: "Do you think scientists are slightly unbalanced people?" The number of students of each calss saying 'yes' and 'no' are:

	Class in College			
	P.U.	1st year	2nd year	3rd year
Yes	15	8	5	2
No	55	42	35	38

Test whether there is any association between opinion and class in College. (Given $\chi^2_{.05} = 7.81$ for 3 d.f.)

71. Random samples from lots supplied by four manufacturers A, B, C, D were examined, and the following defectives were detected:

Manufacturer	A	B	C	D
Sample Size	100	200	150	250
Number of Defectives	20	35	40	45

Do you consider that the lots supplied by the different manufacturers are of the same quality? Given, $\chi^2_{.05} = 7.81$ for 3 degrees of freedom.

72. A Company using door-to-door sales procedure is testing a new sales approach and has the following results on a comparative test under otherwise identical conditions:

	Sales	No Sales
Old Approach	8	12
New Approach	10	10

Use χ^2 test to determine the significane of the observed difference. [Give that the values of χ^2 from statistical tables for 1 degree of freedom are 3.8 and 6.6 at significance levels .05 and .01 respectively.]

73. In an infantile paralysis epidemic, 500 persons contracted the disease. 200 received no serum treatment and of these 75 became paralysed. Of those who did receive serum treatment 65 became paralysed. Was the serum treatment effective? $\chi^2_{.01} = 6.64$ for 1 d.f. *[C.A., Nov. '79]*

74. From the following results regarding eye-colour of fathers and sons, test if the colour of son's eyes is associated with that of father's.

Father's Eye-colour	*Son's Eye-colour* Light	Not-light
Light	47	16
Not-light	14	23

(Given $\chi^2_{.05} = 3.84$ for 1d.f.)

75. Out of 8,000 graduates in a town 800 are females, out of 1,600 graduate employees 120 are females. Use χ^2 test to determine if any distinction is made in appointment on the basis of sex. Value of χ^2 for 5% level for one degree of freedom is 3.84. *[C.A., Nov. '81]*

76. A variable x is normally distributed in the population with mean 20 and standard deviation 5. If a random sample of size 25 is drawn, what is the probability that the sample mean $\bar{x}$ will be greater than 21? (Probability that the normal deviate lies between 0 and 1 is .3413). *[B.U., B.A.(Econ) '71]*
77. The mean of a random sample of size 10 is 12.1. It is known that the population s.d. is 3.2. Can it be concluded that the sample came from a normal population with mean 14.5? Find the 99% confidence limits for the population mean.
78. A sample of 100 iron bars is said to be drawn from a large number of bars whose lengths are normally distributed with mean 4 ft. and s.d. 0.6 ft. If the sample mean is 4.2 ft., can the sample be regarded as a truly random sample? (Null hypothesis and assumptions should be stated clearly). *[I.C.W.A., June '82]*
79. A manufacturer of string has found from past experience that samples of a certain type have a mean breaking strength of 15.6 kg and a s.d. of 2.2 kg. A time-saving change in the manufacturing process of this stirng is tried. A sample of 50 pieces is then taken, for which the mean breaking strength turns our to be 15.5 kg. On the basis of this sample can it be concluded that the new process has a harmful effect on the strength of the string? (Assume that breaking strength of string is normally distriubuted). *[C.U., B.Sc.(Math) '65]*
80. Independent samples of sizes 30 and 55 from two normal populations having a common variance 17.6 were found to have means 23.0 and 21.9 respectively. Test at 5% level of significance whether the populations have also the same mean. Also, find the 95% confidence limits for the difference of population means. Given $P(z > 1.96) = 0.025$.
81. Two independent random samples of sizes 10 and 25 from two normal populations having variances 9.61 and 7.29 were found to have means 23.0 and 20.3 respectively. Test at 1% significance level the hypothesis that mean of the first population is larger. Calculate also the 99% confidence limits for the difference of means.
82. For a random sample of size 10 the mean is 12.1 and the standard deviation is 3.2. Is it reasonable to suppose that this sample came from a normal population with mean 14.5? [Given $P(|t| > 2.26) = .05$ for d.f.]. Also, find 95% confidence interval for mean of the population.
83. A Company has been producing steel tubes of mean inner diameter 2.00 cm. A sample of 10 tubes gives a mean inner diameter of 2.01 cm and a variance of .004 sq. cm. Is the difference in the value of mean significant? (Value of t for d.f. 9 at 5% level = 2.262). *[C.A., Nov. '81]*
84. The heights of 10 males of a given locality are found to be 70, 67, 62, 68, 61, 68, 70, 64, 64, 66 inches. Is it reasonable to believe that the average height is greater than 64 inches? Test at 5% significance level, assuming that for 9 degrees of freedom $P(t > 1.83) = 0.05$. *[C.U. B.Sc. (Math) '78]*
85. It is claimed that students entering a college have an average I.Q. higher than 100. A random sample of 16 is taken and the sample mean is found to be 106. The sample s.d. is 10. Is the claim supportable? (It is assumed that the I.Q. are normally distributed, and given $t_{.01} = 2.82$ for 9 d.f.). *[C.U., B.Sc. (Econ) '81]*

86. The heights of 10 adult males selected at random from a given locality had a mean 63.2 inches and variances 6.25 sq. inches. Test at 5% significance level the hypothesis that the adult males of the given locality are on the average less than 65 inches tall. [Given, for 9 degrees of freedom $P(t > 1.83) = 0.05$]

87. (a) How do you test the significance of the difference between the means of two samples?

(b) Two samples of 6 and 5 items respectively gave the following data:

Mean of the first sample = 40
S.D. of the first sample = 8
Mean of the second sample = 50
S.D. of the second sample = 10

Is the difference of the means significant? (The value of t for 9 d.f. at 5% level is 2.26). *[C.A., May '81]*

88. The means of two independent random samples of sizes 8 and 6 are 38.4 and 33.7 respectively. The sum of the squares of deviations from the respective sample means are 20.8 and 15.7. Do you think that mean of the first population is the larger? (Assume Normal population with the same standard deviation). Also find 99% confidence limits for the difference of the population means. Given $t_{.01} = 2.68$ and $t_{.005} = 3.06$ for 12 d.f.

89. The heights of 6 randomly chosen Sailor are (in inches): 63, 65, 68, 69, 72, 71, and those of 10 Soldiers are 66, 62, 69, 61, 65, 72, 69, 73, 70, 71. In the light of these data, discuss the suggestion that Sailors are on the average taller than Soldiers. (Given, the 5% value of t for 14d.f. is 1.76).

90. Ten soldiers visit a rifle range for two consecutive weeks. Their scores are

Soldier	*A*	*B*	*C*	*D*	*E*	*F*	*G*	*H*	*I*	*J*
1st week	67	24	57	55	63	54	56	68	33	43
2nd week	70	38	58	58	56	67	68	72	42	38

Examine if there is any significant improvement in their performance. Given $t_{.01} = 2.82$ for 9 d.f.

91. Two laboratories carry out independent estimates of a particular chemical in a medicine produced by a certain firm. A sample is taken from each batch, halved and the separate halves sent to the two laboratories. The following data are obtained:

Number of samples	...	10
Mean value of the difference of estimates	...	0.6
Sum of the squares of the differences (from their mean)	...	20

Is the difference significant? (Value of t at 5% level for 9 d.f. is 2.26). *[C.A., May '82]*

92. Nine patients to whom a certain drug was administered, registered the following rise in blood pressure: 3, 7, 4, – 1, – 3, 6, – 4, 1, 5. Test the hypothesis that the drug does not raise the blood pressure at 10% level of significance.

Assume that the sample is from a normal population and $P(t > 1.397) = 0.10$ for 8 degrees of freedom. *[C.U., B.Sc. (Math) '67]*

93. A random sample is drawn from a normal population. The data give sample size and sample variance only. What statistic would you use to test the hypothesis that the population variance has a particular value ? Give reasons. *[C.U., B.Sc. (Econ) '82]*

94. Obtain formulae for 95% confidence limits of the variance of a normal population, when the mean is (i) known, (ii) unknown.

95. Given that x is normally distributed and that a sample of 20 yields a mean of 42 and a variance of 25, test the hypothesis that the population standard deviating is $\sigma = 8$ at 5% level of significance.
(Given the value of χ^2 for d.f. 19 is 30.14 at 5% level.)

96. A random sample of 12 values gave an unbiased estimate s^2 of the population variance equal to 10.62. May the sample be reasonably regarded as drawn from a normal population with variance 7? Calculate the 95% confidence limits for the population s.d. Given that for 11 degrees of freedom, $\chi^2_{.05}$ = 19.68, $\chi^2_{.025}$ = 21.92, $\chi^2_{.975}$ = 3.82.

97. State the assumtions underlying Student's t-test and Snedecor's F-test when applied to both single and two-sample problems. *[C.A., May '80]*

98. The standard deviations calculated from two independent random samples of sized 9and 10 were found to be 2.4 and 1.8 respectively. Can the samples be regarded to have been drawn from equally variable normal populations? (Given that 5% value of F from the tables is 3.23 for degrees of freedom 8 and 9 respectively).

99. A random sample of size 8, drawn from a normal population, shows that the sum of observations is 9.6 and the sum of the squares ofobservations is 60.52. An independent sample of size 11 taken from another normal population, gave these figures as 16.5 and 64.75 respectively. Test whether the two populations may have equal means and variances. Given that $F_{.05}$ = 3.14 for degrees of freedom (7, 10), and $t_{.025}$ = 2.11 for 17 d.f.

100. A correlation coefficient of 0.22 is obtained from random sample of 1024 pairs of observations from a bivariate normal population. Find 95% confidence limits for the correlation coefficient in the population.

101. A random sample of 27 pairs of observations from a bivariate normal population gives a correlation coefficient of 0.42. Is it likely that the variables are uncorrelated inthe population? (Given $t_{.025}$ = 2.06 for 25 d.f.).

102. Find the least value of r in a sample of 27 pairs from a bivariate normal population which is significant at 15 level. (Given $t_{.005}$ = 2.79 for 25 d.f.).

103. Explain Fisher's Z-transformation of correlation coefficient and indicate its uses in test of significance.

104. The value of r obtained from a random sample of 19 pairs of observations from a bivariate normal population is 0.8. Is this value consistent with the hypothesis that the population correlation coefficient is 0.55?

105. The first of two samples consisting of 19 pairs of observation gives a correlation coefficient of 0.5, while an independent sample of 23 pairs shows a correlation coefficient of 0.6. Are the two correlation coefficients significantly different at 1% level?

ANSWERS

3. 33.3 and 17.3 (see Example 14.3).

8. (a) $\frac{(T_1+T_2)}{2}$, $\frac{(T_1-T_2)}{2}$; (b) $3T_2 - T_1$; $T_1 - 2T_2$

11. $\sigma_0^2 = \frac{\Sigma x_i^2}{n}$ 12. $\frac{\bar{x}}{n}$ 13. $\frac{\Sigma x_i}{n}$

17. 63.1 and 64.9 inches. $\left(\text{use} \quad \bar{x} \pm \frac{3\sigma}{\sqrt{n}}\right)$

18. 28.45 to 28.69 gms $\left(\text{use} \quad \bar{x} \pm \frac{1.96S}{\sqrt{n}}\right)$

19. Rs 48.04 to 51.96 (see 14.5.6).
20. 4.60 to 5.00 (see Example 14.12)
21. 3.7% to 8.3% (see Example 14.11)
22. (see Example 14.11)
23. 10%; 5.5% to 14.5% (see Example 14.11)
24. 400 to 1600 (see Example 14.14)
25. 1070 to 1930 (see Example 14.14)
26. (a) Normal population with σ *known*, or Large sample from any population; (b) (see p. 538); Normal population with σ *unknown*.
27. (see 14.5.6 and 14.5.8b)
28. 34.12 to 45.88 (see 14.5.6)
29. 27 and 53 (see 14.5.8b)
30. 51.2 and 62.8 (see 14.5.8b)
31. 1.82 to 13.58 (see 14.5.7)
32. 0.39 to 14.16 (see 14.5.9a), populations have equal s.d.
33. (i) 2.66 to 9.07 (see 14.5.10); (ii) 2.74 to 10.8 (see 14.5.11)
34. 0.20 to 12.0 (see 14.5.13) 37. $\frac{1}{32}$
38. The teaching technique is *not* effective.
39. $\left(\frac{1}{2}\right)^5 = .03125$; $1 - (0.2)5 = 0.99968$
40. .0000,8448; 0.9987,0967
41. No; $|z| = 2.68 > 1.96$, $H_0\left(P = \frac{1}{2}\right)$, $H_1\left(P \neq \frac{1}{2}\right)$; (see Example 14.20).

42. Yes; $|z| = 1.47 < 1.96$, Confidence limits for P : 46% to 54%.
43. Yes; $|z| = 1.70 < 1.96$, $H_0\left(P = \frac{1}{2}\right)$, $H_1\left(P \neq \frac{1}{2}\right)$; (see Example 14.20).
44. No; $|z| = 2.70 > 2.58$, $H_0\left(P = \frac{1}{2}\right)$, $H_1\left(P \neq \frac{1}{2}\right)$; (see Example 14.20).
45. Yes; $z = 1.90 > 1.645$; $H_0\left(P = \frac{1}{3}\right)$, $H_1\left(P > \frac{1}{3}\right)$; (see Example 14.21).
46. No; $z = 2.29 > 1.645$; $H_0(P = .05)$, $H_1(P > .05)$; see Example 14.21.
47. $z = -2.60$; $H_0(P = .95)$, $H_1(P < .95)$; see Example 14.22.
    (i) since $z < -1.645$, claim not justified at .05 level.
    (ii) since $z < -2.33$, claim not justified at .01 level.
48. Not tenable; $z = -3.77 < -1.645$; $H_0(P = 0.9)$, $H_1(P < 0.9)$; see Example 14.22
49. Rs 6,000 to Rs 54,000 (see Example 14.14).
50. Yes; $|z| = 2.11 > 1.96$; $H_0(P_1 = P_2)$, $H_1(P_1 \neq P_2)$; see Example 14.23.
51. Yes; $|z| = 6.38 > 2.58$; H_0(Proportion of students using dot pens equal in two colleges), H_1(Proportion different); see Example 14.23.
52. Yes; $z = 4.94 > 2.33$; $H_0(P_1 = P_2)$, $H_1(P_1 > P_2)$; see Example 14.24
53. No; $z = -1.21 > -1.645$; $H_0(P_1 = P_2)$, $H_1(P_1 < P_2)$; see Example 14.25. Confidence limits for $P_1 - P_2$: – 5.8% to 1.3%.
54. No; $|z| = 2.67 > 2.58$; $H_0(\mu = 16.8)$, $H_1(\mu \neq 16.8)$; see Example 14.26.
55. Not; $|z| = 3.54 > 2.58$; $H_0(\mu = 1 \text{ kg})$, $H_1(\mu \neq 1 \text{ kg})$; see Example 14.26.
56. No; $|z| = 2.5 > 1.96$; $H_0(\mu = 1600)$, $H_1(\mu \neq 1600)$; see Example 14.27.
57. Yes; $|z| = 1.72 > 1.96$; $H_0(\mu = 3.25)$, $H_1(\mu \neq 3.25)$; see Example 14.27.
    Confidence limits for μ: 3.23 to 3.57 (cm).
58. Yes; $z = 2.71 > 2.33$; $H_0(\mu = 80.2)$, $H_1(\mu > 80.2)$.
59. Yes; $z = 3.54 > 2.33$; $H_0(\mu = 1800)$, $H_1(\mu > 1800)$.
60. $\bar{x} = 28.8$, $S = 6.35$; Claim justified; $z = -2.68 < -2.33$;
    $H_0(\mu = 30.5)$, $H_1(\mu < 30.5)$; see Example 14.28.
61. Yes; $|z| = 1.79 < 1.96$; $H_0(\mu_1 = \mu_2)$, $H_1(\mu_1 \neq \mu_2)$; see Example 14.30.
    Confidence limits for $\mu_1 - \mu_2$: – 0.27 to 1.47.
62. Yes; $|z| = 4.73 > 1.96$; $H_0(\mu_1 = \mu_2)$, $H_1(\mu_1 \neq \mu_2)$; see Example 14.29
63. No; $|z| = 1.86 < 1.96$; $H_0(\mu_1 = \mu_2)$, $H_1(\mu_1 \neq \mu_2)$; see Example 14.29
64. Yes; $z = -9.9 < -2.33$; $H_0(\mu_1 = \mu_2)$, $H_1(\mu_1 < \mu_2)$; see Example 14.29.
65. see (14.7.11).
66. Yes; $\chi^2 = 3.7 < 5.99$; H_0(Data agree with model); see Example 14.32
67. Yes; $\chi^2 = 6.3 < 16.92$; H_0(Table is random; i.e., probability of each digit appearing is 1/10); see Example 14.33.
68. Yes; $\chi^2 = 3.0 < 6.25$; H_0(Die is honest; i.e., probabilities of the 4 classes are 1/6, 1/3, 1/3, 1/6 respectively); see Example 14.33.
69. see (14.7.12).

70. No; $\chi^2 = 5.64 < 7.81$; H_0(Attributes are independent); see Example 14.35.
71. (Form a 2×4 contingency table showing 'No. of defectives' and. 'No. of non-defectives' against A, B, C, D). Yes; $\chi^2 = 5.57 < 7.81$; H_0(Lots are of same quality); see Example 14.35.
72. Not significant; $\chi^2 = 0.101 < 3.8$ (apply Yates' correction); H_0 ('Procedure' and 'Sales' independent); see Example 14.36.
73. (Form a 2×2 contingency table showing treatment with 'serum' and 'no serum' against 'paralysed' and 'not paralysed'). Yes; $\chi^2 = 14.1 > 6.64$); H_0(Attributes are not independent); see Example 14.36.
74. Yes; $\chi^2 = 11.74 > 3.84$; H_0 (Attributes are not independent); see Example 14.36
75. (Form a 2×2 table showing 'males' and 'females' against 'employed' and 'unemployed'). Yes; $\chi^2 = 13.54 > 3.84$; H_0(Attributes are not independent); see Example 14.36.
76. $P(\bar{x} > 21) = P(z > 1) = 0.1587$; see Example 13.26 and 27 and 14.38.
77. No; $|z| = 2.37 > 1.96$; $H_0(\mu = 14.5)$, $H_1(\mu \neq 14.5$; σ known; confidence limits for μ: 9.5 and 4.7; see Example 14.37.
78. No; $|z| = 3.33 > 2.58$; H_0(Sample randomly drawn); see Example 14.37.
79. No; $z = -0.32 > -1.645$; $H_0(\mu = 15.6)$, $H_1(\mu < 15.6)$; see Example 14.37.
80. Yes; $|z| = 1.16 < 1.96$; $H_0(\mu_1 = \mu_2)$, $H_1(\mu_1 \neq \mu_2)$; see Example 14.39.Confidence limits for $\mu_1 - \mu_2$: -0.77 and 2.97.
81. Yes, larger; $z = 2.41 > 2.33$; $H_0(\mu_1 = \mu_2)$, $H_1(\mu_1 > \mu_2)$; see Example 14.39. Confidence limits for $\mu_1 - \mu_2$: -0.19 and 5.59.
82. Yes; Student's t-test; $|t| = 2.25 < 2.26$; $H_0(\mu = 14.5)$, $H_1(\mu \neq 14.5)$; see Example 14.40. Confidence limits for μ: 9.69 and 4.51.
83. No; Student's t-test; $|t| = 0.47 < 2.262$; $H_0(\mu = 2.00)$, $H_1(\mu \neq 2.00)$; Assumption: Normal population. See Example 14.40
84. Yes; Student's t-test, $t = 2 > 1.83$; $H_0(\mu = 64)$, $H_1(\mu > 64)$; $\bar{x} = 66$, $S = 3$; see Example 14.41 and 42.
85. No; Student's t-test, $t = 2.32 < 2.82$; $H_0(\mu = 100)$, $H_1(\mu > 100)$; see Example 14.42
86. Yes; Student's t-test, $t = -2.16 < -1.83$; $H_0(\mu = 65)$, $H_1(\mu < 65)$; see Example 14.40 to 42.
87. (a) (i) For large samples, use approximate z-test (14.7.9); (ii) for normal populations with *known s.d.*, use exact z-test (14.8.4); (iii) for normal populations with *unknown s.d.* (but assumed equal), use Fisher's t-test (14.8.9).
    (b) No; Fisher's t-test, $|t| = 1.67 < 2.26$; $H_0(\mu_1 = \mu_2)$, $H_1(\mu_1 \neq \mu_2)$; see Example 14.43.
88. Yes; Fisher's t-test, $t = 4.99 > 2.68$; $H_0(\mu_1 = \mu_2)$, $H_2(\mu_1 > \mu_2)$; see Example 14.43 and 44; $s^2 = (20.8 + 15.7)/12$; Confidence limits for $\mu_1 - \mu_2$: 1.82 and 7.58.
89. No; Fisher's t-test, $t = -0.57 > -1.76$; $H_0(\mu_1 = \mu_2)$, $H_2(\mu_1 < \mu_2)$; $n_1 = 6$, $\bar{x}_1 = 68$, $S_1^2 = 10$, $n_2 = 10$, $\bar{x}_2 = 69$, $S_2^2 = 10$; see Example 14.44.
90. No; Paired t-test, $t = 2.04 < 2.82$; $H_0(\mu_x = \mu_y)$, $H_1(\mu_x > \mu_y)$; (x = final score, $d = x - y$, $\Sigma d = 47$, $\Sigma d^2 = 699$)); see Example 14.45A
91. No; Paired t-test, $|t| = 1.27 < 2.26$; $H_0(\mu_x = \mu_y)$, $H_1(\mu_x \neq \mu_y)$, ($\Sigma d = 6$, $\Sigma(d - d)^2 = 20$); see Example 14.45
92. Yes, it raises; Paired t-test, $t = 1.51 > 1.397$ (8 d.f.); $H_0(\mu_x = \mu_y)$; $H_1(\mu_x = \mu_y)$; (x = final score); see Example 14.45A

93. Statistic: see (14.8.11); Reason: see (13.8.12)
94. see (4.5.10) & (14.5.11)
95. σ may be 8; x^2-test, $x^2 = 7.81 < 30.14$; $H_0(\sigma = 8)$, $H_1(\sigma > 8)$; see Example 14.46–47
96. Yes; χ^2-test, $\chi^2 = 16.7 < 19.68$; use $nS^2 = (n - 1)s^2$; $H_0(\sigma^2 = 7)$; $H_1(\sigma^2 = 7)$;see Example 14.46–47; Confidence limits for σ : 2.31 and 5.53; see Example 14.18
97. (a) Student's *t*-test: (i) In *single sample problem* of testing for a specified mean $H_0(\mu > \mu_0)$, the sample must be drawn *randomly* from *Normal* population, with *unknown s.d.*; see (14.8.6). (ii) In *two sample* problem of testing for equality of two population means, i.e. $H_0(\mu_1 = \mu_2)$, the two samples must be *independent*, and *randomly* drawn from two *Normal populations* with *equal s.d., but unknown*; see (14.8.9)
    (b) *F*-test: (i) Not applicable in *single sample problems*; (ii) In *two sample* problem of testing for equality of two standard deviations $H_0(\sigma_1 = \sigma_2)$, the two samples must be *independent* and drawn *randomly* from *Normal populations* (any or both the means may be known or unknown); see (14.8.12) & (13.8.21 to 24)
98. Yes; *F*-test, $F = 1.8 < 3.23$; $H_0(\sigma_1 = \sigma_2)$, $H_1(\sigma_1 = \sigma_2)$; see Example 14.48–50
99. Yes; *t* and *F*-tests, $F = 1.75 < 3.14$; $H_0(\sigma_1 = \sigma_2)$, $H_1(\sigma_1 = \sigma_2)$; see Example 14.52. Use $S^2 = \Sigma x^2/n - (\Sigma x/n)^2$ and $(n - 1)s^2 = nS^2$. Also $|t| = 0.28 < 2.11$; $H_0(\mu_1 \neq \mu_2)$, $H_1(\mu_1 \neq \mu_2)$.
100. 0.16 & 0.28; see Example 14.53.
101. No; $|t| = 2.31 > 2.06$; $H_0(\rho = 0)$, $H_1(\rho \neq 0)$; see Example 14.54
102. $|r| = 0.487$; see Example 14.55
103. see (14.9.4)
104. Yes; Using Fisher's *Z*-transformation, calculate (14.9.6);
     $|U| = 1.92 < 1.96$; $H_0(\rho = 0.55)$, $H_1(\rho \neq 0.55)$; see Example 14.56
105. No; Using Fisher's *Z*-transformation, calculate (14.9.6):
     $|U| = 0.43 < 1.96$; $H_0(\rho_1 = \rho_2)$, $H_1(\rho_1 \neq \rho_2)$; see Example 14.58

15 ANALYSIS OF VARIANCE

15.1 INTRODUCTION

Analysis of Variance has been defined as the statistical technique for the "separation of variation due to a group of causes from the variation due to other groups". Here we shall discuss the simplest use of this technique, namely testing whether the means of a number of populations are equal. The method is based upon an unusual result that the equality of several population means can be tested by comparing the sample variances using F distribution. It may be recalled that the t statistic (14.8.9) is used for testing whehter *two* population means are equal. The analysis of variance test may there fore be taken as an extension of this test for the case of more than two means.

15.2 DIFFERENT SOURCES OF VARIATION

Consider the data of Example 15.1 (page 619) which show the life-time of motor-car tyres manufactured by 3 companies. We are required to test on the basis of these sample observations whether the means of there populations are equal or not. It will be noticed that there are differences in the observations 'within' each sample or group. This variability occurs due to change, and it reflects the natural differences that occur with each of the populations from each the samples have been taken. If we examine the means of three samples, we see that another kind of variability is present. This variability is due to the combined effect of the natural differences and possible real differences 'between' the different groups.

Our problem is to decide whether the differences among sample means is only due to chance or whether the differences occur because the means of three populations (from which the samples have been drawn) are actually different. Hence we are now concerned with the problem to testing the hypothesis that the means of several populations are equal. Rejection of this hypothesis will lead to be conclusion that at least one of the population means is different from others.

The necessary assumptions in the analysis of variance are:

1. The samples are *independently* drawn;
2. The population are *normally* distributed, with a *common variance.*
3. The effects of various components are *additive.*

If the means of all the populations are equal, then the variability "between" groups would result only from chance and hence would be the same as the variability arising from "within" groups. On the other hand, if the population means are not equal, the variability "between" groups would be more than the variability "within" groups.

The measure of variability used in the analysis of variance is called a "*Mean square*". This is similar to a variance and is defined by

$$\text{Mean square} = \frac{\text{Sum of squared deviation from mean}}{\text{Degree of freedom}}$$

Note that in the *t*-test (14.8.6) for a specified mean, the population variance σ^2 as estimated as the sum of squared deviations from mean divided by the 'sample size minus one'. In the *t*-test (14.8.9) for equality of two means, the common population variance was estimated as the sum of the squared deviations of the two groups of observations from the respective means divided by the 'sum of sample sizes minus two'. These divisors are referred to as "degrees of freedom".

The mean squares appearing here are similar to the estimated variances cited above, but they relate to different 'sources of variation' and are based on different degrees of freedom. One means square is used to measure variability 'within' the groups. This based on the sum of squared deviations of the observations within each group, the deviations being taken from the respective group means, and has degrees of freedom "total sample size minus number of samples". The "*Sum of Squares Within groups*", as it is called, when divided by the number of degrees of freedom provides the "*Mean Square Within groups*". This mean square represents a measure of variability due to chance or 'experimental error'.

The other mean square is used to measure group-effect or possible difference existing, if any, 'between' the groups. This is based on the sum of squared deviations of the individual sample or group means the deviations being taken from the grand mean of all observations being considered as one sample, and has the degrees of freedom "number of samples minus one". Of course, the sample mens are weighted by the respective sample sizes. This "*Sum of Squares Between groups*" when divided by its degrees of freedom provides the "*Mean Square Between groups*". If the means of all populations are equal, there is no group-effect and the mean square between groups will also represent variability due to chance alone. Consequently when the group means in the population are equal, the mean square within groups and the mean square between groups should not be much different, and their ratio should be close to one. Unusually large ratios would indicate that the group means are not equal in the population.

The observations in Example 15.1 are shown in three groups classified according to the brand of tyres, i.e., the "factor of classification" is the brand of tyres. These are therefore known as '*One-way Classified Data*'. Subsequently it will be shown that observations may be classified according to two factors, and these will be known as '*Two-way Classified Data.*'

15.3 TECHNIQUE IN ONE-WAY CLASSIFIED DATA

We have *k independent random* samples (or groups of observations), one group from each of *k Normal populations* with means $\mu_1, \mu_2, \ldots\ldots, \mu_k$ and a *common variance* σ^2

(unknown). On the basis of the data, it is required to test the null hypothesis that the population means are equal.

$$H_0 (\mu_1 = \mu_2 = ... = \mu_k) \tag{15.3.1}$$

against the alternative hypothesis H_1 (all μ_1 are not equal).

The mathematical mode is

$$x_{ij} = \mu_i + e_{i,j} \tag{15.3.2}$$

where x_{ij} denotes the j-th observations in the i-th group;
μ_i denotes mean of the i-th population; and
e_{ij} denotes error due to many unspecified causes.

This is called a *Linear Model,* since x_{ij} is assumed to be made up of the 'sum' of effects due to different components. It is assumed that e_{ij} are independent and identical distributed normal variates with mean 0 and variance σ^2. This model may also be written in the form

$$x_{ij} = \mu + \alpha_i + e_{ij} \tag{15.3.3}$$

where μ denotes the general effect;
α_i denotes effect specifically to the i-th population; and
e_{ij} denotes error component

Then null hypothesis (15.3.1) is equivalent to stating that there is no special effect due to any population, i.e.,

$$H_0 (\alpha_1 = \alpha_2 = ... = \alpha_k = 0) \tag{15.3.4}$$

against the alternative hypothesis H_1 (all α_i are not zero).

We introduce the following notations:

k = Number of independent samples or groups,
n_i = Number of observations in the i-th sample,
N = Total number of observations in all the samples = Σn_i
x_{ij} = The j-th observation in the i-th group,
T_i = Total of the i-th sample = $\sum\limits_i x_{ij}$
T = Grand total of all observations = $\Sigma T_i \ \Sigma\Sigma x_{ij}$
$\bar{x}_i$ = Mean of the i-th sample = $\dfrac{T_i}{n_i}$
$\bar{x}$ = Grand mean of all observations = $\dfrac{T}{N}$

It can be shown algebraically that

$$\sum_i \sum_j (x_{ij} - \bar{x})^2 = \sum_i \sum_j (x_{ij} - \bar{x}_i)^2 + \sum_i n_i (\bar{x}_i - \bar{x})^2 \tag{15.3.5}$$

$$\text{Total } SS = SSW + SSB$$

The expression on the left of (15.3.5) is the Sum of Squares (abbreviated *SS*) of all sample observations about the grand mean and is called '*Total Sum of Squares*' (*Total SS*). The first term on the right is the Sum of Squares of deviations of observations within each group about the respective group means and is called '*Sum of Squares Within groups*' (*SSW*), or '*Sum of Square due to Error*' (*SSE*). The second term is the (weighted) Sum of Squares of group means about the grand mean and is called '*Sum of Squares Between groups*' (*SSB*).

The degrees of freedom can similarly be split up. Thus, we have

$$\text{Sum of Squares} \quad : \quad Total\ SS = SSW + SSB$$
$$\text{Degrees of freedom} : \quad N - 1 \quad = (N - k) + (k - 1)$$

The Mean Squares are now calculated no dividing the Sum of Squares by the corresponding degrees of freedom.

$$\text{Mean Square Within groups } (MSW) = SSW \div (N - k)$$
$$\text{Mean Square Between groups } (MSB) = SSB \div (k - 1)$$

Assuming Normal distribution in the population with a common variance (σ^2), it can be shown that the expected value of *MSW* is σ^2, i.e.,

$$E(MSW) = \sigma^2$$

irrespective of whether the population means are equal or not. But

$$E(ESB) \geq \sigma^2$$

where the sign of equality holds only when the population means are equal.

Thus when the null hypothesis H_0 is true, both the mean squares *MSB* and *MSW* provide independent unbiased estimates of the same population variance (σ^2). Hence, we have the test statistic

$$F = \frac{MSB}{MSW}$$

which under H_0 follows *F* distribution with degrees of freedom ($k - 1$, $N - k$). If the observed value of the statistic equals or exceeds the theoretical *F* value from Statistical Tables at a specified level of significance, we reject the null hypothesis and conclude that all the population means are not equal. On the other hand, if the observed value of the statistic is less than the theoretical value all the population means may be considered to be equal.

The various sums of squares, degrees of freedom, mean squares corresponding to the different sources of variation, and also the *F* values (both observed and theoretical) are shown in the form of a table known as "*Analysis of Variance Table*", as in the Examples below.

15.4 STEPS IN COMPUTATION (ONE-WAY CLASSIFIED DATA)

1. Reduce the sample observations by subtracting a suitable constant.
2. From these reduced figures, obtain
   (i) total (T_i) for each group;
   (ii) grand total ($T = \Sigma T_i$), i.e., sum of group totals T_i,
   (iii) total of the squares of all figures ($\Sigma\Sigma\, x_{ij}^2$)
3. Calculate
   (i) Correction Factor (*CF*) $= \dfrac{T^2}{N}$
   (ii) Total *SS* $= \Sigma\Sigma x_{ij}^2 - CF$
   (iii) *SSB* $= \Sigma\left(\dfrac{T_i^2}{n_i}\right) - CF$
   (iv) *SSW* $= Total\ SS - SSB$

4. Write down the Degrees of Freedom (D.F.)
   (i) D.F. for $SSB = (k-1)$
   (ii) D.F. for $SSW = (N-k)$
5. Calculate the Mean Squares :
   (i) $MSB = SSB \div (k-1)$
   (ii) $MSW = SSW \div (N-k)$
6. Obtain the observed value of F on dividing MSB by MSW :

$$F = MSB \div MSW$$

7. Consult the F-tables and obtain the theoretical values of F at (say) 5% level corresponding to the degrees of freedom $(k-1, N-k)$
8. If the observed value of F at (6) equals or exceeds the theoretical value of F at (7), reject the null hypothesis, and conclude that all the population means are not equal. Otherwise, they may be taken to be equal.

15.5 LOCATING UNEQUAL PAIRS OF MEANS

In the Analysis of Variance, if the observed of F is found to be significant, the conclusion is that all the population means are not equal. This however does not rule out the possibility that some of them are equal, while others are different. In such a case we may be interested to find which pair of means differ. This is really a problem of "test for equality of two population means" using t distribution (see 14.8.9). If the number of observations in each sample is the same, say n, then the appropriate statistic for testing equality of means of μ_i and μ_j is

$$t = \frac{\bar{x}_i - \bar{x}_j}{s\sqrt{\left(\frac{1}{n}+\frac{1}{n}\right)}} = \frac{\bar{x}_i - \bar{x}_j}{s\sqrt{\frac{2}{n}}}$$

where s^2 is the unbiased estimator of the common (but unknown) population variance σ^2. In the analysis of variance tests, σ^2 is estimated by the *Mean Square Error (MSE)*, i.e., *MSW* in one-way classified data.

$$s^2 = MSE$$

The population means μ_i and μ_j will be considered to be equal, at say, 5% leave of significance, if $|t| \le t_{.025}$, where $t_{.025}$ denotes the upper 2.5% value of t distribution with $NM-k$ degrees of freedom. (Note that this is the d.f. of *Error SS*, i.e., the divisor used in *MSE*). Thus

$$\left|\frac{\bar{x}_i - \bar{x}_j}{s\sqrt{\frac{2}{n}}}\right| \le t_{.025}$$

i.e.,

$$\left|\bar{x}_i - \bar{x}_j\right| \le s\sqrt{\frac{2}{n}}\, t_{.025}$$

Multiplying both sides by n,

$$|T_i - T_j| \le s\sqrt{2n}\, t_{.025}$$

where T_i and T_j represent the totals of the i-th and j-th samples respectively.

Therefore, in cases when all the samples are of the same size (n), we calculate the "*Critical Difference*" between sample totals,

$$\text{Critical Difference (C.D.)} = s\sqrt{2n}\, t_{.025} \qquad (15.5.1)$$

The differences between all pairs of samples totals $|T_i - T_j|$ are now compared with C.D. If the difference between any particular pair, say for samples 1 and 3, viz., $|T_1 - T_3|$, exceeds the C.D., then the population means μ_1 and μ_3 are considered to be unequal.

Let us now consider a problem of analysis of variance in one-way classified data with equal number of observations in each sample, which illustrates the theories discussed so far.

Example 15.1 *A random sample of five motor-car tyres is taken from each of 3 brands manufactured by three companies. The lifetime of these tyres (as measured by the mileage run) is show in below. One the basis of the data, test whether the average lifetime of the 3 brands of tyres are equal or not.*

Lifetime of Tyres ('000 miles)			
Brand	*A*	*B*	*C*
	35	32	34
	34	32	33
	34	31	32
	33	28	32
	34	29	33

Also test which pairs of tyres differ in quality.

Solution Each observation is reduced by 30, and shown below:

Table 15.1 Calculations for Analysis of Variance

Samples	1	2	3	
	5	2	4	
	4	2	3	
	4	1	2	
	3	–2	2	
	4	–1	3	
Total:	$T_1 = 20$	$T_2 = 2$	$T_3 = 14$	$T = 36$
Total of Squares:	82	14	42	$\Sigma\Sigma x_{ij}^2 = 138$
Sample Size:	$n_1 = 5$	$n_2 = 5$	$n_3 = 5$	$N = 15$

$$\text{C.F.} = \frac{T^2}{N} = \frac{36^2}{15} = 86.4$$

$$\textit{Total SS} = \Sigma\Sigma x_{ij}^2 - \text{C.F.} = 138 - 86.4 = 51.6$$

$$SSB = \Sigma\left(\frac{T_i^2}{n_i}\right) - \text{C.F.} = \left(\frac{20^2}{5} + \frac{2^2}{5} + \frac{14^2}{5}\right) - 86.4$$

$$= \frac{(400 + 4 + 196)}{5} - 86.4 = 120 - 86.4 = 33.6$$

$$SSE = \textit{Total SS} - SSB = 51.6 - 33.6 = 18.0$$

Table 15.2 Analysis of Variance Table

Source of Variation	S.S	d.f.	M.S.	F Values Observed	F Values Tabulated
Between Groups	33.6	2	16.8	11.2	$F_{.05} = 3.89$
Within Groups (Error)	18.0	12	1.5		$F_{.01} = 6.93$
Total	51.6	14	—		—

[**Note:** (i) Since the total number of observations in all samples is 15, the d.f. for *Total SS* is 15 – 1 = 14. Since there are 3 groups of observations, the d.f. for, *SSB* in 3 – 1 = 2. The d.f. for *SEE* is the difference 14 – 2 = 12.

(ii) The observed value of *F* is *MSB* ÷ *MSE* = 16.8 ÷ 1.5 = 11.2.

(iii) The tabulated values of *F* are obtained from Statistical Tables corresponding to d.f. (2, 12). It may be noticed that the first d.f. is the divisor used in numerator and the second is the divisor used in the denominator for calculation of *F*.]

Since the observed value of *F* (viz. 11.2) exceeds the 1% tabulated value (viz. 6.93), we reject the null hypothesis of equality of population means, and conclude that the average lifetimes of the 3 brands of tyres are not equal.

In order to test which brands of tyres differ in quality, we calculate the critical difference (15.5.1). Here $s = \sqrt{1.5}$, $n = 5$, $t_{.025} = 2.18$ (for 12 d.f.).

$$\text{Critical Difference} = \sqrt{1.5} \times \sqrt{2 \times 5} \times 2.18$$
$$= \sqrt{15} \times 2.18$$
$$= 3.87 \times 2.18 = 8.44$$

The sample totals (of the reduced observations) are $T_1 = 20$, $T_2 = 2$, $T_3 = 14$. We have

$$|T_1 - T_2| = 18$$
$$|T_1 - T_3| = 6$$
$$|T_1 - T_3| = 12$$

Comparing these figures with the Critical Difference 8.44 we find that brands *A* and *B* are of different quality, and so are brands *B* and *C*. Brands *A* and *C* may be taken to be of the same quality.

Example 15.2 *An experimenter wished to study the effect of 4 fertilisers on the yield of crop. He divided the field into 24 plots and assigned each fertiliser at random to 6 plots. Part of his calculations are shown below :*

Source	*df*	*SS*	*MS*	*F*	$F_{5\%}$
Fertilisers	–	*2940*	–	–	*3.10*
Within Group	–	–	–		
Total	–	*6212*			

(a) Complete the above table by filling in the values marked by —

(b) Test at the 5% level to see whether the fertilisers differ significantly.

[W.B.H.S. '78]

Solution (a) [Working **Note:** (i) Since 24 observations are available, one from each plot, 'Total' degrees of freedom (*df*) is (24 – 1) = 23. Again, since 4 fertilisers are being compared, the *df* for Fertilisers is (4 – 1) = 3. Hence, the *df* for 'Within Group' is (23 – 3) = 20.

(ii) Sums of squares (*SS*) are additive. Therefore

$$\text{Within Group } SS = \text{Total } SS - \text{Fertilisers } SS$$
$$= 6212 - 2940 = 3272$$

(iii) Mean squares (*MS*) are obtained on dividing *SS* by *df*:

$$MS \text{ between Fertilisers} = 2940 \div 3 = 980$$
$$MS \text{ within Group} = 3272 \div 20 = 163.6$$

(iv) $$F = \frac{MS \text{ between Fertilisers}}{MS \text{ within Group}} = \frac{980}{163.6} = 5.99]$$

The completed table is as follows:

Table 15.3 Analysis of Variance Table

Source	*df*	*SS*	*MS*	*F*	$F_{5\%}$
Fertilisers	3	2940	980	5.99	3.10
within Group	20	3272	163.6		
Total	23	6212			

(b) Since the observed value of *F* (viz. 5.99) is larger than the 5% tabulated value (viz. 3.10), it is significant at 5% level. We therefore conclude that the fertilisers differ significantly.

Example 15.3 *Each of the following sets of observations is a random sample from a normal population.*

Sets	*Observations*						
I	*249,*	*242,*	*247,*	*250,*	*252*		
II	*251,*	*256,*	*255,*	*258*			
III	*266,*	*261,*	*265,*	*264*			
IV	*262,*	*260*	*263,*	*262,*	*261*	*264,*	*262*

Test whether the population means are equal (Assume that the population standard deviations are same.)

Solution The observations, reduced by 250, and also the preliminary results for calculating the various sum of squares, are shown in Table 15.4.

Table 15.4 Calculations for Analysis of Variance

	Sets				
	I	II	III	IV	
	–1	1	16	12	
	–8	6	11	10	
	–3	5	15	13	
	0	8	14	12	
	2			11	
				14	
				12	
Total	$T_1 = -10$	$T_2 = 20$	$T_3 = 56$	$T_4 = 84$	$T = 150$
Total of Sqs.	78	126	798	1018	$\Sigma\Sigma x_{ij}^2 = 2020$
Sample Size	$n_1 = 5$	$n_2 = 4$	$n_3 = 4$	$n_4 = 7$	$N = 20$

[**Note:** (i) This is a problem of analysis of variance in one-way classified data with unequal numbers of observations in the samples.

(ii) Degrees of freedom under different sources of variation and also the F-values are obtained in the same way as indicated in Example 15.1]

$$\text{C.F.} = \frac{150^2}{20} = 1125$$

$$\text{Total } SS = 2020 - 1125 = 895$$

$$SSB = \left[\frac{(-10)^2}{5} + \frac{20^2}{4} + \frac{56^2}{4} + \frac{84^2}{7}\right] - 1125$$

$$= (20 + 100 + 784 + 1008) - 1125 = 787$$

$$SSE = 895 - 787 = 108$$

The various Sums of Squares (*S.S.*) along with the degree of freedom (*d.f.*) are shown in the following table.

Table 15.5 Analysis of Variance Table

Source of Variation	*S.S*	*d.f.*	*M.S.*	*F value* Observed	*F value* Tabulated
Between Sets	787	3	262.3	38.9	$F_{.05} = 3.24$
Within Sets (Error)	108	16	6.75		$F_{.01} = 5.29$
Total	895	19	—	—	—

Since the observed value F is larger tan the 1% tabulated value corresponding to d.f. (3, 16), we reject the null hypothesis and conclude that the means of normal populations are not equal.

15.6 TECHNIQUE IN TWO-WAY CLASSIFIED DATA

We have a random sample consisting of hk observations, which are classified according to two factors—into h classes according to factor A, and also into k classes according to factor B. There is one observation in each of line hk cells corresponding to a class of factor A and simultaneously a class of factor B. These hk observations may therefore be arranged in the form of a two-way table with h rows and k columns.

The mathematical model is that any observation is made up of the sum 4 components:

$$x_{ij} = \mu + \alpha_i + \beta_j + e_{ij} \tag{15.6.1}$$

where μ is the general effect;

α_i is the effect special to the i-th class according to factor A,

β_i is the effect special to the j-th class according to factor B,

e_{ij} is the error component.

This is called a *Linear Model,* since any observation x_{ij} is supposed to be made up of the '*sum*' of effect of the various components. It is assumed that e_{ij} are independently distributed normal variables each with mean 0 and variance σ^2 (unknown).

In the two-way classified data, the analysis of variance may be used to decide on two types of problems simultaneously, viz. (i) whether there is any differential effect due to classification by factor *A*. Null hypothesis:

$$H_{01}\ (\alpha_1 = \alpha_2 = ... = \alpha_n = 0)$$

and/or (ii) whether there is any differential effect due to classification by factor *B*. Null hypothesis :

$$H_{02}\ (\beta_1 = \beta_2 = ... = \beta_n = 0)$$

We introduce the following notations:

h = Number of classes by factor *A*, (say, number of rows).
k = Number of classes by factor *B* (say, number of columns)
$N = hk$ = Total number of observations.
T_i = Total of the *i*-th row $(i = 1, 2, ..., h)$
T_j = Total of the *j*-th column $(j = 1, 2, ..., k)$
T = Grand total of all observations $= \Sigma T_i = \Sigma T'_j = \Sigma\Sigma x_{ij}$

$$\bar{x} = \frac{T_i}{k} = \text{Mean of the } i\text{-th row.}$$

$$\bar{x}_j = \frac{T_j}{h} = \text{Mean of the } j\text{-th column}$$

$$\bar{x} = \frac{T}{N} = \text{Grand mean of all observations.}$$

Algebraically, it can be shown that

$$\Sigma\Sigma\,(x_{ij} - \bar{x})^2 = k\Sigma\,(\bar{x}_i - \bar{x})^2 + h\Sigma\,(\bar{x}_j - \bar{x})^2 + \Sigma\Sigma\,(x_{ij} - \bar{x}_i - \bar{x}_j + \bar{x})^2$$

i.e., *Total Sum of Square*

= (*SS between* factor *A*) + (*SS between* factor *B*)
+ (*SS* due to *Error*)

or *Total SS* = *SS* + *SSB* + *SSE*

The degrees of freedom (d.f.) for the various *SS* can also be spilt up:

$$hk - 1 = (h-1) + (k-1) + (h-1)\,(k-1)$$

[**Note:** It will be easier to calculate the d.f. as follows: Since there are *hk* observations in all, the d.f. for *Total SS* is one less, viz. $hk - 1$; since there are *h* classes according to factor *A*, the d.f. for *SSA* is one less, viz. $(h - 1)$; similarly the d.f. for *SSB* is one less, viz. $(k - 1)$. Again just as we have *SSE* = *Total SS* – *SSA* – *SSB,* similarly,

(d.f. for *SSE*) = (d.f. for *Total SS*) – (d.f. for *SSA*) – (d.f. for *SSB*)
$= (hk - 1) - (h - 1) - (k - 1) = (h - 1)\,(k - 1)$]

Mean Squares are now calculated as before, on dividing the Sums of Squares by the corresponding degrees of freedom.

Mean Square between classes by factor *A* (*MSA*) = *SSA* ÷ $(h - 1)$
Mean Square between classes by factor *B*(*MSB*) = *SSB* ÷ $(k - 1)$
Mean Square Error (*MSE*) = *SSE* ÷ $(h - 1)\,(k - 1)$

Assuming normal distribution in the population with a constant variance σ^2, it can be shown that the expected value of *MSE* is σ^2.

$$E(MSE) = \sigma^2$$

irrespective of whether the null hypotheses H_{01} and H_{02} are true or not. However, in general

$$E(MSA) \geq \sigma^2 \text{ and } E(MSB) \geq \sigma^2$$

But $E(MSA) = \sigma^2$ only if H_{01} is true; and $E(MSB) = \sigma^2$ only if H_{02} is true. Thus when H_{01} is true the statistic

$$F_1 = \frac{MSA}{MSE}$$

follows F distribution with the appropriate degrees of freedom. Similarly, when H_{02} is true, the statistic

$$F_2 = \frac{MSA}{MSE}$$

follows F distribution with the appropriate degrees of freedom.

15.7 STEPS IN COMPUTATION (TWO-WAY CLASSIFIED DATA)

1. Reduce the observation by subtracting a constant from each.
2. From the reduced figures, obtain
   (i) totals (T_i) for each group classified according to factor A, i.e., for each row.
   (ii) totals (T_j') for each group classified according to factor B, i.e., for each column.
   (iii) grand total ($T = \Sigma T_i = \Sigma T_j'$) for all figures.
   (iv) total of the squares of all figure ($\Sigma\Sigma\, x_{ij}^2$)
3. Calculate the CF and Sums of Squares:
   (i) Correction Factor (CF) $= \dfrac{T^2}{N}$
   (ii) *Totals SS* $= \Sigma\Sigma x_{ij}^2 - CF$
   (iii) SSA $= \Sigma\left(\dfrac{T_i^2}{k}\right) - CF$
   (iv) SSB $= \Sigma\left(\dfrac{T_j'^2}{h}\right) - CF$
   (iv) SSE $=$ *Total SS* $- SSA - SSB$

[**Note:** In the calculate of various SS, we have expression like $\dfrac{T^2}{N}, \dfrac{T_i^2}{k}, \dfrac{T_j'^2}{h}$ in each of which the divisor is the number of observations included in the corresponding totals T, T_i, T_j')

4. Write down the Degrees of freedom (D.F.):
   (i) D.F. for *Total SS* $= hk - 1$
   (ii) D.F. for $SSA = h - 1$
   (iii) D.F. for $SSB = k - 1$
   (iv) D.F. for SSE = (D.F. for *Total SS*) minus D.Fs. for SSA and SSB.
5. Calculate the Mean Squares :
   (i) $MSA = SSA \div$ D.F.
   (ii) $MSB = SSB \div$ D.F.
   (iii) $MSE = SSE \div$ D.F.

6. Obtain the observed values of F on dividing MSA and MSB by MSE :
   (i) $F_1 = MSA \div MSE$
   (ii) $F_2 = MSB \div MSE$
7. Consult F-tables and obtain the theoretical values of F at (say) 5% level for the appropriate degrees of freedom.

[**Note:** Degrees of Freedom for F are: Divisor used for MS in numerator, followed by Divisor used for MS in denominator]

8. (i) If the observed value of F_1 exceeds the theoretical value, we conclude that classification according to factor A has a differential effect on the value of the variable. That is, the means of classes by factor A are significantly different. Otherwise, there is no differential effect.
   (ii) If the observed value of F_2 exceeds the theoretical value, we conclude that classification according to factor B has a differential effect on the values. If not, then there is no differential effect.

Example 15.4 *There experimenters determine the moisture content of samples of power, each man taking a sample from each of 4 consignments. The results are given below. (a) Perform an analysis of variance on these data and discuss whether there is any significant difference between consignments or between experiments.*

Experiment	*Consignment*			
	I	*II*	*III*	*IV*
A	*9*	*10*	*9*	*10*
B	*12*	*11*	*9*	*11*
C	*11*	*12*	*10*	*12*

(b) Also, test at 5% level which pairs of experiments differ signi-ficantly, if any. [Given $F_{.05} = 5.14$ for d.f. (2, 6), $F_{.05} = 4.76$ for d.f. (3, 6), and $t_{.025} = 2.45$ for 6 d.f.]

Solution Each observation is reduced by 10, and shown below:

Table 15.6 Calculations for Analysis of Variance

Experimenter	*Consignment*				*Total* (T_i)
	I	*II*	*III*	*IV*	
A	–1	0	–1	0	–2
B	2	1	–1	1	3
C	1	2	0	2	5
Total (T_j')	2	3	–2	3	$T = 6$

Total of the squares of all figures

$$\Sigma\Sigma x_{ij}^2 = (-1)^2 + 0^2 + (-1)^2 + \ldots + 0^2 + 2^2 = 18$$

$$\text{Correction Factor } (CF) = \frac{T^2}{N} = \frac{6^2}{12} = 3$$

$$Total\ SS = \Sigma\Sigma x_{ij}^2 - CF = 18 - 3 = 15$$

$$SS \text{ between Experimenters} = \frac{(T_1^2 + T_2^2 + T_3^2)}{4} - CF$$

$$= \frac{[(-2)^2 + 3^2 + 5^2]}{4 - 3} = 6.5$$

$$SS \text{ between Consignments} = \frac{(T_1'^2 + T_2'^2 + T_3'^2 + T_4'^2)}{3} - CF$$

$$= \frac{[2^2 + 3^2 + (-2)^2 + 3^2]}{3 - 3} = 5.67$$

SS due to Error = *Total SS* – (SS between experimenters) – (SS between consignments)

= 15 – 6.5 – 5.67 = 2.83

Table 15.7 Analysis of Variance Table

Source of Variation	*df*	*SS*	*MS*	*F(obs.)*	*F (tab.)*
Between Experimenters	2	6.5	3.25	6.91	$F_{.05} = 5.14$
Between Consignments	3	5.67	1.89	4.02	$F_{.05} = 4.76$
Error	6	2.83	0.47	–	–
Total	11	15.0	–	–	–

(a) Since the observed value of F for experimenters (viz., 6.91) is larger than the corresponding tabulated value for d.f. (2, 6), it is significant at 5% level. We therefore conclude that the mean moisture content as determined by the 3 experimenters are not equal; i.e., there are significant difference between experimenters.

The observed value of F for consignments (viz. 4.02) is less than the corresponding tabulated value or d.f. (3, 6), and hence is not significant at 5% level. We conclude that the moisture content of the 4 consignments may not be different from one another, i.e., there are no significant differences between consignments.

(b) Critical Difference (see 15.5.1) between totals for experimenters:

$$\text{C.D.} = \sqrt{0.47} \times \sqrt{2 \times 4} \times 2.45 = 4.8$$

We have $T_1 = -2$, $T_2 = 3$, $T_3 = 5$. The differences between T_1 and T_2, T_1 and T_3 and T_2, and T_3 are respectively 5, 7 and 2. Since 5 and 7 are larger than C.D., experimenters *A* and *B* and *A* & *C* differ significantly.

EXERCISES

1. What is ‘Analysis of Variance’? Describe its usefulness in tests of significance.
2. State the assumptions made in the analysis of variance. Explain the ‘linear model’ one-way classified data.

3. Describe the technique of analysis of variance in one-way classified data, stating clearly the mathematical model and the assumptions you make, by giving analysis of variance table. *[W.B.H.S. '83]*
4. Write notes on:
   (i) Between Group Sum of Squares,
   (ii) Within Group Sum of Squares.
5. The following table shows the retail prices (Rs per kg) of a commodity in some shops selected at random in four cities:

Cities			
A	*B*	*C*	*D*
34	29	27	34
37	33	29	36
32	30	31	38
33	34	28	35

Carry out the analysis of variance to test the significance of the difference between prices of the commodity in the four cities. [Given, $F_{.05} = 3.49$ for (3, 12) degrees of freedom].

6. Prepare the analysis of variance table for the following one-way classified data and comment:

	Weight of Balls (gms)		
	Machine 1	*Machine 2*	*Machine 3*
	2.0	1.8	3.0
	2.2	2.2	2.8
	1.7	2.0	3.2
Total	5.9	6.0	9.0

[W.B.H.S. '81]

[Given $F_{.05} = 5.14$ for (2, 6) degrees of freedom.]

7. The following table gives the yield of three varieties of wheat, each grown on four plots and allocated completely at random. Find out if the variety differences are significant. (Tabulated value of F for d.f. (2, 9) is 8.02 at 1% level).

Varieties	*Plot Yields*			
1	26	27	28	31
2	18	19	21	22
3	16	18	19	19

8. Random samples of electric lamps taken from 4 batches show the following data on life (in hours) of these lamps:

Batch I	1600,	1610,	1660,	1700,	1680
Batch II	1590,	1680,	1640,	1730	
Batch III	1600,	1620,	1570,	1570,	1590
Batch IV	1580,	1600,	1500,	1540,	1550,	1590

Analyse the data and test whether the average life of the lamps in the four batches are equal or not. (Value of $F_{.05}$ from tables at (3, 16) d.f. is 3.24).

9. To test the effect of a small percentage of rubber added to cellulose used for making toys, several batches were mixed under practically identical conditions except for the variation in the percentage of rubber. From each batch four pieces were tested for strength. The results in suitable units are:

Percentage of Rubber		
0.0	0.1	0.2
8	10	12
9	9	8
7	8	9
9	12	11

Examine from an analysis of variance, whether the addition of rubber has any effect on the strength. Given $F_{.05} = 4.26$ for d.f. (2, 9).

10. Twenty observations on the yield of a certain crop grown in plots of equal size are given below. The data are split into five groups of four observations each, according to the fertiliser used.

Fertilizer Used	*Yield (in suitable units)*			
A (No Fertiliser)	67	63	55	58
B	78	71	76	70
C	60	69	60	65
D	69	64	71	70
E	80	70	79	78

(a) Make an analysis of variance of the data stating clearly the appropriate model and the necessary assumptions.

(b) Test if the fertilisers *B* and *C* differ significantly at 5% level. Given $F_{.01} = 4.89$ for d.f. (4, 15), and $t_{.025} = 2.13$ for 15 d.f.

11. The following data show the production of 4 workmen on three machines. Test whether the three workmen differ in respect of mean productivity: (Given $F_{.05} = 6.94$ for (2, 4) degree of freedom)

Machine	*Workmen*		
	A	*B*	*C*
1	38	33	46
2	36	44	42
3	49	42	39

12. Apply the technique of analysis of variance of the two following data, showing the yields of 4 varieties of a crop each from 3 blocks, and test whether the mean yields of the varieties are equal or not. Also test the equality of the block means.

Varieties	Blocks			
	I	*II*	*III*	*IV*
A	4	8	6	8
B	5	5	7	8
C	6	7	9	5

[Given $F_{.05} = 5.143$, $F_{.01} = 10.95$ for d.f. (2, 6); and $F_{.05} = 4.757$, $F_{.01} = 9.779$ for d.f. (3,6). *[C.U., B.Sc, '79]*

13. The following table gives the estimates of acreage of cultivable but not-cultivated land out of 100 acres of total land, as obtained by three investigators in each of three districts. Perform an analysis of variance to test whether there are significant differences between investigators and districts. Given $F_{.05} = 6.94$ for d.f. (2, 4).

Investigator	District		
	I	*II*	*III*
A	23	28	26
B	24	25	27
C	24	22	26

ANSWERS

5.

Source of Variation	*df*	*SS*	*MS*	*F(obs.)*	*F(tab.)*
Between Cities	3	111.5	37.2	9.2	$F_{.05} = 3.49$
Within Cites	12	48.5	4.04		
Total	15	160.0	—		

Conclusion: Prices in the four cities are not equal.

6.

Source of Variation	*df*	*SS*	*MS*	*F(obs.)*	*F(tab.)*
Between Machines	2	2.069	1.034	21.5	$F_{.05} = 5.14$
Within Machines	6	0.287	0.048		
Total	8	2.356			

Comment: Mean weight of ball produced by the 3 Machines are not equal.

7.

Source of Variation	*df*	*SS*	*MS*	*F(obs.)*	*F(tab.)*
Between Varieties	2	224	112	33.6	$F_{.01} = 8.02$
Within Varieties	9	30	3.33		
Total	11	254	—		

Conclusion: Variety differences are significant.

8.

Source of Variation	*df*	*SS*	*MS*	*F(obs.)*	*F(tab.)*
Between batch	3	35,000	11,667	6.91	$F_{.05} = 3.24$
Within batch	16	27,000	1,687.5		
Total	19	62,000			

Conclusion: Average life is not for all batches.

9. *Source of*

Variation	*df*	*SS*	*MS*	*F(obs.)*	*F(tab.)*
Between groups	2	7.17	3.58	1.50	$F_{.05} = 4.26$
Within groups	9	21.50	2.39		
Total	11	28.67			

Conclusion: No effect due to addition of rubber.

10. *Source of*

variation	*df*	*SS*	*MS*	*F(obs.)*	*F(tab.)*
Between fertilisers	4	722.30	180.6	9.7	$F_{.01} = 4.89$
Within fertilisers	15	278.25	18.5		
Total	19	1000.55			

Conclusion: (a) Fertilisers have different effect on yield.

(b) Fertilisers *B* and *C* differ significantly. (*C.D.* = 25.9)

11. *Source of*

Variation	*df*	*SS*	*MS*	*F(obs.)*	*F(tab.)*
Between Machine	2	28.67	14.33	0.35	$F_{.05} = 6.94$
Between Workmen	2	10.67	5.33	0.13	$F_{.05} = 6.94$
Error	4	162.66	40.67		
Total	8	202.00			

Conclusion: Workmen do not differ in respect of mean productivity.

12. *Source of*

Variation	*df*	*SS*	*MS*	*F(obs.)*	*F(tab.)*
Between Varieties	2	0.5	0.25	.09	$F_{.05} = 5.143$
Between Blocks	3	9.67	3.22	1.15	$F_{.05} = 4.757$
Error	6	16.83	2.81		
Total	11	27.00			

Conclusion : (a) Mean yields of varieties may be equal.

(b) Mean yields of blocks may be equal.

13. *Source of*

Variation	*df*	*SS*	*MS*	*F(obs.)*	*F(tab.)*
Between investigators	2	4.67	2.33	0.63	$F_{.05} = 6.94$
Between districts	2	10.67	5.33	1.45	$F_{.05} = 6.94$
Error	4	14.66	3.67		
Total	8	30.00			

Conclusion: No significant difference between investigators, or between districts.

16 TIME SERIES

16.1 MEANING AND NECESSITY OF 'TIME SERIES ANALYSIS'

A series of observations recorded in accordance with the time of occurrence is called "*Time Series*". Such data are of particular interest in economics, business and commerce, where the values of a variable are observed chronologically be days, weeks, months, quarters, or years. Production, consumption sales, profits, bank clearings during successive *periods* of time; and population, price etc. at successive *points* of time are examples of time series. Symbolically, the value of the variable related to time t is denoted by y_i.

Example 16.1 *Explain the necessity of analysing time series data.*

[I.C.W.A., Jan '70; C.A., May '76]

Solution A study of time series data discloses that observed values of the variable are always fluctuating from time to time. The fluctuations are the result of the joint action of various forces, like changes in tastes and habits of people, increase in population, development of new techniques resulting in lower cost of production changes in weather conditions, etc. The forces are everchanging and due to the interaction among them, values of the variable undergo change with the passage of time.

The objects of time series analysis are to isolate and measure the effects of various components. Such an analysis will help in understanding the past behaviour, so that we may be able to predict the future tendency. To business executives who have to plan their production much ahead of sales, the analysis of time series will be of great assistance in planning for the hiring of personnel for peak periods, to accumulate an inventory of raw materials, to ready equipment, and finally in forecasting the future demand of their product.

16.2 COMPONENTS OF TIME SERIES

A graphical representation of time series data (Fig. 12.1, 12.2, 12.4–12.9) reveals the changes over time. However, these changes are not totally haphazard, and a part of it at least can be accounted for. The four components of time series are

(1) Secular trend, or Trend (T)
(2) Seasonal variation (S)
(3) Cyclical fluctuation (C)
(4) Irregular or Random movement (I)

In the classical or traditional approach, it is assumed that there is a multiplicative relationship between the four components, i.e., any particular observation is considered to be the product of the effects of four components:

$$y_t = T \times S \times C \times I \text{ (Multiplicative Model)} \qquad (16.2.1)$$

Another approach is to assume an additive relationship between them,

$$y_t = T + S + C + I \text{ (Additive Model)} \qquad (16.2.2)$$

Although the additive model facilitates easier calculations, it has been found inappropriate in many practical situations, and hence is not generally used.

Example 16.2 *Describe the various components of time series.*
[C.U., B.Com. '82; M.Com. '80; W.B.H.S. '82; I.C.W.A., June '83; C.A., May '83]

Solution There are four components of time series: (i) Secular trend, (ii) Seasonal variation, (iii) Cyclical fluctuation, and (iv) Irregular or Random movement.

Secular trend (or simply *Trend*) of time series is the smooth, regular and long-term movement exhibiting the tendency of growth or decline over a period of time. The trend is that part which the series would have exhibited, had there been no other factors affecting the values. The population growth together with advances in technology and methods of business organisation are the main factors for the growth or upward trend in most of the economic and business data. The decline or downward trend may be due to the decreasing demand of the product, or a substitute taking its place, or difficulty in obtaining raw materials, etc. Many industries, however, initially show a steady growth until a saturation point is reached, and then the trend declines steadily. But sudden or frequent changes are incompatiable with the idea of trend.

Seasonal variation represents a type of *periodic* movement, where the period is not longer than one year. Business activities are found to have brisk and slack periods at different parts of the year. This up-and-down movement of time series, recurring with remarkable regularity year after year, is attributable to the presence of seasonal variations. The factors which cause this type of variation are the climatic changes of the different seasons, such as changes in rainfall, temperature, humidity etc., and the customs and habits which people follow at different parts of the year.

Cyclical fluctuation is another type of *periodic* movement, where the period is more than a year. Such movements are fairly regular and oscillatory in nature. One complete period is called a *cycle.* Cyclical fluctuation is found to exist in most of the business and economic time series, where it is known as *business cycle.* Business cycles are caused by a complex combination of forces affecting the equilibrium of demand and supply. Prosperity, decline, depression and recovery are usually considered to be the four phases of business cycles. The swing from prosperity to recovery and back again to prosperity varies both in time span and intensity. Very often two different time series data are found to follow closely the same pattern of cyclical movement because of inter-relations between them.

Irregular or *Random movements* are such variations which are caused by factors of an erratic nature. These are completely unpredictable or caused by such unforeseen events as war, flood, earthquake, strike and lockout, etc. and may sometimes be the result of many small forces, each of which has a negligible effect, but their combined effect is not negligible. Random movements do not reveal any pattern of the repetitive tendency and may be considered as residual variation.

Example 16.3 *With which characteristic movement of a time series would you associate: (i) a recession, (ii) increasing demand of smaller automobiles, (iii) decline in death rate due to advances in medical science.* [I.C.W.A., Dec. '79]

Solution The characteristic movements associated with the given phenomena are: (i) Cyclical fluctuation, (ii) Secular trend, (iii) Secular trend.

Example 16.4 *Distinguish between 'seasonal' and 'cyclical' fluctuations in time series data.* [I.C.W.A., July '71; C.U., M.Com '71, '73; M.B.A. '77]

Solution The components of time series are of three main types: (i) an overall long-term tendency to change, called 'trend', (ii) a *periodic* movement about the trend, and (iii) an irragular component. Periodic fluctuations are of two kinds—Seasonal and Cyclical.

'Seasonal fluctuations' are comparatively short-range in nature, whose period *does not exceed one year.* The fluctuations are found to maintain a definite periodicity, and reappear almost unerringly at regular interals of time. Examples of seasonal fluctuations are found in passengar traffic during the 24 hours of a day, number of books issued from a library during the 7 days of a week, sales of a departmental store during the 12 months of a year, the quarterly production of a factory, etc. (The periods here are 1 day, 1 week, 1 year and 1 year respectively). The fractors which mainly cause this type of variation in time series are the climatic changes of different seasons, and the social customs and habits of people.

'Cyclical fluctuations' refer to the oscillatory movement of time series with a period *more than a year*. One complete period is called a 'cycle'. Cyclical fluctuations are not as regular as seasonal fluctuations and the period of cycle as well as the intensity of fluctuations very from one cycle to another. The changes occurring in successive periods of time are gradual, each cycle passing through a sequence of 4 phases: prosperity, decline, depression and recovery. The ups-and-downs in business recurring at intervals of time are the effects of cyclical fluctuations. Cyclical fluctuations are caused by the joint in teraction of many factors, and are found to exist almost in all business and economic time series.

16.3 ADJUSTMENTS TO TIME SERIES DATA

Example 16.5 *Discuss some of the adjustments for population changes, calendar variation and price changes, which are necessary to make the time series data homogeneous and comparable.*

Solution Before trying to analyse time series data, it is often necessary to adjust them for calendar variation, population changes and price changes, in order to make them homogeneous and comparable as far as possible.

(i) The adjustment for *calendar variation* is necessary as the number of days in a month, as well as the number of working days, vary from month to month. In such cases we may divide the monthly figures by the corresponding number of days in the month or by the number of working days (giving half weight to half-days), and thus obtain figures on per-day basis.

(ii) Sometimes, the production or consumption figures may show an upward trend indicating greater business activity solely due to *changes in the size of total population.* In this case, the figures may be reduced to per capita basis, dividng by the appropriate population size, and then made comparable.

(iii) Adjustment for *price changes* is made by the process, known as *deflation*. This consists in dividing the original observations by an appropriate index number. For example, in order to compare the volume of sales of a commodity over different periods, the rupee value of the sales may not be sufficient, because of variations in piece. If, however, these rupee values are divided by a price index number, the quantity figures will give a better indication of the volume of sales.

16.4 SECULAR TREND

Example 16.6 *What do you understand by 'Secular Trend' in the analysis of a time series? Explain with examples.* [W.B.H.S. '81]

Solution *'Secular Trend'* or simply *'Trend'* is the smooth, regular and long-term movement of time series exhibiting the basic tendency to growth, decline or stagnation over a period of time. It is the most important component of time series. For example, it we observe the figures of monthly bank deposits published by the Reserve Bank of India for quite a number of years, it may be noticed that apart from fluctuations from month to month, the figures of deposits are on the whole rising as time passes. This means that the series of data have an upward tendency or 'increasing trend'. The position will be clear if we plot the data on a graph paper, as in Fig. 12.1. We may find similar increasing trend in the figures of population, agricultural production, prices, profits, or generation of electricity. A downward tendency or 'decreasing trend' may be observed in the birth rate, or death rate.

The essential feature of trend is its smooth pattern of movement. Sudden or frequent changes are inconsistent with the idea of trend. It should be noted that trend may be computed if data for a sufficiently long period are available. Evaluation of trend from a short period is not advisable, as other components especially cyclical fluctuations may affect the results.

Two important methods of measuring trend are the Moving Average method and the method of Fitting Mathematical Curves. In the former method, trend is obtained by smoothing out fluctuations by the process of averaging. In the latter, a mathematical model is assumed for trend, e.g. linear, exponential, or other type. Measurement of trend is necessary for (i) studying the behaviour of the time series, (ii) for studying the influence of other components like seasonal and cyclical by removing trend from the original series, and (iii) for forecasting the future position.

16.5 MEASUREMENT OF TREND

Example 16.7 *Describe the various methods used in isolating secular trend in time series, explaining the logic behind the methods.*
[C.U., B.Com(Hons) '82, '83; M.Com. '81; W.B.H.S. '83;
I.C.W.A., June '81; C.A. Nov. '80, '81]

Solution There are four methods of isolating secular trend in time series:

(1) Free-hand method,
(2) Semi-average method,
(3) Moving average method, and
(4) Fitting mathematical curves.

(1) Free-hand Method In *Free-hand method,* the given data are plotted as points on a graph paper against time. The time series data (y_t) are shown along the vertical axis and time (t) along the horizontal axis. Then a smooth free-hand curve is drawn through the scatter of the plotted points, which appears to represent their pattern of movement over time. The distance of this line, known as *trend line,* gives the trend value for each time period. The advantages of the method are that a quick estimate of the trend is obtained and that the method can be used to obtain a preliminary knowledge of the nature of trend with a view to applying more refined methods. However, the free-hand method depends too much on individual judgment and different persons will obtain different trend values from the same data (Example 16.8).

(2) Semi-average Method *Semi-average method* consists in dividing the data into two parts, and then finding an average for each part. These averages are plotted as points on a graph paper against the mid-point of the time interval covered by each part. The straight line joining these two points gives the trend line. As before, the distances of trend line from the horizontal axis give the trend values. If the actual trend is a straight line, the method will give quite satisfactory results. Although this method is simple to apply, it may lead to poor' results when used indiscriminately. If the ratios of successive time series data are approximately equal, the method should be applied to the logarithms of values. The trend values will then be obtained by taking antilogarithms of the distances of trend line from the horizontal axis.

(3) Moving Average Method *Moving average method* is very commonly used for the isolation of trend and in smoothing out fluctuations in time series. In this method, a series of arithmetic means of successive observations, known as *moving averages,* are calculated from the given data, and these moving averages are used as trend values. Precisely, moving averages of *period n* are a series of arithmetic means of groups of successive n observations, and are shown against the mid-points of time intervals covered by the respective groups. If the period of moving average is odd, the trend values correspond to the given time. If the period of moving average is even, a two-point moving average of the moving averages so obtained, has to be found out for 'centering' them.

The logic behind the moving average method is that if the period of moving average is exactly equal to the period of cycle (or its multiple) present in the series, then the method will completely eliminate cyclical fluctuations (Example 16.14). The method is very simple and needs no complicated mathematical calculations. Moving averages can also adapt themselves to changing circumstances. A disadvantage of the method is that some trend values at the beginning and at the end of the series cannot be obtained. Also, since moving averages assume no law of change, this method cannot be used for forecasting future trend (Example 16.12).

(4) Fitting Mathematical Curves *Method of fitting mathematical curves* is perhaps the best and most objective method of determining trend. In this method, an appropriate type of mathematical equation is selected for trend, and the constants appearing in the trend equation are determined on the basis of the given time series data. The choice of the appropriate type of equation is facilitated by a graphical representation of the data.

(i) If the plotted data show approximately a straight line tendency on an ordinary graph paper, the equation used is:

$y = a + bx$ (Straight line)

(ii) If they show a straight line on a semi-logarithmic graph paper, the equation used is:

$\log y = a + bx$ (Exponential curve)

(iii) Sometimes a parabola or higher order polynomial may also be fitted.

$y = a + bx + cx^2$ (Parabola)

(iv) Special types of curves are also used in certain cases.

$y = a + bc^x$ (Modified exponential curve)

$1/y = a + bc^x$ (Logistic curve)

$\log y = a + bc^x$ (Gompertz curve)

The constants, appearing in the equations referred to at (i) to (iii) above are obtained by applying the *principle of least squares* (Section 8.4). This states that the values of contants should be such as to make the sum of the squares of vertical distances from the trend line as small as possible. The method of fitting mathermatical curves can be used for forecasting the future trend. This method also involves considerable numerical calculations.

Example 16.8 *With the help of graph paper supplied, obtain the trend values:*

Year	*1962*	*1963*	*1964*	*1965*	*1966*	*1967*	*1968*	*1969*	*1970*	*1971*	*1972*	*1973*	*1974*
Value	*64*	*82*	*97*	*71*	*78*	*112*	*115*	*131*	*88*	*100*	*146*	*150*	*120*

[C.A., May '75]

Solution The given data are plotted as points on the graph paper (Fig. 16.1) and a free-hand curve is drawn through the scatter of the points. This is the 'trend line'. The distances of trend line from the horizontal axis are now read from the graph for all the years, giving 'trend values'. The original data are also shown against each for a comparison.

Year	1962	1963	1964	1965	1966	1967	1968	1969	1970	1971	1972	1973	1974
Value value	75	79	82	86	91	96	102	107	112	118	124	130	136
Original Data	64	82	97	71	78	112	115	131	88	100	146	150	120

[**Note:** (i) The original data are fluctuating up and down (see the zig-zag line in Fig. 16.1), but the trend values show gradual changes over the years.

(ii) Trend values from a free-hand curve, which is drawn by one's individual judgement, may not agree with those obtained by others or by applying other methods.]

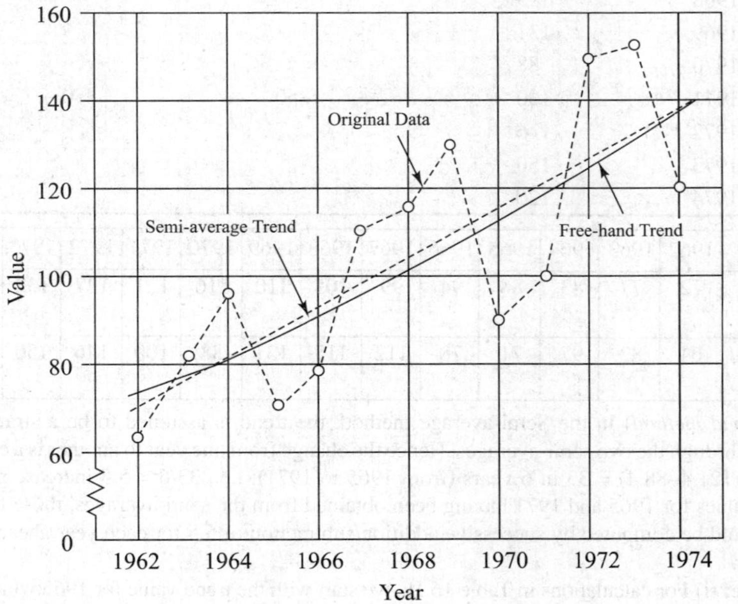

Fig. 16.1 Trend by (i) Free-hand and (ii) Semi-average Methods

Example 16.9 *Apply the Semi-average Method and obtain trend values for all the years from the data of Example 16.8.*

Solution Since data are given for an odd number of years (13 years), we cannot divide them into exactly two equal parts. Let us take two halves of 7 years each—first half from 1962 to 1968, and second half again from 1968 to 1974. The averages for these periods are 88.4 and 121.4 (see Table 16.1a).

These averages (known as 'semi-average') correspond to the middle of the time intervals covered, i.e., 1965 and 1971 respectively.

(*First method*) The semi-averages are plotted on a graph paper against the years 1965 and 1971. A straight line trend is now drawn to join these two points (Fig. 16.1) which is extended at both ends. The distances of the base line at different years from the trend line are now read from the vertical scale of the graph paper. These give the trend values and are shown below along with the original data for a comparison.

Table 16.1a Trend Measurement by Semi-Average Method

Year	*Time Series* (y_i)	*7-year Semi-Totals*	*Semi-Averages*
1962	64		
1963	82		
1964	97		
1965	71	619	619 ÷ 7 = 88.4
1966	78		
1967	112		
1968	115		
1969	131		
1970	88		
1971	100	850	850 ÷ 7 = 121.4
1972	146		
1973	150		
1974	120		

Year	1962	1963	1964	1965	1966	1967	1968	1969	1970	1971	1972	1973	1974
Value value	72	77	83	88	94	99	105	110	116	121	127	132	138
Original Data	64	82	97	71	78	112	115	131	88	100	146	150	120

(*Second method*) In the Semi-average method, the trend is assumed to be a straight line passing through the two semi-averages. Hence the change from one year to another is a constant, namely (121.4–88.4) = 33 in 6 years (from 1965 to 1971); i.e., 33/6 = 5.5 increase per year. Trend values for 1965 and 1971 having been obtained from the semi-averages, those for other years could be computed by successive addition/subtraction of 5.5 for each year ahead/behind any year.

[**Note:** (i) For calculations in Table 16.1b, we start with the trend value for 1965 which is the computed semi-average. Trend values are then calculated successively for 1964, 1963, 1962 backwards each time reducing the previous figure by 5.5. For the years after 1965, we proceed ahead with 1966, 1967, 1968 etc. each time adding 5.5 to the previously computed value.

(ii) The Second Method stated here is generally known by *Method of Selected Points*, the two selected points 88.4 and 121.4 corresponding to 1965 and 1971 computed by the semi-average process. The trend is a straight line with equation $y = a + bx$, the values of a and b being obtained from the selected points. If we start with year 1965 as '*origin*' and time scale 1 year as one '*unit*' of x, then $a = 88.4$ and $b = 5.5$ (i.e., increment per unit of x). The trend values (y) are obtained by putting the appropriate value of x in the trend equation $y = 88.4 + 5.5x$. For instance, in the year 1968 we have $x = 3$ (3 years ahead of origin 1965). Hence trend value is $y = 88.4 + 5.5(3) = 88.4 + 6.5 = 104.9$ which agrees with the figure shown in Table 6.1b. Detailed dicussions will follow in the method of "*Fitting Mathematical Curves*".]

Table 16.1b Trend Measurement by Semi-average Method

Year (t)	*Time Series* (y_i)	*Trend Values* (T)
1962	64	77.4 – 5.5 = 71.9
1963	82	82.9– 5.5 = 77.4
1964	97	88.4 – 5.5 = 82.9
1965	71	88.4 = Semi-average for 1st half
1966	78	88.4 + 5.5 = 93.9
1967	112	93.9 + 5.5 = 99.4
1968	115	99.4 + 5.5 = 104.9
1969	131	104.9 + 5.5 = 110.4
1970	88	110.4 + 5.5 = 115.9
1971	100	121.4 = Semi-aveage for 2nd half
1972	146	121.4 + 5.5 = 126.9
1973	150	126.9 + 5.5 = 132.4
1974	120	132.4 + 5.5 = 137.9

Example 16.10 *Describe the method of moving average and discuss its role in the isolation fo trend and in smoothing time series data.*

[C.U., B.A. (Econ.) '72, '74, '77; I.C.W.A., Dec. '75, June '78; M.B.A. '78]

Solution *Method of Moving Average* is a very convenient tool for ironing out fluctuations in time series. It is generally used in all parts of time-series analysis—for the isolation of trend, and also in connection with seasonal, cyclical and irregular components—by eliminating the oscillatory movements. Even when fluctuations in the data cannot be removed completely moving averages considerably reduce their intensity. This is called "*smoothing*".

Moving averages are a sequence of arithmetic means, each based on a fixed number of successive observations in time series. Suppose, the time series data are given by years. Then, moving averages of *period 3* years, say, (also called *3-year moving averages*) give us a series of arithmetic means each of three successive observations. We start with the first 3 observations and calculate the arithmetic mean. This is placed against the 2nd year. At the next stage, we leave the first observation and calculate the A.M. of the next three, viz. 2nd to 4th, and place it against the 3rd year. The process is repeated until we arrive at the last three observations. Similarly, moving averages of 4, 5, 6, or any period may be calculated.

Sometimes, weighted moving averages are also used. For the calculation of weighted moving average, the successive observations in a group are multiplied by a given set of numbers (called 'weights'), and the weighted sum is divided by the sum of weights (see Example 16.13). The same set of weights is used for the successive observations in each group.

Each moving average is placed against the mid-point of time interval included in the calculation of A.M. Thus when the period is odd, all moving averages will coincide with the given years. However, if the period is even, moving averages will fall mid-way between two successive years. In this case, again two-item moving averages of the moving averages already found, have to be calculated for syncronizing them with the given data. This process is known as *'centering'*.

Moving averages with a period exactly equal to (or a multiple of) the period of cycle present in the series will completely eliminate the cyclical component from a time series and give an estimate of trend. So, the moving average method is used for measurement of trend from a given time series data, by taking the period of moving average exactly equal to the period of cycle present in the series. However, usually the *period* and intensity (called *amplitude*) vary from cycle to cycle. The best results will be obtained by using moving averages with a period equal to the average period of fluctuations, which can be obtained by graphical method. In all cases, however, moving averages will reduce fluctuations present in the series and smooth out the short-term seasonal and irregular movements, even if the cyclical component is not removed altogether.

Merits and Demerits of Moving Average Method

Merits

(i) Moving average method is simple to apply and involves no difficult calculations.

(ii) If the time series contain regular cyclical fluctuations, these fluctuations are automatically removed provided an appropriate period is chosen. Even when the fluctuations are not completely eliminated, moving average process reduces their intensity.

(iii) Moving averages adapt themselves to the general movements of data. The shape of the trend line is thus determined by the data themselves rather than by the statistician's choice of amthematical curve (as in the method of fitting curves).

(iv) Moving average method is flexible in the sense that if some more observations are added to the original series, the entire calculations need not be changed. We get only some more trend values.

Demerits

(i) Trend values for all the given time periods cannot be obtained. Some trend values at the beginning and at the end of the series have to be left out, their number increasing with increase in the period of moving average.

(ii) The period of moving average has to be chosen very carefully. There are no hard and fast rules for the purpose.

(iii) Since moving averages assume no law of change, the method cannot be used for forecasting future trend.

(iv) Moving averages are useful only when the trend is linear or approximately so. If the actual is curvilinear, moving averages may deviate considerably from the trend.

Example 16.11 *(a) Calculate the five-yearly Moving Average of the following:*

Year	1950	1951	1952	1953	1954	1955	1956	1957	1958
Values	105	115	100	90	80	95	85	75	60
	1959	1960	1961	1962	1963	1964	1965	1966	
	65	70	58	55	53	60	52	50	

(b) Illustrate with four examples, how the four-yearly Moving Average can be calculated. [C.A., Nov. '67]

Solution

Table 16.2 Calculations for 5-yearly Moving Average

Year (1)	Value (2)	5-year Moving Total (3)	5-year Moving Average (4)
1950	105	—	—
1951	115	—	—
1952	100	490	98
1953	90	480	96
1954	80	450	90
1955	95	425	85
1956	85	395	79
1957	75	380	76
1958	60	355	71
1959	65	328	65.6
1960	70	308	61.6
1961	58	301	60.2
1962	55	296	59.2
1963	53	278	55.6
1964	60	270	54
1965	52	—	—
1966	50	—	—

[**Note:** (i) Calculations for Moving Totals in Col. (3) may be simplified as follows. Direct calculation gives the sum of the first 5 values (105 + 115 + 100 + 90 + 80) = 490. The next moving total (115 + 100 + 90 + 80 + 95) differs from the former with the new value 95 replacing the 1st value 105, i.e., an increase of 95—105 = – 10; so the new moving total is 490 – 10 = 480. Similarly, the next moving total is 480 + (85 – 115) = 480 – 30 = 450; and so on. The increase or decrease over the preceding moving total may be mentally calculated and the new moving total obtained easily. The accuracy of calculations can be checked by verifying that the last moving total obtained by this process, viz. 278 + (50 – 58) = 270, is exactly equal to the total of the last five values (55 + 53 + 60 + 52 + 50) by direct calculations.

(ii) Col. (4) = Col. (3)¸ 5.]

(b) Let us calculate 4-yearly moving averages from the values for the years 1950–1958. Since the period of moving average is now even, it is necessary to 'centre' the moving averages. The procedure is shown below:

I. ***Direct Method:***

Table 16.3 Calculations for 4-yearly Moving Average

Year (1)	*Value* (2)	*4-year Moving Total (not centered)* (3)	*4-year Moving Average (not centered)* (4)	*2-item Moving Total (centered)* (5)	*4-year Moving Average (centered)* (6)
1950	105	—	—	—	—
1951	115	—	—	—	—
		410	102.5		
1952	100			198.75	99.375
		385	96.25		
1953	90			187.5	93.750
		365	91.25		
1954	80			178.75	89.375
		350	87.5		
1955	95			171.25	85.625
		335	83.75		
1956	85			162.5	81.250
		315	78.75		
1957	75	—	—	—	—
1958	60	—	—	—	—

Col. (4) = Col. (3) ÷ 4 Col. (6) = Col. (5) ÷ 2

II. ***Short-cut Method:***

Table 16.4 Calculations for 4-Yearly Moving Average

Year (1)	*Value* (2)	*4-year Moving Total (not centered)* (3)	*2-item Moving Total of Col (3) (centered)* (4)	*4-year Moving Average (centered)* (5)
1950	105	—	—	—
1951	115	—	—	—
		410		
1952	100		795	99.375
		385		
1953	90		750	93.750
		365		
1954	80		715	89.375
		350		
1955	95		685	85.625
		335		
1956	85		650	81.250
		315		
1957	75	—	—	—
1958	60	—	—	—

Col. (5) = Col. (4) ÷ 8.

Example 16.12 *Find the trend for the following series using a three-year moving averages:*

Year	1	2	3	4	5	6	7
Values	2	4	5	7	8	10	13

[C.U., B.Com. (Hons) '83]

Solution

Table 16.5 Calculations for Moving Average

Year (1)	Values (2)	3-year Moving Total (3)	3-year Moving Average (4)
1	2	—	—
2	4	11	3.67
3	5	16	5.33
4	7	20	6.67
5	8	25	8.33
6	10	31	10.33
7	13	—	—

Col. (4) = Col. (3) ÷ 3

Example 16.13 *Find the trend for the following series using a three-year weighted moving average with weights 1, 2, 1.*

Year	1	2	3	4	5	6	7
Values	2	4	5	7	8	10	13

[I.C.W.A., June '78]

Solution Instead of taking simple averages of the values for three consecutive years, the weighted average is calculated.

Table 16.6 Calculations for Weighted Moving Average

Year (1)	Values (2)	3-year Moving Total* (3)	3-year Moving Average@ (4)
1	2	—	—
2	4	15	3.75
3	5	21	5.25
4	7	27	6.75
5	8	33	8.25
6	10	41	10.25
7	13	—	—

* $2 \times 1 + 4 \times 2 + 5 \times 1 = 15$; $4 \times 1 + 5 \times 2 + 7 \times 1 = 21$; etc.

@ Col. (4) = Col. (3) ÷ (sum of weights; i.e., 4)

Note: Compare 'Simple' and 'Weighted' moving averages from Table 16.5. and 16.6.

Example 16.14 *'If the business cycles are of uniform period (length) and amplitude (height), they can be completely eliminated by making the period of the moving average equal to that of the cycle"—Explain the statement with help of an example.*

Solution The statement implies that if the time series data contain only trend (*T*) and cyclical (*C*) components, then the moving averages calculated from such series will reveal only the trend values, provided cyclical fluctuations are of uniform period and amplitude and that the period of moving average is taken to be equal to that of cycle. This means that the cyclical fluctuations must have been removed by taking moving averages of an appropriate period. The theory provides a method of isolating trend by the moving average process.

Let the trend (*T*) and cyclical (*C*) components of a time series be as shown in cols. (2) and (3) respectively (Table 16.7). From col. (3) it will be noticed that the effects of boom and depression in the cycles repeat exactly after a period of 3 years, i.e., the cycles are of *uniform period* 3 years. Also the magnitude of cyclical fluctuation every third year is the same, indicating a *uniform amplitude* of the cycles. The trend values mixed with the cyclical component give the time series values (y_t) as shown in col. (4). We now calculate moving averages taking a period exactly the same as the period of cycles. It is found from col. (6) that the moving averages coincide with the corresponding trend values shown in col. (2). This verifies that cyclical fluctuations have been completely eliminated from time series by the moving average process.

Table 16.7 Moving Average from Uniformly Cyclical Series

Year (1)	*Trend* (*T*) (2)	*Cyclical Fluctuations* (*C*)* (3)	*Time Series* $y_t = T + C$ (4)	*3-year Moving Total* (5)	*3-year Moving Average* (6)
1950	40	– 4	36	—	—
1951	50	5	55	150	50
1952	60	– 1	59	180	60
1953	70	– 4	66	210	70
1954	80	5	85	240	80
1955	90	– 1	89	270	90
1956	100	– 4	96	300	100
1957	110	5	115	—	—

*The total of fluctuations in a complete cycle is expected to be zero, in the additive model.

Example 16.15 *Compute the trend by the method of moving averages, assuming that a 4-yearly cycle is present in the following series:*

Year	*1958*	*1959*	*1960*	*1961*	*1962*	*1963*	*1964*	*1965*	*1966*	*1967*	*1968*
Annual Value	*54.0*	*40.5*	*47.0*	*48.5*	*42.9*	*42.1*	*36.6*	*42.7*	*45.7*	*45.1*	*37.8*

[I.C.W.A., June '75]

Solution In order to obtain trend values by the moving average method, the period of moving average must be 4 years, i.e., the same as the period of cycle present in the series.

Table 16.8 Calculation of Trend by Moving Averages

Year (1)	Annual Sales (2)	4-year Moving Total (3)	2-item Moving Total of Col. (3) (centered) (4)	4-year Moving Average (centered) (5)
1958	54.0	—	—	—
1959	40.5	—	—	—
		190.0		
1960	47.0		368.9	46.11
		178.9		
1961	48.5		359.4	44.92
		180.5		
1962	42.9		350.6	43.82
		170.1		
1963	42.1		334.4	41.80
		164.3		
1964	36.6		331.4	41.42
		167.1		
1965	42.7		337.2	42.15
		170.1		
1966	45.7		341.4	42.68
		171.3		
1967	45.1	—	—	—
1968	37.8	—	—	—

Col. (5) = Col. (4) ÷ 8

Example 16.15A *Obtain trend values by the moving average method from the data of Example 16.8, by choosing a suitable period. Plot the original data and the moving averages on the same graph paper.*

Solution The data are plotted on a graph paper (Fig. 16.2). It is observed that "peaks" occur in the years 1964, 1969 and 1973—the interval of time between consecutive peaks being 5 and 4 years.

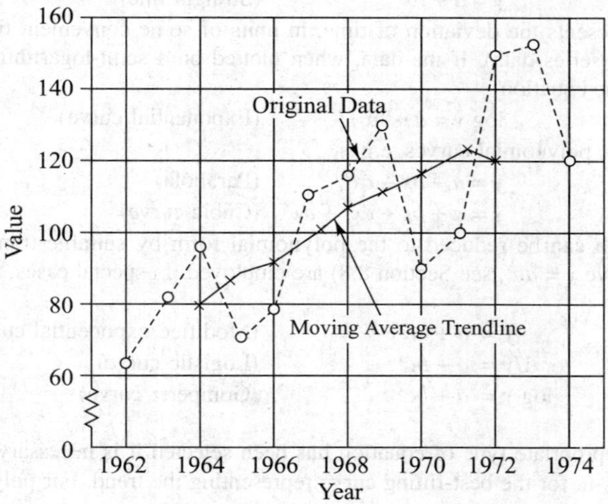

Fig. 16.2 Trend by 5-year Moving Average

There are two "troughs" in the years 1965 and 1970, the time interval between them being 5 years. Thus the average period between con secutive peaks and troughs is $(5 + 4 + 5) \div 3 = 4.7$, i.e., 5 years approximately. Hence, 5 years is taken as the period of the moving average.

Table 16.9 Trend by 5-year Moving Averages

Year	*Value*	*5-year Moving Total*	*5-year Moving Average*
1962	64	—	—
1963	82	—	—
1964	97	392	78.4
1965	71	440	88.0
1966	78	473	94.6
1967	112	507	101.4
1968	115	524	104.8
1969	131	546	109.2
1970	88	580	116.0
1971	100	615	123.0
1972	146	604	120.8
1973	150	—	—
1974	120	—	—

The 5-year moving averages are also plotted on the graph paper (Fig. 16.2).

Example 16.16 *Discuss the method of fitting mathematical curves for determining the trend in time series data.*

Solution The method of fitting mathematical curves is the best and most objective method of determining trend in time series. It is assumed that the trend values follow a mathematical law of a certain type, e.g., straight line, parabola, etc., and the best-fitting curve of this type is obtained for the data. The appropriate type of curve (Section 8.1) is determined by plotting the data on a graph paper. If the points follow a straight line pattern, the equation

$$y = a + bx \qquad \text{(Straight line)}$$

is taken (x represents the deviation of time, in units of some convenient time period, and y represents time series data). If the data, when plotted on a semi-logarithmic paper, show a linear pattern the equation

$$\log y = a + bx \qquad \text{(Exponential curve)}$$

is used. Usually, polynomial curves, e.g.

$$y = a + bx + cx^2 \qquad \text{(Parabola)}$$

$$y = a + bx + cx^2 + dx^2 \qquad \text{(Cubic curve)}$$

or curves which can be reduced to the polynomial form by suitable transformations (e.g. exponential curve $y = ab^x$, see Section 8.8) are employed. In special cases, more complicated curves, e.g.

$$y = a + bc^x \qquad \text{(Modified exponential curve)}$$

$$1/y = a + bc^x \qquad \text{(Logistic curve)}$$

$$\log y = a + bc^x \qquad \text{(Gompertz curve)}$$

are also used.

Once the appropriate type of equation has been selected it is necessary to determine the constant $a, b, c, \ldots$ for the best-fitting curve representing the trend. For polynomial curves or curves reducible to polynomial form, this is done by the *method of least squares* (Section 8.4). According to this method the *sum of the squares of vertical distances of the given points from the best-fitting curve should be the minimum possible.* The problem is tackled by mathematical methods, leading to a set of equations, known as *normal equations* solving which we get the values of *a, b, c*, etc. Special methods are available for fitting modified exponential, logistic, and Gompertz curves. Putting the appropriate values of x in the trend equation trend values for the respective time periods are obtained.

The method of fitting mathematical curves is the most suited for forecasting future trend.

Merits and Demerits of Fitting Mathematical Curves

Merits

(i) The method of fitting mathematical curves complely eliminates personal bias which is inherent in the free-hand method. It has also the advantage of choosing the appropriate type of curve depending on the nature of data (unlike the semi-average method where a linear trend is always assumed).

(ii) Trend values for all the given time periods can be obtained. This is not possible in the moving average method.

(iii) This method enables us to forecast future trend.

Demerits

(i) The choice of the type of curve is rather subjective. One cannot be sure whether a linear or parabolic or any other type of curve will represent the trend best.

(ii) The calculations for this method are more difficult than in the other methods.

(iii) If some more data are available, all the calculations have to be done afresh, which is not necessary in the other methods.

Example 16.17 *In a study of its sales, a motor company obtained the following least squares trend equation:*

$$y = 1600 + 200x:$$

(origin 1950, x units = 1 year, y = total number of units sold annually).

The company has physical facilties to produce only 3600 units a year and it believes that at least for the next decade the trend will continue as before.

(a) What is the average annual increase in the number of units sold? (b) By what year the company's expected sales have equalled its present physical capacity? (c) Estiamte the annual sales for 1965. How much in excess of the company's present physical capacity is this estiamted value?

Solution (a) The given trend equation $y = 1600 + 200x$ represents a straight line (Section 8.2) and its slope (i.e., the coefficient of x on the right hand side) is 200. Since 'slope' represents the change in the value of y for a unit increase in x, hence the average annual increase in the number of units sold is 200.

(b) The trend equation is given by $y = 1600 + 200x$; (origin 1950, x units = 1 year, y = number of units sold yearly). When sales equal the production capacity of 3600 units, we get $y = 3600$. Putting this value in the above trend equation, $3600 = 1600 + 200x$, solving which $x = 2000/200 = 10$. Hence, sales will equal the production capacity after 10 years from origin, i.e., in the year 1960.

(c) Since the origin is at the year 1950, and x units = 1 year, hence for the year 1965,

$$x = \frac{\text{year 1965} - \text{year 1950}}{\text{1 year}} = 15.$$

Putting $x = 15$ in the trend equation $y = 1600 + 200x$, the estimated annual sales for 1965 are given by $y = 1600 + 200 \times 15 = 6400$ units. The present physical capacity being only 3600 units this estiamte is in excess by $(4600 - 3600) = 1000$ units.

Ans. (a) 200; (b) Year 1960; (c) 4600 units; 1000 units.

Example 16.18 *Fit a straight line trend by the least squares method to the following figures of production of a sugar factory:*

Year	*1969*	*1970*	*1971*	*1972*	*1973*	*1974*	*1975*
Production ('000 tons)	*76*	*87*	*95*	*81*	*91*	*96*	*90*

Estiamte the production for 1976. [W.B.H.S. '83]

Solution Let $y = a + bx$ ... (i)

be the equation of the straight line trend with *origin at the year 1972* and x unit = 1 year. By the least squares method, the normal equations for finding the constants a and b are (see Section 8.5)

$$\Sigma y = an + b\Sigma x, \Sigma xy = a\Sigma x + b\Sigma x^2 \quad \text{... (ii)}$$

Table 16.10 Fitting Straight Line Trend

Year	*Production (y)*	x	x^2	xy
1969	76	– 3	9	– 228
1970	87	– 2	4	– 174
1971	95	– 1	1	– 95
1972	81	0	0	0
1973	91	1	1	91
1974	96	2	4	192
1975	90	3	9	270
Total	616	0	28	56

Number of observations $n = 7$. Substituting the values from the table in equation (ii),

$$616 = a(7) + b(0) \quad \text{or} \quad 7a = 616$$

$$56 = a(0) + b(28) \quad 28b = 56$$

Solving, we get $a = 616/7 = 88$ and $b = 56/28 = 2$. Putting these values of a and b in (i), the trend equation is

$$y = 88 + 2x \quad \text{... (iii)}$$

with origin at 1972 and x unit = 1 year.

The value of x for the year 1976 would be 4 (see Table 16.10, co. x). Hence putting $x = 4$ in (iii), the estimate for 1976 is

$$y = 88 + 2(4) = 96 \text{ ('000 tons)}$$

Ans. $y = 88 + 2x$; 96 ('000 tons)

Remarks: The trend line and the original data may be shown on a graph paper as in Fig. 16.3. To draw the straight line trend we need only two points, say trend values for 1969 and 1975. Putting $x = -3$ and 3, we get $y = 88 + 2(-3)$ 82 as trend value for 1969 and $y = 88 + 2(3) = 94$ as the trend value for 1975. These two values are plotted on the graph paper and a straight line is drawn through the points, giving the trend line. If necessary, the straight line may be extended to obtain trend values for future years.

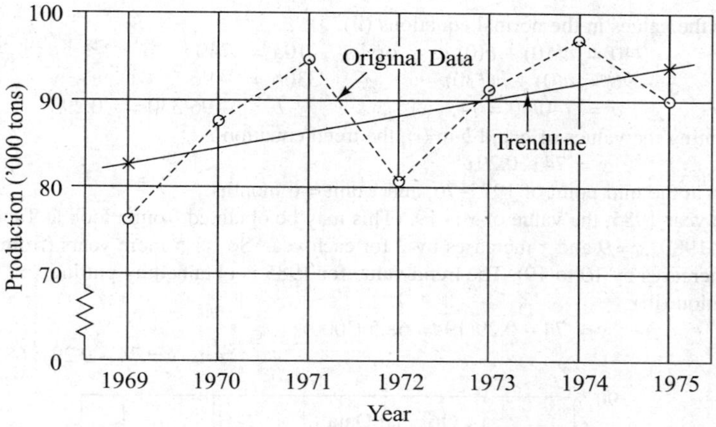

Fig. 16.3 Straight Line Trend (Negative Slope)

Example 16.19 *Fit a straight line trend equation by the method of least squares from the following data and then estimate the trend value for the year 1985:*

Year	*1971*	*1972*	*1973*	*1974*	*1975*	*1976*	*1977*	*1978*	*1979*	*1980*
Value ('000)	*65*	*80*	*84*	*75*	*77*	*71*	*76*	*74*	*70*	*68*

[C.U., B.Com (Hons) '82]

Solution Let $$y = a + bx \quad \text{... (i)}$$
be the equation of the straight line trend with *origin at the mid-point of 1975 and 1976,* and *x unit = 6 months.* (Since data are given for an even number of years, i.e. $n = 10$ is even, the origin and unit of x have been so chosen to make $\Sigma x = 0$). The normal equations for finding the values of a and b are

$$\Sigma y = an + b\Sigma x, \quad \Sigma xy = a\Sigma x + b\Sigma x^2 \quad \text{... (ii)}$$

Table 16.11 Fitting Straight Line Trend

Year	*Value (y)*	*x*	x^2	*xy*	*Trend Values ('000)*
1971	65	−9	81	−585	76.61
1972	80	−7	49	−560	76.03
1973	84	−5	25	−420	75.45
1974	75	−3	9	−225	74.87
1975	77	−1	1	−77	74.29
1976	71	1	1	71	73.71
1977	76	3	9	228	73.13
1978	74	5	25	370	72.55
1979	70	7	49	490	71.97
1980	68	9	81	612	71.39
Total	740	0	330	−96	740.00

Putting the values in the normal equations (ii),

$$740 = a(10) + b(0) \quad \text{or} \quad 10a = 740$$
$$-96 = a(0) + b(330) \qquad 330b = -96$$
$$\therefore \quad a = 740/10 = 74, \qquad b = -96/330 = -0.29$$

Substituting the values of a and b in (i), the trend equation is

$$y = 74 - 0.29x \qquad \text{... (iii)}$$

with origin at the mid-point of 1975–76, and x unit = 6 months.

For the year 1985, the value of x is 19. (This may be obtained from a look at Table 16.11, col. x. For 1980, $x = 9$ and x increases by 2 for each year. So for 5 more years (from 1980 to 1985), x increases by 10 to 19). The trend value for 1985 is obtained by putting $x = 19$ in the trend equation (iii):

$$y = 74 - 0.29(19) = 68.5 \text{ ('000)}$$

Ans. $y = 74 - 0.29x$; 68.5 ('000)

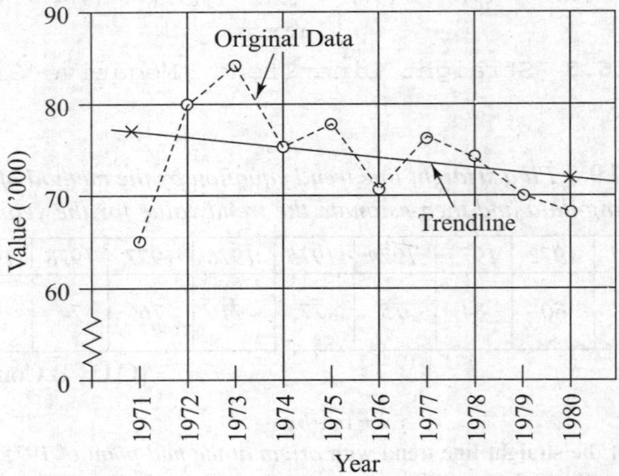

Fig. 16.4 Straight Line Trend (Negative Slope)

[**Note:** (i) The trend line and the original data of Example 16.19 are abown graphically in Fig. 16.4.

It may be observed from Figs 16.3 and 16.4 that the data of Example 16.18 show an 'increasing trend' and the data of Example 16.19 show a 'decreasing trend'. This is also evident from the straight line trend equations: $y = 88 + 2x$ shows a positive slope, viz. + 2 (i.e. y increases as x increases), whereas $y = 74 - 0.29\,x$ shows a negative slope, viz., – 0.29 (i.e. y decreases as x increases).

(ii) If the trend is a straight line, the trend values for successive years change by a constant amount. In Example 16.18 the slope of the straight line is + 2 and since x changes by 1 for each year (i.e. x unit = 1 years) trend values increases annually by 2 units, i.e. 2 thousands tons. In Example 16.19 the slope is – 0.29, and sicne x changes by 2 for each year (i.e. x unit = 6 months), the trend values increase annually by $(-0.29) \times 2 = -0.58$, i.e. decrease by 0.58 units.

With a straight line trend, the trend values of all the given years may therefore by obtained quickly as follows: By direct caculations, trend value for 1971 is $y = 74 - 0.29(-9) = 74 + 2.61 = 76.61$ units, and for each year trend decreases (because slope is negative) by 0.58 units. For successive years, trend values are then $76.61 - 0.58 = 76.03$, $76.03 - 0.58 = 75.45$, etc., as shown in the last column of Table 16.11].

Example 16.20 *Using 1964 as the origin, obtain a straight line trend equation by the method of least squares. Find the trend value of the missing year 1961.*

Year	1960	1962	1963	1964	1965	1966	1969
Value	140	144	160	152	168	176	180

[C.A., Nov. '80]

Solution Let $y = a + bx$ be the straight line trend equation with origin at the year 1964 and x unit = 1 year. By the method of least squares, the values of a and b are obtained by solving the normal equations $\Sigma y = an + b\Sigma x$, $\Sigma xy = a\Sigma x + b\Sigma x^2$.

Table 16.12 Fitting Straight Line Trend

Year	Value (y)	x	x^2	xy
1960	140	– 4	16	– 560
1962	144	– 2	4	– 288
1963	160	– 1	1	– 160
1964	152	0	0	0
1965	168	1	1	168
1966	176	2	4	352
1969	180	5	25	900
Total	1120	1	51	412

Note: Here the given years are not equidistant. Hence col. x should be carefuly obtained.

Substituting the values from the table (here $n = 7$),

$$1120 = a(7) + b(1) \quad \text{or} \quad 7a + b = 1120$$

$$412 = a(1) + b(51) \quad\quad a + 51b = 412$$

Multiplying the second equation by 7 and subtracting from the first equation, we have

$$7a + b = 1120$$

$$7a + 357b = 2884$$

Subtracting $-356b = -1764$ i.e. $b = 1764/356 = 4.96$

Putting the value of $b = 4.96$ in the first equation,

$$7a + 4.96 = 1120; \quad \text{or,} \quad 7a = 1120 - 4.96 = 1115.04$$

$$\therefore \quad a = 1115.04/7 = 159.29$$

The trend equation is therefore $y = 159.29 + 4.96x$ with origin at 1964 and x unit = 1 year. For the missing year 1961, $x = -3$. Hence, the trend value for 1961 is

$$y = 159.29 + 4.96(-3) = 159.29 - 14.88 = 144.41 \text{ units.}$$

Example 16.21 *Fit a second degree polynomial to the following data:*

Year	1882	1883	1884	1885	1886	1887	1888	1889	1890
Price Index	84	82	76	72	69	68	70	72	73

[B.U., B.A. (Econ) '71]

Solution Let $y = a + bx + cx^2$
be the equation of the second degree polynomial, i.e. parabola (see Section 3.14) with *origin of* x at the year 1886, and unit of $x = 1$ year. (Note that data are given for $n = 9$, i.e. an odd number of years). Using the method of least squares, the normal equations for determining the constants a, b, c are (see Section 8.7)

$$\Sigma y = an + b\Sigma x + c\Sigma x^2 \qquad \Sigma xy = a\Sigma x + b\Sigma x^2 + c\Sigma x^3$$
$$\Sigma x^2 y = a\Sigma x^2 + b\Sigma x^3 + c\Sigma x^4$$

Table 16.13 Fitting Second Degree Polynomial

Year	*y*	*x*	x^2	x^3	x^4	*xy*	x^2y
1882	84	– 4	16	– 64	256	– 336	1344
1883	82	– 3	9	– 27	81	– 246	738
1884	76	– 2	4	– 1	16	– 152	304
1885	72	– 1	1	– 1	1	– 72	72
1886	69	0	0	0	0	0	0
1887	68	1	1	1	1	68	68
1888	70	2	4	8	16	140	280
1889	72	3	9	27	81	216	648
1890	73	4	16	64	256	292	1168
Total	666	0	60	0	708	– 90	4622

Substituting the values from the table in the normal equations,

$$666 = a(9) + b(0) + c(60) \quad \text{or,} \quad 666 = 9a + 60c$$
$$-90 = a(0) + b(60) + c(0) \qquad -90 = 60b$$
$$4622 = a(60) + b(0) + c(708) \qquad 4622 = 60a + 708c$$

Solving these there equations, we find $a = 70.06$, $b = -1.50$, $c = 0.59$. The equation of the fitted second degree polynomial is therefore

$$y = 70.06 - 1.50x + 0.59x^2$$

where the origin of x is at the year 1886, and the unit of $x = 1$ year.

Example 16.22 *Fit an exponential trend $y = ab^x$ to the following data by the method of least squares and find the trend values for the years 1944–48:*

Year (x)	*1942*	*1943*	*1944*	*1945*	*1946*	*1947*	*1948*
Sales (y)	*87*	*97*	*113*	*129*	*202*	*195*	*193*

(Logarithmic tables to be supplied).

[C.U., M.Com. '70]

Solution Let us take the origin of x at the year 1945. Taking logarithms of both sides of the equation $y = ab^x$, we have

$$\log y = (\log a) + x (\log b).$$

This can be written in the form of a straight line.

$$Y = A + Bx,$$

where $Y = \log y$, $A = \log a$ and $B = \log b$. Using the method of least squares, the normal equations for determining A and B are

$$\Sigma Y = An + B\Sigma x, \quad \Sigma xY = A\Sigma x + B\Sigma x^2$$

Table 16.14 Fitting Exponential Trend

Year	*Sales(y)*	*x*	*Y = log y*	x^2	*xY*
1942	87	– 3	1.9395	9	– 5.8185
1943	97	– 2	1.9868	4	– 3.9736
1944	113	– 1	2.0531	1	– 2.0531
1945	129	0	2.1106	0	0
1946	202	1	2.3054	1	2.3054
1947	195	2	2.2900	4	4.5800
1948	193	3	2.2856	9	6.5868
Total	—	0	14.9710	28	1.8970

Substituting the values in the normal equations, we get $7A = 14.9710$ and $28B = 1.8970$, i.e. $A = 2.1387$, $B = .0678$. Hence,

a = antilog 2.1387 = 137.6, b = antilog .0678 = 1.169

Putting the values of a and b, the equation of the fitted exponential trend is $y = 137.6\,(1.169)^x$, with the origin of x at the year 1945.

In order to find the trend values, we use the equation $Y = A + Bx$, i.e., $\log y = 2.1387 + .0678x$. Antilog then gives the trend values.

Table 16.15 Calculation of Exponential Trend

Year	*x*	*.0678x*	*log y = 2.1387 + .0678x*	*Trend Value*
1944	– 1	– .0678	2.0709	117.7
1945	0	0	2.1387	137.6
1946	1	.0678	2.2065	160.9
1947	2	.1356	2.2743	188.0
1948	3	.2034	2.3421	219.9

16.6 MONTHLY TREND FROM ANNUAL DATA

In time series, 'annual data' are usually available in two different forms: (i) Monthly Averages for each year (see Example 16.25) and (ii) Annual Totals (see Example 16.26). Separate methods are applicable for treating the cases when the number of years is odd or even.

Case I *Monthly Averages* for odd number of years given:

Since data for an odd number of years are available, we take the middlemost year as origin, x unit = 1 year, and fit the trend equation (see Example 16.25). Let the linear trend fitted to monthly averages, be

$$y = a + bx \tag{16.6.1}$$

(origin: 1950; x unit = 1 year)

The origin is precisely at the mid-point of year 1950, i.e., 30th June, 1950. The slope b represents the increase in monthly average for unit increase in x, i.e. for *12 monhts*. Therefore, the monthly trend increment is $b/12$; and the monthly trend equation is

$$y = a + \frac{b}{12}x$$

(origin: 30th June, 1950; x unit = 1 *month*)

For using the trend equation, we must take some specified month as origin; i.e. precisely the mid-point of a month. If July, 1950 is used as origin (i.e. the origin is shifted half a unit *later*—from 30th June 1950 to 15th July 1950), we have to replace x by $(x + \frac{1}{2})$. Hence, the trend equation for monthly values is

$$y = a + \frac{b}{12}(x + \frac{1}{2}) \tag{16.6.2}$$

(origin: July 1950; x unit = 1 month)

Note: If the origin is shifted half a unit *earlier* to June 1950, x is to be replaced by $x - \frac{1}{2}$ and the trend equation will be $y = a + (b/12)(x - \frac{1}{2})$, with origin at June 1950 and x unit = 1 month.

Case II Monthly Averages for *even* number of years given:

If data for an even number of years are available, we take the mid-point of the two middlemost years as origin and x unit = 6 months (see Example 16.19). For example, suppose that monthly averages in Example 16.25 are given for 8 years from 1946 to 1953. Let the trend equation, fitted to monthly averages, be

$$y = a + bx$$

(origin: 1949–1950; x unit = 6 months)

The origin is at the point midway between 1949 and 1950, i.e. 31st December, 1949. The slope b now represents the increase in monthly average for unit increase in x, i.e. for 6 *months*; hence the monthly trend increment is $b/6$. So, the monthly trend equation is

$$y = a + \frac{b}{6}x$$

(origin: 31st Dec., 1949; x unit = 1 *month*)

For using the monthly trend equation, we must use some specified month as origin; i.e., precisely the mid-point of a month. If January, 1950 is taken as origin (i.e. the origin is shifted half a unit later—from 31st Dec., 1949 to 15th Jan., 1950, x is to be replaced by $(x + \frac{1}{2})$, and the monthly trend equation is now,

$$y = a + \frac{b}{6}(x + \frac{1}{2}) \tag{16.6.3}$$

(origin: *January* 1950; x unit = 1 month)

Case III Annual Totals given:

Let the trend equation fitted to annual totals, given for a number of years (odd or even), be

$$y = A + Bx \tag{16.6.4}$$

If monthly averages were first computed from the given annual totals (dividing each of them by 12), and then a straight line trend fitted to these monthly averages, the equation would be

$$y = \frac{A}{12} + \frac{B}{12}x = a + bx \text{ (say)} \tag{16.6.5}$$

where $a = A/12$, $b = B/12$.

Therefore, the trend equation fitted to monthly averages can be obtained from the equation fitted to annual totals, simply by dividing the right hand side of $y = A + Bx$ by 12 (see Example 16.26). The monthly trend equation may now be obtained by applying the methods shown above, depending on whether the number of years given is odd or even. In fact, this equation would be

$$y = \frac{A}{12} + \frac{B}{144}\left(x + \frac{1}{2}\right) \tag{16.6.6}$$

for odd number of years with origin as at (16.6.2); or

$$y = \frac{A}{12} + \frac{B}{72}\left(x + \frac{1}{2}\right) \tag{16.6.7}$$

for *even* number of years with origin as at (16.6.3).

Example 16.23 *The trend equation fitted to annual average sales is given by y = 230 + 20x, unit of x—one year, origin—30th June, 1960. Adjust the trend equation for finding the monthly trend values, and find trend values for the months of January–March, 1972.* [C.U., M.Com. '73]

Solution It is presumed that 'annual average' refers to "average per *month* for each year". The trend equation is

$$y = 230 + 20x \quad \text{... (i)}$$

(origin: 30 June, 1960; Unit of x = 1 year)

Applying (16.6.2), the equation for monthly trend is

$$y = 230 + \frac{20}{12}\left(x + \frac{1}{2}\right); \quad \text{i.e.} \quad y = 230\frac{5}{6} + \frac{5}{3}x \quad \text{... (ii)}$$

(origin: July, 1960; Unit = 1 month)

The trend values for Jan., Feb., Mar. 1972 are obtained by putting x = 138, 139 and 140 respectively in (ii).

$$y = 230\frac{5}{6} + \frac{5}{3} \times 138 = 460.83 \quad \text{(Trend for Jan. 1972)}$$

$$y = 230\frac{5}{6} + \frac{5}{3} \times 139 = 462.50 \quad \text{(Trend for Feb. 1972)}$$

$$y = 230\frac{5}{6} + \frac{5}{3} \times 140 = 464.17 \quad \text{(Trend for Mar. 1972)}$$

Example 16.24 *Trend equation for certain production data is*

$$y = 240 + 36x$$

where *y = annual production ('000 tons);*
x = time with origin at year 1980, unit 1 year.

Estimate the trend value for March, 1982.

Solution The trend equation for *monthly averages* is (see 16.6.5)

$$y = \frac{240}{12} + \frac{36}{12}x = 20 + 3x$$

where y = monthly production; x = time with origin at 1980 (i.e. 30th June 1980), unit 1 year.

If the unit of x is changed to 1 month, we have

$$y = 20 + \frac{3}{12}x = 20 + 0.25x$$

where y = monthly production; x = time with origin at 30th June 1980, unit 1 *month*.

The origin is shifted to *July,* 1980 (i.e. 15th July 1980). Hence by (16.6.2),

$$y = 20 + 0.25\left(x + \frac{1}{2}\right) = 20.125 + 0.25x$$

Thus, the trend equation for monthly values is

$$y = 20.125 + 0.25x$$

where y = monthly production ('000 tons)

x = time with *origin* at July 1980, *unit* 1 month.

Now, March 1982 is 20 months (i.e. 20 units) later than the origin July 1980. Hence Putting $x = 20$,

$$y = 20.125 + 0.25 \times 20 = 25.125$$

i.e. the estimated trend values for March 1982 is 25.125 ('000 tons).

Example 16.25 *Fit a straight line used trend to the following data and show how you would obtain the monthly trend values from the trend line fitted to the yearly values, and obtain two such monthly values:*

Year	*1946*	*1947*	*1948*	*1949*	*1950*	*1951*	*1952*	*1953*	*1954*
Average Monthly Profits (million Rs)	*6.3*	*7.4*	*9.3*	*7.4*	*8.3*	*10.6*	*9.0*	*8.7*	*7.9*

Solution Let $y = a + bx$ be the equation of the straight line and fitted to the given *yearly* data (origin: 1950; x unit = 1 year; y unit = million Rs). The normal equatins for finding the constants a and b *are* $\Sigma y = an + b\Sigma bx$, $\Sigma xy = a\Sigma x = b\Sigma x^3$

Table 16.16 Fitting Straight Line Trend

Year	*y*	*x*	x^2	*xy*
1946	6.3	– 4	16	– 25.2
1947	7.4	– 3	9	– 22.2
1948	9.3	– 2	4	– 18.6
1949	7.4	– 1	1	– 7.4
1950	8.3	0	0	0
1951	10.6	1	1	10.6
1952	9.0	2	4	18.0
1953	8.7	3	9	26.1
1954	7.9	4	16	31.6
Total	74.9	0	60	12.9

Using the results from Table 16.16 in the normal equations,

$$74.9 = a(9) + b(0) \quad \text{or,} \quad 9a = 74.9$$

$$12.9 = a(0) + b(60) \qquad 60b = 12.9$$

Therefore, $a = 74.9/9 = 8.32$, $b = 12.9/60 = 0.215$; and the trend equation fitted to *yearly* values is

$$y = 8.32 + 0.215x$$

(origin: 1950; x unit = 1 *year*)

Since y represents the *monthly average* for each year and the unit of x is 12 months, we see that the trend of monthly average increases by 0.215 in 12 months, i.e. 0.215/12 per month. So, the trend equation for monthly values is

$$y = 8.32 + \frac{0.215}{12}x; \quad \text{i.e.} \quad y = 8.32 + 0.018x$$

(origin: 30th June, 1950; x unit = 1 *month*)

If July 1950 is chosen as origin, i.e. the origin is shifted half a month later, x should be replaced by $(x + \frac{1}{2})$. The monthly trend equation is therefore,

$$y = 8.32 + 0.018(x + \frac{1}{2}) = 8.32 + 0.018x + 0.009$$

or,
$$y = 8.33 + 0.018x,$$

(origin: July 1950; unit of x = 1 month; unit of y = million Rs)

Let us find the trend values for December 1950 and August 1949.

Since, December 1950 is 5 months (i.e. 5 units) later than the origin, viz. July 1950, putting $x = 5$ in the trend equation

$$y = 8.33 + 0.018 \times 5 = 8.42 \text{ million Rs}$$

Similarly, August 1949 is 11 months earlier than the origin, and hence putting $x = -11$, we get, $y = 8.33 + 0.018\ (-11) = 8.13$ million Rs.

Ans. (a) $y = 8.32 + 0.215x$, origin 1950; unit of x = 1 year;
(b) Trend value for Dec. 1950 = 8.42 million Rs
Trend value for Aug. 1949 = 8.13 million Rs

Example 16.26 *Fit a straight line trend by the method of least squares to the following series of observations:*

Year	*1943*	*1944*	*1945*	*1946*	*1947*	*1948*	*1949*	*1950*
Y	*683*	*687*	*678*	*665*	*656*	*689*	*691*	*698*

Y being the production in thousand tons. From the fitted line, indicate how you would find monthly trend values.

Solution [Note that the given values of Y are *annual totals*.]

Let the equation of straight line trend for the annual totals be $y = a + bx$, with origin at the mid-point of 1946–1947; x unit = 6 months; and $y = Y - 680$. The normal equations for finding the constants a and b are $\Sigma y = an + b\Sigma x$, $\Sigma xy = a\Sigma x + b\Sigma x^2$.

Table 16.17 Fitting Straight Line Trend

Year	*Y*	*y = Y – 680*	*x*	x^2	*xy*
1943	683	3	– 7	49	– 21
1944	687	7	– 5	25	– 35
1945	678	– 2	– 3	9	6
1996	665	– 15	– 1	1	15
1947	656	– 24	1	1	– 24
1948	689	9	3	9	27
1949	691	11	5	25	55
1950	698	18	7	49	126
Total	5447	7	0	168	149

Substituting the values in the normal equations (here $n = 8$),

$$7 = a(8) + b(0) \quad \text{or,} \quad 8a = 7$$
$$149 = a(0) + b(168) \quad\quad 168b = 149$$

Therefore, $a = 0.875$; $b = 0.887$. Hence, the straight line trend is

$$y = a + bx; \quad \text{i.e.,} \quad Y - 680 = 0.875 + 0.887x; \text{ or, } \quad Y = 680.875 + 0.887x$$

(origin: 1946–47; x unit = 6 months; Y unit = thousand tons).

This is the equation of straight line trend fitted to the *annual totals*.

The trend line fitted to *monthly averages* would be

$$Y = \frac{680.875}{12} + \frac{0.887}{12}x; \quad \text{i.e.} \quad Y = 56.74 + 0.074x$$

(origin, 1946–47; x unit = 6 *months*; Y = *monthly average* production).

In order to change the unit of x to 1 month, we divide the coefficient of x by 6. Therefore, the trend line for *monthly production* is

$$Y = 56.74 + \frac{0.074}{6}x; \quad \text{or.} \quad Y = 56.74 + 0.0123\,x$$

(origin: Dec. 1946–Jan. 1947; x unit = 1 *month*; Y unit = *monthly* production in thousand tons).

Now we have to change the origin of x from the middle of two months to a specified month, i.e. Middle of a month. If the origin be Jan. 1947, i.e. it is changed to a point half a month *later*, x is to be replaced by $(x + \frac{1}{2})$. So,

$$Y = 56.74 + 0.0123\,(x + \tfrac{1}{2}); \quad \text{or} \quad Y = 56.75 + 0.0123x$$

(origin: January 1947; x unit = 1 *month*; *Y unit* = *monthly* production in thousand tons). This is the suitable equation for determining monthly trend values.

Let us find trend values for July, 1949 (a period *later* than the origin) and September, 1943 (a period *earlier* than the origin). July, 1949 being 30 months later than the origin, viz. Jan. 1947, we have $x = 30$. Again, Sept. 1943 being 40 months earlier than the origin, $x = -40$. Putting these values in the monthly trend equation:

Trend value for July, 1949 is

$$Y = 56.75 + 0.0123 \times 30 = 57.12 \text{ thousand tons.}$$

Trend value for September, 1943 is

$$Y = 56.75 + 0.0123(-40) = 56.26 \text{ thoudand tons.}$$

Ans. (i) $Y = 680.875 + 0.887x$ (origin: 1946–47; x unit = 6 months); (ii) Trend values for July, 1949 and Sept. 1943 are 57.12 and 56.26 (thousand tons).

16.7 SEASONAL VARIATION

Example 16.27 *Define or describe Seasonal Fluctuations in business or economic time series.* [I.C.W.A., July '72]

Solution *Seasonal fluctuations* in time series refer to a type of periodic movement,where the period does not exceed one year. Variations in passenger traffic during the 24 hours of a day, number of books issued from a library during the seven days of a week, and values of cheque clearance during the 12 months of a year, are examples of seasonal fluctuations. In these cases, the periods are respectively 1 day, 1 week and 1 year. Most of the business and economic activities are found to have brisk and slack periods during some specific parts of a year, and these fluctuations are found to repeat with striking regularity year after year. Such up-and-down movements are the results of seasonal fluctuations. Climatic changes of seasons and the customs and habits of people at different parts of the year are the main factors responsible for seasonal fluctuations.

Example 16.28 *Discuss the various uses of seasonal index in time series analysis.* [M.Com. '73; I.C.W.A., July '72]

Solution There are two major uses of seasonal indices:

(1) Adjusting time series data for seasonal variation,

(2) Forecasting on monthly or quarterly basis.

Seasonal indices give us a clear idea about the relative position of each month or quarter in time series data relating to such matters as sales, production, employment, etc. The indices may, therefore, be used to 'deseasonalise' (i.e. eliminate the seasonal effects of) the series—(i) by expressing the original monthly or quarterly data as percentages of the corresponding seasonal indices, when *multiplicative model* is used, or (ii) by subtracting the seasonal component from original figures, when *additive model* is used. Deseasonalisation is often necessary for studying the trend or cyclical movements (Example 16.36). Moreover, in combining or comparing data that have differing seasonal factors, it is first necessary to eliminate the effects due to seasonal variations (Example 16.30, 16.31)

Seasonal indices may be used for short-term forecasting, which is so necessary for planning the future course of action. For example, in studying the production figures of a company over time, seasonal indices may be used to plan for the hiring of personnel for peak periods, to accumulate an inventory of raw materials, to ready equipment, and to allocate vacation time. Similarly, in order to meet the possible demands of customers during the winter, a department store may utilise seasonal indices of their sales data for optimum allocation of capital in the warm clothings department (Example 16.29, 16.39).

Example 16.29 *A certain company estimates its average monthly sales in a particular year to be Rs 20,00,000. The seasonal indices of the sales data are given below:*

Month	*Jan.*	*Feb.*	*Mar.*	*Apr.*	*May*	*June*	*July*	*Aug.*	*Sep.*	*Oct.*	*Nov.*	*Dec.*
S.I.	*78*	*75*	*100*	*126*	*138*	*121*	*101*	*104*	*99*	*103*	*80*	*75*

Assuming that there is no trend, use the above information to draw up a monthly sales budget for the company. [I.C.W.A., Dec. '82]

Solution Seasonal indices are usually expressed as percentages, their average being 100. (It may be verified that the total of all the given S.I. is 1200). Hence, seasonal indices must be divided by 100, to obtain *seasonal effect.* The average monthly sales being Rs 20,00,000 the estimated monthly sales, for a specified month will be Rs 20,00,000 × (seasonal effect), where

Seasonal Effect = Seasonal Index ÷ 100

Table 16.18 Budget Estimates of Monthly Sales

Month (1)	*Seasonal Index* (2)	*Seasonal Effect* (3)*	*Estimated Sales (Rs Lakh)* (4)**
Jan.	78	.78	15.6
Feb.	75	.75	15.0
Mar.	100	1.00	20.0
Apr.	126	1.26	25.2
May.	138	1.38	27.6

(*Contd*)

Month	*Seasonal Index*	*Seasonal Effect*	*Estimated Sales (Rs Lakh)*
(1)	*(2)*	*(3)**	*(4)***
June	121	1.21	24.2
July	101	1.01	20.2
Aug.	104	1.04	20.8
Sep.	99	.99	19.8
Oct.	103	1.03	20.6
Nov.	80	.80	16.0
Dec.	75	.75	15.0
Total	1200	12.00	240.0

*Col. (3) = Col. (2) ÷ 100 **Col. (4) = Col. (3) × 20.

Examples 16.30 *Deseasonalise the following sales data and interpret them:*

Quarter	*Sales (Rs '000)*	*Seasonal Index*
1st	*23.7*	*0.78*
2nd	*25.2*	*1.24*
3rd	*21.4*	*0.50*
4th	*65.4*	*1.48*

Solution 'Seasonal Index' shown here actually refers to *seasonal effect*. (Note that "seasonal index" is generally used to show 'percentage' position and should be distinguished from "seasonal effect", indicating effect 'per unit'. The average seasonal index is 100, but average seasonal effect is 1.) Since, seasonal indices give ratio changes over the normal value, a multiplicative model is to be assumed for the data, $v_i = T \times S \times C \times I$. In order to deseasonalise (i.e. eliminate the seasonal effect) it is therefore necessary to divide the data by the 'seasonal effects'.

$$\text{Deseasonalised data} = \frac{y_t}{S} = \frac{T \times S \times C \times I}{S} = T \times C \times I$$

Table 16.19 Deseasonalising Time Series Data

Quarter	*Sales (Rs '000)* y_t	*Seasonal Index* S	*Deseasonalised Data* $= y_t/S$
1st	23.7	0.78	30.4
2nd	25.2	1.24	20.3
3rd	21.4	0.50	42.8
4th	65.4	1.48	44.2
Total	—	4.00	—

The significance of deseasonalised data is that sales in the different quarters would have been as shown in the last column of Table 16.19, if they were not affected by seasonal fluctuations (Also see Example 16.36).

Example 16.31 *Sales of a company rose from Rs 39,45,000 to Rs 46,21,000 from second quarter to third quarter. The seasonal indices for these quarters are 103 and 150 respectively. The owner of the company holds that it is a losing concern. Analyse the above information for supporting the owner's view.*

Solution The seasonal index for the second quarter is 103 and the actual sales were Rs 39,45,000. On this basis, the normal quarterly sales would be Rs 39,45,000 ÷ 1.03; and the expected sales during the third quarter, when the seasonal index is 150, would be

$$\frac{\text{Rs } 39,45,000}{1.03} \times 1.50 = \text{Rs } 57,45,146$$

But the actual sales during the third quarter, viz. Rs 46,21,000 are far less than this, justifying the owner's view that it is a losing concern. (Alternatively, we may compare the deseasonalised data for 2nd and 3rd quarters, viz. 38,30,097 and 30,80,667 (Rs), and justify the view).

16.8 MEASUREMENT OF SEASONAL VARIATION

Example 16.32 *Describe briefly but carefully the various methods which can be used for computing a seasonal index for time series data.*
[I.C.W.A., Dec. '72; B.U., B.A.(Econ) '72; C.U., B.Sc. (Econ) '80]

Solution There are four methods of measuring seasonal fluctuations:

(1) Method of (monthly, or quarterly) averages,
(2) Moving average method,
(3) Trend-ratio method
(4) Link relative method.

These are desribed below using quarterly data, although similar procedures are applicable for data given by months, weeks, days or hours.

(1) Method of (Monthly, or Quarterly) Averages This method is applied, when the given time series data do not contain trend or cyclical fluctuations to any appreciable extent. From the quarterly data the totals for each quarter and averages A_1, A_2, A_3, A_4 for the 4 quarters Q_1, Q_2, Q_3, Q_4 are found. The Grand Average

$$G = \frac{1}{4}(A_1 + A_2 + A_3 + A_4)$$

is also calculated. If the additive model is used, the deviations of quarterly averages from the grand average give *seasonal variation* (Example 16.33):

$$S_1 = A_1 - G, S_2 = A_2 - G, S_3 = A_3 - G, S_4 = A_4 - G$$

If the multiplicative model is used, each quarterly average is expressed as a percentage of the Grand Average giving the *seasonal indices* (Example 16.34):

$$S_1 = \frac{A_1}{G} \times 100, \quad S_2 = \frac{A_2}{G} \times 100, \quad S_3 = \frac{A_3}{G} \times 100, \quad S_4 = \frac{A_4}{G} \times 100$$

If monthly figures are given, we find 12 averages $A_1, A_2 \ldots A_{12}$ for the months January, February, ... December respectively, and then proceeding the same way as before, the seasonal index for each month is obtained. The total (or average) seasonal *variation* (in the additive model) is 0, and the average seasonal *index* (in the multipicative model) is 100.

(2) Moving Averages Method From the given quarterly figures the trend is estimated by taking four-quarter moving averages (centered). The effect of trend is then eliminated from the original data. If the additive model is used, the moving average trend values are subtracted from the original data to give us '*derivations from trend*'. Since these deviations do not contain any effect of trend, the method of quarterly averages is applied to these deviations, using additive model. (see Example 16.35).

If the mutliplicative model is taken, then (instead of deviations from moving averages, as in the additive model) we find '*ratios to moving averages*,' expressed as percentages, i.e. the original values are expressed as percentages of the corresponding moving averages values. These percentages are now arranged by quarters and the average for each quarter, P_1, P_2, P_3, P_4 (suppose) are found out. These are adjusted to a total of 400, multiplicative each by $100/P$, where

$$P = \frac{1}{4}(P_1 + P_2 + P_3 + P_4)$$

is the Grand Average. The seasonal indices are

$$S_1 = \frac{P_1}{P} \times 100, \quad S_2 = \frac{P_2}{P} \times 100, \quad S_3 = \frac{P_3}{P} \times 100, \quad S_4 = \frac{P_4}{P} \times 100.$$

respectively for the quarters Q_1, Q_2, Q_3 and Q_4 (see Example 16.37).

(3) Trend-Ratio Method In this method, the multiplicative model is always taken. Trend values are obtained by fitting a mathematical curve, and the original data are expressed as percentages of the corresponding trend. As in the moving average method, these percentages are arranged by quarters and the average trend-ratio for each quarter, viz. P_1, P_2, P_3, P_4 are found out. Each of these is now multiplied by $100/P$, to give the seasonal indices

$$S_1 = \frac{P_1}{P} \times 100, \quad S_2 = \frac{P_2}{P} \times 100, \quad S_3 = \frac{P_3}{P} \times 100, S_4 = \frac{P_4}{P} \times 100$$

corresponding to the quarters Q_1, Q_2, Q_3, Q_4 respectiely, where

$$P = \frac{1}{4}(P_1 + P_2 + P_3 + P_4)$$

The total of the 4 seasonal indices will be 400. (Example 16.38).

(4) Link Relative Method If Quarterly data are given, each value is expressed as a percentage the value for the immediately preceding period. These are known as *Link Relatives* (L.R.). Of course, the link relative for the first quarter (Q_1) of the first year cannot be obtained. The L.R.s are arranged by quarters and the average L.R. for each quarter is found, either by using the arithmetic mean, or median. The average link relatives show the average relation of each quarterly value to the value of the previous quarter (Example 16.40).

From these average L.R.s we find *Chain Relatives* (C.R.) by relating them to a common base, e.g., the first quarter, for which C.R. is taken as 100. The C.R. for any quarter is now obtained on multiplying the L.R. for that quarter by the C.R. for the immediaely preceding quarter, and dividing by 100. Proceeding this way, we find a second C.R. for the first quarter (Q_1) by the relation.

$$\text{Second C.R. for } Q_1 = \frac{(\text{C.R. for } Q_4) \times (\text{L.R. for } Q_1)}{100}$$

Usually, the second C.R. for Q_1 will differ from the originally assumed C.R. 100, owing to the presence of trend. Some adjustment to the C.R.s are therefore necessary.

Let c be the average quarterly deviation of the 2nd C.R. from 100, i.e.

$$c = \frac{1}{4}(\text{Second C.R. for } Q_1 - 100).$$

Subtracting c, $2c$, $3c$ and $4c$ from the C.R.s for Q_2, Q_3, Q_4 and the 2nd C.R. for Q_1, we find that both the C.R.s for Q_1 are now equal to 100. The adjusted C.R.s for Q_1, Q_2, Q_3, Q_4 are now expressed as percentages of their A.M. to give the seasonal indices. The total of these seasonal indices will be 400.

Example 16.33 *Compute the average seasonal movements by the method of quarterly total (average) for the following series of observations:*

Total Production of Paper (thousand tons)				
	Quarters			
Year	*I*	*II*	*III*	*IV*
1951	*37*	*38*	*37*	*40*
1952	*41*	*34*	*25*	*31*
1953	*35*	*37*	*35*	*41*

Solution The calculations are shown below using an *additive model*:

Table 16.20 Calculations for Average Seasonal Movement

	Production ('000 tons)				
Year/Quarter	*I*	*II*	*III*	*IV*	*Total*
1951	37	38	37	40	—
1952	41	34	25	31	—
1953	35	37	35	41	—
Total	113	109	97	112	431
A.M.	37.67	36.33	32.33	37.33	143.66
Average Seasonal Movement	1.75	0.42	– 3.59	1.42	0

Grand Average = 143.66 ÷ 4 = 35.92

[**Note:** Average Seasonal Movements have been obtained by subtracting the Grand Average from the A.M. for each quarter. Slight adjustments are usually necessary to make the total seasonal movement zero. Thus

$37.67 - 35.92 = 1.75$; $36.33 - 35.92 = 0.41$

$32.33 - 35.92 = -3.59$; $37.33 - 35.92 = 1.41$

Arbitrarily, 0.41 and 1.41 have been changed to 0.42 and 1.42 to make the total seasonal movement zero.] *Ans.* 1.75, 0.42, – 3.59, 1.42 ('000 tons) for I, II, III, IV.

Example 16.34 *Calculate the seasonal indices in the case of the following quarterly data in certain units. You may use any method you think appropriate:*

Year	Q_1	Q_2	Q_3	Q_4
1960	*39*	*21*	*52*	*81*
1961	*45*	*23*	*63*	*76*
1962	*44*	*26*	*69*	*75*
1963	*53*	*23*	*64*	*84*

[C.U., M.Com. '65]

Solution *Methods of Quarterly Averages* seems appropriate here, since no appreciable trend is noticed in the given data (Note that the values in any quarter do not reveal any definite tendency to change). The calculations are shown below, using the *multiplicative model* (as an illustrative example).

Table 16.21 Calculations for Seasonal Index

Year	Quarter Q_1	Q_2	Q_3	Q_4	Total
1960	39	21	52	81	—
1961	45	23	63	76	—
1962	44	26	69	75	—
1963	53	23	64	84	—
Total	181	93	248	316	838
A.M.	45.25	23.25	62.00	79.00	209.50
*Seasonal Index	86.4	44.4	118.4	150.8	400.0

Grand Average = 209.50 ÷ 4 = 52.38

[***Note:** (i) Seasonal Index = (A.M. ÷ Grand Average) × 100

(45.25 ÷ 52.38) × 100 = 86.4; (23.25 ÷ 52.38) × 100 = 44.4

(62.00 ÷ 52.38) × 100 = 118.4; (79.00 ÷ 52.38) × 100 = 150.8

(ii) The total of Seasonal Index for the 4 quarters must be 400 with slight adjustments in the last place if necessary.]

Ans. 86.4, 44.4, 118.5, 150.8

Example 16.35 *Obtain seasonal fluctuation from the following time series data:*

Years/Quarters	*Quarterly Output of Coal for 4 years* I	II	III	IV
1928	*65*	*58*	*56*	*61*
1929	*68*	*63*	*63*	*67*
1930	*70*	*59*	*56*	*52*
1931	*60*	*55*	*51*	*58*

Solution We apply the method of moving averages. For this purpose, first find four-quarter moving averages, giving trend. Since col. (6) of Table 16.22 shows figures free from trend, the method of quarterly averages is then applied (see Theory in Example 16.36).

[**Note:** Moving Average method, using the additive model, is the most popular method of obtaining seasonals, and this method should generally be applied unless any particular method is suggested.]

Table 16.22 Moving Averages and Deviations from Trend

Year/ Quarter (1)		Output (2)	4-quarter Moving Total (3)	2-period Moving Total of Col. (3) (4)	4-quarter Moving Average (5)	Deviation from Trend (6)
1928	I	65	—	—	—	—
	II	58	—	—	—	—
			240			
	III	56		483	60.38	– 4.38
			243			
	IV	61		491	61.38	– 0.38
			248			
1929	I	68		503	62.88	5.12
			255			
	II	63		516	64.50	– 1..50
			261			
	III	63		524	65.50	– 2.50
			263			
	IV	67		522	65.25	1.75
			259			
1930	I	70		511	63.88	6.12
			252			
	II	59		489	61.12	– 2.12
			237			
	III	56		464	58.00	– 2.00
			227			
	IV	52		450	56.25	– 4.25
			223			
1931	I	60		441	55.12	4.88
			218			
	II	55		442	55.25	– 0.25
			224			
	III	51	—	—	—	—
	IV	58	—	—	—	—

Col. (5) = Col. (4) ÷ 8 Col. (6) = Col. (2) minus Col. (5)

Table 16.23 Calculations for Seasonal Fluctuations

Year/ Quarter	Deviations from Trend I	II	III	IV	Total
1928	—	—	– 4.38	– 0.38	—
1929	5.12	– 1.50	– 2.50	1.75	—
1930	6.12	– 2.12	– 2.00	– 4.25	—
1931	4.88	– 0.25	—	—	—
Total	16.12	– 3.87	– 8.88	– 2.88	0.49
Average	5.37	– 1.29	– 2.96	– 0.96	0.16
Adjustment	– 0.04	– 0.04	– 0.04	– 0.04	– 0.16
Seasonal	5.33	– 1.33	– 3.00	– 1.00	0

Grand Average = 0.16 ÷ 4 = 0.04

Example 16.36 *Deseasonalise the following production data by the method of moving average:*

	Quarterly Output ('000 tons)			
Quarters	*1940*	*1941*	*1942*	*1943*
I	*30*	*49*	*35*	*75*
II	*49*	*50*	*62*	*79*
III	*50*	*61*	*60*	*65*
IV	*35*	*20*	*25*	*70*

[C.U., M.Com. '73]

Solution [**Theory:** We use the additive model]

$$y_t = T + S + C + I.$$

At first, trend (T) is estimated by taking movin averages of period 4 quarters (since quarterly data are given); it is then eliminated from the original data by subtraction.

$$y_t - T = (T + S + C + I) - T = S + C + I$$

When these 'deviations from trend' are averaged for each quarter, the cyclical (C) and irregular (I) components are removed to a large extent. This leaves only seasonals (S), which are suitably adjusted to a total zero.

While using the additive model, the adjustment is made in the following way:—Calculate (see Table 16.23 and 16.25)

Adjustment = Grand average with the sign reversed.

This is added to the average for each quarter to obtain the 'seasonals'

Seasonals = Average for each quarter + Adjustment.

It must, however, be ensured that the total of all seasonals is zero, in the additive model.

The seasonal for each quarter is *subtracted* from the corresponding quarterly data to give the deseasonalised figures ('Deseasonalisation' means elimination of seasonal component from time series data).

If the multiplicative model is used, (see Example 16.37), the total Seasonal Index will be 400 for quarterly data, and deseasonalised figures are obtained on *dividing* the original data by 'seasonal effects' (see Example 16.30.]

Table 16.24 Calculations for Moving Average Trend

Year/ Quarter (1)		*Output ('000 tons)* (2)	*4-quarter Moving Total* (3)	*2-period Moving Total of Col. (3)* (4)	*4-quarter Moving Average* (5)	*Deviations from Trend* (6)
1940	I	30	—	—	—	—
	II	49	—	—	—	—
			164			
	III	50		347	43.38	6.62
			183			
	IV	35		367	45.88	10.88
			184			
1941	I	49		379	47.38	1.62
			195			
	II	50		375	46.88	3.12
			180			
	III	61		346	43.25	17.75
			166			

(*Contd*)

Year/ Quarter (1)		Output ('000 tons) (2)	4-quarter Moving Total (3)	2-period Moving Total of Col. (3) (4)	4-quarter Moving Average (5)	Deviations from Trend (6)
	IV	20		344	43.00	– 23.00
			178			
1942	I	35		355	44.38	9.38
			177			
	II	62		359	44.88	17.12
			182			
	III	60		404	50.50	9.50
			222			
	IV	25		461	57.62	– 32.62
			239			
1943	I	75		483	60.38	14.62
			244			
	II	79		533	66.62	12.38
			289			
	III	65	—	—	—	—
	IV	70	—	—	—	—

Table 16.25 Calculations for Seasonal Variation

Year/ Quarter	Deviations from Trend [Col. (6) of Table 6.24] I	II	III	IV	Total
1940	—	—	6.62	– 10.88	
1941	1.62	3.12	17.75	– 23.00	
1942	– 9.38	17.12	9.50	– 32.62	
1943	14.62	12.38	—	—	
Total	6.86	32.62	33.87	– 66.50	6.85
Average	2.29	10.87	11.29	– 22.17	2.28
Adjustment*	– 0.57	– 0.57	– 0.57	– 0.57	– 2.28
Seasonal @	1.72	10.30	10.72	– 22.74	0

* Adjustment = Grand average with opposite sign = – (2.28 ÷ 4) = – 0.57

@ Seasonal = Average for each quarter + Adjustment.

Table 16.26 Deseasonalising Time Series Data

Year and Quarter		Output ('000 tons) y_i	Seasonals S	Deseasonalised Data $y_i - S$
1940	I	30	1.72	28.28
	II	49	10.30	38.70
	III	50	10.72	39.28
	IV	35	– 22.74	57.74
1941	I	49	1.72	47.28
	II	50	10.30	39.70
	III	61	10.72	50.28
	IV	20	– 22.74	42.74

(*Contd*)

Year and Quarter		Output ('000 tons) y_i	Seasonals S	Deseasonalised Data $y_i - S$
1942	I	35	1.72	33.28
	II	62	10.30	51.70
	III	60	10.72	49.28
	IV	25	10.72	47.74
1943	I	75	1.72	73.28
	II	79	10.30	68.70
	III	65	10.72	54.28
	IV	70	– 22.74	92.74

Example 16.37 *Find the seasonal indices by the method of moving from the following series of observations:*

	Sales of Woollen Yarn ('000 Rs)		
Quarter/Year	*1959*	*1960*	*1961*
I	*101*	*106*	*110*
II	*93*	*96*	*101*
III	*79*	*83*	*88*
IV	*98*	*103*	*106*

Solutions [**Theory:** If we assume a *multiplicative* mode, then trend is eliminated from original data by taking ratios of moving averages.

$$\frac{y_i}{T} = \frac{T \times S \times C \times I}{T} = S \times S \times \text{I}$$

When these ratios are arranged for each quarter, cyclical-irregular components vanish, leaving only Seasonal Index (S)].

Note: The total seasonal index for the four quarters must be 400. It is therefore necessary to adjust the averages for each quarter proportionately (see Table 16.28).

$$\text{Seasonal Index} = \frac{\text{Average for a quarter}}{\text{Grand Average for all quarters}}$$

Table 16.27 Calculations for Moving Averages

Year and Quarter (1)		y_i (2)	4-quarter Moving Total (3)	2-item Moving Total of Col. (3) (4)	4-quarter Moving Average* (trend) (5)	Ratio to Moving Average @ (6)
1959	I	101	—	—	—	—
	II	93	—	—	—	—
			371			
	III	79		747	93.38	84.6
			376			
	IV	98		755	94.38	103.8
			379			
1960	I	106		762	95.25	111.3
			383			

(*Contd*)

Year and Quarter (1)	y_i (2)	4-quarter Moving Total (3)	2-item Moving Total of Col. (3) (4)	4-quarter Moving Average* (trend) (5)	Ratio to Moving Average @ (6)
II	96		771	96.38	99.6
		388			
III	83		780	97.50	85.1
		392			
IV	103		789	98.62	104.4
		397			
1961 I	110		799	99.88	110.1
		402			
II	101		807	100.88	100.1
		405			
III	88	—	—	—	—
IV	106	—	—	—	—

*Col. (5) = Col. (4) ÷ 8 @ Col. (6) = [Col. (2) ÷ Col. (5)] × 100

The figures in Col. (6) are arranged for each quarter and shown in the following table.

Table 16.28 Calculations for Seasonal Index

Year/Quarter	Ratio to Moving Average I	II	III	IV	Total
1959	—	—	84.6	103.8	—
1960	111.3	99.6	85.1	104.4	—
1961	110.1	100.1	—	—	—
Total	221.4	199.7	169.7	208.2	—
Average	110.7	99.8	84.8	104.1	399.4
*Seasonal Index	111	100	85	104	400

Grand Average = 399.4 ÷ 4 = 99.85

[***Note:** Seasonal Index = (Average ÷ Grand Average) × 100. For example, (110.7 ÷ 99.85) × 100 = 111, (99.8 ÷ 99.85) × 100 = 100 etc. The total of Seasonal index for the four quarters must be 400].

Example 16.38 *The number of traffic accidents in Calcutta in four quarters of a year during the period 1977–79 are given below:*

Year	Quarters I	II	III	IV
1977	165	135	140	180
1978	152	121	127	163
1979	140	100	105	158

Find seasonal indices by Trend-ratio method, assuming a linear trend for the data.

Solution First, we have to find the trend values. Let $y = a + bx$ be the equation of trend (origin : mid-point of quarters II and III, 1978; unit of $x = \frac{1}{2}$ quarter). By the method of least squares, the values of a and b are obtained from the normal equations $\Sigma y = an + b\Sigma x$ and $\Sigma xy = a\Sigma x + b\Sigma x^2$.

Table 16.29 Fitting Linear Trend to Quarterly Data

Year/ Quarter		Time Series (y)	x	x^2	xy	Trend Values (T)	Trend-ratio (y/T) × 100
1977	I	165	– 11	121	– 1815	155.7	106
	II	135	– 9	81	– 1215	152.9	88
	III	140	– 7	49	– 980	150.2	93
	IV	180	– 5	25	– 900	147.4	122
1978	I	152	– 3	9	– 456	144.6	105
	II	121	– 1	1	– 121	141.9	85
	III	127	1	1	127	139.1	91
	IV	163	3	9	489	136.4	120
1979	I	140	5	25	700	133.6	105
	II	100	7	49	700	130.8	76
	III	105	9	81	945	128.1	82
	IV	158	11	121	1739	125.3	126
Total		1686	0	572	– 788	—	—

Substituting the values from the table in the normal equations,

$$a = 1686/12 = 140.5, \qquad b = -788/572 = -1.38$$

The trend equation is therefore $y = 140.5 - 1.38x$ (origin: mid-point of quarters II and III, 1978; unit of $x = \frac{1}{2}$ quarter). Putting appropriate values of x, we get the trend values (see Table 16.29).

The next step is to express the original data as percentages of trend, giving "trend-ratios" (Table 16.29, las column). The trend-ratios are arranged by quarters as in Table 16.30 and the seasonal index is calculated by the methods shown in Example 16.28.

Table 16.30 Calculations for Seasonal Index

Year	I	II	III	IV	Total
1977	106	88	93	122	
1978	105	85	91	120	
1979	105	76	82	126	
Total	316	249	266	368	—
Average	105	83	89	12	400
Seasonal Index	105	83	89	123	400

The seasonal indices are 105, 83, 89, 123 for quarters I and IV respectively.

Example 16.39 *On the basis of quarterly sales (in Rs lakhs) of a certain commodity for the years 1961–65, the following calculations were made:*

Trend $y = 25.0 + 0.6t$, with origin at 1st quarter of 1961,

t = time units (one quarter), and

y = quarterly sales (Rs lakhs),

Seasonal variations:

Quarter	...	*1st*	*2nd*	*3rd*	*4th*
Seasonal Index	...	*90*	*95*	*110*	*105*

Estimate the quarterly sales for the year 1962 (use multiplicative model).

[I.C.W.A., July '72]

Solution Trend equation $y = 25.0 + 0.6t$ has origin of t at Q_1 of 1961, in units of one quarter. The trend values for Q_1, Q_2, Q_3 and Q_4 of 1962 are therefore obtained from this equation by putting $t = 4, 5, 6, 7$ respectively.

Trend for Q_1 1962 : $y = 25.0 + 0.6(4) = 27.4$ (Rs lakhs)
Trend for Q_2 1962 : $y = 25.0 + 0.6(5) = 28.0$ (Rs lakhs)
Trend for Q_3 1962 : $y = 25.0 + 0.6(6) = 28.6$ (Rs lakhs)
Trend for Q_4 1962 : $y = 25.0 + 0.6(7) = 29.2$ (Rs lakhs)

The quarterly sales are obtained on multiplying trend values (T) by the seasonal effects (S): $y_i = T \times S$.

Year/ Quarter	*Trend (T)*	*Seasonal Effects (S)*	*Estimated Sales (Rs lakhs)* $T \times S$
1962 Q_1	27.4	0.90	$27.4 \times 0.90 = 24.66$
Q_2	28.0	0.95	$28.0 \times 0.95 = 26.60$
Q_3	28.6	1.10	$28.6 \times 1.10 = 31.46$
Q_4	29.2	1.05	$29.2 \times 1.05 = 30.66$

Ans. 24.66, 26.60, 31.46, 30.66 (Rs lakhs)

Example 16.40 *Compute Seasonal Indices from the data of Example 16.35, using the method of Link Relatives.*

Solution Each quarterly figure is expressed as a percentage of the figure in the preceding quarter, giving the Link Relatives, shown below:

Table 16.31 Seasonals by Link Relative Method

	Link Relatives				
Year/Quarter	*I*	*II*	*III*	*IV*	*I*
1928	—	89.23	96.55	108.93	—
1929	111.48	92.65	100.00	106.35	—
1930	104.48	84.29	94.92	92.86	—
1931	115.38	91.67	92.73	113.73	—
Total	331.34	357.84	384.20	421.87	—
A.M.	110.45	89.46	96.05	105.47	—
Chain Relative	100	89.46	85.93	90.63	100.10
Adjusted C.R.	100	89.44	85.93	90.56	100.10
Seasonal Index	109	98	94	99	—

Working **Notes:**

(i) *Link Relatives (L.R.):*

$(58 \div 65) \times 100 = 89.23$, $(56 \div 58) \times 100 = 96.55$,
$(61 \div 56) \times 100 = 108.93$, $(68 \div 61) \times 100 = 111.48$, etc.

The A.M. of chain relatives for each quarter shows on an average the percentage value of that quarter in relation to the preceding quarter.

(ii) *Chain Relatives* (*C.R.*):

First C.R. for I = 100 (assumed starting point)

C.R. for II = 100 × 89 ÷ 100 = 89.46

C.R. for III = 89.46 × 96.05 ÷ 100 = 85.93

C.R. for IV = 85.93 × 105.47 ÷ 100 = 90.63

Second C.R. for I = 90.63 × 110.45 ÷ 100 = 100.10

The second C.R. for I, viz. 100.10, differs from the originally assumed C.R. for 1, viz. 100, due to the presence of trend.

Adjustment factor (c) = (100.10 – 100) ÷ 4 = .025

(iii) *Adjusted Chain Relatives:*

for I = 100

for II = 89.46 – .025 = 89.44

for III = 85.93 × 2 × .025 = 85.88

for IV = 90.63 – 3 × .025 = 90.56

for Second I = 100.10 – 4 × .025 = 100.00

The two C.Rs for I now agree.

(iv) Average of "Adjusted C.R." for 4 quarters

= (100 + 89.44 + 85.88 + 90.56) ÷ 4 = 91.47

(v) *Seasonal Index:*

S.I. for I = (100 ÷ 91.47) × 100 = 109

S.I. for II = (89.44 ÷ 91.47) × 100 = 98

S.I. for III = (85.88 ÷ 91.47) × 100 = 94

S.I. for IV = (90.56 ÷ 91.47) × 100 = 99

The total of S.I. for the four quarters must be 400.]

16.9 CYCLICAL FLUCTUATION

Example 16.41 *Explain 'business cycles', and describe a method of isolating the cyclical variation from the time series data.* [C.U., M.Com. '69]

Solution *Business Cycles*—It is a common knowledge that business activities have periods of boom and depression which repeat one after another periodically. These ups and downs in business are attributed to the presence of 'business cycles'. The effects of business cycles are periodic in nature, but they are not as regular in intensity and periodicity as seasonal variations. Business cycles are characterised by oscillatory movements and have four distinct phases—prosperity, decline, depression and recovery. During the period of boom, a peak point is reached after which business activity gradually declines and enters the period of depression. The depression persists for sometime until a very low point is reached, after which business gradually recovers towards prosperity. The time span between two successive periods of boom or depression usually covers a number of years (say, 4 or 5 years). Business cycles are caused by a complex combination of forces affecting the equilibrium of demand and supply. Often two different time series are found to follow closely the same pattern of cyclical fluctuation, because of interrelations between them.

Residual Method for Cyclical Fluctuation

Time series data (y_t) consist of four components, viz., Trend (T), Seasonal variation (S), Cyclical fluctuation (C), and Irregular movement (I). '*Residual Method*' consists in removing the three components, viz. trend, seasonal and irregular, from the original data, in order to isolate the cyclical component. At first, trend and seasonal effects are measured by suitable methods, say, by the Moving Average Method. These two components are then removed from the data,

If the multiplicative model ($y_t = T \times S \times C \times I$) is assumed, in order to get $C \times I$, we remove T and S from y_t by division. We may either divide y_t by T first, and then S; or divide y_t by S first and then by T; or divide y_t by the product $T \times S$, called *normal value*.

$$\frac{y_i}{Normal\ value} = \frac{T \times S \times C \times I}{T \times S} = C \times I$$

If, however, we assume the additive model ($y_i = T + S + C + I$), trend and seasonal components are removed by subtraction :

$$y_t - T - S = (T + S + C + I) - T - S = C + I$$

The 'residual' now consists of a combination of cyclical and irregular components (either $C \times I$, or $C + I$). At the next stage, the irregular component (I) is removed from residuals by smoothing, using moving averages. The appropriate period of the moving average depends on the average duration of irregular movements. We are now left with only the cyclical variations (C).

Residual Method is laborious, but gives accurate results.

16.10 BUSINESS FORECASTING

All businessmen have to make a certain amount of forecasting regarding business conditions. Forecasts in business are necessary for various purposes, e.g. judging future markets, making decisions on production, inventories, selling, pricing, etc. Business forecasting is neither pure guesswork, nor finding the exact figures of business conditions. The scientific methods of forecasting refer to the analysis of past and present conditions with a view to arriving at rough estimates about the future conditions.

Business forecasting is essen tially statistical nature. The study of past conditions is done by analysing time series data into various components—trend, seasonal, cyclical and irregular. Here trend is determined by free-hand method or by fitting a mathematical curve, and is projected into the future. Cyclical fluctuations reveal the periods of boom and depression and are superimposed on trend, previously obtained. The seasonal indices indicate the conditions in the immediate future and are multiplied to the trend-cyclical product. Another important part of the work of business forecasting lies in the construction of index numbers of business activity.

Business forecasting is now-a-days done on scientic principles. Some of the important theories of business forecasting are ...(i) Economic Rhythm theory, (ii) specific Historical Analogy theory, (iii) Action and Reaction theory, (iv) Time-lag or Sequence theory, and (v) Cross-cut Analysis theory. Business forecasting has attained such an important position that in the economically advanced countries of the world, many forecasting agencies. e.g. Harvard Committee on Economic Research (U.S.A.), London an Cambridge Economic Service (U.K.). Swedish Board of Trade, etc. undertake the work of business forecasting on a regular basis.

16.11 EXPONENTIAL SMOOTHING

Exponential Smoothing is a technique for "forecasting" by smoothing out fluctuations in time series data. The forecasts are obtained by weighted moving averages of all observations upto the current period. Thus all past data are involved in computing the forecast. More weight is given to more recent observations and less to those in the past.

In exponential smoothing, the weights used are: α for the current observations, $\alpha(1-\alpha)$ for the immediately preceding observation, $\alpha(1-\alpha)^2$ for the still preceding observation, and so on, where α is a given constant, called the "*smoothing coefficient*", and has a value lying between 0 and 1. It may be noted that the weights of observations from the current period backwards are diminishing with a common ratio $(1-\alpha)$; i.e. the weights form an infinite G.P. series:

$$\alpha,\ \alpha(1-\alpha),\ \alpha(1-\alpha)^2,\ \alpha(1-\alpha)^3,\quad \ldots \quad \ldots$$

and the sum of these weights is 1.

$$\text{Sum of weights} = \alpha + \alpha(1-\alpha) + \alpha(1-\alpha)^2 + \quad \ldots \quad \ldots$$

$$= \frac{\alpha}{1-(1-\alpha)} = 1. \tag{16.11.1}$$

Table 16.32 Weights used in Exponential Smoothing

Time	*Observation*	*Weight*	*Forecast*
⋮	⋮	⋮	⋮
$t-2$	y_{t-2}	$w_{t-2} = \alpha(1-\alpha)^2$	u_{t-2}
$t-1$	y_{t-1}	$w_{t-1} = \alpha(1-\alpha)$	u_{t-1}
t	y_t	$w_t = \alpha$	u_t

The "*exponentially smoothed average*" for time t is the weighted average of observations y_t, y_{t-1}, y_{t-2} ... ... corresponding to time t and backwards with weights w_t, w_{t-1}, w_{t-2}, ..., and is thus given by

$$u_t = \frac{w_t y_t + w_{t-1}\, y_{t-1} + w_{t-2}\, y_{t-2} + \ldots\ldots}{w_t + w_{t-1} + w_{t-2} + \ldots\ldots}$$

$$= w_t y_t + w_{t-1}\, y_{t-1} + w_{t-2}\, y_{t-2} + \ldots\ \ldots, \qquad \text{by (16.11.1)}$$

$$= \alpha y_t + \alpha(1-\alpha) y_{t-1} + \alpha(1-\alpha)^2\, y_{t-2} + \ldots\ \ldots$$

The exponentially smoothed average, which is used as a "forecast" for time t, is thus given by

$$u_t = \alpha y_t + \alpha(1-\alpha) y_{t-1} + \alpha(1-\alpha)^2 y_{t-2} + \ldots\ \ldots \tag{16.11.2}$$

This may also be written as a recurrence relation

$$u_t = \alpha y_t + (1 - \alpha)\, u_{t-1} \tag{16.11.3}$$

where u_t and u_{t-1} are the forecasts at time t and $t - 1$ respectively, y_t is the observation for time t, and α is the smoothing coefficient. Relation (16.11.3) shows that the forecast for a given time may be obtained as the sum of

(i) *Current observation* multiplied by α, and

(ii) *Previous forecast* multiplied by $(1 - \alpha)$.

We may also write (16.11.3) in a different form:

$$u_t = u_{t-1} + \alpha (y_t - u_{t-1}) \tag{16.11.4}$$

i.e.

$$u_t = u_{t-1} + \alpha\, e_t \tag{16.11.5}$$

where $e_t = (y_t - u_{t-1})$ is the "error" or discrepancy of the latest observation from the forecast in the previous period. Thus the exponentially smoothed average at a given time may also be shown as the sum of

(i) *Previous forecast*, and

(ii) *Error* in the latest observation from the previous forecast, multiplied by α.

For numerical computations, relation (16.11.5) may be used with the following steps:

(1) Find the error or discrepancy in the latest observation from the previous forecast: $e_t = y_t - u_{t-1}$

(2) Multiply the error by the smoothing coefficient to obtain the correction (αe_i).

(3) The correction (αe_i) is added to the previous forecast (u_{t-1}) to obtain the forecast (u_t) for the current period t.

In exponential smoothing, the first forecast has to be obtained by some subjective method, or by taking the average of the first few time periods. Subsequent forecasts are then obtained by repeatedly using the relation

$$u_t = u_{t-1} + \alpha . e_t$$

When $\alpha = 0$, no adjustment is necessary to the previous forecast. When $\alpha = 1$, the new forecast will always be the latest observation. Between 0 and 1, the smoothing coefficient α is selected on the basis of experiments with different values. For practical purposes, a value of α between 0.1 and 0.4 is ordinarily chosen.

The method of exponential smoothing provides a convenient and systematic way of obtaining forecast by revising the old forecast on the basis of the latest observation. The computations are simple, depending only on the old forecast (u_{t-1}), the latest observation (y_1) and the smoothing coefficient (α).

Example 16.42 *An old forecast of 39 is given initially and the smoothing coefficient is 0.1. Smooth exponentially the following observations : 41, 41, 34, 39, 35, 40, 36, 41, 33.* [C.A., Nov. '74]

Solution The exponentially smoothed average at time t is

$$u_t = u_{t-1} + \alpha\, e_t$$

where $e_t = y_t - u_{t-1}$ is the "error". Here, we are given $\alpha = 0.1$.

Table 16.33 Calculations for Exponential Smoothing

Time (t)	*Observation* y_t	*Previous Forecast* u_{t-1}	*Error* $e_t = y_t - u_{t-1}$	αe_t	*Current Forecast* u_t
(1)	*(2)*	*(3)*	*(4)*	*(5)*	*(6)*
1	41	39	2	0.2	39.2
2	41	39.2	1.8	0.18	39.38
3	34	39.38	– 5.38	– 0.54	38.84
4	39	38.84	0.16	0.02	38.86
5	36	38.86	– 2.86	– 0.29	38.57
6	35	38.57	– 3.57	– 0.36	38.21
7	40	38.21	1.79	0.18	38.39
8	36	38.39	– 2.39	– 0.24	38.15
9	41	38.15	2.85	0.28	38.43
10	33	38.43	– 5.43	– 0.54	37.89

Steps in computation:

(i) Col. (2) shows the given observations.

(ii) In Col. (3), the first figure 39 is the old forecast (given).

(iii) Col. (4) = Col. (2) minus Col. (3).

(iv) Col. (5) = 0.1 × Col. (4).

(v) Col. (6) = Col. (3) plus Col. (5).

(vi) Figure obtained in Col. (6) is written in Col. (2) against the nest time period.

(vii) All the previous steps are repeated.

EXERCISES

1. What do you mean by a time series? Explain the different components of such a series. *[W.B.H.S. '82]*
2. With what characteristic component of time series should each of the following be associated ?
   (i) An upturn in business activity;
   (ii) Fire loss in a factory;
   (iii) Withdrawal of bank deposits by 15th March for tax payment;
   (iv) General increase in sale of T.V. sets. *[C.A., May '81]*
3. Describe briefly the different components of a time series. With which component of the time series would you associate each of the following? Why?
   (i) The rainfall that occurred in Calcutta for four days in February, 1981;
   (ii) A decline in ice cream sales during November to March;
   (iii) An era of prosperity;
   (iv) Increase in garment sales in October. *[C.U., M.Com. '80]*

4. (a) Describe briefly the various methods of determining trend in a time series.
   (b) Using 3-year moving averages, determine the trend and short term fluctuations. Plot the original and the trend values on the same graph paper:

Year	1968	1969	1970	1971	1972	1973	1974	1975	1976	1977
Production ('000 tons)	21	22	23	25	24	22	25	26	27	26

[C.A., Nov. '81]

5. The net profits of a company for eleven successive years are given below. Find the three-year moving averages.

Year	1956	'57	'58	'59	'60	'61	'62	'63	'64	'65	'66
Profit in Lakhs of Rs	2.7	2.9	3.4	5.2	5.8	6.4	9.3	9.2	9.8	10.2	11.0

[I.C.W.A., July '69]

6. (a) How do you determine the period of a moving average for estimating the trend of a time series? *[C.U., M.Com, '77; C.A., Nov. '74]*
   (b) The following data give sales of a shop observing a five-day week, over four successive weeks. Determine the period of the moving average and calculate the moving averages accordingly.

Day ...	1	2	3	4	5	6	7	8	9	10
Sales ...	26	29	35	47	51	26	32	37	46	53
Day ...	11	12	13	14	15	16	17	18	19	20
Sales ...	28	30	36	46	54	28	31	36	46	54

[C.A., Nov. '74]

7. From the following data calculate the 4-yearly moving average and determine the trend values. Find the short term fluctuations. Plot the original values and the trend on a graph paper.

Year	1958	1959	1960	1961	1962	1963	1964	1965	1966	1967
Value	50.0	36.5	43.0	44.5	38.9	38.1	32.6	41.7	41.1	33.8

[C.A., May '80]

8. Determine trend by the *method of moving averages* from the following figures of quarterly production of a commodity:

	Production (in thousand tons)		
Quarter/Year	1975	1976	1977
I	115	119	149
II	180	189	209
III	108	149	179
IV	99	119	145

[W.B.H.S. '82]

9. Find the quarterly trend values from the following data by the moving average method, using an appropriate period:

Quarter/Year	Quarterly output (million tons)		
	1964	1965	1966
I	52	59	57
II	54	63	61
III	67	75	72
IV	55	65	60

[I.C.W.A., July '71]

10. Assuming a four-yearly cycle, calculate the trend by the method of moving averages from the following data relating to the production of tea in India:

Year	1941	1942	1943	1944	1945	1946	1947	1948	1949	1950
Production (mn. lbs.)	464	515	518	467	502	540	557	571	586	612

[I.C.W.A., Jan. '68]

11. For the following series of observation verify that the 4-year centred moving average is equivalent to a 5-year weighted moving average with weights 1, 2, 2, 2, 1 respectively:

Year	1964	1965	1966	1967	1968	1969	1970	1971	1972	1973	1974
Sales Rs '0000	2	6	1	5	3	7	2	6	4	8	3

[C.A., Nov. '79; I.C.W.A., Dec. '76]

12. Fit a suitable straight line to the following data by the method of least squares:

Year	1959	1960	1961	1962	1963
% of Insured People	11.3	13.0	9.7	10.6	10.7

[Dip. Management, '72]

13. Fit a straight line trend to the following data, and show the original observations and trend values on graph paper.

Year	1965	1966	1967	1968	1969	1970	1971
Gross Ex-factory Value of Output (Rs crores)	672	824	967	1204	1464	1758	2057

[I.C.W.A., Dec. '75]

14. Find the value of the trend ordinates by the method of least squares from the data given below:

Year	1971	1972	1973	1974	1975	1976	1977
Sales (Rs '000)	125	128	133	135	140	141	143

[I.C.W.A., Dec. '80]

15. The following data give the values of sales of a company for the years 1968 – 1978 (Sales in Rs '000):

Year (X)	1968	1969	1970	1971	1972	1973	1974	1975	1976	1977	1978
Sales (Y)	50.0	36.5	43.0	44.5	38.9	38.1	32.6	38.7	41.7	41.1	33.8

Use the method of least squares to fit a straight line trend to the data given above. Compute the trend values for 1971 and 1976. (Take $X = 0$ for the year 1973 and the unit of X is 1 year).

Construct a 5-year moving average and compare the trend values for the years 1971 and 1976. *[I.C.W.A., June '80]*

16. Fit a linear trend equation to the following series on production.

Year	1961	1962	1963	1964	1965	1966
Production (tons)	21	37	48	56	62	69

[M.B.A. '79]

17. Fit a straight line trend to the following series of production data:

Electricity Generated (monthly average) in West Bengal

Year	1951	1952	1953	1954	1955	1956
Electricity Generated (million Kw)	101	107	113	121	136	148

[C.U., M.Com. '80]

18. The annual revenue expenditure (in Rs crores) of Govt. of India is given below for 6 successive years:

Year	1953–54	1954–55	1955–56	1956–57	1957–58	1958–59
Revenue Expenditure	225	238	262	293	399	520

Fit a linear trend by the method of least squares. *[W.B.H.S. '80]*

19. (a) Define trend. Enumerate the methods of determining trend in time series.
(b) Fit a straight line trend equation by the method of least squares and estimate the value for 1969.

Year	1960	1961	1062	1963	1964	1965	1966	1967
Value	380	400	650	720	690	600	870	930

[C.A., May '78]

20. Fit a parabolic curve of second degree ($y = a + bx + cx^2$) to the data given below by the method of least squares:

Year (x)	1973	1974	1975	1976	1977
Import (y) in '000 bales	10	12	13	10	8

(Take 1975 as origin and unit of x as 1 year). *[I.C.W.A., Dec. '81]*

21. Fit a quadratic trend to the following data:

Year	1960	1961	1962	1963	1964	1965	1966
Average Production	37	38	37	40	41	45	50

[C.U., B.A. (Econ) '76]

22. (a) Briefly indicate the procedure of fitting an exponential trend to a time series. *[C.U., M.Com. '79]*

(b) Fit a trend equation $\log y = A + Bx$ to the series of sales data given below:

Year (x)	1943	1944	1945	1946	1947	1948	1949	1950	1951
Sales (y)	97	113	129	202	195	193	192	237	235

[C.U., B.A. (Econ) '71]

23. The following data relate to average monthly number of tourists coming to India in different years. Fit an exponential trend by the method of least squares:

Year	1971	1972	1973	1974	1975
Number of Tourists	25,083	28,579	34,157	35,267	28,773

[W.B.H.S. '81]

24. Fit a straight line trend to the following series of production data:

Year	1960	1961	1962	1963	1964	1965	1966
Y	37	38	37	40	41	45	50

Y values being the average production in thousand tons, what is the monthly trend increment? Find the monthly trend values from the fitted equation for January, March and December of 1961.

25. Determine the linear trend equation that fits the following figures on quarterly consumption of a raw material in some factory. Given that hte seasonal index for the third quarter of a year is 117%, what is the estimated consumption for the third quarter of 1982?

		Consumption (in tons)			
Year	*Quarter*	*1*	*2*	*3*	*4*
1976		28	25	31	39
1977		42	44	48	51
1978		55			

[D.S.W., Nov. '78]

26. (a) Describe the method of fitting a straight line trend to a time series of equispaced observations. Give computational layouts and mention separately the cases where the number of observations is even and when the number of observations is odd.

(b) Suppose we have a series of quarterly production figures (in thousand tons) in an industry for the years 1970 to 1976, and the equation of the linear trend fitted to the annual data is $5\ X_t = 107.2 + 2.93t$ where t = year – 1973 and X_t = annual production in time period t.

Use this equation to estimate the annual production for the year 1977, and for the year 1971.

Suppose now quarterly indices of Seasonal variation are: January-March 125, April-June 105, July-September 87, October-December 83. (The multiplicative model for the time series is assumed. Use these indices to estimate the production during the first quarter of 1977.

[C.U., B.A. (Econ) '78]

27. A large company estimates its average monthly sales in a year to be Rs 2,00,000. The seasonal indices of the sales data are as follows:

Month	*Jan.*	*Feb.*	*Mar.*	*Apr.*	*May*	*Jun.*	*Jul.*	*Aug.*	*Sep.*	*Oct.*	*Nov.*	*Dec.*
Seasonal Index	76	77	98	128	137	122	101	104	100	102	82	73

Using this information, draw up a monthly sales budget for the company. (Assume that there is no trend). *[C.A., Nov. '78]*

28. (a) Explain the meaning of deseasonalising data. What purpose does it serve?
(b) Deseasonalise the following data with the help of the seasonal index given against:

Month	*January*	*February*	*March*	*April*	*May*	*June*
Cash Balance (Rs '000)	360	400	550	360	350	550
Seasonal Index	120	80	110	90	70	100

[C.A., May '83]

29. The following table gives the cash receipts and the seasonal indices for 12 months:

Cash-Receipts	*Jan.*	*Feb.*	*Mar.*	*Apr.*	*May*	*June*
(millions of Rs)	35.1	23.7	20.8	21.1	28.3	22.5
Seasonal Index	1.30	.67	.57	.57	.71	.63
	July	*Aug.*	*Sept.*	*Oct.*	*Nov.*	*Dec.*
	23.1	24.3	41.3	62.1	65.4	71.7
	.71	.71	1.37	1.82	1.45	1.49

Eliminate the seasonal variations in the case-receipts and discuss the significance of such data. *[C.U., B.A.(Econ) '66]*

30. What are the major uses of Seasonal indices in time series analysis? Name four methods by which one can compute a seasonal index from time series data.

The sale of a company rose from Rs 60,000 in the month of August to Rs. 69,000 in the month of September. The seasonal indices for these two months are 105 and 140 respectively. The owner of the company was not at all satisfied with the rise of the sale in the month of September by Rs. 9,000. He expected much more because of the seasonal index for that month. What were his estimate of sales for the month of September? *[I.C.W.A., Dec '77]*

31. Suppose that the secular trend of sales of a company is accurately described by the equation $Y_c = 120{,}000 + 1{,}000x$ where x represents a period of one month and has a value 0 in December, 1981. The seasonal indices for the company's sales are as follows:

January	100	*February*	80	*March*	90	*April*	120
May	115	*June*	95	*July*	75	*August*	70
September	90	*October*	95	*November*	120	*December*	150

Ignoring cyclical and random influences, forecast sales for (i) February 1983, (ii) May 1986, (iii) December 1994. *[C.U., B.Sc. (Econ) '82]*

32. (a) What do you mean by Seasonal variation? Explain with a few examples the utility of such a study.

(b) Calculate the Seasonal Index from the following data using the Average Method:

Year	*1st Qr.*	*2nd Qr.*	*3rd Qr.*	*4th Qr.*
1974	72	68	80	70
1975	76	70	82	74
1976	74	66	84	80
1977	76	74	84	78
1978	78	74	86	82

[C.A., May '79]

33. (a) What is meant by seasonal fluctuations in a time series? How do you distinguish them from other types of fluctuations? Discuss the different methods for determining seasonal fluctuations.

(b) Construct Indices of seasonal variations from the following time series data on consumption of cold drinks, which contains only seasonal and irregular variations.

		Consumption of Cold Drinks ('000 bottles)			
	Quarters	I	II	III	IV
Year	1971	90	75	87	70
	1972	75	80	78	75
	1973	80	75	75	72
	1974	85	82	80	81

[C.U., M.Com. 75; B.Com.(Hons) '81]

34. What is a 'seasonal index'? Briefly explain the various methods of constructing such an index. *[C.U., B.Sc.(Econ) '83]*

35. (a) What do you understand by Seasonal Indices? What methods are used to determine them?

(b) Use the *method of monthly averages* to determine the monthly indices for the following data of production:

	(Production in lakhs of tons)											
Month / *Year*	Jan.	Feb.	Mar.	Apr.	May.	Jun.	Jul.	Aug.	Sep.	Oct.	Nov.	Dec.
1979	12	11	10	14	15	15	16	13	11	10	12	15
1980	15	14	13	16	16	15	17	12	13	12	13	14
1981	16	15	14	16	15	17	16	13	10	10	11	15

[C.A., May '82]

36. Using additive model, estimate the seasonal indices by the method of moving averages from the table given below. Deseasonalise the given production figures with the help of the seasonal indices obtained and explain the significance of the deseasonalised data:

Year/Quarter	Total Production of Paper ('000 tons) I	II	III	IV
1951	37	38	37	40
1952	41	34	25	31
1953	35	37	35	41

[C.U., M.Com. '66]

37. Obtain seasonal indices for the following data:

Season/Year	Output in Thousand Units 1960	1961	1962	1963	1964
Summer	31	42	49	47	51
Rains	39	44	53	51	54
Winter	45	57	65	62	66

[I.C.W.A., (old), Dec. '74]

38. What do you mean "business forecasting" ? Indicate its uses. *[C.A., Nov. '80]*
39. Describe the process of Exponential Smoothing. *[I.C.W.A., Dec. '81]*
40. Using the method of exponential smoothing, find forecasts for the following sales data, taking an initial forecast 25 and a smoothing coefficient 0.4.

Day	1	2	3	4	5	6	7	8
Sales	26	28	23	27	24	30	26	27

ANSWERS

2. (i) Cyclical fluctuation, (ii) Random movement, (iii) Seasonal variation, (iv) Secular trend. (see Example 16.3)
3. (i) *Irregular movement*; since four days' rainfall is unusual in Calcutta during February. (ii) *Seasonal variation*; since this part of the year is the winter season during which ice cream sales come down every year. (iii) *Cyclical fluctuation*; because an era of prosperity will be followed by decline, depression and recovery in course of time. (iv) *Seasonal variation*; since this is a short-term fluctuation noticed every year. (see Example 16.3)
4. (b) Trend: 22.00, 23.33, 24.00, 23.67, 23.67, 24.33, 26.00, 26.33 (in '000 tons) for 1969-1976. Short-term fluctuations are obtained by subtracting trend from original data: 0, –0.33, 1.00, 0.33, –1.67, 0.67, 0, 0.67 (in '000 tons). (see Example 16.11)
5. 3.00, 3.83, 4.80, 5.80, 7.17, 8.30, 9.43, 9.73, 10.3 for 1957–1965. (see Example 16.11)
6. (a) When time series data show a fairly regular periodicity, the period of fluctuation (or its mutliple) is taken as the period of the moving average. In general, however, the data are plotted on a graph paper to observe the periodicity of fluctuation. The average period between consecutive "peaks" and consecutive "troughs" is then taken as the period of moving average (see Example 16.15A).

(b) The data show a regular cycle of 5 days, because every 5th figure is the highest after which there is slump, followed by gradual recovery. Hence 5-day moving averages: 37.6, 37.6, 38.2, 38.6, 38.4, 38.8, 39.2, 38.8, 38.6, 38.6, 38.8, 38.8, 39.0, 39.0, 39.0, 39.0 for days 3 to 18. (see Example 16.15 and 15A)

7. 42.1, 40.9, 39.8, 38.2, 38.1, 37.8 for 1960 to 1965 (see Example 16.12b). Short-term fluctuations: 0.9, 3.6, –0.9, –0.1, –5.5, 3.9.
8. (Take 4-quarter period): 126.0, 127.6, 133.9, 141.5, 147.8, 154.0, 160.2, 167.2 ('000 tons) for 1975-III to 1977-II. (see Example 16.12b)
9. (Take 4-quarter moving averages): 57.9, 59.9, 62.0, 64.2, 65.2, 64.8, 64.1, 63.1 (million tons) for 1964-III to 1966-II. (see Example 16.15)
10. 495.8, 503.6, 511.6, 529.5, 553.0, 572.5 (mn. lbs.) for 1943–1948, (see Example 16.15)
11. Moving averages for 1966 to 1972 are (in Rs '000) 29/8, 31/8, 33/8, 35/8, 37/8, 39/8, 41/8. (see Example 16.12b and 16.13)
12. $y = 11.06 - 0.36x$ (origin: 1961, unit of $x = 1$ year). (see Example 16.18)
13. $y = 1278 + 232.9x$ (origin: 1968, x unit = 1 year). Trend values; 579, 812, 1045, 1278, 1511, 1744, 1977 (Rs crores). (see Example 16.18)
14. $y = 135 + 3.1x$ (origin: 1974, unit of $x = 1$ year). Trend values; 125.7, 128.8, 131.9, 135.0, 138.1, 141.2, 144.3 (Rs '000). (see Example 16.18)
15. $y = 39.9 - 0.77x$ (origin: 1973, unit of $x = 1$ year). Trend values (in Rs '000) by least squares: 41.44 and 37.59 by moving averages: 40.20 and 37.58. (see Example 16.18 and 12a)
16. $y = 48.83 + 4.61x$ (origin: mid-point of 1963–64, unit of $x = 6$ months). (see Example 16.19)
17. $y = 121 + 4.71x$ (origin: mid-point of 1953 and 1954; Unit of $x = 6$ months). (see Example 16.19)
18. $y = 322.8 + 28.4x$ (origin: Mid-point of 1955–56 and 1956-57; Unit of $x = 6$ months). (see Example 16.19)
19. $y = 655 + 35.83x$ (origin: Mid-point of 1963–64; Unit of $x = 6$ months). Estimated value 1049. (see Example 16.19)
20. $y = 12.31 - 0.60x - 0.86x^2$ (origin: 1975; Unit of $x = 1$ year). (see Example 16.21)
21. $y = 39.2 + 2.04x + 0.49x^2$ (origin: 1963; Unit of $x = 1$ year), (see Example 16.21)
22. $\log y = 2.2290 + .0471x$ (origin: 1947, Unit of $x = 1$ year). (see Example 16.22)
23. (Using 4-figure log table) $y = 31990\ (1.114)^x$ (origin: 1973, Unit of $x = 1$ year). (see Example 16.22)
24. $y = 41.14 + 2.04x$ (origin: 1963, Unit of $x = 1$ year); 0.17 thousand tons. Monthly trend $y = 41.23 + 0.17x$ (origin: July 1963, unit of $x = 1$ month); trend values: 36.13, 36.47, 38.00 ('000 tons) when $x = -30, -28, -19$. (see Example 16.23 and 16.25)
25. $y = 40.33 + 3.75x$ (origin: 1st quarter of 1977, Unit of $x = 1$ quarter). Putting $x = 22$, trend value = 122.83. Estimated consumption = $122.83 \times 1.17 = 143.7$ tons. (see Example 16.38 and 16.33)

26. Putting $t = 4$ and -2, annual production (trend) are 118.92 and 101.34 ('000 tons). Trend equation for quartely values $X_t = \frac{107.2}{4} + \frac{2.93}{4 \times 4}\left(t + \frac{1}{2}\right) = 26.89 + 0.183t$ (origin: 3rd quarter of 1973; Unit of t = 1 quarter). Putting $t = 14$, trend value for 1st quarter of 1977 is 29.452. Multiplying by the seasonal effect 1.25 for 1st quarter estimated production = 36.82 ('000 tons). (see Example 16.24 and 16.33).
27. 152, 154, 196, 256, 274, 244, 202, 208, 200, 204, 164, 146 (Rs. '000) for January to December. (see Example 16.29)
28. (b) (Divide 'cash balance' by seasonal effects) 300, 500, 500, 400, 500, 550 ('000 Rs) (see Example 16.30).
29. (Divide case receipts by 'seasonal index') 27.0, 35.4, 36.5, 37.0, 39.9, 35.7, 32.5, 34.2, 30.1, 34.1, 45.1, 48.1 (Rs million) for January to December. (see Example 16.30).
30. Based on the August sales (seasonal index 105), the expected September sales (seasonal index 140) are Rs 80,000. (see Example 16.31).
31. (i) 107,200, (ii) 198,950, (iii) 414,000 (**Hint:** See Example 16.33; Putting $x = 14$, 53 and 156, obtain trend values 134,000, 173,000 and 276,000 and multiply them by seasonal effects 0.80, 1.15 and 1.50).
32. 98, 92, 109, 101 using multiplicative model. (see Example 16.34).
33. 105, 99, 101, 95 for quarters I to IV (using Method of Quarterly Averages, and Mulitplicative Model). Because there is no trend. (see Example 16.34).
35. (b) Using Additive Model: 0.66, –0.33, –1.33, 1.67, 1.67, 2.00, 2.66, –1,00, –2.33, –3.00, –1.67, 1.00 (in lakh tons) for January to December. (see Example 16.33)
36. Seasonal indices: 3.12, 0.69, –4.25, 0.44 ('000 tons) for quarters I to IV. Deduct seasonals from the given figures, and obtain deseasonalised data ('000 tons): 33.88, 37.31, 41.25, 39.56, 37.88, 33.31, 29.25, 30.56, 31.88, 36.31, 39.25, 40.56. (see Example 16.36).
37. –4.49, –2.27, 6.76 ('000 units) for Summer, Rains, Winter, using additive model and moving average method (**Hint:** Use 3-period moving average for trend). (see Example 16.35).
40. 25.40, 26.44, 25.06, 25.84, 25.10, 27.06, 26.64, 26.78.

17 INDEX NUMBERS

17.1 MEANING OF 'INDEX NUMBER'

We begin with a problem. Given the following information regarding the prices of a group of good items in the years 1970 and 1980:

Table 17.1 Prices of Food Items

Commodity	Unit	*Price per unit (Rs)*	
		1970	*1980*
Rice	Quintal	80.00	200.00
Wheat	kg	0.50	1.40
Fish	kg	10.00	23.00
Bread	lb.	0.60	1.35
Milk	litre	1.50	3.00

On the basis of these, answer the following question:

"How does the overall food price in 1980 compare with that in 1970—how many times or what per cent?"

A number which provides an answer to this question will be called an "index number".

Several points invite attention: (i) The proposed index number will give a comparison between 'prices' and hence it will be a 'price index number'. (ii) The index number covers a 'group' of related items, here food items. Hence we shall speak of the 'index number of food prices'. Note that 'food' whose price the index number compares, is not a commodity by itself. (iii) The comparison is made between two 'periods of time'. Here the position in 1980 is compared against that in 1970. (iv) The index number is an 'average' computed from data given in heterogeneous units—Quintal, kg, lb., litre. It compares the overall position of food prices, i.e. an average computed from several items. Again, the prices quoted in the table are some sort of 'average' prices—Rice, for instance, is sold in several qualities or grades whose prices are different. Lastly, did you notice that the price of Rice in 1970 cannot remain a contain throughout the year. It is an average for the whole year. The index number is thus a "special type of average".

Example 17.1 *What are index numbers?*
[C.U., B.Com.(Hons) '82; M.Com. '72; W.B.H.S.'83; D.M. '77; I.C.W.A., June '75, '76; D.S.W., '71; M.B.A. '78]

Solution *Index Numbers* are numerical figures which indicate the relative position in respect of price, or quantity or value of a group of articles at certain periods of time as compared with another period; called *base period*. When the comparison is in respect of price, they are called 'Price Index Numbers' ; similarly we have 'Quantity Index Numbers' and 'Value Index Numbers'. Index number for the base period is always taken as 100. Index number for any other period, called *current period,* shows the overall level of price (or quantity or value) of the group of articles as a percentage of that in the base period.

The statement "*Index Number of Wholesale Prices in India for the year 1977 was 185* (*Base: 1970–71 = 100*)" signifies that as compared with the wholesale prices prevailing during the period 1970–71, the wholesale price of all articles during 1977 was on an average 185%. In other words, prices increased in 1977 by 85% over the 1970–71 prices.

Example 17.2 *State the uses of index number.*
[W.B.H.S. '83; C.U., M.Com. '72; I.C.W.A., June '75]

Solution Index numbers are primarily used to measure the relative position of business and economic conditions. There are many different types of index numbers and the use of an index number depends on its type. Index numbers of wholesale prices, retails prices cost of living, industrial production, quantum of exports and imports, business activity, to name only a few, are useful in their own fields.

Price index numbers are used for various purposes. 'Wholesale price index number' tells us about changes taking place in the value of money. 'Consumer price index number', or 'Cost of living index number' measures changes in the real income of people. It helps in the calculation of dearness allowance, so that the real wage may not decrease. 'Index numbers of stock prices' are used by economists, speculators and bankers in various ways. An economist uses them to measure changes in the purchasing power of money over stocks, a speculator uses them for forecasting the future course of the market, and the insurance company may require the index numbers for estimating future interest rate. Similarly 'index number of industrial production' reveals the comparative position in productivity and 'index number of business activity' throws light on the progress of business conditions.

Index numbers are also used to measure the comparative position in respect of price in different regions at the same period of time, e.g. for comparing the standards of living in several cities.

Example 17.3 *Fill up the blanks—*

(a) *The wholesale price index number of agricultural commodities in a given region at a given date is 280. The percentage rise in prices of agricultural commodities over the base year is ...* [W.B.H.S. '81]

(b) *If now the prices of all commodities in a place have decreased by 35% over the base period prices, then the index number of prices for the place is now (index number of prices of base period = 100) ...* [W.B.H.S. '82]

Solution (a) 180; (b) 65.

17.2 PROBLEMS IN CONSTRUCTION OF INDEX NUMBERS

Example 17.4 *Discuss the various steps and problems involved in the construction of index numbers.* [C.U., B.Com (Hons.) '80; D.S.W., '71, 76; D.M. '78]

Solution The steps in the construction of index numbers are as follows:

1. *Definition of Purpose and Scope*—Before going to construct an index number, a clear statement as to the purpose and its scope is necessary. All index numbers do not serve the same purpose and there is no all-purpose index. The selection of items etc. will depend upon the purpose of construction and the people for whom it is intended. For example, in constructing an index number of wholesale prices, the prices from retailers are unnecessary, just as for a cost of living index number, quotations of cloth price ex-mill or prices of cotton yarn are useless. One must be sure of what the index number is going to measure.
2. *Selection of Items*—For reasons of economy and ease of calculation, *t* is not possible to include all commodities in the construction of an index number. For a price index number, only a few selected items are, therefore, included whose price movements appear to be representative for the whole group of commodities. On the other hand, inclusion of too few items would make the index unrepresentative of the general level. With the passage of time, some items lose importance while some other new items appear to be more useful. The less important items should then be deleted from the list of commodities and replaced by new ones in conformity with their relative importance.
3. *Selection of Sources an Collection of Data*—For a regular source of index numbers, a systematic collection of prices and quantities should be made at regular intervals of time from prominent business firms or standard retail stores located at different important centres. The selected shops should be those which are visited by a large majority of customers. Due care must also be taken in selecting the enumerators, who are entrusted with the collection of data; because, upon their honesty and intelligence will depend the quality and reliability of index numbers.
4. *Choice of Base*—The base period should be chosen with much care and be such a one when abnormal increase or decrease in price was noticed. It is desirable to select a base period of recent past. The base should not be too long or too short a period. Generally, a year is taken as base, preferably a year of some economic importance for the country, e.g. the year 1951, being the first year of India's Five-year Plans. Sometimes a month, or a group of years, is also taken as base.
5. *System of Weighting*—All the commodities includes in the construction of an index number are not of equal importance, in the sense that a change in the price of an item does not affect the price level to the same extent as does the same amount of change in another item. The system of weighting and particularly the allocation of weights to the different items is, therefore, of utmost importance. Price relatives are weighted by values, prices by quantities and quantities by prices. The prices or quantities used as weights may relate either to the base period or to the current period. In Laspeyres' price index formula the base period quantities are used as weights, while in Paasche's price index formula the weights used are the current period quantities. Edgeworth-Marshall's price index formula uses the average of base and current period quanties. In the relative method (e.g. in constructing Cost of Living Index), the weights used for combining group index are the money value devoted to each group, but expressed as percentage of total value.

6. *Form of Average of Use*—Price index number is sometimes computed by averaging the percentage positions in price of the commodities. For the purpose of averaging, arithmetic mean or geometric mean is generally employed. In certain cases, median is also used. The arithmetic mean, due to the simplicity in calculation, is used in a great majority of cases; but since it is highly affected by even a few very large or small values, the geometric mean is preferred in many cases.

17.3 METHODS OF CONSTRUCTION OF INDEX NUMBERS

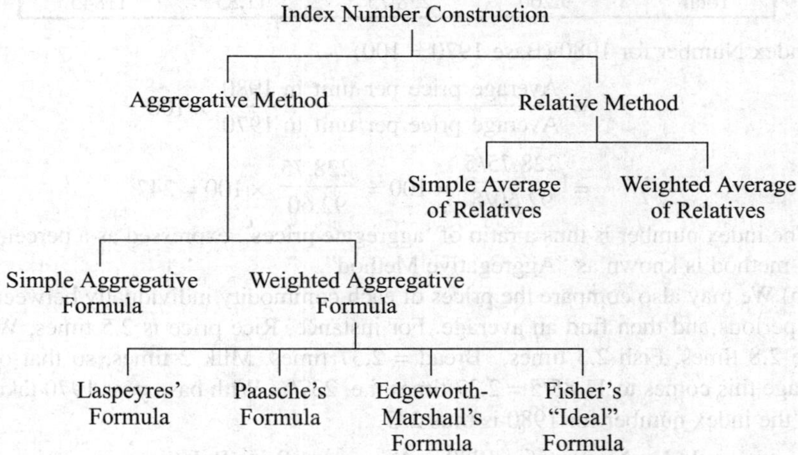

Here, we shall discuss the methods of construction in relation to a "Price Index". The following notations are used:

1. p_0 , p_n denote *price per unit* in base year and current year respectively.
   q_0, q_n denote *quantity* in base year and current year respectively.
   However, when data for several years are available, price (p) and quantity (q) are used with subscripts 0, 1, 2, 3, etc. Thus,
2. p_0 , p_1, p_2 ... denote *price* in the years 0, 1, 2, ... respectively;
   q_0, q_1, q_2 ... denote *quantity* in the years 0, 1, 2, ... respectively;
   The letter I, with appropriate subscripts, is used to denote "index number".
3. I_{0n} denotes *index number for year n with base year 0*
   I_{n0} denotes *index number for year 0 with base year n*
   I_{23} denotes *index number of year 3 with base year 2*
   (The first subscript indicates the base and the second subscript the current year).

In order to devise formulae for index numbers, let us consider the question raised at the beginning of this chapter. Our problem is to find a quantity which will indicate a comparison of the price of food in general in 1980 as against that in 1970. This can be done in two ways:

(a) We may compare the 'average price per unit' of a commodity in 1980 against the same in 1970.

Table 17.2 Construction of Price Index

Commodity	p_o	p_n	p_n/p_o	*Price Relative*
Rice	80.00	200.00	2.5	250
Wheat	0.50	1.04	2.8	280
Fish	10.00	23.00	2.3	230
Bread	0.60	1.35	2.25	225
Milk	1.50	3.00	2.0	200
Total	92.60	228.75	11.85	11845

Index Number for 1980 (Base 1970 = 100)

$$= \frac{\text{Average price per unit in 1980}}{\text{Average price per unit in 1970}} \times 100$$

$$= \frac{228.75/5}{92.60/5} \times 100 = \frac{228.75}{92.60} \times 100 = 247$$

The index number is thus a ratio of 'aggregate prices', expressed as a percentage. This method is known as "Aggregative Method".

(b) We may also compare the prices of each commodity individually between the two periods and then find an average. For instance, Rice price is 2.5 times, Wheat price 2.8 times, Fish 2.3 times, Bread = 2.37 times, Milk 2 times, so that on an average this comes to 11.85/5 = 2.37 times, i.e. 237%. With base year 1970 taken as 100, the index number for 1980 is thus 237.

Index Number for 1980 = Average of Price Relatives
= 1185 ÷ 5 = 237

(Price Relative = Price ratio × 100). This method is known as the "Relative Method".

It may be noted that

(i) Aggregative Method shows "*Relative of averages* (*or aggregates*)"

(ii) Relative Method shows "*Average of relatives*".

The average used may however be either 'simple' or 'weighted'. Thus, we have Simple Aggregative or Weighted Aggregative Index and Simple Average or Weighted Average of Relative Index.

I. Aggregative Method

In this method, the aggregate price of all items in the given year is expressed as a percentage of the same in the base year, giving the index number.

$$\text{Index Number} = \frac{\text{Aggregate Price in the given year}}{\text{Aggregate Price in the base year}} \times 100 \qquad (17.3.1)$$

If simple aggregates of prices are compared, we get

$$\text{Simple Aggregative Index } (I_{0n}) = \frac{\Sigma p_n}{\Sigma p_n} \times 100 \qquad (17.3.2)$$

the summation extending over all items included for the construction of index number.

If, however, weighted aggregates of prices in the two periods are compared, we have

$$\text{Weighted Aggregative Index } (I_{0n}) = \frac{\Sigma p_n w}{\Sigma p_0 w} \times 100 \qquad (17.3.3)$$

where w represents the "weight". It should be noted that the same set of weights must be used both for base year as well as for current year.

In the construction of a Price Index, *quantities* (q) are used as weights. There are several formulae for weighted aggregative index depending on the nature of weights employed:

(i) If the base year quantity (q_0) is used as weight, i.e. $w = q_o$, we get

$$\textbf{Laspeyres's Index } (I_{0n}) = \frac{\Sigma p_n q_0}{\Sigma p_0 q_0} \times 100 \qquad (17.3.4)$$

(ii) If the current year quantity (q_n) is used as weight, i.e. $w = q_n$, we get

$$\textbf{Paasche's Index } (I_{0n}) = \frac{\Sigma p_n q_n}{\Sigma p_0 q_n} \times 100 \qquad (17.3.5)$$

(iii) If the sum of quantities in the base year and the current year is used as weight, i.e. $w = (q_0 + q_n)$, we get

$$\textbf{Edgeworth-Marshall's Index } (I_{on}) = \frac{\Sigma p_n (q_0 + q_n)}{\Sigma p_0 (q_0 + q_n)} \times 100 \qquad (17.3.6)$$

(iv) The geometric mean (i.e. square-root of the product) of Laspeyres' Index and Paasche's index is of special importance, because of certain properties (Example 17.28), and is known as

Fisher's Ideal Index (I_{0n})

$$= \sqrt{(\text{Laspeyres' Index}) \times (\text{Paasche's Index})}$$

$$= \sqrt{\frac{\Sigma p_n q_0}{\Sigma p_0 q_0} \times \frac{\Sigma p_n q_n}{\Sigma p_0 q_n}} \times 100 \qquad (17.3.7)$$

The following index number of the weighted aggregative type are also sometimes used:

(v) The arithmetic mean of Laspeyres' index and Paasche's index is known as **Bowley's Index (I_{0n})**

$$= \frac{1}{2}(\text{Laspeyres' index} + \text{Paasche's index})$$

$$= \frac{1}{2}\left[\frac{\Sigma p_n q_0}{\Sigma p_0 q_0} + \frac{\Sigma p_n q_n}{\Sigma p_0 q_n}\right] \times 100 \qquad (17.3.7a)$$

(vi) If the geometric mean of basic year and current year quantities is used as weight, i.e. $w = \sqrt{q_0 q_n}$, we get

$$\textbf{Walsh's Index } (I_{0n}) = \frac{\Sigma p_n \sqrt{q_0 q_n}}{\Sigma p_0 \sqrt{q_0 q_n}} \times 100 \qquad (17.3.7b)$$

(vii) If the weights used are kept fixed for all periods i.e. weights are constant quantities (q), without any reference to base or current period, we get

$$\textbf{Kelly's Index } (I_{0n}) = \frac{\Sigma p_n q}{\Sigma p_0 q} \times 100 \qquad (17.3.7c)$$

This is also known as "*Aggregative index with fixed weights*".

II. Relative Method

In this method, the price of each item in the current year is expressed as a percentage of the price in the base year. This is called *Price Relative* and is given by the formula

$$\text{Price Relative} = \frac{\text{Price in the given year}}{\text{Price in the base year}} \times 100$$

$$= \frac{p_n}{p_0} \times 100 \qquad (17.3.8)$$

The average of price relatives, which shows the average percentage change for the whole group of items, gives the index number.

$$\text{Price Index} = \text{Average of Price Relatives} \qquad (17.3.8a)$$

Usually A.M. or G.M. is used for averaging the relatives. In special cases, H.M., or median, is also used. Again, the average employed may be either 'simple' or 'weighted'. If a simple average is used, the index number is called *Simple Average of Relatives Index.* If a weighted average is used, it is known as *Weighted Average of Relatives Index.* Thus,

Simple A. M. of Relatives Index (I_{0n})

$$= \Sigma\,(Price\ Relatives) \div k \qquad (17.3.9)$$

where k is the number of items included.

Simple G.M. of Relative Index (I_{0n})

$$= \sqrt[k]{Product\ of\ price\ relatives} \qquad (17.3.10)$$

Weighted A.M. of Relatives Index (I_{0n})

$$= \frac{\Sigma(Price\ Relative) \times w}{\Sigma w} \qquad (17.3.11)$$

The weights (w) employed for averaging price relatives are the values (= price × quantity) of items. In most cases, these values are given not in absolute units, but as percentages of the total value for all the items, i.e. the weights are given as pure numbers [see (17.8.1)]

Aggregative Formulae by Relative Method

It is interesting to note that the weighted average of relatives leads to several index number formulae of the aggregative type, depending on the nature of weights used. Considering price index numbers

1. The A.M. of relatives formula weighted by *base year values* (p_0q_0) gives exactly the same formula as Laspeyres':

$$I_{0n} = \frac{\Sigma\left(\dfrac{p_n}{p_0} \times 100\right) p_0q_0}{\Sigma p_0q_0} = \frac{\Sigma p_nq_0}{\Sigma p_0q_0} \times 100 = \text{Laspeyres' index} \qquad (17.3.12)$$

2. The A.M. of relatives formula weighted by value of *current year quantities at base year prices* (p_0q_n) gives Paasche'e formula:

$$I_{0n} = \frac{\Sigma\left(\dfrac{p_n}{p_0} \times 100\right) p_0q_n}{\Sigma p_0q_n} = \frac{\Sigma p_nq_n}{\Sigma p_0q_n} \times 100 = \text{Paasche's index} \qquad (17.3.13)$$

3. The H.M. of relatives formula weighted by *current year values* (p_0q_n) gives the same formula as Paasche's:

$$I_{0n} = \frac{\Sigma\, p_n q_n}{\Sigma (p_n q_n) / \left(\frac{p_n}{p_0} \times 100\right)} = \frac{\Sigma p_n q_n}{\Sigma p_0 q_n} \times 100 = \text{Paasche's index.} \qquad (17.3.14)$$

Construction of General Index from Group Indices

In the construction of any index number, the items included are usually classified under some broad categories called *Groups*, with similar or related items coming under each group. A separate index number is constructed for each group, and is called *Group Index*. The weighted average (usually A.M.) of group index numbers gives the *General Index*.

$$\textbf{General Index} = \frac{\Sigma IW}{\Sigma W} \qquad (17.3.15)$$

where I represents the *Group Index* and W is the *Group Weight*.

Example 17.5 *Find index numbers by the (a) method of aggregates, and (b) method of relatives (using arithmetic mean), from the following :*

Commodity	*Base Price*	*Current Price*
Rice	*35*	*42*
Wheat	*30*	*35*
Pulse	*40*	*38*
Fish	*107*	*120*

[C.U., B.Com. '73]

Solution (a) Let p_0, p_n denoted the Base Price and Current Price.

Table 17.3 Calculations for Index Number

Commodity	p_0	p_n	*Price Relative* = $\frac{p_n}{p_0} \times 100$
Rice	35	42	$\frac{42}{35} \times 100 = 120.0$
Wheat	30	35	$\frac{35}{30} \times 100 = 116.7$
Pulse	40	38	$\frac{35}{40} \times 100 = 95.0$
Fish	107	120	$\frac{120}{107} \times 100 = 112.1$
Total	212	235	443.8

(a) Simple Aggregative Index:

$$I_{0n} = \frac{\Sigma p_n}{\Sigma p_0} \times 100 = \frac{235}{212} \times 100 = 110.8$$

(b) Simple A.M. of price Relative Index:

$$I_{0n} = \frac{\Sigma(\text{Price Relative})}{k} = \frac{443.8}{4} = 111.0$$

[**Note:** An index number compares current prices as a percentage of base price. Since a number of commodities are to be covered , a measure of this comparison can be given either (i) as the *percentage of average prices* per commodity (the same as percentage of *aggregate* prices), or (ii) as the *average of percentage prices* for each commodity. The former gives Aggregative Index, and the latter gives A.M. of Relatives Index.]

Example 17.6 *Calculate price index numbers from the following data, using (i) weighted aggregative formula, and (ii) weighted arithmetic mean of price relative formula:*

Commodity	*Unit*	*Price (Rs) per unit*		*Weight*
		Base period	*Current period*	
A	*Quintal*	*80*	*110*	*14*
B	*Kg*	*10*	*15*	*20*
C	*Dozen*	*40*	*56*	*35*
D	*Litre*	*50*	*95*	*15*
E	*lb.*	*12*	*18*	*16*

[I.C.W.A., July '17]

Solution [**Note:** Compare this with Example 17.5, where no weights are given].

Table 17.4 Calculations for Index Numbers

Commodity	Base Price (p_0)	Current Price (p_n)	Weight (w)	$p_0 w$	$p_n w$	Price* Relative (I)	Iw
A	80	110	14	1,120	1,540	137.5	1,925
B	10	15	20	200	300	150	3,000
C	40	56	35	1,400	1,960	140	4,900
D	50	95	15	750	1,425	190	2,850
E	12	18	16	192	288	150	2,400
Total	—	—	100	3,665	5513	—	15,075

[***Note:** Price Relative = $(p_n/p_0) \times 100$. For example,
$(110 \div 80) \times 100 = 137.5$, $(15 \div 10) \times 100 = 150$, etc.]

(i) Weighted Aggregative Index

$$= \frac{\Sigma p_n w}{\Sigma p_0 w} \times 100 = \frac{5513}{3662} = 150.55$$

(ii) Weighted Arithmetic Mean of price Relatives Index

$$= \frac{\Sigma Iw}{\Sigma w} = \frac{15075}{100} = 150.75$$

Example 17.7 *From the following price and quantity data, compute Paasche's price index number for 1980 with 1970 as base:*

	Price (Rs per kg)		*Quantities Sold (kg)*	
	1970	*1980*	*1970*	*1980*
Commodity A	*4*	*5*	*95*	*120*
Commodity B	*60*	*70*	*118*	*130*
Commodity C	*35*	*40*	*50*	*70*

Solution Paasche's Price Index is obtained by weighted aggregative formula with current year quantities as weight.

Table 17.5 Calculations for Paasche's Price Index

Commoditiy	p_0	p_n	q_0	q_n	p_0q_n	p_nq_n
A	4	5	95	120	480	600
B	60	70	118	130	7800	9100
C	35	40	50	70	2450	2800
Total	—	—	—	—	10730	12500

$$\text{Paasche's Price Index} = \frac{\Sigma p_n q_n}{\Sigma p_0 q_n} \times 100 = \frac{12500}{10730} \times 100 = 116.$$

Example 17.8 *Construct Fisher's ideal index number for the following data:*

Commodity	*1960 (Base year)*		*1968 (Current year)*	
	Price	*Quantity*	*Price*	*Quantity*
A	*8*	*6*	*12*	*5*
B	*10*	*5*	*11*	*6*
C	*7*	*8*	*8*	*5*

Solution Fisher's Ideal Index = $\sqrt{\text{Laspeyres'} \times \text{Paasche' index}}$

Table 17.6 Calculations for Fisher's Ideal Index

Commodity	p_0	q_0	p_n	q_n	p_0q_0	p_nq_0	p_0q_n	p_nq_n
A	8	6	12	5	48	72	40	60
B	10	5	11	6	50	55	60	66
C	7	8	8	5	56	64	35	40
Total	—	—	—	—	154	191	135	166

$$\text{Laspeyres' index} = \frac{\Sigma p_n q_0}{\Sigma p_0 q_0} \times 100 = \frac{191}{154} \times 100 = 124.0$$

$$\text{Paasche's index} = \frac{\Sigma p_n q_n}{\Sigma p_0 q_n} \times 100 = \frac{166}{135} \times 100 = 123.0$$

$\therefore$ Fisher's Ideal Index = $\sqrt{124.0 \times 123.0} = 123.5$

Example 17.9 *"Marshall-Edgeworth index number is a good approximation to the Fisher's Ideal Index number"—verify the truth of this statement from the following data:*

Year	*Rice*		*Wheat*		*Jowar*	
	Price	*Quantity*	*Price*	*Quantity*	*Price*	*Quantity*
1970	*9.3*	*100*	*6.4*	*11*	*5.1*	*5*
1977	*4.5*	*90*	*3.7*	*10*	*2.7*	*3*

[I.C.W.A. June '80]

Solution Let us take 1970 as Base and 1977 as Current year, Marshall-Edgeworth Price Index

$$= \frac{\Sigma p_n(q_0 + q_n)}{\Sigma p_0(q_0 + q_n)} \times 100 = \frac{\Sigma p_n q_0 + \Sigma p_n q_n}{\Sigma p_0 q_0 + \Sigma p_0 q_n} \times 100$$

$$\text{Fisher's Ideal Price Index} = \sqrt{\frac{\Sigma p_n q_0}{\Sigma p_0 q_0} \times \frac{\Sigma p_n q_n}{\Sigma p_0 q_n}} \times 100$$

Table 17.7 Marshall-Edgeworth's and Fisher's Price Index

Commodity	p_0	p_n	q_0	q_n	p_0q_0	p_0q_n	p_nq_0	p_nq_n
Rice	9.3	4.5	100	90	930	837	450	405
Wheat	6.4	3.7	11	10	70.4	64	40.7	37
Jowar	5.1	2.7	5	3	25.5	15.3	13.5	8.1
Total	—	—	—	—	1025.9	916.3	504.2	450.1

$$\text{Marshall-Edgeworth Price Index} = \frac{504.2 + 450.1}{1025.9 + 916.3} \times 100 = 49.135.$$

$$\text{Fisher's "Ideal" Price Index} = \sqrt{\frac{504.2}{1025.9} \times \frac{450.1}{916.3}} \times 100 = 49.134.$$

The two index numbers are very close to each other. The statement is thus verified.

Example 17.10 *Calculate the price index number for the year 1978 with 1976 as base using Laspeyres' or Paasches' formula, whichever will be applicable, on the basis of the following data:*

Commodity	*Price (in Rs)*		*Money value ('000 Rs)*
	1976	*1978*	*1976*
A	*12.50*	*14.00*	*112.50*
B	*10.50*	*12.50*	*126.00*
C	*15.00*	*14.00*	*105.00*
D	*9.40*	*11.20*	*47.00*

(Here money value means total of a commodity). [W.B.H.S. '82]

Solution We are given p_0, p_n and p_0q_0 (*i.e.* value in "base" year), from which it is possible to find q_0 by the relation

$$q_0 = \frac{p_0q_0}{p_0} = \frac{\text{Money value in 1976 ('000 Rs.)}}{\text{Price in 1976 (Rs.)}}$$

in units of '000. Now using p_0, p_n and q_0 we can only find

$$\text{Laspeyres' Price Index} = \frac{\Sigma p_nq_n}{\Sigma p_0q_0} \times 100$$

(Note that it is not possible to find Paasche's Index, because the formula involves q_n which is not available from the given data).

Table 17.8 Calculations for Laspeyres's Price Index

Commodity (1)	p_o (2)	p_n (3)	p_oq_o (4)	q_o = (4) ÷ (2) (5)	p_nq_o (6)
A	12.50	14.00	112.50	9	126.00
B	10.50	12.00	126.00	12	144.00
C	15.00	14.00	105.00	7	98.00
D	9.40	11.20	47.00	5	56.00
Total	—	—	390.50	—	424.00

$$\text{Laspeyres' Price index} = \frac{\Sigma p_nq_0}{\Sigma p_0q_0} \times 100 = \frac{424.00}{390.50} \times 100 = 109.$$

Example 17.11 *With regard to Laspeyres's and Paasche's price index numbers, it is maintained that "if the price of all the goods change in the same ratio, the two indexes will be equal, for then the weighting system is irrelevant ; or, if the quantities of all the goods change in the same ratio, they will be equal, for then the two weighting systems are the same relatively."*

Verify the above statement. [C.U., M.Com. '81]

Solution Laspeyres' Price Index $(L) = \frac{\Sigma p_nq_0}{\Sigma p_0q_0} \times 100$

Paasche's Price Index $(P) = \frac{\Sigma p_nq_n}{\Sigma p_0q_n} \times 100$

(i) If all prices changes in the same ratio, we have $p_n = k \cdot p_0$, where k is a constant. Then

$$L = \frac{\Sigma(k \cdot p_0)q_0}{\Sigma p_0q_0} \times 100 = \frac{k\Sigma p_0q_0}{\Sigma p_0q_0} \times 100 = 100k$$

$$P = \frac{\Sigma(k \cdot p_0)q_n}{\Sigma p_0q_n} \times 100 = \frac{k\Sigma p_0q_n}{\Sigma p_0q_n} \times 100 = 100k$$

Thus, we find that $L = P$.

(ii) If all quantities change in the same ratio, we put $q_n = k' \cdot q_0$, where k' is a constant. Then

$$L = \frac{k\Sigma p_nq_0}{\Sigma p_0q_0} \times 100$$

$$P = \frac{\Sigma p_n(k' \cdot q_0)}{\Sigma p_0(k' \cdot q_0)} \times 100 = \frac{\Sigma p_nq_0}{\Sigma p_0q_0} \times 100 \text{ ; (because } k' \text{ cancels)}$$

Again we find that $L = P$. The statements are thus verified.

Example 17.12 *Given below are the data on prices of some consumer goods and the weights attached to the various items. Compute price index numbers for the year 1969 (Base: 1968 = 100), using (i) simple average, and (ii) weighted average, of price relatives.*

Item	*Unit*	*Price (Rs)* 1968	*Price (Rs)* 1969	*Weight*
Wheat	*Kg.*	*0.50*	*0.75*	*2*
Milk	*Litre*	*0.60*	*0.75*	*5*
Egg	*Dozen*	*2.00*	*2.40*	*4*
Sugar	*Kg.*	*1.80*	*2.10*	*8*
Shoes	*Pair*	*8.00*	*10.00*	*1*

[I.C.W.A. (old), Dec. '75]

Solution We use formulae (17.3.9) and (17.3.11)

Table 17.9 Calculations for Price Relatives Index

Item	p_0	p_n	*Price Relative* $I = (p_n/p_0) \times 100$	*Weight (w)*	*Iw*
Wheat	0.50	0.75	150	2	300
Milk	0.60	0.75	125	5	625
Egg	2.00	2.40	120	4	480
Sugar	1.80	2.10	117	8	936
Shoes	8.00	10.00	125	1	125
Total	—	—	637	20	2466

$$\text{Simple Average of P.R. Index} = \frac{\Sigma(\text{Price Relative})}{\text{No. of items}} = \frac{637}{5} = 127.4$$

$$\text{Weight Average of P.R. Index} = \frac{\Sigma(\text{Price Relative} \times \text{Weight})}{\Sigma\text{Weight}} = \frac{2466}{20} = 123.3$$

Example 17.13 *On the basis of the following data, compute the wholesale price index number for the 5 groups combined:*

Group	*Weight*	*Index Number for Week Ending 27-9-69 (Basic: 1952–53 = 100)*
Food articles	*50*	*241*
Liquor and tobacco	*2*	*221*
Fuel, power, light and lubricants	*3*	*204*
Industrial raw materials	*16*	*256*
Manufactured commodities	*29*	*179*

[C.U., M.Com. '70]

Solution *Using the weighted arithmetic mean of group indices as the method of combination, the General Index is given by the formula*

$$\text{General Index} = \frac{\Sigma IW}{\Sigma W}$$

where I = Group Index, and W = Group Weight.

Table 17.10 General Index from Group Indices

Group	Weight (W)	Group Index (I)	IW
1. Food articles	50	241	12,050
2. Liquor and tobacco	2	221	442
3. Fuel, power, light and lubricants	3	204	612
4. Industrial raw materials	16	256	4,096
5. Manufactured commodities	29	179	5,191
Total	100	—	22,391

$$\therefore \quad \text{Index number of Wholesale Prices} = \frac{22391}{100} = 223.91.$$

Example 17.14 *Apply the geometric mean to find general index from the following group indices, by assigning the given weights:*

Group	A	B	C	D	E	F
Group Index	118	120	97	107	111	93
Weight	4	1	2	6	5	2

[I.C.W.A., July '67]

Solution *The weighted geometric mean of the group indices will be found by applying logarithms:*

$$\log(\text{General Index}) = \frac{\Sigma(\log I) \times W}{\Sigma W}$$

Table 17.11 General Index using G.M.

Group	Group Index (I)	Weight (W)	log I	(log I) × W
A	118	4	2.0719	8.2876
B	120	1	2.0792	2.0792
C	97	2	1.9868	3.9736
D	107	6	2.0294	12.1764
E	111	5	2.0453	10.2265
F	93	2	1.9685	3.9370
Total	—	20	—	40.6803

Substituting the values from the table,

$$\log(\text{General Index}) = 40.6803 \div 20 = 2.0340$$

$$\therefore \quad \text{General Index} = \text{antilog } 2.0340 = 108.1.$$

Example 17.15 *Prove that Laspeyre's and Paasche's index numbers can be derived as weighted arithmetic averages of the price relatives and specify these weights.*

[C.U., B. A.(Econ) '73]

Solution [**Hint:** See formulae (17.3.12) and (17.3.13)]

[**Note:** The correct spelling is *Laspeyres*' (not *Laspeyre's*), named after the German economist and statistician Etienne Laspeyres.]

Example 17.16 Discuss the importance and use of weights in the construction of index numbers. [I.C.W.A. July '71; C.U., M.Com.'72, '81]

Solution Weights play a very important part in the construction of index numbers. Index numbers of price are calculated either by taking the average of price relatives or by taking the relative of average prices of the items at two periods of time (see Example 17.5). In either case the averaging process is involved, and naturally the question arises whether it should be a simple average or a weighted average. If a simple average is used, it will be assumed that all the items included are equally important. But in almost all cases the cannot be so. All items cannot be considered as equally important in the sense that a change in the price of one of the items does not affect the price level to the same extent as does the same amount of change in the price of another item. For instance, is constructing a wholesale price index number for India, textiles must have greater weight than tobacco. If we ignore weights we shall not get an unweighted index but an inappropriately weighted index.

Since index number should not depend on the units in which the price or quantities are reported, price relatives are weighted by 'values' (= price × quantity), prices by quantities and quantities by prices. The quantity or value used as weight may relate either to the base period or to the current period or to a combination to both. The weights used in some price index formulae are:

(i) *Laspeyres' Index*—Base period quantity ($w = q_0$)
(ii) *Paasche's Index*—Current period quantity ($w = q_n$),
(iii) *Edgeworth-Marshall's Index*—Sum of base and current period quantities ($w = q_0 + q_n$).
(iv) *Simple A.M. of Price Relatives Index*—Number of units of the commodity, that can be purchased in the base period by one unit of money ($w = 1/p_0$).
(v) *Weighted A.M. of Price Relatives Index*—Usually, the base period values are used as weights ($w = p_0q_0$) leading to Laspeyre's index. This has the advantage that the same set of weights calculated from the base year data can be used for a long period of time. If the values of current period quanties at base year prices are used as weights ($w = p_0q_n$), Paasche's index in obtained.

Again, the quantity or value used as weight need not necessarily be the actual physical quantities or values produced or consumed, but their relative magnitudes. Weights are, therefore, as a rule expressed as percentages of total, which is taken as 100 (see Example 17.34).

Example 17.17 *Discuss the advantages of geometric mean in the construction of index numbers.*

Solution Index numbers are designed to measure the 'average' level of any particular factor (e.g. price, quantity or value) from one period to another. Naturally the question arises as to which average to use. For reasons of simplicity in calculation, the arithmetic mean is used in a great majority of cases. But the geometric mean (g.m.) has definite advantages from several standpoints:

(i) The g.m. is useful in averaging ratios, rates and percentages. It is particularly suitable for the construction of index numbers; because index numbers show percentage changes, rather than absolute amounts of change. It also gives equal weight to equal ratios of change.

(ii) Again since the g.m. is less affected than the arithmetic mean by the presence of extremely large or small values, it is considered all the more appropriate in index number construction. An unusual change in the price of a single commodity should not upset the whole index number.

(iii) The g.m. also makes index numbers time-reversible. While the arithmetic mean of relatives index does not satisfy time reversal test (Section 17.5), the g.m. of relatives index satisfies this test. Laspeyre's and Paasche's index numbers do not satisfy either the time reversal or the factor reversal test (Example 17.27), but their g.m., viz. Fisher's ideal index number, satisfies both these tests (Example 17.28), and as such is considered "ideal" from theoretical considerations.

17.4 QUANTITY INDEX NUMBER

Just as price index numbers measure and permit comparison of the price of a group of related items, *quantity Index numbers* similarly measure and permit comparison of the physical quantity of goods produced or consumed or marketed or distributed. Quantity index number formulae may be obtained from the corresponding price index number formulae replacing p and q and q by p.

$$\text{Simple Aggregative Quantity Index} = \frac{\Sigma q_n}{\Sigma q_0} \times 100 \tag{17.4.1}$$

$$\text{Laspeyres' Quantity Index} = \frac{\Sigma q_n p_0}{\Sigma q_0 p_0} \times 100 \tag{17.4.2}$$

$$\text{Paasche's Quantity Index} = \frac{\Sigma q_n p_n}{\Sigma q_0 p_n} \times 100 \tag{17.4.3}$$

$$\text{Edgeworth-Marshall's Index} = \frac{\Sigma q_n (p_0 + p_n)}{\Sigma q_0 (p_0 + p_n)} \times 100 \tag{17.4.4}$$

$$\text{Fisher's Ideal Index} = \sqrt{\frac{\Sigma q_n p_0}{\Sigma q_0 p_0} \times \frac{\Sigma q_n p_n}{\Sigma q_0 p_n}} \times 100 \tag{17.4.5}$$

$$\text{Quantity Relative} = \frac{q_n}{q_0} \times 100 \tag{17.4.6}$$

Simple A.M. of Quantity Relatives Index

$$= \Sigma\,(\text{Quantity Relatives}) \div k \tag{17.4.7}$$

Weighted A.M. of Quantity Relatives Index

$$= \frac{\Sigma(\text{Quantity relative} \times \text{Weight})}{\Sigma(\text{Weight})} \tag{17.4.8}$$

Example 17.18 *Prepare price and quantity index numbers for 1972 with 1961 as base from the following data by using (i) Laspeyre's, (ii) Paasche's and (iii) Fisher's method.*

Commodity	Unit	1961		1972	
		Quantity	*Price (Rs)*	*Quantity*	*Price (Rs)*
A	*Kg*	*5*	*2.00*	*7*	*4.50*
B	*Quintal*	*7*	*2.50*	*10*	*3.20*
C	*Dozen*	*6*	*8.00*	*6*	*4.50*
D	*Kg*	*2*	*1.00*	*9*	*1.80*

[C.U., B. Com. (Hons) '81]

Solution

Table 17.12 Laspeyres's, Paasche's and Fisher's Index

Commodity	q_0	p_0	q_n	p_n	q_0p_0	q_0p_n	q_np_0	q_np_n
A	5	2.00	7	4.50	10.00	22.50	14.00	31.50
B	7	2.50	10	3.20	17.50	22.40	25.00	32.00
C	6	8.00	6	4.50	48.00	27.00	48.00	27.00
D	2	1.00	9	1.80	2.00	3.60	9.00	16.20
Total	–	–	–	–	77.50	75.50	96.00	106.70

(a) *Prince Index Numbers:*

$$\text{Laspeyres' formula} = \frac{\Sigma p_n q_0}{\Sigma p_0 q_0} \times 100 = \frac{75.50}{77.50} \times 100 = 97$$

$$\text{Paasche's formula} = \frac{\Sigma p_n q_n}{\Sigma p_0 q_n} \times 100 = \frac{106.70}{96.00} \times 100 = 111$$

$$\text{Fisher's formula} = \sqrt{\text{Laspeyres'} \times \text{Passche's}} = \sqrt{97 \times 111} = 104$$

(b) *Quality Index Numbers:*

$$\text{Laspeyres' formula} = \frac{\Sigma q_n p_0}{\Sigma q_0 p_0} \times 100 = \frac{96.00}{77.50} \times 100 = 124$$

$$\text{Paasche's formula} = \frac{\Sigma q_n p_n}{\Sigma q_0 p_n} \times 100 = \frac{106.70}{75.50} \times 100 = 141$$

$$\text{Fisher's formula} = \sqrt{\text{Laspeyres'} \times \text{Paasche's}} = \sqrt{124 \times 141} = 132$$

Example 17.19 *Annual production (in million tons) of four commodities are given below:*

Commodity	*Production in Year*			*Weights*
	1950	*1954*	*1955*	
A	*160*	*200*	*216*	*20*
B	*24*	*42*	*45*	*30*
C	*50*	*72*	*68*	*13*
D	*120*	*168*	*156*	*17*

Calculate quantity index numbers for the years 1954 and 1955 with 1950 as base year, using (i) simple arithmetic mean, and (ii) weighted arithmetic mean, of the relatives. [I.C.W.A., July '67]

Solution [Working **Notes:** Quantity Relatives for 1954 (Base 1950)

$$\frac{200}{150} \times 100 = 125; \quad \frac{42}{24} \times 100 = 175; \quad \frac{72}{50} \times 100 = 144; \quad \frac{168}{120} \times 100 = 140$$

Quantity Relatives for 1955 (Base 1950)

$$\frac{216}{160} \times 100 = 135; \quad \frac{45}{24} \times 100 = 187.5; \quad \frac{68}{50} \times 100 = 136; \quad \frac{156}{120} \times 100 = 130]$$

Table 17.13 Calculations for Quantity Index Number

Commodity	*Quantity 1954*	*Relatives 1955*	*Weight*	*(Q.R. for 1954) × (weight)*	*(Q.R. for 1955) × (weight)*
A	125	135	20	2,500	2,700
B	175	187.5	30	5,250	5,625
C	144	136	13	1,872	1,768
D	140	130	17	2,380	2,210
Total	584	588.5	80	12,002	12,303

Using *simple arithmetic mean of quantity relatives,*

Index Number for 1954 = 584 ÷ 4 = 146

Index Number for 1955 = 588.5 ÷ 4 = 147

Using *weighted arithmetic mean of quantity relatives,*

Index Number for 1954 = 12002 ÷ 80 = 150

Index Number for 1955 = 12303 ÷ 80 = 154

Ans. (i) 146, 147; (ii) 150, 154.

Example 17.20 *Calculate a number which will indicate the percentage change in volume of traffic (Oct. 1929 = 100) from October 1929 to October 1930, when account is taken of the relative values of the different types of traffic.*

Type of Traffic	*Tons ('000)*		*Receipts (£ '000)*
	Oct. 1929	*Oct. 1930*	*Oct. 1929*
(a) Merchandies	*1,246*	*1,206*	*776*
(b) Minerals	*1,125*	*981*	*252*
(c) Fuel	*4,794*	*4,229*	*562*

[C.U., M.Com. '70; B.A.(Econ) '70; I.C.W.A., Dec. '73]

Solution *(First method)* We have to find a *quantity index number* for Oct. 1930 with base Oct. 1929. Since 'Receipts in 1929' represents the base period values (p_0q_0), the required quantity index may be obtained as the weighted A.M. of quantity relatives, using these receipts as weights.

Table 17.14 Calculations for Quantity Index Number

Type of Traffic	q_0	q_n	*Weight*	*Quantity Relative* $(q_n/q_0) \times 100$	*(5) × (4)*
(1)	*(2)*	*(3)*	*(4)*	*(5)*	*(6)*
(a)	1246	1206	776	(1206 ÷ 1246) × 100 = 97	75,272
(b)	1125	981	252	(981 ÷ 1125) × 100 = 87	21,924
(c)	4794	4229	562	(4229 ÷ 4794) × 100 = 88	49,456
Total	—	—	1590	—	146,652

$$\text{Quantity Index} = \frac{\Sigma\,(\text{Quantity relative} \times \text{Weight})}{\Sigma\,(\text{Weight})} = \frac{146,652}{1590} = 92$$

(Second method) We are given q_0, q_n and p_0q_0 for each of the three types of traffic, and it is required to find a *quantity index number*. However, we can find p_0 by using the relation

$$p_0 = \frac{p_0q_0}{q_0} = \frac{\text{Receipts in Oct. 1929 (£ '000)}}{\text{Volume of traffic (Tons '000)}}$$

Now, using p_0, q_0 and q_n it is possible to find

$$\text{Laspeyres' Quantity Index} = \frac{\Sigma\, q_n p_0}{\Sigma\, q_0 p_0} \times 100$$

Table 17.15 Calculations for Quantity Index Number

Type of Traffic	q_0	q_n	p_0q_0	$p_0 = \dfrac{\text{Col. (4)}}{\text{Col. (2)}}$	$q_n p_0 =$ *Col. (3) × Col. (5)*
(1)	*(2)*	*(3)*	*(4)*	*(5)*	*(6)*
(a)	1246	1206	776	776 ÷ 1246 = 0.623	1206 × 0.623 = 751.338
(b)	1125	981	252	252 ÷ 1125 = .224	981 × .224 = 219.744
(c)	4794	4229	562	562 ÷ 4794 = .117	4229 × .117 = 494.793
Total	—	—	1590	—	1465.875

$$\therefore \quad \text{Quantity Index} = \frac{1465.875}{1590} \times 100 = 92$$

[**Note:** (i) This proboem is similar to Example 17.10, which relates to *price index number*. (ii) '*First method*' as used here, may also be applied to Example 17.10]

Example 17.21 *From the following data calcuate Paasche's quantity index number for the year 1969, with 1951 as base:*

Commodity	*Quantity 1951*	*Quantity 1969*	*Value 1969*
A	*54*	*250*	*540*
B	*93*	*75*	*825*
C	*18*	*56*	*448*
D	*6*	*8*	*56*
E	*23*	*47*	*141*

[I.C.W.A., Jan., '72]

Solution (**Note:** Here, values in *current* period are given)

$$\text{Paasche's Quantity Index} = \frac{\Sigma\, q_n p_n}{\Sigma\, q_0 p_n} \times 100$$

where q_0, q_n denote quantities in the base year (1951) and current year (1969) respectively; and p_n denotes price in the current year. Since, 'Value' denotes the product of 'price per unit' and 'quantity', i.e. *Value* = (*price* × *quantity*), hence, 'Value in 1969' as shown in the question must be the product $p_n q_n$. Thus, we are given q_0, q_n and $p_n q_n$ for each commodity, and it is required to find Paasche's Quantity Index by combining the data. However, we can find the value of p_n, using the relation $p_n = \dfrac{p_n q_n}{q_n} = \dfrac{\text{Value in 1969}}{\text{Quantity in 1969}}$. Using the values of q_0, q_n and p_n the index can be calculated. (See Example 17.10 and 17.20, where values in *base period* are given).

Table 17.16 Calculations for Paasche's Quantity Index

Commodity	q_0	q_n	$p_n q_n$	$p_n = \dfrac{\text{Col. (4)}}{\text{Col. (3)}}$	$q_0 p_n =$ Col. (2) × col. (5)
(1)	*(2)*	*(3)*	*(4)*	*(5)*	*(6)*
A	54	250	540	540 ÷250 = 2.16	54 × 2.16 = 117
B	93	75	825	825 ÷75 = 11.00	93 × 11.00 = 1023
C	18	56	448	448 ÷56 = 8.00	18 × 8.00 = 144
D	6	8	56	56 ÷ 8 = 7.00	6 × 7.00 = 42
E	23	47	141	141 ÷ 47 = 3.00	23 × 3.00 = 69
Total	—	—	2010	—	1395

$$\therefore \quad \text{Paasche's Quantity Index} = \frac{2010}{1395} \times 100 = 144$$

[**Note:** If value in the base year ($p_0 q_0$) were given, the weighted *arithmetic mean* of relatives, using those values as weight, would lead to Laspeyres' index: $\dfrac{\Sigma\,(p_0 q_0)\left(\dfrac{q_n}{q_0} \times 100\right)}{\Sigma\, p_0 q_0} = \dfrac{\Sigma\, q_n p_0}{\Sigma\, q_0 p_0} \times 100$ = Laspeyres' Quantity Index.

However, since values in the current year ($p_n q_n$) are given, this method will not be applicable. Here, the weighted *harmonic mean* of relatives is the appropriate index, using the current year values as weights (Croxton & Cowden, *Applied General Statistics,* second edition, footnote p. 722). This leads to the same formula as Paasche's (17.4.3).

$$\frac{\Sigma\, p_n q_n}{\Sigma \dfrac{p_n q_n}{\left(\dfrac{q_n}{q_0} \times 100\right)}} = \frac{\Sigma\, q_n p_n}{\Sigma\, q_0 p_n} \times 100 = \text{Paasche's Quantity Index.}$$

In the soltuion given above, Paasche's Quantity index formula has been applied directly without using the harmonic mean of quantity relatives.]

17.5 TESTS OF INDEX NUMBERS

Example 17.22 *Describe the time reversal test, factor reversal test and circular test of index numbers.*

[CU., B.A.(Econ) '72, '73, '75; M.Com. '74, '77, '80; B.Com.(Hons) '82; M.B.A. '78; I.C.W.A., Jan. '72; June '75, '82]

Solution In order to judge the efficiency of an index number formula as a measure of the level of a phenomenon from one period to another, the noted economist Irving Fisher suggested certain tests. The three most important *test of index numbers* are (1) Time Reversal test, (2) Factor Reversal test, and (3) Circular test. These tests are based on the analogy that what is true for an individual item should also hold for a group of items.

(1) Time Reversal Test

According to this test, a good index number formula should work both wasy, forward and backward, with respect to time. In other words, we should get the same picture of change between two points of time, no mater which of the two is taken as base. Consequently, the index number (I_{0n}) for period n with base period 0 should be the reciprocal of the index number (I_{n0}) for period 0 with base period n (omitting the factor 100 from each index). Symbolically,

$$I_{0n} \times I_{n0} = 1 \tag{17.5.1}$$

An index number formula which obeys this relation is said to satisfy the time reversal test.

Time reversal test is satisfied by simple aggregative formula, Marshall-Edgeworth's formula, Fisher's ideal index formula, and simple geometric mean of relatives formula. Weighted aggregative formula and weighted geometric mean of relatives formula also satisfy this test, if constant weights are used which do not depend upon the base or current period.

Time reversal test is based on the following analogy: If the price of a commodity changes from Rs 4 per unit in 1961 to Rs 8 in 1978, the price in 1978 is 200% of (i.e. 2 times) the price in 1961, and the price in 1961 is 50% of (i.e. 0.50 times) the price in 1978. The product of the two price ratios is $2 \times 0.50 = 1$. This is true for each commodity and time reversal test ensures that the same principle holds for an index number, which embraces a group of commodities.

(2) Factor Reversal Test

An index number formula is said to satisfy the factor revesal test, if the product of Price Index (P_{0n}) and Quantity Index (Q_{0n}) gives the true Value Ratio (omitting the factor 100 from each index). In other words, a good index number formula should be such that the price ratio multiplied by the quantity ratio between two points of time gives the ratio of total values. Symbolically,

$$P_{0n} \times Q_{0n} = \frac{\Sigma\, p_n q_n}{\Sigma\, p_0 q_0} \tag{17.5.2}$$

Fisher's ideal index is the only formula which satisfies this test.

Factor reversal test is based on the following analogy: If the price per unit of a commodity changes from Rs 4 in 1961 to Rs 8 in 1978, and the quantity of consumption changes from 60 units to 90 units during the same period, then the price and quantity in 1978 are 200% and 150% respectively of the corresponding factors in 1961. The values (price × quantity) of consumption were Rs 240 in 1961 and Rs 720 in 1978, so that the value ratio is 720/240 = 3. Thus we find that the product of price ratio and quantity ratio equals the value ratio: $2 \times 1.50 = 3$. Factor reversal test ensures that the principle which holds for a single commodity should apply to the index number as a whole.

(3) Circular Test

This is an extension of time reversal test. An index number formula is said to satisfy the circular test, if the time reversal test is satisfied through a number of intermediate years. Symbolically,

$$I_{01} \times I_{12} \times I_{23} \times \ldots \times I_{(n-1),\, n} \times I_{n0} = 1 \qquad (17.5.3)$$

This means that the relation is satisfied in a circular fashion through several years, 0 to 1, 1 to 2, 2 to 3, ..., $(n-1)$ to n, and finally from n back to 0, Simple aggregative formula and the simple geometric mean of relatives formula satisfy this test. Weighted aggregative formula and weighted geometric mean of relatives formula satisfy this test, it constant weights are used for all time periods.

Example 17.23 *Using the following data, verify that Laspeyres' formula does not satisfy Time Reversal Test:*

	1979		*1980*	
Commodity	*Price*	*Quantity*	*Price*	*Quantity*
Rice	*32*	*50*	*30*	*50*
Barley	*30*	*35*	*25*	*40*
Maize	*16*	*55*	*18*	*50*

Solution Using *Laspeyres' Price Index* formula and omitting the factor 100,

Index Number for 1980 with base 1979 $(I_{0n}) = \dfrac{\Sigma\, p_n q_0}{\Sigma\, p_0 q_0}$

Interchanging the suffixes 0 and n,

Index Number for 1979 with base 1980 $(I_{n0}) = \dfrac{\Sigma\, p_0 q_n}{\Sigma\, p_n q_n}$

[Note that we have to calculate all 4 combinations of p and q, viz, p_0q_0, p_nq_0, p_0q_n, p_nq_n.]

Table 17.17 Calculations for Laspeyres' Index

Commodity	p_0	q_0	p_n	q_n	p_0q_0	p_0q_0	p_nq_0	p_nq_n
Rice	32	50	30	50	1600	1600	1500	1500
Barley	30	55	25	40	1050	1200	875	1000
Maize	16	55	18	50	880	800	990	900
Total	—	—	—	—	3530	3600	3365	3400

Substituting the values,

$$I_{0n} = \frac{3365}{3530}, \quad I_{n0} = \frac{3600}{3400}$$

We find that $\quad I_{0n}, I_{n0} = \dfrac{3365}{3530} \times \dfrac{3600}{3400} \neq 1$

This verifies that Laspeyres' formula does not satisfy Time Reversal Test.

Example 17.24 *With the help of the data of Example 17.23, calculate Price Index number using Fisher's formla and show that it satisfies the Time Reversal Test.* [C.U., B.Com(Hons.) '83]

Solution (Calculations are shown in Table 17.17).

(I) *Price Index Number for year 1980 with base 1979:*

$$\text{Laspeyres' Index} = \frac{\Sigma\, p_n q_0}{\Sigma\, p_0 q_0} = \frac{3365}{3530}$$

$$\text{Paasche's Index} = \frac{\Sigma\, p_n q_n}{\Sigma\, p_0 q_n} = \frac{3400}{3600}$$

$$\therefore \quad \text{Fisher's Ideal Index } (I_{0n}) = \sqrt{\text{Laspeyres'} \times \text{Paasche's Index}}$$

$$= \sqrt{\frac{3365}{3530} \times \frac{3400}{3600}} \qquad \text{(i)}$$

(II) Interchanging the suffixes *o* and *n* in the above formulae, *Price Index Number for year 1979 with base 1980:*

$$\text{Laspeyres' Index} = \frac{\Sigma\, p_0 q_n}{\Sigma\, p_0 q_n} = \frac{3600}{3400}$$

$$\text{Paasche's Index} = \frac{\Sigma\, p_0 q_0}{\Sigma\, p_n q_0} = \frac{3530}{3365}$$

$$\therefore \quad \text{Fisher's Ideal Index } (I_{n0}) = \sqrt{\text{Laspeyres'} \times \text{Paasche's Index}}$$

$$= \sqrt{\frac{3600}{3400} \times \frac{3530}{3365}} \qquad \text{(ii)}$$

Multiplying (i) and (ii),
Fisher's Index (I_{0n}) × Fisher's Index (I_{n0})

$$= \sqrt{\frac{3365}{3530} \times \frac{3400}{3600} \times \frac{3600}{3400} \times \frac{3530}{3365}} = \sqrt{1} = 1$$

Using Fisher's formula, we find thta $I_{0n}.I_{n0} = 1$. This verifies that Fisher's formula satisfies Time Reversal Test.

Example 17.25 *Calculate the quantity index number using Fisher's formula for the following data and show that it satisfies the Time Reversal Test.*

Commodity	*1973*		*1974*	
	Price	*Quantity*	*Price*	*Quantity*
A	*6*	*70*	*8*	*120*
B	*8*	*90*	*10*	*100*
C	*12*	*140*	*16*	*280*

[C.U., M.Com. '81]

Solution Let us take 1973 as base year and 1974 as current year.

Table 17.18 Calculations for Quantity Index

Commodity	p_0	q_0	p_n	q_n	q_0p_0	q_0p_n	q_np_0	q_np_n
A	6	70	8	120	420	560	720	960
B	8	90	10	100	720	900	800	1000
C	12	140	16	280	1680	2240	3360	4480
Total	—	—	—	—	2820	3700	4880	6440

(I) *Quantity Index* Number for year 1974 with 1973 as base:

$$\text{Laspeyres's Index} = \frac{\Sigma\, q_n p_0}{\Sigma\, q_0 p_0} = \frac{4880}{2820}$$

$$\text{Paasche's Index} = \frac{\Sigma\, q_n p_n}{\Sigma\, q_0 p_n} = \frac{6440}{3700}$$

$$\text{Fisher's Ideal Index } (I_{0n}) = \sqrt{\frac{4880}{2820} \times \frac{6440}{3700}} \qquad \text{(i)}$$

(II) Interchanging the suffixes 0 and *n*,
Quantity Index Number for year 1973 with 1974 as base:

$$\text{Laspeyres' Index} = \frac{\Sigma\, q_0 p_n}{\Sigma\, q_n p_n} = \frac{3700}{6440}$$

$$\text{Paasche's Index} = \frac{\Sigma\, q_0 p_0}{\Sigma\, q_n p_0} = \frac{2820}{4880}$$

$$\text{Fisher's Ideal Index } (I_{n0}) = \sqrt{\frac{3700}{6400} \times \frac{2820}{4880}} \qquad \text{(ii)}$$

Multiplying (i) and (*ii*),

$$I_{0n}, I_{n0} = \sqrt{\frac{4880 \times 6440}{2820 \times 3700} \times \frac{3700 \times 2820}{6440 \times 4880}} = 1$$

This verifies that the quantity index number using Fisher's formula satisfies Time Reversal test.

[**Note:** Examples 17.24 and 17.25 both verify that Fisher's formula satisfies Time Reversal test. The former uses Price Index, and the latter uses Quantity Index.]

Example 17.26 *(a) Using the following data, show that Paasche's formula does not satisfy the Factor Reversal Test.*

Commodity	*Prince (Rs) per unit*		*Number of units*	
	Base period	*Current period*	*Base period*	*Current period*
A	*6*	*10*	*50*	*56*
B	*2*	*2*	*100*	*120*
C	*4*	*6*	*60*	*60*
D	*10*	*12*	*30*	*24*
E	*8*	*12*	*40*	*36*

(b) Also use the same data to show that Fisher's Ideal formula satisfies the Factor Reversal Test. [I.C.W.A. Dec. '74]

Solution The factor reversal test may be represented in symbols as $P_{0n}.Q_{0n}$ = *Value Ratio.*

Table 17.19 Calculations for Factor Reversal Test

Commodity	p_0	p_n	q_0	q_n	p_0q_0	p_0q_n	p_nq_0	p_nq_n
A	6	10	50	56	300	336	500	560
B	2	2	100	120	200	240	200	240
C	4	6	60	60	240	240	360	360
D	10	12	30	24	300	240	360	288
E	8	12	40	36	320	288	480	432
Total	—	—	—	—	1360	1344	1900	1880

(a) Using *Paasche's formula* and omitting the factor 100,

$$\text{Price Index } (P_{0n}) = \frac{\Sigma p_n q_n}{\Sigma p_0 q_n} = \frac{1880}{1344}$$

Interchanging p and q,

$$\text{Quantity Index } (Q_{0n}) = \frac{\Sigma q_n p_n}{\Sigma q_0 p_n} = \frac{\Sigma p_n q_n}{\Sigma p_n q_0} = \frac{1880}{1900}$$

$$\text{Also,} \quad \text{Value Ratio} = \frac{\Sigma p_n q_n}{\Sigma p_0 q_0} = \frac{1880}{1360}$$

We find that $P_{0n}.Q_{0n} \neq$ *Value Ratio*. This shows that Paasche's formula does not satisfy that Factor Reversal test.

(b) Using *Fisher's Ideal formula* and omitting the factor 100,

$$\text{Price Index } (P_{0n}) = \sqrt{\frac{\Sigma p_n q_0}{\Sigma p_0 q_0} \times \frac{\Sigma p_n q_n}{\Sigma p_0 q_n}} = \sqrt{\frac{1900 \times 1880}{1360 \times 1344}}$$

Interchanging p and q,

$$\text{Quantity Index } (Q_{0n}) = \sqrt{\frac{\Sigma q_n p_0}{\Sigma q_0 p_0} \times \frac{\Sigma q_n p_n}{\Sigma q_0 p_n}} = \sqrt{\frac{1344 \times 1880}{1360 \times 1900}}$$

$$\text{Value Ratio} = \frac{\Sigma p_n q_n}{\Sigma p_0 q_0} = \frac{1880}{1360}$$

$$\text{Now,} \quad P_{0n} \times Q_{0n} = \sqrt{\frac{1900 \times 1880}{1360 \times 1344} \times \frac{1344 \times 1880}{1360 \times 1900}}$$

Cancelling the common factors under the square-root sign and simplifying

$$P_{0n} \times Q_{0n} = \frac{1880}{1360} = \text{Value Ratio}$$

This shows that Fisher's Ideal formula satisfies Factor Reversal Test.

Example 17.27 *Show that neither Laspeyres' formula nor Paasche's formula obeys time reversal or factor reversal tests of index numbers.*

[C.U., M.Com. '70; B.A(Econ) '73, '75; B.U., B.A.(Econ) '70; I.C.W.A., June '74; D.S.W. '70, '73]

Solution

(I) *Time Reversal Test* may be symbolically expressed as $I_{0n} \times I_{n0} = 1$.

(a) Using *Laspeyres' Price Index* formula and omitting the factor 100,

Index Number for year n with base year 0 (I_{0n}) = $\dfrac{\Sigma\, p_n q_0}{\Sigma\, p_0 q_0}$

Interchanging the suffixes 0 and n,

Index Number for year 0 with base year n (I_{n0}) = $\dfrac{\Sigma\, p_0 q_n}{\Sigma\, p_n q_n}$

$$\therefore \quad I_{0n} \times I_{n0} = \frac{\Sigma\, p_n q_0}{\Sigma\, p_0 q_0} \times \frac{\Sigma\, p_0 q_n}{\Sigma\, p_n q_n} \neq 1.$$

Thus, Laspeyres' formula does *not* obey Time Reversal Test.

(b) Using *Paasche's Price Index* formula and omitting the factor, 100,

$$I_{0n} = \frac{\Sigma\, p_n q_n}{\Sigma\, p_0 q_n}$$

Interchanging the suffixes 0 and n, $I_{n0} = \dfrac{\Sigma\, p_0 q_0}{\Sigma\, p_n q_0}$

$$\therefore \quad I_{0n} \times I_{n0} = \frac{\Sigma\, p_n q_n}{\Sigma\, p_0 q_n} \times \frac{\Sigma\, p_0 q_0}{\Sigma\, p_n q_0} \neq 1$$

Thus, Paasche's formula also does *not* obey Time Reversal Test.

(II) *Factor Reversal Test* may be expressed as

$$P_{0n} . Q_{0n} = \frac{\Sigma\, p_n q_n}{\Sigma\, p_0 q_0}$$

(c) Using *Laspeyres'* formula, and omitting the factor 100,

Price Index for year n with base year 0 (P_{0n}) = $\dfrac{\Sigma\, p_n q_0}{\Sigma\, p_0 q_0}$

Interchanging p and q, *Quantity Index* for year n with base year 0

$$Q_{0n} = \frac{\Sigma\, q_n p_0}{\Sigma\, q_0 p_0} = \frac{\Sigma\, p_0 q_n}{\Sigma\, p_0 q_0}$$

Multiplying, we have by Laspeyres' formula

$$P_{0n} \times Q_{0n} = \frac{\Sigma\, p_n q_0}{\Sigma\, p_0 q_0} \times \frac{\Sigma\, p_0 q_n}{\Sigma\, p_0 q_0} \neq \frac{\Sigma\, p_n q_n}{\Sigma\, p_0 q_0}$$

i.e. $$P_{0n} \times Q_{0n} \neq \frac{\Sigma\, p_n q_n}{\Sigma\, p_0 q_0}$$

This proves that Laspeyres' formula does *not* satisfy Factor Reversal Test.

(d) Applying *Paasche's* formula, it will be found that

$$P_{0n} \times Q_{0n} = \frac{\Sigma\, p_n q_n}{\Sigma\, p_0 q_n} \times \frac{\Sigma\, p_n q_n}{\Sigma\, p_0 q_0} \neq \frac{\Sigma\, p_n q_n}{\Sigma\, p_0 q_0}$$

This proves that Paasche's formula does *not* satisfy factor reversal test.

Example 17.28 *Examine whether Fisher's ideal index formula satisfies the Time reversal and Factor reversal tests.*

[I.C.W.A., Dec. '75; C.U., M.Com. '70; B.Com.(Hons.) '82; B.U., B.A.(Econ) '70, '73; C.U., B.A.(Econ). '73, '75, '81]

Solution Using Fisher's "ideal" index formula, Price Index for year *n* with base year 0 is given by (omitting the factor 100)

$$I_{0n} = \sqrt{\frac{\Sigma\, p_n q_0}{\Sigma\, p_0 q_0} \cdot \frac{\Sigma\, p_n q_n}{\Sigma\, p_0 q_n}}$$

Interchanging the suffixes 0 and *n*, Price Index for year 0 with base year *n* is

$$I_{no} = \sqrt{\frac{\Sigma\, p_0 q_n}{\Sigma\, p_n q_n} \cdot \frac{\Sigma\, p_0 q_0}{\Sigma\, p_n q_0}}$$

Multiplying we get

$$I_{0n} \times I_{n0} = \sqrt{\frac{\Sigma\, p_n q_0}{\Sigma\, p_0 q_0} \cdot \frac{\Sigma\, p_n q_n}{\Sigma\, p_0 q_n}} \sqrt{\frac{\Sigma\, p_0 q_n}{\Sigma\, p_n q_n} \cdot \frac{\Sigma\, p_0 q_0}{\Sigma\, p_n q_0}}$$

$$= \sqrt{\frac{\Sigma\, p_n q_0}{\Sigma\, p_0 q_0} \cdot \frac{\Sigma\, p_n q_n}{\Sigma\, p_0 q_n} \cdot \frac{\Sigma\, p_0 q_n}{\Sigma\, p_n q_n} \cdot \frac{\Sigma\, p_0 q_0}{\Sigma\, p_n q_0}} = \sqrt{1} = 1$$

(since all terms cancel one another)

i.e. $I_{0n} \times I_{n0} = 1.$

This shows that Fisher's ideal formula obeys Time Reversal test.

In order to apply Factor Reversal test, we see that Price index by Fisher's ideal formula is

$$P_{0n} = \sqrt{\frac{\Sigma\, p_n q_0}{\Sigma\, p_0 q_0} \cdot \frac{\Sigma\, p_n q_n}{\Sigma\, p_0 q_n}}$$

Interchanging *p* and *q*, Quantity Index is given by

$$Q_{0n} = \sqrt{\frac{\Sigma\, q_n p_0}{\Sigma\, q_0 p_0} \cdot \frac{\Sigma\, q_n p_n}{\Sigma\, q_0 p_n}}$$

$$= \sqrt{\frac{\Sigma\, p_0 q_n}{\Sigma\, p_0 q_0} \cdot \frac{\Sigma\, p_n q_n}{\Sigma\, p_n q_0}}, \text{ (rearranging the factors } p, q)$$

Multiplying $P_{0n} = Q_{0n}$, we have

$$P_{0n} \times Q_{0n} = \sqrt{\frac{\Sigma\, p_n q_0}{\Sigma\, p_0 q_0} \cdot \frac{\Sigma\, p_n q_n}{\Sigma\, p_0 q_n}} \sqrt{\frac{\Sigma\, p_0 q_n}{\Sigma\, p_0 q_0} \cdot \frac{\Sigma\, p_n q_n}{\Sigma\, p_n q_0}}$$

$$= \sqrt{\frac{\Sigma\, p_n q_n}{\Sigma\, p_0 q_0} \cdot \frac{\Sigma\, p_n q_n}{\Sigma\, p_0 q_n} \cdot \frac{\Sigma\, p_0 q_n}{\Sigma\, p_0 q_o} \cdot \frac{\Sigma\, p_n q_n}{\Sigma\, p_n q_0}}$$

$$= \sqrt{\frac{\Sigma\, p_n q_n}{\Sigma\, p_0 q_0} \cdot \frac{\Sigma\, p_n q_n}{\Sigma\, p_0 q_0}} = \frac{\Sigma\, p_n q_n}{\Sigma\, p_0 q_0}$$

i.e. $$P_{0n} \times Q_{0n} = \frac{\Sigma\, p_n q_n}{\Sigma\, p_0 q_0}.$$

This shows that Fisher's ideal index formula obeys Factor Reversal test.

Example 17.29 *Show that the index number obtained by averaging the unweighted price relatives does not satisfy "time reversal test".*

Solution The price relative for year n with base year 0 in given by the formula (omitting the factor 100)

$$\text{Price Relative} = p_n/p_0$$

If there are k items in the series, the unweighted A.M. of Price Relatives Index is given by

$$I_{0n} = \Sigma\,(\text{Price Relatives}) \div k = \Sigma\,(p_n/p_0) \div k$$

Interchanging the suffixes 0 and n,

$$I_{n0} = \Sigma(p_0/p_n) \div k$$

$$\therefore \quad I_{0n} \times I_{n0} = \frac{1}{k^2}\,\Sigma\,(p_n/p_0)\,.\,\Sigma\,(p_0/p_n)$$

$$= \frac{1}{k^2}\left(\frac{p_n'}{p_0'} + \frac{p''}{p_0''} + \dots + \frac{p_n^{(k)}}{p_0^{(k)}}\right)\left(\frac{p_0'}{p_n} + \frac{p_0''}{p_n'} + \dots + \frac{p_0^{(k)}}{p_n^{(k)}}\right) \neq 1$$

where $p_0', p_0'', \dots\dots, p_0^{(k)}$ represent the prices of items 1, 2, ... k respectively in the year o, and $p_n', p_n'', \dots\dots, p_n^{(k)}$ represent the corresponding prices in the year n. Thus

$$I_{0n}\,.\,I_{n0} \neq 1$$

This result shows that the index number obtained by averaging the unweighted price relatives *does not* satisfy the time reversal test.

Example 17.30 *Test whether the index number due to Walsh given by*

$$I = \frac{\Sigma p_1\sqrt{q_0 q_1}}{\Sigma p_0\sqrt{q_0 q_1}} \times 100$$

(symbols have their usual meaning) satisfies time reversal test.

[C.U., M.Com. '80]

Solution Using Walsh's formula and omitting the factor 100. Index Number for year 1 with

base year 0 $(I_{01}) = \dfrac{\Sigma p_1\sqrt{q_0 q_1}}{\Sigma p_0\sqrt{q_0 q_1}}$

Interchanging the suffixes 0 and 1,
Index Number for year 0 with the base year 1 (I_{10})

$$= \frac{\Sigma p_0\sqrt{q_1 q_0}}{\Sigma p_1\sqrt{q_1 q_0}} = \frac{\Sigma p_0\sqrt{q_0 q_1}}{\Sigma p_1\sqrt{q_0 q_1}}$$

Multiplying, we find that $I_{01}, I_{10} = 1$. This shows that Walsh's formula satisfies Time Reversal test.

17.6 CHAIN BASE METHOD

Example 17.31 *Describe the chain base method of construction of index numbers and discuss its advantages and disadvantages as compared with the fixed base method.*

[C.U., M.Com. '79; B.Com. (Hons) '83 ; B.A. (Econ) '72, '75, '78; I.C.W.A., Jan '71, '73; Dec. '74, '75, '76, '81]

Solution There are two methods of construction of index numbers depending on the nature of base period employed: (i) *Fixed Base method,* and (ii) *Chain Base method.* Most of the index numbers in common use are of the fixed base type, where a fixed period is chosen as base and the index number for any given year is calculated by direct reference to this fixed base period. This fixed base index for any year is no, therefore, affected by changes in price or quantity in any other year. "It is however considered that the net changes in any given year are the result of gradual changes that have taken place during the past years. This idea is reflected in Chain Base Index" numbers.

For the construction of index numbers by the chain base method, using an appropriate index number formula (say, Laspeyres' formula), it is first necessary to compute index numbers for all the years, always using the preceding year as base. thesz are known as *Link Index.*

Link Index = Index Number with preceding period as base.

For example, using Laspeyres' formula,

$$\text{Link index for year 1 } (I_{01}) = \frac{\Sigma p_1 q_0}{\Sigma p_0 q_0} \times 100;$$

$$\text{Link index for year 2 } (I_{12}) = \frac{\Sigma p_2 q_1}{\Sigma p_1 q_1} \times 100;$$

$$\text{Link index for year 3 } (I_{23}) = \frac{\Sigma p_3 q_2}{\Sigma p_2 q_2} \times 100;$$

$$\text{Link index for yera 4 } (I_{34}) = \frac{\Sigma p_4 q_3}{\Sigma p_3 q_3} \times 100; \text{ etc.}$$

The link indices I_{01}, I_{12}, I_{23}, I_{34}, ..., are then multiplied successively (called *chaining process*) in order to relate them to a common base. The progressive products, expressed as percentages, give the requred index numbers by the chain base method. These are called *Chain Index Numbers* or *Chain Base Index Numbers.* Thus, a chain index number is the product of several index numbers, each calculated with the preceding period as base.

The chain index numbers with reference to year *o* are (omitting the factor 100 from each index)

$$\begin{aligned} I'_{01} &= I_{01} \\ I'_{02} &= I_{01} \times I_{12} \\ I'_{03} &= I_{01} \times I_{12} \times I_{23} \\ I'_{04} &= I_{01} \times I_{12} \times I_{23} \times I_{34} \end{aligned}$$

(Here I' is used for Chain Index and I for index of the base type).

The chain index number I'_{0n} will not in general be equal to the corresponding fixed base index number I_{0n} unless the formula employed satisfies the circular test of index numbers.

Advantages—

(1) The chain base index is more realistic in nature than the fixed base index, since the effects of all intermediate years are taken into consideration.

(2) The chain base method enables comparison between two adjacent time periods through the link indices. This is far more useful in business and commerce than the indirect comparison through a remote fixed base.

(3) The method also makes it possible the dropping of obsolete items and inclusion of new ones. The necessity of substituting certain items in the existing list is frequently felt when computing a series of index numbers over a long period of time, because of the changing habits of people and new commodities coming into use. If the fixed base method were used, the entire series of index numbers will have to be recalculated, when the list of commodities is altered.

Disadvantages—

(1) The significance of index numbers calculated by the chain base method is diffiult to understand.
(2) The calculations are heavier in the chain base method.
(3) If an error is committed in the calculation of any link index number the entire series of chain base index number will be wrong. Also, if data for even one year are missing, the subsequent chain index numbers cannot be calculated.
(4) Chain base index numbers are really suitable for short periods only. If changes in the list of items are frequent, the index may in the later years reflect quite different movements than the figures in the earlier periods.

Example 17.32 *Given the following information, construct chain index numbers (Base 1962 = 100) for the years 1963–67:*

Year	*1963*	*1964*	*1965*	*1966*	*1967*
Link Index	*103*	*98*	*105*	*112*	*108*

[I.C.W.A., July '72]

Solution

Table 17.20 Calculations for Chain Index

Year	*Link Index*	*Chain Index (Base 1962 = 100)*	
$Y_0 = 1962$	100	100	
$Y_1 = 1963$	$I_{01} = 103$	$I'_{01} = 100 \times 1.03$	= 103
$Y_2 = 1964$	$I_{12} = 98$	$I'_{02} = 100 \times (1.03 \times .98)$	= 101
$Y_3 = 1965$	$I_{23} = 105$	$I'_{03} = 100 \times (1.03 \times .98 \times 1.05)$	= 106
$Y_4 = 1966$	$I_{34} = 112$	$I'_{04} = 100 \times (1.03 \times .98 \times 1.05 \times 1.12)$	= 119
$Y_5 = 1967$	$I_{45} = 108$	$I'_{05} = 100 \times (1.03 \times .98 \times 1.05 \times 1.12 \times 1.08)$	= 128

Ans. 103, 101, 106, 119, 128

17.7 COST OF LIVING INDEX NUMBERS

Example 17.33 *What is a 'cost of living index number'? Describe the steps for its construction, with special reference to the choice of items, and the determination of weights.*

[I.C.W.A., June '78, '79, Dec. '73 and '75 C.U., B.A. (Econ) '73, '76; D.M. (Suppl.) '77; D.S.W., '75, '76, '77, '78]

Solution *Cost of Living Index* numbers are special-purpose index numbers which are designed to measure the relative change in the cost level for maintaining similar standard of living in two different situations. These are generally intended to represent the average changes in prices over time, paid by the ultimate consumer for a specified group of goods and services; and hence are also called *Consumer Price Index* numbers. Generally, the consumption pattern varies with the class of people and the geographical area covered. Hence cost of living index (C.L.I.) numbers must always relate to a specified class of people and a specified geographical area.

The steps in the construction of a Cost of Living Index are as follows:

(1) The first step is to decide on the class of people for whom the index number is intended. It is extremely important to define this in clear terms.

(2) The next step is to conduct a *'family budget enquiry'* in the base period relating to the class of people concerned, by the process of random sampling. This would give us information regarding the nature and quality of goods consumed by an *'average family'* and also enable detemination of weights for computing the index. Only important items among those which are used by the majority of th class of people are included in the construction of a cost of living index.

(3) The items of expenditure are classfied in certain major *Groups*, e.g. (i) Food, (ii) Clothing, (iii) Fuel and light, (iv) Housing, and (v) Miscellaneous. These major groups are further divided into smaller groups and sub-groups, so that the items are individually mentioned. (See Example 17.36).

(4) Arrangements should be made to collect retail prices of the items at regular intervals of time from important local markets. Price quotations are taken at least once a week.

(5) For each item there will be a number of price quotations covering different qualities and markets. The simple average of price relatives of the different quotations is taken as the price relative for the particular item.

(6) A separate index number is then computed for each Group, using Laspeyres' formula in the form of weighted average of price relatives.

$$\text{Group Index } (I) = \frac{\sum w\left(\dfrac{p_n}{p_0} \times 100\right)}{100}, \text{ where } w = \frac{p_0 q_0}{\Sigma p_0 q_0} \times 100$$

Thus, in the construction of a Group Index, the weight (w) of an item is the percentage expenditure of an 'everage family' on that itemex in relation to the total expenditure in the Group, as obtained from the family budget enquiry.

(7) The weighted average of group index numbers gives the final Cost of Living Indx number. (See Example 17.34 and 36).

$$\text{Cost of Living Index} = \frac{\Sigma\, IW}{100}$$

The weight (W) of a group index is the percentage of total expenditure of an average family spent on that group as shown by the family budget enquiry.

(8) Cost of living index numbers are generally constructed for each week. The average of the weekly index numbers is taken as the index number for a month. The average of monthly index numbers gives the cost of living index for the whole year.

Uses—

(i) Cost of Living Index numbers are primarily used for the calculation of dearness allowance, so that the samer standard of living as in the base year can be maintained (Example 17.37).

(ii) The reciprocal of C.L.I. may be used to measure the purchasing power of money (Example 17.43).

(iii) Cost of living index numbers are also used to find "real wage" by the process of 'deflation' (Example 17.42).

Labour Bureau Consumer Price Index Number for Working Class (New Series)

Consumer Price Index Numbers (New Series) for industrial workers are being compiled by the Labour Bureau under the Govt. of India for 50 different centre (e.g. Ahmedabad, Kanpur, Sholapur, Cuttack, etc.) located throughout India with base year 1960 = 100.

For each centre the index is based on about 100 items of expenditure classified under 6 major *groups* and several *sub-groups*. Weekly price quotations are obtained, through State Govts., for each item from some popular retail shops in each of a few selected markets located in the centre. For some items like tea, toilet, soap, etc. quotations are obtained montly and for house rent *ad hoc* surveys are arranged periodically. Laspeyres' formula, in the form of weighted average of price relatives is used for the the construction of consumer price index numbers. The weights have been determined on the basis of *family living surveys* among the working class people conducted at the centres during the period 1958–59. The average of the weekly index numbers gives the monthly index and the average of the monthly indices gives the Consumer Price Index number for a year. The numbers are published in *Indian Labour Journal*, a monthly magazine of Labour Bureau, Govt. of India.

Example 17.34 *From the table of group index numbers and group weights given below, calculate the cost of living index number.*

Group	*Index Number*	*Weight*
Food	*428*	*45*
Clothing	*240*	*15*
Fuel and Light	*200*	*8*
House Rent	*125*	*20*
Miscellaneous	*170*	*12*

[C.U., B. Com. '78]

Solution

Table 17.21 Calculations for C.L.I

Group	*Group Index (I)*	*Weight (W)*	*IW*
Food	428	45	19,260
Clothing	240	15	3,600
Fuel and Light	200	8	1,600
House Rent	125	20	2,500
Miscellaneous	170	12	2,040
Total	—	100	29,000

$$\text{Cost of Living Index} = \frac{\Sigma IW}{\Sigma W} = \frac{29{,}000}{100} = 290.$$

Example 17.35 *In calculating a certain cost of living index number the following weights were used : Food 15, Clothing 3, Rent 4, Fuel and light 2, Miscellaneous 1. Calculate the index for a date when the average percentage increases in prices of items in the various groups over the base period were 32, 54 , 47, 78 and 58 respectively.*

[I.C.W.A., Jan. '73]

Solution

Table 17.22 Calculations for Average Price Increase

Group	*Average % Increase in Price (i)*	*Weight (w)*	*iw*
Food	32	15	480
Clothing	54	3	162
Rent	47	4	188
Fuel and light	78	2	156
Miscellaneous	58	1	58
Total	—	25	1044

Average percentage increase for all groups taken together

$$= \frac{\Sigma iw}{\Sigma w} = \frac{1044}{25} = 41.76$$

Cost of living index = 100 + 41.76 = 141.76.

Example 17.36 *The data below show the percentage increases in price of a few selected food items and the weights attached to each of them. Calculate the index number for the Food group.*

Food Items	*Rice*	*Wheat*	*Dal*	*Ghee*	*Oil*	*Spices*	*Milk*	*Fish*	*Vegetables*	*Refreshments*
Weight	*33*	*11*	*8*	*5*	*5*	*3*	*7*	*9*	*9*	*10*
Percentage Increase in Price	*180*	*202*	*115*	*212*	*175*	*517*	*260*	*426*	*332*	*279*

Using the above Food index and the information given below, calculate the cost of living index number.

Group	*Food*	*Clothing*	*Fuel and Light*	*Rent and Rates*	*Miscellaneous*
Index		*310*	*220*	*150*	*300*
Weight	*60*	*5*	*8*	*9*	*18*

[I.C.W.A., Jan. '72]

Solution The average percentage increase in food price plus 100 will give the index number for Food group.

Table 17.23 Calculations for Food Index

Food Items	*Weight (w)*	*Percentage Increase (i)*	*iw*
Rice	33	180	5,940
Wheat	11	202	2,222
Dal	8	115	920

(*Contd*)

Food Items	Weight (w)	Percentage Increase (i)	iw
Ghee	5	212	1,060
Oil	5	175	875
Spices	3	517	1,551
Milk	7	260	1,820
Fish	9	426	3,834
Vegetables	9	332	2,988
Refreshments	10	279	2,790
Total	100	—	24,000

$$\text{Average Percentage increase in food price} = \frac{\Sigma iw}{\Sigma w} = \frac{24{,}000}{100} = 240$$

$\therefore$ Food Index = 240 + 100 = 340

Table 17.24 Calculation for Cost of Living Index

Group	Index (I)	Weight (W)	IW
Food	340	60	20,400
Clothing	310	5	1,550
Fuel and Light	220	8	1,760
Rent and Rates	150	9	1,350
Miscellaneous	300	18	5,400
Total	—	100	30,460

$$\text{Cost of Living Index} = \frac{\Sigma IW}{\Sigma W} = \frac{30,460}{100} = 304.6.$$

Example 17.37 *The menial class cost of living index numbers for a certain city in two years 1968 and 1972 and 220 and 288 respectively. Suppose a menial worker was getting Rs 90 per month in 1968 and Rs 110 in 1972. Should he receive an extra allowance in 1972 to maintain his standard of living of 1968; and if so, how much?*

Solution In order to maintain the same standard of living, the salary should be proportionate of the C.L.I. This means that in 1972 he should get (288/220) times his salary in 1968, i.e.

Rs 90 × (288/220) = Rs 117.82.

So, he should get an extra allowance of Rs 7.82 over his pay of Rs 110.

Example 17.38 *The relative importance of the following eight groups of family expenditure was found to be—food 348, rent 88, clothing 97, fuel and light 65, housebold durable goods 72 miscellaneous goods 35, services 79, drink and tobacco 217. The corresponding increase in price for October 1960, gave the following percentages —25, 1, 22, 18, 14, 13,? and 4. Calculate the percentage increase in the group—'services', if the percentage for the whole group is 15.278 per cent.*

Solution

Table 17.25 Calculations for Index Number

	Group	Percentage Increase in Price (i)	Weight (w)	iw
1.	*Food*	25	348	8,700
2.	*Rent*	1	88	88
3.	*Clothing*	22	97	2,134
4.	*Fuel and light*	18	65	1,170
5.	*Household durable goods*	14	71	994
6.	*Miscellaneous goods*	13	35	455
7.	*Services*	? = x (say)	79	$79x$
8.	*Drink and tobacco*	4	217	868
	Total	—	1000	$14{,}409 + 79x$

Average percentage increase in price for the whole group = $\frac{\Sigma\, iw}{\Sigma\, w}$

or, $$15.278 = \frac{14{,}409 + 79x}{1{,}000}$$

or, $$14{,}409 + 79x = 15.278 \times 1000 = 15{,}278$$

or, $$79x = 15{,}278 - 14{,}409 = 869$$

$\therefore$ $$x = 869 \div 79 = 11$$ *Ans.* 11

Example 17.39 *When the cost of tobacco was increased by 50%, a certain hardened smoker, who maintained his former scale of consumption, said that the rise has increased his cost of living by 5%. What percentage of his cost of living was due to buying tobacco before the change in price?* [D.S.W. 76; I.C.W.A. June '83]

Solution Let x be the required percentage. The percentage of total expenditure under each group before the change in price (i.e. in the base period) represents the weight of the group. Let us suppose that the items of expenditure are classified under two groups—(1) Tobacco, (2) Other Items. If the toatal weight of the two groups combinded be taken as 100, then the weight for 'Tobacco' will be x and that for 'Other Items' group will be $100 - x$. The data are shown in the following table:

Table 17.26 Calculations for Index Number

Group	Weight (w)	% Increase in Price (i)	wi
(1) Tobacco	x	50	$50x$
(2) Other Items	$100 - x$	0	0
Total	100	—	$50x$

Average percentage increase for all items = $\Sigma iw/\Sigma w$

i.e. $5 = 50x/100$; or, $50x = 500$; $\therefore$ $x = 10$ *Ans.* 10

Example 17.40 *Owing to change in prices the consumer price index of working class in a certain area rose in a month by one quarter of what it was before to 225. The index of food became 252 from 198, that of clothing from 185 to 205, that of fuel and lighting from 175 to 195, and that of miscellaneous from 138 to 212. The index of rent however, remained unchanged at 150. It was known that the weights of clothing, rent, and fuel and lighting were the same. Find out the exact weights of all the groups.*

Solution If I denotes the original consumer price index, then

$$I + \frac{1}{4}I = 225; \quad \text{or,} \quad \frac{5}{4}I = 225; \quad \therefore \quad I = 225 \times \frac{4}{5} = 180.$$

Again, let x represent the common weight of Clothing, Rent, and Fuel and lighting; y represent the weight of Food; and the total weight be 100. Therefore, the weight of Miscellaneous group must be $100 - 3x - y$. The information is shown below:

Table 17.27 Group Indices and Weights for C.L.I.

Group	*Group Index of Preceding Month (I)*	*Group Index of Current Month (I′)*	*Weight (W)*
Food ...	198	252	y
Clothing ...	185	205	x
Fuel and lighting ...	175	195	x
Rent ...	150	150	x
Miscellaneous ...	138	212	$(100 - 3x - y)$
All Items ...	180	225	110

$$\Sigma IW = 198y + 185x + 175x + 150x + 138(100 - 3x - y)$$
$$= 13800 + 96x + 60y.$$
$$\Sigma I'W = 252y + 205x + 195x + 150x + 212(100 - 3x - y)$$
$$= 21200 - 86x + 40y.$$

Consumer Price Index for the preceding month = $\Sigma IW/\Sigma W$

or, $180 = (13800 + 96x + 60y)/100$

or, $18000 = 13800 + 96x + 60y$

or, $96x + 60y = 4200$

$\therefore$ $8x + 5y = 350$ (i)

Consumer Price Index for the current month = $\Sigma I'W/\Sigma W$

or, $225 = (21200 - 86x + 40y)/100$

or, $22500 = 21200 - 86x + 40y$

or, $86x - 40y = -1300$

$\therefore$ $43x - 20y = -650$ (ii)

Solving (i) and (ii), we get $x = 10$, $y = 54$.

From Table 17.27, the weights of different groups are: Food 54, Clothing 10, Fuel and lighting 10, Rent 10, and Miscellaneous 16; Total 100.

Example 17.41 *The following data show that cost of living indices for the groups Food, Clothing, Fuel and light, House rent and Miscellaneous with their respective weights for inustrial workers in an Indian city in March 1972 (Base: 1960 = 100).*

Group	*Group Index*	*Weight*
Food	221.6	44.65
Clothing	164.8	5.47
Fuel and light	202.9	4.26
House rent	187.4	9.16
Miscellaneous	164.5	36.46

Shri Chakravarty's basic pay was Rs 460.00 in March 1972. According to an agreement with the workers in 1969, cost of living index upto 100 was merged with basic pay and for every four point rise in cost of living index there would be a rise of 3% of basic pay as Dearness Allowance. How much was the D.A. of Shri Chakravarty, an industrial worker? (The D.A. was calculated upto the nearest rupee).

[C.U., M.Com. '81]

Solution As in Example 17.34, the cost of living index number was calculated and found to be 194. Beyond 100, the cost of living index increased by 94; i.e. there has been a four-point rise in cost of living index 23 times. Therefore, D.A is now 3 × 23 = 69 per cent of his basic pay. D.A. = 69% of Rs 460 = $\frac{69}{100}$ × Rs 460 = Rs 317.

Example 17.42 *Given below are the average wages in rupees per hour for unskilled workers in a factory during the year 1975–80. Also shown in Consumer price Index for these years (taking 1975 as base year with Price Index 100). Determine the "real wages" of the workers during 1975–80 compared with their wages in 1975.*

Year	*1975*	*1976*	*1977*	*1978*	*1979*	*1980*
Consumer Price Index	*100*	*120.2*	*121.7*	*125.9*	*129.3*	*140.0*
Average Wage (Rs/hour)	*1.19*	*1.94*	*2.13*	*2.28*	*2.45*	*3.10*

[I.C.W.A., June '82]

Solution *Real Wage* in a given period is the equivalent of actual wage in terms of the base period money, i.e. the wage after eliminating the effect of change in consumer price index. It is obtained by "deflating" the actual wage received during the period by the consumer price index of that period.

$$\text{Real Wage} = \frac{\text{Actual Wage}}{\text{Consumer Price Index}} \times 100$$

[**Note:** From the given Consumer Price Index figures, it may be observed that so far as expenditure on cost of living is concerned, Rs 120.2 of 1976 are equivalent to Rs 100 of 1975. As such proportionately, the actual wage Rs 1.94 of 1976 is equivalent to $\frac{1.94}{120.2} \times 100$ = 1.61 Rs of 1975. This amount Rs 1.61 is the "real wage" for 1976 (Base 1975 = 100) corresponding to the actual wage Rs 1.94. The process of obtaining real wage after adjustment by the consumer price index is known as "deflation"]

Table 17.28 Calculation of Real Wages

Year	Consumer Price Index (Base 1975 = 100)	Actual Wages (Rs per hour)	Real Wages $= \frac{\text{Col. (3)}}{\text{Col. (2)}} \times 100$
(1)	(2)	(3)	(4)
1975	100	1.19	1.19
1976	120.2	1.94	$\frac{1.94}{120.2} \times 100 = 1.16$
1977	121.7	2.13	$\frac{2.13}{121.7} \times 100 = 1.75$
1978	125.9	2.28	$\frac{2.28}{125.9} \times 100 = 1.81$
1979	129.3	2.45	$\frac{2.45}{129.3} \times 100 = 1.89$
1980	140.0	3.10	$\frac{3.10}{140.0} \times 100 = 2.21$

Example 17.43 *The index numbers of wholesale prices for the years 1947–54 with 1947–49 = 100 are given below. Determine the wholesale purchasing power of a Rupee in term of 1954 Rupee in each of the given years:*

Year	*1947*	*'48*	*'49*	*'50*	*'51*	*'52*	*'53*	*'54*
Wholesale Price Index (1947–49 = 100)	*96.4*	*104.4*	*99.2*	*103.1*	*114.8*	*111.6*	*110.1*	*110.3*

Solution The significance of the given price index numbers is that on an average the items that could be purchased Rs 96.4 in 1947 cost Rs 110.3 in 1954. In other words, so far as wholesale purchases are concerned Rs 96.4 of 1947 are equivalent to Rs 110.3 of 1954. Hence, the purchasing power of a Rupee of 1947 is Rs (110.3 ÷ 96.4) of 1954. Thus, the purchasing power of a Rupee in any year in terms of 1954-Rupee will be given by

$$\text{Purchasing Power} = \frac{\text{Index Number for 1954}}{\text{Index Number for the year}}$$

Table 17.29 Calculations for Purchasing Power of Rupee

Year	Wholesale Price Index	Purchasing Power in Terms of 1954 Rupee
1947	96.4	110.3 ÷ 96.4 = 1.14
1948	104.4	110.3 ÷ 104.4 = 1.06
1949	99.2	110.3 ÷ 99.2 = 1.11
1950	103.1	110.3 ÷ 113.1 = 1.07
1951	114.8	110.3 ÷ 114.8 = 0.96
1952	111.6	110.3 ÷ 111.6 = 0.99
1953	110.1	110.3 ÷ 110.1 = 1.00
1954	110.3	110.3 ÷ 110.3 = 1.00

Example 17.44 *Explain how index number is used to measure the purchasing power of money.* [I.C.W.A., DEC. '81]

Solution The purchasing power of money (suppose, Rupee) is the relative quantity of goods available with the same amount of money spent in the base period. If the price index number goes down to 50, the significance is that the same amount of money will not be able to purchase double the quantity of goods which could be obtained in the base period. The purchaing power in $\frac{100}{50} = 2$. Again, if the price index goes up to 125, the same amount as spent in the base period will now fetch $\frac{100}{125} = 0.80$ or 80% of the volume of goods as obtainable in the base period. Thus we find that the purchasing power of money is inversely proportional to the price index. The purchasing power of money at a given period with reference to the base period may therefore be measured by the formula:

$$\text{Purchasing Power of money} = \frac{100}{\text{Price index number}}$$

A higher price index number means that the same amount of money can purchase less quantity of goods.

17.8 BIAS IN LASPEYRES' AND PAASCHE'S FORMULAE FOR C.L.I.

The various tests of index numbers viz. time reversal test, factor reversal test and circular test, are not the sole criteria for determining the suitability of a formula, and practical considerations often influence the choice of one formula in preference to another. It has been shown that none of Laspeyres' and Paasches' formulae obeys any of the tests of index number (Example 17.27). However, Laspeyres' formula has the superior advantage that it uses the base period quantities (for price index) as weights; so that the same set of weights can be used over a long period of time, until it becomes necessary to change the base. On the other hand, Paasche's formula, which uses the current period quantities, necessitates determination of weights every time the index number is calculated. It is for this reason that Laspeyres' formula is by far the most widely used in the construction of index numbers, especially in the form of weighted A.M. of price relatives (17.3.12)

$$\text{Laspeyres' Price Index} = \frac{\Sigma\, p_n q_0}{\Sigma\, p_0 q_0} \times 100 = \frac{\Sigma\left(\dfrac{p_n}{p_0} \times 100\right) p_0 q_0}{\Sigma\, p_0 q_0}$$

$$= \frac{\Sigma\left(\dfrac{p_n}{p_0} \times 100\right) \times w}{100} = \frac{\Sigma\,(\text{Price relative}) \times w}{100} \qquad (17.8.1)$$

where $w = \frac{p_0 q_0}{\Sigma\, p_0 q_0} \times 100$ is a pure *number* showing the value $(p_0 q_0)$ of each item in the base period expressed as a percentage of total value $(\Sigma p_0 q_0)$, and $\Sigma w = 100$.

Cost of Living Index number measures the ratio of money values required to maintain equal satisfaction for a particular class of people in two different situations. Let p_0, p_n denote the price per unit of a set of goods and services consumed, and q_0, q_n denote the actual quantities (number of units) consumed in the base and current periods respectively. If $q^{(n)}$ denotes the current period quantities which will produce equal satisfaction relative to base period quantities to consumption q_0, then the total money value of consumption is $\Sigma p_0 q_0$ in the base period and the amount necessary to produce the same satisfaction in the current period is $\Sigma p_n q^{(n)}$ [Note the distribution between q_n and $q^{(n)}$]. The true cost of living index is then

$$I = \frac{\Sigma\ p_n q^{(n)}}{\Sigma\ p_0 q_0} \times 100$$

The practical difficulty of calculating this index lies in determining the exact quantities $q^{(n)}$, which yield equal satisfaction as in the base period, especially because the consumption pattern varies with change in the real income level in the two periods.

$$\text{Laspeyres' formula } L = \frac{\Sigma\ p_n q_0}{\Sigma\ p_0 q_0} \times 100$$

measures roughly the cost of maintaining the base period rate of consumption at current period prices, compared with base period cost.

$$\text{Paasche's formula } P = \frac{\Sigma\ p_n q_n}{\Sigma\ p_0 q_n} \times 100$$

shows a comparison of the cost in the current period relative to what it would have cost if current period quantities were consumed in the base period. None of these two formulae measures the true index. They may only be used as approximations.

It is common experience that when prices increase, relatively smaller quantities are consumed, and cheaper articles are used in larger quantities. It is for this reason that in Laspeyres' formula (L) the numerator is slighly larger than that in the ture index I, making L large than I. Similarly, the denominator in Paasche's formula would be relatively larger than that in the true index I. Thus Laspeyres' formula has a positive bias and Paasche's formula has a negative bias. The concept can be clearly explained with the help of weighted co-efficient of correlation.

Example 17.45 *With the help of the concept of weighted co-efficient of correlation, show that Laspeyres' formula has an upward bias.*

Solution In a bivariate distribution with variables x and y if the pairs of observations (x_1, y_1), (x_2, y_2), ... (x_n, y_n) have weights f_1, f_2, ..., f_n respectively, the coeficient of correlation between the variables is

$$r = \frac{\text{Covariance}}{\sigma_x\ \sigma_y}$$

where

$$\text{Covariance} = \frac{\Sigma fxy}{N} - \left(\frac{\Sigma fx}{N}\right)\left(\frac{\Sigma fy}{N}\right),\ N = \Sigma f,$$

and σ_x, σ_y denote the standard deviations of x and y. Let the variable x represent *price relative*, y represent *quantity relative*, and the weight f represent *value in the base period,* i.e. (omitting the factors 100 from each relative)

$$x = \frac{p_n}{p_0}, \quad y = \frac{q_n}{q_0}, \quad f = p_0 q_0.$$

Then

$$\text{Covariance} = \frac{\Sigma\left(p_0q_0 \cdot \frac{p_n}{p_0} \cdot \frac{q_n}{q_0}\right)}{\Sigma p_0 q_0} - \frac{\Sigma\left(p_0q_0 \cdot \frac{p_n}{p_0}\right)}{\Sigma p_0 q_0} \cdot \frac{\Sigma\left(p_0q_0 \cdot \frac{q_n}{q_0}\right)}{\Sigma p_0 q_0}$$

$$= \frac{\Sigma p_n q_n}{\Sigma p_0 q_0} - \frac{\Sigma p_n q_0}{\Sigma p_0 q_0} \cdot \frac{\Sigma p_0 q_n}{\Sigma p_0 q_0}$$

$$= \frac{\Sigma p_n q_n}{\Sigma p_0 q_n} \frac{\Sigma q_n p_0}{\Sigma q_0 p_0} - \frac{\Sigma p_n q_0}{\Sigma p_0 q_0} \cdot \frac{\Sigma q_n p_0}{\Sigma q_0 p_0}$$

$$= P_p \cdot L_q - L_p \cdot L_q = L_q (P_p - L_p)$$

where L_p = Laspeyres' Price index, L_q = Laspeyres' Quantity index, and P_p = Paasche's Price index (omitting the factor 100 from each index). We have shown that

Convariance between Price Relative and Quantity Relative

$$= L_q(P_p - L_p)$$

Since the sign of correlation coefficient is the same as that of covariance (because in the denominator σ_x and σ_y are always positive), the covariance will be positive or negative according as there is positive or negative correlation between the variables. In the persent case, the variables x and y are price relative and quantity relative respectively, and the nature of correlation between them can be determined from the following considerations. It is a common knowledge that for any commodity the quantity consumed will be relatively less when price is higher, i.e. they move in opposite directions; hence price relative and quantity relative will be negatively correlated. Therefore, the covariance between them, viz. $L_q(P_p - L_p)$, will be negative. Since L_q is always positive, the other factor $(P_p - L_p)$, must be negative, and consequently Laspeyres' Price index L_p is large than Paasche's Price index P_p. In other words, Laspeyres' formula, has an upward bias.

17.9 BASE SHIFTING, SPLICING AND DEFLATION

Base Shifting: This refers to the technique of changing the given base period of a series of index numbers and recasting them to form a new series with reference to a new base period. Base shifting is used in the following situations:

(i) When the base period is too old, it is necessary to shift the base to a more recent period in order that the data are more meaningful.

(ii) Base shifting is also necessary for comparing two or more series of index numbers with different base periods. If all the series are expressed with a common base, the comparison is easier the quicker.

Strictly speaking, base shiting will involve recomputation of the entire series of index numbers by applying the formula already employed. However, this is a difficult job. A relatively simple but approximate method consists in assuming the index number of the new base period as 100 and expressing the old series of index numbers as percentages of the index number for the new base.

Index Number for any year (with new base)

$$= \frac{\text{Old Index Number for the year}}{\text{Old Index Number for the new base year}} \times 100$$

Theoretically, this method will give exact results if the formula employed for index number obeys the Circular Test.

Example 17.46 *The following table shows the Index Number of Industrial Production in India:*

Year	*1973*	*1974*	*1975*	*1976*	*1977*
Index Number (Base 1970 = 100)	*112*	*114*	*119*	*132*	*139*

Shift the base to the year 1973 and recast the data.

Solution

Table 17.30 Base Shifting from 1970 to 1973

Year	*Index Number (1970 = 100)*	*Index Number (1973 = 100)*
1973	112	100
1974	114	$\frac{114}{112} \times 100 = 102$
1975	119	$\frac{119}{112} \times 100 = 106$
1976	132	$\frac{132}{112} \times 100 = 118$
1977	139	$\frac{139}{112} \times 100 = 124$

Example 17.47 *Assume that an index number is 100 in 1968; it rises 3% in 1969, falls 1% in 1970 and rises 2% in 1971 and 3% in 1972; rise and fall begin with respect to the previous year. Calculate the index for the five years, using 1972 as the base year.* [I.C.W.A.(old), Dec., '76]

Solution

Table 17.31 Base Shifting from 1968 to 1972

Year	*Index Number (Base 1968 = 100)*	*Index Number (Base 1972 = 100)*
1968	100	$\frac{100}{107} \times 100 = 93$
1969	$\frac{103}{100} \times 100 = 103$	$\frac{103}{107} \times 100 = 96$
1970	$\frac{99}{100} \times 103 = 102$	$\frac{102}{107} \times 100 = 95$
1971	$\frac{102}{100} \times 102 = 104$	$\frac{104}{107} \times 100 = 97$
1972	$\frac{103}{100} \times 104 = 107$	100

Splicing This is the technique of combining two or more overlapping series of index numbers with different base periods to obtain a single continuous series of index numbers with a common base period. In effect, this is equivalent to shifting the bases of the different series to one fixed base period. Splicing helps comparison among the different years by means of a single continuous series of index numbers. Like base shifting, the technique of splicing will give accurate results only when the formula enployed satisfied the Circular Test.

Example 17.48 *Two series of index numbers are given below:*

	'A' series							
Year	*1940*	*1941*	*1942*	*1943*	*1944*	*1945*	*1946*	*1947*
Index (1939 = 100)	*120*	*130*	*200*	*300*	*350*	*370*	*380*	*400*
	'B' series							
Year	*1947*	*1948*	*1949*	*1950*	*1951*	*1952*	*1953*	*1954*
Index (1947 = 100)	*100*	*110*	*90*	*98*	*101*	*110*	*98*	*96*

For purposes of continuity of records, you are required to construct a combined continuous series 'C', say, on 1939 = 100 basis, and covering records upto 1954. What is this technique called? Suggest any alternative continuous series which can serve the main purpose.

Solution The technique is called "Splicing".

We shall first shift the base of *B*-series to the year 1939, so that the index numbers for all the years from 1940 to 1954 will be comparable. This combined continuous series is shown in Col. (4) of Table 17.32. The alternative continuous series is obtained by shifting the base of the *A*-series to the year 1947. This is shown in Col. (5) of the Table.

Table 17.32 Splicing Two Series of Index Numbers

Year	*A-series Index (1939 = 100)*	*B-series Index (1947 = 100)*	*New Continuous C-series Index (1939 = 100)*	*(1947 = 100)*
(1)	*(2)*	*(3)*	*(4)*	*(5)*
1940	120	—	120	30
1941	130	—	130	32.5
1942	200	—	200	50
1943	300	—	300	75
1944	350	—	350	87.5
1945	370	—	370	92.5
1946	380	—	380	95
1947	400	100	400	100
1948	—	110	440	110
1949	—	90	360	90
1950	—	98	392	98
1951	—	101	404	101
1952	—	110	440	110
1953	—	98	392	98
1954	—	96	384	96

[**Note:** In Col. (4), the figures for 1948 to 1954 are obtained on multiplying the *B*-series by 400/100 = 4. In Col. (5), the figures for 1940 to 1946 are obtained on multiplying the *A*-series by 100/400 = $\frac{1}{4}$]

Deflation This refers to the technique of adjusting the values of a series after making allowance for changes in price level. When prices increase, we can purchase less with the same amount of money spent in earlier years; i.e. the 'purchasing power' of money diminishes. Consequently a rise in the 'money income' will not really mean the same percentage increase in the standard of living; i.e. the 'real income' will be less than the 'money income'. For a true comparsion, it becomes necessary to adjust the money income for changes in the cost of living index numbers. Thus, the real income is determined after 'deflating' the money income by cost of living index.

$$\text{Real Income} = \frac{\text{Money Income}}{\text{Cost of Living Index}} \times 100$$

The technique of deflation is used extensively to find 'real wages' and also to deflate value series, rupee sales, etc. by the corresponding price index numbers.

Example 17.49 *Deflate the per capita income shown in the following table on the basis of the rise in the cost of living index and comment on your results:*

Year	*1965*	*1966*	*1967*	*1968*	*1969*	*1970*	*1971*	*1972*
Cost of Living Index	*100*	*110*	*120*	*130*	*150*	*200*	*250*	*350*
Per Capita Income (Rs)	*65*	*70*	*75*	*80*	*90*	*100*	*110*	*130*

[C.U., M.Com. '73]

Solution

Table 17.33 Deflating Per Capita Income

Year	*Cost of Living Index (1965=100)*	*Actual Per Capita Income (Rs)*	*"Real Income" (Rs)*
(1)	*(2)*	*(3)*	*(4)*
1965	100	65	65.00
1966	110	70	$\frac{70}{110} \times 100 = 63.64$
1967	120	75	$\frac{75}{120} \times 100 = 62.50$
1968	130	80	$\frac{80}{130} \times 100 = 61.54$
1969	150	90	$\frac{90}{150} \times 100 = 60.00$
1970	200	100	$\frac{100}{200} \times 100 = 50.00$
1971	250	110	$\frac{110}{250} \times 100 = 44.00$
1972	350	130	$\frac{130}{350} \times 100 = 37.14$

Comments: It is observed from Col. (2) that although actual income has gradually increased from Rs 65 in 1965 to double its value in 1972, the "real income" has considerably gone down. This indicates that people of the particular category have been hard hit by the substantial rise in the cost of living index.

17.10 ERRORS IN INDEX NUMBERS

All index numbers are affected by mainly three types of errors:

(1) Formula error,
(2) Sampling Error,
(3) Homogeneity error.

Formula error There is no index number formula which can measure the price changes exactly. Each formula in common use has its own defect, and consequently some error is inherent in each index number formula. This is known as 'formula error'. This error can never by eliminated.

Sampling error All index numbers are computed on the basis of price and quantity of some selected commodities. It is expected that this sample of commodities will give a fair picture of the level of price or quantity. However, since many other commodities cannot be taken into consideration, the calculated index number can never represent the changes in the phenomenon accurately. The error thus introduced by selecting a sample of commodities is known as the 'sampling error'. Naturally, the sampling error diminishes with increase in the number of commodities.

Homogeneity error This error arises due to the fact that index numbers are constructed from such commodities which are marketed approximately in the quality both in the base and the current periods. With the passage of time, new commodities replace many of the old commodities and hence the homogeneity in composition of the commodities cannot be strictly maintained. Consequently, the error increases as the gap between the base and current periods increases.

EXERCISES

1. What are index number? How do you choose the base year for constructing an index number series? *[M.B.A. '78]*
2. Explain with example what you meant by a price index number and write down its uses. *[W.B.H.S. '83]*
3. Write down the correct answer:
   If now the prices of all commodities in a place have been decreased by 35% over the base period prices, then the index number of prices for the place is (index number of prices of base period = 100): (i) 100, (ii) 135, (iii) 65, (iv) 35, (v) None of these. *[W.B.H.S. '82]*
4. "Index Numbers are economic barometers"—Explain. *[I.C.W.A., Dec., '79]*
5. Discuss the different steps that have to be taken in the construction of a price index number. *[C.U., B.Com.(Hons) '80]*

6. Write down the well-known formulae for comparing price levels in two time periods, explaining every symbol used. Give two interpretations of Laspeyres' Price Index number. [C.U., B.A.(Econ) '77]
7. Show that both Laspeyres' and Paasche's price index numbers may be regarded as weighted averages of price relatives. [B.U., B.A.(Econ) '80, '82]
8. Find the Simple Aggregative index number from the following data:

Commodity	*Base Price*	*Current Price*
Rice	140	180
Sugar	100	300
Oil	400	550
Wheat	125	150
Pulse	160	200

9. Find by the weighted aggregative method, the index number of the following data:

Commodity	*Base Price*	*Current Price*	*Weight*
Rice	140	180	10
Oil	400	550	7
Sugar	100	250	6
Wheat	125	150	6
Fish	200	300	4

[B.U., B.Com. '77]

10. Calculate the price index number by (a) Paasche's method, (b) Laspeyre's method, (c) Bowley's method, (d) Fisher's ideal formula.

Commodities	*1979*		*1980*	
	Price (Rs)	*Quantity (Kgs)*	*Price (Rs)*	*Quantity (Kgs)*
A	20	8	40	6
B	50	10	60	5
C	40	15	50	10
D	20	20	20	15

[I.C.W.A., Dec. '80]

11. Prepare price index numbers for 1977 with 1975 as base year from the following data, using (i) Laspeyres', (ii) Paasche's, (iii) Fisher's method.

Commodity	*Unit*	*Quantity*	*Price (Rs)*	*Quantity*	*Price (Rs)*
A	kg	5	2.00	7	4.50
B	Quintal	7	2.50	10	3.20
C	Dozen	6	3.00	6	4.50
D	kg	2	1.00	9	1.80

[C.U., M.Com. '76]

12. Using the data given below, calculate price index numbers for the year 1958 by (i) Laspeyres' formula (ii) Paasche's formula, (iii) Fisher's formula, with the year 1949 as base:

Commodity	Price (Rs)		Quantity('000 kg)	
	1949	1958	1949	1958
Rice	9.3	4.5	100	90
Wheat	6.4	3.7	11	10
Pulses	5.1	2.7	5	3

State with reasons one advantage of the Laspeyres' index over the Paasche index in case revisions of an index number are to be made from year to year.

[I.C.W.A., Dec. '78]

13. Given the following data, calculate price index numbers by (i) Laspeyres' formula (ii) Paasche's formula, and (iii) Fisher's formula, with 1927 as base:

Year	Rice		Wheat		Jowar	
	Price	Qty.	Price	Qty.	Price	Qty.
1927	9.3	100	6.4	11	5.1	5
1934	4.5	90	3.7	10	2.7	3

[B.U., B.A.(Econ) '71]

14. Calculate the price index number for 1940 with 1937 as base year by the aggregative method, using (a) base year quantities as weights, and (b) given year quantities as weights, from the following data:

Commodity	1937		1940	
	Quantity ('000 tons)	Price per ton (Rs)	Quantity ('000 tons)	Price per ton (Rs)
A	350	100	400	120
B	200	130	180	200
C	140	50	200	110
D	80	125	100	140

15. The following table gives the change in the price and consumption of three commodities in the workers' consumption basket. Compute Fisher's ideal index number from the data given in the table.

Commodity	1950		1960	
	Price (Rs)	Consumption (units)	Price (Rs)	Consumption (units)
Wheat	100	10	110	6
Rice	150	15	170	18
Cloth	5	50	4	30

[C.U., B.Sc.(Econ) '80]

16. From the data given below, calculate Fisher's Ideal Index number of prices for 1963 with reference to 1960 as base period:

Commodities	Price (Rs)		Quantity ('000 kg)	
	1960	1963	1960	1963
a	4.3	5.2	20	16
b	2.1	3.9	5	4
c	0.8	1.6	11	8
d	3.2	4.8	8	6

17. Find by Arithmetic Mean method the index number from the following data:

Commodity	Base Price	Current Price
Rice	140	180
Sugar	100	300
Oil	400	550
Wheat	125	150
Pulse	160	200

[B.U., B.Com. '76]

18. Find the index number from the following data by the method of Relatives (use A.M.):

Commodity	Rice	Wheat	Fish	Potato	Coal	Pulse
Base Price	30	22	54	20	15	4
Current Price	33	25	64	23	16	5

[C.U., B.Com. '79]

19. Calculate a suitable index number from the data given below:

Commodity	Price Relative	Weight
A	125	5
B	67	2
C	250	2

[D.S.W. '78]

20. Explain the term 'Price Relative'. Find by Arithmetic Mean method the index number from the following:

Commodity	Base Price	Current Price	Weight
Rice	30	52	8
Wheat	25	30	6
Fish	130	150	3
Potato	35	49	5
Oil	70	105	7

[B.U., B.Com. '74]

21. The price quotations of 4 different commodities for 1951 and 1965 are given below. Calculate the index number for 1965 with 1951 as base, by using (i) the simple average of price relatives, (ii) the weighted average of price relatives.

Commodity	Unit	Weight (Rs '000)	Price (Rs) 1951	1965
A	Seer	5	2.00	4.50
B	Maund	7	2.50	3.20
C	Dozen	6	3.00	4.50
D	Seer	2	1.00	1.80

22. An index number of wholesale prices, based on the simple arithmetic mean of price relatives comprises 40 items. There are divided into seven groups. A separate index is published for each group. Find the index number for all the groups combined for 1968, from the following data:

Group	A	B	C	D	E	F	G
No. of Items	10	5	8	4	3	4	6
Group Index for 1968	120	95	115	142	86	100	105

23. In 1976 the average price of a commodity was 20% more than in 1975, but 20% less than in 1974; and moreover it was 50% more than in 1977. Reduce the data to price relatives using 1975 as base (1975 Price Relative = 100).

[I.C.W.A., Dec. '78]

24. Using Paasche's formula, compute the quantity index and the price index numbers for 1970 with 1966 as base year:

Commodity	*Quantity Units*		*Value Rs*	
	1966	1970	1966	1970
A	100	150	500	900
B	80	100	320	500
C	60	72	150	360
D	30	33	360	297

[I.C.W.A., June '75]

25. Using Fisher's 'ideal' formula, calculate the quantity index number from the following data:

Commodity	*Base Year Price (Rs)*	*Base Year Quantity (kg)*	*Current Year Price (Rs)*	*Current Year Quantity (kg)*
A	5	50	10	56
B	3	100	4	120
C	4	60	6	60
D	11	30	14	24
E	7	40	10	36

[I.C.W.A., Jan. '73]

26. Annual production in million tons of three commodities are given:

Commodity	Production in Year		Weights
	1935	1940	
A	160	200	13
B	10	12	21
C	80	100	35

Calculate quantity index number for the year 1940 with 1935 as base year, using simple arithmetic mean and weighted arithmetic mean of the relatives.

27. Explain the terms: Price Relative, Quantity Relative, and Value Relative—with reference to a single commodity and deduce the Factor Reversal property.

In 1970 the price of a commodity increased by 50% over that in 1952 while the production of the quantity decreased by 30%. By what percentage did the total rupee value of the commodity in 1970 increase or decrease with respect to the 1952 value? *[I.C.W.A., June '76]*

28. Explain and give the expressions for Time Reversal test and Factor Reversal test. *[I.C.W.A., Dec. '80]*

29. Using the following data, show that Laspeyres' price index formula does not satisfy the time reversal test:

Commodity	Base Year		Current Year	
	Price	Quantity	Price	Quantity
A	6	50	10	56
B	2	100	2	120
C	4	60	6	60
D	10	30	12	24
E	8	40	12	36

30. From the following data on shoe princes and quantities, show that Laspeyres index does not meet the factor reversal test:

Type of Shoe	P_{1950}	Q_{1950}	P_{1960}	Q_{1960}
Men's	£ 7	36	£ 10	48
Women's	5	50	9	80
Children's	4	18	6	26

[C.U., M.Com. '79]

31. Prove using the following data that the factor reversal test and time reversal test are satisfied by Fisher's Ideal formula for index numbers:

	Rice		Wheat		Jowar	
Year	Price	Quantity	Price	Quantity	Price	Quantity
1949	4	50	3	10	2	5
1959	10	40	8	8	4	4

[D.S.W., Nov. '70]

32. Show that Fisher's 'ideal' price index number satisfies both the time reversal and the factor reversal tests and verify this from the following data:

Commodity	1970		1972	
	Price	*Quantity*	*Price*	*Quantity*
A	6	50	10	56
B	2	100	2	120
C	4	60	6	60
D	10	30	12	24
E	8	40	12	36

[C.U., B.Sc.(Econ) '81]

33. Calculate a suitable price index number for the year 1934 from the following data and verify numerically whether the formula employed satisfies the appropriate test or not.

Commodity		*A*	*B*	*C*	*D*	*E*
Price (Rs)	1927	6	2	4	10	8
	1934	10	2	6	12	12

[I.C.W.A., June '74]

34. What are the tests to be satisfied by a good index number ? Examine how far they are met by Fisher's ideal index number. *[C.U., B.Com.(Hons) '82]*

35. Explain briefly: Time Reversal and Factor Reversal Tests of index number. Indicate whether the following index numbers satisfy one or other of these tests: Laspeyres'. Paasche's, Marshall-Edgeworth's and Fisher's Ideal index numbers. *[C.U., B.A.(Econ) '75; I.C.W.A., June '82]*

36. Show that the simple aggregative type of index number satisfies the time reversal and circular tests, but does not satisfy the factor reversal test. *[I.C.W.A., July '72]*

37. What is the chain base method of construction of index numbers and how does it differ from the fixed base method? Explain. *[C.U., B.Com(Hons) '83]*

38. What do you mean a link index? Discuss the relative merits and demerits of chain base and fixed base index numbers. *[C.U., M.Com. '79. '81]*

39. Find index numbers for the years 1961, 1962, 1963 by the chain base method, with base-year 1960, from the following table:

Year	1960	1961	1962	1963
Link Index	100	110	95.5	109.5

[I.C.W.A, Jan. '69]

40. Compute chain index numbers with 1970 prices as base, from the following table giving the average wholesale prices for the years 1970–74.

	Average Wholesale Prices (Rs)				
Commodity	*1970*	*1971*	*1972*	*1973*	*1974*
A	20	16	28	35	21
B	25	30	24	36	45
C	20	25	30	24	30

[I.C.W.A., Dec. '76]

41. Briefly describe the various steps involved in constructing the cost of living index number. *[D.S.W., '78]*

42. From the table of group index numbers and group expenditures given below calculate the cost of living index number:

Group	*Index Number*	*Percentage of Total Expenditure*
Food	428	45
Clothing	250	15
Fuel and light	220	8
House rent	125	20
Others	175	12

[W.B.H.S. '83]

43. The following are the group index numbers and the group weights of an average working class family's budget. Construct the cost of living index number.

Groups	*Food*	*Fuel and lighting*	*Clothing*	*Rent*	*Miscellaneous*
Index No.	352	220	230	160	190
Weight	48	10	8	12	15

[I.C.W.A., Jan. '70]

44. The following table gives group index numbers and corresponding group weights with regard to cost of living for a given year. Construct the overall cost of living index for the year.

Group	*Index No.*	*Weight*
Food	350	5
Fuel and Lighting	220	1
Clothing	230	1
Rent	160	1
Miscellaneous	190	2

How is the overall index number altered, if

(i) all the group index numbers are changed in the same ratio;

(ii) all the weights and group indices are changed in the same ratio;

(iii) all the group index numbers are increased by 10 and all the weights doubled? *[I.C.W.A., July '68]*

45. The percentage increase in price in 1971 over 1960 in the following groups for middle class people in Calcutta and the percentage of total expenditure spent on those groups are shown below. Calculate the cost of living index number for 1971 with 1960 as base.

Group	*Percentage Increase in Price*	*Percentage of Total Expenditure*
Food	125	45
Clothing	66	6
Fuel and Light	112	5
House Rent and Tax	90	10
Miscellaneous	105	34

[W.B.H. S. '78]

46. Determine the relative importance for the food group, given that the cost of living index number for 1975 with 1970 as base is 175 from the following figures:

Group	*% Increase in Expenditure*	*Weight*
Food	65	
Clothing	90	12
Fuel etc.	20	18
Miscellaneous	70	10
Rent etc.	150	20

[D.S.W. '78]

47. The group indices and the corresponding weights for the working class cost of living index numbers in an industrial city for the years 1976 and 1980 are given below:

Group	*Weight*	*Group Index* 1976	1980
Food	71	370	380
Clothing	3	423	504
Fuel etc.	9	469	336
House rent	7	110	116
Miscellaneous	10	279	283

Compute the cost of living indices for the two years 1976 and 1980. If a worker was getting Rs 300 per month in 1976, do you think that he should be given some extra allowance so that he can maintain his 1976 standard of living? If so, what should be the minimum amount of this extra allowance?

[I.C.W.A, June '81; C.U., B.Com(Hons) '82]

48. An enquiry into the budgets of the middle class families of a certain city revealed that on an average the percentage expenses on the different groups were—Food 45, Rent 15, Clothing 12, Fuel and Light 8, and Miscellaneous

20. The group index numbers for the current year as compared with a fixed base period were respectively 410, 150, 343, 248 and 285. Calculate the consumer price index number for the current year. Mr. *X* was getting Rs. 240 in the base period and Rs 430 in the current year. State how much he ought to have received as extra allowance to maintain his former standard of living.

49. During a certain period the cost of living index number goes up from 110 to 200 and the salary of a worker is also raised from Rs 325 to Rs 500. Does the worker really gain, and if so, by how much in real terms?

[I.C.W.A, Dec. '75]

50. The average weekly wages for all manufacturing industries for a number of months in 1960 are Rs 78.52, 79.71, 78.55, 78.17, 78.99, the corresponding consumer price index numbers are 115, 116, 118, 117, 120. Find the real wages for the different months and calculate the percentage change in the real wages during the period.

51. Monthly wages average in different years is as follows:

Year	1967	1968	1969	1970	1971	1972	1973
Wages (Rs)	200	240	350	360	360	380	400
Price Index	100	150	200	220	230	250	250

Calculate the real wages index numbers. *[I.C.W.A., Dec. '79]*

52. Given the following table, calculate the real wage rates and the purchasing power of the rupee for the years 1947–1954, taking 1947 as the base year:

Year	1947	1948	1949	1950	1951	1952	1953	1954
Wage Rate Per Day (Rs)	1.19	1.33	1.44	1.57	1.75	1.84	1.89	1.94
Consumer Price Index (1947–49 = 100)	95.5	102.8	101.8	102.8	111.0	113.5	114.4	114.8

[B.U., B.A.(Econ) '72]

53. Given below are the average wages in rupees per hour of unskilled workers of a factory during the years 1975–1980. Also shown is Consumer Price index for these years (taking 1975 as base year with Price Index 100). Determine the real wages of the workers during 1975–1980 compared with their wages in 1975.

Year	1975	1976	1977	1978	1979	1980
Consumer Price Index	100	120.2	121.7	125.9	129.3	140
Average Wage (Rs/hour)	1.19	1.94	2.13	2.28	2.45	3.10

How much is the worth of one rupee of 1975 in subsequent years?

[I.C.W.A, June '82]

54. Explain what is precisely mean by saying that Laspeyres' formula has an upward bias while Paasche's has a downward bias.

[B.U., B.A.(Econ) '80, '82]

55. What is meant by (i) base shifting, (ii) splicing, and (iii) deflating of index numbers? Explain with illustrations.

56. The following are index numbers of prices (1969 = 100):

Year	1969	1970	1971	1972	1973	1974	1975	1976	1977	1978
Index	100	120	180	207	243	270	300	360	400	420

Shift the base from 1969 to 1975 and recast the index numbers.

57. The following table shows the Index Number of Wholesale Prices in India (Revised Series) with base 1970–71 = 100:

Year	1971	1973	1974	1975	1976	1977
Index Number	105	132	169	176	172	185

Find the index numbers for these years with base 1973 = 100.

58. An index number is at 100 in 1971. It rises 10% in 1972, falls 4% in 1973, falls 2% in 1974, and again rises 10% in 1975, over the preceding year. Calculate the index number for the 5 years with 1974 as base.

59. Given below are two series of index numbers, one with 1961 as base and the other with 1970 as base:

(a) *Year*	*Index*	(b) *Year*	*Index*
1965	180	1970	100
1966	192	1971	108
1967	208	1972	112
1968	220	1973	125
1969	232	1974	130
1970	250	1975	150

The index number series (a) was discontinued in 1971. Splice the series (a) to the series (b) with 1970 as base.

60. In 1950, a Statistical Bureau started constructing an index number series with 1950 as base.

Year	1950 (Base)	1956	1960
Index	100	140	200

In 1961, the Bureau reconstructed the index number series on a plan with base 1960.

Year	1960 (Base)	1965	1970
Index	100	150	210

In 1971, the Bureau again reconstructed the series on yet another plan with base year 1970.

Year	1970 (Base)	1975	1981
Index	100	180	240

Obtain a continuous series with base 1970, by splicing the three series.

61. Deflate the per capita income shown in the following table on the basis of the rise in the cost of living index and comment on you results:

Year	1965	'66	'67	'68	'69	'70	'71	'72
Cost of Living Index	100	110	120	130	150	200	250	350
Per Capita Income (Rs)	65	70	75	80	90	100	110	130

[*C.U., M.Com. '73*]

62. Discuss the different types of errors that affect a price index number.

[*C.U., M.Com. '79*]

ANSWERS

3. 65 8. 149 9. 145
10. 125.2, 124.7, 125.0, 125.0 11. 159, 162, 160
12. 49.15, 49.12, 49.13 13. 49.15, 49.12, 49.13
14. 139.2, 139.7 15. 111 16. 137
17. 162 18. 115 19. 150.9
20. 145 21. 171, 164 22. 111
23. 150, 100, 120, 80 for 1974-77 [**Hint:** Assume 1976 price to be 100; then 1975 price is $100 \times \frac{100}{120}$, 1974 price is $100 \times \frac{100}{80}$, and 1977 price is $100 \times \frac{100}{150}$. If price relative for 1975 is taken as 100, multiply each by $\frac{120}{100}$.]
24. 131, 119 [**Hint:** As in Example 17.20 and 17.21, find p_0, p_n and then calculate $\Sigma p_0 p_n = 1570$, $\Sigma p_0 q_n = 1726$. Last column give $\Sigma p_n q_n = 2057$]
25. 100.2
26. 123.3, 123.5
27. 5% [**Hint:** Value Ratio = 1.50 × 0.70 = 1.05.]
29. (see Example 17.23)
30. [see Example 17.26(a).]
31. & 32. [see Example 17.24 and 17.26(b).]
33. Simple Aggregative Index 140; formula satisfies T.R. test.
34. (see Example17.28)
35. Marshall-Edgeworth's and Fisher's Ideal index numbers satisfy T.R. test; Fisher's Ideal index only satisfies F.R. test.
39. 110, 105, 115
40. 100, 108, 135, 160, 166 (using A.M. of relatives)
42. 293.7
43. 276.4
44. 274; (i) change in the same ratio; (ii) change in the same ratio as group index numbes; (iii) increase by 10.
45. 210.51

46. 64 (see Example 17.38)
47. 353.20, 351.58. Since index number for 1980 is smaller, no extra allowance need be given.
48. 325, Rs 350.
49. No; Real wage decreases by Rs 45 (from Rs 295 to Rs 250).
50. (Rs) 68.28, 68.72, 66.57, 66.81, 65.82; 3.6% decrease.
51. 100, 80, 88, 82, 78, 76, 80 (base 1967 = 100). [**Hint:** First calculate 'real wages', and then convert them as percentage of real wage for 1967.]
52. Real wages (Rs) 1.25, 1.29, 1.41, 1.53, 1.58, 1.62, 1.65, 1.69

    Purchasing power 1.00, 0.93, 0.94, 0.93, 0.86, 0.84, 0.83, 0.83
53. Real wage (Rs/hr) 1.19, 1.61, 1.75, 1.81, 1.89, 2.21

    * Purchasing power of 1975 rupee 1.00, 1.20, 1.22, 1.26, 1.29, 1.40

    (* **Hint:** Consumer Price Index ÷ 100)
54. (see Section 17.8) 55. (see Section 17.9)
56. 33, 40, 60, 69, 81, 90, 100, 120, 133, 140
57. 80, 100, 128, 133, 130, 140
58. 96.6, 106.3, 102.0, 100, 110.0
59. 72, 76.8, 83.2, 88, 92.8, 100, 108, 112, 125, 130, 150
60. 23.8, 33.3, 47.6, 71.4, 100, 180, 240
61. (Rs) 65, 64, 63, 62, 60, 50, 44, 37. Although 'money wages' have doubled (from Rs 65 to Rs 130) during the period, 'real wages' have practically been reduced to half (from Rs 65 to Rs 37).

18 VITAL STATISTICS

18.1 INTRODUCTION

'*Vital Statistics*' refer to the data or the methods of analysis of data relating to such vital events of human life as birth, death, sickness marriage, etc. The basic data are usually available from two principle sources:

(i) Population census,

(ii) Vital statistics registers.

Population census, conducted every ten years in almost all countries of the world, provides information relating to the number of persons inhabiting a certain area, classified according to age, sex, occupation, etc. at specified 'points' of time. Registration of births, deaths, etc. by the appropriate authorities provides data relating to births classified by sex of the new born child, age of mother, etc., and deaths classified according to causes, ages at death, etc. during different 'periods' of time. Hospital records show various data relating to sickness, number surviving a certain disease, etc.

Here, we shall be concerned with the measurement of some rates of mortality, fertility and overall growth of population. In general, we define,

$$\textit{Rate of a vital event} = \frac{\textit{No. of cases of the event during the period}}{\textit{Total no. of persons exposed to the risk of that event at the middle of the period}} \qquad (18.1.1)$$

Although (18.1.1) is always a fraction, rates of vital events are usually expressed '*per thousand*'. Also, the rate must relate to a particular (i) type of event, e.g. birth-rate, death-rate; (ii) geographical area, e.g. India, Kerala: and (iii) period of time, e.g. year 1971. For example, the birth-rate in India during 1976 was 34.4 per thousands population.

The following rates and ratios have been considered:

Measurement of Mortality

(a) Crude Death Rate,

(b) Specific Death Rate,

(c) Standardised Death Rate.

Measurement of Fertility

(a) Crude Birth Rate,
(b) General Fertility Rate,
(c) Total Fertility Rate.

Measurement of Population Growth

(a) Vital Index,
(b) Gross Reproduction Rate,
(c) Net Reproduction Rate.

18.2 CRUDE DEATH RATE

The simplest way of measuring mortality is to relate the total number of deaths during the period to the total population at the middle of the period.

$$\text{Crude Death Rate} = \frac{\text{Total number of deaths during the period}}{\text{Total population at the middle of the period}}$$

$$= \frac{\Sigma D_x}{\Sigma P_x} \qquad (18.2.1)$$

where D_x = Number of deaths of persons aged x years last birthday;
P_x = Number of persons aged x years last birthday (1.b.d.)

Crude Death Rate (*CDR*) gives the probability that a person chosen randomly from the given population will die during the period from any cause whatsoever. It is simple to calculate. But the drawback of *CDR* is that it is really a crude measure of mortality, and docs not take into account the age and sex composition of the population. The probability of death is not the same at all ages, and also varies for males and females at the same age. As such, *CDR* fails to give a true comparison of the mortality conditions of two communities. Even if the mortality rates at all ages are lower for one region than another, *CDR* for the may be larger. (see Example 18. 1).

18.3 SPECIFIC DEATH RATE

Specific Death Rate is the mortality rate calculated for a specific class or section of the population.

Specific Death Rate

$$= \frac{(\text{No. of deaths in a specific class of population during the period})}{(\text{Total no. of persons in the specific class of population at the middle of the period})} \qquad (18.3.1)$$

Usually death rates are made specific with respect to age and sex, although specificity with respect to race, religion, cause of death, etc. is also possible.

Age-specific Death Rate (m_x) at age x measures the probability that a persons aged x years last birthday will die before reaching his next birthday.

$$\text{Age-specific Death Rate } (m_x) = \frac{D_x}{P_x} \qquad (18.3.2)$$

Specific Death Rates are the best measures of mortality. Using (18.3.2), the crude death rate (18.2.1) may be written as

$$\text{Crude Death Rate} = \frac{\Sigma P_x m_x}{\Sigma P_x} \tag{18.3.3}$$

Crude Death Rate may, therefore, be viewed as a weighted arithmetic mean of age-specific death rates (m_x), using the population (P_x) at different ages as weights.

Example 18.1 *From the following data, find (a) Crude Death Rates, (b) Specific Death Rates for each age group, for towns I and II separately; and comment on the results.*

Age-group (years)	*Town I*		*Town II*	
	Population	*No. of deaths*	*Population*	*No. of deaths*
0–9	*1,500*	*45*	*6,000*	*150*
10–24	*3,000*	*15*	*5,000*	*20*
25–44	*5,000*	*30*	*6,000*	*30*
45 and over	*500*	*12*	*3,000*	*54*
Total	*10,000*	*102*	*20,000*	*254*

Solution

(a) Crude Death Rate for Town I $= \frac{102}{10,000} = .0102 = 10.2$ per thousand

Crude Death Rate for Town II $= \frac{254}{20,000} = .0127$

$= 12.7$ per thousand

(b) Specific Death Rates are shown in Table 18.1.

Comments: From Table 18.1, it is seen that the specific death rates for Town I are higher than those for Town II at each age group. Yet, crude death rate for Town I is the smaller. This shows that Crude Death Rate is highly unreliable for comparing the over-all mortality situation between two places.

Table 18.1 Calculation of Specific Death Rates

Age-group	*Town I*	*Town II*	*Sp. Death Rate (per thousand)*	
			Town I	*Town II*
0–9	45 ÷ 1500 = .030	150 ÷ 6000 = .025	30	25
10–24	15 ÷ 3000 = .005	20 ÷ 5000 = .004	5	4
25–44	30 ÷ 5000 = .006	30 ÷ 6000 = .005	6	5
45 and over	12 ÷ 500 = .024	54 ÷ 3000 = .018	24	18

18.4 STANDARDISED DEATH RATE

Although specific death rates may be used to compare the mortality position for two communities or places at different ages, they cannot provide a comparison of the over-all mortality for these two populations; because, the specific death rates for one community may be higher at certain ages, but lower at other ages than the second community. What we need for the purpose is a single figure, some sort of an average—a weighted average—of age specific death rates for the each community, so that the figures will be directly comparable. If age-specific death rates (m_x) are weighted by the population (P_x) of the community, the average gives Crude Death Rate (18.3.3), which is not at all reliable (see Example 18.1). Even if the age-specific death rates were equal for two communities, the crude death rates will not be equal, unless the age composition of the two are identical, i.e., the corresponding weights P_x are proportional. In order to overcome this difficulty, it is necessary to 'standardise' the death rates.

Standardised Death Rate

Standardised Death Rate of a community is the weighted arithmetic mean of its age-specific death rates, using the figures of a given 'standard population' as weights. The population of a larger community (usually covering all the communities whose mortality situations are to be compared) is taken as a standard for the purpose, and is called '*standard population*'.

$$\textit{Standardised Death Rate} = \textit{Weighted average of } m_x \textit{ using } P'_x \textit{ as weights} = \frac{\Sigma P'_x \cdot m_x}{\Sigma P'_x} \tag{18.4.1}$$

where P'_x = standard population at age x,

m_x = specific death rate of the community at age x.

Standardised Death Rate shows what the crude death rate in the standard population would be, if it experienced the age-specific death rates of the given community. It is used for comparing the mortality situations of two or more places.

Adjusted Death Rate

It the age-specific death rates (m_x) of the community are not available. Standardised Death Rate (18.4.1) cannot be calculated. However, under certain assumptions, an approximate value may be obtained from Crude Death Rate (CDR) on multiplication by an 'adjustment factor'.

$$\textit{Adjustment factor} = \frac{\textit{CDR of standard population}}{\begin{array}{l}(\textit{CDR that would be observed in the given}\\ \textit{community, if it experienced the age-specific}\\ \textit{death rates of standard population})\end{array}}$$

$$= \frac{\textit{CDR of standard population}}{\Sigma P_x . m'_x / \Sigma P_x} \tag{18.4.2}$$

where P_x= population of the community at age x,

m_x= specific death rate of standard population at age x.

$$\textit{Adjusted Death Rate} = \textit{CDR} \times (\textit{Adjustment factor})$$

$$= \frac{(\textit{CDR of community}) \times (\textit{CDR of standard population})}{\textit{Weighted average of } m'_x, \textit{ using } P_x \textit{ as weights}}$$

$$= \frac{(CDR) \times (CDR)'}{\Sigma P_x m'_x / \Sigma P_x} \tag{18.4.3}$$

[Note that we have used prime (′) to indicate the factors in 'standard population'.]

Example 18.2 *Calculate Standardised Death Rates for Towns I and II, from the following data:*

	Specific Death Rate (per thousands)		
Age-group	*Town I*	*Town II*	*'Standard Population'*
0 – 9	*30*	*25*	*2,200*
10 – 24	*5*	*6*	*3,000*
25 – 44	*6*	*5*	*2,700*
45 and over	*24*	*28*	*2,100*
			10,000

Solution Using (18.4.1),

Standardised Death Rate for Town I

= (2200 × 30 + 3000 × 5 + 2700 × 6 + 2100 × 24) ÷ 10,000 = 14.76 (per thousand)

Standardised Death Rate for Town II

= (2200 × 25 + 3000 × 6 + 2700 × 5 + 2100 × 28) ÷ 10,000 = 14.53 (per thousand).

Example 18.3

Age	*Population of City A*	*Rates of Mortality (%) in a Standard Population*
0 – 5	*9,736*	*8.90*
5 – 10	*8,349*	*1.19*
10 – 20	*17,101*	*1.07*
20 – 40	*37,298*	*2.22*
40 – 60	*20,746*	*5.21*
60 – 80	*6,298*	*6.32*
80 –	*472*	*38.98*

Given that the crude death-rate of the standard population in 37.4 per thousand and the same for city A is 10% in excess, find out the adjusted death-rate for city A.

Solution Crude Death Rate of Standard Population = 37.4 per thousand. Since 10% of 37.4 is 3.74, therefore

Crude Death Rate of City A = 37.4 + 3.74 = 41.14 per thousand.

Table 18.2 Calculations for Adjusted Death Rate

Age	*Population of City A* (P_x)	*Mortality (%) of Standard Population* (m'_x)	$P_x m'_x$
0 – 5	9,736	8.90	86,650
5 – 10	8,349	1.19	9,935
10 – 20	17,101	1.07	18,298
20 – 40	37,298	2.22	82,802
40 – 60	20,746	5.21	108,087
60 – 80	6,298	6.32	39,803
80 –	472	38.98	18,399
Total	100,000	—	363,974

$$\therefore \quad \frac{\Sigma P_x m'_x}{\Sigma P_x} = \frac{363{,}974}{100{,}000} = 3.64 \text{ (per cent)} = 36.4 \text{ per thousand}$$

$$\text{Adjustment factor} = \frac{CDR \text{ of standard population}}{\Sigma p_x m'_x / \Sigma P_x} = \frac{37.4}{36.4} \text{ By (18.4.3),}$$

$$\text{Adjusted Death Rate} = 41.14 \times \frac{37.4}{36.4} = 42.27 \text{ (per thousand)}$$

18.5 LIFE TABLE

Life table is a tabular presentation of the mortality experiences of a community. It shows (Table 18.3) how many of, say, 100,000 babies (called '*cohort*' or '*radix*' of the life table) born at the same moment are expected to live at various ages, the probabilities of death within a (or an age-group), the expected life-time after a certain age, etc.

Explanation of Symbols used in Life Table:

(1) x = Age in years.

(2) l_x = Number of persons out of the cohort, who are expected to complete exactly x years of life. Table 18.3 shows that out of 100,000 Indian girls born alive at the same moment 86,174 survive 1 years, 83,055 survive 2 years and so on; i.e. $l_1 = 86{,}174$, $l_2 = 83{,}055$. The *cohart* is $l_0 = 100{,}000$. Note that l_x decreases as x increases.

(3) d_x = Number of persons out of l_x who die before completing age $x + 1$

$$d_x = l_x - l_{x+1} \tag{18.5.1}$$

(4) q_x = Probability that a person who attains age x will be before reaching age $x + 1$.

$$q_x = d_x \div l_x \tag{18.5.2}$$

The values of q_x are the most important in the construction of a life table. If these are known, the life table can be completed with an arbitrary cohort. Usually, q_x is calculated by the relation

$$q_x = \frac{2m_x}{2 + m_x} \tag{18.5.3}$$

where m_x denotes the observed age-specific death rate at age x.

Some life tables show an extra column for

p_x = Probability that a person who attains age x will survive another year.

$$p_x = 1 - q_x \tag{18.5.4}$$

(5) L_x = Total number of years lived, in the aggregate, by the cohort between ages x and $x + 1$. It is called the *life table stationary population* at age x. Approximately

$$L_x = \frac{1}{2}(l_x + l_{x+1}) \tag{18.5.5}$$

Using (18.5.1),
$$L_x = l_x - \frac{1}{2} d_x \tag{18.5.6}$$

Considering the 83,055 (= l_2) girls at age 2, we note that 80,677 (= l_3) of them reached the next birthday each living a complete year, while 2378 (= d_2) died before completing 1 year. Suppose 1 died at age 2 years 1 day, another at age

2 years 4 days, another at age 2 years 6 days, another at 2 years 10 days, etc. Then, the total number of years lived between ages 2 and 3 is L_2 = (80677 years + 1 day + 4 days + 6 days + 10 days + ...) = 80950 years (given). This is obviously the total time lived by the cohort only between ages 2 and 3.

Table 18.3 Extracts From All-India Life Table (Females), 1951-60

x	l_x	d_x	q_x	L_x	T_x	e°_x
0	100000	13826	.1383	89631	4055487	40.55
1	86174	3119	.0362	83390	3965856	46.02
2	83055	2378	.0286	80950	3882466	46.75
3	80677	1797	.0223	79100	3801516	47.12
4	78880	1343	.0170	77708	3722416	47.19
	...		...		...	
10	74600	261	.0035	74470	3266067	43.78
	...		...		...	
30	66235	825	.0125	65823	1845620	27.86
	...		...		...	
50	42690	1274	.0298	42053	745184	17.46
	...		...		...	
70	16599	1167	.0703	16016	154053	9.28
	...		...		...	
90	1325	344	.2596	1153	3303	2.49
	...		...		...	
98	8	6	.7982	5	5	0.63
99	2	2	.9086	—	—	—

(6) T_x = Total number of years lived by the cohort *after* age x years. In fact, this is the total future life time of the l_x persons (until all of them die off). l_x persons (until all of them die off).

$$T_x = L_x + L_{x+1} + L_{x+2} + \dots \qquad (18.5.7)$$

Hence,

$$T_x - L_x = T_{x-1} \qquad (18.5.8)$$

(7) e°_x= Average number of years lived *after* age x, by each of the l_x persons. This is called *expectation of life at age x*.

$$e^\circ_x = T_x \div l_x \qquad (18.5.9)$$

Complete Life Table and Abridged Life Table

The life table which shows the various quantities l_x, d_x, q_x, etc. for each year of age x = 0, 1, 2, 3, ..., is called a *Complete Life Table*. However, if the life table shows these quantities only at some specified ages, say for x = 0, 5, 10, 15, or for age-intervals 0–5, 5–10, 10–15, etc., it is known as *Abridged Life Table*.

Uses of Life Table:

(1) Life Table presents a vivid picture of the mortality experiences of a given community. It may, therefore, be used to compare the mortality position of

two different communities at various ages or age groups. In particular, a comparison of the expectation of the life (e_x°) at various ages may be interesting.

(2) Life Table is useful of measuring population growth, e.g. in the measurement of Net Reproduction Rate (Section 18.12) and in population projection, i.e. in estimating the age distribution of the community at a future date, assuming that it experiences the mortality shown in the life table.

(3) Life Table is useful to the life insurance companies for fixing the rates of premium payable at different ages. Government and private companies may make use of life table for fixing the rates of retirement benefits for their employees.

Method of Construction of Life Table

The basic element in the Life Table is q_x, the probability of dying between one age and the next. Once the values of q_x are known for all ages x, the completion of Life Table is a simple work.

In the given community, let

D_x = Number of deaths of persons between ages x and $x + 1$,

P_x = Population between ages x and $x + 1$ at the middle of the year.

Values of D_x are obtained from Death Registers and those of P_x from Census Reports. For each age,

$$m_x = \frac{D_x}{P_x}$$

gives the observed age-specific death rate. P_x being the number of persons between the ages x and $x + 1$ who are *still alive* at the middle of the year, does not indicate precisely how many started the year of life x to $x + 1$; these persons were, on an average, $x + \frac{1}{2}$ years old at that point. Again, assuming that deaths are evenly spread throughout the age interval, half of the D_x deaths took place before these persons attained the age $x + \frac{1}{2}$. Hence the number of persons who started the age interval x to $x + 1$ is $P_x + \frac{1}{2} D_x$. Therefore, an approximate value of q_x is given by

$$q_x = \frac{D_x}{P_x + \frac{1}{2} D_x} = \frac{P_x m_x}{P_x + \frac{1}{2} P_x m_x} = \frac{2m_x}{2 + m_x} \tag{18.5.10}$$

[**Note:** The assumption that deaths are uniformly distributed over the age interval is not correct for the first few years of life, especially for $x = 0$. The number of deaths in the first 6 months after birth is much higher than that in the next 6 months of life. The application of formula (18.5.10) is therefore not justified in these cases. A more appropriate formula is $q_x = 10m_x/(10 + 7m_x)$].

When the values of q_x have been computed, the Life Table is completed as follows:

(i) *Columns l_x and d_x*—Start with any arbitrary cohort, say $l_0 = 100{,}000$. Then calculate $d_0 = l_0 q_0$, which gives the number of persons expected to die before reaching exact age 1 year. Then

$$l_1 = l_0 - d_0, \qquad d_1 = l_1 q_1$$
$$l_2 = l_1 - d_1, \qquad d_2 = l_2 q_2; \text{ etc.}$$

(ii) *Column* L_x—Using the relation $L_x = l_x - \frac{1}{2}d_x$, now we find

$$L_0 = l_0 - \frac{1}{2}d_0; \qquad L_1 = l_1 - \frac{1}{2}d_1; \text{ and so on.}$$

(iii) *Column* T_x—In order to complete this column, we have to start from the bottom of the Life Table corresponding to the maximum age and proceed backwards, using the relation $T_x = T_{x+1} + L_x$. Thus

$$T_{103} = T_{104} + L_{108}; \qquad T_{102} = T_{103} + T_{102}; \text{ etc.}$$

(iv) *Column* e_x°—It is now easy to complete this column, using the relation $e_x^\circ = T_x \div l_x$. Thus,

$$e_0^\circ = T_0 \div l_0;\ e^\circ_1 = T_1 \div l_1;\ e^\circ_2 = T_2 \div l_2; \text{ etc.}$$

Example 18.4 *In the following table, fill in the blanks which are marked with a query, explaining the meaning of the symbols at the heads of the columns:*

Age	l_x	d_x	p_x	q_x	L_x	T_x	e_x°
20	793,435	?	?	?	?	35,081,126	?
21	760,673					?	?

By (8.5.1), $d_{20} = l_{20} - l_{21} = 793{,}435 - 760{,}673 = 32{,}762$

By (8.5.2), $q_{20} = d_{20} \div l_{20} = 32{,}762 \div 793{,}435 = .04129$

By (8.5.4), $p_{20} = 1 - q_{20} = 1 - .04129 = .95871$

By (8.5.5), $L_{20} = \frac{1}{2}(l_{20} + l_{21}) = \frac{1}{2}(793{,}435 + 760{,}673) = 777{,}054$

By (8.5.8), $T_{21} = T_{20} - L_{20} = 35{,}081{,}126 - 777{,}054 = 34{,}304{,}072$

By (8.5.9), $e^\circ_{20} = T_{20} \div l_{20} = 35{,}081{,}126 \div 793{,}435 = 44.21$

$e^\circ_{21} = T_{21} \div l_{21} = 34{,}304{,}072 \div 760{,}673 = 45.10$

Solution *The completed table is*

Age	l_x	d_x	p_x	q_x	L_x	T_x	e_x°
20	793,435	32,762	.95871	.04129	777,054	35,081,126	44.21
21	760,673					34,304,072	45.10

Example 18.5 *From the life-table we have the following data:*

$$l_{40} = 81{,}685 \qquad l_{50} = 74{,}528 \qquad l_{60} = 62{,}170 \qquad l_{70} = 42{,}091$$

where l_x is the number surviving to exact age x.

(a) Smith, aged 40, enters into partnership with Jones, aged 50. What is the probability that the partnership will continue undissolved by death for 20 years?

(b) In the partnership just considered what is the probability that in the next 20 years Smith will die but Jones survive?

Solution Out of l_{40} = 81,685 persons at exact age 40, the number expected to survive 20 years, i.e. reach age 60, is l_{60} = 62,170. Hence, the probability that Smith survives 20 years is

$$p = l_{60} \div l_{40} = \frac{62{,}170}{81{,}685} = 0.76$$

Similarly, the probability that Jones, aged 50, survives 20 years is

$$p' = l_{70} \div l_{50} = \frac{42,091}{74,528} = 0.56$$

(a) The partnership will continue undissolved, if both Smith and Jones survive 20 years. Considering the events of survival of these two persons as independent, by (11.6.12) at p. 365,
Required probability = $p.p' = 0.76 \times 0.56 = 0.43$

(b) Since the probability that Smith will die in the next 20 years is $1 - p$, therefore by the same formula,
Required probability = $(1 - p)\, p' = (1 - 0.76)\,(0.56) = 0.13$ *Ans.* 0.43, 0.13

18.6 CRUDE BIRTH RATE

$$\text{Crude Birth Rate} = \frac{\text{Total number of live-births during the period}}{\text{Total population at the middle of the period}} \quad (18.6.1)$$

Crude Birth Rate shows the rate at which population increases through births. It is simple to calculate and has the same defects as Crude Death Rate (18.2.1). Furthermore, the 'total population' used as denominator in (18.6.1) cannot be supposed to be experiencing the risk of child-birth. Only the female population, and that too, within a certain age interval in the child-bearing period (say, between 15 and 49 years) is exposed to the risk of child-births

18.7 GENERAL FERTILITY RATE

$$\text{General Fertility Rate} = \frac{\text{Total number of live-births}}{\text{No. of women in the child-bearing ages}} \quad (18.7.1)$$

General Fertility Rate (*GFR*) shows the rate at which females in the child-bearing ages add tc the total population through births. Although *GFR* shows slight improvement over Crude Birth Rate, it is not a satisfactory measure; because the probability of a female bearing a child varies from age to age. As such, even though two communities may have the same age specific fertility rates (Section 18.8) at corresponding ages, *GFR* will be different, unless the ratio of female population in the two communities is also the same at all ages.

Example 18.6 *In 1981, a city had a total of 507 thousand live births, while its total population was 27512 thousand and total female population in the age group 15–49. l.b.d. was 7576 thousand. Obtain the crude birth rate and the general fertility rate.* [B.U., B.A.(Econ) '82]

Solution Using (18.6.1),

$$\text{Crude Birth Rate} = \frac{507,000}{27,512,000} = 0.0184$$
$$= (0.0184 \times 1000) \text{ per thousand}$$
$$= 8.4 \text{ per thousand}$$

Using (18.6.2),

$$\text{General Fertility Rate} = \frac{507{,}000}{7{,}576{,}000} = 0.0669$$

$$= (0.0669 \times 1000) \text{ per thousand}$$

$$= 66.9 \text{ per thousand}$$

18.8 AGE-SPECIFIC FERTILITY RATE

Age-specific Fertility Rate (f_x) at ages x is the fertility rate calculated with respect to women aged x last birth day (l.b.d.).

Age-specific Fertility Rate at age x

$$= \frac{\textit{No. of live-births during the year from women aged x l.b.d.}}{\textit{No. of women aged x l.b.d. at the middle of the year}} \qquad (18.8.1)$$

This measures the probability that a woman aged x years will give birth to a live baby during the year. Alternatively, the age-specific fertility rate may be viewed as the average number of babies born during the year per woman aged x l.b.d. Age-specific Fertility Rates are also calculated for different *age-groups* of females. For instance,

Age-specific Fertility Rate of age-group (15–19) years

$$= \frac{\textit{No. of live-births during the year from women aged 15–19 years}}{\textit{No. of women aged 15–19 years at the middle of the year}} \qquad (18.8.2)$$

This shows the average number of babies born during the year per woman in the age-group (15–19) years.

18.9 TOTAL FERTILITY RATE

Total Fertility Rate = Sum of Age-specific Fertility Rates at each years of mother's are in the child-bearing period.

$$= \sum_{x=15}^{49} f_x \qquad (18.9.1)$$

Total Fertility Rate (*TFR*) shows the rate at which a new-born female would, on an average, add to the total population, if the remained alive and experienced the age-specific fertility rates throughout the child-bearing period. *TFR* is only a hypothetical measure; because *all* the new-born females cannot be expected to remains alive till the end of the child-bearing period.

18.10 VITAL INDEX

The simplest way of measuring the grown of population is by

Crude Rate of Natural Increase

= (*Crude Birth Rate*) – (*Crude Death Rate*) (18.10.1)

Another measure is the birth-death ratio, called

$$\textit{Vital Index} = \frac{\textit{Total live-births}}{\textit{Total deaths}} \tag{18.10.2}$$

18.11 GROSS REPRODUCTION RATE

In the measurement of population growth, it is appropriate to consider only the female births; because, population increases through females.

Gross Reproduction Rate

$$= \sum_{x=15}^{49} \frac{\textit{No. of female births during the year from women aged } x \textit{ l.b.d.}}{\textit{No. of women aged } x \textit{ l.b.d. at the middle of the year}} \tag{18.11.1}$$

Gross Reproduction Rate (*GRR*) measures the rate at which a new born female would, on an average, add to the total *female* population, *if* she remained alive and experienced the age-sepcific fertility rates till the end of the child-bearing period. Like Total Fertility Rate, *GRR* is also a hypothetical figure, because it does not take into account the mortality experiences during the period.

Note the *GRR* (18.11.1) differs from *TFR* (18.9.1) only in the numerator. The former utilises only *female* births, whereas the latter utilises *total* births (male and female) at each year of the mother's age during the child-bearing period. If it is assumed that the sex-ratio at birth (i.e. male births : female births) is a constant at each year of mother's age throughout the child-bearing period, then

$$\textit{Gross Reproduction Rate} = k \times (\textit{Total Fertility Rate}) \tag{18.11.2}$$

where k = (Number of female births) ÷ (Total births).

18.12 NET REPRODUCTION RATE

Gross Reproduction Rate does not take into consideration the fact that some of the new-born females would die before reaching the child-bearing period. Their number also goes on diminishing as they pass through the child-bearing period. For a correct measure of population growth, the factor of mortality experienced by females must also be taken into account.

Net Reproduction Rate

= Average number of females that would be borne, during the child-bearing period per new born female assuming that she experiences the current age-specific mortality and fertility rates of the community. (18.12.1)

$$\textit{Net Reproduction Rate} = \frac{1}{l_0} \sum_{x=15}^{49} L_x f_x \times k \tag{18.12.2}$$

where l_0 = Cohort of the Life Table for *females*,

L_x = stationary life table population for *females* at ages x.

f_x = age-specific fertility rate at age x,

k = proportion of female births to total births.

Net Reproduction Rate (*NRR*) is the best measure of population growth. It can never exceed Gross Reproduction Rate. *NRR* = 1 signifies that a group of new-born females will replace itself in the next generation. If *NRR* is greater than 1, the population is expanding (Example 18.9); while if it is less than 1, the population is diminishing (Example 18.10).

Example 18.7 *Calculate (i) General Fertility Rate, (ii) Specific Fertility Rates, (iii) Total Fertility Rate, and (iv) Gross Reproduction Rate, from the following data. It is given that out of 9.000 births, the number of females is 4,300.*

Age-group (years)	*16–20*	*21–25*	*26–30*	*31–35*	*36–40*	*41–45*	*46–50*
No. of Women ('000)	*25*	*24*	*20*	*16*	*15*	*12*	*8*
No. of Live Births	*1140*	*3000*	*2740*	*1360*	*600*	*150*	*10*

Solution

$$G.F.R. = \frac{\text{Total live-births}}{\text{Total number of women in the child-bearing ages}}$$

$$= \frac{9,000}{120,00} = .075 \text{ (see Table 18.4)}$$

$$S.F.R. = \frac{\text{Number of live-births}}{\text{Number of women}}, \text{(for a particular age-group)}$$

Calculation for Specific Fertility Rates (S.F.R.) in the various age-groups are shown in the Table below.

Table 18.4 Calculations for Fertility Rates

Age-group (1)	*No. of Women (2)*	*No. of Live-births (3)*	*S.F.R. = (3) ÷ (2) (4)*
16–20	25,000	1,140	0.0456
21–25	24,000	3,000	.1250
26–30	20,000	2,740	.1370
31–35	16,000	1,360	.0850
36–40	15,000	600	.0400
41–45	12,000	150	.0125
46–50	8,000	10	.00125
Total	120,000	9.000	0.44635

S.F.R. for the age-group 16–20 is .0456. This implies that a woman in that age-group will on an average give birth to .0456 children (live-births only) *per year*. Therefore, a woman aged exactly 16 years would, by the time she reaches 20, bear .0456 × 5 babies during the 5-year period. Hence a woman is a expected to bear throughout the child-bearing period (here, 16–50 years), a total of

$$.0456 \times 5 + .1250 \times 5 + ... = 0.44635 \times 5 = 2.23175$$

children. It is of course assumed here that the woman remains alive throughout the period and experiences the sates specific fertility rates.

$$T.F.R. = \Sigma(S.F.R. \times 5) = 2.23175$$

The only difference in calculating the Gross Reproduction Rate (*G.R.R.*) from *T.F.R.* is that we take into account only female births (not all births). Assuming that the sex-ratio of children is the same for all age-groups of mothers, the total number of *female* children expected to be borne by a woman throughout the child-bearing period (assuming no death) gives the *G.R.R.*

$$G.R.R. = \frac{\text{No. of female birhts}}{\text{Total births}} \times T.F.R.$$

$$= \frac{4300}{9000} \times 2.23175 = 1.07$$

Ans. (i) .075; (ii) see Col. (4); (iii) 2.23; (iv) 1.07

Example 18.8 *Calculate the Gross Reproduction Rate (G.R.R.) and the Net Reproduction Rate (N.R.R.) from the following data:*

Age-group (years)	*No. of Female Children born to 1000 Women Passing through the Age-group*	*No. out of 1000 New-born Female Babies who Survive upto the Beginning of Age-interval*
15–19	*220*	*750*
20–24	*400*	*710*
25–29	*380*	*690*
30–34	*300*	*660*
35–39	*200*	*610*
44–44	*70*	*550*
45–49	*20*	*500*

Solution

Table 18.5 Calculations for G.R.R. and N.R.R.

Age-group (years) (1)	*No. of Female Children Borne per Woman* (2)	*Survival Rate for New-born Females upto Beginning of Age-interval* (3)	*No. of Female Children Born and Survived (2) × (3)* (4)
15–19	0.22	0.75	0.1650
20–24	.40	.71	.2840
25–29	.38	.69	.2622
30–34	.30	.66	.1980
35–39	.20	.61	.1220
40–44	.07	.55	.0385
45–49	.02	.50	.0100
Total	1.59	—	1.0797

G.R.R. = Average number of female children borne per woman if she remained alive and experienced the fertility rates throughout the child-bearing period (i.e. assuming no deaths).

= Σ(S.F.R. for female births), at all child-beaing ages.

= 1.59

N.R.R. = Average number of female children borne per woman, if she experienced the prevailing mortality rates and fertility rates upto the end to the child-bearing period.

= Σ(S.F.R. for female births) × (Survival rate)

= 1.0797 = 1.08

Ans. 1.59, 1.08

Example 18.9 *From the following table, calculate (i) Total Fertility Rate, (ii) Gross Reproduction Rate, and (iii) Net Reproduction Rate, if it is known that the ratio of females to total births is 0.48:*

Age-group	*15–20*	*20–25*	*25–30*	*30–35*	*35–40*	*40–45*	*45–50*
Birth Rates per 1000 Women	*23.6*	*114.9*	*145.1*	*122.6*	*79.6*	*35.9*	*3.3*
Year Lived by 1000 Live-born Females	*4482*	*4426*	*4359*	*4288*	*4202*	*4100*	*3979*

Solution Assuming that the age-specific fertility rate is the same at each year within an age-group, and since each age-group covers 5 years, hence by (18.9.1).

$$\textit{Total Fertility Rate} = 0.5250 \times 5 = 2.625$$

This shows the average number of children both males and females borne per woman throughout the child-bearing period (here, ages 15–50), if she remained alive till the end of the period.

Table 18.6 Calculations for T.F.R., G.R.R. and N.R.R.

Age-group	*Age Specific Fertility Rate* (f_x)	*Years Lived by 1000 Females* (L_x)	$L_x f_x$
15–20	0.0236	4482	106
20–25	0.1149	4426	509
25–30	0.1451	4359	632
30–35	0.1226	4288	526
35–40	0.0796	4202	334
40–45	0.0359	4100	147
45–50	0.0033	3979	13
Total	0.5250	—	2267

By (18.11.2), *Gross Reproduction Rate* = 0.48 × 2.625 = 1.26

This shows the average number of *females* borne per woman throughout the child-bearing period, assuming no deaths of females aged 15–50 years.

By (18.12.2), *Net Reproduction Rate* = $\frac{t}{1000} \times 2267 \times 0.48 = 1.088$ This gives the average number of potential mothers during the next generation, per new-born female.

Example 18.10 *Calculate the Gross Reproduction Rate and Net Reproduction Rate from the following table:*

Age-range	*Female Population ('000)*	*Female Births*	*Survival Rate*
15–19	*1558*	*18,900*	*0.914*
20–24	*1112*	*71,100*	*0.899*
25–29	*1595*	*96,900*	*0.884*
30–34	*1629*	*64,200*	*0.868*
35–39	*1627*	*34,900*	*0.852*
40–44	*1522*	*10,800*	*0.834*
45–49	*1401*	*800*	*0.813*

[C.U., B.Sc. '72]

Solution Since each age-interval covers 5 years, the total number of females that would be borne per woman throughout the age intervals is

$$G.R.R. = 0.205 \times 5 = 1.025$$

However, it is not expected that all these new-born females would survive. The number of females surviving each year in the given age intervals is shown in the last column of Table 18.7. Thus,

$$N.R.R. = 0.181 \times 5 = 0.905$$

Table 18.7 Calculations for G.R.R. and N.R.R.

Age-range	*No. of Female Births per Woman per Year*	*No. of New-born Females Surviving*
15–19	18,900 ÷ 1558,000 = .012	.012 × .914 = .011
20–24	71,100 ÷ 1112,000 = .064	.064 × .899 = .057
25–29	96,900 ÷ 1595,000 = .061	.061 × 884 = .054
30–34	64,200 ÷ 1629,000 = .039	.039 × .868 = .034
35–39	34,900 ÷ 1627,000 = .021	.021 × .852 = .018
40–44	10,800 ÷ 1522,000 = .007	.007 × .834 = .006
45–49	800 ÷ 1401,000 = .001	.001 × .813 = .001
Total	0.205	0.181

EXERCISES

1. Define *Crude Death Rate* and point out its defects. Discuss the use of *Specific Death Rates* in this connection. *[W.B.H.S. '78]*
2. Why do we need *Standardised Death Rate*? How is this rate constructed? Also describe the method of indirect standardisation. *[B.U., B.A.(Econ) '80]*
3. Calculate (i) Crude Death Rate, (ii) Specific Death Rate for each age-group, and (iii) Standardised Death Rate, from the following data:

Age-group	Population	Number of Deaths in a Year	'Standard Population (thousand)
0–4	5,000	150	110
5–14	7,000	21	210
15–34	14,000	63	360
35–59	16,000	176	240
60 and over	8,000	320	80

4. Confining attention to persons below 25 years only, find out the standardized mortality rates for places *A* and *B*, given the following:

Age	Standard Population ('000)	Mortality Rates (per '000) Place A	Place B
Under 1 year	21	26.9	43.2
1–4	86	1.0	2.1
5–14	161	0.4	0.8
15–24	147	0.8	1.9

[C.U., B.Sc. '74]

5. The following table gives the population and number of deaths for two towns in 1975:

Age	Town I		Town II	
	Population	No. of deaths	Population	No. of deaths
Under 10	24,000	500	32,000	784
10–20	20,000	80	22,000	88
20–60	65,200	540	55,000	385
60 and above	13,500	920	12,000	960

Calculate the standardised death rate of Town II taking Town I as the standard.

6. Calculate (i) Crude Death Rate, and (ii) Standardised Death Rate of District A, from the following data:

Age-group (years)	Population of District A	Deaths in District A	Standard Population
0–5	1,000	50	10,000
5–10	800	20	8,000
10–25	1,200	12	12,000
25–45	3,000	15	30,000
45 and over	4,000	52	40,000
Total	10,000	149	100,00

Are the two rates equal? If so, why? [C.U., B.Sc. '81]

7. Calculate the Crude Death Rate and the Standardized Death Rate for the following data:

Age-group (years)	Population ('000)	Number of Deaths	Standard Age Distribution (per thousand)
0–9	21	350	221
10–24	30	102	298
25–44	37	229	285
45–64	17	354	149
65 and over	5	415	47

8. The age-specific death rates of two countries are given below. Which country would you consider to be more healthy and Why?

Age-group (years)	Death Rate (per 1000)		Standard Population
	Country I	Country II	(lakh)
0–4	20.0	15.0	100
5–14	1.2	1.1	200
15–24	1.4	1.0	190
25–34	2.0	1.6	180
35–44	3.3	2.8	120
45–54	7.5	8.3	100
55–64	15.0	16.0	70
65–74	40.3	55.6	30
75 and above	120.0	134.7	10

9. Refer to data at Question 3 above. If the specific death rates of the standard population are respectively 26, 4, 5, 10, 45 (per thousand) for the different age-groups, find the Adjusted Death Rate.

10. Describe the main columns of a complete life table, explaining their significance. [C.U., B.A.(Econ) '76]

11. In a complete life table $l_{20} = 50{,}000$ and $L_{20} = 49{,}865$. Find the value of q_{20}.

12. Find the values of l_{91}, l_{92} and l_{93}, given that

$$l_{90} = 717,\ d_{90} = 214,\ q_{91} = .33640,\ p_{92} = .62006$$

13. Given $d_{10} = 58$, $q_{10} = 0.00062$, and $T_{10} = 5{,}583{,}559$, find the values L_{10} and e°_{10}. [C.U., B.Sc. '69]

14. In the following table, fill in the blanks which are marked with a query (?):

Age (x)	l_x	d_x	p_x	q_x	L_x	T_x	e°_x
10	74,600	?	?	?	?	3,266,067	?
11	74,340	—	—	—	—	?	?

15. Complete the following table:

x	l_x	d_x	q_x	L_x	T_x	e°_x
83	3,560	?	0.16	?	?	?
84	?	?	0.17	?	11,975	?

16. Supply the missing entries in the following table:

x	l_x	d_x	q_x	L_x	T_x	e°_x
20	71167	...	...	...	...	...
21	70768	...	...	...	...	...
22	70367	403	...	70166	2394139	...

17. From the data given below, calculate the probability that a person aged 50 years will survive at least one year:

Age (x)	50	51	52	53
$1000q_x$	26.35	28.41	30.66	33.12

18. Given the following table:

Age x	90	91	92	93	94	95	96
l_x	16090	11490	8040	5510	3610	2320	1447

The ages of three men *A*, *B* and *C* are 90, 91 and 92 respectively. Find the probabilities that

(i) *A*, *B* and *C* will be alive individually for at least 2 years;
(ii) all three of them will be alive in 2 years' time;
(iii) one at least of the three will be alive in two years' time;
(iv) exactly one of them will be alive in two years' time.

19. Define *Crude Birth Rate* and point out its defects. Discuss the uses of *General Fertility Rate* and *Age-specific Fertility Rate* in this connection. [*W.B.H.S. '83*]

20. The number of births occurring in a country in a certain year is shown below classified according to age of mother, together with the female population in different age-groups in the child-bearing period:

Age-group	*15–19*	*20–24*	*25–29*	*30–34*	*35–39*	*40–44*	*45–49*
Female Population ('000)	85	80	73	76	75	72	67
Number of Births to Mothers in the Age-group ('00)	23	145	167	102	51	14	1

Total population of the country in the year was 2286 thousand. Obtain the Crude Birth Rate and the General Fertility Rate. [*W.B.H.S. '78*]

21. From the following data, compute (i) *General Fertility Rate*, and (ii) *Total Fertility Rate*:

Age-group	*Female Population ('000)*	*No. of Births to Mothers in the Age-group*
15–19	20.5	902
20–24	19.0	2356
25–29	15.8	2054
30–34	12.0	1056
35–39	11.0	572
40–44	9.2	138
45–49	7.5	9

22. Define *Gross Reproduction Rate* (G.R.R.), and *Net Reproduction Rate* (N.R.R.). Show that the former must be larger than the latter.

[B.U., B.A.(Econ) '81]

23. Pick out the correct answer: If you are given taht the Total Fertility Rate is 2006 (per thousand) and sex-ratio at birth is 1 male to 1 female, then the Gross Reproduction Rate will be (i) 1.003, (ii) 1003, (iii) 2.006, (iv) none of these.

24. From the following data, calculate (i) Total fertility rate, and (ii) Gross reproduction rate, assuming that for the given population the ratio of female births to total births is 49%.

Age-group	*15–19*	*20–24*	*25–29*	*30–34*	*35–39*	*40–44*	*45–49*
Fertility Rate per 1000 Women	20	169	252	199	146	64	8

25. Calculate (i) general fertility rate, (ii) gross reproduction rate, from the following data, assuming that sex-ratio at birth is 52 males: 48 females.

Age-group	*Female Population ('000)*	*Total Live-births ('000)*
16–20	1,700	85
21–25	1,660	190
26–30	1,650	206
31–35	1,640	148
36–40	1,600	64
41–45	1,550	20
46–50	1,500	3
	11,300	716

26. The following data show the female population of a city in the child-bearing ages and the number of live-births in a year:

Age-group (years)	*16–20*	*21–25*	*26–30*	*31–35*	*36–40*	*41–45*	*46–50*
Female Population ('000)	20	18	16	14	15	13	12
No. of Live-births	400	1800	2240	1680	1200	650	120

Calculate (i) General Fertility Rate, (ii) Age-specific Fertility Rate for each age group, (iii) Total Fertility Rate, and (iv) Gross Reproduction Rate, assuming that the ratio of male to female births is 52 : 48.

27. Calculate (i) Gross Reproduction Rate, and (ii) Net Reproduction Rate from the following data:

Age-group	*Specific Fertility Rate per Thousand Females*	*Years Lived by 1000 New-born Girls*
15–20	140	3,600
20–25	250	3,500
25–30	240	3,400
30–35	190	3,200
35–40	130	2,900
40–45	50	2,600
45–50	20	2,300

Assume that the sex ratio at birth is 98 females per 100 males.

28. From the information given below, estimate the gross and net reproduction rates (Ratio of male to female births is 1.05 : 1):

Exact-age (years)	*Number of Expected Survivals for Females*	*Specific Fertility Rate per Annum per 1000 Women*
0	1,000	—
15	732	80
20	712	200
25	690	160
30	662	100
35	614	60
40	552	40
45	492	10
50	428	—

29. For a country for a particular year the total fertility rate is found to be 2.893 and the net reproduction rate 1.123. Explain what precisely is meant by these figures. *[C.U.,B.A.(Econ) '80]*

30. Discuss the following statements : "For any community the net reproduction rate is necessarily less than the gross reproduction rate." *[C.U., B.Sc. '78]*

ANSWERS

3. (i) 14.6; (ii) 30, 3, 4.5, 11, 40; (iii) 11.39 (per thousand)
4. 2.01, 3.60 (per thousand)
5. 16.6 (per thousand). [**Hint:** Calculate age-specific death rates for Town II and find their weighted A.M., using population of Town I as weights.]
6. (i) 14.9 per thousand, (ii) 14.9 per thousand. Yes, because the proportion of all people in District *A* is the same as that in the Standard Population, for each age-group.
7. 13.2, 13.5 (per thousand)
8. Country I; because its death rate (7.47 per thousand) is less than that for Country II (7.50 per thousand).
9. 11.22 per thousand   11. 0.0054   12. 503, 334, 207
13. 93,519; 59.69

14.

x	l_x	d_x	p_x	q_x	L_x	T_x	e°_x
10	74,600	260	0.99651	0.00349	74,470	3,266,067	43.78
11	74,340	—	—	—	—	3,191,597	42.93

15.

x	l_x	d_x	q_x	L_x	T_x	e°_x
83	3560	570	0.16	3275	15250	4.28
84	2990	508	0.17	2736	11975	4.01

16.

x	l_x	d_x	q_x	L_x	T_x	e°_x
20	71,167	399	.00561	70,968	2,535,675	35.63
21	70,768	401	.00567	70,568	2,464,707	34.83
22	70,367	403	.00573	70,166	2,394,139	34.02

17. Required probability $= p_{50} = 1 - q_{50} = 0.97365$
18. (i) 0.50, 0.48, 0.45; (ii) 0.108; (iii) 0.857; (iv) 0.392
20. CBR = 22 per thousand; GFR = 97 per thousand
21. (i) 74.6 per thousand; (ii) 2271 per thousand
23. 1.003
24. 4.29, 2.10
25. 0.063, 1.043
26. (i) 0.075, (ii) .02, .10, .14, .12, .08, .05, .01; (iii) 2.6, (iv) 1.248
27. 2.52, 1.66
28. 1.59, 1.05

19 STATISTICAL QUALITY CONTROL

19.1 INTRODUCTION

Statistical Quality Control (S.Q.C.) refers to the statistical techniques employed for the maintenance of uniform quality in a continuous flow of manufactured products. One of its major tools is the '*control chart*', first introduced by W.A. Shewhart, through the application of normal distribution. S.Q.C. has been so widely accepted in modern industry that any manufacturing company that does not use these techniques definitely stands at a disadvantage with respect to its competitors.

In large-scale industrial production the ultimate aim of the manufacturer is to produce articles or components as uniform as possible at the most economical rate. With all the precision of modern engineering, however, it is a common feature in all production processes that no two items passing off the same machine are exactly alike. Variations in "quality" of the product (e.g. length of a screw, diameter of ball bearings, thickness of washers, etc.) are inevitable. If the variation from the specified standard is not much, the product can be accepted as satisfactory. But if the quality is beyond certain fixed limits, called 'tolerance', the article has to be rejected as defective.

One of the methods employed to ensure that the defective are not passed into stock from the factory is to have an inspection department, which will examine each and every item. But this system has two major disadvantages: firstly, the defective is traced only after it has been produced; and secondly, the cost of such inspection is often considerable, increasing the per unit cost of production. Moreover, human nature is such that even 100 per cent inspection cannot guarantee that only satisfactory products will leave the factory.

By the use of '*control charts*', the manufacturer (technically called '*producer*') can however control the manufacturing process ensuring more or less a uniform quality of products with the minimum possible spoilage and rework, which results in a lower cost of production. This is called "*Process Control*". On the other hand, '*sampling inspection plans*' help the purchaser (technically called '*consumer*') of manufactured articless to dispense with 100% inspection and yet guarantee him against accepting lots beyond a specified percentage of defectives in the long run. This is known as "*Lot Control*" or "*Product Control*".

19.2 CHANCE CAUSES AND ASSIGNABLE CAUSES

Variations in the quality of products in a manufacturing process are attributed to two distinct types of causes—(i) *Chance causes* and (ii) *Assignable causes*. It is certain that some small variations are inherent in the production process and cannot be removed altogether, however refined our machinery may be. These are generated by numerous independent factors, which are generally known as 'chance causes'. Since chance causes do not affect the quality of the product to any appreciable extent, it is uneconomic and even impossible to trace them all. The important and larger variations are generated by faults in the production process, which by themselves adversely affect the quality of product. The factors responsible for such large differences in quality, which are known as 'assignable causes', can and must be traced out immediately if the product is to conform to the required specifications.

When assignable causes have been eliminated from the production process, the variations in the quality will be solely due to chance causes and are called 'controlled variations'. It has been observed that the quality of a product under controlled variation conforms to a definite statistical pattern. If therefore, quality measurements in a production process are found to follow a similar pattern, it is assumed that no assignable causes of variation are present. The process is then said to be '*under control*' or '*in a state of statistical control*'. Control charts enable, by simple graphical means to know whether assignable causes of variation are operating in the process, so that remedial measures can be taken to rectify the process before any real damage is done.

19.3 CONTROL CHART—HOW IT WORKS

Control chart is a graphical device for presenting a running record of suitable sample characteristics. In any manufacturing process, it is used to detect lack of control in the process and to give indications of any future trouble. Control charts are usually prepared for Mean, Range, Fraction defective, or Number of defects per article.

The control chart consists of three horizontal lines with a vertical line on the left. The horizontal line at the middle is called *Central Line*, and those above and below it are known as *Upper Control Limit* (*U.C.L.*) and *Lower Control Limit* (*L.C.L.*) respectively. The vertical line shows the scale of measurements and the horizontal scale indicates sample numbers. The distances of Control Limits from the Central Line are determined on the basis of probability. Ordinarily, U.C.L. and L.C.L. are at equal distances from Central Line, this common distance being equal to 3 times the standard deviation (called 'Standard Error' in sampling theory) of the sample characteristic for which the control chart is prepared. Such limits are known as '*three-sigma control limits*' (see Figs 19.1, 19.2, 19.3).

As production goes on, quality measurements are taken at regular intervals of time from a small sample (known as '*rational sub-group*') of consecutive items produced. The characteristic (e.g. mean or range) for which the control chart has been prepared is calculated for each of these samples and plotted as points on the control chart against the corresponding sample number. Since variations in quality due to chance causes follow a probability distribution, so long as the production process is under statistical control, all sample points will fall within the control limits, and no action is called for. But as soon as assignable causes of variation appear in the process, the quality of the product varies widely from the set standard and consequently sample points will fall outside the control limits (either above UCL or below LCL) indicating lack of control of the process. The fact is immediately brought to the notice of the foreman or the engineer in charge who shuts down production, looks for the cause of trouble and finally rectifies the defect before production is resumed.

Control charts are based on the principles of the important probability law known as Normal Distribution. It is assumed that when the process is in a state of control the 'quality' of the product follows a normal distribution. The mean and standard deviation of the sample characteristic (e.g. mean range, etc.), which are expected in the long run are then theoretically obtained from the sampling distribution. The Central Line of the control chart is drawn at the mean and the control limits (UCL and LCL) are drawn at position 3-sigma distances away from the Central Line ('sigma' denoting the standard deviation of the characteristic). This ensures that if the production process is under control, only about 27 points in 10,000 on the average will fall outside the control limits. This is so because in a normal distribution, the probability of a value lying between *Mean+3(s.d.)* and *Mean – 3(s.d.)* is 0.9973 (see page 119); and consequently the probability of a point falling outside these limits is 0.0027.

Rational Sub-group

The success of the control chart technique depends largely upon the efficient grouping of items into samples, such that variation in quality among items within the same sample is small, but variation between one sample and another is as large as possible. Such a sample is known as '*rational sub-group*'. Usually, the order of production is taken as a basis for obtaining rational sub-groups, the consecutive samples being taken from the production line at intervals of time, say about 30 minutes. If articles are produced in more than one machine, the source may also be used for rational sub-grouping.

Why 3-Sigma Control Limits

In the control chart, two kinds of errors may be made—

(i) The sample point may fall outside the control limits, and consequently we any be looking for the presence of assignable causes, when really none is present.

(ii) The sample point may fall within the control limits, so that we may not be aware of the actual trouble, although assignable causes may be operating in the process.

If control limits are set at wider distance from the Central Line, say at plus and minus four times the standard deviation, the possibility of the first kind of error will diminish, and that of the second kind will increase. If control limits are closer, say at two times the standard deviation from the Central Line, the reverse will be the case. Experience has shown that control limits set at about three times the standard deviation provide a good economic balance between these two types of errors. As such, 3-sigma control limits are commonly used unless there are special reasons for using any other limits.

19.4 CONTROL CHARTS FOR VARIABLES AND ATTRIBUTES

Control charts are of two types, depending on the nature of quality which it is desired to control. If quality is stated in terms of a variable, e.g. diameter, tensile strength etc., we have *Control Charts for Variables*. Three such charts are in common use—

(a) Mean chart ($\overline{x}$ chart),
(b) Range chart (R chart),
(c) Standard Deviation chart (σ chart)

Sometimes the quality cannot be measured directly, or it is not measured for economy, and the products are classified by an attribute (see page 3). These are known as *Control Charts for Attributes*. For example, very often articles are inspected or tested in batches and classified either as 'good' (i.e. those which conform to specifications) or 'defective' (i.e. those which are considered as useless rejections). The quality of a product is then measured by "fraction defective" $p = d/n$, where p is a fraction and d is the number of defectives found in a sample of n articles inspected. Again, a single product or a specified size of the product may contain one or more defects, e.g. paper, textiles, metallic sheets, pipes etc. The number of defects c is then used as a quality measure. Two important control charts for attributes are

(d) Fraction Defective Chart (p chart)
(e) Number of Defects chart (c chart)

Statistical Concepts

(I) *Control Charts for Variables*—It is assumed that the quality characteristic x follows a Normal distribution (page 459) with mean $\overline{x}'$ and standard deviation σ', if only chance causes are at work. The mean $\overline{x}$ from random samples of size n, therefore, follows a normal distribution with mean $\overline{x}'$ and s.d $\sigma'/\sqrt{n}$. Consequently, the probability that $\overline{x}$ lies within the interval bounded by

$$\overline{x}' - 3\frac{\sigma'}{\sqrt{n}} \text{ and } \overline{x}' + 3\frac{\sigma'}{\sqrt{n}}$$

is 0.9973 and the chance of lying outside these limits is $1 - 0.9973 = .0027$. In other words, if control limits for mean are drawn at $\overline{x}' \pm 3\sigma'/\sqrt{n}$ on the average only 27 points out of 10,000 will fall outside these limits.

Neither the range (R) nor the standard deviation (σ) from random samples follows a normal distribution. But each has its own probability distribution with a mean and a standard deviation, and almost all sample values of R or σ will lie within the interval bounded by its mean plus and minus three times standard deviation.

(II) *Control Charts for Attributes*—Let us suppose that random samples of a fixed size n are taken from a large population with proportion of defectives p'. Then the number of defectives (d) observed in the sample follows a binomial distribution (page 445) with mean np' and s.d. $\sqrt{np'(1 - p')}$. Consequently, the fraction defective $p = d/n$ observed in large samples approximately follows a normal distribution with mean p' and s.d $\sqrt{p'(1 - p')/n}$ (see Theorem II, page 519).

Again, let us suppose that the number of defects (c) per piece follows Poisson distribution (page 441) with means c'. Since mean and variance for Poisson distribution are equal, therefore its s.d. = $\sqrt{c'}$. Again, we know that when the mean is large, the Poisson distribution can be approximated by a normal distribution. Consequently, if c' is not very small, the number of defects c follows approximately a normal distribution (page 475).

It may be observed that the S.Q.C. theory employs notations usually different from those used in statistical sampling theory (p. 500). A list of these notations is given below:

	S.Q.C. theory	*Sampling theory*
Mean (population)	$\bar{x}'$	m (or μ)
Mean (sample)	$\bar{x}$	$\bar{x}$
Standard deviation (population)	σ'	σ
Standard deviation (sample)	σ	S
Fraction defective (population)	p'	P
Fraction defective (sample)	p	p
Average number of defects (Poisson population)	c'	m
Number of defects (sample)	c	—

19.5 FORMULAE FOR CENTRAL LINE AND CONTROL LIMITS

From the point of view of applications, the calculations of Central Line and the Control Limits (*LCL* and *UCL*) depend on two different situations:

(i) *Standards given*—Sometimes from a long history of the process in statistical control, we may know the *process average* ($\bar{x}'$) and the *process standard deviation* (σ'). Similarly, in control charts for attributes, the fraction defective (p'), or the average number of defects (c') may be specified in advance. These aimed-at-values are then taken as "standards" and use to set control lines for subsequent data.

(ii) *Standards not given*—In most practical applications, no standards are available, but we analyse past data from the process and estimate $\bar{x}'$ and σ' therefrom. Suppose we have k subsgroups each containing n observations. If $\bar{x}_i$ denotes the mean of the i-th subgroup, then x' is estimated by the average of subgroup means:

$$\bar{\bar{x}} = (\bar{x}_1 + \bar{x}_2 + \ldots + \bar{x}_k) \div k$$

There are two ways of estimating the process variability—either by using standard deviation, or range, of subgroups. Let σ_i and R_i denote respectively the standard deviation and range of the i-th subgroup. The average of subgroup standard deviations

$$\bar{\sigma} = (\sigma_1 + \sigma_2 + \ldots + \sigma_k) \div k$$

or the average of ranges (called *Mean Range*)

$$\bar{R} = (R_1 + R_2 + \ldots + R_k) \div k$$

multiplied by the appropriate constant gives an estimate of σ'. In fact, σ' is estimated by $\bar{\sigma}/c_2$ or by $\bar{R}/d_2$, the values of c^2 and d_2 being available from the tables for different n (see Table 19.1).

For attribute charts, p' and c' are estimated by averages obtained from past data. If $d_1, d_2, ..., d_k$ denote the number of defective articles observed in k subgroups of equal size n, the fraction defectives $p_1, p_2, ... p_k$ are given by $p_i = d_i/n$. Again, let $c_1, c_2, ... c_k$ denote the number of defects observed in k single pieces of the product. Assuming that the process is under control, the estimates of p' and c' are respectively.

$$\bar{p} = (p_1 + p_2 + ... + p_k) \div k$$

and

$$\bar{c} = (c_1 + c_2 + ... + c_k) \div k$$

The following formulae are used:

(1) Control Chart for Mean ($\bar{x}$ *chart*):

	Standards given	Standards not given (using s.d.)	(using range)
Central Line	$\bar{x}'$	$\bar{\bar{x}}$	$\bar{\bar{x}}$
LCL	$\bar{r}' - A\sigma'$	$\bar{\bar{x}} - A_1\bar{\sigma}$	$\bar{\bar{x}} - A_2\bar{R}$
UCL	$\bar{x}' + A\sigma'$	$\bar{\bar{x}} + A_1\bar{\sigma}$	$\bar{\bar{x}} + A_2\bar{R}$

(2) Control Chart for Standard Deviation (σ *chart*):

	Standard given	Standard not given
Central Line	$c_2\sigma'$	$\bar{\sigma}$
LCL	$B_1\sigma'$	$B_2\bar{\sigma}$
UCL	$B_2\sigma'$	$B_4\bar{\sigma}$

(3) Control Chart for Range (R *chart*):

	Standard given	Standard not given
Central Line	$d_2\sigma'$	$\bar{R}$
LCL	$D_1\sigma'$	$D_3\bar{R}$
UCL	$D_2\sigma'$	$D_4\bar{R}$

(4) Control Chart for Fraction Defective (p *chart*):

	Standard given	Standard not given
Central Line	p'	$\bar{p}$
LCL	$p' - 3\sqrt{\frac{p'(1-p')}{n}}$	$\bar{p} - 3\sqrt{\frac{\bar{p}(1-\bar{p})}{n}}$
UCL	$p' + 3\sqrt{\frac{p'(1-p')}{n}}$	$\bar{p} + 3\sqrt{\frac{\bar{p}(1-\bar{p})}{n}}$

(5) Control Chart for Number of Defects (c *charts*):

	Standard given	Standard not given
Central Line	c'	$\bar{c}$
LCL	$c' - 3\sqrt{c'}$	$\bar{c} - 3\sqrt{\bar{c}}$
UCL	$c + 3\sqrt{c'}$	$\bar{c} + 3\sqrt{\bar{c}}$

[**Note:** Since 'fraction defective' and 'number of defects' cannot be negative in the p chart or c chart if the formula L.C.L. yields negative values, we must take L.C.L. = 0.]

Example 19.1 *In order to construct a control chart, samples of sizes 10 are taken. The mean sample range is found tube 0.20 and the process average of the machine is set at 8.25. Find the upper and the lower control limits for mean. (Given that the constant of conversion for a sample of size n = 10 is $A_2 = 0.308$).*

[I.C.W.A., Dec. '75]

Solution We are given $n' = 10$, $\overline{R} = 0.20$, $\overline{\overline{x}} = 8.25$ and $A_2 = 0.308$. Control Limits for mean chart are:

$$\overline{\overline{x}} \pm A_2 \overline{R} = 8.25 \pm 0.308 \times 0.20$$

$$= 8.25 \pm 0.0616 = 8.3116 \text{ and } 8.1884$$

Upper Control Limit (U.C.L.) = 8.31

Lower Control Limit (L.C.L.) = 8.19

Table 19.1 Factors used in Control Charts

	Mean Chart			*Standard Deviation Chart*					*Range Chart*				
n	*for Control Limits*			*for Central Line*	*for Control Limits*				*for Central Line*	*for Control Limits*			
	A	A_1	A_2	c_2	B_1	B_2	B_3	B_4	d_2	D_1	D_2	D_3	D_4
2	2.12	3.76	1.88	0.564	0	1.84	0	3.27	1.13	0	3.69	0	3.27
3	1.73	2.39	1.02	.724	0	1.86	0	2.57	1.69	0	4.36	0	2.57
4	1.50	1.88	0.73	.798	0	1.81	0	2.27	2.06	0	4.70	0	2.28
5	1.34	1.60	.58	.841	0	1.76	0	2.09	2.33	0	4.92	0	2.11
6	1.22	1.41	.48	.869	0.03	1.71	0.03	1.97	2.53	0	5.08	0	2.00
7	1.13	1.28	.42	.888	.10	1.67	.12	1.88	2.70	0.21	5.20	0.08	1.92
8	1.06	1.17	.37	.903	.17	1.64	.19	1.81	2.85	.39	5.31	.14	1.86
9	1.00	1.09	.34	.914	.22	1.61	.24	1.76	2.97	.55	5.39	.18	1.82
10	0.95	1.03	.31	.923	.26	1.58	.28	1.72	3.08	.69	5.47	.22	1.78

Example 19.2 *(a) Explain the differences between specification limits tolerance limits, and control limits. What indications about lack of control are provided by a control chart?*

(b) Work out control limits on an $\overline{x}$ chart for samples of size 4 if the process has to meet a lower and an upper specification limit of 79 cms and 104 cms respectively.

[I.C.W.A., June '73]

Solution **Specification Limits**—When an article is proposed to be manufactured, engineers have to decide upon the maximum and the minimum allowable dimensions of some quality characteristic so that the product can be gainfully utilised for which it is intended. If the dimensions are beyond these limits, the product is treated as defective and cannot be used. These maximum and minimum limits of variation of individual items, as mentioned in the product design, are known as 'Specification Limits'.

Tolerance Limits—These are limits of variation of a quality measure of the product between which at least a specified proportion of the product is expected to lie (with a given probability), provided the process is in a state of control. For example, we may claim with a probability of

0.95 that a least 80% of the products will have dimensions between some stated limits. These limits are also known as 'statistical tolerance limits', while the words 'engineering tolerance limits' are used in the sense of specification limits.

Control Limits—These are limits of sampling variation of a statistical measure (e.g. mean, range, or fraction-defective) such that if the production process is under control, the values of the measure calculated from different rational subgroups will lie within these limits. Points falling outside the control limits indicate that the process is not operating under a system of chance causes; i.e. assignable causes of variation are present, which must be eliminated. Control limits are used in Control Charts.

Indications about Lack of Control

Control charts enable us to know whether the production process is under the influence of only chance causes of variation (when it is said to be in a "state of control") or any assignable causes is operating in the process. Control charts also give indications of impending lack of control, so that remedial measures can be taken before any real damage is done.

Principal indications about lack of control in the production process, as given by a control chart, are as follows:

(i) Any point on the control chart falling outside control limits.
(ii) A large majority of points, especially several consecutive points, falling near one of the control limits.
(iii) An unusually large number of points falling on one side of the central line in the control chart.
(iv) A trend exhibited by the plotted points.

(b) We know that, if *a* variable *x* follows Normal distribution with mean m and standard deviation σ, then almost all the values of x (about 99.73%) will lie between the limits $\mu \pm 3\sigma$ (page 471). This means that for a normally distributed variable the upper and the lower limits to the individual observations may be taken as $\mu + 3\sigma$ and $\mu - 3\sigma$. Therefore, for a statistically controlled process with a process average, $\bar{x}'$ and process s.d. σ' (these correspond to l^t and σ in sampling theory), wherein the individual observations are assumed to follow normal distribution, we have

$$\bar{x}' + 3\sigma' = \text{Upper Specification Limit (USL)} = 104$$

$$\bar{x}' - 3\sigma' = \text{Lower Specification Limit (LSL)} = 79$$

Solving these two equation for $\bar{x}'$ and σ', we get

$$\bar{x}' = 91.5 \text{ and } \sigma' = 25/6 = 4.167$$

Again, in random samples of size n from a normal population with mean μ and s.d. σ, the sample mean $\bar{x}$ follows normal distribution with mean μ and standard error (S.E.) $\sigma/\sqrt{n}$, so that the upper and the lower limits for $\bar{x}$, which are 3-sigma distance on either side of the mean, are $\mu \pm 3\sigma/\sqrt{n}$. Hence the 3-sigma control limits for $\bar{x}$-chart are

$$\bar{x}' \pm 3\frac{\sigma'}{\sqrt{n}} = 91.5 \pm \frac{3 \times 4.167}{\sqrt{4}} = 91.5 \pm 6.25 = 97.75 \text{ and } 85.25$$

Example 19.3 *The table below shows the thickness of mica discs produced in a machine, as obtained from 10 samples of size 5 each drawn at regular intervals of half an hour. Draw control charts for Mean and Range, and comment on the state of control.*

	Thickness of Mica Discs (thousandths of an inch)									
Sample No.	*1*	*2*	*3*	*4*	*5*	*6*	*7*	*8*	*9*	*10*
	13	*14*	*11*	*14*	*12*	*10*	*10*	*8*	*13*	*5*
	8	*12*	*10*	*10*	*10*	*12*	*16*	*12*	*8*	*8*
	12	*18*	*8*	*12*	*12*	*8*	*8*	*10*	*11*	*9*
	15	*19*	*16*	*9*	*13*	*10*	*8*	*8*	*14*	*5*
	8	*10*	*9*	*13*	*7*	*8*	*10*	*10*	*7*	*10*

(Conversion factors for n = 5 are $A_2 = 0.577$, $D_3 = 0$, $D_4 = 2.115$)

Solution For drawing control charts, we have to calculate the values of Central Line and Control Limits for each chart.

Table 19.2 Calculations for Mean and Range

Sample No.	*1*	*2*	*3*	*4*	*5*	*6*	*7*	*8*	*9*	*10*	*Total*
	13	14	11	14	12	10	10	8	13	5	
	8	12	10	10	10	12	16	12	8	8	
	12	18	8	12	12	8	8	10	11	9	
	15	19	16	9	13	10	8	8	14	5	
	8	10	9	13	7	8	10	10	7	10	
Total	56	73	54	58	54	48	52	48	53	37	—
Mean ($\bar{x}$)	11.2	14.6	10.8	11.6	10.8	9.6	10.4	9.6	10.6	7.4	106.6
Range ($\bar{R}$)	7	9	8	5	6	4	8	4	7	5	63

For the 10 samples, $\Sigma\bar{x} = 106{,}6$, $\Sigma R = 63$ (see Table 19.2)

Therefore, Process Average ($\bar{\bar{x}}$) = $106.6 \div 10 = 10.66$

Mean Range ($\bar{R}$) = $63 \div 10 = 6.3$

Control Chart for Mean ($\bar{x}$):

$$\text{Central Line} = \bar{\bar{x}} = 10.66$$

$$\text{U.C.L.} = \bar{\bar{x}} + A_2\bar{R} = 10.66 + 0.577 \times 6.3 = 10.66 + 3.64 = 14.30$$

$$\text{L.C.L.} = \bar{\bar{x}} - A_2\bar{R} = 10.66 - 3.64 = 7.02$$

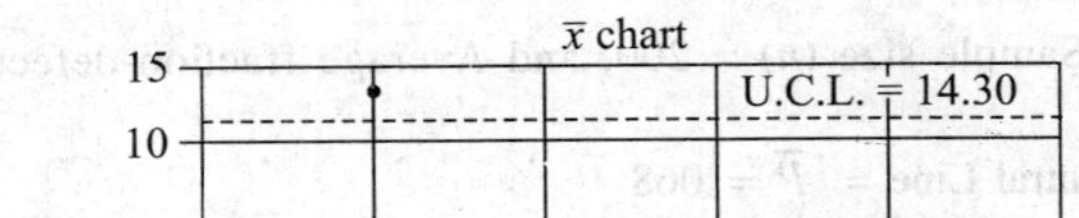

Fig. 19.1 Control Chart for Mean

Control Chart for Range (R):

$$\text{Central Line} = \overline{R} = 6.3$$

$$\text{U.C.L.} = D_4\overline{R} = 2.115 \times 6.3 = 13.32$$

$$\text{L.C.L.} = D_3\overline{R} = 0 \times 6.3 = 0$$

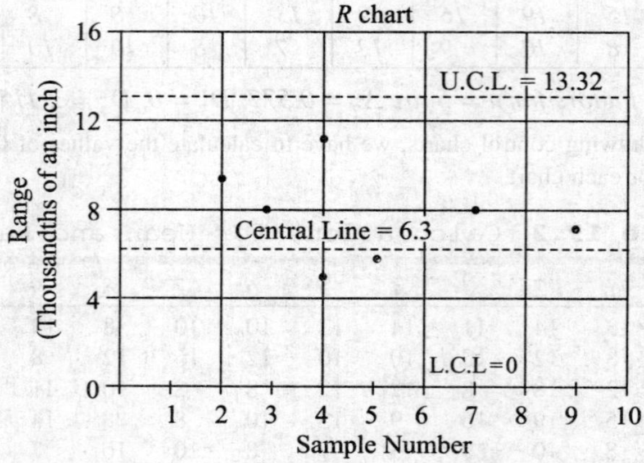

Fig. 19.2 Control Chart for Range

Comments—In the *x* chart (Fig. 19.1), the point for Sample No. 2 lies beyond the U.C.L., and hence the production process is not is a state of control, so far as the 'average' is concerned. In the *R*-chart (Fig. 19.2), all points lie within the control limits and no trend is exhibited by the plotted points; hence the process is in a state of control, so far as the 'variability' is concerned.

Example 19.4 *Based on 15 subgroups each of size 200 taken at intervals of 45 minutes from a manufacturing process, the average fraction-defective was found to be .068. Calculate the values of Central Line, and the Control Limits for a p-chart.*

[I.C.W.A., June '75]

Solution Here Sample size (n) = 200, and Average fraction-defective ($\overline{p}$) = .068.
For a *p*-chart

$$\text{Central Line} = \overline{p} = .068$$

$$\text{Control Limits} = \overline{p} \pm 3\sqrt{\frac{\overline{p}(1-\overline{p})}{n}} = .068 \pm 3\sqrt{\frac{.068 \times .932}{200}}$$

$$= .068 \pm 3\sqrt{.0003169} = .068 \pm 3 \times .0178$$

$$= .068 \pm .0534 = .1214 \text{ and } .0146.$$

Example 19.5 *In a glass factory the task of quality control was done with the help of mean ($\overline{x}$) and standard deviation (σ) chats, 18 samples of 10 items each were chosen and then values Σx and $\Sigma\sigma$ were found to be 595.8 and 8.28 respectively. Determine the 3-σ limits for mean and standard deviation charts. You may use the following control factors for your calculations:*

n	A_1	B_3	B_4
10	1.03	0.28	1.72

[I.C.W.A., June '82]

Solution We are given

No. of samples (k) = 18 $\quad \Sigma \bar{x} = 595.8$

Sample size (n) = 10, $\quad \Sigma \sigma = 8.28$

Hence, $\bar{\bar{x}} = \Sigma \bar{x}/k = 595.8 \div 18 = 33.1$

$\bar{\sigma} = \Sigma \sigma / k = 8.28 \div 18 = 0.46$

(i) *For Mean-Chart,*

$$\text{Central Line} = \bar{\bar{x}} = 33.1$$

$$\text{L.C.L.} = \bar{\bar{x}} - A_1 \bar{\sigma} = 33.1 - 1.03 \times 0.46$$
$$= 33.1 - 0.4738 = 32.6262$$

$$\text{U.C.L.} = \bar{\bar{x}} + A_1 \bar{\sigma} = 33.1 + 0.4738 = 33.5738$$

(ii) *For S.D.-Chart,*

$$\text{Central Line} = \bar{\sigma} = 0.46$$
$$\text{L.C.L.} = B_3 \bar{\sigma} = 0.28 \times 0.46 = 0.1288$$
$$\text{U.C.L.} = B_4 \bar{\sigma} = 1.72 \times 0.46 = 0.7912$$

Example 19.6 *Measurements on averages ($\bar{x}$) and ranges (R) from 20 samples each of size 5 gave the following results:*

$\bar{\bar{x}} = 99.6$, $\bar{R} = 7.0$. Determine the value of the control limits for drawing a Mean Chart. (Given that for n = 5, mean range = 2.32 × Population s.d.)

[I.C.W.A., Dec. '83]

Solution We are given

Mean Range = 2.32 × Population s.d.

or, $\bar{R} = 2.32\ \sigma'$

Since $\bar{R} = 7.0$, we have $\sigma' = 7.0/2.32 = 3.017.$

The standard error (S. E.) of $\bar{x}$ is $\sigma'/\sqrt{n} = 3.017/\sqrt{5}$

Hence, the three-sigma control limits for mean-chart are

$$\bar{\bar{x}} \pm 3\ \frac{\sigma'}{\sqrt{n}} = 99.6 \pm 3 \times \frac{3.017}{\sqrt{5}}$$
$$= 99.6 \pm 4.0477$$
$$= 103.6477 \text{ and } 95.5523.$$

Example 19.7 *The following figures give the number of defectives in 20 samples each containing 2000 items: 425, 430, 216, 341, 225, 322, 280, 306, 337, 305, 356, 402, 216, 264, 126, 409, 193, 326, 280, 390. Calculate the values for Central Line and the Control Limits for p chart (Fraction-Defective chart).*

[I.C.W.A., June '78]

Solution Total number of defectives out of 40,000 items inspected in the 20 samples is

$$\Sigma d = 425 + 430 + \ldots + 389 = 6148$$

The average fraction defective is therefore,

$$\bar{p} = 6148 \div 40{,}000 = 0.1537$$

For *p Chart:*

Central Line = $\bar{p}$ = 0.1537

$$\text{L.C.L.} = \bar{p} - 3\sqrt{\frac{\bar{p}(1-\bar{p})}{n}} = 0.1537 - \sqrt{\frac{0.1537\,(1-0.1537)}{2000}}$$

$$= 0.1537 - 0.0242 = 0.1295$$

$$\text{U.C.L.} = \bar{p} + 3\sqrt{\frac{\bar{p}(1-\bar{p})}{n}} = 0.1537 + 0.0242 = 0.1779$$

Ans. 0.1537; 0.1295, 0.1779

Example 19.8 *Draw control chart for fraction defective (p Chart) for the data given in Example 19.7, and comment.*

Solution *From the given data, we calculate the fraction defective (p) for each sample (here sample size = 2000).*

$$\text{Fraction defective} = \frac{\text{Number of defectives}}{\text{Sample size}}$$

Table 19.3 Calculations for *p* Chart

Sample No.	*Number of Defectives*	*Fraction Defective (p)*	*Sample No.*	*Number of Defectives*	*Fraction Defective (p)*
1	425	0.2125	11	356	0.1780
2	430	0.2150	12	402	0.2010
3	216	0.1080	13	216	0.1080
4	341	0.1705	14	264	0.1320
5	225	0.1125	15	126	0.0630
6	322	0.1610	16	409	0.2045
7	280	0.1400	17	193	0.0965
8	306	0.1530	18	326	0.1630
9	337	0.1685	19	280	0.1400
10	305	0.1525	20	389	0.1945
			Total	6148	3.0740

The average fraction defective is

$$\bar{p} = \Sigma p/k = 3.0740 \div 20 = 0.1537$$

The control limits have been calculated in Example 19.7.

L.C.L. = 0.1295, U.C.L. = 0.1779.

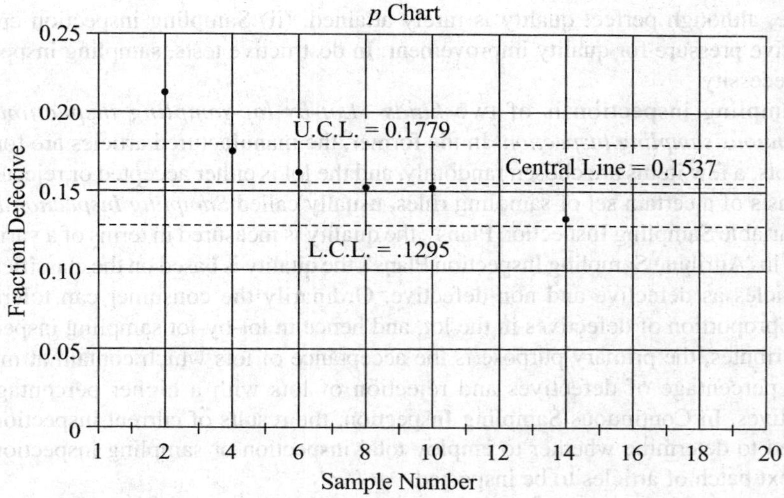

Fig. 19.3 Control Chart for Fraction Defective

Comment—We find that all points do not lie between the control limits. Hence the process is not under control.

Example 19.9 *The following are the number of defects noted in 20 hundred-yard pieces of woollen goods:*

3, 3, 6, 3, 0, 1, 3, 5, 7, 8, 4, 10, 5, 5, 4, 3, 4, 5, 4, 1.

Find $\bar{c}$ and the control limits for the number of defects.

Solution Total number of defects (Σc) = 84

Average number of defects per piece ($\bar{c}$) = 84 ÷ 20 = 4.2

For *c Chart:* Central Line = $\bar{c}$ = 4.2

$$\text{L.C.L.} = \bar{c} - 3\sqrt{c} = 4.2 - 3\sqrt{4.2}$$

$$= 4.2 - 3 \times 2.05 = 4.2 - 6.15, \text{ which is negative.}$$

$$\therefore \quad \text{L.C.L.} = 0$$

$$\text{U.C.L.} = \bar{c} + 3\sqrt{c} = 4.2 + 6.15 = 10.35$$

Ans. 4.2; 0, 10.35.

19.6 SAMPLING INSPECTION

When manufactured articles are submitted by the producer to the consumer for acceptance, the latter is always faced with the problem: whether to inspect the articles 100 per cent or to decide the acceptance or rejection of a lot on the basis of sampling only. There are two main points in favour of sampling inspection—(i) The cost of total inspection is usually very high, thereby considerably increasing the cost per unit article, although perfect quality is rarely attained. (ii) Sampling inspection creates effective pressure for quality improvement. In destructive tests, sampling inspection is a necessity.

Sampling inspection is of two *kinds—Lot-by-lot sampling inspection* and *Continuous sampling inspection.* In the former, the manufactured articles are formed into lots, a few items are chosen randomly, and the lot is either accepted or rejected on the basis of a certain set of sampling rules, usually called *Sampling Inspection Plan.* In 'Variable Sampling Inspection Plans', the quality is measured in terms of a variable, while in 'Attribute Sampling Inspection Plans', the quality is based on the classification of articles as defective and non-defective. Ordinarily the consumer can tolerate a small proportion of defectives in the lot, and hence in lot-by-lot sampling inspection by attributes, the primary purpose is the acceptance of lots which contain at most a fixed percentage of defectives and rejection of lots with a higher percentage of defectives. In Continuous Sampling Inspection, the results of current inspection are utilised to determine whether to employ total inspection or sampling inspection for the next batch of articles to be inspected.

19.7 SINGLE AND DOUBLE SAMPLING INSPECTION PLANS

In lot-by-lot sampling inspection by attributes, two types of sampling plans are commonly used—'Single Sampling Inspection Plan' and 'Double Sampling Inspection Plan'.

Single Sampling Plan—A sample of n items is chosen randomly from a lot of N items. If the sample contains, say; c or fewer defectives, the lot is accepted; while if it contains more than c defectives, the lot is rejected (c is known as *'acceptance number'*). For example, consider the following single sampling plan:

$$N = 150, \qquad n = 10, \qquad c = 1$$

These numbers are interpreted as under: "Take a random sample of 10 items from a lot containing 150 items. If the sample is found to contain at most 1 defective, accept the lot; otherwise reject the lot."

Double Sampling Plan—A sample of n_1 items is first chosen at random from the lot of size N. If the sample contains, say, c_1 or fewer defectives, the lot is accepted; if it contains more than c_2 defectives, the lot is rejected. If, however, the number of defectives in the sample exceeds c_1 but is not more than c_2, a second sample of n_2 items is taken from the same lot. If now the total number of defectively in the two samples together does not exceed c_2, the lot is accepted; otherwise it is rejected. For example, consider the following double sampling inspection plan:

$$N = 1000; \qquad n_1 = 50, \qquad c_1 = 1, \qquad n_2 = 40, \qquad c_2 = 3$$

These numbers are interpreted as under:

(i) Inspect first a random sample of 50 items from a lot containing 1000 items.

(ii) If the number of defectives found in the sample is at most 1, accept the lot; if it is more than 3, reject the lot.

(iii) If the number of defectives in the first sample exceeds 1, but does not exceed 3 (i.e. either 2 or 3), inspect a second sample of 40 items.

(iv) Accept the lot, if the combined sample of 90 items contains at most 3 defectives. Reject the lot, if the number of defectives in the combined sample exceeds 3.

19.8 IMPORTANT TERMS USED IN SAMPLING INSPECTION

Under a sampling inspection plan, a lot is submitted by the 'producer' to the 'consumer' for inspection, and the latter then examines only a sample of items from the lot. On the basis of the findings from the sample the consumer decides either to "accept" or to "reject" the whole lot.

(1) *Producers Risk* and *Consumer's Risk*—Since the composition of a sample depends on pure chance, there are two possible risks in volved in the procedure:

(a) *Rejection* of a lot of acceptable quality. A lot containing actually a small fraction defective may at times be rejected as unsatisfactory by the sampling plan.

(b) *Acceptance* of a lot of unsatisfactory quality. A lot containing actually a large fraction defective may at times be accepted as satisfactory (because the number of defectives in the sample happens to be less than the acceptance number by chance).

The risk of the first type which is detrimental to the interests of the producer is known, as *Producer's Risk*. The risk of the second type which goes against the interests of the consumer is known as *Consumer's Risk*.

(2) *OC (Operating Characteristic)*—The OC of a sampling inspection plan may be defined as the "*probability of acceptance*" of a lot. Naturally the OC depends on the actual fraction defective (p') of the lot. For example, in the single sampling plan: $N = 150$, $n = 10$, $\sigma = 1$, if the fraction defective is $p' = 0.1$, then there are $150 \times 0.1 = 15$ defectives, and the remaining 135 are non-defectives. The lot will be accepted, if the number of defectives in the sample of 10 items is either 0, or 1. Therefore the probability of acceptance of a lot on the basis of the plan is (see 11.7.1, p. 373)

$$P_a = (\text{Prob. of 0 defective}) + (\text{Prob. of 1 defective}).$$

$$= \frac{{}^{15}C_0 \times {}^{135}C_{10}}{{}^{150}C_{10}} + \frac{{}^{15}C_1 \times {}^{135}C_9}{{}^{150}C_{10}}$$

$$= 0.3367 + 0.4009 = 0.7376.$$

It is evident that the OC is a function of p', say

$$P_a = L(p')$$

If the values of P_a are calculated for different values of p', and these are plotted on a graph paper against the corresponding values of p' shown along the horizontal axis, then the curve obtained is known as *OC Curve*.

(3) *AQL (Acceptable Quality Level)*—It is the lot quality (as shown by fraction defective p') at which the probability of rejection has a small specified value (usually .05). If the AQL is denoted by P_1, the Producer's risk corresponding to p_1 is .05; i.e. the probability of acceptance is 0.95.

$$P_a = 0.95 = L(p_1)$$

(4) *LTPD (Lot Tolerance Per cent Defective)*—It is the lot quality (measured as per cent defective, or fraction defective p') at which the probability of acceptance has a small specified value (usually 0.10. If the LTPD is denoted

by p_2, the consumer's risk corresponding to p_2 is 0.10: i.e. the probability of acceptance is 0.10.

$$P_a = 0.10 = L(p_2)$$

(5) *ASN (Average Sample Number)*—In single sampling plans, a predetermined number of items (i.e. sample size *n*) is first examined from the lot. If the number of defectives in the sample does not exceed the acceptance number *c*, no further items need be examined and the whole lot of size *N* is accepted (along with whatever defectives that may remain among the unexamined items). The defectives observed in the sample are however replaced by good items. If the number of defectives in the sample exceeds *c*, the lot is not considered acceptable. In this case it has become customary to examine all the items in the lot (i.e. the remaining *N–n* items also), and get all the defective items replaced by good items.

It is evident that the number of items that has to be examined per lot is in some cases *n* and in some other cases it is *N*. Of course, the proportion of lots that need be examined hundred per cent depends on the fraction defective of the incoming lot. The average number of items examined per lot in the long run is called the *ASN* of the plan. ASN is a function of the fraction defective p'. If ASN is numerically calculated for different values of p' and then plotted on a graph paper the curve obtained is known as *ASH Curve.*

(6) *AOQ (Average Outgoing Quality)*—It is the average fraction defective in all the lots taken together, after the rejected lots have been cleared of all defectives. The AOQ depends on the fraction defective p' of the incoming lot. If the values of AOQ are plotted on a graph paper against the corresponding values of p', the curve obtained is known as *AOQ Curve.*

(7) *AOQL (Average Outgoing Quality Limit)*—It is the maximum value of AOQ, according to the plan, for all values of the fraction defective p'. Thus the AOQL of a sampling inspection plan fixes an upper limit to the proportion of defectives in the accepted lots, whatever be the fraction defective of the incoming lots submitted for inspection.

EXERCISES

1. Explain what you mean by Statistical Quality Control. Distinguish between Process Control and Product Control. *[W.B.H.S. '83; I.C.W.A.,(old) June '74]*
2. Explain what are *chance causes* and *assignable causes* of variation in the quality of manufactured product. *[I.C.W.A., Dec., '82]*
3. Explain the terms: (i) *quality,* and (ii) *control.* Hence discuss the role of Statistics in controlling the quality of a product. *[I.C.W.A. June '81]*
4. Explain the term "rational subgroup", as used in S.Q.C. *[I.C.W.A., Dec. '80; W.B.H.S. '82]*
5. Explain the theoretical background of a Control Chart. *[I.C.W.A., June '74, '75]*
6. Examine each of the following statements and write whether. It is *true* or *false:*

(a) A controlled process means one which allows no variation in the products. *[W.B.H.S. '79]*

(b) A process in control means that all its products are according to the manufacturer's specification. *[W.B.H.S. '78]*

(c) If the production process is under control, most of the values of a statistical measure calculated from different rational subgroups lie within the control limits of that measure. *[W.B.H.S. '82]*

(d) Control charts for mean and range serve the same purpose. So we use one of them and not both. *[W.B.H.S. '78]*

7. Assuming that the characteristic variable follows a normal distribution (mean and standard deviation unknown) give the *control limits* and the *central line* for the mean and range charts. *[I.C.W.A., Dec. '82]*

8. For 20 samples of 5 items each the sum of the sample means was 256.8 and the sum of the sample ranges was 7.84. Compute 3-sigma control limits of the sample mean and range on the basis of the table below:

Sample Sizes	d_2	A_1	A_2	D_1	D_2	D_3	D_4
4	2.059	1.880	0.729	0	4.698	0	2.282
5	2.326	1.596	0.577	0	4.918	0	2.115
6	2.534	1.410	0.483	0	5.078	0	2.004

[C.U., B.Sc. '72]

9. The following data show the values of sample mean ($\bar{x}$) and range (*R*) for ten samples of size 6 each. Calculate the values for Central Line and the Control Limits for mean-chart and range-chart, and comment on the state of control.

Sample No.	*1*	2	3	4	5	6	7	8	9	10
Mean ($\bar{x}$)	43	49	37	44	45	37	51	46	43	47
Range (*R*)	5	6	5	7	7	4	8	6	4	6

(Conversion factors for $n = 6$ are $A_2 = 0.483$, $D_3 = 0$, $D_4 = 2.004$)

10. A machine is set to deliver packets of a given weight. Ten samples of size 5 each were recorded. Below are given relevant data:

Sample No.	*1*	2	3	4	5	6	7	8	9	10
Mean (*x*)	15	17	15	18	17	14	18	15	17	16
Range (*R*)	7	7	4	9	8	7	12	4	11	5

Calculate the values for the central line and the control limits for mean chart and the range chart and then comment on the state of control. (Conversion Factors for $n = 5$ are $A_2 = 0.58$, $D_3 = 0$, $D_4 = 2.11$). *[I.C.W.A., Dec. '80]*

11. Describe the construction of control charts for mean and fraction defective from appropriate data when standards are given and stands are not given. *[W.B.H.S., '81]*

12. A process is designed to yield machine parts known as spacers with mean thickness 0.8470 inch. Successive samples of sizes 5 yield $\bar{x}$ values as follows:

.8464 , .8465, .8477, .8464, .8473, .8469, .8472, .8482, .8478, .8474.

The mean range $\bar{R}$ in these samples is .0005. Construct an $\bar{x}$ chart and determine whether the process is in control.

Factors for determining control limits for $\bar{x}$ chart

n	3	4	5	6
A_2	1.02	0.73	0.58	0.48

[C.U., B.Sc. '75]

13. A machine is designed to produce ball bearings having a mean diameter of 0.574 cm and a standard deviation of 0.008 cm. To determine whether the machine is in proper working order, a sample of 6 ball bearings is taken every two hours on all the working days of the week and the mean diameter is computed from this sample. Design a rule whereby one can be fairly certain that the quality of the products are conforming to the required standards. *[I.C.W.A., June '76]*
14. Measurements of averages ($\bar{x}$) and ranges (R) from 20 samples, each of sizc 5, gave the following results: $\Sigma\bar{x} = 99.6$. SR = 7.0.
    (i) Determine the Central Line for mean-chart. (ii) Compute the 3-sigma upper and lower control limits for mean-chart on the assumption that the production process is in control. (You may make use of the following information:
    Mean range = $d_2 \times$ Population standard deviation
    where d_2 for samplo size 5 equals 2.326).
15. A large sample of a product gave an "average fraction defective" of 0.068. Calculate for a p-chart the values of control limits (upper, lower and central line), if the size of each sample subgroup is 200. *[I.C.W.A., Dec. '81]*
16. The average number of defectives in 22 sampled lots of 2000 rubber belts each was found to be 16%. Indicate how to construct the relevant control chart. *[I.C.W.A., June '83]*
17. The past records of a factory using quality control methods show that on the average 4 articles produced are defective out of a batch of 100. What is the maximum number of defective articles likely to be encountered in a batch of 400 when the production process is in a state of control? *[I.C.W.A., Dec. '74]*
18. During an examination of equal length of cloth the following are the *number of defects* observed: 2, 3, 4, 0, 5, 6, 7, 4, 3, 2. Draw a control chart for the number of defects and comment whether the process is under control or not. *[I.C.W.A., June '80]*
19. The following table shows the number of missing rivets noted at the final inspection of l2 aircrafts. Find the control limits for the number ot defects chart, and comment on the state of control:

Aircraft No.	*1*	*2*	*3*	*4*	*5*	*6*	*7*	*8*	*9*	*10*	*11*	*12*
No. of Missing Rivets	7	15	13	18	10	14	13	10	20	11	22	15

[I.C.W.A., Dec. '78]

20. Explain the concepts of Producer's risk, Consumer's risk, OC and ASN in sampling inspection plans. What are AQL and LTPD?

ANSWERS

1. (see Section 19.1) 2. (see Section 19.2)
3. *Quality*—This refers to some characteristic of an industrial product. The 'quality' may be either a variable (e.g. thickness of metal sheet, diameter of a component, density of a chemical product, etc.) or an attribute (e.g. whether the product is 'defective' or 'non-defective').

   Control—In any manufacturing process. variations in quality from item to item are inevitable. Variations due to 'assignable causes' can be removed by the application of S.Q.C. A production process is said to be in a "state of control", if it is under the inftuence of 'chance causes' only and no assignable causes are present. The quality of the product or the production process is then said to be "controlled".
4. (see Section 19.3) 5. (see Section 19.3)
6. (a) False, (b) False, (c) True, (d) False.
7.

	Central Line	*L.C.L.*	*U.C.L.*
Mean chart	$\bar{\bar{x}}$	$\bar{\bar{x}} - A_2 \bar{R}$	$\bar{\bar{x}} + A_2 \bar{R}$
Range chart	$\bar{R}$	$D_3\bar{R}$	$D_4\bar{R}$

8.

	Central Line	*L.C.L.*	*U.C.L.*
Mean chart	12.84	12.61	13.07
Range chart	0.392	0	0.83

9.

	Central Line	*L.C.L.*	*U.C.L.*
Mean chart	44.2	41.4	47.0
Range chart	5.8	0	11.6

   Comment: The production process is not in a state of control, so far as the 'average' is concerned (because all the sample means *do not* lie between the control limits). The process is, however, in a state of control, so far as the 'variability' is concerned (all sample ranges lie between the control limits).

10.

	Central Line	*L.C.L.*	*U.C.L.*
Mean chart	16.2	11.91	20.49
Range chart	7.4	0	15.61

   Comment: Since all the sample means and all the sample ranges lie between the corresponding control limits, the process is in a state of control.

12. Central Line = 0.8470, L.C.L. = 0.8467, U.C.L. = 0.8473.

    The process is not in a state of control.
13. $\bar{x}$ = 0.574. σ' = .008. The quality conforms to the standards so long as the sample means are found to lie between the control limits: L.C.L.= 0.564 cm and U.C.L. = 0.584 cm.
14. [**Hint:** see Example 19.6] (i) 4.98. (ii) 5.18 and 4.78.
15. Central Line = .068. L.C.L. = .0146. U.C.L. = 0.1214.
16. For *p*-chart, Central Line = 0.16, L.C.L.= 0.135. U.C.L. = 0.185.

17. 28. [**Hint:** For p-chart, with $\overline{p} = .04$ and $n = 400$, the U.C.L. is 0.0694; hence the upper limit to the number of defectives is $0.0694 \times 400 = 27.76$].
18. Central Line = 3.6, L.C.L. = 0, U.C.L. = 9.3. Since none of the observations exceeds U.C.L., the process is in a state of control.
19. Central Line = 14, L.C.L. = 2.8, U.C.L. = 25.2. Yes, the process is in a state of control.
20. (see Section 19.8).

THEORETICAL QUESTIONS

1. Distinguish between a *discrete* variable and a *continuous* variable. *[W.B.H.S. '84]*
2. State which of the following represent discrete data and which represent continuous data:
   (i) Number of inches of rainfall in a city during various months of the year ;
   (ii) Speed of a car in miles per hour;
   (iii) Number of ten-rupee notes circulating in India at any time;
   (iv) Annual incomes of college professors. *[C.U., B.Sc. (Econ) '86]*
3. Define *primary* data and *secondary* data. Discuss the different methods of collecting primary data. *[I.C.W.A., June '86]*
4. Enumerate the important sources of secondary data. What precautions one should take before using such data? *[C.A., Nov. '84]*
5. Write short explanatory notes (with illustrations) on:— Population and Sample. *[C.U., B.Com(H) '86]*
6. Distinguish between the 'census' and 'sampling' methods of collecting data and compare their merits and demerits.
7. What are the advantages of sampling over census as a method of investigation?
8. Explain the different parts of a table. *[W.B.H.S. '83, '87]*
9. Define graphical representation and tabular representation of data and distinguish between them, and write with examples where and why they are suitable. *[W.B.H.S. '88]*
10. Discuss briefly the different types of diagrams generally used to represent numerical data and point out the relative merits and demerits of diagrammatic representation compared to other method used for the purpose. *[C.U., B.Com.(H) '81]*
11. Distinguish between the *natural* scale and *logarithmic* scale used in graphical presentation of data.
12. Explain what is meant by semi-logarithmic chart. Discuss the uses of such a chart. *[C.U., B.Sc.(Econ) '87; W.B.H.S. '87]*
13. What do you mean by *bar diagram?* Describe the different types of bar diagrams, and the situations and techniques of applying them. *[W.B.H.S. '86]*

14. What is a Pie diagram? Describe how it is drawn.
15. Discuss the different considerations to be kept in view in drawing up a frequency distribution from raw data. *[C.U., B.Sc.(Econ) '85]*
16. Explain clearly with illustrations the following term: Class limits, Class boundaries, Frequency density; and discuss their uses with reference to frequency distribution. *[C.U., B.Sc.(Econ) '88]*
17. Explain clearly the distinction between (i) class limits and class boundaries; (ii) relative frequency and density; (iii) histogram and frequency polygon; (iv) less-than and more-than cumulative frequencies.
18. Discuss how would draw a histogram to represent the frequency distribution of a continuous variable. *[C.U., B.Com.(H) '88]*
19. Explain the construction of histogram with class intervals of unequal width. *[I.C.W.A; Dec. '84]*
20. What do you mean by ogives? Describe how they are drawn. *[W.B.H.S. '86]*
21. The point of intersection of the less-than ogive and greater-than ogive corresponds to .... (fill up the blank). *[C.U., B.Com.(H) '83]*
22. Explain what is meant by 'central tendency' of data. *[I.C.W.A., Dec. '78, '80 & June '83 ; W.B.H.S. '79, '85]*
23. Define 'weighted arithmetic mean' of a set of observations. Show that the weighted arithmetic mean is unaffected if all the weights are multiplied by a common factor. *[C.U., B.A.(Econ) '77]*
24. Prove that the sum of the deviations $x_1, x_2, \ldots x_n$ from their mean $\bar{x}$ is equal to zero. *[I.C.W.A. June '83]*
25. Let $x_1, x_2, \ldots, x_n$ be the value of a variable with frequencies $f_1, f_2, \ldots, f_n$ respectively, Prove that $\Sigma f_i (x_i - \bar{x}) = 0$. *[C.U., B.Com.(H) '80]*
26. Prove that the sum of squares of deviations is minimum when deviations are measured from the mean. *[W.B.H.S. '78, '82 ; C.U., B.Com.(H) '81, '83 ; B.Sc.(Econ) '87]*
27. Show that if $\bar{x}_1$ and x_2 are the arithmetic means of two sets of values of x, the mean $\bar{x}$ of the combined set will lie between $\bar{x}_1$ and $\bar{x}_2$. *[C.U., B.Sc. (Econ) '85]*
28. Define A.M., G.M. and H.M. of n positive quantities $x_1, x_2, \ldots, x_n$. Of the three means which has the greatest value? *[I.C.W.A., Dec. '81; C.U., B.Com.(H) '83]*
29. Define geometric mean. Show that the geometric mean of n observations $x_1, x_2, \ldots, x_n$ with frequencies $f_1, f_2, \ldots, f_n$ is the antilogarithm of the weighted arithmatic mean of the logarithms of the set of x's. *[C.U., B.Com.(H) '82]*
30. If A, G and H be the arithmetic, geometric and harmonic mean respectively between x and y, show that $A.H = G^2$. *[C.U., B.Com.(H) '85, '88]*
31. Prove that the geometric mean of two positive numbers is equal to the geometric mean of their arithmetic and harmonic means. *[I.C.W.A., Dec; '81, '83, '85]*
32. If a variety assumes n values $a, ar, ar^2, \ldots, ar^{n-1}$ with $(r < 1)$ equal (constant) frequencies then find the arithmetic mean (A), geometric mean (G) and harmonic mean (H), and show that $AH = G^2$. *[W.B.H.S. '86]*

33. Prove that for any two positive quantities.
A.M. ≥ G.M. ≥ H.M. *[I.C.W.A., Dec. '82, '84]*
34. If x and y are two variables such that $y = a + bx$ and if M is the median of x, then show that the median of y is $a + bM$.
35. What is meant by mode of a distribution? Does it always exist? Is it unique? Give examples to illustrate your answer. *[C.U., B.Com.(H) '82]*
36. If mean and median of a frequency distribution are known, how do you estimate the mode?
37. Compare mean, median and mode as measures of central tendency. *[W.B.H.S. '79, '81, '83, '87, '88]*
38. Indicate the merits and shortcomings of the different measures of central tendency as well as situations where to use each. *[C.U., B.Com.(H) '81]*
39. What are 'quartiles' of a distribution? How do you use them for measuring dispersion and skewness? *[C.U., B.Com.(H) '80]*
40. Explain how by graphically estimating appropriate measures from the ogive of a given frequency distribution on a continuous variable, one can throw light on its central tendency, dispersion and skewness. *[C.U., B.Com.(H) '88]*
41. Explain what your mean by the dispersion of a frequency distribution. *[W.B.H.S. '78, '80, '82, '86]*
42. Show that mean deviation is least when calculated about median. *[W.B.H.S. '79, '81, '84, '87]*
43. Obtain the standard deviation of the first n natural numbers. *[C.U., B.Sc.(Econ) '85]*
44. Prove that the standard deviation calculated from two values x_1 and x_2 of a variable x is equal to half their difference. *[W.B.H.S. '82, '86 ; I.C.W.A., June '80, Dec., '82]*
45. Prove that the standard deviation is independent of any change of origin but is dependent on change of scale. *[W.B.H.S. '82 ; C.U., B.Com.(H) '88 ; I.C.W.A., June '82, '86]*
46. Show that root-mean-square deviation is minimum when measured about the arithmetic mean. *[W.B.H.S. '80]*
47. Find the A.M. and S.D. of y when $y_i = (x_i - \bar{x})/S_x$, where and S_x have usual meaning. *[C.U., B.A.(Econ.) '79]*
48. Let $x_1, x_2, ..., x_n$ be a set of observations. Suppose we compute $y_i = a + bx_i$ $(i = 1, 2, ..., n)$ where a and b are constants. Express the s.d. of y's in terms of the s.d. of x's and comment on the relation between the two. *[C.U., B.A.(Econ.) '78]*
49. $\bar{x}$ is the mean of x_1, x_2 and x_3. If d_1, d_2 and d_3 are the deviations of x_1, x_2 and x_3 from $\bar{x}$ respectively, prove that
$d_1^2 + d_2^2 + d_3^2 = x_1^2 + x_2^2 + x_3^2 - 3\bar{x}^2$ *[C.U., BSc. (Econ.) '88]*
50. Compare range, mean deviation and standard deviation as measures of dispersion. *[W.B.H.S. '78, '80]*
51. Define 'coefficient of variation'. Point out its importance in statistical data analysis. *[I.C.W.A., Dec. '84]*

52. Write short notes on:—Relative merits and demerits of various measures of dispersion. *[C.U., B.Com.(H) '82, '84]*
53. Discuss how different measures of variability are affected when all variety values are changed (i) by the same amount, (ii) in the same proportion. *[C.U., B.Com.(H) '87]*
54. Define 'moments' of a distribution. What are raw and central moments? *[C.U., B.Com.(H) '86]*
55. Express the second, third and fourth central moments in terms of raw moments. *[I.C.W.A., Dec. '76]*
56. Show that central moments are invariant under change of origin but not under change of scale. *[I.C.W.A., Dec. '75]*
57. Let $x_1, x_2, ..., x_n$ be n observations with mean and $\bar{x}$ standard deviation σ. Let m_r denote the r-th central moment of the above observations and let α_r denote the r-th central moment of $z_1, z_2, ..., z_n$ where $z_i = (x_i - \bar{x})/\sigma$. Prove that $\alpha_r = m_r/\sigma^r$. *[I.C.W.A. Dec. '81, June '85]*
58. Explain the terms 'skewness' and 'kurtosis' of a frequency distribution. Write down the different measures of skewness and kurtosis. *[C.U., B.Com.(H) '82, '87]*
59. With the help of rough diagrams show the different types of skewness and kurtosis indicating their names. *[C.U., B.Com.(H) '84, '85]*
60. Show by means of sketches the relative positions of mean, median and mode for frequency curves which are (i) positively skew, (ii) negatively skew, and (iii) symmetrical.
61. What do you understand by bivariate data and a two-way table? *[C.A., May '82]*
62. Define the term 'correlation' between two variables x and y. *[I.C.W.A., June '85, C.U., B.Com.(H) '84]*
63. Write down Karl Pearson's *product moment* formula for measuring correlation between two variables. *[C.U., B.Com.(H) '80 , '87; I.C.W.A., June '82, '85; W.B.H.S. '88]*
64. Prove that the product moment correlation coefficient (r) lies between –1 and +1. *[I.C.W.A., Dec. '81, '83; W.B.H.S. '81, '84, '88; C.U., B.com.(H) '81, 87]*
65. Prove that $-1 \le r \le +1$. Also interpret the cases when $r = -1$, $r = 0$ and $r = 1$. *[I.C.W.A. Dec. '85]*
66. Prove that the correlation coefficient is independent of change of origin and scale. *[I.C.W.A. June '86]*
67. If $u = ax + b$ and $v = cy + d$, where a, b, c and d are constants, express r_{xy} in terms of r_{xy}. *[C.U., B.Com.(H) '88]*
68. Mention the various properties of correlation coefficient. *[C.A., Nov. '78]*
69. If x and y are independent variates, then prove that they are uncorrected. Is the converse true? Explain your answer with the help of an example. *[W.B.H.S. '80; I.C.W.A., June '80, Dec. '81]*
70. Show that $\text{Var}(X - Y) = \text{Var}(X) + \text{Var}(Y) - 2.\ \text{Cov}(X, Y)$, where Var denotes variance and Cov denotes covariance. *[C.U., B.Sc.(Econ) '85]*

71. If $s_y^2 = s_x^2$, show that $(x + y)$ and $(x - y)$ are uncorrected.
[C.U., B.Com.(H) '87]
72. If X_1, X_2 and X_3 are uncorrected variables each having the same standard deviation, obtain the correlation coefficient between $X_1 + X_2$ and $X_2 + X_3$.
[W.B.H.S. '84]
73. Describe the uses and limitations of correlation coefficient.
[I.C.W.A. Dec. '83 ; C.U., B.Com.(H) '87]
74. What is a scatter diagram? Indicate by mean of suitable scatter diagrams the different types of correlation that may exist between the variables in bivariate data. *[C.U., B.Com.(H) '83 ; I.C.W.A., June '82]*
75. Write short notes on—Least squares theory and its uses.
[W.B.H.S. '78, '84, '86, '88]
76. Define the term 'regression'. What is meant by 'regression of x on y'?
[I.C.W.A., Dec. '82, '84, '85; C.U., B.Com.(H) '88]
77. What is the difference between 'correlation' and 'regression'?
[I.C.W.A., Dec. '79]
78. In bivariate data, what are regression lines? *[W.B.H.S. '82, '83]*
79. Explain how you will find the linear regression equation of one variable on another. *[C.U., B.Com.(H) '87]*
80. Explain the concept of regression and write down the equations of regression lines. *[C.A., May '81]*
81. Show that the two regression lines meet at the point . *[C.U., B.Com.(H) '88]*
82. Show that $r^2 = b_{yz} \cdot b_{zy}$, where the symbols have their usual meaning.
[I.C.W.A., Dec. '76]
83. Prove that the arithmetic mean of the two regression coefficients is greater than the correlation coefficient (r) for positive r. *[W.B.H.S. '83]*
84. If the two coefficients of regression are negative, does it mean that their correlation would be positive ? *[I.C.W.A., Dec., '79]*
85. If the two regression lines coincide, show that the correlation coefficient $r = \pm 1$. If $r = 0$, show that the two regression lines are at right angles.
[I.C.W.A., Dec. '79, '84, '85]
86. If the two variables are independent, show that the two regression lines are at right angles. *[I.C.W.A., Dec. '77]*
87. What can you say about the angle between the regression lines when (i) $r = 0$, (ii) $r = \pm 1$, (iii) r increases from 0 to 1 ? *[I.C.W.A., Dec. '76]*
88. When are the regression lines (i) perpendicular to each other, and (ii) coincide ?
[I.C.W.A., June; C.A., Nov. '79]
89. Deduce the expression of regression coefficients by the method of least squares. Point out the important properties of these estimates. *[I.C.W.A., June '83]*
90. If $u = ax + b$ and $v = cy + d$, where a, b, c and d are constants, express b_{uv} in terms of b_{xy}, and b_{vu} in terms of b_{yz}.
91. Why should there be two regression lines?
[C.A., Nov. '79; I.C.W.A., June '84 and Dec. '84, '85; C.U., B.Com.(H) '88]
92. If x and y are two correlated variables with the same variance and the correlation coefficient is r, find the regression coefficient of x on $(x + y)$ and that of $(x + y)$ on x. Hence find the correlation coefficient between x and $(x + y)$.
[W.B.H.S. '85]

93. Are the following statements valid? Justify your answers.
  (a) Positive value of correlation coefficient between x and y implies that if x decreases, y tends to increase.
  (b) Correlation coefficient is independent of the origin of reference.
  (c) Correlation coefficient between x and y turned out to be 1.02.
[I.C.W.A., Dec. '80]

94. Give reasons to justify that the following statements cannot be true:
  (i) $b_{yx} = 0.8$ and $b_{xy} = -0.3$;
  (ii) $b_{yx} = 0.8$ and $b_{xy} = 2.4$. *[C.U., B.Com.(H) '84)*

95. Is it possible for two regression lines to be as follows:
$Y = -1.5X + 7$ and $X = 0.6Y + 9$? Give reasons. *[C.U., B.Com. (H) '87]*

96. For a set of pairs of observations $(x_1, y_1), (x_2, y_2), \ldots, (x_n, y_n)$ prove that

$$\sum_i (y_i - \bar{y})^2 - \sum_i (y_i - y_i)^2 + \sum_i (y_i - \bar{y})^2$$

where y_i represents the values of y for given values of

$x = x_1, x_2, \ldots, x_n$

obtained from the regression equation of y on x. *[C.U., B.Sc.(Econ) '88]*

97. Define the term 'rank correlation'. Write down the formula for Spearman's 'rank correlation coefficient'.
[C.A., May '82; C.U., B.Com.(H) '83; I.C.W.A., Dec., '84]

98. If the difference in ranks in each pair is zero, show that the rank correlation coefficient is + 1. *[I.C.W.A., Dec. '84]*

99. Derive the formula for Spearman's rank correlation coefficient when ther are noties. *[I.C.W.A., Dec. '84; W.B.H.S. '79, '82, '86]*

100. Prove that $-1 \le \rho \le +1$, where ρ is Spearman's rank correlation coefficient in case of no ties. *[W.B.H.S. '82]*

101. Explain the significance of the results; (i) $\rho = +1$, (ii) $\rho = -1$, where ρ Denotes the rank correlation coefficient.

102. Explain the concept of 'multiple regression'. *[I.C.W.A., June '81]*

103. In the case of three variables x_1, x_2 and x_3, write down the formulae for
  (i) Multiple regression equation of x_1 on x_2 and x_3;
  (ii) Partial correlation coefficient $r_{12.3}$;
  (iii) Multiple correlation coefficient $R_{1.23}$;
  (iv) Partial regression coefficient $b_{12.3}$;

104. What are the limits for (i) Partial correlation coefficient, (ii) Multiple correlation coefficient?

105. What is meant by 'interpolation'? *[C.U., B.Com.(H) '82, '84, '86]*

106. Define the two operators Δ and E, and show that $E = 1 + \Delta$. If $U_x = x_3$, evaluate $\Delta^2 U$. *[C.U., B.Com.(H) '83]*

107. State Newton's 'forward' and 'backward' interpolation formulae, and comment on their uses. *[C.U., B.Com.(H) '82, '84, '85, '86]*

108. State Newton's Backward Interpolation formula with assumptions.
[I.C.E.A., June '83, '84]

109. What is 'inverse interpolation'? *[I.C.W.A., June '81. Dec. '82]*

110. State Lagrange's interpolation formula. *[I.C.W.A., June '82, Dec. '83]*

111. Show that the weighted A.M. of first n natural numbers whose weights are equal to the corresponding numbers is equal to $(2n + 1)/3$.

[I.C.W.A., June '81]

112. Let x be a variable assuming the values 1, 2, ..., n and let $F_1, F_2, ..., F_n$ be the corresponding cumulative frequencies of the more-than type. Prove that the A.M. of x is given by

$$\bar{x} = (F_1 + F_2 + ... + F_n)/N$$

where N is the total frequency.

113. Given a set of n values $x_1, x_2, ..., x_n$, show that the square of their arithmetic mean cannot exceed the arithmetic mean of their squares.

114. Show that if a variable takes the values 0, 1, 2, ..., n with frequencies 1, ${}^nC_1$, ${}^nC_2$, ..., ${}^nC_n$ respectively, then the mean square deviation about the origin is $n(n + 1)/4$. Also show that the mean and variance are $n/2$ and $n/4$ respectively.

[I.C.W.A., Dec. '79]

115. Suppose that a variable x assumes the values a and b and $(n - 2)$ other values all equal to $(a + b)/2$. Find
 (i) the mean deviation about mean,
 (ii) the standard deviation.

[B.U., B.A. (Econ.) '82]

[*Ans.* $|a - b|/n$, $|a - b|\sqrt{2n}$].

116. If the observations $x_1, x_2, ..., x_n$ all lie between a and b $(a < b)$, show that their s.d. cannot be larger than $(b - a)/2$.

117. For a given set of observations with mean x, prove that

$$S^2 = \sigma^2 + d^2$$

where S^2 is the mean square deviation from a constant A, σ is the standard deviation and $d = \bar{x} - A$.

118. If in a series of measurements we obtain m_1 values of magnitude x_1, m_2 values of magnitude x_2 and so on, and if x is the of all the measurements, prove that the s.d. is

$$\sqrt{\left[\frac{\Sigma m_r (k - x_r)^2}{\Sigma m_r} - \delta^2\right]}$$

where $x = k + \delta$, and k is a constant. *[I.C.W.A. June '86]*

119. Given n pairs of observations (x_i, y_i) for $i = 1, 2, ..., n$, let $z_i = x_i + y_i$. Prove that

$$\sigma_z^2 = \sigma_z^2 + \sigma_y^2 + 2r_{zy}\,\sigma_x\sigma_y.$$

120. Show that $R_{1.23}^2 = r_{12}^2 + r_3^2$ if $r_{23} = 0$ *[I.C.W.A., June,'87]*

SUPPLEMENT

CHAPTER 1

COLLECTION OF DATA, CLASSIFICATION AND TABULATION

(Summary)

STATISTICS

The word "*Statistics*" has two different meanings. In the plural sense, it is used to denote 'facts and figures' on a particular topic. In the singular sense, Statistics refers to the subject of study which deals with the theories and methods of collection, analysis and interpretation of such numerical figures.

VARIABLE AND ATTRIBUTE

Variable is a quantitative character of an individual or item. *Attribute* is a qualitative character. Two types of variables—*discrete* (or *discontinous*) and *continuous.* 'Height' and 'temperature' are continous variables; 'family size' is a discrete variable. 'Mother-tongue' of an individual is an attribute.

PRIMARY AND SECONDARY DATA

Statistical data which are collected for a specific purpose directly from the field of enquiry are called *primary data.* "Reserve Bank of India Bulletin" published by Reserve Bank of India contains primary data regarding banking and currency in India.

Secondary data are those which have previously been collected by one agency for some purpose and merely compiled from that source for use in a different connection. "Statistical Abstract of the Indian Union" published by Central Statistical Organisation, Government of India, is a source of secondary data.

Methods of Collecting Primary Data—(1) Direct personal observation, (2) Indirect oral investigation, (3) Questionnaires sent through mail, (4) Interview method with schedules.

POPULATION AND SAMPLE

Statistical information on a particular character in respect of each and every unit convered by an enquiry is called *population.* A selected part of the population which is analysed with a view to ascertaining the properties of the population is called a *sample*.

CENSUS AND SAMPLE SURVEY

Collection of statistical data from each and every member of the population is called *census*; e.g. census of industries in West Bengal. Collection of data from only a selected part of the population (i.e. a sample) is called *sample survey*.

Sample survey method has definite advantages over census method of collecting data—(1) It takes less time, labour and money. (2) In sample survey the magnitude of error can be estimated; but in census method this is not possible. (3) In some cases census method is not possible, and sample survey is the only alternative.

QUESTIONNAIRE

This is a proforma containing a sequence of questions relevant to statistical enquiry, and is used for collection of primary data from individuals through their replies to the questions.

Essentials—(1) The questionnaire should be short. (2) Questions should be simple, unambiguous and precise. (3) Questions should be logically arranged. (4) Questions affecting pride and sentiments should not be asked.

CLASSIFICATION AND TABULATION

Classification is the process of arranging the collected information under different categories or classes according to some common characteristic of the individual members.

Tabulation is the logical and systematic organisation of data in rows and columns.

Different Parts of a Table: (1) Title, (2) Stub, (3) Caption and Box-head, (4) Body, (5) Footnote.

ADDITIONAL EXERCISES 1

1. Classify each of the following characters are either a discrete variable or a continuous variable: (i) Monthly income of a person, (ii) Heights of students of a particular school in the year 1978. *[W.B.H.S. '78]*
2. State which of the following represent discrete data and which represent continuous data: (i) Number of inches of rainfall in a city during various months of the year, (ii) Speed of a car in miles per hour, (iii) Number of ten-rupee notes circulating in India at any time, (iv) Annual incomes of college professors. *[C.U., B.Sc.(Econ.) '86]*
3. Distinguish between primary data and secondary data. *[W.B.H.S. '83 ; C.A., May '84]*
4. What do you mean by Primary data? How can they be collected? *[W.B.H.S. '85]*
5. Distinguish between the interview method and the mail questionnaire method of collecting data, state their relative merits and demerits. *[B.U., B.A. (Econ.) '79]*

6. Distinguish between a population and a sample. *[C.U., B.Sc. (Econ.) '83]*
7. Describe the characteristics of a good questionnaire. *[C.A., Nov. '82]*
8. Form a blank table to show the total number of students with males and females reading in 1st year, 2nd year and 3rd year in the streams Arts, Science, and Commerce in 1982 of a particular College under Burdwan University. *[W.B.H.S. '86]*
9. Prepare a model table with different columns and subcolumns to present the statistical data, giving the values of exports and imports through the ports of Calcutta and Bombay in 1971 and 1972 with respect to foodgrains, cash crops, and industrial products. Make provisions for the following in the table: (i) Totals for different columns and subcolumns, (ii) % increase or decrease in trade in 1972 with respect to 1971. *[W.B.H.S '81]*
10. Present the following data in a table:
    The Government grants to a University in the years 1946–47, 1947–48 and 1948–49 are respectively Rs 5,59,685, Rs 7,77,892 and Rs 7,48,000. The receipts by way of fees, endowments and donations during the year 1946–47 are respectively Rs 2,52,220, Rs 2,71,636 and Rs 4,83,934. The receipts for the years 1947–48 and 1948–49 in respect of fees are Rs 3,21,286 and Rs 3,83,153. In 1947–48 the University received by way of donation Rs 3,71,971 and in 1948–49 the University received by way of endowments Rs 3,350. The total receipts in the years 1947–48 and 1948–49 are respectively Rs 14,80,362 and Rs 14,39,025. *[I.C.W.A., June '86 (old)]*
11. Present the following information in a tabular form:
    In 1965 out of a total of 2000 workers of a factory, 1500 workers were members of a trade union. The number of women employed was 150 of which 128 do not belong to a trade union. In 1970 the number of union workers increased to 1620 of which 1582 were men. On the other hand the number of non-union workers fell down to 448 of which 318 were men. *[W.B.H.S. '85]*
12. Present the following data in a tabular form:
    The Orissa Government has taken up a scheme of providing drinking water for each village. During the first four years of the sixth plan the Government has installed 39,664 tubewells. Whereas out of the funds earmarked for natural calamities the Government has sunk 14,072 tubewells during the first four years of the sixth plan.

    Thus out of the sixth plan fund 9,245 and 8,630 tubewells were sunk respectively in 1980-81 and 1981–82. Whereas out of natural calamities fund the number of tubewells installed in 1980–81 and 1981–82 were 4,511 and 637 respectively. The investment for 1980–81 and 1981–82 were Rs 863.41 lakhs and 1185.65 lakhs respectively.

    The number of tubewells installed in 1982–83 were 16,740, out of which 4,800 were installed out of natural calamities fund and the expenditure in respect of sinking of tubewells during 1982–83 were Rs 1411.17 lakhs.

    The number of tubewells installed in 1983–84 were 13,973, out of which 9,849 tubewells were sunk out of fund for sixth plan and the total expenditure during the first four years were Rs 5443.05 lakhs. *[I.C.W.A., Dec, '84]*

ANSWERS

1. (i) Discrete, (ii) Continuous
2. Continuous—(i), (ii); Descrete—(iii), (iv)
8. **Number of Students in a College under Burdwan Univesity, 1982**

Class	*Arts*	*Science*	*Commerce*	*Total*
1st year:				
Males				
Females				
Total				
2nd year:				
Males				
Females				
Total				
3rd year:				
Males				
Females				
Total				
All Classes:				
Males				
Females				
Total				

9. **Exports & Imports through Calcutta and Bombay Ports (1971-72)**

Values (Rs Crores)

Ports/ Items	*Imports*			*Exports*		
	1971	*1972*	*% increase (+) decrease (–)*	*1971*	*1972*	*% increase (+) decrease (–)*
CALCUTTA:						
1. *Foodgrains*						
2. *Cash crops*						
3. *Industrial products*						
Total						
BOMBAY:						
1. *Foodgrains*						
2. *Cash crops*						
3. *Industrial products*						
Total						
Grand Total						

10. **Annual Receipts of a University (1946–49)**

(*Rupees*)

	Source	*1946–47*	*1947–48*	*1948–49*
1.	*Govt. Grant*	5,59,685	7,77,892	7,48,000
2.	*Fees*	2,52,220	3,21,286	3,83,153
3.	*Endowments*	2,71,636	9,213	3,350
4.	*Donations*	4,83,934	3,71,971	3,04,522
	Total	15,67,475	14,80,362	14,39,025

11. **Number of Trade Union Members among Workers in a Factory (Years 1965 & 1970)**

Trade Union	*1965*			*1970*		
	Men	*Women*	*Total*	*Men*	*Women*	*Total*
Member	1478	22	1500	1582	38	1620
Non-member	372	128	500	318	130	448
Total	1850	150	2000	1900	168	2068

12. **Tubewells for Drinking Water in Orissa Villages during Sixth Plan**

Year	*Number of Tubewells sunk*			*Expenditure (Rs lakh)*
	6th Plan Fund	*Natural Calamities Fund*	*Total*	
1980–81	9,245	4,511	13,756	863.41
1981–82	8,630	637	9,267	1,185.65
1982–83	11,940	4,800	16,740	1,411.17
1983–84	9,849	4,124	13,973	1,982.82
Total	39,664	14,072	53,736	5,443.05

CHAPTER 2

CHARTS AND DIAGRAMS

(*Summary*)

CHARTS AND DIAGRAMS

(1) Line Diagram (or Graph),
(2) Bar Diagram,
(3) Pie Diagram,
(4) Pictorgram,
(5) Histogram, Frequency Polygon and Ogives.

LINE DIAGRAM

'Natural scale' or 'Arithmetic scale' shows equal distances for equal amounts of change. 'Ratio scale' or 'Logarithmic scale' shows equal distances for equal ratios of change.

Ordinary graph shows natural scale in both horizontal and vertical directions. Semi-logarithmic graph shows natural scale in the horizontal direction, but ratio scale in the vertical direction.

BAR DIAGRAM

Bar diagram consists of a group of equispaced rectangular diagrams each of equal width. The rectangles start from a common base line and their lengths are proportional to the statistical data.

Bar diagrams may be 'vertical' or 'horizontal'. They may again be (i) Single bar, (ii) Grouped (or Multiple) bar, (iii) Component (or Subdivided) bar, (iv) Paired bar; etc.

PIE DIAGRAM

Pie diagram is a circular diagram whose area is divided proportionately into different sectors by straight lines drawn from the centre to the circumference of the circle. For drawing a Pie diagram, it is first necessary to express each value as a percentage of the total. The percentages are then multiplied by 3.6 each to give the angle of the sector at the centre of the circle. A compass and a protractor are necessary for drawing a Pie diagram.

ADDITIONAL EXERCISES 2

1. Represent the following data by a suitable diagram:

Year	1970	1971	1972	1973	1974	1975
Sale of Steel (thousand tonnes)	8	8.8	9.2	10.2	7.6	12.5

[W.H.B.S. '82]

2. What do you mean by Bar Diagram? Describe the different types of bar diagrams, the situations and techniques of applying them. *[W.B.H.S. '86]*

3. Production of Rubber in 1968 in the principal rubber producing countries:

Country	*Malaysia*	*Indonesia*	*Thailand*	*Sri Lanka*
Production ('000 tonne)	1,042	740	254	146
	India	*Combodia*	*Others*	*Total*
	68	51	294	2,595

Give a proper diagrammatic representation of the above data. *[W.B.H.S. '81]*

4. The percentage of literacy in West Bengal is shown below separately for males and females for 4 years:

Years	1941	1951	1961	1971
Males	27.4	34.1	40.1	44.8
Females	3.9	12.3	17.0	22.1

Represent the data in a suitable diagram. *[W.B.H.S. '80]*

5. Represent the following data of revenue of two Governments in a particular year by a suitable chart:

Sources	*Revenue (Rs Crores)* *Govt. I*	*Govt. II*
Customs	170	110
Excise	480	380
Income Tax	350	320
Corporation Tax	110	140
Others	90	50
Total	1200	1000

[W.B.H.S. '85]

6. The population of India is to be classified Statewise and is to be shown separately for males and females. Suggest suitable methods of diagrammatic representation of the data. *[W.B.H.S. '80]*

7. Draw a (rough) Pie Chart to represent the following data relating to the production cost of manufacture:

Cost of Materials	:	Rs 38,400
Cost of Labour	:	Rs 30,720
Direct Expenses	:	Rs 11,520
Overhead Expenses	:	Rs 15,360

[C.U., B.Com.(Hons) '86]

8. The following table gives the distribution of out-lay in a Five-year Plan of India under the major heads of development expenditure:

Head of Expenditure	*Expenditure (Crores of Rs)*
Agriculture and Community Development	8,000
Irrigation and Power	4,000
Industry and Mining	7,000
Transport and Communication	5,500
Miscellaneous	2,500
Total	27,000

Represent the above information by a Pie Chart. *[I.C.W.A., June '83]*

9. Construct a Pie Diagram for the following data:

Mines	*Coal*	*Iron Ore*	*Copper*	*Manganese*	*Mica*	*Others*
Average Daily Number of Workers Employed ('000)	348	34	4	90	31	84

[C.U., M.Com., '83]

ANSWERS

1. Line diagram
3. Horizontal Bar diagram
4. Grouped Vertical Bar
5. Component Bar
6. Paired Bar (Horizontal)
7. **Hint:** Total 96,000; Percentages 40, 32, 12, 16;
   Angle (degree) 144, 115, 43, 58.
8. **Hint:** Percent. 30, 15, 26, 20, 9;
   Angle (degree.) 108, 54, 94, 72, 32.
9. **Hint:** Total 591; Percent. 59, 6, 1, 15, 5, 14;
   Angle (degree.) 212, 22, 4, 54, 18, 50.

CHAPTER 4

FREQUENCY DISTRIBUTION
(Solved Problems)

Problem 4.1 *Construct the Grouped Frequency Distribution and Cumulative Frequency Distribution from the following:*

(a)	*Marks*	*No. of Students*	*(b)*	*Age*	*No. of Persons*
	Below 10	*15*		*Under 10*	*15*
	Below 20	*35*		*Under 20*	*30*
	Below 30	*60*		*Under 30*	*53*
	Below 40	*84*		*Under 40*	*75*
	Below 50	*106*		*Under 50*	*100*
	Below 60	*120*		*Under 60*	*110*
	Below 70	*125*		*Under 70*	*115*
	(Source: I.C.W.A., June '84)			*Under 80*	*125*
				(Source: I.C.W.A., June '80)	

Solution (a) [**Hint:** 'Marks' obtained by students is a *discrete* variable, which can take the values 0, 1, 2, 3, etc. So, mark "Below 10" can be only the values 0, 1, 2, ..., 9, i.e. upto 9; "Below 20" denotes upto 19, and so on. If marks "Below 10" are separated from those "Below 20", there ramain only the values between 10 and 19. Hence, the class intervals are 0–9, 10–19, etc. The class boundaries between successive classes are then 9.5, 19.5, ..., 69.5. Note that the lower boundary of the first class interval is not stated in Table 2, because a negative class boundary (– 0.5) would be meaningless here. Also see No. 9, page 98.]

Table 1 Grouped Frequency Distribution

Class interval	*Frequency*
0–9	15
10–19	20
20–29	25
30–39	24
40–49	22
50–59	14
60–69	5
Total	125 = *N*

Table 2 Cumulative Frequency Distribution

Class	*Cumulative*
9.5	15
19.5	35
29.5	60
39.5	84
49.5	106
59.5	120
69.5	125 = *N*

(b) [**Hint:** Here the variable 'Age' is *continuous*, which can take any value, even fractional values. "Under 10" can be any value between 0 and a number very close to 10, say 9.9999; no definite upper limit can be stated. Hence, for all practical purposes, age "Under 10" may be assumed to be between 0 and 10, i.e. 0–10. Similarly, "Under 20" can go upto a number which is practically 20. Hence the class intervals are 0–10, 10–20, etc. It should be noted that we cannot speak of a person aged exactly 10 years; because time ticks off in split-seconds.]

Table 3 Grouped Frequency Distribution

Class Interval	Frequency
0–10	15
10–20	15
20–30	23
30–40	22
40–50	25
50–60	10
60–70	5
70–80	10
Total	125 = N

Table 4 Cumulative Frequency Distribution

Class Boundary	Cumulative Frequency
0	0
10	15
20	30
30	53
40	75
50	100
60	110
70	115
80	125 = N

[**Note:** For calculating Mean, Standard Deviation (S.D.), Mean Deviation and Moments from data shown as in Problem 4.1 (a) and (b), it is necessary to obtain *mid-values* of class intervals. Hence Grouped Frequency Distribution is to be constructed.

For calculating Median, Mode, Quartiles, Deciles etc. it is necessary to obtain *class boundaries*. This cannot be done correctly unless one has a clear idea about the nature of Grouped Frequency Distribution. If class intervals are shown by using class limits (see Table 1), we must calculate class boundaries therefrom (see Table 2). If however class intervals are shown by using class boundaries (see Table 3), these class boundaries are shown directly in the Cumulative Frequency Distribution (Table 4). Also see No. 38, Page 171.]

Problem 4.2 *Construct the Grouped Frequency Distribution and Cumulative Frequency Distribution in each of the following cases:*

(a) Value	Frequency
Less than 10	*5*
Less than 20	*15*
Less than 30	*35*
Less than 40	*81*
Less than 50	*90*
Less than 60	*118*
Less than 70	*130*
Less than 80	*140*

(Source : I.C.W.A., Dec, '81)

(b) Annual Sales (Rs '000)	Frequency
Less than 10	*4*
Less than 20	*20*
Less than 30	*35*
Less than 40	*55*
Less than 50	*62*
Less than 60	*67*

(Source : I.C.W.A., June '83)

Solution (a) [**Hint:** If nothing is known about the nature of the variable, whether continuous or discrete, it is assumed to be *continuous*. This should be treated as in Problem 4.1 (b), the class intervals being 0–10, 10–20, etc. and the class boundaries being 0, 10, 20, etc. Also see Example 4.9, page 81.]

(b) [**Hint:** Strictly speaking the variable 'Annual Sales' is *discrete*, because Sales "Less than 10" thousand rupees can be only upto Rs 9999.99 P. However, for practical purposes Sales (in Rs '000) are taken to be 0–10, 10–20, etc.; i.e. the variable is treated as *continuous*. See Problem 4.1 (b) and Problem 5.4]

Problem 4.3 *Construct Cumulative Frequency Distribution and Grouped Frequency Distribution from the following data:*

Mid-value	*10*	*30*	*50*	*70*	*90*
Frequency	*40*	*51*	*64*	*38*	*7*

(Source : I.C.W.A., June '85)

Solution [**Hint:** If a grouped frequency distribution has classes of *equal width*, then the successive

Mid-values have a common difference,
Class boundaries have a common difference,
Lower class limits have a common difference,
Upper class limits hava a common difference,

this common difference being the common width of classes.

Hence, if the successive mid-values are found to have a common difference, the frequencey distribution must have classes of equal width, and the class boundaries will lie exactly midway between the successive mid-values. The common width of classes will be equal to the common difference between mid-values.]

Table 5 Cumulative Frequency Distribution

Class Boundary	*Cumulative Frequency*
0	0
20	40
40	91
60	155
80	193
100	200 = N

Table 6 Grouped Frequency Distribution

Class Interval	*Frequency*
0–20	40
20–40	51
40–60	64
60–80	38
80–100	7
Total	200 = N

[**Note:** For calculating Mean, S. D., M. D. and Moments from data as shown in this problem, the mid-values may be used directly.

For calculating Median, Quartiles, Deciles and Percentiles, Cumulative Frequency Distribution (Table 5) is necessary.

For calculating Mode, Grouped Frequency Distribution (Table 6) is required.]

ADDITIONAL EXERCISES 4

1. Consider the following frequency distribution:

Monthly Wages (Rs)	50.00–59.99	60.00–69.99	70.00–79.99	80.00–89.99
No. of Employees	8	10	16	14
	90.00–99.99	100.00–109.99	110.00–119.99	Total
	10	5	2	65

Determine (i) lower limit of the sixth class;
(ii) class-mark of the third class;
(iii) class boundaries of the fifth class; and
(iv) relative frequency of the third class.

[C.U., B.Sc. (Econ.) '83]

2. Find the class boundaries of the frequency distribution of weight in pounds of a group of boys, where the class limits used are 40.0–42.4, 42.5–44.9, 45.0–47.4, 47.5–49.9, 50.0–52.4, the measurement were made to the nearest $\frac{1}{10}$ th of a pound. *[C.U., B.Sc. (Econ) '86]*

3. Construct (i) Grouped frequency distribution, and (ii) Cumulative frequency distribution, from the following data:

(a) No. of Days Absent	No. of Students	*(b)* Value	Frequency	*(c)* Weekly Wages (Rs)	No. of Workers
Less than 5	29	Below 10	2	Below 10	8
Less than 10	239	Below 20	21	Below 20	18
Less than 15	469	Below 30	56	Below 30	45
Less than 20	584	Below 40	133	Below 40	90
Less than 25	634	Below 50	238	Below 50	113
Less than 30	644	Below 60	303	Below 60	120
Less than 35	650	Below 70	333		
Less than 40	655	Below 80	351	*(Source:*	
		Below 90	359	*(C.U., B.Com. (H) '80)*	
(Source:		Below 100	360		
I.C.W.A. Dec., '83)		*(Source: I.C.W.A. June '85)*			

4. From the following table, find (i) the number of men whose heights are less than 67.2 inches, and (ii) the number of men whose heights lie between 58.5 and 67.2 inches:

Height (inches)	56–57	58–59	60–61	62–63	64–65	66–67	68–69	70–71	72–73
Frequency	2	4	9	30	23	20	9	2	1

[C.U., B.Sc. (Econ) '86]

5. From the following table, find the number of wage earners receiving between Rs 37 and Rs 47 per week:

Weekly Wages (Rs) :	30–	32–	34–	36–	38–	40–	42–	44–	46–	48–50
Frequency :	3	8	24	31	50	61	38	21	12	2

[I.C.W.A., June '82]

6. (a) From the frequency distribution of heights shown in Q. 4, prepare a table showing class bounaries, class marks and cumulative frequencies (less than and more than).
   (b) Draw the histogram for the height distribution. Shade that portion of the histogram which will represent the total number of men whose heights lie between 60.5 and 66.5 inches. *[C.U., B.Com.(H) '84]*

7. Draw histogram of the following frequency distribution and use it to find the total number of wage earners in the age group 19–32 years:

Age Group	14–15	16–17	18–20	21–24	25–29	30–34	35–39
No. of Wage Earners	60	140	150	110	110	100	90

[C.U., B.Com.(H) '81]

8. Draw histogram for the following frequency distribution of heights of students. Hence find the height above which 120 students have their heights.

Height (cm)	126–135	136–145	146–150	151–155	156–160	161–165	166–180
Frequency	11	42	33	41	36	21	16

[W.B.H.S. '86]

9. Construct (a) a frequency polygon and (b) a cumulative frequency polygon, from the frequency distribution shown in Q. 1 above. *[C.U., B.Com.(H) '85]*

10. Draw a suitable cumulative frequency diagram for the following distribution of family size for 100 families:

Family Size	1	2	3	4	5	6	7	8	9	10
Frequency	2	7	12	21	26	17	9	3	2	1

[W.B.H.S. '79]

11. Draw cumulative frequency polygon (less than type) on the basis of the following data:

Mid-value of the Class	7	14	21	28	35	42
Frequency	8	18	32	28	18	6

[W.B.H.S. '85]

ANSWERS

1. (i) Rs 100.00, (ii) Rs 74.995, (iii) Rs 89.995–99.995, (iv) 16/65 = 0.246.
2. (in pound) 39.95–42.45, 42.45–44.95, 44.95–47.45, 47.45–49.95, 49.95–52.45.
3. (a) **Hint:** "No. of days absent" is a discrete variable. See Problem 4.1 (a). (b) **Hint:** See Problem 4.2(a).
   (c) **Hint:** "Weekly wages" is a discrete variable, although it may be treated as continuous. See Problem 4.2(b).

(i) *Class Interval*	0–9.99	10–19.99	20–29.99	30–39.99	40–49.99	50–59.99	Total
Frequency	8	10	27	45	23	7	120

(ii) *Cl. Boundary*	9.995	19.995	29.995	39.995	49.995	59.995
Cum. Freq.	8	18	45	90	113	120

4. (i) 85, (ii) 81 (see Example 4.12, p. 86)
5. 192 (See p. 86)
6. (a) Table showing Class boundaries, Class marks etc.

Class *Lower*	*Boundaries* *Upper*	*Class mark*	*Cumulative* *Less than*	*Frequencies* *More than*
55.5	57.5	56.5	2	100
57.5	59.5	58.5	6	98
59.5	61.5	60.5	15	94
61.5	63.5	62.5	45	85
63.5	65.5	64.5	68	55
65.5	67.5	66.5	88	32
67.5	69.5	68.5	97	12
69.5	71.5	70.5	99	3
71.5	73.5	72.5	100	1

Note: 'Less than' cumulative frequency corresponds to 'Upper' class boundary and 'More than' corresponds to 'Lower' class boundary.

(b) **Hint:** Since all classes are of equal width, the class frequency is taken as height of each rectangle in the histogram. Area under the histogram between the vertical lines drawn at 60.5 and 66.5 (inches) on the horizontal scale, is shaded.

7. **Hint:** The classes are of unequal width. See Example 4.15. page 92. 345 wage earners.
8. Note the unequal width of classes. See Example 4.15, page 92.
9. **Hint:** See Example 4.13, page 89.
10. **Hint:** See Fig. 4.8 page 95.
11. **Hint:** See Problem 4.3 and Fig. 4.4 (page 91).

Class Boundary	3.5	10.5	17.5	24.5	31.5	38.5	45.5
Cum. Frequency	0	8	26	58	86	104	110

CHAPTER 5

MEASURES OF CENTRAL TENDENCY

(Solved Problems)

Problem 5.1 *The average weight of the following frequency distribution is 117 lbs. Find the value of x.*

Weight (in lbs.)	*:*	*100*	*110*	*120*	*x + 25*	*140*	*Total*
No. of Men	*:*	*1*	*4*	*2*	*2*	*1*	*10*

[C.U., B.Com (P) '83]

Solution Mean $\Sigma fX/N$.

Here, $\Sigma fX = (100 \times 1) + (110 \times 4) + (120 \times 2) + (x + 25) \times 2 + (140 \times 1)$

$= 100 + 400 + 240 + (2x + 50) + 140$

$= 2x + 970$ Mean = 117 (given)

$\therefore \quad 117 = \dfrac{2x + 970}{10}$; i.e. $1170 = 2x + 970$

or, $2x = 1170 - 970 = 200$

$\therefore \quad x = 100.$ *Ans.* 100.

Problem 5.2 *The weighted arithmetic mean of the four numbers 26, 18, 12 and 4 is 10.4 If the weight of the first three numbers are 1, 3 and 4 respectively, find the weight of the fourth number.*

Solution Let the required weight be *w*.

x	f	fx
26	1	26
18	3	54
12	4	48
4	w	$4w$
Total	$8 + w$	$128 + 4w$

$$\bar{x} = \Sigma fx/N$$

or, $10.4 = \dfrac{128 + 4w}{8 + w}$ or, $\dfrac{104}{10} = \dfrac{128 + 4w}{8 + w}$

or, $832 + 104w = 1280 + 40w$

or, $64w = 448$

$\therefore \quad w = .448/64 = 7$

Ans. 7.

Problem 5.3 *Prove that for any two real quantities,*
A.M. ≥ G.M. ≥ H.M.

[I.C.W.A., June '79 ; Dec. '82, '84]

Solution Let a and b be two real quantities (both positive).

Then $A.M. = \frac{a+b}{2}$, $G.M. = \sqrt{ab}$, $H.M. = \frac{2ab}{a+b}$

Since $(a-b)^2$ is a perfect square, it cannot be negative.

$$(a-b)^2 \geq 0$$

or, $$(a-b)^2 + 4ab \geq 0 + 4ab$$

or, $$(a+b)^2 \geq 4ab$$

or, $$\left(\frac{a+b}{2}\right)^2 \geq ab$$

or, $$\frac{a+b}{2} \geq \sqrt{ab} \qquad \text{... (i)}$$

or, $$A.M. \geq G.M. \qquad \text{... (ii)}$$

Again, dividing both sides of (i) by ab,

$$\frac{a+b}{2ab} \geq \frac{\sqrt{ab}}{ab}$$

or, $$\frac{a+b}{2ab} \geq \frac{1}{\sqrt{ab}}$$

or, $$(a+b)\sqrt{ab} \geq 2ab$$

or, $$\sqrt{ab} \geq \frac{2ab}{a+b}$$

or, $$G.M. \geq H.M. \qquad \text{... (iii)}$$

From (ii) and (iii), we get $A.M. \geq G.M. \geq H.M.$

Problem 5.4 *Calculate Median and Mode of the following:*

Annual Sales (Rs '000)	*Frequency*
less than 10	*4*
less than 20	*20*
less than 30	*35*
less than 40	*55*
less than 50	*62*
less than 60	*67*

Is it possible to calculate the arithmetic mean? If possible calculate it.

[I.C.W.A., June '83]

Solution [**Note:** Median is calculated from the Cumulative Frequency Distribution and Mode from the Grouped Frequency Distribution. See Problem 4.2 (b), page 359.]

Table 7 Cumulative Frequency Distribution

Class Boundary	*Cumulative Frequency*
10	4
20	20
Median →	← $N/2 = 33.5$
30	35
40	55
50	62
60	$67 = N$

Using simple interpolation,

$$\frac{\text{Median} - 20}{30 - 20} = \frac{33.5 - 20}{35 - 20}$$

or, $$\frac{\text{Median} - 20}{10} = \frac{13.5}{15}$$

or, $$\text{Median} - 20 = \frac{13.5}{15} \times 10 = 9$$

∴ $$\text{Median} = 20 + 9 = 29 \text{ (Rs '000)}$$

Table 8 Grouped Frequency Distribution

Class Interval	*Frequency*
upto 10	4
10–20	16
20–30	15
(30–40)	20
40–50	7
50–60	5
Total	67

The maximum frequency 20 lies in the class interval (30–40), which is the *modal class*.

$$\text{Mode} = l_1 + \frac{d_1}{d_1 + d_2} \times c = 30 + \frac{(20 - 15)}{(20 - 15) + (20 - 7)} \times 10$$

$$= 30 + \frac{5}{18} \times 10 = 30 + 2.78$$

$$= 32.78 \text{ (Rs '000)}$$

It is possible to calculate the Mean if we assume that the class intervals are of equal width, and hence the first class interval is 0–10. (Note that the grouped frequency distribution contains an open-end class.)

Table 9 Calculations for Mean

Class Interval	*Frequency (f)*	*Mid-value (x)*	$y = \dfrac{x - 25}{10}$	*fy*
0–10	4	5	– 2	– 8
10–20	16	15	– 1	– 16
20–30	15	25	0	0
30–40	20	35	1	20
40–50	7	45	2	14
50–60	5	55	3	15
Total	67	–	–	25

If $y = (x - c)/d$, then $\bar{x} = c + d\bar{y}$

$$\text{Mean} = 25 + 10(25/67) = 25 + 3.73 = 28.73 \text{ (Rs '000)}$$

Ans. 29.00, 32.78, 28.73 (Rs '000)

Problem 5.5 *In the following frequency distribution, two class frequencies are missing:*

Intelligence Quotient	*No. of Students*	*Intelligence Quotient*	*No. of Students*
55–64	*2*	*105–114*	*?*
65–74	*19*	*115–124*	*92*
75–84	*78*	*125–134*	*14*
85–94	*?*	*135–144*	*4*
95–104	*301*		

It is however known that the total frequency is 900 and the Median 100.048. Find the two missing frequencies. [C.U., B.Com. (H) '81]

Solution Let us denote the missing frequencies in the fourth and the sixth classes by f_4 and f_6 respectively.

	Class Boundary	*Cumulative Frequency*
	54.5	0
	64.5	2
	74.5	21
	84.5	99
	94.5	$99 + f_4$
Median = 100.048→		←$N/2 = 450$
	104.5	$400 + f_4$
	114.5	$400 + f_4 + f_6$
	124.5	$492 + f_4 + f_6$
	134.5	$506 + f_4 + f_6$
	144.5	$510 + f_4 + f_6 = N = 900$ (given)

[**Note:** Since Median = 100.048 lies between 94.5 and 104.5, $N/2 = 450$ must lie in the corresponding interval.]

Using simple interpolation,

$$\frac{100.048 - 94.5}{104.5 - 94.5} = \frac{450 - (99 + f_4)}{(400 + f_4) - (99 + f_4)}$$

or, $$\frac{5.548}{10} = \frac{450 - 99 - f_4}{400 + f_4 - 99 - f_4}$$

or, $$0.5548 = \frac{351 - f_4}{301}$$

or, $$0.5548 \times 301 = 351 - f_4$$

or, $$167 = 351 - f_4$$

or, $$f_4 = 351 - 167 = 184 \quad \dots (1)$$

Again, Total frequency = 900 (given)

or, $$2 + 19 + 78 + f_4 + 301 + f_6 + 92 + 14 + 4 = 900$$

or, $$510 + f_4 + f_6 = 900$$

or, $$f_4 + f_6 = 390$$

Using (1), we get $$f_6 = 390 - 184 = 206.$$

Ans. $f_4 = 184, f_6 = 206.$

(Also see Example 5.55, page 149).

Problem 5.6 *The median and mode of the following frequency distribution are known to be 27 and 26 respectively. Find the values of a and b.*

Values	*0–10*	*10–20*	*20–30*	*30–40*	*40–50*
Frequency	*3*	*a*	*20*	*12*	*b*

[I.C.W.A., Dec. '85]

Solution

x	*Cum. Freq.*
0	0
10	3
20	$3 + a$
Median = 27→	←$N/2$
30	$23 + a$
40	$35 + a$
50	$35 + a + b = N$

(i) Median = 27 (given) lies between 20 and 30, and so $N/2$ lies between the corresponding cumulative frequencies $3 + a$ and $23 + a$, as indicated by arrow marks. Now using simple interpolation,

$$\frac{27-20}{30-20}=\frac{N/2-(3+a)}{(23+a)-(3+a)}$$

or, $$\frac{7}{10}=\frac{N/2-3-a}{20}=\frac{N-6-2a}{40}$$

or, $$N-6-2a=\frac{7}{10}\times 40=28$$

Putting $N = 35 + a + b$, we have $35 + a + b - 6 - 2a = 28$

or, $a - b = 1$ ... (1)

(ii) Mode = 26 (given) lies in the class (20 – 30). This is the modal class. Applying formula for mode,

$$\text{Mode}=l_1+\frac{d_1}{d_1+d_2}\times C$$

or, $$26=20+\frac{(20-a)}{(20-a)+(20-12)}\times 10$$

or, $$6=\frac{20-a}{28-a}\times 10;\quad \text{Simplifying } a=8$$ ... (2)

Now using (1) and (2), we get $a = 8$, $b = 7$

Problem 5.7 *If $y = a + bx$ and M_o is the mode of x, then show that the mode of y must be $a + bM_o$.* [C.U., B.Sc.(Econ.) '85 ; M.Com. '75]

Solution [Hint: We begin with an example. Let 6, 8, 9, 10 be the values of x with frequencies 2, 5, 4, 1 respectively and $y = 3 + 2x$.

x	f	$y = 3 + 2x$
6	2	15
8	5	19
9	4	21
10	1	23

Since 5 is the largest frequency, mode of x is the corresponding value of x; i.e. $M_o = 8$.

If $y = 3 + 2x$, the values of y are 15, 19, 21, 23 with frequencies 2, 5, 4, 1 respectively. Thus $y = 19$ has the largest frequency. This means that Mode of y is 19. But $19 = 3 + 2 \times 8$; i.e. Mode of $y = 3 + 2$ (Mode of x).]

Let $x_1, x_2, ..., x_n$ be the values of x and $f_1, f_2, ..., f_n$ be the respective frequencies. If the maximum frequency is f_k, then the

x	f	$y = a + bx$
x_1	f_1	$y_1 = a + bx_1$
x_2	f_2	$y_2 = a + bx_2$
⋮	⋮	⋮
x_n	f_k	$y_k = a + bx_k$
⋮	⋮	⋮
x_n	f_n	$y_n = a + bx_n$

mode of x is the corresponding value of x, viz. x_k; i.e.

$$M_o = x_k \quad \text{... (i)}$$

If $y = a + bx$, then the values of y, viz. $y_1, y_2, ..., y_n$ and obtained by putting $x = x_1, x_2, ..., x_n$. These values of y must be all different; because $y = a + bx$ will give different values of x. Hence that value of y has the largest frequency for which $x = x_k$. That is, the mode of y is y_k But $y_k = a + bx_k$; i.e.

$$\text{Mode} = y = a + bM_o. \quad \text{from (i)}$$

SHORT-ANSWER TYPE QUESTIONS

Problem 5.8 *Find the A.M. of the numbers 3, 5, 7, ..., 47.*

Solution [Hint : A.M. of a series of numbers having a common difference, is equal to the A.M. of the 1st and the last numbers]. A.M. = (3 + 47)/2 = 25.

Problem 5.9 *(i) If a variable assumes the values 1, 2, 3, ..., 10 with frequencies 1, 2, 3, ..., 10 respectively, find the A.M.* [W.B.H.S. '80]

(ii) The weighted A.M. of 3, 5 and 10 with weights x, y and z respectively is known to be 8. What is the weighted A.M. if the weights are changed to 2x, 2y and 2z respectively?

(iii) A variable x takes the values 10 and 20 with equal frequencies. Find the A.M. [W.B.H.S. '78]

(iv) A variable x takes all integral values from 1 to 10 with equal frequencies. Find its A.M. [W.B.H.S. '79]

Solution (i) $\text{A.M.} = \dfrac{1 \times 1 + 2 \times 2 + 3 \times 3 + ... + 10 \times 10}{1 + 2 + 3 + ... + 10} = \dfrac{385}{55} = 7$

(ii) If all the weights are multiplied by a constant, the A.M. does not change. A.M. = 8.

(iii) Since the frequencies are equal,

Weighted A.M. = Simple A.M. = (10 + 20)/2 = 15

(iv) Since the frequencies are equal,

Weighted A.M. = Simple A.M.

= A.M. of 1, 2, 3, ..., 10

= (1 + 10)/2, since there is common difference.

= 11/2 = 5.5

Problem 5.10 *If the A.M. of 3, 5, x, 12, 17 be 9, find the value of x.*

Solution $(3 + 5 + x + 12 + 17)/5 = 9$ (given)

or, $37 + x = 45$, i.e. $x = 8$

Problem 5.11 *The algebraic sum of the deviations of 25 observations measured from 45 is – 55. Find the A.M. of the observations.* [C.U., B.Com.(Pass) '85]

Solution [**Hint:** If $y = x - c$, then $\bar{x} = c + \bar{y}$; p. 136]

Here $y = x - 45$ = Deviation from 45; $n = 25$, $\Sigma y = -55$

$\therefore \quad \bar{x} = 45 + (-55)/25 = 45 - 2.2 = 42.8$

Problem 5.12 *(i) Find the geometric mean of 3, 12 and 48.*

[C.U., B.Com(Pass) '83]

(ii) If a variable increase from 20 to 2000 at the same ratio in 10 years, find its value after 5 years. [W.B.H.S. '80]

Solution (i) $G.M. = \sqrt[3]{3 \times 12 \times 48}$

$$= \sqrt[3]{3 \times (3 \times 4) \times (3 \times 4 \times 4)} = 12$$

(ii) [**Hint:** (1) If there is a *common difference* in a series, Middle-most value = A.M. of the 1st and the last values.

(2) If there is *common ratio* in a series, Middle-most value = G.M. of the 1st and the last values.

Since there is a common ratio, and the values after 5 years is the middle-most item,

Required value = G.M. of 20 and 2000

$$= \sqrt{20 \times 2000} = \sqrt{40000} = 200$$

Problem 5.13 *(i) A piece of work was done in 2, 4 and 3 days by 3 men. Find (with the help of harmonic mean) the average rate of working.*

[C.U., B.Com(Hons) '84]

(ii) Find the H.M. of 1, 1/2, 1/3 and 1/4.

Solution (i) Average Rate of working

= *H.M.* of the 3 observations 2, 4, 3

$$= \frac{3}{\frac{1}{2} + \frac{1}{4} + \frac{1}{3}} = \frac{36}{13} = 2.77 \text{ days.}$$

(ii) $$H.M. = \frac{4}{\frac{1}{1} + \frac{1}{\frac{1}{2}} + \frac{1}{\frac{1}{3}} + \frac{1}{\frac{1}{4}}} = \frac{4}{1 + 2 + 3 + 4} = 0.4$$

Problem 5.14 *(i) Find A.M., G.M. and H.M. of 2, 9, 12; and show that A.M. > G.M. > H.M.*

(ii) The A.M. of two observations is 36 and their G.M. is 24. Find the H.M.

(iii) Find A.M., G.M. and H.M. of the following: 2, 9, 2, 12, 9, 12

(iv) Find A.M., G.M, and H.M. of the following:

(a) 3, 3, 3, 3; *(b) 3, – 2, 4, 0, 5.*

Solution (i) $A.M. = \frac{1}{3}(2 + 9 + 12) = \frac{1}{3} \times 23 = 7.67$

$$G.M. = \sqrt[3]{2 \times 9 \times 12} = \sqrt[3]{216} = 6$$

$$H.M. = \frac{1}{\frac{1}{2} + \frac{1}{9} + \frac{1}{12}} = \frac{3}{\frac{18 + 4 + 3}{36}}$$

$$= \frac{108}{25} = 4.32$$

From these results we see that $A.M. > G.M. > H.M.$

(ii) For two observations, $A.M. \times H.M. = (G.M.)^2$

$36 \times H.M. = 24^2; \quad \therefore \quad H.M. = 16$

(iii) Since 2, 9 and 12 occur with equal frequency, therefore

Weighted A.M. = Simple A.M. = 7.67; see (i)

Weighted G.M. = Simple G.M. = 6

Weighted H.M. = Simple H.M. = 4.32

(iv) (a) Since all observations are equal,

$A.M. = G.M. = H.M. = 3$

(b) $A.M. = \frac{3 - 2 + 4 + 0 + 5}{5} = \frac{10}{5} = 2$

G.M. and H.M. cannot be determined, since one of the observations is zero. Moreover, G.M. and H.M. become meaningless, when all observations do not have the same sign.

Problem 5.15 *Find median and mode of the following data:*
4, 7, 10, 7, 9, 15, 12, 7, 9, 6

Solution Arranging: 4, 6, 7, 7, 7, 9, 9, 10, 12, 15.

There are two middlemost observations 5th and 6th. viz. 7 and 9.

Median = (5th item + 6th item)/2 = (7 + 9)/2 = 8

Amongst the given observations, 7 occurs the largest number of times.

Mode = 7.

Problem 5.16 *Find the mode of the following:*

(i) 2, 7, 4, 3, 6

(ii) 2, 7, 4, 3, 6, 7, 4

Solution (i) Mode does not exist; because all values occur with equal frequency.

(ii) There are two modes: 4 and 7; because both of them have the highest frequency.

Problem 5.17 *If mean and median of a frequency distribution are 20 and 17 respectively, find the mode.*

Solution Mean – Mode = 3(Mean – Median)
or, 20 – Mode = 3 (20 – 17)
Solving, Mode = 11.

Problem 5.18 *(a) Suppose, the highest value (97.5) of a set of 100 observations is changed to 197.5. what would happen to its mean, median and mode?*

(b) For the following observations regarding monthly income for 6 persons, which measure of central tendency, you think, would be appropriate? Justify your answer.

Rs 500, 525, 475, 480, 520, 2593

(*c*) In a set of values of a variable *x*, the smallest value is 0.3 and the highest is 1.1. Which of the following is a possible value of the mean? Give reason in support of your answer:

5.2, 1.0, 0.2, – 0.9, [C.U., M.Com. '80]

Solution (a) Mean increases by 1; becauses the total of 100 observations increases by 100 (from 97.5 to 197.5).

Median does not change; because with the change in the highest value, the middlemost observation remains unaffected.

Mode does not change; because usually mode lies near the middle of the given values.

(b) Since one observation is extremely large compared to the others, Median is the appropriate average. Arranging the data, Median = (500 + 520)/2 = 510 Rs.

(c) 1.0; because all measures of central tendency must lie between the smallest and the largest values.

Problem 5.19 *(a) The arithmetic mean of a variable x is 100. Find the mean of the variable 5x – 10.* [W.B.H.S. '78]

(b) Two variables x and y are given y = 2x – 11. If the median of x be 30, find the median of y. [W.B.H.S. '78]

(c) If the relation between two variables x and y is 3y – 2x + 5 = 0, and median of y is 40, find the median of x. [W.B.H.S. '83]

(d) If the relation between two variable x and y is 2y – 6x = 6 and mode of x is 21, find the mode of y. [W.B.H.S. '82]

Solution [**Note:** If $y = a + bx$ ... (i)

then

$$\text{A.M. of } y = a + b.(\text{A.M. of } x)$$
$$\text{Median of } y = a + b.(\text{Median of } x)$$
$$\text{Mode of } y = a + b.(\text{Mode of } x)$$

In general, if two variables *x* and *y* are related by an equation of the form

$$Ax + By + C = 0 \quad \text{... (ii)}$$

where*A*, *B*, *C* are constants, then

$$A.(\text{A.M. of } x) + B.(\text{A.M. of } y) + C = 0$$
$$A.(\text{Median of } x) + B.(\text{Median of } y) + C = 0$$
$$A.(\text{Mode of } x) + B.(\text{Mode of } y) + C = 0$$

Such relations do not however hold for G.M. and H.M. (unless $C = 0$). Note that (i) and (ii) are both equations of straight lines.]

(a) Let $y = 5x - 10$.

Then $\bar{y} = 5\bar{x} - 10 = 5 \times 100 - 10 = 490$.

(b) Since $y = 2x - 11$

∴ Median of y = 2. (Median of x) – 11 = 49

(c) 3. (Median of y) – 2. (Median of x) + 5 = 0

or, 3×40 – 2. (Median of x) + 5 = 0

Solving, Median of $x = 125/2 = 62.5$

(d) Since $2y - 6x = 6$

∴ 2. (Mode of y) – 6. (Mode of x) = 6

Putting Mode of $x = 21$ and solving, se get Mode of $y = 66$

Problem 5.20 *If $2u = 5x$, G.M. of x is 1 and H.M. of x is 0.4, find G.M. and H.M. of u.* [W.B.H.S. '86]

Solution [**Hint:** If $y = kx$, where k is a constant, the (Averate of y) = k. (Average of x)

This is true for hte Averages, vix. A.M., G.M., H.M., Median and Mode.] Here $2u = 5x$, so that $u = (2.5)x$

G.M. of u = 2.5 × (G.M. of x) = 2.5 × 1 = 2.5

H.M. of u = 2.5 × (H.M. of x) = 2.5 × 0.4 = 1

Problem 5.21 *(a) Find the 1st and the 3rd quartiles for the following data: 0, 1, – 1, – 2, 6, 4, 5, 8, 12, 10, 11,* [I.C.W.A., June '86]

(b) If for a symmetrical distribution $Q_1 = 20$ and $Q_3 = 30$, find Median.

(c) How many decides lie below the first quartile? [C.A., Nov. '82]

Solution (a) The data are arranged in *increasing* order and serially numbered (i.e. ranked):

Values	– 2	– 1	0	1	4	5	6	8	10	11	12	
Ranks	1	2	3	4	5	6	7	8	9	10	11	(n = 11)

Q_1 = Value of $(n + 1)/4$th item, i.e. 3rd item = 0

Q_3 = Value of $3(n + 1)/4$th item, i.e. 9th item = 10

(b) For a symmetrical distribution,

Median = $(Q_1 + Q_3)/2 = (20 + 30)/2 = 25$

(c) First decile $D_1 = P_{10}$ and second decile $D_2 = P_{20}$ are the only two deciles which lie below 1st quartile $Q_1 = P_{25}$.

ADDITIONAL EXERCISES 5

1. The weighted arithmetic mean of 3.2, 3.4, 3.8 and 4.6 is known to be 4.656. If the weights of the first observations are respectively 12, 20 and 8, find the weight of 4.6. *[K.U., B.Com.(Pass) '81]*

2. From the following table, calculate the value of mean:

Monthly Wages (Rs)	0–10	10–20	20–30	40–50	50–60	60–70	70–80	80–90
No. of Servants	1	4	10	30	35	10	7	1

[C.U., B.Com.(Hons) '84]

3. A machine depreciates at a rate of 20% for the first year, 10% for the next two years and 5% for the next three years. What is the average rate of depreciation ? *[C.U., M.B.A. '81]*
4. Prove that the logarithm of the geometric mean of n positive values is the arithmetic mean of the logarithms of the values. *[I.C.W.A., June '86]*
5. The total interest paid on each of three different sums of money (for a given period of time) yielding 3%, 4% and 5% simple interest per annum respectively is the same. Which of the three means will you use to find the average yield per cent of the total sum invested? Calculate this average. *[C.U., B.Com.(Hons) '83]*
6. One person purchased 4 types of tea at the prices of Rs 16, Rs 20, Rs 32 and Rs 80 each per kilogram by spending equal amounts of money for each type of tea and mixed them together to sell at a profit of Re 1.00 per kilogram. Find the selling price of the mixed tea. *[W.B.H.S. '83]*
7. Find the arithmetic mean, geometric mean and harmonic mean of the elements of the set $\{1, 2, 3, 4, 5\}$, and verify the relation $A > G > H$. *[C.U., B.Com.(Hons) '85]*
8. Using the formula, find median of the following:

Class Interval	130–134	135–139	140–144	145–149	150–154	155–159	160–164
Frequency	5	15	28	24	17	10	1

[C.U., B.Com.(Hons) '86]

9. Draw Ogive (less than type) from the following data and find Median from the ogive:

Weekly Wages (Rs)	51–60	61–70	71–80	81–90	91–100
No. of Workers	10	15	34	23	11

[K.U., B.Com.(Pass) '83]

10. (a) What is meant by Mode of a distribution? Does it always exist? Is it always unique? Give examples to illustrate your answer. (**Hint:** Illustration 14, page 124.)

(b) Calculate mode of the following frequency distribution:

Marks (x)	5–10	10–15	15–20	20–25	25–30	30–35	35–40	40–45
No. of (f) Students	5	6	15	10	5	4	2	2

[C.U., B.Com.(Hons) '82]

11. Runs scored by individual English batsmen in the 1st innings of the 3rd test match (India vs. England, 1986-series) are as follows: 0, 21, 0, 49, 183 (not out), 44, 38, 17, 18, 8, 0. Find mean, median and mode.

12. Calculate Arithmetic Mean and Median of the frequency distribution given below. Hence calculate Mode using the empirical relation between the three.

Class Limits	1–10	11–20	21–30	31–40	41–50	51–60	61–70
Frequency	8	15	25	20	16	10	6

[C.U., B.Com.(Pass) '82]

13. Calculate Mean and Median from the following data:

Weakly Wages (Rs)	*Number of Workers*
Below 10	8
Below 20	18
Below 30	45
Below 40	90
Below 50	113
Below 60	120

[C.U., B.Com.(Hons) '80]

(**Hint:** No. 3 (c), page 361 and Note on p. 359.)

14. For a certain frequency table with total frequency 150, the mean was found to be 56.47. But while copying out the table, a typist left out two of the class frequencies, way f_3 and f_5, so that the table is given to you in the following form. Find the values of f_3 and f_5.

Weakly Wages in Rs (mid value)	45	50	55	60	65	70	75	Total
Frequency	5	48	f_3	30	f_5	8	6	150

[I.C.W.A., Dec. '83]

15. If the mode of the following frequency distribution which represents the number of minutes spent per week by a group of teenagers in going to movies is 339.5 minutes per week, find the missing frequency. Then find the median of the complete frequency distribution.

Number of minutes/week	0–99	100–199	200–299	300–399	400–499	500 and above
Number of Teenagers	26	32	65	?	60	42

[W.B.H.S. '84]

16. Calculate the quartiles from the following:

Weight (kg)	14–18	18–22	22–26	26–30	30–34	Total
Frequency	15	40	25	13	7	100

[K.U., B.Com.(Pass) '85]

17. If a variate assumes n values $a, ar, ..., ar^{n-1}$ with $(r > 1)$ equal (constant) frequencies then find the arithmetic mean (A), geometric mean (G) and harmonic mean (H) and show that $AH = G^2$. *[WB.H.S. '82]*

SHORT ANSWER TYPE QUESTIONS

18. Find the A.M., G.M. and H.M. of the four number 3, 6, 24 and 48. *[K.U., B.Com.(Pass) '84]*
19. Find the arithmetic mean of two observations if their geometric mean and harmonic mean are 15 and 9 respectively. *[W.B.H.S. '85]*
20. Find mean, median and mode of the following data: 44, 60, 33, 63, 48, 61, 51, 48. *[C.U., B.Com.(Pass) '85]*
21. For some symmetrical distribution $Q_1 = 24$ and $Q_3 = 42$. Find Median. *[C.U., B.Com.(Pass) '85]*
22. If the relation between two variables x and y is $2x + 3y = 7$ and the median of y is 2, find the median of x. *[W.B.H.S. '85]*
23. Examine the validity of the following statements:
    (a) The geometric mean can never be calculated when one of the observations is a negative one.
    (b) One can calculate the harmonic mean for a set of observations which includes 0 (zero) as one of its observations.
    (c) The median of a set of 31 values is the 11th value in the arrangement of the values in an increasing order (or in a decreasing order) of magnitude.
    (d) The mode of a variable is the highest frequency.
    (e) The relation: $\bar{x} - Mode = 3(\bar{x} - Median)$ holds only when the distribution is moderately stewed. *[C.U., M.Com. '82]*
24. Examine the validity of the following:
    (a) A.M. is unit free.
    (b) When all observations are equal, $A.M. = G.M. = H.M.$
    (c) Median can be calculated even for a frequency distribution with open-end classes.
    (d) When all observations are equal, $A.M. \geq G.M. \geq H.M.$
    (e) For a set of observations, one obtains the following:
    $\Sigma | x_i - 10 | /n = 19.35$, $\Sigma | x_i - 5 | /n = 14.35$,
    where Median $(x) = 10$.
    (f) Q_2 = 2nd Quartile = N/2, where N = Total number of observations. *[C.U., M.Com '83]*
25. Fill up the missing parts:
    (a) A.M. of a set of observations is defined as ......
    (b) G.M. of a set of n positive observations is defined as ......
    (c) H.M. of a set of n positive observations is defined as ......
    (d) Median of a set of observations is defined as ......
    (e) Mode of a set of observations is defined as ......

26. Fill up the missing parts:
    (a) Simple and weighted arithmetic means are equal only when ......
    (b) The relations between A.M., G.M. and H.M. are ......
    (c) When all observations are equal, the relations between A.M., G.M. and H.M. are ......
    (d) For two observations *a* and *b*, the expressions for A.M., G.M. and H.M. are respectively ......
    (e) Mean, Median and Mode calculated from two observations x_1 and x_2 are respectively ......
    (f) For two observations only, if A.M. and G.M. are known, then H.M. may be obtained by the relation ......
    (g) The empirical relation connecting Mean, Median and Mode is ......
27. State the appropriate type of average to use in each of the following cases:
    (a) Marks obtained in Mathematics by the students in an examination.
    (b) The size of shoes sold in a shop.
    (c) A frequency distribution having open-end classes.
    (d) Different rates of depreciation charged on diminishing balance.
    (e) Speeds in several journeys when the distance covered is the same in each journey.
    (f) Speeds in several journeys when the time taken is the same in each journey.
28. Comment on the following statements:
    (a) Mean is independent of origin, but depends on scale.
    (b) Median is unaffected by the presence of extreme values.
    (c) Mode is the value of the largest observation in the series.
    (d) The lower quartile of a distribution is $N/4$ and the upper quartile is $3N/4$.
    (e) The quartiles of a distribution are connected by the relation $Q_3 - Q_2 = Q_2 - Q_1$.

ANSWERS

1. 10 2. Rs 46.83 3. 9.3% 5. H.M. = 3.83%
6. Rs 26.60 per kg. 7. 3.00, 2.61, 2.19
8. 144.92 9. Rs 76.82 10. 18.21 11. 34.4, 18, 0
12. 33.0, 31.5, 28.5 13. (Rs) 32.16, 33.33.
14. 41, 12 15. 75, 335.5 min/week
16. (kg) 19.0, 21.5, 25.2
17. $A = \dfrac{a(r^n - 1)}{n(r-1)}, G = ar^{(n-1)/2} \quad H = \dfrac{na(r-1)r^{n-1}}{r^n - 1}$
18. 20.25, 12, 7.11 19. 25
20. 51, 49.5, 48 21. 33

22. 0.5
23. True—(a), (c), (e); False—(b), (d).
24. True (b), (c); False (a), (d), (e) since M.D. is minimum when deviations are taken from median, and (f).
25. (a) the sum of the observations, divided by their number.
    (b) the n-th root of their product.
    (c) the reciprocal of the A.M. of their reciprocals.
    (d) the middlemost observation, when all items are arranged in order of magnitude.
    (e) that observation(s) which has (have) the highest frequency.
26. (a) all weights are equal;
    (b) $A.M. \geq G.M. \geq H.M.$;
    (c) $A.M. = G.M. = H.M.$
    (d) $(a + b)/2$, $\sqrt{ab}$, $2ab/(a + b)$;
    (e) $(x_1 + x_2)/2$, $(x_1 + x_2)/2$, no mode:
    (f) A.M. × H.M. = $(\text{G.M.})^2$;
    (g) Mean – Mode = 3(Mean – Median).
27. (a) A.M., (b) Mode, (c) Median, (d) G.M.,
    (e) H.M., (f) A.M.
28. (a) False; Mean depends on the both origin and scale.
    (b) True.
    (c) False; Mode is the value of the observation which has the largest frequency.
    (d) False; Lower quartile and upper quartile are those values which correspond to cumulative frequencies $N/4$ and $3N/4$ respectively.
    (e) False; Quartiles are connected by $Q_1 \leq Q_2 \leq Q_3$. The given relation holds only for a symmetrical distribution.

CHAPTER 6

MEASURES OF DISPERSION
(Solved Problems)

Problem 6.1 *Find Range from the following data:*

(a) 0, 1, –1, –2, 6, 4, 5, 8, 12, 10, 11.

[I.C.W.A., June'86]

(b)

x	*f*
3	5
4	9
6	10
7	4
11	2
Total	30

(c)

Length (cm)	*Frequency*
20–24	3
25–29	6
30–34	13
35–44	8
46–59	2
Total	32

Solution (a) Maximum value = 12, Minimum value = –2

∴ Range = 12 – (– 2) = 14

(b) Maximum value = 11, Minimum value = 3

∴ Range = 11 – 3 = 8

(c) Max. value = Upper boundary of the class 45 – 59 = 59.5

Min. value = Lower boundary of the class 20 – 24 = 19.5

∴ Range = 59.5 – 19.5 = 40.0 cm.

(**Note:** All absolute measures of dispersion are expressed in the same unit as the observations.)

Problem 6.2 *Find Semi-interquartile Range from the following:*

(a) 0, 1, –1, –2, 6, 4, 5, 8, 12, 10, 11.

[I.C.W.A. June '86]

(b) 391, 384, 591, 407, 672, 522, 777, 733, 2488, 1490.

(c)

x :	0	1	2	3	4	5	6	7	Total
f :	14	21	25	43	51	40	39	12	245

Solution (a) The data are arranged in ascending order of magnitude and ranked:

Value	–2	– 1	0	1	4	5	6	8	10	11	12
Rank	1	2	3	4	5	6	7	8	9	10	11

Rank of $Q_1 = \frac{1}{4}(11 + 1) = 3$; Rank of $Q_3 = \frac{3}{4}(11 + 1) = 9$

Q_1 = Value with rank 3 = 0

Q_5 = Value with rank 9 = 10

$\therefore$ Semi-interquartile Range = $\frac{1}{2}(10-0)=5$

(b) [**Hint:** see Example 5.63, p. 160]. $Q_1 = 403$, $Q_3 = 955.25$.

Semi-interquartile Range = $\frac{1}{2}$ (955.25 – 403) = 276.125

(c) [Ref.: Example 5.57, p. 154] Here $N = 245$

x	f	*Cum. f*
0	14	14
1	21	35
2	25	60
3	43	103
4	51	154
5	40	194
6	39	233
7	12	245 = N
Total	245 = N	

$$\text{Rank of } Q_1 = \frac{1}{4}(N+1) = \frac{1}{4}(245+1) = 61.5$$

$$\text{Rank of } Q_3 = \frac{3}{4}(N+1) = \frac{3}{4}(245+1) = 184.5$$

Q_1 = 61.5th value = 3 (see Note, page 167)

Q_3 = 184.5th value = 5

$\therefore$ Quartile Deviation = $\frac{1}{2}(Q_3 - Q_1) = \frac{1}{2}(5-3) = 1.$

Problem 6.3 *Calculate Mean Deviation about mode in respect of marks (out of 25) obtained by 9 students given below:*

Students :	*A*	*B*	*C*	*D*	*E*	*F*	*G*	*H*	*I*
Marks :	*4*	*7*	*10*	*7*	*9*	*15*	*12*	*7*	*9*

[I.C.W.A., June '86]

Solution Mode = 7 (because it occurs the largest number of times).

Student	*Marks*	*Differences from Mode (= 7)*
A	4	3
B	7	0
C	10	3
D	7	0
E	9	2
F	15	8
G	12	5
H	7	0
I	9	2
	Total	23

M.D. about mode = A.M. of differences from mode

= 23/9 = 2.56

Problem 6.4 *After shift of origin and change of scale a frequency distribution of a continuous variable takes the form as under:*

Step Deviation (u) :	*– 3*	*– 2*	*– 1*	*0*	*1*	*2*	*3*
Frequency (f) :	*5*	*7*	*18*	*3*	*124*	*12*	*3*

If the mean and variance of the distribution are 46 and 189 respectively, find the original frequency distribution. [W.B.H.S.'85]

Solution If the origin of x is shifted to A and the scale changed to h, we write $u = (x - A)/h$, so that $x = A + hu$.

Therefore (i) $\bar{x} = A + h\bar{u}$

(ii) $\sigma_x = h.\sigma_u$ i.e. $\sigma_x^2 = h^2.\sigma_u^2$

We calculate mean (u) and variance (σ_u^2) from the given data and use relations (i) and (ii) to obtain the values of A and h.

u	f	fu	fu^2
– 3	5	– 15	45
– 2	7	– 14	28
– 1	18	– 18	18
0	31	0	0
1	24	24	24
2	12	24	24
3	3	9	27
Total	100	10	190

Hence

$$\bar{u} = \Sigma fu/N = 10/100 = 0.1$$

$$\sigma_u^2 = \Sigma fu^2/N = (\Sigma fu/N)^2 = 190/100 - (10/100)^2$$

$$= 1.9 - 0.01 = 1.89$$

We are given $\bar{x} = 46$, $\sigma_x^2 = 189$.

Putting these results in (i) and (ii) above,

$46 = A + h(0.1)$ and $189 = h^2(1.89)$

Solving, $A = 45$ and $h = 10$.

Hence, $u = (x - 45)/10$; i.e. $x = 45 + 10u$

When $u = -3, -2, -1, ..., 3$, the values of x are respectively 15, 25, 35, 45, ..., 75. These are midvalues of class intervals 10–20, 20–30, 30–40, ..., 70–80. (see Problem 4.3, Page 360). The original frequency distribution is as follows:

Class Interval	10–20	20–30	30–40	40–50	50–60	60–70	70–80	Total
Frequency	5	7	18	31	24	12	3	100

Problem 6.5 *Find the missing figures from the following table:*

	Group I	*Group II*	*Group III*	*Combined*
Number	*50*	*?*	*90*	*200*
Mean	*113*	*?*	*115*	*116*
Standard Deviation	*6*	*7*	*?*	*7.746*

[C.U., B.Com. (Hons.) '86]

Solution From 3 groups, Mean and S.D. of Composite Group are given by the formulae (see Example 6.39, p. 204)

(1) $N = n_1 + n_2 + n_3$

(2) $N\bar{x} = n_1\bar{x}_1 + n_2\bar{x}_2 + n_3\bar{x}_3$

(3) $N\sigma^2 = (n_1\sigma_1^2 + n_2\sigma_2^2 + n_3\sigma_3^2) + (n_1d_1^2 + n_2d_2^2 + n_3d_3^2)$

Using (1), $200 = 50 + n_2 + 90$. $\therefore\ n_2 = 60$

Using (2), $200 \times 116 = 50 \times 113 + 60\bar{x}_2 + 90 \times 115$

or, $23200 = 5650 + 60\bar{x}_2 + 10350$. $\therefore\ \bar{x}_2 = 120$

Using (3), $d_1 = (113 - 116) = -3$, $d_2 = (120 - 116) = 4$,

$d_3 = (115 - 116) = -1$

$200 \times (7.746)^2 = (50 \times 6^2 + 60 \times 7^2 + 90\sigma_3^2) + (50 \times 9 + 60 \times 16 + 90 \times 1)$

or, $12000 = (18000 + 2940 + 90\sigma_3^2) + (450 + 960 + 90)$

Solving $\sigma_3^2 = 64$. $\therefore\ \sigma_3 = 8$

Ans. Number = 60, mean = 120, s.d. = 8

Problem 6.6 *The mean of 5 observations is 4.4 and the variance is 8.24. If 3 of the observations are 4, 6, 9, find the other two.* [I.C.W.A., June' 85]

Solution Let the two other observations be *a* and *b*. We use the formula for Mean and Variance, namely

$\bar{x} = \Sigma x/n, \quad \sigma^2 = \Sigma x^2/n - \bar{x}^2$

$4.4 = (4 + 6 + 9 + a + b)/5. \qquad \therefore\ a + b = 3 \qquad \text{... (i)}$

$8.24 = (4^2 + 6^2 + 9^2 + a^2 + b^2)/5 - (4.4)^2. \qquad \therefore\ a^2 + b^2 = 5 \qquad \text{... (ii)}$

Solving (i) and (ii), we get *a* and *b*.

$(a + b)^2 - (a^2 + b^2) = 3^2 - 5 = 4.$

or, $2ab = 4$

$(a - b)^2 = (a^2 + b^2) - 2ab = 5 - 4 = 1 \qquad \therefore\ a - b = \pm 1 \qquad \text{... (iii)}$

From (i) and (iii), either $a = 2, b = 1$ or, $a = 1, b = 2$

Ans. 1 and 2

Problem 6.7 *Show that M.D. about mean for a set of values of a variable cannot exceed S.D.* [B.U., B.A. (Econ) '82]

Solution [**Hint:** Also see Example 6.45, p. 209]

Let $x_1, x_2, ..., x_n$ be a set of values and $\bar{x}$ their arithmetic mean.

M.D. $= \Sigma \mid x_i - \bar{x} \mid /n$, S.D. $= \sqrt{\Sigma(x_i - \bar{x})^2/n}$

To prove that M.D. ≤ S.D., we use Cauchy-Schwarz inequality (page 70):

$$\Sigma a_i^2 \Sigma b_i^2 \geq (\Sigma a_i b_i)^2$$

Putting $a_i = |x_i - \bar{x}|,\ b_i = 1\ (i = 1, 2, ..., n)$

$\Sigma a_i^2 = \Sigma(|x_i - \bar{x}|)^2 = \Sigma(x_i - \bar{x})^2$

$\Sigma a_i^2 = \Sigma 1^2 = n$

$\Sigma a_i b_i = \Sigma |x_i - \bar{x}|$

Hence, $\Sigma(x_i - \bar{x})^2 \cdot n \geq \{\Sigma |x_i - \bar{x}|\}^2$

Dividing both sides by n^2,

$$\Sigma(x_i - \bar{x})^2/n \geq \{\Sigma |x_4 - \bar{x}|/n\}^2$$

or, $(\text{S.D.})^2 \geq (\text{M.D.})^2$

i.e. S.D. ≥ M.D. *(Proved)*

Problem 6.8 *For the values $x_1, x_2 ..., x_n$ with respective frequenices $f_1, f_2, ..., f_n$, show that*

$$\frac{1}{N}\sum_{i=1}^{n} f_i\left(x_i + \bar{x}\right)^2 \geq \frac{1}{N}\sum_{i=1}^{n} f_i\left(x_i - \bar{x}\right)^2$$

where $\bar{x}$ is the arithmetic mean and $N = \Sigma f_i$.

[C.U., B.Com.(Hons) '82]

Solution We use Cauchy-Schwarz inequality:

$\Sigma a_i^2\ \Sigma a_i^2 \geq (\Sigma a_i b_i)^2$

Putting $a_i = \sqrt{f_i} \mid x_i - \bar{x} \mid,\quad b_i = \sqrt{f_i}$

We have $\Sigma a_i^2 = \Sigma\left\{\sqrt{f_i} \mid x_i - \bar{x} !\right\}^2 = \Sigma f_i\left(x_i - \bar{x}\right)^2$

$\Sigma b_i^2 = \Sigma f_i = N$

$(\Sigma a_i\, b_i)^2 = \{\Sigma f_i \,|x_i - \bar{x}\,|\}^2$

Therefore, $\Sigma f_i (x_i - \bar{x})^2 . N \geq \{\Sigma f_i \,|x_i - \bar{x}\,|\}^2$

Dividing both sides N^2,

$\Sigma f_i (x_i - \bar{x})^2/N \geq \{\Sigma f_i \,|x_i - \bar{x}\,|/N\}^2$ *(Proved)*

SHORT-ANSWER TYPE QUESTIONS

Problem 6.9 *The three quartiles of a variable x are 5, 12 and 17. Find its Quartile Deviation.*

[W.B.H.S. '79]

Solution Q.D. $\frac{1}{2}$ $(Q_3 - Q_1) = \frac{1}{2}$ $(17 - 5) = 6$

Because Q_1 is the smallest and Q_3 is the largest among the 3 quartiles.

Problem 6.10 *(a) Find the Standard Deviation, if in distribution n = 10, Σx = 120, Σx^2 = 1690.* [C.U., B.Com. (Pass)'83]

(b) Calculate the Variance of the sample {2, 5, 8}. [C.U., B.Com. (Pass'82]

(c) Find the S.D. of 1, 2, 3, 4, ..., 10. [C.U., B.Com.(Pass '80]

Solution (a) $\sigma^2 = \Sigma x^2/n - (\Sigma x/n)^2 = 1690/10 - (120/10)^2$
$= 169 - 144 = 25; \quad \therefore \quad \sigma = 5$

[**Note:** Do not write $\sigma = \pm 5$.]

(b) $\Sigma x = 2 + 5 + 8 = 15$

$\Sigma x^2 = 4 + 25 + 64 = 93$ and $n = 3$

Variance $(\sigma^2) = \Sigma x^2/n - (\Sigma x/n)^2 = 93/3 - (15/3)^2 = 6$

(c) Using formulae for the sum and the sum of squares of natural numbers (see Sec. 3.8, p.67) ; here $n = 10$,

$$\Sigma x = \frac{n(n+1)}{2} = \frac{10 \times 11}{2} = 55$$

$$\Sigma x^2 = \frac{n(n+1)(2n+1)}{6} = \frac{10 \times 11 \times 21}{6} = 385$$

$$\sigma^2 = 385/10 - (55/10)^2 = 38.5 - 30.25 = 8.25$$

$$\therefore \quad \sigma = \sqrt{8.25} = 2.9$$

Problem 6.11 *Find the S.D. of x from the following:*
$\Sigma f_i x_i^2/\Sigma f_i = 25, \ \Sigma f_i x_i = 300, \ \Sigma f_i = 100$

[C.U., M.Com.'83]

Solution For a frequency distribution
$\sigma^2 = \Sigma fx^2/N - (\Sigma fx/N)^2 = 25 - (300/100)^2 = 16$

$\therefore \quad \sigma = 4$

Problem 6.12 *(a) A certain student obtained for a set of 25 values of the variable x, $\Sigma x_i = 12.4$, $\Sigma x_i^2 = 34.39$.*

Would you take his computations to be correct ? Give reasons for your answer.

[B.U., B.A.(Econ) '79]

(b) Examine whether the following results of a piece of computation are consistent or not:

Total frequency = 120, $\Sigma fd = -125$, $\Sigma fd^2 = 128$. [I.C.W.A., Dec. '82]

Solution (a) $\sigma^2 = \Sigma(x - \bar{x})^2/n$ can never be negative.

Again, $\sigma^2 = \Sigma x^2/n - (\Sigma x/n)^2$ (Working formula)

Hence, the values of n, Σx^2 will be such that σ^2 does not give negative value.

Here $\sigma^2 = \frac{34.39}{25} - \left(\frac{12.4}{25}\right)^2 = \frac{34.39 \times 25 - (12.4)^2}{625}$

$= \frac{859.75 - 153.76}{625} = \frac{705.99}{625}$ (which is positive)

Thus, the computations may be correct.

(b) For a frequency distribution, the working formula is $\sigma^2 = \Sigma fd^2/N - (\Sigma fd/N)^2$, where $d = x - A$. The values of N, Σfd^2 must be such that σ^2 will not give negative value in any numerical computation.

Here, $\sigma^2 = \frac{128}{120} - \left(\frac{-125}{120}\right)^2 = \frac{128 \times 120 - (-125)^2}{14400} = \frac{-265}{14400}$ (negative). Thus, the given results are not consistent.

Problem 6.13 *If the S.D. of x is 6, find the S.D. of*

(i) $x + 7$,	(ii) $x - 5$
(iii) $3x$	(iv) $-3x$
(v) $x/2$	(vi) $-x/2$
(vii) $3x + 10$	(viii) $3x - 4$
(ix) $10 - 3x$	(x) $100 - 3x$
(xi) $-x$.	

Solution [**Hint:** Addition or Subtraction of a constant does not alter the S.D. of x.]

(i) S.D. of $(x + 7)$ = S.D. of $x = 6$

(ii) S.D. of $(x - 5)$ = S.D. of $x = 6$

[**Hint:** Multiplication or Division by a constant changes the S.D. similarly. The sign in ignored.]

(iii) S.D. of $(3x)$ = 3 (S.D. of x) = 3×6 = 18

(iv) S.D. of $(-3x)$ = 3 (S.D. of x) = 3×6 = 18

(v) S.D. of $(x/2)$ = (S.D. of x)/2 = 6/2 = 3

(vi) S.D of $(-x/2)$ = (S.D. of x)/2 = 6/2 = 3

(vii) S.D. of $(3x + 10)$ = 3 (S.D. of x) = 3×6 = 18

(viii) S.D. of $(3x - 4)$ = 3 (S.D. of x) = 3×6 = 18

(ix) S.D. of $(10 - 3x)$ = 3 (S.D. of x) = 3×6 = 18

(x) S.D. of $(100 - 3x)$ = 3 (S.D. of x) = 3×6 = 18

(xi) S.D. of $(-x)$ = S.D. of x = 6

[**Note:** (1) If $y = a + bx$, where a and b are constants, then S.D.$(y) = |b|$. S.D.(x).

(2) This result holds also for Range, Q.D. and M.D.]

Problem 6.14 *(a) Calculate S.D. of the sample 3, 5, 8, 7, 5, 7. Without any further calculations, write down the S.D. of the following two samples:*

(i) 25, 27, 30, 29, 27, 29

(ii) 6, 10, 16, 14, 10, 14 [C.U., B. Com. (Hons) '83]

(b) The S.D. of a variable x is 10. Find the S.D. of the variable (5x – 10).

[W.B.H.S. '78]

(c) *Find the S.D. of the following data:*

x: 0, 1, –1, – 2, 6, 4, 5, 8, 12, 10, 11

Hence obtain the Variance of the following numbers:

y: –1, 1, –3, –5, 11, 7, 9, 15, 23, 19, 21

[I.C.W.A., June '86]

Solution Since S.D. is unaffected by change of origin, we subtract a constant 5 from each observation. $d = x - 5$.

x	d	d^2
3	– 2	4
5	0	0
8	3	9
7	2	4
5	0	0
7	2	4
Total	5	21

$$\sigma^2 = \Sigma d^2/n - (\Sigma d/n)^2$$
$$= 21/6 - (5/6)^2 = 101/36$$

$$\therefore \quad \sigma = \sqrt{101/6} = 10.05/6 = 1.675$$

(i) We see that these values are obtained by adding a constant 22 to each observation. Since S.D. is unaffected if each observation is increased or decreased by a constant, $\sigma = 1.675$.

(ii) We see that these values are obtained on multiplying each observation by a constant 2. Hence S.D. will also be multiplied by 2.

$$\sigma = 2 \times 1.675 = 3.35$$

(b) If $y = a + bx$, then $\sigma_y = |b| . \sigma_x$

Hence if we write $y = 5x - 10$ (here $a = -10$, $b = 5$)

$$\sigma_y = 5.\sigma_x = 5 \times 10 = 50$$

(c) $n = 11$

$$Sx = 0 + 1 - 1 - 2 + 6 + 4 + 5 + 8 + 12 + 10 + 11 = 54$$
$$\Sigma x^2 = 0 + 1 + 1 + 4 + 36 + 16 + 25 + 64 + 144 + 100 + 121 = 512$$
$$\sigma_x^2 = \Sigma x^2/n - (\Sigma x/n)^2$$
$$= 512/11 - (54/11)^2 = 2716/121 = 22.45$$

$$\therefore \quad \sigma_x = \sqrt{22.45} = 4.74$$

It will be observed that the given values of y are obtained on multiplying x by 2 and then subtracting 1 from it.

$$y = 2x - 1 ; \quad \therefore \quad \text{S.D. } (y) = 2 \times \text{S.D. } (x)$$

Squaring, Variance $(y) = 4 \times$ Variance $(x) = 4 \times 22.45 = 89.8$

Ans. 4.74, 89.8

Problem 6.15 *(a) What is the S.D. of the two values 10 and 7?*

(b) S.D. of 2 quantities is 3. If one of them is 7, find the other.

(c) Write down the S.D. of the six numbers: 5, 5, 5, 7, 7, 7. [W.B.H.S. '80]

(d) For 2 values of a variable x, the mean and S.D. are 10 and 2 respectively. What are the values of x? [B.U., B.A. (Econ.) '80]

Solution (a) S.D. of two values = Half their difference

$$= \frac{1}{2} \times 3 = 1.5$$

(b) If the other value is a, then

$$\frac{1}{2} \text{ (difference of 7 and } a) = 3$$

$\therefore$ Difference of 7 and $a = 6$

(Note that a may be larger or smaller than 7)

$\therefore$ $a = 1$ or 13.

(c) There are 6 observations in which the 2 values 5 and 7 occur with equal frequency. Therefore

S.D. of 5, 5, 5, 7, 7, 7 = S.D. of 5 and 7

$$= \frac{1}{2} \text{ (Difference of 5 and 7)}$$

$$= \frac{1}{2} \times 2 = 1$$

(d) Let the two values be a and b ($b \geq a$)

Then Mean $= \dfrac{a+b}{2} = 10$ (given)

S.D. $= \dfrac{b-a}{2} = 2$ (given), since b is larger, $b - a$ is positive.

$\therefore$ $a + b = 20$, $b - a = 4$

Adding, $2b = 24$. So, $b = 12$. $\therefore$ $a = 8$.

Ans. 8 and 12.

Problem 6.16 *(a) Write down the algebraic expression for root-mean-square-deviation from 10.*

(b) For a set of observations, variance of x turns out to be 5.95. What would be the minimum value of $\Sigma(x_i - A)^2/n$?

(c) When Variance (x) = 0, what can you say about the set of observations?

(d) What is the unit of variance ?

Solution (a) $\sqrt{\Sigma(x_i - 10)^2/n}$

(b) Minimum value = Variance = 5.95,

Because, S.D. = Minimum value of Root-Mean-Square-Deviation

$\therefore$ Variance = Minimum value of Mean-Square-Deviation.

(c) All the values of x are equal.

(d) Square of the unit in which the observations are given. For example, if observations are given in Rs, Variance will be in $(Rs)^2$.

Problem 6.17 *(a) If the relation between two variables x and y is $5x - 7y = 11$ and range of x is 5, find the range of y.* [W.B.H.S., '82]

(b) If the relation between 2 variables x and y is $2x + 3y = 7$, and quartile deviation of x is 3, then write down the Q.D. of y. [W.B.H.S. '83]

(c) If the relation between the variables y and v is $5v + 2y = 17$, and the mean deviation of y about its mean 6 is 5, write down the M.D. of v about its mean. [W.B.H.S.'86]

(d) The mean and S.D. of a variable x are known to be 38 and 6 respectively. What are the means and S.D.s of

(i) $100 - 2x$ (ii) $60 - x/2$ [B.U., B.A. (Econ) '82]

Solution [**Note:** If $Ax + By + C = 0$, then

$|A|$ Dispersion $(x) = |B|$ Dispersion (y) for any absolute measure of Dispersion.]

(a) $5 \times$ Range $(x) = 7 \times$ Range (y)

or, $5 \times 5 = 7 \times$ Range (y); $\therefore$ Range $(y) = 25/7$

(b) $2 \times$ Q.D. $(x) = 3 \times$ Q.D. (y)

or, $2 \times 3 = 3 \times$ Q.D. (y): $\therefore$ Q.D. $(y) = 2$

(c) $5 \times$ M.D. $(v) = 2 \times$ M.D. (y) for M.D. about mean.

$= 2 \times 5 = 10.$ $\therefore$ M.D. $(v) = 2$

(d) If $y = a + bx$, then $\bar{y} = a + b\bar{x}$ and $\sigma_y = |b| \cdot \sigma_x$

Here $\bar{x} = 38, \sigma_x = 6$

(i) $y = 100 - 2x_j$ $\bar{y} = 100 - 2 \times 38 = 24$

$\sigma_y = 2 \times 6 = 12$

(ii) $y = 60 - \frac{1}{2}x;$

$y = 60 - \frac{1}{2} \times 38 = 41$

$\sigma_y = \frac{1}{2} \times 6 = 3$

(Also see Problem 6.13, p. 439)

Problem 6.18 *Are the following statements consistent? Justify your answer.*

(a) When all observations are same, any measure of dispersion would be 1.0 (one).

(b) If $x_i = 5 + y_i$, then $R(x) = 5 + R(y)$, where R is the Range of the variable concerned.

(c) Mean Absolute Deviation from median and mean are 5 and 4 respectively.

(d) $V(x) = \Sigma x_i^2/n - \bar{x}^2 = -30$; x_i, $\bar{x}$ and n are the usual notation.

(e) The unit of Variance is that of the observations. [C.U., M.Com., '82]

Solution (a) No; it is zero.

Range = 0, because maximum and minimum values are equal.

Q.D. = 0, because then $Q_3 = Q_1$ = common observation.

M.D. = 0; because mean is also the same as the common observation, so that $x_i - \bar{x} = 0$.

S.D. = 0; because $x_i - \bar{x} = 0$.

C.V. and all other relative measures contain an absolute measure of dispersion in the numerator, which is zero. So, any relative measure of dispersion = 0.

(b) No; $R(x) = R(y)$. Because Range (in fact, any absolute measure of dispersion) is unaffected by change of origin.

(c) No; because M.D. about median should be smaller than about mean (or any other value).

(d) No; because Variance cannot be negative.

(e) No; the unit of variance is the square of the unit in which observations are given.

Problem 6.19 *The range, standard deviation and coefficient of variation in respect of a set of values are respectively 15, 4.38 and 7.78%. How would these results be affected if each value of the set be increased by 5?* [I.C.W.A., Dec. '85]

Solution Original series: Range = 15, S.D. = 4.38

$$\text{C.V.} = 7.78\%; \quad \text{i.e. } \frac{4.38}{\text{mean}} \times 100 = 7.78;$$

$$\therefore \quad \text{mean} = \frac{4.38 \times 100}{7.78} = 56.3$$

New Series: If each observation is increased by 5,

(i) Range and S.D. do not change (because absolute measures of dispersion are unaffected by change of origin).

$$\text{Range} = 15, \qquad \text{S.D.} = 4.38$$

(ii) Mean increases by 5.Mean = 56.3 + 5 = 61.3

$$\therefore \quad \text{C.V.} = \frac{4.38}{61.3} \times 100 = 7.15$$

Ans. 15, 4.38, 7.15%

Problem 6.20 *If the third quartile is 30 and median 22, find the coefficient of quartile deviation, assuming the distribution to be symmetrical.*

[I.C.W.A., June '86]

Solution For a symmetrical distribution

$$\text{Median} = \frac{1}{2}(Q_1 + Q_3)$$

$$\text{or,} \quad 22 = \frac{1}{2}(Q_1 + 30); \quad \therefore \quad Q_1 = 14$$

$$\text{Hence,} \quad \text{Q.D.} = \frac{1}{2}(Q_3 - Q_1) = \frac{1}{2}(30 - 14) = 8$$

$$\text{Coeff. of Q.D.} = \frac{\text{Q.D.}}{\text{Median}} \times 100 = \frac{8 \times 100}{22} = 36.4$$

ADDITIONAL EXERCISES 6

(Q.D. = Quartile Deviation, M.D. = Mean Deviation,
S.D. = Standard Deviation, C.V. = Coefficient of Variation)

1. Find Range and Q.D. for the following data:

Variable	3	5	7	9	11	13
Frequency	2	7	10	9	6	1

2. Calculate Q.D. from the following:

Weight (kg.)	14–18	18–22	22–26	26–30	30–34	Total
Frequency	15	40	25	13	7	100

[K.U., B.Com. (Pass) '85]

3. Calculate M.D. about median from the following data:

(a) (Rs) 46, 79, 26, 85, 39, 65, 29, 59, 73.

[C.U., B.Com. (Pass) '85]

(b) 79, 85, 99, 46, 56, 72, 65, 32, 26, 36.

[I.C.W.A., June'85]

4. Calculate Median and the M.D. about median from the following:

x	5	6	7	8	9
y	1	4	7	6	2

[K.U., B.Com.(Pass)'80]

5. Find Mean Absolute Deviation for the following:

Height (inches)	60–62	63–65	66–68	69–71	72–74
Frequency	5	18	42	27	8

[C.U., B.Sc. (Econ.) '83]

6. Find S.D. from the following data:

x	10	20	30	40	50	60	Total
f	9	18	25	27	14	7	100

[C.U., B.Com. (Pass) '84, M.Com. '82]

7. Calculate S.D. from the following distribution of weights (in gm) of Zinc coating:

Cell Boundaries	*Frequency*
12.75–13.25	3
13.25–13.75	7
13.75–14.25	15
14.25–14.75	11
14.75–15.25	8
15.25–15.75	4
15.75–16.25	2
Total	50

[C.U., B.Com. (Hons.) '83]

8. Calculate Mean and S.D. from the following data:

Value :	90–99	80–89	70–79	60–69	50–59	40–49	30–39	Total
Frequency :	2	12	22	20	14	2	1	75

[I.C.W.A., Dec.'81]

9. Calculate S.D. of the following frequency distribution showing Wages in a sample of factory workers in Calcutta, 1980:

Weekly Wage (Rs)	*No. of Employees*
30–34.99	3
35–39.99	9
40–44.99	15
45–49.99	27
50–54.99	18
55–59.99	12
60–64.99	9
65–69.99	7

[C.U., B.Sc. (Econ.) '82]

10. After change of origin and scale, a frequency distribution of a continuous variable takes the form as under:

Step Deviation (u) :	– 3	– 2	– 1	0	1	2	3
Frequency (f) :	10	15	25	25	10	10	5

If the mean and variance of the distribution are respectively 31 and 254, find the original frequency distribution. *[I.C.W.A., Dec. 85]*

[**Hint:** Page 382].

11. Calculate S.D. stating the necessary assumption:

Heights in Inches	*No. of Men (hundred)*
Under 62	5
62 and under 63	10
63 and under 64	10
64 and under 65	30
65 and over	45

[I.C.W.A., June'86]

12. Calculate mean and S.D. from the following data:

Age	*No. of Persons Dying*	*Age*	*No. of Persons Dying*
Under 10	15	Under 50	100
Under 20	30	Under 60	110
Under 30	53	Under 70	115
Under 40	75	Under 80	125

[**Hint:** —Table 3, Page 359]. *[I.C.W.A., June '80]*

13. Find the number of items lying within Mean ± σ of following distribution:

Class Interval	*Frequency*	*Class Interval*	*Frequency*
11–12	5	21–22	395
13–14	426	23–24	38
15–16	720	25–26	8
17–18	741	27–28	5
19–20	665	29–30	7

[Hint: Example 6.34, p. 199], *[I.C.W.A., Dec. '82]*

14. For a number of 21 observations, the mean and s.d. are 10 and 2 respectively. It was found after the calculations were made that one of the observations recorded as 7 was incorrect. Find the s.d. of 20 observations, if the incorrect observation is omitted. *[C.U., B.Sc. (Econ) '86]*

15. For a number of $N = 10$ observations it was found that mean was 20 and S.D. $= \sqrt{35}$. After these computations were performed it was found that observations 5 and 15 were incorrect. Find the mean and S.D. if these two incorrect observations are omitted. *[C.U., M.Com. '82]*

16. In a certain factory,
    Number of employees = 100
    Average wages per employee per month = 85 Rs
    S.D. of the wages per employee per month = 4 Rs
    Suppose the wages of an employee were wrongly noted as Rs 120 instead of Rs 100. What would be the correct S.D.? *[W.B.H.S. '84]*

17. For two values, say a and b ($a < b$), of a variable the mean and the s.d. are respectively 17 and 3. Find a and b. *[B.U., B.A. (Econ)'82]*

18. Find the missing figures from the following table:

	No. of Observations	*Mean*	*Variance*
Group I:	—	45	—
Group II:	150	—	144
Combined Group of I and II:	250	51	130

[I.C.W.A., Dec. '85]

19. The mean and variance of 200 values are 60 and 400 respectively. If at the time of calculation, two values were taken as 3 and 67 in place of 13 and 17, find the correct Coefficient of Variation. *[I.C.W.A., June '86]*

20. Using the values of medians and quartiles given below for two distributions A and B, examine which is more variable:
    For A : $Q_1 = 22.65$, Median = 29.45, $Q_3 = 31.70$
    For B : $Q_1 = 192.72$, Median = 242.65, $Q_3 = 292.69$

[I.C.W.A., Dec. '85]

SHORT ANSWERS TYPE QUESTIONS

21. (a) Find Range of the daily wages of 10 persons:

(Rs) 24, 18, 25, 16, 20, 28, 22, 17, 21, 27

[C.U., B.Com(P) '84]

(b) What would be the value of Range, if each observation is

(i) increased by 5, (ii) decreased by 3,
(iii) multiplied by 2, (iv) multiplied by –2,
(v) divided by 10, (vi) multiplied by 2 and then 3 added to it ?

22. Find Semi-interquartile Range from the following data:

4, –1, 8, 0, –3, 18, 7.

23. Find M.D. (i) about 7, (ii) about median, (iii) about mean, from the following data: 4, 0, 8, 6, –3.

24. Find Standard Deviation and Coefficient of Variation from the following:
41, 37, 45, 43, 34.

25. What is the S.D. of two observations (in cm) 53 and 57?

26. Calculate (i) Range, (ii) Quartile Deviation, (iii) Mean Deviation, (iv) Standard Deviation, and (v) Coefficient of Variation, from the following:

x	40	50	60	70	Total
f	3	7	4	1	15

27. Calculate (i) Range, (ii) Quartile Deviation, (iii) Mean Deviation, (iv) Standard Deviation, and (v) Coefficient of Variation, from the following:

Weight (kg)	6–9	10–13	14–17	18–21	Total
Frequency	2	10	5	3	20

28. A student obtained the following results. Do you think that the computations are correct?

(a) $n = 10$, $\Sigma x = 100$, $\Sigma x^2 = 800$

(b) $\Sigma f = 200$, $\Sigma fx = 1000$, $\Sigma fx^2 = 4800$

29. Weekly average wage of workers in a factory increases from Rs 8.00 to Rs 12.00 and s.d. increases from Re 1 to Rs 5. Have the wages become less uniform now? *[I.C.W.A., Dec. '82]*

30. Two samples A and B have the same s.d., but the mean of A is greater than that of B. The Coefficient of Variation of A is: (i) Greater than that of B, (ii) Less than that of B, (iii) Equal to that of B, (iv) None of these. Which is correct? *[C.A., Nov. '82]*

31. If the third quartile is 178 and median is 160, find the coefficient of Quartile Deviation, assuming the distribution to be symmetrical. *[I.C.W.A., June '85]*

32. If the quartiles of a distribution are 28, 35 and 40, find Semi-interquartile Range.

33. In a certain distribution the sum of the two quartiles is 20 and Quartile Deviation is 4. Find the quartiles.

34. The mean and s.d. calculated from 9 observations are 17 and 3 respectively. If an additional observation 27 is included, find the mean and s.d. of all the 10 observations. *[K.U., B.Com(P) '83]*

35. The first of two groups has 100 items with mean 45 and variance 49. If the combined group has 250 items with mean 51 and variance 130, find the mean and s.d. of the second group. *[I.C.W.A., June '84]*

36. If $x_i = 99 + y_i$ and Range $(x) = 99$, find Range (y).

37. If the Quartile Deviation (Q.D.) of x is 3, what is the Q.D. of (i) $5 + 4x$, (ii) $5 - 4x$?

38. (a) If $2x + 3y + 9 = 0$ and range of x is 12, find range of y.
    (b) If two variables x and y are related by $7x - 2y + 5 = 0$ and M.D. (about mean) of x is 4, find M.D. of y.

39. (a) Two variables x and y are given by $y = 11 - 5x$. If the s.d. of x is 4, write down the s.d. of y. *[W.B.H.S. '81]*
    (b) If $x + 3y - 7 = 0$ and s.d. of y is 5, what is the s.d. of x?
    (c) If $x = \frac{3}{5}(y - 10)$ and Variance $(x) = \frac{1}{5}$, find Variance (y).

40. If the s.d. of 1, 2, 3, 4, 5, 6, 7 is known to be 2, what is the s.d. of 2, 4, 6, 8, 10, 12, 14? Hence write down the s.d. of 1, 3, 5, 7, 9, 11, 13.

41. Weights of 10 students are recorded by a machine as under: (kg) 50, 55, 57, 49, 54, 61, 64, 59, 58, 56. If the weighing machine shows weight less by 5 kgs, find the correct values of Range, S.D. and C.V. without calculating the correct weights.
    (Hint: Range and S.D. do not change; but mean increases by 5 kg. See Problem 6.19, page 390) *[I.C.W.A., June '84]*

42. Examine each of the following statements and state whether it is True or False:
    (a) If each observation decreases by 10, Range decreases by 10.
    (b) Q. D. is independent of origin.
    (c) M.D. may be negative
    (d) S.D. does not depend on the scale of measurement.
    (e) Coefficient of Variation is unit-free.
    (f) If each observation is multiplied by 3, coefficient of Q.D. will not change.

43. Fill up the missing parts:
    (a) Range of a given set of observations is defined as ... ...
    (b) Q.D. is defined as ... ...
    (c) M.D. about an arbitrary origin A is defined as ... ...
    (d) S.D. of a given set of observations is defined as ... ...
    (e) Coefficient of Variation is defined as ... .
    (f) Coefficient of Quartile Deviation is defined as ... ...

44. Fill up the missing parts:
   (a) M.D. is minimum when deviations are taken about ....
   (b) Root-Mean-Square-Deviation is minimum when deviations are taken about ...
   (c) S.D. is ... than M.D. about mean.
   (d) S.D. of two observations is equal to ... their differences.
   (e) S.D. is ... when all observations are equal.
   (f) S.D. of one observations is ...

45. Fill up the blanks:
   (a) The quickest measure of dispersion is ...
   (b) The appropriate measure of dispersion from a frequency distribution with open-end class is ...
   (c) The measure of dispersion which is based on all observation and also easy to understand is ...
   (d) The most useful measure of dispersion is ...
   (e) For comparing the variability of two or more distributions whose means differ widely, the appropriate measure of dispersion is ...
   (f) The most widely-used relative measure of dispersion is ...

ANSWERS

1. 10, 2 2. 3.1 kg 3. Rs 18, 20.4 4. 7, 0.8
5. 2.265 inch 6. 13.56 7. 0.073 gm 8. 68.1, 12.5
9. Rs 8.76
10. Class intervals 0 – 10, 10 – 20, ... 60 – 70
11. 1.18 inch (assuming equal width for all classes)
12. 35.16, 19.8 13. 1929 14. 1.93
15. 22.5, 2.5 16. Rs 2.44 17. 14, 20 18. 100, 55, 49
19. 33.6 20. *B* (Coeff. of Q.D. larger)
21. (a) 12 Rs; (b) (Rs) 12, 12, 24, 24, 1.20, 24
22. 4.5 23. 4.4, 3.4, 3.6 24. 4, 10 25. 2 cm
26. 30, 5, 6.67, 8.33, 16 27. 16 kg, 2.6 kg, 2.96 kg, 3.46 kg, 26
28. No (see Problem 6.12) 29. Yes (use C.V.)
30. (ii) 31. 11.25 32. 6 33. 6, 14
34. 18, 4.1 35. 55, 12 36. 99 37. 12, 12
38. 8, 14 39. 20, 15, 5/9 40. 4, 4 41. 15 kg, 4.38 kg, 7.15
42. True—(b), (e), (f); False—(a), (c), (d)

43. (a) the difference between the largest and the smallest items.
    (b) half difference between the upper and lower quartiles.
    (c) the mean of absolute deviations from A.
    (d) the root-mean-square-deviation from mean; (i.e. the square-root of the mean of squared deviations from mean).
    (e) the ratio of S.D., to A.M., multiplied by 100.
    (f) the ratio of Q.D. to Median, multiplied by 100.
44. (a) median, (b) mean, (c) not smaller, (d) half, (e) and (f) zero.
45. (a) Range, (b) Q.D., (c) M.D., (d) S.D., (e) C.V., (f) C.V

CHAPTER 7

MOMENTS, SKEWNESS AND KURTOSIS

(Summary)

Moment Given n observations $x_1, x_2, \ldots, x_n$ the "r-th moment about an arbitrary origin A" is defined by the formula

$$m_r = \frac{1}{n}\,\Sigma(x - A)^r;\ r = 0, 1, 2, 3, \ldots..$$

Thus the r-th moment about A is the mean of the r-th powers of deviations from A.

Moments about zero are called "Raw Moments", and moments about mean are called "Central moments".

r-th raw moment: $m'_r = \frac{1}{n}\,\Sigma\, x^r$

r-th central moment: $m_r = \frac{1}{n}\,\Sigma\,(x - \bar{x})^r$

[**Note:** For a frequency distribution, the formulae can be written down easily: Change n to N, and bring f after Σ sign.]

Important Results

(1) Mean ($\bar{x}$) = 1st raw moment (m'_1)

Variance (σ^2) = 2nd central moment (m_2)

(2) If the Ist and 2nd moments about any origin A are known,

Mean ($\bar{x}$) = $A + m'_1$

S.D. (σ) = $\sqrt{m'_2 - m'^2_1}$

(3) Relation between central moment and non-central moments:

$$m_1 = 0 \text{ (always)}$$
$$m_2 = m'_2 - m'^2_1$$
$$m_3 = m'_3 - 3m'_2 m'_1 + 2m'^3_1$$
$$m_4 = m'_4 - 4m'_3 m'_1 + 6m'_2 m'^2_1 - 3m'^4_1$$

(4) Moments can be used to measure all the four characteristics of statistical data:

(i) Central tendency : Mean = 1st raw moment (m'_1)

(ii) Dispersion : S.D. = $\sqrt{m_2}$

(iii) Skewness : Skewness = $m_3/(\sqrt{m_2})^3$

(iv) Kurtosis : Kurtosis = $m_4/(m_2)^2 - 3$.

(5) Moments of odd order (e.g. 1st moment, 3rd moment) may be positive, negative or zero; but moments of even order (e.g. 2nd moment, 4th moment) can never be negative.

(6) If observations are given in inch, then the 1st moment has unit 'inch', 2nd moment has unit 'square inch', 3rd moment has unit 'cubic inch', and so on.

(7) The central moments m_2, m_3, m_4 etc. are independent of origin, but depend on scale. Thus, if $y = (x - c)/d$, then

$$m_2 = d^2 \cdot \text{(2nd central moment of } y\text{)},$$

$$m_3 = d^3 \cdot \text{(3rd central moment of } y\text{); etc.}$$

Standardized Variable

If the variable x has mean $\bar{x}$ and standard deviation σ, then the variable z defined by

$$z = \frac{x - \bar{x}}{\sigma}$$

is called the 'Standardized variable'. The mean and s.d. of z and always 0 and 1 respectively.

$$r\text{-th moment of } z = m_r/\sigma^r$$

(Note that 3rd and 4th moments of z are used to measure "skewness" and "kurtosis" respectively.)

Skewness

"Skewness" denotes the degree of asymetry (i.e. lack of symmetry) in the data. It is used to indicate the 'shape' of the frequency curve as regards its distortion from symmetry. Skewness may be positive, negative or zero.

Measures of Skewness

(1) Pearson's 1st measure $= \dfrac{\text{Mean} - \text{Mode}}{\text{S.D.}}$

(2) Pearson's 2nd measure $= \dfrac{3\,(\text{Mean} - \text{Median})}{\text{S.D.}}$

(3) Bowley's measure $= \dfrac{Q_3 - 2 \cdot Q_2 + Q_1}{Q_3 - Q_1}$

(4) Moment measure $= \dfrac{m_3}{\sigma^3} = \dfrac{m_3}{(\sqrt{m_2})^3}$

When skewness is positive, distribution is called "positively skew"
When skewness is negative, distribution is called "negatively skew"
When skewness is zero, distribution is called "symmetrical"

Important Results

(a) All measures of skewness are pure numbers (i.e. unit-free)

(b) (i) For symmetrical distributions (i.e. skewness = 0)
Mean = Median = Mode; $Q_2 = (Q_1 + Q_3)/2$; $m_3 = 0$. The frequency curve is symmetrical.

(ii) For positively skew distributions,
Mean > Median > Mode; m_3 positive.
The frequency curve has the longer tail towards the right.

(iii) For negatively skew distribution,

Mean < Median < Mode ; m_3 negative

The frequency curve has the longer tail towards the left.

(c) Bowley's measure always lies between – 1 and + 1.

Kurtosis

"Kurtosis" denotes the degree of 'peakedness' of the frequency curve. It is used to indicate the 'shape' of the frequency curve as regards the sharpness of its peak. Kurtosis may be positive, negative or zero.

The only measure of kurtosis is based on moments.

$$\text{Kurtosis} = \frac{m_4}{\sigma^4} - 3 = \frac{m_4}{(m_2)^2} - 3$$

When kurtosis is positive, distribution is called "leptokurtic"

When kurtosis is negative, distribution is called "platykurtic"

When kurtosis is zero, distribution is called "mesokurtic"

Important Results

(a) The measure of kurtosis is a pure number (i.e. unit-free).

(b) The frequency curve of

"leptokurtic" distribution has a sharp peak;

"platykurtic" distribution has a blunt peak;

"mesokurtic" distribution has a moderate peak

(c) "Normal distribution" (Chapter 12) is symmetrical and mesokurtic (i.e. skewness = 0, kutosis = 0).

SOLVED PROBLEMS

Problem 7.1 *Find the first, second and third moments about 4 of the set of numbers 2, 4, 6, 8.* [I.C.W.A., June' 80]

Solution

x	$x - 4$	$(x - 4)^2$	$(x - 4)^3$
2	– 2	4	– 8
4	0	0	0
6	2	4	8
8	4	16	64
Total	4	24	64

Moments about 4:

$$m_1' = \frac{\Sigma(x - 4)}{n} = \frac{4}{4} = 1$$

$$m_2' = \frac{\Sigma(x - 4)^2}{n} = \frac{24}{4} = 6$$

$$m_3' = \frac{\Sigma(x-4)^3}{n} = \frac{64}{4} = 16 \qquad \textit{Ans. } 1, 6, 16.$$

Problem 7.2 *Find the 1st, 2nd and 3rd raw moments of 2, 4, 6, 8.*

Solution

x	x^2	x^3
2	4	8
4	16	64
6	36	216
8	64	512
20	120	800

Raw moments:

$$m_1' = \frac{\Sigma x}{n} = \frac{20}{4} = 5$$

$$m_2' = \frac{\Sigma x^2}{n} = \frac{120}{4} = 30$$

$$m_3' = \frac{\Sigma x^3}{n} = \frac{800}{4} = 200 \qquad \textit{Ans. } 5, 30, 200.$$

Problem 7.3 *Find the first four moments about mean for the set of numbers 2, 4, 6, 8.* [I.C.W.A., June' 82, '83]

Solution

x	$x-\bar{x}$	$(x-\bar{x})^2$	$(x-\bar{x})^3$	$(x-\bar{x})^4$
2	− 3	9	− 27	81
4	− 1	1	− 1	1
6	1	1	1	1
8	3	9	27	81
20	0	20	0	164

Moments about mean:

$$m_1 = \frac{\Sigma(x-\bar{x})^2}{n} = \frac{0}{4} = 0$$

$$m_2 = \frac{\Sigma(x-\bar{x})^2}{n} = \frac{20}{4} = 5$$

$$m_3 = \frac{\Sigma(x-\bar{x})^3}{n} = \frac{0}{4} = 0$$

$$m_4 = \frac{\Sigma(x-\bar{x})^4}{n} = \frac{164}{4} = 41$$

Mean ($\bar{x}$) = 20/4 = 5. *Ans.* 0, 5, 0, 41.

Problem 7.4 *The first three moments of a distribution about the value 2 of a variable x and 1, 16 and –40. Find the mean, variance and the third central moment.*
[C.U., B.Com. (Hons.) '84]

Solution Give, Moments about $A = 2$,
$m_1' = 1, m_2' = 16, m_3 = -40.$

Mean $(\bar{x}) = A + m_1' = 2 + 1 = 3$

Variance $(m_2) = m_2' - (m_1')^2 = 16 - 1^2 = 15$

3rd Central Moment $(m_3) = m_3' - 3m_2' m_1' + 2(m_1')^3$

$= -40 - 3 \times 16 \times 1 + 2 \times 1^3$

$= -40 - 48 + 2$

$= -86.$ *Ans.* 3, 15, – 86.

Problem 7.5 *Find the first, second and third moments about 7 from the following data:*

Values	2	4	6	8	10	12	*Total*
Frequency	6	9	7	15	7	6	50

Hence find the 1st, 2nd and 3rd central moments. [C.U., B.Com. (Pass) '81]

Solution

x	f	$d = x - 7$	fd	fd^2	fd^3
2	6	– 5	– 30	150	– 750
4	9	– 3	– 27	81	– 243
6	7	– 1	– 7	7	– 7
8	15	1	15	15	15
10	7	3	21	63	189
12	6	5	30	150	750
Total	50	–	2	466	– 46

Moments about 7: Writing $d = x - 7$,

$$m_1 = \frac{\Sigma fd}{N} = \frac{2}{50} = 0.04$$

$$m_2 = \frac{\Sigma fd^2}{N} = \frac{466}{50} = 9.32$$

$$m_3 = \frac{\Sigma fd^3}{N} = \frac{-46}{50} = -0.92$$

Central Moments:

$$m_1 = 0 \text{ (always)}$$
$$m_2 = m_2' - (m_1)^2$$
$$= 9.32 - (0.04)^2 = 9.32 - .0016 = 9.3184$$
$$m_3 = m_3' - 3m_2m_1 + 2(m_1')^3$$
$$= -0.92 - 3 \times 9.32 \times 0.04 + 2(0.04)^3$$
$$= -0.92 - 1.1184 + .000128$$
$$= -2.038,272$$

[**Note:** Odd order moments like m_1', m_3', m_3 may be positive negative or zero; but even order moments like m_2', m_4', m_2, m_4 can never be negative.]

Problem 7.6 *Find the first, second and third moments about 3 from the following distribution:*

Class interval	*0–2*	*2–4*	*4–6*	*6–8*	*Total*
Frequency	*3*	*8*	*5*	*4*	*20*

Also, calculate the third central moment. [K.U., B.Com. (Pass)' 85]

Solution

Midvalue (x)	*frequency (f)*	*d = x – 3*	*fd*	*fd²*	*fd³*
1	3	– 2	– 6	12	– 24
3	8	0	0	0	0
5	5	2	10	20	40
7	4	4	16	64	256
Total	20	–	20	96	272

Moment about 3:

$$m_1 = 20/20 = 1$$
$$m_2' = 96/20 = 4.8$$
$$m_3' = 272/20 = 13.6$$

Third Central Moment:

$$m_3 = m_3' - 3m_2' m_1 + 2m_1'^3$$
$$= 13.6 - 3 \times 4.8 \times 1 + 2 \times 1^3$$
$$= 13.6 - 14.4 + 2$$
$$= 1.2$$

Ans. 1, 4.8, 13.6 ; 1.2

Problem 7.7 *In a distribution the difference of two quartiles is 2.03; their sum is 72.67 and the median is 36.18. Find the coefficient of Skewness.*

[I.C.W.A., June '81 ; C.U., B.Com (Hons) '85]

Solution Since the third quartile (Q_3) is larger than the first quartile (Q_1), we have $Q_3 - Q_1 = 2.03$. Also $Q_3 + Q_1 = 72.67$ and median (Q_2) = 36.18. Using Bowley's formula (7.7.3, p. 234)

$$\text{Skewness} = \frac{Q_3 + Q_1 - 2Q_2}{Q_3 - Q_1} = \frac{72.67 - 2 \times 36.18}{2.03} = +\,0.15.$$

Problem 7.8 *The central moments of a distribution are given by $m_2 = 140$, $m_3 = 148$, $m_4 = 6030$. Calculate the moment measures of skewness and kurtosis, and comment on the shape of the distribution.* [I.C.W.A., Dec. '84]

Solution

$$\text{Skewness} = \frac{m_3}{(\sqrt{m_2})^3} = \frac{148}{(140)^3} = \frac{148}{140\sqrt{140}}$$

$$= \frac{148}{1656.5} = +\,0.089$$

$$\text{Kurtosis} = \frac{m_4}{m_2^2} - 3 = \frac{6030}{19600} - 3 = -\,2.692$$

The distribution is *positively skew* and *playkurtic* (kurtosis negative).

Problem 7.9 *The first four moments of a distribution about origin are are 1, 4, 10 and 46 respectively. Obtain the various characteristics of the distribution on the basis of the information given. Comment upon the nature of the distribution.* [I.C.W.A., June '84]

Solution Given Moments about zero (i.e. Raw Moments):

$$m_1' = 1,\ m_2' = 4,\ m_3' = 10,\ m_4' = 46.$$

The various characteristics, namely (1) Central tendency, (2) Dispersion, (3) Skewness, and (4) Kutrosis, are measured by using 1st raw moment m_1' and central moments m_2, m_3, m_4.

$$m_2 = m_2' - m_1'^2 = 4 - 1^2 = 3$$

$$m_3 = m_3' - 3m_2' m_1' + 2m_1'^3$$

$$= 10 - 3 \times 4 \times 1 + 2 \times 1^3 = 10 - 12 + 2 = 0$$

$$m_4 = m_4' - 4m_3' m_1' + 6m_2' m_1'^2 - 3m_1'^4$$

$$= 46 - 4 \times 10 \times 1 + 6 \times 4 \times 1^2 - 3 \times 1^4$$

$$= 46 - 40 + 24 - 3 = 27$$

Mean = 1st Raw Moment (m_1') = 1

$$\therefore \quad \text{S.D.}\ (\sigma) = \sqrt{m_2} = \sqrt{3} = 1.732$$

$$\text{Skewness} = m_3/\sigma^3 = 0/(\sqrt{3})^3 = 0$$

$$\text{Kurtosis} = m_4/\sigma^4 - 3 = (27/9) - 3 = 0$$

Comment: The distribution is *symmetrical* (skewness = 0) and *mesokurtic* (kurtosis = 0).

Problem 7.10 *The following values were obtained from a frequency distribution: $\Sigma f = 100$, $\Sigma fx = -100$, $\Sigma fx^2 = 5500$, $\Sigma fx^3 = -6250$ and $\Sigma fx^4 = 775000$. Comment on the nature of the distribution after calculating a measure of skewness and a measure of kurtosis.* [I.C.W.A. (New), June '86]

Solution For a frequency distribution, the Raw Moments are

$$m_1' = \Sigma fx/N = -100/100 = -1$$
$$m_2' = \Sigma fx^2/N = 5500/100 = 55$$
$$m_3' = \Sigma fx^3/N = -6250/100 = -62.5$$
$$m_4' = \Sigma fx^4/N = 775000/100 = 7750$$

Using them, we find the Central Moments:

$$m_2 = m_2' - m_1'^2$$
$$= 55 - (-1)^2 = 55 - 1 = 54$$
$$m_3 = m_3' - 3m_2' m_1' + 2m_1'^3$$
$$= -62.5 - 3 \times 55 \times (-1) + 2(-1)^3$$
$$= -62.5 + 165 - 2 = 100.5$$
$$m_4 = m_4' - 4m_3' m_1' + 6m_2' m_1'^2 - 3m_1'^4$$
$$= 7750 - 4(-62.5)(-1) + 6 \times 55\,(-1)^2 - 3\,(-1)^4$$
$$= 7750 - 250 + 330 - 3 = 7827$$

$$\text{Skewness} = \frac{m_3}{(\sqrt{m_2})^3} = \frac{100.5}{396.8} = 0.253$$

$$\text{Kurtosis} = \frac{m_1}{m_2^{\;2}} - 3 = \frac{7827}{2916} - 3 = -0.316$$

The distribution is "positively skew" and "platykurtic".

Problem 7.11 *Find the moment measures of skewness (γ_1) and kurtosis (γ_2) from the following distribution:*

Class Interval	*0–10*	*10–20*	*20–30*	*30–40*	*Total*
Frequency	*1*	*3*	*4*	*2*	*10*

[I.C.W.A.., Dec. '82, '84]

Solution

Mid-value *(x)*	*f*	$y = \frac{x-25}{10}$	*fy*	fy^2	fy^3	fy^4
5	1	– 2	– 2	4	– 8	16
15	3	– 1	– 3	3	– 3	3
25	4	0	0	0	0	0
35	2	1	2	2	2	2
Total	10	—	– 3	9	– 9	21

Raw Moments of *y*:

$$m_1' = \Sigma fy/N = -3/10 = -0.3$$
$$m_2' = \Sigma fy^2/N = 9/10 = 0.9$$
$$m_3' = \Sigma fy^3/N = -9/10 = -0.9$$
$$m_4' = \Sigma fy^4/N = 21/10 = 2.1$$

Centeral Moments of y:

$$m_2 = m_2' - (m_1')^2$$
$$= 0.9 - (-0.3)^2 = 0.9 - 0.09 = 0.81;$$

$\therefore$ $$\sigma = \sqrt{0.81} = 0.9$$

$$m_3 = m_3' - 3m_2'm_1' + 2(m_1')^3$$
$$= -0.9 - 3(0.9)(-0.3) + 2(-0.3)^3$$
$$= -0.9 + 0.81 - 0.054 = -0.144$$
$$m_4 = m_4 - 4m_3' m_1' + 6m_2'(m_1')^2 - 3(m_1')^4$$
$$= 2.1 - 4(-0.9)(-0.3) + 6(0.9)(-0.3)^2 - 3(-0.3)^4$$
$$= 2.1 - 1.08 + 0.486 - 0.0243$$
$$= 1.4817$$

Moment measure of Skewness and Kurtosis are unaffected by changes of origin and scale. Hence using the central moments of y,

$$\text{Skewness} = m_3/\sigma^3 = -0.144/0.729 = -0.198$$
$$\text{Kurtosis} = (m_4/\sigma^4) - 3 = (1.4817/0.6561) - 3 = -0.742.$$

Problem 7.12 *Derive the formula*

$$m_3 = m_3 - 3m_2\, m_1 + 2(m_1')^3$$

where m_1, m_2', m_3' denote moments about an arbitrary origin A.

Solution Let $x_1, x_2, \ldots, x_n$ be the observations. Then

$$m_1' = \Sigma(x - A)/n = \bar{x} - A = d \text{ (suppose)} \qquad \ldots \text{(i)}$$
$$m_2' = \Sigma(x - A)_x^2/n,\ m_3' = \Sigma(x - A)^3/n \qquad \ldots \text{(ii)}$$

3rd central moment $m_3 = \Sigma(x - \bar{x})^3/n$

We can write $(x - \bar{x}) = (x - A) - (\bar{x} - A)$

$$= (x - A) - d, \text{ from (i)}$$

$\therefore$ $(x - \bar{x})^3 = \Sigma(x - A)^3 - 3(x - A)^2 d + 3(x - A)d^2 - d^3$

Applying Σ operator on both sides and then dividing by n,

$$\Sigma(x - \bar{x})^3/n = \Sigma(x - A)^3/n - 3d\,\Sigma(x - A)^2/n + 3d^2\,\Sigma(x - A)/n - nd^3/n$$

or, $$m_3 = m_3' - 3dm_2' + 3d^2 m_1' - d^3, \qquad \text{from (i) and (ii)}$$
$$= m_3' - 3m_1' m_2' + 3(m_1')^2 m_1' - (m_1')^3, \qquad \text{from (i)}$$
$$= m_3' - 3m_2' m_1' + 3(m_1')^3 - (m_1')^3$$
$$= m_3' - 3m_2' m_1' + 2(m_1')^3.$$

ADDITIONAL EXERCISES 7

1. Find the first, second and third moments about 10 from the following data:

   6, 7, 12, 9, 15

2. Find the first three (i) raw moments, and (ii) central moments, from the following:

   2, 5, 8 9.

3. Find the first, second and third moments about 5 from the following:

x	1	3	5	7	9	Total
f	3	5	16	18	8	50

Hence, calculate 2nd moment about mean. *[K.U., B. Com(Pass) '83]*

4. Calculate the first four central moments:

Class Interval	12–13	14–15	16–17	18–19	Total
Frequency	3	10	5	2	20

5. The first two moments of a distribution about the value 3 of a variable are 1 and 10 respectively. Calculate the arithmetic mean and the standard deviation of the distribution. *[C.U., M.B.A. '81]*

6. For a given set of observations, the first three moments about 8 are 2, 17 and 42 respectively. Find (i) the 3rd raw moment, and (ii) and 2nd central moment. (**Hint:** see Example 7.2, p. 226) *[K.U., B.Com(Pass) '81]*

7. If M_2' and M_2 are respectively the second moment about an arbitrary origin A and that about $\bar{x}$, then show that

$$M_2' = M_2 + d^2, \quad \text{where } d = \bar{x} - A.$$ *[C.U., M.Com '83]*

8. Compare the skewness of the following distributions:

	Distr. I	*Distr. II*
Mean	Rs 190	£ 210
Mode	Rs 175	£ 250
S.D.	Rs 25	£ 20

[C.U., M.Com '79]

9. The mean, median and the coefficient of variation based on 100 observations are found to be 90, 84 and 80. Find the coefficient skewness. *[I.C.W.A., Dec. '83]*

10. In a distribution, Mean = 65, Median = 70 and Coefficient of skewness = –0.6. Find Mode and Coefficient of variation. *[C.A., Nov. '78]*

11. The following results were obtained:
   (i) Arithmetic mean = 40;
   (ii) Coefficient of variation = 5;
   (iii) Karl Pearson's measure of skewness = 0.5.

   Using them, find the mode and the s.d. *[K.U., B.Com (Pass) '85]*

12. Calculate with the use of quartiles the coefficient of skewness for the following distribution of age:

Under Years	10	20	30	40	50	60	70	80
No. of Persons	15	32	51	78	97	107	114	120

[**Hint:** Quartiles 18.82, 33.33, 46.32] *[C.U., M.Com. '78]*

13. Calculate the coefficient of skewness for the following data by using quartiles.

Marks Above	0	15	30	45	60	75	90
No. of Students	180	160	130	100	65	20	5

[I.C.W.A., June '83]

14. Particulars relating to the wage distributions of two manufacturing firms are given below:

	Firm A	*Firm B*
Median Wage (Rs)	270	260
Quartile Wages (Rs)	260 and 290	250 and 290

Calculate a measure of skewness for both the firms and on that basis compare the two distributions. *[C.U., M.Com. '82]*

15. In a certain distribution the sum of the two quartiles is 102 and their difference is 12. If the median is 49, find the coefficient of skewness, *[K.U., B,Com. (Pass) '86]*

16. The coefficient of skewness for a certain distribution based upon the quartiles is – 0.8. If the sum of the upper and lower quartiles is 100.7 and the median is 55.35, find the upper and lower quartiles. *[I.C.W.A., June '84]*

17. The first three moments of a distribution about 3 are respectively 4, 65 and 134. Find the arithmetic mean, s.d. and moment measure of skewness. *[C.U., B.Com. (Pass) '80]*

18. The first three moments of a distribution about the value 1 are 2, 25 and 80. Find its mean, S.D. and the moment-measure of skewness. *[C.U., B.Com. (Pass) '83]*

19. The first three moments of a distribution about the value 4 are – 1.5, 17, – 30. Find the values of mean, S.D. and the moment measure of skewness. *[I.C.W.A.(new), Dec. '85]*

20. Comment on the shape of the following two distributions:

Distribution	*Mean*	*Variance*	M_3	M_4
I	3.0	4.0	16.0	48.0
II	3.0	4.0	12.0	56.0

where M_3 and M_4 are the 3rd and 4th central moments. *[C.U., M.Com. ' 83]*

21. In a certain distribution, the first four moments about the point 4 are 1.5, 17, –30 and 108. Find Kurtosis of the frequency curve and comment on its shape. *[I.C.W.A., Dec ' 82]*

22. Calculate Skewness and Kurtosis for the following distribution and comment on the nature of the distribution:

Mid-value (x)	34.5	44.5	54.5	64.5	74.5	84.5	94.5	Total
Frequency	2	3	11	20	32	25	7	100

[C.U., M.Com. '81]

23. Obtain suitable measure of skewness and kurtosis from the following data:

Class Interval	15–19	20–24	25–29	30–34	35–39
Frequency	1	3	7	12	29
Class Interval	40–44	45–49	50–54	55–59	Total
Frequency	20	17	9	2	100

[B.U., B.A.(Econ) '80]

24. In calculating the moments of a frequency distribution based on 100 observations the following results were obtained:

$$\text{Mean} = 9, \text{Variance} = 18, m_3^2/m_2^3 = 0.8$$

where m_2 and m_3 represent the 2nd and 3rd moment respectively. But later it was noticed that one observation 12 was misread as 21. Obtain the correct values of the first three moments about zero. *[I.C.W.A., Dec. '83]*

ANSWERS

1. – 0.2, 11, 8.2
2. 6, 43.5, 343.5; 0, 7.5, – 7.5
3. 0.92, 5.36, 8.48; 4.51
4. 0, 2.84, 1.872, 21.2432
5. 4, 3
6. 1346, 13 (see Example 7.2, p. 226)
8. 0.6, – 2.0; I is positively skew and II negatively skew. But skewness is more is distribution II.
9. 0.25 (**Hint:** Find s.d. from C.V. and use (7.7.2), p. 234)
10. 80, 38 (**Hint:** (i) Use (5.17.1), p. 127; (ii) find s.d. from (7.7.2) and then C.V.
11. 39, 2 (**Hint:** Find S.D. from (i) & (ii); use (7.7.1), p. 234)
12. – 0.055 (Quartiles 18.82, 33.33, 4632) [See—Table 4, p. 359]
13. – 0.11 (Quartiles 28,00, 49.79, 67.17). [**Hint:** Table 2, p. 358 and Example 7.12, p. 236]
14. Bowley's measure 0.33, 0.50; wage distr. for *B* more skew.
15. 0.33
16. 56.6, 44.1
17. 7, 7, – 1.51
18. 3, 4.58, – 0.56
19. 2.5, 3.84, 0.70
20. Skewness 2 and 1.5; Kurtosis 0 and 0.5; I more asymmetrical with longer tail to the right and II has sharper peak.
21. –0.69; platykurtic (**Hint:** $m_2 = 14.75$, $m_4 = 502.3125$)
22. Skewness –0.50, Kurtosis + 0.18; Negatively skew and Leptokurtic.
23. Moment measures –0.209, –0.115.
24. 8.91, 96.03, 1207.975.

CHAPTER 8
CURVE FITTING AND METHOD OF LEAST SQUARES
(Summary)

METHOD OF LEAST SQUARES

Method of Least Squares is a technique employed for finding the equation of a specified type of curve which best fits a given set of observations relating to the variables. This is based on the principle that for the "best-fitting" curve the sum of the squares of differences between the observed and the corresponding estimated values, as obtained from the equation, should be the minimum possible.

FITTING STARTING LINE

Let $(x_1, y_1), (x_2, y_2), \ldots, (x_n, y_n)$ be the given set of observations on two variables x and y.

I. (*When x is the independent variable*) The equation $y = a + bx$ is taken, where a and b are constants. By the method of least squares, these constants are obtained by minimising $\Sigma(y_i - a - bx_i)^2$. This gives the normal equations, solving which we get a and b:

$$\Sigma y = an + b \Sigma x$$
$$\Sigma xy = a \Sigma x + b \Sigma x^2$$

Geometrical interpretation: The observations are plotted as points on a graph paper. Then $y = a + bx$ shows that straight line (among all straight lines) for which the sum of the squares of *vertical distance* from the given points is the smallest possible. (see Fig. 8.3, page 289)

II. (*When y is the independent variable*) The equation of the straight line is now of the form $x = c + dy$, where c and d are constants. The values of these constants are obtained by minimising $\Sigma(x_i - c - dy_i)^2$, giving the normal equations

$$\Sigma x = cn + d \Sigma y$$
$$\Sigma xy = c\Sigma y + d \Sigma y^2$$

Geometrically, this gives the straight line for which the sum of hte squares of *horizontal distances* from the given points is the smallest possible.

Obviously, for the same set of observations $(x_1, y_1), (x_2, y_2), \ldots, (x_n, y_n)$ the two best-fitting straight lines $y = a + bx$ and $x = c + dy$ are usually different.

$y = a + bx$ is used for estimating y when x is given.

$x = c + dy$ is used for estimating x when y is given.

FITTING PARABOLA

The equation is $y = a + bx + cx^2$, where x is the independent variable. The constants a, b, c are obtained by solving the normal equations

$$\Sigma y = an + b \Sigma x + c \Sigma x^2$$
$$\Sigma xy = a\Sigma x + b \Sigma x^2 + c \Sigma x^3$$
$$\Sigma x^2 y = a\Sigma x^2 + b \Sigma x^3 + c \Sigma x^4$$

FITTING EXPONENTIAL CURVE

The equation is $y = ab^x$, where x is the independent variable.

Taking logarithms of both sides, log y = (log a) + (log b) x. This is of the form $Y = A + Bx$, where $Y = \log y$, $A = \log a$, $B = \log b$. The normal equations for finding A and B are

$$\Sigma Y = nA + B\,\Sigma x$$
$$\Sigma xY = A\Sigma x + B\,\Sigma x^2$$

Hence we obtain a = antilog A, b = antilog B.

(**Note:** The computations can be considerably reduced by changes of origin and scale of the independent variable. In particular, if the successive values of the independent variable have a common difference special methods are available for the cases when n is odd or even (see Example 8.4, 8.5, p. 255, 6).

Uses—

(1) Regression lines are straight lines of best-fit obtained by the method of least squares. The regression equation of y on x is a straight line of the form $y = a + bx$, where x is the independent variable. The regression equation of x on y is the straight line of the form $x = c + dy$, where y is the independent variable. (see pages 250, 251).

(2) Method of least squares is also used in Time Series analysis (Chapter 16) for finding the equation of Trend line.

Problem 8.1 *Fit a least square line to the data below, using (i) x as the independent variable, (ii) y as the independent variable. Find y when x = 5 and x = 12. Also find x when y = 7.*

x	3	5	6	8	9	11
y	2	3	4	6	5	8

[C.U., B.Com.(Hons) '83]

Solution (i) When x is the independent variable, the equation of straight line is of the form $y = a + bx$. Normal equations for finding a and b are $\Sigma y = an + b\Sigma x$, $\Sigma xy = a\Sigma x + b\Sigma x^2$.

Fitting Straight Line

x	y	x^2	xy	y^2
3	2	9	6	4
5	3	25	15	9
6	4	36	24	16
8	6	64	48	36
9	5	81	45	25
11	8	121	88	64
42	28	336	226	154

Putting the values,

$$28 = 6a + 42b \qquad \text{...(1)}$$
$$226 = 42a + 336b \qquad \text{...(2)}$$

Multiplying (1) by 7 and subtracting form (2), we get 30 = 42b.

$\therefore$ $$b = 30/42 = 5/7$$

Using this in (1) and solving

$$a = -1/3$$

$\therefore$ The fitted line is $$y = -1/3 + (5/7)\,x \qquad \text{...(3)}$$

(ii) When y is the independent variable, the equation of straight line is of the form $x = c + dy$. Normal equations for finding the constants c and d are

$$\Sigma x = cn + d\Sigma y \quad \text{or,} \quad 42 = 6c + 28d \quad \dots (4)$$

$$\Sigma xy = c\Sigma y + d\Sigma y^2 \quad\quad 226 = 28c + 154d \quad \dots (5)$$

Multiplying (4) by 14 and (5) by 3 and subtracting, we get

$70d = 90$; i.e., $d = 9/7$. Hence, $c = 1$.

The fitted line is $x = 1 + (9/7)y$ ... (6)

To find y when x is given, we must use (3)

When $x = 5$, $y = \frac{1}{3} + \frac{5}{7} \times 5 = \frac{68}{31} = 3.24$

When $x = 12$, $y = -\frac{1}{3} + \frac{5}{7} \times 12 = \frac{178}{21} = 8.24$

To find x when y is given, we must be (6).

When $y = 5$, $x = 1 + \frac{9}{7} \times 7 = 10$

Ans. (i) $y = -1/3 + (5/7)x$, (ii) $x = 1 + (9/7)y$; 3.24, 8.24; 10.

Problem 8.2 *Fit the equatioin $y = 2 + ab^x$ to the following data by the method of least-squares:*

x	2	3	4	5	6
y	14	25	50	98	194

[I.C.W.A., Dec. '85]

Solution (**Note:** This is a Modified Exponential Curve; see page 245). We may write, $y - 2 = ab^x$.

$\therefore \quad \log (y - 2) = (\log a) + x (\log b)$

or, $\quad Y = A + Bx$

where $\quad Y = \log (y - 2)$, $A = \log a$, $B = \log b$. Normal equations:

$$\Sigma Y = nA + B\Sigma x, \quad \Sigma xY = A\Sigma x + B\Sigma x^2$$

Calculations for Curve Fitting

x	y	$(y-2)$	$\log (y-2) = Y$	x^2	xY
2	14	12	1.0792	4	2.1584
3	26	24	1.3802	9	4.1406
4	50	48	1.6812	16	6.7248
5	98	96	1.9823	25	9.9115
6	194	192	2.2833	36	13.6998
20	–	–	8.4062	90	36.6351

Normal equations: $\quad 8.4062 = 5A + 20B \quad \dots (1)$

$$36.6351 = 20A + 90B \quad \dots (2)$$

Multiplying (1) by 4 and subtracting from (2)

$10B = 3.0103$; $\quad B = 0.30103$

Putting this value in (1) and solving, $A = 0.47712$

a = antiolog A = angtilog 0.47712 = 3

b = antilog B = antilog 0.30103 = 2

The fitted equation is $y = 2 + 3 \times 2^x$.

ADDITIONAL EXERCISES 8

1. Fit a straight line by the method of least squares to the following data. What is the value corresponding to $x = 5$?

x	1	2	3	4	5
y	3	4	6	9	10

[I.C.W.A. June '85]

2. Use the method of least squares to determine a and b in $y = a + bx$to fit in the following data:

x	0	1	2	3	4
y	1.8	5.1	8.9	14.1	19.8

[I.C.W.A., June '82]

3. A firm has the following data on the price (x) of its product and the volume of sales (y):

Price per unit (Rs)	10	7	5	4
Volume of Sales ('000) units)	5	7	8	10

Assuming that a linear relationship exits between price and volume of sales, estimate the parameters of this relationship; and hence find the likely volume of sales when price rises to Rs 12 per unit. If price increases by one rupee, how much sales decreases? *[B.U., B.Com.(Hons) '82]*

4. Given the following data on price and quantity demanded, fit a linear demand curve by the method of least squares:

Price (x)	15	12	9	8	7	6
Quantity (y)	75	85	93	115	120	130

[B.U., B.Com.(Hons) '85]

5. Based on the following data, find the linear regression equation of Y on X by the method of least squares. Predict value of Y if $X = 4$.

X	1	2	3	4	5	6
Y	6	4	3	5	4	2

[I.C.W.A., Dec. '82]

6. Fit a least square straight line to the following data taking x as independent variable and y as dependent variable:

x	0	5	10	15	20	25
y	12	15	17	22	24	30

[C.U., B.Com(H) '84]

7. Fit a curve of the $y = AB^x$ to the following data, by the method of the least squares:

x	0	1	2
y	2	6	18

[I.C.W.A., Dec. '86]

8. Given the following data on the sales of a company in different years, fit a linear trend to the data by the method of least squares and estimate the likely sales in the year 1989:

Year	1981	1982	1983	1984	1985
Sales ('000 Rs)	120	130	135	145	165

(a) What is the annual increase in sales?
(b) In what year will sales be Rs 1,80,000?

ANSWERS

1. $y = 0.7 + 1.9x$; 10.2
2. $a = 0.94$, $b = 4.5$
3. $a = 12.45$, $b = -0.76$; 3.33 ('000 units); 760 units.
4. $y = 160.66 - 6.07x$
5. $Y = 5.8 - 0.514X$; 3.74 (Linear regression equation of Y on X is the best-fitting straight line of the form $Y = a + bX$; see page 290).
6. $y = 11.3 + 0.7x$ (**Hint:** Example 8.5 and 8.6, p. 256-257)
7. $y = 2 \times 3^x$ (**Hint:** Example 8.10, p. 263)
8. $y = 139 + 10.5x$, origing in year 1983, x unit 1 year; 202 ('000 Rs); (a) 10,500; (b) Year 1987.
(**Hint:** Example 16.17 and 16.18, pp. 290, Part II).

CHAPTER 9
CORRELATION AND REGRESSION
(Summary)

BIVARIATE DATA

Statistical data relating to the simultaneous measurement of two variables (say, x and y) are called 'bivariate data'. The observations are then paired—(x_1, y_1), (x_2, y_2), ..., (x_n, y_n).

CORRELATION AND REGRESSION

Correlation denotes the strength of association between two variables; i.e. how colsely the variables are related to each other.

Regression denotes prediction or estimation of the most likely value of one variable for a specified value of the other variable.

In *linear correlation* and *linear regression* it is assumed that there is a linear relationship between the variables.

SCATTER DIAGRAM

Scatter diagram is the graphical representation of bivariate data. The paires of observations are plotted as points on a graph paper. This diagram which shows the observations as points on the graph paper is known as 'scatter diagram'. This is used to indicate (i) the type correlation (positive, negative or zero) and (ii) the degree of correlation (high or low).

If the points have more or less a straight line pattern, the variables are correlated. The type of correlation is positive or negative, according as the slope of the straight line is positive or negative. If these is no linear pattern, the variables are 'uncorrelated' ($r = 0$). If all the points lie exactly on a straight line, the variables are 'perfectly correlated' ($r = \pm 1$). The more scattered the points are from the linear path, the less is the degree of correlation.

CORRELATION COEFFICIENT (r)

Karl Pearson's product-moment formula:

$$r = \frac{\text{cov}\,(x,\, y)}{\sigma_x\, \sigma_y} = \frac{\Sigma\,(x - \overline{x})\,(y - \overline{y})}{\sqrt{[\Sigma\,(x - x)^2\; \Sigma(y - \overline{y})^2]}}$$

PROPERTIES OF CORRELATION COEFFICIENT (r)

1. r is a independent of origin and scale of the observations. In particular, if $u = ax + b$, $v = cy + d$, then $r_{uv} = \pm\, r_{xy}$ according as the coefficients of x and y have the same sign or opposite signs.
2. r is a pure number.
3. r lies between -1 and $+1$; i.e. $-1 \leq r \leq +1$.

VARIANCE OF THE SUM (DIFFERENCE)

$$\text{Var}(x + y) = \text{Var}(x) + \text{Var}(y) + 2.\text{Cov}(x, y)$$
$$\text{Var}(x - y) = \text{Var}(x) + \text{Var}(y) + 2.\text{Cov}(x, y)$$

If x and y are uncorrelated (i.e. $r = 0$), then cov(x, y) = 0. Hence

$$\text{Var}(x + y) = \text{Var}(x - y) = \text{Var}(x) + \text{Var}(y)$$

In general,

$$\text{Var}(ax + by) = a^2.\text{Var}(x) + b^2.\text{Var}(y) + 2ab.\text{Cov}(x, y)$$

$$\text{Cov}(ax + by, cx + dy) = ac.\text{Var}(x) + bd.\text{Var}(y) + (ad + bc).\text{Cov}(x, y)$$

Note:

(i) To find Var($ax + by$), think of the square

$$(ax + by)^2 = a^2x^2 + b^2y^2 + 2ab.xy.$$

Replace the square terms x^2, y^2 by Var(x), Var(y) and the product term xy by Cov(x, y).

(ii) To find Cov($ax + by$, $cx + dy$), think of the product $(ax + by)(cx + dy) = ac.x^2 + bd.y^2 + (ad + bc).xy$. As before, replace the square terms by Var and the product term by Cov.

For example, since $(2x - 3y)^2 = 4x^2 + 9y^2 - 12xy$.

$$\text{Var}(2x - 3y) = 4.\text{Var}(x) + 9.\text{Var}(y) - 12.\text{Cov}(x, y)$$

Similarly since $(2x - 3y)(x + 2y) = 2x^3 - 6y^2 + xy$,

$$\text{Cov}(2x - 3y, x + 2y) = 2.\text{Var}(x) - 6.\text{Var}(y) + \text{Cov}(x, y)$$

If x and y are uncorrelated, Cov(x, y) = 0; then

$$\text{Var}(2x - 3y) = 4.\text{Var}(x) + 9.\text{Var}(y)$$

$$\text{Cov}(2x - 3y, x + 2y) = 2.\text{Var}(x) - 6.\text{Var}(y)]$$

LINEAR REGRESSION

The two regression equations are:

(1) Regression equation of y on x (used to estimate y):

$$y - \bar{y} = b_{yx}(x - \bar{x})$$

(2) Regression equation of x on y (used to estimate x):

$$x - \bar{x} = b_{xy}(y - \bar{y})$$

where $b_{yx} = \dfrac{\text{Cov}(x, y)}{\text{Var}(x)} = r\dfrac{\sigma_y}{\sigma_x}$ and $b_{xy} = \dfrac{\text{Cov}(x, y)}{\text{Var}(y)} = r\dfrac{\sigma_x}{\sigma_y}$

Regression equations represent straight lines of best fit obtained by the method of least squares. Regression equation of y on x is of the form $y = a + bx$, where x is the independent variable and y is dependent. Regression equation of x on y is of another form $x = c + dy$, where y is independent and x is dependent (see page 75).

Geometrically, regression line of y on x is that straight line of the scatter diagram for which the sum of the squares of *vertical distances* of the points is minimum. For the regression line of x on y the sum of the squares of *horizontal distances* is minimum. So the two regression lines are usually different.

PROPERTIES OF LINEAR REGRESSION

(1) Correlation coefficient is the geometric mean of the two regression coefficients.

$r = \sqrt{(b_{yx} \times b_{xy})}$.

(2) Correlation coefficient and the two regression coefficients, all have the same sign.

(3) The regression lines always intersect at the point $(\bar{x}, \bar{y})$.

(4) The angle θ between the regression lines is given by

$$\tan \theta = \pm \frac{1 - r^2}{r} \frac{\sigma_x \sigma_y}{\sigma_x^2 + \sigma_y^2}$$

(5) When $r = 0$, the regression lines are perpendicular to each other. When $r = \pm 1$, the regression lines coincide, the angle between them being zero.

(6) When $r = 0$, neither x nor y can be estimated from linear regression equations. When $r = \pm 1$, there is an exact linear relationship between the variables.

(7) Regression coefficients are independent of origin but depend on scale. In particular, if $u = ax + b$, $v = cy + d$, then

$$b_{uv} = \frac{a}{c} b_{xy};\ b_{vu} = \frac{c}{a} b_{yx}$$

EXPLAINED VARIATION

Proportion of Total Variation explained by Regression = r^2

RANK CORRELATION COEFFICIENT (R)

Rank Correlation Coefficient is the product-moment correlation coefficient calculated from *ranks* (i.e. serial numbers indicating the order of preference/magnitude), and not from the actual magnitudes of observations.

Spearman's formula (when there are no ties):

$$R = 1 - \frac{6\Sigma d^2}{n(n^2 - 1)}$$

Rank Correlation Coefficient (R) lies between – 1 and + 1.

$$-1 \leq R \leq +1$$

$R = +1$ when all the individuals have identical ranks in the two charactes $R = -1$ when the individuals have opposite ranks.

Rank correlation coefficient is generally used as a measure of association between two attributes when measurements are either not available or not reliable, but the individuals can be arranged in order of preference. It is therefore applied in educational and psychological studies to find the degree of correspondence between abilities.

MULTIPLE AND PARTIAL CORRELATION (3 VARIABLES)

Multiple Correlation Coefficient of x_2 on x_2 and x_3 measures the strength of the combined influence of x_2 and x_3 on x_1.

$$R_{1\cdot 23} = \sqrt{\frac{r_{12}^2 + r_{133}^2 - 2r_{12}r_{23}r_{13}}{1 - r_{32}^2}}$$

It lies between 0 and 1. $0 \leq R_{1\cdot 23} \leq 1$

Partial Correlation Coefficient of x_1 and x_2, the other variable being x_3, measures the strength of association between x_1 and x_2, after eliminating the effect of x_3 from both.

$$r_{12\cdot 3} = \frac{r_{12} - r_{1n}r_{2n}}{\sqrt{(1 - r_{13}^2)\ (1 - r_{23}^2)}}$$

It lies between – 1 and + 1; $-1 \leq r_{12\cdot 3} \leq +1$

MULTIPLE REGRESSION EQUATION

Regression equation of x_1 on x_2 and x_3 (used to estimate x_1) is:

$$x_1 - \bar{x}_1 = b_{12.3}(x_2 - \bar{x}_2) + b_{13.2}(x_3 - \bar{x}_3)$$

where $$b_{12\cdot3} = \frac{\sigma_1}{\sigma_2}\left(\frac{r_{12} - r_{13}\, r_{23}}{1 - r_{23}^2}\right);\ b_{13\cdot2} = \frac{\sigma_1}{\sigma_2}\left(\frac{r_{13} - r_{12}\, r_{23}}{1 - r_{23}^2}\right)$$

It may be shown that $b_{12\cdot3} \times b_{21\cdot3} = r_{12\cdot3}^2$

$b_{12\cdot3}$, $b_{21\cdot3}$, etc. are known as *Partial Regresstion Coefficients.*

Note:

(1) $r_{12} = r_{21}$; $r_{12\cdot3} = r_{21\cdot3}$; but $r_{12\cdot3} \neq r_{13\cdot2}$

(2) $R_{1\cdot23} = R_{1\cdot32}$; but $R_{1\cdot23} \neq R_{2\cdot13}$

(3) $b_{12\cdot3} \neq b_{21\cdot3}$

SOLVED PROBLEMS

Problem 9.1 *Calculate correlation coefficient from the following results: $n = 10$, $\Sigma x = 140$, $\Sigma y = 150$, $\Sigma(x - 10)^2 = 180$, $\Sigma(y - 15)^2 = 215$, $\Sigma(x - 10)(y - 15) = 60$.*

[C.A., May '80]

Solution Let $u = x - 10$, $v = y - 15$. Then

$$\Sigma u = \Sigma(x - 10) = \Sigma x - 10n = 140 - 10 \times 10 = 40$$

$$\Sigma v = \Sigma(y - 15) = \Sigma y - 15n = 150 - 15 \times 10 = 0$$

Also given, $\Sigma u^2 = 180$, $\Sigma v^2 = 215$, $\Sigma uv = 60$

$$\therefore \quad \sigma_u^2 = \Sigma u^2/n - (\Sigma u/n)^2 = 180/10 - (40/10)^2 = 2$$

$$\sigma_v^2 = \Sigma v^2/n - (\Sigma v/n)^2 = 215/10 - (0/10)^2 = 21.5$$

$$\text{cov}(u, v) = \Sigma uv/n - (\Sigma u/n)(\Sigma v/n) = 60/10 - (40/10)(0/10) = 6$$

$$\therefore \quad r_{uv} = \frac{\text{cov}(u, v)}{\sigma_u\, \sigma_v} = \frac{6}{\sqrt{(2 \times 21.5)}} = +0.91;$$

$$r_{xy} = r_{uv} = +0.91.$$

Ans. 0.91

Problem 9.2 *Given $r = 0.8$, $\Sigma xy = 60$, $\sigma_y = 2.5$, $\Sigma x^2 = 90$, find the number of items.*

[C.A., May '82]

Solution (**Note:** Here x and y denote deviations of X and Y from their means).

$$\text{cov}(X, Y) = \Sigma xy/n = 60/n;\ \sigma_X^2 = \Sigma x^2/n = 90/n$$

$$r = \frac{\text{cov}(X, Y)}{\sigma_X\, \sigma_Y};\ \text{or, } 0.8 = \frac{60/n}{\sqrt{(90/n) \times 2.5}}$$

or, $60/n = 2 \times \sqrt{(90/n)}$; Squaring both sides and solving $n = 10$.

Problem 9.3 *This lines of regression of y on x on y are respectively $y = x + 5$ and $16x = 9y - 94$. Find the variance of x if the variance of y is 16. Also find the covariance of x and y.*

[I.C.W.A., Dec. '86]

Solution Regression equations are $y = x + 5$ and

$$x = (9/16)y - (94/16). \text{ This shows that } b_{yx} = 1,\ b_{xy} = 9/16.$$

$$\therefore \quad r^2 = b_{yx}.b_{xy} = 1 \times \frac{9}{16} = \frac{9}{15}$$

or

$$r = +\tfrac{3}{4} \text{ (positive; because, } b_{yx} \text{ and } b_{xy} \text{ are positive.)}$$

But

$$b_{yx} = r\frac{\sigma_y}{\sigma_x};\ \text{or } 1 = \frac{3}{4} \times \frac{4}{\sigma_x}$$

Solving, $\sigma_x = 3$; or, $\text{var}(x) = 9$

$$\therefore \quad \text{cov}(x, y) = r.\sigma_x\sigma_y = \frac{3}{4} \times 3 \times 4 = 9.$$

Ans. 9; 9

Problem 9.4 *If x and y are two correlated variables with same variance and the correlation coefficient is r, find the regression coefficient of x on (x + y) and that of (x + y) on x. Hence find correlation coefficient between x and (x + y).* [W.B.H.S. '85]

Solution Let $\text{var}(x) = \text{var}(y) = \sigma^2$

Since $r = \dfrac{\text{cov}(x,\ y)}{\sigma_x\ \sigma_y}$, by cross multiplication

$$\text{cov}(x, y) = r.\sigma_x\sigma_y = r.\sigma.\sigma = r\sigma^2$$

If we write $u = x + y$, we have to find

$$b_{xu} = \frac{\text{cov}(x,\ u)}{\text{var}(u)};\ b_{ux} = \frac{\text{cov}(x,\ u)}{\text{var}(x)}$$

Now,

$$\begin{aligned} \text{var}(u) &= \text{var}(x + y) \\ &= \text{var}(x) + \text{var}(y) + 2.\text{cov}(x, y);\ \text{see p. 469} \\ &= \sigma^2 + \sigma^2 + 2_r\cdot\sigma^2 \\ &= 2\sigma^2(1 + r) \end{aligned}$$

Again,

$$\begin{aligned} \text{cov}(x, u) &= \text{cov}(x, x + y);\ \text{see Note, p. 469} \\ &= \text{var}(x) + \text{cov}(x, y) \\ &= \sigma^2 + r.\sigma^2 = \sigma^2(1 + r) \end{aligned}$$

Thus,

$$\text{cov}(x, u) = \sigma^2(1 + r),\ \text{var}(u) = 2\sigma^2(1 + r),\ \text{var}(x) = \sigma^2$$

$$\therefore \quad b_{xu} = \frac{\sigma^2(1 + r)}{2\sigma^2(1 + r)} = \frac{1}{2};\ b_{ux} = \frac{\sigma^2(1 + r)}{\sigma^2} = 1 + r$$

Correlation coefficient between x and u is

$$r_{xu} = \sqrt{b_{xu}\cdot b_{ux}} = \sqrt{(1 + r)/2}$$

Ans. 1/2, $1 + r$; $\sqrt{(1 + r)/2}$

Problem 9.5 *The coefficient of rank correlation of the marks obtained by 10 students in Mathematics and Statistics was found to be 0.5. It was then detected that the difference in ranks in the two subjects for one particular student was wrongly taken to be 3 in place of 7. What should be the correct rank correlation coefficient?*

[I.C.W.A., June '81; C.U., B.Com (Hons) '83]

Solution Putting $R = 0.5$ and $n = 10$ in the formula for Rank Correlation Coefficient (p. 304),

$$0.5 = 1 - \frac{6\Sigma d^2}{990}$$

Solving, $\Sigma d^2 = 82.5$. When one incorrect value $d = 3$ is replaced by 7,

$$\Sigma d^2 = 82.5 - 3^2 + 7^2 = 122.5.$$

$$\therefore \quad R = 1 - \frac{6 \times 122.5}{990} = +0.26$$

Ans. + 0.26

Problem 9.6 *x_1, x_2, x_3 are three variates measured from their means with $N = 10$, $\Sigma x_1^2 = 90$, $\Sigma x_2^2 = 160$, $\Sigma x_3^2 = 40$, $\Sigma x_1x_2 = 60$, $\Sigma x_3x_1 = 40$, $\Sigma x_2x_3 = 60$. Calculate the partial correlation coefficient $r_{31\cdot 2}$ and multiple correlation coefficient $R_{1\cdot 23}$.*

[I.C.W.A., June '86]

Solution Since x_1, x_2, x_3 are measured from their means,

$$\text{var}(x_1) = \Sigma x_1^2/N = 90/10 = 9; \quad \therefore \quad \sigma_1 = 3$$
$$\text{var}(x_2) = \Sigma x_2^2/N = 160/10 = 16; \quad \sigma_2 = 4$$
$$\text{var}(x_3) = \Sigma x_3^2/N = 40/10 = 4; \quad \sigma_3 = 2$$
$$\text{cov}(x_1, x_2) = \Sigma x_1x_2/N = 60/10 = 6$$
$$\text{cov}(x_2, x_3) = \Sigma x_2x_5/N = 60/10 = 6$$
$$\text{cov}(x_3, x_1) = \Sigma x_3x_1/N = 40/10 = 4$$

$$r_{12} = \frac{\text{cov}(x_1, x_2)}{\sigma_1\ \sigma_2} = \frac{6}{3 \times 4} = \frac{1}{2}; \text{ Similarly } r_{23} = \frac{n}{4},\ r_{31} = \frac{2}{3}$$

In the formula for $r_{12\cdot 3}$ (page 308) if we write the subscripts 3, 1, 2 in place of 1, 2, 3 respectively, we get

$$r_{31\cdot 2} = \frac{r_{31} - r_{25}r_{12}}{\sqrt{(1 - r_{23}^2)\ (1 - r_{12}^2)}} = \frac{\frac{2}{3} - \left(\frac{3}{4} \times \frac{1}{2}\right)}{\sqrt{\frac{7}{16} \times \frac{3}{4}}} = \frac{\sqrt{21}}{9} = 0.51$$

Using the formula (page 308)

$$R_{1\cdot 23} = \sqrt{\frac{\frac{1}{5} + \frac{4}{9} - 2 \times \frac{1}{2} \times \frac{3}{4} \times \frac{2}{3}}{1 - \frac{9}{16}}} = \frac{2}{3} = 0.67$$

Ans. 0.51, 0.67

(SHORT-ANSWER TYPE QUESTIONS)

Problem 9.7 *(a) Given cov(x, y) = – 30, $\sigma_x = 6$, $\sigma_y = 10$; find coefficient of corelation bgetween x and y.* [C.U., B.Com(P) '85]

(b) In studying a set of pairs of related variates, the following results are obtained: $N = 100$, $\Sigma x_i = 57$, $\Sigma y_i = 43$, $\Sigma(x_i - \bar{x})^2 = 169$, $\Sigma(y_i - \bar{y})^2 = 64$, $\Sigma(x_i - \bar{x})(y_i - \bar{y}) = 101$. Find the correlation coefficient. [C.U., B.Com(Hons) '84]

Solution (a) $r = \dfrac{\text{cov}(x, y)}{\sigma_x\ \sigma_y} = \dfrac{-30}{6 \times 10} = -0.5$

(b) $r = \dfrac{\Sigma(x - \bar{x})\ (y - \bar{y})}{\sqrt{[\Sigma(x - \bar{x})^2\ \Sigma(y - \bar{y})^2]}} = \dfrac{101}{\sqrt{(169 \times 64)}} = 0.97$

Ans. – 0.5; 0.97

Problem 9.8 *(a) The corresponding values of two variables are:*

x	2	3	5	8	9
y	4	6	10	16	18

What is the correlation coefficient between x and y? [C.A., May '82]

(b) If u + 3x = 5, 2y – v = 7 and the correlation coefficient of x and y is 0.12, find the correlation coefficient of u and v. [W.B.H.S., '83]

Solution

(a) Since each value of y is double the corresponding value of x, we have $y = 2x$. Thus x and y are bound by an exact linear relation with a positive slope. $r = +1$. (Also see Ex. 9.32, p. 342).

(b) $u = 5 - 3x$, $v = 2y - 7$, $r_{xy} = 0.12$. Thus u and v are linear functions of x and y respectively. So r_{uv} is numerically equal to r_{xy}. But since the coefficients of x and y have opposite signs $r_{uv} = -r_{xy} = -0.12$. (Also see Example 9.15, p. 284). *Ans.* +1; – 0.12

Problem 9.9 *(a) Karl Pearson's coefficient of correlation between x and y is 0.52, their covariance is + 7.8. If the variance of x is 16, find the standard deviation of y.* [C.U., B.Com(P) '83]

(b) Two uncorrelated random variables x and y have standard deviations 3 and 4 respectively. Find the s.d. of (x + y).

Solution

(a) $r = \dfrac{\text{cov}(x,\ y)}{\sigma_x\ \sigma_y}$. Putting $r = 0.52$, $\text{cov}(x, y) = 7.8$, $\sigma_x = 4$, and solving $\sigma_y = 15/4 = 3.75$.

(b) If $r = 0$, $\text{var}(x + y) = \text{var}(x) + \text{var}(y)$

$= 3^2 + 4^2 = 25; \quad \therefore \quad \sigma_{x+y} = 5$ *Ans.* 3.75; 5

Problem 9.10 *Are the following statements valid? Justify your answers.*

(i) Positive value of correlation coefficient between x and y implies that if x decreases, y tends to increase.

(ii) Correlation coefficient is independent of the origin of reference but is dependent on the units of measurement.

(iii) Correlation coefficient between x and y turned out to be 1.02.

[I.C.W.A., Dec. '80]

Solution No; the statements are not valid.

(i) Positive value of correlation coefficient implies that if x decreases, y also tends to *decrease*.

(ii) Correlation coefficient is independent of both the origin of reference and the units of measurement.

(iii) Correlation coefficient cannot be larer than 1.

Problem 9.11 *(a) Regression equation of y on x is 3x + 2y = 26, and regression equation of x on y is 6x + y = 31. Find the regression coefficients b_{yx} and b_{xy}.*

(b) The equations of two lines of regression are 2x + 3y – 8 = 0 and x + 2y – 5 = 0. Find the means of x and y.

Solution

(a) Regression equation of y on x may be written as $2y = 26 - 3x$; or, $y = 13 - \frac{3}{2}x$. This shows that

Regression coefficient of y on x $(b_{yx}) = -\frac{3}{2}$

Regression equation of x on y may be written as $6x = 31 - y$; or, $x = \frac{31}{6} - \frac{1}{6}y$. This shows that

Regression coefficient of x on y $(b_{xy}) = -\frac{1}{6}$

Note : Express the *regression equaton of y on x* in the form $y = a + bx$; i.e. L.H.S. is only y and R.H.S. contains no y at all. The coefficient of x on R.H.S. gives b_{yx}. Similarly, express, the *regression equation of x on y* in the form $x = c + dy$, and coefficient of y on R.H.S. gives b_{xy}.]

(b) Solving the two equations we get $x = 1$, $y = 2$. Since the regression lines intersect at $(\bar{x}, \bar{y})$, $\bar{x} = 1$, $\bar{y} = 2$.

Problem 9.12 *(a) If $b_{yx} = -0.3$ and $b_{xy} = -1.2$, find the coefficient of correlation between x and y.*

(b) If $u = 2x + 5$, $v = -3y + 1$, and the regression coefficient of y on x is -1.2, find the regression coefficient of v on u.

Solution

(a) $r^2 = b_{yx}.b_{xy} = (-0.3)(-1.2) = +0.36$; hence $r = \pm 0.6$. But r, b_{yx}, b_{xy} all must have the same sign. $\therefore \quad r = -0.6$

(b) Given $u = 2x + 5$, $v = -3y + 1$, $b_{yx} = -1.2$.

$\therefore \quad b_{vu} = (-3/2)\, b_{yx} = (-3/2) \times (-1.2) = +18.$

[See Problem 9.14(b), Solution, p. 423] *Ans.* -0.6; $+1.8$

Problem 9.13 *(a) Give reasons to justify that the following statement cannot be true: (i) $b_{yx} = 0.8$ and $b_{xy} = -0.3$; (ii) $b_{yx} = 0.8$ and $b_{xy} = 2.4$.*

[C.U., B.Com(Hons) '84]

(b) Criticise: The correlation coefficient between x and y(r_{xy}) is 0.9 and the regression coefficient (b_{xy}) is -1. [I.C.W.A., June '79]

(c) A student obtained the two regression lines as $2x - 5y - 7 = 0$ and $3x + 2y - 8 = 0$. Do you agree with him? [I.C.W.A., Dec. '81]

Solution

(a) (i) Both b_{yx} and b_{xy} must have the same sign,

(ii) The product $b_{yx}.b_{xy}$ must not exceed 1; because $b_{yx}.b_{xy} = r^2$, and r^2 cannot be greater than 1.

(b) The statement cannot be true; because r_{xy}, b_{yx} and b_{xy} all three must have the same sign.

(c) No; because slopes of the two regression lines must be either both positive or both negative.

Problem 9.14 *(a) With $b_{xy} = 0.5$, $r = 0.8$ and variance of y = 16, the standard deviation of x equals to 2.5, 6.4, 10.0, 25.6 —Which is correct?* [C.A., Nov. '83]

(b) Pick out the correct answer: If $4u = 2x + 7$ and $6v = 2y - 15$, and the regression coefficient of y on x is 3, then regression coefficient of v on u is 2, 3, 9/2, – 2, None of these. [W.B.H.S. '82]

Solution

(a) $b_{xy} = r\dfrac{\sigma_x}{\sigma_y}$. Putting the values $b_{xy} = 0.5$, $r = 0.8$, $\sigma_y = 4$ and solving $\sigma_x = 2.5$.

Ans. 2.5

(b) [Note: $\text{var}(ax + b) = a^2\,.\text{var}(x)$;
$\text{cov}(ax + b, cy + d) = ac\,.\text{cov}(x, y)$].

Here, $u = \dfrac{1}{2}x + \dfrac{7}{4}$, $v = \dfrac{1}{3}y - \dfrac{5}{2}$; $b_{yx} = 3$.

$$\text{Now, } b_{vu} = \frac{\text{cov}(u, v)}{\text{var}(u)} = \frac{\frac{1}{2} \times \frac{1}{3} \times \text{cov}(x, y)}{\left(\frac{1}{2}\right)^2 \text{var}(x)} = \frac{2}{3} \times \frac{\cos(x, y)}{\text{var}(x)}$$

$$= \frac{2}{8}\, b_{yx} = \frac{2}{8} \times 3 = 2$$

Ans. 2

Problem 9.15 *(a) If $\Sigma D^2 = 33$ and $N = 10$, find the value of the coefficient of rank correlation, where D represents the difference between the ranks of two series and N is the number of pairs of observations.* [C.U., B.Com(P) '85]

(b) Mention the correct answer: The ranks according to two attributes in a sample are given below:

R_1	1	2	3	4	5
R_2	5	4	3	2	1

The rank correlation between than is 0, + 1, – 1, none of these. [C.A., Nov. '82]

Solution

(a) Putting the values in the formula for rank correlation coefficient

$$R = 1 - \frac{6 \times 33}{10 \times 99} = 0.8$$

Ans. 0.8

(b) Since the ranks are just opposite, $R = -1$.

Ans. – 1

ADDITIONAL EXERCISES 9

1. Find (a) the coefficient of correlation, and (b) the regression equation of x on y, from the following data:
   $\Sigma x = 60$, $\Sigma y = 40$, $\Sigma xy = 1150$, $\Sigma x^2 = 4160$, $\Sigma y^2 = 1720$, $N = 10$
   [C.A., Nov. '85]
2. For 12 pairs of observations the results are: $\Sigma(x_i - \bar{x})^2 = 84.68$, $\Sigma(y_i - \bar{y})^2 =$

38.92, $\Sigma(x_i - \bar{x})(y_i - \bar{y}) = 40.34$. Find the value of product-moment correlation coefficient. *[I.C.W.A., Dec. '84]*

3. Calculate the coefficient of correlation from

x	1	2	3	4	5
y	6	8	11	8	12

[C.U., B.Com(P) '85]

4. Calculate Karl Pearson's coefficient of correlation from the following data, using 20 as the working mean for price, and 70 as the working mean for demand:

Price	14	16	17	18	19	20	21	22	23
Demand	84	78	70	75	66	67	62	58	60

[C.A., May '86]

5. The following data give the heights in inches (x) and the weights in lbs. (y) of 10 students of age 17 years:

x	61	66	68	64	65	70	63	62	64	67
y	112	123	130	115	110	125	100	113	116	126

Calculate the correlation coefficient. *[C.U., B.Com(Hons) '81]*

6. For some bivariate data, the following results were obtained:
   Mean value of the variable X = 53.2
   Mean value of the variable Y = 27.9
   Regression coefficient of Y on X = – 1.5
   Regression coefficient of X on Y = – 0.2
   What is the most likely value of Y when X = 60? What is the Coefficient of correlation between X and Y? *[I.C.W.A., June '84; C.U., B.Com(Hons) '86]*
7. For a bivariate data the mean value of X is 20 and the mean value of Y is 45. The regression coefficient of Y on X is 4 and that of X on Y is 1/9. Find
   (i) the coefficient of correlation,
   (ii) the s.d. of X if the s.d. of Y is 12,
   (iii) the equations of regression lines. *[C.A., Nov, '82]*
8. To study the effect of rain on yield of wheat, the following results were obtained:

	Mean	s.d.
Yield in lbs. per acre	800	12
Rainfall in inches	50	2

$r = 0.80$

Estimate the yield when rainfall is 80 inches. *[I.C.W.A., June '86]*

9. From the following data find the two regression equations:

X	1	2	3	4	5
Y	2	3	5	4	6

Predict the value of Y when X = 2.5. *[C.U., B.Com(P) '83]*

10. The following table shows the age (x) in years, of cars of a certain make and the annual maintenance cost (y) in hundred Rs.

x	1	2	3	4	5	6	7
y	20	19	22	23	21	25	27

Obtain the regression equation of y on x. Estimate the cost of maintaining a six-year old car of the same make. (Answer should be correct upto 2 places of decimal). *[I.C.W.A. June '86 (old)]*

11. The grades of 9 students at the College test (x) and that University examination (y) are as follows:

x	77	50	71	72	81	94	96	99	67
y	82	66	78	34	47	85	99	99	68

(i) Draw the scatter diagram for the data (rough sketch only).
(ii) Calculate the value of r.
(iii) Find a linear regression equation for the data and then estimate the grade of the University examination for a student who received 85 in the College test but was sick at the time of University examination.
[C.U., B.Com(Hons) '83]

12. In the estimation of regression equations of two variables X and Y the following results were obtained:
$\bar{x} = 90$, $\bar{y} = 70$, $N = 10$, $\Sigma x^2 = 6360$, $\Sigma y^2 = 2860$, $\Sigma xy = 3900$. Obtain the two equations. *[C.A., May '83]*
(Note that x and y denote deviations of X and Y from their means)

13. Given that the regression equations of y on x and of x on y are respectively $y = 2x$ and $6x - y = 4$. Find (i) Correlation coefficient between x and y, and (ii) the means of x and y. *[I.C.W.A., June '86]*

14. In a partially destroyed record the following data are available:
Variance of $x = 25$
Regression equation of x on y : $5x - y = 22$
Regression equation of y on x : $64x - 45y = 24$
Find
(i) the mean values of x and y.
(ii) the coefficient of correlation between x and y.
(iii) the standard deviation of y. *[C.A., Nov. '81]*

15. The equations of two regression lines between two variables are expressed as $2x - 3y = 0$ and $4y - 5x - 8 = 0$.
(i) Identify which of the two can be called regression of y on x and which of x on y.
(ii) Find $\bar{x}$ and $\bar{y}$ and correlation coefficient (r) from the equations.
[I.C.W.A., June '83]

16. If the equations of two regres ion lines obtained in a correlation analysis are $3x + 12y = 19$ and $3y + 9x = 46$, determine which one of these is the regression equation of y on x and which one is the regression equation of x on y. Give reasons for your answer.

Find the means, correlation coefficient, and ratio of the variances of x and y. *[W.B.H.S. '83]*

17. Two regression lines of a sample are $x + 6y = 6$ and $3x + 2y = 10$. Find the correlation coefficient. *[C.U., B.Com(Hons) '86]*
18. Marks obtained in Statistics and Economics by 10 M.Com. students are given below:

Student	A	B	C	D	E	F	G	H	I	J
Statistics	61	38	49	57	42	34	93	87	46	35
Economics	31	38	28	35	19	42	45	48	22	25

Calculate Rank Correlation and interpret the result. *[C.U., M.Com. '83]*

19. Ten students were given the following marks in Mathematics and Business Statistics:

Student No.	1	2	3	4	5	6	7	8	9	10
Mathematics (x)	40	20	10	48	6	37	14	8	7	18
B. Statistics (y)	30	35	25	32	15	45	20	3	10	27

Calculate
   (i) the correlation coefficient between x and y.
   (ii) the rank correlation coefficient between x and y.
   (iii) Predict the value of y when $x = 41$. *[I.C.W.A., Dec, '85 (old)]*

20. In a trivariate distribution,
$S_1 = 3, S_2 = S_3 = 5, r_{12} = 0.7, r_{23} = r_{51} = 0.6.$
Find
   (i) Partial correlation coefficient $r_{12\cdot3}$,
   (ii) Multiple correlation coefficient $R_{1\cdot23}$,
   (iii) Regression coefficient $b_{12\cdot3}$. *[I.C.W.A., Dec. '85]*

(SHORT-ANSWER TYPE QUESTIONS)

21. Pick out the correct answer:
   (a) If $u = 2x + 3$ and $v = 3y - 7$ and the correlation coefficient between x and y is 0.86, then the correlation coefficient between u and v is: 0.86, 0.43, – 0.43, – 0.86, none of these. *[W.B.H.S. '81]*
   (b) If relation between two varibles x and y and between other two variables u and v are $2x + 3y = 7$ and $3u + 4v = 5$ respectively, and correlation coefficient of x and u is 0.2, then correlation coefficient of y and v is: – 0.2, 0.2, – 0.4, none of these. *[W.B.H.S.'85]*
22. Pick out the correct answer:
   (a) If $b_{xy} = 0.4$, $b_{yx} = 0.9$, then r equals: 0.36, 0.6, 1.3 *[C.A., May '81]*
   (b) If the relationship between x and u is $u + 2x = 0$ and other two variables y and v is $v = 6 + 3y$, and the regreesion coefficient of x on y is 2.4, then the regression coefficient of u on v is: – 0.4, 0.4, 2.4, none of these. *[W.B.H.S. '84]*
23. If $N = 10$, $\Sigma d^2 = 280$, then the rank correlation coefficient is: + 2.8, + 0.28, – 0.7. Which is correct? *[C.A., May '81]*
24. Which of the following two statements is correct? Give reasons for your answer.
   (i) Two independent variables are always uncorrelated.

(ii) Two uncorrelated variables must be always independent.
[I.C.W.A., June '85]

25. If two regression coefficients are negative, does it mean that their correlation would be positive? *[I.C.W.A., Dec. '79]*
26. State whether 'true' or 'false':
    (a) The correlation coefficient of $3x$ and $-2y$ is the same as the correlation coefficient of x and y.
    (b) If the correlation coefficient of x and y is zero, then $\text{Var}(x+y) = \text{Var}(x-y)$. *[W.B.H.S. '81]*
27. Find the equations of the lines of regression using the following data: $\bar{x} = 4$, $\bar{y} = 5$, $b_{yx} = 0.39$, $b_{xy} = 0.69$. *[C.U., B.Com(P) '82]*
28. (a) If the two regression lines coincide, show that the correlation coefficient $r = \pm 1$.
    (b) If $r = 0$, show that the two regression lines are at right angles. *[I.C.W.A., Dec. '85]*
29. From the following data, calculate the coefficient of correlation between x and y:
    (i) Number of item in x-series or y-series = 12.
    (ii) Sum of the square of deviations from mean: 360 for x-series and 250 for y-series.
    (iii) Sum of the products of deviations of the two series from their respective means = 225. *[I.C.W.A, June '83]*
30. Write down the expression for $r_{12\cdot3}$ and $R_{1\cdot23}$ in terms of r_{12}, r_{23} and r_{31}. What are the limits between which $r_{12\cdot3}$ and $R_{1\cdot23}$ must lie? Also state the relation between $r_{12\cdot3}$, $b_{12\cdot3}$ and $b_{21\cdot2}$. *[I.C.W.A., Dec. '86]*

ANSWERS

1. (a) 0.37, (b) $12x = 7y + 44$ (**Hints:** Example 9.7, p. 278 and Example 9.23, p. 294)
2. 0.70 (**Hint:** Use formula (9.7.2.), p. 276)
3. 0.77 (**Hint:** Find Σx, Σy, Σx^2, Σy^2, Σxy; Example 9.7).
4. – 0.95 (**Hint:** Example 9.9, p. 279).
5. 0.76.
6. 17.7 (**Hint:** formula (9.13.4.), p. 292; $Y = 107.7 - 1.5X$) – 0.55 (**Hint:** formula (9.13.8), p. 292 and property (3))
7. (i) 2/3 (**Hint:** formula (9.13.8). p. 292);
   (ii) 2 (**Hint:** Use (9.13.6) or (9.13.7) p. 292)
   (iii) $Y = 4X - 35$; $X = Y/9 + 15$ (Hint: (9.13.4 and 5).
8. 944 lbs. per acre (**Hint:** Example 9.25, p. 294)
9. $Y = 0.9X + 1.3$; $X = 0.9Y - 0.6$ (**Hint:** Obtain ΣX, ΣY, ΣX^2, ΣY^2, ΣXY; Example 9.23, p. 293); 3.55 (**Hint:** Use Ist equations)
10. $7y = 8x + 125$; 24.71 (hundred Rs.). (**Hint:** Example 9.28, p. 296)
11. (i) **Hint:** see p. 283; (ii) 0.56 (**Hint:** Example 9.9, p. 279)
    (iii) $y = 0.777x + 12.07$; 78 (**Hint:** Example 9.27, p. 295)

12. $Y = 0.6132X + 14.812$; $X = 1.3636Y - 5.452$. (**Hint:** $b_{YX} = \Sigma xy/\Sigma x^2$, $b_{XY} = \Sigma xy/\Sigma y^2$; use (9.13.4 and 5), p. 292).
13. (i) $1/\sqrt{3} = 0.58$, (ii) 1, 2 (**Hint:** In standard form $y = 2x$, $x = (1/6)y + 2/3$; see Example 9.31, p. 299)
14. (i) 6, 8; (ii) 8/15; (iii) 40/3 (**Hint:** Example 9.31 and 9.34)
15. (i) 1st, 2nd; (ii) – 24/7, – 16/7; (iii) 0.73 (**Hint:** Example 9.34, p. 300)
16. Ist, 2nd; because then $r^2 = 1/2$ does not exceed 1, as it must, 5, 1/3; $r = -1/2\sqrt{3}$; 4.3 (**Hint:** Example 9.34, p. 300)
17. – 1/3 (**Hint:** Example 9.34, p, 300)
18. 0.39 (**Hint:** Example 9.40, p. 308). On the whole, a student doing well in Statistics is also found good in Economics.
19. (i) 0.735 (**Hint:** Example 9.9, p. 279)
    (ii) 0.842 (**Hint:** Example 9.40, p. 308)
    (iii) 36.3 (**Hint:** Example 9.27, p. 295; $y = 11.72 + 0.60x$)
20. (i) 0.53125, (ii) 0.73527, (iii) 0.31875 (**Hint:** Use formulae (9.17.5), (9.17.3) and (9.17.2), p. 313)
21. (a) 0.86, (b) 0.2 (**Hint:** see formula (9.8.2). p. 277)
22. (a) 0.6 (**Hint:** (9.13.8), p. 292 & Problem 9.12a)
    (*b*) None of these (**Hint:** Problem 9.12(b); correct – 1.6)
23. – 0.7 (**Hint:** formula (9.16.1), p. 304)
24. (i) is correct; because then Covariance is zero.
    (**Hint:** Example 9.21, p. 289, Example 12.28, p. 128 of Part II).
25. No; is must be negative. (**Hint:** Property (3), p. 292)
26. (a) Flase (**Hint:** Example 9.15, p. 303)
    (b) True (**Hint:** Formula (9.11.5), p. 306)
27. $y = 0.39x + 3.44$; $x = 0.69y + 0.55$ (**Hint:** use (9.13.4 & 5), p. 292)
28. **Hint:** (a) see p. 292 (last para); (b) see p. 292 (para 3).
29. 0.75 (**Hint:** Given $n = 12$, $\Sigma(x - \bar{x})^2 = 360$, $\Sigma(y - \bar{y})^2 = 250$, $\Sigma(x - \bar{x})(y - \bar{y}) = 225$; use formula (9.7.2), p. 276)
30. **Hint:** Formula (9.17.3) to (9.17.6) & (9.17.10), pp. 292–3.

CHAPTER 10

INTERPOLATION

(Solved Problems)

Problem 10.1 *Construct a table of difference. Express y as a function of x and obtain the value of y when x = 22 form the following:*

x	*10*	*15*	*20*	*25*	*30*	*35*
y	*3*	*6*	*11*	*18*	*27*	*38*

[I.C.W.A., June '86 (new)]

Solution [see Example 10.8, p. 324]

Difference able

x	y	Δy	$\Delta^2 y$	$\Delta^3 y$
10	3			
		3		
15	6		2	
		5		0
20	11		2	
		7		0
25	18		2	
		9		0
30	27		2	
		11		
35	38			

The second differences $\Delta^2 y$ are found to be constant. Hence we may assume that y is a 2nd degree polynomial.

Let $\quad y = a + b\,(x - 10) + c\,(x - 10)(x - 15)$ ... (1)

where $\quad a, b, c$ are constants,

Putting $x = 10$ in (1),

$$3 = a + b\,(10 - 10) + c\,(10 - 10)\,(10 - 15)$$

or, $\quad 3 = a + b\,(0) + c \times 0(-5)$

or, $\quad 3 = a + 0 + 0$; $\quad$ or, $a = 3$ ... (2)

Putting $x = 15$ in (1),

$$6 = a + b\,(15 - 10) + c\,(15 - 10)\,(15 - 15);$$

or, $\quad 6 = 3 + b\,(5) + c\,(5)(0)$; $\quad$ since $a = 3$

or, $\quad 5b = 3$; $\quad$ or, $\quad b = 3/5 = 0.6$ ... (3)

Putting $x = 20$ in (1),

$$11 = a + b\,(20 - 10) + c\,(20 - 10)(20 - 15);$$

or, $\quad 11 = 3 + 6 + 50c$; since $a = 3$, $b = 0.6$

or, $\quad 50c = 2$; $\quad$ or, $\quad c = \dfrac{2}{50} = 0.04$ ... (4)

Putting the values of a, b, c in (1),

$$
\begin{aligned}
y &= 3 + 0.6(x - 10) + 0.04(x - 10)(x - 15) \\
&= 3 + 0.6x - 6 + 0.04(x^2 - 25x + 150) \\
&= 3 + 0.6x - 6 + 0.04x^2 - x + 6 \\
&= 3 - 0.4x + 0.04x^2 \qquad \text{... (5)}
\end{aligned}
$$

When $x = 22$, from (5) we get

$$
\begin{aligned}
y &= 3 - 0.4 \times 22 + 0.04 \times 484 \\
&= 3 - 8.8 + 19.36 = 13.56
\end{aligned}
$$

Ans. $y = 3 - 0.4x + 0.04x^2$; 13.56

Problem 10.2 *With the help of the symbolic operator E, find the value of $\log_{10}$ 666 from the following table:*

$\log_{10} 654 = 2.8156$, $\log_{10} 658 = 2.8182$, $\log_{10} 662 = 2.8209$

[I.C.W.A., June '79]

Solution

x	$y = \log_{10}$	x
654	2.8156	(y_0)
658	2.8182	(y_1)
662	2.8209	(y_2)
666	?	(y_3)

Let $y = \log_{10} x$. For *equidistant* values of $x = 654, 658, 662, 666$ let just denote the values of y by y_0, y_1, y_2, y_3 respectively. Since only 3 values of y are known, we assume that y is a 2nd degree polynomial. Hence the 3rd differences are zero. In particular

$$
\begin{aligned}
& \Delta^3 y_0 = 0 \\
\text{or,} \quad & (E - 1)^3 y_0 = 0 \\
\text{or,} \quad & (E^3 - 3E^2 + 3E - 1)y_0 = 0 \\
\text{or,} \quad & y_3 - 3y_2 + 3y_1 - y_0 = 0 \\
\text{or,} \quad & y_3 - 3(2.8209) + 3(2.8182) - 2.8156 = 0 \\
\text{or,} \quad & y_3 - 8.4627 + 8.4546 - 2.8156 = 0 \\
\text{or,} \quad & y_3 - 2.8237 = 0 \\
\therefore \quad & y_3 = 2.8237; \quad \text{i.e. } \log_{10} 666 = 2.8237
\end{aligned}
$$

Problem 10.3 *Using the following results, find $\sqrt[3]{26}$.*

$\sqrt[3]{27} = 3.0000$, $\sqrt[3]{28} = 3.0366$, $\sqrt[3]{29} = 3.0723$, $\sqrt[3]{30} = 3.1072$

Solution

x	$y = \sqrt[3]{x}$	
26	?	(y_0)
27	3.0000	(y_1)
28	3.0366	(y_2)
29	3.0723	(y_3)
30	3.1072	(y_4)

Let $y = \sqrt[3]{x}$. For *equidistant* values of $x = 26, 27, 28, 29, 30$, we denote the values of y by y_0, y_1, y_2, y_3, y_4 respectively. Since only 4 values of the function are given, it is assumed that y

is a 3rd degree polynomial. Hence the 4th differences $\Delta^4 y$ are zero. In particular $\Delta^4 y_0 = 0$. Expanding $(E - 1)^4\, y_0 = 0$

or, $y_4 - 4y_3 + 6y_2 - 4y_1 + y_0 = 0$

or, $3.1072 - 4(3.0723) + 6(3.0366) - 4(3.0000) + y_0 = 0$

Solving, $y_0 = 2.9624$. (Note: The true value is 2.9625).

Problem 10.4 *Obtain estimates of the missing figures in the following table:*

x	2.0	2.1	2.2	2.3	2.4	2.5	2.6
U_x	0.135	–	0.111	0.100	–	0.082	0.074

[I.C.W.A., Dec. '82, '85]

Solution For equidistant values of $x = 2.0, 2.1, ..., 2.6$, let us denote the values of $y = U_x$ by $y_0, y_1, ..., y_6$ respectively. Since only 5 values are given we assume a 4th degree polynomial. Hence 5th differences are zero.

$$\Delta^5 y_0 = 0 \qquad \Delta^5 y_1 = 0$$

or, $$(E-1)^5 y_0 = 0 \qquad (E-1)^5 y_1 = 0$$

Expanding we get (see Ex. 10.1, p. 367)

$$y_5 - 5y_4 + 10y_3 - 10y_2 + 5y_1 - y_0 = 0 \quad \text{... (i)}$$

$$y_6 - 5y_5 + 10y_4 - 10y_3 + 5y_2 - y_1 = 0 \quad \text{... (ii)}$$

[Note that (ii) may be obtained from (i) by increasing each suffix by 1]

or, $0.082 - 5y_4 + 10(0.100) - 10(0.111) + 5y_1 - 0.135 = 0$

$0.074 - 5(0.082) + 10y_4 - 10(0.100) + 5(0.111) - y_1 = 0$

Simplifying, $$-5y_4 + 5y_1 = 0.163 \quad \text{... (iii)}$$

$$10y_4 - y_1 = 0.781$$

Multiplying (ii) by 2 and adding to (iv),

$$9y_1 = 1.107, \quad \therefore \; y_1 = 0.123$$

From (*vi*), $$10y_4 = 0.904 \quad \therefore \; y_4 = 0.0904$$

Problem 10.5 *If $U_x = x^3$, evaluate $\Delta^2 U_x$.* [C.U., B.Com(Hons) '83]

Solution Assuming that the interval of differencing is *unity* (i.e. the successive values of x increase by 1),

$$\Delta U_x = U_{x+1} - U_x, \quad \text{(by definition)}$$

$$= (x+1)^3 - x^3$$

[Note that U_{x+1} is obtained from $U_x = x^3$ on replacing x by $(x + 1)$.]

$$= (x^3 + 3x^2 + 3x + 1) - x^3$$

$$= 3x^2 + 3x + 1$$

$\therefore$ $$\Delta^2 U_x = \Delta U_{x+1} - \Delta U_x, \text{ (by definition)}$$

$$= \{3(x+1)^2 + 3(x+1) + 1\} - \{3x^2 + 3x + 1\}$$

[$\because$ ΔU_{x+1} is obtained from $\Delta U_x = 3x^2 + 3x + 1$ on replacing x by $(x + 1)$.]

$$= 3(x+1)^2 + 3(x+1) + 1 - 3x^2 - 3x - 1$$

$$= 6(x+1), \text{ on simplification.}$$

Ans. $6(x + 1)$

Problem 10.6 *Using an appropriate formula for interpolation, estimate the number of students who obtained marks between 40 and 45 from the following table :*

Marks	*30–40*	*40–50*	*50–60*	*60–70*	*70–80*
Number of Students	*31*	*42*	*51*	*35*	*31*

[C.U., B.Com.(Hons.) '82]

Solution Number of students obtaining marks between 40–45

= (No. of students obtaining less than 45) – (No. of students obtaining less than 40)

= (Cumulative frequency corresponding to 45) – (Cumulative frequency corresponding to 40)

Cumulative frequency corresponding to 40 is 31. Cumulative frequency corresponding to 45 may be obtained from a cumulative frequency table by using *Newton's forward interpolation* formula, because 'marks 45' lies near the beginning of the table.

Marks (x)	*Cum. Freq. (y)*	Δy	$\Delta^2 y$	$\Delta^3 y$	$\Delta^4 y$	$\Delta^5 y$
30	0					
		31				
40	31		11			
		42		−2		
50	73		9		−23	
		51		−25		60
60	124		−16		37	
		35		12		
70	159		−4			
		31				
80	190					

Newton's forward formula (see p. 326).

Here $x_0 = 30, y_0 = 0, x = 45, h = 10.$

$\therefore \quad u = (x - x_0)/h = (45 - 30)/10 - 3/2$

$$y = 0 + \frac{3}{2} \times 31 + \frac{3}{2}\left(\frac{8}{2} - 1\right) \times \frac{11}{2!} + \frac{3}{2}\left(\frac{3}{2} - 1\right)\left(\frac{3}{2} - 2\right) \times \frac{(-2)}{3!}$$

$$+ \frac{3}{2}\left(\frac{3}{2} - 1\right)\left(\frac{3}{2} - 2\right)\left(\frac{3}{2} - 3\right) \times \frac{(-23)}{4!}$$

$$+ \frac{3}{2}\left(\frac{3}{2} - 1\right)\left(\frac{3}{2} - 2\right)\left(\frac{3}{2} - 3\right)\left(\frac{3}{2} - 4\right) \times \frac{60}{5!}$$

$$= 0 + \frac{93}{2} + \frac{3}{2} \times \frac{1}{2} \times \frac{11}{2} + \frac{3}{2} \times \frac{1}{2} \times \left(-\frac{1}{2}\right)\left(-\frac{2}{8}\right)$$

$$+ \frac{3}{2} \times \frac{1}{2} \times \left(-\frac{1}{2}\right)\left(-\frac{3}{2}\right)\left(-\frac{28}{24}\right) + \frac{3}{2} \times \frac{1}{2} \times \left(-\frac{1}{2}\right)\left(-\frac{3}{2}\right)\left(-\frac{5}{2}\right) \times \frac{60}{120}$$

$$= 0 + \frac{93}{2} + \frac{33}{8} + \frac{1}{8} - \frac{69}{128} - \frac{45}{64}$$

$= 0 + 46.5 + 4.125 + 0.125 - 0.539 - 0.703.$

$= 49.508 = 50$ (rounded to the nearest whole number)

$\therefore$ Required number of students $= 50 - 31 = 19$

Problem 10.7 *Using Newton's Backward formula find the annual premium at the age of 29 from the following data:*

Age	*24*	*28*	*32*	*36*	*40*
Annual Premium (Rs)	*28.06*	*30.19*	*32.75*	*34.94*	*40.00*

[C.U., B.Com(Hons) '84]

Solution (For simplicity, we ignore the decimal point while computing the differences).

Age (x)	*Premium (y)*	*Δy*	*Δ²y*	*Δ³y*	*Δ⁴y*
24	28.06				
		213			
28	30.19		43		
		256		– 80	
32	32.75		– 37		404
		219		324	
36	34.94		287		
		506			
40	40.00				

Here $v = (x - x_n)/h = (29 - 40)/4 = -11/4$

Using 2 decimals, and the Backward formula (p. 330),

$$y = 4000 + \left(-\frac{11}{4}\right) \times 506 + \left(-\frac{11}{4}\right)\left(-\frac{11}{4} + 1\right) \times \frac{287}{2\,!}$$

$$+ \left(-\frac{11}{4}\right)\left(-\frac{11}{4} + 1\right)\left(-\frac{11}{4} + 2\right) \times \frac{324}{3\,!}$$

$$+ \left(-\frac{11}{4}\right)\left(-\frac{11}{4} + 1\right)\left(-\frac{11}{4} + 2\right)\left(-\frac{11}{4} + 3\right) \times \frac{404}{4\,!}$$

$$= 4000 + \left(\frac{-11}{4}\right) \times 506 + \left(\frac{-11}{4}\right)\left(\frac{-7}{4}\right) \times \frac{287}{2}$$

$$+ \left(\frac{-11}{4}\right)\left(\frac{-7}{4}\right)\left(\frac{-3}{4}\right) + \left(\frac{-11}{4}\right)\left(\frac{-7}{4}\right)\left(\frac{-3}{4}\right)\left(\frac{1}{4}\right) \times \frac{404}{24}$$

$$= 4000 - \frac{2783}{2} + \frac{22099}{32} - \frac{6237}{32} - \frac{7777}{512}$$

$= 4000 - 1391.5 + 690.6 - 194.9 - 15.2$

$= 3089.0 = 3089$ (rounded)

Adjusting the decimal point, the required value is 30.89 (Rs)

Note: (i) If nothing is mentioned, we must use the most appropriate formula, here Newton's "forward".

(ii) Both Newton's forward and backward formulae applied to the same problem would give exactly the same result. However, by the appropriate formula the successive terms diminish rapidly, so that the first few terms together will give fairly accurate result.

(iii) If we use Newton's forward formula, here

$$u = (x - x_0)/h = (29 - 24)/4 = 5/4$$

Using 2 decimals,

$$y = 2806 + \frac{5}{4} \times 213 + \frac{5}{4}\left(\frac{5}{4} - 1\right) \times \frac{43}{2!} + \frac{5}{4}\left(\frac{5}{4} - 1\right)\left(\frac{5}{4} - 2\right) \times \frac{-80}{3!}$$

$$+ \frac{5}{4}\left(\frac{5}{4} - 1\right)\left(\frac{5}{4} - 2\right)\left(\frac{5}{4} - 3\right) \times \frac{404}{4!}$$

$$= 2806 + \frac{1065}{4} + \frac{5}{4}.\frac{1}{4}.\frac{43}{2} + \frac{5}{4}.\frac{1}{4}.\frac{-3}{4}.\frac{-80}{6} + \frac{5}{4}.\frac{1}{4}.\frac{-3}{4}.\frac{-7}{4}.\frac{404}{24}$$

$$= 2806 + \frac{1065}{4} + \frac{215}{32} + \frac{25}{8} + \frac{3535}{512}$$

$$= 2806 + 266.2 + 6.7 + 3.1 + 6.9$$

$$= 3088.9 = 3089 \text{ (rounded)}$$

Adjusting the decimal point, the required value is 30.89 (Rs), which is exactly the same as obtained by Newton's backward formula.

The first 3 terms by the forward formula together give 30.79 (pretty close to 30.89), whereas by the backward formula we get only 32.99 (far away from 30.89) ; because, in this problem the appropriate formula is Newton's *Forward.*

ADDITIONAL EXERCISES 10

1. Given the following table, find the function $f(x)$ assuming it to be a polynomial of degree 4. Find also $f(5)$.

x	0	1	2	3	4
$f(x)$	1	5	31	121	341

[I.C.W.A., Dec. '79]

2. Obtain an estimate of the missing figure in the following table:

x	100	101	102	103	104
y_z	2.0000	2.0043	–	2.0128	2.0170

[I.C.W.A. Dec. '86]

3. Obtain an estimate of the missing figure in the following:

x	5.1	5.2	5.3	5.4	5.5
y	–	0.110	0.100	0.090	0.082

[I.C.W.A., Dec. '85]

4. Obtain an estimate of the missing figures in the following table:

x	3.0	3.1	3.2	3.3	3.4	3.5	3.6
$f(x)$	0.270	–	0.222	0.200	–	0.164	0.148

[I.C.W.A., Dec '83]

5. Given $U_{50} = 939$, $U_{52} = 907$, $U_{53} = 841$, $U_{55} = 773$, estimate U_{51} and U_{54} under suitable assumptions. [I.C.W.A. Dec. '84]

6. From the following data, obtain the value of y when $x = 9$ by using Newton's Forward Difference Interpolation formula:

x	3	7	11	15	19
y	42	43	47	53	60

[I.C.W.A., Dec. '85]

7. The population of a city for five census years are given below :

Year (x)	1941	1951	1961	1971	1981
Population in Lakhs (y)	46	66	81	93	101

Estimate by using Newton's forward interpolation formula the population of the city for the year 1945. [C.U., B.Com (P) '82]

8. Use Newton's Forward Interpolation formula to find y when $x = 12$ from the following table:

x	10	15	20	25	30	35
y	35.4	32.2	29.1	26.0	23.1	20.4

[C.U., B.Com. (Hon) '86]

9. The following table gives the values of an unknown function $f(x)$ for certain equidistant values of x. Use a suitable interpolation formula to find the value of the function at $x = 24$:

x	18	22	26	30	34
$f(x)$	11.725	14.221	16.958	19.954	23.222

[C.U., B.Com.(Hons) '81]

10. The values of y are 7, 21, 43 and 73 respectively, when the values of x are 2, 4, 6 and 8. Find by a suitable method of interpolation the value of y when x is equal to 7. [C.U., B.Com. (P) '80]

11. Given $\sqrt[3]{5} = 1.710$, $\sqrt[3]{7} = 1.913$, $\sqrt[3]{9} = 2.080$, $\sqrt[3]{11} = 2.224$, find $\sqrt[3]{10}$. [K.U., B.Com, (P) '83]

12. Use a suitable interpolation formula to find y when $x = 33$ from the following data:

x	24	28	32	36	40
y	25.06	27.19	29.75	31.94	37.00

[C.U., B.Com.(Hons.) '85]

13. Given $f(45) = 0.7071, f(50) = 0.7660, f(55) = 0.8192$, and $f(60) = 0.8693$, find $f(59)$ correct to 4 places decimals. *[I.C.W.A., June '79]*

14. Using an appropriate interpolation formula, estimate the number of students who secured marks upto 45 and hence obtain the number so students securing marks between 45 and 50 from the following:

Marks Upto	40	50	60	70	80
No. of Students	31	73	124	159	190

[I.C.W.A., June '86 (old)]

15. Use Newton's Backward formula for interpolation to find the number of workers earning Rs 36 or more both less than Rs 40 per day, from the following table:

Earning per day less than (Rs)	20	25	30	35	40
Number of Workers	20	45	115	210	225

[I.C.W.A., June '83]

16. Given the following table of values, find $f(1.75)$:

x	1.40	160	1.70	1.80
$f(x)$	0.9855	0.9995	0.9917	0.9739

[I.C.W.A., Dec. '84]

17. Given $\log_{10} 654 = 2.8156$, $\log_{10} 658 = 2.8182$, $\log_{10} 659 = 2.8189$, $\log_{10} 661 = 2.8202$, find $\log_{10} 656$ using the interpolation formula available for observations at unequal intervals, say Lagrange's formula.

[I.C.W.A., June '83]

18. Use Lagrange's interpolation formula or otherwise to find $f(5)$ from the following data: $f(3) = 8.71940, f(7) = 9.08914, f(9) = 9.19433$.

[I.C.W.A., June '83]

19. Given the following table, find the value of x corresponding to $y = 2.8198$:

x	654	658	659	661
y	2.8156	2.8182	2.8189	2.8202

[I.C.W.A., June '81]

20. What is the inverse interpolation? Given the following table, find the value of x corresponding to $y = 1.2852$:

x	0.736	0.737	0.738	0.739
y	1.2833	1.2841	1.2849	1.2857

[I.C.W.A., Dec. '82]

ANSWERS

1. $f(x) = x^4 + x^3 + x^2 + x + 1$; 781.
   [**Hint:** Example 10.8 (Second Method), p. 324 and Problem 10.1; write $f(x) = a + bx + cx(x - 1) + dx(x - 1)(x - 2) + ex(x - 1)(x - 2)(x - 3)$.]
2. 2.0086 [**Hint:** Example 10.6, p. 323]
3. 0.118 [**Hint:** Problem 10.3, p. 430]
4. 0.246, 0.1808 [**Hint:** Problem 10.4, p. 431]
5. 950.6, 785.4 [**Hint:** Example 10.5; p. 323]
6. 44.6875 [**Hint:** Example 10.11, p. 326]
7. 58.85 lakh [**Hint:** Example 10.11, p. 326].
8. 34.07 [**Hints:** Example 10.11, p. 326]
9. 15.558 [**Hint:** Newton's forward formula; ignore decimals in computing differences, see Problem 10.7]
10. 57 [**Hint:** Newton's backward formula Example 10.13, p. 329]
11. 2.154 [**Hint:** see Problem 10.3, $y = N x$; use Newton's backward formula, Example 10.13, p. 329]
12. 30.27 [**Hint:** Newton's backward; Example 10.13, p. 329]
13. 0.8594 [**Hint:** Newton's backward; ignore decimals in difference table.]
14. 48, 25 [**Hint:** Problem 10.6]
15. 16 [**Hint:** Problem 10.6; use Newton's backward]
16. 0.9841 [**Hint:** Lagrange's formula; Example 10.15, p. 335]
17. 2.8168 [**Hint:** Write $y = \log_{10} x$; see Problem 10.2]
18. 8.93083 [**Hint:** Example 10 : 15, p. 335 and Example 10.6, p. 332]
19. 660.4 [**Hint:** Example 10.17, p. 334]
20. 0.7384 [**Hint:** Example 10.17, p. 334]

CALCUTTA UNIVERSITY

MODEL SHORT-ANSWER TYPE QUESTIONS— B.COM. (PASS)—STATISTICS

1. (a) Mention, giving reasons, the correct answer from the following: Statistics consists of (i) a single figure: (ii) a group of homogeneous facts and figures; (iii) isolated fact.
   (b) What are the methods of collecting Primary data?
   (c) Which of the following is concerned with Primary data? (i) Annual Statement of the Foreign Trade of India, (ii) Annual Report of the Railway Board, (iii) International Labour Review, (iv) Reserve Bank of India Bulletin.
   (d) What methods would you employ in collection of data considering accuracy, time and cost involved when the field of enquiry is (i) small, (ii) fairly large, (iii) very large.
2. You are given the following distribution:

Height (cm) :	150.0–154.9	155.0–159.9	160.0–164.9	165.0–169.9	170.0–174.9
No. of Students :	5	10	16	12	7

   (a) Write down the class limits and class boundaries.
   (b) What are the frequency densities of the distributions?
   (c) Write down the percentage frequency in each class interval.
   (d) What are the cumulative frequencies from below?
3. State briefly, giving reasons, the kind of diagram you consider most appropriate for use with each of the following data:
   (a) Monthly production of Hind Motor Cars in India for the year 1980.
   (b) Budget allotments for the various departments.
   (c) Different heads of income of a Railway Company during 1979 and 1980.
   (d) Production and sale of electrical energy in West Bengal during the last 5 years.
4. (a) Find the arithmetic mean of 1, 2, 3, ..., n.
   (b) Find the geometric mean of 3, 12 and 48.
   (c) Find the H.M. of 1, 1/2, 1/3, 1/4.
   (d) The A.M. of two observations is 36 and their G.M. is 24; find the H.M.

5. What type of average should be employed in each of the following cases? Give reasons for your answer:
   (a) The marks obtained by 100 students in an examination.
   (b) The size of shoes sold in largest number from a shop.
   (c) A frequency distribution having open-end classes.
   (d) Different rates of depreciation charged on diminishing balances.
6. (a) If the Mean and Median of a frequency distribution be 30 and 27 respectively, find the Mode.
   (b) If the algebraic sum of the deviations of 10 observations measured from 20 is 30, find the Arithmetic Mean.
   (c) If the A.M. of 3, 5, x, 12, 17 be 9, find the value of x.
   (d) If for a symmetrical distribution $Q_1 = 24$, $Q_3 = 32$, find Median.
7. (a) Write down the value of Range of the daily wages of 6 persons given below: (Rs) 18, 25, 15, 20, 28, 22.
   (b) If for a distribution, $Q_1 = 10$, $Q_2 = 22$, $Q_3 = 33$, find Quartile Deviation.
   (c) If the absolute deviations of 7 observations from mean be 31, 35, 29, 63, 55, 72, 37, find Mean Deviation about mean.
   (d) If for a frequency distribution Mean = 48, S.D. = 18.72, write down the value of Coefficient of Variation.
8. (a) If the means of two samples of sizes 40 and 60 are respectively 25 and 30, write down the value of Mean of the Combined Sample when the two samples are pooled together.
   (b) The mean of 10 observations is 20. On checking it was found that two observations were wrongly copied as 15 an 17, while the correct observations were 18 and 19. What is the Correct Mean?
   (c) Write down the value of S.D. of a distribution when Coefficient of Variation is 40 and Mean is 45.
   (d) If for a distribution, $Q_1 = 5$, $Q_2 = 10$, $Q_3 = 15$, write down the value of Coefficient of Quartile Deviation.
9. Comment on the following statements:
   (a) "The median of a distribution is $N/2$, the lower quartile is $N/4$ and the upper quartile is $3N/4$."
   (b) "If Q_1, Q_2, Q_3 be respectively the lower quartile, the median and the upper quartile of a distribution, then $Q_3 - Q_2 = Q_2 - Q_1$."
   (c) "The standard deviation of a distribution is invariant under changes of origin and scale."
   (d) "Coefficient of variation is expressed in the same unit as the original data."
10. State with reasons whether the following statements are correct:
   (a) "The standard deviation (s.d.) of 20 items is 8.5. If every item is increased by 4, the s.d. is increased by 4."
   (b) "If the mean and S.D. of n observations $x_1, x_2, \ldots x_n$ be $\bar{x}$ and σ respectively, then the mean and the s.d. of $-x_1, -x_2, \ldots, x_n$ will be $-\bar{x}$ and $-\sigma$ respectively.
   (c) "For a symmetrical distribution, all central moments of odd order are zero."
   (d) "The 1st moment about origin is the mean and the 2nd moment about origin is the variance of the distribution."

11. (a) Write down the value of Mean, when the first moment of a distribution about the value 5 is 2
(b) Write down the value of skewness of a distribution, when its Mean = 25, Mode = 22, S.D. = 6?
(c) What is the value of Skewness of a distribution, when its Mean = 25, Median = 49, S.D. = 16?
(d) What is the value of Kurtosis of a distribution, when its 4th central moment is 27 and $\sigma = \sqrt{3}$?

12. (a) If for a distribution $Q_1 = 5$, $Q_2 = 10$, $Q_3 = 13$, write down the value of Coefficient of Quartile Deviation.
(b) What is the value of Coefficient of Mean Deviation when Mean Deviation about mean = 24 and Mean = 80 ?
(c) Using Bowley's measure, write down with reasons the value of Skewness of a symmetrical distribution.
(d) What is the value of Skewness of a distribution when the 2nd and the 3rd central moments are respectively 4 and 6?

13. (a) Write down the value of Correlation Coefficient r between x and y when $\text{cov}(x, y) = 9$, $\sigma_x = 3$, $\sigma_x = 4$.
(b) If $r = 0.6$, $\text{cov}(x, y) = 12$, $\sigma_x = 5$, find the value of σ_y.
(c) Write down the value of r when $b_{yx} = -0.4$, $b_{0.6} = -0.9$.
(d) Write down the value of b_{yx} when $\sigma_x = 3$, $\sigma_y = 4$ and $r_{xy} = 0.75$.

14. (a) State in each case whether you would expect to find a positive correlation, a negative correlation: (i) Amount of rainfall and yield of crop, (ii) Shoe size and intelligence, (iii) Insurance companies profits and the number of claims they have to pay, (iv) The ages of husbands and wives.
(b) State with reasons with measure of dispersion would you use to compare the variability of two or more distributions given in different units.
(c) Mention the type of Kurtosis of a frequency distribution when (i) $\beta_2 = 4.56$, and (ii) $\beta_2 = 2.84$, where $\beta_2 = m_4/\sigma^4$.
(d) Write down the value of Rank Correlation Coefficient when $V = 10$, $\Sigma D^2 = 30$, where D = difference between the ranks of two given series.

15. (a) You are given the following values of x and $f(x)$

x :	2	5	8	11	14
$f(x)$:	3	20	56	120	210

State with reasons which formula and would consider most appropriate to compute the value of $f(3.5)$ and $Q(9.3)$.
(b) Write down the value of Coefficient of Skewness, when $Q_1 = 25$, $Q_2 = 36$, $Q_3 = 45$.
(c) Draw the Scatter Diagram when $r = -1$ and $r = 1$.
(d) The coefficient of correlation r of the following data is 0.8

X :	1	2	3	4	5
Y :	3	2	5	4	6

If the values of X are increased by 4 and those of Y decreased by 3, what will be the value of r after the changes? Give reasons for your answer.

ANSWERS

1. (a) (ii), since Statistics must be aggregates of facts and related to some specific field of enquiry (p. 2). (b) See p. 5. (c) (ii) and (iv). (d) (i) Direct personal observation, (ii) Schedules sent through investigators, (iii) Questionnaires sent by mail.
2. (a) *Class limits:* 150.0–154.9, 155.0–159.9 160.0–164.9, 165.0–169.9, 170.0–174.9; Class boundaries: 149.95–154.95, 154.95–159.95, 159.95–164.95, 164.95–169.95, 169.95–174.95. (b) 1.0, 2.0, 3.2, 2.4, 1.4. (c) 10, 20, 32, 24, 14 (since $N = 50$). (d) 0, 5, 15, 31, 43, 50 corresponding to heights (cm) 149.95, 154.95, 159.95, 164.95, 169.95, 174.95 respectively (p. 83).
3. (a) Vertical Bar diagram (or Line diagram); since these are useful for the series data. (b) Pie diagram; since this is used for comparison among different components and also in relations to the total (p. 41). (c) Component Bar diagrams separately for each year; since comparison among different heads of income and also between the two years can be made (p. 39). (d) Grouped Bar diagram, one pair for each year showing Production and Sale; since comparison between them or year-to-year comparison is possible (p. 38).
4. (a) $(n + 1)/2$, (b) 12, (c) 0.4, (d) 16 (Example 5.44).
5. (a) A.M.; since it is simple to understand and based on all observations (b) Mode; since it is representative because of its maximum frequency. (c) Median; since midvalues and widths of open-end classes are not known. (d) G.M.; since it is appropriate for averaging rates of change.
6. (a) 21 (Example 5. 62), (b) 23, (c) 8, (d) 28 (Example 7.15).
7. (a) Rs 13, (b) 11.5, (c) 46, (d) 39.
8. (a) 28 (Example 5.25), (b) 20.5, (c) 18, (d) 50 (p. 216).
9. (a) and (b) See Nos. 50 (a) (b), p. 171, (c) Invariant under change of origin, but depends on scale, (d) Not correct; it is a pure number.
10. (a) Not correct; S.D. is unaffected by change of origin. (b) See No. 21, p. 220. (c) Correct; (see p. 240). (d) 1st part correct, 2nd part not correct; since Variance = 2nd moment about *mean*.
11. (a) 7 (Example 7.1), (b) 0.5, (c) – 0.75,(d) $y_2 = 0$ (p. 241)
12. (a) 40 (p. 209), (b)30, (c) 0, since for a symmetrical distribution $Q_1 - Q_2 = Q_2 - Q_1$. (d) $\gamma_1 = 0.75$ (p. 235).
13. (a) 0.75, (b) 4, (c) – 0.6, (d) 1
14. (a) See Example 9.6, (b)Coefficient of Variation; since absolute measures of dispersion will be in different units and hence not comparable. (c) (i) Leptokurtic, (ii) Platykurtic (p. 242). (d) + 0.82
15. (a) Since the given values of x are equidistant, Newton's Forward formula is appropriate for finding $f(3.5)$, because $x = 3.5$ lies near the beginning of the table, and Newton's Backward formula for $f(9.3)$, because $x = 9.3$ lies near the end of the table. (b) – 0.1,(c) See p. 275, (d) $r = 0.8$. since is unaffected by changes of origin (and also scale).

IMPORTANT FORMULAE AND RESULTS

	Ref.	Page

Measures of Central Tendency

Arithmetic Mean ($\bar{x}$):

1. Simple A.M. $= \frac{1}{n} \Sigma x_i$

   Weighted A.M. $= \frac{1}{N} \Sigma f_i x_i$ — (5.2.3) — 106
2. $\Sigma x_i = n\bar{x}$; $\Sigma f_i x_i = N\bar{x}$ — (5.3.1) — 110
3. $\Sigma(x_i - \bar{x}) = 0$; $\Sigma f_i(x_i - \bar{x}) = 0$ — (5.3.2) — 110
4. If $y = (x - c)/d$, then $\bar{x} = c + d\bar{y}$ — (5.4.2) — 116
5. $N\bar{x} = n_1\bar{x}_1 + n_2\bar{x}_2$ — (5.5.1) — 124

Geometric Mean (G.M.):

6. Simple G.M. $= (x_1 \times x_2 \times \ldots \times x_n)^{1/n}$

   Weighted G.M. $= (x_1^{f_1}, x_2^{f_2}, x_n^{f_n})^{1/N}$ — (5.6.3) — 127
7. $\log$ (simple G.M) $= \frac{1}{n} \Sigma (\log x_i)$

   $\log$ (weighted G.M.) $= \frac{1}{N} \Sigma f_i (\log x_i)$ — (5.6.4) — 127

Harmonic Mean (H.M.):

8. Simple H.M. $= \dfrac{n}{\dfrac{1}{x_1} + \dfrac{1}{x_2} + \ldots + \dfrac{1}{x_n}}$ — (5.8.1) — 133

   Weighted H.M. $= \dfrac{N}{\dfrac{f_1}{x_1} + \dfrac{f_2}{x_2} + \ldots + \dfrac{f_n}{x_n}}$ — (5.8.2) — 133

Relation between A.M., G.M. and H.M.:

9. A.M. $\geq$ G.M. $\geq$ H.M. — (5.10.1) — 140

10.	$\frac{A.M.}{G.M.} = \frac{G.M.}{H.M.}$ (for 2 observations only)	(5.10.2)	137
11.	Median $= l_1 + \frac{N/2 - F}{f_m} \times c$	(5.12.1)	141
12.	$\frac{Median - l_1}{l_2 - l_1} = \frac{(N/2) - F_1}{F_2 - F_1}$	(5.12.2)	141

Mode:

13.	$Mode = l_1 + \frac{d_1}{d_1 + d_2} \times c$ $= l_1 + \frac{f_0 - f_{-1}}{2f_0 - f_{-1} - f_1} \times c$	(5.15.1)	152
14.	*Mean* – *Mode* = 3 (*Mean* – *Median*)	(5.17.1)	157

Quartiles, Deciles and Percentiles:

15.	*Median* $= Q_2 = D_5 = P_{50}$; $Q_1 = P_{25}$; $Q_3 = P_{75}$	(5.18.3)	158
16.	(For simple series) *Median*, Q_1, Q_3, have ranks $(n + 1)/2$, $(n + 1)/4$ and $3(n + 1)/4$	(5.19.1)	159
17.	(For Grouped Frequency distribution) *Median*, Q_1. Q_3 correspond to cumulative frequencies $N/2$, $N/4$, $3N/4$	(5.19.4)	160

Measures of Dispersion

18.	*Range* = *Maximum value* – *Minimum value*	(6.2.1)	179
	Quartile Deviation $= (Q_3 - Q_1)/2$	(6.3.1)	180
	Mean Deviation $= \frac{1}{n} \Sigma \mid x_i - \bar{x} \mid$, or, $\frac{1}{N} \Sigma f_i \mid x_i - \bar{x} \mid$	(6.4.2, 3)	182

Standard Deviation (σ) :

19,	$\sigma^2 = \frac{1}{n} \Sigma (x_i - \bar{x})^2$, or $\frac{1}{N} \Sigma f_i (x_i - \bar{x})^2$	(6.5.2, 3)	185
	$\sigma^2 = \frac{\Sigma x^2}{n} - \left(\frac{\Sigma x}{n}\right)^2$, or $\frac{\Sigma fx^2}{N} - \left(\frac{\Sigma fx}{N}\right)^2$	(6.7.1, 2)	194
20.	If $y = x - c$, then $\sigma_x = \sigma_y$	(6.7.3)	194
	If $y = (x - c)/d$, then $\sigma_x = d.\sigma_y$	(6.7.4)	194
21.	$N\sigma^2 = (n_1\sigma_1^2 + n_2\sigma_2^2) + (n_1 d_1^2 + n_2 d_2^2)$	(6.8.1)	203

Relative Measures:

22. Coeff. of Variation $= 100 \times \frac{\text{S.D.}}{\text{Mean}}$

Coeff. of Quartile Deviation $= 100 \times \frac{\text{Q.D.}}{\text{Mean}}$ 177

Coeff. of Mean Deviation $= 100 \times \frac{\text{M.D.}}{\text{Mean (or Median)}}$

Moments, Skewness and Kurtosis

23. r-th moment $(m_r') = \frac{1}{n}\Sigma(x_i - A)^r$ (7.1.7) 225

r-th central moment $(m_r) = \frac{1}{n}\Sigma(x_i - \bar{x})^r$ (7.1.7) 225

24. $m_1' = \bar{x} - A, \quad (m_1 = 0, m_2 = \sigma^2)$ (7.1.2, 6) 224

$\bar{x} = m_1' + A, \quad \sigma^2 = m_2' - m_1'^2$ (7.1.10) 225

25. $m_2 = m_2' - m_1'^2$

$m_3 = m_3' - 3m_2'm_1' + 2m_1'^3$

$m_4 = m_4' - 4m_3'm_1' + 6m_2'm_1'^2 - 3m_1'^4$ (7.1.9) 225

Beta-coefficients and Gamma-coefficients:

26. $\beta_1 = \frac{m_3^2}{m_2^3}\ ; \quad \beta_2 = \frac{m_4}{m_2^2}$ (7.3.1) 229

$\gamma_1 = \sqrt{\beta_1}\ ; \quad \gamma_2 = \beta_2 - 3$ (7.3.2) 229

Sheppard's Corrections for Grouping:

27. No corrections necessary for Mean and m_3.

m_2 (corrected) $= m_2 - c^2/12$

m_4 (corrected) $= m_4 - \frac{1}{2}c^2 m_2 + 7c^4/240$ 233

Skewness:

28. Pearson's measure $= \frac{\text{Mean} - \text{Mode}}{\text{S.D.}}$

or $\frac{3\,(\text{Mean} - \text{Median})}{\text{S.D.}}$ (7.7.1, 2) 234

Bowley's measure

$= \frac{(Q_3 - Q_2) - (Q_2 - Q_1)}{(Q_3 - Q_2) + (Q_2 - Q_1)} = \frac{Q_3 - 2Q_2 + Q_1}{Q_3 - Q_1}$ (7.7.3) 234

Moment measure $(\gamma_1) = \frac{m_3}{\sigma^3} = \frac{m_3}{(\sqrt{m_2})^3}$ (7.7.4) 234

Kurtosis:

29. Measure of kurtosis $(\gamma_2) = \frac{m_4}{\sigma^4} - 3$ (7.8.1) 241

Curve Fitting and Method of Least Squares

Straight Line:

30. $y = a + bx$; Normal equations $\Sigma y = an + b\Sigma x$

$\Sigma xy = a\Sigma x + b\Sigma x^2$ (8.5.2) 251

Parabola:

31. $y = a + bx + cx^2$;

Normal equations $\Sigma y = an + b\Sigma x + c\Sigma x^2$

$$\Sigma xy = a\Sigma x + b\Sigma x^2 + c\Sigma x^3$$
$$\Sigma xy = a\Sigma x^2 + b\Sigma x^3 + c\Sigma x^4 \qquad (8.7.2) \qquad 259$$

32. When values of x have a common difference,

write, $u = \dfrac{x - \text{central value of } x}{\text{common difference}}$; ($n$ odd) (8.6.3) 252

$u = \dfrac{x - (\text{mean of the two central values})}{\frac{1}{2}(\text{common difference})}$; ($n$ even) (8.6.5) 252

and rewrite Straight line as $y = a + bu$, Parabola as $y = a + bu + cu^2$, with u replacing x in normal equations.

Exponential and Geometric Curves:

33. $y = ab^x$; $y = ax^b$. Take logarithm of both sides, and transform equation into Straight line form with $Y = \log y$, $X = \log x$. 263

Correlation and Regression

Correlation:

34. Correlation coefficient (r) = $\dfrac{\text{cov}(x, y)}{\sigma_x \sigma_y}$ (9.7.1) 276

where $\text{cov}(x, y) = \dfrac{\Sigma xy}{n} - \left(\dfrac{\Sigma x}{n}\right)\left(\dfrac{\Sigma y}{n}\right)$ (9.6.2) 275

$\sigma_x^2 = \dfrac{\Sigma x^2}{n} - \left(\dfrac{\Sigma x^2}{n}\right)^2$; $\sigma_y^2 = \dfrac{\Sigma y^2}{n} - \left(\dfrac{\Sigma y}{n}\right)^2$

35. If $u = (x - c)/d$, $v = (y - c')/d'$, then $r_{xy} = r_{uv}$ (9.8.1) 277

36. $-1 \le r \le +1$ (9.8.3) 277

37. $\sigma_x + y^2 = \sigma_x^2 + \sigma_y^2 + 2r\sigma_x\sigma_y$

$\sigma_{x-y}^{\;2} = \sigma_x^2 + \sigma_y^2 - 2r\sigma_x\sigma_y$ (9.11.3-4) 286

Regression:

38. $y - \bar{y} = b_{yx}(x - \bar{x})$ Regression eqn. of y on x (9.13.4) 292

$x - \bar{x} = b_{xy}(y - \bar{y})$ Regression eqn. of x on y (9.13.5) 292

where $b_{yx} = \dfrac{\text{cov}(x, y)}{\sigma_y^2} = r\dfrac{\sigma_x}{\sigma_y}$ (9.13.6) 292

$b_{xy} = \dfrac{\text{cov}(x, y)}{\sigma_y^2} = r\dfrac{\sigma_x}{\sigma_y}$ (9.113.7) 292

39. $b_{xy} \cdot b_{yx} = r^2$ (9.13.8) 292

40. $\Sigma(y_i - \bar{y})^2 = \Sigma(y_i - y_i')^2 + \Sigma(y_4' - \bar{y})^2$ 301

r^2 = proportion of variation explained by regression

Rank Correlation:

41. Rank correlation coefficient (R) = $1 - \dfrac{6\Sigma d^2}{n^3 - n}$ (9.16.1) 304

Multiple and Partial Correlation:

42. $R_{1.23} = \sqrt{\dfrac{r_{12}^2 + r_{13}^2 - 2r_{12}r_{13}r_{23}}{1 - r_{23}^2}}$ (9.17.3) 309

43. $r_{12.3} = \dfrac{r_{12} - r_{13}r_{23}}{\sqrt{(1 - r_{13}^2)(1 - r_{23}^2)}}$ (9.17.5) 310

44. $0 \le R_{1.23} \le 1;\ -1 \le r_{12.3} \le 1$ (9.17.4, 6) 310

45. $1 - R_{1.23}^2 = (1 - r_{12}^2)\,(1 - r_{13.2}^2)$ (9.17.7) 310

$= (1 - r_{13}^2)\,(1 - r_{12.3}^2)$ (9.17.8) 310

46. $R_{1.23} \ge |r_{12}|;\ r_{1.23} \ge |r_{13}|$ (9.17.8) 310

Interpolation

47. $\Delta y_r = y_{r+1} - y_r;$ (10.2.1) 320

$Ey_r = y_{r+1};\ E = 1 + \Delta$ (10.2.2) 320

Newton's Forward formula: Writing $u = (x - x_0)/h$,

48. $y = y_0 + u\Delta y_0 + \dfrac{u(u-1)}{1 \times 2}\,\Delta^2 y_0 + \dots\dots$

$+ \dfrac{u(u-1)\dots(u-n+1)}{1 \times 2 \times \dots \times n}\,\Delta^n y_0$ (10.4.1) 326

Newton's Backward formula: Writing $v = (x - x_n)/h$,

49. $y = y_n + v\Delta y_{n-1} + \dfrac{v(v+1)}{1 \times 2}\,\Delta^2 y_{n-2} + \dots\dots$

$+ \dfrac{v(v+1)\dots(v+n-1)}{1 \times 2 \times \dots \times n}\,\Delta^n y_0$ (10.5.1) 329

Stirling's formula:

50. $y = y_0 + u \,.\, \dfrac{(\Delta y_0 + \Delta y_{-1})}{2} + \dfrac{u^2}{2!}\,\Delta^2 y_{-1}$

$+ \dfrac{u(u^2 - 1^2)}{3!} \,.\, \dfrac{\Delta^3 y_{-1} + \Delta^3 y_{-2})}{2} + \dots\dots$

where $u = (x - x_0)/h$. (10.6.1) 332

Bessel's formula:

51. $y = \dfrac{(y_0 + y_1)}{2} + v \,.\, \Delta y_0 + \dfrac{(v^2 - \frac{1}{4})}{2!} \,.\, \dfrac{(\Delta^2 y_0 + \Delta^2 y_{-1})}{2}$

$+ \dfrac{v(v^2 - \frac{1}{4})}{3!} \,.\, \Delta^3 y_{-1} + \dots\dots$

where $v = u - \dfrac{1}{2}$ and $u = (x - x_0)/h$. (10.6.2) 332

Lagrange's formula:

52. $y = \dfrac{(x - x_1)\,(x - x_2)\dots(x - x_n)}{(x_0 - x_1)\,(x_0 - x_2)\dots(x_0 - x_n)}\,y_0$

$+ \dfrac{(x - x_0)\,(x - x_2)\dots(x - x_n)}{(x_1 - x_0)\,(x_1 - x_2)\dots(x_1 - x_n)}\,y_1$

$+ \quad \dots \quad \dots \quad \dots$

$+ \dfrac{(x - x_0)\,(x - x_1)\dots(x - x_{n-1})}{(x_n - x_0)\,(x_n - x_1)\dots(x_n - x_{n-1})}\,y_n$ (10.7.1) 333

Theory of Probability

Ref. Page

Classical definition:

1. $P(A) = \dfrac{m}{n}$	(11.5.1)	351

Rules for Combining Probabilities:

2. $P(A+B) = P(A) + P(B) - P(AB)$	(11.6.6)	363
3. $P(AB) = P(A).P(B/A)$	(11.6.10)	364
4. $P(A+B) = P(A) + P(B)$, when *mutually exclusive.*	(11.6.1)	362
5. $P(AB) = P(A).P(B)$, when *independent.*	(11.6.12)	365
6. $P(\bar{A}) = 1 - P(A)$	(11.6.4)	363

Drawing with or without Replacement:

7. $\dfrac{{}^{A}C_a \cdot {}^{B}C_b}{{}^{A+B}C_{a+b}}$ (*without* replacement)	(11.7.1)	373
8. ${}^{n}C_r p^r q^{n-r}$ (*with* replacement)	(11.8.1)	377
9. Bayes' Theorem $P(B_i/A) = \dfrac{P(B_i)\cdot P(A/B_i)}{\Sigma P(B_i)\cdot P(A/B_i)}$	(11.9.1)	379

Mathematical Expectation:

10. $E(x) = \Sigma p_i x_i = m$ (suppose)	(11.10.1)	383
11. $\text{Var}(x) = E(x-m)^2 = \Sigma p_i(x_i - m)^2$	(11.10.2)	383
$= \Sigma p_i x_i^2 - m^2$	(11.10.3)	383

Theorems on Expectation:

12. $E(x+y) = E(x) + E(y)$	(11.10.4)	383
13. $E(x, y) = E(x).E(y)$, when x and y are *independent.*	(11.10.5)	383

Axioms of Probability:

14. Axiom I. For every event A, $P(A) \geq 0$ Axiom II. For the sure event S, $P(S) = 1$ Axiom III. For mutually exclusive events, $P(A \cup B) = P(A) + P(B)$	(11.14.1-3)	396

Conditional Probability and Independence:

15. $P(A/B) = \dfrac{P(A \cap B)}{P(B)}$	(11.17.3)	401
16. $P(A \cap B) = P(A)\cdot P(B)$, if A and B are *independent.*	(11.18.2)	403

Binomial, Normal and Poisson Distributions

Discrete Probability Distribution:

17. Mean $(\mu) = E(x) = \Sigma x \cdot f(x)$	(12.3.3,5)	429
Variance $(\sigma^2) = E(x^2) - \mu^2$	(12.3.6)	429
18. Moments: $\mu'_r = E(x - A)^r$ $\mu_r = E(x-\mu)^r$, where $\mu = E(x)$	(12.3.7)	429

Binomial Distribution:

19. $f(x) = {}^nC_x p^x q^{r-x}$; $(x = 0, 1, 2, \ldots n)$ (12.5.1) 433

Mean = np; S.D. = $\sqrt{npq}$ (12.5.3) 435

Poisson Distribution:

20. $f(x) = \dfrac{e^{-m} m^x}{x!}$; $(x = 0, 1, 2, \ldots \infty)$ (12.6.1) 441

Mean = m; S.D. = $\sqrt{m}$ (12.6.3) 441

Hypergeometric Distribution:

21. $f(x) = \dfrac{{}^{Np}C_x \cdot {}^{Nq}C_{n-x}}{{}^{N}C_n}$ $(x = 0, 1, 2, \ldots m)$ (12.8.2) 448

Mean = np; S.D. = $\sqrt{npq(N-n)/(N-1)}$ (12.8.4) 448

Normal Distribution:

22. $f(x) = \dfrac{1}{\sigma\sqrt{2x}} e^{\frac{-(x-\mu)2}{2\sigma^2}}$ $(-\infty < x < \infty)$ (12.13.1) 468

Mean = μ; S.D. = σ

Sampling Theory

23. **Standard Error** (S.E.) of a statistic is the standard deviation calculated from the sampling distribution of the statistic.

Formulae for Standard Error (S.E.):

24. S.E. of $\overline{x} = \dfrac{\sigma}{\sqrt{n}}$ (13.7.1) 501

25. S.E. of $p = \dfrac{\sqrt{PQ}}{n}$, where $P + Q = 1$. (13.7.2) 501

If the random sample is drawn without replacement from a finite population of size N, the expressions above are multiplied by the Correction Factor $\sqrt{(N-n)/(N-1)}$. If σ (or P) is not known, use S (or P) to find S.E.

Estimation and Test of Significance

Criteria of good Estimators (Point Estimation):

26. (i) Unbiasedness; (ii) Consistency; (iii) Efficiency; (iv) Sufficiency. 529

Estimators of μ and σ^2:

27. $E(\overline{x}) = \mu$, $E(S^2) = \dfrac{n-1}{n}\sigma^2$ (14.3.4-5) 529

28. $E(s^2) = \sigma^2$ where $s^2 = \dfrac{\Sigma(x_i - \overline{x})^2}{n-1} = \dfrac{nS^2}{n-1}$ (14.3.7) 529

Method of Maximum Likelihood: At $\theta = \theta_0$,

29. $\dfrac{\partial}{\partial\theta} \log L = 0, \quad \dfrac{\partial^2}{\partial\theta^2} \log L < 0$ (14.4.2) 534

Interval Estimation:

30. Two statistics t_1 and t_2 are so selected that the probability of the parameter θ being included in the interval (t_1, t_2) has a specified value c.

31. $P(t_1 \le \theta \le t_2) = c$ (14.5.1) 538

Approximate Confidence Limits (large *n*):

	Parameter	95% limits	99% limits	
32.	μ	$\bar{x} \pm 1.96$ (*SE* of $\bar{x}$)	$\bar{x} \pm 2.58$ (*SE* of $\bar{x}$)	
	P	$p \pm 1.96$ (*SE* of p)	$p \pm 2.58$ (*SE* of p)	539

Exact Confidence Limits (any *n*):

	Parameter	95% limits		
33.	$\mu(\sigma$ known)	$\bar{x} \pm 1.96\ (\sigma/\sqrt{3})$	(14.5.6)	542
	$\mu(\sigma$ unknown)	$\bar{x} \pm t_{\cdot 025}\left(S/\sqrt{n-1}\right)$	(14.5.8a)	543
	$\mu_1 - \mu_2(\sigma$'s known)	$(\bar{x}_1 - \bar{x}_2) \pm 1.96\sqrt{\sigma_1^2/n_1 + \sigma_2^2/n_2}$	(14.5.7)	543
	$\mu_1 - \mu_2$ (σ's unknown).	$(\bar{x}_1 - \bar{x}_2) \pm t_{\cdot 025}\left(s\sqrt{1/n_1 + 1/n_2}\right)$	(14.5.9a)	544
	σ^2	$nS^2/\chi^2_{\cdot 025} \le \sigma^2 \le nS^2/\chi^2_{\cdot 975}$	(14.5.11)	545

Large Sample Tests of Significance

Test for a specified proportion: $H_0(P = P_0)$

34. $z = \dfrac{p - P_0}{\text{S.E. of } p}$ (14.7.1) 555

Reject H_0 at 5% level, if $|z| \ge 1.96$, $z \ge 1.645$, or $z \le -1.645$, according as H_1 is $P \ne P_0$, $P > P_0$, or $P < P_0$.

Test for equality of two proportions: $H_0(P_1 = P_2)$

35. $z = \dfrac{p_1 - p_2}{S.E.}$ (14.7.4) 558

where $S.E. = \sqrt{pq\left(\dfrac{1}{n_1} + \dfrac{1}{n_2}\right)}$ and $p = \dfrac{n_1 p_1 + n_2 p_2}{n_1 + n_2}$

Reject H_0 at 5% level, if $|z| \ge 1.96$, $z \ge 1.645$, or $z \le -1.645$, according as H_1 is $P_1 \ne P_1$, $P_1 > P_2$, or $P_1 < P_2$.

Test for a specified mean: $H_0(\mu = \mu_0)$

36. $z = \dfrac{\bar{x} - \mu_0}{\text{S.E. of } \bar{x}}$ (14.7.6) 561

where S.E. of $\bar{x}$ is shown at (24).

Test for equality of two means: $H_0(\mu_1 = \mu_2)$

37. $z = \dfrac{\bar{x}_1 - \bar{x}_2}{S.E.}$ where $S.E. = \sqrt{\dfrac{\sigma_1^2}{n_1} + \dfrac{\sigma_2^2}{n_2}}$ (14.7.8,9) 561

(If σ_1 and σ_2 are unknown, use S_1 and S_2 to calculate $S.E.$).

Test for goodness of fit: H_0 (Data support the theory)

38. $\chi^2 = \sum \dfrac{(f_0 - f_e)^2}{f_e}$, d.f. = (No. of classes – 1). 567

Test for independence of attributes: H_0 (Attributes are independent)

39. $$\chi^2 = \sum \frac{(f_0 - f_e)^2}{f_e},$$ (14.7.12) 569

$d.f.$ = (No. of Rows – 1) × (No. of Columns – 1);

$$f_e = \frac{\text{(Row total)} \times \text{(Column total)}}{\text{Total frequency}}$$

Simplified formula for (2 × 2) table:

40. $$\chi^2 = \frac{N(ad - bc)^2}{R_1 R_2 C_1 C_2};\ \text{d.f.} = 1$$ (14.7.13) 570

Small Sample Tests of Significance

Test for a specified mean (σ known): H_0 $(\mu = \mu_0)$

41. $$z = \frac{\bar{x} - \mu_0}{\sigma/\sqrt{n}}$$ (14.8.1) 572

Test for equality of two means (σ_1, σ_2 known): H_0 $(\mu = \mu_0)$

42. $$z = \frac{\bar{x}_1 - \bar{x}_2}{\sqrt{\dfrac{\sigma_1^2}{n_1} + \dfrac{\sigma_2^2}{n_2}}}$$ (14.8.4) 573

Test for a specified mean (σ known): H_0 $(\mu = \mu_0)$

43. $$t = \frac{\bar{x}_1 - \mu_0}{\dfrac{s}{\sqrt{n}}} = \frac{\bar{x} - \mu_0}{\dfrac{S}{\sqrt{n-1}}}\ \text{d.f.} = (n-1)$$ (14.8.6) 575

where $s^2 = \dfrac{\Sigma(\bar{x}_i - \bar{x})^2}{(n-1)}$, S = sample s.d.

Test for equality of two means ($\sigma_1 = \sigma_2$ unknown): H_0 $(\mu_1 = \mu_2)$

44. $$t = \frac{\bar{x}_1 - \bar{x}_2}{s\sqrt{\dfrac{1}{n_1} + \dfrac{1}{n_2}}} \quad \text{where } s^2 = \frac{n_1 S_1^2 + n_2 S_2^2}{(n_1 + n_2 - 2)}$$ 575

d.f. = $n_1 + n_2 - 2$

Test for a specified s.d. (mean unknown): H_0 ($\sigma = \sigma_0$)

45. $$\chi^2 = \frac{nS^2}{\sigma_0^2}, \text{ d.f.} = (n-1)$$ (14.8.11) 582

Test of equality of two s.d. (means unknown): H_0 ($\sigma_1 = \sigma_2$)

46. $$F = \frac{s_1^2}{s_2^2}, \text{ d.f.} = (n_1 - 1), (n_2 - 1)$$ (14.8.12) 584

Analysis of Variance

One-way Classified Data:

47.

Source of Variation	*S.S.*	*d.f.*	*M.S.*	*F Values*	
				Obs.	*Tab.*
Between Groups	*SSB*	$k-1$	*MSB*	$F = \frac{MSB}{MSW}$	$F_{.05}$
Within Groups	*SSW*	$N-k$	*MSW*	—	—
Total	Total *SS*	$N-1$	—	—	—

If $F > F_{.05}$, reject null hypothesis H_0 (all μ's are equal).

48. **Critical Difference between totals** = $s \cdot \sqrt{2n} \cdot t_{.025}$ (15.5.1) 620

Time Series

Components:

49. (i) Secular trend, or Trend; (ii) Seasonal variation: (ii) Cyclical fluctuation; (iv) Irregular movement. 633

Isolation of Components:

50. Trend: (i) Free-hand method; (ii) Semi-average method; (iii) Moving Average method; (iv) Fitting mathematical curves. 635

51. Seasonal: (i) Method of (monthly or quarterly) Averages; (ii) Moving Average method; (iii) Trend-ratio method; (iv) Link Relative method. 661

Index Numbers

Price Index Formulae:

53. Simple Aggregative Index = $\frac{\Sigma p_n}{\Sigma p_0} \times 100$ (17.3.2) 690

54. Weighted Aggregative Index = $\frac{\Sigma p_n w}{\Sigma p_0 w} \times 100$ (17.3.3) 691

55. Laspeyres' Index = $\frac{\Sigma p_n q_0}{\Sigma p_0 q_0} \times 100$ (17.3.4) 691

56. Paasche's Index = $\frac{\Sigma p_n q_n}{\Sigma p_0 q_n} \times 100$ (17.3.5) 691

57. Edgeworth-Marshall's Index $= \dfrac{\Sigma\, p_n(q_0 + q_n)}{\Sigma\, p_0(q_0 + q_n)} \times 100$ (17.3.6) 691

58. Fisher's Ideal Index $= \sqrt{\text{Laspeyres'} \times \text{Paasche's}}$ (17.3.7)) 691

59. Simple A.M. of Price Relatives Index

$= \Sigma$ (Price Relatives) $\div k$ (17.3.9) 692

where Price Relative $= (p_n \div p_0) \times 100$ (17.3.8) 692

60. Weighted A.M. of the Price Relatives Index

$$= \frac{\Sigma(\text{Price Relatives} \times w)}{\Sigma w}$$ (17.3.11) 692

General Index from Group Indices:

61. General Index $= \dfrac{\Sigma IW}{\Sigma W}$ (17.3.15) 693

Test of Index Numbers:

62. Time Reversal Test: $I_{0n} \cdot I_{n0} = 1$ (17.5.1) 706

63. Factor Reversal Test: $P_{0n} . Q_{0n} = \dfrac{\Sigma p_n q_n}{\Sigma p_0 q_0}$ (17.5.2) 706

64. Circular Test: $I_{01} . I_{12} . I_{23} \ldots I_{(n-1),\,n} . I_{n0} = 1$ (17.5.3) 707

Chain Base Method:

65. Calculate *Link Index* numbers I_{01}, I_{12}, I_{23}, ..., (i.e. with preceding period as base) by the given formula. Multiply them progressively to get *Chain Index* numbers: 714

$$I_{01}' = I_{01}$$
$$I_{02}' = I_{01} . I_{12}$$
$$I_{03}' = I_{01} . I_{12} . I_{23}; \text{ etc.}$$

Cost of Living Index Numbers:

66. For each Group (viz. Food, Clothing, Fuel and light, Housing, Miscellaneous), calculate 716

$$\text{Group Index} = \frac{\Sigma(\textit{Price Relative} \times w)}{\Sigma w}$$

where w = percentage expenditure of an 'average family' on the item;

$$\text{Cost of Living Index} = \frac{\Sigma IW}{100}$$ 716

where I = Group Index, and W = Percentage expenditure of an '*Average family*' on the Group.

Vital Statistics

Crude Death Rate (*CDR*):

67. $CDR = \dfrac{\text{Total number of deaths}}{\text{Total population}} = \dfrac{\Sigma D_x}{\Sigma P_x}$ (18.2.1) 744

Specific Death Rate (m_x):

68. $m_x = D_x \div P_x$ (18.3.2) 744

69. $CDR = \dfrac{\Sigma P_x m_x}{\Sigma P_x}$ (18.3.3) 745

70. **Standardised Death Rate** $= \dfrac{\Sigma P'_x m_x}{\Sigma P'_x}$ (18.4.1) 746

Life Table:

71. l_x = No. of persons out of a given cohort (l_0) who are expected to complete exactly x years of life

$d_x = l_x - l_{x+1}$ (18.5.1) 748

$q_x = d_x \div l_x$ (18.5.2) 748

$p_x = 1 - q_x$ (18.5.4) 748

$L_x = \frac{1}{2}(l_x + l_{x+1}) = l_x - \frac{1}{2} d_x$ (18.5.5-6) 748

$T_x = L_x + L_{x+1} + L_{x+2} + \ldots$ (18.5.7) 749

$e_x^0 = T_x \div l_x$ (18.5.9) 749

Crude Birth Rate (CBR):

72. $CBR = \dfrac{\text{Total number of live-births}}{\text{Total population}}$ (18.6.1) 752

General Fertility Rate (GFR):

73. $GFR = \dfrac{\text{Total number of live-births}}{\text{(Total number of } women \text{ in the child-bearing ages)}}$ (18.7.1) 752

Age-specific Fertility Rate (f_x):

74. $f_x = \dfrac{\text{No. of live-births from women aged } x \text{ 1.b.d.}}{\text{Total No. number of women aged } x \text{ last birthday}}$ (18.8.1) 753

Total Fertility Rate (TFR):

75. $TFR = \sum_{x=15}^{49} f_x$ (18.9.1) 753

Gross Reproduction Rate (GRR):

76. $GRR = \sum_{x=15}^{49} \dfrac{\text{(No. of female live-births from women aged } x \text{ 1.b.d.)}}{\text{Total No. of women aged } x \text{ 1.b.d.}}$

(18.11.1) 754

$GRR = k \times TFR$, **where** k **= (No. of female births) + Total births**

(18.11.2) 754

Net Reproduction Rate (*NRR*):

77. $NRR = \frac{1}{l_0} \sum_{x=15}^{49} L_x f_x \times k$ (18.12.2) 754

Statistical Quality Control

Control Charts (*Standards not given*):

78.

Chart for	Central Line	Control Limits	
Mean ($\overline{x}$)	$\overline{\overline{x}}$	$\overline{\overline{x}} \pm A_2 \overline{R}$	770
Range (R)	$\overline{R}$	$D_3 \overline{R}$, $D_4 \overline{R}$	
Fraction defective (p)	$\overline{p}$	$\overline{p} \pm 3 \sqrt{\frac{\overline{p}(1-\overline{p})}{n}}$	
Number of defects (c)	$\overline{c}$	$\overline{c} \pm 3 \sqrt{c}$	770

APPENDIX

Table I Area Under Standard Normal Curve

(Proportion of area under standard normal curve between the ordinates at z = 0 and given values of z)										
z	*.00*	*.01*	*.02*	*.03*	*.04*	*0.5*	*.06*	*.07*	*.08*	*.09*
0.0	.0000	.0040	.0080	.0120	.0160	.0199	.0239	.0279	.0319	.0359
0.1	.0398	.0438	.0478	.0517	.0557	.0596	.0636	.0675	.0714	.0753
0.2	.0793	.0832	.0871	.0910	.0948	.0987	.1026	.1064	.1103	.1141
0.3	.1179	.1217	.1255	.1293	.1331	.1368	.1406	.1443	.1480	.1517
0.4	.1554	.1591	.1628	.1664	.1700	.1736	.1772	.1808	.1844	.1879
0.5	.1915	.1950	.1985	.2019	.2054	.2088	.2123	.2157	.2190	.2224
0.6	.2257	.2291	.2324	.2357	.2389	.2422	.2454	.2486	.2517	.2549
0.7	.2580	.2611	.2642	.2673	.2704	.2734	.2764	.2794	.2823	.2852
0.8	.2881	.2910	.2939	.2967	.2995	.3023	.3051	.3078	.3106	.3133
0.9	.3159	.3186	.3212	.3238	.3264	.3289	.3315	.3340	.3365	.3389
1.0	.3413	.3438	.3461	.3485	.3508	.3531	.3554	.3577	.3599	.3621
1.1	.3643	.3665	.3686	.3708	.3729	.3749	.3770	.3790	.3810	.3830
1.2	.3849	.3869	.3888	.3907	.3925	.3944	.3962	.3980	.3997	.4015
1.3	.4032	.4049	.4066	.4082	.4099	.4115	.4131	.4147	.4162	.4177
1.4	.4192	.4207	.4222	.4236	.4251	.4265	.4279	.4292	.4306	.4319
1.5	.4332	.4345	.4357	.4370	.4382	.4394	.4406	.4418	.4429	.4441
1.6	.4452	.4463	.4474	.4484	.4495	.4505	.4515	.4525	.4535	.4545
1.7	.4554	.4564	.4573	.4582	.4591	.4599	.4608	.4616	.4625	.4633
1.8	.4641	.4649	.4656	.4664	.4671	.4678	.4686	.4693	.4699	.4706
1.9	.4713	.4719	.4726	.4732	.4738	.4744	.4750	.4756	.4761	.4767
2.0	.4772	.4778	.4783	.4788	.4793	.4798	.4803	.4808	.4812	.4817
2.1	.4821	.4826	.4830	.4834	.4838	.4842	.4846	.4850	.4854	.4857
2.2	.4861	.4864	.4868	.4871	.4875	.4878	.4881	.4884	.4887	.4890
2.3	.4893	.4896	.4898	.4901	.4904	.4906	.4909	.4911	.4913	.4916
2.4	.4918	.4920	.4922	.4925	.4927	.4929	.4931	.4932	.4934	.4936

(C

(Proportion of area under standard normal curve between the ordinates at z = 0 and given values of z)										
z	.00	.01	.02	.03	.04	.05	.06	.07	.08	.09
2.5	.4938	.4940	.4941	.4943	.4945	.4946	.4948	.4949	.4951	.4952
2.6	.4953	.4955	.4956	.4957	.4959	.4960	.4961	.4962	.4963	.4964
2.7	.4965	.4966	.4967	.4968	.4969	.4970	.4971	.4972	.4973	.4974
2.8	.4974	.4975	.4976	.4977	.4977	.4973	.4979	.4979	.4980	.4981
2.9	.4981	.4982	.4982	.4983	.4984	.4984	.4985	.4985	.4986	.4986
3.0	.4987	.4987	.4987	.4988	.4988	.4989	.4989	.4989	.4990	.4990
3.1	.4990	.4991	.4991	.4991	.4992	.4992	.4992	.4992	.4993	.4993
3.2	.4993	.4993	.4994	.4994	.4994	.4994	.4994	.4995	.4995	.4995
3.3	.4995	.4995	.4995	.4996	.4996	.4996	.4996	.4996	.4996	.4997
3.4	.4997	.4997	.4997	.4997	.4997	.4997	.4997	.4997	.4997	.4998

Table II Significance Points of χ^2

[Values of chi-square for given probabilities $P = Prob(\chi^2 > \chi_0^2)$]						
Degrees of Freedom	Probability (P)					
	.99	.975	.95	.05	.025	.01
1	—	—	—	3.84	5.0	6.63
2	0.020	0.051	0.103	5.99	7.38	9.21
3	.115	.216	.352	7.81	9.35	11.34
4	.297	.484	.711	9.49	11.14	13.28
5	.554	.831	1.15	11.07	12.83	15.09
6	.872	1.24	1.64	12.59	14.45	16.81
7	1.24	1.69	2.17	14.07	16.01	18.48
8	1.65	2.18	2.73	15.51	17.53	20.09
9	2.09	2.70	3.33	16.92	19.02	21.67
10	2.56	3.25	3.94	18.31	20.48	23.21
11	3.05	3.82	4.57	19.68	21.92	24.73
12	3.57	4.40	5.23	21.03	23.34	26.22
13	4.11	5.01	5.89	22.36	24.74	27.69
14	4.66	5.63	6.57	23.68	26.12	29.14
15	5.23	6.26	7.26	25.00	27.49	30.58
16	5.81	6.91	7.96	26.30	28.85	32.00
17	6.41	7.56	8.67	27.59	30.19	33.41
18	7.01	8.23	9.39	28.87	31.53	34.81
19	7.63	8.91	10.12	30.14	32.85	36.19
20	8.26	9.59	10.85	31.41	34.17	37.57
21	8.90	10.28	11.59	32.67	35.48	38.93
22	9.54	10.98	12.34	33.92	36.78	40.29
23	10.20	11.69	13.09	35.17	38.08	41.64
24	10.86	12.40	13.85	36.42	39.36	42.98
25	11.52	13.12	14.61	37.65	40.65	44.31

(Contd)

[Values of chi-square for given probabilities $P = Prob(x^2 > x_0^2)$]						
Degrees of	Probability (P)					
Freedom	.99	.975	.95	.05	.025	.01
26	12.20	13.84	15.38	38.89	41.92	45.64
27	12.88	14.57	16.15	40.11	43.19	46.96
28	13.56	15.31	16.93	41.34	44.46	48.28
29	14.26	16.05	17.71	42.56	45.72	49.59
30	14.95	16.79	18.49	43.77	46.98	50.89

Note: These are also known as Percentage Points of chi-square distribution (see page 195). Example: For 10 degrees of freedom (d.f.) the *upper* 5% point is 18.31, and the lower 5% point is 3.94. In symbols, for 10 d.f. $\chi^2_{.05} = 18.31$ and $\chi^2_{.95} = 3.94$.

Table III Significance Points of *t*

[Values of t_0 for given probabilities $P = Prob.(t > t_0) = Prob.\ (t < -t_0)$]				
Degrees of	Probability (P)			
Freedom	.05	.025	.01	.005
1	6.31	12.71	31.82	63.66
2	2.92	4.30	6.97	9.93
3	2.35	3.18	4.54	5.84
4	2.13	2.78	3.75	4.60
5	2.02	2.57	3.37	4.03
6	1.94	2.45	3.14	3.71
7	1.90	2.37	3.00	3.50
8	1.86	2.31	2.90	3.36
9	1.83	2.26	2.82	3.25
10	1.81	2.23	2.76	3.17
11	1.80	2.20	2.72	3.11
12	1.78	2.18	2.68	3.06
13	1.77	2.16	2.65	3.01
14	1.76	2.15	2.62	2.98
15	1.75	2.13	2.60	2.95
16	1.75	2.12	2.58	2.92
17	1.74	2..11	2.57	2.90
18	1.73	2.10	2.55	2.88
19	1.73	2.09	2.54	2.86
20	1.73	2.09	2.53	2.85
21	1.72	2.08	2.52	2.83
22	1.72	2.07	2.51	2.82
23	1.71	2.07	2.50	2.81
24	1.71	2.06	2.49	2.80
25	1.71	2.06	2.49	2.79
26	1.71	2.06	2.48	2.78
27	1.70	2.05	2.47	2.77
28	1.70	2.05	2.47	2.76
29	1.70	2.05	2.46	2.76
30	1.70	2.04	2.46	2.75

Note: These are also known as *Percentage* points of *t* distribution. Example: For 6 degrees of freedom (d.f.), the upper 1% point is 3.14 and the lower 1% point is–3.14 (because of summetry of the *t*-curve. In symbols, for 6 d.f.

$$t_{.01} = 3.14 \quad t_{.99} = -3.14$$

Table IV Significance Points of F

(5% values)

Degree of Freedom (n_2)	Degree of Freedom n_1: 1	2	3	4	5	6	7	8	9	∞
1	161.4	199.5	215.7	224.6	230.2	234.0	236.8	238.9	240.5	254.3
2	18.51	19.00	19.16	19.25	19.30	19.33	19.35	19.37	19.38	19.50
3	10.13	9.55	9.28	9.12	9.01	8.94	8.89	8.85	8.81	8.53
4	7.71	6.94	6.59	6.39	6.26	6.16	6.09	6.04	6.00	5.63
5	6.61	5.79	5.41	5.19	5.05	4.95	4.88	4.82	4.77	4.36
6	5.99	5.14	4.76	4.53	4.39	4.28	4.21	4.15	4.10	3.67
7	5.59	4.74	4.35	4.12	3.97	3.87	3.79	3.73	3.68	3.23
8	5.32	4.46	4.07	3.84	3.69	3.58	3.50	3.44	3.39	2.93
9	5.12	4.26	3.86	3.63	3.48	3.37	3.29	3.23	3.18	2.71
10	4.96	4.10	3.71	3.48	3.33	3.22	3.14	3.07	3.02	2.54
11	4.84	3.98	3.59	3.36	3.20	3.09	3.01	2.95	2.90	2.40
12	4.75	3.88	3.49	3.26	3.11	3.00	2.91	2.85	2.80	2.30
13	4.67	3.80	3.41	3.18	3.02	2.92	2.83	2.77	2.71	2.21
14	4.60	3.74	3.34	3.11	2.96	2.85	2.76	2.70	2.65	2.13
15	4.54	3.68	3.29	3.06	2.90	2.79	2.71	2.64	2.59	2.07
16	4.49	3.63	3.24	3.01	2.85	2.74	2.66	2.59	2.54	2.01
17	4.45	3.59	3.20	2.96	2.81	2.70	2.61	2.55	2.49	1.96
18	4.41	3.55	3.16	2.93	2.77	2.66	2.58	2.51	2.46	1.92
19	4.38	3.52	3.13	2.90	2.74	2.63	2.54	2.48	2.42	1.88
20	4.35	3.49	3.10	2.87	2.71	2.60	2.51	2.45	2.39	1.84
30	4.17	3.32	2.92	2.69	2.53	2.42	2.33	2.27	2.21	1.62
∞	3.84	3.00	2.60	2.37	2.21	2.10	2.01	1.94	1.88	1.00

(Contd)

Significance Points of F

(1% values)

Degree of Freedom (n_2)	Degree of Freedom n_1: 1	2	3	4	5	6	7	8	9	∞
1	4052	4999	5403	5625	5764	5859	5928	5981	6022	6366
2	98.50	99.00	99.17	99.25	99.30	99.33	99.36	99.37	99.39	99.50
3	34.12	30.82	29.46	28.71	28.24	27.91	27.67	27.49	27.35	26.13
4	21.20	18.00	16.69	15.98	15.52	15.21	14.98	14.80	14.66	13.46
5	16.26	13.27	12.06	11.39	10.97	10.67	10.46	10.29	10.16	9.02
6	13.75	10.92	9.78	9.15	8.75	8.47	8.26	8.10	7.98	6.88
7	12.25	9.55	8.45	7.85	7.46	7.19	6.99	6.84	6.72	5.65
8	11.26	8.65	7.59	7.01	6.63	6.37	6.18	6.03	5.91	4.86
9	10.56	8.02	6.99	6.42	6.06	5.80	5.61	5.47	5.35	4.31
10	10.04	7.56	6.55	5.99	5.64	5.39	5.20	5.06	4.94	3.91
11	9.65	7.21	6.22	5.67	5.32	5.07	4.89	4.74	4.63	3.60
12	9.33	6.93	5.95	5.41	5.06	4.82	4.64	4.50	4.39	3.36
13	9.07	6.70	5.74	5.21	4.86	4.62	4.44	4.30	4.19	3.17
14	8.86	6.51	5.56	5.04	4.69	4.46	4.28	4.14	4.03	3.00
15	8.68	6.36	5.42	4.89	4.56	4.32	4.14	4.00	3.89	2.87
16	8.53	6.23	5.29	4.77	4.44	4.20	4.03	3.89	3.78	2.75
17	8.40	6.11	5.18	4.67	4.34	4.10	3.93	3.79	3.68	2.65
18	8.29	6.01	5.09	4.58	4.25	4.01	3.84	3.71	3.60	2.57
19	8.18	5.93	5.01	4.50	4.17	3.94	3.77	3.63	3.52	2.49
20	8.10	5.85	4.94	4.43	4.10	3.87	3.70	3.56	3.46	2.42
30	7.56	5.39	4.51	4.02	3.70	3.47	3.30	3.17	3.07	2.01
∞	6.63	4.61	3.78	3.32	3.02	2.80	2.64	2.51	2.41	1.00

Table V Values of e^{-m}

m	e^{-m}	m	e^{-m}	m	e^{-m}
0.0	1.0000	1.5	0.2231	3.0	0.0498
0.1	0.9048	1.6	.2019	3.2	.0408
0.2	.8187	1.7	.1827	3.4	.0334
0.3	.7408	1.8	.1653	3.6	.0273
0.4	.6703	1.9	.4497	3.8	.0224
0.5	.6065	2.0	.1353	4.0	.0183
0.6	.5488	2.1	.1225	4.2	.0150
0.7	.4966	2.2	.1108	4.4	.0123
0.8	.4493	2.3	.1003	4.6	.0100
0.9	.4066	2.4	.0907	4.8	.00823
1.0	.3679	2.5	.0821	5.0	.00674
1.1	.3329	2.6	.0743	5.5	.00409
1.2	.3012	2.7	.0672	6.0	.00248
1.3	.2725	2.8	.0608	6.5	.00150
1.4	.2466	2.9	.0550	7.0	.00091

Table VI Random Numbers

4062	9228	6347	1043	5824	1560	6507	2405	3832	2993
6489	7404	6051	5954	0008	5231	3311	9324	6679	0345
8163	4275	9177	9796	3253	1562	5746	8523	9118	7332
2956	8728	9335	4461	1315	0647	7177	7246	3114	0813
9203	5790	5225	7478	4920	8721	8044	1806	8271	9724
4895	7266	5890	5011	5478	6475	8887	5014	7672	3442
9388	8173	0357	1623	6416	1466	9510	0343	5816	1909
8924	9121	1570	0374	8156	1949	7720	4802	0133	4496
1829	6798	7836	8856	4001	1638	6113	8612	5795	5846
1798	2280	9290	1379	5472	0460	1544	0965	5563	5652
7306	8564	1301	7910	4952	9390	6210	5867	7018	2032
9642	5428	7002	1835	9928	5132	1453	3939	4201	6703
8194	4347	2268	0425	1026	8375	4640	7178	8317	8489
2380	1000	2259	2701	9227	1349	0647	6078	6595	6619
5491	6250	4713	5947	9580	9438	3403	3513	4613	2690
2752	3281	8909	2901	8215	2067	9589	3465	5757	3417
8767	6707	7561	1191	8583	3705	0934	2916	9749	3936
1959	8451	2771	6219	2276	5028	2750	6224	0656	6126
7943	4347	8451	5713	8193	0313	6989	0088	5867	7992
3966	1653	4509	1921	4973	9488	3410	9540	8314	0808
7056	4276	9813	9777	5686	3428	7604	0591	6542	7752
8805	3993	4048	6604	6948	5519	3769	0403	4575	3098
0298	3634	6429	2478	1197	5680	6168	9324	2506	6616
2828	6589	8721	4082	3463	1859	5918	4605	1848	0348
7973	6207	2960	8675	0536	9518	7974	9749	5723	6888

INDEX

(*The figures indicate page numbers*)

I

J

K

L

M

N

O

P